Environmental Geochemistry

Citations
Please use the following example for citations:
Watts R.J. and Teel A.L. (2003) Groundwater and air contamination: risk, toxicity, exposure assessment, policy and regulation, pp. 1–16. In *Environmental Geochemistry* (ed. B.S. Lollar) Vol. 9 *Treatise on Geochemistry* (eds. H.D. Holland and K.K. Turekian), Elsevier–Pergamon, Oxford.

Cover photo: Bingham Canyon Mine is located 25 miles southwest of Salt Lake City, Utah, in the Oquirrh Mountains. At 3/4 of a mile deep and 2-1/2 miles wide, this is the largest man-made excavation on Earth, and along with the Great Wall of China, is visible with the naked eye by astronauts from outer space. (Photograph provided by M. Logsdon)

Environmental Geochemistry

Edited by

B. Sherwood Lollar
University of Toronto, ON, Canada

TREATISE ON GEOCHEMISTRY
Volume 9

Executive Editors

H. D. Holland
Harvard University, Cambridge, MA, USA

and

K. K. Turekian
Yale University, New Haven, CT, USA

ELSEVIER

2005

AMSTERDAM – BOSTON – HEIDELBERG – LONDON – NEW YORK – OXFORD
PARIS – SAN DIEGO – SAN FRANCISCO – SINGAPORE – SYDNEY – TOKYO

ELSEVIER B.V.
Radarweg 29
P.O. Box 211, 1000 AE Amsterdam
The Netherlands

ELSEVIER Inc.
525 B Street, Suite 1900
San Diego, CA 92101-4495
USA

ELSEVIER Ltd
The Boulevard, Langford Lane
Kidlington, Oxford OX5 1GB
UK

ELSEVIER Ltd
84 Theobalds Road
London WC1X 8RR
UK

First edition 2005

Library of Congress Cataloging in Publication Data
A catalog record is available from the Library of Congress.

British Library Cataloguing in Publication Data
A catalogue record is available from the British Library.

ISBN: 0-08-044643-4 (Paperback)

♾ The paper used in this publication meets the requirements of ANSI/NISO Z39.48-1992 (Permanence of Paper).
Printed in Italy

DEDICATED
TO

CLAIR PATTERSON
(1922–1995)

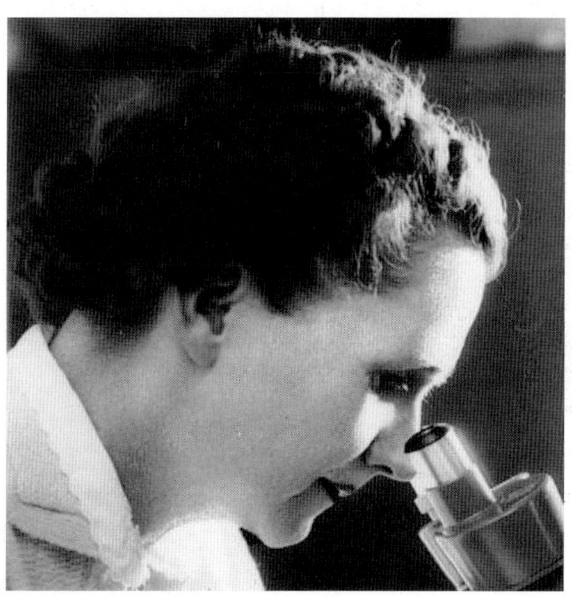

RACHEL CARSON
(1907–1964)

Photograph provided by California Institute of Technology
Photograph provided by The Lear/Carson Collection, Connecticut College

Contents

Executive Editors' Foreword

H. D. Holland
Harvard University, Cambridge, MA, USA
and
K. K. Turekian
Yale University, New Haven, CT, USA

Geochemistry has deep roots. Its beginnings can be traced back to antiquity, but many of the discoveries that are basic to the science were made between 1800 and 1910. The periodic table of elements was assembled, radioactivity was discovered, and the thermodynamics of heterogeneous systems was developed. The solar spectrum was used to determine the composition of the Sun. This information, together with chemical analyses of meteorites, provided an entry to a larger view of the universe.

During the first half of the twentieth century, a large number of scientists used a variety of methods to determine the major-element composition of the Earth's crust, and the geochemistries of many of the minor elements were defined by V. M. Goldschmidt and his associates using the then new technique of emission spectrography. V. I. Vernadsky founded biogeochemistry. The crystal structures of most minerals were determined by X-ray diffraction techniques. Isotope geochemistry was born, and age determinations based on radiometric techniques began to define the absolute geologic timescale. The intense scientific efforts during World War II yielded new analytical tools and a group of people who trained a new generation of geochemists at a number of universities. But the field grew slowly. In the 1950s, a few journals were able to report all of the important developments in trace-element geochemistry, isotopic geochronometry, the exploration of paleoclimatology and biogeochemistry with light stable isotopes, and studies of phase equilibria. At the meetings of the American Geophysical Union, geochemical sessions were few, none were concurrent, and they all ranged across the entire field.

Since then the developments in instrumentation and the increases in computing power have been spectacular. The education of geochemists has been broadened beyond the old, rather narrowly defined areas. Atmospheric and marine geochemistry have become integrated into solid Earth geochemistry; cosmochemistry and biogeochemistry have contributed greatly to our understanding of the history of our planet. The study of Earth has evolved into "Earth System Science," whose progress since the 1940s has been truly dramatic.

Major ocean expeditions have shown how and how fast the oceans mix; they have demonstrated the connections between the biologic pump, marine biology, physical oceanography, and marine sedimentation. The discovery of hydrothermal vents has shown how oceanography is related to economic geology. It has revealed formerly unknown oceanic biotas, and has clarified the factors that today control, and in the past have controlled the composition of seawater.

Seafloor spreading, continental drift and plate tectonics have permeated geochemistry. We finally understand the fate of sediments and oceanic crust in subduction zones, their burial and their exhumation. New experimental techniques at temperatures and pressures of the deep Earth interior have clarified the three-dimensional structure of the mantle and the generation of magmas.

Moon rocks, the treasure trove of photographs of the planets and their moons, and the successful search for planets in other solar systems have all revolutionized our understanding of Earth and the universe in which we are embedded.

Geochemistry has also been propelled into the arena of local, regional, and global anthropogenic problems. The discovery of the ozone hole came as a great, unpleasant surprise, an object lesson for optimists and a source of major new insights into the photochemistry and dynamics of the atmosphere. The rise of the CO_2 content of the atmosphere due to the burning of fossil fuels and deforestation has been and will continue to be at the center of the global change controversy, and will yield new insights into the coupling of atmospheric chemistry to the biosphere, the crust, and the oceans.

The rush of scientific progress in geochemistry since World War II has been matched by organizational innovations. The first issue of *Geochimica et Cosmochimica Acta* appeared in June 1950. The Geochemical Society was founded in 1955 and adopted *Geochimica et Cosmochimica Acta* as its official publication in 1957. The International Association of Geochemistry and Cosmochemistry was founded in 1966, and its journal, *Applied Geochemistry*, began publication in 1986. *Chemical Geology* became the journal of the European Association for Geochemistry.

The Goldschmidt Conferences were inaugurated in 1991 and have become large international meetings. Geochemistry has become a major force in the Geological Society of America and in the American Geophysical Union. Needless to say, medals and other awards now recognize outstanding achievements in geochemistry in a number of scientific societies.

During the phenomenal growth of the science since the end of World War II an admirable number of books on various aspects of geochemistry were published. Of these only three attempted to cover the whole field. The excellent *Geochemistry* by K. Rankama and Th.G. Sahama was published in 1950. V. M. Goldschmidt's book with the same title was started by the author in the 1940s. Sadly, his health suffered during the German occupation of his native Norway, and he died in England before the book was completed. Alex Muir and several of Goldschmidt's friends wrote the missing chapters of this classic volume, which was finally published in 1954.

Between 1969 and 1978 K. H. Wedepohl together with a board of editors (C. W. Correns, D. M. Shaw, K. K. Turekian and J. Zeman) and a large number of individual authors assembled the *Handbook of Geochemistry*. This and the other two major works on geochemistry begin with integrating chapters followed by chapters devoted to the geochemistry of one or a small group of elements. All three are now out of date, because major innovations in instrumentation and the expansion of the number of practitioners in the field have produced valuable sets of high-quality data, which have led to many new insights into fundamental geochemical problems.

At the Goldschmidt Conference at Harvard in 1999, Elsevier proposed to the Executive Editors that it was time to prepare a new, reasonably comprehensive, integrated summary of geochemistry. We decided to approach our task somewhat differently from our predecessors. We divided geochemistry into nine parts. As shown below, each part was assigned a volume, and a distinguished editor was chosen for each volume. A tenth volume was reserved for a comprehensive index:

(i) *Meteorites, Comets, and Planets*: Andrew M. Davis

(ii) *Geochemistry of the Mantle and Core*: Richard Carlson

(iii) *The Earth's Crust*: Roberta L. Rudnick

(iv) *Atmospheric Geochemistry*: Ralph F. Keeling

(v) *Freshwater Geochemistry, Weathering, and Soils*: James I. Drever

(vi) *The Oceans and Marine Geochemistry*: Harry Elderfield

(vii) *Sediments, Diagenesis, and Sedimentary Rocks*: Fred T. Mackenzie

(viii) *Biogeochemistry*: William H. Schlesinger

(ix) *Environmental Geochemistry*: Barbara Sherwood Lollar

(x) *Indexes*

The editor of each volume was asked to assemble a group of authors to write a series of chapters that together summarize the part of the field covered by the volume. The volume editors and chapter authors joined the team enthusiastically. Altogether there are 155 chapters and 9 introductory essays in the Treatise. Naming the work proved to be somewhat problematic. It is clearly not meant to be an encyclopedia. The titles *Comprehensive Geochemistry* and *Handbook of Geochemistry* were finally abandoned in favor of *Treatise on Geochemistry*.

The major features of the Treatise were shaped at a meeting in Edinburgh during a conference on Earth System Processes sponsored by the Geological Society of America and the Geological Society of London in June 2001. The fact that the Treatise is being published in 2003 is due to a great deal of hard work on the part of the editors, the authors, Mabel Peterson (the Managing Editor), Angela Greenwell (the former Head of Major Reference Works), Diana Calvert (Developmental Editor, Major Reference Works),

Bob Donaldson (Developmental Manager), Jerome Michalczyk and Rob Webb (Production Editors), and Friso Veenstra (Senior Publishing Editor). We extend our warm thanks to all of them. May their efforts be rewarded by a distinguished journey for the Treatise.

Finally, we would like to express our thanks to J. Laurence Kulp, our advisor as graduate students at Columbia University. He introduced us to the excitement of doing science and convinced us that all of the sciences are really subdivisions of geochemistry.

Contributors to Volume 9

T. A. Abrajano Jr.
Rensselaer Polytechnic Institute, Troy, NY, USA

P. Adriaens
The University of Michigan, Ann Arbor, MI, USA

H.-J. Albrechtsen
Technical University of Denmark, Lyngby, Denmark

A. L. Baehr
US Geological Survey, W. Trenton, NJ, USA

J. E. Barbash
United States Geological Survey, Tacoma, WA, USA

P. L. Bjerg
Technical University of Denmark, Lyngby, Denmark

D. W. Blowes
University of Waterloo, ON, Canada

C. R. Bryan
Sandia National Laboratories, Albuquerque, NM, USA

E. Callender
US Geological Survey, Westerly, RI, USA

T. Charette
University of Alberta, Edmonton, AB, Canada

T. H. Christensen
Technical University of Denmark, Lyngby, Denmark

I. Cozzarelli
US Geological Survey, Reston, VA, USA

I. M. Cozzarelli
US Geological Survey, Reston, VA, USA

W. F. Fitzgerald
University of Connecticut, CT, USA

F. M. Fordyce
British Geological Survey, Keyworth, Nottingham, UK

C. Gruden
The University of Michigan, Ann Arbor, MI, USA

J. L. Jambor
University of British Columbia, Vancouver, BC, Canada

D. G. Kinniburgh
British Geological Survey, Keyworth, Nottingham, UK

P. Kjeldsen
Technical University of Denmark, Lyngby, Denmark

B. A. Klinck
British Geological Survey, Keyworth, Nottingham, UK

C. H. Lamborg
Woods Hole Oceanographic Institution, MA, USA

M. L. McCormick
Hamilton College, Clinton, NY, USA

S. A. Norton
University of Maine, Orono, ME, USA

V. O'Malley
Enterprise Ireland, Glasnevin, Republic of Ireland

J. A. Plant
British Geological Survey, Keyworth, Nottingham, UK

G. S. Plumlee
US Geological Survey, Denver, CO, USA

E. E. Prepas
Lakehead University, Thunder Bay, ON, Canada

C. J. Ptacek
University of Waterloo, ON, Canada

M. D. Siegel
Sandia National Laboratories, Albuquerque, NM, USA

S. Sillman
University of Michigan, Ann Arbor, MI, USA

P. L. Smedley
British Geological Survey, Keyworth, Nottingham, UK

A. L. Teel
Washington State University, Pullman, WA, USA

A. Vengosh
Ben Gurion University of the Negev, Beer Sheva, Israel

J. Veselý
Czech Geologic Survey, Prague, Czech Republic

R. J. Watts
Washington State University, Pullman, WA, USA

C. G. Weisener
University of Waterloo, ON, Canada

B. Yan
Rensselaer Polytechnic Institute, Troy, NY, USA

T. L. Ziegler
US Geological Survey, Denver, CO, USA

Volume Editor's Introduction

B. S. Lollar
University of Toronto, ON, Canada

Several recurring themes in the evolution of geochemistry during the twentieth century are apparent as one scans the Introductions and Tables of Content of Volumes 1–8 of the *Treatise on Geochemistry*. These are themes that echo throughout Volume 9 as well, and include what the Executive Editors refer to as the broadening of geochemistry beyond older more narrowly defined areas into atmospheric geochemistry, aqueous geochemistry, cosmochemistry and biogeochemistry to name just a few. This expansion presented a major challenge to all the editors—the challenge of designing a *Treatise* that moves beyond the traditional mode of examining one or a small group of elements per chapter as was done in the earlier *Handbook of Geochemistry* (Wedepohl, 1970), to a model in which chapters reflect the increasingly interdisciplinary nature of geochemistry and focus on Earth systems spanning the lithosphere, atmosphere, hydrosphere, and biosphere. During the early organizational phase of the project, the Executive Editors and Volume Editors gave careful attention to designing the scope and dimension of each volume, and within the volumes, for each chapter, to ensure both comprehensive coverage with a minimum of overlap.

This challenge was particularly relevant for Volume 9 of the *Treatise*, with its focus on environmental geochemistry. How best to define the scope of environmental geochemistry vis-à-vis other closely related topics in Volumes 1–8? Two guiding principles were selected. First, while Volume 9 focused on the impact of environmental geochemistry on both human health and on the broader concept of the health of the environment, it was clearly important to incorporate both the impact of natural geochemical processes and of anthropogenic perturbations of natural systems. The impact of natural processes and their importance relative to anthropogenic effects is a critical component of the chapters dealing with the metalloids (Chapter 9.02), heavy metals (Chapter 9.03), and mercury (Chapter 9.04), as well as minerals, dusts, and Earth materials (Chapter 9.07), eutrophication (Chapter 9.08), salinization and saline environments (Chapter 9.09), etc. In contrast, the chapters dealing with acid mine drainage (Chapter 9.05), radioactive elements (Chapter 9.06), acidification (Chapter 9.10), tropospheric ozone and photochemical smog (Chapter 9.11), and the chapters dealing with organic contaminants such as the volatile and high-molecular-weight fuel hydrocarbons (Chapters 9.12, 9.13, and 9.16); halogenated hydrocarbons (Chapter 9.14), and pesticides (Chapter 9.15) focus more specifically on anthropogenic impacts. The second guiding principle was the decision to focus Volume 9 more closely on geochemical impacts at the local and regional than on the global scale. The chapters in Volume 9 were carefully cross-referenced to related topics in other volumes, but global-scale processes such as stratospheric ozone depletion and global climate change are addressed in other volumes in the *Treatise*.

Volume 9 progresses from topics related to inorganic constituents (Chapters 9.02–9.09), to an increasing emphasis on the interface between inorganic and organic geochemistry in Chapters 9.10 and 9.11, to chapters firmly focused on organic environmental pollutants in Chapters 9.12–9.16. In each chapter, the fundamental geochemical processes controlling the source, transport, and transformation of pollutants in the environment are accompanied by an assessment of their impact on the environment and on human health. Watts (Chapter 9.01) provides an overview of the current policy and regulatory framework and of models for evaluating impacts in terms of risk, toxicity, and exposure assessment. This perspective is fitting, as the two scientists to whom this volume is dedicated personified the fusion of fundamental scientific investigation with informed action in the sphere of social policy and regulation.

Clair C. Patterson (1922–1995) was a pioneer across a broad range of geochemical research. He is perhaps best known for his work on the isotopic composition and concentration of lead in terrestrial materials and meteorites to provide the first reliable date for the age of the earth—4.55 Gyr (Patterson, 1956). In the course of his research on expanding models of lead isotope geochemistry, his work on the mass balance of natural background versus anthropogenic lead inputs to the oceans transformed the prevailing thinking about the magnitude of anthropogenic lead pollution (Patterson and Chow, 1962). His landmark work showed that levels of anthropogenic lead above background were orders of magnitude higher than had previously been calculated, because the degree of blank contamination had been seriously underestimated (The National Academies Press Biographical Memoirs, 1998; Casanova, 1998).

While pursuing a very different scientific career path through her work in aquatic biology in the US Fish and Wildlife Service, Rachel Carson (1907–1964) displayed a similar breadth of scientific insight and vision by recognizing that the impact of organic pesticides, most notably DDT, on wildlife and human health through bioaccumulation of pesticides in the food chain had been seriously underestimated. Published in the same year as Patterson's groundbreaking paper on lead, Carson's (1962) book *Silent Spring* had a similar galvanizing impact on environmental geochemistry, government policy, and public perception. Not content to simply publish their findings, both Patterson and Carson embraced the role of challenging science, industry, and government to address the implications of their research in environmental geochemistry through public advocacy and testimony before US government committees and regulatory bodies. Carson's advocacy was a catalyst for the establishment of the US Environmental Protection Agency (EPA) in 1970 and the banning of DDT (US Environmental Protection Agency, 2000; Matthiessen, 2003). Patterson's testimony before the Subcommittee on Air and Water Pollution, the EPA, and the Bureau of Foods is credited with the eventual elimination of lead pipes, lead-soldered cans and lead in gasoline (Casanova, 1998).

As is often the case with visionaries, both Patterson and Carson persevered in their scientific efforts and their public policy efforts in the face of sometimes outspoken criticism. In doing so, they proved themselves pioneers of yet another nature, furnishing inspiring examples of the role of scientific researchers as agents for education and change in society. Their innovative role as leaders in scientific engagement in society is perhaps as important a legacy as their achievements in environmental geochemistry.

In closing, I would like to thank all the authors in Volume 9 for the outstanding commitment they have made to the *Treatise on Geochemistry*. Each author rose to the challenge of integrating the fundamental scientific principles and emerging breakthroughs in their area of expertise with a broad perspective on the past and future impact of environmental geochemistry on human health and the environment. The time, intellectual input, and professionalism they brought to the work are deeply appreciated.

REFERENCES

Carson R. (1962) *Silent Spring*. Fawcett Crest Books, New York, 304p.

Casanova I. (1998) Clair C. Patterson (1922–1995), discoverer of the age of the Earth. *Int. Microbiol.* **1**, 231–232.

Matthiessen P. (2003) *Rachel Carson*, TIME100: Scientists & Thinkers (http://www.time.com/time/time100/scientist/profile/crason03.html).

Patterson C. C. (1956) Age of meteorites and the Earth. *Geochim. Cosmochim. Acta* **10**, 230–237.

Patterson C. C. and Chow T. J. (1962) The occurrence and significance of lead isotopes in pelagic sediments. *Geochim. Cosmochim. Acta* **26**, 263–308.

The National Academies Press Biographical Memoirs (1998) National Academy of Sciences, vol. 74, pp. 266–287 (http://books.nap.edu/books/0309060869/html/266.html).

US Environmental Protection Agency (2000) *People & Profiles (2000)*. The Power of One—Rachel Carbon, Silent Spring.

Wedepohl K. H. (ed.) (1970) *Handbook of Geochemistry* Springer, Berlin.

9.01
Groundwater and Air Contamination: Risk, Toxicity, Exposure Assessment, Policy, and Regulation

R. J. Watts and A. L. Teel

Washington State University, Pullman, WA, USA

9.01.1 INTRODUCTION

The improper disposal of hazardous wastes and subsequent contamination of surface and groundwaters has exposed the public and ecosystems to toxic chemicals that have detrimental consequences. The cost of cleaning up the thousands of hazardous waste sites throughout the world is daunting, and the effort to do so is economically impractical. As a result, some level of contamination will always remain, both locally and globally. The presence of a residual level of contamination carries with it the probability of negative impacts on the world's population; e.g., enhanced risk of cancer or the onset of neurological disorders. Risk is the probability of such events. Risk assessments are routinely performed at contaminated sites and in areas of widespread environmental contamination, such as an entire aquifer, as a means of quantifying the potential threats to public health and to ecosystems.

9.01.2 PRINCIPLES, DEFINITIONS, AND PERSPECTIVES OF HAZARDOUS WASTE RISK ASSESSMENTS

Risk assessment is the attempt to measure the potential for harm. It is a process that aids in site assessments, determining end points in remediation, and evaluating the danger of engaging in potentially hazardous acts such as drinking contaminated groundwater. The use of risk assessment has become commonplace since the promulgation of Comprehensive Environmental Response, Compensation, and Liability Act (CERCLA), or Superfund, and has been important in assessing hazards such as the occurrence of earthquakes, hurricanes, and floods.

Hazardous waste risk assessments are systematic and quantitative; there is a well-established algorithm for conducting the process. However, a significant amount of uncertainty and data gaps are inherent in making risk assessments of contaminated sites and contaminated groundwaters; therefore, the quantitative methodologies are constrained by uncertainty limits. Furthermore, input data for many of the calculations (e.g., the volume of groundwater ingested per individual per day) may be difficult to obtain, or totally unavailable. Because of this, risk assessment teams must have sufficient risk assessment experience to accurately evaluate the inevitable data gaps.

9.01.2.1 Definitions of Hazard and Risk

Risk is the probability of harm or loss and can be considered to be a product of the probability and the severity of specific consequences. Risk, as it relates to hazardous wastes and groundwater contamination, may be defined as the chance that humans or other organisms will sustain adverse effects from exposure to these environmental hazards. Risk is inherent in the life of all organisms—humans, animals, and plants. Tornadoes, landslides, hurricanes, earthquakes, and other natural disasters carry a risk of injury or death to any living thing in their path. Similarly, human-caused risks such as automobile accidents, plane crashes, and nuclear disasters occur with varying levels of severity.

Specific definitions apply to different aspects of risk assessment in hazardous waste management. *Background risk* is the risk to which a population is normally exposed, excluding risks from hazardous chemicals or groundwater contamination. *Incremental risk* is the additional risk caused by hazardous chemicals or the contaminated groundwater. *Total risk* is the background risk plus the incremental risk. For example, the background risk of cancer for the average US citizen is one in four, or 0.25 (Guidotti, 1988). The target incremental risk at Superfund sites for carcinogen exposure to the "most exposed individual," proposed by the Environmental Protection Agency (EPA), is 1×10^{-6}. The target for total lifetime risk for exposure to carcinogenic contaminants at Superfund sites is then 0.25 plus 1×10^{-6}. Analysis of the total risk often involves critical evaluation of the quantitative risk assessment itself, including analysis of the uncertainties of the assessment and the acceptable risk of the hazardous waste.

Hazard is different from risk; it is a descriptive term that characterizes the intrinsic ability of an event or a substance to cause harm. Hazard is one source of risk and is a function of the persistence, mobility, and toxicity of the contaminants.

9.01.2.2 Typical Risks Encountered—Natural and Anthropogenic

Risk assessment and risk management are used widely in scientific, engineering, medical, economic, and even sociological evaluations. Risk from natural hazards, such as storms, floods, hurricanes, tornadoes, and even insect and animal attacks have been studied in detail to evaluate the potential for disaster and economic effects. Structural engineers often evaluate the risk of bridge failures, building failures during earthquakes, and other damage that can result from natural disasters and the aging of structural materials. Epidemiologists often determine risk from disease outbreaks, and other health effects. Many insurance companies and investment firms have risk management departments that use risk models for quantifying economic risks.

9.01.2.3 Risks Associated with Contaminated Sites and Groundwater

More than 600 chemicals have been discovered at Superfund sites. The contaminants that are found most frequently at National Priorities List (NPL) sites are lead (43% of sites), trichloroethylene (42%), chromium (35%), benzene (34%), perchloroethylene (28%), arsenic (28%), and toluene (27%) (ATSDR, 1989). Although chemicals regulated as hazardous wastes are often classified as corrosive, flammable, explosive, or toxic, toxicity is the most common concern in regard to groundwater contaminants. Toxicity, in turn, is classified as acute or chronic. Acute toxicity results from short-term exposure to relatively high contaminant dosages. Chronic toxicity occurs as the result of drinking low contaminant concentrations over decades. The most common concern resulting from the improper disposal of hazardous chemicals and subsequent groundwater contamination is chronic toxicity and resulting effects such as cancer and neurological diseases.

9.01.3 REGULATORY AND POLICY BASIS FOR RISK ASSESSMENT

9.01.3.1 Examples of Contaminated Sites and Potential Risk Exposure Pathways

Before strict regulatory measures were passed that prevented the improper land disposal of hazardous wastes, numerous disposal practices were used that produced thousands of contaminated sites requiring cleanup activities that have lasted for decades. A common disposal practice was to spread waste liquids, especially lubricating oils and other petroleum residues, on soils and unpaved roads. Many industries disposed of waste chemicals by placing them in unlined soil pits and lagoons. Workers simply dug pits into which wastes were poured; the wastes then disappeared by seeping into and through the soil. Many large industries constructed landfills that were used primarily for the land disposal of industrial by-products, such as building materials or out-of-date equipment. Unfortunately, these landfills were also used for the disposal of chemical wastes. Sanitary landfills that were designed to accept newspapers, cans, bottles, and other household wastes also received waste petroleum products, solvents, pesticides, transformer oils, etc. Though liquid hazardous wastes were often disposed of in drums, in some cases they were poured directly into the landfills. Since these sanitary landfills were unlined, the wastes often migrated to surface and groundwater. Waste chemicals stored in 55-gallon drums were often placed on loading docks, concrete pads, or other temporary storage areas awaiting disposal. The drums often accumulated, sometimes to the point where thousands were stored and stacked. Drums stored in this manner eventually corrode and leak, resulting in chemical releases into the underlying soil and groundwater. Underground storage tanks (USTs) that had been buried for decades began to leak in the 1970s resulting in the saturation of soil with leaking chemicals, and the eventual contamination of groundwater.

As a result of these improper disposal practices, sites contaminated with hazardous wastes came to public attention throughout the 1970s and 1980s. The effects of improper hazardous waste disposal may persist for decades or even centuries when the contaminants have low degradation rates and migrate slowly through the subsurface.

The US EPA summarized the results of studies of potential pathways for the release of chemicals from Superfund sites (US EPA, 1988). Migration to groundwater was cited as the primary pathway of contaminants at these hazardous waste sites, a trend confirmed by the data in Table 1; 37% of sites involved releases to groundwater and 23% were responsible for releases to both groundwater and surface water. Other studies document the potential hazards of hazardous waste disposal. The EPA, in a survey of 466 public water supply wells, found that one or more volatile organic

Table 1 Pathways of releases of hazardous chemicals from NPL landfills.

Observed releases from NPL landfills to water and air	*Percent*
Groundwater only	37
Groundwater and surface water	23
None observed	15
Surface water only	9
Groundwater, surface water, air	8
Groundwater and air	3
Surface water and air	3
Air only	2

Source: US EPA (1988).

compounds (VOCs) were detected in 16.8% of small water systems and 28% of large water systems. The VOCs found most often were trichloroethylene and perchloroethylene (Westrick *et al.*, 1983). A survey of 7,000 wells conducted in California from 1984 through 1988 showed that 1,500 contained detectable concentrations of organic chemicals and 400 contained these in concentrations exceeding the state's regulatory requirement or the maximum contaminant level (MCL) prescribed by the Safe Drinking Water Act (SDWA) (MacKay and Smith, 1990). The most common chemicals detected were perchloroethylene, trichloroethylene, chloroform, 1,1,1-trichloroethane, and carbon tetrachloride. The impact of hazardous wastes is serious because of our dependence on groundwater resources; 48% of the US population as a whole receives its drinking water from groundwater; 95% of the rural US population relies on groundwater for domestic use (Patrick, 1983).

9.01.3.2 Risk-based Nature of CERCLA

CERCLA was passed in 1980 to provide a federally supervised system for the mitigation of chronic environmental damage, particularly the cleanup of sites contaminated with hazardous waste. In 1986, CERCLA was amended by the Superfund Amendments and Reauthorization Act (SARA). Each Superfund site has been assessed, characterized, and prioritized based on risk. Potential sites are first screened using a preliminary assessment (PA); sites deemed a significant threat are then evaluated using a hazard ranking system (HRS) to measure the risk of the site relative to that of other potential sites. The most hazardous sites are then placed on the NPL in the order of their potential risk.

Hundreds of chemicals are regulated under CERCLA; they are classified as (i) hazardous substances and (ii) pollutants or contaminants. The definition of a hazardous substance under CERCLA is broad and is based on other environmental regulations. A CERCLA hazardous substance does not need to be a waste or waste material. It can be a commercial chemical, formulation, or product. A CERCLA *hazardous substance* is defined as any chemical regulated under the Clean Water Act, the Clean Air Act, the Toxic Substances Control Act (TSCA), or the Resource Conservation and Recovery Act (RCRA). However, two materials that are excluded from the hazardous substances list are petroleum and natural gas. A CERCLA *pollutant or contaminant* is defined as any other chemical or substance that "will or may reasonably be anticipated to cause harmful effects to human or ecological health." Together, these two categories encompass a broad range of chemicals.

A primary directive of CERCLA is the protection of public health. Because the hazards that exist at Superfund sites tend to be quite variable, it has not been possible to establish specific cleanup criteria for the hazardous substances regulated under CERCLA; potential human health effects must be evaluated by quantitative risk assessment on a site-by-site basis. Each Superfund site is assessed individually to determine *how clean is clean*. The rationale is that the hazard of a contaminant is a function of its potential to reach a receptor (e.g., groundwater, population) and the potential harm to the exposed receptor. The ability of a contaminant to migrate, its potential to degrade, and its distance to a receptor of concern (i.e., the risk), all are site-specific. Only on the basis of such individualized risk assessment is it possible to achieve efficient and cost-effective cleanup of the thousands of hazardous waste sites throughout the US.

9.01.3.3 Risk-based Corrective Actions

Throughout the 1980s, tens of thousands of USTs began to leak due to corrosion. By October 1994, more than 270,000 leaking USTs had been discovered in the United States. Until that time, absolute standards for various petroleum indicators were used as cleanup standards. Although cleanup levels varied from state to state, the most common standard was a soil concentration of 100 mg kg^{-1} of total petroleum hydrocarbons (TPH). Such absolute guidelines were thought to streamline corrective action at UST sites, because minimal exposure and toxicity assessments were required. However, as UST cleanups proceeded, it became apparent to owners and operators of USTs, as well as regulators, that site cleanup to an absolute standard sometimes displaced thousands of cubic meters of soil that posed no risk to human health or to the environment.

Although absolute cleanup standards have been used for a number of contamination problems in addition to those related to leaking USTs, the trend in managing these releases has been toward risk-based decision making. State and local agencies are implementing a process that is based on risk and exposure assessments to evaluate the extent and urgency of needed cleanup actions. The risk-based decision process for the UST corrective action process is called risk-based corrective action (RBCA). The first approach to RBCAs was developed by the American Society of Testing and Materials (ASTM); this has since been implemented by the EPA and state and local agencies.

9.01.3.4 Use of Applicable or Relevant and Appropriate Requirements

The passage of SARA in 1986 resulted in the development of *applicable or relevant and appropriate requirements* (ARARs), which are used as *de facto* values for cleanup end points. The ARARs are usually based on other environmental laws, such as the SDWA or the RCRA.

One of the more common ARARs is the use of SDWA MCLs as an action level for contaminated groundwater. As part of the SDWA, MCLs were established for many contaminants to protect the health of the public over a lifetime of drinking water. The MCLs of many of these common hazardous chemicals are in the low $\mu g\ L^{-1}$ range; this makes analytical sensitivity and quality control a necessity in the chemical analyses of drinking water. The MCLs are based on MCL goals (MCLGs), which are nonenforceable goals based on extremely low risk. The EPA has been given the directive to set MCLs as close to MCLGs as possible. In many cases, MCLs do not correspond to the MCLG level for a cancer risk of 1×10^{-6} (based on an intake of 2 L (0.53 gal) of water per day). Instead, they are set at a pragmatic level dictated by water treatment technologies and analytical detection limits (Travis *et al.*, 1987). Many state and local regulatory authorities use MCLs as *de facto* cleanup criteria for contaminated groundwater.

9.01.3.5 Limited Uses of Absolute Standards

Universal across-the-board cleanup criteria are not commonly used as end points for soil and groundwater cleanup, because of the wide range of risks found at these sites. However, two classes of contaminants have been subject to universal action levels for cleanup: petroleum and polychlorinated biphenyls (PCBs). Most petroleum hydrocarbon action levels are regulated by state and local agencies; the parameters used and their corresponding action levels vary widely from state to state. The specific petroleum parameters that are regulated include total petroleum hydrocarbon–gasoline fraction (TPH-G); total petroleum hydrocarbon–diesel component (TPH-D); benzene, toluene, ethylbenzene, and xylenes (BTEX); and benzene alone. Common regulatory levels include concentrations in soil of $100\ mg\ kg^{-1}$ for TPH-G, $200\ mg\ kg^{-1}$ for TPH-D, and $1\ mg\ kg^{-1}$ for benzene. A state-by-state listing of petroleum standards has been reported by the Association for Environmental Health and Science (Nascarella *et al.*, 2002).

In the United States, PCBs are regulated under the TSCA. The universal standard for total PCBs (i.e., the total of all 209 congeners) in commercial products such as electrical transformers is $50\ mg\ kg^{-1}$. The same regulatory standard is applied to soils and other media contaminated by PCB spills and other environmental releases of PCBs.

9.01.4 THE RISK ASSESSMENT PROCESS

9.01.4.1 Sources, Pathways, and Receptors: The Fundamental Algorithm for Risk Assessments

Hazardous waste problems are frequently generated by mixtures of complex wastes that have been disposed of on land and that have migrated through the subsurface. One approach to assessing the risks of contaminated sites has been to divide the problem into three elements: sources, pathways, and receptors (Watts, 1998) as noted in Table 2. The first step in assessing the risk at a hazardous waste site is to identify the waste components at the *source*, including their concentrations and physical properties such as density, water solubility, and flash point. After the source has been characterized, the *pathways* of the hazardous chemicals are analyzed by quantifying the rates at which the

Table 2 Elements of sources, pathways, and receptors algorithm used in hazardous waste risk assessments.

Sources
- Time since environmental release
- Contaminants potentially present
- Sampling
  - Contaminant concentrations
  - Contaminant locations
- Contaminant properties
  - Water solubility
  - Octanol–water partition coefficient
  - Vapor pressure
  - Henry's law constant

Pathways
- Rate of release from the source
  - Air
  - Groundwater
- Atmospheric transport
  - Wind speed
  - Dispersion
- Groundwater transport
  - Advection–dispersion
  - Sorption
- Distance to receptors

Receptors
- Characteristics of receptor population
- Acute toxicity
- Chronic toxicity
  - Noncarcinogenic
  - Carcinogenic

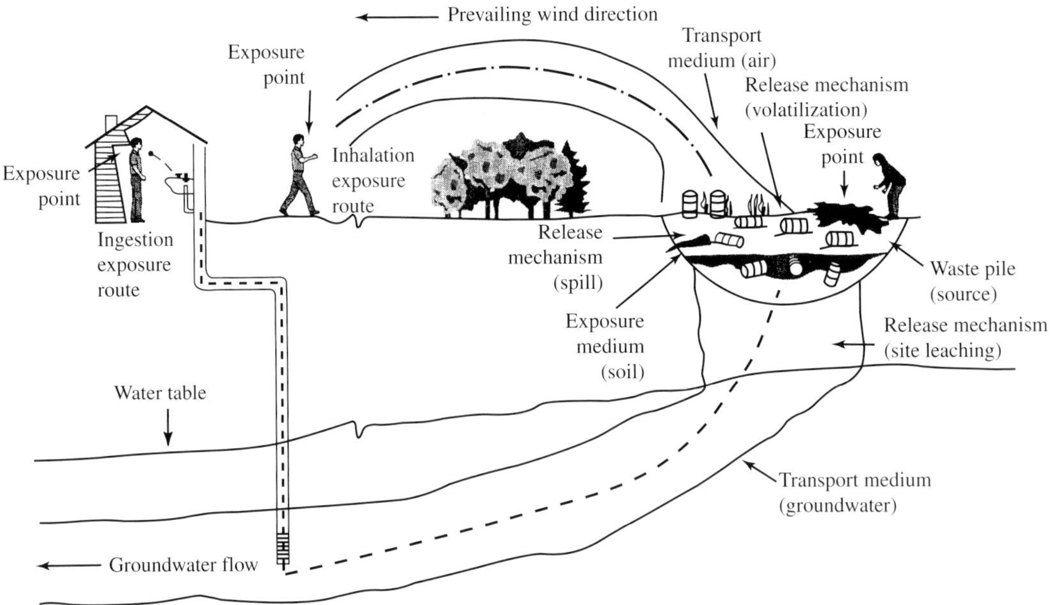

Figure 1 Common pathways for contaminant migration from a hazardous waste source.

waste compounds volatilize, degrade, and migrate from the source (Figure 1). Pathway analysis is built on source information—the identity and nature of the source chemicals must be known in order to quantify their potential to migrate, degrade, or be treated. Pathway analyses may show that the contaminant will be transformed within weeks and cease to be a problem, or they may show that the contaminant persists in the environment and will reach a receptor (such as a drinking water well) long before it is degraded. Finally, if the pathway analysis shows that the contaminant will come into contact with *receptors* (humans, endangered species, etc.), the hazard must be assessed with the aid of toxicological data.

9.01.4.2 The Four-step Risk Assessment Process

The National Academy of Sciences and the EPA have defined four steps in the assessment of risk from hazardous wastes (US EPA, 1989a; NAS, 1983):

(i) *hazard identification* (source analysis; investigating the chemicals present at the site and their characteristics),

(ii) *exposure assessment* (pathway analysis; estimating the potential transport of the chemicals to receptors and levels of intake),

(iii) *toxicity assessment* including the determination of numerical indices of toxicity (receptor analysis), and

(iv) *risk characterization* involving the determination of a number that expresses risk quantitatively, such as one in one hundred (0.01) or one in one million (1×10^{-6}).

9.01.5 HAZARD IDENTIFICATION

9.01.5.1 Determining Contaminant Identity, Concentration, and Distribution

The first step in risk assessment is the evaluation of the identity and properties of the contaminants, their potential for release from the source, and their rate of release. Specific tasks in hazard identification include sampling, installation of monitoring wells, chemical analysis, creating quality assurance/quality control plans, and data analysis. Sampling is usually conducted so as to assess both the identity and the concentration and distribution of contaminants at the source. Based on the sampling and their analyses, a data set is developed (subdivided by type of media–soil, sludge, standing surface water, etc.) for input into exposure assessment models. Hazard identification is usually focused more on contaminants of concern, i.e., those presenting the highest hazard. For example, more detailed source characterization is required for 2,3,7,8-TCDD than for dodecane.

9.01.5.2 Contaminant Surrogate Analysis

Most hazardous waste sites and facilities contain tens to hundreds of chemicals. The process of collecting data can therefore be overwhelming and is sometimes unrealistic. However, risk at a hazardous waste site is often dictated by a few contaminants or one or two pathways; these are the risk drivers. The most common practice in assessing the risks of Superfund sites is to screen the contaminants and pathways using surrogate analysis. The use of surrogates reduces the numbers used in pathway studies and streamlines the computing efforts significantly.

The two most important source characteristics used in screening a large number of contaminants at a hazardous waste site or facility are their concentrations and toxicities. Screening for surrogates has been called *concentration–toxicity screening* by the EPA (US EPA, 1989a). A risk factor or *chemical score* is determined for each chemical based on its concentration and toxicity, and the scores are reported by medium. The units for the chemical score (R) are a function of the medium that is evaluated. The units do not usually matter, as long as they are consistent. If toxicity data are available for both oral and inhalation routes, the most conservative number (i.e., the most toxic) is used in concentration–toxicity screening. After Rs are determined for all of the chemicals, surrogates are chosen that account for 99% of the potential risk.

9.01.6 EXPOSURE ASSESSMENT

Exposure is the contact of an organism with a toxic substance; *exposure assessment* is the estimation of the magnitude, frequency, duration, and route of exposure (Patton, 1993). The primary tasks of exposure assessments include (i) identifying the populations that may be exposed, (ii) identifying the possible exposure pathways, (iii) estimating the concentrations to which the populations are exposed, and (iv) estimating chemical intakes.

9.01.6.1 Potential Exposure Pathways

The rate at which a contaminant is released from a source is a critical parameter in quantifying risk at contaminated sites. Picture two different extremes of release from the source: In the first case, 1,000 kg of benzene (a confirmed human carcinogen) exists in a lined surface impoundment and is volatilizing at a rate of 0.5 kg d^{-1} and is potentially inhaled by a nearby population. In the second case, the same mass of benzene is strongly sorbed to a clay soil rich in organic matter 1 m below the soil surface, from which the volatilization rate is 10^{-7} kg d^{-1}. Obviously, the risk of hazardous chemicals depends on the rate of contaminant release.

Quantifying contaminant release rate is an integral part of source release assessment. Because release rates may vary over space and time, a spatial- and temporal-dependent emission rate is often obtained. Most analytical contaminant transport models are based on simple or constant spatial and temporal release rates; however, actual release rates are often very complex, varying significantly with time and space. Nevertheless, release rates are often approximated using constant rates in space and time.

When the effects of time variations in release rates are included, pulse (or instantaneous) and continuous (plume) emissions are the two most common time-variant inputs in transport models. The classic example of a pulse release is a hazardous waste spill. The steady release of contaminants into groundwater from a subsurface contaminant and the continuous release of volatile solvents from an air-stripping tower are examples of plume emissions.

Environmental releases of hazardous waste from contaminated sites can result in transport through several media. The most common pathways include (i) transport through the subsurface to groundwater and (ii) atmospheric transport after release into the air. Other less common pathways after release include surface waters, and plant and animal uptake.

9.01.6.2 Estimating Exposure Concentrations

The goal of exposure assessment is to determine the concentration of contaminant to which the receptor organism is exposed. This procedure involves two steps: (i) determining the concentration of the contaminant to which the population is exposed, and (ii) quantifying contaminant intake at the point of exposure.

In some cases, direct measurements of contaminant exposure may be made, such as in the assessment of the steady-state release of an established hazardous waste source. Groundwater monitoring wells or air-sampling devices can be used to determine current exposure concentrations for exposed populations. Commonly, however, air, surface water, and groundwater sampling is neither a logical nor a practical choice. Typical situations in which sampling is not feasible include (i) the evaluation of future exposure and risk to potentially exposed populations and (ii) the potential risk from an event that has yet to occur (e.g. a hazardous waste spill).

Contaminant transport models are most commonly used for determining the concentrations of contaminants that reach the exposed population. Such models for the atmosphere, surface waters, and the subsurface have developed substantially; their use has been described by Anderson and Woessner (1992), Watts (1998), and Zheng and Bennett (2002).

9.01.6.3 Identifying Potentially Exposed Populations

An important part of exposure assessment is the identification of the population that may be exposed to hazardous chemicals. Characterizing potentially exposed populations is often an intricate and difficult process (Van Leeuwen and Hermens, 1995). It involves visiting sites, screening populations near the sites, evaluating land use and housing maps, and surveying recreational data. Prospective land use patterns must be considered as hazardous chemical exposures may occur in the future. The daily and seasonal activity patterns of the population must also be evaluated, though these are often difficult to quantify. The demographics of the receptor population may include the total population of a residential area, the identification of sensitive populations (e.g., children, elderly, infirm), or site workers and personnel. In complex systems, the aid of a geographer and geographic information systems (GIS) may be necessary. In the end, however, identifying potential receptor populations is often a qualitative exercise requiring the use of professional judgment.

9.01.6.4 Estimating Chemical Intake

After the receptor population is identified, data about population behavior are gathered. These include duration of exposure, frequency of exposure, mean body weight, and probable future demographics such as population increases or decreases. Intake (I) is estimated using equations consisting of three types of variables: (i) contaminant-related (exposure concentration), (ii) exposed population (contact rate, exposure frequency, duration, and body weight), and (iii) assessment-determined (averaging time). Exposure concentration (C) is the arithmetic mean of the contaminant concentration over the period of exposure. The contact rate (CR) is the amount of contaminant encountered per unit time. The time of exposure is estimated using the site- and population-specific terms, exposure frequency (EF) and exposure duration (ED). Conservative assumptions are normally used for each variable, such as the 95th percentile value for exposure time (US EPA, 1989a). The body weight (BW) value is the average weight of a member of the receptor population over the exposure period. If children are the primary individuals exposed, this value should be the average child's body weight; a BW of 70 kg is commonly used for adult exposures. The averaging time (AT) is a value that is based on the mechanism of toxicity. The most commonly used averaging time for carcinogens is $70\ \text{yr} \times 365\ \text{d yr}^{-1}$; a pathway-specific period of exposure is generally used for noncarcinogenic effects (i.e., the $\text{ED} \times 365\ \text{d yr}^{-1}$).

Contaminant intake may be estimated by using the mean exposure concentration of contaminants in conjunction with the exposed population variables and the assessment-determined variables. The general equation for chemical intake is

$$I = \frac{C \times \text{CR} \times \text{EFD}}{\text{BW}} \times \frac{1}{\text{AT}} \qquad (1)$$

where I is the intake (the amount of chemical at the exchange boundary) ($\text{mg kg}^{-1}\ \text{d}^{-1}$); C the average exposure concentration over the period (e.g., mg L^{-1} for water or mg m^{-3} for air); CR the contact rate, the amount of contaminated medium contacted per unit time (L d^{-1} or $\text{m}^3\ \text{d}^{-1}$); and EFD the exposure frequency and duration. EFD is usually divided into two terms: EF the exposure frequency (d yr^{-1}), and ED, the exposure duration (yr). BW is average body mass over the exposure period (kg) and AT the averaging time (the time period over which the exposure is averaged; d).

Accurate intake data are sometimes difficult to obtain; because exposure frequency and duration vary among individuals, these variables must often be estimated using professional judgment. Daily contamination intake rates from air, contaminated food, drinking water, and through dermal exposure to water while swimming may be estimated by other equations reported by the US EPA (1989b). Drinking contaminated water and breathing contaminated air are two of the most common exposure routes. The intake for ingestion of waterborne chemicals is estimated by the following equation:

$$\text{Intake}\ (\text{mg kg}^{-1}\ \text{d}^{-1}) = \frac{\text{CW} \times \text{IR} \times \text{EF} \times \text{ED}}{\text{BW} \times \text{AT}} \qquad (2)$$

where CW is the chemical concentration in water (mg L^{-1}), IR is the ingestion rate (L d^{-1}), and EF, ED, BW, and AT have the same meaning as in Equation (1).

Some of the values used for the variables in Equation (2) include:

- CW: site-specific measured or modeled value;
- IR: $2\ \text{L d}^{-1}$ (adult, 90th percentile) (US EPA, 1989b), $1.4\ \text{L d}^{-1}$ (adult, average) (US EPA, 1989b);
- EF: pathway-specific value (dependent on frequency of exposure-related activities);
- ED: 70 yr (lifetime; by convention), 30 yr (national upper-bound time (90th percentile) at one residence) (US EPA, 1989b), and 9 yr (national median time (50th percentile) at one residence) (US EPA, 1989b);
- BW: 70 kg (adult, average) (US EPA, 1989b), age-specific values (US EPA, 1985, 1989b); and

- AT: pathway-specific period of exposure for noncarcinogenic effects (i.e., ED × 365 d yr^{-1}), and 70 year lifetime for carcinogenic effects (i.e., 70 yr × 365 d yr^{-1}).

The intake for inhalation of airborne contaminants is

$$\text{Intake (mg kg}^{-1}\text{ d}^{-1}) = \frac{CA \times IR \times ET \times EF \times ED}{BW \times AT} \quad (3)$$

where CA is contaminant concentration in air (mg m^{-3}), IR the inhalation rate (m^3 h^{-1}), ET the exposure time (h d^{-1}), and other symbols have the same meaning as before.

Some of the values used in Equation (3) include:

- CA: site-specific measured or modeled value;
- IR: 30 m^3 d^{-1} (adult, suggested upper bound value) (US EPA, 1989b); 20 m^3 d^{-1} (adult, average) (US EPA, 1989b);
- ET: pathway-specific values (dependent on duration of exposure-related activities);
- EF: pathway-specific value (dependent on frequency of exposure-related activities);
- ED: 70 yr (lifetime; by convention); 30 yr (national upper-bound time (90th percentile) at one residence) (US EPA, 1985); 9 yr (national median time (50th percentile) at on residence) (US EPA, 1989b);
- BW: 70 kg (adult, average) (US EPA, 1989b); age-specific values (US EPA, 1985, 1989b);
- AT: pathway-specific period of exposure for noncarcinogenic effects (i.e., ED × 365 d yr^{-1}), and 70 yr lifetime for carcinogenic effects (i.e., 70 yr × 365 d yr^{-1}).

9.01.7 TOXICITY ASSESSMENT

9.01.7.1 Overview of Human Health Toxicology

Most health effects from environmental toxins are due to a detrimental change in the structure or function of biological molecules in the organism, which lead to a disruption of biochemical and physiological function. Two basic categories of cytological damage are recognized: (i) binding to an enzyme or energy-carrying molecule, resulting in cellular dysfunction or cell death (necrosis); and (ii) binding to or modification of the cell's genetic material (deoxyribonucleic acid; DNA), resulting in abnormal changes in the cell's rate of reproduction and other cellular behavior. Genetic damage to somatic cells may result in cancer, and genetic damage to germ cells can result in teratogenesis (i.e., birth defects). Cellular injury is a detrimental effect in which the extent of toxicity is frequently a function of contaminant dose, while genetic toxicity is classified as stochastic without a safety threshold.

A fundamental principle of hazardous waste risk assessments is the concept of dose response. Except for genetic toxins (e.g., carcinogens, in which the exposure to one molecule can potentially cause a toxic response), every chemical is toxic at some level, and the toxicity is usually directly proportional to the mass ingested. For example, 210 g of table salt is toxic to the average adult, as is 2,080 g of sugar. Hazardous chemicals are, of course, significantly more toxic than sugar. Most exhibit no toxicity below certain concentrations or exposure levels. Furthermore, elements such as chromium, iron, selenium, iodine, and compounds such as vitamins are toxic at high concentrations, but are vital nutrients at lower concentrations.

Toxicity data, which are determined experimentally in studies with laboratory animals, may not be generally applicable due to biological variability. Susceptibility to toxicological effects within a population of organisms is described by a Gaussian distribution. Individuals at one end of the spectrum may be highly susceptible to toxicity, whereas individuals at the other end of the spectrum exhibit more tolerance of the toxic effect. The dose representing the midpoint of the curve is the LD_{50}: the lethal dose to 50% of the population. The values of the LD_{50} vary with the animal species tested, and are generally unavailable for humans; however, data for closely related species, such as monkeys, are considered a close approximation to human responses. The LD_{50} is only useful in cases of acute toxicity. In hazardous waste risk assessment it is applied where a nearby population may be exposed to a toxic cloud or plume.

9.01.7.1.1 Classification of toxic responses

The most important factor that influences toxicity is the dose: it has been said that the dose makes the poison. A second important factor is the time period of exposure. Toxicity is often classified by the number and/or the duration of exposures. For example, a worker exposed to a mixture of sulfuric acid and sodium cyanide might receive a one-time, high dose of this toxic substance that can cause death through the binding of cyanide to energy-transferring molecules known as cytochromes. Repeated exposures are most simply classified as *chronic*. A common chronic exposure to hazardous chemicals is the ingestion drinking water contaminated with trace levels of hazardous chemicals.

Many chemicals that are acutely toxic are not chronically toxic, and vice versa. For example, pure vitamin D exhibits high acute toxicity. However, low repeated doses (such as in the normal intake of milk) are not only nontoxic but also essential to good health. For chemicals that are both acutely and chronically toxic, the mechanisms of the two types of toxicity are often different. For example, acute toxicity from a large dose of chloroform is caused by effects on the central nervous system that cause dizziness and narcosis. However, ingesting water containing trace concentrations of chloroform over a lifetime results in liver damage and cancer (Stewart, 1971).

9.01.7.1.2 Quantifying reversible toxic effects

In hazardous waste management, chronic toxicity is often caused by long-term, low-level exposure to hazardous chemicals. Chronic toxicity is difficult to quantify, because less is known about the long-term effects of chemicals than about their acute toxicity. The lack of toxic response from zero dose to the threshold dose is the result of a biochemical or physiological defense (e.g., detoxification or excretion) that prevents the occurrence of toxicological effects. Toxicity first appears at the threshold dose. As the exposure to the chemical is increased, the detoxification mechanisms are overwhelmed and the effects increase as a function of dose. Eventually, all of the toxicological effects are exhibited. Animal studies are used to evaluate the effects of chronic toxicity. Feeding or inhalation evaluations are carried out for most of the animals' lives. Upon completion of the study, the animals are sacrificed and pathological evaluations are conducted. The results are then extrapolated to humans using biochemically based models for carcinogens and a series of safety factors for noncarcinogens.

9.01.7.2 Quantifying Noncarcinogenic Risk: Reference Dosages

Noncarcinogenic toxicities are detrimental effects caused by chemicals that do not induce cancer. The most common effects are due to interactions between the chemical and the biological molecules in the receptor, especially enzymes. Toxic chemicals can bind to an important enzyme and reduce or eliminate its function. Some of the most important noncarcinogenic interactions between toxic chemicals and biological molecules include the inhibition of acetylcholinesterase by organophosphate ester and carbamate insecticides, the binding of carbon monoxide to hemoglobin, and the binding of cyanide to cytochromes.

Estimates of the toxicity of noncarcinogens are based on the concept of their threshold: the dose below which there are no short- or long-term effects on the organism. The lack of effect below the threshold dose can be understood in terms of its molecular basis. If millions of receptor biomolecules are available for a given function (e.g., nerve transmission, transport of oxygen), the binding of a toxic chemical to a small number of these receptor molecules does not produce a measurable toxic effect. A good example is the inactivation of a very small number of molecules of acetylcholinesterase by an insecticide such as parathion.

9.01.7.2.1 The no observed adverse effect level

An important parameter in the risk assessment of hazardous wastes is the no-effect level to which a population may be exposed. This level, defined as the *no observed effect level* (NOEL), is difficult to measure and also difficult to define accurately. Its values are based on epidemiological data and controlled animal experiments designed to determine the highest dose that will not produce an adverse effect. The *no observed adverse effect level* (NOAEL) is a variant of the NOEL in that it classifies only toxicological effects. Other measures related to the NOEL and the NOAEL are the LOEL (*lowest observed effect level*) and LOAEL (*lowest observed adverse effect level*), a stricter version of the LOEL.

9.01.7.2.2 Acceptable daily intakes and reference doses

The *acceptable daily intake* (ADI) is the level of daily intake of a toxic substance that does not produce an adverse health effect. ADIs are based on NOAELs, but are not considered an absolute physiological threshold; they are based on safety factors that reflect variations in the population. Therefore, values for ADIs are significantly lower than values of corresponding NOAELs (US EPA, 1986, pp. 33992–34003).

Reference doses (RfDs) are a regulatory parameter also based on NOAELs. The RfD is used by the EPA in place of the ADI (which sometimes results in lower values for acceptable intakes). Safety factors are used in the derivation of RfDs to account for hypersensitivity reactions. RfDs usually incorporate two safety factors: (i) a factor of 10 for variation among individuals and (ii) another factor of

10 for extrapolation from experimental animal data to humans (Barnes and Dourson, 1988). Another safety factor that may be used with the RfD is a "modifying factor" ranging from 1 to 10, which is based on professional judgment. The procedure for establishing RfDs is somewhat more detailed than for ADI development, and includes use of the most sensitive species, the appropriate route of exposure, and the most sensitive end point (US EPA, 1986, pp. 34028–34040). RfDs have become a widely used indicator of chronic toxicity, and have been established for oral and inhalation routes. RfDs for the most common hazardous compounds are listed in Table 3.

9.01.7.3 Quantifying Carcinogenic Risk: Slope Factors

Two methods have been used to estimate carcinogenicity in humans: epidemiological studies and animal studies. Of the two, the use of animals is considered more reliable among toxicologists because epidemiological studies are usually based on random data (i.e., without experimental design) and yield a less secure basis for establishing cause–effect relationships. Because of the difficulty of measuring human carcinogenicity, the International Agency for Research on Cancer (IARC), an organization that is part of the World Health Organization

Table 3 Reference doses of some common hazardous compounds.

Compound	*Oral RfD* ($mg\ kg^{-1}\ d^{-1}$)	*Inhalation RfD* ($mg\ kg^{-1}\ d^{-1}$)
Monocyclic aromatic hydrocarbons		
Benzene	0.029	0.029
Toluene	0.2	
Ethylbenzene	0.1	0.286
Xylenes	2.0	
Polycyclic aromatic hydrocarbons		
Anthracene	0.3	
Fluorene	0.04	
Pyrene	0.03	
Nonhalogenated solvents		
Acetone	0.1	
Methyl ethyl ketone	0.6	0.286
Chlorinated solvents		
Carbon tetrachloride	7.0×10^{-4}	
Chloroform	0.01	
PCE	0.01	
Insecticides		
Aldrin	3.0×10^{-5}	
Carbaryl	0.1	
Dieldrin	5.0×10^{-5}	
DDT	5.0×10^{-4}	
Lindane	3.0×10^{-4}	
Malathion	0.02	
Herbicides		
Atrazine	0.035	
2,4-D	0.01	
2,4,5-T	0.01	
Trifluralin	0.0075	
Fungicides		
Pentachlorophenol	0.03	
Industrial intermediates		
Chlorobenzene	0.02	
2,4-Dichlorophenol	0.003	
Hexachlorocyclopentadiene	0.007	
Phenol	0.6	
Explosives		
2,4,6-Trinitrotoluene	5.0×10^{-4}	
Polychlorinated biphenyls		
Aroclor 1016	7.0×10^{-5}	
Aroclor 1254	2.0×10^{-5}	

(WHO) of the United Nations, has developed three categories for evidence of carcinogenicity. These include: (i) sufficient evidence of human cancer definitely caused by exposure, (ii) limited evidence due to inadequate data, and (iii) inadequate evidence (i.e., no data available) (IARC, 1987).

A significant question in the quantification of carcinogen risk is the hazard imposed by exposure to very low concentrations of chemicals. To quantify potential carcinogenesis using the low contaminant concentrations to which the public is typically exposed would require chronic toxicity studies with an enormously large number of test animals. Dose–response relationships can be determined experimentally at higher doses, and these high-dose data can be extrapolated to the lower doses to which populations are typically exposed. Therefore, most animal studies use higher doses than commonly found in the environment. Other guidelines include the use of lifetime studies on at least 50 animals × two sexes × two species × at least three doses (a control and two doses) (IRLG, 1979). A phenomenon that must be considered is the background cancer level, which may be a function of the environment and the genetic makeup of the population being evaluated.

The extrapolation of empirical data requires a careful assessment of the assumptions regarding the mechanisms of carcinogenesis at the lower dosages. Two classes of models have been developed to quantify the probability of cancer as a function of dose, including the extrapolation from higher doses to lower doses. *Tolerance models* (e.g., log-probit, log-logistic) are based on statistics. *Mechanistic models* (e.g., one-hit, gamma, multihit, and Weibull) are based on known biochemical and physiological processes. Mechanistic models are usually favored over tolerance models. Information regarding carcinogenic dose–response models is available in Rai and Van Ryzin (1979) and Gaylor and Kodell (1980).

Two criteria must be considered when evaluating the carcinogenicity of hazardous chemicals: (i) the carcinogenic potential and (ii) the availability of data on the carcinogens. As in all toxicological evaluations, a dose–response relationship is the basis for quantifying a toxicological response. Dose–response evaluations are usually based on animal studies and provide a quantitative relationship between tumor incidence and dose. The procedures for such assays involve giving test animals a dose according to body weight per day (mg kg^{-1} d^{-1}). The response (*y*-axis) is the incidence of cancer corrected for background cancer rates. The cancer incidence in the test animals must then be extrapolated to humans. In linking animal data to humans it is assumed that there is no threshold dose for cancer-causing chemicals because even one molecule can initiate cancer. Therefore, any concentration of a carcinogen carries some degree of risk. The most common value of acceptable risk used in hazardous waste assessments is 1×10^{-6}, or one in one million. However, bioassays for cancer cannot evaluate doses for such a low response by any currently known method. The data collected from animal studies are therefore extrapolated to a 1×10^{-6} risk. Such extrapolation of carcinogen risk uses a *carcinogen potency factor* (CPF), which is the slope of dose–response curves at low exposures. The EPA has adopted the use of carcinogenic dose-response relationship slopes and these have become commonly known as *slope factors* (SFs) with units of (mg kg^{-1} $d^{-1})^{-1}$. As with RfDs, oral and inhalation slope factors have been reported by the EPA (US EPA, 1995). SFs for the most common hazardous chemicals are listed in Table 4.

9.01.8 RISK CHARACTERIZATION

Risk characterization is the calculation of risk for all potential receptors that may be exposed to hazardous wastes. It includes calculating risk for different exposure routes to both noncarcinogenic and carcinogenic hazardous chemicals. Often this requires the use of toxicological data derived from animal studies. Furst (1994) discussed issues in the interpretation of this data, and suggested that animal toxicity data should only be used for risk calculation when experiments employ routes that mimic human exposure (i.e., oral, inhalation, or dermal).

9.01.8.1 Determination of Noncarcinogenic Risk

Noncarcinogenic risk is represented by the hazard index (HI), which is the ratio of the chronic daily intake to the RfD:

$$\mathrm{HI} = \frac{I}{\mathrm{RfD}} \tag{4}$$

where HI, the hazard index, is dimensionless, I is the intake (mg kg^{-1} d^{-1}), and RfD is the reference dose (mg kg^{-1} d^{-1}).

A hazard index of <1.0 represents an acceptable cumulative risk for all contaminants and routes of exposure. In other words, if the hazard index is <1.0, the receptors are exposed to concentrations that do not present a hazard because detoxification and other mechanisms forestall toxic effects that could result from exposure to the contaminant. The quantitative value obtained for the HI is not a value of risk; it does not indicate the probability of harm as the result of exposure.

Table 4 Slope factors of some common hazardous compounds.

Compound	*Oral SF* $(mg\ kg^{-1}\ d^{-1})^{-1}$	*Inhalation SF* $(mg\ kg^{-1}\ d^{-1})^{-1}$
Monocyclic aromatic hydrocarbons		
Benzene	0.029	0.029
Chlorinated Solvents		
Carbon tetrachloride	0.13	0.13
Chloroform	0.0061	0.081
1,1,2-TCA	0.057	0.057
Insecticides		
Aldrin	17	17
Dieldrin	16	16
DDT	0.34	0.34
Hexachlorobenzene	1.6	1.6
Herbicides		
Trifluralin	0.0077	
Fungicides		
Pentachlorophenol	0.12	0.12
Industrial intermediates		
2,4,6-Trichlorophenol	0.011	0.011
Explosives		
2,4,6-Trinitrotoluene	0.03	

The hazard index only provides an indication of the probable presence or absence of effects from exposure to noncarcinogens.

9.01.8.2 Determination of Carcinogenic Risk

Carcinogenic risk is a function of the chronic daily intake (calculated using Equation (1) and the slope factor (SF)):

$$\text{Risk} = \text{CDI} \times \text{SF} \qquad (5)$$

where Risk is the probability of carcinogenic risks (fraction), CDI the chronic daily intake $(mg\ kg^{-1}\ d^{-1})$, and SF the carcinogenic slope factor $(mg\ kg^{-1}\ d^{-1})^{-1}$.

The value for risk is the quantitative end point determined in risk assessment calculations; it is commonly used in regulatory and management decisions regarding hazardous wastes. As with noncarcinogens, the risk term is calculated for each contaminant, each route of exposure, and for all sets of receptor populations; each element of risk is then summed to provide the value of cumulative risk.

9.01.9 SOURCES OF UNCERTAINTIES IN RISK ASSESSMENT

A limitation of both human health and ecological risk assessments is the uncertainty inherent in almost every level of the calculations. Although the risk assessment process results in a relatively straightforward numerical value for the probability of hazard, a significant degree of uncertainty is inherent in that value.

Suter (1990) emphasized that at our current level of knowledge, the detrimental effects of hazardous chemicals on ecosystems cannot be adequately predicted. The current methods can only assess risks in a simplified manner by providing a relative ranking of risk—from chemical-to-chemical or site-to-site. Nonetheless, such relative-risk ranking provides a useful basis for prioritizing environmental hazards, particularly if data are analyzed by qualified risk assessors.

Most risk assessments use data based on assumptions and extrapolations. These generate uncertainties due to the lack of knowledge or data. A degree of caution is used by risk assessors in assigning absolute numbers to risk values and hazard indices because a significant degree of uncertainty is inherent in each of the four steps of risk assessment, which is then compounded into the final risk value.

9.01.9.1 Source Characterization

Uncertainties in source characterization include imprecise source sampling, limitations of the analytical results, and selection of the contaminants used to calculate risk. For example, priority pollutant analyses are often used to screen chemicals at hazardous waste sites. However, contaminants other than the 129 priority pollutants, as well as their metabolites, may be found at these sites, and are not detected by Methods 624, 625, or other standard analytical schemes. The problems of including nondetected chemicals in a health risk assessment were discussed by Seigneur *et al.* (1995). They noted that this practice may lead to estimated risks that exceed regulatory thresholds, because the detection limit or half of the detection limit must be used in risk calculations.

9.01.9.2 Lack of Available Data

Lack of data is often a significant source of uncertainty in risk assessments. Unavailable data may include source concentrations or source contaminants such as those not quantified by standard analyses such as EPA methods 624 and 625. RfDs and SFs are currently only available for fewer than two hundred chemicals; however, since thousands of chemicals are potentially present at contaminated sites and hazardous waste facilities, the paucity of available data may be responsible for a significant amount of uncertainty.

9.01.9.3 Exposure Assessment Models and Methods

Whether predictive models or sampling and analysis of the media are used to determine contaminant concentrations to which receptor organisms are exposed, uncertainties are inevitable. Many risk assessments use exposure assessment models to predict possible future contaminant concentrations at a point of exposure. The uncertainty of such modeling of contaminant concentrations projected into the future is obvious. The possible occurrence of hazardous waste spills and traffic accidents involving hazardous materials is also predicted using stochastic modeling. These predictive models also contain a significant degree of inherent uncertainty. Although most hazardous waste sampling is conducted using strict quality control and quality assurance measures some uncertainty in sampling is inevitable. Furthermore, some standard analytical procedures are subject to false positive and false negative results; for example, natural soil organic matter appears in TPH analyses in the same manner as hydrocarbons, resulting in a false positive error. Some contaminants are not efficiently extracted into organic solvents for analysis from soils and sludge, resulting in a false negative error.

9.01.9.4 Quality of Toxicological Data

ADIs, NOAELs, RfDs, and SFs are based on extrapolation of animal toxicity data to humans using accepted safety factors. These parameters incorporate a degree of variability of up to several orders of magnitude (1,000–10,000), which significantly affects the outcome of any risk assessment. Toxicity data for ecological risk assessments are often subject to even greater uncertainty than human health toxicity data. Furthermore, the toxicological effects on communities via disruption of predator–prey relationships and changes in species diversity and community structure are unknown, and estimates of their magnitude are likely to be very uncertain.

9.01.9.5 Evaluating Uncertainty

One approach to evaluating uncertainty is the use of stochastic modeling. Ünlü (1994) used Monte Carlo, first-order, and point estimate methods to evaluate uncertainties in contaminant concentrations downgradient from a waste pit. A comparison analysis showed that for conservative contaminants, the accuracy of the first-order method is comparable with that of the Monte Carlo method and can be used as an alternative to Monte Carlo analyses. In general, the accuracy of the first-order and point-estimate methods was sensitive to the fate and transport of the contaminants. Uncertainties in groundwater contamination risk have also been evaluated with a model that addressed the effects of source-, chemical-, and aquifer-related uncertainties (Hamed *et al.*, 1995). The application of the method was demonstrated with a nonreactive solute and with *o*-xylene as a reactive contaminant in groundwater. The model results were checked against those from a Monte Carlo simulation method, and were found to be in good agreement except for low probability events. Shevenell and Hoffman (1993) used uncertainty analyses to improve the reliability of risk assessment modeling, a procedure that helps to moderate the sometimes unknown model assumptions and the uncertainty associated with some model parameters. Finally, Doyle and Young (1993) recommended that conclusions about whether the uncertainty in the risk assessment over- or underestimates risk should be included in the assessment report. In most cases, the effect of uncertainties in the potential negative public health effects are addressed by the use of conservative assumptions. For example, conservative estimates of attenuation and lower natural degradation rates than normal may be used to counter the uncertainty of exposure assessment models.

9.01.10 RISK MANAGEMENT AND RISK COMMUNICATION

Risk assessment in itself provides a quantitative value for potential hazards from hazardous waste sites and facilities; however, risk evaluation is usually carried further with risk management. *Risk management* is a decision-making process that uses the quantitative values obtained from risk assessment models along with the insight, experience, and judgment of professionals. Risk assessment and risk management are multifaceted methodologies that require the consideration of quantitative values, qualitative assessments, and professional judgment. The lines between science, engineering, economics, and policy become blurred, and, therefore, the ultimate decisions based on risk analysis have to be balanced by professional judgment.

Scientists and engineers involved in hazardous waste risk assessment often have an innate appreciation of the risks associated with hazardous wastes. However, the public, especially those who live near a hazardous waste site or facility, require information regarding the risk assessment process, its uncertainties, and the value judgments that have been made. This information is conveyed through *risk communication*, an integral part of the risk assessment process.

A hazardous waste risk assessment team must be able to communicate effectively with the public about the risk assessment process and the results of the risk assessment at the site that is of concern to the local population. Some of the specific elements of risk communication include (i) the steps of the risk assessment process, (ii) acceptable levels of risk, and (iii) the uncertainties and value judgments that are inherent in the risk assessment process. The four fundamental steps in the risk assessment process for hazardous wastes (source characterization, exposure assessment, toxicity assessment, and risk estimation) can easily be presented in outline form at citizens' meetings and hearings. Typical acceptable levels of risk, such as the typical Superfund level of 1×10^{-6}, are often discussed in relation to other natural and anthropogenic risks that are part of the world in which we live. In addition to communicating the risk assessment procedure and commonly used levels of risk, the public must be made aware of the uncertainties at the different stages of the risk assessment process, and must know that the quantitative value of risk that is obtained is bounded above and below that number by a range of uncertainties. Furthermore, the public must be aware that these uncertainties are part of quantifying such complex processes, and absolute certainty in such assessments will probably never be realized. The public should also be aware that the experience and value judgments of the risk assessment team are an important and necessary part of the risk assessment process, and that the use of such qualitative tools is a common and effective procedure.

REFERENCES

Anderson M. P. and Woessner W. W. (1992) *Applied Groundwater Modeling*. Academic Press, San Diego, CA.

ATSDR (1989) *ATSDR Biannual Report to Congress: October 17–September 30, 1986*. Agency for Toxic Substances and Disease Registry, US Public Health Service, Atlanta, GA.

Barnes D. G. and Dourson M. (1988) Reference dose (RfD): description and use in health risk assessments. *Regulat. Toxicol. Pharmacol.* **8**, 471–486.

Doyle M. E. and Young J. C. (1993) Human health risk assessments. In *Ecological Assessment of Hazardous Waste Sites* (ed. J. T. Maughan). Van Nostrand Reinhold, New York, chap. 5.

Furst A. (1994) Issues in interpretation of toxicological data for use in risk assessment. *J. Haz. Mater.* **39**, 143–148.

Gaylor D. W. and Kodell R. L. (1980) Linear interpolation algorithm for low dose risk assessment of toxic substances. *J. Environ. Pathol. Toxicol.* **4**, 305–312.

Guidotti T. L. (1988) Exposure to hazard and individual risk: when occupational medicine gets personal. *J. Occupat. Med.* **30**, 571–577.

Hamed M. M., Conte J. P., and Bedient P. B. (1995) Probabilistic screening tool for ground-water contamination assessment. *J. Environ. Eng.* **121**, 767–775.

Interagency Regulator Liaison Group (IRLG) (1979) Scientific bases for identification of potential carcinogens and estimation of risks. *J. Natl. Cancer Inst.* **63**, 241–268.

International Agency on Research of Cancer (IARC) (1987) Overall evaluations of carcinogenicity. In *IARC Monographs on the Evaluation of Carcinogenic Risks to Humans*. An Updating of International Agency on Research of Cancer Monographs, vol. 1–42, suppl. 7.

MacKay D. M. and Smith L. A. (1990) Agricultural chemicals in groundwater: monitoring and management in California. *J. Soil Water Conserv.* **45**, 253–255.

Nascarella M. A., Kostecki P., Calabrese E., and Click D. (2002) AEHS's 2001 survey of states' soil and groundwater cleanup standards. *Contamin. Soil Sedim. Water*, Jan./Feb., 15–68.

National Academy of Sciences (NAS) (1983) *Risk Assessment in the Federal Government: Managing the Process*. National Academy Press, Washington, DC.

Patrick R. (1983) *Groundwater Contamination in the United States*. National Academy Press, Washington, DC.

Patton D. E. (1993) The ABCs of risk assessment. *EPA J.* **19**, 10–15.

Rai K. and Van Ryzin J. (1979) Risk assessment of toxic environmental substances using a generalized multi-hit dose response model. In *Energy and Health* (eds. N. E. Breslow and A. S. Whittemore). Society for Industrial and Applied Mathematics, Philadelphia, PA.

Seigneur C., Constantinou E., Fencl M., Levin L., Gratt L., and Whipple C. (1995) The use of health risk assessment to estimate desirable sampling detection limits. *J. Air Waste Manage. Assoc.* **45**, 823–830.

Shevenell L. and Hoffman F. O. (1993) Necessity of uncertainty analyses in risk assessment. *J. Haz. Mater.* **35**, 369–386.

Stewart R. D. (1971) Methyl chloroform intoxication: diagnosis and treatment. *JAMA* **215**, 1789–1792.

Suter F. W., II (1990) Endpoints for regional ecological risk assessments. *Environ. Manage.* **14**, 9–23.

Travis C. C., Crouch E. A. C., Wilson R., and Klema E. D. (1987) Cancer risk management: a review of 132 federal regulatory cases. *Environ. Sci. Technol.* **21**, 415–420.

Ünlü K. (1994) Assessing risk of ground-water pollution from land-disposed wastes. *J. Environ. Eng.* **120**, 1578–1597.

US EPA (1985) *Development of Statistical Distributions or Ranges of Standard Factors used in Exposure Assessments.* US Environmental Protection Agency Office of Health and Environmental Assesment, US Government Printing Office, Washington, DC.

US EPA (1986) *Guidelines for carcinogen risk assessment.* US Environmental Protection Agency, Federal Register, 51. US Government Printing Office, Washington, DC, pp. 33992–34003; 34028–34040.

US EPA (1988) *Report to Congress: Solid Waste Disposal in the United States, Vol. II,* EPA/530-SW-88-011B. US Environmental Protection Agency, US Government Printing Office, Washington, DC.

US EPA (1989a) *Risk Assessment Guidance for Superfund: Environmental Evaluation Manual* (EPA/540/1-69/001A, OSWER Directive 9285.7-01). US Environmental Protection Agency, US Government Printing Office, Washington, DC.

US EPA (1989b) *Exposure Factors Handbook.* (Publication EPA/600/8-89/043). US Environmental Protection Agency, US Government Printing Office, Washington, DC.

US EPA (1995) *Integrated Risk Information System (IRIS).* US Environmental Protection Agency, US Government Printing Office, Washington, DC.

Van Leeuwen C. J. and Hermens J. L. M. (1995) *Risk Assessment of Chemicals: An Introduction.* Kluwer, Dordrecht, The Netherlands.

Watts R. J. (1998) *Hazardous Wastes: Sources, Pathways, Receptors.* Wiley, New York.

Westrick J. J., Mills J. W., and Thomas R. F. (1983) *The Ground Water Supply Survey: Summary of Volatile Organic Contaminant Occurrence Data.* US EPA, Office of Drinking Water, Cincinnati, OH.

Zheng C. and Bennett G. D. (2002) *Applied Contaminant Transport Modeling.* Wiley, New York.

9.02
Arsenic and Selenium

J. A. Plant, D. G. Kinniburgh, P. L. Smedley, F. M. Fordyce, and B. A. Klinck

British Geological Survey, Keyworth, Nottingham, UK

9.02.1 INTRODUCTION

Arsenic (As) and selenium (Se) have become increasingly important in environmental geochemistry because of their significance to human health. Their concentrations vary markedly in the environment, partly in relation to geology and partly as a result of human activity. Some of the contamination evident today probably dates back to the first settled civilizations which used metals.

Arsenic is in group 15 of the periodic table (Table 1) and is usually described as a metalloid. It has only one stable isotope, ^{75}As. It can exist in the −III, −I, 0, III, and V oxidation states (Table 2).

Selenium is in group 16 of the periodic table and although it has chemical and physical properties intermediate between metals and nonmetals (Table 1), it is usually described as a nonmetal. The chemical behavior of selenium has some similarities to that of sulfur. Formally, selenium can exist in the −II, 0, IV, and VI oxidation states (Table 2). Selenium has six natural stable isotopes, the most important being ^{78}Se and ^{80}Se. Although ^{82}Se is generally also regarded as a stable isotope, it is a β-emitter with a very long half-life (1.4×10^{20} yr). Both arsenic and selenium tend to be covalently bonded in all of their compounds.

Arsenic is 47th and Se 70th in abundance of the 88 naturally occurring elements. Much more has become known about the distribution and behavior of arsenic and selenium in the environment since the 1980s because of the increased application of improved analytical methods such as inductively coupled plasma-mass spectrometry (HG-ICP-MS), inductively coupled plasma-atomic emission spectrometry (HG-ICP-AES), and hydride generation-atomic fluorescence spectrometry (HG-AFS). These methods can detect the low concentrations of arsenic and selenium found in environmental and biological media, and as a result arsenic and selenium are increasingly included in determinand suites of elements during systematic geochemical mapping and monitoring campaigns (Plant *et al.*, 2003).

Arsenic is highly toxic and can lead to a wide range of health problems in humans. It is carcinogenic, mutagenic, and teratogenic (National Research Council, 1999). Symptoms of arsenicosis include skin lesions (melanosis, keratosis) and skin cancer. Internal cancers, notably bladder and lung cancer, have also been associated with arsenic poisoning. Other problems include cardiovascular disease, respiratory problems, and diabetes mellitus. There is no evidence of a beneficial role for arsenic (National Research Council, 1999) and it is unclear whether there is any safe dose for humans. Indeed, the precise nature of the relationship between arsenic dose and carcinogenic effects at low arsenic concentrations remains a matter of much debate (Clewell *et al.*, 1999; Smith *et al.*, 2002).

The principal public health concern with arsenic is from the development of naturally high-arsenic groundwaters (Smedley and Kinniburgh, 2002). Extensive arsenicosis from such sources has been

Table 1 Physical properties of arsenic and selenium.

Name	*Arsenic*	*Selenium*
Symbol	As	Se
Atomic number	33	34
Periodic table group	15	16
Atomic mass	74.9216	78.96
Classification	Metalloid	Nonmetal
Pauling electronegativity	2.18	2.55
Density (kg m^{-3})	5,727	4,808
Melting point (°C)	817 (at high pressure)	220
Boiling point (°C)	614 (sublimes)	685
Natural isotopes and abundance	^{75}As (100%)	^{74}Se (0.87%) ^{76}Se (9.02%) ^{77}Se (7.58%) ^{78}Se (23.52%) ^{80}Se (49.82%) ^{82}Se (9.19%)

Table 2 Chemical forms of arsenic and selenium.

Element and formal oxidation state	*Major chemical forms*
As(−III)	Arsine [H_3As]
As(−I)	Arsenopyrite [FeAsS], loellingite [$FeAs_2$]
As(0)	Elemental arsenic [As]
As(III)	Arsenite [$H_2AsO_3^-$, H_3AsO_3]
As(V)	Arsenate [AsO_4^{3-}, $HAsO_4^{2-}$, $H_2AsO_4^-$, H_3AsO_4]
Organic As (V and III)	Dimethylarsinate [DMA, $(CH_3)_2AsO(OH)$], monomethylarsonate [MMA(V), $CH_3AsO(OH)_2$ or MMA(III), $CH_3As(OH)_2$], arsenobetaine [AsB, $(CH_3)_3As^+ \cdot CH_2COO^-$], arsenocholine [AsC, $(CH_3)_3As^+ \cdot CH_2CH_2OH$]
Se(−II)	Selenide [Se^{2-}, HSe^-, H_2Se]
Se(0)	Elemental selenium [Se]
Se(IV)	Selenite [SeO_3^{2-}, $HSeO_3^-$, H_2SeO_3]
Se(VI)	Selenate [SeO_4^{2-}, $HSeO_4^{2-}$, H_2SeO_4]
Organic Se	Dimethylselenide [DMSe, CH_3SeCH_3]; dimethyldiselinide [DMDSe, $CH_3SeSeCH_3$], selenomethionine [$H_3N^+CHCOO^- \cdot CH_2CH_2SeMe$], selenocysteine [$H_2N^+CHCOO^- \cdot CHSeH$]

reported from Argentina, Bangladesh, Chile, China, Mexico, India, Thailand, and Taiwan. "Blackfoot disease," a form of gangrene arising from excessive arsenic intake, was first described in Taiwan by Tseng *et al.* (1968).

In contrast to arsenic, trace concentrations of selenium are essential for human and animal health. Until the late 1980s, the only known metabolic role for selenium in mammals was as a component of the enzyme glutathione peroxidase (GSH-Px), an anti-oxidant that prevents cell degeneration. There is now growing evidence, however, that a seleno-enzyme is involved in the synthesis of thyroid hormones (Arthur and Beckett, 1989; G. F. Combs and S. B. Combs, 1986). Selenium deficiency has been linked to cancer, AIDS, heart disease, muscular dystrophy, multiple sclerosis, osteoarthropathy, immune system and reproductive disorders in humans, and white muscle disease in animals (Levander, 1986; WHO, 1987, 1996). Selenium deficiency in humans has also been implicated in the incidence of a type of heart disease (Keshan Disease) and an osteoathropathic condition (Kashin-Beck disease) in extensive regions of China. Domestic animals also suffer from white muscle disease in these areas (Tan, 1989). Selenium deficiency has been reported from New Zealand and Finland, and falling concentrations of selenium in the diet are of increasing concern in many western countries (Oldfield, 1999). In Europe, this trend may have been exacerbated by the increasing use of native low-selenium grains rather than the imported selenium-rich grains of North America (Rayman, 2002). Selenium supplementation of livestock is common.

Despite the necessity for selenium, the range of intake between that leading to selenium deficiency ($<40\ \mu g\ d^{-1}$) and that leading to toxicity (selenosis) ($>400\ \mu g\ d^{-1}$) is very narrow in humans (WHO, 1996). Investigations of the relationships between selenium in the environment and animal health were pioneered by Moxon (1937) in the western USA, where selenium accumulator plants are found and both selenium toxicity and deficiency are of concern. Human selenosis at the population level is rare and is generally related to excesses from food rather than from drinking water. It has been reported from China and Venezuela where selenium-rich food is grown and consumed locally (Tan, 1989; WHO, 1996). Cancers of the skin and pancreas have been attributed to high selenium intake (Vinceti *et al.*, 1998). Selenium sulfide is used in antidandruff shampoos and is potentially carcinogenic but is not absorbed through the skin unless there are lesions (WHO, 1987). Selenium toxicity can lead to hair and nail loss and disruption of the nervous and digestive systems in humans and to alkali disease in animals. Chronic selenosis in animals is not common but has been reported from parts of Australia, China, Ireland, Israel, Russia, South Africa, USA, and Venezuela (Oldfield, 1999). Liver damage is a feature of chronic selenosis in animals (WHO, 1987).

It is now recognized that arsenic and selenium interact with each other in various metabolic functions and animal models indicate that each element can substitute for the other to some extent (Davis *et al.*, 2000). This could partly explain the reported protective effect of selenium against some diseases, including some cancers (Shamberger and Frost, 1969). Arsenic was also shown long ago to protect against selenium poisoning in experimental studies with rats (Moxon, 1938). Following the relatively recent discovery of dissimilatory As(V) reduction and the various mechanisms which organisms have developed to deal with the toxicity of As(V) and As(III), there has been a rapid increase in our understanding of the microbial chemistry of arsenic (see Frankenberger (2002) and the individual chapters therein). Selenium chemistry is also closely linked to microbial processes, but these are less well understood.

Contamination as a result of human activity is of increasing concern for both elements but especially for arsenic. In the past, the problem was exacerbated by an absence of waste management strategies. Arsenic concentrations in the natural environment have increased as a result of a number of activities, for example, mining and smelting, combustion of arsenical coals, petroleum recovery (involving the release of production waters), refining, chemical production and use, use of biocides including wood preservatives, use of fertilizers, the manufacture and use of animal feed additives, and the development of high-arsenic groundwater for drinking water and irrigation. Such activities have progressively transferred arsenic from the geosphere into the surface environment and have distributed it through the biosphere, where it poses a potential risk to humans and the wider environment. Arsenicosis caused by the indoor combustion of arsenic-rich coals has also been reported from Guizhou province, China (Aihua *et al.*, 2000; Ding *et al.*, 2000).

Human activities that have increased the concentration of selenium in the environment include the mining and processing of base-metal, gold, coal, and phosphate deposits, the use of rock phosphate as fertilizer, the manufacture of detergents and shampoo, and the application of sewage sludge to land. The increased use of selenium in the pharmaceutical, glazing, photocopying, ceramic, paint, and electronics industries may also be increasing the amount of selenium entering the environment. In certain occupational settings, the principal pathway of arsenic to humans can be through inhalation.

In the following sections, we first review the source and occurrence of arsenic in the environment and then consider its pathways as a basis for an improved understanding of exposure and risk assessment. This should lead to better risk management. A similar format is followed for selenium. In discussing the two elements, and in line with the threats outlined above, emphasis on arsenic is on the behavior of arsenic in water, whereas in the case of selenium it is on the soil–water–plant relationships.

9.02.2 SAMPLING

Selenium and arsenic have been measured in a wide range of environmental media. Here, we describe sampling procedures for rocks, soils, sediments, and natural waters.

9.02.2.1 Rocks, Soils, and Sediments

In the case of rocks, soils, and sediments, sufficient material to be representative of the medium to be analyzed should be collected. Soil and sediment samples should be dried at temperatures $<35\ ^{\circ}C$ to avoid volatilization losses of arsenic or selenium (Rowell, 1994) and ideally freeze-dried (BGS, 1979–2002). Sampling, analysis, and quality control should be carried out with recognized procedures wherever possible (Darnley *et al.*, 1995; Salminen and Gregorauskiene, 2000).

9.02.2.2 Water

9.02.2.2.1 Techniques

As with other solutes, sampling natural waters for arsenic and selenium require: (i) the sample to represent the water body under investigation, and (ii) that no artifacts are introduced during sampling or storage. Sampling methods vary according to whether "dissolved," "particulate," or "total" concentrations are to be determined and whether speciation studies are to be undertaken. Water samples are most commonly analyzed for the "total" concentration of arsenic and selenium. Speciation measurements require additional precautions to ensure preservation of the *in situ* species until separation or measurement.

Since arsenic and selenium are normally present in natural waters in only trace concentrations (less than 10 $\mu g\ L^{-1}$ and frequently much lower), considerable care is required to perform reliable trace analyses. Marine chemists were the first to undertake reliable low-level trace analyses of natural waters and to develop "clean/ultraclean" sampling procedures (Horowitz *et al.*, 1996). Probably the most thorough account of sampling procedures for surface and groundwaters are those given by the USGS (Wilde and Radtke, 1998). Specific procedures for sampling rainwater, lake water, and seawater are also given. The precautions required in sampling for arsenic and selenium are the same as those for other trace elements present in water at $\mu g\ L^{-1}$ concentrations. For example, there should be minimal contact between the sample and metallic substances. Sample bottles should be tested first by analyzing deionized water stored in them to ensure that they do not contaminate the sample. They should also be rinsed thoroughly with the sample water before collection.

Ideally, groundwater should be sampled from purpose-built water-quality monitoring boreholes or piezometers. In practice, existing wells or boreholes are frequently used. It is important as far as practically possible to purge the borehole by pumping at least three borehole volumes to remove standing water before sampling. Low-flow ($<4\ L\ min^{-1}$) pumping is preferred to minimize resuspension of colloidal material.

Several procedures have been devised to obtain water-quality depth profiles in wells and aquifers. These include depth samplers, nested piezometers, strings of diffusion cells, multilevel samplers, and multiple packers. Each method has its advantages and disadvantages with very different costs and sampling logistics. As of early 2000s, few methods have been devised for arsenic and selenium profiling specifically; probably the most detailed profiling has been carried out in ocean sediments (Sullivan and Aller, 1996). Pore water, including water from the unsaturated zone, can be obtained using a high-pressure squeezer or high-speed centrifugation (Kinniburgh and Miles, 1983; Sullivan and Aller, 1996).

There are, as of early 2000s, no methods for the *in situ* determination or continuous monitoring of arsenic and selenium. The diffusive gradient thin-films (DGT) method is a novel sampling method that has been used mainly for cationic metals but may be adaptable for measuring arsenic and selenium as it has been for phosphorus (Zhang *et al.*, 1998). In this method, the normal cation-exchange resin is replaced by an iron oxide (ferrihydrite) gel. Solutes sorbed by the resin or gel are displaced and subsequently analyzed in the laboratory. In principle, the DGT approach is sensitive; detection limits are on the order of $ng\ L^{-1}$. The method also has the advantage that it can measure a wide range of solutes simultaneously with high spatial resolution (at the millimeter scale) and determine the average water quality over relatively long timescales (days or longer).

Most water samples do not require pre-treatment for total elemental analysis, but where organic arsenic or selenium compounds are suspected, pre-treatment by digestion with a strong acid mixture, for example, a 3 min sulfuric acid–potassium persulfate digestion or a nitric acid digestion, is necessary. Where pre-concentration is required, cold trapping of the hydrides or liquid–solid extraction has been used but this is very labor intensive when performed offline.

9.02.2.2.2 Filtered or unfiltered samples

Studies of the behavior of arsenic and selenium usually require the proportions of their dissolved and particulate components to be identified, since this affects their biological availability, toxicity, and transport. It also affects the interpretation of their mineral solubility, adsorption, and redox behavior. Specifications for compliance testing vary with regulatory authority, for example, US EPA specify a 0.45 μm filter while most authorities in developing countries specify (or assume) that water samples are unfiltered. If the water is reducing, it should be filtered before any oxidation occurs. Geochemists typically filter water samples using membrane filters in the range 0.1–0.45 μm, but the effective size of the filter can change as it becomes clogged. There continues to be much discussion about the merits of various filtering strategies (Hinkle and Polette, 1999; Horowitz *et al.*, 1996; Shiller and Taylor, 1996). Small iron-rich particles with adsorbed arsenic, selenium, and other trace elements can pass through traditional filters (Chen *et al.*, 1994; Litaor and Keigley, 1991) and subsequently dissolve when the sample is acidified.

Colloids tend to be most abundant in reducing groundwaters and turbid surface waters. In clear groundwater samples which have been filtered naturally by movement through an aquifer, differences between concentrations in filtered and unfiltered aliquots are often relatively small.

Filtered and unfiltered groundwater samples from high-arsenic areas in Bangladesh were found to have broadly similar arsenic concentrations (within $\pm 10\%$), although larger differences were found occasionally (Smedley *et al.*, 2001b). Similarly, nine out of 10 groundwater samples from arsenic-affected wells in Oregon showed little difference (mostly $<10\%$) between filtered and unfiltered samples (Hinkle and Polette, 1999).

Some studies have reported larger differences. A survey of 49 unfiltered groundwater sources in the USA found that particulate arsenic accounted for more than half of the total arsenic in 30% of the sources (Chen *et al.*, 1999) although the arsenic concentrations were all relatively small.

9.02.2.2.3 Sample preservation and redox stability

For analysis of total arsenic and selenium, samples are normally preserved by adding ultrapure acid (1% or 2% by volume); the choice of acid depends on the analytical procedures to be used. HCl is used before HG-AAS, HG-ICP-AES, and HG-AFS; and HNO_3 before ICP-MS, GF-AAS, and ASV. Acidification also helps to stabilize the speciation (see below), although Hall *et al.* (1999) recommended that nitric acid not be used for acidifying samples collected for speciation. Organic arsenic species are relatively stable, and inorganic As(III) species are the least stable (National Research Council, 1999). As of early 2000s, there are no well-established methods for preserving water samples for arsenic or selenium speciation analysis, although methods are being investigated for arsenic (National Research Council, 1999 ; Rasmussen and Andersen, 2002).

Laboratory observations indicate that the oxidation of As(III) and Se(IV) by air is slow and is often associated with microbial activity. MnO_2(s), which can precipitate following atmospheric oxidation of manganese-rich water, is also known to be a very efficient catalyst for the chemical oxidation of As(III) (Daus *et al.*, 2000; Driehaus *et al.*, 1995; Oscarson *et al.*, 1983). Iron oxides have also been implicated in increasing the abiotic rate of oxidation of As(III), although the evidence for this is somewhat equivocal and probably does not occur in minutes or hours unless some H_2O_2 is present (Voegelin and Hug, 2003). Precipitation of manganese and iron oxides can be minimized by ensuring sufficient acidity (pH 2 or less) and/or adding a reducing/complexing agent such as ascorbic acid, EDTA, or phosphate. Recent studies have demonstrated the efficacy of EDTA (Bednar *et al.*, 2002; Gallagher *et al.*, 2001) and phosphate (Daus *et al.*, 2002) in preserving arsenic speciation.

Arsenic speciation in urine is stable for at least two months without additives at 4 °C (National Research Council, 1999). It is reasonable to conclude that natural water samples probably behave in a similar way. As(III) in samples of Ottawa River water survived oxidation for at least 3 d at ambient temperature and without preservatives (Hall *et al.*, 1999). The lowest rates of oxidation occur under slightly acid conditions (Driehaus and Jekel, 1992) and acidification to pH 3–5 has been found to help stabilize As(III), although it is not always successful (J. Y. Cabon and N. N. Cabon, 2000). HCl normally prevents reduction of As(V) to As(III) and arsenic speciation has been shown to be preserved for many months, even in the presence of high Fe(II) concentrations, if water samples are filtered and acidified in the usual way (1% or 2% HCl) (McCleskey and Nordstrom, 2003). Traces of chlorine in HCl can lead to some long-term oxidation of As(III). One of the critical factors enhancing the oxidation of As(III) is the presence of dissolved Fe(III). The presence of Fe^{2+} or SO_4^{2-}, two species often found in arsenic-rich acid mine drainage (AMD) waters, inhibit the oxidation (McCleskey and Nordstrom, 2003).

Reduction of As(V) can occur in the presence of air if samples contain dissolved organic carbon (DOC), arsenate-reducing bacteria, and no preservatives (Bednar *et al.*, 2002; Hall *et al.*, 1999; Inskeep *et al.*, 2002). Arsenic(V) can then be reduced rapidly, within a few days. Storage at 3–5 °C and in the dark helps to preserve the speciation (Hall *et al.*, 1999; Lindemann *et al.*, 2000). Ideally, speciation studies for arsenic or selenium should involve a minimum of time between sampling and analysis.

An alternative approach to the determination of As(III)/As(V) speciation is to separate the As(V) species in the field using an anion-exchange column (Bednar *et al.*, 2002; Vagliasindi and Benjamin, 2001; Wilkie and Hering, 1998; Yalcin and Le, 1998). At near neutral to acidic pH, typical of most natural waters, uncharged As(III) is not retained by the resin and the retained As(V) can be eluted subsequently with high-purity acid. Providing that total arsenic is known, As(III) can be estimated by difference. Bednar *et al.* (2002) favored an acetate resin because of its high pH buffering capacity. Such anion-exchange methods do not work for selenium speciation, since both the Se(IV) and Se(VI) species are negatively charged and retained by the column.

9.02.3 ANALYTICAL METHODS

9.02.3.1 Arsenic

9.02.3.1.1 Total arsenic in aqueous samples

Laboratory methods. Methods for arsenic analysis in water, food, and biological samples have been reviewed in detail elsewhere (ATSDR, 2000; Irgolic, 1994; National Research Council, 1999; Rasmussen and Andersen, 2002) (Table 3).

Early colorimetric methods for arsenic analysis used the reaction of arsine gas with either mercuric bromide captured on filter paper to produce a yellow-brown stain (Gutzeit method) or with silver diethyl dithiocarbamate (SDDC) to produce a red dye. The SDDC method is still widely used in developing countries. The molybdate blue spectrophotometric method that is widely used for phosphate determination can be used for As(V), but the correction for P interference is difficult. Methods based on atomic absorption spectrometry (AAS) linked to hydride generation (HG) or a graphite furnace (GF) have become widely used. Other sensitive and specific arsenic detectors (e.g., AFS, ICP-MS, and ICP-AES) are becoming increasingly available. HG-AFS, in particular, is now widely used for routine arsenic determinations because of its sensitivity, reliability, and relatively low capital cost.

Conventional ICP-MS has great sensitivity but suffers from serious interferences. Chlorine interference leads to the formation of $^{40}Ar^{35}Cl^+$, which has the same mass/charge ratio as the monoisotopic ^{75}As ($m/z = 75$). Hence, HCl and $HClO_4$ should not be used for preservation or dissolution if ICP-MS is to be used. There may also be significant interference in samples with naturally high Cl/As ratios. A chlorine concentration of 1,000 mg L^{-1} gives an arsenic signal equivalent to ~3–10 μg L^{-1}.

The use of a high-resolution magnetic sector mass spectrometer, which can resolve the small difference in m/z for $^{75}As^+$ at 74.922 from that of $^{40}Ar^{35}Cl^+$ at 74.931, eliminates the chlorine interference. New collision-cell techniques, in which the atomized samples are mixed with a second gas (usually H_2) in a reaction cell, also minimize this interference. Arsenic detection limits of a few ng L^{-1} have been reported in matrices containing 1,000 mg L^{-1} NaCl. The chlorine interference can also be avoided by preseparation using HG, GF, or chromatography.

Field-test kits. A large number of wells need to be tested (and retested) for arsenic worldwide. Hence, there is a need for reliable field-test kits that can measure arsenic concentrations down to 10 μg L^{-1}, the WHO guideline value for arsenic in drinking water. Some of the more recently developed kits based on the Gutzeit method can achieve this semiquantitatively (Kinniburgh and Kosmus, 2002).

9.02.3.1.2 Total arsenic in solid samples

X-ray fluorescence spectrometry (XRF) and instrumental neutron activation analysis (INAA) are commonly used for multi-element analysis of rock, soil, and sediment samples since they do not require chemical dissolution. However, the detection limit for arsenic using XRF is on the order of 5 mg kg^{-1} and is too high for many environmental purposes. Once dissolved, arsenic can be determined using many of the methods described above

Table 3 Recognized methods of arsenic analysis.

Technique	*LoD* (μg L^{-1})	*Sample size* (mL)	*Equipment cost* (US$)	*Analytical throughput per day*	*Comments*	*Accredited procedure*
HG-AAS	0.05	50	20–100,000	30–60	Single element	ISO 11969 SM 3114
GF-AAS	1–5	1–2	40–100,000	50–100		ISO/CD 15586 SM 3113
ICP-AES	35–50	10–20	60–100,000	50–100	Multi-element; requires Ar gas supply. Can reduce LoD with HG	ISO/CD 11885 SM 3120
ICP-MS	0.02–1	10–20	150–400,000	20–100	Multi-element	SM 3125 US EPA 1638
HG-AFS	0.01	40–50	20–25,000	30–60	Single element but can be adapted for Se and Hg	
ASV	0.1	25–50	10–20,000	25–50	Only free dissolved As	US EPA 7063
SDDC	1–10	100	2–10,000	20–30	Simple instrumentation	SM 3500 ISO 6595

Source: Rasmussen and Andersen (2002).
Technique: HG = hydride generation; AAS = atomic absorption spectrometry; GF = graphite furnace; AES = atomic emission spectrometry; MS = mass spectrometry; AFS = atomic fluorescence spectrometry; ASV = anodic stripping voltammetry; SDDC = sodium diethyl dithiocarbamate. Procedures: ISO = International Standards Organization; ISO/CD = ISO Committee Draft; SM = 'Standard Methods'; US EPA = US Environmental Protection Agency; LoD = limit of detection.

for aqueous samples, although the method of digestion must be capable of destroying all solids containing arsenic.

9.02.3.1.3 Arsenic speciation

Aqueous speciation. At its simplest, speciation of arsenic consists of separating the element into its two major oxidation states, As(III) and As(V). This can be achieved on unacidified samples by ion chromatography. More detailed speciation involves determining organic species and less common inorganic species such as sulfide (thio), carbonate, and cyanide complexes, as well as less common oxidation states such as As(−III) and As(0). There is increasing interest in the bioavailability of arsenic. Organic speciation usually involves quantifying the two or three major (mainly the methylated) species present. The oxidation state of arsenic in these organic species can be either As(III) or As(V). Generally, such studies are carried out in research rather than in water-testing laboratories.

A two-stage approach to speciation is often used. This involves preseparation by high-performance liquid chromatography (HPLC) or ion chromatography followed by arsenic detection. The detection methods must be highly sensitive and capable of quantifying inorganic and organic species at the ng L^{-1} to μg L^{-1} level (Yalcin and Le, 1998). Many combinations of separation and detection methods have been used (Bohari *et al.*, 2001; Ipolyi and Fodor, 2000; Lindemann *et al.*, 2000; Martinez-Bravo *et al.*, 2001; National Research Council, 1999; Taniguchi *et al.*, 1999). All of them require expensive instrumentation and highly skilled operators and none has acquired "routine" or accredited status.

A widely used but indirect method of As(III)/As(V) speciation involves no preseparation but involves two separate determinations, with and without prereduction. The rate of AsH_3 production by sodium borohydride ($NaBH_4$) reduction depends primarily on the initial oxidation state of the arsenic in solution and the solution pH. Under typical operating conditions of about pH 6 where the neutral As(III) species, H_3AsO_3, predominates, only As(III) is converted to the hydride (Anderson *et al.*, 1986; Driehaus and Jekel, 1992). For the most part, the negatively charged As(V) species are not converted. For the determination of total arsenic, As(V) to As(III) prereduction can be achieved by adding a mixture of HCl, KI, and ascorbic acid ideally at pH < 1 to ensure full protonation and efficient HG. High concentrations of HCl are particularly effective. Arsenic(V) can then be estimated by difference. AFS or AAS provide sensitive and fairly robust detectors for the arsine gas produced. High concentrations of some metal ions, particularly Fe^{3+} and Cu^{2+}, can interfere with the HG but this can be overcome by adding masking agents such as thiourea (Anderson *et al.*, 1986) or by their prior removal with a cation-exchange resin.

Solid-phase speciation. While most speciation studies have been concerned with redox speciation in solution, speciation in the solid phase is also of interest. Both reduced and oxidized arsenic and selenium species can be adsorbed on minerals, soils, and sediments albeit with differing affinities (see Sections 9.02.5.3 and 9.02.7.2). Such adsorption has been demonstrated on metal oxides and clays and also probably takes place, to some extent, on carbonates, phosphates, sulfides, and perhaps organic matter. Structural arsenic and selenium may also be characterized.

Solid-phase speciation has been measured both by wet chemical extraction and, for arsenic, by instrumental methods principally X-ray absorption near edge structure spectroscopy (XANES) (Brown *et al.*, 1999). La Force *et al.* (2000) used XANES and selective extractions to determine the likely speciation of arsenic in a wetland affected by mine wastes. They identified seasonal effects with As(III) and As(V) thought to be associated with carbonates in the summer, iron oxides in the autumn and winter, and silicates in the spring. Extended X-ray absorption fine structure spectroscopy (EXAFS) has been used to determine the oxidation state of arsenic in arsenic-rich Californian mine wastes (Foster *et al.*, 1998b). Typical concentrations of arsenic in soils and sediments (arsenic < 20 mg kg^{-1}) are often too low for EXAFS measurements, but as more powerful photon beams become available, the use of such techniques should increase.

Classical wet chemical extraction procedures have also been used to assess the solid-phase speciation of arsenic, but care must be taken not to oxidize As(III) during extraction (Demesmay and Olle, 1997). Extractions should be carried out in the dark to minimize photochemical oxidation.

9.02.3.2 Selenium

9.02.3.2.1 Total selenium in aqueous samples

Historically, analysis for selenium has been difficult, partly because environmental concentrations are naturally low. Indeed, selenium analysis still remains problematic for many laboratories at concentrations below 0.01 mg L^{-1}, a relatively high concentration in many environments (Steinhoff *et al.*, 1999). Hence, selenium has often been omitted from multi-element geochemical surveys despite its importance (Darnley *et al.*, 1995). Analytical methods with limits of detection of <0.01 mg L^{-1} include colorimetry, total reflectance-XRF, HG-AFS, gas chromatography

(GC) of organic species, ICP-MS, and HG-ICP-AES. Of these, HG-AFS and ICP-MS are probably the most widely used methods. Like arsenic, there are no generally accepted ways of preserving selenium speciation in water samples, and even fewer studies of the factors controlling the stability of the various species. Many of the precautions for arsenic-preserved species (Section 9.02.2.2.3) are also likely to apply to the preservation of selenium species.

Pre-treatment to destroy organic matter. Organic selenium species are more widespread in the environment than comparable arsenic species. The determination of total selenium by most analytical methods requires samples to be pre-treated to remove organic matter, release selenium, and change its oxidation state.

Wet digestion using mixtures of nitric, sulfuric, phosphoric, and perchloric acids, with or without the addition of hydrogen peroxide, has been used for organic samples and natural waters. Nitric acid reduces foaming and/or charring. The trimethyl-selenonium ion is resistant to decomposition by wet digestion. A long period of digestion is, therefore, required for urine and plant materials, which may contain the ion.

Laboratory methods. Fluorimetry has been used widely for selenium analysis in environmental samples but is being superseded by more sensitive instrumental methods. Some of the instrumental methods used for arsenic speciation and analysis can also be used for selenium. In particular, HPLC and HG can separate selenium into forms suitable for detection by AAS, AFS (Ipolyi and Fodor, 2000), or ICP-AES (Adkins *et al.*, 1995). Only Se(IV) forms the hydride, and so Se(VI) must be prereduced to Se(IV) if total selenium is to be determined. This is normally achieved using warm HCl/KBr followed by co-precipitation with $La(OH)_3$ if necessary (Adkins *et al.*, 1995). KI is not used, since it tends to produce some Se(0), which is not reduced by HG. $La(OH)_3$ collects only Se(IV), so the prereduction step to include the contribution from Se(VI) is required *before* co-precipitation. Other methods of pre-concentration include co-precipitation of Se(IV) with hydrous iron oxide or adsorption onto Amberlite IRA-743 resin (Bueno and Potin-Gautier, 2002).

ICP-MS detection of selenium is now favored because of its sensitivity even without HG or other forms of preconcentration. However, selenium can be seriously affected by matrix interferences when using ICP-MS. The polyatomic Ar_2^+, with a mass of 80, overlaps with the most abundant isotope of selenium (^{80}Se). Even using hydrogen as a collision gas results in the formation of 2–5% of selenium hydride for which a correction must be applied. For routine analysis of selenium, the hydride-interference free, but less abundant isotopes, ^{76}Se and ^{82}Se, are usually determined.

9.02.3.2.2 Selenium in solid samples

Direct analysis of solids for selenium by XRF has a detection limit of ~0.5 mg kg^{-1} and so is often insufficiently sensitive. Rock, sediment, and soil samples can be dissolved using wet chemical methods (HF–HCl–etc.) followed by $La(OH)_3$ co-precipitation to separate hydride-forming elements including selenium. This is present as Se(IV) following acid dissolution (Hall and Pelchat, 1997). The methods described above for aqueous samples can then be used.

Modern thermal ionization mass spectrometry (TIMS) is now sufficiently sensitive and precise to measure individual selenium-isotope abundances (e.g., $^{80}Se/^{76}Se$) in solid samples or residues so that it can be used to study environmental cycling/distributions (Johnson *et al.*, 1999). Microbial reduction of selenate leads to isotopically lighter selenite, i.e., the reduction has a $^{80}Se/^{76}Se$ fractionation factor, ε, of about −5.5‰ (Johnson *et al.*, 1999). INAA has been used to determine different selenium isotopes, especially ^{75}Se in plant tracer studies.

9.02.3.2.3 Selenium speciation

Selenium speciation in waters is poorly understood, although in principle, it can be determined using HG with and without a prereduction step (see Section 9.02.3.2.1). Ion-exchange chromatography is used extensively to determine selenium species in plant extracts, and GC can measure volatile selenium compounds. Recent developments in anion-exchange HPLC and MS techniques (ICP–dynamic reaction cell–MS, TIMS and multiple collector–MS) mean that it is now possible to determine selenium-isotope abundances and concentrations in selenamino acids including selenocysteine and selenomethionine (Gomez-Ariza *et al.*, 2000; Sloth and Larsen, 2000). Electrochemical methods such as cathodic stripping voltammetry (CSV) are highly sensitive and, in principle, can be used for speciation, because only Se(IV) species are electroactive (Lange and van den Berg, 2000). Because of the important role that selenium plays in human nutrition, there is increasing interest in measuring the "bioavailable" amounts especially in foodstuffs using various bioassays.

9.02.3.3 Quality Control and Standard Reference Materials

Although analysts usually determine the precision of their analyses using replicate determinations, the analysis of arsenic and selenium can be affected seriously by contamination and matrix interferences during sampling and analysis. These can be difficult to identify and are best found by sample randomization and the collection of duplicate samples as part of an objective, independent quality-control system (Plant *et al.*, 1975).

The measurement of standard reference materials (SRMs) provides the best method for ensuring that an analytical procedure is producing accurate results in realistic matrices. Many SRMs are available (Govindaraju, 1994; Rasmussen and Andersen, 2002). The most widely used are those supplied by National Institute of Standards and Technology (NIST). Arsenic and selenium concentrations have been certified in a range of natural waters, sediments, and soils (Tables 4 and 5). The SLRS range of certified standards from the National Research Council of Canada also includes several river waters with much lower arsenic concentrations than the NIST standards (~0.2–1 μg L^{-1}). Certified standards for As(III)/As(V) and Se(IV)/Se(VI) speciation are available commercially (e.g., SPEX Certiprep® speciation standards).

The Canadian Certified Reference Materials Project (CCRMP) also provides reference materials for lake sediment, stream sediments, and soils (tills) for arsenic but not for selenium. However, the BCR-320 certified river sediment from the Institute for Reference Materials and Measurements (IRRM), Geel, Belgium is certified at 76.7 mg kg^{-1} arsenic and 0.214 mg kg^{-1} selenium.

The Geological Survey of Japan (GSJ) provides a wide range of rock RMs along with "recommended" arsenic and selenium concentrations. The US Geological Survey (USGS) issues 21 RMs for which it provides "recommended" and "information" (when less than three independent methods have been used) concentrations. Thirteen of these include data for arsenic and two for both arsenic and selenium (the SGR-1 shale and CLB-1 coal samples). Hall and Pelchat (1997) have analyzed 55 geological RMs for arsenic, bismuth, antimony, selenium, and tellurium including RMs from the USGS, IGGE (China), GSJ, CCRMP, and NRC (Canada) programs.

9.02.4 ABUNDANCE AND FORMS OF ARSENIC IN THE NATURAL ENVIRONMENT

9.02.4.1 Abundance in Rocks, Soils, and Sediments

The average crustal abundance of arsenic is 1.5 mg kg^{-1}. The element is strongly chalocophile. Approximately 60% of natural arsenic minerals are arsenates, 20% sulfides and sulfosalts, and the remaining 20% are arsenides,

Table 4 SRMs for natural waters from various suppliers.

SRM	*TMRAIN 95*	*SLRS-4*	*SRM 1640*	*SRM 1643d/e*[a]	*CRM 609*	*CRM 610*	*CRM 403*	*CASS-4*
Supplier	NWRI	NRC	NIST	NIST	IRMM	IRMM	IRMM	NRC
Medium	Rainwater	River water	Natural water	Freshwater	Groundwater	Groundwater	Seawater	Seawater, nearshore
Arsenic (μg L^{-1})	1.07	0.68	26.67	56.02	1.29[b]	10.8[b]	1.46[b]	1.11
Selenium (μg L^{-1})			21.96	11.43				

[a] 1643d is presently out of stock; a new batch (1643e) is being prepared. NWRI = National Water Research Institute, Environment Canada; NIST = National Institute of Standards and Technology, Gaithersburg, Maryland, USA; IRMM = Institute for Reference Materials and Measurements; NRC = National Research Council of Canada. [b] μg kg^{-1}.

Table 5 SRMs for soils, sediment, and sludges from the National Institute of Standards and Technology (NIST).

SRM	*Medium*	*As* (mg kg^{-1})	*Se* (mg kg^{-1})
1646a	Estuarine sediment	6.23	0.193
1944	New York/New Jersey waterway sediment	18.9	1.4
2586	Trace elements in soil containing lead paint	8.7	0.6
2587	Trace elements in soil containing lead paint	13.7	
2709	San Joaquin soil	17.7	1.57
2710	Montana soil (highly elevated traces)	626	
2711	Montana soil (moderately elevated traces)	105	1.52
2781	Domestic sludge	7.82	16.0

arsenites, oxides, alloys, and polymorphs of elemental arsenic. Arsenic concentrations of more than 10^5 mg kg^{-1} have been reported in sulfide minerals and up to 7.6×10^4 mg kg^{-1} in iron oxides (Smedley and Kinniburgh, 2002). However, concentrations are typically much lower. Arsenic is incorporated into primary rock-forming minerals only to a limited extent, for example, by the substitution of As^{3+} for Fe^{3+} or Al^{3+}. Therefore, arsenic concentrations in silicate minerals are typically ~1 mg kg^{-1} or less (Smedley and Kinniburgh, 2002). Many igneous and metamorphic rocks have average arsenic concentrations of 1–10 mg kg^{-1}. Similar concentrations are found in carbonate minerals and rocks.

Arsenic concentrations in sedimentary rocks can be more variable. The highest arsenic concentrations (20–200 mg kg^{-1}) are typically found in organic-rich and sulfide-rich shales, sedimentary ironstones, phosphatic rocks, and some coals (Smedley and Kinniburgh, 2002). Although arsenic concentrations in coals can range up to 3.5×10^4 mg kg^{-1}, concentrations in the range <1–17 mg kg^{-1}are more typical (Gluskoter *et al.*, 1977; Palmer and Klizas, 1997). Evidence for arsenic enrichment in peat is equivocal. Shotyk (1996) found a maximum of 9 mg As kg^{-1} in two 5,000–10,000 yr old Swiss peat profiles. In the profile with the lower ash content, the arsenic content was 1 mg kg^{-1} or lower.

In sedimentary rocks arsenic is concentrated in clays and other fine-grained sediments, especially those rich in sulfide minerals, organic matter, secondary iron oxides, and phosphates. The average concentration of arsenic in shale is an order of magnitude greater than in sandstones, limestones, and carbonate rocks. Arsenic is strongly sorbed by oxides of iron, aluminum, and manganese as well as some clays, leading to its enrichment in ferromanganese nodules and manganiferous deposits.

Alluvial sands, glacial till, and lake sediments typically contain <1–15 mg kg^{-1} arsenic. Stream sediments from England and Wales had a median arsenic concentration of 10 mg kg^{-1} (Webb, 1978). The median arsenic concentration in stream sediments from 20 study areas across the USA collected as part of the National Water-Quality Assessment (NAWQA) program was 6.3 mg kg^{-1} (Rice, 1999).

The arsenic concentration in soils shows a similar range to that found in sediments except where contaminated by industrial or agricultural activity. A survey of 2,600 soils from the Welsh borderlands had a median arsenic concentration of 11 mg kg^{-1} (BGS, 1979–2002). Concentrations of 1,000 mg kg^{-1} or more have been found at contaminated sites close to smelters or industrial sites (Lumsdon *et al.*, 2001).

9.02.4.2 National and International Standards for Drinking Water

National standards for maximum concentrations of arsenic in drinking water have been declining since the early 1980s as the high toxicity of arsenic has become apparent. The 1903 report of the Royal Commission on Arsenic Poisoning in the UK set a standard of 150 μg L^{-1}. In 1942, the US Public Health Service set a drinking-water standard of 50 μg L^{-1} for interstate water carriers and this was adopted nationally by the US EPA in 1975. The WHO guideline value for arsenic in drinking water was reduced from 50 μg L^{-1} to a provisional value of 10 μg L^{-1} in 1993. In most western countries, the limit for arsenic in drinking water is now also 10 μg L^{-1} (Yamamura, 2003). This includes the EC and the USA. The standard in Switzerland remains at 50 μg L^{-1}. While the US EPA maximum contaminant level (MCL) is now 10 μg L^{-1}, they have also set a maximum contaminant level goal (MCLG) of zero for arsenic in drinking water, reflecting the risk to human health.

9.02.4.3 Abundance and Distribution in Natural Waters

Concentrations of arsenic in natural waters vary by more than four orders of magnitude and depend on the source of the arsenic and the local geochemical conditions (Smedley and Kinniburgh, 2002). The greatest range and highest concentrations of arsenic are found in groundwaters, soil solutions, and sediment pore waters because of the presence of favorable conditions for arsenic release and accumulation. Because the range in concentrations of arsenic in water is large, "typical" values are difficult to derive. Concentrations can also vary significantly with time.

9.02.4.3.1 Atmospheric precipitation

Arsenic enters the atmosphere as a result of wind erosion, volcanic emissions, low-temperature volatilization from soils, marine aerosols, and pollution. It is returned to the Earth's surface by wet and dry deposition. The most important pollutant inputs are from smelter operations and fossil-fuel combustion. Concentrations of arsenic in rainfall and snow in rural areas are typically less than 0.03 μg L^{-1} (Table 6), although they are generally higher in areas affected by smelters, coal burning, and volcanic emissions. Andreae (1980) found arsenic concentrations of ~0.5 μg L^{-1} in rainfall from areas affected by smelting and coal burning. Higher concentrations (average 16 μg L^{-1}) have been reported in rainfall 35 km downwind of a copper smelter in Seattle, USA (Crecelius, 1975). Values for Arizona snowpacks

Table 6 Concentration ranges of arsenic in various water bodies.

Water body and location	*Arsenic concentration: average or range* (μg L^{-1})
Rainwater	
Maritime	0.02
Terrestrial (W. USA)	0.013–0.032
Coastal (Mid-Atlantic, USA)	0.1 (<0.005–1.1)
Snow (Arizona)	0.14 (0.02–0.42)
Terrestrial rain	0.46
Seattle rain, impacted by copper smelter	16
River water	
Various	0.83 (0.13–2.1)
Norway	0.25 (<0.02–1.1)
Southeast USA	0.15–0.45
USA	2.1
Dordogne, France	0.7
Po River, Italy	1.3
Polluted European rivers	4.5–45
River Danube, Bavaria	3 (1–8)
Schelde catchment, Belgium	0.75–3.8 (up to 30)
High-As groundwater influenced	
Northern Chile	190–21,800
Northern Chile	400–450
Córdoba, Argentina	7–114
Geothermally influenced	
Sierra Nevada, USA	0.20–264
Waikato, New Zealand	32 (28–36)
	44 (19–67)
Madison and Missouri Rivers, USA	10–370
Mining influenced	
Ron Phibun, Thailand	218 (4.8–583)
Ashanti, Ghana	284 (<2–7,900)
British Columbia, Canada	17.5 (<0.2–556)
Lake water	
British Columbia	0.28 (<0.2–0.42)
Ontario	0.7
France	0.73–9.2 (high Fe)
Japan	0.38–1.9
Sweden	0.06–1.2
Geothermally influenced	
Western USA	0.38–1,000
Mining influenced	
Northwest Territories, Canada	270 (64–530)
Ontario, Canada	35–100
Estuarine water	
Oslofjord, Norway	0.7–2.0
Saanich Inlet, British Columbia	1.2–2.5
Rhône Estuary, France	2.2 (1.1–3.8)
Krka Estuary, Yugoslavia	0.13–1.8
Mining and industry influenced	
Loire Estuary, France	up to 16
Tamar Estuary, UK	2.7–8.8
Schelde Estuary, Belgium	1.8–4.9
Seawater	
Deep Pacific and Atlantic	1.0–1.8
Coastal Malaysia	1.0 (0.7–1.8)
Coastal Spain	1.5 (0.5–3.7)
Coastal Australia	1.3 (1.1–1.6)

(continued)

Table 6 (continued).

Water body and location	*Arsenic concentration: average or range* ($\mu g\ L^{-1}$)
Groundwater	
Various USA aquifers	<1–2,600
Various UK aquifers	<0.5–57
Bengal Basin, West Bengal, Bangladesh	<0.5–3,200
Chaco-Pampean Plain, Argentina	<1–5,300
Lagunera, northern Mexico	8–620
Inner Mongolia, China	<1–2,400
Taiwan	<10 to 1,820
Great Hungarian Plain, Hungary, Romania	<2–176
Red River Delta, Vietnam	1–3,050
Mining-contaminated groundwaters	50–10,000
Geothermal water	<10–50,000
Mineralized area, Bavaria, Germany	<10–150
Herbicide-contaminated groundwater, Texas	408,000
Mine drainage	
Various, USA	<1–850,000
Ural Mountains	400,000
Sediment pore water	
Baseline, Swedish estuary	1.3–166
Baseline, clays, Saskatchewan, Canada	3.2–99
Baseline, Amazon shelf sediments	Up to 300
Mining-contam'd, British Columbia	50–360
Tailings impoundment, Ontario, Canada	300–100,000

After Smedley and Kinniburgh (2002).

(Barbaris and Betterton, 1996) are also slightly above baseline concentrations, probably because of inputs from smelters, power plants, and soil dust. In most industrialized countries, sources of airborne arsenic are limited as a result of air-pollution control measures. Unless significantly contaminated, atmospheric precipitation contributes little arsenic to surface waters.

9.02.4.3.2 River water

Concentrations of arsenic in river waters are also low (typically in the range 0.1–2.0 $\mu g\ L^{-1}$; Table 6). These vary according to bedrock lithology, river flow, the composition of the surface recharge, and the contribution from base flow. The lowest concentrations have been found in rivers draining arsenic-poor bedrocks. Seyler and Martin (1991) reported average concentrations as low as 0.13 $\mu g\ L^{-1}$ in rivers flowing over karstic limestone in the Krka region of Yugoslavia. Lenvik *et al.* (1978) also reported average concentrations of ~0.25 $\mu g\ L^{-1}$ arsenic in rivers draining basement rocks in Norway.

Relatively high concentrations of naturally occurring arsenic in rivers can occur as a result of geothermal activity or the influx of high-arsenic groundwaters. Arsenic concentrations of 10–70 $\mu g\ L^{-1}$ have been reported in river waters from geothermal areas, including the western USA and New Zealand (McLaren and Kim, 1995; Nimick *et al.*, 1998; Robinson *et al.*, 1995). Higher concentrations, up to 370 $\mu g\ L^{-1}$, have been reported in the Madison River in Wyoming and Montana as a result of inputs from the Yellowstone geothermal system. Wilkie and Hering (1998) also found concentrations in the range 85–153 $\mu g\ L^{-1}$ in Hot Creek, a tributary of the Owens River, California. Some river waters affected by geothermal activity show distinct seasonal variations in arsenic concentration. Concentrations in the Madison River are highest during low-flow conditions, reflecting the increased proportion of geothermal water (Nimick *et al.*, 1998). In the Waikato river system of New Zealand, arsenic maxima occur in the summer months, reflecting temperature-controlled microbial reduction of As(V) to the more mobile As(III) species (McLaren and Kim, 1995).

Increased arsenic concentrations are also found in some river waters dominated by base flow in arid areas. Such waters often have a high pH and alkalinity. For example, surface waters from the Loa River Basin of northern Chile (Atacama desert) contain naturally occurring arsenic in the range 190–21,800 $\mu g\ L^{-1}$ (Cáceres *et al.*, 1992). The high arsenic concentrations correlate with high salinity. While geothermal inputs of arsenic are likely to be important, evaporative concentration of the base-flow-dominated river water is also likely to concentrate arsenic in the prevailing arid conditions. High arsenic concentrations (up to 114 $\mu g\,L^{-1}$) have also been reported in alkaline river waters from central Argentina, where regional groundwater arsenic concentrations are high (Lerda and Prosperi, 1996).

Although bedrock influences river-water arsenic concentrations, rivers with typical pH and alkalinity values (ca. pH 5–7, $HCO_3^- < 100$ mg L^{-1}) generally contain lower concentrations of arsenic, even where groundwater concentrations are high, because of oxidation and adsorption of arsenic onto particulate matter in the stream bed and dilution by surface runoff. Arsenic concentrations in the range <0.5–2.7 $\mu g\,L^{-1}$ have been reported for seven river-water samples from Bangladesh, with one sample containing 29 $\mu g\,L^{-1}$ (BGS and DPHE, 2001).

High arsenic concentrations in river waters can also reflect pollution from industrial or sewage effluents. M. O. Andreae and T. W. Andreae (1989) reported arsenic concentrations up to 30 $\mu g\,L^{-1}$ in water from the River Zenne, Belgium which is affected by urban and industrial waste, particularly sewage. The background arsenic concentration was in the range 0.75–3.8 $\mu g\,L^{-1}$. Durum *et al.* (1971) found that 79% of surface waters from the USA had arsenic concentrations below the detection limit of 10 $\mu g\,L^{-1}$. The highest concentration, 1,100 $\mu g\,L^{-1}$, was reported from Sugar Creek, South Carolina, downstream of an industrial complex.

Arsenic can also be derived from mine wastes and tailings. Azcue and Nriagu (1995) reported baseline concentrations of 0.7 $\mu g\,L^{-1}$ in the Moira River, Ontario, upstream of gold-mine tailings, and concentrations up to 23 $\mu g\,L^{-1}$ downstream. Azcue *et al.* (1994) reported concentrations up to 556 $\mu g\,L^{-1}$ (average 17.5 $\mu g\,L^{-1}$) in streams draining mine tailings in British Columbia. Williams *et al.* (1996) and Smedley (1996) noted high arsenic concentrations (typically ~200–300 $\mu g\,L^{-1}$) in surface waters from areas of tin and gold mining, respectively. Such anomalies tend to be localized because of the strong adsorption of arsenic by oxide minerals, especially iron oxide, under oxidizing and neutral to acidic conditions typical of many surface waters. Arsenic concentrations are, therefore, not always very high even in mining areas. For example, stream-water arsenic concentrations from the Dalsung Cu–W mining area of Korea ranged from 0.8 $\mu g\,L^{-1}$ to 19 $\mu g\,L^{-1}$ (Jung *et al.*, 2002).

9.02.4.3.3 Lake water

Arsenic concentrations in lake waters are typically close to or lower than those of river water. Baseline concentrations of <1 $\mu g\,L^{-1}$ have been reported from Canada (Table 6) (Azcue *et al.*, 1995; Azcue and Nriagu, 1995). Higher concentrations in lake waters may reflect geothermal sources or mining activity. Concentrations of 100–500 $\mu g\,L^{-1}$ have been reported in some mining areas and up to 1,000 $\mu g\,L^{-1}$ in geothermal areas. However, arsenic concentrations can be much lower in mining-affected lake waters as a result of adsorption onto iron oxides under neutral to mildly acidic conditions. For example, Azcue *et al.* (1994) reported concentrations in lake waters affected by mining activity in Canada of ~0.3 $\mu g\,L^{-1}$, close to background values.

High arsenic concentrations can also occur in alkaline, closed-basin lakes. Mono Lake, California, USA has dissolved arsenic concentrations of $(10–20) \times 10^4$ $\mu g\,L^{-1}$ with pH values in the range 9.5–10 as a result of the combined influences of geothermal activity, weathering of mineralized volcanic rocks, evaporation of water at the lake surface, and a thriving population of arsenate-respiring bacteria (Maest *et al.*, 1992; Oremland *et al.*, 2000).

Arsenic concentrations show considerable variations in stratified lakes because of changes in redox conditions or biological activity (Aggett and O'Brien, 1985; Hering and Kneebone, 2002). Arsenic concentrations increase with depth in lake waters in Ontario, probably because of an increasing ratio of As(III) to As(V) and an influx of mining-contaminated sediment pore waters at the sediment–water interface (Azcue and Nriagu, 1995). In other cases, seasonal depletion at the surface parallels that of nutrients such as silicate (Kuhn and Sigg, 1993). Concentrations are higher at depth in summer when the proportion of As(III) is greatest, probably reflecting lower oxygen concentrations as a result of biological productivity.

9.02.4.3.4 Seawater and estuaries

Average arsenic concentrations in open seawater are typically ~1.5 $\mu g\,L^{-1}$ (Table 6). Surface depletion, as with nutrients such as silicate, has been observed in some seawater samples but not others. Concentrations in estuarine water are more variable because of different river inputs and salinity or redox gradients but they typically contain less than 4 $\mu g\,L^{-1}$. Peterson and Carpenter (1983) found arsenic concentrations of 1.2–2.5 $\mu g\,L^{-1}$ in waters from Saanich Inlet, British Columbia. Concentrations below 2 $\mu g\,L^{-1}$ were found in Oslofjord, Norway (Abdullah *et al.*, 1995). Higher concentrations reflect industrial or mining effluent (e.g., Tamar, Schelde, Loire Estuaries) or inputs of geothermal water.

Some studies have reported conservative behavior during estuarine mixing. In the unpolluted Krka Estuary of Yugoslavia, Seyler and Martin (1991) observed a linear increase in total arsenic with increasing salinity, ranging from $0.13\ \mu g\ L^{-1}$ in freshwaters to $1.8\ \mu g\ L^{-1}$ offshore. Other studies however, have observed nonconservative behavior in estuaries due to processes such as diffusion from sediment pore waters, co-precipitation with iron oxides, or anthropogenic inputs (M. O. Andreae and T. W. Andreae, 1989; Andreae *et al.*, 1983). The flocculation of iron oxides at the freshwater–saline interface as a result of increase in pH and salinity can lead to major decrease in the arsenic flux to the oceans (Cullen and Reimer, 1989).

9.02.4.3.5 Groundwater

The concentration of arsenic in most groundwaters is $<10\ \mu g\ L^{-1}$ (Edmunds *et al.*, 1989; Welch *et al.*, 2000) and often below the detection limit of routine analytical methods. An analysis of groundwaters used for public supply in the USA showed that only 7.6% exceeded $10\ \mu g\ L^{-1}$ with 64% containing $<1\ \mu g\ L^{-1}$ (Focazio *et al.*, 1999). Nonetheless, naturally high-arsenic groundwaters are found in aquifers in some areas of the world and concentrations occasionally reach the $mg\ L^{-1}$ range (Smedley and Kinniburgh, 2002). Industrially contaminated groundwater can also give rise to very high dissolved arsenic concentrations, but the affected areas are usually localized. For example, Kuhlmeier (1997) found concentrations of arsenic up to $4.08 \times 10^5\ \mu g\ L^{-1}$ in groundwater close to a herbicide plant in Texas.

The physicochemical conditions favoring arsenic mobilization in aquifers are variable, complex, and poorly understood, although some of the key factors leading to high groundwater arsenic concentrations are now known. Mobilization can occur under strongly reducing conditions where arsenic, mainly as As(III), is released by desorption from, and/or dissolution of, iron oxides. Many such aquifers are sufficiently reducing for sulfate reduction, and in some cases for methane generation, to occur (Ahmed *et al.*, 1998). Immobilization under reducing conditions is also possible. Some sulfate-reducing microorganisms can respire As(V) leading to the formation of an As_2S_3 precipitate (Newman *et al.*, 1997a,b). Some immobilization of arsenic may also occur if iron sulfides are formed.

Reducing conditions favorable for arsenic mobilization have been reported most frequently from young (Quaternary) alluvial, deltaic sediments where the interplay of tectonic, isostatic, and eustatic factors have resulted in complex patterns of sedimentation and the rapid burial of large amounts of sediment together with fresh organic matter during delta progradation. Thick sequences of young sediments are quite often the sites of high groundwater arsenic concentrations. The most notable example of these conditions is the Bengal Basin which includes Bangladesh and West Bengal (BGS and DPHE, 2001). Other examples include Nepal, Myanmar, Cambodia, parts of northern China (Luo *et al.*, 1997; Smedley *et al.*, 2003; Wang and Huang, 1994), the Great Hungarian Plain of Hungary and Romania (E. S. Gurzau and A. E. Gurzau, 2001; Varsanyi *et al.*, 1991), the Red River Delta of Vietnam (Berg *et al.*, 2001), and parts of the western USA (Korte, 1991; Welch *et al.*, 2000). Recent groundwater extraction in many of these areas, either for public supply or for irrigation, has induced increased groundwater flow. This could induce further transport of arsenic (Harvey *et al.*, 2002).

High concentrations of naturally occurring arsenic are also found in oxidizing conditions where groundwater pH values are high (ca. >8) (Smedley and Kinniburgh, 2002). In such environments, inorganic As(V) predominates and arsenic concentrations are positively correlated with those of other anion-forming species such as HCO_3^-, F^-, H_3BO_3, and $H_2VO_4^-$. Examples include parts of western USA, for example, San Joaquin Valley, California (Fujii and Swain, 1995), Lagunera region of Mexico (Del Razo *et al.*, 1990), Antofagasta area of Chile (Cáceres *et al.*, 1992; Sancha and Castro, 2001), and the Chaco-Pampean Plain of Argentina (Nicolli *et al.*, 1989; Smedley *et al.*, 2002) (Table 6). These high-arsenic groundwater provinces are usually in arid or semi-arid regions where groundwater salinity is high. Evaporation has been suggested to be an important additional cause of arsenic accumulation in some arid areas (Welch and Lico, 1998).

High concentrations of arsenic have also been found in groundwater from areas of bedrock and placer mineralization which are often the sites of mining activities. Arsenic concentrations of up to $5{,}000\ \mu g\ L^{-1}$ have been found in groundwater associated with the former tin-mining activity in the Ron Phibun area of Peninsular Thailand, the source most likely being oxidized arsenopyrite (FeAsS). Many cases have also been reported from other parts of the world including the USA, Canada, Poland, and Austria. Examples include the Fairbanks mining district of Alaska where arsenic concentrations up to $10^4\ \mu g\ L^{-1}$ have been found in groundwater (Welch *et al.*, 1988), and the Coeur d'Alene district of Idaho where groundwater arsenic concentrations of up to $1{,}400\ \mu g\ L^{-1}$ have been reported (Mok and Wai, 1990).

Groundwater arsenic problems in nonmined mineralized areas are less common, but Boyle *et al.* (1998) found concentrations up to

580 μg L^{-1} in groundwater from the sulfide mineralized areas of Bowen Island, British Columbia. Heinrichs and Udluft (1999) also found arsenic concentrations up to 150 μg L^{-1} in groundwater from a mineralized sandstone aquifer in Bavaria.

9.02.4.3.6 Sediment pore water

Much higher concentrations of arsenic frequently occur in pore waters extracted from unconsolidated sediments than in overlying surface waters. Widerlund and Ingri (1995) reported concentrations in the range 1.3–166 μg L^{-1} in pore waters from the Kalix River estuary, northern Sweden. Yan *et al.* (2000) found concentrations in the range 3.2–99 μg L^{-1} in pore waters from clay sediments in Saskatchewan, Canada.

High concentrations are frequently found in pore waters from geothermal areas. Aggett and Kriegman (1988) reported arsenic concentrations up to 6,430 μg L^{-1} in anoxic pore waters from Lake Ohakuri, New Zealand. Even higher concentrations have been found in pore waters from sediments contaminated with mine tailings or draining ore deposits. McCreadie *et al.* (2000) reported concentrations up to 10^5 μg L^{-1} in pore waters extracted from mine tailings in Ontario, Canada. High pore-water arsenic concentrations probably reflect the strong redox gradients often over a few centimeters below the sediment–water interface. Burial of fresh organic matter and the slow diffusion of oxygen through the sediment lead to reducing conditions with the consequent reduction of As(V) to As(III) and the desorption and dissolution of arsenic from iron and manganese oxides.

There is much evidence for arsenic release into shallow sediment pore waters and overlying surface waters in response to temporal variations in redox conditions. Sullivan and Aller (1996) investigated arsenic cycling in shallow sediments from an unpolluted area of the Amazonian offshore shelf. They found pore-water arsenic concentrations up to 300 μg L^{-1} in anaerobic sediments with nearly coincident peaks of dissolved arsenic and iron. The peaks for iron concentration were often slightly above those of arsenic (Figure 1). The magnitude of the peaks and their depths varied from place to place and possibly seasonally but were typically between 50 cm and 150 cm beneath the sediment–water interface (Sullivan and Aller, 1996). There was no correlation between pore-water arsenic concentrations and sediment arsenic concentrations (Figure 1).

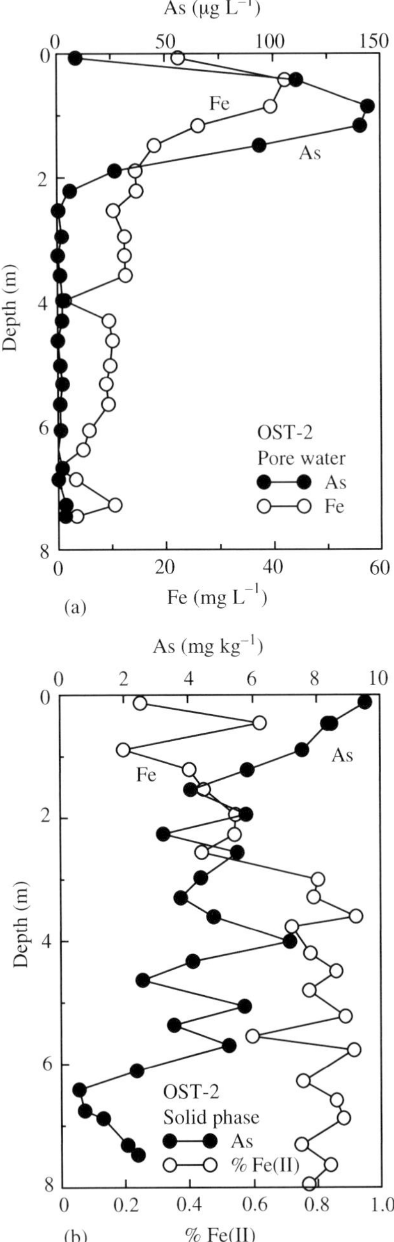

Figure 1 (a) Pore water and (b) cold 6 M HCl-extractable element concentrations in marine sediments from a pristine environment on the River Amazon shelf some 120 km off the coast of Brazil (after Sullivan and Aller, 1996, figure 3).

9.02.4.3.7 Acid mine drainage

Acid mine drainage, which can have pH values as low as −3.6 (Nordstrom *et al.*, 2000), can contain high concentrations of many solutes, including iron and arsenic. The highest reported arsenic concentration, 8.5×10^5 μg L^{-1}, was found in an acid seep in the Richmond mine, California (Nordstrom and Alpers, 1999). Plumlee *et al.* (1999) reported concentrations ranging from <1 μg L^{-1} to 3.4×10^5 μg L^{-1} in 180 samples of mine drainage from the USA, with the highest values from the Richmond mine. Gelova (1977) also reported arsenic concentrations of 4×10^5 $\mu g\ L^{-1}$ in the Ural Mountains. Dissolved arsenic in AMD is rapidly removed as the pH increases and as iron is oxidized and precipitated as hydrous ferric oxide (HFO), co-precipitating large amounts of arsenic.

9.02.4.4 Arsenic Species in Natural Waters

The speciation of arsenic in natural waters is controlled by reduction, oxidation, and methylation reactions that affect its solubility, transport, bioavailability, and toxicity (Hering and Kneebone, 2002). Inorganic speciation is important since the varying protonation and charge of the arsenic species present at different oxidation states has a strong effect on their behavior, for example, their adsorption. While the concentrations of organic arsenic species are low in most natural environments, the methylated and dimethylated As(III) species are now of considerable interest since they have been found to be more cytotoxic, more genotoxic, and more potent enzyme inhibitors than inorganic As(III) (Thomas *et al.*, 2001).

9.02.4.4.1 Inorganic species

Redox potential (Eh) and pH are the most important factors governing inorganic arsenic speciation. Under oxidizing conditions, and pH less than ~6.9, $H_2AsO_4^-$ is dominant, whereas at higher pH, $HAsO_4^{2-}$ is dominant. $H_3AsO_4^0$ and AsO_4^{3-} may be present in extremely acid and alkaline conditions, respectively (Figure 2; Nordstrom and Archer, 2003; Yan *et al.*, 2000). Under reducing conditions where the pH is less than ~9.2, the uncharged arsenite species, H_3AsO_3, predominates. Native arsenic is stable under strongly reducing conditions.

In the presence of high concentrations of reduced sulfur and low pH, dissolved As(III) sulfide species can be formed rapidly by reduction of arsenate by H_2S. There is strong evidence for the existence of the trimer, $As_3S_4(SH)_2^-$, under strongly reducing, acidic and sulfur-rich conditions, with the thioarsenite species, $AsO(SH)_2^-$ appearing at higher pHs (Helz *et al.*, 1995; Nordstrom and Archer, 2003; Rochette *et al.*, 2000; Schwedt and Rieckhoff, 1996). Reducing, acidic, and sulfur-rich conditions also favor precipitation of orpiment (As_2S_3), realgar (AsS), or other arsenic sulfide minerals (Cullen and Reimer, 1989). High concentrations of arsenic are unlikely in acidic waters containing high concentrations of free sulfide (Moore *et al.*, 1988). In more alkaline environments, As(III) sulfides are more soluble and higher dissolved arsenic concentrations could persist. There is some evidence for the existence of As(V) carbonate species (Kim *et al.*, 2000) but their environmental significance

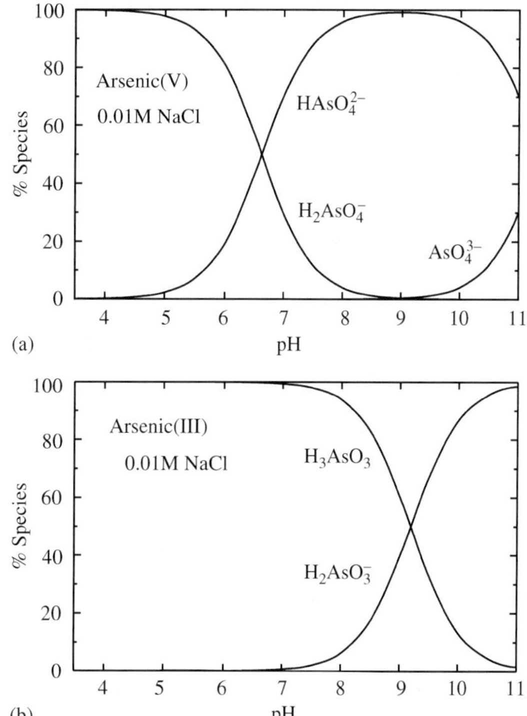

Figure 2 Speciation of: (a) As(V) and (b) As(III) in a 0.01 M NaCl medium as a function of pH at 25 °C (source Smedley and Kinniburgh, 2002).

remains to be understood. Like dissolved hydrogen, arsine is only expected under extremely reducing conditions. Green rusts, complex Fe(II)–Fe(III) hydroxide minerals which form under reducing conditions, have been shown to be able to reduce selenate to selenite abiotically but not arsenate to arsenite (Randall *et al.*, 2001).

The Eh–pH diagram for the As–O–S system is shown in Figure 3. While such diagrams are useful, they necessarily simplify highly complex natural systems. For example, iron is not included despite its strong influence on arsenic speciation. Hence, scorodite ($FeAsO_4 \cdot 2H_2O$), an important arsenic-bearing mineral found under a wide range of near-neutral, oxidizing conditions, is not represented. Neither is the co-precipitation of arsenic with pyrite or the formation of FeAsS under reducing conditions. The relative stability of the various As–S minerals is very sensitive to their assumed free energies of formation and the stability of the various soluble As–S species. The Eh–pH diagram can vary significantly depending on the chosen forms of realgar and As_2S_3, including their crystallinity.

The extent of redox equilibrium in natural waters has been the cause of considerable discussion. In the case of arsenic, Cherry *et al.* (1979) suggested that redox equilibrium was sufficiently rapid for As(V)/As(III) ratios to be

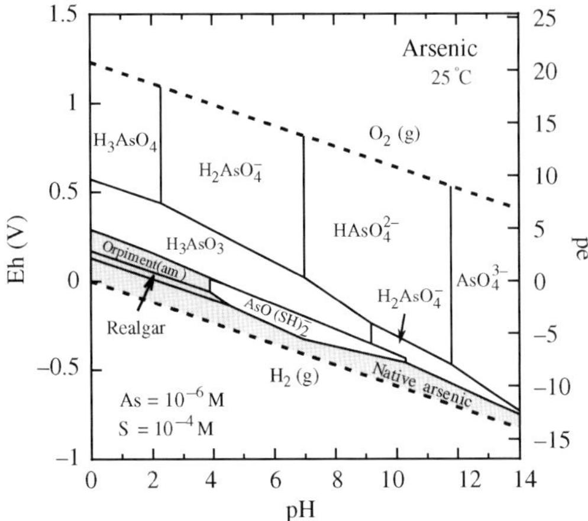

Figure 3 Eh–pH stability diagram for arsenic in the presence of sulfur at 25 °C, 1 bar total pressure. The stability field for water is shown by the dashed lines. The gray area represents a solid phase.

useful indicators of redox state. Subsequent findings have been somewhat equivocal (Welch *et al.*, 1988), although some recent data for pore water As(V)/As(III) ratios and Eh measurements have indicated substantial consistency (Yan *et al.*, 2000). While the rate of oxidation of As(III) in groundwater is difficult to quantify under field conditions, the rates are believed to be slow. However, biological activity in these waters is also generally low making redox equilibrium easier to attain than in more productive environments.

9.02.4.4.2 *Organic species*

Organic-arsenic species are important in food, especially fish and marine invertebrates such as lobsters (AsB and arsenosugars) and in blood and urine (MMA and DMA), although they usually are only a minor component of arsenic in natural waters (Francesconi and Kuehnelt, 2002; National Research Council, 1999). Their concentrations are greatest in organic-rich waters such as soil and sediment pore waters and productive lake waters, and least in groundwaters. The concentrations of organic species are increased by methylation reactions catalyzed by microbial activity, including bacteria, yeasts, and algae. The dominant organic forms found are DMA and MMA. Proportions of these two species are reported to increase in summer as a result of increased microbial activity (Hasegawa, 1997). Organic species may also be more abundant close to the sediment–water interface (Hasegawa *et al.*, 1999).

Small concentrations of trimethylarsonate (TMA), AsB, AsC, and phenylarsonate have been observed occasionally (Florencio *et al.*, 1997). Arsenic can also be bound to humic material, but this association has not been well characterized and may involve ternary complexes with strongly bound cations such as Fe^{3+}.

There have been reports of "hidden" arsenic species in natural waters. These are organic species that do not form arsine gas with $NaBH_4$ and were, therefore, undetected in early speciation studies. Some, though not all, such arsenic species are detected after UV irradiation of samples (Hasegawa *et al.*, 1999; National Research Council, 1999).

9.02.4.4.3 *Observed speciation in different water types*

The oxidation states of arsenic in rainwater vary according to source but are likely to be dominantly as As(III) when derived from smelters, coal burning, or volcanic sources. Organic arsenic species may be derived by volatilization from soils, and arsine ($As(-III)H_3$) may be produced in landfills and reducing soils such as paddy soils and peats. Arsenate may be derived from marine aerosols. Reduced forms undergo oxidation in the atmosphere and reactions with atmospheric SO_2 or O_3 are likely (Cullen and Reimer, 1989).

In oxic seawater, the As(V) species predominates, though some As(III) is invariably present especially in anoxic bottom waters. Arsenic(V) should exist mainly as $HAsO_4^{2-}$ and $H_2AsO_4^-$ in the pH range of seawater (pH ~8.2; Figures 2 and 3) and As(III) mainly as the neutral species H_3AsO_3. In fact, relatively high proportions of H_3AsO_3 occur in surface ocean waters (Cullen and Reimer, 1989) where primary productivity is high, often accompanied by increased concentrations of organic arsenic species as a result of methylation reactions by phytoplankton.

The relative proportions of arsenic species in estuarine waters are more variable because of changes in redox, salinity, and terrestrial inputs (Abdullah *et al.*, 1995; Howard *et al.*, 1988). Arsenic(V) tends to dominate, although M. O. Andreae and T. W. Andreae (1989) found increased proportions of As(III) in the Schelde Estuary of Belgium. The highest values occur in anoxic zones near sources of industrial effluent. Increased proportions of As(III) also occur near sources of mine effluent (M. O. Andreae and T. W. Andreae, 1989). Seasonal variations in concentration and speciation have been reported in seasonally anoxic waters (Riedel, 1993). Peterson and Carpenter (1983) reported a clear crossover in the proportions of the two species with increasing depth in the Saanich Inlet of British Columbia. Arsenic(III) represented only 5% (0.10 $\mu g\ L^{-1}$) of the dissolved arsenic above the redox front but 87% (1.58 $\mu g\ L^{-1}$) below it. In marine and estuarine waters, organic forms of arsenic are usually less abundant but are often detected (Howard *et al.*, 1999; Riedel, 1993). Their concentrations depend on the abundance and species of the biota and on temperature.

In lake and river waters, As(V) is generally dominant (Pettine *et al.*, 1992; Seyler and Martin, 1990), although concentrations and relative proportions of As(V) and As(III) vary seasonally according to changes in input sources, redox conditions, and biological activity. The presence of As(III) may be maintained in oxic waters by biological reduction of As(V), particularly during summer months. Higher proportions of As(III) occur in rivers close to sources of As(III)-dominated industrial effluent (M. O. Andreae and T. W. Andreae, 1989) or where there is a component of geothermal water.

Proportions of As(III) and As(V) are particularly variable in stratified lakes with seasonally variable redox gradients (Kuhn and Sigg, 1993). In the stratified, hypersaline, hyperalkaline Mono Lake (California, USA), As(V) predominates in the upper oxic layer and As(III) in the reducing layer (Maest *et al.*, 1992; Oremland *et al.*, 2000). Oremland *et al.* (2000) measured *in situ* rates of dissimilatory As(V) reduction in the lake and found that this could potentially mineralize 8–14% of the annual pelagic primary productivity during meromixis, a significant amount for a trace element, and ~1/3 of the amount of sulfate reduction. Such reduction does not occur in the presence of NO_3. In fact, NO_3 leads to the rapid, microbial re-oxidation of As(III) to As(V) (Hoeft *et al.*, 2002). Iron(III) acts similarly.

The speciation of arsenic in lakes does not always follow thermodynamic predictions. Recent studies have shown that arsenite predominates in the oxidized epilimnion of some stratified lakes, whereas arsenate may persist in the anoxic hypolimnion (Kuhn and Sigg, 1993; Newman *et al.*, 1998; Seyler and Martin, 1989). Proportions of arsenic species can also vary according to the availability of particulate iron and manganese oxides (Kuhn and Sigg, 1993; Pettine *et al.*, 1992). Sunlight could promote oxidation in surface waters (Voegelin and Hug, 2003).

In groundwaters, the ratio of As(III) to As(V) can vary greatly in relation to changes in the abundance of redox-active solids, especially organic carbon, the activity of microorganisms and the extent of convection and diffusion of oxygen from the atmosphere. Arsenic(III) typically dominates in strongly reducing aquifers in which Fe(III) and sulfate reduction is taking place. Reducing, high-arsenic groundwaters from Bangladesh have As(III)/As_T ratios varying between 0.1 and 0.9 but are typically ~0.5–0.6 (Smedley *et al.*, 2001a). Ratios in reducing groundwaters from Inner Mongolia are typically 0.6–0.9 (Smedley *et al.*, 2003). Concentrations of organic forms of arsenic are generally small or negligible in groundwaters (e.g., Chen *et al.*, 1995; Del Razo *et al.*, 1990).

9.02.4.5 Microbial Controls

The toxicity of arsenic results from its ability to interfere with a number of key biochemical processes. Arsenate can interfere with phosphate biochemistry (oxidative phosphorylation) as a result of their chemical similarity. Arsenite tends to inactivate sulfhydryl groups of cysteine residues in proteins (Oremland *et al.*, 2002; Santini *et al.*, 2002). Microbes have evolved various detoxification strategies for dealing with this (Frankenberger, 2002; Mukhopadhyay *et al.*, 2002; Rosen, 2002). Some microbes have also evolved to use arsenic as an energy source. Certain chemoautotrophs oxidize As(III) by using O_2, nitrate, or Fe(III) as a terminal electron acceptor and CO_2 as their sole carbon source. A select group of organisms grows in anaerobic environments by using As(V) for the oxidation of organic matter or H_2 gas (Newman *et al.*, 1998; Oremland *et al.*, 2002, 2000; Stolz and Oremland, 1999). Such so-called dissimilatory arsenate reduction (DAsR) was only discovered relatively recently (Ahmann *et al.*, 1994). Fourteen species of Eubacteria, including *Sulfurospirullum* species, have so far been shown to be capable of DAsR (Herbel *et al.*, 2002a) as well as two species of hyperthermophiles from the domain Archea.

The bacterium *Thiobacillus* has been shown to have a direct role in precipitating ferric arsenate sulfate (Leblanc *et al.*, 1996). Temporal variations between the proportions of arsenate and arsenite have been observed in the Waikato River, New Zealand and may reflect the reduction of As(V) to As(III) by epiphytic bacteria associated with the alga *Anabaena oscillaroides*. Arsenate reduction does not necessarily take place as an energy-providing (dissimilatory) process (Hoeft *et al.*, 2002). Detoxifying arsenate reductases in the cytoplasm do not provide a means of energy generation. Macur *et al.* (2001) found active As(V) to As(III) reduction under oxic conditions in limed mine tailings which they ascribed to a detoxification rather than an energy-producing, respiratory process. The detoxification process is often combined with an As(III) efflux pump to expel the toxic As(III) from the cell. Purely chemical (abiotic) reduction of As(V) to As(III) has not been documented.

Arsenic can also be released indirectly as a result of other microbially induced redox reactions. For example, the dissimilatory iron-reducing bacterium *Shewanella alga* (strain BrY) reduces Fe(III) to Fe(II) in $FeAsO_4 \cdot 2H_2O$, releasing As(V) but not As(III) (Cummings *et al.*, 1999). This process can be rapid (Langner and Inskeep, 2000).

The rapid oxidation of As(III) has also been observed in the geothermally fed Hot Creek in California (Wilkie and Hering, 1998). Oxidation with a pseudo-first-order half-life of ~0.3 h was found to be controlled by bacteria attached to macrophytes. Where microbial activity is high, there is frequently a lack of equilibrium between the various redox couples, including that of arsenic (Section 9.02.4.4). This is especially true of soils (Masscheleyn *et al.*, 1991).

9.02.5 PATHWAYS AND BEHAVIOR OF ARSENIC IN THE NATURAL ENVIRONMENT

Most high-arsenic natural waters are groundwaters from particular settings such as mineralized, mined and geothermal areas, young alluvial deltaic basins, and inland semi-arid basins (Smedley and Kinniburgh, 2002). The most extensive areas of affected groundwater are found in the low-lying deltaic areas of SE Asia, especially the Bengal Basin, and in the large plains ("pampas") of South America. The sediments in these areas typically have "average" total arsenic concentrations, although concentrations may be higher in iron-oxide-rich sediments. The chemical, microbiological, and hydrogeological processes involved in the mobilization of arsenic in such groundwaters are poorly understood, but they probably involve early diagenetic reactions driven by redox and/or pH changes.

9.02.5.1 Release from Primary Minerals

The arsenic in many natural waters is likely to have been derived naturally from the dissolution of a mineral phase. The most important primary sources are sulfide minerals, particularly arsenic-rich pyrite, which can contain up to 10% arsenic and FeAsS. In one study, the greatest concentrations of arsenic were found in fine-grained (<2 μm) pyrite formed at relatively low temperatures (120–200 °C) (Simon *et al.*, 1999). A variety of other sulfide minerals such as orpiment As_2S_3, and realgar As_2S_2 also occur in association with gold and base-metal deposits. Arsenic is a component of some complex copper sulfides such as enargite (Cu_3As_4) and tennantite ($(Cu,Fe)_{12}As_4S_{13}$). Rarer arsenides are also found in mineralized areas. All of these minerals oxidize rapidly on exposure to the atmosphere releasing the arsenic for partitioning between water and various secondary minerals, particularly iron oxides. Both microbially mediated redox reactions (Section 9.02.4.5) and abiotic processes are involved. The microbial oxidation of arsenic minerals such as FeAsS, Cu_3As_4, and As_2S_3 has been discussed by Ehrlich (2002).

Oxidation of sulfide minerals can occur naturally or as a result of mining activity. Arsenic-rich minerals around mines may, therefore, produce arsenic-rich drainage locally, but this tends to be attenuated rapidly as a result of adsorption of various arsenic species by secondary minerals. Some of the best-documented cases of arsenic contamination occur in areas of sulfide mineralization, particularly those associated with gold deposits.

The oxidation is enhanced by mining, mine dewatering, ore roasting, and the redistribution of tailings in ponds and heaps. In the past this has been the cause of serious environmental damage by leading to high arsenic concentrations in soils, stream sediments, surface waters and some groundwaters, and even the local atmosphere. Although these activities have often had a severe impact on the local environment, the arsenic contamination in surface water and groundwater tends to be restricted to areas within a few kilometers of mine sites.

Oxidation of FeAsS can be described by the reaction.

$$4FeAsS + 13O_2 + 6H_2O = 4Fe^{2+} + 4AsO_4^{3-} + 4SO_4^{2-} + 12H^+$$

which involves the release of acid, arsenic, and sulfate as AMD (see Chapter 9.05). Further acidity is released by the oxidation of the Fe^{2+} and precipitation of HFO or schwertmannite. These minerals readsorb some of the released arsenic, reducing dissolved arsenic concentrations, and may eventually lead to the formation of $FeAsO_4{\cdot}2H_2O$.

Experience with bioleaching of arsenic-rich gold ores has shown that the ratio of pyrite to FeAsS is an important factor controlling the speciation of the arsenic released (Nyashanu *et al.*, 1999). In the absence of pyrite, ~72% of the arsenic released was As(III), whereas in the presence of pyrite and Fe(III), 99% of the arsenic was As(V). It appears that pyrite catalyzed the oxidation of As(III) by Fe(III), since Fe(III) alone did not oxidize the arsenic (Nyashanu *et al.*, 1999).

9.02.5.1.1 Examples of mining-related arsenic problems

Arsenic contamination from former mining activities has been identified in many areas of the world including the USA (Plumlee *et al.*, 1999; Welch *et al.*, 1999, 1988, 2000), Canada, Thailand, Korea, Ghana, Greece, Austria, Poland, and the UK (Smedley and Kinniburgh, 2002).

Groundwater in some of these areas has been found with arsenic concentrations as high as $4.8 \times 10^4\ \mu g\ L^{-1}$. Some mining areas have AMD with such low pH values that the iron released by oxidation of the iron sulfide minerals remains in solution and, therefore, does not scavenge arsenic. The well-documented cases of arsenic contamination in the USA include the Fairbanks gold-mining district of Alaska (Welch *et al.*, 1988; Wilson and Hawkins, 1978), the Coeur d'Alene Pb–Zn–Ag mining area of Idaho (Mok and Wai, 1990), the Leviathan Mine (S), California (Webster *et al.*, 1994), Mother Lode (Au), California (Savage *et al.*, 2000), Summitville (Au), Colorado (Pendleton *et al.*, 1995), Kelly Creek Valley (Au), Nevada (Grimes *et al.*, 1995), Clark Fork River (Cu), Montana (Welch *et al.*, 2000), Lake Oahe (Au), South Dakota (Ficklin and Callender, 1989), and Richmond Mine (Fe, Ag, Au, Cu, Zn), Iron Mountain, California (Nordstrom *et al.*, 2000).

Phytotoxic effects attributed to high concentrations of arsenic have also been reported around the Mina Turmalina copper mine in the Andes, northeast of Chiclayo, Peru (Bech *et al.*, 1997). The main ore minerals involved are chalcopyrite, FeAsS, and pyrite. Arsenic-contaminated groundwater in the Zimapan Valley, Mexico has also been attributed to interaction with Ag–Pb–Zn, carbonate-hosted mineralization (Armienta *et al.*, 1997). Arsenopyrite, $FeAsO_4 \cdot 2H_2O$, and $(Cu,Fe)_{12}As_4S_{13}$ were identified as probable source minerals in this area.

Data for 34 mining localities of different metallogenic types in different climatic settings were reviewed by Williams (2001). He proposed that FeAsS is the principal source of arsenic released in such environments and concluded that *in situ* oxidation generally resulted in the formation of poorly soluble $FeAsO_4 \cdot 2H_2O$ which limited the mobility and ecotoxicity of arsenic. The Ron Phibun tin-mining district of Thailand is an exception (Williams *et al.*, 1996). In this area, FeAsS oxidation products were suggested to have formed in the alluvial placer gravels during mining. Following cessation of mining and pumping, groundwater rebound caused dissolution of the oxidation products. The role of $FeAsO_4 \cdot 2H_2O$ in the immobilization of arsenic from mine workings has been questioned by Roussel *et al.* (2000a), who point out that the solubility of this mineral exceeds drinking-water standards irrespective of pH.

9.02.5.1.2 Modern practice in mine-waste stabilization

Although large international mining companies generally have high environmental standards, mineral working by uncontrolled and disorganized groups (especially for gold) continues to create environmental problems in a number of developing countries.

Modern mining practices, including waste storage and treatment are designed to minimize the risk of environmental impacts (Johnson, 1995). In most countries, environmental impact assessments and environmental management plans are now a statutory requirement of the mining approval process. Such plans include criteria for siting and management of waste heaps and for effluent control. Closure plans involving waste stabilization and capping to limit AMD generation are also required to reduce any legacy of environmental damage (Lima and Wathern, 1999).

Treatment of AMD includes the use of liming, coagulation, and flocculation (Kuyucak, 1998). Other passive technologies include constructing wetlands that rely on sulfate reduction, alkali generation, and the precipitation of metal sulfides. These are often used as the final step in treating discharged water. More recently, permeable reactive barriers (PRBs) have been advocated. For example, Harris and Ragusa (2001) have demonstrated that sulfate-reducing bacteria can be stimulated to precipitate arsenic sulfides by the addition of rapidly decomposing plant material. Monhemius and Swash (1999) investigated the addition of iron to copper- and arsenic-rich liquors to form $FeAsO_4 \cdot 2H_2O$. The arsenic is immobilized by incorporation into a crystalline, poorly soluble compound (Sides, 1995). Swash and Monhemius (1996) have also investigated the stabilization of arsenic as calcium arsenate.

9.02.5.2 Role of Secondary Minerals

9.02.5.2.1 The importance of arsenic cycling and diagenesis

The close association between arsenic and iron in minerals is frequently reflected by their strong correlation in soils and sediments. Iron oxides play a crucial role in adsorbing arsenic species, especially As(V), thereby lowering the concentration of arsenic in natural waters. Manganese oxides play a role in the oxidation of As(III) to As(V) and also adsorb significant quantities although to a much lesser degree than the more abundant iron oxides. HFO is a very fine-grained, high surface area form of iron oxide that is often formed in iron-rich environments in response to rapid changes in redox or pH. It is frequently involved in the cycling of As(III) and As(V). Significant As(V) desorption occurs at pH values of approximately pH 8 and higher (Lumsdon *et al.*, 2001) and this process has been suggested to be important in generating high-arsenic groundwaters (Smedley, 2003; Welch *et al.*, 2000).

Arsenic can also be released under reducing conditions (Section 9.02.5.5).

The mobility of arsenic can also be limited in sulfur-rich, anaerobic environments by its co-precipitation with secondary sulfide minerals, and more generally by clays. The precise behavior of arsenic in sediments is poorly understood, but it is likely that important changes occur during sediment diagenesis. Arsenic adsorbed on mineral surfaces is likely to be sensitive to changes in the mineral properties such as surface charge and surface area. A very small mass transfer from solid to solution can lead to a large change in dissolved arsenic concentration. For example, sediments with average arsenic concentrations of less than 5–10 mg As kg^{-1} can generate mg L^{-1} concentrations of arsenic when only a small fraction of the total arsenic is partitioned into the water.

9.02.5.2.2 Redox behavior

Solid surfaces of many minerals, especially redox-sensitive minerals such as iron and manganese oxides, also play an important role in redox reactions and interactions with microbes (Brown *et al.*, 1999; Grenthe *et al.*, 1992). Solid Mn(IV)O_2, notably birnessite (δ-MnO_2), assists in the oxidation of As(III) to As(V) while itself being partially reduced to Mn(II) (Oscarson *et al.*, 1983; Scott and Morgan, 1995). The rate of oxidation depends on the surface area and surface charge of the MnO_2 and is slightly greater at low pH (pH 4). The Mn(II) and As(V) produced are partially retained or re-adsorbed by the MnO_2 surface, which may lead, in turn, to a deceleration in the rate of As(III) oxidation (Manning *et al.*, 2002). Reactions with birnessite at very high initial As(III) concentrations may lead to the formation of the insoluble mineral, krautite ($MnHAsO_4 \cdot H_2O$) on the birnessite surface (Tournassat *et al.*, 2002). The catalytic role of solid MnO_2 in removing As(III) is used to advantage in water treatment (Daus *et al.*, 2000; Driehaus *et al.*, 1995). TiO_2 minerals and titanium-containing clays may also be able to oxidize As(III). The photocatalytic activity of anatase (TiO_2) has been shown to catalyze the oxidation of As(III) in the presence of light and oxygen (Foster *et al.*, 1998a). Unlike the role of manganese oxides in As(III) oxidation, there is no change in the oxidation state of the surface Ti(IV) atoms.

HFO and other iron oxides may also play a significant role in the oxidation of As(III) in natural waters since the oxidation of As(III) adsorbed by HFO is catalyzed by H_2O_2 (Voegelin and Hug, 2003). This reaction may be significant in natural environments with high H_2O_2 concentrations (1–10 μM) and alkaline pH values, and in water treatment systems where H_2O_2 is used. Similar surface-catalyzed reactions do not occur with aluminum oxides (Voegelin and Hug, 2003).

The reductive dissolution of Fe(III) oxides in reducing sediments and soils (McGeehan *et al.*, 1998) can also lead to the release of adsorbed and co-precipitated arsenic. Reduction and release of arsenic can precede any dissolution of the iron oxides themselves (Masscheleyn *et al.*, 1991). These processes are likely to be the same as those responsible for the development of high-arsenic groundwaters in the Bengal Basin (Bhattacharya *et al.*, 1997; Kinniburgh *et al.*, 2003; Nickson *et al.*, 2000) and other reducing alluvial aquifers (Korte and Fernando, 1991). The release of sorbed arsenic during diagenetic changes of iron oxides including loss of surface area, and changes in surface structure and charge following burial may also be important under both reducing and oxidizing conditions.

9.02.5.3 Adsorption of Arsenic by Oxides and Clays

Metal ion oxides are often important in minimizing the solubility of arsenic in the environment in general and more specifically for localizing the impact of arsenic contamination near contaminated sites, especially old mines (La Force *et al.*, 2000; Roussel *et al.*, 2000b; Webster *et al.*, 1994). Organic arsenic species tend to be less strongly sorbed by minerals than inorganic species.

There have been many laboratory studies of the adsorption of arsenic species by pure minerals, especially iron and aluminum oxides and clays (Goldberg, 1986; Inskeep *et al.*, 2002). The general features of the processes involved have been established. Dzombak and Morel (1990) critically reviewed the available laboratory data for the adsorption of a wide range of inorganic species, including those for arsenic, by HFO and fitted the most reliable data to a surface complexation model—the diffuse double-layer model. This model, and the accompanying thermodynamic database, are now incorporated into several general-purpose geochemical speciation and transport models, including PHREEQC2 (Parkhurst and Appelo, 1999) and The Geochemist's Workbench (GWB) (Bethke, 2002). These software packages enable rapid calculations of the possible role of arsenic adsorption by HFO to be made. Critically, in PHREEQC2 and GWB, this adsorption behavior can also be automatically linked to the dissolution/precipitation of HFO. The results of such calculations demonstrate the important role of both oxidation states (arsenate versus arsenite) and pH (Figure 4).

The oxidized and reduced species of arsenic behave very differently on HFO (Figure 4). This, along with the pH dependence of adsorption,

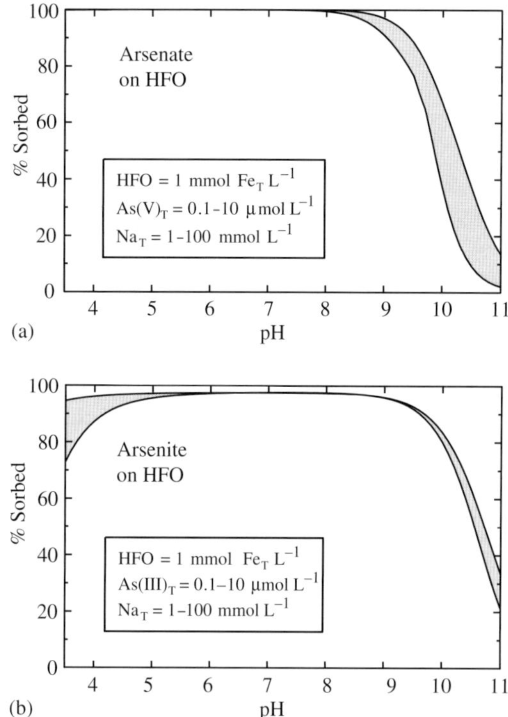

Figure 4 Calculated percent adsorption of: (a) oxidized and (b) reduced arsenic species by HFO. Infilled areas show the adsorption for a range of total As concentrations (0.1–10 μmol L^{-1}) and ionic strengths (1–100 mmol L^{-1}).

accounts, at least in part for their different behavior with oxides and clays and hence their behavior in natural waters. Arsenic(V) is very strongly adsorbed by HFO, especially at low pH and low concentrations, but is desorbed as the pH increases as a result of the increasingly strong electrostatic repulsion on the negatively charged HFO surface. The adsorption isotherm for arsenate is, therefore, highly nonlinear and can be approximated by a pH-dependent Freundlich isotherm, i.e., the slope of the adsorption decreases markedly with increasing arsenic concentration (the K_d varies with concentration). In contrast, As(III) in the pH range 4–9 is present mainly in solution as the neutral $As(OH)_3$ species and so electrostatic interactions are not nearly so important. Therefore, arsenite is adsorbed over a wide range of pH and because the adsorbed species is uncharged, arsenite adsorption tends to follow a Langmuir isotherm, i.e., the isotherm has an adsorption maximum and approaches linearity at low concentrations. It is also almost independent of pH. Organic arsenic species are weakly adsorbed by oxides; their formation can, therefore, increase arsenic mobility.

In oxidizing environments, arsenate is more strongly adsorbed than arsenite in neutral to acidic conditions, and especially at low concentrations. The weaker adsorption of As(V) at high pH has important environmental consequences. The precise pH where this desorption occurs depends on several other factors (e.g., the total arsenic concentration, and the concentrations of other competing anions), but it is in the region pH 8–9. Under these conditions, arsenite may be more strongly bound.

The adsorption of arsenic species also depends, to some extent, on competition from other anions. In reducing groundwaters these include phosphate, silicate, bicarbonate, and fulvic acids (Appelo *et al.*, 2002; Hiemstra and van Riemsdijk, 1999; Jain and Loeppert, 2000; Meng *et al.*, 2002; Wang *et al.*, 2001; Wijnja and Schulthess, 2000). Arsenic(V) and phosphorus sorption on HFO are broadly similar although there is usually a slight preference for phosphorus (Jain and Loeppert, 2000). Not surprisingly, As(V) is much more strongly affected by phosphate competition than As(III) (Jain and Loeppert, 2000). Cations, such as Ca^{2+} and Fe^{2+}, may increase arsenic adsorption (Appelo *et al.*, 2002). Once the oxides have an adsorbed load, any change in their surface chemistry or the solution chemistry can lead to the release of adsorbed arsenic, thereby increasing groundwater concentrations. The extremely high solid/solution ratio of soils and aquifers makes them very sensitive to such changes, and redox changes are likely to be particularly important (Meng *et al.*, 2001; Zobrist *et al.*, 2000).

Adsorption by aluminum and manganese oxides and clays has been studied much less (Inskeep *et al.*, 2002). Arsenic(III) binds strongly to amorphous $Al(OH)_3$ over the range pH 6–9.5, a somewhat greater range than found for HFO. It also binds significantly but somewhat less strongly to montmorillonitic and kaolinitic clays (Manning and Goldberg, 1997). Arsenic (V) shows the same declining affinity for clays at high pH as shown by HFO, but in the case of the clays this decline may begin to occur above pH 7.

9.02.5.4 Arsenic Transport

There are few observations of arsenic transport in aquifers, and its rate of movement is poorly understood. The transport of arsenic, as that of many other chemicals, is closely related to adsorption–desorption reactions (Appelo and Postma, 1993). Arsenate and arsenite have different adsorption isotherms. They, therefore, travel through aquifers at different velocities, and tend to be separated.

Gulens *et al.* (1979) used breakthrough experiments with columns of sand (containing 0.6% iron and 0.01% manganese) and various groundwaters pumped continuously from piezometers to study As(III) and As(V) mobility over a range of Eh and pH conditions. Radioactive ^{74}As (half-life = 17.7 d) and ^{76}As (half-life = 26.4 h) were used to monitor the breakthrough of arsenic. The results showed that: (i) As(III) moved 5–6 times faster than As(V) under oxidizing conditions at a pH in the range 5.7–6.9; (ii) As(V) moved much faster at the lowest pH but was still slower than As(III) under reducing groundwater conditions; and (iii) with a pH of 8.3, both As(III) and As(V) moved rapidly through the column but when the amount of arsenic injected was substantially reduced, the mobility of As(III) and As(V) was greatly reduced. This chromatographic effect (used to advantage in analytical chemistry to speciate arsenic) may account, in part, for the variable As(III)/As(V) ratios found in many reducing aquifers. Chromatographic separation of arsenic and other species during transport destroys the original source characteristics, for example, between arsenic and iron, further complicating the interpretation of well water analyses.

Few field-based investigations have been carried out on natural systems which allow the partition coefficient (K_d) or retardation factor of arsenic species to be determined directly. However, the work of Sullivan and Aller (1996) indicates that K_d values calculated for sediment profiles on the Amazon Shelf are in the approximate range of 11–5,000 L kg^{-1}. High-arsenic pore waters were mostly found in zones with low K_d values (typically <100 L kg^{-1}). Evidence from various studies also suggests low K_d values (<10 L kg^{-1}) for arsenic in aquifers in which there are high arsenic concentrations (Smedley and Kinniburgh, 2002). Factors controlling the partition coefficients are poorly understood, and they involve the chemistry of groundwater, and the surface chemistry and stability of the solid phases.

9.02.5.5 Impact of Changing Environmental Conditions

Arsenic moves between different environmental compartments (rock–soil–water–air–biota) from the local to the global scale partly as a result of pH and redox changes. Being a minor component in the natural environment, arsenic responds to such changes rather than creating them. These changes are driven by the major (bio)geochemical cycles.

9.02.5.5.1 Release of arsenic at high pH

High arsenic concentrations can develop in groundwaters as As(V) is released from oxide minerals and clays at high pH. High pH conditions frequently develop in arid areas as a result of extensive mineral weathering with proton uptake. This is especially true in environments dominated by sodium rather than by calcium, since $CaCO_3$ minerals restrict the development of high pH values.

9.02.5.5.2 Release of arsenic on reduction

Flooding of soils generates anaerobic conditions and can lead to the rapid release of arsenic (and phosphate) to the soil solution (Deuel and Swoboda, 1972; Reynolds *et al.*, 1999). Similarly, arsenic can be released to pore water in buried sediments. The concentration of dissolved arsenic in some North Atlantic pore waters varies inversely with the concentration of easily leachable arsenic in the solid phase and directly with increasing concentrations of solid phase Fe(II) (Sullivan and Aller, 1996). This reflects a strong redox coupling between arsenic and iron, whereby oxidized arsenic is associated with iron oxides in surface sediments and is subsequently reduced and released into pore water on burial. Upward diffusion and reworking of sediment releases the dissolved arsenic to the water column or releases it for re-adsorption in surface sediments as HFO is formed (Petersen *et al.*, 1995). Some reducing, iron-rich aquifers also contain high concentrations of arsenic (Korte, 1991), but there are also many iron-rich groundwaters with low arsenic concentrations.

9.02.5.6 Case Studies

9.02.5.6.1 The Bengal Basin, Bangladesh, and India

In terms of the numbers of people at risk, the high-arsenic groundwaters in the alluvial and deltaic aquifers of Bangladesh and West Bengal represent the most serious threat to public health. Health problems from this source were first identified in West Bengal in the 1980s but remained unrecognized in Bangladesh until 1993. Concentrations of arsenic in groundwaters from the affected areas have a very large range from <0.5 μg L^{-1} to ca. 3,200 μg L^{-1} (Kinniburgh *et al.*, 2003). In a survey of Bangladesh groundwater by BGS and DPHE (2001), 27% of shallow (<150 m) tubewells in Bangladesh were found to contain more arsenic than the national standard in drinking water, 50 μg L^{-1}. Groundwater surveys indicate that the most affected area is in southeast Bangladesh (Figure 5), where more than 60% of the wells in some districts have concentrations 50 μgL^{-1}. Approximately, 30–35 million people in Bangladesh and six million in West Bengal are at risk from arsenic concentrations in excess of 50 μg L^{-1} in their drinking water (BGS and DPHE, 2001).

The affected aquifers of the Bengal Basin are generally shallow (less than 100–150 m deep),

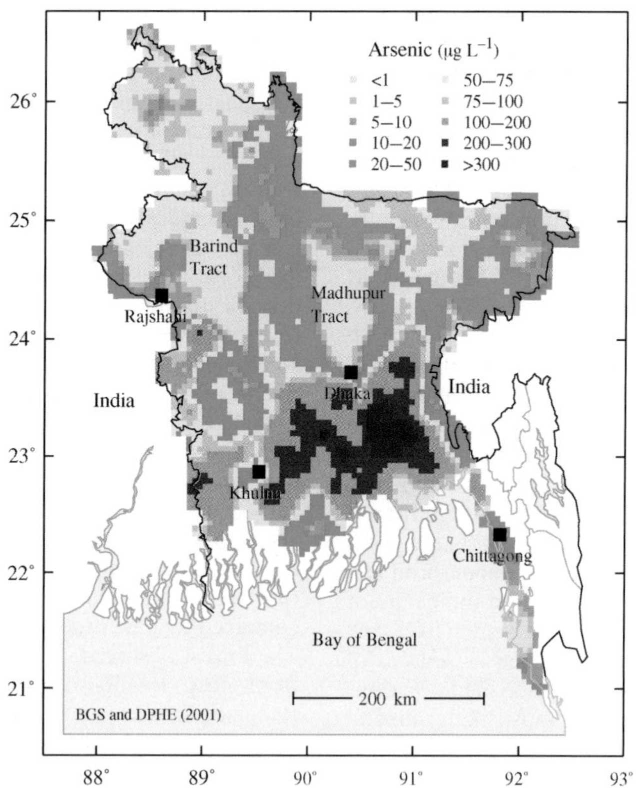

Figure 5 Map showing the distribution of arsenic in shallow (<150 m) Bangladesh groundwaters based on some 3,200 groundwater samples (source BGS and DPHE, 2001).

and consist mainly of Holocene micaceous sands, silts, and clays associated with the Ganges, Brahmaputra, and Meghna river systems. In West Bengal, the area east of the Hoogli River is affected. The sediments are derived from the Himalayan highlands and Precambrian basement complexes in northern and western West Bengal. In most of the areas with high-arsenic groundwater, alluvial and deltaic aquifer sediments are covered by surface horizons of fine-grained overbank deposits. These restrict the entry of air to underlying aquifers and, together with the presence of reducing agents such as organic matter, facilitate the development of strongly reducing conditions in the affected aquifers. Mobilization of arsenic probably reflects a complex combination of redox changes in the aquifers resulting from the rapid burial of the alluvial and deltaic sediments, reduction of the solid-phase As to As(III), desorption of arsenic from iron oxides, reductive dissolution of the oxides, and changes in iron oxide structure and surface properties in the ambient reducing conditions (BGS and DPHE, 2001). Some workers have also suggested that, in parts of Bangladesh at least, enhanced groundwater flow and redox changes may have been imposed on the shallow aquifer as a result of recent irrigation pumping (Harvey *et al.*, 2002).

Deep tubewells (>150–200 m depth), mainly in the southern coastal region, and wells in older Plio-Pleistocene sediments from the Barind and Madhupur Tracts of northern Bangladesh almost invariably have arsenic concentrations below 5 $\mu g\ L^{-1}$ and usually less than 0.5 $\mu g\ L^{-1}$ (BGS and DPHE, 2001). It is fortunate that Calcutta and Dhaka draw their water from these older sediments and do not have drinking-water arsenic problems. Dhaka is sited at the southern tip of the Madhupur Tract (Figure 5). Shallow open dug wells also generally have low arsenic concentrations, usually <10 $\mu g\ L^{-1}$ (BGS and DPHE, 2001).

The high-arsenic groundwaters of the Bengal Basin typically have near-neutral pH values. They are strongly reducing. Measured redox potentials are usually less than 100 mV (BGS and DPHE, 2001). The source of the organic carbon responsible for the reducing conditions has been variously attributed to dispersed sediment carbon (BGS and DPHE, 2001), peat (McArthur *et al.*, 2001), or soluble carbon brought down by a combination of surface pollution and irrigation (Harvey *et al.*, 2002). High concentrations of iron (>0.2 mg L^{-1}), manganese (>0.5 mg L^{-1}), bicarbonate (>500 mg L^{-1}), ammonium (>1 mg L^{-1}), and phosphorus (>0.5 mg L^{-1}), and low concentrations of nitrate (<0.5 mg L^{-1}) and sulfate

(<1 mg L^{-1}) are also typical of the high-arsenic areas. Some Bangladesh groundwaters are so reducing that methane production has been observed (Ahmed *et al.*, 1998; Harvey *et al.*, 2002). Positive correlations between arsenic and iron in the groundwaters have been reported in some studies at the local scale (e.g., Nag *et al.*, 1996), although the correlations are generally poor on a regional scale (Kinniburgh *et al.*, 2003). Arsenic(III) typically dominates the dissolved arsenic load, although As(III)/As(V) ratios are variable (BGS and DPHE, 2001).

The arsenic-affected groundwaters in the Bengal Basin are associated with alluvial and deltaic sediments with total arsenic concentrations in the range <2–20 mg kg^{-1}. These values are close to world-average concentrations for such sediments. However, even though the arsenic concentrations are low, there is a significant variation both regionally and locally, and the sediment iron and arsenic concentrations appear to be indicators of the concentration of dissolved arsenic (BGS and DPHE, 2001). The mineral source or sources of the arsenic are not well established. Various workers have postulated the most likely mineral sources as iron oxides (BGS and DPHE, 2001; Bhattacharya *et al.*, 1997; Nickson *et al.*, 1998), but pyrite or FeAsS (Das *et al.*, 1996) and phyllosilicates (Foster *et al.*, 2000) have also been cited as possible sources. The solid-solution mass transfers involved are small so that it is difficult to identify, or even eliminate, any particular sources using mass balance considerations alone.

The reasons for the differing arsenic concentrations in the shallow and deep groundwaters of the Bengal Basin are not completely understood. They could reflect different absolute arsenic concentrations in the aquifer sediments, different oxidation states, or differences in the arsenic-binding properties of the sediments. The history of groundwater movement and aquifer flushing in the Bengal Basin may also have contributed to the differences. Older, deeper sediments will have been subjected to longer periods of groundwater flow, aided by greater hydraulic heads during the Pleistocene period when sea level during glacial periods was up to 130 m lower than today (Umitsu, 1993). These sediments will therefore have undergone a greater degree of flushing and removal of labile solutes than Holocene sediments at shallower depths.

Isotopic evidence suggests that groundwater in some parts of the Bengal Basin has had a variable residence time. At a site in western Bangladesh (Chapai Nawabganj), tritium was found to be present at 2.5–5.9 TU (tritium units) in two shallow piezometer samples (10 m or less) indicating that they contain an appreciable component of post-1960s recharge (Smedley *et al.*, 2001b). At this site and in two others in south and central Bangladesh (Lakshmipur and Faridpur, respectively), groundwater from piezometers between 10 m and 30 m depth had tritium concentrations in the range <0.1–9.6 TU indicating a variable proportion of post-1960s recharge. Some of the low-tritium wells contain high arsenic concentrations, suggesting that the arsenic was released before the 1960s, i.e., before the recent rapid increase in groundwater abstraction for irrigation and water supply. Groundwater from piezometers at 150 m depth in central and south Bangladesh contained <1 TU also indicative of pre-1960s water.

Radiocarbon dating has a longer time frame than tritium and provides evidence for water with ages on the scale of thousands of years. Radiocarbon dating of groundwater sampled from the above piezometers in the 10–40 m depth range typically contained 65–90% modern carbon (pmc), whereas below 150 m the groundwater contained 51 pmc or less (Smedley *et al.*, 2001b). The lowest observed ^{14}C activities were in water from deep (>150 m) piezometers in southern Bangladesh. Here, activities of 28 pmc or less suggested the presence of palaeowater with ages of the order of 2,000–12,000 yr.

Taken together with the tritium data, these results indicate that the water below 31 m or so tends to have ages of between 50 yr and 2,000 yr. Broadly similar results and conclusions were reported by Aggarwal (2000). However, Harvey *et al.* (2002) drew the opposite conclusion from data for their field site just south of Dhaka. They found that a water sample from 19 m depth contained DIC with a ^{14}C composition at bomb concentrations and was, therefore, less than 50 yr old. This sample contained ~200 μg L^{-1} arsenic. These authors proposed that the rapid expansion of pumping for irrigation water has led to an enhanced inflow of organic carbon and that this has either initiated the enhanced reduction and release of arsenic, or has led to the displacement of arsenic by carbonate. However, a sample from 31 m depth which had a lower ^{14}C DIC activity and an estimated age of 700 yr also contained a high As concentration (~300 μg L^{-1}). This predates modern irrigation activity. Whether, in general, irrigation has had a major impact on arsenic mobilization in the Bengal aquifers is a matter of recent debate.

9.02.5.6.2 Chaco-Pampean Plain, Argentina

The Chaco-Pampean Plain of central Argentina covers ~1 million km^2 and perhaps is one of the largest regions of high-arsenic groundwaters. High concentrations of arsenic have been documented from Córdoba, La Pampa, Santa Fe, Buenos Aires, and Tucumán Provinces. Symptoms typical of chronic arsenic poisoning, including skin lesions and some internal cancers, have been recorded in these areas (Hopenhayn-Rich *et al.*, 1996). The climate is temperate with aridity increasing toward the west. The high-arsenic groundwaters occur in Quaternary deposits of loess (mainly silt) with intermixed rhyolitic or dacitic volcanic ash (Nicolli *et al.*, 1989; Smedley *et al.*, 2002), often situated in closed basins. The sediments display abundant evidence of postdepositional diagenetic changes under semi-arid conditions. Calcrete is common.

Many investigations of groundwater quality have identified variable and often extremely high arsenic concentrations. Nicolli *et al.* (1989) found arsenic concentrations in groundwaters from Córdoba in the range 6–11,500 $\mu g\ L^{-1}$ (median 255 $\mu g\ L^{-1}$). Smedley *et al.* (2002) found concentrations for groundwaters in La Pampa Province in the range <4–5,280 $\mu g\ L^{-1}$ (median 145 $\mu g\ L^{-1}$), and Nicolli *et al.* (2001) found concentrations for groundwaters in Tucuman Province in the range 12–1,660 $\mu g\ L^{-1}$ (median 46 $\mu g\ L^{-1}$). A map showing the distribution of arsenic in groundwaters from northern La Pampa is shown in Figure 6.

The geochemistry of the high-arsenic groundwaters of the Chaco-Pampean Plain is quite distinct from that of the deltaic areas of the Bengal Basin. The Argentine groundwaters often have high salinity and the arsenic concentrations are generally highly correlated with other anionic and oxyanionic species of F, V, HCO_3, B, and Mo. Many of the waters exceed the WHO guideline value for fluoride in drinking water (1.5 mg L^{-1}) as well as for arsenic, molybdenum, boron, and uranium. Arsenic is dominantly present as As(V) (Smedley *et al.*, 2002).

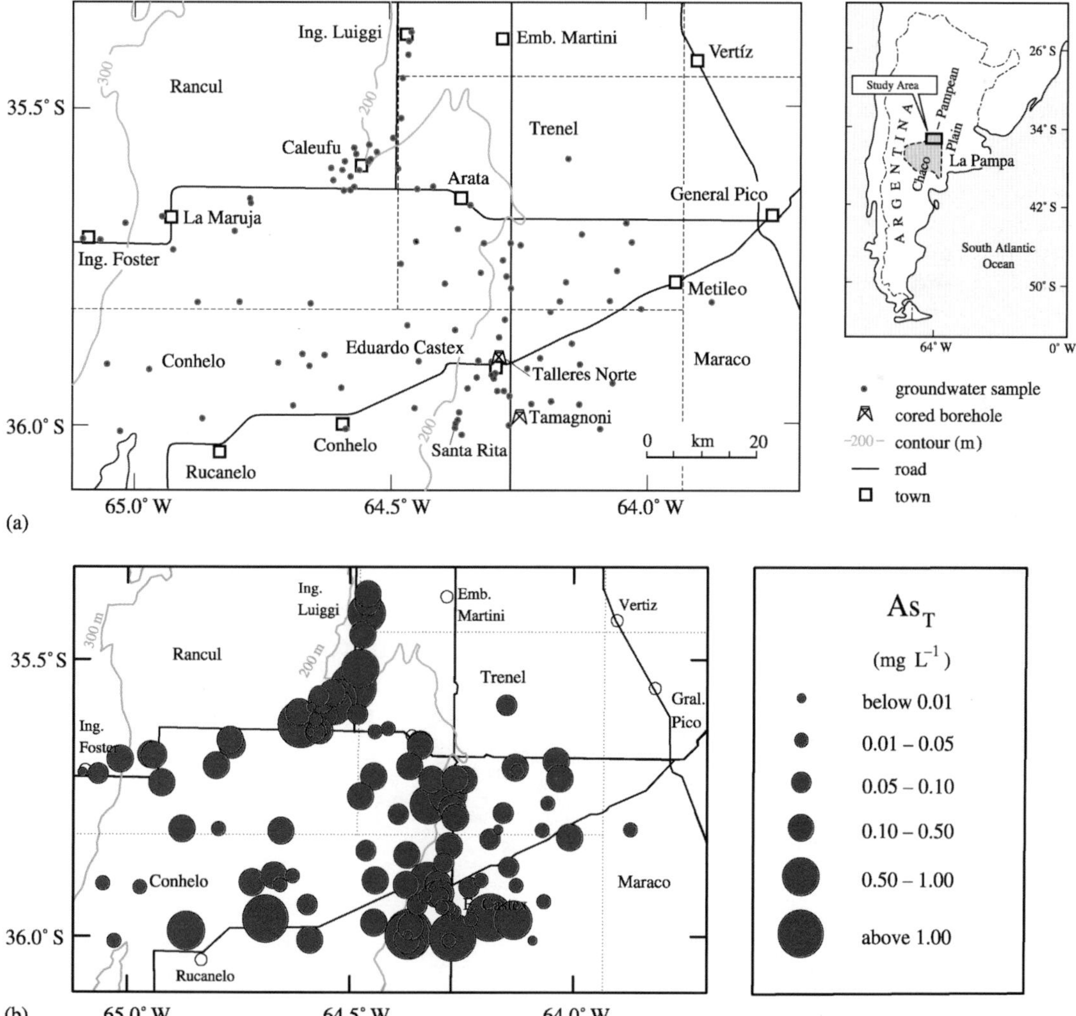

Figure 6 Map showing: (a) groundwater sampling locations and (b) observed arsenic concentrations in the Chaco-Pampean Plain of central Argentina (source Smedley *et al.*, 2002).

The groundwaters are also predominantly oxidizing with low dissolved iron and manganese concentrations. There is no indication of reductive dissolution of iron oxides or of pyrite oxidation. Under the arid conditions, silicate and carbonate weathering reactions are pronounced and the groundwaters often have high pH values. Smedley *et al.* (2002) found pH values typically in the range of 7.0–8.7.

While the reasons for these high arsenic concentrations are unclear, metal oxides in the sediments (iron, manganese, and aluminum) are thought to be the main source of dissolved arsenic, although the direct dissolution of volcanic glass has also been cited as a potential source (Nicolli *et al.*, 1989). The arsenic is believed to be desorbed under the high-pH conditions (Smedley *et al.*, 2002). A change in the surface chemistry of the iron oxides during early diagenesis may also be an important factor in arsenic desorption. The arsenic released tends to accumulate where natural groundwater movement is slow, especially in low-lying discharge areas. Evaporative concentration is also a factor, but the lack of correlation between arsenic and chloride concentrations in the groundwaters suggests that it is not the dominant control (Smedley *et al.*, 2002).

9.02.5.6.3 Eastern Wisconsin, USA

The analysis of some 3.135×10^4 groundwaters throughout the USA indicates that ~10% exceed the current $10\ \mu g\ L^{-1}$ drinking-water MCL (Welch *et al.*, 2000). At a broad regional scale (Figure 7), arsenic concentrations exceeding $10\ \mu g\ L^{-1}$ are more frequently observed in the western USA than in the east. The Mississippi delta shows a locally high pattern but is not exceptional when viewed nationally. Arsenic concentrations in groundwater from the Appalachian Highlands and the Atlantic Plain are generally very low ($<1\ \mu g\ L^{-1}$). Concentrations are somewhat greater in the Interior Plains and the Rocky Mountains and recently, areas in New England, Michigan, Minnesota, South Dakota, Oklahoma, and Wisconsin have been shown to have groundwaters with arsenic concentrations exceeding $10\ \mu g\ L^{-1}$, sometimes appreciably so. Eastern Wisconsin is one such area.

The St. Peter Sandstone (Ordovician) aquifer of eastern Wisconsin (Brown, Outagamie, Winnebago Counties) is a locally important source of water for private supplies. Arsenic contamination was first identified at two locations in 1987, and subsequent investigations showed that 18 out of 76

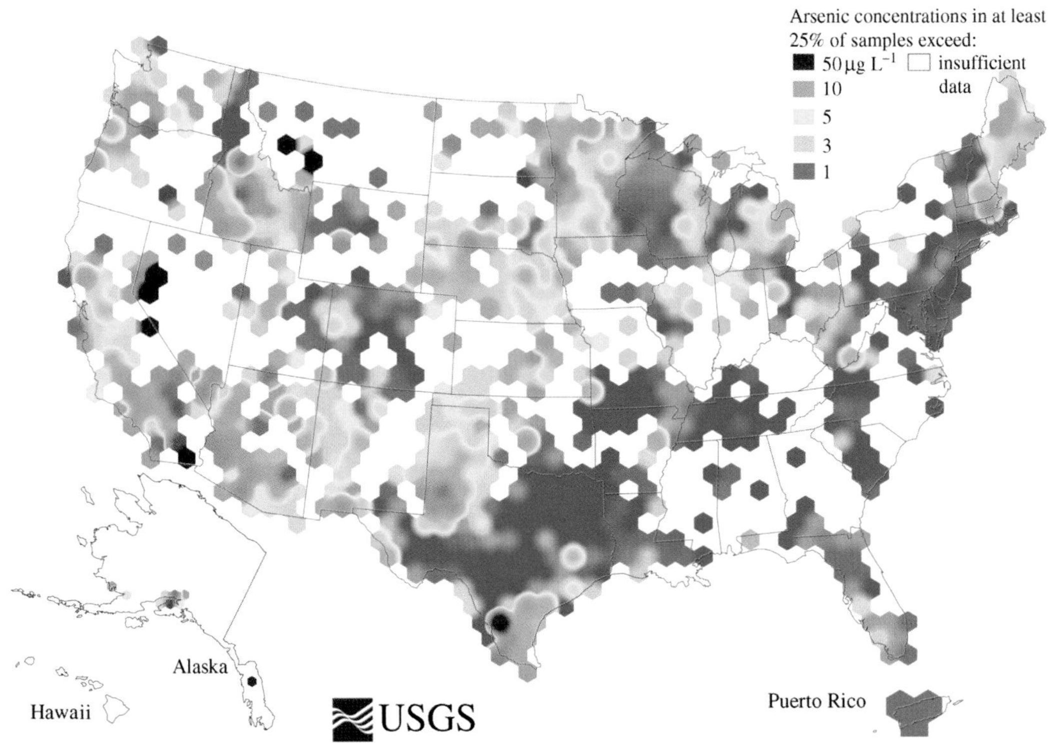

Figure 7 Map of the USA showing the regional distribution of arsenic in wells (from http://webserver.cr.usgs.gov/trace/pubs/geo_v46n11/fig3.html after Ryker, 2001). This shows where 25% of water samples within a moving 50 km radius exceed a certain arsenic concentration. It is computed from ~3×10^4 water samples and updated from the results presented by Welch *et al.* (2000).

sources (24%) in Brown County, 45 out of 1,116 sources (4.0%) in Outagamie County, and 23 out of 827 sources (2.8%) in Winnebago County exceeded the then current MCL for arsenic of 50 $\mu g\ L^{-1}$ (Burkel and Stoll, 1999). The highest arsenic concentration found was 1,200 $\mu g\ L^{-1}$. A depth profile in one of the affected wells showed that most of the groundwater was slightly acidic (pH 5.2–6.6) and in some places very acidic (pH <4). There were also high concentrations of Fe, Cd, Zn, Mn, Cu, and SO_4, and it was concluded that the As and other elements were released following the oxidation of sulfide minerals (pyrite and marcasite) present in a cemented horizon at the boundary between the Ordovician Sinnipee Group and the underlying unit, either the St. Peter Sandstone or the Prairie du Chien Group. The low pH values are consistent with iron sulfide mineral oxidation.

Subsequent detailed studies in the Fox River Valley towns of Algoma and Hobart confirmed the importance of the sulfide-rich cement horizon as a probable source of the arsenic (Schreiber *et al.*, 2000) (Figure 8). In the town of Algoma, one well contained $1.2 \times 10^4\ \mu g\ L^{-1}$ arsenic. There was, however, much apparently random spatial variation. Two wells close to the high-arsenic well contained much lower arsenic concentrations (12 $\mu g\ L^{-1}$ and 34 $\mu g\ L^{-1}$). The highest arsenic concentration found in wells in the town of Hobart was 790 $\mu g\ L^{-1}$.

The oxidation of sulfide minerals appears to have been promoted by groundwater abstraction which has led to the lowering of the piezometric surface at a rate of ~0.6 m yr^{-1} since the 1950s, leading to partial dewatering of the confined aquifer. The high arsenic concentrations occur where the piezometric surface intersects, or lies close to, the sulfide cement horizon (Schreiber *et al.*, 2000).

9.02.6 ABUNDANCE AND FORMS OF SELENIUM IN THE NATURAL ENVIRONMENT

9.02.6.1 Abundance in Rocks, Soils, and Sediments

The average crustal abundance of selenium is 0.05 mg kg^{-1} (Jacobs, 1989). Like arsenic, selenium is strongly chalcophile and is partitioned into sulfides and rare selenides, such as

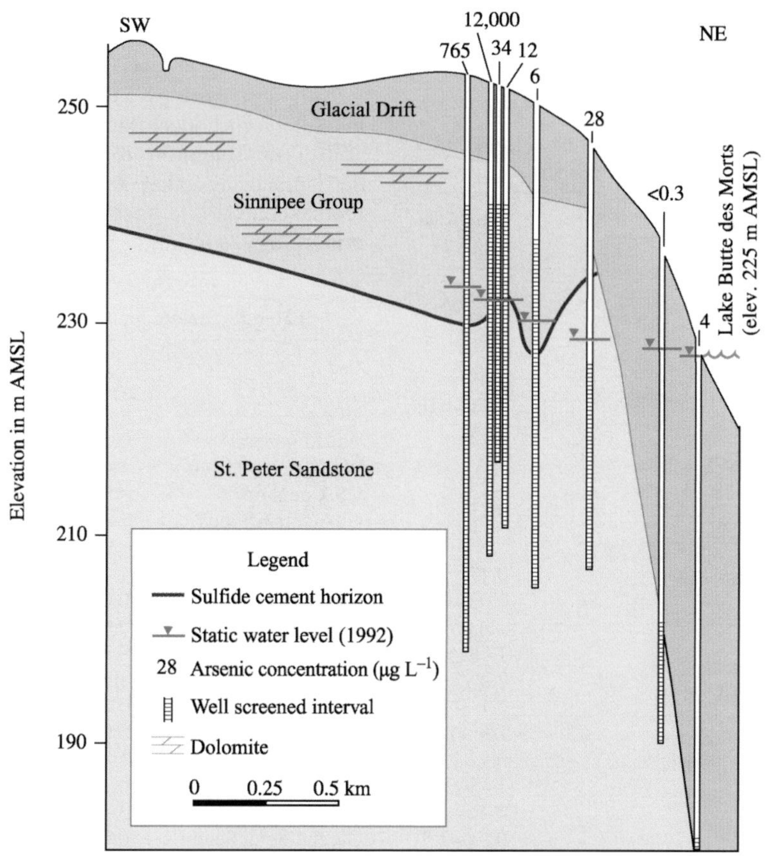

Figure 8 Hydrogeological section through part of Algoma, Winnebago County, Wisconsin showing the arsenic concentration in various wells in relation to the cemented sulfide-rich horizon and the static water level (after Schreiber *et al.*, 2000).

crooksite and clausthalite. Selenium generally substitutes for sulfur in sulfide minerals, but elemental (native) selenium has also been reported (Alloway, 1995; Davies, 1980; Tokunaga *et al.*, 1996).

Selenium concentrations in coal and other organic-rich deposits can be high and typically range from 1 mg kg^{-1} to 20 mg kg^{-1}. The average selenium concentration in coals from the USA is 4.1 mg kg^{-1} (Swanson *et al.*, 1976). Large concentrations of selenium, like arsenic, are often associated with the clay fraction of sediments because of the abundance of free iron oxides and other strong sorbents. Selenium concentrations are generally larger in shales than in limestones or sandstones (Table 7). Selenium concentrations in the excess of 600 mg kg^{-1} are found in some black shales. Selenium is present in these shales as organo-selenium compounds or adsorbed species (Jacobs, 1989). Concentrations exceeding 300 mg Se kg^{-1} have also been reported in some phosphatic rocks (Jacobs, 1989).

Generally, there is a strong correlation between the selenium content of rocks and the sediments and soils derived from them. Soil selenium concentrations are typically in the range 0.01–2 mg kg^{-1} with a world average of 0.4 mg kg^{-1}. Extremely high concentrations (up to 1,200 mg kg^{-1}) have been found in some organic-rich soils derived from black shales in Ireland (Table 8). Soils from England derived from black shales had an average concentration of 3.1 mg kg^{-1} compared with an overall average of 0.48 mg kg^{-1} for a range of more typical English soils (Thornton *et al.*, 1983). Concentrations of 6–15 mg kg^{-1} have been reported in volcanic soils such as those of Hawaii (Jacobs, 1989). High concentrations tend to be found in soils from mineralized areas and in poorly drained soils. The median selenium concentration in stream sediments from 20 study areas across the USA was 0.7 mg kg^{-1} (Rice, 1999) and 0.5 mg kg^{-1} in 1.9×10^4 stream sediments from Wales (BGS, 1979–2002). Relatively low selenium concentrations are found in well-drained soils derived from limestones and coarse sands.

Selenium-rich plants, including the selenium-indicating vetches (*Astragalus* sp.), are widespread in South Dakota and Wyoming, USA. The vegetation grows on soils developed over black shales and sandstones with high selenium concentrations (Moxon, 1937). Selenium toxicity was first documented in 1856 near Fort Randall, where a physician in the US Cavalry reported horses experiencing hair, mane and tail loss and sloughing of hooves. Forage that contains 2–5 mg Se kg^{-1} poses a marginal threat to livestock, and acute effects are likely to occur above 5 mg Se kg^{-1}.

Although geology is the primary control on the selenium concentration of soil, the bioavailability of selenium to plants and animals is determined by other factors including pH and redox conditions, speciation, soil texture and mineralogy, organic matter content, and the

Table 7 Selenium concentrations in selected rock types.

Material	*Selenium* (mg kg^{-1})
Earth's Crust	0.05
Igneous rocks	
Ultramafic rocks	0.05
Mafic rocks	0.05
Granite	0.01–0.05
Volcanic rocks	0.35
Volcanic rocks, USA	<0.1
Volcanic rocks, Hawaii	<2.0
Volcanic tuffs	9.15
Sedimentary rocks	
Marine carbonates	0.17
Limestone	0.03–0.08
Sandstone	<0.05
W. USA shale	1–675
Wyoming shale	2.3–52
S. Korean shale	0.1–41
Carbon-shale China	206–280
Mudstone	0.1–1,500
Phosphate	1–300
USA Coal	0.46–10.7
Australian coal	0.21–2.5
Chinese stone-coal	<6,500
Oil	0.01–1.4

Sources: Jacobs (1989), Fordyce *et al.* (2000b), WHO (1987), Oldfield (1999), Alloway (1995), and Davies (1980).

Table 8 Selenium concentrations in soils.

Soil	*Total Se* (mg kg^{-1})	*Water-soluble Se* (mg kg^{-1})
World general	0.4	
World seleniferous	1–5,000	
USA general	<0.1–4.3	
USA seleniferous	1–10	
England/Wales general	<0.01–4.7	0.05–0.39
Ireland seleniferous	1–1,200	
China general	0.02–3.8	
China Se deficient	0.004–0.48	0.00003–0.005
China Se adequate	0.73–5.7	
China seleniferous	1.49–59	0.001–0.25
Finland	0.005–1.25	
India Se deficient	0.025–0.71	0.019–0.066
India seleniferous	1–20	0.05–0.62
Sri Lanka Se deficient	0.11–5.2	0.005–0.043
Norway	3–6	
Greece Se deficient	0.05–0.10	
Greece Se adequate	>0.2	
New Zealand	0.1–4	

Sources: Davies (1980), Thornton *et al.* (1983), Jacobs (1989), WHO (1987), Alloway (1995), Oldfield (1999), Fordyce *et al.* (2000a,b).

presence of competing ions. Even soils with relatively high total selenium concentrations can give rise to selenium deficiency if the selenium is not bioavailable. The first map of the selenium status of soil and vegetation in relation to animal deficiency and toxicity was prepared by Muth and Allaway (1963).

Several techniques are available to assess selenium bioavailability in soils. The most widely used is the water-soluble concentration (Fordyce *et al.*, 2000b; Jacobs, 1989; Tan, 1989). In most soils, only a small proportion of the total selenium is in dissolved form (0.3–7%), and water-soluble selenium contents are generally <0.1 mg kg^{-1} (Table 8).

Selenium is also added to soils as a trace constituent of phosphate fertilizers and in selenium-containing pesticides and fungicides, as well as by the application of sewage sludge and manure (Alloway, 1995; Frankenberger and Benson, 1994; Jacobs, 19891). Sewage sludge typically contains ~1 mg Se kg^{-1} dry wt. Precautionary limits are set for several chemical elements likely to be increased by the application of sewage sludge to land. For example, in the EU the banning of the discharge of sewage sludge to the sea since 1999 has increased its application to land. The maximum admissible concentration of selenium in sewage sludge in the UK is 25 mg kg^{-1} and that in soil after application is 3 mg kg^{-1} in the UK and 10 mg kg^{-1} in France and Germany (ICRCL, 1987; Reimann and Caritat, 1998). In the USA, the limit in soils is 100 mg kg^{-1}.

9.02.6.2 National and International Standards in Drinking Water

The WHO guideline value for selenium in drinking water is currently 10 μg L^{-1} and this standard has been adopted by the EC, Australia, Japan, and Canada. The US EPA primary drinking-water standard is 50 μg L^{-1}.

9.02.6.3 Abundance and Distribution in Natural Waters

The selenium concentration in most natural waters is very low, often less than 1 μg L^{-1} and frequently just a few ng L^{-1}. Hence, selenium from drinking water only constitutes a health hazard in exceptional circumstances (Fordyce *et al.*, 2000a; Vinceti *et al.*, 2000). However, occasionally much greater concentrations are found. Groundwaters containing up to 275 μg L^{-1} have been reported from aquifers in China and 1,000 μg L^{-1} selenium from seleniferous aquifers in Montana, USA (Table 9). Selenium concentrations of up to 2,000 μg L^{-1} or more have also been reported in lakes from saline, seleniferous areas. Such areas are rare but include some arid parts of the USA, China, Pakistan, and Venezuela. In general, data on selenium concentrations in water are scarce. Reported ranges from the literature are summarized in Table 9.

Waters containing 10–25 μg L^{-1} selenium may have a garlic odor, whereas waters containing 100–200 μg L^{-1} selenium also have an unpleasant taste. Groundwaters generally contain higher selenium concentrations than surface waters, because water–rock interactions have been more extensive (Frankenberger and Benson, 1994; Jacobs, 1989).

9.02.6.3.1 Atmospheric precipitation

Selenium in rainfall is derived principally from earth-surface volatilization, volcanic sources, fossil-fuel combustion (especially coal), and the incineration of municipal wastes. Few determinations of selenium in atmospheric precipitation have been reported, but concentrations are usually very low. Hashimoto and Winchester (1967) found concentrations in the range 0.04–1.4 μg L^{-1} (Table 9).

9.02.6.3.2 River and lake water

Selenate (Se(VI)) is only weakly adsorbed by oxides and clays at near-neutral pH. Hence, oxidation of Se(IV) to Se(VI) enhances selenium mobility and persistence in natural waters. High concentrations of selenate can occur in agricultural drainage waters in arid areas. Seleniferous soils, especially those derived from black shales, are common in the central and western USA, and irrigation can give rise to selenate concentrations of several hundred μg L^{-1} in drainage water. Further concentration can occur in lakes by evapotranspiration. Well-documented cases of such situations include examples from California (Kesterson Reservoir, Richmond Marsh, Tulare Basin, and Salton Sea), North Carolina (Belews Lake, Hyco Reservoir), Texas (Martin Reservoir), and Wyoming (Kendrick Reclamation Project) in the USA. Problems of selenium toxicity are also found in other semi-arid areas. In the Soan-Sakesar Valley of Punjab, Pakistan, average selenium concentrations were 302 μg L^{-1} ($n = 13$) in streams and springs and 297–2,100 μg L^{-1} in lake water (three lakes) (Afzal *et al.*, 2000). The highest concentrations were reported from low-lying, salinized areas.

The Colorado River catchment, Utah, USA is also a seleniferous area. Median selenium concentrations in the Colorado River and its major tributaries are in the range 1–4 μg L^{-1} (Engberg, 1999), although values up to 400 μg L^{-1} have

Table 9 Concentration ranges of Se in various water bodies.

Water body and location	*Se concentration and range* ($\mu g\,L^{-1}$)	*Reference(s)*
Rain water		
Various	0.04–1.4	Hashimoto and Winchester (1967)
Polar ice	0.02	Frankenberger and Benson (1994)
River and lake water		
Jordan River, Jordan	0.25	Nishri *et al.* (1999)
River Amazon, Brazil	0.21	Jacobs (1989)
Colorado River, USA	<1–400	NAS (1976), Engberg (1999)
Mississippi River, USA	0.14	Jacobs (1989)
Lake Michigan, USA	0.8–10	Jacobs (1989)
Gunnison River, USA	10	Jacobs (1989)
Cienaga de Santa Clara wetland, Mexico	5–19	García-Hernández *et al.* (2000)
Seawater and estuaries		
Seawater	0.09; 0.17	Hem (1992), Thomson *et al.* (2001)
San Francisco Bay, USA	0.1–0.2	Cutter (1989)
Carquinez Strait, San Francisco Bay, USA	0.07–0.35	Zawislanski *et al.* (2001a)
Groundwater		
East Midlands Triassic Sandstone, UK	<0.06–0.86	Smedley and Edmunds (2002)
Chaco-Pampean Plain, loess aquifer, Argentina	<2–40	Nicolli *et al.* (1989), Smedley *et al.* (2002)
Bengal Basin alluvial aquifer, Bangladesh	<0.5	BGS and DPHE (2001)
Soan-Sakesar Valley alluvial aquifer, Punjab, Pakistan	Avg 62	Afzal *et al.* (2000)
Colorado River catchment, USA	up to 1,300	Engberg (1999)
Coast Range alluvial aquifer, San Joaquin Valley, California, USA	<1–2,000	Deverel *et al.* (1994)
Sierra Nevada alluvial aquifer, San Joaquin Valley, California	<1	Deverel *et al.* (1994)
Central Barents groundwater, Norway	0.01–4.82	Reimann *et al.* (1998)
Slovakian groundwater	0.5–45	Rapant *et al.* (1996)
Pore water		
Baseline, estuarine Lake Macquarie, Australia	<0.2	Peters *et al.* (1999)
Smelter and power-station-impacted, Lake Macquarie, Australia	0.3–5.0	Peters *et al.* (1999)

been reported (NAS, 1976). Irrigation is believed to have been responsible for ~70% of the selenium reaching Lake Powell (Engberg, 1999). Selenium concentrations in the Cienega de Santa Clara wetlands on the east side of the Colorado River Delta, Mexico are also in the range 5–19 $\mu g\,L^{-1}$ (García-Hernández *et al.*, 2000). However, high selenium concentrations do not occur in all rivers in arid areas. For example, concentrations in the Jordan River average only 0.25 $\mu g\,L^{-1}$ (Nishri *et al.*, 1999). Selenium concentrations in surface waters may be increased locally near sources of waste, including sewage effluent.

9.02.6.3.3 Seawater and estuaries

Selenium concentrations in estuarine water and seawater are also generally low. An average concentration of 0.17 $\mu g\,L^{-1}$ was estimated for seawater by Thomson *et al.* (2001). Dissolved concentrations in the range 0.1–0.2 $\mu g\,L^{-1}$ have been reported in San Francisco Bay (Cutter, 1989). Zawislanski *et al.* (2001b) reported concentrations of 0.07–0.35 $\mu g\,L^{-1}$ in the nearby Carquinez Strait. Much of the selenium is thought to have been derived from industrial sources including historical releases from oil refineries.

During low-flow conditions, oil refineries contribute up to 75% of the total selenium load entering San Francisco Bay. Refineries processing oil derived from the neighboring San Joaquin Valley, California, produce effluent containing selenium concentrations an order of magnitude greater than those in refinery effluent from Alaskan North Slope crude oil (Zawislanski and Zavarin, 1996).

9.02.6.3.4 Groundwater

As in the case of surface waters, the concentrations of selenium in groundwater are usually low and commonly below analytical detection limits. Concentrations tend to be higher in oxidizing groundwaters, because the dominant form present, Se(VI), is less prone to adsorption by metal oxides than Se(IV). Elemental selenium is also unstable under oxidizing conditions. High selenium concentrations have been found under oxidizing conditions in groundwaters in some arid and semi-arid areas as a result of evaporation. Extremely high concentrations (up to 1,300 $\mu g\,L^{-1}$) have been reported from shallow wells in the upper reaches of the Colorado River catchment, Utah (Engberg, 1999).

Deverel and Fujii (1988) also reported concentrations in the range of <1–2,000 $\mu g\,L^{-1}$ (Table 9) in shallow groundwater from Coast Range alluvial fan sediments near Kesterson Reservoir, California. Concentrations of <20 $\mu g\,L^{-1}$ were found in the middle fan deposits but reached several hundreds of $\mu g\,L^{-1}$ in the lower fan deposits. Concentrations increased with groundwater salinity, probably as a result of leaching of soil salts by irrigation and subsequent evaporation. Deverel and Fujii (1988) found low concentrations of selenium in groundwater from the eastern side of the San Joaquin Valley in alluvial sediments of the Sierra Nevada Formation. Values were generally less than 1 $\mu g\,L^{-1}$, probably as a result of reducing conditions in which selenium occurred in less mobile forms, notably Se(IV).

Selenium-rich groundwaters are also found in the semi-arid regions of Argentina (Table 9). Nicolli *et al.* (1989) found concentrations up to 24 $\mu g\,L^{-1}$ in oxidizing groundwater from Córdoba Province. Concentrations correlated positively with salinity. Smedley *et al.* (2002) also found selenium concentrations in the range <2–40 $\mu g\,L^{-1}$ ($n = 34$) in oxidizing groundwaters from the neighboring province of La Pampa, with the highest concentrations in high-salinity shallow groundwaters in which selenium was concentrated by evaporation. No speciation studies were carried out, although selenate is likely to dominate. Eleven of the groundwater samples analyzed (32%) in the Smedley *et al.* (2002) study exceeded the WHO guideline value for selenium in drinking water of 10 $\mu g\,L^{-1}$.

In the Soan-Sakesar Valley of the Punjab, Pakistan, the average selenium concentration in groundwater was 62 $\mu g\,L^{-1}$ ($n = 29$) (Afzal *et al.*, 2000). Again there was a positive correlation between salinity and selenium concentration. Most of the selenium in the groundwater was present as Se(VI).

Selenium concentrations in reducing groundwaters are very low or undetectable as a result of reduction to Se(IV). Concentrations in samples of the strongly reducing high-arsenic groundwaters of Bangladesh were <0.5 $\mu g\,L^{-1}$ (BGS and DPHE, 2001). In the Triassic Sandstone aquifer of the English East Midlands, selenium concentrations varied from less than 0.06 $\mu g\,L^{-1}$ to 0.86 $\mu g\,L^{-1}$ (Table 9). Concentrations were highest in the unconfined oxidizing part of the sandstone aquifer and fell abruptly to less than 0.06 $\mu g\,L^{-1}$ at and beyond the redox boundary (Smedley and Edmunds, 2002).

9.02.6.3.5 Sediment pore water

Few data are available for the selenium content of pore waters. However, Peters *et al.* (1999) reported concentrations of up to 5 $\mu g\,L^{-1}$ in estuarine pore waters from Mannering Bay (Lake Macquarie), New South Wales, Australia. Investigations followed concerns during the 1990s about high selenium concentrations in marine organisms from the area. Concentrations were highest in the upper 5 mm of the profile, and were substantially higher throughout the profile than from nearby Nord's Wharf, where concentrations were typically <0.2 $\mu g\,L^{-1}$ selenium (i.e., below the detection limit). Although redox controls influenced the trends with depth, the high selenium concentrations in the uppermost sediments were thought to reflect contamination from smelter and power station inputs.

9.02.6.3.6 Mine drainage

Since selenium substitutes for sulfur in the structure of sulfide minerals, drainage from mineralized and mined areas may have high dissolved selenium concentrations. Acid seeps derived from oxidation of sulfide minerals draining the Moreno Shale in the Coast Ranges, USA, have selenium concentrations up to 420 $\mu g\,L^{-1}$ with concentrations of aluminum, manganese, zinc, and nickel in the mg L^{-1} range (Presser, 1994).

9.02.6.4 Selenium Species in Water, Sediment, and Soil

The behavior of selenium in the environment is similar in many respects to that of arsenic.

Importantly, it also occurs naturally in several oxidation states and is, therefore, redox sensitive. Methylation and hydride formation are important, and sulfur and iron compounds play an important role in the cycling of selenium. Microbiological volatilization of organic selenium, particularly dimethyl selenide, is known to be an important factor in the loss of selenium from some selenium-rich soils and waters (Frankenberger and Arshad, 2001; Oremland, 1994). Phytoplankton can also promote the production of gaseous selenium compounds in the marine environment (Amouroux *et al.*, 2001).

Selenium occurs in natural waters principally in two oxidation states, Se(IV) (selenite) and Se(VI) (selenate). Elemental selenium, Se(0) (red and black forms), and selenide, Se(−II), are essentially insoluble in water $Se(VI)O_4^{2-}$ occurs mainly in oxidizing waters while $HSe(IV)O_3^-$ and $Se(IV)O_3^{2-}$ dominate under reducing conditions (Figure 9). The concentration ratio of Se(IV) to Se(VI) species in natural waters does not necessarily follow that of other redox couples (e.g., Fe^{2+}/Fe^{3+}). This reflects the slow kinetics involved (White and Dubrovsky, 1994).

Elemental selenium, selenides, and selenium sulfide salts are only stable in reducing, acidic conditions and are largely unavailable to plants and animals. Zawislanski *et al.* (2001b) found a strong positive correlation between selenium and organic carbon in suspended particulate matter from San Francisco Bay possibly reflecting reduction of selenium by organic matter. Strong positive correlations of particulate selenium with particulate iron and aluminum were also noted. The oxidation and reduction of selenium is related to microbial activity. For example, the bacterium *Bacillus megaterium* can oxidize elemental selenium to selenite.

It has been estimated that up to 50% of the selenium in some soils may be present as organic compounds, although few such compounds have been isolated and identified (Jacobs, 1989). Selenomethionine has been extracted from soils and is 2–4 times more bioavailable to plants than inorganic selenite, although selenocysteine is less bioavailable than selenomethionine (Alloway, 1995; Davies, 1980; Frankenberger and Benson, 1994; Jacobs, 1989). The bioavailability of the different selenium species in soils can be summarized as

$$\text{selenate} > \text{selenomethionine} > \text{selenocysteine} > \text{selenite} > \text{elemental selenium} > \text{selenide}$$

In general, selenate is more available and more mobile than selenite in the environment, so that selenium is much more bioavailable under oxidizing alkaline conditions.

An Eh–pH diagram for the system Se–O–S is given in Figure 10 as a guide but, as in the case of arsenic, it is necessarily an oversimplification of a complex natural system. Se_2S_3 is not shown as no thermodynamic data are available for this species. It would probably displace elemental selenium (black) as the dominant phase under strongly reducing sulfur-rich conditions. Also the stability of Fe–Se minerals and the effects of selenium adsorption by metal oxides are not represented in the diagram.

Selenium is more readily reduced than arsenic. In soils and sediments, elemental selenium dominates under strongly reducing conditions. The gas H_2Se is important under acid, strongly reducing conditions. Organic selenides occur in biological materials. Some of these selenides are highly volatile.

The most detailed studies of selenium distribution and speciation have been carried out for seawater. G. A. Cutter and L. S. Cutter (2001) found that selenate had generally higher concentrations in marine waters from the southern

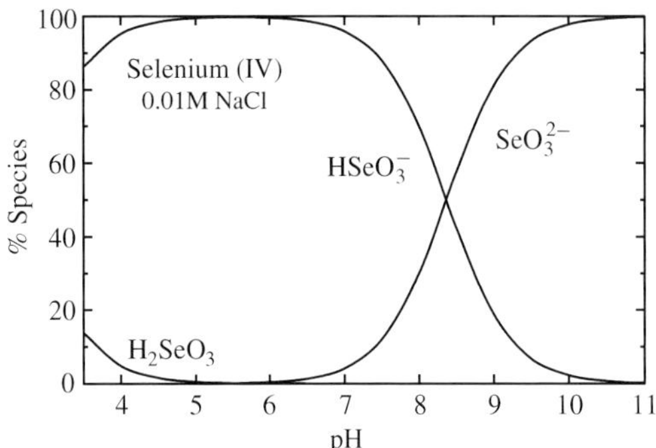

Figure 9 Speciation of selenium in a 0.01 M NaCl medium as a function of pH at 25 °C. The plot for Se(VI) is not shown since this is always dominated by SeO_4^{2-}.

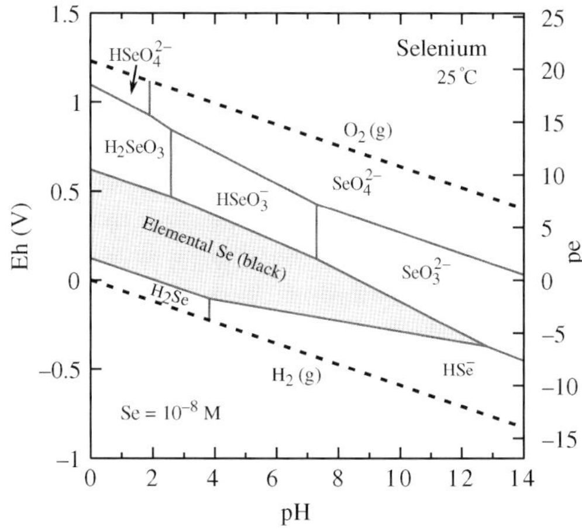

Figure 10 Eh–pH stability diagram for selenium at 25 °C, 1 bar total pressure. The stability field for water is shown by the dashed lines.

($0.019\ \mu g\ L^{-1}$) than the northern hemisphere ($0.014\ \mu g\ L^{-1}$). In contrast, selenite had low concentrations in seawater from the southern hemisphere ($0.005\ \mu g\ L^{-1}$) with the highest concentrations in the equatorial region and below the Intertropical Convergence Zone ($0.009\ \mu g\ L^{-1}$). Depth profiles of total dissolved selenium, selenite, and selenate in Atlantic seawater all showed surface-water depletion and deep-water enrichment, characteristic of nutrient-like behavior. In North Atlantic Deep Water, the Se(IV)/Se(VI) ratios were generally similar to those found in the eastern Atlantic and North Pacific (0.7), but waters originating in the southern polar regions were enriched in selenate and had low Se(IV)/Se(VI) ratios (~0.4). Organic selenide was found in surface ocean waters but was not detected in mid- or deep waters.

Selenium profiles in sediments from the northeast Atlantic Ocean indicate concentrations of ~0.2–0.3 mg kg^{-1} in the oxic zone, and typically 0.3–0.5 mg kg^{-1} below the redox boundary reflecting immobilization under reduced conditions (Thomson *et al.*, 2001). Similar increases for cadmium, uranium, and rhenium have also been observed in the suboxic zone.

As with arsenic, microbiological processes are important in the reduction of selenium, principally through the microbial reduction of Se(IV) and Se(VI) (Oremland *et al.*, 1990). Oremland (1994) found that the areal rate of dissimilatory selenium reduction in sediments from an agricultural evaporation pond in the San Joaquin Valley was ~3 times lower than for denitrification and 30 times lower than for sulfate reduction. Stable-isotope studies of water in the Tulare Lake Drainage district wetland, California, indicated little selenium-isotope fractionation (Herbel *et al.*, 2002b). This suggested that the primary source of reduced selenium was its assimilation by plants and algae followed by deposition and mineralization rather than the direct bacterial reduction of Se(VI) or Se(IV).

9.02.7 PATHWAYS AND BEHAVIOR OF SELENIUM IN THE NATURAL ENVIRONMENT

9.02.7.1 Release from Primary Minerals

As noted above, the principal natural sources of selenium in water are likely to be sulfides or metal oxides containing adsorbed selenium, especially Se(IV). Coal can be an additional primary source of selenium either directly through oxidation or indirectly via atmospheric precipitation following combustion. Selenium is readily oxidized during the weathering of minerals. Seleniferous groundwater areas such as those in the USA and Pakistan are most common where underlain by selenium- and organic-rich shales, which release selenium on weathering. Selenium-rich groundwaters tend to be found in semi-arid areas under irrigation. Examples are found in the central and western USA (Deverel *et al.*, 1994) and in parts of Pakistan (Afzal *et al.*, 2000).

9.02.7.2 Adsorption of Selenium by Oxides and Clays

In contrast to arsenic, the reduced form of selenium, Se(IV), is very strongly adsorbed by HFO. This may account, in part, for the very low selenium concentration in many strongly reducing environments. Furthermore, also in contrast with arsenic, the oxidized form of selenium, Se(VI), is less strongly adsorbed to HFO than the reduced species. These differences, also reflected by other oxide-based sorbents including clays, account for the markedly different behavior of arsenic and selenium in natural waters.

The behavior of selenium in soils mirrors that of the pure oxides (Goldberg, 1985). In acid soils, selenium is likely to occur mainly as Se(IV) strongly adsorbed to iron oxides. Less commonly, Se(IV) may form highly insoluble iron compounds such as ferric selenite ($Fe_2(OH)_4SeO_3$) or iron selenide (FeSe). In alkaline, oxidized and selenium-rich soils, most of the selenium is likely to be present as Se(VI) which is very weakly adsorbed. Furthermore, there are no common insoluble selenate minerals. Hence, selenate accumulates in soluble form particularly in arid and semi-arid areas where evaporation tends to concentrate selenium along with other soluble salts (Deverel *et al.*, 1994).

The strong affinity of iron oxides for Se(IV) has been well documented (Dzombak and Morel, 1990) and calculations based on the Dzombak and Morel (1990) diffuse double-layer model and default HFO database show the principal response to pH and redox speciation changes (Figure 11). The selenate species is less strongly adsorbed by iron oxides at near-neutral pH than the selenite species (Figure 11). Clay minerals (Bar-Yosef and Meek, 1987) also adsorb Se(IV).

The iron oxide and clay content of soils and sediments can affect the bioavailability of selenium markedly. The strong pH dependence of adsorption is an important control. Maximum adsorption occurs between pH 3 and pH 5 and decreases as the pH rises. Organic matter also removes selenium from soil solution, possibly as a result of the formation of organometallic complexes. Addition of PO_4 to soils increases selenium uptake by plants, because the PO_4^{3-} ion displaces selenite from soil particles making it more bioavailable. Conversely, increasing the concentrations of PO_4 in soils can

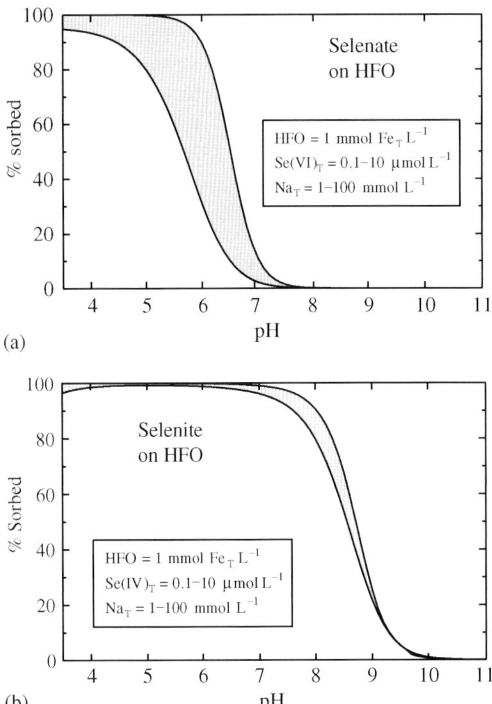

Figure 11 Calculated percent adsorption of: (a) oxidized and (b) reduced selenium species by HFO. Infilled areas show the adsorption for a range of total Se concentrations (0.1–10 μmol L^{-1}) and ionic strengths (1–100 mmol L^{-1}).

dilute the selenium content of vegetation by inducing increased plant growth (Frankenberger and Benson, 1994; Jacobs, 1989).

9.02.7.3 Selenium Transport

The transport of selenium is related strongly to its speciation. The weak adsorption of selenate by soil and aquifer materials, especially in the presence of high SO_4^{2-} concentrations, means that it is relatively unretarded by groundwater flow (Kent *et al.*, 1995). There is also little likelihood that insoluble metal selenates, such as $CaSe(VI)O_4{\cdot}2H_2O$, will limit Se(VI) solubility under oxidizing conditions (White and Dubrovsky, 1994). By contrast, the strong adsorption tendency of Se(IV) and low solubility of Se(0) and Se(−II) species mean that transport of selenium is strictly limited under reducing conditions.

The strong contrast in selenium mobility between reducing and oxidizing conditions means that changes in redox conditions in soils, sediments, or aquifers can result in significant changes in selenium concentrations in water and crops. For example, the accumulation of selenium in the reduced bottom sediments of the Salton Sea, California could be mobilized if engineered changes to transfer water out of the Salton Sea Basin lead to oxidation of the sediments (Schroeder *et al.*, 2002). Such a process has already occurred at Kesterson Reservoir (Section 9.02.7.4.1). The change of land use, for example, from wet paddy soils to dryland agriculture, could also lead to an increase in uptake of selenium by crops (Yang *et al.*, 1983).

9.02.7.3.1 Global fluxes

Selenium is dispersed through the environment and is cycled by biogeochemical processes involving the rock weathering, rock–water interactions, and microbiological activity. Estimates of the selenium fluxes through the atmosphere, land, and oceans indicate that the anthropogenic flux now exceeds the marine flux, the principal natural pathway (Table 10).

The global flux of selenium from land to the oceans via rivers has been estimated as 1.538×10^4 t yr^{-1} (Haygarth, 1994). The cycling of selenium from land to water is poorly understood, but ~85% of the selenium in rivers is thought to be in particulate rather than dissolved form.

Typical concentrations of selenium in seawater are ~0.1–0.2 μg L^{-1} (Table 9) with an estimated mean residence time of 70 yr in the mixed layer and 1,100 yr in the deep ocean. The oceans are, therefore, an important sink for selenium (Haygarth, 1994; Jacobs, 1989). Biogenic volatilization of selenium from seawater to the atmosphere is estimated to be 5,000–8,000 t yr^{-1}. Amouroux *et al.* (2001) have demonstrated that biotransformation of dissolved selenium in seawater by blooms of phytoplankton in the spring is a major pathway for the emission of gaseous selenium to the atmosphere. Hence, oceans are an important part of the selenium cycle.

9.02.7.3.2 Selenium fluxes in air

Volatilization of selenium from volcanoes, soils, sediments, the oceans, microorganisms, plants, animals, and industrial activity all contribute to selenium in the atmosphere. Natural background concentrations of selenium in nonvolcanic areas are only around 0.01–1 ng m^{-3}, but the short residence time, usually a matter of weeks, makes the atmosphere a rapid transport route for selenium. Volatilization of selenium into the atmosphere results from microbial methylation of selenium from soil, plant, and water, and is affected by the availability of selenium, the presence of an adequate carbon source, oxygen availability, and temperature (Frankenberger and Benson, 1994; Jacobs, 1989).

Most gaseous selenium is thought to be in the dimethylselenide form. Terrestrial biogenic

Table 10 Global selenium fluxes.

Source	*Pathways*	*Se flux* ($\times 10^4 t yr^{-1}$)
Anthropogenic releases	Mining, anthropogenic releases to atmosphere, water, land, and oceans	7.6–8.8
Marine loss	Volatilization, sea salt suspension, into marine biota, sediment transfer to land	3.825
Terrestrial loss	Volatilization, particle resuspension, dissolved and suspended load to oceans	1.538
Atmospheric loss	Wet and dry deposition to the oceans and land	1.53

Source: Haygarth (1994).

sources probably contribute 1,200 t yr^{-1} of selenium to the atmosphere. Atmospheric dusts derived from volcanoes and wind erosion of the Earth's surface (180 t yr^{-1}) and suspended sea salts (550 t yr^{-1}) from the oceans are also significant sources of atmospheric selenium. Particle-bound selenium can be transported several thousands of kilometers before deposition. Wet deposition from rain, snow, and other types of precipitation is thought to contribute 5,610 t ha^{-1} yr^{-1} selenium to the terrestrial environment. For example, in the UK wet deposition has been shown to account for 76–93% of the total with >70% in soluble form. Near to point sources of selenium, for example, from industry, atmospheric deposition can account for 33–82% of the selenium present on the leaves of plants (Frankenberger and Benson, 1994; Jacobs, 1989).

9.02.7.3.3 Soil–water–plant relationships

There is little evidence that selenium is essential for plants. The selenium concentrations in plants generally reflect those of the soils in which they are grown. An important factor, which may determine whether or not selenium-related health problems affect man and animals, is the variable capacity of different plant species to accumulate selenium (Alloway, 1995; Frankenberger and Benson, 1994; Jacobs, 1989; Oldfield, 1999).

Rosenfield and Beath (1964) classified plants into three groups based on their selenium uptake from seleniferous soils: (i) selenium accumulator plants which can contain >1,000 mg kg^{-1} selenium and grow well on high-selenium soils; (ii) secondary selenium absorbers with concentrations in the range of 50–100 mg kg^{-1}; and (iii) others, which include grains and grasses that can contain up to 50 mg kg^{-1} selenium. Most plants contain less than 10 mg kg^{-1}. Selenium concentrations can range from 0.005 mg kg^{-1} up to 5,500 mg kg^{-1} in selenium accumulators. Accumulators belonging to the plant genera *Astragalus*, *Haplopappus*, and *Stanleya* are commonly found in the semi-arid seleniferous environments of the western USA and elsewhere and are used as indicators of high - selenium environments, although other species of these genera are nonaccumulators (Alloway, 1995; Jacobs, 1989).

The exclusion of selenium from the proteins of accumulator plants is thought to be the basis for their selenium tolerance. Their selenium metabolism is based mainly on water-soluble nonprotein forms such as selenium methylselenomethionine (Jacobs, 1989). The "garlic" odor characteristic of selenium-accumulator plants reflects the volatile organic compounds dimethylselenide and dimethyldiselenide. Plants can suffer selenium toxicity as a result of selenium competition with essential metabolites for biochemical sites, replacement of essential ions by selenium, mainly major cations, selenate occupation of the sites of essential groups such as phosphate and nitrate, or selenium substitution in essential sulfur compounds.

Experimental evidence suggests that there is a negative correlation between very high soil selenium concentrations and plant growth. Alfalfa yields have been shown to decline when extractable selenium in soil exceeds 500 mg kg^{-1}. Yellowing, black spots, and chlorosis of plant leaves and pink root tissue can occur (Frankenberger and Benson, 1994; Jacobs, 1989). Phytotoxicity in nature has only been reported from China, where high selenium concentrations in soil caused discoloration of maize corn-head embryos and also affected the growth and yield of wheat and pea crops, respectively (Yang *et al.*, 1983).

Food crops generally have a low selenium tolerance, but most crops have the potential to accumulate selenium in quantities toxic to animals and humans (Jacobs, 1989). In general, root crops contain the highest selenium concentrations (Table 11) with plant leaves containing a higher concentration than the tuber. For example, Yang *et al.* (1983) noted that selenium concentrations in vegetables (0.3–81.4 mg kg^{-1}) were generally higher than in cereal crops (0.3–28.5 mg kg^{-1} in rice and maize) in seleniferous regions of China. Turnip leaves had particularly high concentrations with an average of 460 mg kg^{-1} ranging up to 2.5×10^4 mg kg^{-1} compared to an average of

Table 11 Examples of Se contents in various crops grown in the USA.

USA crop type	*Average Se* ($mg\ kg^{-1}$ dry wt.)
Roots and bulbs	0.407
Grains	0.297
Leafy vegetables	0.110
Seed vegetables	0.066
Vegetable fruits	0.054
Tree fruits	0.015

Source: Jacobs (1989).

12 $mg\ kg^{-1}$ in the tuber. Brassicas are unable to distinguish selenium from sulfur and so tend to accumulate selenium.

In moderate- to low-selenium environments, alfalfa (*Medicago* sp.) has been shown to take up more selenium than other forage crops. Crop species grown in low-selenium soils generally show little difference in selenium uptake so that changing the type of crop grown makes little impact. An exception has been reported from New Zealand, where changing from white clover to the grass *Agrostis tenuis* increased the selenium content of fodder (Davies and Watkinson, 1966).

9.02.7.4 Case Studies

9.02.7.4.1 Kesterson reservoir, USA

One of the best-documented cases of selenium toxicity in animals occurred at Kesterson Reservoir, California, USA (Jacobs, 1989; Wu *et al.*, 2000). Soil irrigation in the western San Joaquin Valley of California began in the late 1800s and accelerated, particularly in the 1930s–1940s. Irrigation water was taken from both surface water and groundwater (Deverel and Fujii, 1988). During the 1970s, flow into the reservoir was mainly surface water, but over the period 1981–1986 almost all the inflow was from shallow agricultural drainage for which the reservoir acted as a set of evaporation ponds. This inflow contained 250–350 $\mu g\ L^{-1}$ selenium, mostly present as bioavailable selenate (Se(VI)).

The primary source of the selenium is believed to have been pyrite in shales, particularly the Upper Cretaceous–Paleocene Moreno Shale and the Eocene–Oligocene Kreyenhagen Shale. Concentrations of selenium in these formations range up to 45 $mg\ kg^{-1}$, median concentrations being 6.5 $mg\ kg^{-1}$ and 8.7 $mg\ kg^{-1}$, respectively (Presser, 1994). The concentration of selenium in the surface sediments (0–0.3 m depth) of the old playas is in the range 1–20 $mg\ kg^{-1}$ reflecting the historical accumulation of selenium from the selenium-rich drain water. Deeper sediments typically contain much lower selenium concentrations of 0.1–1 $mg\ kg^{-1}$ (Tokunaga *et al.*, 1994).

Between 1983 and 1985, the USA Wildlife Service compared the biological impact of the high selenium in the Kesterson Reservoir region to that of the adjacent Volta Wildlife area, which was supplied with water containing normal selenium concentrations. The research showed that the high concentrations of selenium in the irrigation waters were having a detrimental effect on the health of fish and wildlife (Tokunaga *et al.*, 1994). Health effects on birds in the Kesterson Reservoir area were very marked. Twenty-two percent of the eggs contained dead or deformed embryos. The developmental deformities included missing or abnormal eyes, beaks, wings, legs, and feet as well as hydrocephaly. It has been estimated that at least 1,000 adult and young birds died between 1983 and 1985 as a result of consuming plants and fish containing 12–120 times the normal amount of selenium. No overt adverse health effects were noted in reptile or mammalian species, but the concentrations of selenium present were of concern in terms of bioaccumulation through the food chain.

These findings led the US Bureau of Reclamation to halt the discharge of agricultural drainage to the reservoir. The reservoir was dewatered, and the lower parts were infilled to prevent groundwater rising to the soil surface. Bioremediation based on microbial reduction of selenite and selenate to insoluble Se(0) or methylation of these species to dimethylselenide was used to immobilize the selenium. Field trials demonstrated that microorganisms, particularly *Enterobacter cloacea*, were effective in reducing selenium to insoluble Se(0) and that the process was stimulated by the addition of organic matter (Wu *et al.*, 2000). The area was also planted with upland grass. Biological monitoring has demonstrated that selenium concentrations in the water and vegetation at Kesterson are now much lower and largely within safe limits. By 1992, concentrations of selenium in surface pools that formed after periods of rainfall were between 3 $\mu g\ L^{-1}$ and 13 $\mu g\ L^{-1}$.

9.02.7.4.2 Enshi, China

Human selenosis has been reported from the Enshi district, Hubei Province, China. Between 1923 and 1988, 477 cases of human selenosis were reported, 338 resulting in hair and nail loss and disorders of the nervous system. In one small village, the population was evacuated after 19 of 23 people suffered nail and hair loss, and all the livestock had died from selenium poisoning. Cases of selenosis in pigs reached a peak between 1979 and 1987, when 280 out of 2,238 pigs were affected in one village.

No human cases of selenium toxicity have been reported in recent years, but animals continue to show health problems as a result of high concentrations of selenium in the environment (Fordyce *et al.*, 2000b; Yang *et al.*, 1983). Yang *et al.* (1983) were the first to compare the concentration of selenium in soil, crops, drinking water, human urine, blood, nail, and hair samples from the Enshi area with other regions of China. They demonstrated that the endemic selenium poisoning was related to the occurrence of selenium-enriched Permian coal deposits. These contained selenium concentrations up to 6,470 mg kg^{-1}. Selenium concentrations in the soil, food, and human samples from areas underlain by these rocks were up to 1,000 times higher than in samples from nearby low-selenium areas, and dietary intakes of selenium greatly exceeded the recommended international and Chinese thresholds (Table 12). Locally grown crops constituted 90% of the diet in the Enshi area. Cereal crops (rice and maize) accounted for 65–85% of the selenium intake. In addition to exposure through the food chain, villagers were exposed to selenium because they mined the coal for fuel and used the residues as a soil conditioner.

Concentrations of selenium in soils and foodstuffs can vary markedly from deficient to toxic in the same village depending on the outcrop of the coal-bearing strata (Fordyce *et al.*, 2000b). Villagers were advised to avoid cultivating areas underlain by the coal or using coal ash to condition the soil. The outbreaks of human selenosis in the late 1950s and early 1960s coincided with drought and failure of the rice crop leading to an increased dependence on locally produced vegetables and maize. As the source of food crops diversified, the incidence of the disease diminished.

9.02.7.4.3 Soan-Sakesar Valley, Pakistan

The Soan-Sakesar Valley is situated in the center of the Salt Range mountains in the Punjab, northeast Pakistan. The geochemistry of the waters in the area has been studied extensively by Afzal *et al.* (2000, 1999). The average altitude of the Soan-Sakesar Valley is 762 m. The mean annual rainfall (1984–1994) is 613 mm. The average summer temperature is 33 °C and the average winter temperature is 3 °C, with periods below freezing. Average evaporation is estimated to be ~950 mm yr^{-1}. The area is essentially a closed basin, although ephemeral streams and rivers, including the River Soan, flow seasonally westwards towards the River Indus.

The area is covered by sedimentary rocks, mainly of Tertiary age. The valley lies between two parallel east–west ridge systems. Wheat and maize are grown in the area. Three quite large brackish-saline lakes (3–14 km^2) occur within synclinal structures formed by the folding of Eocene rocks. The largest and most saline lake, Lake Uchhali, has a TDS of ~36 g L^{-1}, a nitrate concentration of 28 mg L^{-1}, and a boron concentration of nearly 1 mg L^{-1}. It also contained a selenium concentration of 2.1 mg L^{-1}. The major-element chemistry is dominated by Na–Mg–Cl–SO_4. All surface waters in the region exceeded the WHO guideline value of 10 μg L^{-1} for selenium in drinking water (Table 13).

The Soan-Sakesar aquifer consists of two major formations, a freshwater Sakesar limestone (Chharat Group) of Eocene age and a brackish formation (Rawalpindi Group). Groundwater recharge is mainly from infiltration through the alluvial fans during times of stream flow. Groundwater has been used extensively for irrigation and has led to substantially altered groundwater flow patterns. The water table is generally 4–7 m bgl and varies by 1–2 m seasonally.

The groundwaters of the area also contain a high salt content. Selenium concentrations are in excess of the WHO guideline value for drinking water and the FAO guideline value for irrigation water (20 μg L^{-1}). The median concentrations of nitrate and boron in groundwaters from the area were 27.5 mg L^{-1} and 0.52 mg L^{-1}, respectively. The boron concentration in many of the groundwaters exceeded 0.5 mg L^{-1}, the revised (1998) provisional WHO guideline value for boron in drinking water, and sometimes exceeded 1 mg L^{-1}, the maximum acceptable concentration for irrigation water. The Sakesar Formation is dominated by shale. This is probably the ultimate source of the selenium. Selenium concentrations, as well as the overall dissolved salt concentration, were greatest in low-lying areas, where a shallow water table existed and where intense evaporation of soil water had occurred.

Selenium speciation confirmed that all of the waters were dominated by selenate with 10–20% selenite (Afzal *et al.*, 2000). A small percentage of the selenium was present as organic selenium in the surface waters but was absent from the groundwaters. Volatilization of selenium from the lakes was suspected but not proven. It is likely that selenate was reduced to selenite or elemental selenium in the anoxic sediments.

9.02.7.4.4 Selenium deficiency, China

In contrast, selenium deficiency has been implicated in several human diseases, most notably Keshan disease (KD) and Kashin-Beck disease (KBD). KD is an endemic selenium-responsive cardiomyopathy that mainly affects children and women of child-bearing age and is named after

Table 12 Deficiency and toxicity thresholds for selenium in various media.

Medium	*Units*	*Deficient*	*Marginal*	*Moderate*	*Adequate*	*Toxic*	*Criterion*	*Reference(s)*
Soils								
Worldwide	mg kg^{-1}	0.1–0.6					Animal health	Various
Chinese soils	mg kg^{-1}	0.125	0.175	0.400		>3	Human health	Tan (1989)
Chinese soil water-soluble	mg kg^{-1}	0.003	0.006	0.008		0.020	Human health	Tan (1989)
Vegetation								
Worldwide	mg kg^{-1}	<0.1			0.1–1.0	3–5	Animal health	Jacobs (1989), Levander (1986)
Chinese cereals	mg kg^{-1}	0.025	0.040	0.070		>1	Human health	Tan (1989)
Animals								
Food, chronic exposure	mg kg^{-1}	<0.04			0.1–3	3–15	Animal health	Jacobs (1989), Mayland (1994)
Cattle and sheep liver	mg kg^{-1}	0.21					Animal health	WHO (1987)
Cattle and sheep blood	mg kg^{-1}	<0.04	0.05–0.06		0.07–0.10		Animal health	Mayland (1994)
Humans								
Chinese human hair	mg kg^{-1}	0.200	0.250	0.500		>3	Human health	Tan (1989)
Urinary excretion rate	μg day^{-1}				10–200		Human health	Oldfield (1999)
Food	mg kg^{-1}	<0.05				2–5	Human health	WHO (1996)
Ref. Dose US EPA	mg kg^{-1} day^{-1}				0.005		Human health	US EPA (2002)
Human diet (WHO)	μg day^{-1}	<40			55–75	>400	Human health	WHO (1996)
Drinking water (WHO)	μg L^{-1}					>10	Maximum admissible concentration	WHO (1993)

Table 13 Average concentrations of groundwater and stream/spring samples from the Soan-Sakesar Valley, Pakistan.

Parameter	*Lake Uchhali n = 3*	*Groundwaters n = 29* ($mg\ L^{-1}$)	*Streams/springs n = 13*
Ca	159	22	64
Mg	1,770	62	104
Na	9,890	130	323
K	254	14	25
Alkalinity	584	234	577
Cl	9,530	89	216
SO_4	14,300	243	551
NO_3	28	30	20
Li	2.86	0.3	0.65
B	0.96	0.61	0.36
P	0.48	0.17	0.52
SiO_2	5.1	3.7	5.2
Se	2.10	0.062	0.302
Mo	0.02	0.021	0.019

Source: Afzal *et al.* (2000).

Keshan County in NE China (Wang and Gao, 2001). During the years of peak KD prevalence (1959–1970), 8,000 cases and up to 3,000 deaths were reported annually (Tan, 1989). The disease occurred in a broad belt from NE to SW China, where subsistence farmers depended on local food supplies. White muscle disease in animals occurred in the same areas. Grain crops in the affected areas contained <0.04 mg Se kg^{-1} which led to extremely low dietary intakes (10–15 μg Se d^{-1}) in the local population. A very low selenium level in the affected population was indicated by selenium concentrations in hair of <0.12 mg kg^{-1} (Tan, 1989; Xu and Jiang, 1986; Yang and Xia, 1995). Supplementation with 50 μg Se d^{-1} prevented the condition but had no effect on those already showing signs of the disease.

The precise biological function of selenium in KD is unclear, and seasonal variations in prevalence suggested the involvement of a virus. High levels of the Coxsackie B virus were found in KD patients (Li *et al.*, 2000). Work by Beck (1999) showed that a normally benign strain of the Coxsackie B3 virus becomes virulent in selenium- or vitamin E-deficiency conditions.

The incidence of the disease has fallen in recent years as a result of selenium supplementation and improved economic conditions in China generally (Burk, 1994). The evidence of viral mutogeny in the presence of selenium deficiency has important implications for many infections. Selenium deficiency may increase the likelihood of dying from HIV-related diseases (Baum *et al.*, 1997) and may have exacerbated the incidence of AIDS in parts of Africa.

The relationship between selenium deficiency and KD in Zhangjiakou district, China has been described in detail by Johnson *et al.* (2000). Soils in the villages with a high prevalence of KD were found to be black or dark brown with a high organic matter content and lower pH than other soils in the region. Although the soil in the KD-affected areas contained highest selenium concentration, the selenium was strongly bound by soil organic matter and it was not in a bioavailable form. Water-soluble selenium concentrations in the villages with a high prevalence of KD were lower than deficiency threshold values (geometric mean, 0.06 μg L^{-1}; threshold, 3 μg L^{-1}).

The study concluded that when the bioavailability of selenium is low, any factor that further reduces its bioavailability and mobility may be critical. Adding selenium fertilizer to crops rather than to soils was recommended to increase the selenium concentrations in local diets. No cases of KD have been reported since 1996, as the diet has become more diversified as a result of improvements in economic conditions and transport.

KBD, named after the two Russian scientists who first described it in the late 1800s, is an endemic osteoarthropathy, which causes deformity of the joints. It is characterized by impaired movement, commonly with shortened fingers and toes, and in extreme cases, dwarfism (Levander, 1986; Tan, 1989; WHO, 1996). In China, the distribution of the disease is similar to KD in the north but the links with selenium deficiency are less clear. Iodine supplementation of the diets of children and nursing mothers, together with 0.5–2.0 mg sodium selenite a week for six years, reduced the disease prevalence from 42% to 4% in children aged 3–10 yr (WHO, 1987). As with KD, other factors have been implicated in the pathogenesis of KBD. These include drinking water high in humic acids, greater fungal (mycotoxin) contamination of grain, and iodine deficiency (Peng *et al.*, 1999; Suetens *et al.*, 2001). KBD also occurs in Siberia, North Korea and possibly parts of Africa.

9.02.8 CONCLUDING REMARKS

The recent surge of research into the behavior of arsenic in the environment has followed the discovery of human health problems linked to high concentrations of the element in some groundwaters, soils, and contaminated land. Rather fewer studies have been undertaken of selenium, although serious health problems related to selenium toxicity or deficiency have been reported. There have also been many studies of the distribution of both arsenic, and more recently selenium, in metalliferous exploration, especially for gold.

This chapter has outlined the main effects of arsenic and selenium on human and animal health, their abundance and distribution in the environment, sampling and analysis, and the main factors controlling speciation and cycling. Such information should help to identify aquifers, water resources and soils at risk from high concentrations of arsenic and selenium, and areas of selenium deficiency. Human activity has had, and is likely to continue to have, a major role in releasing arsenic and selenium from the geosphere and in perturbing the natural distribution of these and other elements over the Earth's surface.

Arsenic and selenium demonstrate many similarities in their behavior in the environment. Both are redox sensitive and occur in several oxidation states under different environmental conditions. Both partition preferentially into sulfide minerals and metal oxides and are concentrated naturally in areas of mineralization and geothermal activity. Both elements occur as oxyanions in solution and, depending on redox status, are potentially mobile in the near-neutral to alkaline pH conditions that typify many natural waters. However, there are also some major differences. Selenium is immobile under reducing conditions while the mobility of arsenic is less predictable and depends on a range of other factors. Selenium also appears to partition more strongly with organic matter than arsenic.

While concern with arsenic in the environment relates principally to toxic conditions, concerns with selenium relate to both deficiency and toxicity. The optimum range of ingested selenium concentrations for health is narrow. In certain environments, high intakes of arsenic or selenium or very low intakes of selenium can occur, potentially leading to a wide range of diseases, not all of which are well understood and some of which may not have been recognized.

Many of the health and environmental problems caused by arsenic and selenium were not predicted until recently because the distribution and behavior of arsenic and selenium in the environment were not sufficiently well known. Recent improvements have been aided by cost-effective analytical techniques and more powerful data-processing techniques, which have made it easier to prepare high-resolution geochemical maps. Also, modern digital data sets of geochemical and hydrochemical data are used increasingly in studies designed to estimate element speciation, bioavailability, and risk.

The discovery of high concentrations of arsenic in groundwater from parts of the Bengal Basin of Bangladesh and West Bengal and elsewhere after several years of groundwater development has highlighted the need to analyse a wide range of water-quality parameters before using these sources. With the advent of modern multi-element analytical techniques capable of measuring arsenic and selenium at environmentally relevant concentrations on an almost routine basis, it should be possible to include arsenic and selenium in more geochemical and hydrochemical surveys and thereby acquire a much better picture of their distribution, behavior, and role in the environment.

ACKNOWLEDGMENTS

The authors thank Kirk Nordstrom for an early copy of his review of the thermodynamics of the As–O–S system and also an anonymous reviewer for constructive comments about the microbiology of arsenate reduction. This chapter is published with the permission of the Executive Director of the British Geological Survey (NERC).

REFERENCES

Abdullah M. I., Shiyu Z., and Mosgren K. (1995) Arsenic and selenium species in the oxic and anoxic waters of the Oslofjord, Norway. *Mar. Pollut. Bull.* **31**, 116–126.

Adkins R., Walsh N., Edmunds W., and Trafford J. M. (1995) Inductively coupled plasma atomic emission spectrometric analysis of low levels of selenium in natural waters. *Analyst* **120**, 1433–1436.

Afzal S., Younas M., and Ali K. (2000) Selenium speciation studies from Soan-Sakesar valley, Salt Range, Pakistan. *Water Int.* **25**, 425–436.

Afzal S., Younas M., and Hussain K. (1999) Selenium speciation of surface sediments from saline lakes of the Soan-Sakesar Valley Salt-Range, Pakistan. *Water Qual. Res. J. Can.* **34**, 575–588.

Aggarwal P. K. (2000) Preliminary Report of Investigations—Isotope Hydrology of Groundwater in Bangladesh: Implications for Characterization and Mitigation of Arsenic in Groundwater. Report BGD/8/016, International Atomic Energy Agency, Vienna.

Aggett J. and O'Brien G. A. (1985) Detailed model for the mobility of arsenic in lacustrine sediments based on measurements in Lake Ohakuri. *Environ. Sci. Technol.* **19**, 231–238.

Aggett J. and Kriegman M. R. (1988) The extent of formation of arsenic (III) in sediment interstitial waters and its release to hypolimnetic waters in Lake Ohakuri. *Water Res.* **22**(4), 407–411.

Ahmann D., Roberts A. L., Krumholz L. R., and Morel F. M. M. (1994) Microbe grows by reducing arsenic. *Nature* **371**, 750.

Ahmed K. M., Hoque M., Hasan M. K., Ravenscroft P., and Chowdhury L. R. (1998) Occurrence and origin of water well methane gas in Bangladesh. *J. Geol. Soc. India* **51**, 697–708.

Aihua Z., Xiaoxin H., Xianyao J., Peng L., Yucheng G., and Shouzheng X. (2000) The progress of study on endemic arsenism due to burning arsenic containing coal in Guizhou province. *Metal Ions Biol. Med.* **6**, 53–55.

Alloway B. J. (1995) *Heavy Metals in Soils*. Blackie, Glasgow.

Amouroux D., Liss P. S., Tessier E., Hamren-Larsson M., and Donard O. F. X. (2001) Role of oceans as biogenic sources of selenium. *Earth Planet. Sci. Lett.* **189**, 277–283.

Anderson R. K., Thompson M., and Culbard E. (1986) Selective reduction of arsenic species by continuous hydride generation: Part II. Validation of methods for application to natural waters. *Analyst* **111**, 1153–1157.

Andreae M. O. (1980) Arsenic in rain and the atmospheric mass balance of arsenic. *J. Geophys. Res.* **85**, 4512–4518.

Andreae M. O. and Andreae T. W. (1989) Dissolved arsenic species in the Schelde estuary and watershed, Belgium. *Estuar. Coast. Shelf Sci.* **29**, 421–433.

Andreae M. O., Byrd T. J., and Froelich O. N. (1983) Arsenic, antimony, germanium and tin in the Tejo estuary, Portugal: modelling of a polluted estuary. *Environ. Sci. Technol.* **17**, 731–737.

Appelo C. A. J. and Postma D. (1993) *Geochemistry Groundwater and Pollution*. AA Balkema.

Appelo C. A. J., Van der Weiden M. J. J., Tournassat C., and Charlet L. (2002) Surface complexation of ferrous iron and carbonate on ferrihydrite and the mobilization of arsenic. *Environ. Sci. Technol.* **36**, 3096–3103.

Armienta M. A., Rodriguez R., Aguayo A., Ceniceros N., Villasenor G., and Cruz O. (1997) Arsenic contamination of groundwater at Zimapan. *Mexico. Hydrogeol. J.* **5**, 39–46.

Arthur J. R. and Beckett G. T. (1989) Selenium deficiency and thyroid hormone metabolism. In *Selenium in Biology and Medicine* (ed. A. Wendell). Springer, Berlin, pp. 90–95.

ATSDR (2000) Toxicological profile for arsenic: Chapter 6. Analytical methods. Agency for Toxic Substances and Disease Registry US Department of Health and Human Services, Public Health Service, pp. 301–312.

Azcue J. M. and Nriagu J. O. (1995) Impact of abandoned mine tailings on the arsenic concentrations in Moira Lake, Ontario. *J. Geochem. Explor.* **52**, 81–89.

Azcue J. M., Mudroch A., Rosa F., and Hall G. E. M. (1994) Effects of abandoned gold mine tailings on the arsenic concentrations in water and sediments of Jack of Clubs Lake, BC. *Environ. Technol.* **15**, 669–678.

Azcue J. M., Mudroch A., Rosa F., Hall G. E. M., Jackson T. A., and Reynoldson T. (1995) Trace elements in water, sediments, porewater, and biota polluted by tailings from an abandoned gold mine in British Columbia, Canada. *J. Geochem. Explor.* **52**, 25–34.

Barbaris B. and Betterton E. A. (1996) Initial snow chemistry survey of the Mogollon Rim in Arizona. *Atmos. Environ.* **30**, 3093–3103.

Bar-Yosef B. and Meek D. (1987) Selenium sorption by kaolinite and montmorillonite. *Soil Sci.* **144**, 12–19.

Baum M. K., Shor-Posner G., Lai S. H., Zhang G. Y., Fletcher M. A., Sauberlich H., and Page J. B. (1997) High risk of HIV-related mortality is associated with selenium deficiency. *J. AIDS Human Retrovirol.* **15**, 370–374.

Bech J., Poschenreider C., Llugany M., Barcelo J., Tume P., Tobias F. J., Barranzuela J. L., and Vasquez E. R. (1997) Arsenic and heavy metal contamination of soil and vegetation around a copper mine in Northern Peru. *Sci. Tot. Environ.* **203**, 83–91.

Beck M. A. (1999) Selenium and host defense towards viruses. *Proc. Nutrit. Soc.* **58**, 707–711.

Bednar A. J., Garbarino J. R., Ranville J. F., and Wildeman T. R. (2002) Preserving the distribution of inorganic arsenic species in groundwater and acid mine drainage samples. *Environ. Sci. Technol.* **36**, 2213–2218.

Berg M., Tran H. C., Nguyen T. C., Pham H. V., Schertenleib R., and Giger W. (2001) Arsenic contamination of groundwater and drinking water in Vietnam: a human health threat. *Environ. Sci. Technol.* **35**, 2621–2626.

Bethke C. (2002) *The Geochemist's Workbench: User Guide*, University of Illinois.

BGS (1979–2002) *Regional Geochemical Atlas Series*. British Geological Survey, Keyworth (formerly Institute of Geological Sciences).

BGS and DPHE (2001) Arsenic contamination of groundwater in Bangladesh (four volumes). In *BGS Technical Report WC/00/19* (eds. D. G. Kinniburgh and P. L. Smedley). British Geological Survey, Keyworth (see www.bgs.ac.uk/Arsenic/Bangladesh).

Bhattacharya P., Chatterjee D., and Jacks G. (1997) Occurrence of arsenic-contaminated groundwater in alluvial aquifers from delta plains, Eastern India: options for safe drinking water supply. *Water Resour. Develop.* **13**, 79–92.

Bohari Y., Astruc A., Astruc M., and Cloud J. (2001) Improvements of hydride generation for the speciation of arsenic in natural freshwater samples by HPLC-HG-AFS. *J. Analyt. Atom. Spectrom.* **16**, 774–778.

Boyle D. R., Turner R. J. W., and Hall G. E. M. (1998) Anomalous arsenic concentrations in groundwaters of an island community, Bowen Island, British Columbia. *Environ. Geochem. Health* **20**, 199–212.

Brown G. E., Henrich V. E., Casey W. H., Clark D. L., Eggleson C., Felmy A., Goodman D. W., Gratzel M., Marciel G., McCarthy I. M., Nealson K. H., Sverjensky D. A., Toney M. F., and Zachara J. M. (1999) Metal oxide surfaces and their interactions with aqueous solutions and microbial organisms. *Chem. Rev.* **99**, 77–174.

Bueno M. and Potin-Gautier M. (2002) Solid-phase extraction for the simultaneous preconcentration of organic (selenocystine) and inorganic Se(IV), Se(VI) selenium in natural waters. *J. Chromatogr. A* **963**, 185–193.

Burk R. F. (1994) *Selenium in Biology and Human Health*. Springer, Berlin.

Burkel R. S. and Stoll R. C. (1999) Naturally occurring arsenic in sandstone aquifer water supply wells of northeastern Wisconsin. *Ground Water Monitor. Remed.* **19**, 114–121.

Cabon J. Y. and Cabon N. (2000) Determination of arsenic species in seawater by flow injection hydride generation *in situ* collection followed by graphite furnace atomic absorption spectrometry: stability of As(III). *Analyt. Chim. Acta* **418**, 19–31.

Cáceres L., Gruttner E., and Contreras R. (1992) Water recycling in arid regions: Chilean case. *Ambio* **21**, 138–144.

Chen H. W., Frey M. M., Clifford D., McNeill L. S., and Edwards M. (1999) Arsenic treatment considerations. *J. Am. Water Works Assoc.* **91**, 74–85.

Chen S. L., Dzeng S. R., Yang M. H., Chlu K. H., Shieh G. M., and Wal C. M. (1994) Arsenic species in groundwaters of the Blackfoot disease areas, Taiwan. *Environ. Sci. Technol.* **28**, 877–881.

Chen S. L., Yeh S. J., Yang M. H., and Lin T. H. (1995) Trace element concentration and arsenic speciation in the well water of a Taiwan area with endemic Blackfoot disease. *Biol. Trace Element Res.* **48**, 263–274.

Cherry J. A., Shaikh A. U., Tallman D. E., and Nicholson R. V. (1979) Arsenic species as an indicator of redox conditions in groundwater. *J. Hydrol.* **43**, 373–392.

Clewell H. J., Gentry P. R., Barton H. A., Shipp A. M., Yager J. W., and Andersen M. E. (1999) Requirements for a biologically realistic cancer risk assessment for inorganic arsenic. *Int. J. Toxicol.* **18**, 131–147.

Combs G. F. and Combs S. B. (1986) *The Role of Selenium in Nutrition*. Academic Press, New York.

Crecelius E. A. (1975) The geochemical cycle of arsenic in Lake Washington and its relation to other elements. *Limnol. Oceanogr.* **20**, 441–451.

Cullen W. R. and Reimer K. J. (1989) Arsenic speciation in the environment. *Chem. Rev.* **89**, 713–764.

Cummings D. E., Caccavo F., Fendorf S., and Rosenzweig R. F. (1999) Arsenic mobilization by the dissimilatory Fe(III)-reducing bacterium Shewanella alga BrY. *Environ. Sci. Technol.* **33**, 723–729.

Cutter G. A. (1989) The estuarine behavior of selenium in San-Francisco Bay. *Estuar. Coast. Shelf Sci.* **28**, 13–34.

Cutter G. A. and Cutter L. S. (2001) Sources and cycling of selenium in the western and equatorial Atlantic Ocean. *Deep-Sea Res.: Part I. Topical Studies Oceanogr.* **48**, 2917–2931.

Darnley A. G., Björklund A., Bølviken B., Gustavsson N., Koval P. V., Plant J. A., Steenfelt A., Tauchid M., and Xuejing Xie X. (1995) A global geochemical database for environmental and resource management. *Recommendations for International Geochemical Mapping. Earth Science Report* **19**, UNESCO Publishing, Paris.

Das D., Samanta G., Mandal B. K., Chowdhury T. R., Chanda C. R., Chowdhury P. P., Basu G. K., and Chakraborti D. (1996) Arsenic in groundwater in six districts of West Bengal, India. *Environ. Geochem. Health* **18**, 5–15.

Daus B., Mattusch J., Paschke A., Wennrich R., and Weiss H. (2000) Kinetics of the arsenite oxidation in seepage water from a tin mill tailings pond. *Talanta* **51**, 1087–1095.

Daus B., Mattusch J., Wennrich R., and Weiss H. (2002) Investigation on stability and preservation of arsenic species in iron rich water samples. *Talanta* **58**, 57–65.

Davies B. E. (1980) *Applied Soil Trace Elements*. Wiley, Chichester.

Davies E. and Watkinson J. (1966) Uptake of native and applied Se by pasture species. *NZ J. Agri. Res.* **9**, 317–324.

Davis C., Uthus E., and Finley J. (2000) Dietary selenium and arsenic affect DNA methylation *in vitro* in Caco-2 cells and in vivo in rat liver and colon. *J. Nutrit.* **130**, 2903–2909.

Del Razo L. M., Arellano M. A., and Cebrian M. E. (1990) The oxidation states of arsenic in well-water from a chronic arsenicism area of northern Mexico. *Environ. Pollut.* **64**, 143–153.

Demesmay C. and Olle M. (1997) Application of microwave digestion to the preparation of sediment samples for arsenic speciation. *Fresen. J. Analyt. Chem.* **357**, 1116–1121.

Deuel L. E. and Swoboda A. R. (1972) Arsenic solubility in a reduced environment. *Soil Sci. Soc. Am. Proc.* **36**, 276–278.

Deverel S. J. and Fujii R. (1988) Processes affecting the distribution of selenium in shallow groundwater in agricultural areas, Western San Josquin Valley, California. *Water Resour. Res.* **24**, 516–524.

Deverel S., Fio J., and Dubrovsky N. (1994) Distribution and mobility of selenium in groundwater in the western San Joaquin valley of California. In *Selenium in the Environment* (eds. W. T. Frankenberger and S. Benson). Dekker, New York, pp. 157–183.

Ding Z., Finkelman R., Belkin H., Aheng B., Hu T., and Xie U. (2000) The mode of occurrence of arsenic in high arsenic coals from endemic arsenosis areas in Southwest Guizhou province, China. *Metal Ions Biol. Med.* **6**, 56–58.

Driehaus W. and Jekel M. (1992) Determination of As(III) and total inorganic arsenic by on-line pretreatment in hydride generation atomic absorption spectrometry. *Fresen. J. Analyt. Chem.* **343**, 352–356.

Driehaus W., Seith R., and Jekel M. (1995) Oxidation of arsenate(III) with manganese oxides in water-treatment. *Water Research* **29**, 297–305.

Durum W. H., Hem J. D., and Heidel S. G. (1971) Reconnaissance of selected minor elements in surface waters of the United States. *Geol. Soc. Circular* **643**, 1–49.

Dzombak D. A. and Morel F. M. M. (1990) *Surface Complexation Modelling—Hydrous Ferric Oxide*. Wiley, New York.

Edmunds W. M., Cook J. M., Kinniburgh D. G., Miles D. L., and Trafford J. M. (1989) Trace-element occurrence in British groundwaters. Research Report SD/89/3. British Geological Survey.

Ehrlich H. (2002) Bacterial oxidation of As(III) compounds. In *Environmental Chemistry of Arsenic* (ed. W. Frankenberger). Dekker, pp. 313–342, chap. 7.

Engberg R. A. (1999) Selenium budgets for Lake Powell and the upper Colorado River Basin. *J. Am. Water Resour. Assoc.* **35**, 771–786.

Farmer J. G. and Lovell M. A. (1986) Natural enrichment of arsenic in Loch Lomond sediments. *Geochim. Cosmochim. Acta* **50**, 2059–2067.

Ficklin W. H. and Callender E. (1989) Arsenic geochemistry of rapidly accumulating sediments, Lake Oahe, South Dakota. In *US Geological Survey Water Resources Investigations Report 88-4420. US Geological Survey Toxic Substances Hydrology Program. Proceedings of the Technical Meeting, Phoenix, Arizona, September 26–30, 1988* (eds. G. E. Mallard and S. E. Ragone). US Geological Survey, Denver, pp. 217–222.

Florencio M. H., Duarte M. F., Facchetti S., Gomes M. L., Goessler W., Irgolic K. J., van't Klooster H. A., Montanarella L., Ritsema R., Boas L. F., and de Bettencourt A. M. M. (1997) Identification of inorganic, methylated and hydride-refractory arsenic species in estuarine waters. Advances by electrospray, ES-MS, pyrolysis-GC-MS and HPLC-ICP/MS. *Analysis* **25**, 226–229.

Focazio M. J., Welch A. H., Watkins S. A., Helsel D. R., and Horng M. A. (1999) *A Retrospective Analysis of the Occurrence of Arsenic in Groundwater Resources of the United States and Limitations in Drinking Water Supply Characterizations*. US Geological Survey, Water Resources Investigation Report 99–4279, 11pp.

Fordyce F. M., Johnson C. C., Navaratna U. R. B., Appleton J. D., and Dissanayake C. B. (2000a) Selenium and iodine in soil, rice and drinking water in relation to endemic goitre in Sri Lanka. *Sci. Tot. Environ.* **263**, 127–141.

Fordyce F. M., Zhang G., Green K., and Liu X. (2000b) Soil, grain and water chemistry in relation to human selenium-responsive diseases in Enshi District, China. *Appl. Geochem.* **15**, 117–132.

Foster A. L., Brown G. E., and Parks G. A. (1998a) X-ray absorption fine-structure spectroscopy study of photocatalyzed, heterogeneous As(III) oxidation on kaolin and anatase. *Environ. Sci. Technol.* **32**, 1444–1452.

Foster A. L., Brown G. E., Tingle T. N., and Parks G. A. (1998b) Quantitative arsenic speciation in mine tailings using X-ray absorption spectroscopy. *Am. Mineral.* **83**, 553–568.

Foster A. L., Breit G. N., Whitney J. W., Welch A. H., Yount J. C., Alam M. K., Islam M. N., Islam M. S., Karim M., and Manwar A. (2000) X-ray absorption fine structure spectroscopic investigation of arsenic species in soil and aquifer sediments from Brahmanbaria, Bangladesh. *4th Annual Arsenic Conference, San Diego, June 18–22, 2000.*

Francesconi K. A. and Kuehnelt D. (2002) Arsenic compounds in the environment. *Environmental Chemistry of Arsenic* (ed. W. Frankenberger). Dekker, New York, Chap. 3, pp. 51–94.

Frankenberger W. T. (2002) *Environmental Chemistry of Arsenic*. Dekker, 391pp.

Frankenberger W. T. and Arshad M. (2001) Bioremediation of selenium-contaminated sediments and water. *Biofactors* **14**, 241–254.

Frankenberger W. T. and Benson S. (eds.) (1994) *Selenium in the Environment*. Dekker, New York, 456pp.

Fujii R. and Swain W. C. (1995) Areal distribution of selected trace elements, salinity, and major ions in shallow ground water, Tulare Basin, Southern San Joaquin Valley, California. US Geological Survey Water-resources Investigations Report 95-4048. US Geological Survey.

Gallagher P. A., Schwegel C. A., Wei X. Y., and Creed J. T. (2001) Speciation and preservation of inorganic arsenic in drinking water sources using EDTA with IC separation and ICP-MS detection. *J. Environ. Moniter.* **3**, 371–376.

García-Hernández J., Glenn E. P., Artiola J., and Baumgartner D. J. (2000) Bioaccumulation of selenium (Se) in the Cienega de Santa Clara wetland, Sonora, Mexico. *Ecotoxicol. Environ. Safety* **46**, 298–304.

Gelova G. A. (1977) *Hydrogeochemistry of Ore Elements*. Nedra, Moscow.

Gluskoter H. J., Ruch R. R., Miller W. G., Cahill R. A., Dreher G. B., and Kuhn J. K. (1977) *Trace Elements in Coal—Occurrence and Distribution*, Circular 499, Illinois State Geological Survey, Urbana, 154p.

Goldberg S. (1985) Chemical modeling of anion competition on goethite using the constant capacitance model. *Soil Sci. Soc. Am. J.* **49**(4), 851–856.

Goldberg S. (1986) Chemical modeling of arsenate adsorption on aluminum and iron oxide minerals. *Soil Sci. Soc. Am. J.* **50**, 1154–1157.

Gomez-Ariza J. L., Velasco-Arjona A., Giraldez I., Sanchez-Rodas D., and Morales E. (2000) Coupling pervaporation-gas chromatography for speciation of volatile forms of selenium in sediments. *Int. J. Environ. Analyt. Chem.* **78**, 427–440.

Govindaraju K. (1994) Compilation of working values and description for 383 geostandards. *Geostand. Newslett.* **18**, 1–158.

Grenthe I., Stumm W., Laaksuharju M., Nilsson A. C., and Wikberg P. (1992) Redox potentials and redox reactions in deep groundwater systems. *Chem. Geol.* **98**, 131–150.

Grimes D. J., Ficklin W. H., Meier A. L., and McHugh J. B. (1995) Anomalous gold, antimony, arsenic, and tungsten in ground water and alluvium around disseminated gold deposits along the Getchell Trend, Humboldt County, Nevada. *J. Geochem. Explor.* **52**, 351–371.

Gulens J., Champ D. R., and Jackson R. E. (1979) Influence of redox environments on the mobility of arsenic in ground water. In *ACS Symposium Series 93. Chemical Modelling in Aqueous Systems* (ed. E. A. Jenne). American Chemical Society, Washington, DC, pp. 81–95.

Gurzau E. S. and Gurzau A. E. (2001) Arsenic exposure from drinking groundwater in Transylvania, Romania: an overview. In *Arsenic Exposure and Health Effects IV* (eds. W. R. Chappell, C. O. Abernathy, and R. L. Calderon). Elsevier, Amsterdam, pp. 181–184.

Hall G. and Pelchat J.-C. (1997) Determination of As, Bi, Sb, Se, and Te in fifty five reference materials by hydride generation ICP-MS. *Geostand. Newslett.* **21**, 85–91.

Hall G. E. M., Pelchat J. C., and Gauthier G. (1999) Stability of inorganic arsenic(III) and arsenic(V) in water samples. *J. Analyt. Atom. Spectrom.* **14**, 205–213.

Harris M. A. and Ragusa S. (2001) Bioremediation of acid mine drainage using decomposable plant material in a constant flow bioreactor. *Environ. Geol.* **40**, 1192–1204.

Harvey C. F., Swartz C. H., Badruzzaman A. B. M., Keon-Blute N., Yu W., Ali M. A., Jay J., Beckie R., Niedan V., Brabander D., Oates P. M., Ashfaque K. N., Islam S., Hemond H. F., and Ahmed M. F. (2002) Arsenic mobility and groundwater extraction in Bangladesh. *Science* **298**, 1602–1606.

Hasegawa H. (1997) The behavior of trivalent and pentavalent methylarsenicals in Lake Biwa. *Appl. Organometal. Chem.* **11**, 305–311.

Hasegawa H., Matsui M., Okamura S., Hojo M., Iwasaki N., and Sohrin Y. (1999) Arsenic speciation including 'hidden' arsenic in natural waters. *Appl. Organometal. Chem.* **13**, 113–119.

Hashimoto Y. and Winchester J. W. (1967) Selenium in the atmosphere. *Environ. Sci. Technol.* **1**, 338–340.

Haygarth P. M. (1994) Global importance and global cycling of selenium. In *Selenium in the Environment* (eds. W. T. Frankenberger and S. Benson). Dekker, New York, chap. 1, pp. 1–27.

Heinrichs G. and Udluft P. (1999) Natural arsenic in Triassic rocks: a source of drinking-water contamination in Bavaria. *Germany. Hydrogeol. J.* **7**, 468–476.

Helz G. R., Tossell J. A., Charnock J. M., Pattrick R. A. D., Vaughan D. J., and Garner C. D. (1995) Oligomerization in As(III) sulfide solutions-theoretical constraints and spectroscopic evidence. *Geochim. Cosmochim. Acta* **59**, 4591–4604.

Hem J. (1992) *Study and Interpretation of Chemical Characteristics of Natural Water*, USGS Water Supply Paper no. 2254 3rd edn. US Geological Survey, Washington, DC.

Herbel M. J., Blum J. S., Hoeft S. E., Cohen S. M., Arnold L. L., Lisak J., Stolz J. F., and Oremland R. S. (2002a) Dissimilatory arsenate reductase activity and arsenate-respiring bacteria in bovine rumen fluid, hamster feces, and the termite hindgut. *Fems Microbiol. Ecol.* **41**, 59–67.

Herbel M. J., Johnson T. M., Tanji K. K., Gao S. D., and Bullen T. D. (2002b) Selenium stable isotope ratios in California agricultural drainage water management systems. *J. Environ. Qual.* **31**, 1146–1156.

Hering J. and Kneebone P. E. (2002) Biogeochemical controls on arsenic occurrence and mobility in water supplies. In *Environmental Chemistry of Arsenic* (ed. W. Frankenberger). Dekker, New York, chap. 7, pp. 155–181.

Hiemstra T. and van Riemsdijk W. H. (1999) Surface structural ion adsorption modeling of competitive binding of oxyanions by metal (Hydr)oxides. *J. Colloid Interface Sci.* **210**, 182–193.

Hinkle S. and Polette D. (1999) Arsenic in groundwater of the Willamette Basin, Oregon. Water-Resources Investigation Report 98-4205. US Geological Survey.

Hoeft S. E., Lucas F., Hollibaugh J. T., and Oremland R. S. (2002) Characterization of microbial arsenate reduction in the anoxic bottom waters of Mono Lake, California. *Geomicrobiol. J.* **19**, 23–40.

Hopenhayn-Rich C., Biggs M. L., Fuchs A., Bergoglio R., Tello E. E., Nicolli H., and Smith A. H. (1996) Bladder-cancer mortality associated with arsenic in drinking water in Argentina. *Epidemiology* **7**, 117–124.

Horowitz A., Lum K., Garbarino J., Hall G., Lemieux C., and Demas C. (1996) Problems associated with using filtration to define dissolved trace element concentrations in natural water samples. *Environ. Sci. Technol.* **30**, 954–963.

Howard A. G., Apte S. C., Comber S. D. W., and Morris R. J. (1988) Biogeochemical control of the summer distribution and speciation of arsenic in the Tamar estuary. *Estuar. Coast. Shelf Sci.* **27**, 427–443.

Howard A. G., Hunt L. E., and Salou C. (1999) Evidence supporting the presence of dissolved dimethylarsinate in the marine environment. *Appl. Organometal. Chem.* **13**, 39–46.

ICRCL (1987) *Interdepartmental Committee on the Redevelopment of Contaminated Land. ICRCL guidance note 59/83* 2nd edn. Department of the Environment, HMSO, London.

Inskeep W. P., McDermott T., and Fendorf S. (2002) Arsenic (V)/(III) cycling in soils and natural waters: chemical and microbiological processes. In *Environmental Chemistry of Arsenic* (ed. W. Frankenberger). Dekker, New York, chap. 8, pp. 183–215.

Ipolyi I. and Fodor P. (2000) Development of analytical systems for the simultaneous determination of the speciation of arsenic As(III), methylarsonic acid, dimethylarsinic acid, As(V) and selenium Se(IV), Se(VI). *Analyt. Chim. Acta* **413**, 13–23.

Irgolic K. (1994) Determination of total arsenic and arsenic compounds in drinking water. In *Arsenic: Exposure and*

Health (eds. W. R. Chappell, C. O. Abernathy, and C. R. Cothern). Science and Technology Letters, Northwood, pp. 51–60.

Jacobs L.W. (1989) Selenium in Agriculture and the Environment. In *Soil Science Society of America, Special Publication 23*. Soil Science Society of America, Madison.

Jain A. and Loeppert R. H. (2000) Effect of competing anions on the adsorption of arsenate and arsenite by ferrihydrite. *J. Environ. Qual.* **29**, 1422–1430.

Johnson C. C., Ge X., Green K. A., and Liu X. (2000) Selenium distribution in the local environment of selected villages of the Keshan Disease belt, Zhangjiakou District, Hebei Province, People's Republic of China. *Appl. Geochem.* **15**, 385–401.

Johnson M. S. (1995) Environmental management in metalliferous mining: the past, present and future. In *Mineral Deposits: from their Origin to their Environmental Impact* (ed. K. Zak). Balkema, pp. 655–658.

Johnson T. M., Herbel M. J., Bullen T. D., and Zawislanski P. T. (1999) Selenium isotope ratios as indicators of selenium sources and oxyanion reduction. *Geochim. Cosmochim. Acta* **63**, 2775–2783.

Jung M. C., Thornton I., and Chon H. T. (2002) Arsenic, Sb and Bi contamination of soils, plants, waters and sediments in the vicinity of the Dalsung Cu-W mine in Korea. *Sci. Tot. Environ.* **295**, 81–89.

Kent D. B., Davies J. A., Anderson L. C. D., and Rea B. A. (1995) Transport of chromium and selenium in a pristine sand and gravel aquifer: role of adsorption processes. *Water Resour. Res.* **31**, 1041–1050.

Kim M. J., Nriagu J., and Haack S. (2000) Carbonate ions and arsenic dissolution by groundwater. *Environ. Sci. Technol.* **34**, 3094–3100.

Kinniburgh D. G. and Kosmus W. (2002) Arsenic contamination in groundwater: some analytical considerations. *Talanta* **58**, 165–180.

Kinniburgh D. G. and Miles D. L. (1983) Extraction and chemical analysis of interstitial water from soils and rocks. *Environ. Sci. Technol.* **17**, 362–368.

Kinniburgh D. G., Smedley P. L., Davies J., Milne C., Gaus I., Trafford J. M., Burden S., Huq S. M. I., Ahmad N., and Ahmed K. M. (2003) The scale and causes of the ground water arsenic problem in Bangladesh. In *Arsenic in Groundwater: Occurrence and Geochemistry* (eds. A. H. Welch and K. G. Stollenwerk). Kluwer, Boston, pp. 211–257.

Korte N. (1991) Naturally occurring arsenic in groundwaters of the midwestern United States. *Environ. Geol.: Water Sci.* **18**, 137–141.

Korte N. E. and Fernando Q. (1991) A review of arsenic(III) in groundwater. *Critical Rev. Environ. Control* **21**, 1–39.

Kuhlmeier P. D. (1997) Partitioning of arsenic species in fine-grained soils. *J. Air Waste Manage. Assoc.* **47**, 481–490.

Kuhn A. and Sigg L. (1993) Arsenic cycling in eutrophic Lake Greifen, Switzerland: influence of seasonal redox processes. *Limnol. Oceanogr.* **38**, 1052–1059.

Kuyucak N. (1998) Mining, the environment and the treatment of mine effluents. *Int. J. Environ. Pollut.* **10**, 315–325.

La Force M. J., Hansel C. M., and Fendorf S. (2000) Arsenic speciation, seasonal transformations, and co-distribution with iron in a mine waste-influence palustrine emergent wetland. *Environ. Sci. Technol.* **34**, 3937–3943.

Lange B. and van den Berg C. M. G. (2000) Determination of selenium by catalytic cathodic stripping voltammetry. *Analyt. Chim. Acta* **418**, 33–42.

Langner H. W. and Inskeep W. P. (2000) Microbial reduction of arsenate in the presence of ferrihydrite. *Environ. Sci. Technol.* **34**, 3131–3136.

Leblanc M., Achard B., Othman D. B., Luck J. M., Bertrand-Sarfati J., and Personne J. C. (1996) Accumulation of arsenic from acidic mine waters by ferruginous bacterial accretions (stromatolites). *Appl. Geochem.* **11**, 541–554.

Lenvik K., Steinnes E., and Pappas A. C. (1978) Contents of some heavy metals in Norwegian rivers. *Nordic Hydrol.* **9**, 197–206.

Lerda D. E. and Prosperi C. H. (1996) Water mutagenicity and toxicology in Rio Tercero (Córdoba, Argentina). *Water Res.* **30**, 819–824.

Levander O. A. (1986) Selenium. In *Trace Elements in Human and Animal Nutrition* (ed. W. Mertz). Academic Press, Orlando, pp. 209–266.

Li Y., Peng T., Yang Y., Niu C., Archard L. C., and Zhang H. (2000) High prevalence of enteroviral genomic sequences in myocardium from cases of endemic cardiomyopathy (Keshan disease) in China. *Heart* **83**, 696–701.

Lima H. M. and Wathern P. (1999) Mine closure: a conceptual review. *Mining Eng.* **51**, 41–45.

Lindemann T., Prange A., Dannecker W., and Neidhart B. (2000) Stability studies of arsenic, selenium, antimony and tellurium species in water, urine, fish and soil extracts using HPLC/ICP-MS. *Fresen. J. Analyt. Chem.* **368**, 214–220.

Litaor M. I. and Keigley R. B. (1991) Geochemical equilibria of iron in sediments of the Roaring river alluvial fan, Rocky Mountain National Park, Colorado. *Earth Surf. Process. Landforms* **16**, 533–546.

Lumsdon D. G., Meeussen J. C. L., Paterson E., Garden L. M., and Anderson P. (2001) Use of solid phase characterisation and chemical modelling for assessing the behaviour of arsenic in contaminated soils. *Appl. Geochem.* **16**, 571–581.

Luo Z. D., Zhang Y. M., Ma L., Zhang G. Y., He X., Wilson R., Byrd D. M., Griffiths J. G., Lai S., He L., Grumski K., and Lamm S. H. (1997) Chronic arsenicism and cancer in Inner Mongolia-consequences of well-water arsenic levels greater than 50 mg l^{-1}. In *Arsenic Exposure and Health Effects* (eds. C. O. Abernathy, R. L. Calderon, and W. R. Chappell). Chapman and Hall, London, pp. 55–68.

Macur R. E., Wheeler J. T., McDermott T. R., and Inskeep W. P. (2001) Microbial populations associated with the reduction and enhanced mobilization of arsenic in mine tailings. *Environ. Sci. Technol.* **35**, 3676–3682.

Maest A. S., Pasilis S. P., Miller L. G., and Nordstrom D. K. (1992) Redox geochemistry of arsenic and iron in Mono Lake, California, USA. *Proceedings of the Seventh International Symposium on Water–Rock Interaction.* AA Balkema, Rotterdam, pp. 507–511.

Manning B. A. and Goldberg S. (1997) Adsorption and stability of arsenic(III) at the clay mineral-water interface. *Environ. Sci. Technol.* **31**, 2005–2011.

Manning B. A., Fendorf S. E., Bostick B., and Suarez D. L. (2002) Arsenic(III) oxidation and arsenic(V) adsorption reactions on synthetic birnessite. *Environ. Sci. Technol.* **36**, 976–981.

Martinez-Bravo Y., Roig-Navarro A. F., Lopez F. J., and Hernandez F. (2001) Multielemental determination of arsenic, selenium and chromium(VI) species in water by high-performance liquid chromatography-inductively coupled plasma mass spectrometry. *J. Chromatogr. A* **926**, 265–274.

Masscheleyn P. H., DeLaune R. D., and Patrick W. H. (1991) Effect of redox potential and pH on arsenic speciation and solubility in a contaminated soil. *Environ. Sci. Technol.* **25**, 1414–1419.

Mayland H. (1994) Selenium in plant and animal nutrition. In *Selenium in the Environment* (eds. W. T. Frankenberger and S. Benson). Dekker, New York, chap. 2, pp. 29–45.

McArthur J. M., Ravenscroft P., Safiulla S., and Thirwall M. F. (2001) Arsenic in groundwater: testing pollution mechanisms for sedimentary aquifers in Bangladesh. *Water Resour. Res.* **37**, 109–117.

McCleskey R. B. and Nordstrom D. K. (2003) Arsenic (III/V) preservation procedures for water samples: new data and an evaluation of the literature. *Appl. Geochem.* (in review).

McCreadie H., Blowes D. W., Ptacek C. J., and Jambor J. L. (2000) Influence of reduction reactions and solid phase composition on porewater concentrations of arsenic. *Environ. Sci. Technol.* **34**, 3159–3166.

McGeehan S. L., Fendorf S. E., and Naylor D. V. (1998) Alteration of arsenic sorption in flooded-dried soils. *Soil Sci. Soc. Am. J.* **62**, 828–833.

McLaren S. J. and Kim N. D. (1995) Evidence for a seasonal fluctuation of arsenic in New Zealand's longest river and the effect of treatment on concentrations in drinking water. *Environ. Pollut.* **90**, 67–73.

Meng X. G., Korfiatis G. P., Jing C. Y., and Christodoulatos C. (2001) Redox transformations of arsenic and iron in water treatment sludge during aging and TCLP extraction. *Environ. Sci. Technol.* **35**, 3476–3481.

Meng X. G., Korfiatis G. P., Bang S. B., and Bang K. W. (2002) Combined effects of anions on arsenic removal by iron hydroxides. *Toxicol. Lett.* **133**, 103–111.

Mok W. and Wai C. M. (1990) Distribution and mobilization of arsenic and antimony species in the Coeur D'Alene River, Idaho. *Environ. Sci. Technol.* **24**, 102–108.

Monhemius A. J. and Swash P. M. (1999) Removing and stabilising arsenic from copper refining circuits by hydrothermal processing. *J. Minerals Metals Materials Soc.* **51**, 30–33.

Moore J. N., Ficklin W. H., and Johns C. (1988) Partitioning of arsenic and metals in reducing sulfidic sediments. *Environ. Sci. Technol.* **22**, 432–437.

Moxon A. L. (1937) Alkali disease or selenium poisoning. In *South Dakota State College Bulletin*, vol. 311, 99p.

Moxon A. L. (1938) The effect of arsenic on the toxicity of seleniferous grains. *Science* **88**, 81.

Mukhopadhyay R., Rosen B. P., Pung L. T., and Silver S. (2002) Microbial arsenic: from geocycles to genes and enzymes. *Fems Microbiol. Rev.* **26**, 311–325.

Muth O. H. and Allaway W. H. (1963) The relationship of white muscle disease to the distribution of naturally occurring selenium. *J. Am. Vet. Med. Assoc.* **142**, 1379–1384.

Nag J. K., Balaram V., Rubio R., Alberti J., and Das A. K. (1996) Inorganic arsenic species in groundwater: a case study from Purbasthali (Burdwan), India. *J. Trace Elements Med. Biol.* **10**, 20–24.

NAS (1976) *Selenium*. National Academy of Sciences, Washington, DC, 203pp.

National Research Council (1999) *Arsenic in Drinking Water*. National Academy Press, Washington, DC.

Newman D. K., Beveridge T. J., and Morel F. M. M. (1997a) Precipitation of arsenic trisulfide by *Desulfotomaculum auripigmentum*. *Appl. Environ. Microbiol.* **63**, 2022–2028.

Newman D. K., Kennedy E. K., Coates J. D., Ahmann D., Ellis D. J., Lovley D. R., and Morel F. M. M. (1997b) Dissimilatory arsenate and sulfate reduction in Desulfotomaculum auripigmentum sp. nov. *Arch. Microbiol.* **168**, 380–388.

Newman D. K., Ahmann D., and Morel F. M. M. (1998) A brief review of microbial arsenate respiration. *Geomicrobiol. J.* **15**, 255–268.

Nickson R., McArthur J., Burgess W., Ahmed K. M., Ravenscroft P., and Rahman M. (1998) Arsenic poisoning of Bangladesh groundwater. *Nature* **395**, 338.

Nickson R. T., McArthur J. M., Ravenscroft P., Burgess W. G., and Ahmed K. M. (2000) Mechanism of arsenic release to groundwater, Bangladesh and West Bengal. *Appl. Geochem.* **15**, 403–413.

Nicolli H. B., Suriano J. M., Peral M. A. G., Ferpozzi L. H., and Baleani O. A. (1989) Groundwater contamination with arsenic and other trace-elements in an area of the Pampa, province of Córdoba, Argentina. *Environ. Geol.: Water Sci.* **14**, 3–16.

Nicolli H. B., Tineo A., García J., Falcón C., and Merino M. (2001) Trace-element quality problems in groundwater from Tucumán, Argentina. *Water–Rock Interaction*. AA Balkema, Rotterdam, pp. 993–996.

Nimick D. A., Moore J. N., Dalby C. E., and Savka M. W. (1998) The fate of geothermal arsenic in the Madison and Missouri Rivers, Montana and Wyoming. *Water Resour. Res.* **34**, 3051–3067.

Nishri A., Brenner I. B., Hall G. E. M., and Taylor H. E. (1999) Temporal variations in dissolved selenium in Lake Kinneret (Israel). *Aquat. Sci.* **61**, 215–233.

Nordstrom D. K. and Alpers C. N. (1999) Negative pH, efflorescent mineralogy, and consequences for environmental restoration at the Iron Mountain superfund site, California. *Natl. Acad. Sci.* **96**, 3455–3462.

Nordstrom D. K. and Archer D. G. (2003) Arsenic thermodynamic data and environmental geochemistry. In *Arsenic in Ground Water: Geochemistry and Occurrence* (ed. K. G. Stollenwerk). Kluwer, Boston, pp. 1–25.

Nordstrom D. K., Alpers C. N., Ptacek C. J., and Blowes D. W. (2000) Negative pH and extremely acidic mine waters from Iron Mountain, California. *Environ. Sci. Technol.* **34**, 254–258.

Nyashanu R., Monhemius A., and Buchanan D. (1999) The effect of ore mineralogy on the speciation of arsenic in bacterial oxidation of refractory arsenical gold ores. In *International Biohydrometallurgy of Symposium IBS 99, Madrid* (eds. R. Amils and A. Ballester). Elsevier, Oxford, pp. 431–441.

Oldfield J. E. (1999) *Selenium World Atlas*. Selenium-Tellurium Development Association.

Oremland R., Newman D., Kail B., and Stolz J. (2002) Bacterial respiration of arsenate and its significance in the environment. In *Environmental Chemistry of Arsenic* (ed. W. Frankenberger). Dekker, New York, chap. 11, pp. 273–295.

Oremland R. S. (1994) Biogeochemical transformations of selenium in anoxic environments. In *Selenium in the Environment* (eds. W. T. Frankenberger and S. Benson). Dekker, New York, chap. 16, pp. 389–419.

Oremland R. S., Steinberg N. A., Maest A. S., Miller L. G., and Hollibaugh J. T. (1990) Measurement of *in situ* rates of selenate removal by dissimilatory bacterial reduction in sediments. *Environ. Sci. Technol.* **24**, 1157–1164.

Oremland R. S., Dowdle P. R., Hoeft S., Sharp J. O., Schaefer J. K., Miller L. G., Blum J. S., Smith R. L., Bloom N. S., and Wallschlaeger D. (2000) Bacterial dissimilatory reduction of arsenate and sulphate in meromictic Mono Lake, California. *Geochim. Cosmochim. Acta* **64**, 3073–3084.

Oscarson D. W., Huang P. M., Liaw W. K., and Hammer U. T. (1983) Kinetics of oxidation of arsenite by various manganese dioxides. *Soil Sci. Soc. Am. J.* **47**, 644–648.

Palmer C. A. and Klizas S. A. (1997) The chemical analysis of Argonne premium coal samples. US Geological Survey Bulletin 2144. US Geological Survey.

Parkhurst D. and Appelo C. (1999) User's Guide to PHREEQC (Version 2)—A Computer Program for Speciation, Batch-Reaction, One-Dimensional Transport, and Inverse Geochemical Calculations. Water-Resources Investigations Report 99-4259. US Geological Survey.

Pendleton J. A., Posey H. H., and Long M. B. (1995) Characterizing Summitville and its impacts: setting and scene. In *Proceedings: Summitville Forum '95; Colorado Geological Survey Special Publication 38* (eds. H. H. Posey, J. A. Pendleton, and D. Van Zyl). US Geological Survey, Denver, pp. 1–12.

Peng A., Wang W. H., Wang C. X., Wang Z. J., Rui H. F., Wang W. Z., and Yang Z. W. (1999) The role of humic substances in drinking water in Kashin-Beck disease in China. *Environ. Health Perspect.* **107**, 293–296.

Peters G. M., Maher W. A., Krikowa F., Roach A. C., Jeswani H. K., Barford J. P., Gomes V. G., and Reible D. D. (1999) Selenium in sediments, pore waters and benthic infauna of Lake Macquarie, New South Wales, Australia. *Mar. Environ. Res.* **47**, 491–508.

Petersen W., Wallmann K., Li P. L., Schroeder F., and Knauth H. D. (1995) Exchange of trace-elements at the sediment-water interface during early diagenesis processes. *Mar. Freshwater Res.* **46**, 19–26.

Peterson M. L. and Carpenter R. (1983) Biogeochemical processes affecting total arsenic and arsenic species distributions in an intermittently anoxic Fjord. *Mar. Chem.* **12**, 295–321.

Pettine M., Camusso M., and Martinotti W. (1992) Dissolved and particulate transport of arsenic and chromium in the Po River (Italy). *Sci. Tot. Environ.* **119**, 253–280.

Plant J. A., Smith D., and Reeder S. (2003) Environmental geochemistry on a global scale. In *Geology and Health: Closing the Gap* (ed. H. C. W. Skinner, A. R. Berger, *et al.*). Oxford University Press, Oxford, chap. 20, pp. 129–134.

Plant J. A., Jeffery K., Gill E., and Fage C. (1975) The systematic determination of accuracy and precision in geochemical exploration data. *J. Geochem. Explor.* **4**, 467–486.

Plumlee G. S., Smith K. S., Montour M. R., Ficklin W. H., and Mosier E. L. (1999) Geologic controls on the composition of natural waters and mine waters draining diverse mineral-deposit types. In *Reviews in Economic Geology Vol. 6B. Environmental Geochemistry of Mineral Deposits. Part B: Case studies* (eds. L. H. Filipek and G. S. Plumlee). Society of Economic Geologists, Littleton, Colorado, chap. 19, pp. 373–432.

Presser T. (1994) Geologic origin and pathways of selenium from the California coast ranges to the west-central San Joaquin Valley. In *Selenium in the Environment* (eds. W. T. Frankenberger and S. Benson). Dekker, New York, chap. 6, pp. 139–155.

Randall S. R., Sherman D. M., and Ragnarsdottir K. V. (2001) Sorption of As(V) on green rust (Fe-4(II)Fe-2(III)(OH)(12)-SO43H(2)O) and lepidocrocite (gamma-FeOOH): surface complexes from EXAFS spectroscopy. *Geochim. Cosmochim. Acta* **65**, 1015–1023.

Rapant S., Vrana K., and Bodis D. (1996) *Geochemical Atlas of Slovakia: Part 1. Groundwater*. Geological Survey of Slovak Republic.

Rasmussen L. and Andersen K. (2002) Environmental health and human exposure assessment (draft). *United Nations Synthesis Report on Arsenic in Drinking Water*. WHO, chap. 2.

Rayman M. P. (2002) *Selenium Brought to Earth*. Chemistry in Britain, October, pp. 28–31.

Reimann C., Ayras M., Chekushin V., Bogatyev I., Boyd R., Caritat P., Dutter R., Finne T. E., Halleraker J. H., Jaeger O., Kashulina G., Lehto O., Niskavaara H., Pavlov V., Raisanen M. L., Strand T., and Volden T. (1998) *Environmental Geochemical Atlas of the Central Barents Region*. Geological Survey of Norway.

Reimann C. and Caritat P. (1998) *Chemical Elements in the Environment*. Springer, Berlin.

Reynolds J. G., Naylor D. V., and Fendorf S. E. (1999) Arsenic sorption in phosphate-amended soils during flooding and subsequent aeration. *Soil Sci. Soc. Am. J.* **63**, 1149–1156.

Rice K. C. (1999) Trace-element concentrations in streambed sediment across the conterminous United States. *Environ. Sci. Technol.* **33**, 2499–2504.

Riedel G. F. (1993) The annual cycle of arsenic in a temperate estuary. *Estuaries* **16**, 533–540.

Robinson B., Outred H., Brooks R., and Kirkman J. (1995) The distribution and fate of arsenic in the Waikato River System, North Island, New Zealand. *Chem. Speciat. Bioavail.* **7**, 89–96.

Rochette E. A., Bostick B. C., Li G. C., and Fendorf S. (2000) Kinetics of arsenate reduction by dissolved sulfide. *Environ. Sci. Technol.* **34**, 4714–4720.

Rosen B. P. (2002) Biochemistry of arsenic detoxification. *Febs Lett.* **529**, 86–92.

Rosenfield I. and Beath O. A. (1964) *Selenium, Geobotany, Biochemistry, Toxicity and Nutrition*. Academic Press.

Roussel C., Bril H., and Fernandez A. (2000a) Arsenic speciation: involvement in evaluation of environmental impact caused by mine wastes. *J. Environ. Qual.* **29**, 182–188.

Roussel C., Bril H., and Fernandez A. (2000b) Arsenic speciation: Involvement in evaluation of environmental impact caused by mine wastes. *J. Environ. Qual.* **29**, 182–188.

Rowell D. L. (1994) *Soil Sci.: Methods and Applications*. Longman.

Ryker S. J. (2001) Mapping arsenic in groundwater: a real need, but a hard problem—why was the map created? *Geotimes* **46**, 34–36.

Salminen R. and Gregorauskiene V. (2000) Considerations regarding the definition of a geochemical baseline of elements in the surficial materials in areas differing in basic geology. *Appl. Geochem.* **15**, 647–653.

Sancha A. M. and Castro M. (2001) Arsenic in Latin America: occurrence, exposure, health effects and remediation. In *Arsenic Exposure and Health Effects IV* (eds. W. R. Chappell, C. O. Abernathy, and R. L. Calderon). Elsevier, Oxford, pp. 87–96.

Santini J. M., vanden Hoven R. N., and Macy J. M. (2002) Characteristics of newly discovered arsenite-oxidizing bacteria. In *Environmental Chemistry of Arsenic* (ed. W. Frankenberger). Dekker, New York, chap. 14, pp. 329–342.

Savage K. S., Bird D. K., and Ashley R. P. (2000) Legacy of the California Gold Rush: environmental geochemistry of arsenic in the southern Mother Lode Gold District. *Int. Geol. Rev.* **42**, 385–415.

Schreiber M. E., Simo J. A., and Freiberg P. G. (2000) Stratigraphic and geochemical controls on naturally occurring arsenic in groundwater, eastern Wisconsin, USA. *Hydrogeol. J.* **8**, 161–176.

Schroeder R. A., Orem W. H., and Kharaka Y. K. (2002) Chemical evolution of the Salton Sea, California: nutrient and selenium dynamics. *Hydrobiologia* **473**, 23–45.

Schwedt G. and Rieckhoff M. (1996) Analysis of oxothio arsenic species in soil and water. *J. Praktische Chemie-Chemiker-Zeitung* **338**, 55–59.

Scott M. J. and Morgan J. J. (1995) Reactions at oxide surfaces: 1. Oxidation of As(III) by synthetic birnessite. *Environ. Sci. Technol.* **29**, 1898–1905.

Seyler P. and Martin J. M. (1990) Distribution of arsenite and total dissolved arsenic in major French estuaries: dependence on biogeochemical processes and anthropogenic inputs. *Mar. Chem.* **29**, 277–294.

Seyler P. and Martin J. M. (1991) Arsenic and selenium in a pristine river-estuarine system: the Krka (Yugoslavia). *Mar. Chem.* **34**, 137–151.

Seyler P. and Martin J.-M. (1989) Biogeochemical processes affecting arsenic species distribution in a permanently stratified lake. *Environ. Sci. Technol.* **23**, 1258–1263.

Shamberger R. and Frost D. (1969) Possible inhibitory effect of selenium on human cancer. *Can. Med. Assoc. J.* **100**, 682.

Shiller A. and Taylor H. (1996) Comment on "Problems associated with using filtration to define dissolved trace element concentrations in natural water samples.". *Environ. Sci. Technol.* **30**, 3397–3398.

Shotyk W. (1996) Natural and anthropogenic enrichments of As, Cu, Pb, Sb, and Zn in ombrotrophic versus minerotrophic peat bog profiles, Jura Mountains, Switzerland. *Water Air Soil Pollut.* **90**, 375–405.

Sides A. D. (1995) Can gold mining at Mokrsko (Czech Republic) be environmentally acceptable? In *Mineral Deposits: from their Origin to their Environmental Impact* (eds. J. Pasava, B. Kribek, and K. Zak). AA Balkema, Rotterdam, pp. 701–703.

Simon G., Kesler S. E., and Chryssoulis S. (1999) Geochemistry and textures of gold-bearing arsenian pyrite, Twin Creeks, Nevada: implications for deposition of gold in Carlin-type deposits. *Econ. Geol. Bull. Soc. Econ. Geol.* **94**, 405–421.

Sloth J. J. and Larsen E. H. (2000) The application of inductively coupled plasma dynamic reaction cell mass spectrometry for measurement of selenium isotopes, isotope ratios and chromatographic detection of selenoamino acids. *J. Analyt. Atom. Spectrom.* **15**, 669–672.

Smedley P. L. (1996) Arsenic in rural groundwater in Ghana. *J. African Earth Sci.* **22**, 459–470.

Smedley P. L. (2003) Arsenic in groundwater—south and east Asia. In *Arsenic in Ground Water: Geochemistry and Occurrence* (eds. A. H. Welch and K. G. Stollenwerk). Kluwer, Boston, pp. 179–209.

Smedley P. L. and Edmunds W. M. (2002) Redox patterns and trace-element behavior in the East Midlands Triassic Sandstone Aquifer, UK. *Ground Water* **40**, 44–58.

Smedley P. L. and Kinniburgh D. G. (2002) A review of the source, behaviour and distribution of arsenic in natural waters. *Appl. Geochem.* **17**, 517–568.

Smedley P. L., Kinniburgh D. G., Huq I., Luo Z. D., and Nicolli H. B. (2001a) International perspective on arsenic problems in groundwater. In *Arsenic Exposure and Health Effects IV* (eds. W. R. Chappell, C. O. Abernathy, and R. L. Calderon). Elsevier, Oxford, pp. 9–26.

Smedley P. L., Kinniburgh D. G., Milne C. J., Trafford J. M., Huq S. I., and Ahmed K. M. (2001b) Hydrogeochemistry of three Special Study Areas. In *Arsenic Contamination of Groundwater in Bangladesh*, Volume: 2. Final Report. British Geological Survey Report WC/00/19 (eds. D. G. Kinniburgh and P. L. Smedley). British Geological Survey, Keyworth, chap 7, pp. 105–149.

Smedley P. L., Nicolli H. B., Macdonald D. M. J., Barros A. J., and Tullio J. O. (2002) Hydrogeochemistry of arsenic and other inorganic constituents in groundwaters from La Pampa, Argentina. *Appl. Geochem.* **17**, 259–284.

Smedley P. L., Zhang M.-Y., Zhang G.-Y., and Luo Z.-D. (2003) Mobilisation of arsenic and other trace elements in fluviolacustrine aquifers of the Huhhot Basin, Inner Mongolia. *Appl. Geochem.* **18**, 1453–1477.

Smith A., Lopipero P., Bates M., and Steinmaus C. (2002) Arsenic epidemiology and drinking water standards. *Science* **296**, 2145–2146.

Steinhoff P. J., Smith B. W., Warner D. W., and Moller G. (1999) Analysis of interlaboratory performance in the determination of total selenium in water. *J. AOAC Int.* **82**, 1466–1473.

Stolz J. F. and Oremland R. S. (1999) Bacterial respiration of arsenic and selenium. *Fems Microbiol. Rev.* **23**, 615–627.

Suetens C., Moreno-Reyes R., Chasseur C., Mathieu F., Begaux F., Haubruge E., Durand M. C., Neve J., and Vanderpas J. (2001) Epidemiological support for a multifactorial aetiology of Kashin-Beck disease in Tibet. *Int. Orthop.* **25**, 180–187.

Sullivan K. A. and Aller R. C. (1996) Diagenetic cycling of arsenic in Amazon shelf sediments. *Geochim. Cosmochim. Acta* **60**, 1465–1477.

Swanson V. E., Medlin J. H., Hatch J. R., Coleman S. L., Woodruff S. D., and Hildebrand R. T. (1976) Collection, chemical analysis, and evaluation of 799 coal samples in 1975. US Geological Survey Open-File Rept. 76-468, 503p. United States Geological Survey.

Swash P. M. and Monhemius A. J. (1996) Characteristics of calcium arsenate compounds relevant to disposal of arsenic from industrial processes. *Min. Metals Environ. II*, 353–361.

Tan J. (1989) *The Atlas of Endemic Diseases and their Environments in the People's Republic of China.* Science Press, Beijing.

Taniguchi T., Tao H., Tominaga M., and Miyazaki A. (1999) Sensitive determination of three arsenic species in water by ion exclusion chromatography-hydride generation inductively coupled plasma mass spectrometry. *J. Analyt. Atom. Spectrom.* **14**, 651–655.

Thomas D. J., Styblo M., and Lin S. (2001) The cellular metabolism and systemic toxicity of arsenic. *Toxicol. Appl. Pharmacol.* **176**, 127–144.

Thomson J., Nixon S., Croudace I. W., Pedersen T. F., Brown L., Cook G. T., and MacKenzie A. B. (2001) Redox-sensitive element uptake in north-east Atlantic Ocean sediments (Benthic Boundary Layer Experiment sites). *Earth Planet. Sci. Lett.* **184**, 535–547.

Thornton I., Kinniburgh D. G., Pullen G., and Smith C. A. (1983) Geochemical aspects of selenium in British soils and implications to animal health. In *Trace Substances in Environmental Health: XVII* (ed. D. D. Hemphill). University of Missouri, Columbia, pp. 391–398.

Tokunaga T., Pickering I., and Brown G. (1996) Selenium transformations in ponded sediments. *Soil Sci. Soc. Am. J.* **60**, 781–790.

Tokunaga T., Zawislanski P., Johannis P., Lipton D., and Benson S. (1994) Field investigations of selenium speciation, transformation, and transport in soils from Kesterson Reservoir and Lahontan valley. In *Selenium in the Environment* (eds. W. T. Frankenberger and S. Benson). Dekker, New York, chap. 5, pp. 119–138.

Tournassat C., Charlet L., Bosbach D., and Manceau A. (2002) Arsenic(III) oxidation by birnessite and precipitation of manganese(II) arsenate. *Environ. Sci. Technol.* **36**, 493–500.

Tseng W. P., Chu H. M., How S. W., Fong J. M., Lin C. S., and Yeh S. (1968) Prevalence of skin cancer in an endemic area of chronic arsenicism in Taiwan. *J. Natl. Cancer Inst.* **40**, 453–463.

US EPA (2002) Air Toxics Web-site: Selenium and Compounds (see http://www.epa.gov/ttn/atw/hlthef/selenium.html). US Environmental Protection Agency.

Umitsu M. (1993) Late quaternary sedimentary environments and landforms in the Ganges Delta. *Sedim. Geol.* **83**, 177–186.

Vagliasindi F. G. A. and Benjamin M. M. (2001) Redox reactions of arsenic in As-spiked lake water and their effects on As adsorption. *J. Water Supply Res. Technology-Aqua* **50**, 173–186.

Varsányi I., Fodr Z., and Bartha A. (1991) Arsenic in drinking water and mortality in the southern Great Plain, Hungary. *Environ. Geochem. Health* **13**, 14–22.

Vinceti M., Rothman K. J., Bergomi M., Borciani N., Serra L., and Vivoli G. (1998) Excess melanoma incidence in a cohort exposed to high levels of environmental selenium. *Cancer Epidemiol. Biomark. Prevent.* **7**, 853–856.

Vinceti M., Nacci G., Rocchi E., Cassinadri T., Vivoli R., Marchesi C., and Bergomi M. (2000) Mortality in a population with long-term exposure to inorganic selenium via drinking water. *J. Clinic. Epidemiol.* **53**, 1062–1068.

Voegelin A. and Hug S. (2003) Catalyzed oxidation of arsenic(III) by hydrogen peroxide on the surface of ferrihydrite: an *in situ* ATR-FTIR study. *Environ. Sci. Technol* **37**, 972–978.

Wang H. C., Wang P. H., Peng C. Y., Liu S. H., and Wang Y. W. (2001) Speciation of As in the blackfoot disease endemic area. *J. Synchr. Radiat.* **8**, 961–962.

Wang L. and Huang J. (1994) Chronic arsenism from drinking water in some areas of Xinjiang, China. In *Arsenic in the Environment: Part II. Human Health and Ecosystem Effects* (ed. J. O. Nriagu). Wiley, New York, pp. 159–172.

Wang Z. J. and Gao Y. X. (2001) Biogeochemical cycling of selenium in Chinese environments. *Appl. Geochem.* **16**, 1345–1351.

Webb J. (1978) *The Wolfson Geochemical Atlas of England and Wales.* Clarendon Press, Oxford.

Webster J. G., Nordstrom D. K., and Smith K. S. (1994) Transport and natural attenuation of Cu, Zn, As, and Fe in the acid mine drainage of Leviathan and Bryant creeks. *ACS Symp. Ser.* **550**, 244–260.

Welch A. H. and Lico M. S. (1998) Factors controlling As and U in shallow ground water, southern Carson Desert, Nevada. *Appl. Geochem.* **13**, 521–539.

Welch A. H., Lico M. S., and Hughes J. L. (1988) Arsenic in groundwater of the Western United States. *Ground Water* **26**, 333–347.

Welch A. H., Helsel D. R., Focazio M. J., and Watkins S. A. (1999) Arsenic in ground water supplies of the United States. In *Arsenic Exposure and Health Effects* (eds. W. R. Chappell, C. O. Abernathy, and R. L. Calderon). Elsevier, Oxford, pp. 9–17.

Welch A. H., Westjohn D. B., Helsel D. R., and Wanty R. B. (2000) Arsenic in ground water of the United States: occurrence and geochemistry. *Ground Water* **38**, 589–604.

White A. and Dubrovsky N. (1994) Chemical oxidation-reduction controls on selenium mobility in groundwater systems. In *Selenium in the Environment* (eds. W. T. Frankenberger and S. Benson). Dekker, New York, chap. 8, pp. 185–221.

WHO (1987) *Environmental Health Criterion 58: Selenium*. World Health Organization.

WHO (1993) *Guidelines for Drinking Water Quality*. World Health Organisation.

WHO (1996) *Trace Elements in Human Nutrition and Health*. World Health Organization.

Widerlund A. and Ingri J. (1995) Early diagenesis of arsenic in sediments of the Kalix River estuary, Northern Sweden. *Chem. Geol.* **125**, 185–196.

Wijnja H. and Schulthess C. P. (2000) Interaction of carbonate and organic anions with sulfate and selenate adsorption on an aluminum oxide. *Soil Sci. Soc. Am. J.* **64**, 898–908.

Wilde F. and Radtke D. (1998) Field Measurements. In *Book 9. Handbooks for Water-resources Investigations. National Field Manual for the Collection of Water-quality Data*. US Geological Survey, chap. A6.

Wilkie J. A. and Hering J. G. (1998) Rapid oxidation of geothermal arsenic(III) in streamwaters of the eastern Sierra Nevada. *Environ. Sci. Technol.* **32**, 657–662.

Williams M. (2001) Arsenic in mine waters: an international study. *Environ. Geol.* **40**, 267–279.

Williams M., Fordyce F., Paijitprapapon A., and Charoenchaisri P. (1996) Arsenic contamination in surface drainage and groundwater in part of the southeast Asian tin belt, Nakhon Si Thammarat Province, southern Thailand. *Environ. Geol.* **27**, 16–33.

Wilson F. H. and Hawkins D. B. (1978) Arsenic in streams, stream sediments and ground water, Fairbanks area, Alaska. *Environ. Geol.* **2**, 195–202.

Wu L., Banuelos G., and Guo X. (2000) Changes of soil and plant tissue selenium status in an upland grassland contaminated by selenium-rich agricultural drainage sediment after ten years transformed from a wetland habitat. *Ecotoxicol. Environ. Safety* **47**, 201–209.

Xu G. and Jiang Y. (1986) Se and the prevalence of Keshan and Kaschin-Beck diseases in China. In *Proceedings of the First International Symposium on Geochemistry and Health* (ed. I. Thornton). Science Reviews Ltd., Northwood, pp. 192–205.

Yalcin S. and Le X C. (1998) Low pressure chromatographic separation of inorganic arsenic species using solid phase extraction cartridges. *Talanta* **47**, 787–796.

Yamamura S. (2003) Drinking water guidelines and standards. In *Arsenic, Water, and Health: the state of the art* (eds. H. Hashizume and S. Yamamura). World Health Organisation, chap. 5.

Yan X. P., Kerrich R., and Hendry M. J. (2000) Distribution of arsenic(III), arsenic(V) and total inorganic arsenic in porewaters from a thick till and clay-rich aquitard sequence, Saskatchewan, Canada. *Geochim. Cosmochim. Acta* **64**, 2637–2648.

Yang G. and Xia M. (1995) Studies on human dietary requirements and safe range of dietary intakes of selenium in China and their application to the prevention of related endemic diseases. *Biomed. Environ. Sci.* **8**, 187–201.

Yang G., Wang S., Zhou R., and Sun S. (1983) Endemic selenium intoxication of humans in China. *Am. J. Clinic. Nutrit.* **37**, 872–881.

Zawislanski P. T. and Zavarin M. (1996) Nature and rates of selenium transformations in Kesterson Reservoir soils: a laboratory study. *Soil Sci. Soc. Am. J.* **60**, 791–800.

Zawislanski P. T., Chau S., Mountford H., Wong H. C., and Sears T. C. (2001a) Accumulation of selenium and trace metals on plant litter in a tidal marsh. *Estuar. Coast. Shelf Sci.* **52**, 589–603.

Zawislanski P. T., Mountford H. S., Gabet E. J., McGrath A. E., and Wong H. C. (2001b) Selenium distribution and fluxes in intertidal wetlands, San Francisco Bay, California. *J. Environ. Qual.* **30**, 1080–1091.

Zhang H., Davison W., Gadi R., and Kobayashi T. (1998) *In situ* measurement of dissolved phosphorus in natural waters using DGT. *Analyt. Chem. Acta* **370**, 29–38.

Zobrist J., Dowdle P. R., Davis J. A., and Oremland R. S. (2000) Mobilization of arsenite by dissimilatory reduction of adsorbed arsenate. *Environ. Sci. Technol.* **34**, 4747–4753.

9.03
Heavy Metals in the Environment—Historical Trends

E. Callender

US Geological Survey, Westerly, RI, USA

9.03.1 INTRODUCTION

9.03.1.1 Metals: Pb, Zn, Cd, Cr, Cu, Ni

These six metals, commonly classified as heavy metals, are a subset of a larger group of trace elements that occur in low concentration in the Earth's crust. These heavy metals were mined extensively for use in the twentieth century Industrial Society. Nriagu (1988a) estimated that between 0.5 (Cd) and 310 (Cu) million metric tons of these metals were mined and ultimately deposited in the biosphere. In many instances, the inputs of these metals from anthropogenic sources exceed the contributions from natural sources (weathering, volcanic eruptions, forest fires) by several times (Adriano, 1986). In this chapter, heavy metals (elements having densities greater than 5) and trace elements (elements present in the lithosphere in concentrations less than 0.1%) are considered synonymous.

It has been observed in the past that the rate of emission of these trace metals into the atmosphere is low due to their low volatility. However, with the advent of large-scale metal mining and smelting as well as fossil-fuel combustion in the twentieth century, the emission rate of these metals has increased dramatically. As most of these emissions are released into the atmosphere where the mammals live and breathe, we see a great increase in the occurrence of health problems such as lead (Pb) poisoning, cadmium (Cd) Itai-itai disease, chromium (Cr), and nickel (Ni) carcinogenesis.

In this chapter, the author has attempted to present a synopsis of the importance of these metals in the hydrocycle, their natural and anthropogenic emissions into the environment, their prevalent geochemical form incorporated into lacustrine sediments, and their time-trend distributions in watersheds that have been impacted by urbanization, mining and smelting, and other anthropogenic activities. These time trends are reconstructed from major–minor–trace–element distributions in age-dated sediment cores, mainly from reservoirs where the mass sedimentation rates (MSRs) are orders of magnitude greater than those in natural lakes, the consequences of which tend to preserve the heavy-metal signatures and minimize the metal diagenesis (Callender, 2000). This chapter focuses mainly on the heavy metals in the terrestrial and freshwater environments whilst the environmental chemistry of trace metals in the marine environment is discussed in Volume 6, Chapter 3 of the Treatise on Geochemistry.

The data presented in Tables 2–5 are updated as much as possible, with many of the references postdate the late 1980s. Notable exceptions are riverine particulate matter chemistry (Table 2), some references in Table 3, and references concerning the geochemical properties of the six heavy metals discussed in this chapter. There appears to be no recent publication that updates the worldwide average for riverine particulate matter trace metal chemistry (Martin and Whitfield, 1981; Martin and Windom, 1991). This is supported by the fact that two recent references (Li, 2000; Chester, 2000) concerning marine chemistry still refer to this 1981 publication. As for references in Table 3, there is a very limited data available concerning the pathways of heavy-metal transport to lakes. Some of the important works have been considered and reviewed in this chapter. In addition, the analytical chemistry of the sedimentary materials has changed little over the past 30 years until the advent and use of inductively coupled plasma/mass spectrometry (ICP/MS) in the late 1990s. Extensive works concerning the geochemical properties of heavy metals have been published during the past 40 years and to the author's knowledge these have survived the test of time.

9.03.1.2 Sources of Metals

There are a variety of natural and anthropogenic sources of these heavy metals (Pb, Zn, Cd, Cr, Cu, Ni) in the environment.

9.03.1.2.1 Natural

The principal natural source of heavy metals in the environment is from crustal material that is either weathered on (dissolved) and eroded from (particulate) the Earth's surface or injected into the Earth's atmosphere by volcanic activity. These two sources account for 80% of all the natural sources; forest fires and biogenic sources, account for 10% each (Nriagu, 1990b). Particles released by erosion appear in the atmosphere as windblown dust. In addition, some particles are released by vegetation. The natural emissions of the six heavy metals are 12,000 (Pb); 45,000 (Zn); 1,400 (Cd); 43,000 (Cr); 28,000 (Cu); and 29,000 (Ni) metric tons per year, respectively (Nriagu 1990b). Thus, we can conclude that an abundant quantity of metals are emitted into the atmosphere from natural sources. The quantity of anthropogenic emissions of these metals is given in the next section.

9.03.1.2.2 Anthropogenic

There are a multitude of anthropogenic emissions in the environment. The major source of these metals is from mining and smelting. Mining releases metals to the fluvial environment as tailings and to the atmosphere as metal-enriched dust whereas smelting releases metals to the atmosphere as a result of high-temperature refining processes. In the lead industry, Pb–Cu–Zn–Cd are released in substantial quantities; during Cu and Ni smelting, Co–Zn–Pb–Mn as well as Cu–Ni are released; and in the Zn industry, sizeable releases of Zn–Cd–Cu–Pb occur (Adriano, 1986). Table 1 shows that the world metal production during the 1970s and the 1980s has remained relatively constant except for Cr production that substantially increased during the 1980s due to the technological advances and increased importance (Faust and Aly, 1981).

Table 1 Global primary production and emissions of six heavy metals during the 1970s and the 1980s.

Metal	*Metal production*		*Emissions to air*		*Emissions to soil*	*Emissions to water*
	1970s	*1980s*	*1970s*	*1980s*	*1980s*	*1980s*
Pb	3,400	3,100	449	332	796	138
Zn	5,500	5,200	314	132	1,372	226
Cd	17	15	7.3	7.6	22	9.4
Cr	6,000	11,250	24	30	896	142
Cu	6,000	7,700	56	35	954	112
Ni	630	760	47	56	325	113

Source: Nriagu (1980a), Pacyna (1986), and Nriagu and Pacyna (1988).
All values are thousand metric tons.

Much of the demand for Cr was due to steel and iron manufacturing and the use of Cr in pressure-treated lumber (Alloway, 1995). Table 1 also shows that anthropogenic emissions to the atmosphere, to which mining and smelting are major contributors, are in the interval of two times (Cu, Ni), five times (Zn, Cd), and 33 times (Pb) greater than the natural emissions of metals to the atmosphere. Anthropogenic atmospheric emissions decreased substantially from the 1970s to the 1980s for Pb, Zn, and Cu (Table 1). On the other hand, Cd and Cr have remained the same and Ni emissions have increased in the 1980s. In addition, anthropogenic emissions of Cr are only about one-half of those from the natural sources. The major contributor of Cr to natural atmospheric emissions is windblown dust (Nriagu and Pacyna, 1988).

Other important sources of metals to the atmosphere include fossil-fuel combustion (primarily coal), municipal waste incineration, cement production, and phosphate mining (Nriagu and Pacyna, 1988). Important sources of metals to the terrestrial and aquatic environment include discharge of sewage sludges, use of commercial fertilizers and pesticides, animal waste and wastewater discharge (Nriagu and Pacyna, 1988). Table 1 shows that metal emissions to soil are several times those to air, suggesting that land disposal of mining wastes, chemical wastes, combustion slags, municipal wastes, and sewage sludges are the major contributors of these emissions. Emissions to water are only about twice those relative to air (except for Pb and Cd) suggesting that direct chemical and wastewater releases to the aquatic environment are the only additional inputs besides the atmospheric emissions (Table 1).

Table 2 gives a comparison of the six heavy-metal contents of a variety of natural earth materials that annually impact atmospheric, terrestrial, and aquatic environments. The primary data of metals are also normalized with respect to titanium (Ti). Titanium is a very conservative element that is associated with crustal rock sources. Normalization with respect to Ti compensates for the relative percentage of various diluents (non-crustal rock sources) and allows one to see more clearly metal enrichment due to anthropogenic inputs. For instance, in Table 2, recent lacustrine sediment is clearly enriched in metal content relative to pre-Industrial lacustrine sediment.

It is obvious that there is a progressive enrichment in the metal content of the earth materials as one migrates from the Earth's upper crust to the soils to river mud to lacustrine sediments, and finally to the river particulate matter. This is especially true for Zn and Cd. If we consider the recent lacustrine sediments, then Pb, Zn, Cd, and Cu are all highly enriched compared to the upper crust and soils. Chromium and Ni, on the other hand, are not especially enriched when compared to the crust and soils (Table 2). The metal content of the river particulate matter is also highly enriched in relation to the crust and soils. It is obvious that anthropogenic activities have a pronounced effect on the particulate matter chemistry of lakes and rivers. It is also obvious that much of the enriched portion of the riverine particulates are deposited near river mouths and in the coastal zone (continental shelf) as the Ti-normalized metals for estuarine sediments and hemipelagic mud are less enriched than riverine particulates but still enriched relative to the crust and soils. Table 2 also shows the effect of diagenetic remobilization and reprecipitation of ferromanganese oxides in surficial pelagic clays as both Cu and Ni (major accessory elements in ferromanganese nodules) are significantly enriched in these marine deposits relative to the precursor earth materials. Finally, Table 2 shows the effects of high-temperature combustion on the enrichment of metals in coals as they are concentrated in fly ash. This is especially true for Pb, Zn, Cr, Cu, and Ni.

9.03.1.3 Source and Pathways

The two main pathways for heavy metals to become incorporated into air–soil–sediment–water are transport by air (atmospheric) and

Table 2 Average concentration of six heavy metals in natural earth materials.

Material	*Pb* (ppm)	*Zn* (ppm)	*Cd* (ppm)	*Cr* (ppm)	*Cu* (ppm)	*Ni* (ppm)	*Ti* (wt.%)	*References*
Upper crust	17(52)	67(203)	0.1(0.30)	69(209)	39(118)	55(167)	0.33	Li (2000)
Average soils	26(68)	74(195)	0.1(0.26)	61(160)	23(60)	27(71)	0.38	Li (2000)
River mud	23(42)	78(142)	0.6(2.0)	85(155)	32(58)	32(58)	0.55	Govindaraju (1989)
Pre-industrial, baseline lacustrine sediment	22(69)	97(303)	0.3(0.55)	48(150)	34(106)	40(125)	0.32	Shafer and Armstrong (1991); Forstner (1981); Heit *et al.* (1984); Mudroch *et al.* (1988); Eisenreich (1980); Kemp *et al.* (1976, 1978); Wren *et al.* (1983); Wahlen and Thompson (1980)
Recent lacustrine sediment	102(316)	207(640)	2.2(6.8)	63(195)	60(186)	39(121)	0.32	Above references plus: Dominik *et al.* (1984); Rowell (1996); Mecray *et al.* (2001)
River particulate matter	68(120)	250(446)	1.2(2.1)	100(178)	100(178)	90(161)	0.56	Martin and Windom (1991); Martin and Whitfield (1981)
Estuarine sediment	54(108)	136(272)	1.2(2.4)	94(188)	52(104)	35(70)	0.50	Alexander *et al.* (1993); Coakley and Poulton (1993); Anikiyev *et al.* (1994); Hanson (1997)
Hemipelagic mud	23(49)	111(236)	0.2(0.44)	79(168)	43(91)	44(94)	0.47	Li (2000); Chester (2000)
Pelagic clay	80(174)	170(370)	0.4(0.9)	90(196)	250(543)	230(500)	0.46	Li (2000)
Coal	15(24)	53(84)	0.4(0.6)	27(43)	16(25)	17(27)	0.63	Tillman (1994); Adriano (1986)
Fly ash	43(70)	149(245)	0.5(0.8)	115(189)	56(92)	84(137)	0.61	Hower *et al.* (1999); Adriano (1986)

Values in parentheses are Ti-normalized.

water (fluvial). In the previous section it was shown that heavy-metal emissions to air and water (Table 1) are a significant percentage of the amounts of metals that are extracted from the Earth's crust by mining. Ores are refined by smelting thus releasing large amounts of metal waste to the environment (primary source). Relatively pure metals are incorporated into a multitude of technological products which, when discarded, produce a secondary, but important, source of metals to the environment. Metals are also incorporated naturally and technologically into foodstuffs which, when consumed and discarded by man, result in an important metal source to the aquatic environment (sewage wastewater), soils, and sediments (sewage sludge).

We can see from Table 3 that except for Pb in the terrestrial environment and Cd in the marine environment, metal transport to the lakes and to the oceans via water (fluvial) is many times greater (2–10) than that by air (atmospheric). This undoubtedly reflects the prevalence of wastewater discharges from sewage–municipal–industrial inputs that are so common in our industrialized society. The prevalence of Pb atmospheric emissions is probably due to the burning of leaded gasoline which was phased out in North America and Western Europe by the early 1990s but is still occurring in the Third World countries. Natural atmospheric emissions of Cd (volcanoes) are most likely the cause of substantial atmospheric Cd fluxes to the marine environment (Nriagu, 1990b).

9.03.2 OCCURRENCE, SPECIATION, AND PHASE ASSOCIATIONS

9.03.2.1 Geochemical Properties and Major Solute Species

9.03.2.1.1 Lead

Lead (atomic no. 82) is a bluish-white metal of bright luster, is soft, very malleable, ductile, and a poor conductor of electricity. Because of these properties and its low melting point (327 °C), and resistance to corrosion, Pb has been used in the manufacture of metal products for thousands of years. In fact, the ancient world technology for smelting Pb–Ag alloys from PbS ores was developed 5,000 years ago (Settle and Patterson, 1980). Lead has a density of $11.342\ \mathrm{g\ cm^{-3}}$, hence finds extensive use as a shield for radiation; its atomic weight is 207.2. Lead has two oxidation states, +2 and +4. The tetravalent state is a powerful oxidizing agent but is not common in the Earth's surficial environment; the divalent state, on the other hand, is the most stable oxidation level and most Pb^{2+} salts with

Table 3 Relative percentage of atmospheric (%A) and fluvial (%F) inputs of six heavy metals to lakes, a coastal zone, and the ocean.

Lake/Ocean	*Metal*												*References*
	Pb		*Zn*		*Cd*		*Cr*		*Cu*		*Ni*		
	%A	*%F*	*%A*	*%F*	*%A*	*%F*	*%A*	*%F*	*%A*	*%F*	*%A*	*%F*	
Lake IJsselmeer	NA	NA	7	93	2	98	0.1	99.9	6	94	1	99	Salomons (1983)
Southern Lake Michigan	47	53	22	78	NA	NA	NA	NA	13	87	NA	NA	Dolske and Sievering (1979)
Lake Michigan	60	40	35	65	10	90	41	59	15	85	NA	NA	Eisenreich (1980)
Lake Erie	40	60	12	88	NA	NA	NA	NA	9	91	NA	NA	Nriagu *et al.* (1979)
South Atlantic Bight	2	98	1	99	41	59	NA	NA	7	93	9	91	Chester (2000)
Ocean	15	85	5	95	17	83	3	97	2	98	5	95	Chester (2000)

NA = not available.

naturally-occurring common anions are only slightly soluble. It is composed of four stable isotopes ($^{208}Pb = 52\%$) and several radioisotopes whose longest half-life is 15 Myr (Reimann and de Caritat, 1998). Lead belongs to group IVa of the periodic table which classifies it as a heavy metal whose geochemical affinity is chalcophilic (associated with sulfur).

In a simple freshwater system, exposed to atmospheric CO_2 and containing 10^{-3} M Cl^-, 10^{-4} M SO_4^{-2}, and 10^{-6} M HPO_4^{-2}, it is predicted that Pb will be complexed by the carbonate species $Pb(CO_3)_2^{-2}$ in the pH range of 6–8 (Hem and Durum, 1973). The complex $PbSO_4^0$ is stable below pH 6 (or in low sulfate waters Pb^{2+}) and the complex $Pb(OH)_2$ is stable above pH 8 (Hem, 1976). In oxygenated stream and lake environments the concentration of dissolved Pb is less than 1 $\mu g\ L^{-1}$ over the pH range of 6–8 (Reimann and de Caritat, 1998) while its average concentration in world river water is 0.08 $\mu g\ L^{-1}$ (Gaillardet *et al.*, 2003). The dissolved Pb concentration in ocean water (0.002 $\mu g\ L^{-1}$) is an order of magnitude lower than that in river water (Chester, 2000).

Adsorption and aggregation-complexation with organic matter appear to be the most important processes that transform dissolved Pb to particulate forms in freshwater systems. Krauskopf (1956) originally suggested that the concentration of Pb, as well as certain other trace metals, could be controlled by adsorption onto the ferric and manganese oxyhydroxides–clay mineral–organic matter. The extent of Pb adsorption onto hydrous Fe and Mn oxides is influenced by the physical characteristics of the adsorbent (specific surface, crystallinity, etc.) and the composition of the aqueous phase (pH, Eh, complexation, competing cations). In a recent study of Fe and Pb speciation, reactivity, and cycling in a lacustrine environment, Taillefert *et al.* (2000) determined that Pb is entrained during the formation of Fe-exocellular polymeric substances (EPS) that aggregate in a water column near the chemocline. It is not yet clear whether the metal is complexed to the EPS or adsorbed directly to the Fe oxide. However, extraction data from lake sediments suggest that the $Pb–FeO_x$ phase is available to chemical attack (see below).

The average concentration of Pb in the lithosphere is about 14 $\mu g\ g^{-1}$ and the most abundant sources of the metal are the minerals galena (PbS), anglesite ($PbSO_4$), and cerussite ($PbCO_3$). The most important environmental sources for Pb are gasoline combustion (presently a minor source, but in the past 40 years a major contributor to Pb pollution), Cu–Zn–Pb smelting, battery factories, sewage sludge, coal combustion, and waste incineration.

9.03.2.1.2 Zinc

Zinc (atomic no. 30) is a bluish-white, relatively soft metal with a density of 7.133 g cm^{-3}. It has an atomic weight of 65.39, a melting point of 419.6 °C, and a boiling point of 907 °C. Zinc is divalent in all its compounds and is composed of five stable isotopes ($^{64}Zn = 49\%$) and a common radioisotope, ^{65}Zn, with a half-life of 245 days. It belongs to group IIb of the periodic table which classifies it as a heavy metal whose geochemical affinity is chalcophilic.

In freshwater, the uncomplexed Zn^{2+} ion dominates at an environmental pH below 8 whereas the uncharged $ZnCO_3^0$ ion is the main species at higher pH (Hem, 1972). Complexing of Zn with SO_4^{2-} becomes important at high sulfate concentrations or in acidic waters. Hydrolysis becomes significant at pH values greater than 7.5; hydroxy complexes of $ZnOH^-$ and $Zn(OH)_2^0$ do not exceed carbonate species at typical environmental concentrations of 15 $\mu g\ L^{-1}$ for world stream water (Reimann and de Caritat, 1998). More recent data of Gaillardet *et al.* (2003) places the concentration of dissolved Zn in average world river water at 0.60 $\mu g\ L^{-1}$. Significant complexing with organic ligands may occur in stream and lake waters with highly soluble organic carbon concentrations. The concentration of Zn in ocean water is 0.39 $\mu g\ L^{-1}$ (Chester, 2000), which is close to its value in world river water.

There are several factors that determine the relative abundance of dissolved and particulate Zn in natural aquatic systems. These include media pH, biogeochemical degradation processes that produce dominant complexing ligands, cation exchange and adsorption processes that control the chemical potential of solid substrates, and the presence of occluded oxyhydroxide compounds (Adriano, 1986). At pH values above 7, aqueous complexed Zn begins to partition to particulate Zn as a result of sorption onto iron oxyhydroxide. The clay mineral montmorillonite is particularly efficient in removing Zn from solution by adsorption (Krauskopf, 1956; Farrah and Pickering, 1977).

The average Zn content of the lithosphere is ~80 $\mu g\ g^{-1}$ and the most abundant sources of Zn are the ZnS minerals sphalerite and wurtzite and to a lesser extent smithsonite ($ZnCO_3$), willemite (Zn_2SiO_4), and zincite (ZnO) (Reimann and de Caritat, 1998). The smelting of nonferrous metals and the burning of fossil fuels and municipal wastes are the major Zn sources contributing to air pollution.

9.03.2.1.3 Cadmium

Cadmium has an atomic number of 48, an atomic weight of 112.40 consisting of eight stable isotopes ($^{112,114}Cd$ are most abundant), and a density of 8.65 g cm^{-3} (Nriagu, 1980a). In several aspects Cd is similar to Zn (it is a neighbor of Zn in the periodic table); in fact it is almost always associated with Zn in mineral deposits and other earth materials. Cadmium is a soft, silvery white, ductile metal with a faint bluish tinge. It has a melting point of 321 °C and a boiling point of 765 °C. It belongs to group IIb of elements in the periodic table and in aqueous solution has the stable 2+ oxidation state. Cadmium is a rare element (67th element in order of abundance) with a concentration of ~0.1 μg g^{-1} in the lithosphere and is strongly chalcophilic, like Zn.

In a natural, aerobic freshwater aquatic system with typical Cd–S–CO_2 concentrations (Hem, 1972), Cd^{2+} is the predominant species below pH 8, $CdCO_3^0$ is predominant from pH 8 to 10, and $Cd(OH)_2^0$ is dominant above pH 10. The solubility of Cd is minimum at pH 9.5 (Hem, 1972). The speciation of Cd is generally considered to be dominated by dissolved forms except in cases where the concentration of suspended particulate matter is high such as "muddy" rivers and reservoirs and near-bottom benthic boundary layers, and underlying bottom sediments in rivers and lakes (Li *et al.*, 1984). The distribution coefficient between the particulate and the dissolved Cd is remarkably consistent for a wide range of riverine and lacustrine situations (Lum, 1987). The sorption of Cd on particulate matter and bottom sediments is considered to be a major factor affecting its concentration in natural waters (Gardiner, 1974). Pickering (1980) has quantitatively evaluated the role clay minerals, humic substances, and hydrous metal oxides in Cd adsorption and concludes that some fraction of the particle-bound Cd is irreversibly held by the solid substrate. The concentration of dissolved Cd in average world river water is 0.08 μg L^{-1} (Gaillardet *et al.*, 2003). This concentration is identical to that of Cd in ocean water (0.079 μg L^{-1}; Chester, 2000).

9.03.2.1.4 Chromium

Chromium has an atomic number of 24, an atomic weight of 51.996 consisting of four stable isotopes ($^{52}Cr = 84\%$), and a density of 7.14 g cm^{-3} (Adriano, 1986). Crystalline Cr is steel-gray in color, lustrous, hard metal that has a melting point of 1,900 °C and a boiling point of 2,642 °C. It belongs to group VIb of the transition metals and in aqueous solution Cr exists primarily in the trivalent (+3) and hexavalent (+6) oxidation states. Chromium, as well as Zn, are the most abundant of the "heavy metals" with a concentration of about 69 μg g^{-1} in the lithosphere (Li, 2000).

In most natural waters at near neutral pH, Cr^{III} is the dominant form due to the very high redox potential for the couple Cr^{VI}/Cr^{III} (Rai *et al.*, 1989). Chromium(III) forms strong complexes with hydroxides. Rai *et al.* (1987) report that the dominant hydroxo species are $CrOH^{2+}$ at pH values 4–6, $Cr(OH)_3^0$ at pH values from 6 to 11.5, and $Cr(OH)_4^-$ at pH values above 11.5. The OH^- ligand was the only significant complexer of Cr^{III} in natural aqueous solutions that contain environmental concentrations of carbonate, sulfate, nitrate, and phosphate ions. The only oxidant in natural aquatic systems that has the potential to oxidize Cr^{III} to Cr^{VI} is manganese dioxide. This compound is common on Earth's surface and thus one can expect to find some Cr^{VI} ions in natural waters. The predominant Cr^{VI} species at environmental pH is CrO_4^{2-} (Hem, 1985). The principal Cr^{III} solid compound that is known to control the solubility of Cr^{III} in nature is $Cr(OH)_3^0$. However, Sass and Rai (1987) have shown that Cr/$Fe(OH)_3$ has an even lower solubility. This compound is a solid solution and thus its solubility is dependent on the mole fraction of Cr; the lower the mole fraction, the lower the solubility (Sass and Rai, 1987). Most Cr^{VI} solids are expected to be relatively soluble under environmental conditions. In the absence of solubility-controlling solids, Cr^{VI} aqueous concentrations under neutral pH conditions will primarily be controlled by adsorption/desorption reactions (Rai *et al.*, 1989). Under environmental conditions, iron oxides are the predominant adsorbents of chromate (Cr^{VI}) in acidic to neutral pH range and oxidizing environments. The Cr concentration in average world river water is 0.7 μg L^{-1} (Gaillardet *et al.*, 2003) and that in ocean water is 0.21 μg L^{-1} (Chester, 2000).

Chromium occurs in nature mainly in the mineral chromite; Cr also occurs in small quantities in many minerals in which it replaces Fe^{3+} and Al^{3+} (Faust and Aly, 1981). The metallurgy industry uses the highest quality chromite ore whilst the lower-grade ore is used for refractory bricks in melting furnaces. Major atmospheric emissions are from the chromium alloy and metal producing industries. Smaller emissions come from coal combustion and municipal incineration. In the aquatic environment, the major sources of Cr are electroplating and metal finishing industries. Hexavalent Cr^{VI} is a potent carcinogen and trivalent Cr^{III} is an essential trace element (Krishnamurthy and Wilkens, 1994).

9.03.2.1.5 Copper

Copper has an atomic number of 29, an atomic weight of 63.546, consists of two stable isotopes ($^{63}Cu = 69.2\%$; $^{65}Cu = 30.8\%$), and has a density of 8.94 g cm^{-3} (Webelements, 2002). Metallic Cu compounds (sulfides) are typically brassy yellow in color while the carbonates are a variety of green- and yellow-colored. The metal is somewhat malleable with a melting point of 1,356 °C and a boiling point of 2,868 °C. It belongs to group Ib of the transition metals and in aqueous solution Cu exists primarily in the divalent oxidation state although some univalent complexes and compounds of Cu do occur in nature (Leckie and Davis, 1979). Copper is a moderately abundant heavy metal with a concentration in the lithosphere of about 39 μg g^{-1} (Li, 2000).

Chemical models for the speciation of Cu in freshwater (Millero, 1975) predict that free $Cu^{2+}(aq)$ is less than 1% of the total dissolved Cu and that $Cu(CO_3)_2^{2-}$ and $CuCO_3^0$ are equally important for the average river water. Leckie and Davis (1979) showed that the $CuCO_3^0$ complex is the most important one near the neutral pH. At pH values above 8, the dihydroxo–Copper(II) complex predominates. The chemical form of Cu is critical to the behavior of the element in geochemical and biological processes (Leckie and Davis, 1979). Cupric Cu forms strong complexes with many organic compounds.

In the sedimentary cycle, Cu is associated with clay mineral fractions, especially those rich in coatings containing organic carbon and manganese oxides. In oxidizing environments ($Cu–H_2O–O_2–S–CO_2$ system), Cu is likely to be more soluble under acidic than under alkaline conditions (Garrels and Christ, 1965). The mineral malachite is favored at pH values above 7. Under reducing conditions, Cu solubility is greatly reduced and the predominant stable phase is cuprous sulfide (Cu_2S) (Leckie and Nelson, 1975). In natural aquatic systems, some of the Cu is dissolved in freshwater streams and lakes as carbonate and organic complexes; a larger fraction is associated with the solid phases. Much of the particulate Cu is fixed in the crystalline matrix of the particles (Gibbs, 1973). Some of the riverine reactive particulate Cu may be desorbed as the freshwater mixes with seawater. The biological cycle of Cu is superimposed on the geochemical cycle. Copper is an essential element for the growth of most of the aquatic organisms but is toxic at levels as low as 10 μg L^{-1} (Leckie and Davis, 1979). Copper has a greater affinity, than most of the other metals, for organic matter, organisms, and solid phases (Leckie and Davis, 1979) and the competition for Cu between the aqueous and the solid phases is very strong. Krauskopf (1956) noted that the concentration of copper in natural waters, 0.8–3.5 μg L^{-1} (Boyle, 1979), is far below the solubility of known solid phases. Davis *et al.* (1978) found that the adsorption behavior of Cu in natural systems is strongly dependent on the type and concentration of inorganic and organic ligands. Recent data of Gaillardet *et al.* (2003) places the concentration of dissolved Cu in average world river water at 1.5 μg L^{-1} and that in ocean water at 0.25 μg L^{-1} (Chester, 2000).

The most common Cu minerals, from which the element is refined into the metal, are Chalcocite (Cu_2S), Covellite (CuS), Chalcopyrite ($CuFeS_2$), Malachite and Azurite (carbonate compounds). It is not surprising that Cu is considered to have a chalcophillic geochemical affinity. In the past, the major source of Cu pollution was smelters that contributed vast quantities of Cu–S particulates to the atmosphere. Presently, the burning of fossil fuels and waste incineration are the major sources of Cu to the atmosphere and the application of sewage sludge, municipal composts, pig and poultry wastes are the primary sources of anthropogenic Cu contributed to the land surface (Alloway, 1995).

9.03.2.1.6 Nickel

Nickel has an atomic number of 28, an atomic weight of 58.71 consisting of five stable isotopes of which ^{58}Ni (67.9%) and ^{60}Ni (26.2%) are the most abundant, and a density of 8.9 g cm^{-3} (National Science Foundation, 1975). Nickel is a silvery white, malleable metal with a melting point of 1,455 °C and a boiling point of 2,732 °C. It has high ductility, good thermal conductivity, moderate strength and hardness, and can be fabricated easily by the procedures which are common to steel (Nriagu, 1980b). Nickel belongs to group VIIIa and is classified as a transition metal (the end of the first transition series) whose prevalent valence states are 0 and 2+. However, the majority of nickel compounds are of the Ni^{II} species.

Morel *et al.* (1973) showed that the free aquo species (Ni^{2+}) dominates at neutral pH (up to pH 9) in most aerobic natural waters; however, complexes of naturally occurring ligands are formed to a minor degree ($OH^- > SO_4^{2-} > Cl^- > NH_3$). Under anaerobic conditions that often occur in the bottom sediments of lakes and estuaries, sulfide controls the solubility of Ni. Under aerobic conditions, the solubility of Ni is controlled by either the co-precipitate $NiFe_2O_4$ (Hem, 1977) or $Ni(OH)_{2(s)}$ (Richter and Theis, 1980). The latter authors performed laboratory adsorption experiments for Ni in the presence of silica, goethite, and amorphous manganese oxide and found that manganese oxide removed 100% of the Ni over the pH range 3–10. The iron oxide began to adsorb Ni at pH 5.5, the oxide's zero point of charge. Hsu (1978) found that Ni was associated with both amorphous iron and manganese oxides that coated silica sand grains.

In 1977, Turekian noted that the calculated theoretical concentrations of Ni and other trace metals in seawater were in orders of magnitude higher than the measured values. Turekian (1977) hypothesized that the role of particulate matter was most important in sequestering reactive elements and transporting them from the continents to the ocean floor. For lakes, Allan (1975) demonstrated that atmospheric inputs were responsible for Ni concentrations in sediments from 65 lakes surrounding a nickel smelter. As Jenne (1968) and Turekian (1977) note, hydrous iron and manganese oxides have a large capacity for sorption or co-precipitation with trace metals such as Ni. These hydrous oxides exist as coatings on the particles, particularly clays, and can transport sequestered metals to great distances (Snodgrass, 1980). In the major rivers of the world, Ni transport is divided into the following phases (Snodgrass, 1980): 0.5% solution, 3.1% adsorbed, 47% as precipitated coating, 14.9% complexed by organic matter, and 34.4% crystalline material.

The concentration of Ni in the lithosphere is 55 $\mu g\ g^{-1}$ (Li, 2000) and the concentration of dissolved Ni in stream water is 2 $\mu g\ L^{-1}$ (Turekian, 1971). More recent data on the concentration of dissolved Ni in average world river water indicates the value to be 0.8 $\mu g\ L^{-1}$ (Gaillardet *et al.*, 2003) and the Ni concentration in ocean water to be 0.47 $\mu g\ L^{-1}$ (Chester, 2000). Natural emissions of Ni to the atmosphere are dominated by windblown dusts while anthropogenic sources that represent 65% of all emission sources are dominated by fossil-fuel combustion, waste incineration and nonferrous metal production (Nriagu, 1980b). Major uses of Ni include its metallurgical use as an alloy (stainless steel and corrosion-resistant alloys), plating and electroplating, as a major component of Ni–Cd batteries, and as a catalyst for hydrogenating vegetable oils (National Science Foundation, 1975).

9.03.2.2 Occurrence in Rocks, Soils, Sediments, Anthropogenic Materials

Table 4 presents the average concentration of six heavy metals (Pb, Zn, Cd, Cr, Cu, Ni) in a variety of earth materials, soils, sediments, and natural waters. For Pb it can be seen that the solid-phase concentration increases little along the transport gradient from the Earth's crust to world soils to lake sediments ($14 < 22 < 23\ \mu g\ g^{-1}$; Table 4). However, stream sediment and particularly riverine particulate matter is substantially enriched (50–68 $\mu g\ g^{-1}$) suggesting that anthropogenic inputs from the past use of leaded gasoline, the prevalent burning of fossil fuels and municipal waste, and land disposal of sewage sludge are mobilized from soils and become concentrated in transported particulate matter. The Pb content of soils in England and Wales (UK) is much higher (74 $\mu g\ g^{-1}$) than that (12 $\mu g\ g^{-1}$) found in remote soils of the USA (Alloway, 1995). This is due in part to the more densely populated regions of the UK that were sampled and the inclusion of metalliferous mining areas. Shallow marine sediments appear not to be

Table 4 Heavy metals in the Earth's crustal materials, soils, freshwater sediments, and marine sediments.

Material	*Pb*	*Zn*	*Cd*	*Cr*	*Cu*	*Ni*	*References*
Crust	14.8	65	0.10	126	25	56	Wedepohl (1995)
Granite	18, 17	40, 50	0.15, 0.13	20, 10	15, 20	8, 10	Adriano (1986); Drever (1988)
Basalt	8, 6	100, 105	0.2, 0.2	220, 170	90, 87	140, 130	"
Shale	23, 20	100, 95	1.4, 0.3	120, 90	50, 45	68, 68	"
Sandstone	10, 7	16, 16	<0.03,	35, 35	2, 2	2, 2	"
Limestone	9, 9	29, 20	0.05, 0.03	10, 11	4, 4	20, 20	"
Soils (general)	19	60	0.35	54	25	19	Adriano (1986)
Soils (World)	30	66	0.06	68	22	22	Kabata-Pendias (2000)
Soils, UK	74	97	0.8	41	23	25	Alloway (1995)
Soils, USA	12	57	0.27		30	24	"
Stream sediments	51 ± 28	132 ± 67	1.57 ± 1.27	67 ± 24	39 ± 13	44 ± 19	Various sources[a]
Lake sediment	22	97	0.6	48	34	40	Table 2
River particulates	68	250	1.2	100	100	90	"
Shallow marine sediment	23	111	0.2	79	43	44	Li (2000); Chester (2000)
Deep-sea clay	80	170	0.4	90	250	230	Li (2000)
Streams	1	30	0.01	1	7	2	Drever (1988)
Ocean	0.03	2	0.05	0.2	0.5	0.5	Drever (1988)

Units are $\mu g\ g^{-1}$ dry weight. Dissolved metal data for streams and ocean water are expressed in units $\mu g\ L^{-1}$.
[a] Various Sources: Dunnette (1992), Aston *et al.* (1974), Presley *et al.* (1980), Olade (1987), Mantei and Foster (1991), Zhang *et al.* (1994), Osintsev (1995), Chiffoleau *et al.* (1994), Borovec *et al.* (1993), Gocht *et al.* (2001).

enriched in Pb related to source materials (crustal rocks and world soils; Table 4) and deep-sea sediments appear to be the final repository of Pb that becomes concentrated in a variety of authigenic phases.

Zinc and Cd show a similar pattern with riverine particulate Zn (250 $\mu g\ g^{-1}$) greatly exceeding average Zn in terrestrial earth materials (68 ± 32 $\mu g\ g^{-1}$) and world soils (66 ± 17 $\mu g\ g^{-1}$), and particulate Cd (1.2 $\mu g\ g^{-1}$) greatly exceeding the terrestrial earth materials and soil concentrations (0.14 ± 0.08 and 0.23 ± 0.15 $\mu g\ g^{-1}$). As for Pb, UK soils are significantly greater in Zn and Cd concentrations relative to USA soils, a fact that reflects the urban and metalliferous character of the UK soils (Alloway, 1995). Stream–lake–shallow marine sediments are all more concentrated in Zn (113 ± 18 $\mu g\ g^{-1}$) than crustal rocks and soils (64 ± 3 $\mu g\ g^{-1}$). As in the case of Pb, deep-sea clays are the ultimate repository for Zn also.

Chromium has the highest concentration of all the six heavy metals in the Earth's crust (Table 4), mainly due to a very high concentration in basalt and shale. Average crustal rocks (72 ± 75 $\mu g\ g^{-1}$) are similar in Cr concentration to world soils (73 ± 19 $\mu g\ g^{-1}$) and the average Cr concentration in stream sediment–riverine particulates–lake sediment–shallow marine sediment (74 ± 22 $\mu g\ g^{-1}$). Only deep-sea clay is slightly enriched relative to all the other earth materials (Table 4). From these data it is apparent that natural Cr concentrations of various earth materials that constitute the weathering-transport continuum from continent to oceans have not been seriously altered by man's activities. As has been seen before, this is not the case for Pb, Zn, and Cd. These metals, along with Cu and Ni, are the backbone of the world's metallurgical industry and thus man's mining and smelting activities that have gone on for centuries have greatly altered the natural cycles.

The Cu concentration of crustal rocks (32 ± 34 $\mu g\ g^{-1}$) is approximately equivalent to that for average soils (25 ± 4 $\mu g\ g^{-1}$). However, as the earth material is weathered and transported to streams–lakes–shallow marine sediments there is a minimal enrichment in Cu concentration (39 ≈ 34 ≈ 43 $\mu g\ g^{-1}$) (Table 4). And, as for Pb–Zn–Cd, riverine particulate matter is greatly enriched (100 $\mu g\ g^{-1}$) relative to the other sedimentary materials. While the Pb–Zn–Cd concentrations of deep-sea clay are enriched 1.5 times that of the continental sedimentary materials, Cu is enriched approximately five times. The substantial enrichment of Cu in oceanic pelagic clay relative to terrestrial earth materials is due to the presence of ubiquitous quantities of ferromanganese oxides in surficial ocean sediments (Drever, 1988).

The Ni concentration of crustal rocks (58 ± 53 $\mu g\ g^{-1}$) is substantially greater than the average world soils (23 ± 3 $\mu g\ g^{-1}$), but essentially equal to continental sedimentary materials (49 ± 13 $\mu g\ g^{-1}$). Riverine particulate matter (90 $\mu g\ g^{-1}$) is nearly twice the Ni concentration of these continental sedimentary materials and deep-sea clay is nearly three times (230 $\mu g\ g^{-1}$) that concentration. As noted for Cu, the substantial Ni enrichment of deep-sea clays is due to the presence of ferromanganese micronodules in the oxidized surficial sediment column (Drever, 1988).

Table 5 gives the average concentration of six heavy metals in anthropogenic by-products; that is, materials refined from natural materials such as fly ash from coal and smelting of metal ores or by-products from man's use such as sewage sludge and animal waste. It is evident that smelting of the metal ores is a major contributor to the environmental pollution caused by atmospheric transport of heavy metals (Table 5). However, fly ash emissions from coal-fired power plants is probably a more important source of atmospheric heavy-metal pollution due to the fact that these power plants are the main sources of electricity for much of the world's population. In addition, sewage sludge is a major contributor of heavy-metal pollution in soils as land disposal of human waste becomes the only practical solution. It is not surprising that riverine particulates are so enriched in Pb, Zn, Cd, Cu, and Ni as soils polluted with atmospheric emissions from mining and smelting activities, and those altered by the addition of sewage sludge are swept into streams and rivers that eventually empty into the ocean.

9.03.2.3 Geochemical Phase Associations in Soils and Sediments

Not all metals are equally reactive, toxic, or available to biota. The free ion form of the metal is thought to be the most available and toxic (Luoma, 1983). With regards to reactivity, it is generally thought that different metal ions display differing affinities for surface binding sites across the substrates (Warren and Haack, 2001). The speciation or dissolved forms of a metal in solution is of primary importance in determining the partitioning of the metal between the solid and solution phases. Mineral surfaces, especially those of Fe oxyhydroxides, have been studied well by aquatic chemists. This is due to their ubiquitous and abundant nature and their proven geochemical affinity (Honeyman and Santschi, 1988). Metals can be incorporated into solid minerals by a number of processes; nonspecific and specific adsorption, co-precipitation, and precipitation of discrete oxides and hydroxides (Warren and

Table 5 Average concentration of six heavy metals in anthropogenic by-products.

By-Product	*Pb*	*Zn*	*Cd*	*Cr*	*Cu*	*Ni*	*References*
Coal	15	53	0.4	27	16	17	Tillman (1994), Adriano (1986)
Fly ash	43	144	0.5	115	56	84	Hower *et al.*(1999), Adriano (1986)
Soils down-wind of smelters	28, 2200	61, 3000	25, 91		184	306	Adriano (1986), Alloway (1995)
Fertilizers	235	288, 371	32, 35	151, 60	18, 84	36, 20	"
Sewage sludge	1049, 820	3025, 2490	72, 18	1221, –	1085, –	319, –	"
Animal waste	45, 11	93, 130	0.36, 0.55	16, 30	20, 31	29, 19	Adriano (1986), Kabata-Pendias and Pendias (2001)

Units are $\mu g\ g^{-1}$ dry weight.

Haack, 2001). Furthermore, Fe and Mn oxyhydroxides form surface coatings on other types of mineral surfaces such as clays, carbonates, and grains of feldspar and quartz. The three most common environmental solid substrates are Fe-oxides, Mn-oxides, and natural organic matter (NOM) (Warren and Haack, 2001).

Sediments are an important storage compartment for metals that are released to the water column in rivers, lakes, and oceans. Because of their ability to sequester metals, sediments can reflect water quality and record the effects of anthropogenic emissions (Forstner, 1990). Particles as substrates of pollutants originate from two sources; (a) particulate materials transported from the watershed that are mostly related to soils and (b) endogenic particulate materials formed within the water column. Since adsorption of metal pollutants onto air- and waterborne particles is the primary factor in determining the transport, deposition, reactivity, and potential toxicity of these metals, analytical techniques should be related to either the chemistry of the particle surface or to the metal species that is highly enriched on the particle surface (Forstner, 1990). In the absence of highly-sophisticated solid-state techniques, chemical methods have been devised to characterize the reactivity of metal-rich phases adsorbed to solid particle surfaces. Single leaching and combined sequential extraction schemes have been developed to estimate the relative phase associations of sedimentary metals in various aquatic environments (Pickering, 1981). The most widely applied extraction scheme was developed by Tessier *et al.* (1979) in which the extracted components were defined as exchangeable, carbonates, easily-reducible Mn oxides, moderately-reducible amorphous Fe oxides, sulfides and organic matter, and lithogenic material.

Partition studies on river sediments were first reported by Gibbs (1973) for suspended loads of the Amazon and Yukon rivers. Nickel was the main heavy metal bound to hydroxide coatings while a lithogenic crystalline phase concentrated the Cr and Cu. Salomons and Forstner (1980), in an extraction study of river sediments from different regions of the world, found that less polluted or unpolluted river systems exhibit an increase in the relative amount of the metals' lithogenic fraction and that the excess of metal contaminants released to the aquatic environment by man's activities exist in relatively unstable chemical associations such as exchangeable and reducible. With the exception of Cd and Mn, the amount of heavy metals in exchangeable positions is generally low (Salomons and Forstner, 1984). In addition to this, Zn is often concentrated in the easily reducible phase (amorphous Fe/Mn oxyhydroxides), and Fe–Pb–Cu–Cr are concentrated in the moderately reducible phase (crystalline Fe/Mn oxyhydroxides) (Salomons and Forstner, 1984). As can be seen later, for reservoir and lake sediments, Pb is almost completely extracted by the mildly acidic hydroxylamine hydrochloride but Zn is only partially extracted by this chemical that defines the easily-reducible phase.

In a series of landmark papers by Tessier and coworkers, the role of hydrous Fe/Mn oxides in controlling the heavy-metal concentrations in natural aquatic systems has been defined by careful field and laboratory studies by comparing with theory (Tessier *et al.*, 1985). They concluded that the adsorption of Cd, Cu, Ni, Pb, and Zn onto Fe-oxyhydroxides is an important mechanism in the lowering of heavy-metal concentrations in oxic pore waters of Canadian-Shield lakes. These heavy-metal concentrations were below the concentrations prescribed by equilibrium solubility models. In a more recent study, Tessier *et al.* (1989) concluded that Zn is sorbed onto Fe oxyhydroxides and that their field data fit reasonably well into a simple model of surface complexation. They also concluded that other substrates (Mn oxyhydroxides, organic matter, clays) can sorb Zn. Also, removal of Zn by phytoplankton has been shown to be an important mechanism for controlling the dissolved Zn concentrations in the eutrophic Lake Zurich (Sigg, 1987). Finally, Tessier *et al.* (1996) expanded their studies to include adsorbed organic matter. Their results strongly suggest that pH plays an important role in determining which types of particle surface binding sites predominate in the sorption of heavy metals in lakes. In circumneutral lakes metals are bound directly to hydroxyl groups of the Fe/Mn oxyhydroxides, and in acidic lakes metals are bound indirectly to these oxyhydroxides via adsorption of metals complexed by NOM.

Some words of caution should be included concerning these "solid speciation" sediment extraction techniques. Kersten and Forstner (1987) noted that "useful information on solid speciation influencing the mobility of contaminants in biogeochemically reactive sediments by the chemical leaching approach requires proper and careful handling of the anoxic sediment samples." Martin *et al.* (1987) showed that the specificity and reproducibility of the extraction method greatly depends on the chemical properties of the element and the chemical composition of the samples. They state that "these methods provide, at best, a gradient for the physicochemical association strength between trace elements and solid particles rather than their actual speciation." The problem of post-extraction readsorption of As, Cd, Ni, Pb, and Zn has been addressed by Belzile *et al.* (1989) who found that by using the "Tessier method" (Tessier *et al.*, 1979) on trace-element spiked natural sediments it is possible to recover the added trace elements within the limits of experimental error.

In a recent study of extraction of anthropogenic trace metals from sediments of US urban reservoirs, Conko and Callender (1999) showed that Pb had the highest anthropogenic content accounting for 80–90% of the total metal concentration. Three extractions were used: (i) easily-reducible 0.25 M Hydroxylamine HCL in 0.25 N HCl (Chao, 1984); (ii) weak-acid digest (Hornberger *et al.*, 1999); and (iii) Pb-isotope digest of 1 N HNO_3 + 1.75 N HCl (Graney *et al.*, 1995). Chao (1984) extraction, originally thought to extract only amorphous Mn oxyhydroxides, is now considered to be an acid-reducible extraction that solubilizes amorphous hydrous Fe and Mn oxides (Sutherland and Tack, 2000). Hornberger *et al.*'s (1999) weak-acid digest (0.6 N HCl) is thought to represent the bioavailable fraction of the metal (Hornberger *et al.*, 2000). Graney *et al.*'s (1995) HNO_3 + HCl acid digest is the most aggressive of the three extracts and has been shown to represent, using Pb isotopes, the anthropogenic fraction of Pb in lacustrine sediments. A plot of extractable Pb and total Pb for a 1997 sediment core from the suburban Lake Anne watershed in Reston, Virginia (Callender and Van Metre, 1997) is presented in Figure 1(a). It can be seen that between 85 and 95% of the total Pb is extracted by these chemicals. In general, the Chao extraction recovered 95% of the total Pb and since this technique is thought to specifically extract amorphous Fe and Mn oxyhydroxides, Conko and Callender (1999) postulated that most of the Pb is bound by these amorphous oxides. Figure 1(b) is a plot of various extractions and total Zn in the same Lake Anne core. Only 70% of the total Zn was extracted by any of the three techniques mentioned before; thus the remaining Zn must be associated with other sedimentary phases. Conko and Callender (1999) suggest Zn fixation by 2 : 1 clay minerals (i.e., montmorillonite), whereby sorbed Zn is fixed in the alumina octahedral layer, is an important phase (Pickering, 1981). An additional phase could be biotic structures that are postulated by Webb *et al.* (2000) as substrates where Zn occurs in intimate combination with Fe and P. While Lake Anne sediment is a typical siliclastic material, Lake Harding (located south of Atlanta, Georgia) sediment is reddish in color and consists of appreciably more iron and aluminum oxides. Much of the iron oxides are undoubtedly crystalline in character and may not be attacked by the mild extraction techniques listed above (especially the Chao extraction). Figure 2 shows a plot of the extractable and total Pb in a sediment core from Lake Harding. Contrary to the Lake Anne Pb data where 95% is extractable, only 75% of the total Pb in Lake Harding is extractable. The Chao easily-reducible

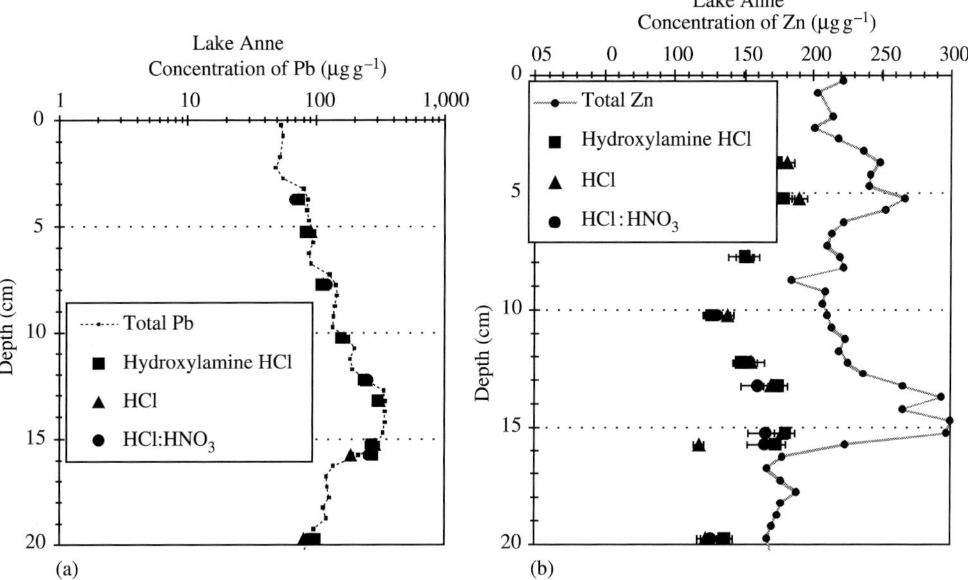

Figure 1 Temporal distribution of total and extractable lead (a) and zinc (b) in a sediment core from Lake Anne, Reston, Virginia, USA.

extraction yielded the lowest extraction efficiences. It is clear from these data that the type and nature of phase components that comprise natural aquatic sediments are most important for understanding the efficiency of any extraction scheme. Very little is known about the relationship between easily-extracted phases removed by sequential extraction (Tessier *et al.*, 1979) and those liberated by single leaches. Sutherland (2002) compared the two approaches using soil and road deposited sediment in Honolulu, Hawaii. The results indicated that the dilute HCl leach was slightly more aggressive than the sequential procedure but that there was no significant difference between the Pb and Zn concentrations liberated by the two approaches. Further, the data also indicated that a dilute HCl leach was a valuable, rapid, cost-effective analytical tool for contamination assessment. The Hawaii data also indicated that between 75% and 80% of the total Pb is very labile and anthropogenically enhanced (Sutherland, 2002; Sutherland and Tack, 2000). On the other hand, while labile Zn comprises 75% of the total, it is equally distributed between acid extractable and reducible (Sutherland *et al.*, 2000). The extractable Pb data agrees well with the lacustrine Pb data presented in Figure 1(a). The single HCl leach method on Hawaii sediments (Sutherland, 2002) extracts about 50% of the total Zn; a figure even lower than the 70% for lake sediments. Unfortunately, no information was available concerning the phase distribution of Zn in these sediments.

An important reason for testing "selective" leach procedures on sediments that are subjected

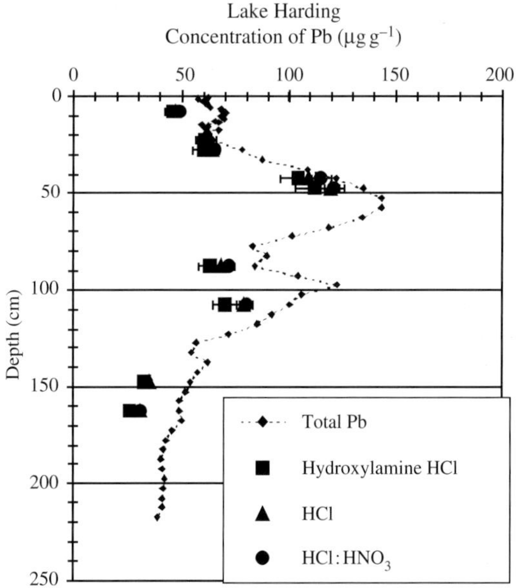

Figure 2 Temporal distribution of total and extractable lead in a sediment core from Lake Harding, Atlanta, Georgia, USA.

to anthropogenic influence is to determine whether such leaches can be used to measure the anthropogenic metal content of sedimentary materials. For Lake Anne sediment, Conko and Callender (1999) calculated the anthropogenic Pb and Zn content by subtracting the background metal concentrations from the total metal content. These were then compared to the "anthropogenic" leach concentrations. For both Pb and Zn, there was essentially no difference between the two procedures. These techniques were applied to several other lake sediments with similar successes suggesting that a mild acid leach might be used to estimate the labile, anthropogenic metal content of a variety of sedimentary materials.

Terrestrial materials (river sediments, lake sediments, and urban particulate matter) appear to have between 50% and 70% exchangeable Pb and Zn while marine sediments contain very little exchangeable metal but appreciably more reducible and much more residual Pb and Zn (Kersten and Forstner, 1995). This may not be too surprising as exchangeable metals are released once freshwater mixes with salt water and redistribution in the marine environment results in some precipitated phases (carbonates, Fe/Mn oxyhydroxides) and the relative increase in the lithogenic fraction. In future, the solid-phase identification techniques should be used to classify the sediments that are to be subjected to "selective" extraction techniques for the purpose of understanding the heavy metal phase associations.

9.03.3 ATMOSPHERIC EMISSIONS OF METALS AND GEOCHEMICAL CYCLES

Both natural and perturbed geochemical cycles include several subcycle elements, not the least of which is the emission of metals into the atmosphere. Atmospheric metals deposited on the land and ocean surface are a part of the runoff from land into the ocean and become incorporated in marine sediments. Thus, the two major pathways whereby heavy metals are injected into the natural geochemical cycles are atmospheric and fluvial. Considering the land surface, atmospheric emissions from stationary and mobile facilities and aqueous emissions from manufacturing and sewage disposal facilities are the primary sources of heavy metal contamination. As for the ocean, atmospheric deposition and continental runoff are the primary inputs. Duce *et al.* (1991) summarized the global inputs of metals to the ocean for the 1980s and these data are presented in Table 6. Riverine inputs are substantially greater than the atmospheric inputs, especially particulate riverine inputs that account for 95% of the total (Chester, 2000). For Pb and Zn, riverine inputs are 20 and 30 times greater than the corresponding atmospheric inputs. For Cd the factor is only 5, while for Cu and Ni the factors are 45 and 30, respectively. Global atmospheric inputs to land and ocean for the same time period are substantially greater (2–3 times, Table 6) than atmospheric deposition to the ocean. This is due to the presence of major pollution sources (mining, smelting, fossil-fuel combustion, waste incineration, manufacturing facilities) on the land masses. In fact, for Pb during the 1970s and 1980s, the use of leaded gasoline in vehicles resulted in the emission of four times the metal to the land surface compared to the ocean (Table 6).

Table 6 Global deposition of metals to the ocean for the 1980s.

Pb	*Zn*	*Cd*	*Cu*	*Ni*
Atmospheric				
90	137	3.1	34	25
Riverine				
1,602	3,906	15.3	1,510	1,411
World atmosphere				
342	177	8.9	63	86

Source: Duce *et al.* (1991).
All deposition values are thousand metric tons per year.

From the above data it is obvious that atmospheric emissions on land are a major source of heavy-metal contamination to our natural environment. In the following sections the focus will be on these emissions due to the fact that there are numerous data available to construct emission estimates (Nriagu and Pacyna, 1988) and that historical atmospheric emissions have been archived in continental ice accumulations (Greenland and Antarctica). The metal emission estimates of Nriagu and Pacyna (1988) are the most complete, and recent data are available for worldwide metal emissions.

9.03.3.1 Historical Heavy Metal Fluxes to the Atmosphere

Claire Patterson and his co-workers have pioneered the study of natural earth materials to uncover the "secrets of the ages". As early as 1963, Tatsumoto and Patterson (1963) related the high concentrations of Pb in surface seawater off Southern California to automotive aerosol fallout. In the United States it was found that Pb in gasoline was the largest single source of air pollution. Aerosols account for about one-third of the industrial Pb added to the oceans (Patterson *et al.*, 1976). Murozumi *et al.* (1969) provided a very convincing argument that airborne Pb particulates can be transported over vast distances in their classic study of the Greenland ice sheet. Their data indicated that before 1750 the concentration of Pb in the atmosphere began to increase above "natural" levels and that this was mainly due to the lead smelters, and that the sharp increase in the atmospheric Pb occurred around 1950 due to the burning of Pb alkyls in gasoline after 1940.

More recently, Claude Boutron and his coworkers in France have published high-quality data for Pb–Zn–Cd–Cu in Greenland snows (Boutron *et al.*, 1991; Hong *et al.*, 1994; Candelone *et al.*, 1995). Table 7 gives heavy metal deposition fluxes for the Summit Central Greenland Icesheet sampling locality. Lead increased dramatically between BP 7760 and AD 1773 (Industrial Revolution), and subsequently through 1850–1960 (Pb alkyl additives to automobile gasoline) (Nriagu, 1990a). Candelone *et al.* (1995) have successfully extended the uncontaminated metal record in ice from Central Greenland. Besides the above Pb record, for Zn, Cd, and Cu there is a clear increasing trend from 1773 to the 1970s (Table 7). However, between BP 7760 and AD 1773, there is essentially no change in metal flux. In fact, Zn decreased slightly; there is no change in Cd; and Cu increased slightly (Table 7). Over the past 200 years, Zn fluxes started to increase but it was not until 1900s that the increase became more rapid. On the other hand, for Cd and Cu, it was not until after the 1850s that their atmospheric concentrations and fluxes increased substantially (Candelone *et al.*, 1995). The maximum remote atmospheric concentrations of Zn occurred around 1960 while those for Cd and Cu occurred around 1970 (Candelone *et al.*, 1995). Finally, 1992 icesheet data indicate that the remote atmospheric Pb fluxes (Table 7) decreased by 6.5-fold in response to the banning of leaded gasoline throughout most of the world. Zinc and Cu decreased only 1.5 times while Cd decreased about 2.5 times (Table 7). These large increases in historical metal fluxes to the remote atmosphere are undoubtedly related to the major changes in the large scale anthropogenic emissions to the atmosphere in the northern hemisphere.

There is a wealth of data available on the world production of heavy metals during the past century or so (Nriagu, 1990b). Candelone *et al.* (1995) present historical Zn–Cd–Cu concentrations in snow/ice deposited at Summit, Central Greenland from 1773 to 1992. If one assumes that the 1773 concentrations are the result of natural atmospheric emissions, then the ratio of 1980s concentrations to 1773 concentrations are a measure of anthropogenic contamination of the remote atmosphere to that date. These ratios are 4, 7, and 3 for Zn, Cd, and Cu, respectively (figure 3 in Candelone *et al.*, 1995). Compare this to the 1983 total emissions divided by the natural emissions (Nriagu, 1990b). These values are 3.9, 6.4, and 2.3 for Zn, Cd, and Cu, respectively. It appears that historical changes in Zn–Cd–Cu deposition in the Greenland icesheet are consistent with the estimates of metal emissions to the global atmosphere (Candelone *et al.*, 1995). These emissions are primarily a result of smelting/refining, manufacturing processes, fossil-fuel combustion, and waste incineration (Nriagu, 1990b).

A similar analysis for Pb yields the following results: icesheet concentration ratio is about 15 and atmospheric emission ratio is about 20. While this is not too bad a comparison, it is not as good as for the other three metals (Zn–Cd–Cu). It is clear that most of the Pb increase in snow/ice samples from Greenland is due to the use of leaded gasoline after the 1950s (Murozumi *et al.*, 1969).

Going back to the Holocene era (BP 7760 years) where dated ice cores give metal concentrations that reflect a time when man's impact on the global environment was minimal, Candelone *et al.* (1995) measured Zn–Cd–Cu concentrations that were comparable to values for ice dated at AD 1773. Even by AD 1900 the concentrations of Zn and Cu were only 1.3 and 1.5 times those recorded for the AD 1773 date (Candelone *et al.*, 1995). Cadmium concentrations had increased more than four times during this period and Pb concentrations had increased nearly 10 times. In fact for Pb, the concentrations recorded in the icesheet have increased at least 30 times between BP 7760 and AD 1900. (Candelone *et al.*, 1995). It is obvious that much Pb was emitted to the atmosphere long before the Industrial Revolution and that some Cd was emitted during the early stages of the Industrial Revolution. It is possible that Cd was a by-product of the Pb mining and smelting during the Greco-Roman civilization (2500–BP 1700 years).

Table 7 Heavy metal deposition fluxes at Summit, Central Greenland.

Age	*Pb*	*Zn*	*Cd*	*Cu*
BP 7760	1.3	53	0.6	3.9
1773	18	37	0.6	5.0
1850	35	70	0.6	5.3
1960s–1970s	250	200	4.1	22
1992	39	120	1.8	17

Source: Candelone *et al.* (1995).
All values are in picograms per cm^2 per year.

9.03.3.2 Perturbed Heavy Metal Cycles

In this discussion of heavy metals, geochemical cycles are treated in a simple manner; emissions from land and oceans to the global atmosphere and subsequent deposition on the land and ocean surface, and runoff from the land to the ocean and eventual deposition in marine sediments. Only two components of this simple cycle will be discussed due to the availability of relatively accurate and complete data; deposition of metals from the atmosphere to the land and ocean surface, and continental runoff to the ocean.

Figure 3 presents these data for two simple scenarios: minimal human disturbances and maximum human disturbances. For the minimal human disturbances senario, it was assumed that deposition from the atmosphere was due to natural sources (Nriagu, 1990b) and that there was minimal anthropogenic impact on the Earth's surface. The continental runoff (riverine) data was taken from Bertine and Goldberg (1971) who calculated the amounts of metals entering into the world's oceans as a result of the weathering cycle. They accounted for both the dissolved and particulate phases by using the marine rates of sedimentation. It can be seen from Figure 3 that for Pb–Zn–Cr–Cu–Ni, continental runoff was 5–10 times greater than the natural atmospheric inputs. For Cd, the atmospheric fluxes are greater than the continental runoff suggesting that continental rocks are depleted in Cd or that there are poor quality Cd data for these two sources. The latter explanation seems to be the most likely.

For the maximum human disturbances scenario, riverine inputs were calculated with the data of Martin and Whitfield (1981). As can be seen from Table 2 in this chapter, Pb–Zn–Cd–Cu–Ni are strongly enriched (by man's activities) when compared to the average Earth's upper crust and soils and the Cr enrichment is found to be only somewhat enriched (Table 2). Atmospheric input data was computed as the average of global emissions data for the 1970s and 1980s (Garrels *et al.*, 1973; Lantzy and Mackenzie, 1979; Nriagu and Pacyna, 1988; and Duce *et al.*, 1991), and was assumed to be the time of maximum anthropogenic emissions to the atmosphere. For the maximum human disturbances scenario, riverine

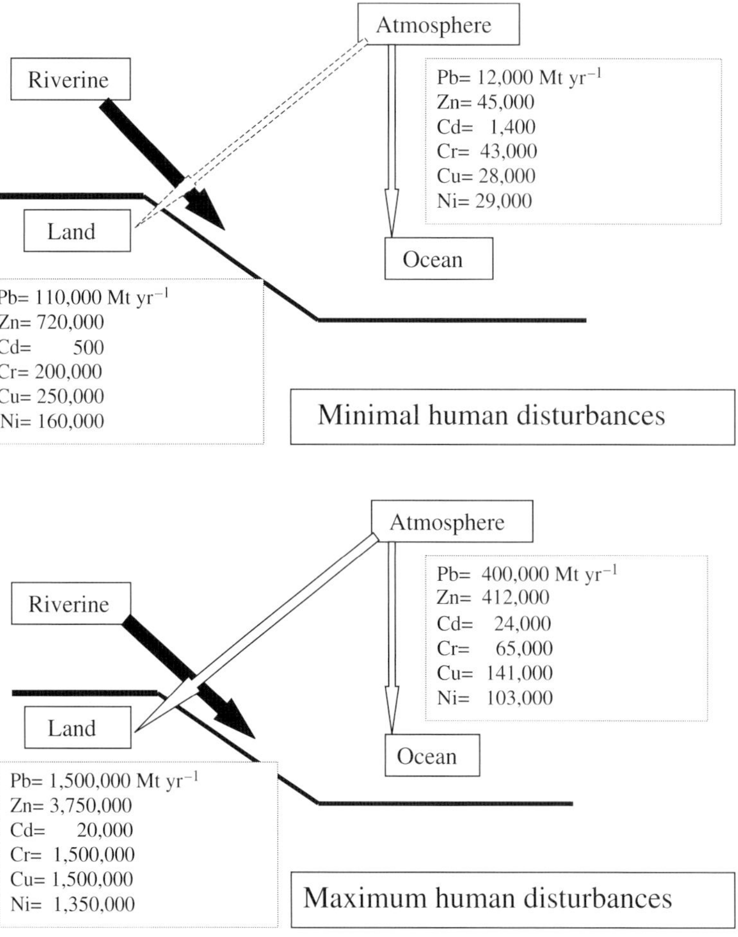

Figure 3 Schematic diagrams of perturbed heavy metal cycles representing prehistoric times of minimal human disturbances and modern times of maximum human disturbances. Data sources for minimum human disturbances: Nriagu (1990b), Bertine and Goldberg (1971). Data sources for maximum human disturbances: Martin and Whitfield (1981), Garrels *et al.* (1973), Lantzy and Mackenzie (1979), Nriagu and Pacyna (1988), Duce *et al.* (1991).

inputs are still larger than the atmospheric inputs except that Pb–Cr are only three times greater, Zn is six times greater, Cu is 10 times greater, Ni is 13 times greater, and Cd is about equal. These differences are undoubtedly due to the magnitude of different source functions.

A calculation of maximum/minimum ratio from the atmospheric input data in Figure 3 yields the following results: Pb = 33, Zn = 9, Cd = 17, Cr = 1.5, Cu = 5, Ni = 4. We know that the burning of leaded gasoline is responsible for the large increase of Pb. Enormous metal production of Zn and Cd ores as well as refuse incineration are responsible for the increases of these metals. In addition, marine aerosols are an important source of Cd (Li, 1981). Obviously, Cu–Ni production from ores increased during this period but not nearly as much as for Zn–Cd. Also, combustion of fossil fuels contributed somewhat to the increase of Cu and Ni. The main source of Cr is steel and iron manufacturing which appears to not be as important an impact on the atmospheric environment as sources for the other metals. The pollution sources of Cr are minimal as reflected in the balance between riverine input and marine sediment output (Li, 1981).

A similar calculation for the riverine inputs (Figure 3) yields the following results: Pb = 14, Zn = 5, Cd = 40, Cr = 7.5, Cu = 6, Ni = 8. With the exception of Pb and Cd, the increases for Zn–Cr–Cu–Ni are similar. Smelting wastes and coal fly ash releases are the common sources of these four metals. Gasoline residues are an obvious source of the Pb increases and urban refuse incineration is a major source of the Cd increase (Nriagu and Pacyna, 1988).

9.03.3.3 Global Emissions of Heavy Metals

Table 8 presents the data on the global emissions of heavy metals to the atmospheric and terrestrial environments for the 1970s and 1980s. The atmospheric and riverine input (weathering mobilization) data are the same as that used for the minimum and maximum human disturbances to the geochemical cycling of the heavy metals presented in the previous section. Total industrial discharges of heavy metals are the calculated discharges into soils and water minus the emissions to the atmosphere (Nriagu, 1990b). Only a fraction of the heavy-metal production from mines is released into the atmosphere in the same year (Nriagu, 1990b). For instance for Pb, in the year 1983, about 30% of the metal produced from mining is used for metal production, other sources, and is wasted as industrial discharges (Table 8): Zn 27%, Cd 190%, Cr 16%, Cu 14%, and Ni 57%. It is not surprising that the price of base metals fluctuate so widely in that there

Table 8 Global emissions of heavy metals to the atmosphere and terrestrial environment during 1970s and 1980s.

Element	*Atmospheric input*[a]	*Weathering mobilization*[b]	*Total industrial discharges*[c]	*Production from mines*[d]	*World Metal production* (Atmos.)[e]	*Other sources* (Atmos.)[f]	*Emissions H_2O, soil* (Atmos.)[g]	*Global natural emissions*[h]
Pb	400,000	295,000	565,000	3,077,000	83,800	292,000	875,000	12,000
Zn	412,000	1,390,000	1,427,000	6,040,000	125,800	67,000	2,083,000	45,000
Cd	24,000	15,000	24,000	19,000	8,500	3,500	43,000	1,400
Cr	65,000	1,180,000	1,010,000	6,800,000	28,500	25,000	1,397,000	43,000
Cu	141,000	635,000	1,048,000	8,114,000	35,400	15,500	1,428,000	28,000
Ni	103,000	540,000	356,000	778,000	15,900	71,000	614,000	29,000

Units are metric tons per year. Atmospheric Input (a) = World Metal Production (e) + Other Sources to the Atmosphere (f) + Natural Emissions to the Atmosphere (h). Pb: 400,000 = 388,000; Zn: 412,000? 238,000; Cd: 24,000 ≅ 13,400; Cr: 65,000 ≅ 96,500; Cu: 141,000 ≅ 79,000; Ni: 103,000 ≅ 116,000.
[a] Source: Lantzy and Mackenzie (1979), Garrels *et al.* (1973), Nriagu and Pacyna (1988), Duce *et al.* (1991). [b] Bertine and Goldberg (1971). [c,d,h] Nriagu (1990b). [e–g] Nriagu and Pacyna (1988).

appears to be a substantial excess of supply over demand (Table 8). This is not the case for Cd; the data presented in Table 8 suggest that there may be a deficit in the supply of Cd. It appears unlikely but it may be that there is a sufficient demand for Cd that can just about balance the mine production. Another explanation is that the estimate of Cd from industrial discharges might be in error. Other discrepancies in Cd estimates have also been noted and it is reasonable to think that since the concentrations of Cd are so low in natural earth materials, the analytical data may not be good.

In order to assess the internal consistency of the emissions, as shown in Table 8, a calculation was made whereby the mean atmospheric input was equated to the world metal production emitted to the atmosphere plus natural emissions and other sources to the atmosphere. With the exceptions of Cu and Zn, the quantities of emissions balance rather well. There is no obvious reason why Cu is out of balance by nearly a factor of 2 (atmospheric input $>$ sources). For Zn, with an imbalance of 1.7 for atmospheric input $>$ sources, there is an obvious problem with other sources in that the impact of rubber tire wear. This source term will be addressed in the next section. However, even with this term, the right side of the equation would increase to a maximum emissions figure of 300,000 t yr^{-1} (Table 8). It is possible that maximum Cu and Zn emissions to the atmosphere have been overestimated but there is no way to check this with the available data.

9.03.3.4 US Emissions of Heavy Metals

While there is a reasonable amount of data pertaining to global emissions of heavy metals during the last half of the twentieth century, there is a wealth of data available for emissions of heavy metals to the US atmosphere. Most of this has been calculated from USEPA and US Bureau of Mines materials production data combined with emission factors for a variety of source functions (Pacyna, 1986). In this section data plots will be presented to show the calculated emissions of several heavy metals to the US atmosphere over a decade of time. Some of the data, such as that for Pb, are from the published literature. On the other hand, much of the data for Zn has been calculated by the author and his colleagues and is presented for the first time.

9.03.3.4.1 Lead

With the scientific realization that Pb had contaminated the global atmosphere (Murozumi *et al.*, 1969), scientists set out to identify the major sources of this contamination. The late Claire Patterson, formerly of the California Institute of Technology, was the leader in this field. In an earlier paper concerning Pb contamination and its effect on human beings, Patterson (1965) wrote "the industrial use of lead is so massive today that the amount of lead mined and introduced into our relatively small urban environments each year is more than 100 times greater than the amount of natural lead leached each year from soils by streams and added to the oceans over the entire earth". This conclusion was reached by Chow and Patterson (1962) in their landmark study of Pb isotopes in pelagic sediments. This information, coupled with the well-known health impacts of Pb (USEPA, 2000a,b), arose the interest of toxicologists worldwide and prompted detailed studies of the cycling of this element in the environment. Ingested Pb (food, water, soil, and dust) damages organs, affects the brain and nerves, the heart and blood, and particularly affects young children and adults (USEPA, 2000a,b). With the use of leaded gasoline that began in the 1930s (Nriagu, 1990a), the public outcry about the outbreak of severe lead poisoning, and the drastic increase in the US in automobile miles traveled, the US Congress passed an amendment to the Clean Air Act (Callender and Van Metre, 1997) banning the use of leaded gasoline. The USEPA (2000a,b), in their most recent air pollutant emission trends report, showed that since 1973 the quantity of Pb emitted to the environment (Table 9) has decreased drastically from about 200,000 t to about 500 t in 1998. As a comparison, European Pb emissions for 1979/1980 were released at a rate of 80,800 t yr^{-1} (Pacyna and Lindgren, 1997) while those for the US were 66,600 tons per year (USEPA, 2000a,b).

Figure 4 presents the important EPA emissions data on a five-year time scale from 1970 to 1995. In the 1970s and 1980s, it is clear that leaded gasoline consumption was the overwhelming

Table 9 Total US emissions of lead (Pb) to the atmosphere.

Source category	*1970*	*1975*	*1980*	*1985*	*1990*	*1995*
Waste incineration	1,995	1,447	1,097	790	729	552
Fossil fuel combustion	9,269	9,385	3,899	469	454	446
Metals processing	21,971	9,000	2,745	1,902	1,968	1,864
Gasoline consumption	164,800	123,657	58,688	17,208	1,086	475

Source: USEPA (1998, 2000a,b).
Units are metric tons.

emitter of Pb to the environment with metal processing a far second. Presently, the total amount of Pb emitted to the environment is a paltry 2,500 t (USEPA, 2000a,b), with metal processing and waste disposal being the main emitters. While the US consumption of leaded gasoline has all but stopped, it is not the case for the rest of the world. As of 1993 when leaded gasoline consumption in North America (mostly Mexico) emitted 1,400 t of Pb to the atmosphere, the rest of the world emitted 69,000 tons of Pb to the atmosphere (Thomas, 1995).

9.03.3.4.2 Zinc

The US atmospheric emissions data for Zn are somewhat sparse. Nriagu (1979) published data on the worldwide anthropogenic emission of Zn to the atmosphere during 1975 (Table 10). The author has taken this report as a model for the type of Zn emissions that appear to be important and has added several categories such as cement and fertilizer production and automobile rubber tire wear.

The reason why the emission data for Zn are sparse is that until recently it was thought that Zn was not harmful to the environment and that health risks were minimal compared to other heavy metals. Zinc is an essential micronutrient and plays a role in DNA polymerization (Sunda, 1991) and nervous system functions (Yasui *et al.*, 1996). Zinc is generally less toxic than other heavy metals (Nriagu, 1980a); however, it is known to cause a variety of

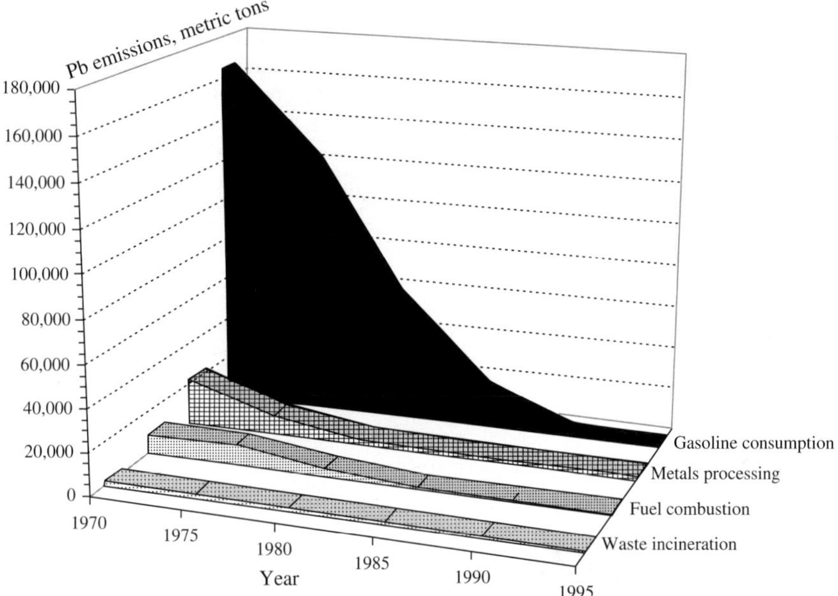

Figure 4 Three-dimensional plot of lead emissions to the US atmosphere for the period 1970–1995. Data from USEPA (2000a).

Table 10 Total US Emissions of zinc (Zn) to the atmosphere.

Source category	*1960*	*1965*	*1970*	*1975*	*1980*	*1985*	*1990*	*1995*
Cement production[a]	617	716	729	667	735	754	752	846
Fertilizer production[b]	1,000	1,000	1,054	1,329	1,632	1,525	1,390	1,365
Copper mining[c]	1,035	1,163	1,200	983	915	795	1,185	1,448
Iron and steel[d]	1,628	2,160	2,236	1,952	1,682	1,223	1,342	1,374
Fossil Fuel combustion[e]	1,532	1,719	2,141	1,878	1,916	1,984	1,658	1,298
Rubber tire wear[f]	3,747	4,901	5,503	5,044	5,983	7,258	8,329	8,847
Waste incineration[g]	6,367	7,280	5,920	5,006	3,232	5,659	7,298	7,941
Zinc mining[h]	101,500	126,280	111,440	55,580	47,600	36,540	36,820	32,480

Units are metric tons.
[a] Source: Nriagu and Pacyna (1988), http://minerals.usgs.gov/minerals/pubs/commodity/cement/stat/tbl1.txt. [b] Source: Nriagu and Pacyna (1988), http://minerals.usgs.gov/minerals/pubs/commodity/phosphate_rock/stat/tbl1.txt. [c] Source: Nriagu and Pacyna (1988), http://minerals.usgs.gov/minerals/pubs/commodity/of01-006/copper. [d] Source: Nriagu and Pacyna (1988), http://minerals.usgs.gov/minerals/pubs/commodity/of01-006/ironand steel [e] Source: Pacyna (1986), Statistical Abstracts of the United States (1998) (Coal and Oil production data). [f] Source: Councell *et al.* (2003). [g] Source: Pacyna (1986), USEPA (1998). [h] Source: Nriagu and Pacyna (1988)http://minerals.usgs.gov/minerals/pubs/commodity/of01-006/zinc.

acute and toxic effects in aquatic biota. Several studies have established links between human activities and environmental Zn enrichment (Pacyna, 1996).

Figure 5 is a plot of second tier Zn emissions to the atmosphere for the period 1960–1995 in five-year time intervals. The only important Zn emission category not included in Figure 5 is Zn mining. This is and has been the largest Zn emission category with 102,000 t in 1960 to 112,000 in 1970, declining to 48,000 in 1980, and stabilizing at about 32,000 t in the 1990s (Nriagu and Pacyna, 1988; www.minerals.usgs.gov). Obviously, emissions from Zn mining and smelting are the overwhelming sources. Total US Zn emissions for the 1980s amount to approximately 60,000 t yr^{-1} while European Zn emissions total 43,000 t yr^{-1} (Pacyna and Lindgren, 1997). Mining–smelting emissions overwhelm others that are important but it is difficult to plot these clearly on Figure 5 if Zn mining is also included.

Of the five Zn emission categories plotted in Figure 5, waste incineration and rubber tire wear are the most important. Note that in general these emissions have increased during the last 40 years such that the second-tier emissions total approximately one-half of the Zn mining–smelting emissions.

9.03.3.4.3 Cadmium

Cadmium has received a wide variety of uses in American industries with the largest being electroplating and battery manufacture. Its emission from natural sources (erosion and volcanic activity) are negligible. The dominant sources of Cd emissions to the atmosphere are primary metals smelting (Cu and Pb), secondary metals production, fossil-fuel combustion, waste incineration, iron and steel production, and rubber tire wear. Figure 6 is a plot of Cd emissions to the US atmosphere for five-year time periods from 1970 to 1990. Between 1970 and 1980, primary metals smelting was the primary source of Cd emissions to the atmosphere. Then fossil-fuel combustion became the primary emitter (60%) with Cu–Pb smelting accounting for much of the remainder (30%) (Wilber *et al.*, 1992). These emissions were concentrated in the central part of the US (Wilber *et al.*, 1992). By 1990, fossil fuel emissions decreased significantly, a fact that is probably related to the increased efficiency of stack emission controls; and secondary metal production became the major source of Cd to the US atmosphere (Figure 6). In the 1980s, Cd emissions to the US atmosphere amounted to 650 t yr^{-1} (Table 11). European emissions were nearly double this amount, i.e., 1,150 t yr^{-1} (Pacyna and Lindgren, 1997).

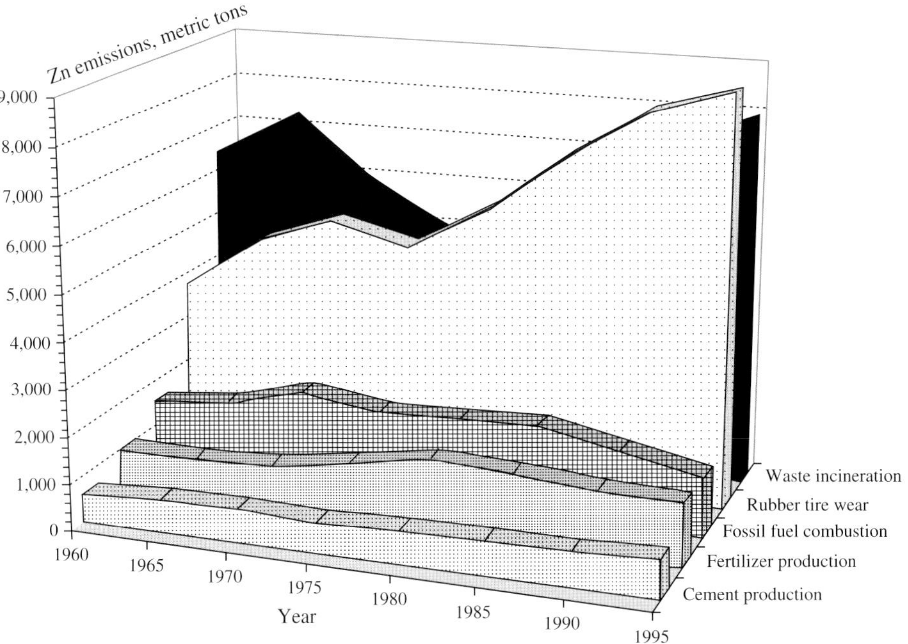

Figure 5 Three-dimensional plot of second tier zinc emissions to the US atmosphere for the period 1960–1995. Data from Councell *et al.* (2003), Nriagu and Pacyna (1988), minerals.usgs.gov/minerals/pubs/commodity/cement/stat/tbl1.txt, minerals.usgs.gov/minerals/pubs/commodity/phosphate_rock/stat/tbl1.txt, Pacyna (1986), USEPA (1998), Statistical Abstracts of the United States (1998), author's calculations.

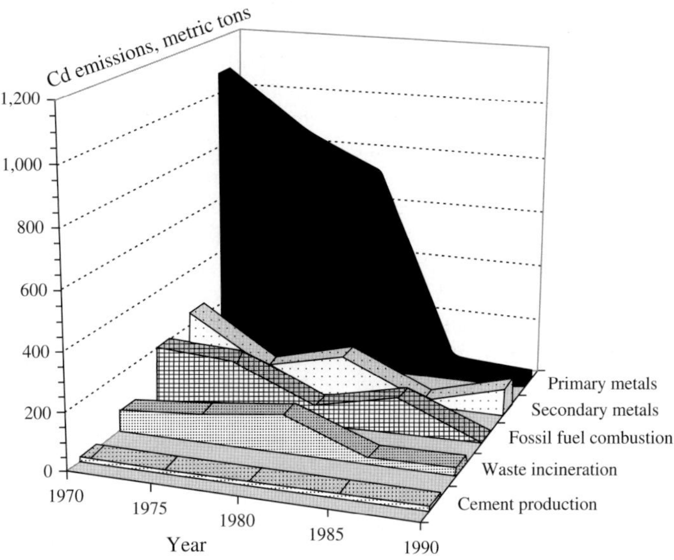

Figure 6 Three-dimensional plot of cadmium emissions to the US atmosphere for the period 1970–1990. (sources Davis and Associates, 1970; USEPA, 1975a,b, 1976, 1978; Wilber *et al.*, 1992).

Table 11 Total U.S. emissions of cadmium (Cd) to the atmosphere.

Source category	*1970*[a]	*1975*[b]	*1980*[c]	*1985*[d]	*1990*[e]
Rubber tire wear	6	6	5	5	5
Cement production	14	13	14	14	15
Waste incineration	72	97	131	21	28
Fossil fuel combustion	200	186	60	121	7
Secondary metals	245	75	143	15	96
Primary metals	1,075	860	728	54	32

Units are metric tons.
[a] Source: Davis and Associates (1970). [b] Source: USEPA (1975a,b, 1976). [c] Source: USEPA (1978). [d] Source: Wilber *et al.* (1992). [e] Source: USEPA (2000b).

9.03.4 HISTORICAL METAL TRENDS RECONSTRUCTED FROM SEDIMENT CORES

9.03.4.1 Paleolimnological Approach

Many governmental agencies collect the data routinely for the assessment of the quality of rivers, streams, lakes, and coastal oceans. Water-quality monitoring involves temporal sampling of water resources that are affected by natural random events, seasonal phenomena, and anthropogenic forces. Testing water quality, monitoring data at regular intervals has become a common feature and an important exercise for water managers who are interested in checking whether the investment of large sums of money has improved the water quality of various water resources during the past 30 years. In addition, trends available by such monitoring can provide a warning of degradation of water quality.

Historical water quality databases suffer from many limitations such as lack of sufficient data, changing sampling and analytical methods, changing detection limits, missing values and values below the analytical reporting level. With regard to trace metal data, the problem is even more difficult in that many metals occur at environmental concentrations so low (parts per billion or less) that current (up until the early 1990s) routine analytical methodology was unable to detect ambient concentrations with adequate sensitivity and precision. Even the best statistical techniques, when applied to questionable data, can produce misleading results. For example, Pb in the Trinity River, south of Dallas, TX, USA. Abundant dissolved Pb data for the period 1977–1992 (Van Metre and Callender, 1996) indicate that there were no trends in Pb; in fact, the concentrations were scattered from 0 to 5 ppb. On the other hand, Pb in sediment cores from Lake Livingston, downstream from the Trinity River sampling station, showed a decline in Pb concentration from 1970 to 1993 (Van Metre and Callender, 1996). Thus, from the core data, one can conclude that there is a declining trend in Pb in the Trinity River.

For these very reasons, the US Congress supported the US Geological Survey's National Water Quality Assessment Program with its goals to describe the status and trends in water quality of our Nation's surface and groundwater, and to provide an understanding of the natural and human factors that affect the observed conditions and trends. The US public is eager to know whether the water quality of US rivers and lakes has benefited from the expenditure of billions of dollars since the passage of the Clean Air and Clean Water Acts in the 1970s.

An alternate approach to statistical analyses of historical water-quality data is to use metal distributions in dated sediment cores to assess the past trends in anthropogenic hydrophobic constituents that impact watersheds (paleolimnological approach). It is well known that marine and lacustrine sediments often record natural and anthropogenic events that occur in drainage basins, local and regional air masses, or are forced upon the aquatic system (Valette-Silver, 1993). A good example of the former is the increase in erosional inputs to lakes in response to anthropogenic activities in the drainage basin (Brush, 1984). Atmospheric pollution resulting from the cultural and industrial activities (Chow *et al.*, 1973) is a compelling example of the latter. Thus, aquatic sediments are archives of natural and anthropogenic change. This is especially true for hydrophobic constituents such as heavy metals.

Because of their large adsorption capacity, fine-grained sediments are a major repository for the contaminants and a record of the temporal changes in contamination. Thus, sediments can be used for historical reconstruction. To guarantee a reliable age dating, and, therefore, to be useful in the historical reconstruction, the core sediment must be undisturbed, fine-grained, and collected in an area with a relatively fast sedimentation rate. These conditions are often found in lakes where studies in the 1970s by Kemp *et al.* (1974) and Forstner (1976) used lake sediments to understand the pollution history of several Laurentian Great Lakes and some European lakes. However, there is a serious limitation in using sediment cores from many natural lakes in that the sedimentation rate is generally too slow in providing the proper time resolution to discern modern pollution trends. Lacustrine sediments usually accumulate at rates less than 1 cm yr^{-1} (Krishnaswami and Lal, 1978) and often at rates less than 0.3 cm yr^{-1} (Johnson, 1984). Thus, there may be sufficient time for early diagenesis, such as microbiologically-mediated reactions, to occur. On the other hand, in lacustrine environments where sediments accumulate at rates exceeding 1 and may exceed 5–10 cm yr^{-1} (Ritchie *et al.*, 1973), such as in surface-water reservoirs, rapid sedimentation exerts a pronounced influence on sedimentary diagenesis.

A brief discussion of sedimentary diagenesis is warranted as post-depositional chemical, and physical stability is probably the most important factor in preserving heavy-metal signatures that may be recorded in aquatic sediments. For sediments to provide a historical record of pollution, the pollutant must have an affinity for the sedimentary particles. It is well known that most of the metals, and certainly the heavy metals discussed in this chapter, are hydrophobic in nature and allow partition to the solid phase. Once deposited in the sediment, the pollutants should not undergo chemical mobilization within the sediment column nor should the sediment column be disturbed by physical and biological processes.

Natural lacustrine and estuarine sediments whose accumulation rates are low, generally below 0.25 cm yr^{-1}, often do not satisfy the above requirements. The biophysical term bioturbation refers to surficial sediments mixed by the actions of deposit feeders, irrigation tube dwellers, and head-down feeders (Boudreau, 1999). In general, these bioturbation processes do not occur in reservoirs where sediment accumulation rates exceed 1 and often 5 cm yr^{-1} (Callender, 2000). At these rates, the sediment influx at the water–sediment interface is too great for benthic organisms to establish themselves.

On the other hand, geochemical mobility affects every sedimentary environment; varying in degree, from slowly accumulating natural lacustrine and estuarine sediments to rapidly accumulating reservoir sediments. The major authigenic solid substrates for adsorption and co-precipitation of heavy metals in aquatic sediments are the hydrous oxides of iron (Fe) and manganese (Mn) (Santschi *et al.*, 1990). These primary metal oxides sorb/co-precipitate Pb–Cr–Cu (Fe oxyhydroxides) and Zn–Cd–Pb (Mn oxyhydroxides) (Santschi *et al.*, 1990). Manganese oxides begin to dissolve in mildly oxidizing sediments while Fe oxides are reduced in anoxic sediments (Salomons and Forstner, 1984). In the mildly oxidizing zone, Mn^{2+} diffuses upward and precipitates as Mn oxide in the stronger oxidizing part of the sediment column. At greater sediment depths, Fe oxide reduction to Fe^{2+} begins and ferrous iron diffuses upward and precipitates as Fe oxide in the mildly oxidizing part of the sediment sequence (Salomons and Forstner, 1984).

An example from a slowly-accumulating (0.01–0.1 cm yr^{-1}) sediment profile in a freshwater lake in Scotland (Williams, 1992) should suffice to illustrate the formation of diagenetic metal profiles. Early diagenetic processes, such as those described before, have promoted extensive metal enrichment immediately beneath the water–sediment interface. The oxic conditions, near the water–sediment interface, that promote metal precipitation and enrichment (Mn, Fe, Pb, Zn, Cu, Ni) are entirely confined to strata of post-industrial age (Williams, 1992).

Callender (2000) extensively studied the geochemical effects of rapid sedimentation in aquatic systems and postulated that rapid sedimentation exerts a pronounced influence on early sedimentary diagenesis. The following are two case studies that illustrate this point. The Cheyenne River Embayment of Lake Oahe, one of the several impoundments on the upper Missouri River, accumulates sediment at an average rate of 9 cm yr^{-1} (Callender and Robbins, 1993). Three interstitial-water Fe profiles from the same site taken over a three-year period (August 1985, August 1986, June 1987), when superimposed on the same depth axis, show the effects of interannual variations in sediment inputs such that in 1986 a rapid input of oxidized material suppressed the dissolved Fe concentration to less than 0.1 mg L^{-1} to a depth of 8 cm. In 1985 when there was a drought and sediment inputs were reduced substantially, near-surface sediment became nearly anoxic and the interstitial Fe concentration rose to a very high 26 mg L^{-1} (Callender, 2000). In Pueblo Reservoir on the upper Arkansas River in central Colorado, cores of bottom sediments showed distinct reddish-brown layers that indicate rapid transport and sedimentation of Fe-rich colloids formed by the discharge of acid-mine waters from abandoned mines upstream (Callender, 2000). The amorphous sedimentary Fe profile from a sediment core near the river mouth shows two peaks at depths that correspond to the dates of heavy metal releases from the mines. Although the amorphous Fe oxyhydroxide concentrations are only 10% of the total Fe concentrations, they are adequate to adsorb Pb (Fergusson, 1990) and produce the anthropogenic Pb concentrations found in the core (Callender, 2000). Copper and Zn show similar distributions in this core whose sedimentation rate is 5 cm yr^{-1}.

In these examples as well as for most aquatic sediments, the principal diagenetic reactions that occur in these sediments are aerobic respiration and the reduction of Mn and Fe oxides. Under the slower sedimentation conditions in natural lakes and estuaries, there is sufficient time (years) for particulate organic matter to decompose and create a diagenetic environment where metal oxides may not be stable. When faster sedimentation prevails, such as in reservoirs, there is less time (months) for bacteria to perform their metabolic functions due to the fact that the organisms do not occupy a sediment layer for any length of time before a new sediment is added (Callender, 2000). Also, sedimentary organic matter in reservoir sediments is considerably more recalcitrant than that in natural lacustrine and estuarine sediments as reservoirs receive more terrestrial organic matter (Callender, 2000).

The author hopes that this discussion of sedimentary diagenesis, as it applies to heavy-metal signatures in natural lacustrine and reservoir sediments, will help the reader interpret the results presented in the following sections on reconstructed metal trends from age-dated reservoir sediment cores.

The approach that Callender and Van Metre (1997) have taken is to select primarily reservoir lakes that integrate a generally sizeable drainage basin that is impacted by a unique landuse such as agriculture, mining, stack emissions, suburban "sprawl", or urban development with some commercial and light industrial activity. Sediment cores are taken to sample the post-impoundment section as much as possible and to penetrate the pre-impoundment material. Core sampling is accomplished with a variety of coring tools (box cores, push cores, piston cores) in order to recover a relatively undisturbed sediment section. The recovered sediment is sampled on approximately an annual sediment thickness and samples are preserved (chilled, then frozen) for future analytical determinations. In the laboratory, sediment samples are weighed, frozen, freeze-dried, weighed again, and ground to a fine powder. Elemental concentrations are determined on concentrated acid digests (nitric and hydrofluoric in microwave pressure vessels) by inductively coupled plasma-atomic emission spectrometry (ICP/AES) or by graphite furnace atomic adsorption spectrometry (GF/AAS).

For reservoirs to be a good medium for detecting the trends in heavy metals, several conditions need to be satisfied. First, the site sampled should be continuously depositional over the life of the reservoir. This condition is most easily satisfied by sampling in the deeper, lacustrine region of the reservoir where sedimentation is slower but more uniform and the sediments predominantly consist of silty clay material. The second condition is that the sediments sampled should not be subject to significant physical and chemical diagenesis; that is, mobilization of chemical constituents after deposition. Callender (2000) has written an extensive paper indicating that rapid sedimentation promotes minimal diagenesis and preserves historical metal signatures. The third condition is that the chemical quality of reservoir bottom sediments should be related to the water quality of the influent river and that the influent water quality be representative of the drainage basin.

9.03.4.2 Age Dating

In general, reservoir sediments can be dated by several techniques. In one technique, the sediment surface is dated by the time of coring while in the other the date is derived from a visual inspection of the cored sediment column which often penetrates the pre-impoundment surface. The primary age dating tool for reservoir sediments is by counting the radioactive isotope ^{137}Cs which has a half-life period of 30 years (Robbins and Edgington, 1975; McCall *et al.*, 1984). The ^{137}Cs activity of freeze-dried sediment samples is measured by counting the gamma activity in fixed geometry with a high-resolution, intrinsic germanium detector gamma-spectrometer (Callender and Robbins, 1993). Depending on the penetration depth of the core and the age of the reservoir, ^{137}Cs can provide one or two date markers and can be used to evaluate the relative amount of postdepositional mixing or sediment disturbance (Van Metre *et al.*, 1997). The peak ^{137}Cs activity in the sediment core is assigned a date of 1964, consistent with the peak in atmospheric fallout levels of ^{137}Cs for 1963–1964. In reservoirs constructed prior to or around 1950, the first occurrence of ^{137}Cs, if it did not appear to have been effected by postdepositional sediment mixing, was assigned a date of 1953 which is consistent with the generally accepted date of 1952 for the first large-scale atmospheric testing of nuclear weapons by the US in Nevada (Beck *et al.*, 1990). This is also the date of the first globally-detectable levels of ^{137}Cs in the atmosphere. In some cases dates for samples between the known date-depth markers were assigned using constant mass accumulation rates (MARs), and in other cases the MARs were varied.

In natural lacustrine and slowly-accumulating reservoir sediments, core dating with the isotope ^{210}Pb has been used extensively (Schell and Barner, 1986). Appleby and Oldfield (1983) found that the constant rate of ^{210}Pb supply model (CRS) provides a reasonably accurate sedimentation chronology. The basic assumption of the CRS model is that the rate of supply of excess ^{210}Pb to the lake is constant. This model, thus, assumes that the erosive processes in the catchment are steady and give rise to a constant rate of sediment accumulation (MAR) (Appleby and Oldfield, 1983). In practice, for reservoirs, this assumption is rarely met because, for example, an increase in the MAR caused by land disturbances, such as those associated with the urban development, transports additional surficial soils and sediments to the lake. This additional erosion increases the MAR and also increases the rate of supply of ^{210}Pb to the lake. In general, because excess ^{210}Pb is an atmospheric fallout radionuclide, the model works better in low sedimentation rate, atmospherically dominated lakes with undisturbed watersheds, than in high sedimentation rate, fluvially dominated urban lakes and reservoirs.

Another problem with age dating of reservoir sediment is the concept of sediment focusing. This concept was developed to correct for postdepositional resuspension and redistribution of sediment in parts of the lake (Hermanson, 1991). A common focus correction factor is derived from the inventory of ^{137}Cs in the sediment column compared to the estimated total ^{137}Cs fallout at the sampling site (Hermanson, 1991). The same concept was found to not work well for lakes and reservoirs where the catchment area far exceeds the lake area. Such is the case for most reservoirs (Van Metre *et al.*, 2000). In these cases where the catchment area is 10–100 times the lake area, sediment focusing in the lake basin is overwhelmed by the concentration effect of atmospheric fallout over the catchment area being funneled into the lake or reservoir. The catchment area focus corrections are calculated the same way as lake basin focus corrections except that there may be some variation in the ^{137}Cs flux to large catchment areas and that there will almost always be a correction factor greater than 1. These focus corrections must be calculated in order to compare the contaminant fluxes between the sites within a lake basin and between lake basins.

9.03.4.3 Selected Reconstructed Metal Trends

9.03.4.3.1 Lead and leaded gasoline: consequence of the clean air act

Of the six heavy metals discussed in this chapter, Pb has been studied extensively with respect to the environmental effects. Clair Patterson, the father of environmental Pb studies, in one of his many major publications concerning the global Pb cycle (Patterson and Settle, 1987), noted that during pre-industrial times Pb in the troposphere originated from soil dusts and volcanic gases. In modern times (1950–1980) the proportion of natural Pb in the atmosphere is overwhelmed by the industrial sources of smelter emissions and automobile exhausts. Lead air pollution levels measured near our Nation's roadways decreased 97% between 1976 and 1995 due to the consequence of the Clean Air Act that eliminated leaded gasoline which interfered with the performance of catalytic converters.

For remote locations on a more global scale, Boyle *et al.* (1994) showed that the stable Pb concentration in North Atlantic waters decreased at least three-fold from 1979 to 1988. Wu and Boyle (1997) confirmed and extended this time series to 1996 whereby the concentration of stable Pb apparently stabilized at 50 pmol kg^{-1} in surface waters near Bermuda. Shen and Boyle (1987) presented a 100-year record of Pb concentration in corals from Bermuda and the Florida Straits showing that Pb peaked in the 1970s and declined thereafter. Veron *et al.* (1987) found high Pb concentrations in northeast Atlantic surficial sediments and noted that the quantity of Pb stored in these sediments is of the same order of magnitude as the amount of pollutant Pb present in the water column.

On a more local level, man's activities in the urban/suburban environment have produced a strong imprint of Pb on the land surface. In the US, automobile and truck travel are the primary means of moving people and goods around the continent. With the introduction of leaded gasoline in the 1950s, the mean annual atmospheric concentration of Pb nearly tripled in value, especially near population centers (Eisenreich *et al.*, 1986). A substantial proportion of these atmospheric emissions of Pb have been deposited relatively close to the source. Figures 7(a)–(c) presents the age-dated sedimentary Pb profiles for reservoirs and lakes from urban–suburban–rural localities. One can see that the peak concentrations decrease from 700 to 300 to 100 μg g^{-1} as the distance from urban centers increase. All but two of the Pb peak concentrations date between 1970 and 1980, and

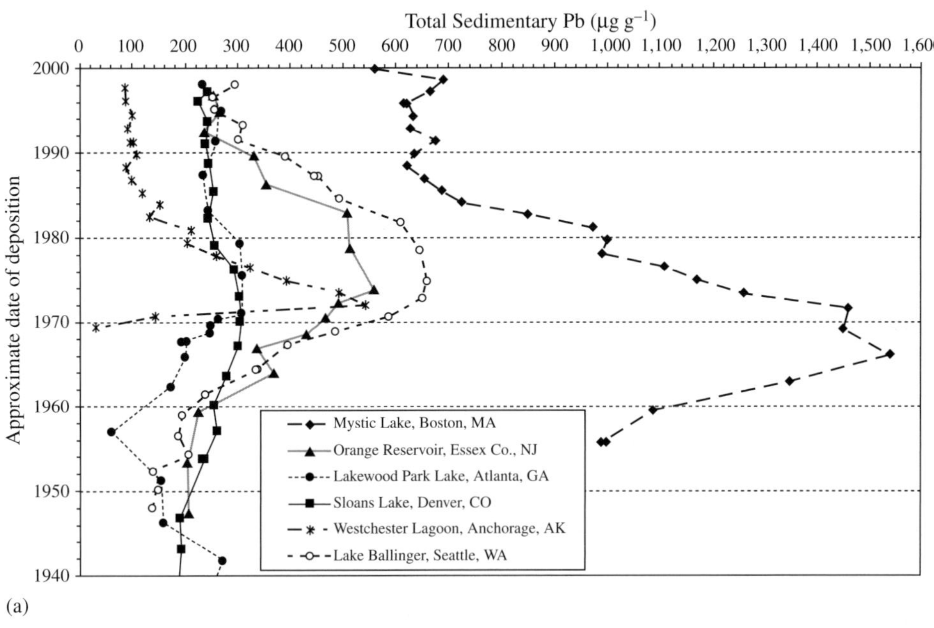

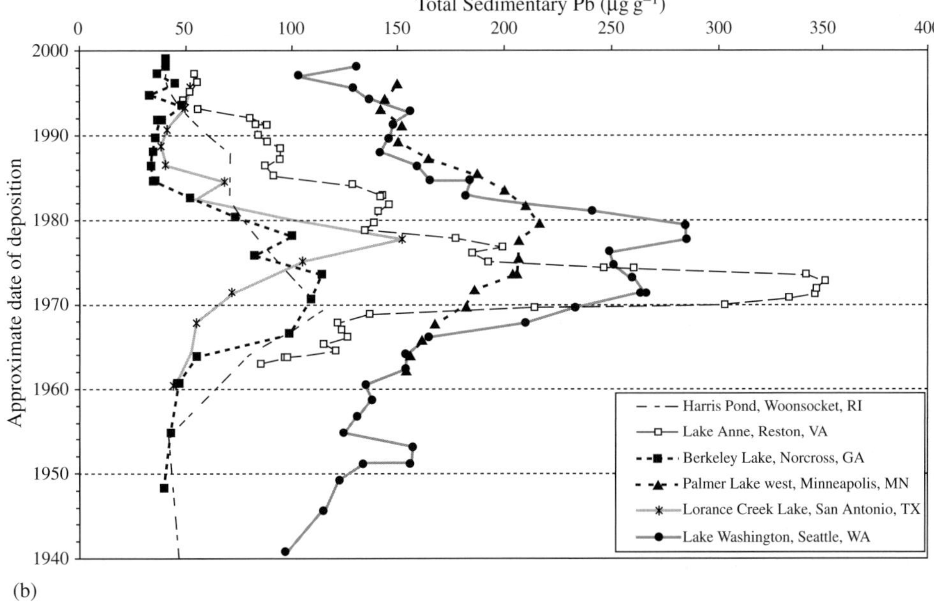

Figure 7 Temporal distribution of total sedimentary lead in sediment cores from (a) urban reservoirs, (b) suburban reservoirs and lakes, and (c) atmospheric reference site reservoirs.

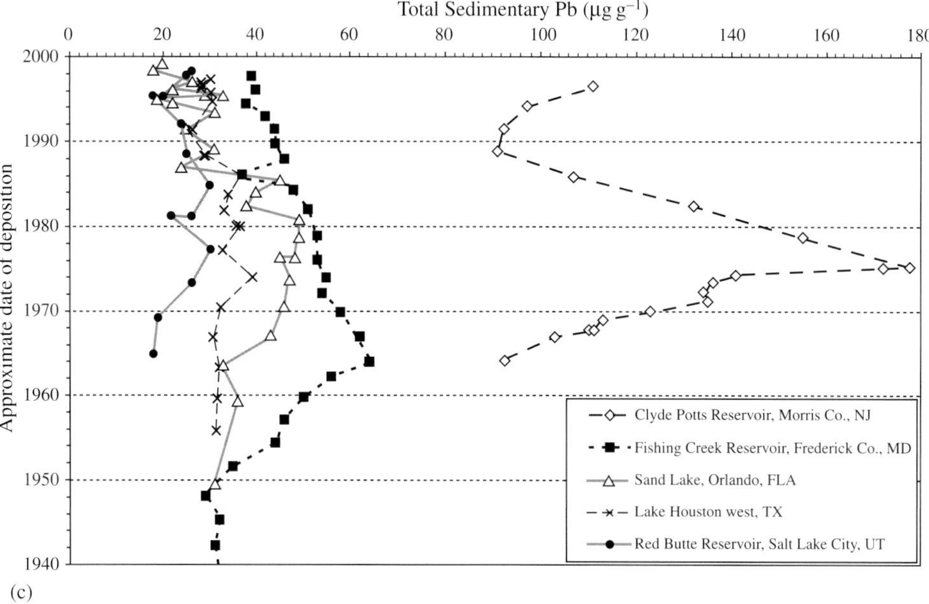

Figure 7 (continued).

are consistent with the decline in US atmospheric Pb concentrations following the ban of unleaded gasoline in 1972 (Callender and Van Metre, 1997). Sedimentary Pb data from many urban centers around the US (Boston, New York, New Jersey, Atlanta, Orlando, Minneapolis, Dallas, Austin, San Antonio, Denver, Salt Lake City, Las Vegas, Los Angeles, Seattle, and Anchorage) have been subjected to statistical trend analysis (B.J. Mahler, personal communication, 2002) and the results plotted in Figure 8. It is obvious that essentially all urban reservoirs and lake records show a very significant decline in Pb since 1975 and that this trend is most probably a result of the ban on leaded gasoline that was instituted in 1972.

9.03.4.3.2 Zinc from rubber tire wear

Contrary to the distribution of Pb in sediment cores whereby peak concentrations occurred during the 1970s, the concentration of sedimentary Zn often increases to the 1990s. It was observed in the atmospheric emissions of metals that waste incineration was one of the major contributors to the second tier of Zn emissions to the US atmosphere. Figure 9(a) presents age-dated sedimentary Zn data from a spectrum of urban/suburban sites around the US It is obvious that the general trend of sedimentary Zn is one of increasing concentrations from 1950s to 1990s. However, the general increasing trend for Zn is not as prevalent as that for Pb and at a few of the urban/suburban/reference sites noted for the Pb trend map there is no significant trend in sedimentary Zn concentration (B.J. Mahler, personal communication, 2002). Figure 5 shows that rubber tire wear is the most important and increasing contributor to the second tier of Zn emissions to the US atmosphere. Tire tread material has a Zn content of about 1% by weight. A significant quantity of tread material is lost to road surfaces by abrasion prior to tire replacement on a vehicle. In Figure 9(b) the anthropogenic Zn data for urban/suburban core sites is regressed against the mass sedimentation rate (MSR) for each core site. When MSR-normalized anthropogenic Zn is plotted against average annual daily traffic (AADT) data for the various metropolitan areas shown in Figure 9(a), a significant regression results (Figure 9(c)) suggesting that there is a causal relationship between anthropogenic Zn and vehicle traffic. Councell *et al.* (2003) produced data that estimates the magnitude of the Zn releases to the environment from rubber tire abrasion. Two approaches, wear rate (g km^{-1}) and tread geometry (abrasion to wear bars), were used to assess the magnitude of this nonpoint source of Zn in the US for the period 1936–1999. For 1999, the quantity of Zn released by tire wear in the US is estimated to be between 10,000 and 11,000 t.

Two specific case studies focused on the impact of vehicle tire wear to the Zn budget of watersheds in the Washington, DC metropolitan area. For Lake Anne, a suburban watershed located 40 km southwest of Washington, DC, the wet deposition atmospheric flux of Zn was 8 μg cm^{-2} yr^{-1} (Davis and Galloway, 1981) and the flux of Zn

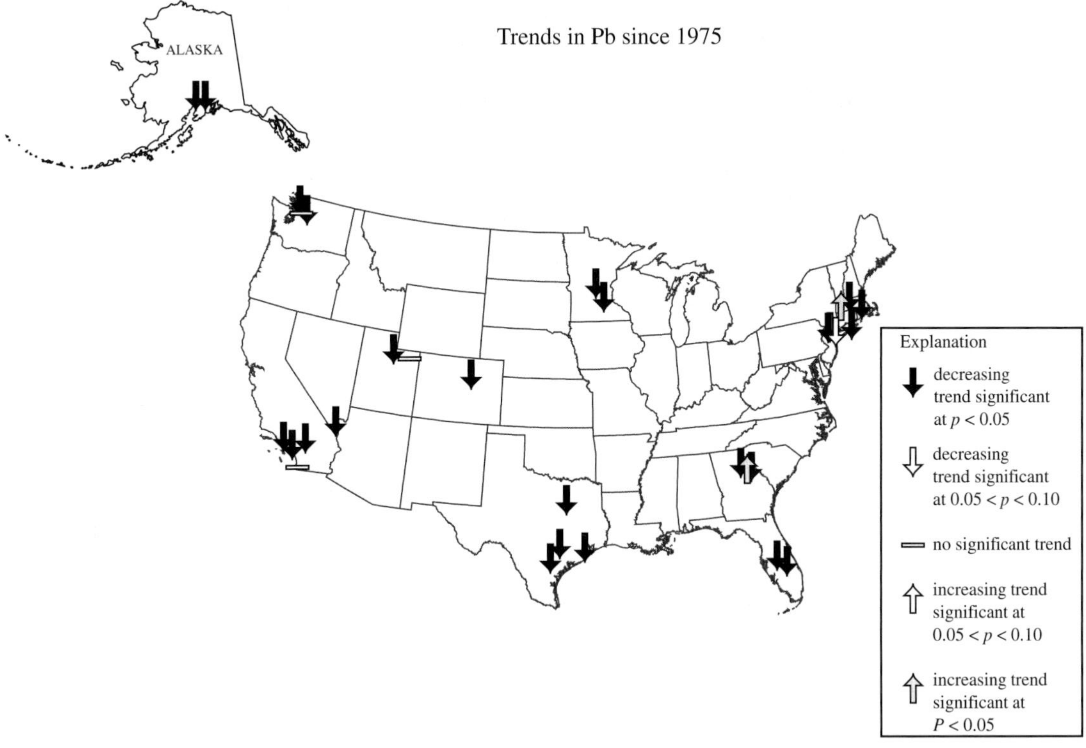

Figure 8 Map showing statistical trends in lead from sediment cores located throughout the US.

estimated from tire wear was 31 μg cm^{-2} yr^{-1} (Landa *et al.* 2002). The measured accumulation rate of Zn in age-dated sediment cores from Lake Anne is 41 μg cm^{-2} yr^{-1} (Landa *et al*, 2002) suggesting that tire-wear Zn inputs to suburban watersheds can be significantly greater than atmospheric inputs. In a rural/atmospheric reference site watershed, located ~90 km northwest of Washington, DC, the atmospheric Zn flux is 12 μg cm^{-2} yr^{-1} (Davis and Galloway, 1981) and that from tire wear is only 1 μg cm^{-2} yr^{-1} (Landa *et al.*, 2002). There are only dirt roads leading to cabins in this protected watershed and it is obvious that vehicle tire wear is only a minor component of the Zn flux in this remote watershed. One conclusion drawn from these case studies and the substantial set of age-dated sediment core Zn profiles is that those watersheds that are impacted by vehicular traffic receive significant amounts of Zn via tire abrasion and that this Zn-enriched particulate matter is fluvially transported to lakes and reservoirs.

9.03.4.3.3 Metal processing and metal trends in sediment cores

While the relationship between Pb and Zn distributions in reservoir and lake sediment cores and environmental forcing functions (leaded gasoline use and vehicular traffic) are clear, the same is not true in the case of other metals such as Cd, Cr, Cu, and Ni. These metals do have one common, major source: nonferrous and ferrous metal production. Approximately 70% of the Cd and Cr anthropogenic emissions, 50% of the Cu, and 21% of the Ni anthropogenic emissions come from mining–smelting–metal processing. In an attempt to interpret metal trend maps for these four metals, historical US metal mining and production statistics have been shown in Figure 10. Copper is the only metal whose production increased during the past 25 years. Starting in 1985, the primary production of Cu has increased from about one million metric tons to a maximum of about two million metric tons in 1998 (Figure 10). Arizona has the largest Cu mining production followed by Utah, New Mexico, and Montana. Thus, all the major point sources of Cu exist in the western states. Figure 11 shows the statistical trends (since 1975) for Cu in sediment cores from 30 reservoir and lake sites around the US (B.J. Mahler, personal communication, 2002). Increasing or decreasing trends in the upper part of sediment cores were tested statistically for eight trace metals. Significant trends in sediments deposited since 1975 were tested using a Spearman's rank correlation

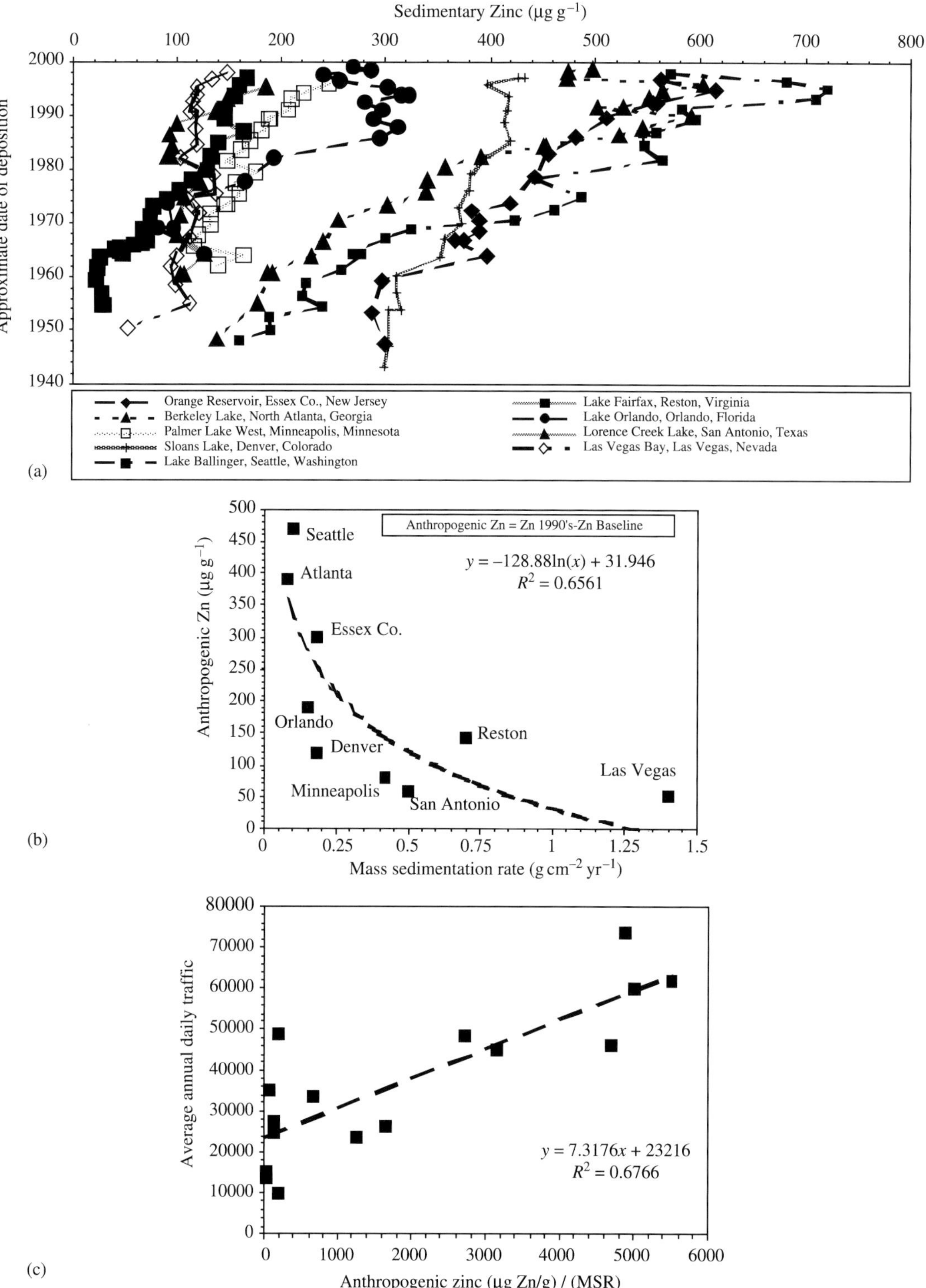

Figure 9 (a) Temporal distribution of total sedimentary zinc in US reservoir sediment cores, (b) regression of anthropogenic zinc versus MSR for US reservoir sediment cores, (c) MSR-normalized anthropogenic zinc versus average annual daily traffic for urban and suburban watersheds throughout the US

(Helsel and Hirsch, 1992). Trends were determined to be significantly increasing or decreasing based on a p value of less than 0.05 (B.J. Mahler, personal communication, 2002). The Cu trend indicators in Figure 11 show that there are increasing trends significant at $p < 0.05$ at the core sites in Washington–California–Nevada and that there are no

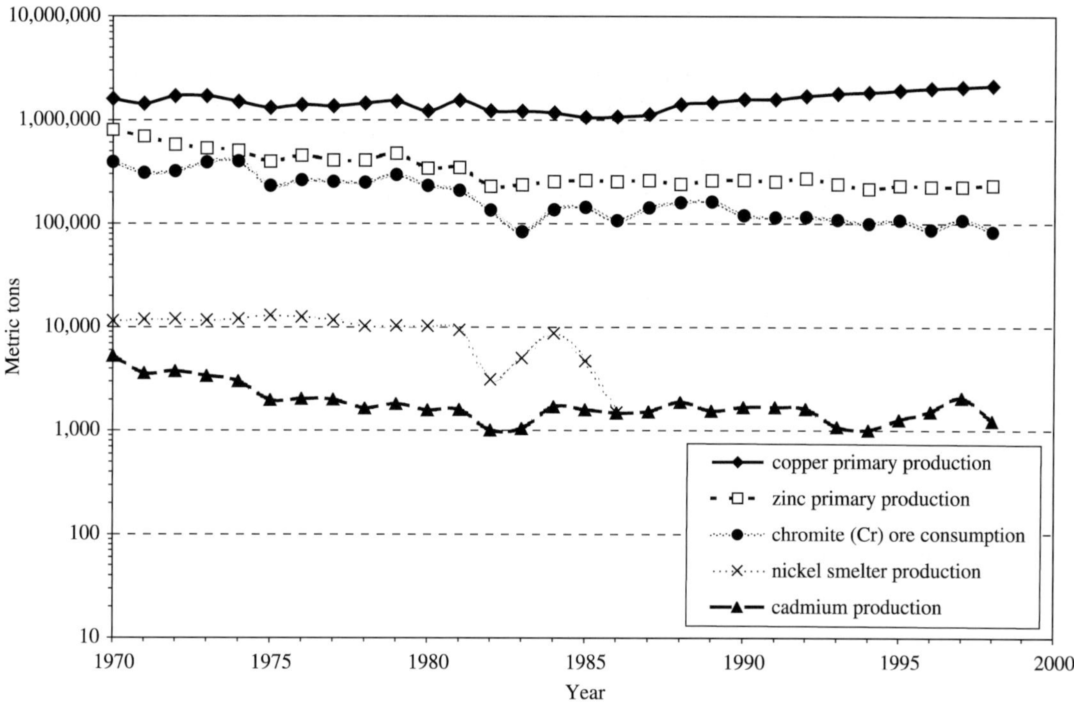

Figure 10 Historical production of nonferrous metals in the US Data from US Geological Survey, Mineral Resources Program.

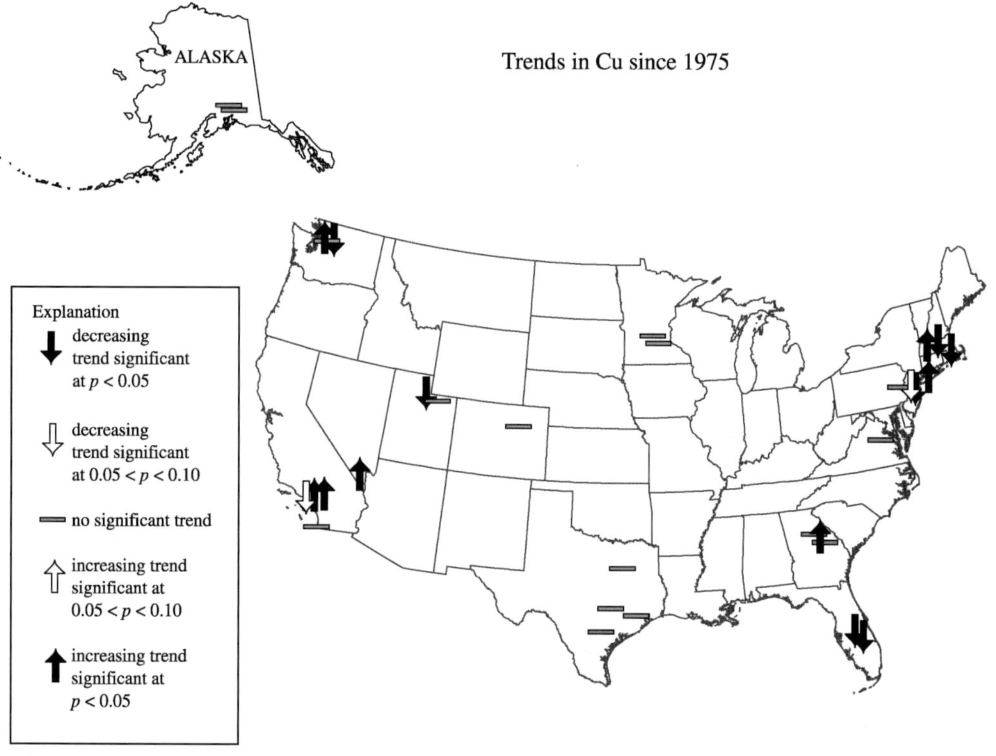

Figure 11 Map showing statistical trends in copper from sediment cores located throughout the US.

significant trends in Cu at core sites in the Midwest and Southwest. Such a pattern suggests that an increase in Cu mining in the Rocky Mountain States may cause airborne emissions that impact the western US but that these emissions are not transported east to the mid-continent. Along the Atlantic coast, from Georgia to Massachusetts, there are some sites that show an increasing trend in sedimentary Cu. It should be noted that many of the east coast sites are small ponds used for water supply and recreation and that some have been treated with $CuSO_4$ to control the algae. In addition, many of these sites are located in the east coast urban/suburban corridor and obviously receive a multitude of anthropogenic contaminants.

Limited space does not allow for the presentation of all metal trend maps such as those presented for Pb and Cu. Of the three remaining metals (Cd, Cr, Ni), the trends in Ni since 1975 is representative of the other metals as well. For the western half of the US, there are nine sites where there is a significant decreasing trend in Ni (Figure 12). This corresponds to the decrease in Ni smelter production for the 1970s and 1980s (Figure 10). There are a few increasing Ni trends along the east coast of the US, a pattern that may reflect the location of many nickel consumption facilities in Pennsylvania, West Virginia, and New Jersey. The trends in Cr since 1975 are very similar to those for Ni; for the western half of the US there are eight sites where there is a significant decreasing trend. Along the east coast of the US there is only one site in the Boston area that shows an increasing trend. Thus, the overall decreasing Cr trend for the US reflects the four-fold decrease in chromite (Cr ore) consumption by metallurgical and chemical firms in the US since 1970 (Figure 10). The trends in Cd since 1975 are not as strongly skewed toward decreasing trends as those for Ni and Cr. This is probably due to the fact that much of the Cd in the US is recovered by the processing of Zn ore and as one can see from Figure 10, Zn production has leveled out in the 1980s and 1990s. In fact, the preponderance of coring sites in the US (20 out of 30) show no significant trend in Cd.

9.03.4.3.4 Reduction in power plant emissions of heavy metals: clean air act amendments and the use of low sulfur coal

Only recently have electric utility power plant emissions been included on the US Environmental Protection Agency's (USEPA) Toxic Release Inventory which reported that electric utilities ranked highest for industrial toxic air emissions in 1998. These emissions were likely to be an important component of toxic air releases in the past, particularly prior to the passage of the Clean Air Act of 1970.

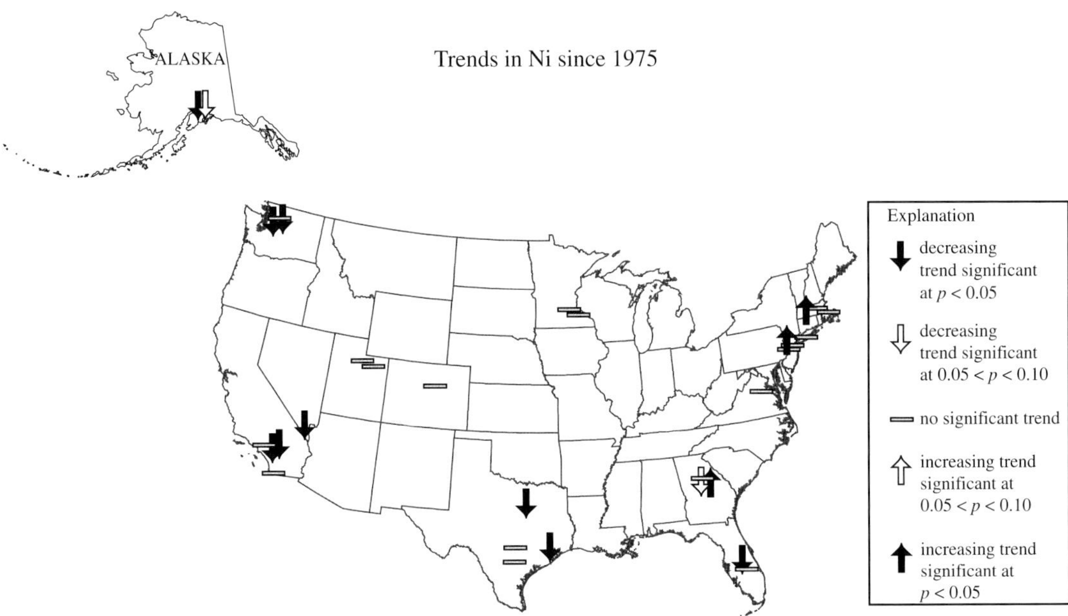

Figure 12 Map showing the statistical trends in nickel from sediment cores located throughout the US.

A sediment core from one reservoir (Mile Tree Run Reservoir) located in southwestern West Virginia was selected for the purpose of identifying particulate signatures of coal-fired power plant emissions from the nearby Ohio River Valley region. Arsenic (As) and some Pb and Zn releases to the environment may reflect coal-related geochemical processes. Stream sediments from the Appalachian Basin Region are particularly enriched in As compared to the sediments outside the Basin. While some As enrichment may come from the weathering of As-rich coal, this area of West Virginia is underlain primarily by sandstone which is low in trace-element composition. Thus, the elevated As contents are not likely to have an origin in the regional country rock but might have originated from numerous large coal-fired power plants situated along the Ohio River.

Figure 13(a) shows the temporal distribution of Ti-normalized As, Pb, and Zn in the sediment core from Mile Tree Run Reservoir. These Ti-normalized metal peaks date back to 1987, 1966, and 1946, respectively (Figure 13(a)). Figure 13(b) shows the temporal distribution of Ti-normalized sulfur (S), isothermal remnant magnetization (IRM) (a magnetite proxy), and Fe in the same core. Peaks in these constituents correspond to the aforementioned dates. These dates match the maximum values in combined coal production for the states of West Virginia, Pennsylvania, and Ohio (M.B. Goldhaber, personal communication, 2002). The temporal profile for the magnetic property IRM that is indicative of the mineral magnetite is shown in Figure 13(b) (M.B. Goldhaber, personal communication, 2002).

Figures 14 (a) and (b) are scatter plots between Zn and coal production, and Zn and IRM (magnetite). Note the excellent correlations suggesting that Zn relates to atmospheric input (fly ash) from power plants. A glance at Table 2 shows that Zn in fly ash is substantially enriched compared to average soils. One can also see this relationship for Mile Tree Run watershed soils in Figure 13(a). Arsenic also has a very strong positive correlation with IRM in the Mile Tree Run Reservoir core. This element is known to be strongly associated with magnetite, an important fly ash mineral that is formed in the high temperature combustion of coal (Locke and Bertine, 1986). Such a geochemical association is not surprising in that it is thought that magnetite is formed from the pyrite in coal that is subjected to high temperature combustion. It is also well known that As is a minor element associated with pyrite.

The simultaneous decline after 1985 of the correlated Fe–As–Pb–Zn and magnetite peaks (Figures 13(a) and (b) suggests that the amount of power plant particulate emissions decreased since the 1977 amendment of the Clean Air Act that mandated reduced amounts of sulfur in the feed coal (Hower *et al.*, 1999). It appears that this action resulted in lower metal and magnetite quantities in the fly ash combusted residue. Despite this decrease, soil samples from the Mile Tree Run Reservoir watershed are strongly enriched in As, Pb, and Zn when compared to average soils (Table 2) and local bedrock (Callender *et al.*, 2001), indicating a regional power plant emissions impact on the geochemical landscape.

9.03.4.3.5 European lacustrine records of heavy metal pollution

Much of the recent literature pertaining to European studies of heavy metal pollution using sediment cores to track time trends focused on Pb. Petit *et al.* (1984) used the stable isotope geochemistry to identify Pb pollution sources and to evaluate the relative importance of anthropogenic sources to total Pb fluxes in a semi-rural region of western Europe. Thomas *et al.* (1984) showed that metal enrichment factors for Pb–Zn–Cd were significantly above unity, indicating a diffuse contribution through atmospheric transport from industrialized areas. A comparison with atmospheric fluxes showed good agreement for diffuse atmospheric supply of Pb, Zn, and Cd in the lake sediments (Thomas *et al.*, 1984).

Much work has been done on Lake Constance situated on the borders of Germany, Austria, and Switzerland. The lake has undergone extensive cultural eutrophication due to the surrounding population, industrial activity, and the input of River Rhine. Muller *et al.* (1977) found that sedimentary Pb and Zn concentrations peaked around 1965. They also noted that there was a strong positive correlation between heavy-metal content and PAHs in the sediment core. Muller *et al.* (1977) suggested that coal burning was the source of this relationship. In a more recent study of Lake Constance sediments, Wessels *et al.* (1995) also noted high concentrations of Pb and Zn that began around 1960. However, they postulated that the origin of the Pb increase was emissions by regional industry and the origin of the Zn increase was a combination of urban runoff and coal burning.

Two groups, one in Switzerland and the other in Sweden, have used age-dated sediment cores from peat bogs and natural lake sediments to record the history of atmospheric Pb pollution dating back to several thousand years.

In Switzerland, Shotyk and co-workers (Shotyk *et al.*, 1998) rebuilt the history of atmospheric Pb deposition over the last 12,000 years. They cored ombrotrophic peat bogs that are hydrologically isolated from the influence of local groundwaters and surfacewaters, and receive their inorganic

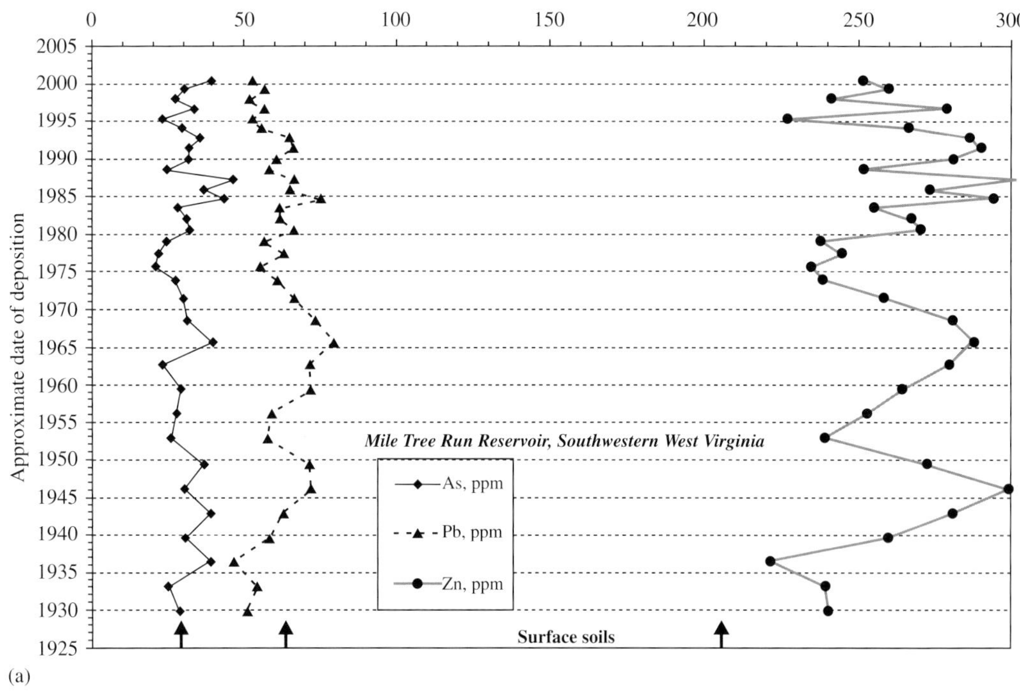

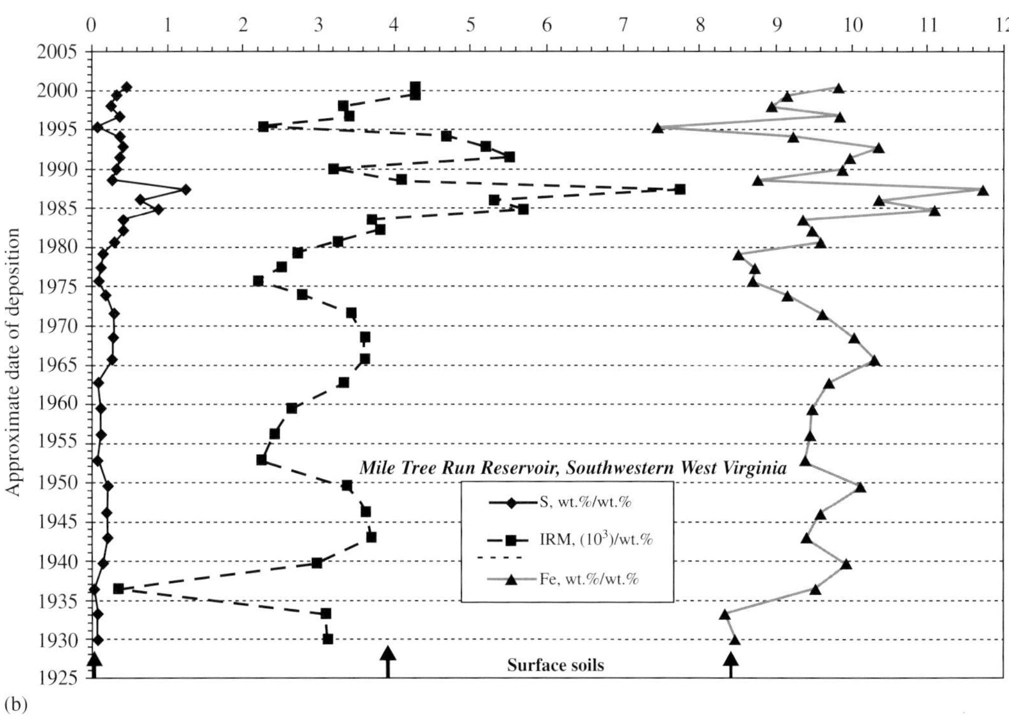

Figure 13 (a) Temporal distribution of Ti-normalized As–Pb–Zn (ppm/wt.%) in a sediment core from Mile Tree Run Reservoir, Southwestern West Virginia, USA; (b) Temporal distribution of Ti-normalized S–IRM–Fe (wt.%/wt.%) in the Mile Tree Run Reservoir sediment core.

solids exclusively by atmospheric deposition. Whereas slowly-accumulating natural lake sediments appear to be affected by chemical diagenesis, studies have shown that peat bogs provide a reliable record of changes in atmospheric metal deposition (Roos-Barraclough and Shotyk, 2003). Their radiocarbon dated core profiles of stable Pb and $^{206}Pb/^{207}Pb$ isotopic

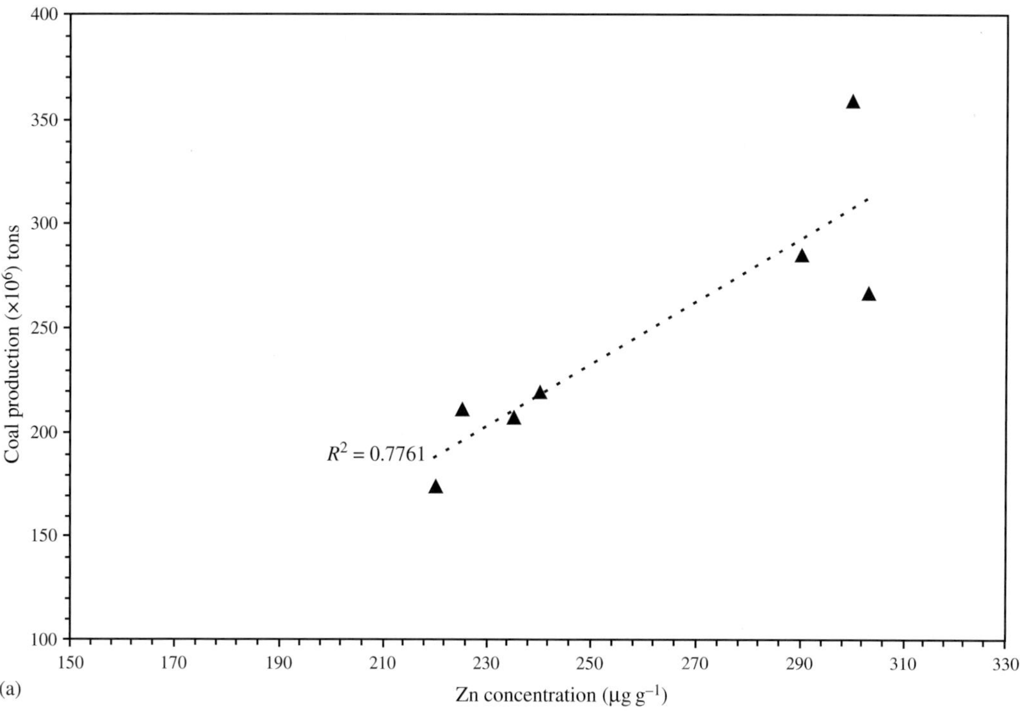

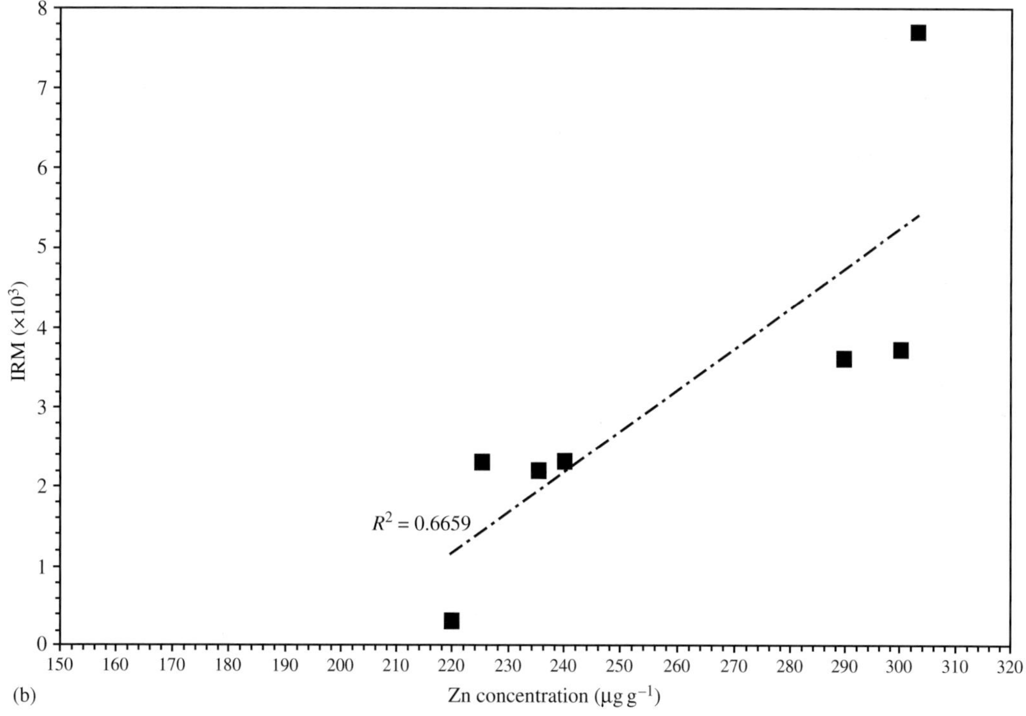

Figure 14 (a) Scatter plot of Appalachian Basin temporal coal production versus temporal concentration of Zn in a Mile Tree Run Reservoir sediment core, (b) Scatter plot of the temporal concentration of IRM (magnetite) versus the temporal concentration of Zn in the Mile Tree Run Reservoir sediment core.

ratios indicate that enhanced fluxes of Pb were caused by climatic changes between 10,500 and 8,250 years before present (BP). Soil erosion, caused by forest clearing and agricultural tillage, increased Pb deposition subsequent to this time. Beginning 3,000 yr BP, Pb pollution from mining and smelting was recorded by a significant increase in normalized Pb concentrations (Pb/Sc) and a decreasing $^{206}Pb/^{207}Pb$ ratio. Around BP 2100, Roman Pb mining became the most important source of atmospheric Pb pollution; and in AD 1830 the effects of the Industrial Revolution were recorded by a very large peak in Pb enrichment. The $^{206}Pb/^{207}Pb$ ratio declined significantly around AD 1940 indicating the use of leaded gasoline which was subsequently discontinued in AD 1979 (Shotyk *et al.*, 1998). In an earlier paper, Shotyk *et al.* (1996) noted that there were significant enrichments in As and Sb as well as for Pb dating back to Roman times (BP 2100). These enrichments in As and Sb were thought to be related to Pb mining and smelting.

In Sweden, Brannvall and his colleagues published several papers culminating in a summary paper (Brannvall *et al.*, 2001) describing four thousand years of atmospheric Pb pollution in northern Europe. They cored 31 lakes throughout Sweden; some were age-dated with radiocarbon while others had varved sediments. Their stable Pb and $^{206}Pb/^{207}Pb$ isotope data indicate that the first influx of noncatchment atmospheric Pb occurred between 3,500 and 3,000 years ago. The large world production of Pb ($80{,}000\ t\ yr^{-1}$) during Greek and Roman times 2,000 years ago caused widespread atmospheric Pb pollution. There was a decline in the atmospheric Pb flux between AD 400 and 900. Brannvall *et al.* (2001) note that the Medieval period, rather than the Industrial Revolution, was the real beginning of the contemporary Pb pollution era. This era extended from AD 1000 to 1800. Lead peaked in the mid-twentieth century in Sweden (1950s–1970s) due to the use of leaded gasoline and fossil-fuel combustion. The recent decline in atmospheric Pb deposition since 1980 is very steep and significant. Johansson (1989) analyzed the heavy-metal content of some 54 lakes in central and northern Sweden, and noted that the Pb content of surface sediment was 50 times greater than the background concentration and that this Pb enrichment decreased substantially from south to north. Ek *et al.* (2001) studied the environmental effects of one thousand years of Cu production in Central Sweden. Metal analyses of the lake sediments showed that Cu pollution was restricted to a smaller area near the emission sources and that Pb, Zn, and Cd pollution was widespread. Sedimentary metal enrichments began about AD 1000 and peaked in the seventeenth century when Central Sweden produced two-thirds of the world's Cu supply.

Sedimentary lacustrine records of heavy metal pollution in Lough Neagh (northern Ireland) and Lake Windermere (England) suggest that there have been two periods of metal disturbance since the AD 1600. Both lakes are situated within rural catchments. In Lough Neagh (Rippey *et al.*, 1982), a change in the catchment erosion regime during the seventeenth century produced an increase in the sedimentary Cd–Cu–Pb concentrations. This change in erosion was a result of widespread and comprehensive forest clearance. A second and larger change occurred about AD 1880 when the concentrations of Cr, Cu, Zn, and Pb increased toward the sediment surface. Sediments from Lake Windermere also show a pronounced increase in Cu–Pb–Zn concentrations within the upper part of the sediment column (Hamilton-Taylor, 1979). Since the catchments are not proximal to the local anthropogenic sources such as mining, smelting, or wastewater inputs, it is possible that a substantial part of these metal enrichments are due to a more regional or even global atmospheric input that was generated by many anthropogenic processes related to the Industrial Revolution.

REFERENCES

Adriano D. C. (1986) *Trace Elements in the Terrestrial Environment*. Springer, New York.

Alexander C. R., Smith R. G., Calder F. D., Schropp S. J., and Windom H. L. (1993) The historical record of metal enrichment in two Florida estuaries. *Estuaries* **16**, 627–637.

Allan R. J. (1975) Natural versus unnatural heavy metal concentrations in lake sediments in Canada. *Proc. Int. Conf. Heavy Metals in the Environ.* **2**, 785–808.

Alloway B. J. (1995) *Heavy Metals in Soils*, 2nd edn. Blackie Academic and Professional, London, UK.

Anikiyev V. V., Perepelitsa S. A., and Shumilin Ye. N. (1993) Effects of man-made and natural sources on the heavy-metal patterns in bottom sediments in the Gulf of Peter the Great, Sea of Japan. *Geokhimiya* **9**, 1328–1340.

Appleby P. G. and Oldfield F. (1983) The assessment of ^{210}Pb data from sites with varying sediment accumulation rates. *Hydrobiologia* **103**, 29–35.

Aston S. R., Thornton I., Webb J. S., Purves J. B., and Milford B. L. (1974) Stream sediment composition: an aid to water quality assessment. *Water Air Soil Pollut.* **3**, 321–325.

Beck H. L., Helfer I. K., Bouville A., and Deicer M. (1990) Estimates of fallout in the continental US from Nevada weapons testing based on gummed-film monitoring data. *Health Phys.* **59**, 565–576.

Belzile N., Lecomte P., and Tessier A. (1989) Testing readsorption of trace elements during partial chemical extractions of bottom sediments. *Environ. Sci. Technol.* **23**, 1015–1020.

Bertine K. K. and Goldberg E. D. (1971) Fossil fuel combustion and the major sedimentary cycle. *Science* **173**, 233–235.

Borovec Z., Tolar V., and Mraz L. (1993) Distribution of some metals in sediments of the central part of the Labe (Elbe) River, Czech Republic. *Ambio* **22**, 200–205.

Boudreau B. P. (1999) Metals and models: diagenetic modeling in freshwater lacustrine sediments. *J. Paleolimnol.* **22**, 227–251.

Boutron C. F., Gorlach U., Candelone J.-P., Bolshov M. A., and Deimas R. J. (1991) Decrease in anthropogenic lead, cadmium and zinc in Greenland snows since the late 1960s. *Nature* **353**, 153–156.

Boyle E. A. (1979) Copper in natural waters. In *Copper in the Environment, Part I: Ecological Cycling* (ed. J. O. Nriagu). Wiley, New York, pp. 77–88.

Boyle E. A., Sherrell R. M., and Bacon M. P. (1994) Lead variability in the western North Atlantic Ocean and central Greenland ice: implications for the search for decadal trends in anthropogenic emissions. *Geochim. Cosmochim. Acta* **58**, 3227–3238.

Brannvall M. L., Bindler R., Emteryd O., and Renberg I. (2001) Four thousand years of atmospheric lead pollution in northern Europe: a summary from Swedish lake sediments. *J. Paleolimnol.* **25**, 421–435.

Brush G. S. (1984) Patterns of recent accumulation in Chesapeake Bay (Virginia-Maryland, USA) tributaries. *Chem. Geol.* **44**, 227–242.

Callender E. (2000) Geochemical effects of rapid sedimentation in aquatic systems: minimal diagenesis and the preservation of historical metal signatures. *J. Paleolimnol.* **23**, 243–260.

Callender E., Goldhaber M. B., Reynolds R. L., and Grosz A. (2001) Geochemical signatures of power plant emissions as revealed in sediment cores from West Virginia reservoirs. *Ann. Meeting Geol. Soc. Am.* (Abstracts).

Callender E. and Robbins J. A. (1993) Transport and accumulation of radionuclides and stable elements in a Missouri River reservoir. *Water Resour. Res.* **29**, 1787–1804.

Callender E. and Van Metre P. C. (1997) Reservoir sediment cores show US lead declines. *Environ. Sci. Technol.* **31**, 424A–428A.

Candelone J.-P., Hong S., Pellone C., and Boutron C. F. (1995) Post-industrial revolution changes in large-scale atmospheric pollution of the northern hemisphere by heavy metals as documented in central Greenland snow and ice. *J. Geophys. Res.* **100**, 16,605–16,616.

Chao T. T. (1984) Use of partial dissolution techniques in geochemical exploration. *J. Geochem. Explor.* **20**, 101–135.

Chester R. (2000) *Marine Geochemistry*, 2nd edn. Oxford, Malden.

Chiffoleau J.-F., Cossa D., Auger D., and Truquet I. (1994) Trace metal distribution, partition and fluxes in the Seine estuary (France) in low discharge regime. *Mar. Chem.* **47**, 145–158.

Chow T. and Patterson C. C. (1962) The occurrence and significance of lead isotopes in pelagic sediments. *Geochim. Cosmochim. Acta* **26**, 263–308.

Chow T. J., Bruland K. W., Bertine K. K., Soutar A., Koide M., and Goldberg E. D. (1973) Lead pollution: records in southern California coastal sediments. *Science* **181**, 551–552.

Coakley J. P. and Poulton D. J. (1993) Source-related classification of St. Lawrence Estuary sediments based on spatial distribution of adsorbed contaminants. *Estuaries* **16**, 873–886.

Conko K. M. and Callender E. (1999) Extraction of anthropogenic trace metals from sediments of two US urban reservoirs. *Abstracts, 1999 Fall Meeting Am. Geophys. Union*.

Councell T. B., Duckenfield K. U., Landa E. R., and Callender E. (2003) Tire-wear particles as a source of zinc to the environment. *Environ. Sci. Technol.* (submitted).

Davis W. E. and Associates (1970) *National Inventory of Sources and Emissions of Cadmium*. Nat'l Tech. Information Service Rept. PB 192250, Springfield, VA.

Davis A. O. and Galloway J. N. (1981) Atmospheric lead and zinc deposition in lakes in the eastern United States. In *Atmospheric Pollutants in Natural Waters* (ed. S. J. Eisenreich). Ann Arbor Science, Ann Arbor, pp. 401–408.

Davis J. A., James R. O., and Leckie J. O. (1978) Surface ionization and complexation at the oxide/water interface: 1. Computation of electrical double layer properties in simple electrolytes. *J. Colloid Interface Sci.* **63**, 480.

Dolske D. A. and Sievering H. (1979) Trace Element loading of southern Lake Michigan by dry deposition of atmospheric aerosol. *Water Air Soil Pollut.* **12**, 485–502.

Dominik J., Mangini A., and Prosi F. (1984) Sedimentation rate variations and anthropogenic metal fluxes into Lake Constance sediments. *Environ. Geol.* **5**, 151–157.

Drever J. I. (1988) *The Geochemistry of Natural Waters*, 2nd edn. Prentice-Hall, New York.

Duce R. A., Liss P. S., Merrill J. T., Atlas E. L., Buat-Menard P., Hicks B. B., Miller J. M., Prospero J. M., Arimoto R., Church T. M., Ellis W., Galloway J. N., Hansen L., Jickells T. D., Knap A. H., Reinhardt K. H., Schneider B., Soudine A., Tokos J. J., Tsunogai S., Wollast R., and Zhou M. (1991) The atmospheric input of trace species to the world ocean. *Global Biogeochem. Cycles* **5**, 193–259.

Dunnette D. A. (1992) Assessing global river quality, overview and data collection. In *The Science of Global Change: the Impact of Human Activities on the Environment*, Am. Chem. Soc. Symposium 483 (eds. D. A. Dunnette and R. J. O'Brien). ACS, pp. 240–286.

Eisenreich S. J. (1980) Atmospheric input of trace elements to Lake Michigan. *Water Air Soil Pollut.* **13**, 287–301.

Eisenreich S. J., Metzer N. A., Urban N. R., and Robbins J. A. (1986) Response of atmospheric lead to decreased use of lead in gasoline. *Environ. Sci. Technol.* **20**, 171–174.

Ek A. S., Lofgren S., Bergholm J., and Qvarfort U. (2001) Environmental effects of one thousand years of copper production at Falun Central Sweden. *Ambio* **30**, 96–103.

Farrah H. and Pickering W. F. (1977) Influence of clay-solute interactions an aqueous heavy metal ion levels. *Water Air Soil Pollut.* **8**, 189–197.

Faust S. D. and Aly O. M. (1981) *Chemistry of Natural Waters*. Ann Arbor Science, Ann Arbor.

Fergusson J. E. (1990) *The Heavy Elements: Chemistry, Environmental Impact, and Health Effects*. Pergamon, Oxford, 614pp.

Forstner U. (1976) Lake sediments as indicators of heavy-metal pollution. *Naturwissenschaften* **63**, 465–470.

Forstner U. (1981) Recent heavy metal accumulation in limnic sediments. In *Handbook of Strata-Bound and Stratiform Ore Deposits* (ed. K. H. Wolf). Elsevier, pp. 179–269.

Forstner U. (1990) Inorganic sediment chemistry and elemental speciation. In *Sediments: Chemistry and Toxicity on In-Place Pollutants* (eds. R. Baudo, J. P. Giesy, and H. Mantau). Lewis, pp. 61–105.

Gaillardet J., Viers J., and Dupre B. (2003) Trace Elements in River Waters. In *Treatise on Geochemistry*. Elsevier, Amsterdam, vol. 5, Chapter 6.

Gardiner J. (1974) The chemistry of cadmium in natural water: II. The adsorption of cadmium on river muds and naturally occurring solids. *Water Res.* **8**, 157–164.

Garrels R. J. and Christ C. L. (1965) *Solutions, Minerals and Equilibria*. Harper and Row, New York.

Garrels R. M., Mackenzie F. T., and Hunt C. (1973) *Chemical Cycles and the Global Environment: Assessing Human Influences*. William Kaufmann, Los Altos.

Gibbs R. J. (1973) Mechanisms of trace metal transport in rivers. *Science* **180**, 71–73.

Gocht T., Moldenhauer K.-M., and Puttmann W. (2001) Historical record of polycyclic aromatic hydrocarbons (PAH) and heavy metals in floodplain sediments from the Rhine River (Hessisches Ried, Germany). *Appl. Geochem.* **16**, 1707–1721.

Govindaraju K. (1989) 1989 compilation of working values and sample description for 272 geostandards. *Geostand. Newslett.* **13**, 1–113.

Graney J. R., Halliday A. N., Keeler G. J., Nriagu J. O., Robbins J. A., and Norton S. A. (1995) Isotopic record of lead pollution in lake sediments from the northeastern United States. *Geochim. Cosmochim. Acta* **59**, 1715–1728.

Hamilton-Taylor J. (1979) Enrichments of Zinc Lead, and Copper in recent sediments of Windermere, England. *Environ. Sci. Technol.* **13**, 693–697.

Hanson P. J. (1997) Response of hepatic trace element concentrations in fish exposed to elemental and organic contaminants. *Estuaries* **20**, 659–676.

Heit M., Klusek C., and Baron J. (1984) Evidence of deposition of anthropogenic pollutants in remote Rocky Mountain lakes. *Water Air Soil Pollut.* **22**, 403–416.

Helsel D. R. and Hirsch R. M. (1992). *Statistical Methods in Water Resources, Studies in Environmental Science*. **49**, Elsevier, Amsterdam.

Hem J. D. (1972) Chemistry and occurrence of cadmium and zinc in surface water and ground water. *Water Resour. Res.* **8**, 661–679.

Hem J. D. (1976) Geochemical controls on lead concentrations in stream water and sediments. *Geochim. Cosmochim. Acta* **40**, 599–609.

Hem J. D. (1977) Reactions of metal ions at surfaces of hydrous iron oxide. *Geochim. Cosmochim. Acta* **41**, 527–538.

Hem J. D. (1985) *Study and Interpretation of the Chemical Characteristics of Natural Water*. US Geol. Survey Water-Supply Paper 2254, US Govt. Printing Office.

Hem J. D. and Durum W. H. (1973) Solubility and occurrence of lead in surface water. *Am. Water Works Assoc. J.* **65**, 562–568.

Hermanson M. H. (1991) Chronology and sources of anthropogenic trace metals in sediments from small, shallow arctic lakes. *Environ. Sci. Technol.* **25**, 2059–2064.

Honeyman B. D. and Santschi P. H. (1988) Metals in aquatic systems. *Environ. Sci. Technol.* **22**, 862–871.

Hong S., Candelone J.-P., Patterson C. C., and Boutron C. F. (1994) Greenland ice evidence of hemispheric lead pollution two millennia ago by Greek and Roman civilizations. *Science* **265**, 1841–1843.

Hornberger M. I., Luoma S. N., Van Geen A., Fuller C. C., and Anima R. J. (1999) Historical trends of metals in the sediments of San Francisco Bay, California. *Mar. Chem.* **64**, 39–55.

Hornberger M. I., Luoma S. N., Cain D. J., Parchaso F., Brown C. L., Bouse R. M., Wellise C., and Thompson J. K. (2000) Linkage of bioaccumulation and biological effects to changes in pollutant loads in south San Francisco Bay. *Environ. Sci. Technol.* **34**, 2401–2409.

Hower J. C., Robl T. L., and Thomas G. A. (1999) Changes in the quality of coal combustion by-products produced by Kentucky power plants, 1978 to 1997: consequences of Clean Air Act directives. *Fuel* **78**, 701–712.

Hsu C. L. (1978) Heavy metal uptake by soils surrounding a fly ash pond. MS Thesis, University of Notre Dame.

Jenne E. A. (1968) Controls on Mn, Fe, Co, Ni, Cu and Zn concentrations in soils and water: the significant role of hydrous Mn and Fe oxides. In *Trace Inorganics in Water*, Am. Chem. Soc. Adv. Chem. Ser. 73, ACS, pp. 337-387.

Johansson K. (1989) Metals in sediment of lakes in Northern Sweden. *Water Air Soil Pollut.* **47**, 441–455.

Johnson T. C. (1984) Sedimentation in large lakes. *Ann. Rev. Earth Planet. Sci.* **12**, 179–204.

Kabata-Pendias A. (2000) *Trace Elements in Soils and Plants*, 3rd edn. CRC Press, Boca Raton.

Kabata-Pendias A. and Pendias H. (1992) *Trace Elements in Soils and Plants*, 2nd edn. CRC Press, Boca Raton.

Kabata-Pendias A. and Pendias H. (2001) *Trace Elements in Soils and Plants*, 3rd edn. CRC Press, Boca Raton.

Kemp A. L. W., Anderson T. W., Thomas R. L., and Mudrochova A. (1974) Sedimentation rates and recent sediment history of Lake Ontario, Erie, and Huron. *J. Sed. Petrol.* **44**, 207–218.

Kemp A. L. W. and Thomas R. L. (1976) Impact of man's activities on the chemical composition in the sediments of Lakes Ontario, Erie, and Huron. *Water Air Soil Pollut.* **5**, 469–490.

Kemp A. L. W., Williams J. D. H., Thomas R. L., and Gregory M. L. (1978) Impact of man's activities on the chemical composition of the sediments of Lakes Superior and Huron. *Water Air Soil Pollut.* **10**, 381–402.

Kersten M. and Forstner U. (1987) Effects of sample pretreatment on the reliability of solid speciation data of heavy metals-implications for the study of early diagenetic processes. *Mar. Chem.* **22**, 299–312.

Kersten M. and Forstner U. (1995) Speciation of trace metals in sediments and combustion waste. In *Chemica Speciation in the Environment* (eds. A. M. Ure and C. M. Davidson). Blackie Academic and Professional, London, pp. 234–275.

Krauskopf K. B. (1956) Factors controlling the concentration of thirteen rare metals in sea-water. *Geochim. Cosmochim. Acta* **9**, 1–32.

Krishnamurthy S. and Wilkens M. M. (1994) Environmental chemistry of chromium. *Northeastern Geol.* **16**, 14–17.

Krishnaswami S. and Lal D. (1978) Radionuclide limnochronol. In *Lakes-Chemistry, Geology, Physics* (ed. A. Lerman). Springer-Verlag, New York, pp. 153–177.

Landa E. R., Callender E., Councell T. B., and Duckenfield K. V. (2002) Where the rubber meets the soil: tire wear particles as a source of zinc to the environment. *Abstracts, Ann. Meeting Soil Sci. Soc. America.*

Lantzy R. J. and Mackenzie F. T. (1979) Atmospheric trace metals: global cycles and assessment of man's impact. *Geochim. Cosmochim. Acta* **43**, 511–525.

Leckie J. O. and Davis J. A. (1979) Aqueous environmental chemistry of copper. In *Copper in the Environment* (ed. J. O. Nriagu). Wiley, New York, pp. 90–121.

Leckie J. O. and Nelson M. B. (1975) Role of natural hetrerogeneous sulfide systems in controlling the concentration and distribution of heavy metals. Paper presented at the Second International Symposium on Environmental Biogeochemistry, Ontario, Canada.

Li Y.-H. (1981) Geochemical cycles of elements and human perturbation. *Geochim. Cosmochim. Acta* **45**, 2073–2084.

Li Y.-H. (2000) *A Compendium of Geochemistry*. Princeton University Press, Princeton.

Li Y.-H., Burkhardt L., and Teraoka H. (1984) Desorption and coagulation of trace elements during estuarine mixing. *Geochim. Cosmochim. Acta* **48**, 1879–1884.

Locke G. and Bertine K. K. (1986) Magnetite in sediments as an indicator of coal combustion. *Appl. Geochem.* **1**, 345–356.

Lum R. R. (1987) Cadmium in freshwaters: the Great Lakes and St. Lawrence River. In *Cadmium in the Aquatic Environment* (eds. J. O. Nriagu and J. B. Sprague). Wiley, New York, pp. 35–50.

Luoma S. N. (1983) Bioavailability of trace metals to aquatic organisms: a review. *Sci. Total Environ.* **28**, 1–23.

Mantei E. J. and Foster M. V. (1991) Heavy metals in stream sediments: effects of human activities. *Environ. Geo. Water Sci.* **18**, 95–104.

Martin J.-M. and Meybeck M. (1979) Elemental mass balance of material carried by major world rivers. *Mar. Chem.* **7**, 173–206.

Martin J.-M., Nirel P., and Thomas A. J. (1987) Sequential extraction techniques: promises and problems. *Mar. Chem.* **22**, 313–341.

Martin J.-M. and Whitfield M. (1981) The significance of the river input of chemical elements to the ocean. In *Trace Elements in Seawater* (eds. C. S. Wong, E. Boyle, K. W. Bruland, J. D. Burton, and E. D. Goldberg). Plenum Press, New York, pp. 265–296.

Martin J.-M. and Windom H. L. (1991) Present and future roles of ocean margins in regulating marine biogeochemical cycles of trace elements. In *Ocean Margin Process in Global Change* (eds. R. F. C. Mantoura, J.-M. Martin, and R. Wollast). Wiley-Interscience, New York, pp. 45–67.

McCall P. L., Robbins J. A., and Matisoff G. (1984) ^{137}Cs and ^{210}Pb transport and geochronologies in urbanized reservoirs with raoidly increasing sedimentation rates. *Chem. Geol.* **44**, 36–65.

Mecray E. L., King J. W., Appleby P. G., and Hunt A. S. (2001) Historical trace metal accumulation in the sediments of an urbanized region of the Lake Champlain watershed Burlington, Vermont. *Water Air Soil Pollut.* **125**, 201–230.

Millero F. J. (1975) The physical chemistry of estuaries. In *Marine Chemistry in the Coastal Environment*, ACS Symposium Series 18 (ed. T. Church). American Chemical Society, pp. 25–55.

Morel F. M. M., McDuff R. E., and Morgan J. J. (1973) Interactions and chemostasis in aquatic chemical systems: role of pH, pE, solubility, and complexation. In *Trace Metals and Metal-Organic Interactions in Natural Waters* (ed. P. C. Singer). Ann Arbor Science, Ann Arbor, pp. 157–200.

Mudroch A., Sarazin L., and Lomas T. (1988) Summary of surface and background concentrations of selected elements in the Great Lakes sediments. *J. Great Lakes Res.* **14**, 241–251.

Muller G., Grimmer G., and Bohnke H. (1977) Sedimentary record of heavy metals and polycyclic aromatic hydrocarbons in Lake Constance. *Naturwissenschaften* **64**, 427–431.

Murozumi M., Chow T. J., and Patterson C. C. (1969) Chemical concentrations of pollutant lead aerosols, terrestrial dusts and sea salts in Greenland and Antarctic snow strata. *Geochim. Cosmochim. Acta* **33**, 1247–1294.

National Science Foundation (1975) *Nickel*. National Academy of Sciences, Washington, DC.

Nriagu J. O. (1979) Global inventory of natural and anthropogenic emissions of trace metals to the atmosphere. *Nature* **279**, 409–411.

Nriagu J. O. (1980a) Global cadmium cycle. In *Cadmium in the Environment, Part I: Ecological Cycling* (ed. J. O. Nriagu). Wiley, New York, pp. 1–12.

Nriagu J. O. (1980b) Global cycle and properties of nickel. In *Nickel in the Environment* (ed. J. O. Nriagu). Wiley, New York, pp. 1–26.

Nriagu J. O. (1988a) A silent epidemic of environmental metal poisoning? *Environ. Pollut.* **50**, 139–161.

Nriagu J. O. (1990a) The rise and fall of leaded gasoline. *Sci. Total Environ.* **921**, 13–18.

Nriagu J. O. (1990b) Global metal pollution. *Environment* **32**, 7–33.

Nriagu J. O. and Pacyna J. M. (1988) Quantitative assessment of worldwide contamination of air, water, and soils by trace metals. *Nature* **33**, 134–139.

Nriagu J. O., Kemp A. L. W., Wong H. K. T., and Harper N. (1979) Sedimentary record of heavy metal pollution in Lake Erie. *Geochim. Cosmochim. Acta* **43**, 247–258.

Olade M. A. (1987) Dispersion of Cadmium, Lead, and Zinc in soils and sediments of a humid tropical ecosystem in Nigeria. In *Lead, Mercury, Cadmium and Arsenic in the Environment* (eds. T. C. Hutchinson and K. M. Meema). Wiley, New York, SCOPE 31, pp. 303–312.

Osintsev S. P. (1995) Heavy metals in the bottom sediments of the Katun' River and the Ob' upper reaches. *Water Resour.* **22**, 42–49.

Pacyna J. M. (1986) Atmospheric trace elements from natural and anthropogenic sources. In *Toxic Metals in the Atmosphere* (eds. J. O. Nriagu and C. I. Davidson). Wiley, New York, pp. 33–50.

Pacyna J. M. (1996) Monitoring and assessment of metal contaminants in the air. In *Toxicology of Metals* (ed. L. W. Chang). CRC Press, Boca Raton, pp. 9–28.

Pacyna J. M. and Lindgren E. S. (1997) Atmospheric transport and deposition of toxic compounds. In *The Global Environment* (eds. D. Brune, D. V. Chapman, M. D. Gwynne, and J. M. Pacyna). Wiley, New York, pp. 386–407.

Patterson C. C. (1965) Contaminated and natural lead environments of man. *Arch. Environ. Health* **11**, 344–360.

Patterson C. C. and Settle D. M. (1987) Review of data on eolian fluxes of industrial and natural lead to the lands and seas in remote regions on a global scale. *Mar. Chem.* **22**, 137–162.

Patterson C. C., Settle D., Schaule B., and Burnett M. (1976) Transport of pollutant lead to the ocean and within ocean ecosystems. In *Marine Pollution Transfer* (eds. H. l. Windon and R. A. Duce). Heath, pp. 23–38.

Petit D., Mennessier J. P., and Lamberts L. (1984) Stable lead isotopes in pond sediments as a tracer of past and present atmospheric lead pollution in Belgium. *Atmos. Environ.* **18**, 1189–1193.

Pickering W. F. (1980) Cadmium retention by clays and other soil or sediment components. In *Cadium in the Environment,* Part I: Ecological Cycling (ed. J. O. Nriagu). Wiley, New York, pp. 365–397.

Pickering W. F. (1981) Selective chemical extraction of soil components and bound metal species. *CRC Critical Rev. Anal. Chem. Nov.*, 233–266.

Presley B. J., Trefry J. H., and Shokes R. F. (1980) Heavy metal inputs to Mississippi Delya sediments. *Water Air Soil Pollut.* **13**, 481–494.

Rai D., Eary L. E., and Zachara J. M. (1989) Environmental chemistry of chromium. *Sci. Tot. Environ.* **86**, 15–23.

Rai D., Sass B. M., and Moore D. A. (1987) Chromium (III) hydrolysis constants and solubility of chromium (III) hydroxide. *Inorg. Chem.* **26**, 345–349.

Reimann C. and de Caritat P. (1998) *Chemical Elements in the Environment*. Springer, Berlin.

Richter R. O. and Theis T. L. (1980) Nickel speciation in a soil/ water system. In *Nickel in the Environment* (ed. J. O. Nriagu). Wiley, New York, pp. 189–202.

Rippey B., Murphy R. J., and Kyle S. W. (1982) Anthropogenically derived changes in the sedimentary flux of Mg Ni, Cu, Zn, Hg, Pb, and P in Lough Neagh, Northern Ireland. *Environ. Sci. Technol.* **16**, 23–30.

Ritchie J. C., McHenry J. R., and Gill A. C. (1973) Dating recent reservoir sediments. *Limnol. Oceanogr.* **18**, 254–263.

Robbins J. A. and Edgington D. N. (1975) Determination of recent sedimentation rates in Lake Michigan using Pb-210 and Cs-137. *Goechim. Cosmochim. Acta* **39**, 285–304.

Roos-Barraclough F. and Shotyk W. (2003) Millennial scale records of atmospheric mercury deposition obtained from ombrotrophic and minerotrophic peatlands in the Swiss Jura Mountains. *Environ. Sci. Technol.* **37**, 235–244.

Rowell H. C. (1996) Paleolimnology of Onondaga Lake: the history of anthropogenic impacts on water quality. *Lake Reserv. Mgmt.* **12**, 35–45.

Salomons W. (1983) Trace metal cycling in a polluted lake (Ijsselmeer, the Netherlands). *Delft Hydraulics Laboratory Rept.* S 357, 50pp.

Salomons W. and Forstner U. (1980) Trace metal analysis on polluted sediments: II. Evaluation of environmental impact. *Environ. Technol. Lett.* **1**, 506–517.

Salomons W. and Forstner U. (1984) *Metals in the Hydrocycle*. Springer-Verlag, New York.

Santschi P., Hohener P., Benout G., and Buchholtz-ten Brink M. (1990) Chemical processes at the sediment-water interface. *Mar. Chem.* **30**, 269–315.

Sass B. M. and Rai D. (1987) Solubility of amorphous chromium (II)-iron (III) hydroxide solid solutions. *Inorg. Chem.* **26**, 2228–2232.

Schell W. R. and Barner R. S. (1986) Environmental isotope and anthropogenic tracers of recent lake sedimentation. In *Handbook of Environmental Isotope Geochemistry, The Terrestrial Environment* (eds. J. C. Fontes and P. Fritz). Elsevier, Amsterdam, Netherlands, pp. 169–206.

Scudlark J. R. and Church T. M. (1997) Atmospheric deposition of trace elements to the mid-Atlantic bight. In *Atmospheric Deposition of Contaminants to the Great Lakes and Coastal Waters* (ed. J. E. Baker). SETAC Press, pp. 195–208.

Settle D. M. and Patterson C. C. (1980) Lead in albacore: guide to lead pollution in Americans. *Science* **207**, 1167–1176.

Shafer M. M. and Armstrong D. E. (1991) Trace element cycling in southern Lake Michigan: role of water column particle components. In *Organic Substances and Sediments in Water* (ed. R. A. Baker). Lewis Publishers, Chelsea, vol. 2, pp. 15–47.

Shen G. T. and Boyle E. A. (1987) Lead in corals: reconstruction of historical industrial fluxes to the surface ocean. *Earth Planet. Sci. Lett.* **82**, 289–304.

Shotyk W., Cheburkin A. K., Appleby P. G., Frankhauser A., and Kramers J. D. (1996) Two thousand years of atmospheric arsenic, antimony, and lead deposition recorded in an ombrotrophic peat bog profile, Jura Mountains, Switzerland. *Earth Planet. Sci. Lett.* **145**, E1–E7.

Shotyk W., Weiss D., Appleby P. G., Cheburkin A. K., Frei R., Gloor M., Kramers J. D., Reese S., and Van Der Knaap W. O. (1998) History of atmospheric lead deposition since 12,370 ^{14}C yr BP from a peat bog, Jura Mountains, Switzerland. *Science* **281**, 1635–1640.

Sigg L. (1987) Surface chemical aspects of the distribution and fate of metal ions in lakes. In *Aquatic surface Chemistry* (ed. W. Stumm). Wiley, New York, pp. 319–349.

Snodgrass W. J. (1980) Distribution and behavior of nickel in the aquatic environment. In *Nickel in the Environment* (ed. J. O. Nriagu). Wiley, New York, pp. 203–274.

Sunda W. G. (1991) Trace metal interactions with marine phytoplankton. *Biol. Oceanogr.* **6**, 411–442.

Sutherland R. A. (2002) Comparison between non-residual Al Co, Cu, Fe, Mn, Ni, Pb and Zn released by a three-step sequential extraction procedure and a dilute hydrochloric acid leach for soil and road deposited sediment. *Appl. Geochem.* **17**, 353–365.

Sutherland R. A. and Tack F. M. G. (2000) Metal phase associations in soils from an urban watershed, Honolulu, Hawaii. *Sci. Total Environ.* **256**, 103–113.

Sutherland R. A., Tack F. M. G., Tolosa C. A., and Verloo M. G. (2000) Operationally defined metal fractions in road deposited sediment Honolulu, Hawaii. *J. Environ. Qual.* **29**, 1431–1439.

Taillefert M., Lienemann C.-P., Gaillard J. F., and Perret D. (2000) Speciation, reactivity, and cycling of Fe and Pb in a meromictic lake. *Geochim. Cosmochim. Acta* **64**, 169–183.

Tatsumoto M. T. and Patterson C. C. (1963) The concentration of common lead in some Atlantic and Mediterranean waters and in snow. *Nature* **199**, 350–352.

Tessier A., Cambell P. G. C., and Bisson M. (1979) Sequential extraction procedure for the speciation of particulate trace metals. *Anal. Chem.* **51**, 844–851.

Tessier A., Carignan R., Dubreuil B., and Rapin F. (1989) Partitioning of zinc between water column and the oxic sediments in lakes. *Geochim. Cosmochim. Acta* **53**, 1511–1522.

Tessier A., Fortin D., Belzile N., DeVitre R. R., and Leppard G. G. (1996) Metal sorption to diagenetic iron and manganese oxyhydroxides and associated organic matter: narrowing the gap between field and laboratory measurements. *Geochim. Cosmochim. Acta* **60**, 387–404.

Tessier A., Rapin F., and Carignan R. (1985) Trace metals in oxic lake sediments: possible adsorption onto iron oxyhydroxides. *Geochim. Cosmochim. Acta* **49**, 183–194.

Thomas V. M. (1995) The elimination of lead in gasoline. *Ann. Rev. Energy Environ.* **20**, 301–324.

Thomas M., Petit D., and Lamberts L. (1984) Pond sediments as historical record of heavy metals fallout. *Water Air Soil Pollut.* **23**, 51–59.

Tillman D. A. (1994) *Trace Metals in Combustion Systems*. Academic Press, New York.

Turekian K. K. (1971) Rivers, tributaries, and estuaries. In *Impingement of Man on the Oceans* (ed. D. W. Hood). Wiley, New York, pp. 9–74.

Turekian K. K. (1977) The fate of metals in the oceans. *Geochim. Cosmochim. Acta* **41**, 1139–1144.

Turekian K. K. and Wedepohl K. H. (1961) Distribution of the elements in some major units of the Earth's crust. *Bull. Geol. Soc. Am.* **72**, 175–192.

United States Environmental Protection Agency (1975a) *Scientific and Technical Assessment Report on Cadmium*. EPA-600/6-6-75-003, US Govt. Printing Office.

United States Environmental Protection Agency (1975b) *Technical and Microanalysis of Cadmium and its Compounds*. EPA-560/3-75-005, US Govt. Printing Office.

United States Environmental Protection Agency (1976) *Cadmium: Control Strategy Analysis*. EPA-GCA-TR-75-36-G, US Govt. Printing Office.

United States Environmental Protection Agency (1978) *Sources of Atmospheric Cadmium*. EPA-68-02-2836, US Govt. Printing Office.

United States Environmental Protection Agency (1998) *Characterization of Municipal Solid Waste in the United States: 1997 Update*. EPA530-R-98-007.

United States Environmental Protection Agency (2000a) *National Air Pollutant Emission Trends, 1900-1998*. EPA-454/R-00-002, US Govt. Printing Office.

United States Environmental Protection Agency (2000b) *Deposition of air Pollutants to the Great Waters*. EPA-453/R-00-005, US Govt. Printing Office.

United States Department of Commerce (1998) *Statistical Abstracts of the United States 1998*. US Govt. Printing Office.

Valette-Silver N. J. (1993) The use of sediment cores to reconstruct historical trends in contamination of estuarine and coastal sediments. *Estuaries* **16**, 577–588.

Van Metre P. C. and Callender E. (1996) Identifying water-quality trends in the Trinity River Texas, USA, 1969–1992, using sediment cores from Lake Livingston. *Environ. Geol.* **28**, 190–200.

Van Metre P. C., Callender E., and Fuller C. C. (1997) Historical trends in organochlorine compounds in river basins identified using sediment cores from reservoirs. *Environ. Sci. Technol.* **31**, 2339–2344.

Van Metre P. C., Mahler B. J., and Furlong E. T. (2000) Urban sprawl leaves its PAH signature. *Environ. Sci. Technol.* **34**, 4064–4070.

Veron A., Lambert C. E., Isley A., Linet P., and Grousset F. (1987) Evidence of recent lead pollution in deep north-east Atlantic sediments. *Nature* **326**, 278–281.

Wahlen M. and Thompson R. C. (1980) Pollution records from sediments of three lakes in New York State. *Geochim. Cosmochim. Acta* **44**, 333–339.

Warren L. A. and Haack E. A. (2001) Biogeochemical controls on metal behavior in freshwater environments. *Earth-Sci. Rev.* **54**, 261–320.

Webb S. M., Leppard G. G., and Gaillard J.-F. (2000) Zinc speciation in a contaminated aquatic environment: characterization of environmental particles by analytical electron microscopy. *Environ. Sci. Technol.* **34**, 1926–1933.

WebElements. http://www.webelements.com/webelements/elements/text/Cu.html (accessed June 16, 2002).

Wedepohl K. H. (1968) Chemical fractionation in the sedimentary environment. In *Origin and Distribution of the Elements* (ed. L. H. Ahrens). Pergamon Press, New York, pp. 999–1015.

Wedepohl K. H. (1995) The composition of the continental crust. *Geochim. Cosmochim. Acta* **59**, 1217–1232.

Wessels M., Lenhard A., Giovanoli F., and Bollhofer A. (1995) High resolution time series of lead and zinc in sediments of Lake Constance. *Aquatic Sci.* **57**, 291–304.

Wilber G. G., Smith L., and Malanchuk J. L. (1992) Emissions inventory of heavy metals and hydrophobic organics in the Great Lakes Basin. In *Fate of Pesticides and Chemicals in the Environment* (ed. J. L. Schnoor). Wiley, New York, pp. 27–50.

Williams T. M. (1992) Diagenetic metal profiles in recent sediments of a Scottish freshwater loch. *Environ. Geol. Water Sci.* **20**, 117–123.

Wren C. D., Maccrimmon H. R., and Loescher B. R. (1983) Examination of bioaccumulation and biomagnification of metals in a Precambrian shield lake. *Water, Air, Soil Pollut.* **19**, 277–291.

Wu J. and Boyle E. A. (1997) Lead in the western North Atlantic Ocean: completed response to leaded gasoline phaseout. *Geochim. Cosmochim. Acta* **61**, 3279–3283.

Yasui M., Strong M. J., Ota K., and Verity M. A. (1996) *Mineral and Metal Neurotoxicology*. CRC Press, Boca Raton.

Zhang J., Huang W. W., and Wang J. H. (1994) Trace-metal chemistry of the Huanghe (Yellow River), China—Examination of the data from *in situ* measurements and laboratory approach. *Chem. Geol.* **114**, 83–94.

9.04
Geochemistry of Mercury in the Environment

W. F. Fitzgerald
University of Connecticut, CT, USA

and

C. H. Lamborg
Woods Hole Oceanographic Institution, MA, USA

9.04.1 INTRODUCTION

Mercurial, the metaphor for volatile unpredictable behavior, aptly reflects the complexities of one of the most insidiously interesting and scientifically challenging biogeochemical cycles at the Earth's surface. Elemental mercury is readily recognized as a silvery liquid at room temperature. Its gas phase is important geochemically, since mercury and some of its compounds have relatively high vapor pressures. Mercury (Hg, from the Latin *hydrargyrum* or "watery silver") is sulfur loving (i.e., chalcophilic) and extremely active biologically. It is mobilized tectonically, and significant deposits are found in mineralized regions characterized by subduction zones and deep-focus earthquakes (Schlüter, 2000). Many of the major deposits are shown in Figure 1 (Kesler, 1994).

The remarkable and useful qualities of mercury and its major mineralized form (cinnabar, HgS) have been well known for thousands of years. The Almadén mine in Spain, for example, the "richest known single source of cinnabar and quicksilver," has been active for more than 2,500 yr (Goldwater, 1972). Mercury products (e.g., thermostats, batteries, switches, fluorescent lighting) and applications (e.g., chlor-alkali production, dentistry, pharmaceuticals, catalysis) have been a practical part of modern life, while the wastes have been quite detrimental. Today, mercury emissions associated with fossil fuel burning, especially coal, and high-temperature combustion processes (e.g., municipal waste incineration) represent the primary sources of pollutant mercury to the environment on a global scale. As a result, mercury emissions since the mid-nineteenth century appear to have been in step with increases in emissions of CO_2 (Lamborg *et al.*, 2002a). During the past decade, emissions, discharges, and nonpoint source inputs of mercury appear to have peaked and may be diminishing in many developed countries. Unfortunately, and on a global scale, declines are countered by increases in anthropogenic mercury releases from developing nations, particularly in Asia (E. G. Pacyna and J. M. Pacyna, 2002). Additionally, environmental gains through pollution controls, remediation, and regulations are tempered by the large reservoir of historic mercury, the pollution "legacy," residing in watersheds and sediments of many terrestrial, freshwater, and marine environments.

Although mercury has been used in "spring tonics," as a "cure" for syphilis, and a panacea for other afflictions, it is now recognized as a highly toxic trace metal that concentrates in aquatic food webs. According to Clarkson (1997), the principal human exposure to inorganic mercury species is from elemental mercury vapor, which is derived principally from industries such as gold and silver mining and chlor-alkali plants, and from dental amalgams. Deleterious health effects (e.g., "Mad Hatters Disease") have been known since ancient times, and as Clarkson states, "severe exposure results in a triad of symptoms, erethism, tremor, and gingivitis." Today, however, the principal mercury-related human health concern is associated with exposure to the highly neurotoxic organomercury species, monomethylmercury, MMHg. This exposure is due almost entirely to consumption of fish and fish products (Fitzgerald and Clarkson, 1991; National Research Council, 2000). Inorganic mercury, whether natural or pollution derived, can be readily methylated in aquatic systems. Mercury methylation appears to be predominately biotic, although some abiotic production is likely in natural waters (Benoit *et al.*, 2003; Gårdfeldt *et al.*, 2003). *In situ* methylation of "reactive" or bioavailable mercury by sulfate-reducing bacteria has been documented to result in the accumulation of MMHg in freshwater food-webs and fish (e.g., Westöö, 1966; Wiener *et al.*, 1990a; Gilmour and Henry, 1991; Watras and Bloom, 1994; Watras *et al.*, 1994; Hall *et al.*, 1998). This linkage is quite evident in the elevated MMHg in fish from reservoirs created by dam construction and subsequent flooding of landscapes (e.g., Cox *et al.*, 1979; Bodaly *et al.*, 1984; Tremblay *et al.*, 1998). A similar process is thought to occur in the marine water column, possibly first through the formation of dimethylHg (DMHg) followed by decomposition to MMHg (Mason and Fitzgerald, 1993; Mason *et al.*, 1998; Mason and Sullivan, 1999). Microbially mediated methylation continues to amplify the insidiousness of current and historic mercury pollution and health risks to wildlife and humans. Indeed, toxicologically, methylation of inorganic mercury is the most important transformation affecting the behavior and fate of mercury in aquatic systems. Its accumulation in freshwater and marine fish can reach levels that not only pose a threat to human health (Grandjean *et al.*, 1997; Davidson *et al.*, 2000), but can reduce the reproductive success of piscivorous wildlife (e.g., Scheuhammer, 1991; special section of *Environmental Toxicology and Chemistry*, 1998, **17**/2: 137–227, 12 papers, M. W. Meyer, editor) and the fish themselves (Wiener *et al.*, 1990b; Wiener and Spry, 1996; Hammerschmidt *et al.*, 2002). Wiener *et al.* (2002) presented an up-to-date review of mercury toxicology in a number of different animal species.

MMHg poisoning is known as "Minamata disease." Unfortunately, and between 1950 and 1975, major industrially related mass poisonings, severe debilitation, and many deaths occurred in Minamata and Niigata, Japan and in Iraq. The Japanese poisonings resulted from consumption of locally caught fish and seafood that had been

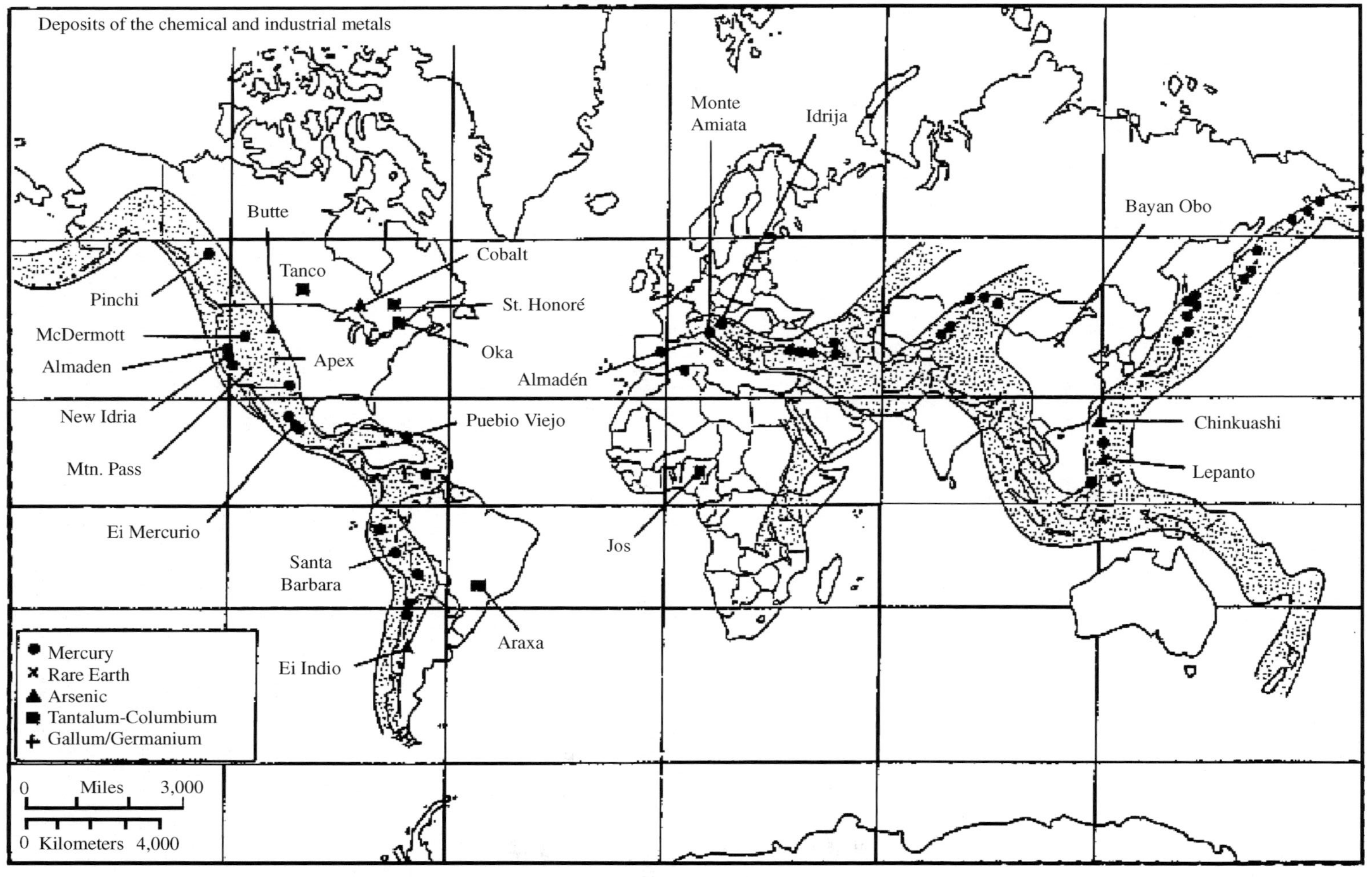

Figure 1 The global Hg belt (source Kesler, 1994).

contaminated principally by MMHg discharged with wastewater from factories making acetaldehyde (Chisso Co. Ltd and Showa Denko Co. Ltd.). The MMHg was synthesized abiotically as a by-product during the production of acetaldehyde (inorganic mercury was used as a catalyst). In the Iraqi tragedy, the source was contaminated bread, which had been made with flour, unknowingly milled from wheat treated with MMHg as a fungicide (Bakir *et al.*, 1973). There is an extensive scientific, medical, and general literature, as well as news accounts of these tragedies. The following works on the Japanese poisonings provide an overview and useful starting point for a more in-depth examination: Smith (1975), Japan Public Health Association (2001), The Social Scientific Study Group on Minamata Disease (2001) and George (2001). There are other examples of severely mercury contaminated sites and the interested reader is directed to the volume entitled *Mercury Contaminated Sites—Characterization, Risk Assessment and Remediation*, edited by Ebinghaus *et al.* (1998), which also contains descriptions of the Minamata situation.

In the mid-twentieth century, Goldschmidt (1954) cryptically summarized knowledge of the environmental cycling of mercury as, "not much information is available concerning the geochemistry of mercury." Moreover, and as he emphasized for that period, "most of the modern analytical data are due to A. Stock and coworkers," which were derived from their pre-World War II studies (e.g., Stock and Cucuel, 1934). In contrast, and at the beginning of the twenty-first century, dozens of environmentally related mercury publications appear yearly. There is an abundance of distributional data, an enhanced knowledge of the biogeochemical cycling of mercury and the impact of anthropogenic activities, a fuller appreciation of the utility, dangers, and complexities characterizing the mobilization, interactions, and fate of this biologically active element, and an awareness of daunting challenges inherent in studying a metal that includes a gas phase as a major feature of its biogeochemistry at the Earth's surface. The potential risks of human exposure to MMHg, especially prenatally, and the potential deleterious ecological consequences from localized to global-scale mercury pollution have given much impetus to mercury studies and regulatory activities internationally. There have been six international conferences on "Mercury as a Global Pollutant," since 1990. The city of Minamata, the venue for the sixth conference in 2001, provided conferees with poignant evidence of the tragic legacy of mercury contamination. The abstracts and publications from these broadly based meetings, which include basic biogeochemistry, environmental and pollution-related studies, ecological and human toxicology, and analytical developments, chronicle the rapid worldwide expansion of mercury research and knowledge.

This chapter focuses principally on the low-temperature environmental biogeochemistry of mercury. We are presenting current understanding of mercury cycling at the Earth's surface (soils, sediments, natural waters, and the atmosphere). Our coverage, as appropriate, will include the anthropogenic interferences (i.e., mercury pollution) and biological mediation, which affect significantly the speciation, behavior, and fate of mercury in the environment. The reader will be referred to other chapters for complementary information on mercury geochemical cycling in and among the various earthly reservoirs.

9.04.1.1 The Global Mercury Cycle

Major features of the global mercury cycle have been illustrated using the relatively simple three-reservoir mass balance developed by Mason, Fitzgerald, and Morel in 1994 ("MFM"). Refinements are considered in Section 9.04.7. The MFM model provides estimates of natural and anthropogenic fluxes and an assessment of impact from human-related mercury emissions on the natural cycle for 1990 (Figure 2(a)). The pre-industrial situation is shown in Figure 2(b). It is evident that the atmosphere and oceans play important roles in the distribution and redistribution of mercury at the Earth's surface. Indeed, atmospheric mercury deposition to the oceans greatly exceeds riverine inputs. The MFM model also suggests that the natural cycle of mercury has been severely perturbed. Human-related mercury emissions dominate the cycle such that most of the mercury in the atmosphere and surface oceans is anthropogenic. Moreover, the integrated estimate for total loadings from globally dispersed, anthropogenic mercury emissions between 1890 and 1990 is 1,000 Mmol (200 kt). The MFM simulation predicts that ~95% of this mercury is sequestered in terrestrial systems, with the remainder in the oceans and atmosphere. The MFM analysis suggested that about half of the modern pollution-related emissions (20 Mmol yr^{-1}) enter the global cycle while the other half is deposited near sources. The MFM estimate for the globally dispersed anthropogenic contribution is similar to the most recent estimate of 9.6 Mmol yr^{-1} (for 1995) reported by Pacyna and Pacyna (2002).

Elemental mercury, a monatomic gas, is the dominant atmospheric form and has a long residence time in the troposphere (>1 yr; Fitzgerald *et al.*, 1981; Slemr *et al.*, 1981; Lindqvist *et al.*, 1991; Lamborg *et al.*, 2000, 2002a). Such longevity allows emissions of mercury to the atmosphere from natural and

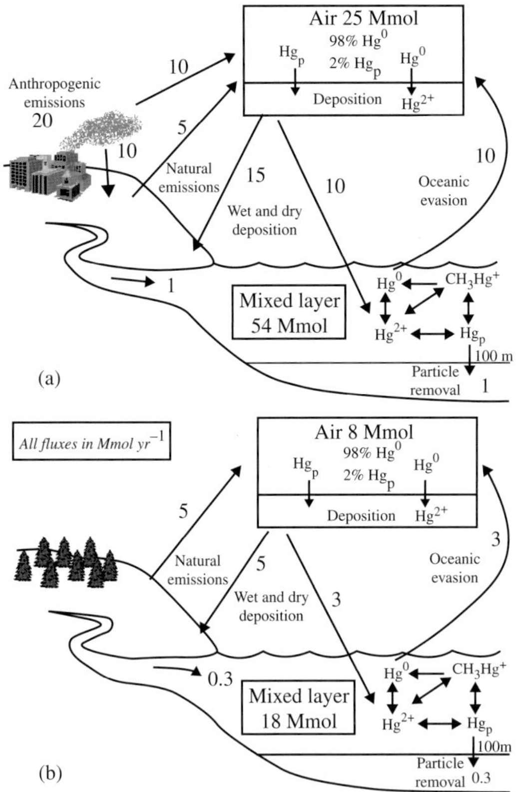

Figure 2 Fluxes of Hg on a global scale. a) The current condition and (b) the pre-industrial condition. All fluxes in Mmol of Hg yr^{-1}. Note the balance of deposition to and evasion from the ocean, making soils the primary sink on decade/century timescales. One result of human activity has been a tripling of Hg in the atmosphere and surface ocean (after Mason *et al.*, 1994).

anthropogenic sources to be dispersed widely across the Earth's surface. The redox couple of Hg^0 with the stable mercuric ion (Hg^0/Hg^{2+}; $E° = 0.85$ V) is similar to that of the Fe(II)/Fe(III) couple, thereby providing the potential for dynamic oxidation and reduction cycling in the range of common environmental redox (i.e., pe) conditions. A unique and important example of this is the biologically and abiotically mediated reduction of Hg^{2+} resulting in widespread supersaturation of Hg^0 in freshwater and saltwater and subsequent evasion (Kim and Fitzgerald, 1986; Mason and Fitzgerald, 1993; Amyot *et al.*, 1997; Rolfhus, 1998). Field and laboratory observations in aqueous systems have documented the important influence of photochemical and bacterial reactions on the *in situ* reduction of Hg^{2+} (Mason and Fitzgerald, 1993; Amyot *et al.*, 1997; Rolfhus, 1998; Rolfhus and Fitzgerald, 2001), and therefore link Hg^0 production and evasion from the ocean to the complex biogeochemical cycles of carbon, nitrogen, phosphorus, and sulfur. Through evasion, the ocean prolongs the residence of mercury at the Earth's surface and exacerbates the human perturbation of the global mercury cycle. The MFM analysis shows such oceanic emission of Hg^0 to be substantial (10 Mmol) and comparable to the atmospheric input of mercury to the oceans.

Although the marine biogeochemical cycling of mercury is greatly simplified in the MFM simulation, the importance of the ocean as both a sink for atmospheric deposition as well as a substantial source of mercury to the atmosphere (Kim and Fitzgerald, 1986; Nriagu, 1989; Lindqvist *et al.*, 1991; Mason *et al.*, 1994) is consistent with models that use more realistic physical and biogeochemical oceanic dynamics (Hudson *et al.*, 1995; Mason *et al.*, 2001; Lamborg *et al.*, 2002a).

The oceans actively and significantly participate in the transport and transformation of this toxic metal. Thus, there is an obvious need to increase knowledge and understanding concerning the biogeochemical cycling of mercury and MMHg and the impact of anthropogenically related inputs in the marine environment. A cursory examination of papers presented at the international mercury conferences is sufficient to show that there has been much interest in and support for studies examining linkages between the cycling of mercury in the atmosphere, anthropogenic emissions/discharges, deposition to terrestrial systems, and the bioaccumulation of MMHg in freshwaters (e.g., *Mercury in Temperate Lakes Program*—Watras *et al.*, 1994; *METAALICUS Project*; USEPA Mercury Study Report to Congress, 1997; Ebinghaus and Krüger, 1996; Petersen *et al.*, 1995; *The Nordic Network*—Lindqvist *et al.*, 1991; Iverfeldt, 1991a). In contrast, mercury cycling in the oceans has received scant attention. This is unfortunate as there is compelling evidence, for example, that the mercury content in fish has been increasing over the past several decades in the North Atlantic (e.g., Monteiro and Furness, 1997). Moreover, in 2001, growing recognition of human exposure to MMHg from marine fish and fish products prompted the US Food and Drug Administration (USFDA) to place four pelagic marine fish (tilefish, king mackerel, swordfish, and shark) on their consumer advisory list (USFDA Consumer Advisory, 2001). The advisory emphasizes the need for women of childbearing age to limit their consumption of marine fish to 12 ounces per week.

A more stringent fish consumption advisory reference dose (Schober *et al.*, 2003) was issued by the US Environmental Protection Agency (USEPA) in 2001. Women of childbearing age were cautioned to limit their MMHg intake to 0.1 μg kg^{-1} of body weight per day. It is illuminating to translate this recommendation into fish consumption. A summary of the MMHg content in some major marine fish and shellfish is presented in Table 1. Using fresh tuna with an average MMHg concentration of 0.32 $\mu g\ g^{-1}$ wet weight (Table 1) as an example, the recommended weekly consumption for a 50–70 kg woman of childbearing age would be 35–49 μg or the equivalent of ~110–155 g (~4–5.5 oz, half a typical can) of fresh tuna. This is much smaller than the USFDA recommendation, which is equivalent to 0.5 $\mu g\ kg^{-1}$ of body weight per day (however, a maximum of 12 oz of fish per week is recommended). The current provisional advisory from the Joint Food and Agricultural Organization/World Health Organization Expert Committee on Food Additives (JECFA, 2000) is 0.5 $\mu g\ kg^{-1}$ of body weight per day. For further information regarding MMHg and fish consumption, the reader is referred to the useful and informative publication from the US National Academy of Sciences entitled "Toxicological Effects of Methylmercury" (NAS, 2000). The examining committee from the National Research Council of the NAS reached the following consensus: "the value of EPA's current reference dose for MeHg, 0.1 $\mu g\ kg^{-1}$ per day is scientifically justifiable for the protection of public health."

In summary, mercury entering the marine environment may continue to actively participate in aquatic chemistry, while much of the mercury deposited from the atmosphere to terrestrial systems is sequestered. Given the importance of the oceans as a whole in global mercury cycling, and international concerns and issues regarding human exposure to MMHg through marine fish and seafood consumption, the current situation is one of insufficient study and undersampling. Indeed, there have been relatively few open-ocean cruises to examine mercury biogeochemistry (e.g., Kim and Fitzgerald, 1986; Gill and Bruland, 1987; Gill and Fitzgerald, 1988; Mason and Fitzgerald, 1993; Dalziel and Yeats, 1985; Dalziel, 1992; Cossa *et al.*, 1997; Mason *et al.*, 1998; Mason and Sullivan, 1999; Lamborg *et al.*, 1999; Leermakers *et al.*, 2001), and almost no longer-term, seasonally oriented mid-ocean studies. Even in the more accessible near-shore zones, the biogeochemistry of mercury is understudied. In this regard, estuaries and adjacent coastal waters, as regions of high biological productivity, MMHg production and fishery activity, are of special interest. They are major repositories for natural and pollutant mercury (see Section 9.04.6).

Table 1 Levels of total mercury ($\mu g\ g^{-1}$ wet weight) in seafood (USFDA, 2001). Most (>95%) of the total mercury in edible fish tissue is MMHg).

Fish species	*Mean (range)*	*n*[a]
Tilefish	1.45 (0.65–3.73)	60
Swordfish	1.00 (0.10–3.22)	598
King Mackerel	0.73 (0.30–1.67)	213
Shark	0.96 (0.05–4.54)	324
Tuna (fresh or frozen)	0.32 (ND–1.30)	191
Tuna (canned)	0.17 (ND–0.75)	248
Atlantic cod	0.19 (ND–0.33)	11
Pollock	0.20 (ND–0.78)	107
Mahi mahi	0.19 (0.12–0.25)	15
American lobster	0.31 (0.05–1.31)	88

Sources: Grieb *et al.* (1990), Bloom (1992), and Hammerschmidt *et al.* (1999). ND denotes that the mercury level was not detectable.
[a] Number of samples analyzed.

9.04.2 FUNDAMENTAL GEOCHEMISTRY

In this section, we consider aspects of the fundamental geochemistry and biogeochemistry of mercury used later in the chapter. These topics include: (i) solid Earth abundance and distribution, (ii) isotopic composition and recent advances in mercury cosmochemistry, (iii) the formation and distribution of minable mercury deposits, and (iv) mercury in fossil fuels.

9.04.2.1 Solid Earth Abundance and Distribution

A summary of mercury data from the report of Turekian and Wedepohl (1961) is shown in Table 2. Due to the chalcophilic nature of its associations, mercury is found in higher abundances in intrusive magmatic rocks and locations of subaerial and submarine volcanism. Peak concentrations in these rocks may be as high as several percent in ore-grade minerals (e.g., 35% mercury in sphalerite; Ozerova, 1996). Also as a result of this association, the distribution of highest mercury concentrations in rocks at the surface and near surface mirrors regions of current and past tectonic activity and has been described as the "global mercury belt" (e.g., Gustin *et al.*, 2000). As noted in greater detail in Section 9.04.3, mercury concentrations in soils weathered from

Table 2 Mercury content of selected rocks and sediments.

Rock type	*Hg content* (ppm)
Ultrabasic igneous	0.0X[a]
Basaltic rocks	0.09
High- and low-Ca granites	0.08
Syenites	0.0X
Shales	0.4
Sandstones	0.03
Carbonates	0.04
Deep-sea carbonate	0.0X
Sediment	
Deep-sea clays	0.X

Source: Turekian and Wedepohl (1961).
[a] The X notation indicates order of magnitude estimate.

this material can be very high as well (e.g., Steamboat Springs, NV, USA: 1.2–14.6 ppm; Gustin *et al.*, 2000) and represent a potentially significant source of mercury to the atmosphere at a variety of spatial scales through low-temperature volatilization. These rocks and their weathered products are rich in other metals as well, and emission of mercury from soils and rock has been used as a tool for large-scale ore and petroleum exploration as well as an indicator of tectonic activity (e.g., McCarthy, 1968; Varekamp and Buseck, 1983; Klusman and Jaacks, 1987). These general characterizations are significant then, as we later consider the cause for concentrations of mercury in soils removed from the mercury belt that are also elevated above the crustal abundances.

Also indicated in Turekian and Wedepohl's compilation is a relatively high concentration of mercury in sedimentary material rich in organic carbon, such as shales. Mercury associations with organic matter are considered in later sections (Sections 9.04.5 and 9.04.8). Here, we stress that such associations can lead not only to higher concentrations of mercury in these types of rock units but also in the transport of mercury away from sites of sediment accumulation as a result of petroleum movement (White, 1967; Fein and Williams-Jones, 1997). The mercury content of major rock types has not been systematically revisited since Turekian and Wedepohl's report, and there is evidence that some of their values may be overestimates. As an example, the recent recovery and analysis of glacial till in Glacier Bay National Park by Engstrom and Swain (1997) indicated much lower concentrations of mercury (<10 ppb). It is reasonable to assume, however, that mercury concentration trends across rock types suggested by Turekian and Wedepohl and the geochemistry they suggest are valid.

9.04.2.2 Isotopic Distributions

Mercury has a relatively even distribution of its seven stable isotopes (196, 0.15%; 198, 10.0%; 199, 16.7%; 200, 23.2%; 201, 13.2%; 202, 29.8%; 204, 6.8%; Friedlander *et al.*, 1981; Lauretta *et al.*, 2001). This pattern presented cosmochemists with a formidable task when mercury isotopic distributions in meteorites were examined (e.g., Jovanovic and Reed, 1976; Thakur and Goel, 1989). Analytical difficulties apparently resulted in inaccurate determinations of the bulk abundance and isotopic composition of some meteorites, leading to the so-called "mercury problem"; examined meteorites did not show the same bulk abundance and isotopic distribution as terrestrial material (Grevesse, 1970; Lauretta *et al.*, 1999). Subsequent advances in mass spectrometry, and especially the development of multi-collectors, have shown that the isotopic distributions of mercury in terrestrial and extraterrestrial material are very similar (e.g., Allende meteorite: 0.03–0.3 ppm; Lauretta *et al.*, 2001).

With so many isotopes from which to choose, one might expect examination of mercury isotopic fractionation in natural media to be a profitable area of research as it has been, for example, with lead, light metals, and nonmetals (e.g., Alleman *et al.*, 2001; Richter *et al.*, 1992; Hoefs, 1980). To date, this has not been widely explored, but there is little indication thus far of either biological fractionation in aquatic systems useful in tracing biogeochemistry or of geogenic fractionation leading to isotopic signatures of particular ores (Klaue *et al.*, 2000; Evans *et al.*, 2001). The wide range of stable isotopes is being used in deliberate addition experiments ranging from benchscale to whole-watershed-scale (e.g., Hintelmann and Evans, 1997; Hintelmann *et al.*, 2002). These advances will be highlighted in Section 9.04.8.

The radioisotope ^{203}Hg has played an important role in laboratory investigations of mercury biogeochemistry (e.g., Gilmour and Riedel, 1995; Stordal and Gill, 1995; Costa and Liss, 1999). Continued production of ^{203}Hg-enriched material has been curtailed recently, and thus future mechanistic studies will instead feature the use of stable isotopes.

9.04.2.3 Minable Deposits

As mentioned, higher mercury concentrations in rock and soil are associated with the global mercury belt. However, the occurrence of minable deposits is not continuous along this belt. In addition to Almadén (Spain), the most productive mercury mines included Idrija (Slovenia), New Almaden (California, USA), and Huancavelica (Peru). Very high concentrations of mercury have been reported to be associated with oceanic hydrothermal sulfide chimneys and their weathered remains (e.g., Koski *et al.*, 1994; Ozerova, 1996; Stoffers *et al.*,1999). In all of these cases, mercury occurs almost exclusively as cinnabar (red HgS), with smaller amounts of metacinnabar (black HgS) and elemental mercury (often as inclusion with HgS in minerals such as sphalerite and chalcopyrite). HgS is extremely insoluble (log K_{so} (cinnabar) $= -36.8$; Martell *et al.*, 1998) under typical surface water conditions, and thus transport of mercury from these mineral rich environments at the Earth's surface generally involves sediment transport (e.g., Ganguli *et al.*, 2000). The fate of cinnabar in anoxic sediments is addressed in Section 9.04.5. Transport of mercury to and from ore bodies invariably involves hydrothermal systems in the subsurface, with HgS solubility strongly controlled by fluid pH, temperature, chloride, sulfide, and organic carbon contents (White, 1967; Varekamp and Buseck, 1984).

9.04.2.4 Occurrence of Mercury in Fossil Fuels

Even with the stated stability of cinnabar, mercury is mobile in the surface environment. This is indicated by the relatively high concentration of mercury in organic-rich deposits, such as fossil fuels and shales. As we will explore later, mercury has high affinities for organic carbon as well as sulfide. Interest in fossil fuel recovery and burning as a source of mercury to the environment has led to a few published studies in this area and we summarize some of these data in Table 3. With notable exceptions, concentrations in coal appear higher than those in oil, suggesting preferential burial of mercury in terrestrial and coastal systems rather than in pelagic marine environments. Though mercury concentrations are not as high in various refined petroleum materials as in coal, the oil and gas recovery process often liberates large amounts of mercury leading to localized contamination of marine sediments (e.g., Grieb *et al.*, 2001). The concentration of mercury is sufficiently high in some crude petroleum materials to also pose a corrosion hazard to the drilling and transportation apparatus and represents a significant engineering problem (e.g., Wilhelm, 1999; Bloom, 2000).

9.04.3 SOURCES OF MERCURY TO THE ENVIRONMENT

There have been few well-designed studies to constrain mercury emission estimates from natural sources and allow global extrapolations. Indeed, some work has been extraordinarily misleading. For example, in flawed studies based on the accumulation of mercury and other metals in glacial ice, Jaworowski *et al.* (1981) estimated the annual mercury flow into the atmosphere at 1,000 Mmol with an anthropogenic contribution at 50 Mmol! Low- and high-temperature volatilization processes were offered by Jaworowski *et al.* as a potential explanation for the huge natural fluxes of mercury. Such inaccuracies as well as the paucity of reliable investigations begs the question "What is the appropriate flux range for global mercury emissions from sources such as volcanism, biomass burning, and low-temperature volatilization from natural waters and soils?" The MFM simulation (Mason *et al.*, 1994) of the global mercury cycle estimates natural terrestrial emissions at 5 Mmol annually. Such emissions would include inputs from subaerial volcanism under erupting and non-erupting conditions, and the pre-industrial mercury fluxes from mineralized regions, forest fires, biological activities, and natural waters. Today, volcanism and volatilization from mineralized regions may be the only purely natural sources of mercury, because, and as illustrated in Figure 2, anthropogenic mercury contamination is present throughout the atmosphere, biosphere, the terrestrial realm, and the hydrosphere. Thus, a portion of the emissions from these secondary reservoirs represents recycled pollutant mercury; this component has often been overlooked when the source strengths from "natural sources" have been compared and assessed relative to modern anthropogenic mercury inputs.

Table 3 Mercury content of fossil fuels.

Sample type	*Total Hg* ($ng\ g^{-1}$)
Coal	
China[a]	220
Std. Ref. Mat.[b]	77.4–433.2
Global estimate[c]	20–1,000
Unrefined petroleum[d]	
Crude oil	<d.l. to >7,000
Condensate	<d.l. to >12,000
Refined petroleum[d]	
Light distillates	1 ± 3
Utility fuel oil	1 ± 1
Asphalt	0.3 ± 0.3
Gasoline	0.2–3
Diesel	0.4–3
Kerosene	0.04
Heating oil	0.59
Naphtha	3–60
Petroleum coke	0–250

[a] Wang *et al.* (2000). [b] Long and Kelly (2002). [c] Pacyna and Pacyna (2002). [d] Wilhelm (2001) and the references therein.

9.04.3.1 Volcanic Mercury Emissions

The determination of global volcanic mercury fluxes from direct measurements is at best a hazardous, expensive, labor-intensive, and, perhaps, impossible task. Reasonably well-constrained global estimates, however, can be achieved through use of elemental ratios and a geochemical indexing approach. Sulfur provides an appropriate index, since it is a major component of volcanic emanations, and there is agreement on its global flux to the atmosphere. An example of this approach is the work of Patterson and Settle (1987), who combined Pb/S, Tl/S, and Bi/S ratios measured in volcanic gases collected under eruptive and quiescent (fumarolic) conditions with the global volcanic sulfur flux to "approximate global volcanic emissions of these three metals to the atmosphere." One of the present authors (WFF), measured mercury during the Patterson and Settle study. Results from this research were presented at two conferences (Fitzgerald, 1981, 1996), and Fitzgerald's estimate of 40 t yr^{-1} appears in the paper by Patterson and Settle (1987). Given the potential importance of volcanic mercury inputs, and the lack of investigations, a detailed description of the mercury study is included here.

Patterson, Settle, Buat-Menard (University of Bordeaux), Fitzgerald, and colleagues (see Acknowledgments section) evaluated volcanic metal fluxes using an experimental design based on the hypothesis that metal volatilization would be dependent on temperature, sulfur and halogen composition of magma and mobilization by volcanic gases. Therefore, volcanoes and fumaroles were selected to provide a range of temperatures and S/Cl ratios. The characteristics of the volcanic gas samples are tabulated in Table 4. The four volcanoes examined were Kilauea (Hawaii, USA), Mt. Etna (Sicily, Italy), Vulcano (Aeolian Islands, Italy), and White Island (New Zealand).

Plumes were sampled for mercury using a simple gas train consisting of a pre-blanked (pyrolyzed) glass fiber prefilter stage for particulate phases, followed by two gold traps arranged in tandem to collect gaseous mercury (Fitzgerald and Gill, 1979; Bloom and Fitzgerald, 1988). The fumarolic collections were made with a gas sampling train designed by Clair Patterson and modified for mercury (Patterson and Settle, 1987; shown here in Figure 3). Any mercury associated with aerosols or the gas phase that escaped the two cold traps was collected by a gas-sampling train analogous to the plume samplers. The gas was pumped through this system at $\sim 1 \text{ L min}^{-1}$.

In general, and as illustrated for the 1977 study of 100 °C fumarole at Kilauea, essentially all of the mercury and the other metals are trapped in the 0° condensate (Table 5). Thus, fumarolic gas collectors for metal studies can be simplified. The observed Hg/S weight ratio in the Kilauea condensate was 0.9×10^{-6}. The Hg/S ratios from the other investigations are summarized in Table 4. Values range from 0.7 to 14×10^{-6} with a suggestion that Hg/S may increase as the S/Cl decreases. Dedeurwaerder *et al.* (1982) were part of the Mt. Etna study and their results for the Bocca Nuevo plume are included in the summary. Their plume-sampling apparatus was based on the Fitzgerald technique and similar to that described above. The sulfur-flux weighted mean volcanic Hg/S from all locations is 5×10^{-6}.

Estimates of global volcanic sulfur emissions are summarized in Table 6. We have chosen a value of $9 \times 10^6 \text{ t S yr}^{-1}$ as representative of the recent estimates. Therefore, by applying the determined Hg/S ratio, a global mercury flux from subaerial volcanism is estimated to be $\sim 45 \text{ t yr}^{-1}$, or 0.23 Mmol annually. These average emissions are only 5% of the natural flux of 5 Mmol yr^{-1} estimated by Mason *et al.* (1994). Thus, and under long-term mean conditions, other types of terrestrial volatilization processes for mercury would dominate. Given this conclusion, it is important to place additional constraints on the validity of the 45 t yr^{-1} estimate for subaerial volcanic mercury emissions.

First, Varekamp and Buseck (1986) reported an average Hg/S weight ratio of 7.4×10^{-6} for emissions under non-erupting conditions from Mts. Hood, Shasta, and St. Helens in the United States, and from Mt. Etna. In 1981, these investigators reported a very high Hg/S weight

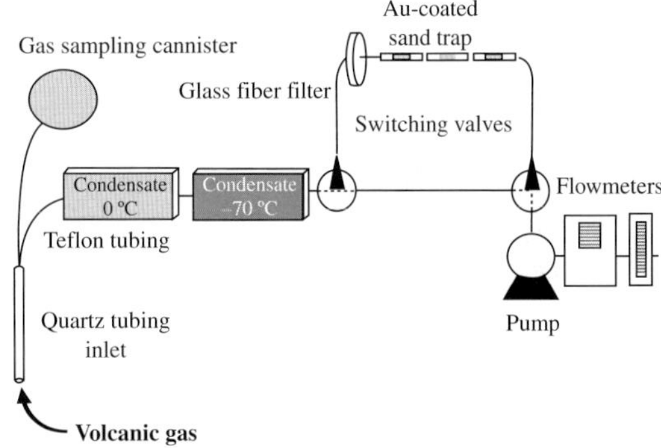

Figure 3 Hg sampling apparatus used to collect fumarole gas samples (after Patterson and Settle, 1987).

Table 4 Volcano sampling locations and gas characteristics.

Volcano	*Date*	*Type of gas*	*Temp.* (°C)	*S* *(wet gas)* (mg S m^3)	*S/Cl* (weight)	*Cl/F* (weight)	*Hg/S* (weight)	*Hg/Bi* (weight)
Kilauea Caldera	6/23/77	Fumarole	100	22,800	40	2	0.9	
Kilauea Pu'u O'o	5/12/84	Eruptive plume	1,140	110	9		0.7	0.0019
Mt. Etna Bocca Nuova[a]	6/7/80	Eruptive plume	1,100	4.3	0.7	8	11–14	
Mt. Etna SE Crater	6/8/80	Eruptive plume	1,100	2.5	0.7	4	<13	
Volcano Site A	6/11/80	Fumarole	280	5,500	0.3	90	0.9	
White Island Site A	7/22/83	Fumarole	180	33,000	45	30	0.9	
White Island Site B	7/22/83	Fumarole	590	54,000			~0.44	0.024
White Island Site C	7/22/83	Fumarole[b]	650			180		0.0015

After Patterson and Settle, 1987.
[a] Dedeurwaerder *et al.* (1982). [b] Collected using plume filter device.

Table 5 Mercury and sulfur concentrations in volcanic gas at sulfur fumarole site, Kilauea caldera.

Collection	*Hg concentration*	*Total Hg* (ng)	*Total S* (g)	*Hg/S ratio* (weight)
0 °C condensate	120 ng g^{-1}	59,200	4.0	14.8×10^{-6}
−70 °C condensate	10 ng g^{-1}	150	0.1	1.5×10^{-6}
Volcanic gas free of water	0.88 ng g^{-1}	230	60	3.9×10^{-9}
Total		59,600	64	9.3×10^{-7}
Ambient air	10.2 ng m^{-3}			

Table 6 Estimates of the global annual sulfur flux from volcanic activity.

Method	*S Flux* (10^6 t yr^{-1})	*References*
Rate of S loss per volcano	5.0	Stoiber and Jepson (1973)
	3.8	Cadle (1975)
	4–5	Friend (1973)
S measurements from selected volcanoes	9.4 Nonexplosive (4.5) Explosive (4.9)	Stoiber *et al.* (1987)
Literature review	9	Spiro *et al.* (1992)
Satellite survey of SO_2 emissions	6.5 Nonexplosive (4.5; Stoiber *et al.*, 1987) Explosive (2.0)	Bluth *et al.* (1993)
Satellite and modeling	7.5–10.5	Halmer *et al.* (2002)

ratio of $6,000 \times 10^{-6}$ for Mt. St. Helens under erupting conditions (Varekamp and Buseck, 1981). However, the mercury and sulfur measurements were not measured simultaneously and, according to Varekamp (personal communication), should not be used for a global-scale analysis of this kind. Thus, we suggest that a mean Hg/S ratio of 5×10^{-6} and the annual emission estimate of 45 t is of the appropriate magnitude. Nriagu and Becker (2003), using the same Hg/S ratio and a more detailed volcanic SO_2 emissions inventory, derived much the same conclusion (~95 t yr^{-1}; 60% eruptive, 40% degassing).

Second, the scale of volcanic mercury fluxes can be approximated indirectly. For example, using the Hg/Bi ratios listed in Table 4 as well as the Lambert *et al.* (1988) estimate for annual global bismuth emissions of 1,500 t, we obtain a range for the global volcanic mercury flux of 2–36 t yr^{-1}. Hinkley *et al.* (1999) suggests that the Lambert *et al.* values are a factor of 3 or 4 too high.

A comparable value of 18 t annually is obtained using Olafsson's (1975) estimate of mercury emissions (7×10^5 g Hg/6×10^{14} g ejecta) for the volcanic eruption at Heimay, Iceland, and an estimate of 6 km^3 (~15×10^{15} g) for annual, subaerial lava production (~20% of total annual lava production being subaerial; Crisp, 1984; Varekamp, personal communication).

Thus, annual global volcanic mercury emissions estimated or measured in several ways are less than 0.5 Mmol (100 t), and our work yielded an average value of 0.23 Mmol (45 t). This scaling serves two purposes: (i) it provides a framework for further, needed studies of mercury cycling associated with volcanism, and (ii) it provides a reasonably well constrained estimate for global emissions for modeling and assessment of human perturbations of the natural mercury cycle. It is clear that average volcanic mercury inputs are small relative to estimates for modern anthropogenic mercury fluxes (~9.6 Mmol; Pacyna and Pacyna, 2002) to the global mercury cycle. Indeed, and using an atmospheric residence time of 1.25 yr (Lamborg *et al.*, 2002a), the average yearly terrestrial deposition of mercury from volcanism would be ~0.07 μg m^{-2}. Since volcanic eruptions vary in time and space, large-scale eruptions might be apparent in natural archives such as lake sediments (Chapter 9.03), and ice cores. Carefully collected and dated lake sediments, however, do not reveal unusually high accumulations in strata coincident with large explosive aperiodic volcanic eruptions (e.g., Pinatubo in 1991, Krakatau in 1883 or Tambora in 1815). In contrast, and according to Schuster *et al.* (2002), volcanic mercury depositional signals may be evident in a recent study of the Fremont Glacier in Wyoming, USA.

For example, their results suggest an average global mercury depositional peak increase of 16 $\mu g\ m^{-2}\ yr^{-1}$ from the 1883 Krakatau eruption, which released ~21 km^3 of volcanic material (Rampino and Self, 1984). As the pre-eruption mercury deposition was ~2 $\mu g\ m^{-2}\ yr^{-1}$, the Krakatau event is extraordinarily prominent at ~8 × the background. In contrast, a mercury signal associated with the June 1991 Mount Pinatubo eruption in the Philippines, which released ~5 km^3 ash and pumice and ~50×10^{12} g SO_2 or 25×10^{12} g S (USGS Fact Sheet 113-97), is not evident in the Fremont Glacier ice core. If Mt. Pinatubo were assumed to be analogous to Krakatau according to the Fremont ice core, then the predicted mercury deposition would be ~4 × less or ~4 $\mu g\ m^{-2}\ yr^{-1}$. The anthropogenically enhanced background mercury deposition for 1991 in the ice core was considerably larger at ~11 $\mu g\ m^{-2}\ yr^{-1}$. Thus, and given the large pollution-derived mercury deposition, the uncertainty inherent in reconstructed fluxes and the assumptions, the lack of a mercury signal from Mt. Pinatubo could reasonably be expected.

We suggest, that the work of Schuster *et al.* (2002), while somewhat speculative and not well constrained, is provocative and stimulating. Moreover, there is reasonable coherence with the Hg/S for volcanic emissions shown in Table 4. For example, the Hg/S of 5×10^{-6} when combined with the 25×10^{12} g sulfur emission estimates for the Mt. Pinatubo eruption yields a mercury input of 125×10^6 g mercury or ~0.6 Mmol. This is just 6% of the 10 Mmol anthropogenic mercury emissions estimated by the MFM simulation for the global contribution in 1990. Therefore, the measured Hg/S ratio suggests the Mt. Pinatubo eruption would not be detectable as well. The Krakatau eruption would be predicted to be ~2.4 Mmol, which would result in an average enhancement of global mercury deposition of only 1 $\mu g\ m^{-2}\ yr^{-1}$. It must be noted, however, that work from long peat cores has suggested that large aperiodic volcanic eruptions may have left detectable signals in that archiving medium as well (e.g., Martínez-Cortizas *et al.*, 1999; Roos-Barraclough *et al.*, 2002). The lack of identifiable volcanic signals in lacustrine sediments is likely due to the "smoothing" effect that a several-year residence time of mercury within a lake and its watershed would exert on such short-duration signals.

In summary, it has been demonstrated that Hg/S ratios measured for a variety of volcanic plumes and fumaroles, when indexed to estimates of global sulfur emissions from volcanism, yield a mean volcanic mercury flux of 0.23 Mmol (45 t), which is consistent with other estimates and observations. Accordingly, average yearly mercury emission from volcanoes is small relative to natural terrestrial fluxes to the atmosphere (5 Mmol) and modern pollution mercury entering the global cycle (9.6 Mmol). Over the 100 yr time period used in the MFM simulation, anthropogenic mercury inputs to the global atmosphere were 1,000 Mmol, while average mercury emissions from volcanoes would be 23 Mmol. Periodic large eruptions, such as Tambora, Krakatau, Mt. St. Helens, and Mt. Pinatubo, would add significantly to this flux but for very short periods.

9.04.3.2 Mercury Input to the Oceans via Submarine Volcanism

The few attempts to determine mercury concentrations in oceanic hydrothermal fluids have not been successful. As a consequence, we must estimate the fluxes of mercury associated with submarine volcanic activity indirectly. First, we assume that the Hg/S of 5×10^{-6} established for subaerial volcanism (Table 4) can be applied to the hydrothermal inputs associated with submarine tectonic activity, and second, that oceanic lava production is ~5 × as large as the amounts formed subaerially (e.g., Crisp, 1984). Accordingly, the submarine inputs would be ~1.3 Mmol annually. An upper estimate of 1.8 Mmol yr^{-1} was proposed by Fitzgerald *et al.* (1998), who used oceanic mercury profiles and an estimate for the rate of vertical mixing in the water column. The agreement for these estimates negates the extraordinarily large fluxes (20—40 ×) suggested by heat flow calculations (e.g., Rasmussen, 1994), and provides additional support for an average Hg/S ratio for volcanic emissions of ~5×10^{-6}. An input of 1–2 Mmol yr^{-1} is significant and comparable to worldwide river input as estimated by MFM (1994). It is probable, however, that only a small fraction of the tectonically associated marine mercury inputs mixes with bulk ocean water. It is likely that mercury will be removed from the hydrothermal fluids through co-precipitation of metal sulfides and scavenging by precipitating hydrous oxides of manganese and iron as hydrothermal fluids mix with the oxygenated seawater near their entry points. Moreover, elevated levels of mercury are present in the hydrothermally derived metal-rich deposits found on the East Pacific Rise (Bostrom and Fisher, 1969) and the Gorda Ridge (Koski *et al.*, 1994). Stoffers *et al.* (1999) observed some nascent elemental mercury around the sulfide chimneys of the White Island (New Zealand) complex, but given the scaling arguments above it would appear that submarine volcanism does not represent a significant source of mercury to the global ocean.

9.04.3.3 Low-temperature Volatilization

As noted in the previous section, direct low-temperature weathering inputs from mineralized mercury deposits to aquatic environments occur primarily through sediment transport of cinnabar-containing material. Volatilization is an additional form of low-temperature weathering in which mercury is unparalleled by other metals. The volatility of elemental mercury is well documented, and to the extent that mercury-containing materials possess some fraction of their burden in the elemental form, weathering by volatilization will occur. Other mercury species are somewhat volatile as well (Table 7), but most are less so by orders of magnitude than elemental mercury. Volatilization of mercury from soils and rock to the atmosphere has only recently received significant attention. Unlike air–water gas exchange, air–soil/rock gas exchange has not been described in theoretical terms; thus, all that is known of mercury volatilization is from direct flux measurements and the results of soil manipulation experiments. Several others (e.g., Poissant and Casimir, 1998 and references therein) have noted from measured fluxes and their temperature dependence, the similarity of estimated and theoretical activation energies of vaporization ($\sim$60 kJ mol^{-1} and 85 kJ mol^{-1}, respectively). These observations hint that a more rigorous theoretical description of this process may be forthcoming.

Measurements using flux chambers and micrometeorological techniques are the most numerous (see special section of *J. Geophys. Res.* **104**(D17), 1999). As noted in Section 9.04.2, Klusman and colleagues (e.g., Klusman and Jaacks, 1987) attempted to develop a tracer approach based on measurements of ^{222}Rn/He/Hg to estimate the flux of mercury from soils indexed to the fluxes of the other gases. Their work, however, has not been extended beyond small-scale applications. Finally, isotope addition experiments including those of Schlüter (2000) using radioactive ^{203}Hg additions and those of Lindberg and colleagues using stable isotopes in the METAALICUS program (see Section 9.04.8) are proving very insightful.

Table 7 Henry's law constants for selected mercury species (at STP).

Equilibrium	*H* (M atm^{-1})
$Hg^0_{(g)} \leftrightarrow Hg^0_{(aq)}$	0.11
$Hg(OH)_{2(g)} \leftrightarrow Hg(OH)_{2\ (aq)}$	1.2×10^4
$HgCl_{2(g)} \leftrightarrow HgCl_{2\ (aq)}$	1.4×10^6
$(CH_3)_2Hg_{(g)} \leftrightarrow (CH_3)_2Hg_{(aq)}$	0.13
$CH_3HgCl_{(g)} \leftrightarrow CH_3HgCl_{(aq)}$	2.2×10^3

Sources: Sanemasa, (1975), Iverfeldt and Lindqvist (1982), and Lindqvist and Rodhe (1985).

The results from some volatilization measurements over a number of substrates are shown in Table 8, and vary widely. In general terms, the various studies indicate that higher concentrations of mercury in the soil/rock substrates lead to higher evasional fluxes. Other factors are strongly influential as well. These include temperature, light, wind speed, and soil moisture (e.g., Gustin *et al.*, 1999). It is clear that evasion of mercury from mineralized areas can be significant; however, the results from other substrates are currently limited by the large uncertainties and variability inherent in making such difficult measurements. Gustin and colleagues have made efforts to scale up their measurements, made primarily from Nevada, to the western US and Mexico (10 Mg yr^{-1}; Gustin *et al.*, 2000). Thus, this important area of research is still developing and should be active in the future.

In his review of soil volatilization experiments, Schlüter (2000) also highlighted the importance of the dissolved organic carbon concentration in the soil fluids, with higher concentrations of fulvic acids, for instance, leading to an enhancement of mercury reduction and evasion by generating Hg(I) and then aiding the disproportionation reaction of (2Hg(I) = Hg(0) + Hg(II)) through sequestration of Hg(II). The source of the reducing equivalents in soils appears to be species generated indirectly through photoreduction of some kind (e.g., organic carbon and Fe(II); Schlüter, 2000). The flux from nonenriched soils, though, is substantially lower than that from the mineralized areas and may average around 0.2 μg m^{-2} yr^{-1} (Schlüter, 2000).

Combining estimates for volcanic and low-temperature inputs of mercury from mineralized

Table 8 Some examples of measured fluxes over natural soils.

Location	*Method(s)*	*Soil conc.* (ng g^{-1})	*Evasional flux* (ng m^{-2} h^{-1})	*References*
Sweden	FC	NA	−2 to 2	Xiao *et al.* (1991)
Tennessee, USA	FC	61–469	−1.81 to 54.94	Carpi and Lindberg (1998)
Quebec, Canada	FC	NA	0.62–8.29	Poissant and Casimir (1998)
Nevada, USA	FC and MM	1,200–14,600	50–600	Gustin *et al.* (1999)
Nova Scotia, Canada	FC	NA	−1.4 to 4.3	Boudala *et al.* (2000)

FC = flux chamber; MM = micrometeorology (Bowen ratio).

areas and non-enriched soils to the atmosphere allows an estimate of the total amount of natural terrestrial emissions to be made. The volcano work benefits from the existence of tracing species such as sulfur that make tractable the scaling of individual measurements to the global scale. In the case of low-temperature volatilization, however, no such index has yet been developed. Therefore, translating values such as those of Table 8 into global fluxes is difficult.

Using the data from Nevada (Gustin *et al.*, 2000; 0.011 Mmol yr^{-1}; $1.8 \times 10^{11} m^2$ area) an emission rate of $\sim 10\ \mu g\ m^{-2}\ yr^{-1}$ for the global mercury belt areas can be estimated. Further assuming that these enriched areas represent no more than ~15% of the continental area suggests a maximum contribution for volatilization from these areas of ~5.6 Mmol yr^{-1}. The addition of the small volcanic contribution suggests that natural emissions of mercury to the atmosphere are <5.8 Mmol yr^{-1} and that subaerial and submarine emissions combined are <7.1 Mmol yr^{-1}.

The volatilization estimates are crude extrapolations, as they are based on assumptions of soil concentration distributions and understanding of driving forces behind volatilization. They do suggest however, that the emissions measured and estimated in some of the work cited are consistent in the first order with Mason *et al.* (1994) and that natural land-based sources of mercury to the atmosphere are consistent and likely to be ~5 Mmol yr^{-1}. It has also been noted that a flux 5 Mmol yr^{-1} for natural sources is consistent with the rate of atmospheric deposition in the pre-industrial past indicated by analysis of lake sediments (e.g., Lamborg *et al.*, 2002b). Finally, it must be noted that emissions from soils removed from natural enrichments likely contain a significant fraction of mercury initially mobilized by anthropogenic activities that was subsequently deposited to soils (see the section on mixed sources below).

9.04.3.4 Anthropogenic Sources

The human-related sources of mercury to the environment are numerous and widespread. Most direct inputs of mercury from point sources to aquatic systems have largely been contained in most developed countries. Inputs of mercury to the environment via the atmosphere are of the greatest concern. These emissions, coupled with long-distance transport of elemental mercury, have resulted in elevated concentrations of mercury in fish from locations that are removed from anthropogenic sources (e.g., open-ocean, and semi-remote regions in the United States, Canada, Scandinavia; Wiener *et al.*, 2002). A summary of the fluxes from major sources (for 1995) is shown in Table 9 (total of 9.6 Mmol yr^{-1}). High temperature processes, principally coal and municipal waste burning, dominate anthropogenic inputs of mercury to the atmosphere. The emission of anthropogenic mercury is higher in the northern hemisphere, as a result of greater industrial activity and population density. Between 1990 and 1995 the emissions from developed economies in North America and Europe have declined substantially. Unfortunately, they have almost been completely replaced by emissions from countries, especially in Asia, that have rapidly developing economies that are coal-driven. Accordingly, Asian sources of mercury currently constitute 56% of all anthropogenic emissions. Based on the Pacyna and Pacyna inventory and the natural source strength suggested by Mason *et al.* (1994), human activity contributes ~2/3 of the mercury emitted from land-based sources each year. Similarly, these estimates suggest that the emission and deposition fluxes of mercury are currently 3 × what they were in the pre-human environment. Such estimates are now widely supported by the reconstruction of mercury deposition from remote lakes worldwide (more below; e.g., Fitzgerald *et al.*, 1998).

9.04.3.5 Mining

Mining has been a long-standing and continuing source of environmental mercury contamination. Indeed, a partial analog to the alchemist's quest to transmute base metals into gold is contained in the *patio* process in which naturally occurring but trace amounts of gold and silver are amalgamated (concentrated) using large amounts of liquid mercury. The dense amalgam can be separated from the crushed, parent rock or from placer and alluvial deposits, often with much loss of mercury to air and aquatic systems. The gold or silver is recovered by heating the amalgam and vaporizing the mercury. This technology has been

Table 9 Major classes of anthropogenic emissions of mercury to the atmosphere in 1995.

Source type[a]	*1995 flux* (Mmol yr^{-1})
Stationary combustion	7.4
Non-ferrous metal production	0.8
Cement production	0.7
Waste disposal	0.6
Pig iron and steel production	0.1
Total	9.6

After: Pacyna and Pacyna, 2002.
[a] Stationary combustion includes fossil fuel burning power plants, while waste disposal includes municipal waste combustion.

employed broadly and often crudely since its introduction by Bartolome de Medina in 1557 (Nriagu, 1979). Historically, uses of mercury in gold and silver mining were especially significant in the Americas from the mid-1500s to the turn of the twentieth century. This unhealthy and ecologically damaging practice continues today, and on a large scale in many countries (e.g., China, Brazil, Philippines, Kenya, and Tanzania). In their review of current gold mining, Lacerda and Salomons (1998) found that environmental losses of mercury are large, 1–1.7 kg per kg gold recovered. Much of the pollutant mercury accumulates in the surrounding lands, watersheds, waterways, and mine tailings, and the associated environmental and human-health concerns are primarily local and regional. However, there are global worries as well, because a portion of the mercury is emitted to the atmosphere (Porcella *et al.*, 1997).

Mercury losses occur not only with the processing and recovery of gold and silver, but in the mining and production of mercury. For example, the nineteenth century "gold rush" in the western United States was fueled by mercury mining in California. Egleston (1887) reports that between 1850 and 1889, the mercury yield from California mines, especially from the New Almaden operation (85%), was 1,518,380 flasks (~34.5 kg/flask). This was comparable to the combined output of the two other major mines, the Almadén (Spain) and Idrija, Austria (now Slovenia), which produced 1,291,636 and 347,586 flasks respectively over the same time period. Moreover, and as Egleston emphasizes, "according to the best California authorities, the loss in the best constructed furnaces as near as can be approximated is not less than 15–20%, and in many of the works the losses will probably amount to double that." Mercury mining activities continue today in Spain though with a higher sensitivity to preventing mercury releases to the environment. Despite this, mercury mines remain significant sources of mercury to watersheds and coastal marine systems including inoperative sites such as Idrija and Clear Lake, CA that supply mercury from abandoned tailings.

9.04.3.6 Biomass Burning, Soil and Oceanic Evasion—Mixed Sources

There are three prominent processes that release mercury of mixed natural and anthropogenic origin to the atmosphere. These three include biomass burning (deliberate and natural) and the evasion of mercury from soils and the ocean. The general factors controlling emission of mercury from soils have been discussed in the section on low-temperature volatilization. The mercury released from soils that are not naturally enriched (unlike some of the mineralized substrates described) is mercury derived principally from atmospheric deposition and is released from the upper horizon pool (Schlüter, 2000 and references therein). As the mercury that is deposited from the atmosphere is of mixed origin, so is the mercury emitted to the atmosphere from these soils. Therefore, though mercury may be released from completely natural and undisturbed soils as regulated by ambient biogeochemical processes, the current flux is not entirely natural. In the case of non-enriched soils, this is not very significant as these materials are net sinks of atmospheric mercury deposition (see Section 9.04.6). In the case of biomass burning and especially oceanic evasion, however, the fluxes to the atmosphere may be very important in the global mercury cycle. As with soils though, these processes mobilize both natural and anthropogenic mercury and represent sources of mixed origin. In this way, these media act to recycle mercury in the environment, extending the residence time of mercury at the Earth's surface.

Recent measurements of mercury in biomass burning plumes from research aircraft suggest that this process releases substantial amounts of mercury. Brunke *et al.* (2001) and Friedli *et al.* (2001) used CO and CO_2 as indexing species to establish fluxes of mercury of ~1–5 Mmol yr^{-1}. A first-order estimate, based on the relative strengths of truly anthropogenic emissions and truly natural emissions, suggests that some two-thirds of the mercury released by biomass burning was initially released by human activities (i.e., 0.67–3.4 Mmol yr^{-1} anthropogenic; 0.33–1.6 Mmol yr^{-1} natural).

Oceanic evasion is a major component of the mercury cycle. As described in greater detail in Section 9.04.5 below, there are a number of processes that may lead to evasion of elemental mercury from the ocean. Mason *et al.* (1994) found that evasion from the ocean had tripled in magnitude in concert with the increase in anthropogenic activities. Therefore, as with biomass burning, nearly two-thirds of the mercury currently evading from the ocean is anthropogenic.

9.04.3.7 Watersheds and Legacy Mercury

Watersheds are sources of mercury to the aquatic environment. However, and similar to biomass burning and evasion, the mercury released from watersheds is of mixed origin. Because the residence time of mercury within watersheds is fairly long (see Section 9.04.6), the potential for the buildup of "legacy" mercury exists. This feature is relevant when considering how rapidly a system might respond to decreased mercury loadings. Therefore, though decreases in mercury deposition to a watershed may occur, the watershed will contribute more mercury to its receiving waters than enters it each year. Legacy mercury, however, was released to the environment as a result of human activity and should be viewed as an anthropogenic source term.

9.04.4 ATMOSPHERIC CYCLING AND CHEMISTRY OF MERCURY

Mercury is found in the atmosphere in both gas and particle phases. Greater than 95% of mercury exists as gas-phase elemental mercury (Fitzgerald and Gill, 1979; Bloom and Fitzgerald, 1988; Iverfeldt, 1991a). There is growing evidence to suggest that some chemical form of Hg^{2+} also exists in the gas phase (the so-called "reactive gaseous mercury" or RGM; Stratton and Lindberg, 1995; Sheu and Mason, 2001; Landis *et al.*, 2002). Concentrations of total gaseous mercury (TGM; including elemental, ionic and gaseous alkylated forms such as dimethylHg) in remote areas are typically in the range of 1–2 ng (as mercury) m^{-3}. Concentrations below 1 ng m^{-3} are to be found under certain conditions (more below) and higher values are often observed in urban/suburban locations. Some selected concentration and deposition data are shown in Table 10. Particle-phase atmospheric mercury appears to be largely Hg^{2+} and comprises a few percent of total atmospheric mercury in the troposphere (Iverfeldt, 1991a; Fitzgerald *et al.*, 1991). There is only one published report of mercury in the stratosphere to our knowledge and no measurements in other regions of the upper atmosphere (Murphy *et al.*, 1998). Not surprisingly, the authors found the concentration of particulate mercury increased above the tropopause as a result of enhanced oxidation of elemental mercury by ozone and the condensation of the less volatile Hg^{2+} onto ambient particles.

The vertical profile of mercury in the troposphere has been determined in a few cases (Banic *et al.*, 2003; Landis and Stevens, 2001). In most situations, it appears that there is little change in the mercury mixing ratio with altitude, indicating thorough vertical mixing and an atmospheric residence time that is long enough to make this possible. There have been some suggestions that elemental mercury decreases with altitude, while reactive gaseous mercury increases, creating a gradient for atmospheric deposition on large scales of the more soluble and surface active Hg^{2+}. For example, Landis and Stevens (2001) have suggested that the vertical RGM gradient is on the order of 400 pg m^{-3} over the troposphere (6,340 m, isobaric). This is a rather large gradient, representing ~25% of the total atmospheric mercury. Applying a vertical eddy diffusivity of 1 $m^2\ s^{-1}$ (Seinfeld, 1986) to this gradient provides an estimate of the potential rate of removal of RGM on a large scale of 2 $\mu g\ m^{-2}\ yr^{-1}$ (or ~10–30% of the observed flux at most locations). It must be noted that these datasets are only now being developed and therefore the flux estimate made above is highly speculative.

The situation of horizontal profiling is better developed. Figure 4 illustrates data from a number of sampling locations worldwide, showing a small but discernible interhemispheric gradient in TGM. Values for TGM in the northern hemisphere (NH; ~1.7 ng m^{-3}) are larger than in the south (~1.2 ng m^{-3}) as a result of the NH representing a greater proportion of land-based natural and anthropogenic emissions. Horizontal gradients on smaller scales (i.e., plumes) have also been observed including continental-scale, urban plumes, and single industries (Fitzgerald, 1995 and the references therein; Lamborg *et al.*, 2002a).

Assuming little vertical variation in the mixing ratio of total mercury in the troposphere and using the available horizontal surface-based measurements, Mason *et al.* (1994) estimated the total

Table 10 Atmospheric deposition concentration data.

Location	*TGM* (ng m^{-3})	*Hg_T in precip.* (ng L^{-1})	*Deposition* ($\mu g\ m^{-2}\ yr^{-1}$)	*Calculated lifetime* (yr)	*Reference(s)*
Florida, USA	1.4–3.1	13–23	15–28	0.3–0.7	Guentzel *et al.* (1995) and Gill *et al.* (1995)
Tennessee, USA	5.8 ± 3.6	3	30	1.2	Lindberg *et al.* (1992)
Michigan, USA	2.0	10	9 ± 3	1.4	Hoyer *et al.* (1995)
S. Atlantic Ocean	1.4	4	6	1.5	Lamborg *et al.* (1999)
Wisconsin, USA	1.6 ± 0.4	6	7	1.5	Lamborg *et al.* (1995)
Alert, CAN	1.2	~15	5	1.5	Schroeder *et al.* (1998) and Schroeder, pers. comm.
Global average	1.6	NA	5.6	1.8	Lamborg *et al.* (2002a)
Eq. Pacific	1.3	3	4	2	Mason and Fitzgerald (1993)
Sweden	2.9 ± 0.7	10	13 ± 12	2 ± 1	Lindqvist *et al.* (1991)

After Lamborg *et al.* (2001).

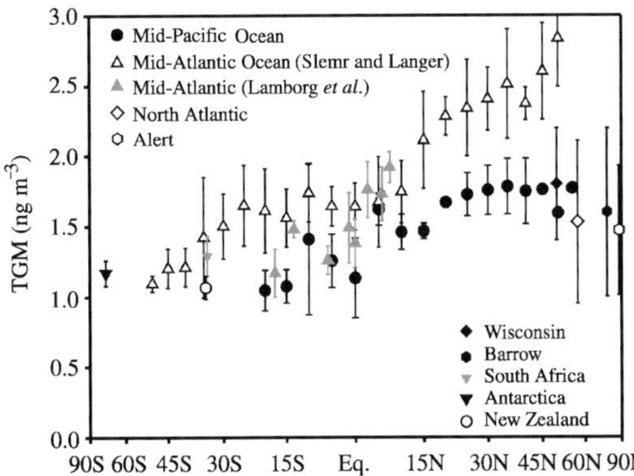

Figure 4 Total gaseous Hg in the atmosphere at several locations. Notice the discernible interhemispheric gradient, resulting from greater emissions of Hg to the atmosphere in the more industrialized northern hemisphere (data from Fitzgerald (1995)—Mid-Pacific (filled circles); Slemr and Langer (1992)—Mid-Atlantic (open triangles); Lamborg *et al.* (1999)—Mid-Atlantic (shaded triangles); Mason *et al.* (1998)—North Atlantic (open diamonds); Schroeder *et al.* (1998)—Alert, N.W.T. (open hexagons); Lamborg *et al.* (1995)—rural Wisconsin, USA (filled diamonds); Lindberg *et al.* (2002)—Point Barrow, Alaska, USA (filled hexagons); Ebinghaus *et al.* (2002)—Cape Town, R.S.A. (shaded inverted triangles) and Antarctica (filled inverted triangles); Fitzgerald (1989)—Ninety Mile Beach, NZ (open circles); figure from Lamborg *et al.*, 2002a).

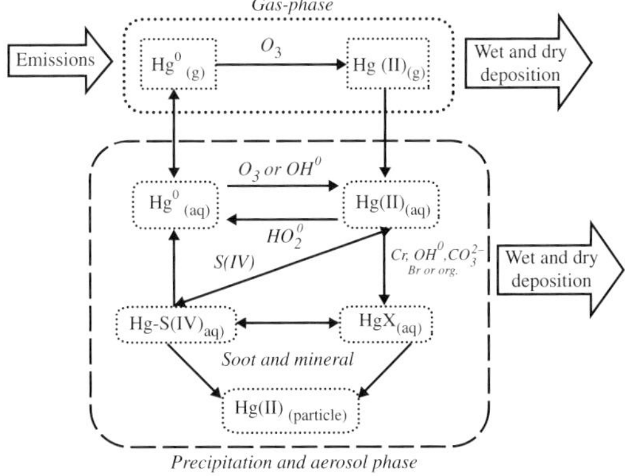

Figure 5 Summary of some of the important physical and chemical transformations of mercury in the atmosphere figure style adapted from Shia *et al.*, 1999; reactions from Shia *et al.* (1999) and Lin and Pehkonen (1999) and the references therein).

atmospheric burden of mercury to be 25 Mmol (5 kt; Figure 2(a)). Accordingly, the average tropospheric air column mercury burden is ~10 μg m^{-2}. If the emissions of mercury as outlined in Section 9.04.3 equal 20 Mmol yr^{-1}, then the average residence time of mercury in the atmosphere is ~1.25 yr. Such a residence time is relatively long on atmospheric timescales (e.g., the mixing time for the hemispheres is ~1.3 yr; Geller *et al.*, 1997), and thus we should expect to find mercury reasonably well mixed vertically and horizontally as we do. This global-scale value is within the range of similar estimates made from various specific locations (Table 10).

It was quickly realized that the mercury species to be found in greatest abundance in precipitation was ionic mercury (e.g., Fogg and Fitzgerald, 1979). Some typical values of total mercury in precipitation are shown in Table 10. Extensive databases of precipitation mercury concentrations are available from monitoring networks in the US, Canada, and Nordic countries (e.g., US Mercury Deposition Network: http://nadp.sws.uiuc.edu/mdn). The discrepancy between the dominant gas and precipitation phase species implied a process of oxidation of elemental mercury in the atmosphere and its subsequent scavenging as being a major component of the mercury cycle. Since the initial work, and partially in response to increased governmental interest in long-range atmospheric transport of pollutant mercury, there has been an extraordinary increase in research on the atmospheric chemistry of mercury. Many mechanisms for elemental mercury oxidation in the atmosphere have been proposed and a few have been studied in detail through laboratory experiments (e.g., Munthe, 1992; Hall *et al.*, 1995; Tokos *et al.*, 1998; Lin and Pehkonen, 1999; Sommar *et al.*, 2001). These include homogeneous gas-phase and heterogeneous-phase reactions occurring in cloud-water/precipitation and aerosols. The principal constraint on gas-phase oxidation is that the overall reaction rate must be similar to the residence time of mercury. The rate constants for oxidation by ozone and hydroxyl radical are consistent with this constraint. In heterogeneous phase reactions, oxidation may be partially balanced by reduction, leading to complex cycling of mercury species within cloud-water or aerosols that includes influences by sorbent surfaces such as soot (Pleijel and Munthe, 1995). Some of the proposed mechanisms are shown in Figure 5. The ongoing challenge for those studying atmospheric mercury is to identify which of these mechanisms is actually influential in the atmosphere and under what conditions. Lamborg *et al.* (2002a), based on strong rainwater correlations between Hg and ^{210}Pb, have suggested that a mechanism including gas-phase oxidation followed by particle and precipitation scavenging is the dominant overall process for the removal of mercury from the global atmosphere. The atmospheric chemistry embedded in the transport and deposition model of Shia *et al.* (1999), which included many of these reactions, estimated the atmospheric residence of mercury to be ~1.7 yr. Their simulation highlighted the importance of aqueous-phase reduction reactions in controlling the atmospheric residence time of mercury. In more urbanized environments and close to sources, dry deposition of particulate mercury emitted from sources or forming shortly after emission is likely to be the dominant removal mechanism (e.g., Keeler *et al.*, 1995; Chiaradia and Cupelin, 2000).

There are, however, new findings from the Arctic and from mid-ocean that suggest this is not the entire story. One of the more dramatic observations in mercury biogeochemistry in recent years is the so-called "spring time depletion" of mercury in high latitudes. Schroeder *et al.* (1998) were the first to observed this phenomenon at Alert on the northern tip of Ellesmere Island (Canada). Total gaseous mercury at this location shows a fairly steady value of 1.5 ng m^{-3}, typical of the northern hemispheric "background." However, at the advent of polar sunrise and lasting for several weeks, the concentration begins to fluctuate between the baseline value and near zero, with the depletion episodes lasting hours to days. There is also a high correlation between depletion of mercury and ozone, thus forging a connection between mercury and the chemistry and physics of Arctic haze formation and breakdown (Figure 6; Barrie and Hoff, 1985; MacDonald *et al.*, 2000). Further work by Bill Schroeder and his colleagues as well as others in the Arctic (e.g., Lu *et al.*, 2001;

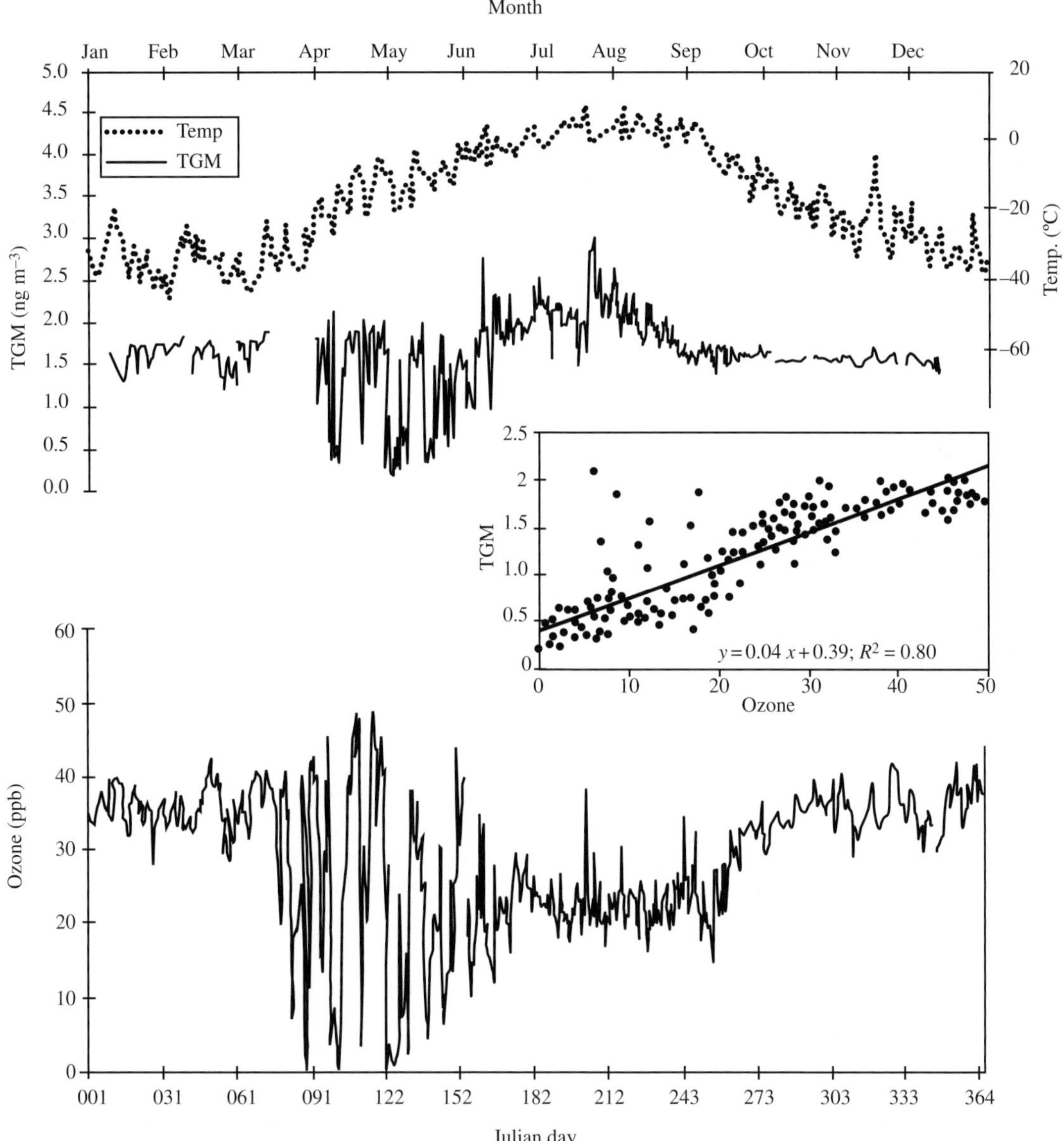

Figure 6 Depletion of Hg° in the atmosphere during Arctic spring as observed by Schroeder *et al.* (1998) at Alert. The traces, from top to bottom, are surface air temperature, total gaseous mercury, and ozone. The inset illustrates the correlation between mercury and ozone during the spring depletion period (source Schroeder *et al.*, 1998).

Munthe and Berg, 2001 and references therein; Lindberg *et al.*, 2002) and Antarctic (Ebinghaus *et al.*, 2002) confirms this phenomenon as recurrent, seasonal, and occurring in both polar regions (though perhaps more dramatically in the Arctic). Working at mid-ocean, Mason *et al.* (2001) have observed a rapid oxidation of mercury near the air–sea interface resulting from reactions with sea-spray halogens. As Lindberg *et al.* (2002) and others have observed, the reaction in the Arctic is also coincident with a buildup of reactive bromine compounds in the polar atmosphere, pointing to a possible mechanistic similarity between the Arctic and mid-ocean phenomena. In both cases, the generation of significant concentrations of oxidized mercury in the gas-phase may lead to significant dry depositional fluxes in addition to fluxes associated with precipitation. The assessment made by Schroeder *et al.* (1998) when describing the depletion events initially was that perhaps 0.5 Mmol of mercury was deposited to the Arctic as a result of this phenomenon (a boundary layer of 500 m over 2×10^5 km^2 containing 1.84 ng m^{-3} emptied of its mercury 5 times: 4.6 μg m^{-2} yr^{-1}). This is significant but not an enormous unanticipated sink for mercury on a global-scale. However, the flux could still be quite significant for delicate Arctic ecosystems and their human inhabitants. This bears further scrutiny. The polar and mid-ocean works also imply that such a mechanism should be looked for in other atmospheric environments and may be of broad significance.

Research on pathways of mercury dry deposition in addition to particle phase and RGM suggests that plants (especially trees) may take up elemental mercury from the atmosphere into their leaves above a certain "compensation point" concentration (Hanson *et al.*, 1995; Benesch *et al.*, 2001; Rea *et al.*, 2002). Elemental mercury absorbed in this way could then be deposited to soils in the form of litterfall (e.g., Johnson and Lindberg, 1995; Grigal *et al.*, 2000; Lee *et al.*, 2000; St. Louis *et al.*, 2001). Forest foliage may also act as a particle interceptor, effectively increasing the dry deposition velocity of particles (throughfall; Iverfeldt, 1991b). Throughfall and litterfall studies suggest that the removal of mercury from the atmosphere in forests might be as much as 3 × more than "open-field" deposition.

Long-term monitoring datasets of mercury in the troposphere are now being developed at several locations. Some studies have suggested secular increases in TGM of as much as 1.6% yr^{-1}, though this has not been widely reported (Slemr and Langer, 1992; Ebinghaus *et al.*, 2001; Baker *et al.*, 2002). Such research efforts are being performed in a social context of decreased mercury emissions from a number of countries (see Section 9.04.3). It is therefore difficult to gauge the current direction of secular change of mercury in the atmosphere. However, as suggested, it may be that the reduced emissions from relatively uncontrolled sources in Eastern Europe, for example, will be more than made up for by increased emissions associated with developing economies such as China.

However, assuming for the moment that the atmosphere is near a steady state between emissions and deposition (Mason *et al.*, 1994), the emissions estimates of Section 9.04.3 can be used to gauge the magnitude of the depositional flux (20 Mmol yr^{-1}). This flux is not uniform, with low latitudes receiving more mercury per unit area than high latitudes and continental regions more than oceanic areas. These general trends are the result of several factors. If wet deposition tends to be more important in the removal of mercury than dry processes, wetter regions such as the tropics can be expected to have higher overall fluxes of mercury. This trend is evident even on regional scales, as for instance the flux of mercury to lakes on the wet side of New Zealand's South Island are much higher than on the dry side (Lamborg *et al.*, 2002b). The difference between mid-continent and mid-ocean fluxes of mercury from the atmosphere appears to be one driven principally by transport from the continents, where sources to the atmosphere are strong, a process recently traced using ^{222}Rn (Lamborg *et al.*, 1999).

There remains an intriguing inconsistency between experiments related to the mechanisms for mercury removal. Many lab, field, and model efforts indicate that the lifetime of mercury in the atmosphere must be 1–2 yr, but there exist a number of plausible removal mechanisms (such as foliar mercury uptake followed by litterfall) that suggest the flux from the atmosphere is more consistent with lifetimes that are less than 1 yr. The likely resolution of this problem lies in the observation that majority of the Earth's surface is covered by areas that are not temperate or boreal forests, including the open ocean and tropical regions. The deposition to the ocean is consistent with an atmospheric residence time in excess of 1 yr, while the mercury cycling within tropical forests is understudied.

9.04.5 AQUATIC BIOGEOCHEMISTRY OF MERCURY

A brief overview of the marine biogeochemical cycle of mercury was presented in the introduction. Here, a broader picture of the reactions and species-specific interactions involving mercury in natural waters appears in Figure 7. This mechanistic scheme, taken from Fitzgerald and Mason (1997), is derived in part from the simulation of

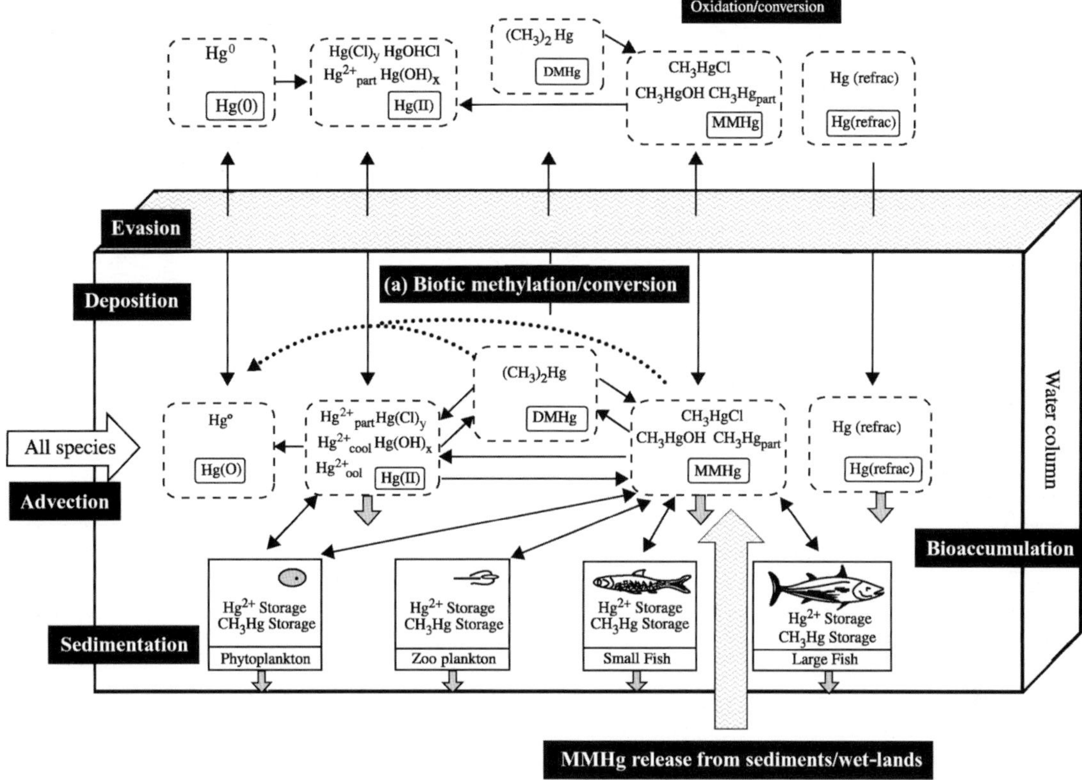

Figure 7 Generalized view of mercury biogeochemistry in the aquatic environment. Prominent processes are labeled (Hudson *et al.*, 1994).

the biogeochemistry of mercury in temperate lakes (Hudson *et al.*, 1994). Hudson and colleagues (1994) as part of the successful and scientifically influential Mercury in Temperate Lakes Program (MTL) conducted in northern Wisconsin, USA, developed a mercury cycling model (MCM). Comparable models have been developed for other freshwater environments such as the Florida Everglades (Beals *et al.*, 2002) and Onondaga Lake, a highly mercury-contaminated and USEPA designated "Superfund" site (Gbondo-Tugbawa and Driscoll, 1998). We anticipate, as information increases, analogous biogeochemical MCMs will be developed and applied for marine systems. Accordingly, we have chosen to illustrate major features of the aquatic cycling of mercury using the near-shore environment as a generalized working analogue for aquatic systems.

Although organic and inorganic ligands and organisms differ in fresh and salty environments, much of the biogeochemical processing and movement of mercury are expected to be similar. Representative distribution and speciation data for mercury in natural waters are presented in Table 11. MMHg was first determined in freshwaters by N. Bloom (Bloom, 1989; Watras *et al.*, 1994), and is now well documented (e.g., Verta and Matilainen, 1995; Mason and Sullivan, 1997). In 1990, Mason and Fitzgerald reported finding methylated mercury species, including DMHg, in the open-ocean waters of the equatorial Pacific Ocean (Mason and Fitzgerald, 1990, 1993). Their presence has been confirmed for the Atlantic Ocean, Mediterranean Sea and estuaries (Cossa *et al.*, 1994; Mason *et al.*, 1998; Mason and Sullivan, 1999). Likewise, the production and supersaturation of Hg^0 is well documented in fresh and salt waters since the initial papers were published by Vandal *et al.*, 1991 for lakes and Kim and Fitzgerald, 1986 for the equatorial Pacific Ocean. As Figure 7 shows, the cycling of Hg^0 and MMHg is intimately linked by their competing and critical roles in the aquatic biogeochemistry of mercury (substrate hypothesis). Notice that, in general, the speciation, transformation pathways, reactions and processes can be connected to reactive mercury. This reactant or "substrate" should be viewed broadly to encompass labile inorganic and organically associated mercury species. Sources include atmospheric deposition/ exchange, watersheds, riverine inputs, sewage and other human-related discharges.

Using Table 11 as background and Figure 7 as a guide, important features associated with mercury cycling in all natural waters but especially

Table 11 Mercury species concentrations in a variety of natural waters. All data in pM, except where noted.

Location	*Dissolved total Hg*	*Particulate total Hg*	*Dissolved reactive Hg*	*Dissolved MMHg*	*Particulate MMHg*	*Dissolved DMHg*	*Dissolved Hg^0*	*Reference(s)*
Freshwaters								
Lake Michigan, USA	1.6	0.58	NA	0.025–0.05	0.01–0.015	NA	0.140 ± 0.085	Mason and Sullivan (1997)
Lake Superior, USA/Canada	ca. 0.5–5[a]	0.04–0.43[b]	0.08–0.57	0.008–0.064	0.00005–3.9[b]	NA	0.03–0.17	Rolfhus *et al.* (2003)
Lake Hoare, Antarctica[a]	2.7–6.8	NA	0.4–1.2	<0.4–1.2	NA	NA	NA	Vandal *et al.* (1998)
Everglades	5–10[a]	NA	0.15–0.5[a]	0.25–2.5[a]	NA	NA	0.025–0.225	Hurley *et al.* (1998)
Wisconsin Lakes, USA	3–6	1–2	NA	0.1–0.9	0.15–0.35	<0.003	0.035–0.355	Watras *et al.* (1994) and Fitzgerald *et al.* (1991)
Estuaries/coastal								
San Francisco Bay, USA	0.4–174	0.3–439	NA	0–1.6	0–1.92	NA	0.043–9.8	Conaway *et al.* (2003)
Long Island Sound, USA	1.6–13.1	<0.1–24.1	<0.1–7.6	0–3.3	<0.01–2.91	NA	0.037–0.89	Vandal *et al.* (2002) and Rolfhus and Fitzgerald (2001)
North Sea and Scheldt Estuary	0.5–14	0.1–6[b]	NA	0.05–1.37	0.0009–0.0435[b]	NA	0.06–0.8	Baeyans and Leermakers (1998) and Leermakers *et al.* (2001)
Siberian Estuaries	0.7–17	0.15–9.4	NA	NA	NA	NA	NA	Coquery *et al.* (1995)
Loire and Seine Estuaries	1–6	0.42–13.3[b]	<0.4–2.1	NA	<0.0015–0.0296[b]	NA	<0.05–0.454	Coquery *et al.* (1997)
Chesapeake Bay, USA[a]	~3–40	NA	NA	~0.05–0.8	NA	NA	~0.1	Mason *et al.* (1999)
Pettaquamscutt R., USA	~1–25	~0–18	0.4–8[a]	<0.05–4	<0.05–6.88	NA	<0.025–0.4	Mason *et al.* (1993)
Brazilian Lagoons	18.5–55.2	18–230	0.18–0.43	NA	NA	NA	NA	Lacerda and Gonçalves (2001)
Open ocean								
Mediterranean Sea	0.8–6.4[a]	NA	<0.2–0.97[a]	<0.15[a]	NA	<0.13–0.29	<0.02–0.39	Cossa *et al.* (1997)
Eq. Pacific Ocean[a]	NA	0.11–5.87	0.4–6.9	<0.05–0.58	NA	<0.005–0.67	0.015–0.69	Mason and Fitzgerald (1993)
North Atlantic	2.4 ± 1.6	0.035 ± 0.02	0.8 ± 0.44	1.04 ± 1.08	NA	0.08 ± 0.07	0.48 ± 0.31	Mason *et al.* (1998)
South Atlantic	2.9 ± 1.7[a]	0.1 ± 0.05	1.7 ± 1.2[a]	<0.05–0.15	NA	<0.01–0.1	1.2 ± 0.8	Mason and Sullivan (1999)

[a] These samples were unfiltered. NA = not available. [b] Units of nmol Hg gm^{-1} of suspended material, dry weight.

seawater are reemphasized and highlighted in the following summary:

(i) The principal source of the toxic species, MMHg, in marine and many freshwater aquatic systems is *in situ* biologically mediated conversion of labile reactive mercury. As discussed, sulfate-reducing bacteria (SRB) have been implicated as the primary synthesizers (Compeau and Bartha, 1985; Winfrey and Rudd, 1990; Gilmour and Henry, 1991). The bioamplification of MMHg in the aquatic food chain yields concentrations in fish that are often more than a million times greater than its levels in water. Many freshwater systems also receive significant inputs of MMHg from their watersheds, particularly wetlands (e.g., Hultberg *et al.*, 1994; Hurley *et al.*, 1995; St. Louis *et al.*, 1996). Salt marshes can also be prolific generators of MMHg, but do not appear to be larger sources than coastal sediments (Langer *et al.*, 2001). They are, however, important nurseries for many aquatic organisms and could represent locations of significant MMHg accumulation in the early life stages of some fish species.

(ii) Hg^0 is an important species in air and water, and *in situ* direct reduction of labile reactive mercury by biotic (i.e., bacterial) and abiotic (i.e., photochemical) means is a principal pathway for its aqueous production (Amyot *et al.*, 1994, 1997; Rolfhus, 1998; Costa and Liss, 1999); biological demethylation mechanisms (see Figure 7) yield small amounts of Hg^0 (Mason *et al.*, 1993). The mechanisms of reduction are unclear and the focus of current study. The reverse reaction, oxidation of Hg^0, also occurs (e.g., Amyot *et al.*, 1997; Lalonde *et al.*, 2001). Thus, ambient Hg^0 concentrations can be expected to vary in space and time in response to changes in the forces that drive the reduction and oxidation reactions (e.g., bacterial activity, light, temperature, DOC, total mercury). Examples of diel and seasonal variations in Hg^0, consistent with this view, are becoming more common in the literature (e.g., Lindberg *et al.*, 2000; Rolfhus and Fitzgerald, 2001; Amyot *et al.*, 2001; O'Driscoll *et al.*, 2003; Tseng *et al.*, 2003; Tseng *et al.*, in press).

(iii) Reiterating, *in situ* Hg^0 production (natural waters are generally supersaturated) and emissions to the atmosphere are major processes; they exert a first-order (primary) control on the overall biogeochemistry and bioavailability of mercury in aqueous systems, and the water–air fluxes must be considered in global/regional atmospheric and aquatic biogeochemical models of the mercury cycle; the reduction reactions (leading to aqueous emissions of Hg^0) are recycling mercury derived from both natural and anthropogenic sources of mercury, and thereby, extending the lifetime of pollutant and natural mercury in active reservoirs.

(iv) The aqueous production of Hg^0 competes for reactant (i.e., labile reactive mercury) with the *in situ* biological synthesis of MMHg; thus, water bodies with a large production of Hg^0 will have less bioavailable mercury, smaller amounts of MMHg in biota and reduced mercury accumulation in the sediment (Wiener *et al.*, 1990b; Rada and Powell, 1993; Fitzgerald *et al.*, 1991).

(v) Given the affinity of Hg^{2+} for sulfur (i.e., sulfhydryl groups) and its ability to form very stable organomercury chelates, organic complexation will exert an important control on the bioavailability of mercury.

(vi) Inorganic complexation with sulfur is a primary reaction in reducing environments. This reaction may occur in oxygenated waters where, for example, microenvironments develop such that oxygen is depleted and sulfate reduction takes place. This is one possible explanation for the presence of MMHg in ocean surface waters. There is a triad of competing ligands for free mercury in most aqueous systems. Organic ligands compete with chloride in saltwater and with hydroxide in freshwater, while sulfur becomes especially competitive as oxygen levels decline, and SRB activity increases.

As a point of analytical and environmental interest, Hg^0 is more readily measured in natural waters than MMHg. Since the *in situ* production of MMHg and Hg^0 is proportional to the supply of reactive mercury, a comprehensive understanding of the aqueous Hg^0 cycle and its temporal and spatial patterns may provide a means to constrain and improve predictive models for the aquatic and atmospheric biogeochemistry of mercury and MMHg in natural waters. For a sense of the potential geochemical benefits from automated Hg^0 measurements, the reader is referred to some recent field studies of Hg^0 (e.g., Lindberg *et al.*, 2000; Amyot *et al.*, 2001; Balcom *et al.*, 2000).

9.04.5.1 Environmental Mercury Methylation

Given the importance of *in situ* synthesis of MMHg through conversion of less toxic mercury species and its prominent role in the aquatic cycling of mercury (Figure 7), aqueous mercury methylation merits added consideration. As outlined, bacterial mediation enhances the rates at which mercury, a "soft acid," can form alkylated species in aquatic environments. This extraordinary interaction and its potential consequences have provided the rationale for much environmentally related mercury research over the past three decades. Indeed, the biologically mediated synthesis of alkylated mercury species can readily account for most of the MMHg accumulating in biota, especially large fish, in most marine and freshwaters (Wiener *et al.*, 1990a; Fitzgerald and Watras, 1989; Watras *et al.*, 1994; Rolfhus and Fitzgerald, 1995; Benoit *et al.*, 2003).

9.04.5.1.1 Near-shore regions

The near-shore environment provides a useful biogeochemical framework for outlining current knowledge regarding mercury methylation in aqueous systems. Mechanistically, recent work in freshwaters and near-shore sediments has not only pointed to SRB as methylating agents, but transition regions between oxygenated and anoxic conditions (e.g., low oxygen/hypoxic) as the principal sites of MMHg production (e.g., Gilmour and Henry, 1991; Watras *et al.*, 1994; Langer *et al.*, 2001). While mercury methylation does occur in the water column and is especially important throughout most of the oceans (i.e., pelagic regions, e.g., Topping and Davies, 1981), the major sites for production are associated with particles and depositional environments such as lake and coastal/estuarine sediments, wetlands, and marshes (Watras *et al.*, 1994; Gilmour *et al.*, 1998; Langer *et al.*, 2001; Hammerschmidt *et al.*, in preparation). Microbial production of MMHg in sediment is influenced by a number of environmental factors that affect either the activity of methylating organisms (i.e., SRB) or the availability of inorganic mercury for methylation. For example, recent studies of MMHg levels in bulk surface sediment (e.g., Benoit *et al.*, 1998a; Gilmour *et al.*, 1998; Krabbenhoft *et al.*, 1999; Mason and Lawrence, 1999; Hammerschmidt *et al.*, in preparation) have shown dependencies on inorganic mercury, organic matter, and sulfide. In marine and estuarine sediments, where seawater provides ample sulfate, rates of sulfate reduction are influenced mostly by availability of organic matter and temperature (Skyring, 1987). King *et al.* (1999, 2000, 2001) have demonstrated that the rate of mercury methylation is closely related to that of sulfate-reduction.

Estuarine/marine systems that are highly productive or receive autochthonous inputs of organic matter are prime locales for enhanced rates of mercury methylation and ecosystem exposure to MMHg. However, recent studies have illustrated that although significant mercury methylation occurs in such environs, production of MMHg is attenuated by accumulation of sulfide, the metabolic by-product of sulfate reduction (Figure 8; Gilmour and Henry, 1991). In estuarine and marine sediments, where activity of SRB is high and largely independent of sulfate, sulfide inhibition of mercury methylation is clearly demonstrated by the inverse relationship with sulfate. In contrast, mercury methylation in freshwater systems is directly related to sulfate, which limits SRB metabolism. Hence, maximum mercury methylation occurs in sediments where organic matter and sulfate are sufficiently high as to stimulate SRB metabolism, but not so high as to cause accumulation of sulfide that inhibits the availability of mercury for methylation

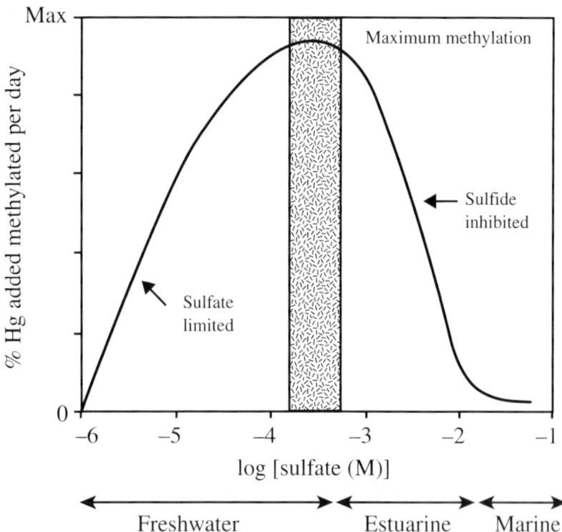

Figure 8 Sulfate/Sulfide controls on mercury methylation in aquatic environments — the "Gilmour curve." At relatively low sulfate concentrations (most freshwaters), methylation of mercury is limited by the rate of sulfate reduction. At higher sulfate concentrations (saltwaters), sulfide buildup from relatively high rates of sulfate reduction results in decreased bioavailability of mercury (figure from Langer *et al.* (2001); after Gilmour and Henry (1991).

(Gilmour and Henry, 1991, Figure 8, the "Gilmour Curve").

A mechanism by which sulfide affects methylation of mercury was recently proposed by Benoit *et al.* (1999a,b, 2001a,b). Sulfide affects the chemistry of inorganic mercury in sediments by precipitating it as solid mercuric sulfide and forming dissolved mercury-sulfide complexes, including HgS^0, HgS_2^{2-}, and $HgHS_2^-$. HgS^0 is a major dissolved mercury-sulfide complex when sulfide is less than 10^{-5} M and charged complexes, mainly as $HgHS_2^-$, are dominant at greater levels (Figure 9; Benoit *et al.*, 1999a). The mechanism for uptake of inorganic mercury by methylating bacteria is not known, though the research of Benoit *et al.* points to diffusion of neutrally charged HgS^0 through the cellular membrane as the key factor. As a result, maximum rates of mercury methylation occur in sediments where SRB activity is significant but the accumulation of sulfide is minimized, thereby favoring speciation of dissolved Hg–S complexes as HgS^0.

Sulfide oxidation occurs both microbially and abiotically. In coastal sediments that are not subject to water column anoxia, burrowing animals mix the upper few centimeters of sediment (i.e., bioturbation), homogenizing the sedimentary solid-phase and pore water constituents (e.g., Gerino *et al.*, 1998). In doing so, underlying anoxic (i.e., sulfidic) sediments are mixed with overlying oxic sediments; thereby minimizing accumulation of sulfide via dilution and abiotic and microbially

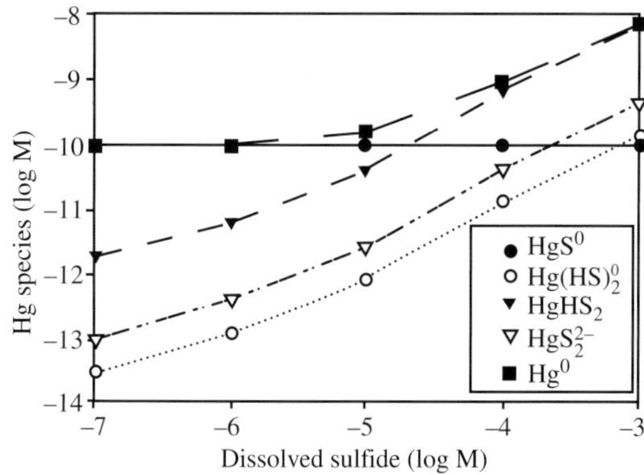

Figure 9 Dissolved mercury speciation in sediment pore waters as a function of sulfide concentration. Note that the most bioavailable form, HgS°, is the dominant chemical form at log S <, ~(−4.7) (source Benoit *et al.*, 1999a).

mediated oxidation reactions. Sulfide-oxidizing bacteria (SOB) are chemolithotrophs that use sulfide as a source of energy and reducing power. Bioturbation also can stimulate SRB activity by translocating organic matter from surface sediments to depth (Hines and Jones, 1985; Skyring, 1987; Gerino *et al.*, 1998). Hence, bioturbation of estuarine sediments likely stimulates mercury methylation by both enhancing SRB activity and minimizing accumulation of sulfide. Biologically mediated reworking of coastal/estuarine sediments, in general, keeps some portion of the historic (buried) inventory of anthropogenic mercury ("mercury pollution legacy") active. Given that legacy mercury can be methylated and mobilized in the coastal zone (unlike in lakes), a significant delay is likely between reductions in modern loadings and expected declines in MMHg in the fish stock. This unfortunate expectation for marine systems must be emphasized when considering the expected and observable benefits from "zero mercury use" environmental legislation and remediation efforts.

In sediments that are less bioturbated, SOB promote mercury methylation by minimizing accumulation of sulfide. These bacteria proliferate in redox transition zones overlying SRB. By consuming sulfide, SOB minimize its accumulation, promoting speciation of mercury-sulfide complexes as HgS^0 and facilitating uptake of inorganic mercury by proximal SRB. The well-defined relationships between redox transition zones, rates of mercury methylation, and MMHg distributions in salt marsh sediments are illustrated in Figure 10 (Langer *et al.*, 2001). These results are consistent with those predicted by the hypotheses of Benoit *et al.* (1999a,b). Clearly, the diverse chemistry and microbiology of the redox transition zone makes it an important location for MMHg synthesis in sediments. Although SRB appear to the principal methylators, there is also evidence for mercury methylation by iron reducing bacteria (Gilmour, personal communication).

Demethylation in the water column and sediments is receiving increasing attention. Both abiotic (e.g., Sellers *et al.*, 1996, 2001) and biotic (e.g., Pak and Bartha, 1998; Marvin-Dipasquale and Oremland, 1998; Marvin-Dipasquale *et al.*, 2000; Hintelmann *et al.*, 2000) processes are implicated. The result is that MMHg accumulation in aquatic systems represents a balance between methylation, bioaccumulation, and the demethylation processes. In sediments, MMHg decomposition is particularly important, and it is possible that some sediments represent net sinks, rather than net sources, for MMHg in the water column.

9.04.5.1.2 Open-ocean mercury cycling

In contrast to the nearshore, mercury methylation in the water column is the primary source of MMHg in the open ocean, which represents ~90% of the marine environment. As shown in Table 11, DMHg is found in seawater, but has not been observed in common freshwaters (Mason *et al.*, 1993; Mason and Fitzgerald, 1993; Mason and Sullivan, 1999). Indeed, MMHg is the predominant alkylated species in temperate lakes, while DMHg, and (to a lesser extent) MMHg are common constituents of the dissolved mercury pool in ocean waters. The unique presence of DMHg in seawater prompted Mason and Fitzgerald (1993, 1996) to propose the following reaction sequence:

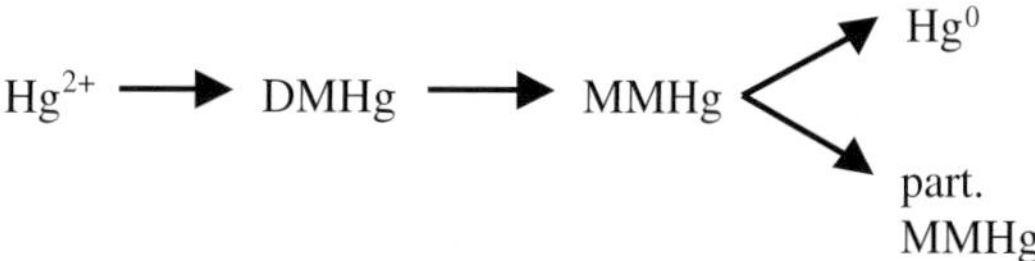

where DMHg would be the principal product from the methylation of inorganic mercury with MMHg and Hg^0 derived from the decomposition reactions. The primary source of Hg^0 in aqueous systems, however, remains the *in situ* direct reduction of labile reactive mercury by biotic (i.e., bacterial) and abiotic (i.e., photochemical) processes (Figure 7). Decomposition is the principal loss term for DMHg, while particulate scavenging and decomposition are important sinks for dissolved MMHg. This view of mercury cycling in the ocean is very speculative. DGM formation has been little studied, and the organisms responsible for methylation in the open ocean are not known (nor have the pathways been elucidated).

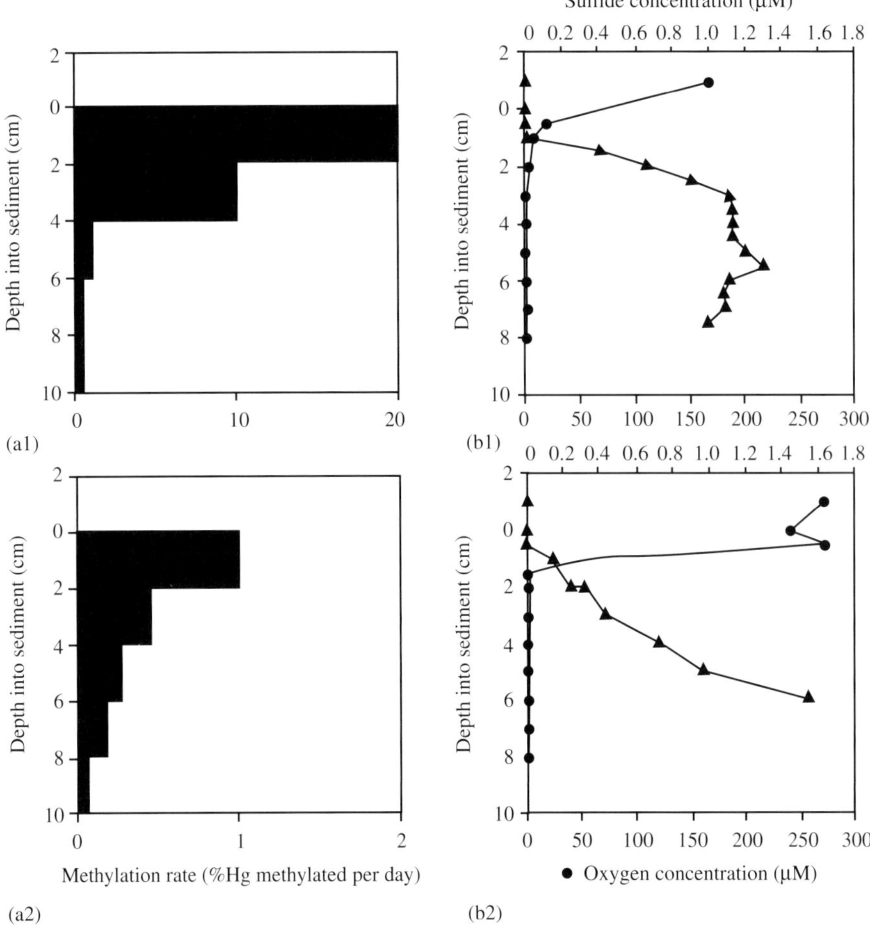

Figure 10 Vertical profiles of methylation rates (a1, a2) and sulfide/oxygen concentrations (b1, b2) in the sediments of sandy (1) and muddy (2) sites in a salt marsh (Barn Island, Connecticut, USA). Maximum methylation occurs in the top 2 cm of these sediments and is coincident with the redox transition zone. Note also that the rates are an order of magnitude faster in the sandy sediments (source Langer *et al.*, 2001).

9.04.5.1.3 Open-ocean mercury profiles

An appreciation of the challenges of ultra-trace metal investigations can be gained from a consideration of mercury speciation and distributional information for open ocean. A classic vertical distributional profile for mercury in the northeast Pacific Ocean is shown in Figure 11 (Gill and Bruland, 1987). As expected with a biogeochemically active element, mercury shows a nonconservative distribution, one that is not governed by simple mixing of ocean waters. Total mercury concentrations range from 1.8 pM at the surface to 0.3 pM in the upper ocean minimum. The atmospheric mercury enrichment in the near-surface waters was captured due to recent rains and short-term stratification. The minimum is indicative of mercury removal via particulate scavenging processes that are biologically mediated. The gradual increase in mercury at depth may reflect regeneration processes. Today, and given the analytical improvements since the

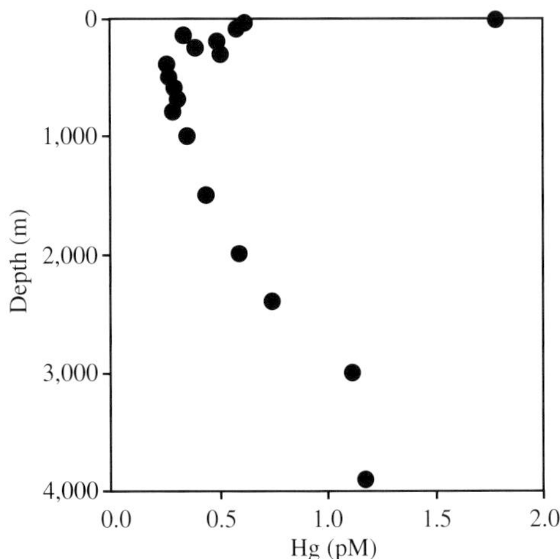

Figure 11 Vertical profile of total dissolved mercury in the northeast Pacific Ocean (after Gill and Bruland (1987), from the Vertex VII cruise, station T7, August 6–10, 1987).

Gill and Bruland report, a more comprehensive suite of mercury species can be determined, along more insightful biogeochemical reaction-based explanations for the distributional patterns in space and time. A summary of such extensive mercury measurements appears in Table 11.

An example of recent data comes from the Third IOC Baseline Trace Metal Cruise, which took place during May–June of 1996 in the equatorial and South Atlantic (Mason and Sullivan, 1999). We selected vertical profiles for mercury species for one station (#8 at 17° S, 25° W) from among six others examined as part of the Mason and Sullivan Program. These data are presented in Figure 12. DMHg, Hg^0 (total dissolved gaseous mercury—the small contribution of DMHg), and total mercury are plotted versus depth. In addition, and as a reference, the distributions of dissolved silicon and salinity are shown. First, the results confirm the presence of methylated mercury species, especially DMHg, and Hg^0 in oxygenated ocean waters. Here, however, and in contrast to the North Atlantic (Table 11), MMHg levels were at the detection level (0.05 pM). This suggests that MMHg is either decomposing or scavenged more rapidly than it is formed. The prominence of Hg^0 (>50%) relative to the total mercury present (average of 2.4 ± 1.4 pM) is a most striking feature. Its abundant presence at depth is consistent with the hypothesis outlined above where MMHg, which is produced as DMHg decomposes. A portion of the MMHg is scavenged by particulate matter and Hg^0 produced as the relatively stable product (under dark conditions) of the decomposition of the MMHg.

Most DMHg is produced in the near surface waters, but, as illustrated in Figure 12, little is found in the euphotic zone because DMHg is readily decomposed photochemically. DMHg accumulates in the intermediate depths (see profile) above 1,500 m. Below 1,500 m, small but significant concentrations (0.02–0.03 pM) occur, but they are considerably smaller than values (0.16 ± 0.8 pM) in the source region of the North Atlantic Deep Water (NADW; Mason *et al.*, 1998). While the DMHg decline in the modestly advecting NADW (southward travel time ~100 yr to reach the study regions of South Atlantic and equatorial Atlantic) is significant, decomposition rate estimates for DMHg in advecting deep ocean waters imply that sufficient production must be occurring to yield measurable concentrations (Mason and Sullivan, 1999).

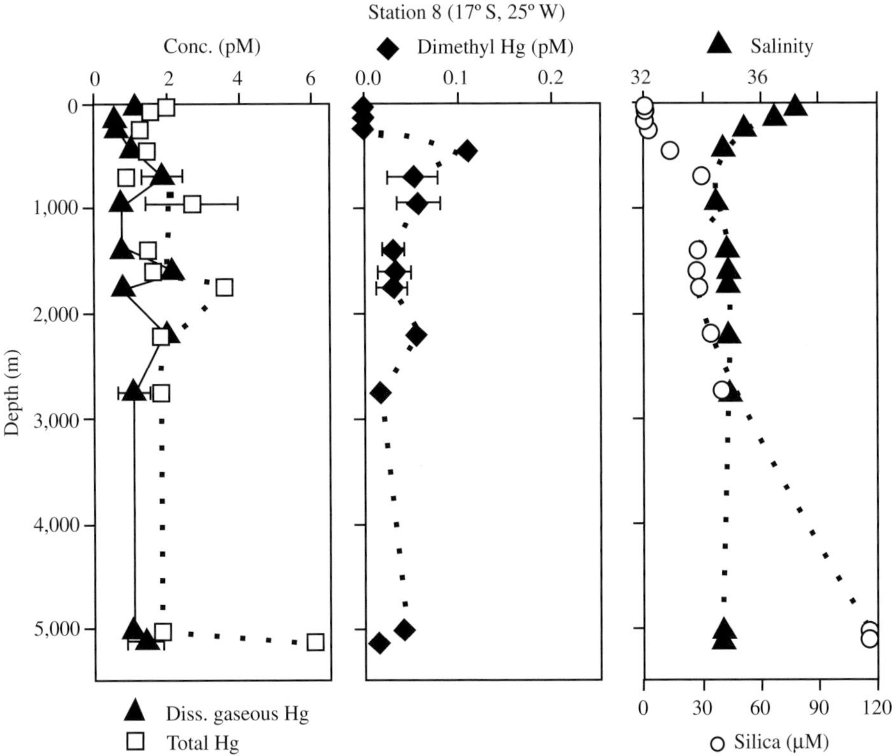

Figure 12 Vertical profiles of total dissolved mercury and mercury species in the South Atlantic, from station 8 of the IOC Baseline South Atlantic Cruise. MonomethylHg was below detection at all depths, while measurable dimethylHg was found within and below the thermocline. Dissolved gaseous mercury, dominated by Hg°, represents the majority of mercury at many depths at this location (source Mason and Sullivan, 1999).

This production is presumably fueled by the very small transport of carbon from the surface regions to below 1,500 m.

In summary, mercury speciation studies in oceanic systems are revealing the complex interactions of this biologically active and reactive element. However, and as emphasized, the spatial and temporal coverage is sparse and many of the biogeochemical insights and hypotheses attempting to explain the behavior and fate of in-marine systems are speculative. Indeed, the challenging and complicated marine biogeochemistry of mercury beckons the curious and innovative. There is much to discover.

9.04.6 REMOVAL OF MERCURY FROM THE SURFICIAL CYCLE

The work summarized in Section 9.04.3 notwithstanding, most studies indicate that soils and terrestrial sediments act as net sinks of mercury on timescales of centuries. Such a statement is supported by a review of the mercury content of soils from around the world and supporting data that allow determinations of mercury inventories in various soil horizons (Table 12). For comparison, we have estimated the "excess mercury" inventories resulting from anthropogenic activities using the integrated

Table 12 Concentrations of mercury in soils and terrestrial sediments and estimates of the anthropogenic mercury inventory (xsHg).

Location(s)	*Soil/Sed. Hg Conc.* (ng g^{-1})	*Soil/Sed. Hg Inventory* (mg m^{-2})	*References*
Soils			
Cedar Creek, MN, USA	140 ± 30	0.3 ± 0.2	Grigal *et al.* (1994)
Organic layer			
0–10 cm	36 ± 7	3.4 ± 0.5	
10–50 cm	11 ± 4	7 ± 4	
Fr. Guiana (rain forest)	122–318		Roulet and Lucotte (1995)
0–30 cm			
Sweden O Horizon	320 ± 10	~1.5	Alriksson (2001)
B Horizon	43 ± 3		
C Horizon	13 ± 1		
Nevada, USA	100–15,000		Gustin *et al.* (1999)
Lake sediments			
Northern Quebec, Canada	25–450		Lucotte *et al.* (1995)
N.S. and N.Z. lakes	10–300	0.32–0.95[a]	Lamborg *et al.* (2002b)
Ice cores			
Fremont glacier, USA	0.002–0.035	~1[a]	Schuster *et al.* (2002)
Model estimate			
Est. avg. anthrop. signal		0.6–2.6[a]	this work; Mason *et al.* (1994)
Coastal sediments			
Long Island Sound, USA	<30–>600	~30–170[a]	Varekamp *et al.* (2000)
Northern Adriatic Sea	20–230	~13[a]	Fabbri *et al.* (2001)
Chesapeake Bay, USA	60–>1,000	NA	Mason *et al.* (1999)
Gulf of Cadiz, Spain	~50–250	~75[a]	Cossa *et al.* (2001)
Gulf of Trieste, Adriatic Sea	100–23,300	~12,000[a]	Covelli *et al.* (2001)
Florida Bay, USA	~10–236	~1.6[a]	Kang *et al.* (2000)
San Francisco Bay, USA	20–700 (MMHg: 0–3.5)	NA	Conaway *et al.* (2003)
S. Baltic Sea	2–340	1.2[a]	Pempkowiak *et al.* (1998)
S. China Sea, Malaysia	20–127	NA	Kannan and Falandysz (1998)
Anadyr est., Bering Sea, Russia	77–2,100	NA	Kannan and Falandysz (1998)
Mouth of St. Lawrence est., Canada	~50–100	~4.5[a]	Gobeil and Cossa (1993)
Greenland	~6–275	NA	Asmund and Nielsen (2000)
Santa Barbara Basin, USA	60–160	NA	Young *et al.* (1973)
Pelagic sediments			
Arctic Ocean	10–116	NA	Gobeil *et al.* (1999)
Kara Sea	75–2045	NA	Seigel *et al.* (2001)
W. Mediterranean	~80	NA	Cossa *et al.* (1997)
Laptev Sea	25–140	NA	Cossa *et al.* (1996)

[a] Includes just the anthropogenic, or excess Hg.

human emission inventory over the last 100 yr estimated by Mason *et al.* (1994; 947 Mmol total, 447 Mmol globally uniform, 500 Mmol near source). As the continental area of the globe is $\sim 1.5 \times 10^8$ km^2 and assuming uniform distribution and no soil loss, we expect to find from 0.6 mg m^{-2} in remote areas to 2.6 mg m^{-2} in "non-remote" areas (taken to be one-third of the continental area). Most of the lake sediment data cited fit this range reasonably well. The soil data also display agreement depending on the assumed depth of penetration of the modern mercury signal. Matilainen *et al.* (2001) found that additions of radioactive ^{203}Hg were quantitatively retained in the organic humus layer of their test soils (O-horizon), and thus we might expect little vertical penetration of the modern signal. This leads to the conclusion that most of the mercury that has been deposited to soils in the last $\sim$100 yr is still present in active biogeochemical zones in these systems.

These estimates suggest that human-related contributions to soil loadings of mercury (the excess mercury) are a significant if not always dominant contributor to the total soil column mercury inventory. The natural weathering inputs of mercury to soils may be examined if we assume that the residence time of mercury within soils is similar to that of organic carbon, citing the strong associations between mercury and humic materials already mentioned. Estimates of organic carbon residence within soils is wide ranging and dependent on ecological setting, but may average $\sim$100 yr. This approach is supported by the observation that agricultural fields, which experience net losses of organic carbon as a result of human use, show lower mercury inventories than undisturbed soils (Grigal *et al.*, 1994). As 100 yr is the approximate timescale over which major human perturbation has taken place, we may subtract the estimated excess mercury loading from the total inventory to arrive at an estimate of the weathering input (Table 12). In most instances, the natural and excess loadings are comparable as was the ratio of natural to human-related emissions inventories during that period. Thus, it appears that on millennial timescales, soils are not net sinks for mercury, and release their burden to natural waters through runoff and erosion and to the atmosphere by volatilization and from fires.

Such bulk, integrative studies are supported by short-term studies of mercury loadings to watersheds and subsequent runoff in surface streams. A number of studies have indicated that 70–95% of the mercury deposited under these conditions is retained within the soils of the watersheds. Retention estimates made in this way could be biased low due to the slow release of "legacy mercury" deposited over the previous decades that increases in magnitude as the total watershed inventory increases (e.g., Mierle, 1990; Aastrup *et al.*, 1991; Swain *et al.*, 1992; Krabbenhoft *et al.*, 1995; Scherbatskoy *et al.*, 1998; Lawson and Mason, 2001; Kamman and Engstrom, 2002; Lamborg *et al.*, 2002b).

The flux of mercury from the continents to the ocean in river runoff has been estimated to be $\sim$1–5 Mmol yr^{-1} (Mason *et al.*, 1994; Cossa *et al.*, 1996). While some of this material is volatilized as outlined in Section 9.04.5, most is buried in coastal sediments (e.g., Fitzgerald *et al.*, 2000). Deeper coastal sediments or remote, surface lacustrine sediments are of the order of 10–50 ng g^{-1} dry weight. Typical surface sediments, even from systems not receiving direct inputs from industrial activities can be 10 $\times$ the background value. Extreme examples, such as the near-shore sediments of the Gulf of Trieste (NE Adriatic Sea), which receives the river discharge from the mercury mining and mineralized area of Idrija, are also to be found (Table 12). This enrichment of surface coastal sediments is not likely to be due to sediment diagenesis as the flux of mass and mercury to the sediment overwhelms upward diffusion (Gobeil and Cossa, 1993). Thus, the excess mercury associated with the enrichments may be used to gauge the impact of human activity on the coastal zone, as shown in Table 12. Many of the values estimated here are larger than that expected from release of mercury from watersheds following atmospheric deposition. This estimate was made by assuming $\sim$25% of the continental deposition (250 Mmol) was delivered to 10% of the ocean area (3.6×10^7 km^2) giving 4 mg m^{-2}. This difference is the result of periods when significant non-atmospheric point- and area-source inputs to coastal systems and their watersheds were in effect. Such a condition exists currently, for example, in the case of relatively unrestricted gold mining in Amazonia resulting in large documented accumulations of mercury in the river sediments and presumably in the coastal sediments of the western equatorial Atlantic (e.g., Lechler *et al.*, 2000). The residence time of mercury in this pool is difficult to estimate given our relatively sparse data, but is likely to be similar to the residence time of the sedimentary material. For example, Herut *et al.* (1996) found that mercury contamination of the sediments of Haifa Bay, Israel, was disappearing in a manner consistent with sediment remobilization, and could have a half-life in excess of 100 yr. These sediments then represent a significant sink for mercury on the global scale. This pool should also be viewed as an environmental and public health concern for the future because this material is concentrated in a relatively small area of the ocean and one that is biologically productive and commercially important (see Section 9.04.5).

Finally, open-ocean sediments receive a portion of the mercury deposited to the ocean. Here, the data are particularly sparse and only a handful of studies are available to guide our discussion. Concentrations of mercury in these materials can be found in Table 12. Given the slow rate of accumulation and rapid recycling of material in the upper ocean, it is to be expected that little of this mercury is of anthropogenic origin, though one study claims to observe surface enrichments consistent with pollution inputs in a relatively nearshore and heavily impacted region (Seigel *et al.*, 2001). Gobeil *et al.* (1999) found evidence for diagenetic remobilization of mercury along with Fe/Mn and Mercone *et al.* (1999) have proposed coupled Hg/Se diagenesis and mineral formation under the slow sediment accumulation conditions present in the open ocean. They argued that this remobilization could result in surface enrichments not associated with anthropogenic inputs. We estimate the total amount of mercury associated with marine sediments to be 6×10^7 Mmol ($\sim$10 ng Hg g^{-1} of sediments, 1% organic carbon content and 12×10^{15} t organic carbon total; Lalli and Parsons, 1993). With inputs of mercury to open-ocean sediments on the order of 1 Mmol yr^{-1} (Mason *et al.*, 1994), the sedimentary pool should be expected to represent the primary sink for mercury on million year timescales.

9.04.7 MODELS OF THE GLOBAL CYCLE

Secular change in the global mercury cycle as a result of human activity is one of the major themes in this chapter and a focus of research currently. One of the most insightful research activities in pursuit of this theme has been the development of historical archives of mercury change. As of early 2000s, this development effort has focused on three archives: peat bogs, lake sediments, and ice cores. All are used to reconstruct historical changes in the flux of mercury from the atmosphere.

Peat cores are generally collected from bogs dominated by *Sphagnum* moss species and from systems that are ombrotrophic ("rain fed"). Under these conditions, it is assumed that the moss is only able to receive mercury from the atmosphere as these systems are isolated from groundwater and geologic inputs other than dust. Mercury delivered to the moss is incorporated in the live material at the surface and remains associated even as the moss continues to grow vertically. Therefore, collection and sectioning of a peat core, followed by dating and mercury analysis, should reveal any changes in the rate of incorporation of mercury into the moss. As discussed by Benoit *et al.* (1998b), the source of the mercury archived could include mercury associated with rain and dust as well as elemental mercury taken up from the gas phase by the moss. A number of studies have made successful use of the peat archive (e.g., A. Jensen and A. Jensen, 1990; Norton *et al.*, 1997; Benoit *et al.*, 1998b; Martínez-Cortizas *et al.*, 1999). There are, however, other reports of variable element mobilities, including the dating isotope ^{210}Pb (e.g., Damman, 1978; Urban *et al.*, 1990; Norton *et al.*, 1997). In our recent work in Nova Scotia, we found evidence for mobility of lead but not mercury (Lamborg *et al.*, 2002a). We therefore submit that peat archives, or sections of archives, that are dated with ^{210}Pb may yield inaccurate estimates of mercury deposition, while other dating techniques may result in reliable data. For example, in Nova Scotia, we dated surface peat using a co-occurring moss species *Polytrichum* that develops annual markings on the male gametophytes. The results of that analysis when compared to lake sediments and rain measurements indicated that dry deposition of dust or uptake of elemental mercury by the *Sphagnum* may contribute some of the mercury archived in peat, but that the contribution was likely small.

Sediments collected from remote lakes have been the most profitable of the archives studied to date. In most cases, lake sediments are free of diagenetic mobility that plagues the use of peat, allowing reliable dating and more precise reconstructions (see Chapter 9.03). The lake approach does have drawbacks, however, as suitable lakes (small watersheds, simple morphology, little in/outflow) are often difficult to find in a location of interest. In the higher latitudes of the northern hemisphere, where extensive glaciation was in effect, small seepage lakes tend to be more common, and thus much of this reconstruction work has been performed in Scandinavia, Canada and the northern US (e.g. Swain *et al.*, 1992; Lockhart *et al.*, 1998; Bindler *et al.*, 2001). Suitable lakes are being investigated currently in additional locations such as in the tropics, temperate, and subpolar southern hemisphere. Extension of lake sediment studies to these new regions should bolster the conclusions from the northern studies. The general picture emerging from the use of lake sediments (and peat) is one of widespread, global-scale 2–4 times increase in the amount of mercury delivered from the atmosphere since the advent of the industrial revolution. The similarity in timing and scale of the increased mercury deposition across these varied systems leaves little doubt that the signal is real and the result of the substantial impact that human activity has had on the global mercury cycle. Application of this information to modeling efforts (see below) represents an important step forward in our understanding of how the mercury cycle operates.

Ice core archives are receiving increasing attention. In general, these archives are not as useful as sediments due to the relatively low rate of accumulation (usable ice cores are generally to be found in dry high latitudes and accumulate at slow rates). Furthermore, recent work by Marc Amyot and colleagues (Lalonde *et al.*, 2002) has suggested that mercury deposited in snow is photoactive and may be lost from the media following reduction to elemental mercury. However, and described in Section 9.04.3, Schuster *et al.* (2002) analyzed ice core samples from the relatively rapidly accumulating ($\sim$70 cm yr^{-1}) Fremont Glacier (Wyoming, USA) to reconstruct deposition over the last 270 yr. As indicated in Table 12, their results are consistent with predictions made for the inventory of anthropogenic mercury made here based on the model of Mason *et al.* (1994). Therefore, mountain glaciers may represent an important archive for future development. We noted that this glacial core also apparently preserved signals associated with volcanic eruptions that have not been found in lake sediment archives, but that these signals are difficult to interpret as they appear not to be self-consistent. In a more conventional location for ice core studies, Boutron *et al.* (1998) used snow blocks to estimate changes in mercury deposition to Greenland over the last 40 yr. Their work illustrates other fundamental difficulties associated with studying Hg in ice cores (small sample sizes and exceptionally low levels, $<$1 pg g^{-1}). Stretching even further back, Vandal *et al.* (1993) used core material from Dome C in Antarctica to reconstruct mercury accumulation at that location to before the last glacial maximum ($\sim$34 ka Bp). This unique study found evidence that increased deposition of mercury to Antarctica during that last glacial maximum was attributable to increased evasion from the ocean and suggests that mercury in ice cores may be a useful paleoproductivity proxy.

Seabird feathers may provide a proxy for oceanic mercury secular change (Monteiro and Furness, 1997). This unique dataset includes feathers retrieved from museum specimens of exclusively pelagic, piscivorous seabirds such as shearwaters and petrels. While there is scatter associated with the values, as should be expected from such a natural archive, an increase of $\sim 3\times$ between 1885 and 1994 can be seen in the mercury concentration of the feathers. Furness, Monteiro and co-workers have extended their work to include studies on the movement of mercury within living birds and the relationship between prey concentrations of mercury and those observed in the bird's feathers (Monteiro *et al.*, 1998; Monteiro and Furness, 2001), which aid in the interpretation of the feather record. This approach has so far been applied to the NE Atlantic (Azores), and should be extended to other ocean regions.

A number of mathematical models have been developed to test the consistency of a variety of environmental mercury data sets as well as aid in generating testable hypotheses concerning those aspects of the global mercury cycle that are difficult to observe directly (Lantzy and MacKenzie, 1979; Millward, 1982; Mason *et al.*, 1994; Hudson *et al.*, 1995; Shia *et al.*, 1999; Bergan *et al.*, 1999; Bergan and Rodhe, 2001; Lamborg *et al.*, 2002a). The earliest efforts were limited by the general lack of high-quality data available at the time. Mason *et al.* (1994; MFM) therefore can be said to be the first global mercury model capable of providing meaningful insight into the whole cycle. The findings of MFM, including their reconstruction of the cycle in its pre-industrial and modern forms have been illustrated (Figure 2) and discussed in Section 9.04.1. As demonstrated, the atmospheric deposition predictions from MFM can be compared to anthropogenic mercury inventories in soils and sediments and appear to accurately predict the flux of mercury from the atmosphere to these archives. Additional insights from MFM include an atmosphere/ocean system that is nearly balanced, indicating the influential role of oceanic evasion in redistributing and sustaining the lifetime of mercury at the Earth's surface. Furthermore, MFM found that anthropogenic activities should have tripled the current atmospheric deposition of mercury, resulting in a tripling of the amount of mercury entering terrestrial and marine ecosystems. As indicated by the lake sediment archives, this is what is found. Finally, the MFM model estimated that only 50% of the mercury emitted to the atmosphere as a result of human activities is able to participate in the global cycle, and that the remaining mercury is removed from the atmosphere close to the source. Thus, the MFM treatment describes the current condition of the mercury cycle as one significantly perturbed by human activity on local, regional, and global scales.

As with any modeling effort, MFM was partially supported by a number of assumptions based on, in some cases, scant information. Since MFM, several other modeling efforts have sought to address particular hypotheses concerning some of these assumptions. For example, Hudson *et al.* (1995) included additional reservoirs to their simulated ocean than were used in MFM in an effort to gauge the importance of ocean mixing in modulating mercury increases in the environment. They also introduced a more complex emissions source function indexed to CO_2 that released less mercury overall than in MFM, and did so in a nonlinear fashion. The conclusions from this work were quite similar to MFM but did indicate that ocean mixing should be considered an important process in the global-scale biogeochemistry of mercury. Perhaps most importantly, ocean mixing increases the size of the oceanic pool and as a result increases the predicted residence time of mercury in the ocean. An increased oceanic residence time could result in an increased overall rate of mercury methylation, particularly as much of this time would be spent in the lower oxygen/microbially active region of the permanent oceanic thermocline. Additional consideration of the oceanic biogeochemistry of mercury by Mason and Fitzgerald (1996) support this model view, particularly as an explanation for local imbalances in the air–sea exchange of mercury in certain regions such as the equatorial Pacific. Hudson *et al.* (1995) were also the first authors to directly compare their results to those provided by the lake sediments and noted, among other findings, that emissions from gold and silver mining prior to 1900 predicted to be significant were not to be found in the archives.

Three other recent global models (Shia *et al.*, 1999; Bergan *et al.*, 1999; Bergan and Rodhe, 2001) focused on the atmospheric aspects of the global cycle and have been able to place important constraints on which of the several chemical mechanisms proposed for elemental mercury oxidation are likely to be important at this scale. Particularly noteworthy are the efforts by Bergan and colleagues, who have exploited the data on interhemispheric concentrations of elemental mercury in the air. Exchange between the two hemispheres is relatively slow (~1.3 yr) due to physical forcings, but yet the north/south gradient of elemental mercury is relatively weak (Slemr and Langer, 1992; Fitzgerald, 1995; Lamborg *et al.*, 1999). As the emissions of mercury in the northern hemisphere are estimated to be larger than in the south, Bergan and colleagues used gradient and mixing time information to place constraints on the atmospheric lifetime of mercury. More recently, we have extended this approach through inverse box modeling (Lamborg *et al.*, 2002a), which used the gradient and lake sediment data explicitly to constrain both the atmospheric lifetime as well as aspects of the oceanic cycle including evasion, particle scavenging, and burial. This particular model (the global/regional interhemispheric mercury model, or GRIMM) suggests that evasion from the ocean is less than that estimated in MFM and that the ocean is a net sink.

Our inverse, or data assimilation, model underscores the importance of developing new data sets to improve our understanding of mercury biogeochemistry. As mentioned here, models such as Hudson *et al.*, Bergan *et al.*, and Lamborg *et al.* are principally driven by atmospheric and atmospherically related data. As these efforts are using essentially the same data, it is not surprising that the conclusions from these efforts are fairly similar. A more rigorous test of the current view of the global mercury cycle will require wholly new information, particularly from the ocean.

9.04.8 DEVELOPMENTS IN STUDYING MERCURY IN THE ENVIRONMENT ON A VARIETY OF SCALES

9.04.8.1 Acid Rain and Mercury Synergy in Lakes

As noted, sulfate reduction is thought to be the driving biogeochemical process behind mercury methylation in many ecosystems. In freshwaters that are relatively low in sulfate, recent increases in sulfate deposition associated with the acid rain phenomenon are hypothesized to result in increased rates of sulfate reduction and subsequently in mercury methylation (e.g., Gilmour and Henry, 1991; see Chapter 9.10, for details on acid rain geochemistry). Thus, not only are lakes receiving more total mercury than in the preindustrial past, they may be methylating a greater proportion of this load. Swain and Helwig (1989), for example, have noted that the concentration of mercury in Minnesota fish (usually dominated by MMHg), has increased by 10 ×, while deposition of mercury has increased by only 3 ×. Recent work by Branfireun *et al.* (1999, 2001) have recreated this phenomenon through sulfate additions to microcosms in the Experimental Lakes Area of Ontario, and there are ongoing efforts to expand this research to watershed/whole-lake scales (D. R. Engstrom, personal communication).

9.04.8.2 METAALICUS

METAALICUS is a project titled *Mercury Experiment To Assess Atmospheric Loading In Canada and the United States*. As described in this chapter, the atmosphere is the principal avenue for the mobilization of mercury in the environment and anthropogenic mercury emissions to the atmosphere, especially from coal combustion, have contaminated all the major reservoirs. We noted the strong interest and support for studies examining linkages between the cycling of mercury in the atmosphere, anthropogenic emissions, deposition to terrestrial systems, and the bioaccumulation of MMHg in freshwaters. We have suggested that there has been much emphasis on freshwater fish contamination at the expense of marine studies, given that the major exposure of humans to MMHg is through the consumption of marine fish and seafood products. Given our concerns as a caveat, we note that many scientists and "stake holders" think that an unequivocal connection between direct atmospheric mercury deposition and the levels of MMHg in fish in freshwater or marine systems has not been established. In other words, if mercury deposition is reduced, through, for example, controls on mercury emissions from coal-burning power plants, would MMHg concentrations in fish decline? The METAALICUS team of scientists from the US and Canada is now attempting empirically to address questions of linkages between mercury deposition and the bioaccumulation of MMHg in fish. They are conducting a whole ecosystem experiment in which stable isotopes of mercury are added to a lake, its upland watershed and adjoining wetland, and mercury in fish as well as the food chain are examined. The project, which began in 1999 and is funded through 2004, is being conducted in the Experimental Lakes region.

The objectives of METAALICUS are as follows (taken from the METAALICUS web page: http://www.umanitoba.ca/institutes/fisheries/METAALICUS1.html):

(i) "To determine the relationship between the rate of atmospheric loading of mercury to lakes and the rate of mercury accumulation in fish.

(ii) To determine the response time of mercury levels in fish to changes in atmospheric loading of mercury.

(iii) To determine the relative importance of different sources of mercury to fish—the upland watersheds, adjacent wetlands, and the lake surface, by the addition of three different stable isotopes of mercury.

(iv) To calibrate and parameterize a mechanistic model that will be used to predict reductions in fish mercury concentrations following controls on mercury emissions."

The use of stable isotopes of mercury to track pathways of inputs and uptake is innovative. The additions will be increased over time to levels comparable to the current mercury deposition in the northeastern US. The work is ongoing and, as of 2002, one paper has been published (Hintelmann *et al.*, 2002). We wish to reemphasize that in productive near-shore regions of marine ecosystems, the legacy of pollution derived mercury in the surficial sediments is likely to predominate over "new mercury" as a substrate for methylation. The intense bioturbation in coastal marine sediments can keep much historical mercury active, relative to the more quiescent sediments of lakes. Unfortunately, the mechanistic predictions for declines in fish mercury levels following controls on mercury emissions, derived from the anticipated successful METAALICUS program, will not be applicable to the marine environment.

9.04.8.3 Ambient Isotopic Studies of Mercury—Biological Fractionation of Mercury Isotopes

With improvements in ICPS-MS detectors, there is great interest in the possibility that there may be biologically induced fractionation of the stable isotopes of mercury. As noted, there have only recently been meaningful measurements of the isotopic composition of mercury in media other than rocks and meteorites. This is largely due to the difficulty in measuring isotopic ratios with suitable accuracy for such a heavy element. However, the range of isotopic masses presented by the mercury system (196–204) at the extreme offers a 4% difference in mass which approaches that of elements whose isotopic composition are routinely used to study cycling in the environment (e.g., $^{14}N-^{15}N$: 7%). It is theoretically possible that biological and physical processes could fractionate mercury by mass.

If fractionation of mercury isotopes does occur, then their possible uses are obviously extraordinary. Isotopic variations would provide biogeochemical tracers for the mercury cycle. For example, the mercury incorporated into MMHg should be isotopically lighter than the bulk mercury of sediment pore waters. Watersheds/rivers may have a unique isotopic patterns, while wastewater effluents another. Coals from different regions may show variations that would characterize the mercury emissions, as has been documented for lead. Perhaps, pre-industrial atmospheric deposition will show a mercury isotopic signature distinct from the modern distribution. The possibilities are many, but evidence for biologically induced fractionation is lacking. Interest was spurred by a recent controversial report (Jackson, 2001a; Hintelmann *et al.*, 2001; Jackson, 2001b) indicating substantial fractionation of mercury isotopes in a lacustrine food chain. There is ongoing research in several laboratories (e.g., Amouroux *et al.*, 2003; Krupp *et al.*, 2003), but few publications as we go to press. Indeed, and to reiterate, as of early 2000s, to our knowledge, no study has been published that convincingly demonstrates ambient fractionation of mercury. The instrumental capability is young, and improvements in technologies and sensitivity in this area of research will be very active in the future.

9.04.8.4 Tracing Atmospheric Mercury with ^{210}Pb and Br

As noted in Section 9.04.4, an ongoing effort among researchers studying the atmospheric chemistry of mercury is the determination of which reaction mechanisms are most influential in the field. Tracers are exceptionally powerful tools in this regard, as they greatly simplify the process of scaling up individual measurements to a wide range of scales. They are often integrative as well, so that discrete measurements of mercury and tracers in a particular medium at a particular time can be easily generalized to a whole year or area.

Two tracers have recently shown some promise in understanding the large-scale behavior of mercury in the atmosphere and in complementary ways: ^{210}Pb and Br.

Lamborg *et al.* (1999, 2000) have demonstrated a strong correlation between Hg and ^{210}Pb in precipitation from samples collected in remote continental (Wisconsin) and mid-ocean (tropical South Atlantic) locations. The relationship from these collections was very similar and indicates that ^{210}Pb may be a useful tracer of mercury in precipitation. The geochemistry of ^{210}Pb is well described and this tracer has been useful for atmospheric studies of numerous kinds. The correlation between these two species implicates, by analogy, a homogeneous gas-phase oxidation of mercury that is first order as ^{210}Pb is generated from the homogeneous-phase first-order radioactive decay of ^{222}Rn. As the flux of ^{222}Rn to the atmosphere is known and constant, the current flux of mercury to and from the atmosphere can be estimated from $Hg/^{210}Pb$ ratios, and the universality of the ratio used to examine regionality of mercury deposition.

Two recent findings suggest that bromine may also be an important tracer and clue in the atmospheric chemistry of mercury. Lindberg *et al.* (2002) have noted that the advent of mercury depletion events in the Arctic is coincident with the buildup of reactive bromine compounds (such as BrO) in the polar atmosphere. Halogen compounds are potent oxidizers of mercury, stabilizing the Hg(II) created by forming halogen complexes. For example, BrCl is routinely used to oxidize solutions for mercury analysis in the laboratory. It has also been noted that such reactive compounds can be created *in situ* from sea salt in aerosols in the marine boundary layer (e.g., Disselkamp *et al.*, 1999), supporting the hypothesis contained in Mason *et al.* (2001) of a rapid oxidation of elemental mercury over the ocean. Roos-Barraclough *et al.* (2002) recently presented findings from long peat cores that showed a strong correlation between higher deposition of mercury coincident with higher deposition of bromine over relatively long time-scales (centuries). The cause of this correlation is still unknown and may have as much to do with sources of mercury to the atmosphere as well as the atmospheric chemistry of mercury.

9.04.8.5 Mercury and Organic Matter Interactions

Mercury, the quintessential soft metal, forms exceptionally strong associations with natural organic matter (e.g., Mantoura *et al.*, 1978). This behavior has been recognized as influential in a number of aspects of mercury biogeochemistry. For example, lakes have often been shown to have higher mercury in higher dissolved organic carbon (DOC) waters (e.g., Vaidya *et al.*, 2000). This indicates that watersheds (the principal supply of the DOC in lakes) can contribute significant amounts of mercury as well, and that the DOC mobilization from uplands mobilizes mercury and MMHg. In addition, higher concentrations of DOC provide enhanced complexation capacity for mercury in the water column of lakes, enhancing mercury solubility and increasing mercury residence within the lake (e.g., Ravichandran *et al.*, 1999). There are also competing effects of DOC on the reduction of mercury within natural waters. Recently, Amyot *et al.* (1997), Rolfhus (1998), and Costa and Liss (1999) have indicated that DOC may act to both enhance and inhibit mercury reduction (and therefore mercury evasion) in both freshwater and salt water depending on the concentration of organic carbon in the water. The mechanisms for these intricate redox processes are not well known, but DOC is clearly an important master variable.

The magnitude of mercury–organic interactions (strength of the complexes and the abundance of the complexing agents) has been studied through partitioning experiments of various kinds. Mantoura *et al.* (1978), a frequently cited report, made use of a form of size exclusion chromatography to separate free mercury from organically complexed mercury. Other important work includes that of Hintelmann *et al.* (1997), who employed dialysis membranes to estimate MMHg partitioning between organic and inorganic complexes, and the recent work of Benoit *et al.* (2001c), who used calibrated octanol–water partitioning behavior of mercury species to study mercury-binding ligands in sediment pore waters from the Everglades. For other trace metals, electrochemical techniques have often been used, and one such study has been published for mercury (Wu *et al.*, 1997). Watson *et al.* (1999) have suggested recently that the gold stripping electrodes used for this determination do not quantitatively release mercury, making this approach somewhat suspect. Lamborg *et al.* (2003), who developed a wet chemical analogue to the electrochemical approach, found affinity results similar to those of Benoit *et al.* and the high end of Mantoura *et al.* from bulk water samples (most of the other studies have been performed on isolated DOC fractions). The general assumption has been that it is the reduced sulfur moieties (thiols) in the macromolecules of natural water DOC that are the sites for binding, and recent spectroscopic evidence supports this (Xia *et al.*, 1999; Hesterberg *et al.*, 2001). However, Lamborg *et al.* (2003) have found ligand concentrations, when normalized to DOC, which suggest mercury-binding functional group abundances on the order of parts per million. This is well below the abundance of reduced sulfur, which is generally found to be in the parts per thousand range. It therefore appears that the locations for mercury binding are rare in the DOC pool but still present in 10–1,000× excess of mercury.

In saltwater, organic matter complexation of mercury may not compete with the more abundant chloride ion (see Section 9.04.5). In estuaries, ligand exchange was observed by Rolfhus and Fitzgerald (2001) and by Tseng *et al.* (2001). One result of the exchange of mercury from organic to inorganic complex forms is a general increase in the reactivity of mercury within estuaries. The change in complexation can result in dramatic changes in the reactivity of the mercury as a result (e.g., enhanced Hg° production, Rolfhus and Fitzgerald, 2001).

9.04.8.6 Bioreporters—A New Technique for Ultra-trace Determinations of Mercury

Modern biotechnology offers the potential for the development of highly sensitive reagents and methods to determine mercury species such as bioreactive or bioavailable mercury and MMHg. In one of the earliest studies on the subject, Barkay and colleagues (Selifonova *et al.*, 1993) suggested that mercury-specific luminescent reagents could be bioengineered (i.e., biologically synthesized and amplified with molecular techniques), for use in environmental studies. They reported the successful development of three "biosensors" for Hg(II). These were created by combining or fusing the "the well-understood Tn21 mercury resistance operon (*mer*) with promoterless *lux-CDABE* from the marine bacterium, *Vibrio fischer*." Light production via *luxCDABE* is controlled by the *mer* regulatory gene and promoter, which is in turn activated by reactive mercury species. These combined reagents are commonly referred to as "bioreporters," where the light can be detected photonically and quantified. Aspects of the techniques are patented (e.g., Rosson, 1996, patent states "the lux operon complex comprises *luxC*, *luxD*, *luxA*, *luxB*, and *luxE* genes but is free of (1) a promoter for the complex and (2) an inducible regulatory gene for the complex." The Sanseverino *et al.* (2002) US patent application indicates that "an exemplary bioreporter is an *E. coli* that has been modified to respond to mercury II as a result of incorporation of a *merRop/lux* gene cassette into its genome."

The first applications were encouraging. However, the mercury "bioreporters" were semiquantitative and not sensitive at environmentally useful levels (i.e., pM to fM). Progress in lowering detection limits has been swift. For example, in 1993, Selifonova and co-workers could work in the 1 nM range, and Tescione and Belfort (1993) at 10 nM. Virta *et al.* (1995) reported a laboratory detection limit of 0.1 fM for mercury using "firefly luciferase gene as a reporter, the mer promoter from transposon Tn21, and *Escherichia coli* (*E. coli*) MC1061 as the host organism." Today, the methods, in general, are quite advanced and the sensitivity appears suitable for studies even in pristine ecosystems such as the open ocean and remote lakes (see recent reports by Kelly *et al.*, 2001; Scott *et al.*, 2001; Barrocas *et al.*, 2001).

Quantification, the operational nature of "bioavailability" and the need to add a suspension of non-standardized genetically engineered bacteria as an integral part of the analysis are limitations to broad application of the "bioreporter" techniques to environmental studies of mercury in aqueous systems. However, variations on this bacterial recombinant DNA technological approach are promising. That is, mercury activated organic molecules can be bioengineered, isolated, and purified to provide reagents for mercury investigations. Wylie *et al.* (1991, 1992) produced monoclonal antibodies specific for ionic mercury and applied the immunoassay to water. Their experiments were conducted in the nM range, concentrations too high for investigating mercury in most natural aquatic environments. However, it is likely that the detection limits can be lowered significantly, and the mercury-specific antibody reagent technique optimized for natural waters.

9.04.9 SUMMARY

In this chapter, we have summarized some of the gains made in understanding the environmental biogeochemistry of mercury since Goldschmidt's groundbreaking work. Much of this advancement has come since the early 1970s, and the growth in mercury research continues at breakneck pace. This is fortunate as there is a need for urgency, we believe, in these endeavors. While human activity has perturbed the mercury cycle by a smaller degree than, for example, lead, the implications for continued perturbation on human and ecological health are enormous.

The way forward will be a fascinating and challenging one. As we have summarized, this is because the biogeochemistry of mercury operates at a variety of time and space scales and in many environmental media. Due to the complexity of the processes and the minute quantities of material often encountered in the environment, future research will also require new hypotheses and new instrumentation. Similarly, and as with so many environmental research efforts, new collaborations among scientific disciplines will be required.

ACKNOWLEDGMENTS

Foremost, we thank the editor, Barbara Sherwood-Lollar. Rob Mason and Gary Gill provided some graphs, Todd Hinkley and Johan Varekamp provided guidance for the volcano section. The volcano study was aided greatly by F. LeGuern, CNRS and P. Buat-Menard, University of Bordeaux, France; W. Giggenbach (deceased), DSIR, New Zealand; J. Spedding (deceased), University of Auckland, New Zealand; L.P. Greenland, USGS, C. C. Patterson (deceased) and D. Settle (retired). Chad Hammerschmidt assisted with the sections on methylation. Dan Engstrom, Don Porcella, Jim Wiener, and Kris Rolfhus graciously read drafts and provided many useful suggestions.

REFERENCES

Aastrup M., Johnson J., Bringmark E., Bringmark L., and Iverfeldt Å. (1991) Occurrence and transport of mercury within a small catchments area. *Water Air Soil Pollut.* **56**, 155–168.

Alleman L. Y., Church T. M., Veron A. J., Kim G., Hamelin B., and Flegal A. R. (2001) Isotopic evidence of contaminant lead in the South Atlantic troposphere and surface waters. *Deep-Sea Res. II* **48**(13), 2811–2827.

Alriksson A. (2001) Regional variability of Cd, Hg, Pb and C concentrations in different horizons of Swedish forest soils. *Water Air Soil Pollut. Focus* **1**, 325–341.

Amouroux D., Donard O. F., Krupp E., Pecheyran C., Besson T., and Fitzgerald W. F. (2003) Isotopic fractionation of mercury species in top predators fishes (tuna and whale) from world ocean: preliminary results. In *Abstracts of ASLO Aquatic Sciences Meeting, Salt Lake City, UT, USA*, 8–14 February.

Amyot M., Mierle G., Lean D. R. S., and McQueen D. J. (1994) Sunlight-induced formation of dissolved gaseous mercury in lake waters. *Environ. Sci. Technol.* **28**(13), 2366–2371.

Amyot M., Gill G. A., and Morel F. M. M. (1997) Production and loss of dissolved gaseous mercury in coastal seawater. *Environ. Sci. Technol.* **31**, 3606–3611.

Amyot M., AuClair J. C., and Poissant L. (2001) *In situ* high temporal resolution analysis of elemental mercury in natural waters. *Anal. Chim. Acta* **447**, 153–159.

Asmund G. and Nielsen S. P. (2000) Mercury in dated Greenland marine sediments. *Sci. Tot. Environ.* **245**, 61–72.

Baeyans W. and Leermakers M. (1998) Elemental mercury concentrations and formation rates in the Scheldt estuary and the North Sea. *Mar. Chem.* **60**, 257–266.

Baker P. G. L., Brunke E. G., Slemr F., and Crouch A. M. (2002) Atmospheric mercury measurements at Cape Point, South Africa. *Atmos. Environ.* **36**(14), 2459–2465.

Bakir F., Damluji S. F., Amin-Zaki L., Murtadha M., Khalidi A., al-Rawi N. Y., Tikriti S., Dahahir H. I., Clarkson T. W., Smith J. C., and Doherty R. A. (1973) Methylmercury poisoning in Iraq. *Science* **181**(96), 230–241.

Balcom P. H., Fitzgerald W. F., Lamborg C. H., Rolfhus K. R., Vandal G. M., Langer C. S., and Tseng C.-M. (2000) Mercury cycling in and emissions from Long Island Sound. Presented at 11th International Conference on Heavy Metals in the Environment. Ann Arbor, MI, USA, August.

Banic C. M., Beauchamp S. T., Tordor R. J., Schroeder W. H., Steffen A., Anlauf K. A., Wong H. K. T. (2003) Vertical distribution of gaseous elemental mercury in Canada. *J. Geo. Res.* **108**(D9), Art no. 4264.

Barrie L. A. and Hoff R. M. (1985) Five years of air chemistry observations in the Canadian Arctic. *Atmos. Environ.* **19**, 1995–2010.

Barrocas P. R. G., Landing W. M., and Procter L. M. (2001) A bacterial biosensor for aquatic mercury(II) speciation and bioavailability. Abstracts 6th International Conference on Mercury as a Global Pollutant, Minamata, Japan, 130p.

Beals D. I., Harris R. C., and Pollman C. (2002) Predicting fish mercury concentrations in Everglades marshes: Handling uncertainty in the Everglades mercury cycling model (E-MCM) with a Monte Carlo approach. Abstracts of papers of the American Chemical Society 223: 200-ENVR, Part 1 APR 7 2002.

Benesch J. A., Gustin M. S., Schorran D. E., Coleman J., Johnson D. A., and Lindberg S. E. (2001) Determining the role of plants in the biogeochemical cycling of mercury on an ecosystem level. Presented at the 6th International Conference on Mercury as a Global Pollutant. Minamata, Japan, October.

Benoit J. M., Gilmour C. C., Mason R. P., Riedel G. S., and Riedel G. F. (1998a) Behavior of mercury in the Patuxent River estuary. *Biogeochemistry* **40**, 249–265.

Benoit J. M., Fitzgerald W. F., and Damman A. W. (1998b) The biogeochemistry of an ombrotrophic bog: evaluation of use as an archive of atmosphere mercury deposition. *Environ. Res.* **78**(2), 118–133.

Benoit J. M., Gilmour C. C., Mason R. P., and Heyes A. (1999a) Sulfide controls on mercury speciation and bioavailability to methylating bacteria in sediment pore waters. *Environ. Sci. Technol.* **33**, 951–957.

Benoit J. M., Mason R. P., and Gilmour C. C. (1999b) Estimation of mercury-sulfide speciation in sediment pore waters using octanol-water partitioning and implications for availability to methylating bacteria. *Environ. Toxicol. Chem.* **18**, 2138–2141.

Benoit J. M., Gilmour C. C., and Mason R. P. (2001a) Aspects of bioavailability of mercury for methylation in pure cultures of *Desulfobulbus propionicus* (1pr3). *Appl. Environ. Microbiol.* **67**, 51–58.

Benoit J. M., Gilmour C. C., and Mason R. P. (2001b) The influence of sulfide on solid-phase mercury bioavailability for methylation by pure cultures of *Desulfobulbus propionicus* (1pr3). *Environ. Sci. Technol.* **35**, 127–132.

Benoit J. M., Mason R. P., Gilmour C. C., and Aiken G. R. (2001c) Constants for mercury binding by dissolved organic matter isolates from the Florida Everglades. *Geochim. Cosmochim. Acta* **65**(24), 4445–4451.

Benoit J. M., Gilmour C. C., Heyes A., Mason R. P., and Miller C. L. (2003) Geochemical and biological controls over methylmercury production and degradation in aquatic ecosystems. In *Biogeochemistry of Environmentally Important Trace Elements*. ACS Symposium Series #835 (eds. Y. Chai and O. C. Braids). American Chemical Society, Washington, DC, pp. 262–297.

Bergan T. and Rodhe H. (2001) Oxidation of elemental mercury in the atmosphere: constraints imposed by global scale modeling. *J. Atmos. Chem.* **40**, 191–212.

Bergan T., Gallardo L., and Rodhe H. (1999) Mercury in the global troposphere: a three-dimensional model study. *Atmos. Environ.* **33**, 1575–1585.

Bindler R., Renberg I., Appleby P. G., Anderson N. J., and Rose N. L. (2001) Mercury accumulation rates and spatial patterns in lake sediments from West Greenland: a coast to ice margin transect. *Environ. Sci. Technol.* **35**(9), 1736–1741.

Bloom N. (1989) Determination of picogram levels of methylmercury by aqueous phase ethylation, followed by cryogenic gas chromatography with cold vapor atomic fluorescence detection. *Can. J. Fish. Aquat. Sci.* **46**(7), 1131–1140.

Bloom N. S. (1992) On the chemical form of mercury in edible fish and marine invertebrate tissue. *Can. J. Fish. Aquat. Sci.* **49**, 1010–1017.

Bloom N. S. (2000) Analysis and stability of mercury speciation in petroleum hydrocarbons. *Fres. J. Anal. Chem.* **366**, 438–443.

Bloom N. S. and Fitzgerald W. F. (1988) Determination of volatile mercury species at the picogram level by low temperature gas chromatography with cold vapor atomic fluorescence detection. *Anal. Chim. Acta* **208**, 151–161.

Bluth G. J. S., Schnetzler C. C., Krueger A. J., and Walter L. S. (1993) The contribution of explosive volcanism to global atmospheric sulfur dioxide concentrations. *Nature* **366**, 327–329.

Bodaly R. A., Hecky R. E., and Fudge R. J. P. (1984) Increases in fish mercury levels in lakes flooded by the Churchill River Diversion, northern Manitoba. *Can. J. Fish. Aquat. Sci.* **41**, 682–691.

Bostrom K. and Fisher D. E. (1969) Distribution of mercury in East Pacific sediments. *Geochim. Cosmochim. Acta* **33**(6), 743–745.

Boudala F. S., Folkins I., Beauchamp S., Tordon R., Neima J., and Johnson B. (2000) Mercury flux measurements over air and water in Kejimkujik National Park, Nova Scotia. *Water Air Soil Pollut.* **122**, 183–202.

Boutron C. F., Vandal G. M., Fitzgerald W. F., and Ferrari C. P. (1998) A forty year record of mercury in central Greenland snow. *Geophys. Res. Lett.* **25**(17), 3315–3318.

Branfireun B. A., Roulet N. T., Kelly C. A., and Rudd J. W. M. (1999) *In situ* sulphate stimulation of mercury methylation in a boreal peatland: toward a link between acid rain and methylmercury contamination in remote environments. *Global Biogeochem. Cycles* **13**(3), 743–750.

Branfireun B. A., Bishop K., Roulet N. T., Granberg G., and Nilsson M. (2001) Mercury cycling in boreal ecosystems: the long-term effect of acid rain constituents on peatland pore water methylmercury concentrations. *Geophys. Res. Lett.* **28**(7), 1227–1230.

Brunke E. G., Labuschagne C., and Slemr F. (2001) Gaseous mercury emissions from a fire in the Cape Peninsula, South Africa, during January 2000. *Geophys. Res. Lett.* **28**(8), 1483–1486.

Cadle R. D. (1975) Volcanic emissions of halides and sulfur compounds to the troposphere and stratosphere. *J. Geophys. Res.* **80**(12), 1650–1652.

Carpi A. and Lindberg S. E. (1998) Application of a Teflon dynamic flux chamber for quantifying soil mercury flux: test and results over background soil. *Atmos. Environ.* **32**(5), 873–882.

Chiaradia M. and Cupelin F. (2000) Gas-to-particle conversion of mercury, arsenic and selenium through reactions with traffic-related compounds (Geneva)? Indications from lead isotopes. *Atmos. Environ.* **34**(2), 327–332.

Clarkson T. W. (1997) The toxicology of mercury. *Crit. Rev. Clin. Lab. Sci.* **34**(4), 369–403.

Compeau G. C. and Bartha R. (1985) Sulfate-reducing bacteria: principal methylators of mercury in anoxic estuarine sediment. *Appl. Environ. Microbiol.* **50**(2), 498–502.

Conaway C. H., Squire S., Mason R. P., and Flegal A. R. (2003) Mercury speciation in the San Francisco Bay estuary. *Mar. Chem.* **80**, 199–225.

Coquery M., Cossa D., and Martin J.-M. (1995) The distribution of dissolved and particulate mercury in 3 Siberian estuaries and adjacent arctic coastal waters. *Water Air Soil Pollut.* **80**(1–4), 653–664.

Coquery M., Cossa D., and Sanjuan J. (1997) Speciation and sorption of mercury in two macro-tidal estuaries. *Mar. Chem.* **58**, 213–227.

Cossa D., Sanjuan J., and Noel J. (1994) Mercury transport in waters of the Strait of Dover. *Mar. Pollut. Bull.* **28**(6), 385–388.

Cossa D., Coquery M., Gobeil C., and Martin J.-M. (1996) Mercury fluxes at the ocean margins. In *Global and Regional Mercury Cycles: Sources, Fluxes and Mass Balances*, NATO ASI: Series 2. Environment (eds. W. Baeyens, R. Ebinghaus, and O. Vasiliev). Kluwer, Boston, vol. 21, pp. 229–248.

Cossa D., Martin J.-M., Takayanagi K., and Sanjuan J. (1997) The distribution and cycling of mercury species in the western Mediterranean. *Deep-Sea Res. II* **44**(3–4), 721–740.

Cossa D., Elbaz-Poulichet F., and Nieto J. M. (2001) Mercury in the Tinto-Odiel estuarine system (Gulf of Cádiz, Spain): sources and dispersion. *Aquat. Geochem.* **7**, 1–12.

Costa M. and Liss P. S. (1999) Photoreduction of mercury in sea water and its possible implications for Hg-0 air–sea fluxes. *Mar. Chem.* **68**(1–2), 87–95.

Covelli S., Faganeli J., Horvat M., and Brambati A. (2001) Mercury contamination of coastal sediments as the result of long-term cinnabar mining activity (Gulf of Trieste, northern Adriatic Sea). *App. Geochem.* **16**, 541–558.

Cox J. A., Carnahan J., Dinunzio J., McCoy J., and Meister J. (1979) Source of mercury in new impoundments. *Bull. Environ. Contamin. Toxicol.* **23**, 779.

Crisp J. A. (1984) Rates of magma emplacement and volcanic output. *J. Volcanol. Geotherm. Res.* **20**, 177–211.

Dalziel J. A. (1992) Reactive mercury on the Scotian Shelf and in the adjacent Northwest Atlantic Ocean. *Mar. Chem.* **37**(3–4), 171–178.

Dalziel J. A. and Yeats P. A. (1985) Reactive mercury in the central North Atlantic Ocean. *Mar. Chem.* **15**(4), 357–361.

Damman A. W. H. (1978) Distribution and movement of elements in ombrotrophic peat bogs. *Oikos* **30**, 480–495.

Davidson P. W., Palumbo D., Myers G. J., Cox C., Shamlaye C. F., Sloane-Reeves J., Cernichiari E., Wilding G. E., and Clarkson T. W. (2000) Neurodevelopmental outcomes of Seychellois children from the pilot cohort at 108 months following prenatal exposure to methylmercury from a maternal fish diet. *Environ. Res.* **84**(1), 1–11.

Dedeurwaerder H., Decadt G., and Baeyens W. (1982) Estimations of mercury fluxes emitted by Mount Etna volcano. *Bull. Volcanol.* **45**(3), 191–196.

Disselkamp R. S., Chapman E. G., Barchet W. R., Colson S. D., and Howd C. D. (1999) BrCl production in NaBr/NaCl/HNO_3/O^3 solutions representative of sea-salt aerosols in the marine boundary layer. *Geophys. Res. Lett.* **26**(4), 2183–2186.

Ebinghaus R. and Krüger O. (1996) Emission and local deposition estimates of anthropogenic mercury in North Western and Central Europe. In *Global and Regional Mercury Cycles: Sources, Fluxes and Mass Balances*. NATO ASI: Series 2. Environment (eds. W. Baeyens, R. Ebinghaus, and O. Vasiliev). Kluwer, Boston, vol. 21, pp. 135–159.

Ebinghaus R., Turner R. R., Lacerda L. D., Vasiliev B. O., and Salomons W. (eds.) (1998) *Mercury Contaminated Sites—Characterization, Risk Assessment and Remediation* Springer, Regensburg, Germany, 548pp.

Ebinghaus R., Kock H. H., and Schmolke S. R. (2001) Measurements of atmospheric mercury with high time resolution: recent applications in environmental research and monitoring. *Fres. J. Anal. Chem.* **371**(6), 806–815.

Ebinghaus R., Kock H. H., Temme C., Einax J. W., Lowe A. G., Richter A., Burrows J. P., and Schroeder W. H. (2002) Antarctic springtime depletion of atmospheric mercury. *Environ. Sci. Technol.* **36**(6), 1238–1244.

Egleston T. (1887) *The Metallurgy of Silver, Gold, and Mercury in the United States*. Wiley, NY.

Engstrom D. R. and Swain E. B. (1997) Recent declines in atmospheric mercury deposition in the upper Midwest. *Environ. Sci. Technol.* **31**(4), 960–967.

Evans R.D., Hintelmann H. and Dillon P.J. (2001) Natural variation in mercury isotope ratios of coals and cinnabars. *Proc. 6th Int. Conf. Mercury Glob. Pollut.* Minimata, Japan, Oct. 15–19.

Fabbri D., Gabbianelli G., Locatelli C., Lubrano D., Trombini C., and Vassura I. (2001) Distribution of mercury and other heavy metals in core sediments of the northern Adriatic Sea. *Water Air Soil Pollut.* **129**, 143–153.

Fein J. B. and Williams-Jones A. E. (1997) The role of mercury-organic interactions in the hydrothermal transport of mercury. *Econ. Geo.* **92**, 20–28.

Fitzgerald W. F. (1981) Volcanic mercury emissions and the global mercury cycle. *Programs and Abstracts, Symposium on the Role of the Ocean in Atmospheric Chemistry.* IAMAP 3rd Scientific Assembly, Hamburg, Germany. August 17–28, 134p.

Fitzgerald W. F. (1989) Atmospheric and oceanic cycling of mercury. In *Chemical Oceanography* (eds. J. P. Riley and R. Chester). Academic Press, New York, chap. 57, vol. 10, pp. 151–186.

Fitzgerald W. F. (1995) Is mercury increasing in the atmosphere? The need for an atmospheric mercury network (AMNET). *Water Air Soil Pollut.* **80**, 245–254.

Fitzgerald W. F. (1996) Mercury emissions from volcanos. In *Abstr. 4th Int. Conf. on Mercury as a Global Pollutant.* Hamburg, Germany, August, pp. 4–7.

Fitzgerald W. F. and Clarkson T. W. (1991) Mercury and monomethylmercury: present and future concerns. *Environ. Hlth. Perspect.* **96**, 159–166.

Fitzgerald W. F. and Gill G. A. (1979) Subnanogram determination of mercury by two-stage gold amalgamation applied to atmospheric analysis. *Anal. Chem.* **51**, 1714–1720.

Fitzgerald W. F. and Mason R. P. (1997) Biogeochemical cycling of mercury in the marine environment. *Metal Ions Biol. Sys.* **34**, 53–111.

Fitzgerald W. F. and Watras C. J. (1989) Mercury in surficial waters of rural Wisconsin lakes. *Sci. Tot. Environ.* **87**(88), 223–232.

Fitzgerald W. F., Gill G. A., and Hewitt A. D. (1981) Mercury a trace atmospheric gas. Presented at the symposium on the role of the ocean in atmospheric chemistry, IAMAP 3rd Scientific Assembly, Hamburg, Germany, August 17–28.

Fitzgerald W. F., Mason R. P., and Vandal G. M. (1991) Atmospheric cycling and air–water exchange of mercury over mid-continental lacustrine regions. *Water Air Soïl Pollut.* **56**, 745–767.

Fitzgerald W. F., Engstrom D. R., Mason R., and Nater E. A. (1998) The case for atmospheric mercury contamination in remote areas. *Environ. Sci. Technol.* **32**(1), 1–7.

Fitzgerald W. F., Vandal G. M., Rolfhus K. R., Lamborg C. H., and Langer C. S. (2000) Mercury emissions and cycling in the coastal zone. *J. Environ. Sci.* **12**, 92–101.

Fogg T. R. and Fitzgerald W. F. (1979) Mercury in southern New England rains. *J. Geophys. Res.* **84**, 6987–6988.

Friedlander G., Kennedy J. W., Macias E. S., and Miller J. M. (1981) *Nuclear and Radiochemistry*, 3rd edn. Wiley, New York, pp. 640–641.

Friedli H. R., Radke L. R., and Lu J. Y. (2001) Mercury in smoke from biomass fires. *Geophys. Res. Lett.* **28**(17), 3223–3226.

Friend J. P. (1973) Global sulfur cycle. In *Chemistry of the Lower Atmosphere* (ed. S. I. Rusool). Plenum, New York, pp. 177–201.

Ganguli P. M., Mason R. P., Abu-Saba K. E., Anderson R. S., and Flegal A. R. (2000) Mercury speciation in drainage from the New Idrija mercury mine, California. *Environ. Sci. Technol.* **34**(22), 4773–4779.

Gårdfeldt K., Munthe J., Strömberg D., and Lindqvist (2003) A kinetic study on the abiotic methylation of divalent mercury in the aqueous phase. *Sci. Total Environ.* **304**, 127–136.

Gbondo-Tugbawa S. and Driscoll C. T. (1998) Application of the regional mercury cycling model to predict the fate and remediation of mercury in Onondaga Lake, New York. *Water Air Soil Pollut.* **105**, 417–426.

Geller L. S., Elkins J. W., Lobert J. M., Clarke A. D., Hurst D. F., Butler J. H., and Myers R. C. (1997) Tropospheric SF6: observed latitudinal distribution and trends, derived emissions and interhemispheric exchange time. *Geophys. Res. Lett.* **24**(6), 675–678.

George T. S. (2001) *Minamata: Pollution and the Struggle for Democracy in Postwar Japan* (Harvard East Asian Monographs, No 194). Harvard University Press, Cambridge, MA, 385pp.

Gerino M., Aller R. C., Lee C., Cochran J. K., Aller J. Y., Green M. A., and Hirschberg D. (1998) Comparison of different tracers and methods used to quantify bioturbation during a spring bloom: 234-thorium, luminophores, and chlorophyll *a*. *Estuar. Coast. Shelf Sci.* **46**, 531–547.

Gill G. A. and Bruland K. W. (1987) Mercury in the northeast Pacific. *EOS* **68**, 1763.

Gill G. A. and Fitzgerald W. F. (1988) Vertical mercury distributions in the oceans. *Geochim. Cosmochim. Acta* **52**, 1719–1728.

Gill G. A., Guentzel J. L., Landing W. M., and Pollman C. D. (1995) Total gaseous mercury measurements in Florida: The FAMS project (1992–1994). *Water, Air Soil Poll.* **80**, 235–244.

Gilmour C. C. and Henry E. A. (1991) Mercury methylation in aquatic systems affected by acid deposition. *Environ. Pollut.* **71**(2–4), 131–169.

Gilmour C. C. and Riedel G. S. (1995) Measurement of Hg methylation in sediments using high specific-activity ^{203}Hg and ambient incubation. *Water Air Soil Pollut.* **80**, 747–756.

Gilmour C. C., Riedel G. S., Ederington M. C., Bell J. T., Benoit J. M., Gill G. A., and Stordal M. C. (1998) Methylmercury concentrations and production rates across a trophic gradient in the northern Everglades. *Biogeochemistry* **40**, 326–346.

Gobeil C. and Cossa D. (1993) Mercury in sediments and sediment pore water in the Laurentian Trough. *Can. J. Fish. Aquat. Sci.* **50**, 1794–1800.

Gobeil C., MacDonald R. W., and Smith J. N. (1999) Mercury profiles in sediments of the Arctic Ocean basins. *Environ. Sci. Technol.* **33**, 4194–4198.

Goldschmidt V. M. (1954) *Geochemistry*. Oxford University Press, Oxford, pp. 274–280.

Goldwater L. J. (1972) Mercury. In *A History of Quicksilver*. York Press, Baltimore, 318pp.

Grandjean P., Weihe P., White R. F., Debes F., Araki S., Yokoyama K., Murata K., Sorensen N., Dahl R., and Jorgensen P. J. (1997) Cognitive deficit in 7-year-old children with prenatal exposure to methylmercury. *Neurotox. Terat.* **19**(6), 417–428.

Grevesse N. (1970) Solar and meteoritic abundances of mercury. *Geochim. Cosmochim. Acta.* **34**, 1129–1130.

Grieb T. M., Driscoll C. T., Gloss S. P., Schofield C. L., Bowie G. L., and Porcella D. B. (1990) Factors affecting mercury accumulation in fish in the Upper Michigan peninsula. *Environ. Toxicol. Chem.* **9**, 919–930.

Grieb T. M., Petcharuttana Y., Roy S., Bloom N., and Brown K. (2001) Mercury studies in the central Gulf of Thailand. *Proc. 6th Int. Conf. Mercury Global Pollut.*, Minimata, Japan, Oct. 15–19.

Grigal D. F., Nater E. A., and Homann P. S. (1994) Spatial distribution patterns of mercury in an east-central Minnesota landscape. In *Mercury Pollution: Integration and Synthesis* (eds. C. Watras and J. Huckabee). Lewis Pubs., Ann Arbor, MI, chap. III. 2, pp. 305–312.

Grigal D. F., Kolka R. K., Fleck J. A., and Nater E. A. (2000) Mercury budget of an upland-peatland watershed. *Biogeochemistry* **50**(1), 95–109.

Guentzel J. L., Landing W. M., Gill G. A., and Pollman C. D. (1995) Atmospheric deposition of mercury in Florida: the FAMS project (1992–1994). *Water, Air, Soil Poll.* **80**, 393–402.

Gustin M. S., Lindberg S., Marsik F., Casimir A., Ebinghaus R., Edwards G., Hubble-Fitzgerald C., Kemp R., Kock H., Leonard T., London J., Majewski M., Montecinos C., Owens J., Pilote M., Poissant L., Rasmussen P., Schaedlich F., Schneeberger D., Schroeder W., Sommar J., Turner R., Vette A., Wallschlaeger D., Xiao Z., and Zhang H. (1999) Nevada STORMS project: measurement of mercury emissions from naturally enriched surfaces. *J. Geo. Res.* **104**(D17), 21831–21844.

Gustin M. S., Lindberg S. E., Austin K., Coolbaugh M., Vette A., and Zhang H. (2000) Assessing the contribution of natural sources to regional atmospheric mercury budgets. *Sci. Tot. Environ.* **259**, 61–71.

Hall B., Schager P., and Ljungstrom E. (1995) An experimental-study on the rate of reaction between mercury-vapor and gaseous nitrogen-dioxide. *Water Air Soil Pollut.* **81**(1–2), 121–134.

Hall B. D., Rosenberg D. M., and Wiens A. P. (1998) Methyl mercury in aquatic insects from an experimental reservoir. *Can. J. Fish. Aquat. Sci.* **55**, 2036–2047.

Halmer M. M., Schmincke H. U., and Graf H. F. (2002) The annual volcanic gas input into the atmosphere, in particular into the stratosphere: a global data set for the past 100 years. *J. Volcanol. Geotherm. Res.* **115**(3–4), 511–528.

Hammerschmidt C. R., Sandheinrich M. B., Wiener J. G., and Rada R. G. (2002) Effects of dietary methylmercury on reproduction of fathead minnows. *Environ. Sci. Technol.* **36**(5), 877–883.

Hanson P. J., Lindberg S. E., Tabberer T., Owens J. G., and Kim K.-H. (1995) Foliar Exchange of Mercury Vapor: evidence for a Compensation Point. *Water Air Soil Pollut.* **80**, 373–382.

Herut B., Hornung H., Kress N., and Cohen Y. (1996) Environmental relaxation in response to reduced contaminant input: the case of mercury pollution in Haifa Bay, Israel. *Mar. Pollut. Bull.* **32**(4), 366–373.

Hesterberg D., Chou J. W., Hutchison K. J., and Sayers D. E. (2001) Bonding of Hg(II) to reduced organic sulfur in humic acid as affected by S/Hg ratio. *Environ. Sci. Technol.* **35**(13), 2741–2745.

Hines M. E. and Jones G. E. (1985) Microbial biogeochemistry in the sediments of Great Bay, New Hampshire. *Estuar. Coast. Shelf Sci.* **20**, 729–742.

Hinkley T. K., Lamothe P. J., Wilson S. A., Finnegan D. L., and Gerlach T. M. (1999) Metal emissions from Kilauea, and a suggested revision of the estimated worldwide metal output by quiescent degassing of volcanoes. *Earth Planet. Sci. Lett.* **170**(3), 315–325.

Hintelmann H. and Evans R. D. (1997) Application of stable isotopes in environmental tracer studies-Measurement of monomethylmercury (CH_3Hg^+) by isotope dilution ICP-MS and detection of species transformation. *Fres. J. Anal. Chem.* **358**, 378–385.

Hintelmann H., Welbourn P. M., and Evans R. D. (1997) Measurement of complexation of methylmercury(II) compounds by freshwater humic substances using equilibrium dialysis. *Environ. Sci. Technol.* **31**(2), 489–495.

Hintelmann H., Keppel-Jones K., and Evans R. D. (2000) Constants of mercury methylation and demethylation rates in sediments and comparison of tracer and ambient mercury availability. *Environ. Tox. Chem.* **19**(9), 2204–2211.

Hintelmann H., Dillon P., Evans R. D., Rudd J. W. M., and Bodaly R. A. (2001) Comment: variations in the isotope composition of mercury in a freshwater sediment sequence and food web. *Can. J. Fish. Aquat. Sci.* **58**(11), 2309–2311.

Hintelmann H., Harris R., Heyes A., Hurley J. P., Kelly C. A., Krabbenhoft D. P., Lindberg S., Rudd J. W. M., Scott K. J., and St. Louis V. L. (2002) Reactivity and mobility of new and old mercury deposition in a boreal forest ecosystem during the first year of the METAALICUS study. *Environ. Sci. Technol.* **36**(23), 5034–5040.

Hoefs J. (1980) *Stable Isotope Geochemistry*. Springer, Heidelburg.

Hoyer M., Burke J., and Keeler G. (1995) Atmospheric sources, transport and deposition of mercury in Michigan: two years of event precipitation. *Water Air Soil Pollut.* **80**, 192–208.

Hudson R. J. M., Gherini S. A., Watras C. J., and Porcella D. B. (1994) Modeling the biogeochemical cycle of mercury in lakes: the mercury cycling model (MCM) and its application to the MTL study lakes. In *Mercury Pollution: Integration and Synthesis* (eds. C. Watras and J. Huckabee). Lewis Publishers, Ann Arbor, MI, chap. V.1, pp. 473–526.

Hudson R. J. M., Gherini S. A., Fitzgerald W. F., and Porcella D. B. (1995) Perturbation of the global mercury cycle: a model-based analysis. *Water Air Soil Pollut.* **80**, 192–208.

Hultberg H., Iverfeldt A., and Lee Y.-H. (1994) Methylmercury input/output and accumulation in forested catchments and critical loads for lakes in southwestern Sweden. In *Mercury Pollution: Integration and Synthesis* (eds. C. Watras and J. Huckabee). Lewis Publishers, Ann Arbor, MI, chap. III. 3, pp. 313–322.

Hurley J. P., Benoit J. M., Babiarz C. L., Shafer M. M., Andren A. W., Sullivan J. R., Hammond R., and Webb D. A. (1995) Influences of watershed characteristics on mercury levels in Wisconsin rivers. *Environ. Sci. Technol.* **29**(7), 1867–1875.

Hurley J. P., Krabbenhoft D. P., Cleckner L. B., Olson M. L., Aiken G. R., and Rawlik P. S., Jr. (1998) System controls on the aqueous distribution of mercury in the northern Florida Everglades. *Biogeochemistry* **40**, 293–310.

Iverfeldt Å. (1991a) Occurrence and turnover of atmospheric mercury over the Nordic countries. *Water Air Soil Pollut.* **56**, 251–265.

Iverfeldt Å. (1991b) Mercury in forest canopy throughfall water and its relation to atmospheric deposition. *Water Air Soil Pollut.* **56**, 553–564.

Iverfeldt Å. and Lindqvist O. (1982) Distribution equilibrium of methyl mercury chloride between water and air. *Atmos. Environ.* **16**, 2917–2925.

Jackson T. A. (2001a) Variations in the isotope composition of mercury in a freshwater sediment sequence and food web. *Can. J. Fish. Aquat. Sci.* **58**(1), 185–196.

Jackson T. A. (2001b) Reply: variations in the isotope composition of mercury in a freshwater sediment sequence and food web. *Can. J. Fish. Aquat. Sci.* **58**(11), 2312–2316.

Japan Public Health Association (2001) *Methylmercury Poisoning in Minamata and Niigata, Japan* (eds. Okizawa and Osame). Tokyo, 154p.

Jaworowski Z., Bysiek M., and Kownacka L. (1981) Flow of metals into the global atmosphere. *Geochim. Cosmochim. Acta* **45**(11), 2185–2199.

JECFA (2000) *Safety Evaluation of Certain Food Additives and Contaminants WHO Food Additives Series: 44; Methylmercury*. Prep. by 53rd Joint FAO/WHO Expert Committee on Food Additives (JECFA), World Health Organization, Geneva, Switzerland. available at: http:// www. inchem. org/documents/jecfa/jecmono/v44jec13.htm

Jensen A. and Jensen A. (1990) Historical deposition rates of mercury in Scandinavia estimated by dating and measurement of mercury in cores of peat bogs. *Water Air Soil Pollut.* **56**, 769–777.

Johnson D. W. and Lindberg S. E. (1995) The biogeochemical cycling of Hg in forests—alternative methods for quantifying total deposition and soil emission. *Water Air Soil Pollut.* **80**(1–4), 1069–1077.

Jovanovic S. and Reed G. W., Jr. (1976) Interrelations among isotopically anomalous mercury fractions from meteorites and possible cosmological inferences. *Science* **193**, 888–891.

Kamman N. C. and Engstrom D. R. (2002) Historical and present fluxes of mercury to Vermont and New Hampshire lakes inferred from ^{210}Pb dated sediment cores. *Atmos. Environ.* **36**, 1599–1609.

Kang W. J., Trefry J. H., Nelsen T. A., and Wanless H. R. (2000) Direct atmospheric inputs versus runoff fluxes of mercury to the lower Everglades and Florida Bay. *Environ Sci. Technol.* **34**(19), 4058–4063.

Kannan K. and Falandysz J. (1998) Speciation and concentrations of mercury in certain coastal marine sediments. *Water Air Soil Pollut.* **103**, 129–136.

Keeler G., Glinsorn G., and Pirrone N. (1995) Particulate mercury in the atmosphere—its significance, transport, transformation and sources. *Water Air Soil Pollut.* **80**(1–4), 159–168.

Kelly C. A., Rudd J. W. M., Golding G., Scott K., Holoka M., Barkay T., and Bloom N. (2001) The use of genetically-engineered bioreporters to study mercury availability at trace concentrations. In *Abstr. 6th Int. Conf. on Mercury as a Global Pollutant, Minamata, Japan.* 129p.

Kesler S. E. (1994) *Mineral Resources, Economics and the Environment.* MacMillan, NY.

Kim J. P. and Fitzgerald W. F. (1986) Sea-air partitioning of mercury over the equatorial Pacific Ocean. *Science* **231**, 1131–1133.

King J. K., Saunders F. M., Lee R. F., and Jahnke R. A. (1999) Coupling mercury methylation rates to sulfate reduction rates in marine sediments. *Environ. Toxicol. Chem.* **18**, 1362–1369.

King J. K., Kostka J. E., Frischer M. E., and Saunders F. M. (2000) Sulfate-reducing bacteria methylate mercury at variable rates in pure culture and in marine sediments. *Appl. Environ. Microbiol.* **66**, 2430–2437.

King J. K., Kostka J. E., Frischer M. E., Saunders F. M., and Jahnke R. A. (2001) A quantitative relationship that demonstrates mercury methylation rates in marine sediments are based on the community composition and activity of sulfate-reducing bacteria. *Environ. Sci. Technol.* **35**, 2491–2496.

Klaue B., Kesler S. E., Blum J. D. (2000) Investigation of natural fractionation of stable mercury isotopes by inductively coupled plasma mass spectrometry. In *Proc. Int. Conf. Heavy Metals Environ.* Ann Arbor, MI, August 6–10.

Klusman R. W. and Jaacks J. A. (1987) Environmental influences upon mercury, radon and helium concentrations in soil gases at a site near Denver, Colorado. *J. Geochem. Explor.* **27**, 259–280.

Koski R. A., Benninger L. M., Zierenberg R. A., and Jonasson I. R. (1994) Composition and growth history of hydrothermal deposits in Escanaba Trough, southern Gorda Ridge. In *Geologic, Hydrothermal, and Biologic Studies at Escanaba Trough, Gorda Ridge, Offshore Northern California.* US Geological Survey Bulletin 2022 (eds. J. L. Morton, R. A. Zierenberg, and C. A. Reiss), pp. 293–324.

Krabbenhoft D. P., Benoit J. M., Babiarz C. L., Hurley J. P., and Andren A. W. (1995) Mercury cycling in the Allequash Creek watershed, northern Wisconsin. *Water Air Soil Pollut.* **80**, 425–433.

Krabbenhoft D. P., Wiener J. G., Brumbaugh W. G., Olson M. L., DeWild J. F., and Sabin T. J. (1999) A national pilot study of mercury contamination of aquatic ecosystems along multiple gradients. In *US Geological Survey Toxic Substances Hydrology Program*, Proceeding of the Technical Meeting, Volume 2 (eds. D. W. Morganwalp and H. T. Buxton). Contamination of Hydrologic Systems and Related Ecosystems: USGS, Water-resources Investigations Report 99-4018B. United States Geological Survey, Reston, VA, pp. 147–160.

Krupp E. M., Pecheyran C., and Donard O. F. X. (2003) Investigation on the mercury isotope distribution of mercury species (MMHg and $Hg2^+$) in different fish tissue samples. In *Abstracts of the 2003 European Winter Conference on Plasma Spectrochemistry, Garmisch–Partenkirchen, Germany*, January, 12–17.

Lacerda L. D. and Gonçalves G. O. (2001) Mercury distribution and speciation in waters of the coastal lagoons of Rio de Janeiro, SE Brazil. *Mar. Chem.* **76**, 47–58.

Lacerda L. D. and Salomons W. (1998) *Mercury from Gold and Silver Mining: a Chemical Time Bomb?* Springer, New York, 146pp.

Lalli C. M. and Parsons T. R. (1993) *Biological Oceanography: An Introduction.* Heinemann, Oxford, UK, 147p.

Lalonde J. D., Amyot M., Kraepiel A. M. L., and Morel F. M. M. (2001) Photooxidation of Hg(0) in artificial and natural waters. *Environ. Sci. Technol.* **35**, 1367–1372.

Lalonde J. D., Poulain A. J., and Amyot M. (2002) The role of mercury redox reactions in snow on snow-to-air mercury transfer. *Environ. Sci. Technol.* **36**(2), 174–178.

Lambert G., Le Cloaree M.-F., and Pennisi M (1988) Volcanic output of SO_2 and trace metals: a new approach. *Geochim. Cosmochim. Acta* **52**(1), 39–42.

Lamborg C. H., Fitzgerald W. F., Vandal G. M., and Rolfhus K. R. (1995) Atmospheric mercury in northern Wisconsin: sources and species. *Water Air Soil Pollut.* **80**, 198–206.

Lamborg C. H., Rolfhus K. R., Fitzgerald W. F., and Kim G. (1999) The atmospheric cycling and air-sea exchange of mercury species in the south and equatorial Atlantic Ocean. *Deep-Sea Res. II.* **46**(5), 957–977.

Lamborg C. H., Fitzgerald W. F., Graustein W. C., and Turekian K. K. (2000) An examination of the atmospheric chemistry of mercury using ^{210}Pb and ^{7}Be. *J. Atmos. Chem.* **36**, 325–338.

Lamborg C. H., Fitzgerald W. F., O'Donnell J., and Torgersen T. (2002a) A non-steady-state compartmental model of global-scale mercury biogeochemistry with interhemispheric atmospheric gradients. *Geochim. Cosmochim. Acta* **66**(7), 1105–1118.

Lamborg C. H., Fitzgerald W. F., Damman A. W. H., Benoit J. M., Balcom P. H., and Engstrom D. R. (2002b) Modern and historic atmospheric mercury fluxes in both hemispheres: global and regional mercury cycling implications. *Glob. Biogeochem. Cycles.* **16**(4), 1104 doi: 10.1029/2001GB001847.

Lamborg C. H., Tseng C.-M., Fitzgerald W. F., Balcom P. H., and Hammerschmidt C. R. (2003) Determination of the mercury complexation characteristics of dissolved organic matter in natural waters with "reducible Hg" titrations. *Environ. Sci. Technol.* **37**, 3316–3322.

Landis M. S. and Stevens R. K. (2001) Preliminary results from the USEPA mercury speciation network and aircraft measurements campaigns. Presented at: 6th International Conference on Mercury as a Global Pollutant, Minamata, Japan, October.

Landis M. S., Stevens R. K., Schaedlich F., and Prestbo E. (2002) Development and characterization of an annular denuder methodology for the measurement of divalent inorganic reactive gaseous mercury in ambient air. *Environ. Sci. Technol.* **36**(13), 3000–3009.

Langer C. S., Fitzgerald W. F., Vandal G. M., and Visscher P. T. (2001) Biogeochemical cycling of methylmercury at Barn Island Salt Marsh, Stonington, CT, USA. *Wetlands Ecol. Manage.* **9**, 295–310.

Lantzy R. J. and MacKenzie F. T. (1979) Atmospheric trace metals: global cycles and assessment of man's impact. *Geochim. Cosmochim. Acta* **43**, 511–525.

Lauretta D. S., Devouard B., and Buseck P. R. (1999) The cosmochemical behavior of mercury. *Earth Planet. Sci. Lett.* **171**, 35–47.

Lauretta D. S., Klaue B., Blum J. D., and Buseck P. R. (2001) Mercury abundances and isotopic compositions in the Murchison (CM) and Allende (CV) carbonaceous chondrites. *Geochim. Cosmochim. Acta* **65**(16), 2807–2818.

Lawson N. M. and Mason R. P. (2001) Concentration of mercury, methylmercury, cadmium, lead, arsenic, and selenium in the rain and stream water of two contrasting watersheds in western Maryland. *Water Res.* **35**(17), 4039–4052.

Lechler P. J., Miller J. R., Lacerda L. D., Vinsond D., Bonzongo J.-C., Lyons W. B., and Warwick J. J. (2000) Elevated mercury concentrations in soils, sediments, water, and fish of the Madeira River basin, Brazilian Amazon: a function of natural enrichments? *Sci. Tot. Environ.* **260**, 87–96.

Lee Y. H., Bishop K. H., and Munthe J. (2000) Do concepts about catchment cycling of methylmercury and mercury in boreal catchments stand the test of time? Six years of atmospheric inputs and runoff export at Svartberget, northern Sweden. *Sci. Tot. Environ.* **260**(1–3), 11–20.

Leermakers M., Galletti S., De Galan S., Brion N., and Baeyens W. (2001) Mercury in the southern North Sea and Scheldt estuary. *Mar. Chem.* **75**(3), 229–248.

Lin C. J. and Pehkonen S. O. (1999) Aqueous phase reactions of mercury with free radicals and chlorine: implications for atmospheric mercury chemistry. *Chemosphere* **38**(6), 1253–1263.

Lindberg S. E., Meyers T. P., Taylor G. E., Jr., Turner R. R., and Schroeder W. H. (1992) Atmospheric surface exchange of mercury in a forest: results of modeling and gradient approaches. *J. Geol. Res.* **97**(2), 2519–2528.

Lindberg S. E., Vette A. F., Miles C., and Schaedlich F. (2000) Mercury speciation in natural waters: measurement of dissolved gaseous mercury with a field analyzer. *Biogeochemistry* **48**, 237–259.

Lindberg S. E., Brooks S., Lin C. J., Scott K. J., Landis M. S., Stevens R. K., Goodsite M., and Richter A. (2002) Dynamic oxidation of gaseous mercury in the Arctic troposphere at polar sunrise. *Environ. Sci. Technol.* **36**(6), 1245–1256.

Lindqvist O. and Rodhe H. (1985) Atmospheric mercury-a review. *Tellus* **27B**, 136–159.

Lindqvist O., Johansson K., Aastrup M., Andersson A., Bringmark L., Hovsenius G., Haakanson L., Iverfeldt Å., Meili M., and Timm B. (1991) Mercury in the Swedish environment—recent research on causes, consequences and corrective methods. *Water Air Soil Pollut.* **55**(1–2), 1–261.

Lockhart W. L., Wilkinson P., Billeck B. N., Danell R. A., Hunt R. V., Brunskill G. J., Delaronde J., and St. Louis V. (1998) Fluxes of mercury to lake sediments in central and northern Canada inferred from dated sediment cores. *Biogeochemistry* **40**(2–3), 163–173.

Long S. E. and Kelly W. R. (2002) Determination of mercury in coal by isotope dilution cold-vapor generation inductively coupled plasma mass spectrometry. *Anal. Chem.* **74**, 1477–1483.

Lu J. Y., Schroeder W. H., Barrie L. A., Steffen A., Welch H. E., Martin K., Lockhart L., Hunt R. V., Boila G., and Richter A. (2001) Magnification of atmospheric mercury deposition to polar regions in springtime: the link to tropospheric ozone depletion chemistry. *Geophys. Res. Lett.* **28**(17), 3219–3222.

Lucotte M., Mucci A., Hillaire-Marcel C., Pichet P., and Grondin A. (1995) Anthropogenic mercury enrichment in remote lakes of northern Quebec (Canada). *WASP* **80**, 467–476.

MacDonald R. W., Barrie L. A., Bidleman T. F., Diamond M. L., Gregor D. J., Semkin R. G., Strachan W. M. J., Li Y. F., Wania F., Alaee M., Alexeeva L. B., Backus S. M., Bailey R., Bewers J. M., Gobeil C., Halsall C. J., Harner T., Hoff J. T., Jantunen L. M. M., Lockhart W. L., Mackay D., Muir D. C. G., Pudykiewicz J., Reimer K. J., Smith J. N., Stern G. A., Schroeder W. H., Wagemann R., and Yunker M. B. (2000) Contaminants in the Canadian Arctic: 5 years of progress in understanding sources, occurrence and pathways. *Sci. Tot. Environ.* **254**, 93–234.

Mantoura R. F. C., Dickson A., and Riley J. P. (1978) The complexation of metals with humic materials in natural waters. *Estuar. Coast. Mar. Sci.* **6**, 387–408.

Martell A. E., Smith R. M., and Motekaitis R. J. (1998) *NIST Critically Selected Stability Constants of Metal Complexes Data Base*. NIST Stand. Ref. Database no. 46. US Dept. of Commerce, Gaithersburg, MD.

Martínez-Cortizas A., Potevedra-Pombal X., García-Rodeja E., Nóvoa Muñoz J. C., and Shotyk W. (1999) Mercury in a Spanish peat bog: archive of climate change and atmospheric metal deposition. *Science* **284**, 939–942.

Marvin-Dipasquale M. and Oremland R. S. (1998) Bacterial methylmercury degradation in Florida Everglades peat sediment. *Environ. Sci. Technol.* **32**, 2556–2563.

Marvin-Dipasquale M., Agee J., McGowan C., Oremland R. S., Thomas M., Krabbenhoft D., and Gilmour C. C. (2000) Methyl-mercury degradation pathways: a comparison among three mercury impacted ecosystems. *Environ. Sci. Technol.* **34**, 4908–4916.

Mason R. P. and Fitzgerald W. F. (1990) Alkylmercury species in the equatorial Pacific. *Nature* **347**, 457–459.

Mason R. P. and Fitzgerald W. F. (1993) The distribution and biogeochemical cycling of mercury in the equatorial Pacific Ocean. *Deep-Sea Res.* **40**, 1897–1924.

Mason R. P. and Fitzgerald W. F. (1996) Sources, sinks and biogeochemical cycling of mercury in the ocean. In *Global and Regional Mercury Cycles: Sources, Fluxes and Mass Balances*. NATO ASI: Series 2. Environment (eds. W. Baeyens, R. Ebinghaus, and O. Vasiliev). Kluwer, Boston, vol. 21, pp. 249–272.

Mason R. P. and Lawrence A. L. (1999) Concentration, distribution, and bioavailability of mercury and methylmercury in sediments of Baltimore Harbor and Chesapeake Bay, Maryland, USA. *Environ. Toxicol. Chem.* **18**, 2438–2447.

Mason R. P. and Sullivan K. A. (1999) Mercury in Lake Michigan. *Environ. Sci. Technol.* **31**, 942–947.

Mason R. P. and Sullivan K. A. (1999) The distribution and speciation of mercury in the south and equatorial Atlantic. *Deep-Sea Res. II.* **46**, 937–956.

Mason R. P., Fitzgerald W. F., Hurley J. P., Hanson A. K., Jr., Donaghay P. L., and Sieburth J. M. (1993) Mercury biogeochemical cycling in a stratified estuary. *Limnol. Oceanogr.* **38**, 1227–1241.

Mason R. P., Fitzgerald W. F., and Morel F. M. M. (MFM) (1994) The biogeochemical cycling of elemental mercury: anthropogenic influences. *Geochim. Cosmochim. Acta* **58**, 3191–3198.

Mason R. P., Rolfhus K. R., and Fitzgerald W. F. (1998) Mercury in the North Atlantic. *Mar. Chem.* **61**, 37–53.

Mason R. P., Lawson N. M., Lawrence A. L., Leaner J. J., Lee J. G., and Sheu G.-R. (1999) Mercury in Chesapeake Bay. *Mar. Chem.* **65**, 77–96.

Mason R. P., Lawson N. M., and Sheu G. R. (2001) Mercury in the Atlantic Ocean: factors controlling air–sea exchange of mercury and its distribution in the upper waters. *Deep-Sea Res. II.* **48**(13), 2829–2853.

Matilainen T., Verta M., Korhonen H., Uusi-Rauva A., and Niemi M. (2001) Behavior of mercury in soil profiles: impact of increased precipitation, acidity, and fertilization on mercury methylation. *Water Air Soil Pollut.* **125**, 105–119.

McCarthy J. H. (1968) Experiments with mercury in soil gas and air applied to mineral exploration. *Min. Eng.* **20**, 46.

Mercone D., Thomson J., Croudace I. W., and Troelstra S. R. (1999) A coupled natural immobilization mechanism for mercury and selenium in deep-sea sediments. *Geochim. Cosmochim. Acta* **63**(10), 1481–1488.

Mierle G. (1990) Aqueous inputs of mercury to Precambrian shield lakes in Ontario. *Environ. Toxicol. Chem.* **9**, 843–851.

Millward G. E. (1982) Nonsteady state simulations of the global mercury cycle. *J. Geol. Res.* **87**, 8891–8897.

Monteiro L. R. and Furness R. W. (1997) Accelerated increase in mercury contamination in North Atlantic mesopelagic food chains as indicated by time series of seabird feathers. *Environ. Toxicol. Chem.* **16**(12), 2489–2493.

Monteiro L. R. and Furness R. W. (2001) Kinetics, dose-response, and excretion of methylmercury in free-living adult Cory's shearwaters. *Environ. Sci. Technol.* **35**(4), 739–746.

Monteiro L. R., Granadeiro J. P., and Furness R. W. (1998) Relationship between mercury levels and diet in Azores seabirds. *Mar. Ecol. Prog. Ser.* **166**, 259–265.

Munthe J. (1992) The aqueous oxidation of elemental mercury by ozone. *Atmos. Environ.* **26A**(8), 1461–1468.

Munthe J., and Berg T. (2001) Reply to comment on "Atmospheric mercury species in the European Arctic: measurement and modeling" by Berg *et al. Atmos. Environ.* **35**(31), 5379–5380.

Murphy D. M., Thomson D. S., and Mahoney M. J. (1998) *In situ* measurements of organics, meteoritic material, mercury, and other elements in aerosols at 5 to 19 kilometers. *Science* **282**, 1664–1669.

National Academy of Sciences (NAS) (2000) In *Toxicological Effects of Methylmercury*. Committee on the toxicological effects of methylmercury (ed. R. A. Goyer, chair). National Academy Press, Washington, DC, 344pp.

National Research Council Committee on the Toxicological Effects of Methylmercury (2000) *Toxicological Effects of Methylmercury*. National Academy Press, Washington, DC.

Nriagu J. O. (1979) Production and uses of mercury. In *The Biogeochemistry of Mercury in the Environment* (ed. J. O. Nriagu). Elsevier/North-Hollan Biomedical Press, Amsterdam, pp. 23–40.

Nriagu J. O. (1989) A global assessment of natural sources of atmospheric trace metals. *Nature* **338**, 47–49.

Nriagu J. O. and Becker C. (2003) Volcanic emissions of mercury to the atmosphere: global and regional inventories. *Sci. Tot. Environ.* **304**, 3–12.

Norton S. A., Evans G. C., and Kahl J. S. (1997) Comparison of Hg and Pb fluxes to hummocks and hollows of ombrotrophic Big Heath Bog and to nearby Sargent Mt. Pond, Maine, USA. *Water Air Soil Pollut.* **100**, 271–286.

O'Driscol N. J., Siciliano S. D., and Lean D. R. S. (2003) Continuous analysis of dissolved gaseous mercury in freshwater lakes. *Sci. Tot. Environ.* **304**, 285–294.

Olafsson J. (1975) Volcanic influence on sea water at Heimaey. *Nature* **255**, 138–141.

Ozerova N. A. (1996) Mercury in geological systems. In *Global and Regional Mercury Cycles: Sources, Fluxes and Mass Balances*. NATO ASI: Series 2. Environment (eds. W. Baeyens, R. Ebinghaus, and O. Vasiliev). Kluwer, Boston, vol. 21, pp. 463–474.

Pacyna E. G. and Pacyna J. M. (2002) Global emission of mercury from anthropogenic sources in 1995. *Water Air Soil Pollut.* **137**, 149–165.

Pak K.-R. and Bartha R. (1998) Mercury methylation and demethylation in anoxic lake sediments and by strictly anaerobic bacteria. *Appl. Environ. Microbiol.* **64**(3), 1013–1017.

Patterson C. C. and Settle D. M. (1987) Magnitude of lead flux to the atmosphere from volcanoes. *Geochim. Cosmochim. Acta* **51**(3), 675–681.

Pempkowiak J., Cossa D., Sikora A., and Sanjuan J. (1998) Mercury in water and sediments of the southern Baltic Sea. *Sci. Tot. Environ.* **213**, 185–192.

Petersen G., Iverfeldt Å., and Munthe J. (1995) Atmospheric mercury species over central and northern Europe: model calculations and comparison with the observations from the Nordic air and precipitation network for 1987 and 1988. *Atmos. Environ.* **29**(1), 47–67.

Pleijel K. and Munthe J. (1995) Modeling the atmospheric mercury cycle-chemistry in fog droplets. *Atmos. Environ.* **29**(12), 1441–1457.

Poissant L. and Casimir A. (1998) Water–air and soil–air exchange rate of total gaseous mercury measured at background sites. *Atmos. Environ.* **32**(5), 883–983.

Porcella D. B., Ramel C., and Jernelov A. (1997) Global mercury pollution and the role of gold mining: an overview. *Water Air Soil Pollut.* **97**(3–4), 205–207.

Rada R. G. and Powell D. E. (1993) Whole-lake burdens and spatial distribution of mercury in surficial sediments in Wisconsin seepage lakes. *Can. J. Fish. Aquatic. Sci.* **50**(4), 865–873.

Rampino M. R. and Self S. (1984) Sulfur-rich volcanic eruptions and stratospheric aerosols. *Nature* **310**, 677–679.

Rasmussen P. (1994) Current methods of estimating atmospheric mercury fluxes in remote areas. *Environ. Sci. Technol.* **28**(13), 2233–2241.

Ravichandran M., Aiken G. R., Ryan J. N., and Reddy M. M. (1999) Inhibition of precipitation and aggregation of metacinnabar (mercuric sulfide) by dissolved organic matter isolated from the Florida Everglades. *Environ. Sci. Technol.* **33**, 1418–1423.

Rea A. W., Lindberg S. E., Scherbatskoy T., and Keeler G. J. (2002) Mercury accumulation in foliage over time in two northern mixed-hardwood forests. *Water Air Soil Pollut.* **133**(1–4), 49–67.

Richter F. M., Rowley D. B., and DePaolo D. J. (1992) Sr isotope evolution of seawater: the role of tectonics. *Earth Planet. Sci. Lett.* **109**, 11–23.

Rolfhus K. R. (1998) The production and distribution of elemental Hg in a coastal marine environment. PhD Dissertation, University of Connecticut.

Rolfhus K. R. and Fitzgerald W. F. (1995) Linkages between atmospheric mercury deposition and the methylmercury content of marine fish. *Water Air Soil Pollut.* **80**, 291–297.

Rolfhus K. R. and Fitzgerald W. F. (2001) The evasion and spatial/temporal distribution of mercury species in Long Island Sound, CT–NY. *Geochim. Cosmochim. Acta* **65**(3), 407–418.

Rolfhus K. R., Sakamoto H. E., Cleckner L. B., Stoor R. W., Babiarz C. L., Back R. C., Manolopoulos H., and Hurley J. P. (2003) Distribution and fluxes of total and methylmercury in Lake Superior. *Environ. Sci. Technol.* **37**(5), 865–872.

Roos-Barraclough F., Martinez-Cortizas A., Garcia-Rodeja E., and Shotyk W. (2002) A 14,500 year record of the accumulation of atmospheric mercury in peat: volcanic signals, anthropogenic influences and a correlation to bromine accumulation. *Earth Planet. Sci. Lett.* **202**(2), 435–451.

Rosson R. (1996) Luciferase operon-containing plasmid engineered for detecting mercury in water. US Patent Application #5571722.

Roulet M. and Lucotte M. (1995) Geochemistry of mercury in pristine and flooded ferralitic soils of a tropical rain forest in French Guiana, South America. *WASP* **80**, 1079–1088.

Sanemasa I. (1975) The solubility of elemental mercury vapor in water. *Bull. Chem. Soc. Japan.* **48**, 1975–1978.

Sanseverino J., Sayler G. S., and Ripp S. A. (2002) Bioluminescent methods for direct visual detection of environmental compounds (US patent # 0214551). *PCT Int. Appl.* 51pp.

Scherbatskoy T., Shanley J. B., and Keeler G. J. (1998) Factors controlling mercury transport in an upland forested catchment. *Water Air Soil Pollut.* **105**, 427–438.

Scheuhammer A. M. (1991) Effects of acidification on the availability of toxic metals and calcium to wild birds and mammals. *Environ. Pollut.* **71**, 329–375.

Schlüter K. (2000) Review: evaporation of mercury from soils. An integration and synthesis of current knowledge. *Environ. Geol.* **39**(3–4), 249–271.

Schober S. E., Sinks T. H., Jones R. L., Bolger P. M., McDowell M., Osterloh J., Garrett E. S., Canady R. A., Dillon C. F., Sun Y., Joseph C. B., and Mahaffey K. R. (2003) Blood mercury levels in US children and women of childbearing age, 1999–2000. *J. Am. Med. Assoc.* **289**, 1667–1674.

Schroeder W. H., Anlauf K. G., Barrie L. A., Lu J. Y., Steffen A., Schneeberger D. R., and Berg T. (1998) Arctic springtime depletion of mercury. *Nature* **394**, 331–332.

Schuster P. F., Krabbenhoft D. P., Naftz D. L., Cecil D., Olson M. L., DeWild J. F., Susong D. D., Green J. R., and Abbott M. L. (2002) Atmospheric mercury deposition during the last 270 years: a glacial ice core record of natural and anthropogenic sources. *Environ. Sci. Technol.* **36**, 2303–2310.

Scott K. J., Hudson R. J. M., Kelly C. A., Rudd J. W. M., and Barkay T. (2001) Effects of chemical speciation on the bioavailability of Hg(II) in defined media and natural water as measured with a *mer-lux* bioreporter. In *Abstr. 6th Int. Conf. on Mercury as a Global Pollutant, Minamata, Japan.* 130p.

Seigel F. R., Kravitz J. H., and Galasso J. J. (2001) Arsenic and mercury contamination in 31 cores taken in 1965, St. Anna

Trough, Kara Sea, Arctic Ocean. *Environ. Geol.* **40**(4–5), 528–542.

Seinfeld J. H. (1986) *Atmospheric Chemistry and Physics of Air Pollution*. Wiley-Interscience, New York, pp. 594–597.

Selifonova O., Burlage R., and Barkay T. (1993) Bioluminescent sensors for detection of bioavailable Hg(II) in the environment. *Appl. Environ. Microbiol.* **59**(9), 3083–3090.

Sellers P., Kelly C. A., Rudd J. W. M., and MacHutchon A. R. (1996) Photodegradation of methymercury in lakes. *Nature* **380**, 694–697.

Sellers P., Kelly C. A., and Rudd J. W. M. (2001) Fluxes of methylmercury to the water column of a drainage lake: the relative importance of internal and external sources. *Limnol. Oceanogr.* **46**(3), 623–631.

Sheu G.-R. and Mason R. P. (2001) An examination of methods for the measurements of reactive gaseous mercury in the atmosphere. *Environ. Sci. Technol.* **35**, 1209–1216.

Shia R.-L., Seigneur C., Pai P., Ko M., and Sze N. D. (1999) Global simulation of atmospheric mercury concentrations and deposition fluxes. *J. Geol. Res.* **104**(D19), 23747–23760.

Skyring G. W. (1987) Sulfate reduction in coastal ecosystems. *Geomicrobiol. J.* **5**, 295–373.

Slemr F. and Langer E. (1992) Increase in global atmospheric concentrations of mercury inferred from measurements over the Atlantic Ocean. *Nature* **355**, 434–437.

Slemr F., Seiler W., and Schuster G. (1981) Latitudinal distribution of mercury over the Atlantic Ocean. *J. Geol. Res.* **80**, 1159.

Smith W. E. (1975) *Minamata*. Henry Holt and Company, New York.

Sommar J., Gårdfeldt K., Stromberg D., and Feng X.-B. (2001) A kinetic study of the gas-phase reaction between the hydroxyl radical and atomic mercury. *Atmos. Environ.* **35**(17), 3049–3054.

Spiro P. A., Jacob D. J., and Logan J. A. (1992) Global inventory of sulfur emissions with $1° \times 1°$ resolution. *J. Geophys. Res.* **97/D5**, 6023–6036.

St. Louis V. L., Rudd J. W. M., Kelly C. A., Beaty K. G., Flett R. J., and Roulet N. T. (1996) Production and loss of methylmercury and loss of total mercury from boreal forest catchments containing different types of wetlands. *Environ. Sci. Technol.* **30**(9), 2719–2729.

St. Louis V. L., Rudd J. W. M., Kelly C. A., Hall B. D., Rolfhus K. R., Scott K. J., Lindberg S. E., and Dong W. (2001) Importance of the forest canopy to fluxes of methyl mercury and total mercury to boreal ecosystems. *Environ. Sci. Technol.* **35**(15), 3089–3098.

Stock A. and Cucuel F. (1934) The quantitative determination of microamounts of mercury. *Naturwissenchaften* **22**, 390.

Stoffers P., Hannington M., Wright I., Herzig P., and de Ronde C. (1999) Elemental mercury at submarine hydrothermal vents in the Bay of Plenty, Taupo volcanic zone, New Zealand. *Geology* **27**(10), 931–934.

Stoiber R. E. and Jepsen A. (1973) Sulfur dioxide contributions to the atmosphere by volcanoes. *Science* **182**, 577–578.

Stoiber R. E., Williams S. N., and Huebert B. (1987) Annual contribution of sulfur dioxide to the atmosphere by volcanoes. *J. Volcanol. Geotherm. Res.* **33**(1–3), 1–8.

Stordal M. C. and Gill G. A. (1995) Determination of mercury methylation rates using a 203-Hg radiotracer technique. *Water Air Soil Pollut.* **80**, 725–734.

Stratton W. J. and Lindberg S. E. (1995) Use of a refluxing mist chamber for measurement of gas-phase mercury(II) species in the atmosphere. *Water Air Soil Pollut.* **80**, 1269–1278.

Swain E. B. and Helwig D. D. (1989) Mercury in fish from northeastern Minnesota lakes: historical trends, environmental correlates, and potential sources. *J. Minn. Acad. Sci.* **55**, 103–109.

Swain E. B., Engstrom D. R., Brigham M. E., Henning T. A., and Brezonik P. L. (1992) Increasing rates of atmospheric mercury deposition in midcontinental North America. *Science* **257**, 784–787.

Tescione L. and Belfort G. (1993) Construction and evaluation of a metal ion biosensor. *Biotechnol. Bioeng.* **42**(8), 945–952.

Thakur A. N. and Goel P. S. (1989) Huge variations in the isotopic ratio $^{196}Hg/^{202}Hg$ in some acid-insoluble resides of Sikhote Alin and other iron meteorites. *Earth Planet. Sci. Lett.* **96**, 235–246.

The Social Scientific Study Group on Minamata Disease (2001) *In the Hope of Avoiding Repetition of a Tragedy of Minamata Disease (What we have learned from the experience)*. National Institute for Minamata Disease, Minamata, Japan, 140p.

Tokos J. J. S., Hall B., Calhoun J. A., and Prestbo E. (1998) Homogeneous gas-phase reactions of Hg° with H_2O_2, O_3, CH_3I and $(CH_3)_2S$: implications for atmospheric Hg cycling. *Atmos. Environ.* **32**(5), 823–828.

Topping G. and Davies I. M. (1981) Methylmercury production in the marine water column. *Nature* **290**, 243–244.

Tremblay A., Lucotte M., and Schetagne R. (1998) Total mercury and methylmercury accumulation in zooplankton of hydroelectric reservoirs in northern Quebec (Canada). *Sci. Tot. Environ.* **213**(1–3), 307–315.

Tseng C. M., Amouroux D., Abril G., Tessier E., Etcheber H., and Donard O. F. X. (2001) Speciation of mercury in a fluid mud profile of a highly turbid macrotidal estuary (Gironde, France). *Environ. Sci. Technol.* **35**(13), 2627–2633.

Tseng C.-M., Balcom P. H., Lamborg C. H., and Fitzgerald W. F. (2003) Dissolved elemental mercury investigations in Long Island Sound using on-line Au amalgamation-flow injection analysis. *Environ. Sci. Technol.* **37**, 1183–1188.

Tseng C.-M., Lamborg C. H., Fitzgerald W. F., and Engstrom D. R. Cycling of dissolved elemental mercury in arctic Alaskan lakes. *Geochim. Cosmochim. Acta* (in press).

Turekian K. K. and Wedepohl K. H. (1961) Distribution of the elements in some major units of the Earth's crust. *Geol. Soc. Am. Bull.* **72**, 175–192.

US Environmental Protection Agency (1997) Mercury Study Report to Congress. EPA 452/R-97-003 available at: http:// www. epa. gov/oar/mercury. html

USAD Consumer Advisory (2001) Mercury levels in seafood species. Center for Food Safety and Applied Nutrition, Office of Seafood.

USFDA (2001) Mercury levels in seafood species. Center for Food Safety and Applied Nutrition, Office of Seafood.

Urban N. R., Eisenreich S. J., Grigal D. F., and Schurr K. T. (1990) Mobility and diagenesis of Pb and ^{210}Pb in peat. *Geochim. Cosmochim. Acta* **54**, 3329–3346.

Vaidya O. C., Howell G. D., and Leger D. A. (2000) Evaluation of the distribution of mercury in lakes in Nova Scotia and Newfoundland (Canada). *Water Air Soil Pollut.* **117**(1–4), 353–369.

Vandal G. M., Fitzgerald W. F., and Mason R. P. (1991) Cycling of volatile mercury in temperate lakes. *Water Air Soil Pollut.* **56**, 791–803.

Vandal G. M., Fitzgerald W. F., Boutron C. F., and Candelone J. P. (1993) Variations in mercury deposition to the Antarctic over the last 34,000 years. *Nature* **362**, 621–623.

Vandal G. M., Mason R. P., McKnight D., and Fitzgerald W. (1998) Mercury speciation and distribution in a polar desert lake (Lake Hoare, Antarctica) and two glacial meltwater streams. *Sci. Tot. Environ.* **213**, 229–237.

Vandal G. M., Fitzgerald W. F., Rolfhus K. R., Lamborg C. H., Langer C. S., and Balcom P. H. (2002) Sources and cycling of mercury and methylmercury in Long Island Sound. Final Report to CTDEP. 105pp.

Varekamp J. C. and Buseck P. R. (1981) Mercury emissions from Mount St. Helens during September 1980. *Nature* **293**, 555–556.

Varekamp J. C. and Buseck P. R. (1983) Mercury anomalies in soils: a geochemical exploration method for geothermal areas. *Geothermics* **12**(1), 29–47.

Varekamp J. C. and Buseck P. R. (1984) The speciation of mercury in hydrothermal systems, with applications to ore deposition. *Geochim. Cosmochim. Acta* **48**(1), 177–185.

Varekamp J. C. and Buseck P. R. (1986) Global mercury flux from volcanic and geothermal sources. *Appl. Geochem.* **1**(1), 65–73.

Varekamp J. C., Buchholdtz ten Brink M. R., Mecray E. L., and Kreulen B. (2000) Mercury in Long Island Sound sediments. *J. Coast. Res.* **16**(3), 613–626.

Verta M. and Matilainen T. (1995) Methylmercury distribution and partitioning in stratified Finnish forest lakes. *Water Air Soil Pollut.* **80**(1–4), 585–588.

Virta M., Lampinen J., and Karp M. (1995) A luminescence-based mercury biosensor. *Anal. Chem.* **67**, 667–669.

Wang Q., Shen W., and Ma Z. (2000) Estimation of mercury emission from coal combustion in China. *Environ. Sci. Technol.* **34**, 2711–2713.

Watras C. J. and Bloom N. S. (1994) The vertical distribution of mercury species in Wisconsin lakes: accumulation in plankton layers. In *Mercury Pollution: Integration and Synthesis,* Chap. I.11 (eds. C. Watras and J. Huckabee). Lewis Publishers, Ann Arbor, MI, pp. 137–152.

Watras C. J., Bloom N. S., Hudson R. J. M., Gherini S., Munson R., Claas S. A., Morrison K. A., Hurley J., Wiener J. G., Fitzgerald W. F., Mason R., Vandal G., Powell D., Rada R., Rislov L., Winfrey M., Elder J., Krabbenhoft D., Andren A. W., Babiarz C., Porcella D. B., and Huckabee J. W. (1994) Sources and fates of mercury and methylmercury in Wisconsin lakes. In *Mercury Pollution: Integration and Synthesis,* chap. I.12 (eds. C. Watras and J. Huckabee). Lewis Publishers, Ann Arbor, MI, pp. 153–177.

Watson C. M., Dwyer D. J., Andle J. C., Bruce A. E., and Bruce M. R. M. (1999) Stripping analyses of mercury using gold electrodes: irreversible adsorption of mercury. *Anal. Chem.* **71**, 3181–3186.

Westöö G. (1966) Determination of methylmercury compounds in foodstuffs: I. Methylmercury compounds in fish, identification and determination. *Acta Chem. Scand.* **20**(8), 2131–2137.

White D. E. (1967) Mercury and base-metal deposits with associated thermal and mineral waters. In *Geochemistry of Hydrothermal Ore Deposits* (ed. H. L. Barnes). Holt, Rinehart and Winston, New York, chap. 13, pp. 575–631.

Wiener J. G. and Spry D. J. (1996) Toxicological significance of mercury in freshwater fish. In *Environmental Contamination of Wildlife* (eds. W. N. Beyer, G. H. Heinz, and A. W. Redmon-Norwood). Lewis Publication, Boca Raton, FL, USA. pp. 297–339.

Wiener J. G., Fitzgerald W. F., Watras C. J., and Rada R. G. (1990a) Partitioning and bioavailability of mercury in an experimentally acidified Wisconsin lake. *Environ. Toxicol. Chem.* **9**(7), 909–918.

Wiener J. G., Martini R. E., Sheffy T. B., and Glass G. E. (1990b) Factors influencing mercury concentrations in walleyes in northern Wisconsin lakes. *Trans. Am. Fisher. Soc.* **119**, 862–870.

Wiener J. G., Krabbenhoft D. P., Heinz G. H., and Scheuhammer A. M. (2002) Ecotoxicology of Mercury. In *Handbook of Ecotoxicology*, 2nd edn. (eds. D. J. Hoffman and B. A. Rattner). Lewis Publishers, Boca Raton, FL, USA.

Wilhelm S. M. (1999) Design mercury removal systems for liquid hydrocarbons. *Hydrocarb. Process.* **78**(4), 61.

Wilhelm S. M. (2001) Estimate of mercury emissions to the atmosphere from petroleum. *Environ. Sci. Technol.* **35**(24), 4704–4710.

Winfrey M. R. and Rudd J. W. M. (1990) Environmental factors affecting the formation of methylmercury in low pH lakes. *Environ. Toxicol. Chem.* **9**(7), 853–869.

Wu Q., Apte S. C., Batley G. E., and Bowles K. C. (1997) Determination of the mercury complexation capacity of natural waters by anodic stripping voltammetry. *Anal. Chim. Acta* **350**, 129–134.

Wylie D. E., Carlson L. D., Carlson R., Wagner F. W., and Schuster S. M. (1991) Detection of mercuric ions in water by ELISA with a mercury-specific antibody. *Anal. Biochem.* **194**, 381–387.

Wylie D. E., Lu D., Carlson L. D., Carlson R., Babacan K. F., Schuster S. M., and Wagner F. W. (1992) Monoclonal antibodies specific for mercuric ions. *Proc. Natl. Acad. Sci. USA* **89**, 4104–4108.

Xia K., Skyllberg U. L., Bleam W. F., Bloom P. R., Nater E. A., and Helmke P. A. (1999) X-ray absorption spectroscopic evidence for the complexation of Hg(II) by reduced sulfur in soil humic substances. *Environ. Sci. Technol.* **33**, 257–261.

Xiao Z. F., Munthe J., Schroeder W. H., and Lindqvist O. (1991) Vertical fluxes of volatile mercury over forest soil and lake surfaces in Sweden. *Tellus* **43B**, 267–279.

Young D. R., Johnson J. N., Soutar A., and Isaacs J. D. (1973) Mercury concentrations in dated varved marine sediments collected off southern California. *Nature* **244**, 273–275.

9.05
The Geochemistry of Acid Mine Drainage

D. W. Blowes and C. J. Ptacek
University of Waterloo, ON, Canada
J. L. Jambor
University of British Columbia, Vancouver, BC, Canada
and
C. G. Weisener
University of Waterloo, ON, Canada

9.05.1 INTRODUCTION

9.05.1.1 Scale of the Problem

Mine wastes are the largest volume of materials handled in the world (ICOLD, 1996). The generation of acidic drainage and the release of water containing high concentrations of dissolved metals from these wastes is an environmental problem of international scale. Acidic drainage is caused by the oxidation of sulfide minerals exposed to atmospheric oxygen. Although acid drainage is commonly associated with the extraction and processing of sulfide-bearing metalliferous ore deposits and sulfide-rich coal, acidic drainage can occur wherever sulfide minerals are excavated and exposed to atmospheric oxygen. Engineering projects, including road construction, airport development, and foundation excavation are examples of civil projects that have resulted in the generation of acidic drainage. On United States Forest Service Lands there are $(2-5) \times 10^4$ mines releasing acidic drainage (USDA, 1993). Kleinmann *et al.* (1991) estimated that more than 6,400 km of rivers and streams in the eastern United States have been adversely affected by mine-drainage water. About $(0.8-1.6) \times 10^4$ km of streams have been affected by metal mining in the western United States. The annual worldwide production of mine wastes exceeded 4.5 Gt in 1982 (ICOLD, 1996). Estimated costs for remediating mine wastes internationally total in the tens of billions of dollars (Feasby *et al.*, 1991).

9.05.1.2 Overview of the Mining Process and Sources of Low-quality Drainage

The recovery of metals from sulfide-rich ore bodies proceeds through a series of steps: from mining to crushing to mineral recovery, followed typically by smelting of the sulfide ores, and thence to metal refining; although the nature of the ore body dictates the processes used to extract metals from ore, each of these steps generates a waste stream. The volumes of the waste streams can be large. For example, production of 1 t of copper typically requires the excavation and processing of 100 t of rock. Each of the steps of metal production can lead to the generation of low-quality water.

9.05.1.2.1 Mine workings and open pits

Minerals are typically excavated by underground mining, strip mining, or open-pit mining. The selection of the mine design is dictated by the physical structure and value of the ore body and by the characteristics of the adjacent geological materials. Although open-pit mines and underground mines are the two most common mining strategies, placer mining and solution mining also have been used for mineral extraction. Placer mining involves excavation of river or stream sediments and the separation of valuable minerals by gravity, selective flotation, or by chemical extraction. Most solution mining is by heap leaching, in which the extractant solution is trickled over broken ore on the surface or in underground workings; less common is injection into underground aquifers. The consequence of the excavation of open pits and other mining-related disturbances is that sulfide minerals previously isolated from the atmosphere are exposed to oxygen. Oxidation of sulfide minerals ensues.

9.05.1.2.2 Waste rock

Open-pit and underground mining result in the excavation of large volumes of rock to gain access to ore bodies. After the ore body is accessed, ore for processing is separated from the host rock on the basis of economic cutoff values. Rock of higher metal grade is processed, and rock below the cutoff grade is put to waste. Frequently, ore is segregated into high-grade and low-grade ore stockpiles. Ore is material that will yield a profit; thus, the metal contents of the discrimination between high-grade and low-grade ores will vary with the costs of mining activities and the value of the metals extracted. The waste rock from mine operations may be used in construction activities at the mine site. Excess waste rock is deposited in waste-rock piles whose composition differs greatly from mine to mine because of variations in ore-deposit and host-rock mineralogy, and because of differences in the processing techniques and ore-grade cutoff values. Daily production of waste rock in Canada is estimated to be 1×10^6 t (Government of Canada, 1991). Because of the large volume of rock excavated in open-pit operations, waste-rock piles may be tens of hectares in area and tens of meters in height (Ritchie, 1994).

9.05.1.2.3 Mill tailings

The ore extracted in most nonferrous, metalliferous mining operations is rich in base or precious metals, but the ore minerals are generally too dilute for direct processing using metallurgical techniques. Thus, most ores are processed through concentration steps that involve costs—crushing, grinding, and milling to a fine grain size—for beneficiation. The grain size of the milled rock is dictated by the process used for mineral recovery. Typical grain sizes range from 25 μm to 1.0 mm. At many plants, differential flotation is used to separate the valuable sulfide minerals containing base or precious metals from others (e.g., pyrite [FeS_2] or pyrrhotite [$Fe_{1-x}S$]) that have little commercial value. A flotation concentrator may contain several circuits for the selective recovery of a variety of metal sulfides and the production of a series of metal-sulfide concentrates. The concentrate from the flotation step is retained for further metallurgical processing. Mill tailings are the residual material, including sulfide gangue minerals, that is discharged to tailings impoundments, typically as a slurry of water and finely ground rock.

The ratio of tailings to concentrate can be very large, particularly at gold and precious-metal mines, at which the concentrate may represent only a small fraction of 1% of the ore processed. The mining industry produces immense masses of mine tailings. The mass of tailings produced daily in Canada is estimated to be 9.5×10^5 t (Government of Canada, 1991). Tailings impoundments may be very large. For example, the Inco Ltd. Central Tailings Disposal Area covers an area of 25 km^2 with tailings up to 50 m in depth, and the ultimate capacity is more than 725 Mt (Puro *et al.*, 1995).

Mill tailings are typically retained in impoundments. The retaining dams of many impoundments are constructed of coarse-grained tailings or of tailings combined with waste rock. These types of impoundments are designed to drain, thereby enhancing their structural integrity, but resulting in the development of a thick zone of only partial saturation. The entry of gas-phase oxygen into the unsaturated tailings results in sulfide-mineral oxidation and the release of low-quality drainage.

9.05.2 MINERALOGY OF ORE DEPOSITS

9.05.2.1 Coal

Coal is an organic rock-like natural product (Speight, 1994) whose beginnings were as the remains of flora that accumulated as peat. The accumulation occurred in submerged conditions, thereby preventing complete decay of the organic material to $CO_2 + H_2O$ during the early stage of maturation. Reflecting its origin, coal occurs in beds—maximum thicknesses ~90 m, and typical mining thicknesses 1–4 m.

The carbon content of coal varies from about 70 wt.% to 95 wt.%, with most of the remainder consisting of oxygen, hydrogen, nitrogen, and sulfur. The oxygen content generally ranges from about 2% to 20%, and the major change that occurs during coalification is a decrease in the oxygen content and an increase in the carbon content. With this change, the physical properties and thermal yield per unit weight also change, and various classifications have been devised to reflect those properties. A common commercial subdivision is into "brown coal" and "hard coal," which in turn is an indication of the degree of induration. At the lowest end is brown coal, which has the lowest carbon and the highest moisture contents, and is most commonly classified as lignite. The upward progression is to sub-bituminous, bituminous, and anthracite, with the last the most highly evolved and of highest metamorphic rank. Most brown coal is lignite, with some overlap into the sub-bituminous category, whereas the others are hard coal. Most mined coal is of the bituminous variety.

Current world production is ~3,500 Mt (million metric tonnes) of hard coal and ~900 Mt of brown coal. Germany, at ~160 Mt, is the largest producer of brown coal, with Russia a distant second. China's production of 900–1,000 Mt of hard coal is the world's highest, with >90% obtained from underground mines. US production is a close second to that of China, whereas India ranks third and produces <300 Mt. In recent years, the US has undergone a pronounced shift to mining of western coal, predominantly sub-bituminous, which has a lower average content of sulfur than coals from Appalachia. Wyoming alone now accounts for about a third of all US production.

Environmental concerns have been focused on the gaseous and particulate emissions, on the environmental quality of the ash and slag residues, and on acid drainage that may ensue as a consequence of the exposure of mining-related wastes to atmospheric weathering. Many of the environmentally least desirable aspects concerning the utilization of coal are related to the presence of mineral matter, especially iron disulfides. The disulfides are a principal source for SO_2 emissions during combustion, and in mine wastes the oxidation of FeS_2 is the principal cause of the development of acidic drainage.

Stach *et al.* (1982) list more than 40 minerals that have been identified as occurring in coal, and recent observations have expanded the total to more than a hundred. Finkelman (1980a,b) concluded that coals yielding >5 wt.% ash have had the bulk of their minerals derived by detrital processes.

Table 1 summarizes and presents an interpretation of the occurrence of the principal nondetrital minerals in coal. The chief detrital minerals are quartz and clay minerals (including K–Al micas), and these minerals commonly form up to 90% of the mineral matter in coals. The bulk of the remainder typically consists of carbonates and pyrite. Renton (1982) observed that most discrete mineral grains observed in coal are ~20 μm in diameter, and few exceed 100 μm. Exceptions are concretions, nodules, and "balls" that typically contain one or more of pyrite, marcasite [FeS_2], calcite [$CaCO_3$], and siderite [$FeCO_3$], and which may be many centimeters in diameter. As well, aggregates of pyrite and marcasite occur within coal and as fracture (cleat) mineralization. The most common cleat-filling minerals are calcite, pyrite, and kaolinite [$Al_2Si_2O_5(OH)_4$] (Renton, 1982). Vassilev *et al.* (1996) observed that higher-rank coals are enriched in elements associated with probable detrital minerals, whereas lower-rank coals are enriched in elements associated with probable authigenic minerals and organic material. The magnitudes of the concentrations of trace elements in coal are summarized in Table 2.

Table 1 Some minerals in coal deposits.[a]

Mineral group	*Inherent*	*Extraneous source*
Clay minerals	Kaolinite, illite-sericite,[b] mixed-layer clays, smectite	Illite-sericite,[b] chlorite
Carbonates	Calcite, siderite, dolomite-ankerite	Calcite, dolomite-ankerite
Sulfides	Pyrite, marcasite, sphalerite, galena, chalcopyrite, pyrrhotite, greigite	Pyrite, marcasite, sphalerite, galena, chalcopyrite
Oxides	Quartz, Fe oxyhydroxides, hematite	Quartz, goethite, lepidocrocite
Phosphates	Apatite, crandallite-group minerals, vivianite	Apatite
Others	Gypsum, halite	Sulfates, chlorides, nitrates

[a] Interpreted partly from data in Mackowsky (1968), Renton (1982), Stach *et al.* (1982), Harvie and Ruch (1986), Birk (1989), Ward (1989), Speight (1994), and Spears (1997). The extraneous-source minerals typically form after consolidation has progressed to the state at which the coal can sustain fracturing, and the minerals occur as fracture fillings and in cavities. The minerals in each group are listed in approximate decreasing order of abundance. [b] Sericite is a general term for fine-grained, mica-like minerals. Illite has been assigned a specific composition by the International Mineralogical Association (Rieder *et al.*, 1998), but illite and sericite are used here only to designate mica-type minerals. The mineral ankerite has formula Fe > Mg, but the name is commonly used for ferroan dolomite [$Ca(Mg,Fe)(CO_3)_2$].

Table 2 The magnitude of the trace-element contents (ppm) of coal and coal ash.

Element	*Coal*		*Coal ash*	
	Average[a]	*Range*[b]	*Average*[c]	*Range*[d]
Antimony	3.0	0.05–10		200?
Arsenic	5.0	0.5–80	500	100–500?
Barium	500	20–1,000		300–900
Beryllium	3	0.1–15	45	1–10
Bismuth	5.5		20	20–50
Boron	75	5–400	600	
Cadmium	1.3	0.1–3	5	5?
Chlorine	1,000	50–2,000		
Chromium	10	0.5–60		100–400
Cobalt	5	0.5–30	300	300
Copper	15	0.5–50		20–200
Fluorine		20–500		
Gallium	7	1–20	100	100?
Germanium	5	0.5–50	500	50–500
Lead	25	2–80	100	5–50?
Lithium	65	1–80		
Manganese	50	5–300		
Mercury	0.01	0.02–1		0.02–0.5?
Molybdenum	5	0.1–10	50	100–200
Nickel	15	0.5–50	700	50–800
Phosphorus	500	10–3,000		
Scandium	5	1–10	60	
Selenium	3	0.2–1.4		60?
Silver	0.50		2	1–5
Strontium	500	15–500		80–170?
Thallium		<0.2–1		1?
Thorium		0.5–10		
Tin		1–10		16–200?
Titanium	500	10–2,000		
Uranium	1.0	0.5–10	400	
Vanadium	25	2–100		100–1,000
Zinc	50	5–300		100–1,000?
Zirconium		5–200		100–500?

[a] From the US National Committee for Geochemistry, as cited in Valković (1983). [b] Swaine (1990). [c] Mason (1958). [d] Krauskopf (1955).

Modern technology utilizes fluidized bed boilers, pressurized fluidized bed boilers, and gasification of coal to improve energy extraction and minimize potential pollutants. The lower temperatures of those processes affect the fate of the individual trace elements that are emitted or are associated with the residues from the consumed coal (Table 2; Clarke, 1993).

9.05.2.2 Base-metal Deposits

Base metal is a wide-ranging term that refers either to metals inferior in value to those of gold and silver, or alternatively, to metals that are more chemically active than gold, silver, and the platinum metals (AGI, 1957). Accordingly, a review of base-metal mineralogy would encompass much of the world's metal production and geology. Usage of the term "base metal" in the minerals industry is rather loose, but a common application is to the nonferrous ore metals that include copper, lead, and zinc. Thus, e.g., Kesler (1994) grouped manganese, nickel, chromium, silicon, cobalt, molybdenum, vanadium, tungsten, niobium, and tellurium as ferroalloy metals, and copper, lead, zinc, and tin as base metals. Among the latter four metals, tin is by far the least significant in terms of volumes consumed and monetary value.

World mine production of copper is currently in the range of 13–14 Mt, about a third of which is from Chile. Other large producers are the United States, followed closely by Indonesia and Australia. The most important ore mineral is chalcopyrite [$CuFeS_2$], and also significant are bornite [Cu_5FeS_4] and chalcocite [Cu_2S]. The first two are primary minerals, whereas chalcocite forms principally by their weathering and subsequent reprecipitation of the solubilized copper as enriched "blankets" of chalcocite ore beneath the oxidation zone.

Copper ore is predominantly derived from porphyry copper deposits, with lesser but significant contributions from massive sulfide, skarn, and other types of deposits. The host rocks for porphyry copper deposits are felsic granitoid intrusions, and in skarn deposits the intrusions penetrate limestone and associated sedimentary-derived assemblages. The deposits are typically large (commonly hundreds of millions of tonnes) and of low grade (commonly $<1\%$ copper), with successful exploitation dependent mainly on open-pit access and on daily large-tonnage extraction and processing.

World mine production of zinc is ~9 Mt, with almost all of it derived from sphalerite [(Zn,Fe)S], which is also the principal primary source of cadmium and several other metals, such as germanium and indium. China and Australia are the largest producers, but several other countries mine significant amounts. About half of the annual consumption is for the manufacture of galvanized products to resist corrosion, primarily in the automotive and construction industries.

Mineral-deposit sources of sphalerite are diverse. Large production is obtained chiefly from skarn (e.g., Atamina, Peru), from volcanogenic massive sulfide deposits in which pyrite is the predominant mineral (e.g., Kidd Creek and Brunswick No. 12, Canada) from sedimentary-exhalative (SEDEX) deposits in which layers of lead, zinc, and iron sulfides were deposited in fine-grained clastic sedimentary rocks (e.g., Broken Hill and Mt. Isa, Australia), and from Mississippi Valley-type deposits, in which sphalerite and galena [PbS] were deposited in large amounts in cavities, breccias, and as replacements of calcareous sedimentary rocks consisting predominantly of limestone (e.g., Viburnum Trend, USA).

Whereas sphalerite is the principal mineral source of zinc, galena is the main source of lead. Annual world mining production of lead is ~3 Mt, and annual consumption is >6 Mt, with the difference made up by recycling. The largest primary producers are China and Australia, and the largest consumers are the European Union and the United States. About 75% of lead consumption is for the manufacture of lead-acid automotive batteries, which are also the principal source of recycled scrap. Unlike zinc, which is an essential biological trace element, lead has no similar function and is an important environmental hazard (Kesler, 1994). Processing of lead and zinc concentrates is almost totally by conventional pyrometallurgical smelting, but the most abundant anthropogenic sources of lead have been coal combustion and gasoline additives (Kesler, 1994).

9.05.2.3 Precious-metal Deposits

The precious-metal group consists of gold, silver, and the platinum-group elements (PGE). The world's leading producer of gold is South Africa, followed by the United States and Australia. Most mining of gold is done specifically for that metal rather than for a polymetallic assemblage, and most gold is produced from auriferous quartz veins. However, appreciable amounts of gold are recovered from the processing of base-metal ores, especially copper deposits. A characteristic feature of all types of deposits is that nearly all of the gold occurs as the native metal, commonly with silver in solid solution and less commonly with small but economically important amounts of gold in solid solution in arsenical pyrite. The traditional method of recovering gold from arsenical pyrite, a type of association that is referred to as refractory gold, is high-temperature roasting to vaporize the sulfur and arsenic, followed by cyanide leaching of the oxidized residue. Environmental concerns about high-arsenic emissions, or the disposal of the As_2O_3 that precipitates from the condensed gases, have led to the increased use of pressurized autoclaves to effect the oxidation step. Another environmental concern has been the use of mercury to recover gold by amalgamation. This practice has been largely discontinued because the effects of mercury poisoning are well known, but a legacy of pollution remains in many areas, and amalgamation on a small scale is still practiced by artisan miners in countries such as Brazil and Indonesia.

In recent years gold has found increased markets in electrical and electronic applications, but these account for <5% of annual consumption. About 90% of the annual production is utilized for jewelry and arts purposes.

Mexico, Peru, and the United States are the largest producers of silver, whose main usage is in photography, plating, jewelry, and electronic and electrical applications. More than three-quarters of the world annual production of silver is obtained from deposits in which base metals or gold are the principal product. For example, the world's largest silver producer is the metamorphosed, stratabound Cannington deposit in Australia. The deposit is of the Broken Hill type and contains ~44 Mt grading 11.6% lead, 4.4% zinc, and 538 g t^{-1} silver (Walters and Bailey, 1998). The dominant sulfide assemblage is galena–sphalerite–pyrrhotite, and the high silver content is related mainly to the presence of argentiferous galena and freibergite [$(Ag,Cu,Fe)_{12}Sb_4S_{13}$].

Mexico, the leading silver producer, obtains about half of its output from mines in which silver is the principal ore metal. Many of the mines are epithermal fissure veins, and most host a polymetallic assemblage whose exploitation is economically dependent on the high silver values. Although acanthite [Ag_2S] and native silver predominate in some veins, in others much of the silver occurs in silver sulfosalts and as silver substitutions in tetrahedrite [$(Cu,Fe,Ag)_{12}Sb_4S_{13}$] and other minerals.

The platinum-group metals consist of ruthenium, rhodium, palladium, osmium, iridium, and platinum. Each of the metals occurs naturally in its native form, and in economically exploitable deposits the elements occur overwhelmingly as individual platinum-group mineral (PGM) species. Mutual substitution of the various PGE is common, but substitutions in other minerals, such as base-metal sulfides, typically occur to only a limited extent. A comprehensive review of PGM and PGE geochemistry is given by Cabri (2002).

The platinum-group metals are generally grouped with gold and silver as precious-metal commodities, but the platinum-group metals have little in common with the other precious metals in terms of their primary geological host-rock associations. The world's largest producer of platinum and rhodium is South Africa, with most of the metal obtained from mines that exploit thin (centimeters rather than meters), PGM-rich layers (averaging <10 g t^{-1} PGE) in the Bushveld complex, a layered mafic intrusion that is also a principal source of chromium and vanadium. Platinum and palladium account for all but a very small percentage of world PGE production. Whereas the Bushveld complex accounts for more than a quarter of world palladium production, more than double that amount is obtained as a byproduct from Cu–Ni mines in layered intrusive complexes such as those at Sudbury, Canada, and Noril'sk–Talnakh, Russia; the latter is the world's largest primary source of palladium.

Braggite [(Pt,Pd)S], cooperite [PtS], sperrylite [$PtAs_2$], and Pt–Fe alloys are among the principal sources of PGE in the Bushveld complex. Michenerite [PdBiTe], moncheite [$PdTe_2$], and sperrylite are common sources of PGE in some of the Sudbury ores, and all of the above minerals and many more have been reported to occur in the Noril'sk deposits (Cabri, 1981).

The principal consumption of PGE is as a catalyst, especially the use of platinum, or the more favored palladium because of its superior high-temperature performance, in catalytic converters in motor vehicles. Among the diverse other chief uses are electrical and electronic applications, jewelry, fabrication of laboratory equipment, and dental repairs.

9.05.2.4 Uranium Deposits

Canada and Australia are the world's largest producers of uranium. All Canadian production is from rich deposits in the Athabasca basin of northern Saskatchewan; among those is the McArthur River mine, which has the world's largest high-grade deposit, estimated at 1.52×10^5 t of uranium from ore grading 15–18% uranium. These "unconformity"-type Saskatchewan deposits, which are also the principal deposit-type for Australian uranium production, contain mainly uraninite [UO_2] with associated coffinite [$U(SiO_4)_{1-x}(OH)_{4x}$] and brannerite [$(U,Ca,Y,Ce)(Ti,Fe)_2O_6$] (Plant *et al.*, 1999). The chief uses of uranium are in nuclear power plants and weaponry.

9.05.2.5 Diamond Deposits

World annual production of natural diamonds, the cubic form of carbon, is about 110 million carats (1 carat = 200 mg). Almost all is derived from kimberlite or its weathered remnants, but Australian production is from the Argyle mine, at which the host rock is lamproite. Kimberlites are olivine- and volatile-rich potassic ultrabasic rocks of variable geological age that typically form near-vertical carrot-shaped "pipes" intruded into Archean cratons. The volatile-rich component is predominantly CO_2 in the carbonate minerals calcite and dolomite, and the texture is characteristically inequigranular, with large grains (macrocrysts), usually of olivine [Mg_2SiO_4], in a fine-grained, olivine-rich matrix.

Australia, Botswana, Russia, and the Democratic Republic of Congo, in decreasing order, account for ~80% of the carats produced annually. In terms of value, however, Australia is surpassed by several countries; for example, South African production in carats is ranked fifth and is only 40% of that of Australia, but the value is three times that of Australia's diamonds.

9.05.2.6 Other Deposits

Table 3 summarizes data on the principal sources and uses of numerous other metals. The listing is not intended to be comprehensive.

9.05.3 SULFIDE OXIDATION AND THE GENERATION OF OXIDATION PRODUCTS

A principal environmental concern associated with mine wastes results from the oxidation of sulfide minerals within the waste materials and mine workings, and the transport and release of oxidation products. The principal sulfide minerals in mine wastes are pyrite and pyrrhotite, but others are susceptible to oxidation, releasing elements such as aluminum, arsenic, cadmium, cobalt, copper, mercury, nickel, lead, and zinc to the water flowing through the mine waste.

9.05.3.1 Pyrite Oxidation

Pyrite is commonly associated with coal and metal ore deposits. Pyrite oxidation and the factors affecting the kinetics of oxidation (O_2, Fe^{3+}, temperature, pH, E_h, and the presence or absence of microorganisms) have been the focus of extensive study because of their importance in both environmental remediation and mineral separation by flotation (Buckley and Woods, 1987; Brown and Jurinak, 1989; Evangelou and Zhang, 1995; McKibben and Barnes, 1986; Moses *et al.*, 1987; Nordstrom, 1982; Wiersma and Rimstidt, 1984; Williamson and Rimstidt, 1994; Luther, 1987; Sasaki *et al.*, 1995). Reviews of pyrite oxidation and the formation of acid mine drainage are given by Lowson (1982), Evangelou (1995), Evangelou and Zhang (1995), Nordstrom and Southam (1997), and Nordstrom and Alpers (1999a).

The oxidation of pyrite can occur when the mineral surface is exposed to an oxidant and water, either in oxygenated or anoxic systems, depending on the oxidant. The process is complex and can involve chemical, biological, and electrochemical reactions. The chemical oxidation of pyrite can follow a variety of pathways involving surface interactions with dissolved O_2, Fe^{3+}, and other mineral catalysts (e.g., MnO_2). Oxidation of pyrite

Table 3 Principal "mineral" sources and usage of various metals.

	Principal "mineral" sources	*Principal usage*
Aluminum	Gibbsite [$Al(OH)_3$], böhmite [$AlO(OH)$]	Transportation, packaging
Antimony	Stibnite [Sb_2S_3]; byproduct from Pb sulfides	Flame-retardant chemical; hardener for Pb in batteries
Arsenic	By-product as As_2O_3	Wood preservatives
Beryllium	Beryl [$Be_3Al_2Si_6O_{18}$] from pegmatite; bertrandite [$Be_4Si_2O_7(OH)_2$] from tuff	Be–Cu alloys (telecommunications)
Bismuth	By-product, mainly from galena	Pharmaceuticals, chemicals
Cadmium	By-product from sphalerite	Batteries
Chromium	Chromite [$FeCr_2O_4$] in mafic-ultramafic intrusions	Stainless steel
Cobalt	Laterites; Ni–Cu sulfide ores; linnaeite [$Co^{2+}Co_2^{3+}S_4$] from sedimentary-hosted Cu–Co deposits	Superalloys; alloys with steel
Gallium	By-product from sphalerite	GaAs in electronic devices
Germanium	By-product from sphalerite	Fiber-optic systems
Indium	By-product from sphalerite	Electronics, LCD screens
Iron	Hematite [Fe_2O_3], goethite [FeOOH], magnetite [$Fe^{2+}Fe_2^{3+}O_4$]	Iron and steel
Magnesium	Brines, seawater; magnesite [$MgCO_3$]	Al alloys; die casting
Manganese	Mn oxides, hydroxides	Alloys with steel
Mercury	Cinnabar [HgS]	Electrolysis; batteries
Molybdenum	Molybdenite [MoS_2] from porphyry Mo and Cu deposits	Alloys with iron, steel
Nickel	Laterite; pentlandite [$(Fe,Ni)_9S_8$] in mafic-ultramafic intrusions	Steel and nonferrous alloys
Niobium	Pyrochlore [$(Ca,Na)_2Nb_2O_6(OH,F)$] from carbonatites	Alloys with steel; superalloys
Rare-earth elements (REEs)	Bastnäsite-Ce [$(Ce,La)(CO_3)F$] from carbonatites	Chemical catalyst; automotive catalytic converters; glass polishing, ceramics
Rhenium	By-product from molybdenite	Pt–Rh catalysts; superalloys
Scandium	By-product	Al alloys; halide lighting additive
Selenium	By-product from Cu anode slimes	Glass; metallurgical additive; electronics
Silicon	Quartz [SiO_2]	Additive to steel
Strontium	Celestine [$SrSO_4$]	Television faceplate glass; ceramics
Tantalum	Tantalite-columbite [$(Fe,Mn,Mg)Ta_2O_6$–$(Fe,Mn,Mg)Nb_2O_6$] from pegmatites	Electronic components
Tellurium	By-product from refining Cu ores	Steel and copper additive
Thallium	By-product from Cu–Zn–Pb sulfides ores	Semiconductor materials; electronics
Thorium	Monazite [$(REE,Th)PO_4$] byproduct from heavy-mineral sands	Refractory applications; catalyst
Tin	Cassiterite [SnO_2] in placers	Plating on cans and containers; solder
Titanium	Ilmenite [$FeTiO_3$] from heavy-mineral sands	TiO_2 pigment
Tungsten	Scheelite [$CaWO_4$] from skarns; ferberite [$Fe^{2+}WO_4$] from veins	Tungsten carbide
Vanadium	Magnetite [$Fe^{2+}(Fe^{3+},V^{3+})_2O_4$] in mafic-ultramafic intrusions	Steel additive
Yttrium	By-product from bastnäsite REE production	Phosphorus
Zirconium	Zircon [$ZrSiO_4$] from heavy-mineral sands	Refractory facings and bricks

by atmospheric oxygen produces one mole of Fe^{2+}, two moles of SO_4^{2-} and two moles of H^+ for every mole of pyrite oxidized (Nordstrom, 1982):

$$FeS_2(s) + \tfrac{7}{2}O_2 + H_2O \rightarrow Fe^{2+} + 2SO_4^{2-} + 2H^+ \quad (1)$$

The Fe(II) thus released may be oxidized to Fe(III):

$$Fe^{2+} + \tfrac{1}{4}O_2 + H^+ \leftrightarrow Fe^{3+} + \tfrac{1}{2}H_2O \quad (2)$$

Fe(III) oxyhydroxides such as ferrihydrite (nominally $5Fe_2O_3{\cdot}9H_2O$) may precipitate:

$$Fe^{3+} + 3H_2O \leftrightarrow Fe(OH)_3 + 3H^+ \quad (3)$$

where $Fe(OH)_3$ is a surrogate for ferrihydrite. Adding Equations (1)–(3) yields the overall reaction (4):

$$\begin{aligned} &FeS_2(s) + \tfrac{15}{4}O_{2(aq)} + \tfrac{7}{2}H_2O_{(aq)} \\ &\rightarrow 2SO_4^{2-} + Fe(OH)_{3(s)} + 4H^+_{(aq)} \end{aligned} \quad (4)$$

This overall reaction results in the release of 4 mol of H^+ for each mole of pyrite oxidized.

9.05.3.1.1 *Mechanism of pyrite oxidation by Fe^{3+} and O_2*

Initially, pyrite oxidation involves the adsorption of O_2 and water to the partly protonated pyrite surface by bonding to Fe^{2+} (Fornasiero *et al.*, 1994). Various iron oxyhydroxide intermediate products can form on the pyrite surface, depending on pH. Singer and Stumm (1970) suggested that, under acidic conditions, the major oxidant of pyrite is Fe^{3+}, whereas O_2 becomes the predominant oxidant at circumneutral pH because of the diminished solubility of Fe^{3+}. Pyrite oxidation by Fe^{3+} at circumneutral pH has also been observed (Brown and Jurinak, 1989; Evangelou and Zhang, 1995; Moses *et al.*, 1987), but the reaction cannot be sustained without the presence of dissolved O_2 to perpetuate the oxidation to Fe^{3+}. When O_2 is the oxidant under near-neutral pH conditions, one oxygen atom in the sulfate is derived from dissolved O_2, with the remainder derived from H_2O. Under acidic conditions, all four oxygen atoms in sulfate are derived from H_2O (Reedy *et al.*, 1991). Although both Fe^{3+} and oxygen can bind chemically to the surface, the more rapid oxidation rates for Fe^{3+} compared to those for O_2 are due to a more efficient electron transfer for Fe^{3+} (Luther, 1987). A molecular orbital model proposed by Luther (1987) is consistent with pyrite oxidation data obtained by McKibben and Barnes (1986), Moses *et al.* (1987) and Wiersma and Rimstidt (1984).

Rate data from the literature for the reaction of pyrite with dissolved O_2 were compiled by Williamson and Rimstidt (1994). They produced a rate law that is applicable for more than four orders of magnitude in O_2 concentration and over a pH range of 2–10:

$$R = 10^{-8.19(\pm 0.04)} \frac{(m_{DO})^{0.5(\pm 0.04)}}{(m_{H^+}^{0.11})^{(\pm 0.01)}} \quad (5)$$

where R is the rate of pyrite dissolution in units of mol m^{-2} s^{-1}.

A series of batch and mixed flow reactor experiments was performed at pH <3 to determine the effect of SO_4^{2-}, Cl^-, ionic strength, and dissolved O_2 on the rate of pyrite oxidation by Fe^{3+}. Of these, only dissolved O_2 had any appreciable affect on the rate of pyrite oxidation in the presence of Fe^{3+}. Williamson and Rimstidt (1994) combined their experimental results with kinetic data reported from the literature to formulate rate laws that are applicable over a range spanning six order of magnitude in Fe^{3+} and Fe^{2+} concentrations, and for a pH range of 0.5–3.0, when fixed concentrations of dissolved O_2 are present:

$$R = 10^{-6.07(\pm 0.57)} \frac{(m_{Fe^{3+}}^{0.93})^{(\pm 0.07)}}{(m_{Fe^{2+}}^{0.40})^{(\pm 0.06)}} \quad (6)$$

where R is the rate of pyrite dissolution in units of mol m^{-2} s^{-1}.

A wide variation in empirical rate laws has been developed to describe pyrite oxidation. The wideness of the range could be due to several factors, among which are differences in sample preparation, different ratios of surface area to volume, and the presence of impurities in the pyrite or in solution. Activation energies determined for pyrite oxidation range from 50 kJ mol^{-1} for pH 2, to 4–92 kJ mol^{-1} for pH 6–8, regardless of whether dissolved O_2 or Fe(III) is used as the oxidant (Nicholson, 1994; Wiersma and Rimstidt, 1984). Table 4 provides a summary of the proposed rate expressions for the dissolution of pyrite in solutions containing dissolved O_2 and Fe(III). Activation energies are observed to be higher for pH values in the range of 6–8 than in the range of 2–4. Casey and Sposito (1991) suggested that the proton adsorption/desorption reactions can contribute up to 50 kJ mol^{-1} to the experimental activation energy of dissolution reactions for silicate minerals. The hydrogen-ion activity or pH, therefore, may play an important role in the observed activation energy for the oxidation of sulfide minerals. Regardless, the high activation energies observed indicate that the rate-limiting step in pyrite oxidation is related to electron transfer at the pyrite surface.

Holmes and Crundwell (2000) studied the kinetics of pyrite oxidation and reduction

Table 4 Summary of proposed rate expressions for the dissolution of pyrite in solutions containing dissolved oxygen and ferric iron.

Source	*Rate expression: dissolved oxygen*	*Rate expression: dissolved iron*
Garrels and Thompson (1960)		$r_{FeS_2} = \frac{k[Fe^{3+}]}{\sum[Fe]}$
Mathews and Robins (1972, 1974)	$r_{FeS_2} = k[O_2]^{0.81}$	$r_{FeS_2} = \frac{k[Fe^{3+}]}{\sum[Fe][H^+]^{0.44}}$
Lowson (1982)		$r_{FeS_2} = \frac{k[Fe^{3+}][Fe^{2+}]}{\sum[Fe]}$
McKibben and Barnes (1986)	$r_{FeS_2} = k[O_2]^{0.5}$	$r_{FeS_2} = \frac{kFe^{3+}]^{0.58}}{[H^+]^{0.5}}$
Williamson and Rimstidt (1994)	$r_{FeS_2} = k[O_2]^{0.5}[H^+]^{-0.11}$	$r_{FeS_2} = \frac{k[Fe^{3+}]^{0.3}}{[Fe^{2+}]^{0.47}[H^+]^{0.32}}$
Holmes and Crundwell (2000)	$r_{FeS_2} = k[H^+]^{-0.18}[O_2]^{0.5}$	$r_{FeS_2} = k[H^+]^{-0.5}\left(\frac{k_{Fe^{3+}}[Fe^{3+}]}{k_{Fe^{2+}}[Fe^{2+}]k_{FeS_2}[H^+]^{-0.5}}\right)^{0.5}$

After Holmes and Crundwell (2000).

independently using electrochemical techniques. The kinetics of the half reactions are related to the overall dissolution reaction assuming no accumulation of charge on the surface. This assumption was used to derive expressions for the mixed potential and rate of dissolution, which agreed with those obtained by McKibben and Barnes (1986) and Williamson and Rimstidt (1994). The results showed that the electrochemical reaction steps occurring at the mineral–solution interface control the rate of dissolution.

9.05.3.2 Pyrrhotite Oxidation

Pyrrhotite is a common iron-sulfide mineral. Although there have been numerous studies of the oxidation of pyrite, fewer studies have focused on pyrrhotite oxidation (Buckley and Woods, 1985; Jones *et al.*, 1992; Nicholson and Scharer, 1994; Pratt *et al.*, 1994a,b; Thomas *et al.*, 1998; Janzen *et al.*, 2000). The pyrrhotite structure is based on hexagonal close packing, but is disordered (i.e., NiAs-type structure), giving rise to nonstoichiometric and stoichiometric compositions in which x in the formula $Fe_{1-x}S$ can vary from 0.125 (Fe_7S_8) to 0 (FeS). The iron vacancies within the structure may be charge-compensated by Fe^{3+} (Vaughan and Craig, 1978) or an approximation thereof. Analyses of cleaved pyrrhotite surfaces under vacuum showed Fe(III)–S interactions on the pyrrhotite surface (Pratt *et al.*, 1994a). The deficiency in iron within the pyrrhotite structure can result in a symmetry that varies from monoclinic (Fe_7S_8) to hexagonal ($Fe_{11}S_{12}$), with the composition progressing to stoichiometric troilite (FeS). Orlova *et al.* (1988) examined the reaction rates for monoclinic and hexagonal pyrrhotite and concluded that the hexagonal form was the more reactive.

The deficiency of iron in the pyrrhotite structure may affect the oxidation behavior. Nicholson and Scharer (1994) observed a dependence of activation energy on pH. The energy ranged from 52 kJ mol^{-1} to 58 kJ mol^{-1} at pH 2–4, and almost doubled to 100 kJ mol^{-1} at circumneutral pH (i.e., 6). These values are similar to activation noted for pyrite, suggesting a chemical-controlled reaction. Orlova *et al.* (1988) observed a range of activation energies for both monoclinic and hexagonal varieties ranging from 50 kJ mol^{-1} to 46 kJ mol^{-1}, respectively. It was argued that the lower activation energy was significant to the hexagonal variety. Janzen *et al.* (2000) did not observe consistent trends between activation energy and crystal structure.

9.05.3.2.1 *Chemical oxidation by O_2 and Fe^{3+}*

Pyrrhotite dissolution can proceed through oxidative or nonoxidative reactions. Oxidative dissolution can be at least 10^3 times slower than nonoxidative reactions (Thomas *et al.*, 1998). Dissolved O_2 and Fe^{3+} can be important oxidants of pyrrhotite. When oxygen is the primary oxidant, the overall reaction may be written as

$$\begin{aligned} &Fe_{1-x}S + (2 - \tfrac{1}{2}x)O_2 + xH_2O \\ &\quad \rightarrow (1-x)Fe^{2+} + SO_4^{2-} + 2xH^+ \end{aligned} \quad (7)$$

The production of protons is linked to the mineral stoichiometry. Up to one-quarter mole of the protons produced are derived from the oxidation of one mole of the iron-deficient form ($x = 0.125$), whereas no protons are produced from the stoichiometric form, which is troilite ($x = 0$). The release of protons can also result from the oxidation of the dissolved iron resulting from the precipitation of ferric hydroxide:

$$Fe^{2+} + \frac{1}{4}O_2 + \frac{5}{2}H_2O \rightarrow Fe(OH)_{3(s)} + 2H^+ \quad (8)$$

In other circumstances, the oxidation reactions may not proceed to completion. Partial oxidation may result in only a small proportion of the sulfur being transformed to sulfate, with the remainder accumulating as reduced sulfur species (polysulfides and elemental sulfur) at the mineral surface (Janzen *et al.*, 2000):

$$Fe_{1-x}S_{(s)} + \left(\frac{1-x}{2}\right)O_2 + 2(1-x)H^+ \rightarrow (1-x)Fe^{2+} + S_x^{n<0} + (1-x)H_2O \quad (9)$$

The rates of oxidation of both pyrite and pyrrhotite at 25 °C and standard atmospheric oxygen indicate that pyrrhotite can react 20–100 times faster than pyrite. During oxidation of a particle of pyrrhotite, iron diffuses to the exposed surface, thereby creating a sulfur-enriched inner zone that contains disulfide and polysulfide-like species (Mycroft *et al.*, 1995).

9.05.3.2.2 Nonoxidative mechanism

Nonoxidative dissolution of pyrrhotite occurs in acidic solutions when predominant S^{2-} surface species are exposed. The reaction occurs as

$$Fe_{(1-x)}S_{(s)} + 2H^+ \rightarrow (1-x)Fe^{2+} + H_2S \quad (10)$$

Jones *et al.* (1992) observed restructuring of sulfur-rich pyrrhotite surfaces in deoxygenated acid, resulting in the development of a surface dominated by a discontinuous layer of a tetragonal intermediate Fe_2S_3 structure. Janzen *et al.* (2000) showed a significant release of Fe^{2+} from pyrrhotite in acidic solutions in which oxygen was not present. Although Fe^{2+} concentrations increased linearly with time, sulfate values remained unchanged, with sulfur from pyrrhotite dissolution remaining in a reduced state (S^{2-}). Thomas *et al.* (1998, 2001) proposed a dissolution mechanism that allows two distinct pathways: (i) iron leaves the surface, with no additional electrons released from the structure, and (ii) after a critical accumulation of charge, reduction of polysulfide to sulfide occurs, resulting in the release of negative charge from the surface in the form of HS^-. The significant feature of this process is the delay between the release of Fe^{2+} and HS^-.

9.05.3.3 Oxidation of Other Metal Sulfides

9.05.3.3.1 Sphalerite

The oxidation of sphalerite is dependent on a number of factors, among which are the concentration of oxidants, such as dissolved O_2 or Fe(III) in solution, the temperature, and the pH (Bobeck and Su, 1985; Crundwell, 1988; Olanipekun, 1999; Perez and Dutrizac, 1991; Rimstidt *et al.*, 1994). For sphalerite, Vaughan and Craig (1978) reported a solubility product of $K_{sp} = 1 \times 10^{-20.6}$ at 25 °C in water. Other researchers have reported similar values (Daskalakis and Helz, 1993).

For sphalerite in dilute Fe(III) solutions, Rimstidt *et al.* (1994) obtained a dissolution rate of 7.0×10^{-8} mol m^{-2} s^{-1} with a corresponding activation energy of 27 kJ mol^{-1} over a range of 25–60 °C. The concentration of Fe(III) used was 10^{-3} M, which is similar to dissolved iron concentrations ($(2-9) \times 10^{-3}$ M) typically observed in acidic mine waters (Lin, 1997). The overall oxidation reaction for pure sphalerite, assuming that all sulfur is oxidized to sulfate, is

$$ZnS_{(s)} + 2O_{2(aq)} \rightarrow Zn^{2+}_{(aq)} + SO^{2-}_{4(aq)} \quad (11)$$

XPS examination of oxidized sphalerite showed the development of a surface layer of metal-deficient sulfide (Buckley *et al.*, 1989) whose formation in acid solution is described by

$$ZnS \rightarrow Zn_{1-x}S + xZn^{2+} + 2xe^- \quad (12)$$

Weisener (2002) observed increased rates of oxidation and increased acid consumption as a function of the amount of solid-solution iron in sphalerite [(Zn,Fe)S]. Apparent activation energies of 21–28 kJ mol^{-1} obtained at 25–85 °C are similar to the values reported by Rimstidt *et al.* (1994). Weisener *et al.* (2003) suggested that the production of polysulfide species results in a lower diffusion gradient at the mineral surface, thus leading to lower reactivity with potential oxidants and to diffusion-limited release of zinc and iron from the bulk mineral. Elemental sulfur was not observed to limit the reactivity of the mineral surface. The accumulation of polysulfides and S^0 on the sphalerite surface under oxygenated conditions can affect the acid-neutralization capacity because the polysulfides and S^0 consume acid when pH is <3. The resulting formation of a sulfur-enriched surface slows the subsequent rate of dissolution of sphalerite in the absence of bacteria. Under these conditions, S^0 does not passivate the surface (Weisener, 2002).

9.05.3.3.2 Galena and chalcopyrite

Galena and chalcopyrite are commonly associated with acid-generating minerals, such as pyrite and pyrrhotite. Acid ferric sulfate solutions, generated through the oxidation of iron sulfides can enhance the oxidation of lead- and copper-bearing sulfide minerals. The oxidation of galena has been studied by Buckley and Woods (1984a), Tossell and Vaughan (1987), Fornasiero *et al.* (1994),

Kim *et al.* (1995), Prestidge *et al.* (1995), Basilio *et al.* (1996), Kartio *et al.* (1996, 1998), Chernyshova and Andreev (1997), Jennings *et al.* (2000), Nowak and Laajalehto (2000), Shapter *et al.* (2000), and others. XPS studies showed that S^0 formed when galena was oxidized in a hydrogen peroxide solution, and that metal-deficient surfaces resulted from oxidation by dilute acetic acid solutions (Buckley and Woods, 1984a). In natural oxygenated environments, galena will weather to anglesite, which is weakly soluble below pH 6 (Lin, 1997; Shapter *et al.*, 2000):

$$PbS_{(s)} + 2O_{2(aq)} \rightarrow Pb^{2+}_{(aq)} + SO_4^{2-}{}_{(aq)} \quad (13)$$

$$Pb^{2+}_{(aq)} + SO^{2-}_{4(aq)} \leftrightarrow PbSO_{4(s)} \quad (14)$$

Galena may also be oxidized by Fe(III) under acidic conditions (Rimstidt *et al.*, 1994):

$$PbS + 8Fe^{3+} + 4H_2O \rightarrow 8H^+ + SO_4^{2-} + Pb^{2+} + 8Fe^{2+} \quad (15)$$

The oxidation of galena in air may result in the formation of lead hydroxide and lead oxide (Evans and Raftery, 1982; Buckley and Woods, 1984a; Laajalehto *et al.*, 1993). Oxidation in aqueous solutions may lead to the formation of lead oxides and lead sulfate surface products (Fornasiero *et al.*, 1994; Kartio *et al.*, 1996; Kim *et al.*, 1995; Nowak and Laajalehto, 2000). In the absence of oxygen, both lead and sulfide ions are released to solution in the form of free lead ions and hydrogen sulfide (Fornasiero *et al.*, 1994). Jennings *et al.* (2000) showed that galena was not acid generating when exposed to accelerated oxidation using hydrogen peroxide. This reaction resulted in the accumulation of anglesite on the mineral surface.

XPS study by Buckley and Woods (1984b) showed that freshly fractured chalcopyrite surfaces exposed to air formed a ferric oxyhydroxide overlayer with an iron-deficient region composed of CuS_2. Acid-treated surfaces of fractured chalcopyrite showed an increase in the thickness of the CuS_2 layer and the presence of elemental sulfur. Hackl *et al.* (1995) suggested that dissolution of chalcopyrite is passivated by a thin (<1 μm) copper-rich surface layer that forms as a result of solid-state changes. The passivating surface layer consists of copper polysulfide, CuS_n, where $n > 2$. Hackl *et al.* (1995) described the dissolution kinetics as a mixed diffusion and chemical reaction whose rate is controlled by the rate at which the copper polysulfide is leached. The oxidation of chalcopyrite in the presence of ferric ions under acidic conditions can be expressed as

$$CuFeS_2 + 4Fe^{3+} \rightarrow 5Fe^{2+} + Cu^{2+} + 2S^0 \quad (16)$$

Hiroyoshi *et al.* (1997) monitored the oxygen consumption, sulfur formation, total iron, and Fe(II) concentrations at different pH levels during the oxidation of chalcopyrite. On the basis of the reaction products formed, it was concluded that ferrous ions catalyzed the oxidation by dissolved oxygen in acidic media:

$$CuFeS_2 + 4H^+ + O_2 \rightarrow Cu^{2+} + Fe^{2+} + 2S^0 + 2H_2O \quad (17)$$

The dissolution of chalcopyrite can be also influenced strongly by galvanic effects. The presence of pyrite or molybdenite in association with chalcopyrite can cause accelerated rates of chalcopyrite dissolution (Dutrizac and MacDonald, 1973), whereas the presence of iron-rich sphalerite and galena can slow the dissolution.

9.05.3.3.3 Arsenic and mercury sulfides

The oxidation of arsenopyrite [FeAsS] releases both sulfur and arsenic. Buckley and Walker (1988) studied the oxidation of arsenopyrite in alkaline and in acidic aqueous solutions. In air, the mineral reacted rapidly, and the oxidation of arsenic to As(III) was more rapid than the oxidation of iron on the same surface. Only a small amount of sulfur oxidation occurred. Under acidic conditions, the mineral formed sulfur-rich surfaces.

Nesbitt *et al.* (1995) conducted a detailed study of the oxidation of arsenopyrite in oxygenated solutions. Arsenic and sulfur were observed to exist in multiple oxidation states near the pristine surface. After reaction with air-saturated distilled water, Fe(III) oxyhydroxides formed the dominant iron surface species, and As(V), As(III), and As(I) were as abundant as As($-$I) surface species. An appreciable amount of sulfate was observed on the mineral surface. Arsenic was more readily oxidized than sulfur, and similar rates of the oxidation of As($-$I) and $Fe(II)^+$ surface species were observed. Nesbitt *et al.* (1995) concluded that continued diffusion of arsenic to the surface under these conditions can produce large amounts of As^{3+} and As^{5+}, promoting rapid selective leaching of arsenites and arsenates.

Rimstidt *et al.* (1994) measured the rate of reaction of arsenopyrite under conditions typical of acid mine-drainage environments. Arsenopyrite was observed to be more reactive than pyrite, chalcopyrite, galena, and sphalerite. Oxidation of arsenopyrite led to the formation of scorodite on the surface. The activation energies for arsenopyrite oxidation varied from 18 kJ mol^{-1} at 0–25 °C, to a slightly negative E_a of -6 kJ mol^{-1} at 25–60 °C. Rimstidt *et al.* (1994) attributed the negative activation energies to competition between the dominant reactions that occur at all temperatures and the less vigorous side-reactions that contribute to rate-limiting behavior at higher temperatures.

An investigation of the surface composition and chemical state of three naturally weathered arsenopyrite samples exposed for periods ranging from 14 d to 25 yr showed that the arsenopyrite surface has an effective passivating layer that protects the mineral from further oxidation (Nesbitt and Muir, 1998). The same samples were then reacted with mine-waste waters, which caused extensive leaching of the arsenopyrite surface below the oxidized overlayers. The acidic nature of the solution caused dissolution of the previously accumulated ferric arsenite and arsenate salts.

Foster *et al.* (1998a,b) provided a detailed comparison of the oxidized speciation and local geometry of arsenic for three California mine-waste materials: fully oxidized tailings, partly oxidized tailings, and roasted sulfide ore. The fully oxidized tailings showed As(V) species sorbed primarily on the ferric oxyhydroxides and aluminosilicates. The partly oxidized tailings contained equal amounts of arsenopyite and arsenian pyrite that accounted for 20% of the total arsenic, and the remaining 80% consisted of As^{5+} as a precipitate of scorodite. The roasted sulfide ore contained an As(V) species that had substituted for sulfate in jarosite $[KFe_3(SO_4)_2(OH)_6]$ and was sorbed onto the surfaces of hematite and ferric oxyhydroxide grains.

Cinnabar [HgS], the principal ore of mercury, is the most thermodynamically stable form at low temperature (Benoit *et al.*, 1999; Barnett *et al.*, 2001). The presence of trace impurities, such as zinc, selenium, and iron, can impede the conversion of metacinnabar, which is the high-temperature form, to cinnabar. The trace impurities decrease the inversion temperature and thus retard the conversion (Barnett *et al.*, 1997). Therefore, in some environments the availability of impurities will favor the *in situ* formation of metacinnabar over cinnabar. The formation of HgS is favored under reducing conditions, in part because of the high affinity of mercury compounds for sulfur. Cinnabar is kinetically resistant to oxidation, and can remain in soil and tailing impoundments even under oxidizing conditions (Barnett *et al.*, 1997). Although little has been published on its oxidation rates in soils, the observed persistence of HgS in soils at mine sites suggests that its weathering is slow under typical oxidative environments (Barnett *et al.*, 1997; Gray *et al.*, 2000). See Fitzgerald and Lamberg (Chapter 9.04) for additional information on the geochemical behavior of mercury in the environment.

9.05.3.4 Bacteria and Sulfide-mineral Oxidation

9.05.3.4.1 Organisms that catalyze sulfide oxidation

Microorganisms contribute to a vast array of biogeochemical cycles, among which are oxygenic photosynthesis, ammonia oxidation, sulfur oxidation/reduction, methanogenesis, fermentation, and respiration. This diversity allows microorganisms to survive in environments that have finite nutrients and low pH, as occur in acidic mine effluents. Bacteria are among the few forms of life that can tolerate these extreme environments. Chemolithotrophs are microorganisms that derive their energy requirements from inorganic sources in the presence of oxygen. Some chemolithotrophs can substitute nitrate or ferric iron as the electron acceptor if oxygen is unavailable. The genera *Thiobacillus*, *Acidithiobacillus*, and *Leptospirillum* contain numerous species that can utilize various sulfur compounds (Table 5). This group of organisms is largely responsible for the oxidation of sulfide minerals and includes iron- and sulfur-oxidizing bacteria. A detailed historical review of the isolation and identification of these microorganisms is provided by Nordstrom and Alpers (1999a).

Sulfur can exist in several oxidation states in nature, the most common being sulfide (-2), polysulfides $(-2 < x < 0)$, elemental sulfur (0), and sulfate $(+6)$. Sulfur oxidizers commonly occur within interfaces of low-energy environments, where anoxic waters mix slowly with oxic conditions above. Acidophilic sulfide-oxidizing bacteria are commonly isolated from acid mine-drainage environments in which abundant sources of sulfur from sulfide-rich material, such as waste rocks and tailing impoundments, are available. The availability of oxygen in conjunction with an abundant source of sulfur supports bacterial growth and activity. Bacterial catalysis of the available sulfide minerals accelerates the transformation of the reduced compounds to sulfate. The subsequent low pH produced, typically <4, allows the acidophilic organisms to thrive where other organisms are excluded. Neutral-pH oxidizers of sulfur include several *Thiobacillus* species. *Thiobacillus thioparus* is a mesophillic bacterium with an optimum pH range between 6.0 and 8.0. *T. thioparus* can oxidize sulfide, thiosulfate, and other reduced species of sulfur (Gould, 1994; Gould and Kapoor, 2003).

Acidithiobacillus thiooxidans, formerly known as *Thiobacillus thiooxidans*, is an acidophilic bacterium that oxidizes S^0 and thiosulfate, but not iron. Due to the rapid kinetics involved in sulfide oxidation by dissolved oxygen, some sulfide-oxidizing bacteria are in continuous competition with the chemical oxidation mechanism. *Acidithiobacillus ferrooxidans*, formerly known

Table 5 Members of the bacteria genera *Thiobacillus*, *Leptospirillum*, *Sulfobacillus* and four additional *Archaea* spp. that have been observed in acid mine waters and are associated with pyrite oxidation. The species in bold type are groups of acidophiles.

Chemolithotrophs	*Inorganic energy source*
Thiobacillus denitrificans	H_2S, S(0), $S_2O_3^{2-}$, $S_4O_6^{2-}$
Thiobacillus delicatus	S(0), $S_2O_3^{2-}$, $S_4O_6^{2-}$
Thiobacillus albertis	H_2S, $S_2O_3^{2-}$
Thiobacillus acidophilus	S(0), $S_2O_3^{2-}$, $S_3O_6^{2-}$, $S_4O_6^{2-}$
Thiobacillus halophilus	S(0)
Thiobacillus intermedius	S(0), $S_2O_3^{2-}$, $S_4O_6^{2-}$
Thiobacillus neapolitanus	H_2S, sulfide minerals, S(0), $S_2O_3^{2-}$, $S_4O_6^{2-}$
Thiobacillus novellus	$S_2O_3^{2-}$, $S_4O_6^{2-}$
Thiobacillus perometabolis	S(0), $S_2O_3^{2-}$, $S_4O_6^{2-}$
Thiobacillus tepidarius	H_2S, S(0), $S_2O_3^{2-}$, $S_3O_6^{2-}$, $S_4O_6^{2-}$
Thiobacillus thermophilica	H_2S, sulfide minerals, S(0)
Acidithiobacillus thioxidans	S(0), $S_2O_3^{2-}$, $S_4O_6^{2-}$
Thiobacillus thioparus	H_2S, sulfide minerals, S(0), $S_2O_3^{2-}$, $S_4O_6^{2-}$
Thiobacillus versutus	H_2S, $S_2O_3^{2-}$
Acidithiobacillus ferrooxidans	H_2S, sulfide minerals, S(0), $S_2O_3^{2-}$, $S_4O_6^{2-}$, Fe(II)
Leptospirillum ferrooxidans	Fe(II), sulfide minerals
Leptospirillum thermoferrooxidans	Fe(II), sulfide minerals
Sulfobacillus thermosulfidooxidans	Fe(II), S(0), sulfide minerals
Archaea spp.	
Acidianus brierleyi	Fe(II), S(0), sulfide minerals
Sulfolobus solfataricus	S(0)
Sulfolobus ambivalens	S(0)
Sulfolobus acidocaldarius	Fe(II), S(0)

After Nordstrom and Southam (1997).

as *Thiobacillus ferrooxidans*, is capable of utilizing Fe(II), S^0, or sulfide minerals as a source of energy for metabolic activity and plays an important role in the biogeochemical cycling of iron and sulfur. *Leptospirillum ferrooxidans* is metabolically similar to *A. ferrooxidans* but morphologically differs in having a spherical structure and a preferred pH range of 1.5–2.1. *L. ferrooxidans* lacks the capacity to oxidize sulfur compounds, preferring the selective oxidation of Fe(II), thus making it less sensitive to Fe(III) inhibition and allowing this organism to thrive at higher redox potentials than *A. ferrooxidans* (Breed and Hansford, 1999).

9.05.3.4.2 Conditions for optimal growth of bacterial species

The mesophiles *A. thiooxidans* and *A. ferrooxidans* have been found in equal proportions in mine effluents regardless of geological and climatological conditions. The temperature optimum for *A. ferrooxidans* and *A. thiooxidans* lies between 15 °C and 35 °C. Both are acidophiles and can tolerate a pH range of 1.5–6.0, with optimum growth between pH 2.0 and 2.5. The ability of these microorganisms to adapt and tolerate different pH and metal concentrations makes them effective catalysts for the oxidation of sulfides in mine wastes. *A. ferrooxidans* is an obligate aerobe and has a strict requirement of CO_2 as a source of carbon for growth. The concentration of CO_2 that effectively supports the maximal growth rate has been found to be 7–8%. An increase above 8% results in inhibited growth characteristics (Barron and Lueking, 1990). Both *L. ferrooxidans* and *A. ferrooxidans* occur in similar environments, but *L. ferrooxidans* is more tolerant to extremes in temperature and pH (Edwards *et al.*, 1999).

9.05.3.4.3 Role of bacteria in generation of acidic drainage

The pre-1991 research involving microbial oxidation of 29 sulfide minerals of iron, copper, arsenic, antimony, gallium, zinc, lead, nickel, and mercury was compiled by Nordstrom and Southam (1997). The importance of microbially mediated sulfide oxidation has been recognized for several decades (Nordstrom and Southam, 1997). Bacteria catalyze the oxidative dissolution of sulfide minerals, increasing the production of acidity in mine wastes. In the absence of bacteria, the rate of sulfide oxidation stabilizes as the pH decreases below 3.5 (Singer and Stumm, 1970).

Sulfide-mineral oxidation by microbial populations has been postulated to proceed via direct or indirect mechanisms (Tributsch and Bennett, 1981a,b; Boon and Heijnen, 2001; Fowler, 2001; Sand *et al.*, 2001; Tributsch, 2001). In the direct mechanism, it is assumed that the action taken by the attached cell or bacterium on a metal sulfide will solubilize the mineral surface through direct enzymatic oxidation reactions. The sulfur moiety on the mineral surface is oxidized to sulfate without the production of any detectable intermediates. The indirect mechanism assumes that the cell or bacteria do not act directly on the sulfide-mineral surface, but catalyze reactions proximal to the mineral surface. The products of these bacterially catalyzed reactions act on the mineral surfaces to promote oxidation of the dissolved Fe(II) and S^0 that are generated via chemical oxidative processes. Ferrous iron and S^0, present at the mineral surface, are biologically oxidized to Fe(III) and sulfate. Physical attachment is not required for the bacterial catalysis to occur. The resulting catalysis promotes chemical oxidation of the sulfide-mineral surface, perpetuating the sulfide oxidation process (Figure 1).

The following equations summarize the two mechanisms using pyrite as an example:

Direct interaction

$$FeS_2 + \tfrac{7}{2}O_2 + H_2O \rightarrow Fe^{2+} + 2H^+ + 2SO_4^{2-} \quad (18)$$

$$2Fe^{2+} + \tfrac{1}{2}O_2 + 2H^+ \rightarrow 2Fe^{3+} + H_2O \quad (19)$$

Indirect interaction

$$FeS_2 + 14Fe^{3+} + 8H_2O \rightarrow 15Fe^{2+} + 16H^+ + 2SO_4^{2-} \quad (20)$$

$$MS + 2Fe^{3+} \rightarrow M^{2+} + S^0 + 2Fe^{2+} \quad (21)$$

$$S^0 + \tfrac{3}{2}O_2 + H_2O \rightarrow 2H^+ + SO_4^{2-} \quad (22)$$

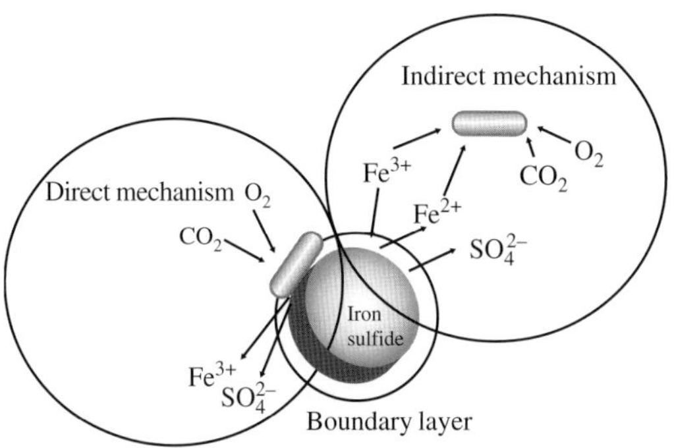

Figure 1 Comparison of the microbial direct versus indirect oxidation mechanism for sulfide (source Sand *et al.*, 2001).

Dissolved oxygen or Fe(III) ions are considered to be the primary oxidants in either chemical or biologically mediated processes which occur in the environment. The ability of *A. ferroxidans* to oxidize metal sulfides in the absence of iron ions has been used as supporting evidence for a direct mechanism (Tributsch and Bennett, 1981a,b). However, *A. ferrooxidans* in the absence of iron ions can behave in a manner similar to that of *A. thiooxidans*, utilizing the oxidation of sulfur as the metabolic energy source (Figure 2; Sand *et al.*, 2001).

Sand *et al.* (2001) suggested that Fe(III) ions or protons (H^+) are the only chemical agents that dissolve a metal sulfide, and that bacteria have functions to regenerate these ions and to concentrate them at the mineral/water or the mineral/bacterial cell interface. The chemical reactions require the presence of surrounding bacterial cells. The concentration of reactants in this nanometer-thick layer causes the observed acceleration of rates of oxidative dissolution of sulfide minerals.

Tributsch (2001) suggested that bacterially mediated oxidation proceeds through mechanisms of cooperative dissolution and contact leaching. Contact leaching replaces the term direct leaching because this definition would require an enzymatic interaction between the bacterial membrane and the cell at the sulfide surface, which is not observed. The production by *A. ferrooxidans* of discrete etch-pit chain sequences observed on pyrite surfaces (Tributsch, 2001) is consistent with the hypothesis of contact leaching. Integrated microbiological, mineralogical, and geochemical studies conducted at the Kidd Creek tailings impoundment, Timmins, Ontario, suggest that the early stages of sulfide oxidation are catalyzed by neutrophilic bacteria, including *T. thioparus* and related species. After acidic conditions are attained, acidophillic species, including *A. thiooxidans* and related species, and *A. ferrooxidans* and related species, become established (Blowes *et al.*, 1995).

Recent studies at the Iron Mountain acid mine-drainage site in California suggest that the main role of *A. ferrooxidans* is to oxidize iron downstream from the principal acid-generating site, and that the primary effect is to enhance the precipitation of iron oxyhydroxides (Banfield and Welch, 2000). Other iron-oxidizing bacteria (e.g., *L. ferrooxidans*) and archeal (e.g., *Thermoplasmales*) species have been observed proximal to the sulfide ore (Edwards *et al.*, 2000). The utilization of energy derived from iron or sulfur oxidation in other prokaryotes remains unclear, but it can be surmised that the mechanism involved could be broadly similar to that determined for *A. ferrooxidans* (Banfield and Welch, 2000).

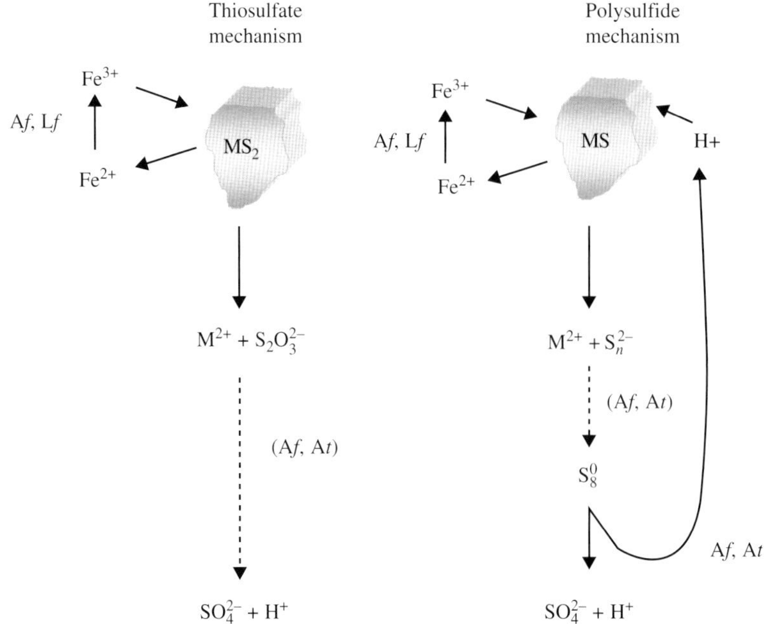

Figure 2 The sulfate and polysulfide mechanisms in microbial leaching of metal sulfides. MS_2 = metal disulfides, MS = metal sulfides; M^{2+} = metal ion; $S_2O_3^{2-}$ = thiosulfate, S_n^{2-} = polysulfide, S_8^0 = elemental sulfur; A*f*, L*f*, A*t* = without brackets refers to the enzymatic reaction by *A. ferrooxidans*, *L. ferrooxidans*, and/or *A. thiooxidans*; (A*f*, A*t*) = with brackets refers to a possible enzymatic reaction (after Sand *et al.*, 2001).

9.05.3.4.4 Impact of bacteria on the rate of sulfide-mineral oxidation

The chemical oxidation of metal sulfides is controlled in part by the dissolution of sulfide minerals under acidic conditions and by the presence of oxidants (DO, Fe^{3+}) that lead to the disruption of sulfide chemical bonds. Bacteria can have a significant effect on the rate of oxidative dissolution of sulfide minerals by controlling mineral solubility and surface reactivity. Metal-enriched waters and solutions rich in sulfuric acid that form in association with mining can be directly linked to microbial activity. The majority of studies to date have focused on the reactivity and kinetics of sulfide minerals in the presence of *A. ferrooxidans* and *L. ferrooxidans*, and in some cases *A. thiooxidans* (Singer and Stumm, 1970; Tributsch and Bennett, 1981a,b; Sand *et al.*, 1992, 2001; Nordstrom and Southam, 1997; Sasaki *et al.*, 1998; Edwards *et al.*, 1998, 1999, 2000; Nordstrom and Alpers, 1999a; Banfield and Welch, 2000; Tributsch, 2001). Additional studies have been conducted on other species of bacteria and archea (Edwards *et al.*, 1998, 1999, 2000).

Singer and Stumm (1970) determined that in the presence of *A. ferrooxidans* the oxidation of Fe(II) from pyrite increased from 3×10^{-12} mol L^{-1} s^{-1} to $\sim 3 \times 10^{-7}$ mol L^{-1} s^{-1}, almost five orders of magnitude. Nordstrom (1985) measured the iron-oxidation rates and stream velocities in downstream effluents from mine tailings. The oxidation rates were estimated to be between 2×10^{-7} mol L^{-1} s^{-1} and 8×10^{-7} mol L^{-1} s^{-1}, with the range attributed to climatic variations. Table 6 provides a compilation of abiotic, microbial, and field rate-data for pyrite, sphalerite, galena, and chalcopyrite.

In the review by Nordstrom and Southam (1997), insightful correlations among the relationships of grain size, surface area, and microbial leaching rates were provided (taken from Olson, 1991; Southam and Beveridge, 1992). Olson (1991) conducted an inter-laboratory comparison of pyrite bioleaching rates by *A. ferrooxidans* using a standard set of conditions. Evaluation of data collected by Olson determined that the average microbial rate of oxidation for pyrite is $\sim 8.8 \times 10^{-8}$ mol m^{-2} s^{-1}. Nordstrom and Alpers (1999a) concluded that, because this rate fell between the abiotic oxidation of pyrite by Fe(III) and the microbial oxidation of Fe(II), little difference was effected by *A. ferrooxidans*. The lower rate of microbial pyrite oxidation compared to the oxidation rate of Fe(II) by *A. ferrooxidans* suggests that the heterogeneous reaction is the rate-determining step (Nordstrom and Alpers, 1999a). Although uncertainties exist regarding rates and the natural variation in the oxidation of Fe(II), the rates are large enough that there is no significant difference. Consequently, the kinetics of pyrite oxidation may be controlled by the rate involved in the production of aqueous ferric

Table 6 Comparison of sulfide–mineral oxidation rates from abiotic, microbial, and field measurements.

Process	*Rates of oxidation pH <2, 25 °C*			*Source*
	Abiotic	*Microbial*	*Field*	
Pyrite oxidation by $[Fe^{2+}]$	3×10^{-12} mol L^{-1} s^{-1}	3×10^{-7} mol L^{-1} s^{-1}	5×10^{-7} mol L^{-1} s^{-1}	Singer and Stumm (1968, 1970) Nordstom (1985)
Pyrite oxidation by $[Fe^{3+}]$	1×10^{-3} to 2×10^{-8} mol m^{-2} s^{-1} 2.7×10^{-7} mol m^{-2} s^{-1}			McKibben and Barnes (1986), Rimstidt *et al.* (1994)
Pyrite oxidation by [DO]	0.3×10^{-9} to 3×10^{-9} mol m^{-2} s^{-1} 1.2×10^{-9} mol m^{-2} s^{-1} 0.5×10^{-9} mol m^{-2} s^{-1}	8×10^{-8} mol m^{-2} s^{-1}		McKibben and Barnes (1986), Olson (1991), Weisener (2002), Nicholson (1994)
Pyrite nonoxidative Dissolution	1.9×10^{-10} mol m^{-2} s^{-1}			Weisener (2002)
Oxidation of waste dump			0.03×10^{-8} mol m^{-2} s^{-1}	Ritchie (1994)
Oxidation in high relative humidity	$\sim 10^{-8}$ mol m^{-2} s^{-1}			Jerz and Rimstidt (2000)
Pyrrhotite oxidation by [DO]	1.3×10^{-8} mol m^{-2} s^{-1} 4×10^{-9} mol m^{-2} s^{-1}			Nicholson and Scharer (1994), Janzen *et al.* (2000)
Pyrrhotite oxidation by $[Fe^{3+}]$	3.5×10^{-8} mol m^{-2} s^{-1}			Janzen *et al.* (2000)
Pyrrhotite nonoxidative dissolution	5×10^{-10} mol m^{-2} s^{-1} 6×10^{-7} mol m^{-2} s^{-1}			Janzen *et al.* (2000), Thomas *et al.* (1998, 2001)
Sphalerite oxidation by $[Fe^{3+}]$	3×10^{-7} to 7×10^{-7} mol m^{-2} s^{-1}			Rimstidt *et al.* (1994)
Sphalerite oxidation by [DO]	1×10^{-8} to 3×10^{-8} mol m^{-2} s^{-1}			Weisener (2002)
Chalcopyrite oxidation by $[Fe^{3+}]$	9.6×10^{-9} mol m^{-2} s^{-1}			Rimstidt *et al.* (1994)
Galena oxidation by $[Fe^{3+}]$	1.6×10^{-6} mol m^{-2} s^{-1}			Rimstidt *et al.* (1994)

iron from ferrous iron via microbial catalysis (Nordstrom and Alpers, 1999a).

9.05.4 ACID NEUTRALIZATION MECHANISMS AT MINE SITES

The oxidation of sulfide minerals in mine-waste rock piles and tailings impoundments generates acidic waters containing high concentrations of SO_4, Fe(II), and other metals. The water affected by sulfide oxidation is displaced into underlying or adjacent geological materials, or it is discharged directly to the adjacent surface-water flow system. Geochemical reactions with the gangue minerals result in progressive increases in the pore-water pH and the attenuation of some dissolved metals. A sequence of geochemical reactions occurring in the mine wastes and in underlying aquifers results in profound changes in the concentrations of dissolved constituents and in the mineralogy and physical properties of the mine waste and aquifer materials. Low-pH conditions promote the dissolution of many metal-bearing solids and the desorption of metals from solid surfaces. Increases in pH through acid-neutralization reactions can lead to pronounced declines in the concentrations of dissolved metals released from mine wastes.

9.05.4.1 Mechanisms of Acid Neutralization

The acid generated through sulfide oxidation reacts with the nonsulfide gangue minerals within the mine wastes. The most significant pH-buffering reactions in mine settings are the dissolution of carbonate minerals, aluminum hydroxide and ferric oxyhydroxide minerals, and aluminosilicate minerals.

9.05.4.1.1 Carbonate-mineral dissolution

The most abundant carbonate minerals in mine wastes are calcite [$CaCO_3$], dolomite [$CaMg(CO_3)_2$], ankerite [$Ca(Fe,Mg)(CO_3)_2$], siderite [$FeCO_3$] or mixtures thereof. The dissolution of calcite can be described as

$$CaCO_3 + H^+ \leftrightarrow Ca^{2+} + HCO_3^- \qquad (23)$$

or, at low pH, as

$$CaCO_3 + 2H^+ \leftrightarrow Ca^{2+} + CO_2 + H_2O \qquad (24)$$

Dissolution of carbonate minerals has the potential to raise the pH of the pore water to near neutral. Dissolution of carbonate minerals releases calcium, magnesium, manganese, iron, and other cations that are present as solid-solution substitutions or as impurities, and increases the alkalinity of the water. At many sites, the mass of carbonate minerals contained in the mine wastes exceeds that of the sulfide minerals, and the rapid dissolution of carbonate minerals is sufficient to maintain neutral pH conditions through the mines or their waste piles. Neutral pH conditions have been observed in tailings impoundments derived from processing carbonate-bearing vein-type gold deposits (Blowes, 1990; McCreadie *et al.*, 2000), and in open pits derived from gold mining (Shevenell *et al.*, 1999).

Many mines and mine wastes derived from massive sulfide ore deposits contain a sulfide content in excess of the carbonate neutralization capacity. At these sites, the available carbonate content is completely consumed, with the most soluble and most reactive carbonate minerals depleted first. Calcite is depleted initially, followed successively by dolomite-ankerite and siderite (Blowes and Ptacek, 1994). As the carbonate content of the waste dissolves, the pH is buffered to near neutral. The depletion of carbonate minerals through consumption by acid-neutralization reactions has been documented for a number of settings, including base-metal mine tailings impoundments (Dubrovsky *et al.*, 1984; Dubrovsky, 1986; Blowes and Jambor, 1990; Johnson *et al.*, 2000), aquifers impacted by drainage from mine wastes (Morin *et al.*, 1988), and coal-mine spoils (Cravotta, 1994). In these studies, the greatest depletion occurs at the advancing acid front. An abrupt increase in solid-phase carbonate concentration is observed, suggesting that the dissolution of carbonate minerals is rapid. Measurements of the carbonate content of mine wastes, in front of and behind the carbonate dissolution front, indicate that carbonate depletion occurs over a small distance along the path of groundwater flow, and that carbonate minerals are absent behind the acid front (Dubrovsky *et al.*, 1984; Johnson *et al.*, 2000). Detailed laboratory column studies conducted by Jurjovec *et al.* (2002, 2003) confirmed that the depletion of carbonate minerals in response to an advancing acid front is rapid. Jurjovec *et al.* (2002, 2003) observed stable pH conditions while alkalinity and cations were produced through carbonate dissolution, followed by a pronounced drop in pH after carbonate minerals were depleted (Figure 3).

The solubility products for the dissolution of calcite, dolomite-ankerite, and siderite vary, with calcite having the highest solubility product and siderite the lowest (Table 7). Field observations suggest that dissolution of the most soluble of these minerals will proceed first, followed by dissolution of the next most soluble mineral. As iron-sulfide minerals oxidize, Fe(II) can be released to anoxic pore waters and can react with the HCO_3^- produced through calcite dissolution, thereby forming siderite. Observations from field sites, coupled with geochemical modeling results, suggest that

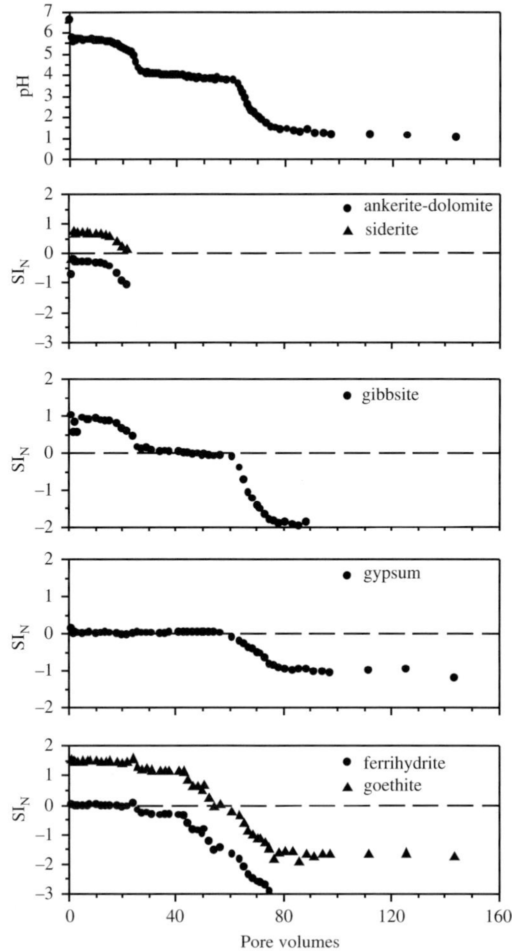

Figure 3 The pH-buffering sequence observed in mine tailings: results of laboratory column experiment with 0.1 N H_2SO_4 input solution (source Jurjovec *et al.*, 2002). SI_N is the saturation index normalized per mole of reactant.

secondary siderite does form through this reaction (e.g., Ho *et al.*, 1984; Morin *et al.*, 1988; Blowes and Jambor, 1990; Ptacek and Blowes, 1994). Al *et al.* (2000) identified secondary siderite on calcite surfaces, consistent with theoretical predictions. In laboratory studies conducted by Jurjovec *et al.* (2002, 2003), dissolution of various carbonate minerals, including calcite, dolomite-ankerite, and siderite could not be distinguished. Under the near-neutral pH conditions maintained by carbonate-mineral dissolution, the precipitation of crystalline and amorphous metal hydroxides is favored. Cations released through sulfide oxidation and the dissolution of carbonate and aluminosilicate minerals can lead to the formation of secondary hydroxide phases that accumulate in the waste solids. These phases may subsequently dissolve as additional acid is produced through sulfide-mineral oxidation reactions. Dissolution of these secondary minerals contributes to acid neutralization.

9.05.4.1.2 Dissolution of hydroxide minerals

After the carbonate minerals are depleted, the pH falls until equilibrium with the most soluble secondary hydroxide mineral is attained. An abrupt increase in the concentration of dissolved aluminum has been observed to coincide with carbonate depletion at several mine-tailings impoundments. These observations suggest that the initial hydroxide mineral to dissolve is an aluminum hydroxide. Although a discrete aluminum-bearing phase has not been isolated, it is generally assumed to be amorphous aluminum hydroxide [$Al(OH)_3$] or gibbsite [crystalline $Al(OH)_3$]. Dissolution of this aluminum-bearing phase buffers the pH in the region of 4.0–4.5

Table 7 Solubility products for selected phases which form at mine drainage sites.

Mineral	*Reaction*	*log K*
Oxides and hydroxides		
Goethite	$FeOOH_{(s)} + 3H^+_{(aq)} \leftrightarrow Fe^{3+}_{(aq)} + 2H_2O_{(l)}$	−1.0
Ferrihydrite	$Fe(OH)_{3(s)} + 3H^+_{(aq)} \leftrightarrow Fe^{3+}_{(aq)} + 3H_2O_{(l)}$	3.0 to 5.0
Gibbsite	$Al(OH)_{3(s)} + 3H^+_{(aq)} \leftrightarrow Al^{3+}_{(aq)} + 3H_2O_{(l)}$	7.94
Böhmite	$AlOOH_{(s)} + 3H^+ \leftrightarrow Al^{3+}_{(aq)} + 2H_2O_{(l)}$	7.83
Sulfates		
Gypsum	$CaSO_4{\cdot}2H_2O_{(s)} \leftrightarrow Ca^{2+}_{(aq)} + SO_{4(aq)} + 2H_2O_{(l)}$	−4.58
Celestine	$SrSO_{4(s)} \leftrightarrow Sr^{2+}_{(aq)} + SO^{2-}_{4(aq)}$	−6.62
Barite	$BaSO_4 \leftrightarrow Ba^{2+}_{(aq)} + SO^{2-}_{4(aq)}$	−0.97
Jarosite	$KFe_3(SO_4)_2(OH)_{6(s)} + 6H^+_{(aq)} \leftrightarrow K^+_{(aq)}$ $+ 3Fe^{3+}_{(aq)} + 2SO^{2-}_{4(aq)} + 6H_2O_{(l)}$	−11.0
Carbonates		
Calcite	$CaCO_{3(s)} \leftrightarrow Ca^{2+}_{(aq)} + CO_{3(aq)}$	−8.48
Dolomite	$CaMg(CO_3)_{2(s)} \leftrightarrow Ca^{2+}_{(aq)} + Mg^{2+}_{(aq)} + 2CO^{2-}_{3(aq)}$	−17.09
Siderite	$FeCO_{3(s)} \leftrightarrow Fe^{2+}_{(aq)} + CO^{2-}_{3(aq)}$	−10.89

Sources: Nordstrom *et al.* (1990), Nordstrom and Munoz (1994), and Baron and Palmer (1996).

(Dubrovsky, 1986; Blowes and Jambor, 1990; Johnson *et al.*, 2000). Although repeated attempts have been made, the presence of these aluminum-bearing phases has not been confirmed through mineralogical studies. After the secondary aluminum phases present in the tailings are depleted, the pH falls until equilibrium with an iron-oxyhydroxide mineral, typically ferrihydrite (nominally $9Fe_2O_8 \cdot 5H_2O$) or goethite (α-FeOOH), has been attained. Dissolution of ferric oxyhydroxide minerals typically maintains pH values in the range of 2.5–3.5.

9.05.4.1.3 Dissolution of aluminosilicate minerals

Gangue minerals associated with sulfide-rich tailings include various aluminosilicate minerals such as chlorite, smectite, biotite, muscovite, plagioclase, amphibole, and pyroxene. Their contribution to acid-neutralization reactions has been examined in a number of field studies and through controlled laboratory experiments. During the period of carbonate- and hydroxide-mineral dissolution, aluminosilicate minerals may also dissolve, thus consuming H^+ and releasing H_4SiO_4, Al^{3+}, and other cations, including potassium, calcium, magnesium, and manganese. Aluminosilicate dissolution is generally not rapid enough to buffer the pore water to a specific pH. Such reactions, however, do consume H^+ and contribute to the overall acid-neutralization potential of mine wastes. In addition to H^+ consumption, aluminum and other metals released from aluminosilicate minerals may accumulate in secondary products such as amorphous $Al(OH)_3$ or gibbsite that act as secondary pH buffers. As with other phases contributing to acid neutralization, the mass and composition of aluminosilicates vary from site to site and within an individual site.

Detailed field studies conducted at the Heath Steele tailings impoundment, New Brunswick, Canada, illustrate the role of aluminosilicate-mineral dissolution in contributing to acid neutralization. The tailings contain in excess of 85 wt.% sulfide minerals, and partial oxidation has led to the production of low-pH pore waters (pH < 0.5) and the extensive dissolution of aluminosilicate minerals. Blowes (1990) noted increased concentrations of aluminum and silicon (up to 2,500 mg L^{-1} and 200 mg L^{-1}, respectively) due to the dissolution of the aluminosilicate minerals. In the deeper tailings, a diverse silicate-oxide gangue was observed, including chlorite, muscovite, quartz, and an iron-bearing amphibole. In the upper 30 cm of the tailings, a less diverse silicate gangue, extensively depleted in chlorite and amphibole, was observed. The order of depletion of the aluminosilicate minerals followed closely the order predicted on the basis of the zero point of charge (ZPC) of the aluminosilicate minerals.

Johnson *et al.* (2000) documented the depletion of aluminosilicates at the Nickel Rim mine-tailings impoundment, where the solid-phase aluminum and biotite are depleted in the upper oxidized portion of the tailings, whereas biotite is abundant in the underlying unoxidized tailings. At the Sherridon tailings impoundment, Manitoba, Moncur *et al.* (2003) observed extensive depletion of biotite, chlorite, and, possibly, smectite in the upper 1 m of the tailings. At this site, there is evidence of secondary replacement of other aluminosilicates, namely, plagioclase, cordierite, and amphibole. At Sherridon, the first appearance of biotite occurs at 72 cm, which coincides with the first appearance of secondary sulfides (marcasite). Chlorite and smectite first appear at a 90 cm depth, indicating the greater tendency for dissolution. In a controlled laboratory experiment using a 0.1 M H_2SO_4 solution, Jurjovec *et al.* (2002) observed extensive depletion of chlorite and minor depletion of biotite as a result of acid leaching of the tailings.

9.05.4.1.4 Development of pH-buffering sequences

The sequence of pH-buffering reactions observed within tailings impoundments results in a progressive increase in the pore-water pH along the groundwater flow path. These changes in pH occur as long zones of relatively uniform pH, which are dominated by a single-pH buffering reaction, separated by fronts or relatively sharp changes in pH as a buffering phase is depleted (Figure 3). Field data from the Nickel Rim tailings impoundment, as were presented by Johnson *et al.* (2000), illustrate a pH-buffering sequence. At the Nickel Rim site, the pH of the pore water at the base of the impoundment is near neutral, varying from 6.5 to 7.0. Geochemical calculations suggest that the pore water in this zone approaches or attains equilibrium with respect to calcite. Overlying the calcite-buffered zone, calcite is absent from the tailings and the pH is buffered by the dissolution of siderite and possibly by the dissolution of dolomite. The complete depletion of carbonate minerals is accompanied by a sharp decline in pH from 5.8 to 4.5. Immediately above the depth of carbonate-mineral depletion, the pH is relatively uniform, varying from 4.0 to 4.5, the dissolved aluminum concentrations increase sharply, and the pore water approaches or attains equilibrium with respect to gibbsite. Near the surface of the impoundment, the pH decreases sharply from 4.0 to 3.5, then more gradually from 3.5 to ~2.5. Throughout this decrease in pH, the pore water is undersaturated with respect to ferrihydrite, and approaches equilibrium with respect to goethite.

9.05.5 GEOCHEMISTRY AND MINERALOGY OF SECONDARY MINERALS

In mine-waste settings, the secondary minerals are those that form after disposal of the wastes. Thus, the mined material may contain both the primary, fresh minerals and the oxidation products that were formed through geological processes. In rare cases, a mineral deposit will consist wholly of oxidized material, but in the acid drainage setting all of these minerals are deemed to constitute the primary assemblage because they were formed prior to disposal as wastes. After final discharge, or even during temporary storage of wastes and low-grade ores, crystallization of secondary minerals occurs *in situ* principally as a result of oxidation of the primary sulfide assemblage. However, upon removal from its disposal site, as occurs during a sampling program, a waste may form additional minerals as the material dries; such products, which are predominantly water-soluble salts that crystallize during evaporation of the pore water, are classified as "tertiary" minerals (Jambor, 1994; Jambor and Blowes, 1998).

9.05.5.1 Soluble Sulfates: Iron Minerals

The most common secondary iron-sulfate minerals associated with oxidized mine wastes are of the type $Fe^{2+}SO_4{\cdot}nH_2O$, wherein n ranges from 7 to 1. The minerals of most frequent occurrence and abundance are melanterite ($n = 7$), rozenite ($n = 4$), and szomolnokite ($n = 1$); the pentahydrate (siderotil) and hexahydrate (ferrohexahydrite) have also been reported, but are much less widespread and occur sparingly. The divalent salts seem to form in the earliest stages of sulfide-mineral oxidation, but they are observed more commonly as superficial evanescent coatings that appear during dry periods and are susceptible to solubilization during rainfall events. Melanterite has been reported from several localities as a precipitate from greenish pools of iron-rich solutions that drained from pyritiferous deposits, and the mineral has also been noted to form a hardpan layer in pyrite-rich tailings at the Heath Steele mine, New Brunswick, Canada (Blowes *et al*., 1992). The tetrahydrate, rozenite, more typically forms as a dehydration product of melanterite.

Although the soluble salts of ferric iron may crystallize on a micro scale in the earliest stages of iron-sulfide oxidation, in field settings the paragenesis indicates that the progression is from the early formation of the divalent salts of iron, to minerals containing both ferrous and ferric iron, and thence to the trivalent iron minerals (Jambor *et al*., 2000a,b). For minerals devoid of other cations, the most commonly occurring of the mixed-valence type is copiapite $[Fe^{2+}Fe^{3+}_4(SO_4)_6(OH)_2{\cdot}20H_2O]$, and less common is römerite $[Fe^{2+}Fe^{3+}_2(SO_4)_4{\cdot}14H_2O]$. Among the trivalent-iron minerals, the most common are coquimbite $[Fe^{3+}_2(SO_4)_3{\cdot}9H_2O]$, ferricopiapite $[Fe^{3+}_{2/3}Fe^{3+}_4(SO_4)_6(OH)_2{\cdot}20H_2O]$, and rhomboclase $[(H_5O_2)^{1+}Fe^{3+}(SO_4)_2{\cdot}2H_2O]$. Also common is fibroferrite $[Fe^{3+}(SO_4)(OH){\cdot}5H_2O]$, which, however, is relatively insoluble except in acidic conditions.

A notable locality for acid drainage is Iron Mountain, California, where oxidation of pyritiferous massive sulfide deposits has resulted in the formation of many of the aforementioned soluble salts in the underground workings. The precipitation order of the salts at Iron Mountain follows the divalent to trivalent trend (Nordstrom and Alpers, 1999b).

9.05.5.2 Soluble Sulfates: Other Elements

Other than the soluble sulfates of iron, only those of magnesium and aluminum are of moderately common occurrence in association with acid drainage. The magnesium salts are chiefly of the type $MgSO_4{\cdot}nH_2O$, with epsomite $[MgSO_4{\cdot}7H_2O]$ and hexahydrite $[MgSO_4{\cdot}6H_2O]$ the predominant minerals. Occurrences of the pentahydrate (pentahydrite) and the tetrahydrate (starkeyite), which are known from a few localities, could be categorized as rare, and the monohydrate (kieserite), although identified as a weathering product of pyritiferous shales, has not yet been observed specifically in association with acid drainage.

Whereas the oxidation of pyrite provides the principal source for the iron analogues of the simple magnesium sulfates, the chief source of the latter is dolomite $[CaMg(CO_3)_2]$. Dissolution of dolomite attenuates the acidity generated by pyrite oxidation, thereby releasing calcium and magnesium to form gypsum $[CaSO_4{\cdot}2H_2O]$, simple magnesium salts, and more complex ones such as magnesiocopiapite $[MgFe^{3+}_4(SO_4)_6(OH)_2{\cdot}20H_2O]$. Thus, both gypsum and the magnesium sulfates may form without the development of acid drainage if dolomite is sufficiently abundant to neutralize the acidity generated by oxidation of the associated sulfides.

The formation of soluble salts containing aluminum is paragenetically later than that of the simple magnesium salts because aluminum is sourced from aluminosilicates, the dissolution of which is typically slow and requires low-pH conditions to accelerate the process. The most common soluble sulfates of aluminum are members of the halotrichite–pickeringite series [$Fe^{2+}Al_2(SO_4)_4 \cdot 22H_2O - MgAl_2(SO_4)_4 \cdot 22H_2O$] and of less widespread occurrence are aluminocopiapite [$Al_{2/3}Fe_4^{3+}(SO_4)_6(OH)_2 \cdot 20H_2O$] and alunogen [$Al_2(SO_4)_3 \cdot 17H_2O$].

Various soluble salts, such as kalinite [$KAl(SO_4)_2 \cdot 11H_2O$], blödite [$Na_2Mg(SO_4)_2 \cdot 4H_2O$], and bilinite [$Fe^{2+}Fe_2^{3+}(SO_4)_4 \cdot 22H_2O$] occur sparingly but may be locally prominent. The most common of the copper salts is chalcanthite [$CuSO_4 \cdot 5H_2O$]. Both goslarite [$ZnSO_4 \cdot 7H_2O$] and gunningite [$ZnSO_4 \cdot H_2O$] have been reported as secondary minerals in sphalerite-bearing wastes, but the salts occur sparingly.

9.05.5.3 Less Soluble Sulfate Minerals

Gypsum has a moderate solubility in water, whereas jarosite [$KFe_3(SO_4)_2(OH)_6$] is relatively insoluble (Table 7). The two minerals are the most abundant and widespread of the secondary minerals associated with acid drainage. Early formation of gypsum is possible because of cation availability through the dissolution of carbonate minerals. Bassanite [$2CaSO_4 \cdot H_2O$] has also been identified as a secondary mineral in tailings impoundments, but the quantities are minute and are insignificant relative to those of gypsum. Jarosite is, by definition, K-dominant, but as a mine-drainage precipitate it invariably is oxonium-bearing, commonly with $K > H_3O > Na$. In the early stages of acid generation the bulk of the potassium at most localities is derived by incongruent alteration of trioctahedral mica. Various other members of the jarosite supergroup occur in acid-drainage settings; other members detected are natrojarosite [$NaFe_3(SO_4)_2(OH)_6$], hydronium jarosite [$(H_3O)Fe_3(SO_4)_2(OH)_6$], plumbojarosite [$PbFe_6(SO_4)_4(OH)_{12}$], alunite [$KAl_3(SO_4)_2\ (OH)_6$], and As-bearing species, but all are of rare occurrence relative to jarosite.

Among the aluminum-sulfate salts, jurbanite [$Al(SO_4)(OH) \cdot 5H_2O$] is commonly used in geochemical modeling, but the mineral itself is of extremely rare occurrence; the most common aluminum hydroxysulfates associated with acid drainage correspond to crystalline and amorphous felsöbányaite (basaluminite) [$Al_4(SO_4)(OH)_{10} \cdot 4H_2O$] and hydrobasaluminite [$Al_4(SO_4)(OH)_{10} \cdot 15H_2O$].

A poorly crystalline hydroxysulfate, schwertmannite, whose crystal structure has not yet been resolved and which is thought to have the ideal formula $Fe_8O_8(OH)_6SO_4 \cdot n\,H_2O$, may be the most common direct precipitate of iron from acidic effluents at pH 2–4 (Bigham and Nordstrom, 2000). The mineral is difficult to identify because of its poor crystallinity and the almost invariable presence of associated iron oxyhydroxides and jarosite, but numerous occurrences have been reported in the recent literature. Detailed studies of the oxidation of pyrite and pyrrhotite indicate that schwertmannite is not an early-formed mineral, and the occurrences suggest an origin in which the requisite iron and SO_4 have undergone solubilization, transportation, and subsequent precipitation rather than maintaining an intimate association with precursor sulfides.

The principal ore mineral of lead is galena [PbS], and the predominant sulfates derived from it are anglesite [$PbSO_4$] and plumbojarosite. Partial substitution of lead for the alkali-site cations also occurs in other members of the jarosite supergroup. An unusual sink for copper and aluminum is the mineral hydrowoodwardite [$Cu_{1-x}Al_x(OH)_2(SO_4)_{x/2}(H_2O)_n$], recently discovered as a supergene product at several old mines in Saxony, Germany (Witzke, 1999). The same mineral was almost simultaneously reported to occur as bluish coatings on weathered waste rocks and in stream sediments at two former mines in northern Italy (Dinelli *et al.*, 1998).

9.05.5.4 Metal Oxides and Hydroxides

The oxidation of iron sulfides invariably leads to the formation of goethite [α-FeOOH] both in intimate association with the sulfides and in distal precipitates. Other oxyhydroxide minerals identified as secondary precipitates are lepidocrocite [γ-FeOOH] and akaganéite [β-FeO(OH,Cl)], but occurrences are generally insignificant relative to the abundance and distribution of goethite. However, lepidocrocite has been observed to be important at some sites; e.g., Roussel *et al.* (2000) determined the mineral to be the principal component of suspended particulate matter in drainage from an arsenic-rich tailings impoundment in France, and Bowell and Bruce (1995) observed lepidocrocite to be a significant component of ochres that precipitated from low-pH (<5) waters at the Levant mine in Cornwall, England.

Ferrihydrite, a poorly crystalline mineral of widespread occurrence, is equivalent to the compound that is commonly referred to as "amorphous $Fe(OH)_3$." Precipitation of ferrihydrite is favored when oxidation of Fe^{2+} and hydrolysis occur rapidly. The mineral is unstable with respect to goethite, and precipitation occurs at a higher pH than that for schwertmannite. The presence of ferrihydrite in precipitates from mine-waste effluents is commonly reported, but unequivocal identifications of its presence in mine-waste solids, such as tailings, have been fewer. For the latter, because the crystallinity is poor and the mineral forms only a small proportion of the total waste solids, recourse is generally made to selective dissolution whereby the fraction that is more soluble in certain reagents is inferred to represent ferrihydrite rather than goethite, which is more dissolution resistant.

Although hematite [α-Fe_2O_3] is a common product of the oxidation of sulfide-bearing mineral deposits, only a few unequivocal occurrences are known in mine-drainage settings (e.g., Hochella *et al.*, 1999; Boulet and Larocque, 1996). Various iron oxides and oxyhydroxides, including hematite, maghemite [γ-Fe_2O_3], and ilmenite [$Fe^{2+}TiO_3$] have been observed as microcrystalline aggregates that have formed within acidophilic microorganisms in sediments affected by effluents from tailings at Elliot Lake, Ontario, Canada (Mann and Fyfe, 1989). Minerals of the spinel group, which includes magnetite [$Fe^{2+}Fe_2^{3+}O_4$], have been detected, but most associations are on a minute scale. However, the presence of extensive accumulations of magnetite derived by the anoxic reduction of transported iron in mining-impacted lake-bottom sediments has been reported by Cummings *et al.* (2000).

Manganese oxides–oxyhydroxides have been observed at a few acid-drainage localities. For example, Benvenuti *et al.* (2000) observed pyrolusite [MnO_2] and pyrochroite [$Mn(OH)_2$] in weathered jig tailings and waste rock at a site in Tuscany, Italy. Hochella *et al.* (1999) identified hydrohetaerolite [$Zn_2Mn_4^{3+}O_8{\cdot}H_2O$] among the secondary minerals in stream sediments at an acid drainage site in Montana, and nsutite [γ-MnO_2], birnessite [$Na_4Mn_{14}O_{27}{\cdot}9H_2O$], and minerals in the ranciéite–takanelite series [$(Ca,Mn^{2+})Mn_4^{4+}O_9{\cdot}3H_2O$–$(Mn^{2+},Ca)Mn_4^{4+}O_9{\cdot}H_2O$] were among the various manganese minerals detected in precipitates in the streambed of Pinal Creek, Arizona (Lind and Hem, 1993). At Pinal Creek, the acidic groundwater plume reacted with calcite-bearing alluvium, initially forming "amorphous iron oxide" as pH increased, and subsequently precipitating the manganese-bearing assemblage farther downstream, where the pH of the stream water was near neutral. Similar relationships were observed by Hudson-Edwards *et al.* (1996), who identified cesarolite [$PbH_2Mn_3O_8$], coronadite [$Pb(Mn^{4+},Mn^{2+})_8O_{16}$], woodruffite [$(Zn,Mn^{2+}Mn_3^{4+}O_7{\cdot}12H_2O$], and hydrohetaerolite in contaminated stream sediments of the Tyne River catchment, northeastern England, and noted the disappearance of these minerals in downstream areas in which the pH of the water was lower.

The presence of plumboferrite [$\sim PbFe_4O_7$] in the Tyne River sediments was reported by Hudson-Edwards *et al.* (1996), and the mineral has also been identified as a secondary product in tailings (Morin *et al.*, 1999). Although aluminum hydroxides play a prominent role in geochemical modeling, few occurrences of such material have been described. Bove *et al.* (2000) reported the presence of an X-ray amorphous aluminum hydroxide precipitate in a stream channel that drains an acid-sulfate hydrothermal system near Silverton, Colorado, and Berger *et al.* (2000) reported the occurrence of a poorly crystalline aluminum hydroxide in particulate matter that was filtered from seeps below an abandoned waste-rock dump in New Mexico; however, detailed characterizations were not published. Diaspore [AlOOH] has been reported to occur as a secondary mineral in a waste-rock dump in Tuscany, Italy (Benvenuti *et al.*, 1997). In the study by Hochella *et al.* (1999), a nearly pure amorphous Si–Al oxyhydroxide (Si : Al = 2 : 1) rather than $Al(OH)_3$ was identified as the major sink for aluminum in the local drainage system.

The occurrence of secondary $SiO_2{\cdot}nH_2O$ has been observed in several tailings impoundments, predominantly as amorphous hydrated pseudomorphs after biotite. Cristobalite [SiO_2] has also been detected in the pseudomorphs. Partial replacement of various other minerals by amorphous silica has also been observed.

An extensive suite of supergene minerals, including chlorides, sulfates, and arsenates occurs at the Levant mine, Cornwall, England, where ingress of seawater has influenced the variety of minerals in the suite. In stagnant pools of mine drainage at the mine, cuprite [Cu_2O] and native copper have precipitated within the ochres (Bowell and Bruce, 1995).

9.05.5.5 Carbonate Minerals

Like the precipitation of secondary $Al(OH)_3$, that of secondary siderite [$FeCO_3$] is widely employed in geochemical models. Secondary siderite and associated iron oxyhydroxides were determined to have formed as coatings, up to 1,000 μm thick, on ankerite-dolomite in the Kidd Creek tailings impoundment at Timmins, Ontario, Canada (Al *et al.*, 2000). At Elliot Lake, Ontario, nodules of siderite, each <10 μm across, and calcite in aggregates up to 400 μm across, occur as secondary minerals in a small tailings impoundment (Paktunc and Davé, 2002). The precipitation of the carbonate minerals at the Elliot Lake site was interpreted to result from high alkalinity related to the presence of decaying organic debris.

At the Matchless pyritiferous deposit in Namibia, huntite [$CaMg_3(CO_3)_4$] occurs within high-pH zones in an oxidized tailings impoundment (Dill *et al.*, 2002). The high-pH zones are sandwiched between low-pH layers, with phyllosilicate-rich layers acting as aquitards that impede vertical movement and homogenization of the pore waters.

Wastes from lead-rich mineral deposits typically form anglesite in sulfate-dominant environments, but in limestone-dominated host rocks and in gangue containing abundant carbonates, both cerussite [$PbCO_3$] and hydrocerussite [$Pb_3(CO_3)_2(OH)_2$] have been reported as secondary minerals in mining-related wastes. Several carbonate and hydroxycarbonate minerals of copper and zinc were reported by Hudson-Edwards *et al.* (1996) as secondary products in stream sediments in the Tyne Basin, England.

9.05.5.6 Arsenates and Phosphates

Secondary phosphates occur rarely in acid-drainage settings. Pyromorphite [$Pb_5(PO_4)_3Cl$] has been identified in tailings (Morin *et al.*, 1999) and in mining-related stream sediments (Hudson-Edwards *et al.*, 1996). Brushite [$CaHPO_4{\cdot}2H_2O$] was reported by Dill *et al.* (2002) to be present in the oxidized tailings of the Matchless mine, Namibia, but the absence of an adequate source of PO_4 in the primary mineral assemblage indicated that the anion was most likely derived from a flotation reagent.

Arsenates are much more common than phosphates in mine-drainage settings, mainly because of the presence of arsenopyrite [FeAsS] and arsenical pyrite as primary sources of arsenic in metalliferous deposits. For many years, the processing of arsenical gold ores involved roasting of the sulfide concentrates, and the off-gases and particulate emissions were dispersed to the atmosphere and surrounding countryside. Subsequent technological improvements led to recovery of the arsenic-rich emissions, principally by condensation as As_2O_3 (arsenolite), the disposal of which has in turn created environmental problems. For example, more than 2×10^5 t of As_2O_3 are temporarily stored in the underground workings of the gold mines at Yellowknife, Canada, and permanent disposal is a major environmental concern. At a site near Marseille, France, wastes from the processing of arsenical ores contain pyrite, arsenopyrite, and arsenolite, and runoff waters are arsenic-rich and low in pH (2.2). The flows subsequently interact with a limestone substratum, thereby raising the pH and precipitating white crusts that contain gypsum, pharmacolite [$CaHAsO_4{\cdot}2H_2O$], haidingerite [$CaHAsO_4{\cdot}H_2O$], weilite [$CaHAsO_4$], and picropharmacolite [$H_2Ca_4Mg(AsO_4)_4{\cdot}11H_2O$]; scorodite [$FeAsO_4{\cdot}2H_2O$] was also identified in the filtered particulate matter (Juillot *et al.*, 1999).

Scorodite and pharmacosiderite [$KFe_4^{3+}(AsO_4)_3(OH)_4{\cdot}6{-}7H_2O$] have been observed in the wastes at abandoned mines (e.g., Brown *et al.*, 1990). At the Mole River mine, Australia, which ceased the processing of arsenopyrite ore in the 1930s and is known as a source of acid drainage and arsenic contamination, Ashley and Lottermoser (1999) identified scorodite in the regionally contaminated soils. The stream sediments contain arsenolite, and at the mine and roaster is a mixed assemblage that includes pharmacolite and clinoclase [$Cu_3(AsO_4)(OH)_3$]. Gypsum, pharmacolite, and krautite [$MnH(AsO_4){\cdot}H_2O$] occur as efflorescences at the furnace.

Acidic drainage (pH 2.2–4) from a former Pb–Zn mine in Gard, France, was determined by Leblanc *et al.* (1996) to average 250 mg L^{-1} arsenic, and ochreous downstream sediments contain up to 20 wt.% arsenic in stromatolite-like accreted material up to 20 cm thick. Precipitation of the accreted material was attributed to direct or indirect microbial action. X-ray study of the material showed it to contain scorodite, poorly crystalline hematite, bukovskýite [$Fe_2^{3+}(AsO_4)(SO_4)(OH){\cdot}7H_2O$], angelellite [$Fe_4^{3+}(AsO_4)_2O_3$], and beudantite [$PbFe_3(AsO_4, SO_4)(OH)_6$]. At Cobalt, Ontario, minerals in the erythrite–annabergite series [$Co_3(AsO_4)_2{\cdot}8H_2O{-}Ni_3(AsO_4)_2{\cdot}8H_2O$] have formed by oxidation of the tailings from the silver deposits, which are rich in Co–Ni–Fe arsenides. Seeps from the tailings contain up to 20 mg L^{-1} dissolved arsenic and precipitate mainly gypsum and thenardite [Na_2SO_4], but pharmacolite and brassite [$MgHAsO_4{\cdot}4H_2O$] are also present (Percival *et al.*, 1996).

Despite the large number of crystalline arsenates that have been recognized as sinks for arsenic, numerous occurrences of amorphous Fe(III) arsenates and the substitution of arsenic in minerals such as those of the jarosite supergroup (e.g., Savage *et al.*, 2000; Dutrizac and Jambor, 2000) are also known. Sorption of arsenic on schwertmannite has been reported to displace a substantial part of that mineral's SO_4 (Carlson *et al.*, 2002). By far the most important sink for arsenic, however, is sorption to iron- and manganese-oxyhydroxides. The literature is replete with examples and relevant laboratory studies; a few of the many recent or pertinent papers are those by Fuller *et al.* (1993), Nriagu (1994), Kimball *et al.* (1995), Manceau (1995), Waychunas *et al.* (1996), Lutzenkirchen and Lovgren (1998), Manning *et al.* (1998), Raven *et al.* (1998), Webster *et al.* (1998), Ding *et al.* (2000), Jain and Loeppert (2000), and Farquhar *et al.* (2002). Sorption of arsenic on clays is discussed by Manning and Goldberg (1996), Foster *et al.* (1998a,b), Lin and Puls (2000), Garcia-Sanchez *et al.* (2002), and in references therein.

9.05.5.7 Secondary Sulfides

The principal sulfides that have been observed in oxidized mine wastes are marcasite [FeS_2] and covellite [CuS], which have distinctly different origins. The initial oxidation of pyrrhotite results in the solid-state diffusion of iron toward grain boundaries, at which the iron is oxidized (Pratt *et al*., 1994a,b; Pratt and Nesbitt, 1997). Upon further iron migration and the consequent enhanced sulfur enrichment of the original pyrrhotite, replacement by marcasite is effected. Hence, pseudomorphism of the pyrrhotite by marcasite and other oxidation products, including native sulfur, is common if leaching rates are not rapid.

In contrast to the spatial restriction of marcasite to replacement of pyrrhotite, most covellite in mine wastes results from redeposition of solubilized copper that is typically derived from primary chalcopyrite. Sorption of copper on iron oxyhydroxides is common, but redeposition as a sulfide occurs where reductive conditions are present. Such conditions seem to be available locally on a micro scale in proximity to altering pyrrhotite, but on a broader scale the formation of covellite is predominant at the interface between the oxidized and reduced zones of a waste body, thus emulating the supergene enrichment process that takes place in sulfide deposits, especially those of porphyry copper. In mine wastes, other copper sulfides that resemble covellite in reflected light may also be present, but small grain size has impeded specific identification.

In permeable reactive barriers, chalcopyrite, bornite [Cu_5FeS_4], greigite [$Fe^{2+}Fe_2^{3+}S_4$], and abundant framboidal pyrite have been observed as secondary precipitates within the reactive medium emplaced to treat waters laden with heavy metals and sulfate derived from oxidized tailings and sulfide concentrates (unpublished data). Organic carbon is added to the barrier mixture to promote sulfate reduction. The occurrence of mackinawite [Fe_9S_8] at one of the barriers was described by Herbert *et al*. (1998).

Nanometer-scale secondary sulfides associated with sulfide-reducing bacteria were reported by Fortin and Beveridge (1997) to have formed within the pyritiferous tailings at Kidd Creek, Timmins, Ontario. The sulfides were identified as mackinawite, amorphous FeS, and pyrite. A bacterially associated secondary Ag_2S phase, possibly acanthite, was observed by Davis (1997) in the Kidd Creek tailings.

9.05.6 ACID MINE DRAINAGE IN MINES AND MINE WASTES

9.05.6.1 Underground Workings

Access to deeply buried ore bodies is attained through underground mining. Workings are excavated to provide access to the ore body (Figure 4) and to mine the ore. These operations typically involve the removal of rock with no or little valuable metal content. Mining and economic limitations frequently result in large amounts of residual sulfide minerals being left adjacent to the excavated stopes. The extent of oxidation of these residual sulfides is primarily dependent on the surface area of exposed sulfide minerals and the duration of exposure to oxygen (Morin and Hutt, 1997). After a mine is decommissioned,

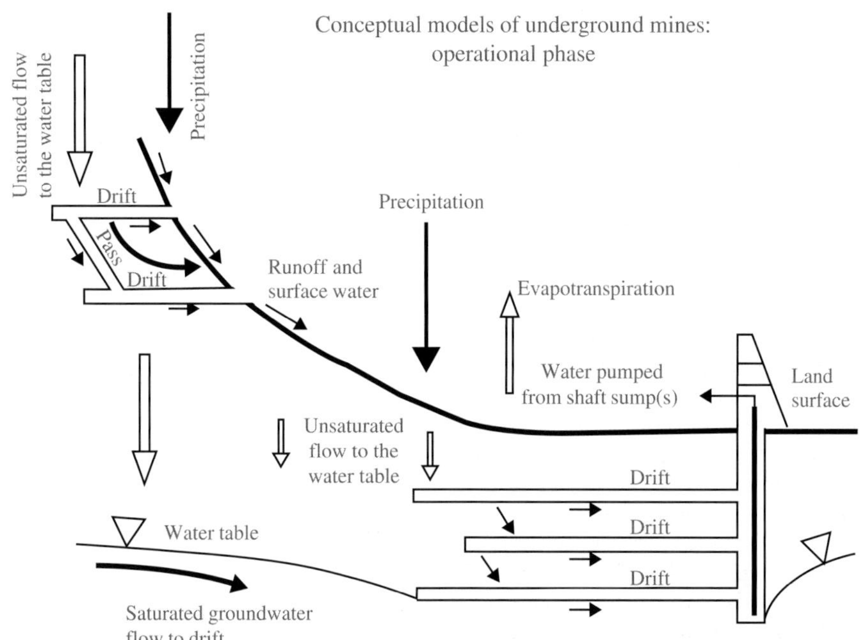

Figure 4 Schematic diagram showing underground mine workings (after Morin and Hutt, 1997).

the weathering products of these exposed sulfides may become a source of acidity, sulfate and dissolved metals.

Many underground mines are allowed to flood shortly after mining ceases, thus limiting the extent of sulfide-mineral oxidation. In areas where the water table is deep, or where adits have been excavated to drain the mine workings, extensive oxidation of the wall rock may persist for years or decades after mining ends. The extent of sulfide oxidation and the quality of the water derived from a mine site is dependent on the duration of exposure to atmospheric O_2, and on the local geological conditions and methods of mining used.

Nordstrom and Alpers (1999b) and Nordstrom *et al.* (2000) described the generation of extremely acidic water at the Iron Mountain mine near Redding, California, where a series of tunnels had been excavated to allow access to deep portions of the ore body. The excavation of drainage tunnels from these sites had the effect of exposing large volumes of sulfide minerals, principally pyrite with lesser amounts of chalcopyrite and sphalerite, to oxidation. Large accumulations of secondary sulfate solids, such as römerite, rhomboclase, and Zn–Cu-bearing melanterite, have been observed as efflorescence on the mine walls and as stalagmites and stalactites (Nordstrom and Alpers, 1999b). Water dripping from the mine walls and in pools contained extremely high concentrations of dissolved solids (up to 7.6×10^5 mg L^{-1} SO_4, 8.6×10^4 mg L^{-1} Fe, and very high concentrations of other metals) and had a pH as low as −3.6 (Nordstrom *et al.*, 2000; Table 8). During dry periods, secondary minerals form within the mine workings, and the dissolved metals in pools become concentrated from evaporative effects. After rainfall events, flushing of the accumulated oxidation products results in effluents that have very high metal concentrations. The concentrations of dissolved metals associated with the Iron Mountain site are much higher than those observed in drainage from other base-metal and gold mines, such as the Carlton Mine and the Roosevelt mine in Colorado (Table 8).

At the WISMUT Königstein mine, near Dresden, Germany, acidic solutions were used in an *in situ* leaching procedure to extract uranium from a sandstone aquifer. During mining, zones of the mine were isolated and acidic leaching solutions were percolated through blocks of aquifer material. The residual sulfuric acid was collected at the base of the block. The blocks were blasted to enhance permeability. In addition to acid used in the leaching process, acidity was released by the oxidation of sulfide minerals contained in the sandstone (Biehler and Falck, 1999). The concentrations of dissolved constituents vary in the vicinity of the mine (Table 8).

Table 8 Example water chemistry observed in mine workings.

	Königstein Mine[a] *groundwater in fourth aquifer*	*Königstein Mine*[a] *unsaturated unleached blocks*	*Königstein Mine*[a] *unsaturated leached blocks*	*Königstein Mine*[a] *flooded leached blocks*	*Richmond Mine portal*[b]	*Carlton Mine tunnel*[c]	*Roosevelt Mine tunnel*[d]
pH (SU)	5.99	1.88	2.92	1.88	0.5–1.0	7.f81	7.69
Eh (mV)	307	747	807	651			
TDS (mg L^{-1})	1,736	12,322	3,827	13,296			
SO_4 (mg L^{-1})	33	8,220	2,090	8,800	20,000–108,000	1,292	849
As (mg L^{-1})					34–59		
Cd (mg L^{-1})					4–19		
Cr_T (mg L^{-1})	<0.002	0.97	0.072	1.34			
Cu (mg L^{-1})					120–650		
Fe_T (mg L^{-1})	1.51	1,171	15.32	1,570	13,000–19,000	0.006	0.012
Pb (mg L^{-1})	0.013	2.1	0.010	1.43			
Zn (mg L^{-1})	<0.01	132	24	164	700–2,600	0.044	0.22
Ra (Bq kg^{-1})	104 ± 7	0.520 ± 0.047	0.0073 ± 0.0016	2.74 ± 0.24			
Th (Bq kg^{-1})	NA	1,333 ± 100	49 ± 6	1,051 ± 95			
U (mg L^{-1})	<0.02	12.3	18.1	50			

[a] Königstein mine, Germany (Biehler and Falck, 1999). [b] Richmond mine portal, California, USA (Nordstrom *et al.*, 2000). [c] Carlton mine tunnel, Colorado, USA (Eary *et al.*, 2003). [d] Roosevelt mine tunnel, Colorado, USA (Eary *et al.*, 2003).

Low concentrations of sulfate and dissolved metals are observed in the groundwater. Higher concentrations are observed in the mined blocks, with the maximum concentrations observed in blocks that had been leached and flooded. WISMUT plans to decommission the Königstein mine. The fate of dissolved metals and radionuclides in the flooding water is an important factor in the development of decommissioning plans (Bain *et al.*, 2001).

The effects of gold mining on groundwater quality were assessed at the Cripple Creek Mining District, Cripple Creek, Colorado, using field measurements of water quality, mineralogical analyses, and geochemical modeling techniques (Eary *et al.*, 2003). The underground workings were dewatered using a series of drainage adits. These have not been plugged, allowing the workings to continue to be exposed to atmospheric O_2. The gold ore occurs as disseminated native gold and is generally associated with pyrite. Oxidation reactions have produced low-pH conditions (pH < 3), with elevated concentrations of SO_4 ($>2{,}000$ mg L^{-1}), Fe (>200 mg L^{-1}), and other dissolved metals in the shallow groundwater. Deeper drainage waters are near-neutral in pH (>7), and have low concentrations of dissolved metals. The upper host rocks were depleted in carbonate minerals, suggesting that the increase in pH was a result of carbonate-mineral dissolution reactions. Removal of iron was attributed to formation of ferrihydrite, manganese to formation of manganese oxyhydroxides and rhodochrosite, and zinc to formation of zinc silicates and/or zinc substitutions in calcite. Predictions suggest that the neutralizing capacity of the rock will result in continued improvements in water quality for the foreseeable future (Eary *et al.*, 2003).

9.05.6.2 Open Pits

Open pits are excavated to extract ore from near-surface ore bodies. Many open pits are adjacent or connected to underground workings. Open-pit mining has increased substantially over the past two decades as improved metallurgical techniques and the increased scale of mechanical operations have made it possible to extract metals from low-grade ores (Miller *et al.*, 1996). These improvements have made large-scale open-pit mining of low-grade copper and gold ores economically viable. Open pits are excavated through a series of benches that provides access to the ore and maintains the stability of the pit walls. In areas where the water table is high, it may be necessary, during the mining operation, to dewater the pits and surrounding materials by using pumping or diversion techniques (Figure 5).

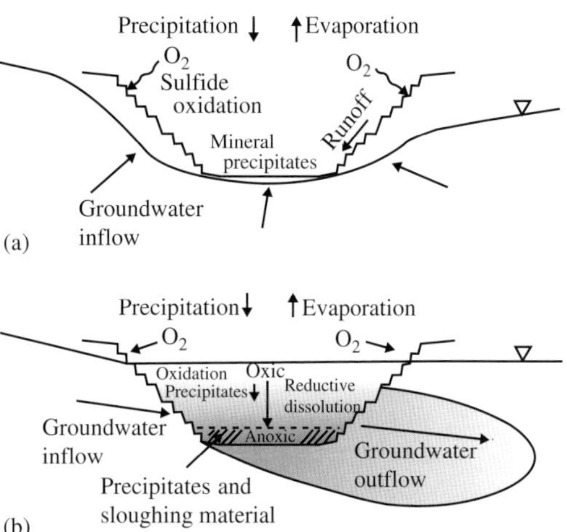

Figure 5 Schematic diagram of an open-pit mine during and after operation (after Miller *et al.*, 1996).

Excavation of the open pit and dewatering expose portions of the ore body and country rock, which were previously buried or submerged, to atmospheric oxygen. Oxidation of sulfide minerals within the wall rock results in the release of acidity, sulfate, dissolved metals, and other trace elements (Davis and Ashenberg, 1989). The extent of sulfide oxidation is dependent on the surface area of the exposed sulfide minerals and the duration of mining. In upper portions of the pit, where benches may be pushed back only occasionally, sulfide minerals may oxidize for a prolonged period of time. The quantity of sulfide minerals exposed is dependent on their abundance and on the surface area of the pit walls. The pit-wall surface area is considerably greater than would be estimated from an ideal three-dimensional model because of irregularities in the pit walls, the presence of fractures and faults intersecting the pit walls, and the accumulation of rubble zones (Morin and Hutt, 1997). The reactive surface of a pit wall has been estimated to be 27–161 times larger than the ideal three-dimensional surface area (MEND, 1995). The total reactive rock-surface area for an open pit can be large. For example, the reactive rock-surface area at the Island Copper mine, near Port Hardy, British Columbia, Canada, was estimated to be 2.44×10^8 m^2, with a 161 : 1 ratio of reactive to modeled planar surfaces (MEND, 1995). Many pits are backfilled with sulfide-bearing waste rock, tailings and other materials, some of which may have oxidized extensively prior to placement in the pit. The presence of these backfill materials may also contribute to the generation of low-quality effluent.

After mining is complete, pumping is terminated and the pits are allowed to flood. The principal components of water gain are precipitation, groundwater inflow, and surface-water inflow. The final elevation of the pit lake will depend on the local hydrological conditions and the geological properties of the surrounding rocks. The principal components of water loss are evaporation, and groundwater and surface-water outflow. A water balance based on these principal components can be constructed to estimate the relative magnitude of each component and to assist in the prediction of pit-water quality. For the Spenceville mine at Spenceville, California, Levy *et al.* (1997) estimated that the principal inflow components were surface runoff > precipitation > groundwater inflow, and the principal outflow components were direct evaporation > groundwater outflow as the pit reflooded. On this basis, it was concluded that two years had been required to reflood the pit. After the pit had filled, outflow of surface water commenced and the magnitude of the groundwater and surface-water outflow components increased. In contrast, the Berkeley Pit at Butte, Montana, is expected to fill over a period of 26 yr (Davis and Ashenberg, 1989), thus allowing a more prolonged exposure of sulfide minerals on the pit walls, and the possible accumulation of secondary minerals. These secondary minerals may dissolve into precipitation runoff, or into inflowing groundwater or during pit refilling. After the pit floods, further oxidation of sulfide minerals by atmospheric oxygen below the pit lake surface will be limited because of the low solubility of O_2 in water. The sulfide minerals above the surface of the pit lake will continue to oxidize long after mining has ceased.

The chemistry of pit lakes depends on a number of factors, among which are the geology of the pit walls and surrounding geological materials, the duration of sulfide oxidation, and the local hydrologic conditions (Table 9; Miller *et al.*, 1996; Shevenell *et al.*, 1999). Some pit lakes have near-neutral pH and contain low concentrations of dissolved constituents. These lakes may be suitable for recreational use (Shevenell *et al.*, 1999). Other lakes, however, are acidic and contain high concentrations of dissolved metals and may be hazardous to waterfowl (Miller *et al.*, 1996; Morin and Hutt, 1997). Miller *et al.* (1996) and Shevenell *et al.* (1999) compared the water chemistry of several pit lakes in the western United States. Pit lakes associated with sulfide ores and carbonate-deficient rocks generated acidic water containing high concentrations of dissolved metals, whereas pits excavated in country rock containing abundant carbonates were neutral in pH and contained low concentrations of dissolved metals. Some elements, including arsenic, selenium, and mercury, were present at high concentrations under the neutral pH conditions that are predominant in carbonate-rich terrains.

The relative depth, which is the ratio of the maximum pit depth to the mean diameter of the pit lake expressed as a percentage, is greater than for natural lakes, resulting in less mixing. The large relative depth typical of pit lakes, combined with the high concentrations of dissolved solids in some pit-lake waters, results in a stable stratification; hence seasonal turnover and metal cycling are limited. For example, Ramstedt *et al.* (2003) observed a halocline, redoxcline, and thermocline at different depths in the Udden pit lake, northern Sweden. The limited mixing that persists in some pit lakes can result in the development of a permanently anoxic zone at the base of a lake. The anoxic zone limits Fe(II) oxidation and provides an environment suitable for anaerobic bacteria, including iron-reducing bacteria and sulfate-reducing bacteria. In lakes derived from metal mining, concentrations of organic carbon may be low enough to limit the extent of iron and sulfate reduction, whereas open pits derived from coal mines may have sufficient organic carbon to promote extensive sulfate reduction (Blodau and Pfeiffer, 2003; Ramstedt *et al.*, 2003). In the oxic portion of the water column, oxidation of ferrous iron and precipitation of ferric (oxy)hydroxide solids may occur. These solids can scavenge trace metals, and can settle through the water column. The upper, oxic portion of a lake may therefore contain lower concentrations of dissolved constituents than the deeper waters. The distribution of dissolved constituents in the pit will affect the quality of water discharged from the pit into the adjacent groundwater system and surface-water outflow. For example, very high concentrations of dissolved zinc have been observed in groundwater adjacent to the Brunswick No. 6 open pit at Bathhurst, New Brunswick (Morin and Hutt, 1997). Eary (1999) provides a summary of secondary minerals that may form in open pits.

9.05.6.3 Waste-rock Piles

Waste rock, the large volume of broken rock that cannot be processed economically, is commonly deposited in large piles that are typically 30–100 m high, but range up to 500 m in height and several square kilometers in area (Figure 6). Waste-rock piles may be deposited in successive lifts or benches, or the waste rock may be end dumped from the top of the pile, enhancing sorting and segregation according to particle size. The selection of a disposal technique will depend on the site conditions, economic considerations, and environmental policy.

Waste-rock materials vary in grain size from fine-grained sand- and gravel-size materials, to

Table 9 Example water chemistry observed in existing pit lakes.

Constituent[a]	*EPA drinking water standard*	*Berkeley Pit, Butte, MT 10/16/87*	*Robinson District*		*Yerington Pit Yerington, NV 1991*	*Getchell Mine*			*Cortez Pit, Cortez, NV 1992–1993*
			Liberty Pit 1993 (0.5 m)	*Kimberley Pit 1993 (0.5 m)*		*South Pit*[c] *4/28/82*	*Center Pit*[c] *4/28/92*	*North Pit*[c] *4/28/92*	
pH	6.5–8.5 (s)	2.8	3.21	7.61	8.45	5.96	5.27	7.67	8.07
TDS[b]	100 (s)		6,240	3,580	631	2,110	2,140	2,420	432
Cl	250 (s)	9	48.9	286	36	34.4	30.2	25.7	24.4
F	1.4–2.4		18.5	3.01	1.4	2.4	2.4	1.6	2.4
NO_3 as N	10		<0.04	<0.02	0.67	0.01	0.01	0.01	0.207
SO_4	250 (s)	5,740	3,700	1,800	270	1,380	1,410	1,570	90.2
As	0.5	0.05	<0.005	<0.005	0.003	0.009	0.008	0.38	0.038
Ba	1		<0.002	<0.002	0.034				0.06
Cd	0.01	1.3	0.647	<0.005	<0.001	<0.005	<0.005	<0.005	
Cr	0.05		0.107	0.059	<0.005	<0.02	<0.02	<0.02	
Cu	1 (s)	156	37.1	0.06	0.16	0.04	0.04	<0.005	
Fe	0.3 (s)	386	62.2	<0.05	0.01	0.8	2.1	0.16	0.134
Pb	0.05		<0.005	<0.005	<0.005	<0.05	<0.05	<0.05	0.0043
Mn	0.05 (s)	95	116	0.17	0.32	1.8	4.3	0.13	0.0017
Hg	0.002		<0.0002	<0.0002	<0.0005	<0.2	<0.2	<0.02	0.00046
Se	0.01		<0.005	<0.005	0.13	<0.002	<0.002	0.003	
Ag	0.05		0.022	0.021	<0.01	<0.005	<0.005	<0.005	
Zn	5 (s)	280	52.1	1.81	0.01	0.33	0.4	0.02	0.002
Ca		462	506	605	93	401	438	530	45.4
Mg		201	344	156	15	79.3	61.9	72.7	18.1
K		10	63	11.4	6.9	7.85	6.36	10.82	11.7
Na		72	53.6	95.3	76	56.3	40	38.1	68.6
Total alk.			0	189	24				

Source Eary *et al.* (1996).
[a] Amounts are in mg L^{-1}; pH is given in pH units. [b] Total dissolved solids. [c] Filtered samples; composites of samples from various depths.
(s) represents a secondary standard.

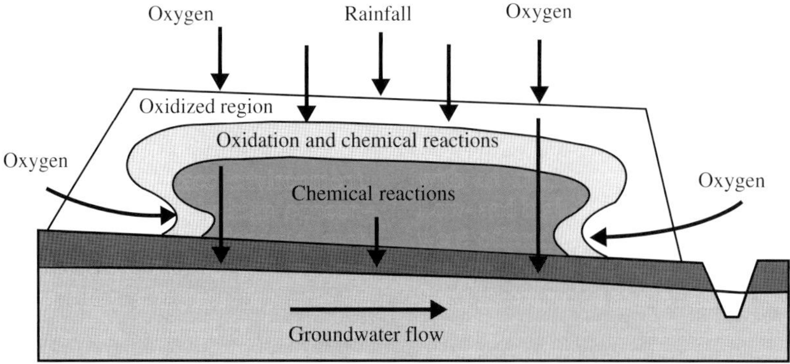

Figure 6 Schematic diagram of a waste-rock dump (after Ritchie, 1994).

Table 10 Physical properties of a typical dump of mine waste.

Symbol	*Definition*	*Value*	*Units*
L	Dump height	15	m
A	Dump area	25	ha
ρ_r	Bulk density of dump material	1,500	kg m^{-3}
ρ_{rs}	Sulfur density as pyrite	30 (2%)	kg m^{-3}
Q_w	Infiltration rate	0.5	m yr^{-1}
ε_g	Porosity of the dumped material	0.4	
ε_w	Water filled porosity at specified infiltration	0.1	
K_s	Saturated hydraulic conductivity of dump	10	m d^{-1}
D	Oxygen diffusion coefficient in the dump	5×10^{-6}	m^2 s^{-1}
C_o	Oxygen concentration in air	0.265	kg m^{-3}
ε	Mass of oxygen consumed per unit mass sulfur oxidized	1.75	
S^*	Mass of oxygen consumed per unit volume and unit time	1×10^{-8}	kg m^{-1} s^{-1}
ρ_c	Carbonate density	0.6 (0.04%)	kg m^3

Source: Ritchie (1994).

large blocks up to several meters in diameter (Ritchie, 1994). The hydrology of waste-rock piles is an active area of research (Gelinas *et al.*, 1992; MEND, 1994; Ritchie, 1994; Nichol *et al.*, 2002). Measurements made on waste-rock piles and in laboratory columns suggest that the coarse nature of waste rock leads to a relatively large gas-filled porosity and high permeability. Ritchie (1994) summarized the characteristics of waste-rock piles (Table 10). This summary indicates that the travel time for vertical transport of water from the surface to the base of a 15 m high waste-rock pile is approximately three years. These calculations assume that water transport through the waste rock can be described in the same manner as water flow through an unsaturated soil. Recent studies of waste materials at a mine site in northern Saskatchewan suggest that more than one flow regime may be present in waste-rock piles, with a portion of the water moving along rapid flow paths (Nichol *et al.*, 2002). At many locations, waste-rock piles are constructed on permeable geological materials (Ritchie, 1994). At these locations, pore water affected by geochemical reactions within the waste-rock pile may be displaced into the underlying geological materials. Travel through the base of a waste-rock pile, 25 ha in area and under saturated flow conditions, is anticipated to require five years or more (Ritchie, 1994). Under conditions that prevail in most waste-rock piles, the supply of oxygen limits the rate and extent of sulfide-mineral oxidation (Ritchie, 1994). Initially, the oxygen contained in a waste-rock pile upon deposition is consumed. This oxygen is gradually replenished by oxygen from the surface of the pile via four gas-transport mechanisms: diffusion, advection, convection, and barometric pumping, with the last probably the least significant (Figure 6).

The rate of oxygen diffusion is proportional to the diffusivity of the waste-rock pile. Although the diffusivity of waste rock is high because of the low moisture content of the waste rock, diffusive transport of oxygen is sufficiently slow that this process limits the rate of sulfide oxidation. Advective transport of oxygen results from changes in gas pressure between the waste-rock pile and the atmosphere. Wind has the potential to drive oxygen deeper into the pile than would occur under diffusive transport mechanisms alone (Ritchie, 1994; Figure 6). The oxidation of pyrite is exothermic, and the progressive accumulation of heat can result in high temperatures in a waste-rock pile undergoing intense oxidation. Temperatures in excess of 60 °C have been measured in oxidizing waste-rock piles. These increased temperatures induce the convective transport of atmospheric oxygen into the pile (Figure 7).

Convective transport of oxygen results in the penetration of oxygen deep into the waste-rock pile, accelerating the rate of oxidation of sulfide minerals and reinforcing the development of the convection cell (Cathles, 1979, 1994). The accelerated rate of oxidation near the pile margins both increases the short-term release of contaminants and decreases the duration of sulfide oxidation at those sites. Measurements of temperature and gas-phase oxygen concentrations in a waste-rock pile at the Rum Jungle mine site in Australia illustrate the regimes of gas transport within the pile (Harries and Ritchie, 1985). On the surface of the pile, away from the pile margins, the temperature remains relatively uniform and the concentration of oxygen in the pore gas decreases from atmospheric levels (20.9%) to <1% within 10 m. Gas transport in this region of the pile is dominated by diffusion. At its western margin, temperatures of up to 50 °C were observed within the pile. High O_2 concentrations coincide with these high temperatures, indicating that convective transport of gas is drawing oxygen-rich air from the margin of the pile to ~150 m into the interior of the pile. The convective transport of O_2 in this region accelerates the rate of sulfide oxidation at the pile margin relative to the pile surface.

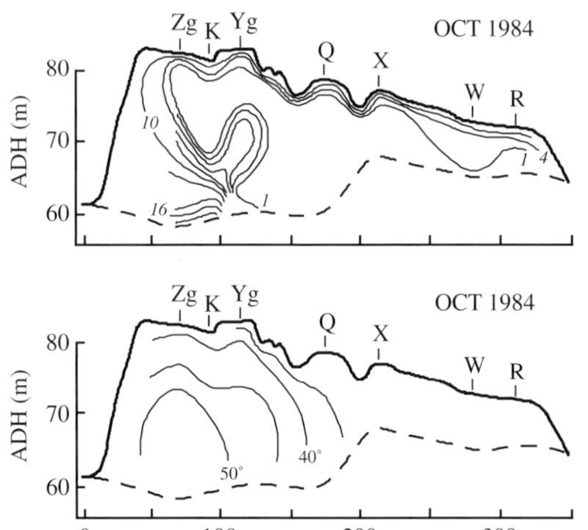

Figure 7 Measurements of temperature and oxygen concentration in the Rum Jungle mine waste-rock pile (after Harries and Ritchie, 1985).

Although seemingly dramatic in its effects, the convective transport of atmospheric gases is of relatively limited importance because a zone of only ~100 m inward from the pile margins is affected (Ritchie, 2003). Based on model calculations, Ritchie (1994) estimated that ~150 years would be required to oxidize all of the pyrite in a typical waste-rock pile containing 2 wt.% pyrite. The duration of oxidation is longer if the pile has a greater sulfide content. The rapid oxidation of sulfide minerals in waste-rock piles can generate low-pH conditions and release very high concentrations of dissolved constituents to the pore water that flows through the waste-rock pile. Because of the difficulty associated with obtaining water samples from unsaturated waste rock, few measurements of water chemistry from within waste-rock piles are available. Measurements made within the unsaturated zone at the Mine Doyon (Québec, Canada) waste-rock pile show low-pH waters containing very high concentrations of dissolved SO_4, Fe, and Al (Table 11; Gelinas *et al.*, 1992; MEND, 1994).

9.05.6.4 Coal-mine Spoils

One of the most serious environmental concerns associated with coal mining is the production of acid mine drainage. Coal mining exposes sulfur-bearing minerals to atmospheric oxygen and water. Pyrite is the principal source of acid production in coal spoils (Rose and Cravotta, 1998). Concerns associated with acidic coal-mine

Table 11 Concentrations of dissolved constituents in samples collected in lysimeters installed into the Mine Doyon waste-rock pile.

Sample	*92-1 L3A*	*92-1 L4A*	*92-1 L5A*	*92-2 L3A*	*92-2 L4A*	*92-2 L5A*
Depth (m)	1.67	2.42	4.05	1.67	2.54	4.07
pH (SU)	6.81	6.81	6.97	1.77	2.03	1.9
Cond. (μS)	1,242	1,625	2,315	21,185	17,588	22,532
E_h (mV)	249	250	226	514	484	432
Fe_T (mg L^{-1})	0	0	0	16,614	2,878	7,888
SO_4 (mg L^{-1})	629	1,550	0	63,029	43,210	40,750
Al (mg L^{-1})				2,324	2,412	2,634

Source: MEND (1994).

drainage include sedimentation of chemical precipitates, soil erosion, and loss of aquatic habitats in contact with waters with high metal loads (Williams *et al.*, 2002). A bimodal distribution of coal-mine drainage has been observed, with acidic (pH 3–5) and near-neutral (pH 5–7) pH values (Brady *et al.*, 1997). Figure 8 illustrates the range of pH values observed for both bituminous and anthracite coals of the eastern United States. Acidic drainage associated with coal spoils develops where carbonate minerals, such as calcite and dolomite, or other calcareous strata that could neutralize acid production are insufficient to neutralize the acidity released by sulfide oxidation. Coal-mine effluents can range in composition from acidic to alkaline, depending on the host-rock geology. Effluents can have pH values as low as 2 and high concentrations of SO_4, iron, manganese, and aluminum, along with common elements such as calcium, sodium, potassium, and magnesium. The latter constituents are commonly elevated due to aggressive acidic dissolution of carbonate, oxide, and aluminosilicate minerals along flow paths that are downgradient from the sources of oxidizing pyrite (Cravotta, 1994).

In cases where neutral or alkaline mine drainage predominates, problems may arise because of elevated concentrations of SO_4, iron, manganese, and other solutes that are derived from sulfide oxidation or from reactions with carbonate or aluminosilicate minerals. Dissolved iron and aluminum may precipitate as the pH increases, and these precipitates can act as substrates for adsorption and co-precipitation (Stumm and Sulzberger, 1992; Foos, 1997; Brake *et al.*, 2001). The dissolution of siderite, which is commonly associated with coal spoils, followed by the precipitation of Fe(III) (oxy)hydroxides, generates no *net* alkalinity (Rose and Cravotta, 1998).

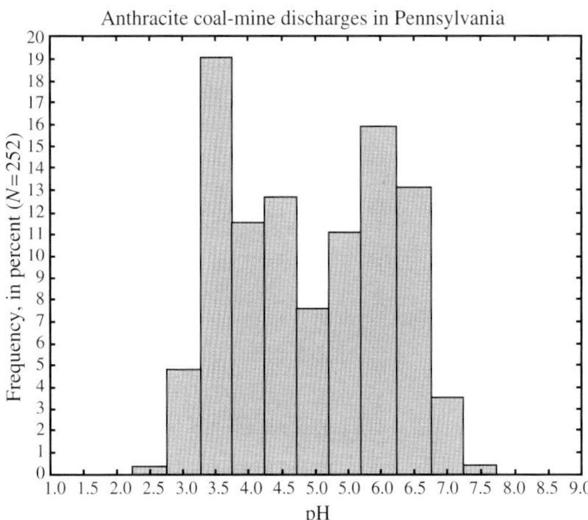

Figure 8 The bimodal distribution of pH in coal-mine drainage, where approximately half the discharges from bituminous and anthracite coal mines are acidic with a pH less than 5 (source http://pa.water.usgs.gov/projects/amd/index.html).

High sulfate concentrations are not dependent on pH conditions and can pose a significant problem in both acidic and alkaline conditions (Rose and Cravotta, 1998). The formation of hydrous sulfate minerals in coal spoils can be significant sources of "stored acidity" (Alpers *et al.*, 1994). This stored acidity can be released when the soluble secondary minerals are dissolved during periods of recharge or runoff and when iron or aluminum components of the minerals undergo hydrolysis (Rose and Cravotta, 1998). An example of this effect is the dissolution of halotrichite and coquimbite, respectively:

Halotrichite

$$FeAl_2(SO_4)_4{\cdot}22H_2O + \tfrac{1}{4}O_2 \leftrightarrow Fe(OH)_3 + 2Al(OH)_3 + 4SO_4^{2-} + 8H^+ + 13H_2O \quad (25)$$

Coquimbite

$$Fe_2(SO_4)_3{\cdot}9H_2O \leftrightarrow 2Fe(OH)_3 + 6H^+ + 3H_2O + 3SO_4^{2-} \quad (26)$$

The storage and release of acidity by these mechanisms can cause considerable temporal variability in water quality and can cause acid drainage to continue even after pyrite oxidation has ceased.

9.05.6.5 Tailings Impoundments

Mine tailings are the finely ground residue from ore extraction. The grain size of the tailings depends on the nature of the ore and the milling process. Size measurements (Robertson, 1994) of tailings from four mines in Ontario, Canada, indicated the tailings materials to be predominantly silt and fine to medium sand with $<10\%$ clay content. Tailings are transported from the mill and are discharged into an impoundment as a slurry containing ~ 30 wt.% solids. The method of deposition affects the distribution of tailings particles within the impoundment. Discharge commonly takes place at elevated perimeter dams; hence there is potential for extensive hydraulic sorting, with coarser fractions settling near the discharge point and finer fractions settling in distal portions of the impoundment (Robertson, 1994). At some sites, tailings are thickened to >60 wt.% solids prior to deposition. Thickening the tailings allows a more rapid settling of the solids, which therefore reduces the potential for hydraulic sorting, resulting in a more uniform grain-size distribution than is observed in conventional tailings areas (Robinsky, 1978; Al, 1996).

During tailings disposal, water is continuously added to the impoundment and the water table remains near the impoundment surface. After tailings deposition ceases, precipitation becomes the dominant source of recharge to the impoundment. The water table falls to an equilibrium position controlled by the rate of precipitation, the rate of evapotranspiration, and the hydraulic properties of the tailings and the underlying materials (Dubrovsky *et al.*, 1984; Blowes and Jambor, 1990).

The fine grain size of mine tailings results in a high moisture-retaining potential for these materials, which is a situation distinctly different from that in waste-rock piles. Whereas waste-rock piles commonly have a large open and free-draining porosity, mine tailings drain slowly, maintaining a large residual moisture content under gravity drainage. Measured moisture contents of conventional tailings impoundments vary from 10% to 100% saturation (Smyth, 1981; Blowes, 1990). The residual moisture content of thickened tailings is greater than that observed for conventional tailings (Robinsky *et al.*, 1991; Al and Blowes, 1996). The high residual moisture content of mine tailings results in a low gas-filled porosity, and in rapid changes in hydraulic gradient in response to precipitation (Blowes and Gillham, 1988; Al and Blowes, 1996).

Precipitation that falls on the impoundment surface migrates downward and laterally through the tailings impoundment into underlying geological materials (Figure 9). Groundwater velocities in tailings impoundments are relatively low. Coggans *et al.* (1999) estimated that the groundwater vertical velocity ranged from $0.2\ \mathrm{m\ a^{-1}}$ to $1.0\ \mathrm{m\ a^{-1}}$ at the Inco Ltd. Copper Cliff Central Tailings area in Sudbury, Ontario, whereas horizontal velocities were on the order of $10-15\ \mathrm{m\ a^{-1}}$. At the Nickel Rim tailings impoundment, also near Sudbury, Johnson *et al.* (2000) estimated groundwater vertical and horizontal velocities were in the range $0.1-0.5\ \mathrm{m\ a^{-1}}$ and $1-16\ \mathrm{m\ a^{-1}}$.

The surface areas of tailings impoundments vary from less than 10 ha to several square kilometers, and the thicknesses of the tailings deposits vary from a few meters to more than 50 m. The relatively low groundwater velocities and the large areal extent of tailings impoundments result in long time intervals between the time of groundwater infiltration and the time of groundwater discharge to an underlying aquifer or to the surface-water environment (Figure 10). These long travel times result in the delay of measurable environmental degradation at the groundwater discharge point until long into the life of the impoundment. The severity of the negative environmental effects associated with tailings impoundments may not be evident until long after mine closure and decommissioning of the impoundments. The subsequent prevention and remediation of low-quality drainage waters are more difficult than during the active mining. The long travel distances and low groundwater velocities result not only in the potential for prolonged release of contaminants from the tailings impoundment, but also in large long-term treatment costs. For example, Coggans (1992) combined estimates of the rate of sulfide

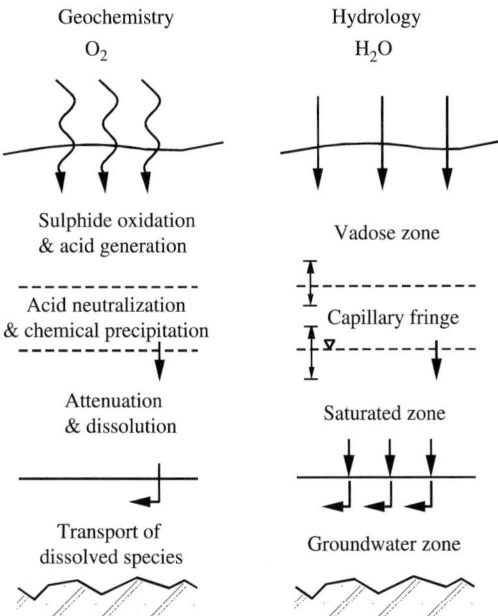

Figure 9 Schematic diagram showing the hydrology and geochemistry of a decommissioned mine tailings impoundment (source Blowes and Ptacek, 1994).

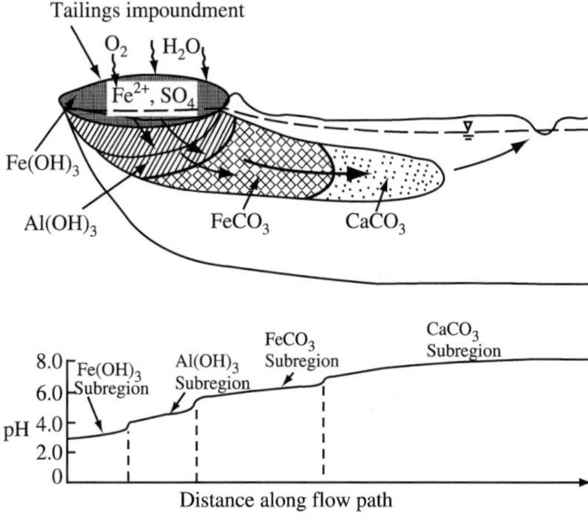

Figure 10 Schematic diagram of tailings impoundment and underlying aquifer, and pH buffering regions (after Blowes and Ptacek, 1994).

oxidation with estimates of groundwater velocity at the Inco Copper Cliff Central Tailings area in Sudbury, Ontario, and predicted (i) that the peak release of sulfide oxidation products will occur ~50 yr after the impoundment is decommissioned, and (ii) that high concentrations of oxidation products will persist for ~400 yr thereafter.

In most tailings impoundments, gaseous diffusion is the most significant mechanism for oxygen transport. The rate of diffusion of oxygen gas is dependent on the concentration gradient and the diffusion coefficient of the tailings material. The diffusion coefficient of tailings is dependent on the air-filled porosity of the tailings; the coefficient increases as the air-filled porosity increases, and the coefficient decreases as the moisture content increases. Several empirical relationships have been developed to describe the dependence of the gas diffusion coefficient on the tailings moisture content (e.g., Reardon and Moddle, 1985). These relationships indicate a maximum diffusion coefficient at low moisture contents, with a gradual decline in diffusion coefficient as moisture content increases to ~70% saturation, followed by a more rapid decline as the moisture content increases further. The relationship between moisture content and diffusion coefficient results in rapid oxygen diffusion in the shallow portion of the vadose zone of a tailings impoundment, where the moisture content is low. The rapid diffusion of oxygen in this zone replenishes oxygen consumed by the oxidation of sulfide minerals. As the sulfide minerals in the shallow portion of the tailings are depleted, the rate of sulfide oxidation decreases due to the longer diffusion distance and the higher moisture content of the deeper tailings.

In many tailings impoundments a variety of sulfide minerals is present. Jambor (1994) reported a general sequence of sulfide-mineral reactivity observed in several tailings impoundments, from the most readily attacked to the most resistant, to be pyrrhotite → galena-sphalerite → pyrite-arsenopyrite → chalcopyrite → magnetite. Blowes and Jambor (1990) observed systematic variations in sulfide-mineral alteration versus depth at the Waite Amulet tailings impoundment, Rouyn-Noranda, Québec. On the basis of the observations, the degree of alteration was classified into a numerical scale as shown in Table 12. The sulfide alteration index indicates the relative degree of alteration of sulfides. Because pyrrhotite is the sulfide mineral most susceptible to alteration, the extent of its replacement forms the basis for the alteration index. When plotted versus depth on a vertical axis, the alteration index estimates made at the Sherridon Mine, Manitoba, correlated well with geochemical parameters measured in adjacent drill-holes, and with gas-phase O_2 concentrations (Figure 11).

The microbially mediated oxidation of sulfide minerals within mine-tailings impoundments generates acidic conditions and releases high concentrations of dissolved metals. Mill tailings

Table 12 Alteration-index criteria for sulfides in the Waite Amulet (Québec) oxidized tailings.

Index	*Alteration*
10	Almost complete oxidation of sulfides; traces of chalcopyrite ± pyrite
9	Only sparse pyrite and chalcopyrite; no pyrrhotite or sphalerite
8	Pyrite and chalcopyrite common, but chalcopyrite proportion higher than normal possibly because of pyrite dissolution; no pyrrhotite or sphalerite
7	Pyrite and chalcopyrite proportions normal; pyrrhotite absent but sparse sphalerite present
6	Pyrrhotite absent but sphalerite common
5	Pyrrhotite represented by marcasite pseudomorphs
4	First appearance of pyrrhotite, but only as remnant cores
3	Cores of pyrrhotite abundant
2	Well-developed cores of pyrrhotite, with narrower alteration rims; replacement by marcasite decreasing, pseudomorphs absent
1	Alteration restricted to narrow rims on pyrrhotite

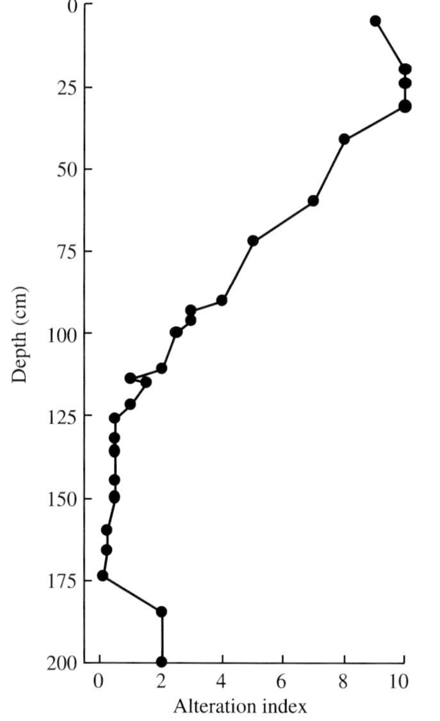

Figure 11 Sulfide alteration index (SAI) for oxidized tailings at the Sherridon mine, Manitoba.

at the Heath Steele mine in New Brunswick contain up to 85 wt.% sulfide minerals (Blowes *et al.*, 1991; Boorman and Watson, 1976). Pore-water pH values as low as 1.0, and concentrations of dissolved SO_4 up to 8.5×10^4 mg L^{-1} were observed in the shallow pore water of the tailings impoundment (Figure 12; Blowes *et al.*, 1991). This water also contained up to 4.8×10^4 mg L^{-1} Fe, 3,690 mg L^{-1} Zn, 70 mg L^{-1} Cu, and 10 mg L^{-1} Pb. The shallow pore waters at the Waite Amulet tailings impoundment in northwestern Québec contain 2.1×10^4 mg L^{-1} SO_4, 9.5×10^3 mg L^{-1} Fe, 490 mg L^{-1} Zn, 140 mg L^{-1} Cu, and 80 mg L^{-1} Pb. The pH of this water varies from 2.5 to 3.5 (Blowes and Jambor, 1990). High concentrations of dissolved zinc (48 mg L^{-1}), copper (30 mg L^{-1}), nickel (2.8 mg L^{-1}), and cobalt (1.5 mg L^{-1}) were observed in the shallow groundwater at the inactive Laver copper mine in northern Sweden (Holmström *et al.*, 1999). These low-pH conditions and high concentrations of dissolved metals occur within the shallowest portions of the tailings impoundment. As this water is displaced downward through the tailings, or through adjacent aquifer materials, the pH gradually rises, and many metals are removed from solution by precipitation, co-precipitation or adsorption reactions. High concentrations of Fe(II) and SO_4, however, move down through the tailings and aquifer sediments relatively unattenuated (Dubrovsky *et al.*, 1984; Johnson *et al.*, 2000). As this groundwater discharges from the tailings impoundment, Fe(II) oxidizes and precipitates as ferric (oxy)hydroxide and ferric hydroxysulfate minerals. These reactions release H^+, generating acidic conditions within surface waters. The transport of Fe(II) along the groundwater flow path, therefore, provides the vehicle for transporting acidity long distances from the oxidation zone to the surface-water flow system.

9.05.7 METHODS OF PREDICTION

9.05.7.1 Laboratory Static Procedures

Static tests are intended to predict whether a sample, and the rock or soil that it represents, will be acid producing after exposure to weathering. The distinguishing characteristic of a static test is that it is a one-time determination, whereas kinetic tests involve repeated cycles in which dosages of humidity or aqueous solutions are applied over a period of time. Thus, kinetic tests can provide information on weathering rates and the abundances of ions in the leachates, data for which are not obtainable from a static test. Nevertheless, static tests are widely used because they have the advantages of being rapid and simple to perform, and commensurate with those advantages is a relatively low cost per determination. Hence, static tests are commonly performed on large numbers of samples for an individual project, and for potentially exploitable mineral and coal deposits the results are commonly used to provide guidance as to which rocks may merit further study, such as by kinetic tests. At active mines, static tests may be used to monitor the potential of various wastes, such as overburden and barren or sulfide-bearing low-grade rocks that host the ores, to generate acidic drainage. The results may be used to govern how and where those wastes are disposed.

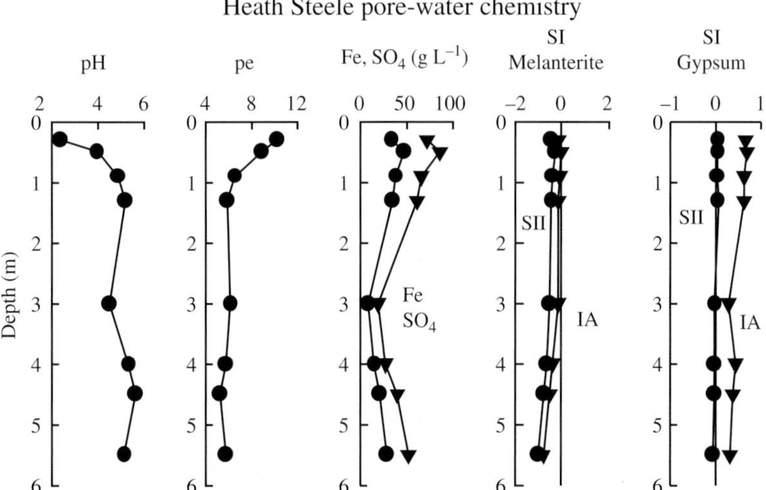

Figure 12 Pore-water chemistry and saturation indices versus depth at the tailings site of the Heath Steele mine. IA represents saturation indices calculated using an ion-association model, and SII represents saturation indices calculated using a specific ion-interaction model (after Ptacek and Blowes, 2000).

Numerous types of static tests have been developed, and some of the laboratory procedures and the variety of tests or their individual developments have been described in various publications, such as those by MEND (1991), Lawrence and Wang (1997), Morin and Hutt (1997), White *et al.* (1999), and Jambor (2003). By far the most widely adopted static test, both for metalliferous and coal deposits, is the acid–base account (ABA) of Sobek *et al.* (1978). As is implied by its name, ABA involves a determination of the acid-producing potential (AP) of a sample, and a determination of the base that is potentially releasable; the latter is generally referred to as the neutralization potential (NP). The two chemically determined values therefore provide a net accounting of the expected behavior during weathering. A common form of expressing the result is to obtain the net NP (NNP) by simple subtraction of the two chemically determined values:

$$\mathrm{NNP} = \mathrm{NP} - \mathrm{AP} \quad (27)$$

If NP > AP, the resulting value for NNP will be positive, thus indicating that the sample should have some acid-neutralizing capacity; the opposite, with AP > NP, is taken as an indication that the sample will be acid generating.

The value for AP is obtained by measuring the sample's weight percentage of sulfur (or wt.% $S_{sulfide}$ in some jurisdictions). It is assumed that the sulfur is present as pyrite [FeS_2], and that the pyrite will weather in accordance with the reaction

$$FeS_2 + \tfrac{15}{4}O_2 + \tfrac{7}{2}H_2O \Rightarrow Fe(OH)_3 + 2SO_4^{2-} + 4H^+ \quad (28)$$

Hence, each mole of sulfur produces $2H^+$ or, stated alternatively, each mole of pyrite produces four moles of H^+. Most nonsulfide minerals will react with acid to some extent, and if the effect of the mineral dissolution is to decrease the acidity of the original solution, then the mineral contributes NP, the amount of which is relative to the acidity consumption effected by calcite in the reaction

$$CaCO_3 + 2H^+ \Rightarrow Ca^{2+} + CO_2 + H_2O \quad (29)$$

The $2H^+$ produced per mole of sulfur in Equation (28) can be neutralized by 1 mol of $CaCO_3$, as in Equation (29). Therefore, 1 mol of $CaCO_3$ is equivalent to 1 mol of sulfur, and their approximate molecular-weight values are 100 and 32, respectively; hence $100 \div 32 = 3.125$, and 1 g of sulfur is equivalent to 3.125 g of $CaCO_3$. If S is in kg t^{-1}, which is a unit commonly employed, the AP of a sample is its wt.% S multiplied by 31.25 to obtain a value expressed in kg t^{-1} of $CaCO_3$ equivalence. The NP of calcite, because it is in g t^{-1}, is 1,000. The NP of material is obtained by the addition of acid to a 2 g mass of a sample to determine how much of the acid it consumes (neutralizes). The step-by-step procedures and calculations are given by Sobek *et al.* (1978), MEND (1991), and Morin and Hutt (1997).

The relationship between the result from a static test and what may occur during weathering is strictly empirical. Many assumptions are involved in the derivation of the NP and AP values, and static tests are a surrogate for weathering, not an emulation of it. Nonetheless, those aspects aside, probably the chief arguments concerning interpretations of static-test results have been focused on which value constitutes an environmentally "safe" NNP, on which minerals will provide NP to prevent ARD rather than just mitigate it, and on which minerals contribute to static-test NP even though their effects on ARD are minimal except over long time periods.

Part of the answer to the arguments has been the recent measurement of the NP values of specific minerals (Jambor *et al.*, 2002), and the replacement, by some regulatory agencies, of the absolute value of NNP, adopting in its place the ratio of NP/AP for environmental assessments. Like other environmental parameters, the ratio deemed to be appropriate is different in various jurisdictions. Whether NNP values or NP/AP ratios are used, however, commonly they are but one part of a more comprehensive, site-specific environmental evaluation that includes kinetic (dynamic) laboratory tests.

9.05.7.2 Laboratory Dynamic Procedures

Laboratory dynamic tests are conducted by exposing small volumes of rock to repeated weathering cycles. These tests are commonly referred to as kinetic tests and are intended to assess the potential for acid generation and metal leaching under flow conditions that are designed to emulate field conditions. The objectives of dynamic testing programs vary from confirming the hypothesis developed through static testing, to estimating the rates of sulfide oxidation and metal release, to assessing the potential for secondary mineral formation and metal attenuation. Various testing procedures have been developed (DIAND, 1992), among which are well-standardized ones (ASTM, 2001). The test apparatus ranges from humidity cells, which contain ~1 kg of sample (ASTM, 2001) to columns containing up to 100 kg of sample (Bennett *et al.*, 1999). The operational steps of the tests vary from procedure to procedure, and the steps may dictate the rate, duration, and volume of sample irrigation, the rate and duration of sample aeration, and the frequency of collection of supernatant samples.

The interpretation of dynamic test data depends on the objective and design of the testing procedure. Although some tests are designed to estimate individual parameters, such as the rate of sulfide-mineral oxidation (Bennett *et al.*, 1999), other tests are intended to provide estimates of the rate of release of dissolved metals and the rate of depletion of sulfide minerals and acid-neutralizing carbonate minerals (DIAND, 1992; ASTM, 2001). It is important that the test design address the objectives. Tests that accelerate the rate of water washing of the sample may not accelerate the rate of sulfide oxidation if that reaction is dependent on the exposed surface area and the extent of bacterial catalysis, but independent of the rate of irrigation. Furthermore, many of the products of sulfide oxidation are sparingly soluble at the neutral or moderately acidic pH values that may be observed in the early stages of the test program. Care must be taken to incorporate the masses of these phases into sulfide-oxidation or metal-leaching calculations.

9.05.7.3 Geochemical Models

9.05.7.3.1 Geochemical modeling approaches

A summary of the computer models available to describe inorganic geochemistry in static batch systems was prepared by Alpers and Nordstrom (1999). This summary indicates that a diverse series of models was developed in the 1970s and 1980s, and by the late 1990s this diversity had decreased to a smaller number of well-supported models. Figure 13 illustrates the current evolution of coupled chemical–hydrological–speciation modeling programs.

Geochemical speciation models and geochemical speciation mass-transfer models are used widely to describe water chemistry in mine-drainage environments. Most of these models are based on the ion-association theory (Garrels and Thompson, 1962), and use the extended Debye–Hückel model or derivative models, such as the Davies or WATEQ equation, for activity corrections (Nordstrom and Munoz, 1994). A variety of geochemical speciation mass-transfer models has been developed (Figure 13; Parkhurst *et al.*, 1985; Ball and Nordstrom, 1987; Allison *et al.*, 1990; Parkhurst, 1998; Alpers and Nordstrom, 1999). These models have become highly sophisticated and are able to describe increasingly complex geochemical reactions. Activities derived from the speciation calculations are used to determine mineral saturation indices, and predict the extent of mass transfer among aqueous, gaseous, and solid phases, over a range in temperature. Processes described by these models include geochemical speciation, acid–base equilibrium, redox equilibrium, precipitation/dissolution, and adsorption/desorption. Some versions of these models are able to incorporate descriptions of kinetically controlled reactions and descriptions of one-dimensional solute transport (e.g., Parkhurst, 1998).

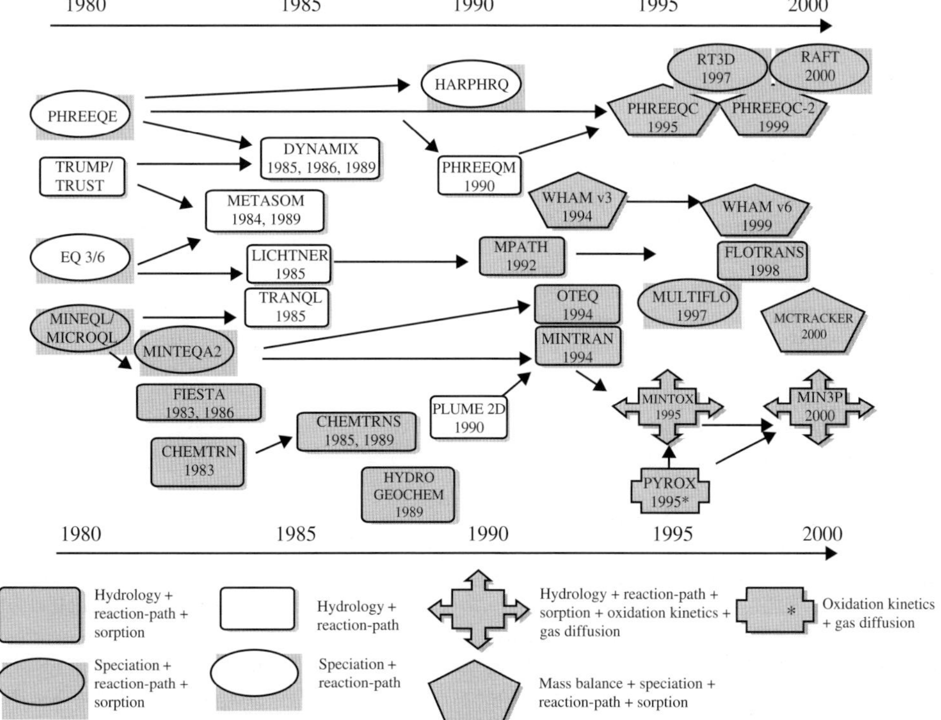

Figure 13 The evolution of coupled chemical–hydrological–speciation modeling programs (after Alpers and Nordstrom, 1999).

Models based on the ion-pair approach are generally for application to dilute waters, such as are observed in relatively uncontaminated lakes or groundwater systems. Efforts to improve predictions of ion activities and mineral saturation indices for mine-drainage settings (e.g., Ball and Nordstrom, 1987; Nordstrom *et al.*, 1990) have included the addition of extensions to the Debye–Hückel equation to cover a wider concentration range, inclusion of a large number of ion-association constants and complexation constants specific to mine-drainage settings, and the addition of mineral-solubility constants for phases typically encountered in mine-drainage environments. As a result, the ion-association approach for predicting metal-speciation and mineral-stability fields is generally accurate for mine waters characterized by relatively high concentrations of dissolved solids (ionic strengths up to 1 M) and a relatively broad range in temperature (e.g., Ball and Nordstrom, 1987; Nordstrom *et al.*, 1990).

9.05.7.3.2 *Application of geochemical speciation mass-transfer models*

Geochemical speciation and geochemical speciation mass-transfer models are used to assist in the interpretation of data collected in mines, deposits of mine wastes, and bodies of receiving water. These models can be used to infer the geochemical reactions that affect the concentrations of major ions and dissolved metals in mine waters and pore waters of mine wastes, to assess the stability of secondary minerals present in mine wastes, and to predict the concentrations of dissolved metals that can be anticipated in mine and mine-waste discharge. The application of these models can be constrained by other observations, particularly characterization of the primary and secondary mineralogy, by measurements of groundwater and surface-water flow directions and velocity, by water-balance calculations, and by measurements of environmental isotopes (e.g., ^{2}H, ^{3}H, ^{18}O, and ^{34}S). Geochemical speciation calculations have been used to assist in the interpretation of water chemistry associated with underground mine workings (Eary *et al.*, 2003), open pits (Davis and Ashenberg, 1989; Eary, 1999; Temple *et al.*, 2000; Ramstedt *et al.*, 2003), tailings impoundments (Blowes and Jambor, 1990; Blowes *et al.*, 1992; Johnson *et al.*, 2000), and rivers (Runkel *et al.*, 1996; Runkel and Kimball, 2002).

Waters associated with mine sites may contain high concentrations of dissolved solids as a result of intense sulfide-oxidation and evaporative processes. Specialized models are required for calculation of ion activities and mineral saturation indices for sites characterized by these high concentrations. The most widely accepted approach for prediction of ion activities and mineral solubilities in complex concentrated electrolyte mixtures is based on the Pitzer ion-interaction formalism (Pitzer, 1973) as summarized in Pitzer (1991). Models based on the Pitzer formalism have been developed for application to various geochemical systems, from acidic to basic, dilute to concentrated, over a range in temperature, total pressure, and partial pressures of component gases (Clegg and Whitfield, 1991; Pitzer, 1991). The first applications of Pitzer-based models to geochemical systems involved calculations of ion-activity and mineral-solubility relations in waters containing seawater components (e.g., Weare, 1987; Plummer *et al.*, 1988; Clegg and Whitfield, 1991). These efforts have been expanded to include development of ion-interaction-based models for a wide range of chemical compositions, including those derived from sulfide-oxidation processes (Reardon and Beckie, 1987; Baes *et al.*, 1993; Ptacek and Blowes, 2000). Mine drainage containing high concentrations of dissolved metals, sulfate, and acid have been reported for sites in Canada (in excess of 3×10^{5} mg L^{-1}; Blowes *et al.*, 1991; Moncur *et al.*, 2003) and in the US (in excess of 9×10^{5} mg L^{-1}; Nordstrom and Alpers, 1999b).

Geochemical models based on the Pitzer approach have been applied to describe mineral stability relationships in concentrated mine-drainage waters. At the Heath Steele site in New Brunswick, Canada, high concentrations in pore waters were observed in shallow tailings as a result of sulfide oxidation. Mineral saturation indices calculated using a Pitzer-based model for melanterite, gypsum, and siderite were close to 0.0 where these minerals were observed to occur, whereas saturation index values calculated using the conventional ion-pair model indicated supersaturated conditions (Blowes *et al.*, 1991; Ptacek and Blowes, 1994). Tailings pore waters should have achieved equilibrium with respect to melanterite and gypsum, and possibly siderite, within the observed residence time in the tailings. These results suggest that the Pitzer-based models have the potential to provide highly accurate predictions of mineral stability relationships at mine-drainage sites with concentrated pore waters.

9.05.7.4 Reactive Solute-transport Models

Reactive solute-transport models couple the equations that describe physical transport processes with equations that describe geochemical reactions. These models can be divided into three basic categories: (i) equilibrium models, (ii) partial equilibrium models, and (iii) kinetic models. The three are differentiated by the approach used to solve the equations for chemical reactions. Equilibrium-based models apply the local equilibrium assumption (LEA), in which all chemical reactions are assumed to proceed to completion, or attain equilibrium, within each time step. This assumption facilitates coupling between the physical transport step and the geochemical reaction step. In most models the iteration between the transport step and the chemical step assures convergence between the two modules (Walter *et al.*, 1994a,b). In some cases, however, the two steps proceed sequentially with no iteration between them. A comparison conducted by Walter *et al.* (1994a) suggests that the error arising from use of the sequential approach is modest. Partial equilibrium models identify a reaction, or a series of reactions, for which the equilibrium assumption is unrealistic. Kinetic expressions are then applied to describe these reactions. The remaining geochemical reactions are described using the local equilibrium assumption (Wunderly *et al.*, 1996). Kinetic models provide the opportunity to consider the rates of all chemical reactions that occur. In several of these models, the chemical reactions are directly substituted into the transport equations and are solved simultaneously (Lichtner, 1993; Steefel and Lasaga, 1994; Steefel and van Capellan, 1998; Mayer *et al.*, 2002). These models provide a comprehensive description of the geochemical system.

Reactive solute-transport models have been used to describe sulfide oxidation and transport of its products through mine wastes. Liu and Narasimhan (1989a,b) used a reactive-transport model to describe the transport of dissolved constituents at a uranium mine site. Sulfide oxidation and transport of oxidation products at the inactive Nordic uranium tailings impoundment were described using reactive-transport models (Walter *et al.*, 1994b; Wunderly *et al.*, 1996). The transport of metal contaminants at Pinal Creek, Arizona, was modeled by Brown *et al.* (1998). Sulfide oxidation and transport of oxidation products through the Nickel Rim tailings impoundment, Sudbury, Ontario, was simulated by Bain *et al.*, (2001) and by Mayer *et al.* (2002).

9.05.8 BIOACCUMULATION AND TOXICITY OF OXIDATION PRODUCTS

Metals released from mines and mine sites, and in some cases from the natural weathering of ore deposits, can harm aquatic biota in adjacent water bodies (Borgmann *et al.*, 2001). Metals are present in aqueous environments in a variety of species with various toxicities and potential for bioaccumulation. Transformations among these species depend on the physical and chemical characteristics of the water body. Metal toxicity can be acute or chronic. For example, aluminum can coordinate or precipitate to form species that result in sudden killing of fish in waters near mine sites. Other metals, such as mercury, can bioaccumulate, leading to chronic toxicity (Domagalski, 2001). The extent of bioaccumulation and the toxicity of metals within natural environments are controlled by a number of factors, among which are the pH, oxidation–reduction potential, the organic carbon content, concentrations and compositions of other dissolved species, and the composition of the sediment (Warren and Haak, 2001).

9.05.8.1 Uptake and Bioaccumulation

Bioaccumulation and biomagnification are two terms commonly used for metal toxicity. Bioaccumulation refers to how pollutants (metals) enter a food chain and relates to the accumulation of contaminants, in biological tissues by aquatic organisms, from sources such as water, food, and particles of suspended sediment (Wang and Fisher, 1999). Bioaccumulation involves, relative to the ambient value, an increased concentration of a metal in a biological organism over time. Accumulation in living things can occur whenever metals are taken up and stored faster than they are metabolized or excreted (Markich *et al.*, 2001). Understanding the dynamic processes of bioaccumulation can have important ramifications in protecting human beings and other organisms from the adverse effects of metal exposure, and hence, bioaccumulation is an important consideration in the regulation and treatment of metals associated with acid mine drainage.

First some terminology: in conjunction with bioaccumulation, we define uptake, bioconcentration, and biomagnification. Uptake describes the entrance of a chemical into an organism such as by breathing, swallowing, or absorbing it through the skin without regard to subsequent storage, metabolism, and excretion. Bioconcentration is the specific bioaccumulation process by which the concentration of a chemical in an organism becomes higher than its concentration in the air or water around the organism. Although the process is the same for natural and anthropogenic chemicals, the term bioconcentration usually refers to chemicals foreign to the organism. For fish and other aquatic animals, bioconcentration after uptake through the gills or, in some circumstances, through the skin, is usually the most important bioaccumulation process. Biomagnification refers to the tendency of pollutants to concentrate as they move from one trophic level to the next. The process occurs when a chemical or metal becomes increasingly concentrated as it moves up through a food chain, i.e., the dietary linkages between single-celled plants and increasingly larger animal species. The natural bioaccumulation process is essential for the growth and nurturing of organisms (Heikens *et al.*, 2001). Bioaccumulation of substances to harmful levels, however, may also occur.

Acid and alkaline mine waters commonly contain high concentrations of dissolved metals and metal-oxide particulates. The acidification of wetlands can elevate the concentrations of metals and increase the potential bioavailability in aquatic plants and freshwater biota (Albers and Camardese, 1993), and can influence the uptake of metals in both submerged and rooted plants (Sparkling and Lowe, 1998). Arsenic concentrations in freshwater macrophytes affected by effluents from a gold mine were examined by Dushenko *et al.* (1995). Macrophytes concentrated arsenic relative to sediment concentrations, with submerged species containing much higher levels of arsenic than those in air-exposed plants. The differences observed were attributed to growth form and the ability of plants to exclude arsenic with increasing sediment concentrations. Plants in the vicinity of high arsenic values showed clear indications of necrosis of leaf tips and reduced micronutrient levels of copper, manganese, and zinc in root tissues. A study of arsenic contamination in wood mice proximal to abandoned mine sites has shown that the extent of accumulation depends on the level of habitat contamination (Erry *et al.*, 2000). Rai *et al.* (2002) observed that metals correlated positively with metal concentrations in adjacent water and sediments, which had been impacted by domestic and industrial discharges.

9.05.8.2 Toxicity of Oxidation Products

Metals such as iron, copper, cadmium, chromium, lead, mercury, selenium, and nickel can produce reactive oxygen species, resulting in lipid peroxidation, DNA damage, depletion of sulfhydryls, and calcium homeostasis (Stohs and Bagchi, 1995). The inherent toxicities produced by the oxidation of these metals generally involve symptoms of neurotoxicity and hepatotoxicity. Metal reactions can be influenced by oxidation–reduction reactions, which often occur in aqueous environments impacted by mine-waste effluents. Species that contain more than one oxidation state in natural waters are inherently more mobile, reactive, and will exhibit differences in toxicity (Ahmann *et al.*, 1997; Brown *et al.*, 1999a,b; Ledin and Pedersen, 1996; Lin, 1997; Nordstrom and Alpers, 1999a; Warren and Haak, 2001). Depending on the metal concentrations entering the environment, most oxidation products in excess of natural requirements can produce toxic responses to aquatic biota.

Iron is an essential element for metabolic systems, but in iron-rich solutions toxicity can develop in both fish and biota. Iron toxicity has occurred in aquatic plants exposed to iron-rich groundwater (Lucassen *et al.*, 2000). Iron species can also affect the gill performance in fish, causing acute toxicity and accumulation on the gills (Dalzell and Macfarlane, 1999). In mine-waste discharge, Fe(III)-sulfate and (oxy)hydroxide precipitates can accumulate on the gill epithelium, resulting in clogging and damage, and decreasing the available surface area and increasing the diffusion distance for respiratory exchange (Dalzell and MacFarlane, 1999).

Arsenic in aquatic environments is usually more concentrated in sediments and pore water than in the overlying water column (Ahmann *et al.*, 1997; Smedley and Kinniburgh, 2002; Williams, 2001). The most abundant forms of arsenic are arsenate [As(V)] and arsenite [As(III)], but methylated forms can occur in mine-impacted environments (i.e., methylarsenic acid and dimethylarsenic acid) (Smedley and Kinniburgh, 2002); see Chapter 9.02. The principal pathway of arsenic toxicity is through dietary exposure to sediment and suspended particulates by fish, followed by human consumption. Environmental exposure to arsenic is a causal factor in human carcinogenous and other related health issues. Chronic exposure symptoms in humans include hyperkeratosis, hyperpigmentation, skin malignancies, and peripheral arteriosclerosis. Water provides the dominant pathway for arsenic exposure in humans (Williams, 2001).

Mercury speciation is dependent on the available oxygen, pH, and dominant redox conditions, which often are site specific. Mercury can be present as either elemental mercury or mercuric phases (i.e., Hg^{2+}, HgS, $HgCl_2$) associated with reduced anoxic environments. Under these conditions, mercury is considered to be relatively insoluble and is less toxic to biota. Under more oxidative conditions, such as those associated with roasting by-products (calcines) from separation procedures, mercury can form soluble sulfates and oxychlorides. Further reduction by sulfate-reducing bacteria can cause the inadvertent methylation of the dissolved mercury (Rytuba, 2000). The methylation of mercury and adsorption of mercury and methylmercury onto iron (oxy)-hydroxides are important processes which control the fate and transport of mercury species in waters impacted by mercury-containing mine drainage. The primary mechanisms controlling the accumulation of methylmercury and inorganic mercury in aquatic food chains are not sufficiently understood, but it is speculated that bacteria in anoxic sedimentary environments associated with the reduction of SO_4^{2-} and S_2^{2-} are responsible (Domagalski, 2001; Rytuba, 2000). The formation of methylmercury within sediments and suspended particulate matter has the potential to increase bioaccumulation across all trophic levels, resulting in biomagnification up the food web. Mercury concentrations in fish are ultimately determined by methylmercury accumulation at the bottom of the food chain, which is governed by water chemistry, primarily pH, and chloride concentration (Mason *et al.*, 1996).

Accumulation of methylmercury in fish is a consequence of the greater trophic transfer efficiency of methylmercury than of inorganic mercury. For example, methylmercury concentrations in phytoplankton accumulate in the cell cytoplasm, and assimilation by zooplankton is four times more efficient than occurs for inorganic mercury, which is bound in cellular membranes (Mason *et al.*, 1996). The toxicity of methylmercury is high because of its increased stability and its affinity for lipid-based compounds, and because its ionic properties lead to an increased ability to penetrate the membranes of living organisms. Because methylmercury is lipid-soluble, it can penetrate the blood–brain barrier. This penetration can affect the central nervous systems of most vertebrates by concentrating in the cerebellum and cerebral cortex, binding tightly to sulfhydryl groups. Developing fetuses are subject to risk exposure because methylmercury can cross the placental barrier (Domagalski, 2001; US EPA, 2000).

9.05.8.3 Assessment of Toxicity

9.05.8.3.1 Predictive models

Models are important tools for the prediction of metal toxicity in aquatic systems. Models relate metal toxicity to site-specific differences in the chemical composition of surface waters. For example, Di Toro *et al.* (2001) and Santore *et al.* (2001) summarized the biotic-ligand model (BLM) approach to account for compositional effects on the acute toxicity of metals to aquatic organisms. The model is based on the premise that mortality occurs when the metal biotic ligand complex reaches a critical concentration. The biotic ligand in fish is either known or suspected to be the calcium- or sodium-channel proteins within the gill surface that facilitate the ionic composition of the blood. The biotic ligand will therefore interact with metal cations in solution. In natural systems, the amount of metal that binds to the gill surface is determined by the competition between the toxic metal and other metals such as calcium. The model, which is a generalization of the free-ion activity model, relates toxicity to the concentration of the divalent metal ions in solution. The difference in this model is the presence of competitive binding at the biotic ligand, which accounts for the protective effects of other cations, and direct influences from pH. The model is applied using the Windemere humic aqueous model (WHAM) (Tipping, 1994) to describe metal complexation to organic matter, in conjunction with normal chemical speciation models such as MINTEQA2 (Allison *et al.*, 1990).

9.05.8.3.2 Biological sensors

Metal contaminants can be monitored using biological sensors. These sensors can be divided into two groups: active and passive. Active monitoring utilizes well-defined species under controlled conditions, whereas passive monitoring refers to direct observation or chemical analysis of indigenous plant and wildlife. Widespread health concerns have led to the adoption and development of a variety of methods for rapid toxicity assessment. These methods include biosensor devices that incorporate biological whole cells on electrode substrates (e.g., cyanobacteria, microalgae, and fish cells) and substrate monitoring (e.g., bivalves, fish parasites, plants, and mosses).

Natural organisms can provide information pertaining to the chemical state within an environment, not through their presence or absence, but through their ability to concentrate heavy metals within tissues. For example, sentinel organisms, which include bivalves, have been used to monitor the concentrations of bioavailable metals and toxicity in aquatic ecosystems (Lau *et al.,* 1998; Hall *et al.*, 2002; Byrne and O'Halloran, 2001). Bivalves have been used to monitor heavy-metal pollutants from gold-mine operations in Sarawak Malaysia (Lau *et al.*, 1998).

Plants and algae species have also been used as biosensors to detect high concentrations of metals from contaminated aquatic ecosystems. Long-term evaluations of zinc and cadmium concentrations using two species of brown algae were conducted in Sepetiba Bay, Rio de Janeiro, Brazil (Amado Filho *et al.*, 1999).

9.05.8.3.3 Molecular tools

Molecular tools can be used for the determination of genotoxicity resulting from contaminated sediments. Molecular tools are based upon the measurement of the relative amount of damage to DNA induced from soils contaminated with metals. Amplified fragment length polymorphism (AFLP) and flow cytometry (FCM) are among the tools applied by molecular biologists to determine changes to DNA.

9.05.9 APPROACHES FOR REMEDIATION AND PREVENTION

Extensive degradation of surface waters by the effluents from abandoned mines and mine wastes has made the treatment of effluents imperative in many mining districts throughout the world. Because the degradation of surface-water resources is obvious, often the remediation of surface waters is the first step in remediation. However, prevention of oxidation in mine wastes and treatment of contaminated groundwater have become areas of active research in mine rehabilitation efforts since the early 1980s. An understanding of the geochemical and hydrological processes that result in the release and transport of the products of sulfide oxidation guides the development and implementation of remedial technologies at many mine sites. Among the various remedial strategies developed for implementation at mine sites are the collection and treatment of contaminated surface water and groundwater, passive treatment of contaminated surface water using constructed wetlands, treatment of contaminated groundwater using permeable reactive barriers or other *in situ* remedial approaches, and the emplacement of covers on tailings impoundments and waste-rock piles to prevent oxygen ingress or infiltration of precipitation. Mine-site remediation may focus on the application of one of these strategies, or more commonly, the integration of a combination of these approaches. The selection of appropriate technologies depends on the physical conditions of the site, the mineralogy of the wastes, and the extent of existing sulfide oxidation.

9.05.9.1 Collection and Treatment

The most common approach to remediate sulfide-bearing mine wastes is collection of the effluent and treatment by pH neutralization and metal precipitation. At many mine sites, water is collected from ponds and ditches and is conveyed to a central treatment facility. The most common approach to water treatment is through pH neutralization using lime, with the subsequent precipitation of metals as a ferric hydroxide sludge. Since the early 1990s, substantial improvements have been made in treatment efficiency, particularly in the development of high-density sludge systems that both remove greater amounts of dissolved metals and produce lower volumes of waste sludges. The wastes derived from these treatment plants may be disposed off in dedicated facilities, or may be co-disposed with mill tailings. Little research has focused on the stability of treatment sludges within tailings impoundments. Zinck (1999) assessed the stability of sludges from wastewater treatment plants from several mine sites in Canada. Although the results suggested that these sludges are stable under the conditions that prevail in many mine-tailings impoundments, the stability of co-disposal waste products from metal refineries seems to differ. For example, Al *et al.* (1996) observed that jarosite residue, derived from the Kidd Creek metallurgical plant at Timmins, Ontario, was unstable in the tailings impoundment at that site. At the Campbell mine at Red Lake, Ontario, As-bearing hematite and maghemite residues derived from the roaster circuit in the Au refinery were mixed with flotation tailings and were co-discharged to the tailings impoundment. The biologically mediated reductive dissolution of the roaster residue was determined to release As to the tailings pore water, which has been subsequently displaced into an underlying aquifer (McCreadie *et al.*, 2000; Stichbury *et al.*, 2000).

Alternative water-treatment technologies have recently been developed and applied for the treatment of mine-site effluents. Biologically mediated systems that reduce sulfate and promote the precipitation of insoluble metal sulfides have been developed for treatment of mine-waste streams. Reverse-osmosis systems have been applied for treatment of mine-waste effluents, or for polishing the effluent from facilities that use lime treatment.

9.05.9.2 Controls on Sulfide Oxidation

9.05.9.2.1 Physical barriers

Figure 14 shows a schematic diagram of a mine-waste impoundment. The generation of acidic drainage results from the exposure of sulfide minerals in the vadose zone of the tailings impoundment to atmospheric O_2. Oxygen penetrates the sulfide-bearing wastes by downward and inward movement. Limiting O_2 transport into mine wastes has the potential to limit the extent of sulfide oxidation and the rate of contaminant release. Various physical barriers have been applied or proposed to prevent oxygen ingress; the most commonly used barriers are covering the wastes with water, which limits oxygen transport because of its low diffusion coefficient, emplacement of covers composed of soil materials designed to maintain high degrees of saturation and low gas-phase diffusivities, and emplacement of synthetic covers, which can be specified to maintain low rates of oxygen transport. In addition to covers that impede oxygen transport, covers composed of oxygen-consuming materials, such as wood wastes, have been applied to prevent oxidation of underlying sulfide minerals.

To be most effective, physical barriers must be applied either at the time that mine-waste deposition ceases or shortly thereafter. The rate of sulfide oxidation is greatest immediately after mine-waste deposition ceases, whereupon the moisture content declines. Measurements of concentrations of dissolved constituents in inactive tailings impoundments, and modeling of the rate of sulfide oxidation in these wastes, indicate that extensive oxidation occurs during the first decade after tailings disposal is complete, and that the rate of sulfide oxidation rapidly declines thereafter as the zone of active oxidation migrates deeper into the impoundment and the path for oxygen diffusion lengthens (Johnson *et al.*, 2000; Mayer *et al.*, 2002). These observations suggest that, for sites where mine-waste oxidation has progressed for more than a decade, the emphasis of remedial programs should shift from preventing sulfide oxidation to managing the fate of dissolved oxidation products.

(i) *Subaqueous disposal*. The solubility of O_2 in water is low, ~8–13 mg L^{-1}, at normal surficial temperatures. Furthermore, the diffusion coefficient of O_2 in water (2×10^{-9} m^2 s^{-1}) is much lower than the diffusion coefficient of O_2 in air (1.78×10^{-5} m^2 s^{-1}). These characteristics have been exploited to limit the rate of sulfide-mineral oxidation by covering mine wastes with water. Subaqueous disposal can be achieved by depositing the wastes in natural water bodies, including lakes or marine systems, or in constructed impoundments designed to maintain the water cover. The extent of sulfide oxidation in tailings submerged in lakes in northern Manitoba was observed to be much less extensive than oxidation in adjacent subaereal tailings impoundments (Pedersen, 1993). The development of reduced conditions in the sediments overlying the tailings resulted in the formation of authigenic sulfide minerals.

The environmental implications of dedicating a lake to mine-waste disposal can be significant. Mining companies proposing subaqueous disposal in natural water bodies face regulatory hurdles, and in some jurisdictions, disposal of mine wastes in natural water bodies is not permitted. In addition, mine sites may be remote from water bodies that have the size or depth to accommodate the mass of mine wastes generated through

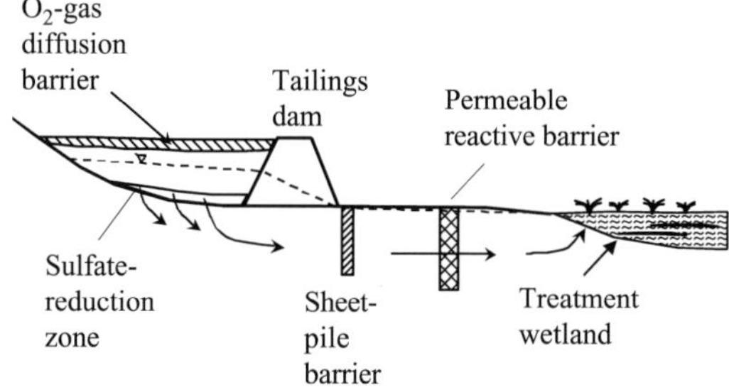

Figure 14 Schematic diagram of a mine-waste impoundment with a combined remediation approach, including a cover to prevent O_2 and water ingress, *in situ* mixing to geochemically stabilize waste, permeable reactive barrier in aquifer to treat subsurface drainage, and wetland for surface treatment of drainage.

mineral exploitation. In these settings, surface repositories with sufficient retention capacity to maintain water covers over wastes have been constructed (Davé and Vivyurka, 1994). To maintain structural stability of the repository, it is desirable to minimize the depth of the free-water cover. Water covers of 1 m or less have been placed on inactive tailings impoundments (Yanful and Verma, 1999; Vigneault *et al.*, 2001). Shallow water covers may be susceptible to resuspension of tailings by wind and wave action (Yanful and Verma, 1999). Furthermore, where sunlight penetrates the water cover, a periphyton layer may develop on the surface of the exposed tailings, producing oxygen at the tailings surface (Vigneault *et al.*, 2001). As a result of these processes, complete prevention of sulfide oxidation may not be attained in repositories with shallow water covers (Yanful and Verma, 1999; Vigneault *et al.*, 2001). For example, Martin *et al.* (2001) observed more extensive oxidation in shallow lakebed sediments than in similar sediments in deeper portions of a lake in Peru. Although, shallow water covers do not completely prevent sulfide oxidation, these covers do substantially lessen the rate of oxidation relative to rates observed under subaereal conditions. Vigneault *et al.* (2001) estimated that tailings submerged beneath a 0.3 m water cover oxidized at a rate ~2,000 times less than the rate observed for moist tailings exposed to oxygen in humidity-cell experiments.

(ii) *Dry covers*. Retaining dams constructed at many mine-waste disposal facilities are composed of coarse-grained mill tailings or mill tailings combined with waste rock. Many of these structures were designed to retain tailings during deposition, and to drain and consolidate after the deposition is complete. Consolidation enhances the stability of the retaining dam, and is a component of the impoundment design. Alternative cover designs may include layers that maintain high moisture contents under negative pressures. These dry covers maintain a near-saturated layer several meters above the water table. Dry covers typically are composed of layers of soil materials with variable grain-size characteristics (Nicholson *et al.*, 1989; Holmström *et al.*, 2001). In their simplest form, these covers can be composed of a single layer of fine-grained material. In 1988, a fine-grained cover was applied to the surface of waste-rock piles at the Rum Jungle mine site in the Northern Territory, Australia (Harries and Ritchie, 1985, 1987). The performance of the Rum Jungle cover has been monitored regularly since installation (Harries and Ritchie, 1985, 1987; Timms and Bennett, 2000). Analysis and modeling of the field data indicate that the cover decreased the rate of oxidation of the underlying sulfide minerals by one-third to one-half of the initial rate, and that recent data indicate that both the oxidation rate and the rate of infiltration have increased (Timms and Bennett, 2000).

More complex covers may be comprised of several layers, each with different soil characteristics. A cover installed at the Bersbo site in Sweden included layers of fine-grained clay, coarse-grained aggregate material, and a protective layer of natural till (Lundgren, 2001).

Dry covers may also be installed with the objective of preventing the infiltration of precipitation into the underlying mine wastes. These covers are designed to retain moisture during wet seasons, and promote evapotranspiration of this moisture during dry periods (Bews *et al.*, 1997; Durham *et al.*, 2000).

(iii) *Synthetic covers*. Covers composed of synthetic materials have been emplaced at mine-waste sites to prevent ingress of both oxygen and water. Synthetic materials used include polyethylene, concrete, and asphalt. A complex cover, including a sealing layer consisting of concrete stabilized with fly-ash, was installed at the Bersbo site in Sweden in 1988. This cover includes a granular stabilization layer, the concrete sealing layer, and a 2-m-thick protective till cover. Instruments were installed through the barrier at the time of implementation, and monitoring since then has shown that, although the cover formed a suitable oxygen barrier, changes in barometric pressure cause migration of O_2 into the waste material (Lundgren, 2001).

To prevent the release of contaminants from an inactive Cu–Zn mine-tailings impoundment near Joutel, Québec, a geomembrane and a protective clay cover overlain by 0.5–1.5 m of till were placed on the impoundment surface (Lewis *et al.*, 2000). The performance of this barrier is the subject of a long-term monitoring program.

(iv) *Oxygen-consuming materials*. Covers containing organic materials have been applied to the surface of mine wastes (Tassé *et al.*, 1994). The principal objective of these covers is to prevent oxygen entry into the mine wastes by intercepting and consuming oxygen in the cover material. Wood wastes and other waste forms of organic carbon have been evaluated to assess their potential as cover materials. Reardon and Poscente (1984) evaluated the potential application of wood wastes as an oxygen-consuming barrier for sulfide-bearing mine tailings. The conclusion was that these materials may provide an effective oxygen barrier, but the life span of the organic carbon cover would be insufficient for this approach to be viable. Tassé *et al.* (1994) described the performance of a wood-waste cover installed at a mine site in Québec. At this site, the cover provided an effective barrier to the ingress of O_2 into the underlying tailings.

Organic materials, such as wood waste or waste materials derived from pulp-and-paper operations, have the potential to leach organic acids, which subsequently move into the underlying tailings. For freshly deposited tailings, in which the oxidation of sulfide minerals has been minimal, the presence of these organic acids is not anticipated to have deleterious environmental effects. If an organic cover is emplaced over weathered sulfide-bearing mine tailings, however, accumulations of ferric-iron-bearing secondary minerals may be reductively dissolved by the organic acids released from the cover materials. Extraction experiments conducted by Ribet *et al.* (1995) indicated that most of the metals present in the shallow portions of the Nickel Rim tailings impoundment, Sudbury, Ontario, were in the oxidized form and would be susceptible to release by reductive dissolution.

9.05.9.2.2 Chemical treatments

A variety of methods has been developed to prevent sulfide-mineral oxidation at the particle scale. Chemical treatments have been proposed to encapsulate sulfide minerals in inert materials, thereby preventing oxidation at the sulfide-mineral surface. Huang and Evangelou (1994) proposed encapsulation with silicate coatings. Silicate binds with ferric iron on the sulfide-mineral surface, rendering the surface inert. Huang and Evangelou (1994) also proposed using phosphate to bind on sulfide surfaces in a similar manner. Laboratory experiments indicated that the rate of oxidation at the sulfide-mineral surface was much less in the presence of phosphate and silicate anions. Silicate coating of sulfide grains was also evaluated by Georgopoulou *et al.* (1995) and by Fytas and Bousquet (2002). All of these studies indicated that coatings formed on the surfaces of sulfide minerals substantially decreased the rate of sulfide oxidation relative to that of pristine sulfide minerals. In addition to chemical treatments that incorporate the addition of anionic reagents, accelerated oxidation of sulfide particles by using strong oxidants, such as permanganate, has been proposed. The objective is to oxidize the surface of the sulfide grain, creating a protective armor layer.

9.05.9.2.3 Bactericides

The rate of abiotic pyrite oxidation declines as the pH decreases below 3.5. Under these conditions, bacterially mediated pyrite oxidation predominates. Numerous researchers have evaluated the use of bactericides, principally anionic surfactants, to prevent bacterial activity and limit the rate of sulfide oxidation. Bactericides can be applied either directly to mine-waste surfaces or as intimate mixtures with the mine wastes (Ericson and Ladwig, 1985). Preliminary results using this approach were encouraging (Sobek, 1987). Slow release of surfactant from rubber-based pellets have been developed to extend the life span of these surfactants in field applications. Stichbury *et al.* (1995) proposed the use of thio-blocking agents to inhibit bacteria involved in the early stages of sulfide oxidation. In shake-flask cultures and in column experiments, Stichbury *et al.* (1995) observed reduced sulfide oxidation that persisted over a one-year study. In a one-year field study, the inhibitor mixture completely prevented the growth of *A. ferrooxidans* and lessened the growth of *A. thiooxidans* and *T. thioparus* (Lortie *et al.*, 1999). Lower concentrations of sulfate, iron, and dissolved metals were observed in the treated field plots than were observed in untreated control plots. The effectiveness of the inhibitor was observed to decline following the first year after application. This observation suggests that inhibitors may be best applied in settings where prevention of sulfide oxidation is required for limited periods only. Examples are stockpiling of waste rock before it is placed in underground workings, stockpiles of mill tailings that will be subsequently used for mine backfill, or waste-rock and tailings piles during periods of temporary closure.

9.05.9.3 Passive Remediation Techniques

9.05.9.3.1 Constructed wetlands

Constructed wetlands have been used for water treatment since the early 1950s (Hedin and Nairn, 1994). The interest in using constructed wetlands for the remediation of mine drainage stems from their ability to filter particulate matter, thus reducing suspended solids, their ability to reduce biological oxygen demand (BOD) and remove and store nutrients and heavy metals, and their natural buffering capacity (Machemer and Wildeman, 1992; Perry and Kleinmann, 1991; Greenway and Simpson, 1996; Hsu and Maynard, 1999). However, wetland systems can loose their effectiveness if the influent pH is too low (Hsu and Maynard, 1999). If properly designed and maintained, constructed wetlands can result in the improvement of the quality of drainage water derived from mines (Walton-Day, 1999). The mechanisms that promote acid neutralization and metal retention include dissolution of primary carbonate minerals, formation and precipitation of metal hydroxides, microbially mediated sulfate reduction, the formation of metals sulfides, complexation of metals by organic carbon, ion-exchange, and direct uptake by plants (Bendell-Young and Pick, 1996). Wetlands can be designed to function under aerobic or anaerobic conditions. Aerobic wetlands consist of *Typha* and other wetland vegetation, planted in shallow relatively impermeable sediments comprised of soil, mine spoil, or clay. The vegetation in anaerobic wetlands is planted in deep (>30 cm) permeable sediments consisting of peat, soil, organically enriched compost, sawdust, etc., which are sequenced between crushed limestone or are mixed (http://www.wvu.edu/~agexten/landrec/passtr/passtrt.htm).

Anaerobic constructed wetlands promote bacterially mediated sulfate reduction. These reactions have the potential to remove metals, as well as high concentrations of sulfate, from mine drainage. A growing concern associated with the use of wetlands to attenuate metals is the potential for the wetland soils to become latent sinks of toxic metals (Gopal, 1999). Heal and Salt (1999) monitored a constructed-wetland system impacted by acidic metal-rich drainage from ironstone mine-spoils. Mine waters (pH 2.7, 247 mg L^{-1} total iron) passing through the wetland were monitored for 12 months. After installation, acidity declined by 33%, and iron, manganese, and aluminum concentrations declined by 20–40%. The performance of the constructed wetland showed favorable response to metal loads during the summer months but the rates of metal removal decreased significantly in the winter months (Heal and Salt, 1999). Sobolewski (1996) monitored the performance of a constructed peat-based system for long-term treatment of mine drainage at the former Bell Copper mine near Smithers, BC, Canada. Copper concentrations were used to provide estimates of the long-term potential for removal of copper from the mine-drainage site. Two copper species were identified: copper sulfides and copper bound to organic material. Removal from mine effluent containing low concentrations of copper at near-neutral pH was through sulfide formation. Removal from acidic drainage containing higher copper concentrations was by complexation to organic matter. The effectiveness of the constructed-wetland system to remove metals under low-pH conditions and with high metal concentrations was insufficient to be suitable for long-term remediation.

The definition of appropriate operating conditions is required to assure success of wetland treatment systems. Acidic drainage containing high concentrations of dissolved metals discharged from abandoned underground mines in Kentucky, USA, was intercepted by a wetland constructed in 1989 to reduce the concentration of metals (Barton and Karathanasis, 1998). The initial attempt to treat the drainage water using the wetland failed after six months because of insufficient utilization of the treatment area, inadequate alkalinity production, and high metal loadings. A renovation added two anoxic limestone drains and a series of anaerobic subsurface compartments to promote vertical flow through a successive alkalinity-producing system. The renovated system, monitored over 19 months, showed improved performance in metal removal. Iron concentrations decreased from 787 mg L^{-1} to 39 mg L^{-1}, the pH increased from 3.38 to 6.46, and the acidity decreased from 2,244 mg L^{-1} to 199 mg L^{-1} (carbonate equivalent) (Barton and Karathanasis, 1998). The mass loading of metals was reduced by an average of 98% for aluminum, 95% for iron, 55% for sulfate, and 49% for manganese.

Improving the efficiency of metals removal using wetlands has proved to be difficult. Detailed assessments of long-term performance of wetland systems with a focus on metal-retention mechanisms, eutrophication, and flushing rates, and their effects on downgradient ecosystems are required if wetland treatment is to be applied in temperate regions.

9.05.9.3.2 Permeable reactive barriers

Permeable reactive barriers are used for treatment and prevention of mine drainage waters within aquifers receiving discharge from mine wastes. These systems are constructed by excavating a portion of the aquifer downgradient from the waste disposal area, and filling the excavation with a permeable mixture composed of reactive components (Blowes *et al.*, 2000). Among the mixtures designed for treating mine-drainage waters have been organic carbon in the form of composted municipal wastes, wood wastes, and byproducts from pulp-and-paper manufacturing (Blowes *et al.*, 1994; Benner *et al.*, 1999), zero-valent iron (Blowes *et al.*, 2000; Naftz *et al.*, 2002; Morrison *et al.*, 2002), limestone, and phosphate-based adsorbent materials (Conca *et al.*, 2002).

Reactive mixtures containing organic carbon are designed to support the bacterially mediated reduction of sulfate and the precipitation of metal sulfides (Blowes *et al.*, 1994; Waybrant *et al.*, 1998, 2002; Cocos *et al.*, 2002). Sulfate-reducing bacteria oxidize organic carbon by using SO_4 as the electron acceptor, thereby generating H_2S and releasing dissolved inorganic carbon:

$$2(CH_2O)_x(NH_3)_y(H_3PO_4)_z + xSO_4^{2-} \Rightarrow 2xHCO_3^- + xH_2S + 2yNH_3 + 2zH_3PO_4 \quad (30)$$

where "$(CH_2O)_x(NH_3)_y(H_3PO_4)_z$" represents organic matter undergoing oxidation, and x, y, z are stoichiometric coefficients. The H_2S produced through sulfate reduction combines with metal cations to form metal sulfides:

$$Me^{2+} + S^{2-} \Rightarrow MeS \quad (31)$$

These metal sulfides are stable below the water table within the reactive barrier. Sulfate reduction also releases dissolved inorganic carbon, which neutralizes the pH and favors the precipitation of metal carbonate minerals, e.g., $FeCO_3$ and $MnCO_3$ (Waybrant *et al.*, 2002).

The first full-scale reactive barrier for treatment of mine drainage was installed at the Nickel Rim mine site, near Sudbury, Ontario in 1995 (Benner *et al.*, 1997). The barrier contains a reactive mixture composed of 50 vol.% organic carbon, 49 vol.% pea gravel, and 1 vol.% limestone. The organic carbon used was a composted waste of leaf mulch and wood chips derived from a municipal recycling program. Semi-annual monitoring of the reactive barrier for five years after installation showed consistent removal of dissolved sulfate and dissolved metals, and increases in pH and alkalinity (Benner *et al.*, 1999). Enumeration of sulfate-reducing bacteria within the barrier indicated that a large population of sulfate-reducing bacteria had become established within a year following installation (Benner *et al.*, 2000). Measurements of the sulfur isotopic ratio indicated enrichment of $^{34}S-SO_4$ in the barrier effluent, which is consistent with the hypothesis that the sulfate removal was through bacterially mediated sulfate reduction. Measurements of the solid-phase sulfide content indicated accumulations of acid-volatile sulfides within the barrier material, and lesser accumulations of total reducible sulfide (Herbert *et al.*, 2000).

The aqueous geochemistry data obtained from the Nickel Rim site indicates that the rate of sulfate reduction is dependent on temperature, with sulfate reduction more rapid during the warm summer months than during the cooler winter months. The relationship between temperature and reaction rate could be explained using an Arrhenius-type relationship (Benner *et al.*, 2002).

A pilot-scale reactive barrier for treating metals and sulfate was installed in 1997 at an industrial site in Vancouver, Canada (Ludwig *et al.*, 2002; McGregor *et al.*, 2002). Monitoring of the barrier between 1997 and 2001 showed that initial concentrations of up to 8,570 μg L^{-1} copper were decreased to 17 μg L^{-1} after passage through the barrier, zinc decreased from 2,780 μg L^{-1} to <140 μg L^{-1} and cadmium and nickel declined from 21 μg L^{-1} and 210 μg L^{-1} to concentrations <0.2 μg L^{-1} and <6 μg L^{-1}, respectively (McGregor *et al.*, 2002). The performance of the pilot-scale barrier led to the installation, in 2002, of a full-scale reactive barrier that is 400 m long and 17 m deep, with a maximum thickness of 5 m.

Permeable reactive barriers have also been installed at uranium mine-tailings impoundments to remove uranium and other metals. A reactive barrier containing three reactive components, as part of a side-by-side comparison, was installed at the Fry Canyon mine site, Utah. The reactor systems included hydrous ferric oxide for the adsorption of uranium, zero-valent iron to induce uranium reduction and precipitation, and bone char, which is rich in phosphate to induce the precipitation of uranium-bearing phosphates. Monitoring showed removal of uranium in all of these systems (Blowes *et al.*, 2000; Fuller *et al.*, 2002; Naftz *et al.*, 2002). A full-scale permeable barrier using zero-valent iron for reduction and precipitation of uranium was installed in 1999 in Monticello, Utah (Morrison *et al.*, 2002). This barrier treats uranium, molybdenum, selenium, and vanadium.

Permeable reactive barriers have significant advantages over the conventional approaches for groundwater remediation. The barriers are passive systems whose effectiveness can persist for several years to decades. Moreover, the reactive material can be adjusted to target specific contaminants from a mine site and the contaminants precipitated within the barrier are isolated from the surface-water environment and biota (Blowes *et al.*, 2000).

9.05.10 SUMMARY AND CONCLUSIONS

The exposure of sulfide minerals contained in mine wastes to atmospheric oxygen results in the oxidation of these minerals. The oxidation reactions are accelerated by the catalytic effects of iron hydrolysis and sulfide-oxidizing bacteria. The oxidation of sulfide minerals results in the depletion of minerals in the mine waste, and the release of H^+, SO_4, Fe(II), and other metals to the water flowing through the wastes. The most abundant solid-phase products of the reactions are typically ferric oxyhydroxide or hydroxysulfate minerals. Other secondary metal sulfate, hydroxide, hydroxy sulfate, carbonate, arsenate, and phosphate precipitates also form. These secondary phases limit the concentrations of dissolved metals released from mine wastes.

At many sites the acid produced by sulfide oxidation is consumed by the pH-buffering reactions of the nonsulfide gangue minerals contained in the wastes. As these acid-neutralization reactions proceed, the pH of the pore water progressively increases, enhancing the potential for attenuation of dissolved metals by adsorption and precipitation reactions. The quality of aqueous effluents from tailings impoundments and waste-rock piles is dependent on the extent of the acid-producing sulfide-oxidation reactions versus the acid-consuming pH-buffering reactions. Regardless of the degree of acid neutralization, the effluent water from the waste material contains increased concentrations of dissolved constituents.

Oxidation products transported from the mine waste can enter streams, lakes, and oceans. A series of further reactions occurs upon the discharge of this mine-drainage water, resulting in acidification and metal release. Complex geochemical models and reactive solute-transport models have been developed and are important tools in the prediction of the environmental impacts of mine wastes and in developing remedial alternatives. The remedial technologies developed and refined since the 1980s are being applied increasingly, thereby decreasing the magnitude of the negative effects of mining operations.

REFERENCES

AGI (1957) *Glossary of Geology and Related Sciences*. American Geological Institute, Washington DC.

Ahmann D., Krumholz L. R., Lovley D. R., and Morel F. M. M. (1997) Microbial mobilization of arsenic from sediments of the Aberjona watershed. *Environ. Sci. Technol.* **31**, 2923–2930.

Al T. A. (1996) The hydrology and geochemistry of thickened, sulfide-rich tailings, Kidd Creek Mine, Timmins, Ontario. PhD Thesis, University of Waterloo.

Al T. A. and Blowes D. W. (1996) Storm-water hydrograph separation of run off from a mine-tailings impoundment formed by thickened tailings discharge at Kidd Creek, Timmins, Ontario. *J. Hydrol.* **180**, 55–78.

Al T. A., Martin C. J., and Blowes D. W. (2000) Carbonate-mineral/water interactions in sulfide-rich mine tailings. *Geochim. Cosmochim. Acta* **64**, 3933–3948.

Albers P. H. and Camardese M. B. (1993) Effects of acidification on metal accumulation by aquatic plants and invertebrates: 1. Constructed wetlands. *Environ. Toxicol. Chem.* **12**, 959–967.

Alpers C. N. and Nordstrom D. K. (1999) Geochemical modeling of water–rock interactions in mining environments. In *The Environmental Geochemistry of Mineral Deposits* (eds. G. S. Plumlee and M. J. Logsdon). Rev. Econ. Geol. **6A**, Society of Economic Geologists Inc. Littleton, co, pp. 289–323.

Alpers C. N., Blowes D. W., Nordstrom D. K., and Jambor J. L. (1994) Secondary minerals and acid mine water chemistry. In *The Environmental Geochemistry of Sulfide Mine-Wastes* (eds. J. L. Jambor and D. W. Blowes). Mineralogical Association of Canada, Nepean, ON, vol. **22**, pp. 247–270.

Allison J. D., Brown D. S., and Novo-Gradac K. J. (1990) *MINTEQA2/PRODEFA2, A Geochemical Assessment Model for Environmental Systems: Version 3.0 User's Manual*. US Environmental Protection Agency Report EPA/600/3-91/021.

Amado Filho G. M., Andrade L. R., Karez C. S., Farina M., and Pfeifferx W. C. (1999) Brown algae species as biomonitors of Zn and Cd at Sepetiba bay, Rio de Janeiro, Brazil. *Mar. Environ. Res.* **48**, 213–224.

Ashley P. M. and Lottermoser B. G. (1999) Arsenic contamination at the Mole River mine, northern New South Wales. *Austral. J. Earth Sci.* **46**, 861–874.

ASTM (2001) Standard test method for accelerated weathering of solid materials using a modified humidity cell. Method D5744–96.

Baes C. F., Jr., Reardon E. J., and Moyer B. A. (1993) Ion interaction model applied to the $CuSO_4$–H_2SO_4–H_2O system at 25°C. *J. Phys. Chem.* **97**, 12343–12348.

Bain J. G., Blowes D. W., Robertson W. D., and Frind E. O. (2001) Modelling of sulfide oxidation with reactive transport at a mine drainage site. *J. Contamin. Hydrol.* **41**, 23–47.

Bain J. G., Mayer K. U., Blowes D. W., Frind E. O., Molson J. W. H., Kahnt R., and Jenk U. (2001) Modelling the closure-related geochemical evolution of groundwater at a former uranium mine. *J. Contamin. Hydrol.* **52**, 109–135.

Ball J. W. and Nordstrom D. K. (1987) *WATEQ4F—a Personal Computer Fortran Translation of the Geochemical Model WATEQ2 with Revised Data Base*. US Geol. Surv. Open-File Report, 50–87.

Banfield J. F. and Welch S. A. (2000) Microbial controls on mineralogy of the environment. In *Environmental Mineralogy*, European Mineralogical Union (eds. D. J. Vaughan and R. A. Wogelius). Budapest University Press, Budapest, vol. 2, pp. 173–196.

Barnett M. O., Harris L. A., Turner R. R., Stevenson R. J., Henson T. J., Melton R. C., and Hoffman D. P. (1997) Formation of mercuric sulfide in soil. *Environ. Sci. Technol.* **31**, 3037–3043.

Barnett M. O., Turner R. R., and Singer P. C. (2001) Oxidative dissolution of metacinnabar (β-HgS) by dissolved oxygen. *Appl. Geochem.* **16**, 1499–1512.

Barron J. L. and Lueking D. R. (1990) Growth and maintenance of *Thiobacillus ferrooxidans* cells. *Appl. Environ. Microbiol.* **56**, 2801–2806.

Baron D. and Palmer C. D. (1996) Solubility of jarosite at 4–35 degrees. *C. Geochim Cosmochim. Acta* **60**, 185–195.

Barton C. D. and Karathanasis A. D. (1998) Renovation of a failed constructed wetland treating acid mine drainage. *Environ. Geol.* **39**, 39–50.

Basilio C. I., Kartio I. J., and Yoon R.-H. (1996) Lead activation of sphalerite during galena flotation. *Min. Eng.* **9**(8), 869–879.

Bendell-Young L. and Pick F. R. (1996) Base cation composition of pore water, peat and pool water of fifteen Ontario peatlands: implications for peatland acidification. *Water Air Soil Pollut.* **96**, 155–173.

Benner S. G., Blowes D. W., and Ptacek C. J. (1997) A full-scale porous reactive wall for prevention of acid mine drainage. In *Ground Water Monitoring and Remediation, Fall*, vol. 17(4), pp. 99–107.

Benner S. G., Blowes D. W., Gould W. D., Herbert R. B., Jr., and Ptacek C. J. (1999) Geochemistry of a permeable reactive barrier for metals and acid mine drainage. *Environ. Sci. Technol.* **33**, 2793–2799.

Benner S. G., Gould W. D., and Blowes D. W. (2000) Microbial populations associated with the generation and treatment of acid mine drainage. *Chem. Geol.* **169**, 435–448.

Benner S. G., Blowes D. W., Ptacek C. J., and Mayer K. U. (2002) Rates of sulfate removal and metal sulfide precipitation in a permeable reactive barrier. *Appl. Geochem.* **17**, 301–320.

Bennett J. W., Comarmond M. J., Clark N. R., Carras J. N., and Day S. (1999) Intrinsic oxidation rates of coal reject measured in the laboratory. *Proc. Sudbury '99 Mining Environ.* **1**, 9–17.

Benoit J. M., Gilmour C. C., Mason R. P., and Heyes A. (1999) Sulfide controls on mercury speciation and bioavailability to methylating bacteria in sediment pore waters. *Environ. Sci. Technol.* **33**, 951–957.

Benvenuti M., Mascaro I., Corsini F., Lattanzi P., Parrini P., and Tanelli G. (1997) Mine waste dumps and heavy metal pollution in abandoned mining district of Boccheggiano (southern Tuscany, Italy). *Environ. Geol.* **30**, 238–243.

Benvenuti M., Mascaro I., Corsini F., Ferrari M., Lattanzi P., Parrini P., Costagliola P., and Tanelli G. (2000) Environmental mineralogy and geochemistry of waste dumps at the Pb(Zn)–Ag Bottino mine, Apuane Alps, Italy. *Euro. J. Min.* **12**, 441–453.

Berger A. C., Bethke C. M., and Krumhansl J. L. (2000) A process model of natural attenuation in drainage from a historic mining district. *Appl. Geochem.* **15**, 655–666.

Bews B. E., O'Kane M. A., Wilson G. W., Williams D., and Currey N. (1997) The design of a low flux cover system, including lysimeters, for acid generating waste rock in semi-arid environments. In *Proc. 4th Int. Conf. Acid Rock Drainage*. MEND, Natural Resources Canada, Ottawa, ON, vol. **2**, pp. 747–762.

Biehler D. and Falck W. E. (1999) Simulation of the effects of geochemical reactions on groundwater quality during planned flooding of the Königstein uranium mine, Saxony, Germany. *Hydrogeol. J.* **7**, 284–293.

Bigham J. M. and Nordstrom D. K. (2000) Iron and aluminum hydroxysulfates from acid sulfate waters. In *Sulfate Minerals—Crystallography, Geochemistry, and Environmental Significance*, Rev. Min. Geochem. (eds. C. N. Alpers, J. L. Jambor and D. K. Nordstrom). Mineralogical Society of America, Washington DC, vol. **40**, pp. 352–403.

Birk D. (1989) Quantitative coal mineralogy of the Sydney Coalfield, Nova Scotia, Canada, by scanning electron microscopy, computerized image analysis, and energy-dispersive X-ray spectrometry. *Can. J. Earth Sci.* **27**, 163–179.

Blodau C. and Peiffer S. (2003) Thermodynamics and organic matter: constraints on neutralization processes in sediments of highly acidic waters. *Appl. Geochem.* **18**, 25–36.

Blowes D. W. (1990) The geochemistry, hydrogeology and mineralogy of decommissioned sulfide tailings. A Comparative Study. PhD Thesis, University of Waterloo.

Blowes D. W. and Gillham R. W. (1988) The generation and quality of streamflow on inactive uranium tailings near Elliot Lake, Ontario. *J. Hydrol.* **97**, 1–22.

Blowes D. W. and Jambor J. L. (1990) The pore-water geochemistry and the mineralogy of the vadose zone of sulfide tailings, Waite Amulet, Québec, Canada. *Appl. Geochem.* **5**, 327–346.

Blowes D. W. and Ptacek C. J. (1994) Acid-neutralization mechanisms in inactive mine tailings. In *The Environmental Geochemistry of Sulfide Mine-Wastes* (eds. J. L. Jambor and D. W. Blowes). Mineralogical Association of Canada, Nepean, ON, vol. 22, pp. 271–292.

Blowes D. W., Reardon E. J., Cherry J. A., and Jambor J. L. (1991) The formation and potential importance of cemented layers in inactive sulfide mine tailings. *Geochim. Cosmochim. Acta* **55**, 965–978.

Blowes D. W., Jambor J. L., Appleyard E. C., Reardon E. J., and Cherry J. A. (1992) Temporal observations of the geochemistry and mineralogy of a sulfide-rich mine-tailings impoundment, Heath Steele mines, New Brunswick. *Explor. Mining Geol.* **1**, 251–264.

Blowes D. W., Ptacek C. J., and Jambor J. L. (1994) Remediation and prevention of low-quality drainage from mine wastes. In *The Environmental Geochemistry of Sulfide Mine-Wastes* (eds. D. W. Blowes and J. L. Jambor). Mineralogical Association of Canada, Nepean, ON, vol. 22, pp. 365–380.

Blowes D. W., Lortie L., Gould W. D., Jambor J. L., and Hanton-Fong C. J. (1995) Geochemical, mineralogical and microbiological characterization of a sulfide-bearing, carbonate-rich, gold mine tailings impoundment, Joutel, Québec. *Appl. Geochem.* **31**, 687–705.

Blowes D. W., Ptacek C. J., Benner S. G., McRae C. W. T., and Puls R. W. (2000) Treatment of dissolved metals and nutrients using permeable reactive barriers. *J. Contamin. Hydrol.* **45**, 120–135.

Bobeck G. E. and Su H. (1985) The kinetics of dissolution of sphalerite in ferric chloride solution. *Met. Trans. B* **16b**, 413–424.

Boon M. and Heijnen J. J. (2001) Solid–liquid mass transfer limitation of ferrous iron in the chemical oxidation of FeS_2 at high redox potential. *Hydrometallurgy* **62**, 57–66.

Boorman R. S. and Watson D. M. (1976) Chemical processes in abandoned sulfide tailings dumps and environmental implications for northeastern New Brunswick. *CIM Bull.* **69**(772), 86–96.

Borgmann U., Norwood W. P., Reynoldson T. B., and Rosa F. (2001) Identifying cause in sediment assessments: bioavailability and sediment quality triad. *Can. J. Fish Aquat. Sci.* **58**, 950–960.

Boulet M. P. and Larocque A. C. L. (1996) A comparative mineralogical and geochemical study of sulfide mine tailings at two sites in New Mexico, USA. *Environ. Geol.* **33**, 130–142.

Bove D. J., Mast M. A., Wright W. G., Verplanck P. L., Meeker G. P., and Yager D. B. (2000) Geologic control on acidic and metal-rich waters in the southeast Red Mountains area, near Silverton, Colorado. In *Proc. 5th Int. Conf. Acid Rock Drainage*. Society for Mining, Metallurgy, and Exploration, vol. 1, pp. 523–533.

Bowell R. J. and Bruce I. (1995) Geochemistry of iron ochres and mine waters from the Levant mine, Cornwall. *Appl. Geochem.* **10**, 237–250.

Brady K. B. C., Hornberger R. J., and Fleeger G. (1998) Influence of geology on postmining water quality: northern Appalachian basin. In *Coal Mine Drainage Prediction and Pollution Prevention in Pennsylvania* (eds. K. B. C. Brady, M. W. Smith, and J. Schueck). The Pennsylvania Department of Environmental Protection, Bureau of Mining and Reclamation, Harrisburg, PA, pp. 8-1–8-92.

Brady B. C., Rose A. W., and Cravotta C. A., III (1997) Bimodal distribution of pH coal-mine drainage (abstr.).

Brake S. S., Connors K. A., and Romberger S. B. (2001) A river runs through it: impact of acid mine drainage on the geochemistry of West Little Sugar Creek pre- and post-reclamation at the Green Valley coal mine, Indiana, USA. *Environ. Geol.* **40**, 1471–1481.

Breed A. W. and Hansford G. S. (1999) Effects of pH on ferrous-iron oxidation kinetics of *Leptospirillium ferrooxidans* in continuous culture. *Biochem. Eng. J.* **3**, 193–201.

Brown A. D. and Jurinak J. J. (1989) Pyrite oxidation in aqueous mixtures. *J. Environ. Qual.* **18**, 545–550.

Brown L. J., Gutsa E. H., Cashion J. D., Fraser J. R., and Coller B. A. W. (1990) Mössbauer effect study of ferric arsenates from an abandoned gold mine. *Hyperfine Interact.* **57**, 2159–2166.

Brown J. G., Bassett R. L., and Glynn P. D. (1998) Analysis and simulation of reactive transport of metal contaminants in ground water in Pinal Creek Basin, Arizona. In *Special Issue—Reactive Transport Modeling of Natural Systems* (eds. C. I. Steefel and P. van Cappellen). *J. Hydrol.* **209**, 225–250.

Brown D. A., Sherriff B. L., Sawicki J. A., and Sparling R. (1999a) Precipitation of iron minerals by a natural microbial consortium. *Geochim. Cosmochim. Acta* **63**, 2163–2169.

Brown G. E., Jr., Henrich V. E., Casey W. H., Clark D. L., *et al.* (1999b) Metal oxide surfaces and their interactions with aqueous solutions and microbial organisms. *Chem. Rev.* **99**, 77–174.

Buckley A. N. and Walker W. (1988) The surface composition of arsenopyrite exposed to oxidizing environments. *Appl. Surf. Sci.* **35**, 227–240.

Buckley A. N. and Woods R. W. (1984a) An X-ray photoelectron spectroscopic study of the oxidation of galena. *Appl. Surf. Sci.* **17**, 401–414.

Buckley A. N. and Woods R. W. (1984b) An X-ray photoelectron spectroscopic study of the oxidation of chalcopyrite. *Austral. J. Chem.* **37**, 2403–2413.

Buckley A. N. and Woods R. W. (1985) X-ray photoelectron spectroscopy of oxidized pyrrhotite surfaces. *Appl. Surf. Sci.* **22/23**, 280–287.

Buckley A. N. and Woods R. W. (1987) The surface oxidation of pyrite. *Appl. Surf. Sci.* **27**, 437–452.

Buckley A. N., Wouterlood H. J., and Woods R. (1989) The surface composition of natural sphalerites under oxidative leaching conditions. *Hydrometallurgy* **22**, 39–56.

Byrne P. A. and O'Halloran J. (2001) The role of bivalve molluscs as tools in estuarine sediment toxicity testing: a review. *Hydrobiologia* **465**, 209–217.

Cabri L. J. (1981) Relationship of mineralogy to the recovery of platinum-group elements from ores. In *Platinum-Group Elements: Mineralogy, Geology, Recovery* (ed. L. J. Cabri). Canadian Institute of Mining, Metallurgy and Petroleum, Montreal, Québec, pp. 233–250.

Cabri L. J. (ed.) (2002) *The Geology Geochemistry, Mineralogy, and Beneficiation of Platinum-group Elements*. Canadian Institute of Mining, Metallurgy and Petroleum.

Carlson L., Bigham J. M., Schwertmann U., Kyek A., and Wagner F. (2002) Scavenging of As from acid mine drainage by schwertmannite and ferrihydrite: a comparison with synthetic analogues. *Environ. Sci. Technol.* **26**, 1712–1719.

Casey W. H. and Sposito G. (1991) On the temperature dependence of mineral dissolution rates. *Geochim. Cosmochim. Acta* **56**, 3825–3830.

Cathles L. M. (1979) Predictive capabilities of a finite difference model of copper leaching in low grade industrial sulfide waste dumps. *Math. Geol.* **11**, 175–191.

Cathles L. M. (1994) Attempts to model the industrial-scale leaching of copper-bearing mine waste. In *Environmental Geochemistry of Sulfide Oxidation* (eds. C. N. Alpers and D. W. Blowes). American Chemical Society, vol. 550, pp. 123–131.

Chernyshova I. V. and Andreev S. I. (1997) Spectroscopic study of galena surface oxidation in aqueous solutions. *Appl. Surf. Sci.* **108**, 225–236.

Clarke L. B. (1993) The fate of trace elements during combustion and gasification: an overview. *Fuel* **72**, 731–736.

Clegg S. L. and Whitfield M. (1991) Activity coefficients in natural waters. In *Activity Coefficients in Electrolyte Solutions* (ed. K. S. Pitzer). 2nd edn. CRC Press, Boca Raton, FL, pp. 279–434.

Cocos I. A., Zagury G. J., Clément B., and Samson R. (2002) Multiple factor design for reactive mixture selection for use in reactive walls in mine drainage treatment. *Water Res.* **32**, 167–177.

Coggans C. J. (1992) Hydrogeology and geochemistry of the Inco Ltd. Copper Cliff, Ontario, mine tailings impoundments. MSc Thesis, University of Waterloo.

Coggans C. J., Blowes D. W., Robertson W. D., and Jambor J. L. (1999) The hydrogeochemistry of a nickel-mine tailings impoundment, Copper Cliff, Ontario. *Rev. Econ. Geol.* **6B**, 447–465.

Conca J., Strietelmeier E., Lu N., Ware S. D., Taylor T. P., Kaszuba J., and Wright J. (2002) Treatability study of reactive materials to remediate groundwater contaminated with radionuclides, metals, and nitrates in a four-component permeable reactive barrier. In *Handbook of Groundwater Remediation Using Permeable Reactive Barriers—Applications to Radionuclides, Trace Metals, and Nutrients* (eds. D. L. Naftz, S. J. Morrison, J. A. Davis, and C. C. Fuller). Academic Press, San Diego, CA, pp. 221–252.

Cravotta C. A., III (1994) Secondary iron sulfate minerals as sources of sulfate and acidity-the geochemical evolution of acidic ground water at a reclaimed surface coal mine in Pennsylvania. In *Environmental Geochemsitry of Sulfide Oxidation* (eds. C. N. Alpers and D. W. Blowes). American Chemical Society, Washington, DC, vol. 550, pp. 345–364.

Crundwell F. K. (1988) Effect of iron impurity in zinc sulfide concentrates on the rate of dissolution. *Am. Inst. Chem. Eng. J.* **34**, 1128–1134.

Cummings D. E., March A. W., Bostick B., Spring S., Caccavo F., Jr., Fendorf S., and Rosenzweig R. F. (2000) Evidence for microbial Fe(III) reduction in anoxic, mining-impacted lake sediments (Lake Coeur d'Alene, Idaho). *Appl. Environ. Microbiol.* **66**, 154–162.

Dalzell D. J. B. and Macfarlane N. A. A. (1999) The toxicity of iron to brown trout and effects on the gills: a comparison of two grades of iron sulfate. *J. Fish Biol.* **55**, 301–315.

Daskalakis K. D. and Helz G. R. (1993) The solubility of sphalerite (ZnS) in sulfidic solutions at 25 °C and 1 atm pressure. *Geochim. Cosmochim. Acta* **57**, 4923–4931.

Davé N. K. and Vivyurka A. J. (1994) Water cover on acid generating uranium tailings—laboratory and field studies. In *Proceedings of the International Land Reclamation and Mine Drainage Conference and the Third International Conference on the Abatement of Acidic Drainage*, US Bureau of Mines, SP064A-94, vol. 1, pp. 297–306.

Davis B. S. (1997) Geomicrobiology of the oxic zone of sulfide mine tailings. In *Biological—Mineralogical Interactions* (eds. J. M. McIntosh and L. A. Groat). Mineralogical Association of Canada, vol. 25, pp. 93–112.

Davis A. and Ashenberg D. (1989) The aqueous geochemistry of the Berkeley Pit, Butte, Montana, USA. *Appl. Geochem.* **4**, 23–36.

Di Toro D. M., Allen H. E., Bergman H. L., Meyer J. S., Paquin P. R., and Santore R. C. (2001) Biotic ligand model of acute toxicity of metals: 1. Technical basis. *Envron. Toxicol. Chem.* **20**, 2383–2396.

DIAND (1992) *Guidelines for Acid Rock Prediction in the North.* Canada Department of Indian Affairs and Northern Development.

Dill H. G., Pöllmann H., Bosecker K., Hahn L., and Mwiya S. (2002) Supergene mineralization in mining residues of the Matchless cupreous pyrite deposit (Namibia)— a clue to the origin of modern and fossil duricrusts in semiarid climates. *J. Geochem. Explor.* **75**, 43–70.

Ding M., DeJong B. H. W. S., Roosendaal S. J., and Vredenberg A. (2000) XPS studies on the electronic structure of bonding between solid and solutes: adsorption of arsenate, chromate, phosphate, Pb^{2+}, and Zn^{2+} ions on amorphous black ferric oxyhydroxide. *Geochim. Cosmochim. Acta* **64**, 1209–1219.

Dinelli E., Morandi N., and Tateo F. (1998) Fine-grained weathering products in waste disposal from two sulphide mines in the northern Apennines, Italy. *Clay Min.* **33**, 423–433.

Domagalski J. (2001) Mercury and methylmercury in water and sediment of the Sacramento river basin, California. *Appl. Geochem.* **16**, 1677–1691.

Dubrovsky N. M. (1986) Geochemical evolution of inactive pyritic tailings in the Elliot Lake uranium district. PhD Thesis, University of Waterloo.

Dubrovsky N. M., Morin K. A., Cherry J. A., and Smyth D. J. A. (1984) Uranium tailings acidification and subsurface contaminant migration in a sand aquifer. *Water Pollut. Res. J. Can.* **19**, 55–89.

Durham A. J. P., Wilson G. W., and Currey N. (2000) Field performance of two low infiltration cover systems in semi arid environment. In *Proc. 5th Int. Conf. on Acid Rock Drainage*. Society for Mining, Metallurgy, and Exploration, vol. 2, pp. 1319–1326.

Dushenko W. T., Bright D. A., and Reimer K. J. (1995) Arsenic bioaccumulation and toxicity in aquatic macrophytes exposed to gold-mine effluent: relationships with environmental partitioning, metal uptake and nutrients. *Aquat. Bot.* **50**, 141–158.

Dutrizac J. E. and Jambor J. L. (2000) Jarosites and their application in hydrometallurgy. In *Sulfate Minerals—Crystallography, Geochemistry, and Environmental Significance* (eds. C. N. Alpers, J. L. Jambor, and D. K. Nordstrom). *Rev. Min. Geochem.* Mineralogical Society of America, vol. 40, pp. 406–452.

Dutrizac J. E. and MacDonald R. J. C. (1973) The effect of some impurities on the rate of chalcopyrite dissolution. *Can. Metall. Quart.* **12**(4), 409–420.

Eary L. E. (1999) Geochemical and equilibrium trends in mine pit lakes. *Appl. Geochem.* **14**, 963–987.

Eary L. E., Runnells D. D., and Esposito K. J. (2003) Geochemical controls on ground water composition at the cripple creek mining district, Cripple Creek, Colorado. *Appl. Geochem.* **18**, 1–24.

Edwards K. J., Schrenk M. O., Hamers R. J., and Banfield J. F. (1998) Microbial oxidation of pyrite: experiments using microorganisms from extreme acidic environment. *Am. Min.* **83**, 1444–1453.

Edwards K. J., Gihring T. M., and Banfield J. F. (1999) Geomicrobiology of pyrite (FeS_2) dissolution: a case study at Iron Mountain, California. *Geomicrobiol. J.* **16**, 155–179.

Edwards K. J., Bond P. L., Druschel G. K., McGuire M. M., Hamers R. J., and Banfield J. F. (2000) Geochemical and biological aspects of sulfide mineral dissolution: lessons from Iron Mountain, California. *Chem. Geol.* **169**, 383–397.

Ericson P. M. and Ladwig J. (1985) *Control of Acid Formation by the Inhibition of Bacteria and by Coating Pyrite Surfaces.* Final Report to the West Virginia Dept. of Natural Resources, 68.

Erry B. V., Macnair M. R., Meharg A. A., and Shore R. F. (2000) Arsenic contamination in wood mice (*Apodemus sylvaticus*) and bank voles (*Clethrionomys glareolus*) on abandoned mine sites in southwest Britain. *Environ. Pollut.* **110**, 179–187.

Evangelou V. P. (1995) *Pyrite Oxidation and its Control: Solution Chemistry, Surface Chemistry, Acid Mine Drainage (AMD), Molecular Oxidation Mechanisms, Microbial Role, Kinetics, Control, Ameliorates and Limitations, Microencapsulation.* CRC Press, Boca Raton, FL.

Evangelou V. P. and Zhang Y. L. (1995) A review: pyrite oxidation mechanisms and acid mine drainage prevention. *Crit. Rev. Environ. Sci. Technol.* **25**(2), 141–199.

Evans S. and Raftery E. (1982) Electron spectroscopic studies of galena and its oxidation by microwave-generated oxygen species and by air. *Chem. Soc. Faraday Trans.* **78**, 3545–3560.

Farquhar M. L., Charnock J. M., Livens F. R., and Vaughan D. J. (2002) Mechanisms of arsenic uptake from aqueous solution by interaction with goethite, lepidocrocite, mackinawite, and pyrite: an X-ray absorption spectroscopy study. *Environ. Sci. Technol.* **36**, 1757–1762.

Feasby D. G., Blanchette M., and Tremblay G. (1991) The mine environment neutral drainage program. In *2nd Int. Conf. Abatement of Acidic Drainage*. MEND Secretariat, Tome, vol. 1, pp. 1–26.

Finkelman R. B. (1980a) Modes of occurrence of trace elements and minerals in coal: an analytical approach. In *Atomic and Nuclear Methods in Fossil Energy Research* (eds. R. H. Filby, B. S. Carpenter, and R. C. Ragaini). Plenum, New York, pp. 141–149.

Finkelman R. B. (1980b) Modes of occurrence of trace elements in coal. PhD Thesis, University of Maryland.

Foos A. (1997) Geochemical modelling of coal drainage, Summit County, Ohio. *Environ. Geol.* **31**, 205–210.

Fornasiero D., Li F., Ralston J., and Smart R. St. C. (1994) Oxidation of galena surfaces. *J. Coll. Interface Sci.* **164**, 333–344.

Fortin D. and Beveridge T. J. (1997) Microbial sulfate reduction within sulfidic mine tailings: formation of diagenetic Fe sulfides. *Geomicrobiol. J.* **14**, 1–21.

Foster A. L., Brown G. E., Jr., and Parks G. A. (1998a) X-ray absorption fine-structure spectroscopy study of photocatalyzed, heterogeneous As(III) oxidation in kaolin and anatase. *Environ. Sci. Technol.* **32**, 1444–1452.

Foster A. L., Brown G. E., Jr., Ingle T. N., and Parks G. A. (1998b) Quantitative arsenic speciation in mine tailings using x-ray absorption spectroscopy. *Am. Min.* **83**, 553–568.

Fowler T. A. (2001) On the kinetics and mechanism of the dissolution of pyrite in the presence of *Thiobacillus ferrooxidans*. *Hydrometallurgy* **59**, 257–270.

Fuller C. C., Davis J. A., and Waychunas G. A. (1993) Surface chemistry of ferrihydrite: Part 2. Kinetics of arsenate adsorption and co-precipitation. *Geochim. Cosmochim. Acta* **57**, 2271–2282.

Fuller C. C., Piana M. J., Bargar J. R., Davis J. A., and Kohler M. (2002) Evaluation of apatite materials for use in permeable reactive barriers for the remediation of uranium-contaminated groundwater. In *Handbook of Groundwater Remediation Using Permeable Reactive Barriers—Applications to Radionuclides, Trace Metals, and Nutrients* (eds. D. L. Naftz, S. J. Morrison, J. A. Davis, and C. C. Fuller). Academic Press, San Diego, CA, pp. 255–280.

Fytas K. and Bousquet P. (2002) Silicate micro-encapsulation of pyrite to prevent acid mine drainage. *CIM Bull.* **95**(1063), 96–99.

Garcia-Sanchez A., Alvarez-Ayuso E., and Rodriguez-Martin F. (2002) Sorption of As(V) by some oxyhydroxides and clay minerals: application to its immobilization in two polluted mining soils. *Clay Min.* **37**, 187–194.

Garrels R. M. and Thompson M. E. (1960) Oxidation of pyrite by iron sulfate solutions. *Austral. J. Sci.* **258A**, 57–67.

Garrels R. M. and Thompson M. E. (1962) A chemical model for seawater at 25°C and one atmosphere total pressure. *Am. J. Sci.* **260**, 57–66.

Gelinas P., Lefebvre R., and Choquette M. (1992) Monitoring of acid mine drainage in a waste-rock dump. In *Environmental Issues and Waste Management in Energy and Minerals Production* (eds. R. K. Singhal, A. K. Mehrotra, A. K. Fytas, and J. L. Collins). Balkema, pp. 747–756.

Georgopoulou Z. J., Fytas K., Soto H., and Evangelou B. (1995) Pyrrhotite coating to prevent oxidation. *Proc. Sudbury '95 Mining Environ.* **1**, 7–17.

Gopal B. (1999) Natural and constructed wetlands for wastewater treatment: potentials and problems. *Water Sci. Technol.* **40**(3), 27–35.

Gould W. D. (1994) The nature and role of microorganisms in the tailings environment. In *The Environmental Geochemistry of Sulfide Mine-Wastes* (eds. J. L. Jambor and D. W. Blowes). Mineralogical Association of Canada, Nepean, ON vol. 22, pp. 163–184.

Gould W. D. and Kapoor A. (2003) The microbiology of acid mine drainage. In *Environmental Aspects of Mine Wastes* (eds. J. L. Jambor, D. W. Blowes and A. I. M. Ritchie). Short Course Series, **31**, Mineralogical Association of Canada, Vancouver, BC.

Government of Canada (1991) *State of Canada's Environment*. Ministry of Supply and Services, Ottawa.

Gray J. E., Theodorakos P. M., Bailey E. A., and Turner R. R. (2000) Distribution, speciation and transport of mercury in stream sediment, stream-water, and fish collected near abandoned mercury mines in southwestern Alaska USA. *Sci. Tot. Environ.* **260**, 21–33.

Greenway M. and Simpson J. S. (1996) Artificial wetlands for wastewater treatment, water reuse and wildlife in Queensland Australia. *Water Sci. Technol.* **33**(10–11), 221–229.

Hackl R. P., Dreisinger D. B., Peters E., and King J. A. (1995) Passivation of chalcopyrite during oxidative leaching in sulfate media. *Hydrometallurgy* **39**, 25–48.

Hall L. W., Jr., Anderson R. D., and Alden R. W., III (2002) A ten year summary of concurrent ambient water column and sediment toxicity tests in the Chesapeake Bay water shed: 1990–1999. *Environ. Monitor. Assess.* **76**, 311–352.

Harries J. R. and Ritchie A. I. M. (1985) Pore gas composition in waste-rock dumps undergoing pyritic oxidation. *Soil Sci.* **140**, 143–152.

Harries J. R. and Ritchie A. I. M. (1987) The effect of rehabilitation on the rate of oxidation of pyrite in a mine waste-rock dump. *Environ. Geochem. Health* **9**, 27–36.

Harvie R. D. and Ruch R. R. (1986) Mineral matter in Illinois and other U. S. coals. In *Mineral Matter and Ash in Coals* (ed. K. S. Vorres). American Chemical Society, Washington, DC, vol. 301, pp. 10–40.

Heal K. V. and Salt C. A. (1999) Treatment of acidic metal-rich drainage from reclaimed ironstone mine spoil. *Water Sci. Technol.* **39**(12), 141–148.

Hedin R. S. and Nairn R. W. (1994) Passive treatment of coal mine drainage. US Bureau of Mines, pp. 2–43.

Heikens A., Peihnenburg W. J. G. M., and Hendricks A. J. (2001) Bioaccumulation of heavy metals in terrestrial invertebrates. *Environ. Pollut.* **113**, 385–393.

Herbert R. B., Jr., Benner S. G., Pratt A. R., and Blowes D. W. (1998) Surface chemistry and morphology of poorly crystalline iron sulfides precipitated in media containing sulfate-reducing bacteria. *Chem. Geol.* **144**, 87–97.

Herbert R. B., Jr., Benner S. G., and Blowes D. W. (2000) Solid phase iron-sulfur geochemistry of a reactive barrier for treatment of mine drainage. *Appl. Geochem.* **15**, 1331–1343.

Hiroyoshi N., Hirota M., Hirajima T., and Tsunekawa M. (1997) A case of ferrous sulfate addition enhancing chalcopyrite leaching. *Hydrometallurgy* **47**, 37–45.

Ho G. E., Murphy P. J., Platell N., and Wajon J. E. (1984) Iron removal from TiO_2-plant acidic wastewater. *J. Environ. Eng.* **110**, 828–846.

Hochella M. F., Jr., Moore J. N., Golla U., and Putnis A. (1999) A TEM study of samples from acid mine drainage systems: metal—mineral association with implications for transport. *Geochim. Cosmochim. Acta* **63**, 3395–3406.

Holmes P. R. and Crundwell F. K. (2000) The kinetics of the oxidation of pyrite by ferric ions and dissolved oxygen: an electrochemical study. *Geochim. Cosmochim. Acta* **64**, 263–274.

Holmström H., Ljungberg J., Ekström M., and Öhlander B. (1999) Secondary copper enrichment in tailings at the Laver mine, northern Sweden. *Environ. Geol.* **38**, 327–342.

Holmström H., Salmon U. J., Carlsson E., Petrov P., and Öhlander B. (2001) Geochemical investigations of sulfide-bearing tailings at Kristineberg, northern Sweden, a few years after remediation. *Sci. Tot. Environ.* **273**, 111–133.

Hsu S. C. and Maynard J. B. (1999) The use of sulfur isotopes to monitor the effectiveness of constructed wetlands in controlling acid mine drainage. *Environ. Eng. Policy* **1**, 223–233.

Huang X. and Evangelou V. P. (1994) Suppression of pyrite oxidation rate by phosphate addition. In *Environmental Geochemistry of Sulfide Oxidation* (eds. C. N. Alpers and D. W. Blowes). American Chemical Society, Washington, DC 550, pp. 562–573.

Hudson-Edwards K. A., Macklin M. G., Curtis C. D., and Vaughan D. J. (1996) Processes of formation and distribution of Pb-, Zn-, Cd-, and Cu-bearing minerals in the Tyne Basin, northeast England: implications for metal-contaminated river systems. *Environ. Sci. Technol.* **30**, 72–80.

ICOLD (1996) *A Guide to Tailings Dams and Impoundments: Design, Construction, Use and Rehabilitation*. International Commission on Large Dams, Bulletin (United Nations Environment Programme) no. 106, 239pp.

Jain A. and Loeppert R. H. (2000) Effect of competing anions on the adsorption of arsenate and arsenite by ferrihydrite. *J. Environ. Qual.* **29**, 1422–1430.

Jambor J. L. (1994) Mineralogy of sulfide-rich tailings and their alteration products. In *The Environmental Geochemistry of Sulfide Mine-wastes* (eds. J. L. Jambor and D. W. Blowes). Mineralogical Association of Canada, Nepean, ON, vol. 22, pp. 59–102.

Jambor J. L. (2003) Mine-waste mineralogy and mineralogical perspectives of acid—base accounting. In *Environmental Aspects of Mine Wastes* (eds. J. L. Jambor, D. W. Blowes, and A. I. M. Ritchie). Mineralogical Association of Canada, vol. 31, 117–145.

Jambor J. L. and Blowes D. W. (1998) Theory and applications of mineralogy in environmental studies of sulfide-bearing mine wastes. In *Modern Approaches to Ore and Environmental Mineralogy* (eds. L. J. Cabri and D. J. Vaughan). Mineralogical Association of Canada, vol. 27, pp. 367–401.

Jambor J. L., Blowes D. W., and Ptacek C. J. (2000a) Mineralogy of mine wastes and strategies for remediation. In *European Mineralogical Union Notes in Mineralogy*, Environmental Mineralogy (eds. D. J. Vaughan and R. A. Wogelius). Budapest University Press, chap. 7, vol. 2, pp. 255–290.

Jambor J. L., Nordstrom D. K., and Alpers C. N. (2000b) Metal-sulfate salts from sulfide oxidation. In *Sulfate Minerals—Crystallography, Geochemistry, and Environmental Significance* (eds. C. N. Alpers, J. L. Jambor, and D. K. Nordstrom). *Rev. Mineral. Geochem.* Mineralogical Society of America, Washington, DC, vol. 40, pp. 303–350.

Jambor J. L., Dutrizac J. E., Groat L. A., and Raudsepp M. (2002) Static tests of neutralization potentials of silicate and aluminosilicate minerals. *Environ. Geol.* **43**, 1–17.

Janzen M. P., Nicholson R. V., and Scharer J. M. (2000) Pyrrhotite reaction kinetics: reaction rates for oxidation by oxygen, ferric iron and nonoxidative dissolution. *Geochim. Cosmochim. Acta* **64**, 1511–1522.

Jennings S. R., Dollhopf D. J., and Inskeep W. P. (2000) Acid production from sulfide minerals using hydrogen peroxide weathering. *Appl. Geochem.* **15**, 247–255.

Jerz J. K. and Rimstidt J. D. (2000) A reactor to measure pyrite oxidation in air. In *Proc. 5th Int. Conf. Acid Rock Drainage*. Society for Mining, Metallurgy, and Exploration, vol. 1, pp. 55–60.

Johnson R. H., Blowes D. W., Robertson W. D., and Jambor J. L. (2000) The hydrogeochemistry of the Nickel Rim mine tailings impoundment, Sudbury, Ontario. *J. Contamin. Hydrol.* **41**, 49–80.

Jones C. F., LeCount S., Smart R. St. C., and White T. J. (1992) Compositional and structural alteration of pyrrhotite surfaces in solution: XPS and XRD studies. *Appl. Surf. Sci.* **55**, 65–85.

Juillot F., Ildefonse Ph., Morin G., Calas G., de Kersabiec A. M., and Benedetti M. (1999) Remobilization of arsenic from buried wastes at an industrial site: mineralogical and geochemical control. *Appl. Geochem.* **14**, 1031–1048.

Jurjovec J., Ptacek C. J., and Blowes D. W. (2002) Acid neutralization mechanisms and metal release in mine tailings: a laboratory column experiment. *Geochim. Cosmochim. Acta* **66**, 1511–1523.

Jurjovec J., Ptacek C. J., Blowes D. W., and Jambor J. L. (2003) The effect of natrojarosite addition to mine tailings. *Environ. Sci. Technol.* **37**, 158–164.

Kartio I., Laajalehto K., Kaurila T., and Suoninen E. (1996) A study of galena (PbS) surfaces under controlled potential in pH 4.6 solution by synchrotron radiation excited photoelectron spectroscopy. *Appl. Surf. Sci.* **93**, 167–177.

Kartio I. J., Laajalehto K., Suoninen E., Buckley A. N., and Woods R. (1998) The initial products of the anodic oxidation of galena in acidic solution and the influence of mineral stoichiometry. *Coll. Surf. A: Physiochem. Eng. Aspects* **133**, 303–311.

Kesler S. E. (1994) *Mineral Resources, Economics and the Environment*. Macmillan, Boca Raton, FL.

Kim B. S., Hayes R. A., Prestidge C. A., Ralston J., and Smart R. St. C. (1995) Scanning tunneling microscopy studies of galena: the mechanisms of oxidation in aqueous solution. *Langmuir* **11**, 2554–2562.

Kimball B. A., Callender E., and Axtmann E. V. (1995) Effects of colloids on metal transport in a river receiving acid mine drainage, Upper Arkansas River, Colorado, USA. *Appl. Geochem.* **10**, 285–306.

Kleinmann R. P. L., Edenborn H. M., and Hedin R. S. (1991) Biological treatment of mine water—an overview. In *2nd Int. Conf. Abatement of Acidic Drainage*. MEND Secretariat, Tome, vol. **1**, pp. 27–42.

Krauskopf K. B. (1955) Sedimentary deposits of rare metals. In *Econ. Geol. Fiftieth Anniversary Volume*, Econ. Geol. (ed. A. M. Bateman) Society of Economic Geologists, Littleton, CO, pp. 411–463.

Laajalehto K., Smart R. St. C., Ralston J., and Suoninen E. (1993) STM and XPS investigation of reaction of galena in air. *Appl. Surf. Sci.* **64**, 29–39.

Lau S., Mohamed M., Tan Chi Yen A., and Su'ut S. (1998) Accumulation of heavy metals in freshwater molluscs. *Sci. Tot. Environ.* **214**, 113–121.

Lawrence R. W. and Wang Y. (1997) Determination of neutralization potential in the prediction of acid rock drainage. In *Proc. 4th Int. Conf. Acid Rock Drainage*. MEND, Natural Resources Canada, Ottawa, ON, vol. 1, 451–464.

Leblanc M., Achard B., Othman D. B., Luck J. M., Bertrand-Sarfati J., and Personné J. Ch. (1996) Accumulation of arsenic from acid mine waters by ferruginous bacterial accretions (stromatolites). *Appl. Geochem.* **11**, 541–554.

Ledin M. and Pedersen K. (1996) The environmental impact of mine wastes—roles of microorganisms and their significance in treatment of mine wastes. *Earth Sci. Rev.* **41**, 67–108.

Levy D. B., Custis K. H., Casey W. H., and Rock P. A. (1997) The aqueous geochemistry of the abandoned Spenceville copper pit, Nevada County, California. *J. Environ. Qual.* **26**, 233–243.

Lewis B. A., Gallinger R. D., and Wiber M. (2000) Poirier site reclamation program. In *Proc. 5th Int. Conf. Acid Rock Drainage*. Society for Mining, Metallurgy, and Exploration, Littleton, Co vol. 2, pp. 959–968.

Lichtner P. C. (1993) Scaling properties of time-space kinetic mass transport equations and the local equilibrium assumption. *Am. J. Sci.* **293**, 257–296.

Lin Z. (1997) Mineralogical and chemical characterization of wastes from a sulfuric acid industry in Falun Sweden. *Environ. Geol.* **30**, 153–162.

Lin Z. and Puls R. W. (2000) Adsorption, desorption, and oxidation of arsenic affected by clay minerals and aging process. *Environ. Geol.* **39**, 753–759.

Lind C. J. and Hem J. D. (1993) Manganese minerals and associated fine particulates in the streambed of Pinal Creek, Arizona, USA: a mining-related acid drainage problem. *Appl. Geochem.* **8**, 67–80.

Liu C. W. and Narasimhan T. N. (1989a) Redox-controlled multiple-species reactive chemical treatment: 1. Model development. *Water Resour. Res.* **25**, 868–882.

Liu C. W. and Narasimhan T. N. (1989b) Redox-controlled multiple-species reactive chemical treatment: 2. Model development. *Water Resour. Res.* **25**, 883–910.

Lortie L., Gould W. D., Stichbury M., Blowes D. W., and Thurel A. (1999) Inhibitors for the prevention of acid mine drainage (AMD). In *Proc. Sudbury '99 Mining and the Environment II Conf., September 13–17, 1999. Sudbury, Ontario*, Laurentian University, Sudbury, ON, pp. 1191–1198.

Lowson R. T. (1982) Aqueous oxidation of pyrite by molecular oxygen. *Chem. Rev.* **82**, 461–497.

Lucassen E. C., Smolders A. J. P., and Roelofs J. G. M. (2000) Increased ground levels cause iron toxicity in Glyceria fluitans (L). *Aquat. Bot.* **66**, 321–327.

Ludwig R. D., McGregor R. G., Blowes D. W., Benner S. G., and Mountjoy K. (2002) A permeable reactive barrier for the treatment of dissolved metals. *Ground Water* **40**, 59–66.

Lundgren T. (2001) The dynamics of oxygen transport into soil covered mining waste deposits in Sweden. *J. Geochem. Explor.* **74**, 163–173.

Luther I. G. W. (1987) Pyrite oxidation and reduction: molecular orbital theory considerations. *Geochim. Cosmochim. Acta* **51**, 3193–3199.

Lutzenkirchen J. and Lovgren L. (1998) Experimental study of arsenite adsorption to goethite. *Min. Mag.* **62A**, 927–928.

Machemer S. D. and Wildeman T. R. (1992) Adsorption compared with sulfide precipitation as metal removal processes from acid mine drainage in a constructed wetland. *J. Contamin. Hydrol.* **9**, 115–131.

Mackowsky M.-Th. (1968) Mineral matter in coal. In *Coal and Coal-bearing Strata* (eds. D. Murchison and T. S. Westoll). Elsevier, New York, pp. 309–321.

Manceau A. (1995) The mechanism of anion adsorption on iron oxides: evidence for the bonding of arsenate tetrahedra on free $Fe(OOH)_6$ edges. *Geochim. Cosmochim. Acta* **59**, 3647–3653.

Mann H. and Fyfe W. S. (1989) Metal uptake and Fe-, Ti-oxide biomineralization by acidophilic microorganisms in mine-waste environments, Elliot Lake, Canada. *Can. J. Earth Sci.* **26**, 2731–2735.

Manning B. A. and Goldberg S. (1996) Modeling arsenate competitive adsorption on kaolinite, montmorillonite and illite. *Clays Clay Min.* **44**, 609–623.

Manning B. A., Fendorf S. E., and Goldberg S. (1998) Surface structures and stability of arsenic(III) on goethite: spectroscopic evidence for inner-sphere complexes. *Environ. Sci. Technol.* **32**, 2383–2388.

Markich S. J., Brown P. L., and Jeffree R. A. (2001) Divalent metal accumulation in freshwater bivalves: an inverse relationship with metal phosphate solubility. *Sci. Tot. Environ.* **275**, 27–41.

Martin A. J., McNee J. J., and Pedersen T. F. (2001) The reactivity of sediments impacted by metal-mining in Lago Junin, Peru. *J. Geochem. Explor.* **74**, 175–187.

Mason B. (1958) *Principles of Geochemistry*. Wiley, New York.

Mason R. P., Reinfelder J. R., and Morel F. M. M. (1996) Uptake, toxicity, and trophic transfer of mercury in a coastal diatom. *Environ. Sci. Technol.* **30**, 1835–1845.

Mathews C. T. and Robins R. G. (1972) The oxidation of ferrous disulfide by ferric sulfate. *Austral. Chem. Eng.* **13**, 21–25.

Mathews C. T. and Robins R. G. (1974) Aqueous oxidation of iron disulfide by molecular oxygen. *Austral. Chem. Eng.* **15**, 19–24.

Mayer K. U., Frind E. O., and Blowes D. W. (2002) Multicomponent reactive transport modeling in variably saturated porous media using a generalized formulation for kinetically controlled reactions. *Water Resour. Res.* **38**, 1174–1195.

McCreadie H., Blowes D. W., Ptacek C. J., and Jambor J. L. (2000) The influence of reduction reactions and solids composition on pore-water arsenic concentrations. *Environ. Sci. Technol.* **34**, 3159–3166.

McGregor R., Benner S., Ludwig R., Blowes D., and Ptacek C. (2002) Sulfate reduction permeable reactive barriers to treat acidity, cadmium, copper, nickel, and zinc: two case studies. In *Handbook of Groundwater Remediation Using Permeable Reactive Barriers—Applications to Radionuclides, Trace Metals, and Nutrients* (eds. D. L. Naftz, S. J. Morrison, J. A. Davis, and C. C. Fuller). Academic Press, San Diego, CA, pp. 495–522.

McKibben M. A. and Barnes H. L. (1986) Oxidation of pyrite in low temperature acidic solutions: rate laws and surface textures. *Geochim. Cosmochim. Acta* **50**, 1509–1520.

MEND (1991) *Acid Rock Drainage Prediction Manual*. MEND Project 1.16.1b, report by Coastech Research. MEND, Natural Resources Canada.

MEND (1994) *Monitoring and Modeling of Acid Mine Drainage from Waste-rock Dumps*. Natural Resources Canada MEND Report 1.14.2g.

MEND (1995) *MINEWALL 2.0 Literature Review and Conceptual Models*. MEND Project 1.15.2b. Natural Resources of Canada.

Miller G. C., Lyons W. B., and Davis A. (1996) Understanding the water quality of pit lakes. *Environ. Sci. Technol.* **30**, 118–123A.

Moncur M. C., Ptacek C. J., Blowes D. W., and Jambor J. L. (2003) Fate and transport of metals from an abandoned tailings impoundment after 70 years of sulfide oxidation. In *Proc. Sudbury '03, Mining and the Environment III*. Sudbury, Ontario, pp. 238–247.

Morin K. A. and Hutt N. M. (1997) *Environmental Geochemistry of Minesite Drainage: Practical Theory and Case Studies*. MDAG Publishing, Vancouver, BC.

Morin K. A., Cherry J. A., Dave N. K., Lim T. P., and Vivyurka A. J. (1988) Migration of acidic groundwater seepage from uranium-tailings impoundments: 1. Field study and conceptual hydrogeochemical model. *J. Contamin. Hydrol.* **2**, 271–303.

Morin G., Ostergren J. D., Juillot F., Ildefonse P., Calas G., and Brown G. E., Jr. (1999) XAFS determination of the chemical form of lead in smelter-contaminated soils and mine tailings: importance of adsorption processes. *Am. Min.* **84**, 420–434.

Morrison S. J., Carpenter C. E., Metzler D. R., Bartlett T. R., and Morris S. A. (2002) Design and performance of a permeable reactive barrier for containment of uranium, arsenic, selenium, vanadium, molybdenum, and nitrate at Monticello, Utah. In *Handbook of Groundwater Remediation Using Permeable Reactive Barriers—Applications to Radionuclides, Trace Metals, and Nutrients* (eds. D. L. Naftz, S. J. Morrison, J. A. Davis, and C. C. Fuller). Academic Press, San Diego, CA, pp. 371–399.

Moses C. O., Nordstrom D. K., Herman J. S., and Mills A. L. (1987) Aqueous pyrite oxidation by dissolved oxygen and ferric iron. *Geochim. Cosmochim. Acta* **51**, 1561–1571.

Mycroft J. R., Nesbitt H. W., and Pratt A. R. (1995) X-ray photoelectron and Auger electron spectroscopy of air oxidized pyrrhotite: distribution of oxidized species with depth. *Geochim. Cosmochim. Acta* **59**, 721–733.

Naftz D. L., Fuller C. C., Davis J. A., Morrison S. J., Feltcorn E. M., Freethey G. W., Rowland R. C., Wilkowske C., and Piana M. (2002) Field demonstration of three permeable reactive barriers to control uranium contamination in groundwater, Fry Canyon, Utah. In *Handbook of Groundwater Remediation Using Permeable Reactive Barriers—Applications to Radionuclides, Trace Metals, and Nutrients* (eds. D. L. Naftz, S. J. Morrison, J. A. Davis, and C. C. Fuller). Academic Press, San Diego, CA, pp. 401–434.

Nesbitt H. W. and Muir I. J. (1998) Oxidation states and speciation of secondary products on pyrite and arsenopyrite reacted with mine waste waters and air. *Min. Petrol.* **62**, 123–144.

Nesbitt H. W., Muir I. J., and Pratt A. R. (1995) Oxidation of arsenopyrite by air and air-saturated, distilled water, and implications for mechanism of oxidation. *Geochim. Cosmochim. Acta* **59**, 1773–1786.

Nichol C., Beckie R., and Smith L. (2002) Characterization of unsaturated flow at different scales in waste rock. In *Groundwater 2002—Proceedings of the International Association of Hydrologists Conference*. International Association of Hydraulic Engineering and Research, Madrid, Spain.

Nicholson R. V. (1994) Iron-sulfide oxidation mechanisms: laboratory studies. In *The Environmental Geochemistry of Sulfide Mine-wastes* (eds. J. L. Jambor and D. W. Blowes). Mineralogical Association of Canada, Nepean, ON, vol. 22, pp. 164–183.

Nicholson R. V. and Scharer J. M. (1994) Pyrrhotite oxidation kinetics. In *Environmental Geochemistry of Sulfide Oxidation* (eds. C. N. Alpers and D. W. Blowes). American Chemical Society, Washington, DC, vol. 550 pp. 14–30.

Nicholson R. V., Gillham R. W., Cherry J. A., and Reardon E. J. (1989) Reduction of acid generation in mine tailings through the use of moisture-retaining layers as oxygen barriers. *Can. Geotech. J.* **26**, 1–8.

Nordstrom D. K. (1982) Aqueous pyrite oxidation and the consequent formation of secondary iron minerals. In *Acid Sulfate Weathering* (eds. J. A. Kittrick, D. F. Fanning, and L. R. Hossner). Soil Sci. Soc. Am. Spec. Publ. 10, pp. 37–56.

Nordstrom D. K. (1985) The rate of ferrous iron oxidation in a stream receiving acid mine effluent. In *Selected Papers in the Hydrogeological Sciences*. US Geol. Surv. Water-supply, paper 2270, pp. 113–119.

Nordstrom D. K. and Alpers C. N. (1999a) Geochemistry of acid mine waters. In *The Environmental Geochemistry of Mineral Deposits* (eds. G. S. Plumlee and M. J. Logsdon).

Society of Economic Geologists Inc., Littleton, DC, vol. 6A, pp. 133–157.

Nordstrom D. K. and Alpers C. N. (1999b) Negative pH, efflorescent mineralogy, and consequences for environmental restoration at the Iron Mountain Superfund site, California. *Proc. Natl. Acad. Sci. USA* **96**, 3455–3462.

Nordstrom D. K. and Munoz J. L. (1994) *Geochemical Thermodynamics*. Blackwell, Boston, MA.

Nordstrom D. K. and Southam G. (1997) Geomicrobiology of sulfide mineral oxidation. In *Geomicrobiology: Interactions between Microbes and Minerals* (eds. J. F. Banfield and K. H. Nealson). Mineralogical Society of America, Washington, DC, vol. 35, pp. 361–385.

Nordstrom D. K., Plummer L. N., Langmuir D., Busenberg E., May H. M., Jones B. F., and Parkhurst D. L. (1990) Revised chemical equilibrium data for major water—mineral reactions and their limitations. In *Chemical Modeling of Aqueous Systems II* (eds. D. C. Melchior and R. L. Bassett). Am. Chem. Soc. Symp. Ser. Washington, DC, vol. 416, pp. 398–413.

Nordstrom D. K., Alpers C. N., Ptacek C. J., and Blowes D. W. (2000) Negative pH and extremely acidic mine waters from Iron Mountain, California. *Environ. Sci. Technol.* **34**, 254–258.

Nowak P. and Laajalehto K. (2000) Oxidation of galena surface—an XPS study of the formation of sulfoxy species. *Appl. Surf. Sci.* **157**, 101–111.

Nriagu J. O. (ed.) (1994) *Arsenic in the Environment: Part I. Cycling and Characterization*. Advances in Environmental Science and Technology, Wiley, New York, vol. 26.

Olanipekun E. O. (1999) Kinetics of sphalerite leaching in acidic ferric chloride solutions. *Trans. Indian Inst. Metall.* **52**(2–3), 81–86.

Olson G. J. (1991) Rate of pyrite bioleaching by *Thiobacillus ferrooxidans*-results of an interlaboratory comparison. *Appl. Environ. Microbiol.* **57**, 642–644.

Orlova T. A., Stupnikov V. M., and Krestan A. L. (1988) Mechanism of oxidative dissolution of sulfides. *Zhurnal Prikladnoi Khimii* **61**, 2172–2177.

Paktunc D. and Davé N. K. (2002) Formation of secondary pyrite and carbonate minerals in the Lower Williams Lake tailings basin, Elliot Lake, Ontario, Canada. *Am. Min.* **87**, 593–602.

Parkhurst D. L. (1998) PHREEQC website (http://wwwbrr.cr.usgs.gov/projects/GWC_coupled/phreeqc/).

Parkhurst D. L., Thorstenson D. C., and Plummer L. N. (1985) PHREEQE—a computer program for geochemical calculations. *US Geol. Surv. Water Res. Invest. Rept.* 80–96.

Pedersen T. F. (1993) The early diagenesis of submerged sulphide-rich mine tailings in Anderson Lake, Manitoba. *Can. J. Earth Sci.* **30**, 1099–1109.

Percival J. B., Dumaresq C. G., and Kwong Y. T. J. (1996) Transport and attenuation of arsenic in an alkaline environment. *Prog. Abstr. Geol. Assoc. Can.: Min. Assoc. Can.* **21**, A74.

Perez I. P. and Dutrizac J. E. (1991) The effect of iron content of sphalerite on its rate of dissolution in ferric sulphate and ferric chloride media. *Hydrometallurgy* **26**, 211–232.

Perry A. and Kleinmann R. L. P. (1991) The use of constructed wetlands in the treatment of acid mine drainage. *Nat. Resour. Forum* **15**(3), 178–184.

Pitzer K. S. (1973) Thermodynamics of electrolytes: I. Theoretical basis and general equations. *J. Phys. Chem.* **77**, 2300–2308.

Pitzer K. S. (1991) Ion interaction approach: theory and data correlation. In *Activity Coefficients in Electrolyte Solutions* (ed. K. S. Pitzer). CRC Press, pp. 75–153.

Plant J. A., Simpson P. R., Smith B., and Windley B. F. (1999) Uranium ore deposits—products of the radioactive earth. In *Uranium: Mineralogy, Geochemistry and the Environment* (eds. P. C. Burns and R. Finch). Mineralogical Society of America Washington, DC, *Rev. Min.* **38**, 255–319.

Plummer L. N., Parkhurst D. L., Fleming G. W., and Dunkle S. A. (1988) A computer program incorporating Pitzer's equations for calculation of geochemical reactions in brines. *US Geol. Surv. Water Res. Invest. Rept.* 88–4153.

Pratt A. R. and Nesbitt H. W. (1997) Pyrrhotite leaching in acid mixtures of HCl and H_2SO_4. *Am. J. Sci.* **297**, 807–828.

Pratt A. R., Muir I. J., and Nesbitt H. W. (1994a) X-ray photoelectron and Auger electron spectroscopic studies of pyrrhotite and mechanisms of air oxidation. *Geochim. Cosmochim. Acta* **58**, 827–841.

Pratt A. R., Nesbitt H. W., and Muir I. J. (1994b) Generation of acids from mine waste: oxidative leaching of pyrrhotite in dilute H_2SO_4 solutions at pH 3.0. *Geochim. Cosmochim. Acta* **58**, 5147–5159.

Prestidge C. A., Skinner W. M., Ralston J., and Smart R. St. C. (1995) The interaction of iron(III) species with galena surfaces. *Coll. Surf.* **105**, 325–339.

Ptacek C. J. and Blowes D. W. (1994) Influence of siderite on the pore-water chemistry of inactive mine-tailings impoundments. In *Environmental Geochemistry of Sulfide Oxidation* (eds. C. N. Alpers and D. W. Blowes). American Chemical Society, Washington, DC, vol. 550, pp. 172–189.

Ptacek C. J. and Blowes D. W. (2000) Prediction of sulfate mineral solubility in concentrated waters. Sulfate minerals: crystallography, geochemistry, and environmental significance. In *Reviews in Mineralogy and Geochemistry* (eds. C. N Alpers, J. L. Jambor, and D. K. Nordstrom). Mineralogical Society of America, Washington DC, vol. 40, pp. 513–540.

Puro M., Kipkie W. B., Knapp R. A., MacDonald T. J., and Stuparyk R. A. (1995) Inco's Copper Cliff tailings area. In *Proc. Sudbury '95*, Canmur, Natural Resources, Ottawa, Ontario, Canada, *Mining and the Environment*, vol. 1, pp. 181–191.

Rai U. N., Tripathi R. D., Vajpajee P., Vidyanath J., and Ali M. B. (2002) Bioaccumulation of toxic metals (Cr, Cd, Pb, and Cu) by seeds of Euryale ferox Salisb. *Chemosphere* **46**, 267–272.

Ramstedt M., Carlsson E., and Lövgren L. (2003) Aqueous geochemistry in the Udden pit lake, northern Sweden. *Appl. Geochem.* **18**, 97–108.

Raven K. P., Jain A., and Loeppert R. H. (1998) Arsenite and arsenate adsorption on ferrihydrite: kinetics, equilibrium, and adsorption envelopes. *Environ. Sci. Technol.* **32**, 344–349.

Reardon E. J. and Beckie R. D. (1987) Modelling chemical equilibria of acid-mine drainage: the $FeSO_4$–H_2SO_4–H_2O system. *Geochim. Cosmochim. Acta* **51**, 2355–2368.

Reardon E. J. and Moddle P. M. (1985) Gas diffusion coefficient measurements on uranium mill tailings: implications to cover layer design. *Uranium* **2**, 111–131.

Reardon E. J. and Poscente P. J. (1984) A study of gas compositions in sawmill waste deposits: evaluation of the use of wood waste in close-out of pyritic tailings. *Reclam. Reveg. Res.* **3**, 109–128.

Reedy B. J., Beattie J. K., and Lowson R. T. (1991) A vibrational spectroscopic ^{18}O study of pyrite oxidation. *Geochim. Cosmochim. Acta* **55**, 1609–1614.

Renton J. J. (1982) Mineral matter in coal. In *Coal Structure* (ed. R. A. Meyers). Academic Press, New York, pp. 283–326.

Ribet I., Ptacek C. J., Blowes D. W., and Jambor J. L. (1995) The potential for metal release by reductive dissolution of weathered mine tailings. *J. Contamin. Hydrol.* **17**, 239–273.

Rieder M., Cavazzini G., D'Yakonov Y. S., Frank-Kamenetskii V. A., Gottardi G., Guggenheim S., Koval P. V., Muller G., Neiva A. M. R., Radoslovich E. W., Robert J. L., Sassi F. P., Takeda H., Weiss Z., and Wones D. R. (1998) Nomenclature of the micas. *Can. Min.* **36**, 905–912.

Rimstidt J. D., Chermak J. A., and Gagen P. M. (1994) Rates of reaction of galena, sphalerite, chalcopyrite, and arsenopyrite. In *Environmental Geochemistry of Sulfide Oxidation* (eds. C. N. Alpers and D. W. Blowes). American Chemical Society, Washington, DC, vol. 550, pp. 2–13.

Ritchie A. I. M. (1994) Sulfide oxidation mechanisms: controls and rates of oxygen transport. In *The Environmental Geochemistry of Sulfide Mine-wastes* (eds. J. L. Jambor, and D. W. Blowes). Mineralogical Association of Canada, Nepean, ON, vol. 22, pp. 201–246.

Ritchie A. I. M. (2003) Oxidation and gas transport in piles of sulfidic material. In *Environmental Aspects of Mine Wastes*, Short Course Series (eds. J. L. Jambor, D. W. Blowes, and A. I. M. Ritchie). Mineralogical Association of Canada, Vancouver, BC, vol. 31, pp. 73–94.

Robertson W. D. (1994) The physical hydrogeology of mill-tailings impoundments. In *The Environmental Geochemistry of Sulfide Mine-wastes* (eds. J. L. Jambor and D. W. Blowes). Mineralogical Association of Canada, Nepean, ON, vol. 22, pp. 1–77.

Robinsky E. (1978) Tailings disposal by the thickened discharge method for improved economy and environmental control. In *Tailings Disposal Today, Proc. 2nd Int. Tailings Symp.*, Miller Freeman San Francisco, CA, vol. **2**, pp. 75–92.

Robinsky E., Barbour S. L., Wilson G. W., Bordin D., and Fredlund D. G. (1991) Thickened sloped tailings disposal: an evaluation of seepage and abatement of acid drainage. In *2nd Int. Conf. Abatement of Acidic Drainage*. MEND Secretariat, vol. 1, pp. 529–550.

Rose A. W. and Cravotta C. A., III (1998) Geochemistry of coal mine drainage. In *Coal Mine Drainage Prediction and Pollution in Pennsylvania* (eds. K. B. C. Brady, M. W. Smith, and J. Schueck). Bureau of Mining and Reclamation, Pennsylvania Department of Environmental Protection, Harrisburg, PA, pp. 1–22.

Roussel C., Bril H., and Fernandez A. (2000) Arsenic speciation: involvement in evaluation of environmental impact caused by mine wastes. *J. Environ. Qual.* **29**, 182–188.

Runkel R. L. and Kimball B. A. (2002) Evaluating remedial alternatives for an acid mine drainage stream–application of a reactive transport model. *Environ. Sci. Technol.* **36**, 1093–1101.

Runkel R. L., Bencala K. E., and Broshears R. E. (1996) An equilibrium-based simulation model for reactive solute transport in small streams. In *US Geol. Surv. Water Resour. Invest. Rept.* 94-4014, pp. 775–780.

Rytuba J. J. (2000) Mercury mine drainage and processes that control its environmental impact. *Sci. Tot. Environ.* **260**, 57–71.

Sand W., Rohde K., Sobotke B., and Zenneck C. (1992) Evaluation of *Leptospirrillium ferrooxidans* for leaching. *Appl. Environ. Microbiol.* **58**, 85–92.

Sand W., Gehrke T., Jozsa P. G., and Schippers A. (2001) (Bio)chemistry of bacterial leaching-direct vs. indirect bioleaching. *Hydrometallurgy* **59**, 159–175.

Santore R. C., Di Toro D. M., Paquin P. R., Allen H. E., and Meyer E. (2001) Biotic ligand model of the acute toxicity of metals: 2. Application to acute copper toxicity in freshwater fish and Daphina. *Environ. Toxicol. Chem.* **20**, 2397–2402.

Sasaki K., Tsunekawa M., Ohtsuka T., and Konno H. (1995) Confirmation of a sulfur-rich layer on pyrite after oxidative dissolution by Fe(III) ions around pH 2. *Geochim. Cosmochim. Acta* **59**, 3155–3158.

Sasaki K., Tsunekawa M., Ohtsuka T., and Konno H. (1998) The role of sulfur oxidizing bacteria *Thiobacillus thiooxidans* in pyrite weathering. *Coll. Surf.* **133**, 269–278.

Savage K. S., Tingle T. M., O'Day P. A., Waychunas G. A., and Bird D. K. (2000) Arsenic speciation in pyrite and secondary weathering phases, Mother Lode gold district, Tuolumne County, California. *Appl. Geochem.* **15**, 1219–1244.

Shapter J. G., Brooker M. H., and Skinner W. M. (2000) Observation of oxidation of galena using Raman spectroscopy. *Int. J. Min. Process.* **60**, 199–211.

Shevenell L., Connors K. A., and Henry C. D. (1999) Controls on pit lake water quality at sixteen open-pit mines in Nevada. *Appl. Geochem.* **14**, 669–687.

Singer P. C. and Stumm W. (1968) Kinetics of the oxidation of ferrous iron. In *2nd Symp. Coal Mine Drainage Research.* National Coal Association/Bituminous Coal Research, pp. 12–34.

Singer P. C. and Stumm W. (1970) Acid mine drainage-rate determining step. *Science* **167**, 1121–1123.

Smedley P. L. and Kinniburgh D. G. (2002) A review of the source, behaviour and distribution of arsenic in natural waters. *Appl. Geochem.* **17**, 517–568.

Smyth D. J. A. (1981) Hydrogeological and geochemical studies above the water table in an inactive uranium tailings impoundment near Elliot Lake, Ontario. MSc Thesis, University of Waterloo.

Sobek A. A. (1987) The use of surfactants to prevent AMD in coal refuse and base metal tailings. In *Proc. Acid Mine Drainage Workshop*, Halifax, Nova Scotia, March 1987. Environment Canada, Ottawa, pp. 357–390.

Sobek A. A., Schuller W. A., Freeman J. R., and Smith R. M. (1978) Field and laboratory methods applicable to overburdens and minesoils. **EPA-600/2-78-954**. US Environmental Protection Agency.

Sobolewski A. (1996) Metal species indicate the potential of constructed wetlands for long term treatment of metal mine drainage. *Ecol. Eng.* **6**, 259–271.

Southam G. and Beveridge T. J. (1992) Enumeration of *Thiobacilli* within pH-neutral and acidic mine tailings and their role in the development of secondary mineral soil. *Appl. Environ. Microbiol.* **58**, 1904–1912.

Sparkling D. W. and Lowe T. P. (1998) Metal concentrations in aquatic macrophytes as influenced by soil and acidification. *Water Air Soil Pollut.* **108**, 203–221.

Spears D. A. (1997) Environmental impact of minerals in UK coals. In *European Coal Geology and Technology* (eds. R. Gayer and J. Pešek). Geol. Soc. Spec. Publ. London, UK, vol. 125, pp. 287–295.

Speight J. G. (1994) *The Chemistry and Technology of Coal.* Marcel Dekker, NY.

Stach E., Mackowsky M.-Th., Teichmüller M., Taylor G. H., Chandra D., and Teichmüller R. (1982) *Stach's Textbook of Coal Petrology*. Gebrüder Borntraeger, Berlin, Germany.

Steefel C. I. and Lasaga A. C. (1994) A coupled model for transport of multiple chemical species and kinetic precipitation/dissolution reactions with application to reactive flow in single phase hydrothermal systems. *Am. J. Sci.* **294**, 529–592.

Steefel C. I. and van Capellan P. (1998) Reactive transport modelling of natural systems. *J. Hydrol.* **209**, 1–7.

Stichbury M., Bechard G., Lortie L., and Gould W. D. (1995) Use of inhibitors to prevent acid mine drainage. *Proc. Sudbury '95 Mining Environ.* **2**, 613–622.

Stichbury M.-L., K., Bain J. G., Blowes D. W., and Gould W. D. (2000) Microbially-mediated reductive dissolution of arsenic bearing minerals in a gold mine tailings impoundment. *Proc. 5th Int. Conf. Acid Rock Drainage* **1**, 97–103.

Stohs S. J. and Bagchi D. (1995) Oxidative mechanisms in the toxicity of metal ions. *Free Rad. Biol. Med.* **18**, 321–336.

Stumm W. and Sulzberger B. (1992) The cycling of iron in natural environments: considerations based on laboratory studies of heterogeneous redox processes. *Geochim. Cosmochim. Acta* **56**, 3233–3257.

Swaine D. J. (1990) *Trace Elements in Coal.* Butterworths, Toronto, ON.

Tassé N., Germain M. D., and Bergeron M. (1994) Composition of interstitial gases in wood chips deposited on reactive tailings—consequences for their use as an oxygen barrier. In *Environmental Geochemistry of Sulfide Oxidation* (eds. C. N. Alpers and D. W. Blowes). American Chemical Society, Washington, DC, vol. 550, pp. 631–644.

Tempel R. G., Shevenell L. A., Lechler P., and Price J. (2000) Geochemical modeling approach to predicting arsenic concentrations in a mine pit lake. *Appl. Geochem.* **15**, 475–492.

Thomas J. E., Jones C. F., Skinner W. M., and Smart R. St. C. (1998) The role of surface sulfur species in the inhibition of pyrrhotite dissolution in acid conditions. *Geochim. Cosmochim. Acta* **62**, 1555–1565.

Thomas J. E., Skinner W. M., and Smart R. St. C. (2001) A mechanism to explain sudden changes in rates and products for pyrrhotite dissolution in acid solution. *Geochim. Cosmochim. Acta* **65**, 1–12.

Timms G. P. and Bennett J. W. (2000) The effectiveness of covers at Rum Jungle after fifteen years. In *Proc. 5th Int. Conf. Acid Rock Drainage*. Society for Mining, Metallurgy, and Exploration, vol. 2, pp. 813–818.

Tipping E. (1994) WHAM—a chemical equilibrium model and computer code for waters, sediments, and soils incorporating a discrete site/electrostatic model of ion-binding by humic substances. *Comput. Geosci.* **20**, 973–1023.

Tossell J. A. and Vaughan D. J. (1987) Electronic structure and the chemical reactivity of the surface of galena. *Can. Min.* **25**, 381–392.

Tributsch H. (2001) Direct versus indirect bioleaching. *Hydrometallurgy* **59**, 177–185.

Tributsch H. and Bennett J. C. (1981a) Semiconductor-electrochemical aspects of bacterial leaching: II. Survey of rate controlling metal sulfide properties. *J. Chem. Technol. Biotechnol.* **31**, 627–635.

Tributsch H. and Bennett J. C. (1981b) Semiconductor-electrochemical aspects of bacterial leaching: I. Oxidation of metal. *J. Chem. Technol. Biotechnol.* **31**, 565–577.

USDA (1993) Acid drainage from mines on the National Forest: a management challenge. *Forest Serv. Publ.* **1505**, 1–12.

US EPA (2000) Bioaccumulation testing and interpretation for the purpose of sediment quality assessment. 1–475, EPA 823-R00-001.

Valković V. (1983) *Trace Elements in Coal: 1*. CRC Press, Boca Raton, FL.

Vassilev S. V., Kitano K., and Vassileva C. G. (1996) Some relationships between coal rank and chemical and mineral composition. *Fuel* **75**, 1537–1542.

Vaughan D. J. and Craig J. R. (1978) *Mineral Chemistry of Metal Sulfides*. Cambridge University Press, London.

Vigneault B., Campbell P. G. C., Tessier A., and de Vitre R. (2001) Geochemical changes in sulfidic mine tailings stored under a shallow water cover. *Water Res.* **35**, 1066–1076.

Walter A. L., Frind E. O., Blowes D. W., Ptacek C. J., and Molson J. W. (1994a) Modelling of multicomponent reactive transport in groundwater: 1. Model development and testing. *Water Resour. Res.* **30**, 3137–3148.

Walter A. L., Frind E. O., Blowes D. W., Ptacek C. J., and Molson J. W. (1994b) Modelling of multicomponent reactive transport in groundwater: 2. Metal mobility in aquifers impacted by acidic mine tailings discharge. *Water Resour. Res.* **30**, 3149–3158.

Walters S. and Bailey A. (1998) Geology and mineralization of the Cannington Ag–Pb–Zn deposit: an example of Broken Hill-type mineralization in the Eastern succession, Mount Isa inlier, Australia. *Econ. Geol.* **93**, 1307–1329.

Walton-Day K. (1999) Geochemistry of the processes that attenuate acid mine drainage in wetlands. *Rev. Econ. Geol.* **6A**, 215–228.

Wang W. X. and Fisher N. S. (1999) Delineating metal accumulation pathways for marine invertebrates. *Sci. Tot. Environ.* **237/238**, 459–472.

Ward C. R. (1989) Minerals in bituminous coals of the Sydney basin (Australia) and the Illinois basin (USA). *Int. J. Coal Geol.* **13**, 455–479.

Warren L. A. and Haak E. A. (2001) Biogeochemical controls on metal behavior in freshwater environments. *Earth Sci. Rev.* **54**, 261–320.

Waybrant K. R., Blowes D. W., and Ptacek C. J. (1998) Prevention of acid mine drainage using in situ porous reactive walls: selection of reactive mixtures. *Environ. Sci. Technol.* **32**, 1972–1979.

Waybrant K. R., Ptacek C. J., and Blowes D. W. (2002) Treatment of mine drainage using permeable reactive barriers: column experiments. *Environ. Sci. Technol.* **36**, 1349–1356.

Waychunas G. A., Fuller C. C., Rea B. A., and Davis J. A. (1996) Wide angle X-ray scattering (WAXS) study of 'two-line' ferrihydrite structure: effect of arsenate sorption and counterion variation and comparison with EXAFS results. *Geochim. Cosmochim. Acta* **60**, 1765–1781.

Weare J. H. (1987) Models of mineral solubility in concentrated brines with application to field observations. In *Thermodynamic Modeling of Geological Materials: Minerals, Fluids and Melts* (eds. I. S. E. Carmichael and H. P. Eugster). Mineralogical Society of America, Washington, DC, Rev. Min. **17**, 143–176.

Webster J. G., Swedlund P. J., and Webster K. S. (1998) Trace metal adsorption onto an acid drainage iron(III) oxy hydroxy sulfate. *Environ. Sci. Technol.* **23**, 1361–1368.

Weisener C. G. (2002) The reactivity of iron and zinc sulfide mineral surfaces: adsorption and dissolution mechanisms. PhD Thesis, University of South Australia.

Weisener C. G., Smart R. St. C., and Gerson A. (2003) Spectroscopic characterisation of leached sphalerite surfaces as a function of temperature at pH1. *Geochim. Cosmochim. Acta* **67**, 823–830.

White W. W., III, Lapakko K. A., and Cox R. L. (1999) Static-test methods most commonly used to predict acid-mine drainage: practical guidelines for use and interpretation. In *The Environmental Geochemistry of Mineral Deposits: Part A. Processes, Techniques, and Health Issues.* (eds. G. S. Plumlee and M. J. Logsdon). *Rev. Econ. Geol.* **6A**, pp. 325–338.

Wiersma C. L. and Rimstidt J. D. (1984) Rates of reaction of pyrite and marcasite with ferric iron at pH 2. *Geochim. Cosmochim. Acta* **48**, 85–92.

Williams M. (2001) Arsenic in mine waters: an international study. *Environ. Geol.* **40**, 267–278.

Williams D. J., Bigham J. M., Cravotta C. A., III, Traina S. J., Andersen J. E., and Lyon J. G. (2002) Assessing mine drainage pH from color and spectral reflectance of chemical precipitates. *Appl. Geochem.* **17**, 1273–1286.

Williamson M. A. and Rimstidt J. D. (1994) The kinetics and electrochemical rate-determining step of aqueous pyrite oxidation. *Geochim. Cosmochim. Acta* **58**, 5443–5454.

Witzke T. (1999) Hydrowoodwardite, a new mineral of the hydrotalcite group from Königswalde near Annaberg, Saxony/Germany and other localities. *Neues Jahrb. Mineral. Mh.*, 75–86.

Wunderly M. D., Blowes D. W., Frind E. O., and Ptacek C. J. (1996) A multicomponent reactive transport model incorporating kinetically controlled pyrite oxidation. *Water Resour. Res.* **32**, 3173–3187.

Yanful E. K. and Verma A. (1999) Oxidation of flooded mine tailings due to resuspension. *Can. Geotech. J.* **36**, 826–845.

Zinck J. M. (1999) Stability of lime treatment sludges. In *Proc. 101st CIM Annual Gen. Mtg., Calgary, May, 1999*, Canadian Institute of Mining.

9.06 Environmental Geochemistry of Radioactive Contamination

M. D. Siegel and C. R. Bryan
Sandia National Laboratories, Albuquerque, NM, USA

NOMENCLATURE

a_{H^+}	activity of the hydrogen ion in solution
$a_{UO_2^{2+}}$	activity of uranyl in solution
$A'_{colloid}$	specific surface area of the colloid, in a colloid/rock system
A'_{IP}	specific surface area of the rock matrix, in a colloid/rock system
C	colloid concentration in solution
C_i	initial concentration of contaminant in solution
C_f	final concentration of contaminant in solution
C_L	concentration of a contaminant in solution
C_S	concentration of a contaminant sorbed onto solid
e	fundamental electrical charge
F	the ratio of the specific surface areas of the colloid and the rock matrix, in a colloid/rock system
k	Boltzmann's constant
K_d	distribution coefficient for a contaminant in a water/rock system
$K_{i,j,k}$	intrinsic formation constant for an aqueous species (i, j, k describe stoichiometry)
m	mass of substrate present in a batch system
R	retardation factor for a contaminant
R_d	sorption coefficient for a contaminant in a water/rock system
$R_{F,eff}$	effective retardation factor (including the effects of transport on colloids)
T	absolute temperature, in K
V_i	initial solution volume in a batch system
V_f	final solution volume in a batch system
β^{cat}	intrinsic surface-complexation constant for a cationic species
ρ	bulk density of a porous medium
ϕ	bulk porosity of a porous medium
ψ_0	electrical potential of the inner (o) surface plane

9.06.1 INTRODUCTION

Psychometric studies of public perception of risk have shown that dangers associated with radioactive contamination are considered the most dreaded and among the least understood hazards (Slovic, 1987). Fear of the risks associated with nuclear power and associated contamination has had important effects on policy and commercial decisions in the last few decades. In the US, no new nuclear power plants were ordered between 1978 and 2002, even though it has been suggested that the use of nuclear power has led to significantly reduced CO_2 emissions and may provide some relief from the potential climatic changes associated with fossil fuel use. The costs of the remediation of sites contaminated by radioactive materials and the projected costs of waste disposal of radioactive waste in the US dwarf many other environmental programs. The cost of disposal of spent nuclear fuel at the proposed repository at Yucca Mountain will likely exceed \$10 billion. The estimated total life cycle cost for remediation of US Department of Energy (DOE) weapons production sites ranged from \$203–247 billion dollars in constant 1999 dollars, making the cleanup the largest environmental project on the planet (US DOE, 2001). Estimates for the cleanup of the Hanford site alone exceeded \$85 billion through 2046 in some of the remediation plans.

Policy decisions concerning radioactive contamination should be based on an understanding of the potential migration of radionuclides through the geosphere. In many cases, this potential may have been overestimated, leading to decisions to clean up contaminated sites unnecessarily and exposing workers to unnecessary risk. It is important for both the general public and the scientific community to be familiar with information that is well established, to identify the areas of uncertainty and to understand the significance of that uncertainty to the assessment of risk.

9.06.1.1 Approach and Outline of Chapter

This chapter provides an applications-oriented summary of current understanding of environmental radioactive contamination by addressing three major questions:

(i) What are the major sources of radioactive contamination on the planet?

(ii) What controls the migration of radioactive contaminants in the environment?

(iii) How can an understanding of radionuclide geochemistry be used to facilitate environmental remediation or disposal of radioactive wastes, and how can we assess the associated risks?

The chapter starts with an overview of the nature of major sites of radioactive environmental contamination. A brief summary of the health effects associated with exposure to ionizing radiation and radioactive materials follows. The remainder of the chapter summarizes current knowledge of the properties of radionuclides as obtained and applied in three interacting spheres of inquiry and analysis: (i) experimental studies and theoretical calculations, (ii) field studies, and (iii) predictions of radionuclide behavior for remediation and waste disposal. Recent studies of radionuclide speciation, solubility, and sorption are reviewed, drawing upon the major US and European nuclear waste and remediation programs. Examples are given of the application of that information to understanding the behavior of radionuclides as observed at sites of natural and anthropogenic radioactive contamination. Finally, the uses of that information in remediation of radioactive contamination and in predicting the potential behavior of radionuclides released from proposed nuclear waste repositories are described.

9.06.1.2 Previous Reviews and Chapter Scope

Information describing the environmental geochemistry of radionuclides is being gathered at a rapid rate due to the high interest and pressing need to dispose of nuclear wastes and to remediate radioactively contaminated areas. Several important books have been written on the subject of actinide chemistry in the last few decades. Notable ones include Seaborg and Katz (1954), Katz *et al.* (1986), Ivanovich (1992), and Choppin *et al.* (1995). This chapter updates similar reviews that have been published in the last few decades, such as those by Allard (1983), Krauskopf (1986), Choppin and Stout (1989), Hobart (1990), Fuger (1992), Kim (1993), Silva and Nitsche (1995), and chapters in Barney *et al.* (1984), Langmuir (1997a), and Zhang and Brady (2002). National symposia dealing with the disposal of nuclear waste and remediation of radioactive environmental contamination have been held annually by the Material Research Society (e.g., McGrail and Cragnolino, 2002) and Waste Management Symposia (e.g., WM Symposia, 2001) since the 1980s. Proceedings of these conferences should continue to be valuable sources of the results of current research and work carried out after the publication of this chapter.

This chapter focuses on the interactions of radionuclides with geomedia in near-surface low-temperature environments. Due to the limitations on the chapter length, this review will not describe the mineralogy or economic geology of uranium deposits; the use of radionuclides as environmental tracers in studies of the atmosphere, hydrosphere, or lithosphere, the nature of the nuclear fuel cycle or processes involved in nuclear weapons production. Likewise, radioactive contamination associated with the use of atomic weapons during World War II, the contamination of the atmosphere, hydrosphere, or lithosphere related to nuclear weapons testing, and concerns over the contamination of the Arctic Ocean by the Soviet nuclear fleet are not discussed. The interested reader is advised to turn to other summaries of these topics and included references. The nuclear fuel cycle is summarized in publications of the US DOE (1997a,b). Eisenbud (1987) provides a comprehensive overview of environmental radioactivity from natural, industrial, and military sources. A recent publication of the Mineralogical Society of America included a series of review articles describing the mineralogy and paragenesis of uranium deposits and the environmental geochemistry of uranium and its decay products (Burns and Finch, 1999). Mahara and Kudo (1995) and Kudo *et al.* (1995) review the environmental behavior of plutonium released by the Nagasaki atomic bomb blast. A recent summary of the extent of contamination from atmospheric nuclear testing is found in Beck and Bennett (2002). Descriptions of radioactive contamination of the Arctic Ocean are found in Salbu *et al.* (1997), Aarkrog *et al.* (1999), and in publications of the Arctic Monitoring and Assessment Programme (AMAP, 2002; http://www.amap.no/).

9.06.2 THE NATURE AND HAZARDS OF RADIOACTIVE ENVIRONMENTAL CONTAMINATION

The relationships between radioactive contamination and the risks to human health have scientific and regulatory dimensions. From the scientific perspective, the risks associated with radioactive contamination depend systematically on the magnitude of the source, the type of radiation, exposure routes, and biological susceptibility to the effects of radiation damage. The effects of radiation on human health can be ascertained from the health status of people exposed to different levels of radiation in nuclear explosions or occupational settings. The most extreme exposures include those experienced by survivors of the atomic bomb blasts in Hiroshima and Nagasaki in World War II. Researchers have not observed an increase in cancer frequency for Japanese bomb survivors below an external dose of 0.2 Gy (20 rad), thus providing a lower limit for human biological susceptibility to the effects of radiation. In contrast, the regulatory perspective addresses estimates of an upper limit to biological susceptibility and involves a *di minimus* approach to exposure limits. Radiation protection standards are based on a "zero threshold" dose-response relationship and exposure for the public is limited to 1 mSv yr^{-1} (0.1 rem yr^{-1}). The following sections summarize basic information about the scientific and regulatory aspects of this issue. The different types of radioactivity are characterized, natural and anthropogenic sources of radioactivity are described, levels of exposure to these sources are estimated and compared to regulatory exposure limits, and the effects of radiation on human health are discussed.

9.06.2.1 Sources of Radioactivity

9.06.2.1.1 Radioactive processes

Only certain combinations of protons and neutrons result in stable atomic nuclei. Figure 1 shows a section of the chart of the nuclides, on which nuclides are plotted as a function of their proton number (Z) and neutron number (N). The radioactive decay chain for ^{238}U is indicated; only ^{206}Pb has a stable combination of protons and neutrons. At low atomic numbers (below $Z = 20$), isotopes with proton: neutron ratios of ~1 are stable, but a progressively higher proportion of neutrons is required to produce stability at higher atomic numbers. Unstable nuclei undergo radioactive decay—spontaneous transformations involving emission of particles and/or photons, resulting in changes in Z and N, and transformation of that atom into another element. Several types of radioactive decay may occur:

- β^--decay—a negatively charged beta particle (electron) is emitted from the nucleus of the atom, and one of the neutrons is transformed into a proton. Z increases by 1 and N decreases by 1.
- β^+-decay—a positively charged beta particle (positron) is emitted from the nucleus, and a proton is transformed into a neutron. Z decreases by 1 and N increases by 1.
- Electron capture—an unstable nucleus may capture an extranuclear electron, commonly a K-shell electron, resulting in the transformation of a proton to a neutron. This results in the same change in Z and N as β^+ decay; commonly, nuclides with a deficiency of neutrons can decay by either mechanism.
- α-decay—nuclei of high atomic number (heavier than cerium), and a few light nuclides, may decay by emission of an α-particle, a ^{4}He nucleus consisting of two protons and two neutrons. Z and N both decrease by 2.

In each case, the daughter nucleus is commonly left in an excited state and decays to the ground state by emission of γ-rays. If there is a significant delay between the two processes, the γ-emission is considered a separate event. Decay by γ-emission, resulting in no change in Z or N, is called an isomeric transition (e.g., decay of ^{99m}Tc to ^{99}Tc).

Many radioactive elements decay to produce unstable daughters. The radioactivity of many forms of radioactive contamination is due primarily to daughter products with short half-lives. The longest such decay chains that occur naturally are those for ^{238}U, ^{235}U, and ^{232}Th, which decay through a series of intermediate daughters to ^{206}Pb, ^{207}Pb, and ^{208}Pb, respectively (see Table 1). The decay of ^{238}U to ^{206}Pb results in the production of eight α-particles and six β-particles; that of

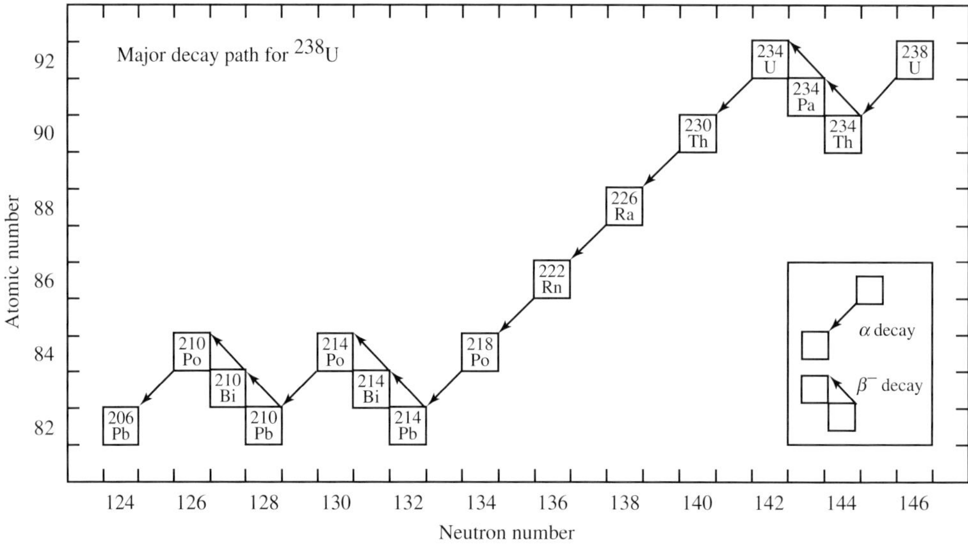

Figure 1 The major decay path for ^{238}U. A small fraction of decays will follow other possible decay paths—for example, decaying by β^- emission from ^{218}Po to ^{218}At, and then by α emission to ^{214}Bi—but ^{206}Pb is the stable end product in all cases.

Table 1 The major decay paths for several important actinides (isotope, half-life, and decay mode).

^{238}U decay series			^{235}U decay series			^{232}Th decay series			^{237}Np decay series			^{239}Pu decay series			^{238}Pu decay series			^{241}Am decay series		
^{238}U	4.47×10^{9} yr	α	^{235}U	7.04×10^{8} yr	α	^{232}Th	1.40×10^{10} yr	α	^{237}Np	2.14×10^{6} yr	α	^{239}Pu	2.44×10^{4} yr	α	^{238}Pu	87.7 yr	α	^{241}Am	433 yr	α
^{234}Th	24.1 d	β^-	^{231}Th	1.06 d	β^-	^{228}Ra	5.76 yr	β^-	^{233}Pa	27.0 d	β^-	^{235}U	As above		^{234}U	As above		^{237}Np	As above	
^{234}Pa	1.17 m	β^-	^{231}Pa	3.28×10^{4} yr	α	^{228}Ac	6.15 h	β^-	^{233}U	1.59×10^{5} yr	α									
^{234}U	2.46×10^{5} yr	α	^{227}Th	21.8 yr	β^-	^{228}Th	1.91 yr	α	^{229}Th	7.30×10^{3} yr	α									
^{230}Th	7.54×10^{4} yr	α	^{223}Ra	18.7 d	α	^{224}Ra	3.66 d	α	^{225}Ra	14.9 d	β^-									
^{226}Ra	1.6×10^{3} yr	α	^{219}Rn	11.4 d	α	^{220}Rn	55.6 s	α	^{224}Ac	10.0 d	α									
^{222}Rn	3.82 d	α	^{215}Po	4.0 s	α	^{216}Po	0.15 s	α	^{221}Fr	4.8 m	α									
^{218}Po	3.10 m	α	^{211}Pb	1.78×10^{-3} s	α	^{212}Pb	10.6 h	β^-	^{217}At	32 ms	α									
^{214}Pb	27.0 m	β^-	^{214}Pb	36.1 m	β^-	^{212}Bi	60.6 m	β^-	^{213}Bi	45.6 m	β^-									
^{214}Bi	19.9 m	β^-	^{211}Bi	2.14 m	α	^{212}Po	45 s	α	^{213}Po	4 μs	α									
^{214}Po	1.64×10^{-4} s	α	^{207}Tl	4.78 m	β^-	^{208}Pb	Stable		^{209}Pb	3.25 h	β^-									
^{210}Pb	22.6 yr	β^-	^{207}Pb	Stable					^{209}Bi	Stable										
^{210}Bi	5.01 d	β^-																		
^{210}Po	138 d	α																		
^{206}Pb	Stable																			

Source: Parrington *et al.* (1996).

^{235}U to ^{207}Pb, seven α- and four β-particles; and that of ^{232}Th to ^{208}Pb, six α- and four β-particles. Thus, understanding the geochemistry of radioactive contamination requires consideration of the chemistries of both the abundant parents and of the transient daughters, which have much lower chemical concentrations. After several half-lives of the longest-lived intermediate daughter, a radioactive parent and its unstable daughters will reach secular equilibrium; the contribution of each nuclide to the total activity will be the same. Thus, a sample of ^{238}U will, after about a million and a half years, have a total α-activity that is ~8 times that of the uranium alone.

Heavy nuclei can also decay by *fission*, by splitting into two parts. Although some nuclei can spontaneously fission, most require an input of energy. This is most commonly accomplished by absorption of neutrons, although α-particles, γ-rays and even X-rays may also induce fission. Fission is usually asymmetric—two unequal nuclei, or *fission products*, are produced, with atomic weights ranging from 66 to 172. Fission product yields vary with the energy of the neutrons inducing fission; under reactor conditions, fission is highly asymmetric, with production maxima at masses of ~95 and ~135. High-energy neutrons result in more symmetric fission, with a less bimodal distribution of products.

Fission products generally contain an excess of neutrons and are radioactive, decaying by successive β^- emissions to stable nuclides. The high radioactivity of spent nuclear fuel and of the wastes generated by fuel reprocessing for nuclear weapons production is largely due to fission products, and decreases rapidly over the first few tens of years. In addition to the daughter nuclei, neutrons are released during fission (2.5–3.0 per fission event for thermal neutrons), creating the potential for a fission chain reaction—the basis for nuclear power and nuclear weapons.

The major decay paths for the naturally occurring isotopes of uranium and thorium are shown in Table 1. Other actinides of environmental importance include ^{237}Np, ^{238}Pu, ^{239}Pu, and ^{241}Am. These have decay series similar to and overlapping those of uranium and thorium. Neptunium-237 ($t_{1/2} = 2.14 \times 10^{6}$ yr, α) decays to ^{209}Bi through a chain of intermediates, emitting seven α- and four β^--particles. Plutonium-238 ($t_{1/2} = 86$ yr, α) decays into ^{234}U, an intermediate daughter on the ^{238}U decay series. Plutonium-239 ($t_{1/2} = 2.44 \times 10^{4}$ yr, α) decays into ^{235}U. Americium-241 ($t_{1/2} = 458$ yr, α) decays into ^{237}Np.

The basic unit of measure for radioactivity is the number of atomic decays per unit time. In the SI system, this unit is the becquerel (Bq), defined as one decay per second. An older, widely used measure of activity is the curie (Ci). Originally defined as the activity of 1 g of ^{226}Ra (1 Ci = 3.7×10^{10} Bq). The units used to describe the dose, or energy absorbed by a material exposed to radiation, are dependent upon the type of radiation and the material. X-ray or γ-radiation absorbed by air is measured in Roentgens (R). The dose absorbed by any material, by any radiation, is measured in rad (radiation-absorbed-dose), where 1 rad corresponds to 100 erg g^{-1} (1 erg = 10^{-7} J) of absorbed energy. The SI equivalent is the gray (Gy), which is equal to 100 rad.

Different types of radiation affect biological materials in different ways, so a different unit is needed to describe the dose necessary to produce an equivalent biological damage. Historically, this unit is the rem (roentgen-equivalent-man). The dose in rem is equal to the dose in rad multiplied by a quality factor, which varies with the type of radiation. For β^--, γ-, and X-ray radiation, the quality factor is 1; for neutrons, it is 2–11, depending upon the energy of the particle; and for α-particles, the quality factor is 20. The SI unit for equivalent dose is the sievert (Sv), which is equivalent to 100 rem.

9.06.2.1.2 Natural sources of radioactivity

Radioactive materials have been present in the environment since the accretion of the Earth. The decay of radionuclides provides an important source of heat that drives many large-scale planetary processes. The most abundant naturally occurring radionuclides are ^{40}K, ^{232}Th, and ^{238}U and ^{235}U. The bulk of the natural global inventory of actinide radioactivity in the upper 100 m of the lithosphere (~10^{22} Bq or 2.7×10^{11} Ci) is due to activity of uranium and thorium isotopes (Santschi and Honeyman, 1989; Ewing, 1999). This is about equal to the total activity of ^{40}K in the world ocean. Average crustal concentrations of uranium (mostly 238) and thorium (mostly 232) are 2.7 $\mu g\ g^{-1}$ and 9.6 $\mu g\ g^{-1}$, respectively. Both elements are enriched in silica-rich igneous rocks (4.4 $\mu g\ g^{-1}$ and 16 $\mu g\ g^{-1}$, respectively in granites) and are highly enriched in zircons (2,000 $\mu g\ g^{-1}$ and 2,500 $\mu g\ g^{-1}$, respectively).

In groundwater, average uranium concentrations range from <0.1 $\mu g\ L^{-1}$ (reducing) to 100 $\mu g\ L^{-1}$ (oxidizing) (Langmuir, 1997a). The average thorium concentration in groundwater is <1 $\mu g\ L^{-1}$ and is not affected by solution redox conditions. Other naturally occurring radionuclides include actinium, technetium, neptunium, and protactinium. Small amounts of actinides (^{237}Np and ^{239}Pu) are present from neutron capture reactions with ^{238}U. Natural ^{99}Tc is a product of ^{238}U spontaneous fission (Curtis *et al.*, 1999). For comparison, Table 2 provides examples of large-scale sources of natural and anthropogenic radioactivity in the environment throughout the world.

9.06.2.1.3 Nuclear waste

It is estimated that the inventory of nuclear reactor waste in the US will reach 1.3×10^{21} Bq by 2020 (Ewing, 1999). Decay of the radionuclides from a reference inventory over 10^{10} yr is shown in Figure 2. In this figure, the change of ingestion toxicity of radionuclides important for

Table 2 Examples of sources of radioactivity in the environment.

Location	*Source of radioactivity*	*Major radionuclides*	*Amount of radioactivity*	*Ref.*
Global	Top 100 m of lithosphere	$^{238,235}U$, ^{232}Th	1.0×10^{22} Bq	1,2
HLW geologic repository	70 kt spent fuel (proposed)	^{137}Cs, ^{90}Sr	1.0×10^{22} Bq	1,2
Atmospheric testing	220 Megaton yield	^{131}I and ^{3}H	2.0×10^{20} Bq	1,2
		$^{239,240}Pu$	1.0×10^{17} Bq	
Mayak, Russia	Nuclear production	Various HLW	3.6×10^{19} Bq	4
		^{90}Sr, ^{137}Cs	2.1×10^{19} Bq	
US weapons complex	High level waste/ 100 million gallons (3.8×10^5 m^3)	Short-lived ($t_{1/2} < 50$ yr): ^{137}Cs, ^{90}Sr, ^{90}Y, ^{137m}Ba, ^{241}Pu	3.3×10^{19} Bq	6
		Longer lived ($t_{1/2}$ = 50–500 yr): ^{238}Pu, ^{131}Sm, ^{241}Am	1.1×10^{17} Bq	
		Long lived ($t_{1/2}$ = 500–50,000 yr): ^{239}Pu, ^{240}Pu, ^{14}C	3.3×10^{15} Bq	
		Longest lived ($t_{1/2}$ >50,000 yr): ^{99}Tc, ^{135}Cs, ^{233}U	2.0×10^{15} Bq	
Chernobyl	Reactor accident in 1986	^{131}I, $^{134,137}Cs$, $^{103,106}Ru$	1.2×10^{19} Bq	1,2,5
US	U mining and milling	^{226}Ra	1.9×10^{15} Bq	1

References: 1. Ewing (1999), 2. Santschi and Honeyman (1989), 3. NRC (2001), 4. Cochran *et al.* (1993), 5. IAEA (1996), and 6. US DOE (1997a).

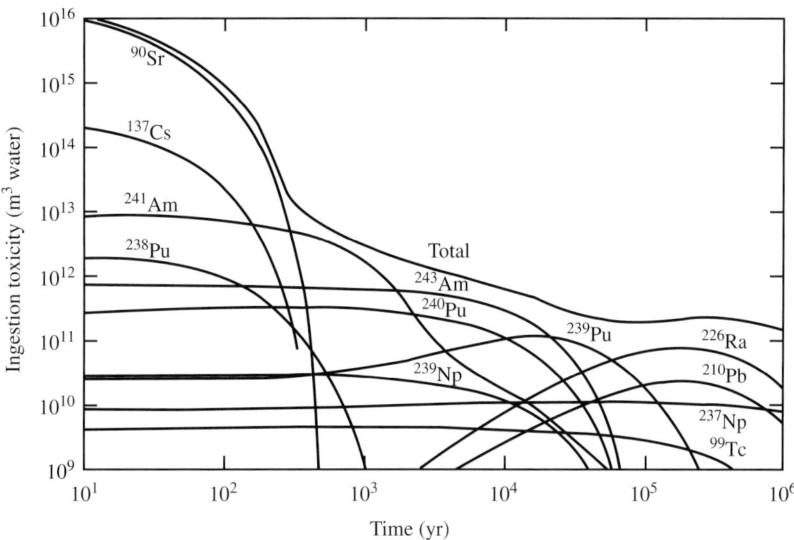

Figure 2 Ingestion toxicity for HLW as a function of decay time (Campbell *et al.* (1978); reproduced by permission of Nuclear Regulatory Commission from *Risk Methodology for Geologic Disposal of Nuclear Waste*, **1978**).

disposal of high-level wastes (HLWs) is shown. Ingestion toxicity for a given radioisotope is defined as the isotopic quantity (in microcuries) divided by the maximum permissible concentration in water (in microcuries per cubic meters) for that isotope (Campbell *et al.*, 1978). Plots of time-dependent thermal output from the radionuclides show similar trends. It can be seen that initially the bulk of the radioactivity is due to short-lived radionuclides ^{137}Cs and ^{90}Sr. After 1,000 yr, the bulk of the hazard is due to decay of ^{241}Am, ^{243}Am, ^{239,240}Pu, and ^{237}Np, and during the longest time periods, a mixture of the isotopes ^{99}Tc, ^{210}Pb, and ^{226}Ra dominates the small amount of radioactivity that remains. US Department of Energy (1980) shows that the relative toxicity (hazard) index (toxicity associated with ingesting a given weight of material) of spent fuel (SF) is about the same as uranium ore (0.2% U ore) after 10^5 yr.

Most regulations focus on the time period up to 10^4 yr and 10^5 yr after emplacement when radioactivity is dominated by the decay of americium, neptunium, and plutonium. Disposal of nuclear waste in the US is regulated by the Environmental Protection Agency (EPA) and the Nuclear Regulatory Commission (NRC). There are several classes of nuclear waste; each type is regulated by specific environmental regulations and each has a preferred disposal option, as described below.

Spent Fuel (SF) consists of irradiated fuel elements removed from commercial reactors or special fuels from test reactors. It is highly radioactive and generates a lot of heat; therefore, remote handling and heavy shielding are required. It is considered a form of HLW because of the uranium, fission products, and transuranics that it contains. HLW includes highly radioactive liquid, calcined or vitrified wastes generated by reprocessing of SF. Both SF and HLW from commercial reactors will be entombed in the geological repository at Yucca Mountain ~100 mile (1 mile = 1.609344 km) northwest of Las Vegas, Nevada. Disposal of spent nuclear fuel and HLW in the US is regulated by 40 CFR Part 191 (US EPA, 2001) and 10 CFR Part 60 (US NRC, 2001). It is discussed in more detail in a later section of this chapter.

Transuranic waste (TRU) is defined as waste contaminated with α-emitting radionuclides of atomic number greater than 92 and half-life greater than 20 yr in concentrations greater than 100 nCi g^{-1} (3.7×10^3 Bq g^{-1}). TRU is primarily a product of the reprocessing of SF and the use of plutonium in the fabrication of nuclear weapons. In the US, the disposal of TRU at the Waste Isolation Pilot Plant in southeastern New Mexico is regulated by 40 CFR Part 194 (US EPA, 1996). It is also discussed in more detail in a later section of this chapter.

Uranium mill tailings are large volumes of radioactive residues that result from the processing of uranium ore. In the US, the DOE has the responsibility for remediating mill tailing surface sites and associated groundwater under the Uranium Mill Tailings Radiation Control Act (UMTRCA) of 1978 and its modification in 1988. *Low-level wastes* (LLWs) are radioactive wastes not classified as HLW, TRU, SF, or uranium mill tailings. They are generated by institutions and facilities using radioactive materials and may include lab waste, towels, and lab coats contaminated during normal operations. Disposal of LLW is governed by agreements between states through state compacts at several facilities in the continental US. Geochemical data, conceptual models, and performance assessment methodologies relevant to LLW are summarized in Serne *et al.* (1990).

9.06.2.1.4 Sites of radioactive environmental contamination

In the US, radioactive contamination is of particular importance in the vicinity of US nuclear weapons production sites, near proposed or existing nuclear waste disposal facilities and in areas where uranium mining was carried out. Contamination from mining was locally significant in areas within the Navajo Nation in the southwestern US. Abdelouas *et al.* (1999) and Jove-Colon *et al.* (2001) provide concise reviews of the environmental problems associated with uranium mill tailings. These large volume mining and milling residues contain ~85% of the radioactivity of the unprocessed uranium ore, primarily in the form of U radioisotopes, ^{230}Th, Ra (^{226}Ra and ^{222}Ra), and Rn isotopes. In the US, more than 230 Mt of tailings are stored at 24 inactive tailing sites; in Canada more than 300 Mt of tailings have accumulated (Abdelouas *et al.*, 1999). By 1999, remediation of all 24 Title I UMTRA uranium mill tailings sites in the US had been completed through either in-place stabilization or relocation to more favorable sites. Other UMTRA sites (Title II sites) still had their mill tailings in place as of 2002 and have relatively long plumes that could release uranium to nearby aquifers (Brady *et al.*, 2002).

Both surface and subsurface processes are important for environmental contamination from uranium mill tailings. Surface soil/water contamination occurs by erosion and wind dispersion of contaminated soil; air pollution occurs by radon emission. Leaching and subsequent leaking of radioactive and hazardous metals (cadmium, copper, arsenic, molybdenum, lead, and zinc) from mill tailings contaminate groundwater. Concentrations of uranium in mill tailings and tailings pore waters in deposits reviewed by Abdelouas *et al.* (1999) were as high as 1.4 g kg^{-1} and 0.5 mg L^{-1}, respectively. Abdelouas *et al.* (1999) provide good examples of the relationships among environmental contamination, radiological hazards, and remediation activities in their discussion of the tailings piles sites at Tuba City, Arizona, at Rayrock, Northwest Territories, Canada, and at the complex of solution mining sites, open pits and tailings piles managed by the WIZMUT company in the uranium mining districts of East Germany.

The US DOE estimates that nuclear weapons production activities have led to radioactive contamination of ~63 Mm^3 of soil and 1,310 Mm^3 of groundwater in the US (US DOE, 1997a,b). The contamination is located at 64 DOE environmental management sites in 25 states. The Hanford Reservation in Washington State provides a good example of the diverse sources of radioactive contamination associated with weapons production. At this site, nine plutonium production reactors were built in the 100 Area; HLWs are stored in buried tanks in the 200 Area where SF was processed; and nuclear fuel was fabricated in the 300 Area. A complex mixture of radioactive and hazardous wastes is either stored in aging underground tanks or has been discharged to seeps or trenches and to the vadose zone in surface impoundments. The total inventory at the Hanford Site is estimated to be (1.33–1.37) × 10^{19} Bq (360–370 MCi); between 8.1 × 10^{15} Bq and 2.4 × 10^{17} Bq (0.22–6.5 MCi) has been released to the ground (National Research Council, 2001).

In Europe and the former Soviet Union (FSU), large areas have been contaminated by nuclear weapons production and uranium mining. The extent of contamination in the FSU is greater than in the US because of fewer controls on the disposal of nuclear wastes and less strict mining regulations. Extensive contamination has occurred due to solution mining of uranium ore deposits. In solution mining, complexing agents (lixiviants) are added to groundwater and reinjected into a uranium ore body. In the US, nearly all of uranium solution mining was carried out using carbonated, oxidizing solutions to produce mobile uranyl carbonato complexes. These practices did not lead to significant contamination. In contrast, more aggressive acid leach techniques using large amounts of sulfuric and nitric acids were used in solution mining operations in the FSU. These operations created a legacy of widespread contamination comprised of large volumes of groundwater contaminated with acid and leached metals. In addition, a large number of open pit and underground mines, tailing ponds, waste rock, and low-grade ore piles are sources of potential radioactive contamination in Central and eastern European countries.

The Straz deposit in the Hamr District of the Czech Republic provides a good example of this problem (Slezak, 1997). Starting in 1968, more than 4 Mt of sulfuric acid, 3 × 10^5 t of nitric acid, and 1.2 × 10^5 t of ammonia were injected into the subsurface to mine uranium ore. Now, ~266 Mm^3 in the North Bohemian Cretaceous Cenomanian and Turonian aquifers are contaminated with uranium, radium, and manganese and other solutes. The contaminated area is more than 24 km^2 and threatens the watershed of the Plucnice River.

Large-scale contamination in regions of the Ural Mountains of Russia is severe due to weapons production at the Chelyabinsk-65 complex near the city of Kyshtym. Cochran *et al.* (1993) and Shutov *et al.* (2002) provide overviews of the extent of the contamination at the site, drawing upon a number of Russian source documents. The main weapons production facility, the Mayak Chemical Combine, produced over 5.5×10^{19} Bq of radioactive wastes from 1948 through 1992. Over 4.6×10^{18} Bq of long-lived radioactivity were discharged into open lake storage and other sites. During the period 1948–1951, $\sim 9.3 \times 10^{16}$ Bq of medium level β^--activity liquid waste was discharged directly into the nearby Techa River; from March 1950 to November 1951, the discharge averaged 1.6×10^{14} Bq d^{-1}. During this period, $\sim 1.24 \times 10^5$ people living downstream from the facility were exposed to elevated levels of radioactivity. In 1951, the medium level waste was discharged into nearby Lake Karachay. Through 1990, the lake had accumulated 4.4×10^{18} Bq of long-lived radionuclides (primarily ^{137}Cs (3.6×10^{18} Bq) and ^{90}Sr (7.4×10^{17} Bq)). By 1993, seepage of radionuclides from the lake had produced a radioactive groundwater plume that extended 2.5–3.0 mile from the lake. In 1993, the total volume of contaminated groundwater was estimated to be more than 4 Mm^3 containing at least 1.8×10^{14} Bq of long-lived (>30 yr half-life) fission products.

Several large-scale nuclear disasters have occurred at the Chelyabinsk-65 complex. Two of these events led to the spread of radioactive contamination over large areas, exposure to large numbers of people and to relocation of entire communities. In 1957, during the so-called "Kyshtym Disaster," an explosion of an HLW storage tank released 7.4×10^{17} Bq of radioactivity into the atmosphere (Medvedev, 1979; Trabalka *et al.*, 1980). About 90% of the activity fell out in the immediate vicinity of the tank. Approximately 7.8×10^{16} Bq formed a kilometer-high radioactive cloud that contaminated an area greater than 2.3×10^4 km^2 to levels above 3.7×10^9 Bq km^{-2} of ^{90}Sr. This area was home to $\sim 2.7 \times 10^5$ people; many of the inhabitants were evacuated after being exposed to radioactive contamination for several years.

A second disaster at the site occurred in 1967, after water in Lake Karachay evaporated during a hot summer that followed a dry winter. Dust from the lakeshore sediments was blown over a large area, contaminating 1,800–2,700 km^2 with a total of 2.2×10^{13} Bq of ^{137}Cs and ^{90}Sr. Approximately 4.1×10^4 people lived in the area contaminated at levels of 3.7×10^9 Bq km^{-2} of ^{90}Sr or higher. According to Botov (1992), releases of radioactive dust from the area continued through at least 1972. Since 1967, a number of measures have been taken to reduce the dispersion of radioactive contamination from Lake Karachay as described in Cochran *et al.* (1993) and source documents cited therein.

9.06.2.2 Exposure to Background and Anthropogenic Sources of Radioactivity

The amounts of radioactivity listed in Table 2 provide some idea of the maximum amount of radioactivity potentially available for exposure to the public. However, such inventories by themselves provide little information about the actual risk associated with the radioactive hazards. Although the exposure has been very high for those unfortunate few that have been involved in nuclear accidents, the exposure of the general public to ionizing radiation is relatively low. The 2×10^5 workers involved in the cleanup of Chernobyl nuclear power plant after the 1986 accident received average total body effective doses of 100 mSv (10 rem) (Ewing, 1999). Current US regulations limit whole-body dose exposures (internal + external exposures) of radiological workers to 50 mSv yr^{-1} (5 rem yr^{-1}).

The calculated effective whole-body γ-radiation doses on tailings piles at the Rayrock mill tailings site in Northwest Territories, Canada, ranged from 42 mSv yr^{-1} to 73 mSv yr^{-1}. This is $\sim$24 times the annual dose from average natural radiation in Canada (Abdelouas *et al.*, 1999). Exposure to natural sources of radon (produced in the decay chain of crustal ^{238}U) averages 2 mSv yr^{-1} while other natural sources account for 1 mSv yr^{-1} (National Research Council, 1995). Total exposure to anthropogenic sources averages $\sim$0.6 mSv yr^{-1}, with medical X-ray tests accounting for $\sim$2/3 of the total. The average exposure related to the nuclear fuel cycle is estimated to be less than 0.01 mSv yr^{-1} and is comparable to that associated with the release of naturally occurring radionuclides from the burning of coal in fossil-fuel plants (Ewing, 1999; McBride *et al.*, 1978).

As shown previously, radioactive hazards associated with SF and HLW decrease exponentially over time (see Figure 2). After 10^4–10^5 yr, the risk to the public of a nuclear waste disposal vault approaches that of a high-grade uranium ore deposit and is less than the time invariant toxicity risk of ore deposits of mercury and lead (Langmuir, 1997a). The new EPA standard for nuclear waste repositories seeks to limit exposures from all exposure pathways for the reasonably maximally exposed individual living 18 km from a nuclear waste repository to 0.15 mSv yr^{-1} (15 mrem yr^{-1}) (US EPA, 2001). For comparison, the radiation dose on the shores of Lake Karachay near the Chelyabinsk-65 complex is ~0.2 Sv h^{-1} (Cochran *et al.*, 1993).

9.06.2.3 Health Effects and Radioactive Contamination

9.06.2.3.1 Biological effects of radiation damage

A summary of the nature of radioactive contamination would be incomplete without some mention of the human health effects related to radioactivity and radioactive materials. Several excellent reviews at a variety of levels of detail have been written and should be consulted by the reader (ATSDR, 1990a,b,c, 1999, 2001; Harley, 2001; Cember, 1996; BEIR V, 1988). The subject is extremely complex, with a number of important controversies that are beyond the scope of this chapter. Some general principles, however, are summarized below.

Ionizing radiation loses energy by producing ion pairs when passing through matter. These can damage biological material directly or produce reactive species (free radicals) that can subsequently react with biomolecules. External and internal exposure pathways are important for human health effects. External exposure occurs from radiation sources outside the body such as soil particles on the ground, surface water, and from particles dispersed in the air. It is most important for γ-radiation that has high penetrating potential; however, except for large doses, most of the radiation will pass through the body without causing significant damage. External exposure is less important for β- and α-radiation, which are less penetrating and will deposit their energy primarily on the outer layer of the skin.

Internal exposures of α- and β-particles are important for ingested and inhaled radionuclides. Dosimetry models are used to estimate the dose from internally deposited radioactive particles. The amount and mode of entry of radionuclides into the body, the movement and retention of radionuclides within various parts of the body, and the amount of energy absorbed by the tissues from radioactive decay are all factors in the computed dose (BEIR IV, 1988). The penetrating power of α-radiation is low; therefore, most of the energy from α-decay is absorbed in a relatively small volume surrounding an ingested or inhaled particle in the gut or lungs. This means that the chance that damage to DNA or other cellular material will occur is greater and the associated human health risk is higher for radioactive contamination composed of α-emitters than for the other forms of radioactive contamination. As mentioned above, weighting parameters that take into account the radiation type, the biological half-life, and the tissue or organ at risk are used to convert the physically absorbed dose in units of gray (or rad) to the biologically significant committed equivalent dose and effective dose, measured in units of Sv (or rem).

There is considerable controversy over the shape of the dose-response curve at the chronic low dose levels important for environmental contamination. Proposed models include linear models, nonlinear (quadratic) models, and threshold models. Because risks at low dose must be extrapolated from available data at high doses, the shape of the dose-response curve has important implications for the environmental regulations used to protect the general public. Detailed description of dosimetry models can be found in Cember (1996), BEIR IV (1988), and Harley (2001).

The health effect of radiation damage depends on a combination of events on the cellular, tissue, and systemic levels. Exposure to high doses of radiation (>5 Gy) can lead to direct cell death before division due to interaction of free radicals with macromolecules such as lipids and proteins. At acute doses of 0.1–5.0 Gy, damage to organisms can occur on the cellular level through single strand and double strand DNA breaks. These led to mutations and/or cellular reproductive death after one or more divisions of the irradiated parent cell. The dose level at which significant damage occurs depends on the cell type. Cells that reproduce rapidly, such as those found in bone marrow or the gastrointestinal tract, will be more sensitive to radiation than those that are longer lived, such as striated muscle or nerve cells. The effect of high radiation doses on an organ depends on the various cell types that it contains.

Cancer is the major effect of low radiation doses expected from exposure to radioactive contamination. Laboratory studies have shown that α-, β-, and γ-radiation can produce cancer in virtually every tissue type and organ in animals that have been studied (ATSDR, 2001). Cancers observed in humans after exposure to radioactive contamination or ionizing radiation include cancers of the lungs, female breast, bone, thyroid, and skin. Different kinds of cancers have different latency periods; leukemia can appear within 2 yr after exposure, while cancers of the breast, lungs, stomach, and thyroid have latency periods greater than 20 yr. Besides cancer, there is little evidence of other human health effects from low-level radiation exposure (ATSDR, 2001; Harley, 2001).

9.06.2.3.2 Epidemiological studies

The five large epidemiological studies that provide the majority of the data on the effects of radiation on humans are reviewed by Harley (2001). These include radium exposures by radium dial painters, atom bomb survivors, patients irradiated with X-rays for ankylosing spondylitis and ringworm (tinea capitis), and uranium miners exposed to radon. The first four studies examine health effects due to external exposures of high doses of ionizing radiation. The studies of uranium miners are more relevant to internal exposures. There have been 11 large follow-up studies of underground miners who were exposed to high concentrations of radon (^{222}Rn) and radon decay products. The carcinogens are actually the short-term decay products of ^{222}Rn (^{218}Po, ^{214}Po), which are deposited on the bronchial airways during inhalation and exhalation. Because of their short range, the α-particles transfer most of their energy to the thin layer of bronchial epithelium cells. These cells are known to be involved in induction of cancer, and it is clear that even relatively short exposures to the high levels possible in mines lead to excess lung cancers.

The results of the studies on miners have been used as a basis for estimating the risks to the general public from exposures to radon in homes. There is considerable controversy over this topic. Although the health effects due to the high radon exposures experienced by the miners have been well established, the risks at the lower exposure levels in residences are difficult to establish due to uncertainties in the dose-response curve and the confounding effects of smoking and urbanization. The reader is referred to extensive documentation by the National Academy of Sciences (1998) and the National Institute of Health (1994) for more information.

Epidemiological studies of populations in the FSU exposed to fallout from the 1986 nuclear reactor explosion at Chernobyl and releases from the Chelyabinsk-65 complex demonstrate the health effects associated with exposure to radioactive iodine, strontium, and caesium. A study of 2.81×10^4 individuals exposed along the Techa River, downstream from Chelyabinsk-65, revealed that a statistically significant increase in leukemia mortality arose between 5 yr and 20 yr after the initial exposure (37 observed deaths versus 14–23 expected deaths; see Cochran *et al.* (1993) and cited references and comments). There has been a significant increase of thyroid cancers among children in the areas contaminated by fallout from the Chernobyl explosion (Harley, 2001; UNSCEAR, 2000). The initial external exposures from Chernobyl were due to ^{131}I and short-lived isotopes. Subsequently, external exposures to ^{137}Cs and ^{134}Cs and internal exposures to radiocaesium through consumption of contaminated foodstuffs were important.

9.06.2.3.3 Toxicity and carcinogenicity

Toxicity and carcinogenicity of radioactive materials are derived from both the chemical properties of the radioelements and the effects of ionizing radiation. The relative importance of the radiological and chemical health effects are determined by the biological and radiological half-lives of the radionuclide and the mechanism of chemical toxicity of the radioelement. Ionizing radiation is the main source of carcinogenicity of the radionuclides. Many of the damaging effects of radiation can be repaired by the natural defenses of cells and the body. However, the longer a radionuclide is retained by the body or localized in a specific organ system, the greater the chances that the damage to DNA or proteins will not be repaired and a cancer will be initiated or promoted. The biological half-life of a radioelement, therefore, is an important determinant of the health risk posed by a radionuclide. This is determined by the chemical and physical form of the radioelement when it enters the body, the metabolic processes that it participates in and the routes of elimination from the body. For example, ^{90}Sr substitutes for calcium and accumulates on the surfaces of bones. It has a long biological half-life (50 yr), because it is recycled within the skeletal system. In young children, it is incorporated into growing bone where it can irradiate both the bone cells and bone marrow. Consequently, high exposures to ^{90}Sr can lead to bone cancer and cancers of the blood such as leukemia.

In contrast, the chemical toxicity of uranium is more important than its radiological hazard. In body fluids, uranium is present as soluble U(VI) species and is rapidly eliminated from the body (60% within 24 h; Goyer and Clarkson (2001)). It is rapidly absorbed from the gastrointestinal tract and moves quickly through the body. The uranyl carbonate complex in plasma is filtered out by the kidney glomerulus, the bicarbonate is reabsorbed by the proximule tubules, and the liberated uranyl ion is concentrated in the tubular cells. This produces systemic toxicity in the form of acute renal damage and renal failure.

Because there are few data on the results of human exposure to actinides, the health effects of these radioelements are more uncertain than those discussed above for ionizing radiation, radon, and fission products. Americium accumulates in bones and will likely cause bone cancer due to its radioactive decay. Animal studies suggest that plutonium will cause effects in the blood, liver, bone, lung, and immune systems. Other potential mechanisms of chemical toxicity and carcinogenicity of the actinides are similar to those of heavy metals and include: (i) disruption of transport pathways for nutrients and ions; (ii) displacement of essential metals such as Cu^{2+}, Zn^{2+}, and Ni^{2+}

from biomolecules; (iii) modification of protein conformation; (iv) disruption of membrane integrity of cells and cell organelles; and (v) DNA damage. More details can be found in the references cited at the beginning of this section.

9.06.3 EXPERIMENTAL AND THEORETICAL STUDIES OF RADIONUCLIDE GEOCHEMISTRY

Analysis of the risk from radioactive contamination requires consideration of the rates of release and dispersion of the contaminants through potential exposure pathways. Prediction of the release and dispersion of radionuclides from nuclear waste sites and contaminated areas must consider a series of processes including: (i) contact of the waste with groundwater, degradation of the waste, and release of radioactive aqueous species and particulate matter; (ii) transport of aqueous species and colloids through the saturated and vadose zone; and (iii) uptake of radionuclides by exposed populations or ecosystems. The geochemistry of the radionuclides will control migration through the geosphere by determining solubility, speciation, sorption, and the extent of transport by colloids. These are strong functions of the compositions of the groundwater and geomedia as well as the atomic structure of the radionuclides. These topics are the main focus of the sections that follow.

General predictions of radionuclide mobility are difficult to make; instead, site-specific measurements and thermodynamic calculations for the site-specific conditions are needed to make meaningful statements about radionuclide behavior. However, for the purposes of this chapter, some underlying themes are described. Trends in solubility and mobility are described from several perspectives: (i) by identity of the radionuclide and its dominant oxidation state under near-surface environmental conditions; (ii) by composition of the groundwater or pore waters (i.e., Eh, pH, and nature of complexing ligands and competing solutes); and (iii) by composition of the geomedia in contact with the solutions (i.e., mineralogy and organic matter).

9.06.3.1 Principles and Methods

9.06.3.1.1 Experimental methods

A wide variety of experimental techniques are used in radiochemical studies; a review of this subject is beyond the scope of this chapter. The interested reader should refer to the reviews and textbooks of actinide chemistry listed in the Section 9.06.1.2 above. Some general points, which should be considered in evaluating available data relevant to environmental radioactive contamination, are made below.

Solubility and speciation. Minimum requirements for reliable thermodynamic solubility studies include: (i) solution equilibrium conditions; (ii) effective and complete phase separation; (iii) well-defined solid phases; and (iv) knowledge of the speciation/oxidation state of the soluble species at equilibrium. Ideally, radionuclide solubilities should be measured in both "oversaturation" experiments, in which radionuclides are added to a solution until a solid precipitates, and "undersaturation" experiments, in which a radionuclide solid is dissolved in aqueous media. Due to the difference in solubilities of crystalline versus amorphous solids and different kinetics of dissolution, precipitation, and recrystallization, the results of these two types of experiments rarely agree. In some experiments, the maximum concentration of the radionuclide source term in specific water is of interest, so the solid that is used may be SF or nuclear waste glass rather than a pure radionuclide solid phase.

In addition, the maximum concentrations measured in laboratory experiments and the solubility-limiting solid phases identified are often not in agreement with the results of theoretical thermodynamic calculations. This discrepancy could be due to differences in the identity or the crystallinity of solubility-limiting solids assumed in the calculation or to errors in the thermodynamic property values used in the calculations. Thus, although theoretical thermodynamic calculations are useful in summarizing available information and in performing sensitivity analyses, it is important also to review the results of empirical experimental studies in site-specific solutions.

Good summaries of accepted experimental techniques can be found in the references that are cited for individual radionuclides in the sections below. Nitsche (1991) provides a useful general summary of the principles and techniques of solubility studies. A large number of techniques have been used to characterize the aqueous speciation of radionuclides. These include potentiometric, optical absorbance, and vibrational spectroscopy. Silva and Nitsche (1995) summarize the use of conventional optical absorption and laser-based photothermal spectroscopy for detection and characterization of solution species and provide an extensive citation list. A recent review of the uses of Raman and infrared spectroscopy to distinguish various uranyl hydroxy complexes is given by Runde *et al.* (2002b).

Extraction techniques to separate oxidation states and complexes are often combined with radiometric measurements of various fractions. A series of papers by Choppin and co-workers provides good descriptions of these techniques (e.g., Caceci and Choppin, 1983; Schramke *et al.*, 1989). Cleveland and co-workers used a variety of extraction techniques to characterize the speciation of plutonium, neptunium, and americium in natural waters (Rees *et al.*, 1983; Cleveland *et al.*, 1983a,b; and Cleveland and Rees, 1981).

A variety of methods have been used to characterize the solubility-limiting radionuclide solids and the nature of sorbed species at the solid/water interface in experimental studies. Electron microscopy and standard X-ray diffraction techniques can be used to identify some of the solids from precipitation experiments. X-ray absorption spectroscopy (XAS) can be used to obtain structural information on solids and is particularly useful for investigating noncrystalline and polymeric actinide compounds that cannot be characterized by X-ray diffraction analysis (Silva and Nitsche, 1995). X-ray absorption near edge spectroscopy (XANES) can provide information about the oxidation state and local structure of actinides in solution, solids, or at the solution/solid interface. For example, Bertsch *et al.* (1994) used this technique to investigate uranium speciation in soils and sediments at uranium processing facilities. Many of the surface spectroscopic techniques have been reviewed recently by Bertsch and Hunter (2001) and Brown *et al.* (1999). Specific recent applications of the spectroscopic techniques to radionuclides are described by Runde *et al.* (2002b). Rai and co-workers have carried out a number of experimental studies of the solubility and speciation of plutonium, neptunium, americium, and uranium that illustrate combinations of various solution and spectroscopic techniques (Rai *et al.*, 1980, 1997, 1998; Felmy *et al.*, 1989, 1990; Xia *et al.*, 2001).

Sorption studies. Several different approaches have been used to measure the sorption of radionuclides by geomedia. These include: (i) the laboratory batch method, (ii) the laboratory flow-through (column) method, and (iii) the *in situ* field batch sorption method. Laboratory batch tests are the simplest experiments; they can be used to collect distribution coefficient (K_d) values or other partitioning coefficients to parametrize sorption and ion-exchange models. The different sorption models are summarized in Section 9.06.3.1.2. The term *sorption* is often used to describe a number of surface processes including adsorption, ion exchange, and co-precipitation that may be included in the calculation of a K_d. For this reason, some geochemists will use the term sorption ratio (R_d) instead of distribution coefficient (K_d) to describe the results of batch sorption experiments. In this chapter, both terms are used in order to be consistent with the terminology used in the original source of information summarized.

Batch techniques. Descriptions of the batch techniques for radionuclide sorption and descriptions of calculations used to calculate distribution coefficients can be found in ASTM (1987), Park *et al.* (1992), Siegel *et al.* (1995a), and US EPA (1999a). The techniques typically involve the following steps: (i) contacting a solution with a known concentration of radionuclide with a given mass of solid; (ii) allowing the solution and solid to equilibrate; (iii) separating the solution from the solid; and (iv) measuring the concentration of radionuclide remaining in solution.

In batch systems, the distribution or sorption coefficient (K_d or R_d) describes the partitioning of a contaminant between the solid and liquid phases. The K_d is commonly measured under equilibrium or at least steady-state conditions, unless the goal of the experiment is to examine the kinetics of sorption. It is defined as follows:

$$K_d(\text{mL g}^{-1}) = \frac{C_S}{C_L} \tag{1}$$

where C_S is the concentration of the contaminant on the solid and C_L is the concentration in solution. In practice, the concentration of the contaminant on the solid is rarely measured. Rather, it is calculated from the initial and final solution concentrations, and the operative definition for the K_d becomes

$$K_d(\text{mL g}^{-1}) = \frac{(C_i V_i - C_f V_f)/m}{C_f} \tag{2}$$

where C_i and C_f are the initial and final concentrations of contaminant in solution, respectively; V_i and V_f are the initial and final solution volumes; and m is the mass of the substrate added to the system.

Measured batch K_d values can be used to calculate a retardation factor (R), which describes the ratio of the groundwater velocity ν_m to the velocity of radionuclide movement ν_r:

$$R = \frac{\nu_m}{\nu_r} = 1 + \frac{K_d \rho}{\phi} \tag{3}$$

where ρ is the bulk density of the porous medium and ϕ is the porosity. This equation can be rearranged, and contaminant retardation values measured from column breakthrough curves can be used to calculate K_d values.

Many published data from batch sorption measurements are subject to a number of limitations as described by Siegel and Erickson (1984, 1986), Serne and Muller (1987), and by US EPA (1999a). These include a solution: solid ratio that is much higher than that present in natural conditions, an inability to account for multiple sorbing species, an inability to measure different adsorption and desorption rates and affinities, and an inability to distinguish between adsorption and co-precipitation.

Batch methods are also used to collect data to calculate equilibrium constants for the surface-complexation models (SCMs) which are described in more detail in the next section. Commonly for these models, sorption is measured as function of pH and data are presented as pH–sorption edges similar to the one shown in Figure 3. Sorption is strongly affected by the surface charge of the geomedia. The proton is the surface potential determining ion (PDI) in metal oxyhydroxides and of the high-energy edge sites in aluminosilicates. In the absence of significant sorption of metal ions (i.e., low surface loading or site occupancy), the surface charge is determined primarily by the difference between the surface concentrations of positively charged, protonated sites, and negatively charged, deprotonated sites.

Considerable data have been collected describing the influence of pH on actinide sorption. Sorption edges are most commonly (and usefully) measured for single oxidation states of the radionuclide. Figure 3 shows a plot of a sorption/desorption edge for uranium under oxidizing conditions and illustrates reversible sorption. The effect of competition between protons and the radionuclide is illustrated by the sorption edge. Appreciable sorption can occur at a relatively low pH for radionuclides that form strong bonds with the surface, resulting in a low pH_{50} value (the pH at which 50% of the radionuclide is adsorbed). Radionuclides that form weaker surface complexes can only sorb appreciably when the concentration of competing protons is low (high pH) and therefore have high pH_{50} values. Comparisons of sorption edges for different radionuclides on the same substrate or for a single radionuclide on several substrates can be made by referring to their pH_{50} values. For example, Kohler *et al.* (1992) show that goethite strongly sorbs Np(V) from $NaClO_4$ solutions, while quartz only weakly sorbs Np(V). The pH_{50} sorption values for Np(V) increase in the order goethite < hematite < gibbsite < albite < quartz.

Other techniques. Laboratory column tests are more difficult to perform but overcome some of the limitations of the batch tests. Proper design, descriptions of experimental procedures, and methods of data interpretation for column tests can be found in US EPA (1999a), Relyea (1982), Van Genuchten and Wierenga (1986), Triay *et al.* (1992, 1996, 1997), Torstenfelt (1985a,b), Siegel *et al.* (1995b), Sims *et al.* (1996), and Gabriel *et al.* (1998). In these experiments, a solution containing a known concentration of radionuclide is introduced into a column of packed soil or rock at a specified flow rate. The concentration of the radionuclide in the column effluent is monitored to obtain a breakthrough curve; the shape of the curve provides information about sorption equilibrium and kinetics and other properties of the crushed rock or intact rock column. Limitations of this technique are: (i) complex, time-consuming and expensive experimental procedures are required; (ii) symmetric breakthrough curves are rarely obtained and a number of ad hoc assumptions may be required to interpret them; and (iii) the sorption parameters are dependent on the hydrodynamics of the specific experiment and are not applicable to other conditions.

In situ (field) batch sorption tests use measurements of the radionuclide contents of samples of rock cores and consanguineous pore water obtained at a field site. The advantage of this approach is that the water and rock are likely to be

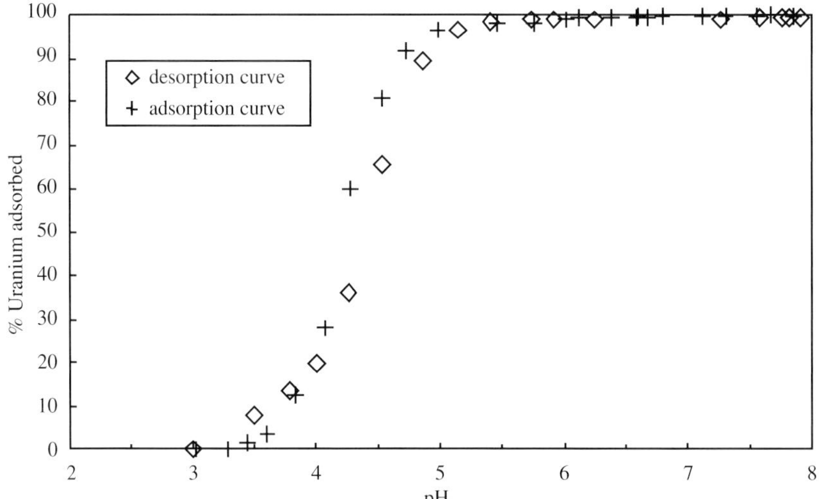

Figure 3 Sorption and desorption edges for uranyl on goethite. Each sample contained 60 $m^2 L^{-1}$ goethite and 100 $\mu g\ mL^{-1}$ uranium. The sorption edge was measured 2 d after uranium addition; desorption samples were contacted with the uranyl solution at pH 7 for 5 d, the pH was adjusted to cover the range of interest, and the samples were re-equilibrated for 2 d prior to sampling (see also Bryan and Siegel, 1998).

in chemical equilibrium and that the concentrations of any cations or anions competing for sorption sites are appropriate for natural conditions. Disadvantages of this technique are associated with limitations in obtaining accurate measurements of the concentration of radionuclides on the rock surface in contact with the pore fluid. Applications of this technique are described in Jackson and Inch (1989), McKinley and Alexander (1993), Read *et al.* (1991), Ward *et al.* (1990), and Payne *et al.* (2001).

Surface analytical techniques. A variety of spectroscopic methods have been used to characterize the nature of adsorbed species at the solid–water interface in natural and experimental systems (Brown *et al.*, 1999). Surface spectroscopy techniques such as extended X-ray absorption fine structure spectroscopy (EXAFS) and attenuated total reflectance Fourier transform infrared spectroscopy (ATR-FTIR) have been used to characterize complexes of fission products, thorium, uranium, plutonium, and uranium sorbed onto silicates, goethite, clays, and microbes (Chisholm-Brause *et al.*, 1992, 1994; Dent *et al.*, 1992; Combes *et al.*, 1992; Bargar *et al.*, 2000; Brown and Sturchio, 2002). A recent overview of the theory and applications of synchrotron radiation to the analysis of the surfaces of soils, amorphous materials, rocks, and organic matter in low-temperature geochemistry and environmental science can be found in Fenter *et al.* (2002).

Before the application of these techniques became possible, the composition and predominance of adsorbed species was generally inferred from information about aqueous species. In some cases, these models for adsorbed species have been shown to be incorrect. For example, spectroscopic studies demonstrated that U(VI)-carbonato complexes were the predominant adsorbed species on hematite over a wide pH range (Bargar *et al.*, 2000). This was contrary to expectations based on analogy to aqueous U(VI)-carbonato complexes, which are present in very low concentration at near-neutral and acidic pH. As discussed later, the use of different sorption models will produce alternate stoichiometries for surface species when data are fit to sorption edges. The information obtained from surface spectroscopy helps to constrain the interpretation of the results of batch sorption tests by revealing the stoichiometry of the sorbed species. For example, Redden and Bencheikh-Latmar (2001) demonstrate how data from an EXAFS study helped to elucidate the interactions among UO_2^{2+}, citrate, and goethite over the pH range 3.5–5.5. Previous studies of sorption equilibria (Redden *et al.*, 1998) showed that whereas citric acid reduced the sorption of uranium by gibbsite and kaolinite, the presence of citrate led to enhanced uranyl sorption at high citrate: UO_2^{2+} ratios. The EXAFS spectra were consistent with the existence of two principal surface species: an inner-sphere uranyl–goethite complex and an adsorbed uranyl–citrate complex, which displaces the binary uranyl–geothite complex at high citrate: UO_2^{2+} ratios. Other examples of the use of these techniques to understand the nature of radionuclide sorption can be found in a recent review by Brown and Sturchio (2002).

9.06.3.1.2 Theoretical geochemical models and calculations

Aqueous speciation and solubility. Several geochemical codes are commonly used for calculations of radionuclide speciation and solubilities. Recent reviews of the codes can be found in Serne *et al.* (1990), Mangold and Tsang (1991), NEA (1996), and US EPA (1999a). Extensive databases of thermodynamic property values and kinetic rate constants are required for these codes. Several databases of thermodynamic properties of the actinides have been developed since the early 1970s. Of historical importance are the compilations and reviews of Lemire and Tremaine (1980), Phillips *et al.* (1988), and Fuger *et al.* (1990). The more recent comprehensive and consistent databases have been based on compilations produced by the Nuclear Energy Agency (NEA) for plutonium and neptunium (Lemire *et al.*, 2001), americium (Silva *et al.*, 1995), uranium (Grenthe *et al.*, 1992), and technetium (Rard *et al.*, 1999). These books contain suggested values for ΔG, ΔH, C_p, and log K_f for formation reactions of radionuclide species. More recent publications often use these compilations as reference and add more recent property values or correct errors. For example, Langmuir (1997a) updates the 1995 NEA database for uranium (Grenthe *et al.*, 1995; Silva *et al.*, 1995) and provides results of solubility, speciation, and sorption calculations using the MINTEQ2A code (Allison *et al.*, 1991) as described later in this chapter.

The more recent compilations have been *internally self-consistent*. The process of compiling an internally self-consistent database consists of several steps: (i) compilation of process values such as equilibrium constants for reactions involving the element of interest; (ii) extrapolation of equilibrium constants to reference conditions (usually zero ionic strength and 25 °C), and (iii) calculation of property values such as free energies of formation of the products through reaction networks (Wagman *et al.*, 1982; Grenthe *et al.*, 1992). Calculated thermodynamic constants from different experimental studies will be incompatible if different reference states or reaction networks are used. Caution must be exercised in combining constants from different compilations, because they are dependent on the methods used for ionic strength correction and the values used for auxiliary species such as OH^- or SO_4^{2-}. Thus, strictly speaking, as new species are identified or suspect ones eliminated and as constants of previously recognized species are revised, the entire reaction network must be used to rederive all of the constants in order to maintain internal consistency. Software is available to recalculate the reaction networks to ensure that internal consistency is maintained with the NEA Thermochemical Database (TDB) Project (http://www.nea.fr/html/dbtdb/cgi-bin/tdbdocproc.cgi). In practice, however, these recalculations are not commonly done, especially if only minor changes in values of equilibrium constants are expected.

For most solubility and speciation studies, calculations of the activity coefficients of aqueous species are required. For waters with relatively low ionic strength (0.01–0.1 *m*), simple corrections such as the Debye–Hückel equation are used (Langmuir, 1997a, p. 127). This model accounts for the electrostatic, nonspecific, long-range interactions between water and the solutes. At higher ionic strengths, short-range, nonelectrostatic interactions must be taken into account. The NEA has developed a database based on the specific interaction theory (SIT) approach of Bronsted (1922), Scatchard (1936), and Guggenheim (1966). In this model, activity coefficients are calculated from a set of virial coefficients obtained from experimental data in simple solutions (Ciavatta, 1980). This method is assumed to be valid for ionic strengths up to 3.0 *m* and had been used to extrapolate experimental data to zero ionic strength to obtain equilibrium constants and free energies for the NEA databases (Lemire *et al.*, 2001; Silva *et al.*, 1995; Grenthe *et al.*, 1992; Rard *et al.*, 1999).

The US DOE has adopted the more complex Pitzer model (Pitzer, 1973, 1975, 1979) for calculations of radionuclide speciation and solubility in its Nuclear Waste Management Programs. This model includes concentration-dependent interaction terms and is valid up to ionic strengths greater than 10 *m*. However, because it requires three parameters instead of the single interaction parameter of the SIT model, this method requires more extensive experimental data. Pitzer coefficients and coefficients from the Harvie–Moller–Weare model (Harvie *et al.*, 1984) are used in several geochemical codes such as PHRQPITZ (Plummer *et al.*, 1988), EQ3/6 (Wolery, 1992a), and REACT (Bethke, 1998). Considerably more experimental data are needed for the radionuclides, and the Pitzer coefficient database is currently being developed by the US DOE. Data are available for some radionuclides (Felmy and Rai, 1999) and have been used in the Waste Isolation Pilot Plant (WIPP) program as described in a later section.

Sorption overview. Both empirical and mechanistic approaches have emerged since the 1970s to describe interactions between radionuclides and geomedia. These are based on "conditional" constants, which are valid for specific experimental conditions, or more robust "intrinsic" constants, which are valid over a wider range of conditions. The "empirical approach" involves measurements of conditional radionuclide distribution or sorption coefficients (K_ds or R_ds) in site-specific water–rock systems using synthetic or natural ground waters and crushed rock samples. Mechanistic-based approaches produce intrinsic, thermodynamic surface-complexation constants for simple electrolyte solutions with pure mineral phases.

The different approaches to describing sorption can be discussed in order of increasing model complexity and the robustness of their associated constants:

(i) linear sorption (K_d or R_d);
(ii) nonlinear sorption (Freundlich and other isotherms);
(iii) constant-charge (ion-exchange) model;
(iv) constant-capacitance model;
(v) double layer or diffuse layer model (DLM); and
(vi) triple-layer model (TLM).

Figure 4 compares several of these models with respect to the nature of the constants that each uses. The simplest model (linear sorption or K_d) is the most empirical model and is widely used in contaminant transport models. K_d values are relatively easy to obtain using the batch methods described above. The K_d model requires a single distribution constant, but the K_d value is conditional with respect to a large number of variables. Thus, even if a batch K_d experiment is carefully carried out to avoid introduction of extraneous effects such as precipitation, the K_d value that is obtained is valid only for the particular conditions of the experiment. As Figure 4 shows, the radionuclide concentration, pH, major and minor element composition, rock mineralogy, particle size and solid-surface-area/solution volume ratio must be specified for each K_d value.

Ion-exchange models are commonly used to describe radionuclide sorption onto the fixed-charged sites of materials like clays. Ion exchange will be strongly affected by competition with monovalent and divalent ions such as Na^+ and Ca^{2+}, whereas it will be less dependent on pH over the compositional ranges common for natural waters. Many studies of strontium and caesium sorption by aluminosilicates (e.g., Wahlberg and Fishman, 1962; Tamura, 1972) have been carried out within the framework of ion-exchange theory. Early mechanistic studies

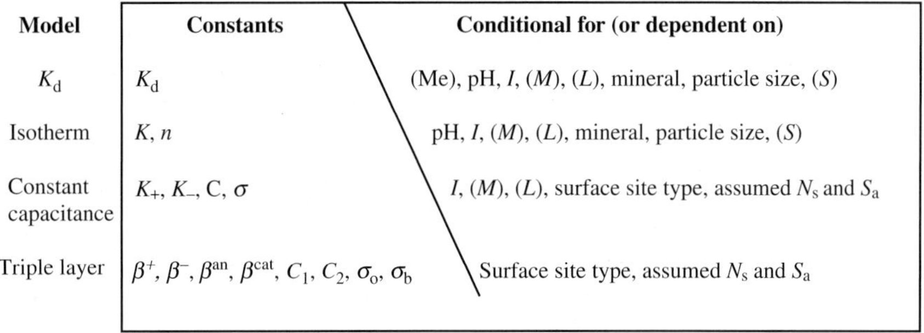

Model	Constants	Conditional for (or dependent on)
K_d	K_d	(Me), pH, *I*, (*M*), (*L*), mineral, particle size, (*S*)
Isotherm	*K*, *n*	pH, *I*, (*M*), (*L*), mineral, particle size, (*S*)
Constant capacitance	K_+, K_-, C, σ	*I*, (*M*), (*L*), surface site type, assumed N_s and S_a
Triple layer	β^+, β^-, β^{an}, β^{cat}, C_1, C_2, σ_o, σ_b	Surface site type, assumed N_s and S_a

Figure 4 Comparison of sorption models. Several commonly used sorption models are compared with respect to the independent *constants* they require. These *constants* are valid only under specific conditions, which must be specified in order to properly use them. In other words, the constants are *conditional* with respect to the *experimental* variables described in the third column of the figure. K_d is the radionuclide distribution constant; *K* and *n* are the Freundlich isotherm parameters; β^+ and β^- are surface complexation constants for protonation and deprotonation of surface sites; K_+, K_-, β^{an}, β^{cat} are surface complexation constants for sorption of cations and anions in the constant capacitance model and TLM, respectively; *C*, C_1, and C_2 are capacitances for the electrical double layers; σ, σ_o, and σ_b are surface charges at different surface planes; (Me) and (S) are concentrations of the sorbing ions and the surface sites, (*M*), (*L*) are concentrations of other cations and ligands in solution, respectively; *I* is the ionic strength of the background electrolyte; N_s and S_a are the site density and specific surface of the substrate, respectively. The requirements of the DLM are similar to those of the constant capacitance model.

of uranium sorption were carried out within an ion-exchange framework (e.g., Tsunashima *et al.*, 1981); however, more recent studies relevant to environmental conditions have used SCMs (e.g., Davis, 2001).

The TLM (Davis and Leckie, 1978) is the most complex model described in Figure 4. It is an example of an SCM. These models describe sorption within a framework similar to that used to describe reactions between metals and ligands in solutions (Kent *et al.*, 1988; Davis and Kent, 1990; Stumm, 1992). Reactions involving surface sites and solution species are postulated based on experimental data and theoretical principles. Mass balance, charge balance, and mass action laws are used to predict sorption as a function of solution chemistry. Different SCMs incorporate different assumptions about the nature of the solid–solution interface. These include the number of distinct surface planes where cations and anions can attach (double layer versus triple layer) and the relations between surface charge, electrical capacitance, and activity coefficients of surface species.

Aqueous radionuclide species and other solutes can sorb to mineral surfaces by forming chemical bonds directly with the amphoteric sites or may be separated from the surface by a layer of water molecules and be bound through longer-range electrostatic interactions. In the TLM, complexes of the former type are often called "inner-sphere" complexes; those of the latter type are called "outer-sphere" complexes (Davis and Kent, 1990). The TLM includes an inner plane (o-plane), an outer plane (β-plane), and a diffuse layer that extends from the β-plane to the bulk solution. Sorption via formation of inner-sphere complexes is often referred to "chemisorption" or "specific sorption" to distinguish it from ion exchange at fixed charged sites or outer-sphere complexation that is dominantly electrostatic in nature.

The sorption of uranium to form the sorbed species ($>FeOH-UO_2^+$) at the o-plane of an iron oxyhydroxide surface (represented as $>FeOH$) can be represented by a surface reaction in a TLM as

$$>FeOH + UO_{2\,(s)}^{2+} \rightarrow\ > FeO-UO_2^+ + H_{(s)}^+ \quad (4)$$

where $UO_{2(s)}^{2+}$ and $H_{(s)}^+$ are the aqueous uranyl ion and proton at the surface, respectively. The mass action law (equilibrium constant) for the reaction using the TLM is

$$\beta^{UO_2^{2+}} = \frac{\{>FeO-UO_2^+\} \times a_{H^+} \times \exp[e\psi_0/kT]}{\{>FeOH\} \times a_{UO_2^{2+}}} \quad (5)$$

where $\beta^{UO_2^{2+}}$ is the intrinsic surface-complexation constant for the uranyl cation; $\{>FeOH\}$ and $\{>FeO-UO_2^+\}$are the activities of the uncomplexed and complexed surface sites, respectively; a_{H^+} and $a_{UO_2^{2+}}$ are activities of the aqueous species in the bulk solution; ψ_0 is the electrical potential for the inner (o) surface plane; and *k*, *T*, and *e* are the Boltzmann constant, absolute temperature, and the fundamental charge, respectively. The exponential term describes the net change in electrostatic energy required to exchange the divalent uranyl ion for the proton at the mineral surface. (The activities of the uranyl ion and proton at the surface differ from their activities in the bulk solution: $\{UO_{2(s)}^{2+}\} = a_{UO_2^{2+}} \times \exp(-2e\psi_0/kT)$ and $\{H_{(s)}^+\} = a_{H^+} \times \exp(-e\psi_0/kT)$. Equation (5) can be derived from the equilibrium constant for Equation (4), $\beta^{UO_2^{2+}} = [\{>FeO-UO_2^+\}/\{>FeOH\}] \times [\{H_{(s)}^+\}/\{UO_{2(s)}^{2+}\}]$, by substitution.)

In natural waters, other surface reactions will be occurring simultaneously. These include protonation and deprotonation of the >FeOH site at the inner o-plane and complexation of other cations and anions to either the inner (o) or outer (β) surface planes. Expressions similar to Equation (5) above can be written for each of these reactions. In most studies, the activity coefficients of surface species are assumed to be equal to unity; thus, the activities of the surface sites and surface species are equal to their concentrations. Different standard states for the activities of surface sites and species have been defined either explicitly or implicitly in different studies (Sverjensky, 2003). Sverjensky (2003) notes that the use of a hypothetical 1.0 M standard state or similar convention for the activities of surface sites and surface species leads to surface-complexation constants that are directly dependent on the site density and surface area of the sorbent. He defines a standard state for surfaces sites and species that is based on site occupancy and produces equilibrium constants independent of these properties of the solids. For more details about the properties of the electrical double layer, methods to calculate surface speciation and alternative models for activity coefficients for surface sites, the reader should refer to the reference cited above and other works cited therein.

The TLM contains eight adjustable constants (identified in the caption of Figure 4) that are valid over the ranges of pH, ionic strength, solution composition, specific areas, and site densities of the experiments used to extract the constants. The surface-complexation constants, however, must be determined for each type of surface site of interest and should not be extrapolated outside the original experimental conditions. Although the TLM constants are valid under a wider range of conditions than are K_ds, considerably more experimental data must be gathered to obtain the adjustable parameters. An important advantage of surface-complexation constant models is that they provide a structured way to examine experimental data obtained in batch sorption studies. Application of such models may ensure that extraneous effects such as precipitation have not been introduced into the sorption experiment.

Between the simplicity of the K_d model and the complexity of the TLM, there are several other sorption models. These include various forms of isotherm equations (e.g., Langmuir and Freundlich isotherms) and models that include kinetic effects. The generalized two-layer model (Dzombak and Morel, 1990) (also referred to as the DLM) recently has been used to model radionuclide sorption by several research groups (Langmuir, 1997a; Jenne, 1998; Davis, 2001). Constants used in this model are dependent upon the concentration of background electrolytes and are thus less robust than those of the TLM. Reviews by Turner (1991), Langmuir (1997b), and US EPA (1999a) provide concise descriptions of many of these models.

Several researchers have illustrated the interdependence of the adjustable parameters and the nonunique nature of the SCM constants by fitting the same or similar sorption edges to a variety of alternate SCM models (Westall and Hohl, 1980; Turner, 1995; Turner and Sassman, 1996). Robertson and Leckie (1997) systematically examined the effects of SCM model choice on cation binding predictions when pH, ionic strength, cation loading, and proposed surface complex stoichiometry were varied. They show that although different models can be used to obtain comparable fits to the same experimental data set, the stoichiometry of the proposed surface complex will vary considerably between the models. In the near future, it is possible that the actual stoichiometry of adsorbed species can be determined using combinations of the spectroscopic techniques discussed in a previous section and molecular modeling techniques similar to those described in Cygan (2002).

There is no set of reference surface-complexation constants corresponding to the reference thermodynamic property values contained in the NEA thermodynamic database described in the previous section (Grenthe *et al.*, 1992; Silva *et al.*, 1995; Rard *et al.*, 1999; Lemire *et al.*, 2001). Wang *et al.* (2001a,b) used the DLM with original experimental data to obtain a set of internally consistent surface-complexation constants for Np(V), Pu(IV), Pu(V), and Am(III), I^-, IO_3^-, and TcO_4^- sorption by a variety of synthetic oxides and geologic materials in low-ionic-strength waters (<0.1 M). Turner and Sassman (1996) and Davis (2001) provide databases for uranium sorption also using the DLM. Langmuir (1997b) compiles surface-complexation constants for actinides and fission products based on several different SCMs. Other compilations are based on the TLM; these include those of Hsi and Langmuir (1985), Tripathi (1983), McKinley *et al.* (1995), Turner *et al.* (1996), and Lenhart and Honeyman (1999) for uranium; Girvin *et al.* (1991) and Kohler *et al.* (1999) for neptunium; Laflamme and Murray (1987), Quigley *et al.* (1996), and Murphy *et al.* (1999) for thorium; and Sanchez *et al.* (1985) for plutonium. Langmuir (1997b), Davis (2001), Wang *et al.* (2001a,b), and Turner *et al.* (2002) provide references to a large number of other surface-complexation studies of radionuclides.

Representation of sorption of radionuclides under natural conditions. Several approaches have been used to represent variability of sorption under natural conditions. These include: (i) sampling K_d values from a probability distribution function (PDF); (ii) calculating a K_d using empirical relations based on measurements over a range of experimental conditions, solution compositions, and mineral properties; and (iii) calculating aqueous and surface speciation using a thermodynamically based surface-complexation model. The first approach is used most commonly in risk assessment and remediation design calculations. Because of the diversity of solutions, minerals, and radionuclides that could be present at contaminated sites and potential waste disposal repository sites, a large body of empirical radionuclide sorption data has been generated. Databases of K_d values that can be used to estimate PDFs for various geologic media are summarized by Barney (1981a,b), Tien *et al.* (1985), Bayley *et al.* (1990), US DOE (1988), McKinley and Scholtis (1992), Triay *et al.* (1997), the US EPA (1999a,b), and Krupka and Serne (2000). Methods used to specify PDFs for K_ds for use with sampling techniques such as Latin Hypercube Sampling (Iman and Shortencarier, 1984; Helton and Davis, 2002) have been described by Siegel *et al.* (1983, 1989), Wilson *et al.* (1994), and Rechard (1996).

Serne and Muller (1987) describe attempts to find statistical empirical relations between experimental variables and the measured sorption ratios (R_ds). Mucciardi and Orr (1977) and Mucciardi (1978) used linear (polynomial regression of first-order independent variables) and nonlinear (multinomial quadratic functions of paired independent variables, termed the Adaptive Learning Network) techniques to examine effects of several variables on sorption coefficients. The dependent variables considered included cation-exchange capacity (CEC) and surface area (SA) of the solid substrate, solution variables (Na, Ca, Cl, HCO_3), time, pH, and Eh. Techniques such as these allow modelers to construct a narrow probability density function for K_ds.

The dependence of a K_d on the composition of the groundwater can also be described in terms of more fundamental thermodynamic parameters. This can be illustrated by considering the sorption of uranyl (UO_2^{2+}) onto a generic surface site (>SOH) of a mineral

$$> SOH + UO_2^{2+} \rightarrow SO{-}UO_2^{+} + H^{+} \quad (6)$$

with an equilibrium sorption binding constant β^{cat} defined for the reaction. The concentration of UO_2^{2+} available to complex with the surface site will be affected by complexation reactions with other ligands such as carbonate. The K_d in a system containing the uranyl ion and its hydroxo and carbonato complexes can be calculated as

$$K_d = \frac{\beta^{cat} \times \{SOH\} \times C}{\{H^+\} \times [1 + \sum_{ijk} K_{i,j,k} \{UO_2^{2+}\}^{i-1} \{OH\}^j \{CO_3^{2-}\}^k]} \quad (7)$$

For simplicity, in this equation, we have assumed that activities are equal to concentrations and brackets refer to activities. C is a units conversion constant $= V_v\, m^{-1}$, relating void volume V_v (mL) in the porous media and the mass m (g) of the aquifer material in contact with the volume V_v; $K_{i,j,k}$ is the formation constant for an aqueous uranyl complex, and the superscripts i, j, k describe the stoichiometry of the complex. The form that the sorption binding constant β^{cat} takes is different for the different sorption models shown in Figure 4 (e.g., see Equation (5)). Leckie (1994) derives similar expressions for more complex systems in which anionic and cationic metal species form polydentate surface complexes. Equation (7) can be derived from the following relationships for this system:

(i) K_d = total sorbed uranium/total uranium in solution;

(ii) total sorbed uranium $= \{>SO{-}UO_2^{2+}\} = \beta^{cat} \times \{>SOH\} \times \{UO_2^{2+}\}/\{H^+\}$;

(iii) total uranium in solution $= \{UO_2^{2+}\} + \{UO_2CO_3\}$ + other uranyl complexes;

(iv) $\{UO_2CO_3\} = K_{UO_2CO_3} \times \{UO_2^{2+}\} \times \{CO_3^{2-}\}$; and

(v) similar expressions can be written for other uranyl species.

Substituting (v) and (iv) into (iii), and then substituting (iii) and (ii) into (i), yields Equation (7) after some manipulation. Note that activity coefficients of all species are assumed to be equal to 1.0.

Expressions like Equation (7) can be solved using computer programs such as HYDRAQL. Using a spreadsheet program for postprocessing of the results, K_d values can easily be calculated over ranges of solution compositions. Using this approach, the effects of relatively small changes in the composition of the groundwaters can be shown to result in order-of-magnitude changes in the K_d. Figure 5 shows that the calculated K_d of uranium in systems containing several competing ligands can be sensitive to the concentration of other cations such as Ca^{2+}. Leckie (1995) provides examples of this methodology and produces multidimensional K_d response surfaces. Approaches to using thermodynamic sorption models to predict, interpret, or guide the collection of K_d data are summarized by the NEA (2001).

K_ds, whether sampled from probability distribution functions or calculated by regression equations or surface-complexation models, can be used in many contaminant transport models. Alternate forms of the retardation factor equation that use a K_d (Equation (3)) and are appropriate for porous media, fractured porous media, or discrete fractures have been used to calculate contaminant velocity and discharge (e.g., Erickson, 1983; Neretnieks and Rasmuson, 1984). An alternative approach couples chemical speciation calculations

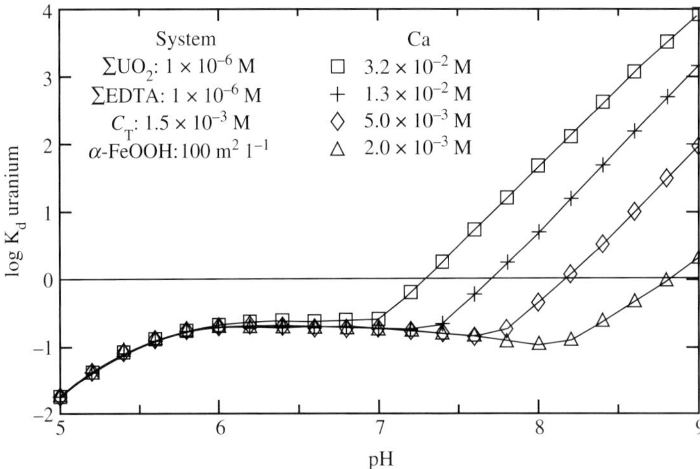

Figure 5 Calculated theoretical K_d for sorption of uranyl onto a goethite substrate as a function of pH at fixed total carbon concentration in the presence of a sequestering agent (EDTA). K_ds are shown for several levels of calcium concentration. Surface area of the substrate is 100 $m^2 L^{-1}$; total carbon is fixed at 1.5×10^{-3} M and total uranium content is 10^{-6} M.

to transport equations. Such models of *reactive transport* have been developed and demonstrated by a number of researchers including Parkhurst (1995), Lichtner (1996), Bethke (1997), Szecsody *et al.* (1998), Yeh *et al.* (1995, 2002), and others reviewed in Lichtner *et al.* (1996), Steefel and Van Cappellen (1998), and Browning and Murphy (2003). Uses of such models to simulate radionuclide transport of uranium in one-dimensional (1D) column experiments are illustrated by Sims *et al.* (1996) and Kohler *et al.* (1996). Glynn (2003) models transport of redox sensitive elements neptunium and plutonium in a 1D domain with spatially variant sorption capacities. Simulations of 2D reactive transport of neptunium and uranium are illustrated by Yeh *et al.* (2002) and Criscenti *et al.* (2002), respectively. Such calculations demonstrate that the results of reactive transport simulations differ markedly from those obtained in transport simulations using constant K_d, Langmuir or Freundlich sorption models. Routine use of the reactive transport codes in performance assessment calculations, however, is still limited by the substantial computer simulation time requirements.

Sorptive properties of mineral assemblages and soils. An important question for the prediction of radionuclide migration is whether sorption in the geomedia can be predicted from the properties of the constituent minerals. Attempts by researchers to use sorption models based on weighted radionuclide K_d values of individual component minerals ("sorptive additivity") have met with limited success (Meyer *et al.*, 1984; Jacquier *et al.*, 2001). Tripathi *et al.* (1993) used a "competitive-additivity" model based on surface-complexation theory to model the pH-dependent sorption of lead by goethite/Ca-montmorillonite mixtures. They used complexation constants obtained from single sorbent systems and predicted the sorption behavior of mineral mixtures from the proportion of the two sorbents and their respective affinities for the metals. Davis *et al.* (1998) describe the component additivity (CA) model, a similar approach in which the wetted surface of a complex mineral assemblage is assumed to be composed of a mixture of one or more reference minerals. The surface properties of the individual phases are obtained from independent studies in monomineralic model systems and then are applied to the mineral assemblage without further fitting, based on the contributions of the individual minerals to the total surface area of the mixture. Applications of this approach to radionuclides are described by McKinley *et al.* (1995), Waite *et al.* (2000), Prikryl *et al.* (2001), Arnold *et al.* (2001), Davis (2001), and Davis *et al.* (2002). Strongly sorbent minerals such as clays or goethite are produced by the alteration of host rocks and line the voids of porous geomedia. In these cases, the sorption behavior of the mineral assemblage can be approximated by using the properties of one or two of minerals even though they constitute a small fraction of the rock mass (Davis and Kent, 1990; Ward *et al.*, 1994; Barnett *et al.*, 2002).

The generalized composite (GC) approach is an alternative approach in which surface-complexation constants are obtained by fitting experimental data for the natural mineral assemblage directly (Koβ, 1988; Davis *et al.*, 1998). A simplified form of this approach fits the pH-dependent sorption of the radionuclide without representation of the electrostatic interaction terms found in other SCMs. The disadvantages of this approach are: (i) the constants obtained are site specific and (ii) it is difficult to apply it to carbonate-rich mineral assemblages. However, it can be used to calibrate simpler sorption models that are used in performance assessment codes.

9.06.3.2 Results of Radionuclide Solubility, Speciation, and Sorption Studies

In this section, results of studies of the geochemistry of fission products and actinides are summarized. The chemistry of the fission products is described as a group first because their behavior is relatively simple compared to the actinides. Next, general trends and then site-specific environmental chemistry of the actinides are summarized.

9.06.3.2.1 Fission products

Fission products of uranium and other actinides are released to the environment during weapons production and testing, and by nuclear accidents. Because of their relatively short half-lives, they commonly account for a large fraction of the activity in radioactive waste for the first several hundred years. Important fission products are shown in Table 3. Many of these have very short half-lives and do not represent a long-term hazard in the environment, but they do constitute a significant fraction of the total released in a nuclear accident. Only radionuclides with half-lives of several years or longer represent a persistent environmental or disposal problem. Of primary interest are ^{90}Sr, ^{99}Tc, ^{129}I, and ^{137}Cs, and to a lesser degree, ^{79}Se and ^{93}Zr; all are β^--emitters.

While fission product mobility is mostly a function of the chemical properties of the element, the initial physical form of the contamination can also be important. For radioactive contaminants

Table 3 Environmentally important fission products.

Fission product	$t_{1/2}$ (yr)
^{79}Se	6.5×10^5
^{90}Sr	28.1
^{93}Zr	1.5×10^6
^{99}Tc	2.12×10^5
^{103}Ru	0.11
^{106}Ru	0.56
^{110m}Ag	0.69
^{125}Sb	2.7
^{129}I	1.7×10^7
^{134}Cs	2.06
^{137}Cs	30.2
^{144}Ce	0.78

released as particulates—"hot particles"—radionuclide transport is initially dominated by physical processes, namely, transport as aerosols (Wagenpfeil and Tschiersch, 2001) or as bedload/suspended load in river systems. At Chernobyl, the majority of fission products were released in fuel particles and condensed aerosols. Fission products were effectively sequestered—for example, little downward transport in soil profiles and little biological uptake—until dissolution of the fuel particles occurred and the fission products were released (Petryaev *et al.*, 1991; Konoplev *et al.*, 1992; Baryakhtar, 1995; Konoplev and Bulgakov, 1999). Thus, fuel particle dissolution kinetics controlled the release of fission products to the environment (Kruglov *et al.*, 1994; Kashparov *et al.*, 1999, 2001; Uchida *et al.*, 1999; Sokolik *et al.*, 2001).

^{90}Sr. Strontium occurs in only one valence state, (II). It does not form strong organic or inorganic complexes and is commonly present in solution as Sr^{2+}. The concentration is rarely solubility limited in soil or groundwater systems because the solubility of common strontium phases is relatively high (Lefevre *et al.*, 1993; US EPA, 1999b). The concentration of strontium in solution is commonly controlled by sorption and ion-exchange reactions with soil minerals. Parameters affecting strontium transport are CEC, ionic strength, and pH (Sr^{2+} sorption varies directly with pH, presumably, due to competition with H^+ for amphoteric sites). Clay minerals—illite, montmorillonite, kaolinite, and vermiculite—are responsible for most of the exchange capacity for strontium in soils (Goldsmith and Bolch, 1970; Sumrall and Middlebrooks, 1968). Zeolites (Ames and Rai, 1978) and manganese oxides/hydroxides also exchange or sorb strontium in soils. Because of the importance of ion exchange, strontium K_ds are strongly influenced by ionic strength of the solution, decreasing with increasing ionic strength (Mahoney and Langmuir, 1991; Nisbet *et al.*, 1994); calcium and natural strontium are especially effective at competing with ^{90}Sr. Strontium in soils is largely exchangeably bound; it does not become fixed with time (Serne and Gore, 1996). However, co-precipitation with calcium sulfate or carbonate and soil phosphates may also contribute to strontium retardation and fixation in soils (Ames and Rai, 1978).

^{137}Cs. Caesium, like strontium, occurs in only one valence state, (I). Caesium is a very weak Lewis acid and has a low tendency to interact with organic and inorganic ligands (Hughes and Poole, 1989; US EPA, 1999b); thus, Cs^+ is the dominant form in groundwater. Inorganic caesium compounds are highly soluble, and precipitation/co-precipitation reactions play little role in limiting caesium mobility in the environment.

Retention in soils and groundwaters is controlled by sorption/desorption and ion-exchange reactions.

Caesium is sorbed by ion exchange into clay interlayer sites, and by surface complexation with hydroxy groups comprised of broken bonds on edge sites, and the planer surfaces of oxide and silicate minerals. CEC is the dominant factor in controlling caesium mobility. Clay minerals such as illite, smectites, and vermiculite are especially important, because they exhibit a high selectivity for caesium (Douglas, 1989; Smith and Comans, 1996). The selectivity is a function of the low hydration energy of caesium; once it is sorbed into clay interlayers, it loses its hydration shell and the interlayer collapses. Ions, such as magnesium and calcium, are unable to shed their hydration shells and cannot compete for the interlayer sites. Potassium is able to enter the interlayer and competes strongly for exchange sites. Because it causes collapse of the interlayers, caesium does not readily desorb from vermiculite and smectite and may in fact be irreversibly sorbed (Douglas, 1989; Ohnuki and Kozai, 1994; Khan *et al.*, 1994). Uptake by illitic clay minerals does not occur by ion exchange but rather by sorption onto frayed edge sites (Cremers *et al.*, 1988; Comans *et al.*, 1989; Smith *et al.*, 1999), which are highly selective for caesium. Although illite has a higher selectivity for caesium, it has a much lower capacity than smectites because caesium cannot enter the interlayer sites.

Caesium mobility increases with ionic strength because of competition for exchange sites (Lieser and Peschke, 1982). Potassium competes more effectively than calcium or magnesium. Since caesium is rapidly and strongly sorbed by soil and sediment particles, it does not migrate downward rapidly through soil profiles, especially forest soils (Bergman, 1994; Rühm *et al.*, 1996; Panin *et al.*, 2001). Estimated downward migration rates for Cs released by the Chernobyl accident are on the order of $0.2-2\ \mathrm{cm\ yr^{-1}}$ in soils in Bohemia (Hölgye and Malú, 2000), Russia (Sokolik *et al.*, 2001), and Sweden (Rosén *et al.*, 1999; Isaksson *et al.*, 2001).

^{99}Tc. Technetium occurs in several valence states, ranging from −1 to +7. In groundwater systems, the most stable oxidation states are (IV) and (VII) (Lieser and Peschke, 1982). Under oxidizing conditions, Tc(VII) is stable as pertechnetate, TcO_4^-. Pertechnetate compounds are highly soluble, and being anionic, pertechnetate is not sorbed onto common soil minerals and/or readily sequestered by ion exchange. Thus, under oxidizing conditions, technetium is highly mobile. Significant sorption of pertechnetate has been seen in organic-rich soils of low pH (Wildung *et al.*, 1979), probably due to the positive charge on the organic fraction and amorphous iron and aluminum oxides, and possibly coupled with reduction to Tc(IV).

Under reducing conditions, Tc(IV) is the dominant oxidation state because of biotic and abiotic reduction processes. Technetium(IV) is commonly considered to be essentially immobile, because it readily precipitates as low-solubility hydrous oxides and forms strong surface complexes on iron and aluminum oxides and clays.

Technetium(IV) behaves like other tetravalent heavy metals and occurs in solution as hydroxo and hydroxo-carbonato complexes. In carbonate-containing groundwaters, $TcO(OH)_{2(aq)}$ is dominant at neutral pH; at higher pH values, $Tc(OH)_3CO_3^-$ is more abundant (Erikson *et al.*, 1992). However, the solubility of Tc(IV) is low and is limited by precipitation of the hydrous oxide, $TcO_2{\cdot}n\,H_2O$. The number of waters of hydration is traditionally given as $n = 2$ (Rard, 1983) but has more recently been measured as 1.63 ± 0.28 (Meyer *et al.*, 1991)). In systems containing H_2S or metal sulfides, the solubility-limiting phase for technetium may be Tc_2S_7 or TcS_2 (Rard, 1983).

Retention of pertechnetate in soil and groundwater systems usually involves reduction and precipitation as Tc(IV)-containing hydroxide or sulfide phases. Several mineral phases have been shown to fix pertechnetate through surface-mediated reduction/co-precipitation. These include magnetite (Haines *et al.*, 1987; Byegård *et al.*, 1992; Cui and Erikson, 1996) and a number of sulfides, including chalcocite, bournonite, pyrrhotite, tetrahedrite, and, to a lesser extent, pyrite and galena (Strickert *et al.*, 1980; Winkler *et al.*, 1988; Lieser and Bauscher, 1988; Huie *et al.*, 1988; Bock *et al.*, 1989). Sulfides are most effective at reducing technetium if they contain a multivalent metal ion in the lower oxidation state (Strickert *et al.*, 1980). Technetium sorption by iron oxides is minimal under near-neutral, oxidizing conditions but is extensive under mildly reducing conditions, where Fe(III) remains stable. It is minimal on ferrous silicates (Vandergraaf *et al.*, 1984).

In addition, technetium may be fixed by bacterially mediated reduction and precipitation. Several types of Fe(III)- and sulfate-reducing bacteria have been shown to reduce technetium, either directly (enzymatically) or indirectly through reaction with microbially produced Fe(II), native sulfur, or sulfide (Lyalikova and Khizhnyak, 1996; Lloyd and Macaskie, 1996; Lloyd *et al.*, 2002).

^{129}I. Iodine can exist in the oxidation states − 1, 0, +1, +5, and +7. However, the +1 state is not stable in aqueous solutions and disporportionates into −1 and +5. In surface- and groundwaters at near-neutral pH, IO_3^- (iodate) is the dominant form in solution, while under acidic conditions, I_2 can form. Under anoxic conditions, iodine is

present as I^- (iodide) (Allard *et al.*, 1980; Liu and van Gunten, 1988).

Iodide forms low-solubility compounds with copper, silver, lead, mercury, and bismuth, but all other metal iodides are quite soluble. As these metals are not common in natural environments, they have little effect on iodine mobility (Couture and Seitz, 1985). Retention by sorption and ion exchange appears to be minor (Lieser and Peschke, 1982). However, significant retention has been observed by the amorphous minerals imogolite and allophane (mixed Al/Si oxides–hydroxides, with SiO_2/Al_2O_3 ratios between 1 and 2). These minerals have high surface areas and positive surface charge at neutral pH and contribute significantly to the anion-exchange capacity in soils (Gu and Schultz, 1991). At neutral pH, aluminum and iron hydroxides are also positively charged and contribute to iodine retention, especially if iodine is present as iodate (Couture and Seitz, 1985). Sulfide minerals containing the metal ions which form insoluble metal iodides strongly sorb iodide, apparently through sorption and surface precipitation of the metal iodide. Iodate is also sorbed, possibly because it is reduced to iodide on the metal/sulfide surfaces (Allard *et al.*, 1980; Strickert *et al.*, 1980). Lead, copper, silver, silver chloride, and lead oxides/hydroxides and carbonates can also fix iodine through surface precipitation (Bird and Lopato, 1980; Allard *et al.*, 1980). None of these minerals are likely to be important in natural soils but may be useful in immobilizing iodine for environmental remediation.

Organic iodo compounds are not soluble and form readily through reaction with I_2 and, to a lesser extent, I^- (Lieser and Peschke, 19821; Couture and Seitz, 1985); retention of iodine in soils is mostly associated with the organic matter (Wildung *et al.*, 1974; Muramatsu *et al.*, 1990; Gu and Schultz, 1991; Yoshida *et al.*, 1998; Kaplan *et al.*, 2000). Several studies have suggested that fixation of iodine by organic soil compounds appears to be dependent upon microbiological activity, because sterilization by heating or radiation commonly results in much lower iodine retention (Bunzl and Schimmack, 1988; Koch *et al.*, 1989; Muramatsu *et al.*, 1990; Bors *et al.*, 1991; Rädlinger and Heumann, 2000).

9.06.3.2.2 Uranium and other actinides [An(III), An(IV), An(V), An(VI)]

General trends in solubility, speciation, and sorption. Actinides are hard acid cations (i.e., comparatively rigid electron clouds with low polarizability) and form ionic species as opposed to covalent bonds (Silva and Nitsche, 1995; Langmuir, 1997a). Several general trends in their chemistry can be described (although there are exceptions). Due to similarities in ionic size, coordination number, valence, and electron structure, the actinide elements of a given oxidation state have either similar or systematically varying chemical properties (David, 1986; Choppin, 1999; Vallet *et al.*, 1999). For a given oxidation state, the relative stability of actinide complexes with hard base ligands can be divided into three groups in the order: CO_3^{2-}, $OH^- > F^-$, HPO_4^{2-}, $SO_4^{2-} > Cl^-$, NO_3^-. Within these ligand groups, stability constants generally decrease in the order $An^{4+} > An^{3+} \approx AnO_2^{2+} > AnO_2^+$ (Lieser and Mohlenweg, 1988; Silva and Nitsche, 1995). In addition, the same order describes the decreasing stability (increasing solubility) of actinide solids formed with a given ligand (Langmuir, 1997a).

These trends have allowed the use of an oxidation analogy modeling approach, in which data for the behavior of one actinide can be used as an analogue for others in the same oxidation state. An oxidation state analogy was used for the WIPP to evaluate the solubility of some actinides and to develop a more complete set of modeling parameters for actinides included in the repository performance calculations. The results are assumed to be either similar to the actual case or can be shown to vary systematically (Fanghänel and Kim, 1998; Neck and Kim, 2001; Wall *et al.*, 2002). The similarities in chemical behavior extend beyond the actinides to the lanthanides—Nd(III) is commonly used as a nonradioactive analogue for the +III actinides. For instance, complexation and hydrolysis constants and Pitzer ion interaction parameters used in modeling Am(III) speciation and solubility for the WIPP were extracted from a suite of published experimental studies involving not only Am(III) but also Pu(III), Cm(III), and Nd(III) (US DOE, 1996).

Oxidation state. Differences among the potentials of the redox couples of the actinides account for much of the differences in their speciation and environmental transport. Detailed information about the redox potentials for these couples can be found in numerous references (e.g., Hobart, 1990; Silva and Nitsche, 1995; Runde, 2002). This information is not repeated here, but a few general points should be made. Important oxidation states for the actinides under environmental conditions are described in Table 4. Depending on the actinide, the potentials of the III/IV, IV/V, V/VI, and/or IV/VI redox couples can be important under near-surface environmental conditions. When the redox potentials between oxidation states are sufficiently different, then one or two redox states will predominate; this is the case for uranium, neptunium, and americium (Runde, 2002). The behavior of uranium is controlled by the predominance of U(VI) species under

Table 4 Important actinide oxidation states in the environment.

Actinide element	*Oxidation states*			
Thorium		IV		
Uranium		IV		VI
Neptunium		IV	V	
Plutonium		IV	V	VI
Americium	III			
Curium	III			

oxidizing conditions and U(IV) under reducing conditions. In the intermediate Eh range and neutral pH possible under many settings, the solubility of neptunium is controlled primarily by the Eh of the aquifer and will vary between the levels set by $Np^{IV}(OH)_{4(s)}$ (10^{-8} M under reducing conditions) and $Np_2^VO_{5(s)}$ (10^{-5} M under oxidizing conditions). Redox potentials of plutonium in the III, IV, V, and VI states are similar (~1.0 V); therefore, plutonium can coexist in up to four oxidation states in some solutions (Langmuir, 1997a; Runde, 2002). However, Pu(IV) is most commonly observed in environmental conditions and sorption of plutonium is strongly influence by reduction of Pu(V) to Pu(IV) at the mineral–water interface. More discussions of these behaviors will be found in the individual sections for each actinide that follow.

Complexation and solubility. In dilute aqueous systems, the dominant actinide species at neutral to basic pH are hydroxy and carbonato complexes. Similarly, solubility-limiting solid phases are commonly oxides, hydroxides, or carbonates. The same is generally true in high-ionic-strength brines, because common brine components—Na^+, Ca^{2+}, Mg^{2+}, Cl^-, SO_4^{2-}—do not complex as strongly with actinides. However, weak mono-, bis-, and tris-chloro complexes with hexavalent actinides (U(VI) and Pu(VI)) can contribute significantly to the solubility of these actinides in chloride-rich brines. Runde *et al.* (1999) measured shifts in the apparent solubility product constants for uranyl and plutonyl carbonate of nearly one log unit as chloride concentrations increased to 0.5 M. Carbonate complexes are important for radionuclides; thorium, plutonium, neptunium, and uranium all have strong carbonate complexes under environmental conditions. Carbonate complexation also leads to decreased sorption by forming strong anionic complexes that will not sorb to negatively charged mineral surfaces. The potential importance of carbonate complexes with respect to increasing actinide solubility and decreasing sorption influenced a decision by the DOE to use MgO as the engineered barrier in the WIPP repository. MgO and its hydration products sequester CO_2 through formation of carbonates and hydroxycarbonates, as well as buffering the pH at neutral to moderately basic values, where actinide solubilities are at a minimum.

Dissolved organic carbon may be present as strong complexing ligands that increase the aqueous concentration limits of actinides (Olofsson and Allard, 1983). In environments with high organic matter from natural or anthropogenic sources, complexation of actinides with ligands such as EDTA and other organic ligands may decrease the extent of sorption onto rocks. Langmuir (1997a) suggests that to a first approximation, complexation of An^{4+}, AnO_2^+, and AnO_2^{2+} with humic and fulvic acids can be ignored, because the actinides form such strong hydroxyl and carbonato-complexes in natural waters. In contrast, however, although An^{3+} species form OH^- and or CO_3^{2-} complexes, important actinide/humic–fulvic complexation does occur. Conditions under which actinide–humic interactions are important are discussed in more detail in Section 9.06.3.3.1.

Sorption. In general, actinide sorption will decrease in the presence of ligands that complex with the radionuclide (most commonly humic or fulvic acids, CO_3^{2-}, SO_4^{2-}, F^-) or cationic solutes that compete with the radionuclide for sorption sites (most commonly Ca^{2+}, Mg^{2+}). In general, sorption of the (IV) species of actinides (Np, Pu, U) is greater than of the (V) species.

As discussed previously (Section 9.06.3.1.1), plots of pH sorption edges (see Figure 3) are useful in summarizing the sorption of radionuclide by substrates that have amphoteric sites (i.e., SOH, SO^-, SOH_2^+). The pH sorption edges of actinides are similar for different aluminosilicates (quartz, α-alumina, clinoptilolite, montmorillonite, and kaolinite). For example, Np(V) and U(VI) exhibit similar pH-dependent sorption edges that are independent of specific aluminosilicate identity (Bertetti *et al.*, 1998; Pabalan *et al.*, 1998). Under similar solution conditions, the amount of radionuclide adsorbed is primarily a function of the surface area. This observation has led several workers to propose that the amount of actinide sorption onto natural materials can be predicted from the surface site density and surface area rather the specific molecular structure of the surface (Davis and Kent, 1990; Turner and Pabalan, 1999).

Carroll *et al.* (1992), Stout and Carroll (1993), Van Cappellen *et al.* (1993), Meece and Benninger (1993), Brady *et al.* (1999), and Reeder *et al.* (2001) summarize empirical data and theoretical models of actinide-carbonate mineral interactions. The surface PDI on carbonate minerals may be Ca^{2+} or Mg^{2+}. Increased solution concentration of Ca^{2+} will lead to decreased actinide sorption, which then leads to complex sorption behavior if the carbonate concentration and pH of the solution are varied. Carroll *et al.* (1992) studied the uptake of Nd(III), U(VI), and Th(VI) by pure calcite in dilute $NaHCO_3$ solutions using a combination of surface analysis techniques. They found that U(VI) uptake was limited to monolayer sorption and uranium–calcium solid solution was minimal even in solutions supersaturated with rutherfordine (UO_2CO_3). In contrast, they found that surface precipitation and carbonate solid solution was extensive for thorium and neodymium. Similarly, irreversible sorption and surface precipitation of americium onto carbonates were observed by Shanbhag and Morse (1982) and Higgo and Rees (1986).

Attempts to propose representative K_d values for actinides have met with controversy. For example, Silva and Nitsche (1995) suggested average K_d values for actinides in the order $An^{4+} > An^{3+} > AnO_2^{2+} > AnO_2^{+}$, as 500, 50, 5, and 1, respectively. This order corresponds to the order of the pH_{50} values of sorption edges for Th(IV), Am(III), Np(V), and Pu(V) in studies of sorption by γ-Al_2O_3 (Bidoglio *et al.*, 1989); and of Pu(IV), U(VI), and Np(V) in studies of sorption by α-FeOOH (Turner, 1995). Calculated or measured element-specific K_ds for natural soils and geomedia for many environmental sites, however, are quite different from these values. For example, a recent compilation listed the following suggested general ranges for soil/mineral K_ds, in mL g^{-1}: Pu, 11–300,000; U, 10.5–4,400; Am,1–47,230 (Krumhansl *et al.*, 2002). These wide ranges exist because sorption of radionuclides is very dependent on the radionuclide oxidation state, groundwater composition, and nature of rock surface, all of which may be variable and/or poorly characterized along the flow path. The databases of K_d values described previously (Section 9.06.3.1.2) should be used to obtain K_ds for site-specific conditions instead of using broad "generic" ranges whenever possible.

Table 5 Compositions of low ionic strength reference waters used in speciation and solubility calculations.

Component	*SKI-90 Stripa* (mM)[a]	*J-13 YM* (mM)[b]
Na^+	1.39	1.96
K^+	0.0256	0.14
Ca^{2+}	0.5	0.29
Mg^{2+}	0.0823	0.07
Fe (total)	0.00179	
SiO_2	0.0682	1.07
Cl^-	0.423	0.18
SO_4^{2-}	0.417	0.19
F^-	0.142	0.11
PO_4^{3-}	3.75e − 5	
HCO_3^-	2.0	2.81
PH[c]	8.2	6.9
Eh (mV)	−0.3	0.34 − 0.7[c]

[a] SKI (1991). [b] Ogard and Kerrisk (1984). [c] Range of Eh used in different works.

Table 6 Compositions of brines used in WIPP speciation and solubility calculations (US DOE, 1996).

Component	*Salado brine* (mM)	*Castile brine* (mM)
Na^+	1,830	4,870
K^+	770	97
Ca^{2+}	20	12
Mg^{2+}	1,440	19
Fe (total)		
SiO_2		
Cl^-	5,350	4,800
SO_4^{2-}	40	170
F^-		
PO_4^{3-}		
Br^-	10	11
$B_4O_7^{2-}$	5	16
pH[a]	8.7	9.2
pC_{H^+}[a,b]	9.4	9.9
P_{CO_2}, atm[a]	$10^{-5.5}$	$10^{-5.5}$
Eh		
Total dissolved solids	306,000	330,000
Ionic strength	6,990	5,320

[a] In equilibrium with brucite and hydromagnesite.
[b] pC_{H^+} = negative log of the molar concentration of H^+.

9.06.3.2.3 *Site-specific geochemistry of the actinides*

Introduction: actinide solubilities in reference waters. In this section, the environmental chemistry of the actinides is examined in more detail by considering three different geochemical environments. Compositions of groundwater from these environments are described in Tables 5 and 6. These include: (i) low-ionic-strength reducing waters from crystalline rocks at nuclear waste research sites in Sweden; (ii) oxic water from the J-13 well at Yucca Mountain, Nevada, the site of a proposed repository for high-level nuclear waste in tuffaceous rocks; and (iii) reference brines associated with the WIPP, a repository for TRU in the Permian Salt beds of SE New Mexico. These last brines are model solutions produced by the reaction of the Permian formation waters with the components of the engineered barrier at the WIPP as discussed below.

The Swedish repository science program has investigated crystalline rock as a host rock for the disposal of radioactive waste and has measured the composition of granitic groundwaters Andersson, (1990). Much of this was done at the Stripa site, an abandoned iron mine located in a granitic intrusion in south-central Sweden. At Stripa, shallow groundwaters are dilute, carbonate-rich, pH neutral, oxidizing waters of meteoric origin; naturally occurring uranium is present in concentrations of 10–90 ppb. Once below this zone, the waters are slightly more saline (up to 1.3 g L^{-1} total dissolved solids), more basic (up to pH 10.1), and the Eh is lower—groundwater uranium concentrations are less than 1 ppb (Andrews *et al.*, 1989; Nordstrom *et al.*, 1989). The trace amounts of sulfide and ferrous iron in the groundwater have little capacity for maintaining reducing conditions, and groundwater interactions with radioactive waste, waste containers, or repository backfill materials are likely to govern the redox conditions in a real repository (Nordstrom *et al.*, 1989). The dilute, near-neutral, mildly reducing groundwater composition given in Table 5 is a composite of analyses from several Swedish sites and is a suggested reference composition (Andersson, 1990) for deep granitic groundwaters.

The proposed nuclear waste repository at Yucca Mountain, Nevada, would be located in a thick sequence of Tertiary volcanic tuffs. The range of groundwater compositions sampled at the site is discussed by Perfect *et al.* (1995). Numerous geochemical studies have been carried out in high-Eh waters from the alluvium and tuffaceous rocks (e.g., UZ-TP-7) from the unsaturated zone, high-Eh waters from the saturated zone (e.g., J-13) within tuffaceous rocks, and in lower-Eh waters from a deeper Paleozoic carbonate aquifer (e.g., UE252p-1) (Tien *et al.*, 1985; Triay *et al.*, 1997). Table 5 describes the composition of water from the J-13 well, which has been used as a reference water in systematic studies of sorption, transport, and solubility (Nitsche *et al.*, 1992). Its composition is controlled by a number of processes including dissolution of vitric and devitrified tuff, precipitation of secondary minerals, and ion exchange (Tien *et al.*, 1985; Triay *et al.*, 1997).

Table 7 contains the results of actinide solubility and speciation calculations for the J-13 and SKI-90 reference waters carried out using the MINTEQ2A code (Allison *et al.*, 1991) as described by Langmuir (1997a). The MINTEQ2A thermodynamic database of Turner *et al.* (1993) was used with revised data for americium (Silva *et al.*, 1995) and modifications for uranium described in Langmuir (1997a). Langmuir (1997a) used formation constants that effectively eliminated the influence of $Np(OH)_5^-$ and $Pu(OH)_5^-$ complexes and assumed that the most soluble amorphous hydroxides and mixed carbonato-hydroxide phases controlled the solubility. These concentrations should be considered maximum soluble concentrations that might be important for short-term behavior of the radionuclides. Over longer time periods, the solubilities are likely controlled by more crystalline phases at levels that are several orders of magnitude lower than those listed in Table 7. These calculations should be considered as a set of baseline calculations for low-ionic-strength solutions; they illustrate the effect of redox potential on speciation and solubility.

WIPP is an underground repository for the permanent disposal of defense-related TRU wastes (NAS, 1996). The facility is located in the US in southeastern New Mexico in a thick, bedded salt, the Salado Formation, at a depth of 655 m. The Castile Formation is an evaporite sequence below the Salado that may serve as a brine source if the repository is breached by human activities in the future. Brines from both formations are a mixture of Na^+, Mg^{2+}, K^+, Ca^{2+}, Cl^-, and SO_4^{2-} (see Table 6) and are saturated with respect to halite (NaCl) and anhydrite ($CaSO_4$). For the discussion below, the pH, P_{CO_2} and radionuclide solubilities and speciation were calculated assuming that the brines were in equilibrium with halite, anhydrite, and minerals produced by hydration and carbonation of the MgO engineered barrier (brucite and hydromagnesite). The Eh of the reference brines was assumed to be controlled by the metallic iron in the waste and waste packages. Calculated solubilities in the two brines are grossly similar. The dominant aqueous species and solubility-limiting phases contain hydroxide and/or carbonate, and solubility differences in the two brines are largely due to differences in the pH (8.7 for the Salado, and 9.2 for the Castile) and CO_3^{2-} activities (3.81×10^{-7} M and 4.82×10^{-6} M, for the Salado and Castile brines, respectively).

Table 7 Solubility-limiting solids and range of solubility-limited concentrations (M concentration) for low-ionic strength geochemical environments.[a]

Element	*Environment*	
	YMP[b]	*Stripa*[c]
Tc	None	$TcO_2{\cdot}2H_2O \leq 3.3 \times 10^{-8}$
Th	$Th(OH)_{4(am)} \leq 6.0 \times 10^{-7}$	$Th(OH)_{4(am)} \leq 5.7 \times 10^{-7}$
U	$Ca(H_3O)_2(UO_2)_2(SiO_4)_2{\cdot}3H_2O_{(cr)}$ 5.4×10^{-9}	$UO_{2\,(am)} \leq 1.4 \times 10^{-8}$
Np	$NaNpO_2CO_3{\cdot}3.5H_2O_{(cr)}$ 8.9×10^{-4}	$Np(OH)_{4(am)} \leq 1.6 \times 10^{-9}$
Pu	$Pu(OH)_{4(am)} \leq 6.6 \times 10^{-8}$	$Pu(OH)_{4(am)} \leq 1.7 \times 10^{-9}$
Am	$AmOH(CO_3)_{(cr)}$ 5.6×10^{-8}	$AmOH(CO_3)_{(cr)} \leq 1.4 \times 10^{-7}$

[a] Based on Langmuir (1997a, table 13.11). [b] J-13 reference water (Ogard and Kerrisk, 1984). [c] SKI-90 reference water (SKI, 1991).

Because of the high ionic strength of the brines, the calculations were carried out using a Pitzer ion interaction model (US DOE, 1996) for the activity coefficients of the aqueous species (Pitzer, 1987, 2000). Pitzer parameters for the dominant non-radioactive species present in WIPP brines are summarized in Harvie and Weare (1980), Harvie *et al.* (1984), Felmy and Weare (1986), and Pitzer (1987, 2000). For the actinide species, the Pitzer parameters that were used are summarized in the WIPP Compliance Certification Application (CCA) (US DOE, 1996). Actinide interactions with the inorganic ions H^+, Na^+, K^+, Mg^{2+}, Cl^-, and HCO_3^-/CO_3^{2-} were considered.

Americium. The low solubilities and high sorption affinity of thorium and americium severely limit their mobility under environmental conditions. However, because each exists in a single oxidation state—Th(IV) and Am(III)—under environmentally relevant conditions, they are relatively easy to study. In addition, their chemical behaviors provide valuable information about the thermodynamic properties of trivalent and tetravalent species of uranium, neptunium, and plutonium.

Silva *et al.* (1995) provide a detailed summary of experimental and theoretical studies of americium chemistry as well as a comprehensive, self-consistent database of reference thermodynamic property values. Solubility and speciation experiments with Am(III) indicate that the mixed hydroxy-carbonate $AmOHCO_{3(cr)}$ is the solubility-limiting solid phase under most surface and subsurface conditions. At neutral pH, $AmOH^{2+}$ or $AmCO_3^+$ can be the dominant solution species depending on the carbonate concentration. Langmuir (1997a) calculated a solubility of 5.6×10^{-8} M for J-13 water with the MINTEQA2 code using a revised formation constant $\log K_{sp} = 7.2$ for $AmOHCO_3$ (compared to NEA $\log K_{sp} = 8.605$ of Silva *et al.* (1995)); see Table 7. This is similar to the value of 1.2×10^{-9} M measured by Nitsche *et al.* (1993) in solubility experiments in J-13 water. The americium solubility calculated by Langmuir (1997a) for reducing water from crystalline rock in Table 7 ($\leq 1.4 \times 10^{-7}$ M) is similar to the range calculated by Bruno *et al.* (2000) using the EQ3NR (Wolery, 1992b) code for slightly basic, reducing groundwaters in granite at Äspö and Gideå, Sweden.

Fanghänel and Kim (1998) evaluated the solubility of trivalent actinides in brines, using Cm(III) as a representative analogue, and found that An(III) hydroxy and carbonato complexes are the most stable aqueous complexes. Multiple-ligand complexes with a high negative charge are more stable in brines than in dilute solutions, apparently because of the high cation concentrations. Chloride and sulfate complexes, although very weak, may be important aqueous species in some brines, especially at low pH.

In the WIPP speciation and solubility calculations, the solubility-controlling solid phase for americium, and by analogy, for all +III actinides under WIPP conditions, was $Am(OH)CO_{3(cr)}$ (Novak, 1997; US EPA, 1998a,b,c,d; Wall *et al.*, 2002). $Am(OH)_2^+$ was the most abundant aqueous species, and estimated americium solubilities in the reference Salado and Castile brines (Table 8) were 9.3×10^{-8} M and 1.3×10^{-8} M, respectively (Novak, 1997; US EPA, 1998d).

Americium is strongly sorbed by tuffaceous rocks from Yucca Mountain in waters of low ionic strength (Triay *et al.*, 1997). In a compilation by Tien *et al.* (1985), americium K_ds obtained with tuff in J-13 water ranged from 130 mL g^{-1} to 13,000 mL g^{-1}. Average values for devitrified, vitric, and zeolitized tuff were 2,975 mL g^{-1}, 1,430 mL g^{-1}, and 1,513 mL g^{-1}, respectively. Turin *et al.* (2002) measured K_ds ranging from 410 mL g^{-1} to 510 mL g^{-1} using similar waters and tuffaceous rocks from Busted Butte on the Nevada Test Site. They also provide Freundlich isotherm parameters from the sorption measurements. A K_d range of 500–50,000 ml g^{-1} is reported for crystalline rocks by McKinley and Scholtis (1992); a value of 5,000 mL g^{-1} is recommended for performance assessment.

Data are sparse for americium sorption in high-ionic-strength solutions. In experimental studies with near-surface sediments from the Gorleben site, Lieser *et al.* (1991) showed that americium sorption did not vary ($K_d \sim 1{,}000$ mL g^{-1}) over a range of NaCl concentrations of 0–2 M, at a pH 7.5. They concluded that

Table 8 Calculated actinide solubilities in WIPP brines.

Actinide	*Solubility* (M)		*Solubility-limiting phase*	*Dominant aqueous phases*
	Salado	*Castile*		
An(VI)[a]	8.7×10^{-6}	8.8×10^{-6}		
An(V)	1.2×10^{-7}	4.8×10^{-7}	$KnpO_2CO_3{\cdot}2H_2O_{(s)}$	$NpO_2CO_3^-$
An(IV)	1.2×10^{-8}	4.1×10^{-8}	$ThO_{2(am)}$	$Th(OH)_3CO_3^-$
An(III)	9.3×10^{-8}	1.3×10^{-8}	$Am(OH)CO_{3(cr)}$	$Am(OH)_2^+$

Source: US EPA (1998d).
[a] Estimated from literature data (values as listed in US DOE, 1996).

americium sorption was not sensitive to ionic strength, because at this pH americium is nearly completely hydrolyzed. Thus, ion-exchange reactions did not contribute to americium sorption, and competing ion concentrations had little effect on sorption K_ds. *In situ* studies of radionuclide transport through brackish bay sediments in Sweden (~seawater solution compositions) measured K_ds of 10^3–10^4 mL g^{-1} (Andersson *et al.*, 1992).

Thorium. Experimental and theoretical studies of thorium speciation, solubility, and sorption in low-ionic-strength waters are described by Langmuir and Herman (1980), Laflamme and Murray (1987), Östhols *et al.* (1994), Östhols (1995), and Quigley *et al.* (1996). Langmuir and Herman (1980) provide a critically evaluated thermodynamic database for natural waters at low temperature that is widely used. However, it does not contain information about important thorium carbonate complexes, and the stability of phosphate complexes may be overestimated (US EPA, 1999b).

In both low-ionic-strength groundwaters and in the WIPP brines, the solubility limiting phase is $ThO_{2(am)}$. In seawater, waters from Yucca Mountain, and reducing waters in crystalline rocks, the dominant aqueous species are $Th(OH)_{4(aq)}$ and mixed hydroxy carbonato complexes. In alkaline lakes and other environments with high carbonate concentrations, thorium carbonate complexes are dominant (Laflamme and Murray, 1987; Öthols *et al.*, 1994). In organic-rich stream waters, swamps, soil horizons, and sediments, organic thorium complexes may predominate (Langmuir and Herman, 1980). Calculated thorium solubilities in waters from Yucca Mountain and crystalline rocks in Table 7 are similar (~6.0×10^{-7} M) but are much higher than those calculated by Bruno *et al.* (2000) for waters from Äspö and Gideå using EQ3NR (~2×10^{10} M).

The solubility of $ThO_{2(am)}$ increases with increasing ionic strength; above pH 7 in 3.0 M NaCl solutions, the solubility is approximately three orders of magnitude higher than that measured in 0.1 M $NaClO_4$ solution (Felmy *et al.*, 1991). Rai *et al.* (1997) describe solubility studies and a thermodynamic model for Th(IV) speciation and solubility in concentrated NaCl and $MgCl_2$ solutions. A Pitzer ion-interaction model was used to obtain a solubility product of $\log K_{sp} = -45.5$ for $ThO_{2(am)}$. In the speciation and solubility calculations for the WIPP performance assessment (Table 8), the only important aqueous species was $Th(OH)_3(CO_3)^-$; the corresponding estimated Th(IV) solubilities were 1.2×10^{-8} M in the Salado brine and 4.1×10^{-8} M in the Castile brine (Novak, 1997; US EPA 1998d).

Thorium sorbs strongly to iron oxyhydroxides and humic matter (Nash and Choppin, 1980; Hunter *et al.*, 1988; Murphy *et al.*, 1999) and weakly to silica at neutral to basic pH (Östhols, 1995). Thorium sorption is sensitive to carbonate alkalinity due to the formation of negatively charged aqueous mixed hydroxy-carbonato complexes (Laflamme and Murray, 1987); at alkalinities of 100 meq L^{-1}, thorium sorption by goethite decreases markedly. However, at the relatively low alkalinities measured at Yucca Mountain, this effect is not important for the proposed repository site (Triay *et al.*, 1997). Measured thorium sorption ratios in J-13 water from Yucca Mountain for devitrified, vitric, and zeolitized tuff ranged from 140 mL g^{-1} to 2.38×10^4 mL g^{-1} (Tien *et al.*, 1985; Thomas, 1987). Other compilations contain representative K_d values for thorium in crystalline rock that range from 100 mL g^{-1} to 5,000 mL g^{-1} (McKinley and Scholtis, 1992) and from 20 mL g^{-1} to 3×10^5 mL g^{-1} for low-temperature geochemical environments (US EPA, 1999b).

Thorium sorption at high ionic strength was examined using uranium-series disequilibrium techniques by Laul (1992). Laul measured thorium retardation in saline groundwaters from the Palo Duro Basin, Texas, and determined sorption K_ds of ~2,100 mL g^{-1}. Because tetravalent actinides are strongly sorbed by mineral colloids and have a strong tendency to form intrinsic colloids, increases in ionic strength may have more effect on An(IV) transport through destabilization and flocculation of colloidal particles (Lieser and Hill, 1992), rather than through changes in the degree of sorption.

Uranium. Uranium, neptunium, and plutonium are probably the most important actinides in assessment of the environmental risks posed by radioactive contamination. Uranium contamination is present at numerous sites contaminated by uranium mining, milling, and solution mining as described in previous sections. It is highly mobile and soluble under near-surface oxidizing conditions and thus presents an exposure hazard to humans and ecosystems.

Under oxidizing near-surface conditions, U(VI) is the stable oxidation state. Figure 6 shows the aqueous speciation of U(VI) and the solubility of crystalline schoepite (β-$UO_2(OH)_2$) under atmospheric conditions ($P_{CO_2} = 10^{-3.5}$) over the pH 4–9. The calculations were carried out with the HYDRAQL code (Papelis *et al.*, 1988) using a thermodynamic database described by Davis *et al.* (2001), which is based primarily on the compilation of Grenthe *et al.* (1992). The figure shows that over the pH 6–8, the mixed hydroxy-carbonato binuclear complex $(UO_2)_2CO_3(OH)_3^-$ is predicted to predominate; that at lower pH, the uranyl ion (UO_2^{2+}) is most important; and that at higher pH, the polycarbonate $UO_2(CO_3)_3^{4-}$ has the highest concentration. In Figure 6, the total

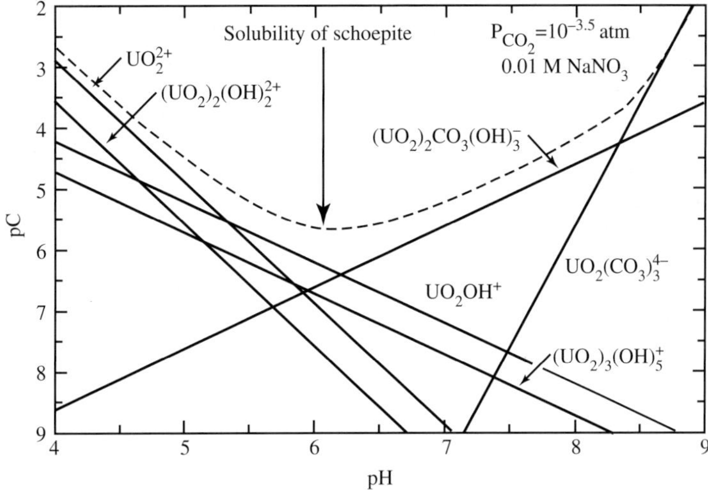

Figure 6 Aqueous speciation of uranium(VI) and the solubility of crystalline schoepite (β-$UO_2(OH)_2$) under atmospheric conditions ($P_{CO_2} = 10^{-3.5}$ atm (Davis (2001); reproduced by permission of Nuclear Regulatory Commission from *Complexation Modeling of Uranium(VI) Adsorption on Natural Mineral Assemblages NUREG/CR- 6708*, **2001**, p.12).

concentration of uranium is limited by the solubility of schoepite; therefore, it varies as a function of pH and can exceed 10^{-3} M. The relative importance of the multinuclear uranyl complexes are different from those shown in Figure 6 at different fixed total uranium concentrations. For example, if the total uranium concentration is limited to $<10^{-8}$ M (e.g., by slow leaching of uranium from a nuclear waste form), then the species $UO_2(OH)_{2(aq)}$ predominates at pH 6–7. At higher total uranium concentrations (e.g., 10^{-4} M), the multinuclear uranyl hydroxy complex $(UO_2)_3(OH)_5^+$ predominates at pH = 5–6 (Davis *et al.*, 2001).

Uncertainties in the identity and solubility product of the solubility-limiting uranium solid in laboratory studies lead to considerable uncertainty in estimates of the solubility under natural conditions. If amorphous $UO_2(OH)_2$ is assumed to limit the solubility in solutions open to the atmosphere, then the value of the minimum solubility and the pH at which it occurs change. If crystalline schoepite (β-$UO_2(OH)_2$) controls the solubility, then the minimum solubility of 2×10^{-6} M occurs at about pH 6.5. If amorphous $UO_2(OH)_2$ controls the solubility, the solubility minimum of 4×10^{-5} M occurs closer to pH 7.0 if the solubility product of Tripathi (1983) is assumed (Davis *et al.*, 2001).

In water of low Eh, such as crystalline rock environments studied in the European nuclear waste programs, uranium solubility is controlled by saturation with UO_2 and coffinite ($USiO_4$) (Langmuir, 1997a). Langmuir (1997a, pp. 501–502) describes some of the controversy surrounding estimation of the solubility of UO_2. Estimates for the log K_{sp} of UO_2 range from −51.9 to −61.0, corresponding to soluble concentrations (as $U(OH)_{4(aq)}$) ranging from 10^{-8} M (measured by Rai *et al.*, 1990) to $10^{-17.1}$ M (computed by Grenthe *et al.* (1992)). The reason for the wide range lies in the potential contamination of the experimental systems by O_2 and CO_2 and the varying crystallinity of the solid phase. Contamination and the presence of amorphous rather than crystalline UO_2 would lead to higher measured solubilities.

In the WIPP performance assessment calculations, it was assumed that redox conditions would be controlled by the presence of metallic iron and that U(VI) would be reduced to U(IV). An upper bound for the solubility of U(IV) was estimated from that of Th(IV), using the oxidation state analogy. Calculations by Wall *et al.* (2002) suggest that this is a conservative assumption—that the solubility of U(IV) is much lower than that of Th(IV). This is because the solubility product constant of $ThO_{2(am)}$, the solubility limiting phase in the An(IV) model, is several orders of magnitude greater than that of $UO_{2(am)}$.

Reed *et al.* (1996) examined An(VI) stability in WIPP brines under anoxic conditions (1 atm H_2 gas) and found that U(VI) was stable as a carbonate complex in Castile brine at pH 8–10. Xia *et al.* (2001) also observed that U(VI) may be stable under some WIPP-relevant conditions, finding that while U(VI) was rapidly reduced to U(IV) by Fe^0 in water and 0.1 M NaCl, it was not reduced in the Castile brine, at pC_H^+ 8–13, over the course of a 55-day experiment. The possible occurrence of U(VI) was considered in WIPP performance assessment calculations—solubility-limited concentrations for U(VI) in the Salado and Castile brines were estimated from literature values to be 8.7×10^{-6} M and 8.8×10^{-6} M, respectively (US DOE, 1996).

In areas affected by uranium solution mining using sulfuric acid, $UO_2SO_{4(aq)}$ will be important. In alkaline waters, carbonate complexes will dominate. Bernhard *et al.* (1998) studied uranium speciation in water from uranium mining districts in Germany (Saxony) using laser spectroscopy, and found that $Ca_2UO_2(CO_3)_{3(aq)}$ was the dominant species in neutral pH carbonate- and calcium-rich mine waters; $UO_2(CO_3)_3^{4-}$ was the dominant aqueous species in basic (pH = 9.8), carbonate-rich, calcium-poor mine waters; and $UO_2SO_{4(aq)}$ dominated in acidic (pH = 2.6), sulfate-rich mine waters.

A large number of studies of uranium sorption have been carried out in support of the nuclear waste disposal programs and the uranium mill tailings program (UMTRA). Park *et al.* (1992) and Prasad *et al.* (1997) describe studies of sorption of uranyl ion by corrensite, the clay mineral lining many fractures in the fractured Culebra Dolomite member of the Rustler Formation above the WIPP in SE New Mexico. The studies were carried out in dilute and concentrated NaCl (0.1–3 M) solutions in the presence of Ca^{2+}, Mg^{2+}, carbonate, and citrate. Binding constants for the TLM were fit to the sorption edges. They found that the adsorption edges were typical of cation adsorption on mineral surfaces; the uranium was nearly completely bound to the surface at neutral and near-neutral pH values. Neither the background electrolyte (NaCl) nor Ca^{2+} or Mg^{2+} ions (at 0.05 M) influenced the adsorption, suggesting that uranyl binds at pH-dependent edge sites on the corrensite surface as an inner-sphere complex. Both carbonate and citrate reduced the adsorption of uranyl on corrensite in near neutral solutions. Redden *et al.* (1998) carried out similar studies of uranium sorption by goethite, kaolinite, and gibbsite in the presence of citric acid. Davis (2001) and Jenne (1998) provide good summaries of studies of uranium sorption by synthetic and natural aluminosilicates and iron oxyhydroxides. Qualitative features of the sorption edges for these minerals are similar: U(VI) sorption at higher pH is typically low and likely is controlled by the predominance of the negatively charged uranyl-carbonate solution species. By analogy, sorption of U(VI) by aluminosilicates is predicted to be low in waters sampled at Yucca Mountain (Turner *et al.*, 1998; Turner and Pabalan, 1999).

Luckscheiter and Kienzler (2001) examined uranyl sorption onto corroded HLW glass simulant in deionized water, 5.5 M NaCl and 5.0 M $MgCl_2$, and found that sorption was greatly inhibited by the magnesium-rich brine, while the NaCl brine had little effect. Uranyl sorption at high ionic strength was also studied by Vodrias and Means (1993), who examined uranyl sorption onto crushed impure halite and limestone from the Palo Duro Basin, Texas, in a synthetic Na–K–Mg–Ca–Cl brine ($I = 10.7$ M). They measured K_ds of 1.3 mL g^{-1} on the halite and 4–7 mL g^{-1} on the limestone. This is in contrast to a K_d of 2,100 mL g^{-1} determined from uranium-series disequilibrium measurements on formation brines from the same region (Laul, 1992). The isotopic ratios suggest that naturally occurring uranium was more strongly sorbed because it was present as U(IV).

Neptunium. Neptunium and plutonium are the radioelements of primary concern for the disposal of nuclear waste at the proposed repository at Yucca Mountain. This is due to their long half-lives, radiotoxicity, and transport properties. Neptunium is considered to be the most highly mobile actinide because of its high solubility and low potential for sorption by geomedia. Its valence state (primarily Np(V) or Np(IV)) is the primary control of its environmental geochemistry. Oxide, hydroxide, and carbonate compounds are the most important solubility-limiting phases in natural waters. In low-ionic-strength, carbonate-free systems, $NpO_2(OH)$ and Np_2O_5 are stable Np(V) solids, while in brines, Np(V) alkaline carbonate solids are stable. Under reducing conditions, $Np(OH)_{4\ am}$ and NpO_2 are the stable Np(IV) solids. Under most near-surface environmental conditions, the dominant complexes of neptunium are those of the pentavalent neptunyl species (NpO_2^+). Neptunium(IV) aqueous species may be important under reducing conditions possible at some underground nuclear waste research facilities such as the WIPP and Stripa.

Kaszuba and Runde (1999)compiled thermodynamic data for neptunium relevant to Yucca Mountain. They updated the database of Lemire (1984) with recent experimental data and used the SIT to calculate ion activity coefficients. Their report has an extensive reference list and list of interaction parameters. Kaszuba and Runde (1999) used the EQ3NR (Wolery, 1992b) and the Geochemist Workbench (Bethke, 1998) codes to calculate solubility and speciation in the J-13 and UE25p#1 well waters that span the expected geochemical conditions for the proposed HLW repository at Yucca Mountain. They predicted that $Np(OH)_{4(aq)}$ is the dominant aqueous complex in neutral solutions at Eh <0 mV, while under oxidizing conditions, NpO_2^+ and $NpO_2CO_3^-$ are predominant at pH <8 and pH 8–13, respectively.

Although the calculations of Kaszuba and Runde (1999) indicate that $NpO_{2(s)}$ is the thermodynamically stable solid for most Eh–pH conditions of environmental interest, that phase has never been observed to precipitate in solubility experiments in natural waters; instead, $Np_2O_{5(s)}$ and amorphous $Np(OH)_{4(s)}$ precipitate. Figure 7 shows that if $Np_2O_{5(s)}$ controls the solubility

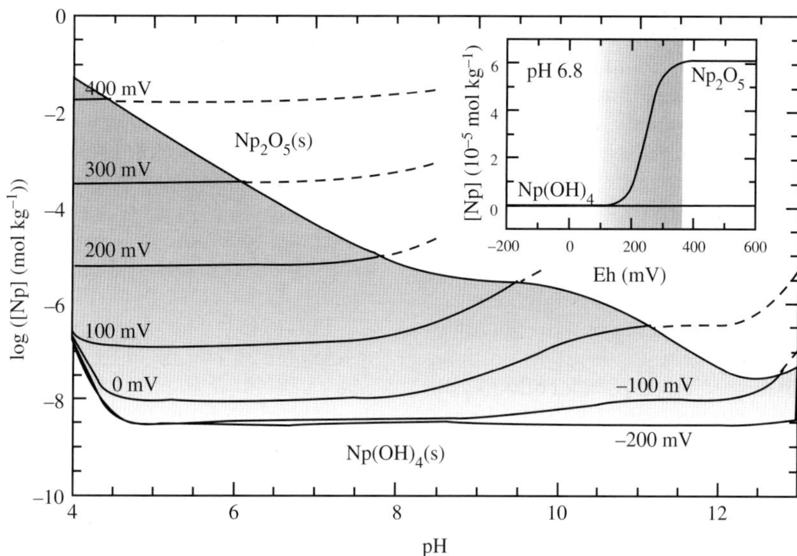

Figure 7 Calculated Np solubilities as a function of pH and Eh in J-13 groundwater variants (Table 5). $Np_2O_{5(s)}$ and $Np(OH)_{4(s)}$ were assumed to be the solubility-limiting phases. Inset shows regions of solubility control versus redox control (shaded area) (Kaszuba and Runde, 1999) (reproduced by permission of American Chemical Society from *Environmental Science and Technology* **1999**, *33*, 4433).

under oxidizing conditions (Eh $>$ 0.25 V), then the calculated solubility of neptunium decreases from $\sim 10^{-3.5}$ M at pH = 6 to 10^{-5} M at pH = 8. If amorphous $Np(OH)_{4(s)}$ controls the solubility under reducing conditions (Eh < -0.10 V), the solubility is $\sim 10^{-8}$ M over the same pH range. In the intermediate Eh range and neutral pH conditions possible under many environmental settings, the solubility of neptunium is controlled primarily by the Eh of the aquifer and will vary between the levels set by the solubilities of $Np(OH)_{4(s)}$ and $Np_2O_{5(s)}$ (Figure 7). The inset in the figure illustrates that at pH = 6.8, at Eh = -0.10 V, the concentration of neptunium in solution is approximately equal to that of the Np(IV) species and is controlled by the solubility of $Np(OH)_{4(s)}$. As the redox potential increases, Np(IV) in solution is oxidized to Np(V) and the aqueous concentration of neptunium increases. Phase transformation of $Np(OH)_{4(s)}$ to $Np_2O_{5(s)}$ occurs at about Eh = 0.25 V and then the solubility of the Np(V) oxide controls the aqueous neptunium concentration at higher Eh values.

Neptunium is expected to be present in the WIPP in either the IV or V oxidation state. For the WIPP CCA speciation and solubility calculations, an upper bound for the solubility of Np(IV) was estimated from that of Th(IV), using the oxidation state analogy. As with U(IV), calculations suggest that this assumption is conservative (Wall *et al.*, 2002). Modeling for the WIPP project suggested that Np(V) solubility in the reference Salado and Castile brines is limited by $KNpO_2CO_3{\cdot}2H_2O_{(s)}$ and is 1.2×10^{-7} M and 4.8×10^{-7} M, respectively (Novak, 1997; US EPA, 1998d). The most abundant aqueous species in both brines is $NpO_2(CO_3)^-$. Experimental measurements of the solubility of Np(V) in laboratory solutions representing unaltered Salado brine yielded a value of 2.4×10^{-7} M, after allowing the brine systems to equilibrate for up to 2 yr (Novak *et al.*, 1996). The solubility-limiting phase was identified as $KNpO_2CO_3{\cdot}n\,H_2O_{(s)}$, in agreement with the results of the WIPP performance assessment modeling.

Aqueous neptunium species including bishydroxo and mixed hydroxy-carbonato species may be important, though not dominant, at higher pH and carbonate concentrations. Such conditions may exist at the Hanford Waste tanks, where $MNpO_2CO_3{\cdot}n\,H_2O$ and $M_3NpO_2(CO_3)_2$ ($M = Na^+, K^+$) are predicted to be stable phases. The solubilities are two to three orders of magnitude higher than in waters in which Np_2O_5 is stable. Under conditions expected in the near field of HLW geologic repositories in saline groundwater environments such as the salt domes and the bedded salts in Europe, Np(VI) species like $NpO_2(CO_3)_3^{4-}$ might be important due to radiolysis.

In general, sorption of Np(V) by aluminosilicates is expected to be low in waters at Yucca Mountain (Turner and Pabalan, 1999; Turner *et al.*, 1998). K_ds for sorption of neptunium by zeolites and tuff particles were typically less than 10 mL g^{-1} in waters from that site (Tien *et al.*, 1985; Runde, 2002). The low neptunium sorption is due to the relative dominance of the poorly sorbed hydrolyzed species $NpO_2(OH)_{(aq)}$ and the anionic $NpO_2CO_3^-$ species in solution. In contrast, the average K_ds for Np(V) uptake by colloidal hematite, montmorillonite, and silica were 880 mL g^{-1}, 150 mL g^{-1}, and 550 mL g^{-1}, respectively, in Yucca Mountain J-13 water (Efurd *et al.* (1998), probably due to the high surface area of the particles.

Similarly, McCubbin and Leonard (1997) reported neptunium K_ds of 10^3–10^4 mL g^{-1} for particulates in seawater, but the oxidation state was uncertain. Like other tetravalent actinides, Np(IV) has a strong tendency to polymerize and form colloids and is strongly sorbed. Neptunium(IV) migration is likely to occur as intrinsic colloids or sorbed species on pseudocolloids, and changes in ionic strength are likely to impact mobility mostly through destabilization of colloidal particles. Neptunium(V) intrinsic colloids are not expected at neutral pH (Tanaka *et al.*, 1992) and uptake by carrier colloids occurs by ion-exchange and surface complexation. Competition for sorption sites between Np(V) species and other ions, especially Ca^{2+} and Mg^{2+}, could be significant (Tanaka and Muraoka, 1999; McCubbin and Leonard, 1997).

Plutonium. Plutonium chemistry is complicated by the fact that it can exist in four oxidations states over an Eh range of −0.6–1.2 V and a pH range of 0–14. In the system $Pu–O_2–H_2O$, four triple points exist (Eh–pH where three oxidation states may coexist) and thus disproportionation reactions can occur in response to radiolysis or changes in Eh, pH, or the concentrations of other chemical species (Langmuir, 1997a). The most important of these reactions are disproportionation of PuO_2^+ to PuO_2^{2+} and Pu^{4+} or disproportionation of plutonium facilitated by humic acid (Guillaumont and Adloff, 1992) and radiolysis (Nitsche *et al.*, 1995).

Langmuir's Eh–pH calculations (1997a) show that in systems containing only Pu, H_2O, and carbonate/bicarbonate (10^{-2} M), the stability field for the species Pu^{4+} is nearly nonexistent (limited to high Eh and very low pH) but the field for $Pu(OH)_{4(aq)}$ is extensive at pH > 5 and Eh < 0.5. Carbonato complexes of Pu(VI) and Pu(V) are important at pH > 5 and higher Eh. Plutonium solubilities are generally low over most environmental conditions (<10^{-8} M); the solubility fields for $PuO_{2(cr)}$ and $Pu(OH)_{4(am)}$ cover the Eh–pH field over the pH > 5 at all Eh and substantial portions of the Eh–pH field at lower pH where Eh > 0.5 V. As discussed below, the system is different when other ligands, cations, and higher concentrations of carbonate are present.

Runde *et al.* (2002a) compiled an internally consistent database to calculate solubility and speciation of plutonium in more complex low-ionic-strength waters. A specific interaction model (Grenthe *et al.*, 1992) was used for ionic strength corrections. The reader is referred to that work for details of the data sources and methods used for extrapolation and interpolation. Where reliable data for plutonium species were unavailable, thermodynamic constants were estimated from data for analogous americium, curium, uranium, and neptunium species. The most important solution species of plutonium are the aqueous ions, hydroxides, carbonates, and fluoride complexes. Important solids include oxides and hydroxides ($Pu(OH)_3$, PuO_2, $PuO_2 \cdot nH_2O$ [or $Pu(OH)_4$], PuO_2OH, $PuO_2(OH)_2$), and the carbonate PuO_2CO_3. The dominant solid phases and species are shown in an Eh–pH diagram for J-13 water variants in Figure 8. Note that in this system, the only triple point occurs at a pH of 2.4 where species in the IV, V, and VI oxidation states are calculated to be in equilibrium. In reference J-13 water (pH = 7, Eh = 0.43 V), $Pu(OH)_{4(aq)}$ dominates solution speciation, and $Pu(OH)_{4(s)}$ is the solubility-limiting phase. Under certain environments affected by interactions of groundwater and nuclear waste forms, Pu(V) or Pu(VI) could be produced by radiolysis or Pu(III) species could be produced by reduction.

Runde *et al.* (2002a) demonstrate that significant changes in plutonium solubility can occur due to the formation of Pu(V) and Pu(VI) species at pH > 6 or due to the formation of Pu(III) species at pH < 6. Using either $Pu(OH)_{4(s)}$ or the more crystalline $PuO_{2(s)}$ as the solubility controlling solids, Runde *et al.* (2002a) calculated plutonium solubilities over ranges of pH (3–10), Eh (0–0.6 V), and total carbonate concentration (0.1–2.8 mmol). For conditions typical of groundwater environments (pH 6–9 and Eh 0.05–0.45 V), $Pu(OH)_{4(aq)}$ is the dominant aqueous species. Under alkaline conditions, solubility increases with Eh due to formation of Pu(V) and Pu(VI) solution species. At pH > 8 and Eh > 0.4 V, carbonate species are dominant. At pH < 7, the solubility increases with decreasing Eh due to the stability of $Pu(OH)_3^+$. The calculated solubilities and speciation are sensitive to changes in both Eh and pH. For systems in which $Pu(OH)_{4(s)}$ is the stable solid, they ranged from > 10^{-1} M at pH = 4 and Eh = 0 V to 10^{-11} M at pH > 5 and Eh < 0.4 V. Calculated solubilities were about four orders of magnitude lower when $PuO_{2(s)}$ was the stable solid.

Experimentally measured solubilities over this range of solution compositions are typically two orders of magnitude higher than calculated values (Runde *et al.*, 2002a), presumably due to the presence of Pu(IV) colloids (Capdevila and Vitorge, 1998; Efurd *et al.*, 1998; Knopp *et al.*, 1999). In experimental studies in J-13 water, at ambient temperature, plutonium solubility decreased from 5×10^{-8} M at pH = 6, to 9×10^{-9} M at pH = 9 (Efurd *et al.*, 1998).

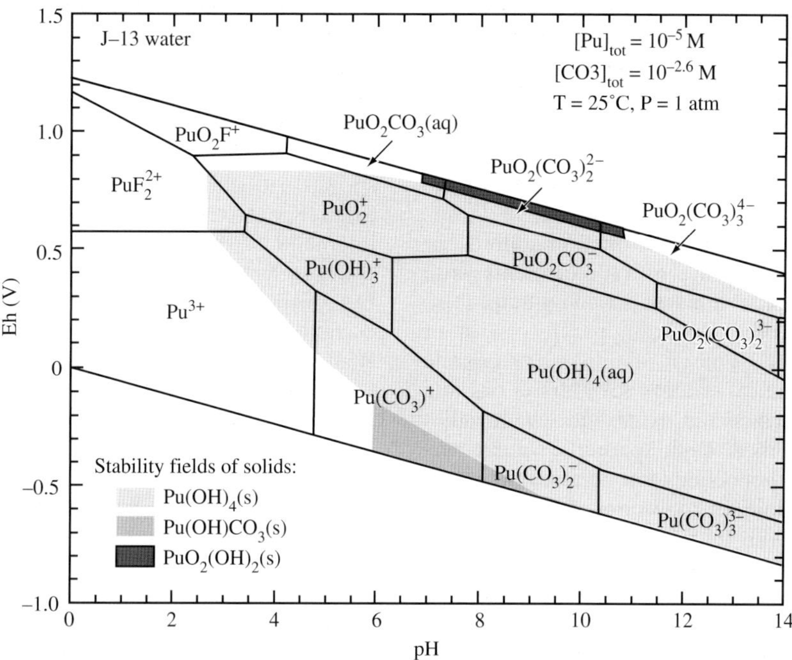

Figure 8 Eh–pH diagram showing dominant plutonium solid phases and species for J-13 water variants at 25 °C as calculated by Runde *et al.* (2002a). Solid lines indicate dominant solution species and shaded areas indicate solids supersaturated in 10^{-5} M Pu solutions. Precipitation of PuO_2 was suppressed in the calculations; see Runde *et al.* (2002a) for details. (reproduced by permission of Elsevier from *Appl. Geochem.*, **2002**, *17*, 844).

Because of the presence of Fe^0 and Fe(II), Pu(VI) is not expected to be stable under WIPP repository conditions. Plutonium(IV) is expected to be the dominant oxidation state, although Pu(III) was also considered to be a possibility in the WIPP CCA. In the WIPP performance assessment speciation and solubility calculations, Th(IV) was used as an analogue for Pu(IV) (US DOE, 1996). Wall *et al.* (2002) evaluated the appropriateness of the analogy and found that this assumption was highly conservative and that predicted solubilities for Pu(IV) in Salado and Castile brines were 10–11 orders of magnitude lower than those for Th(IV). Similarly, Am(III) was used to estimate the solubility of Pu(III).

Studies of the sorption of plutonium are complicated by the high redox reactivity of plutonium. Sorption of Pu(V) by pure aluminosilicates and oxyhydroxide phases is usually characterized by initial rapid uptake followed by slow irreversible sorption and may represent a reductive uptake mechanism catalyzed by the electrical double layer of the mineral surface (Turner *et al.*, 1998; Runde *et al.*, 2002a). In Yucca Mountain waters, the K_d ranges for Pu(V) uptake by hematite, montmorillonite, and silica colloids were 4.9×10^3 mL g^{-1} to 1.8×10^5 mL g^{-1}; 5.8×10^3 mL^{-1}; and 8.1×10^3 mL g^{-1}, respectively. These are much higher than those observed for Np(V) in the same waters as described previously. High surface redox reactivity for plutonium and possible disproportionation of Pu(V) to Pu(VI) and Pu(IV) were observed in sorption studies using goethite by Keeney-Kennicutt and Morse (1985) and Sanchez *et al.* (1985). Desorption by plutonium was typically less from hematite than from aluminosilicates in studies with J-13 water described by Runde *et al.* (2002a).

9.06.3.3 Other Topics

9.06.3.3.1 Colloids

Introduction. Colloidal suspensions are defined as suspensions of particles with a mean diameter less than 0.45 μm, or a size range from 1 nm to 1 μm. They represent potentially important transport vectors for highly insoluble or strongly sorbing radionuclides in the environment. Colloids are important in both experimental systems and natural settings. In the former, unrecognized presence of colloids may lead to overestimation of the solubility and underestimation of the sorption of radionuclides if they are included in the estimation of the concentration of radionuclide solution species. In natural systems, they may provide an important transport mechanism for radionuclides not filtered out by the host rock. In fractured rock, local transport of radionuclides by colloids may be important.

Useful reviews of the behavior of colloids in natural systems and their potential role in transporting contaminants include those of Moulin and Ouzounian (1992), Ryan and Elimelech (1996), Kretzschmar *et al.* (1999), and Honeyman and Ranville (2002).

Two types of colloids are recognized in the literature. Intrinsic colloids (also called "true" colloids, type I colloids, precipitation colloids, or "Eigencolloids") consist of radioelements with very low solubility limits. Carrier colloids (also known as "pseudocolloids," type II colloids or "Fremdkolloides") consist of mineral or organic phases (in natural waters primarily organic complexes, silicates and oxides) to which radionuclides are sorbed. Both sparingly soluble and very soluble radionuclides can be associated with this type of colloid. In addition, radionuclides can be associated with microbial cells and be transported as biocolloids.

Natural carrier colloids exist in most groundwaters; they include mineral particles, alteration products of mineral coatings, humic substances, and bacteria. In nuclear waste repositories, carrier colloids will be produced by degradation of engineered barrier materials and waste components: iron-based waste package materials can produce iron oxyhydroxide colloids, degradation of bentonite backfills can produce clay colloids, and alteration of HLW glass can produce a variety of silicate particulates. Intrinsic colloids potentially could be produced by direct degradation of the nuclear waste or by remobilization of precipitated actinide compounds (Avogadro and de Marsily, 1984; Bates *et al.*, 1992; Kim, 1994).

Two different processes could be important for the initiation of radionuclide transport by carrier colloids: (i) reversible sorption of radionuclides from solution onto pre-existing colloids and (ii) detachment of colloids from the host rock with high concentrations of previously sorbed irreversibly bound colloids. In both cases, radionuclides that sorb strongly to the rock matrix and would normally migrate very slowly will travel at a rapid rate while they are bound to colloids.

Naturally occurring colloids. Degueldre *et al.* (2000) and Honeyman and Ranville (2002) summarize modern techniques used to sample colloids from groundwater and to characterize particle concentration and size distributions. Naturally occurring colloids and radionuclide-colloid associations have been characterized at several natural analogue sites for nuclear waste repositories. These include: the Cigar Lake Uranium deposit in altered sandstone (Vilks *et al.*, 1993); the altered schist at the Koongarra Uranium deposit (Payne *et al.*, 1992); altered volcanic rock sites in Pocos de Caldas, Brazil (Miekeley *et al.*, 1991); shallow freshwater aquifers above the salt-hosted Gorleben repository test site in Germany (Dearlove *et al.*, 1991); the Grimsel test site in the Swiss Alps (Degueldre *et al.*, 1989); the Whiteshell Research area in fractured granite in Canada (Vilks *et al.*, 1991); the El Borrocal site in weathered fractured granite near Madrid, Spain (Gomez *et al.*, 1992); and 24 springs and wells near or within the Nevada Test Site (Kingston and Whitbeck, 1991). Major international studies of the occurrence of natural colloids and their potential importance to the European nuclear waste disposal program were carried out by the MIRAGE 2 (Migration of Radionuclides in the Geosphere) project and the Complex Colloid Group of the Commission of European Communities; these are reviewed in Moulin and Ouzounian (1992).

Although locally and globally there are wide variations in colloid concentration and size distribution, several general trends can be observed. Many of the observed particle concentrations fall within the range 0.01–5 mg L^{-1}; however, concentrations of >200 mg L^{-1} have been observed. There is an inverse correlation between particle concentration and particle size. Degueldre (1997) summarized the occurrence of colloids in groundwater from 17 different sites. In a marl aquifer near a proposed Swiss repository site for low-level nuclear waste, the concentration was found to be independent of flow rate, and colloid generation was caused only by resuspension/detachment of the rock clay fraction. Degueldre *et al.* (2000) expanded the scope of that study and found that colloid stability can be parsimoniously described as a function of groundwater chemistry while colloid composition is a function of rock composition. They found that at certain ionic strengths, the concentration of colloids was inversely correlated to the concentration of alkali metals and alkaline earth elements (below about 10^{-4} M and 10^{-2} M, respectively). Large concentrations of organics and the process of water mixing enhance colloid stability and concentration.

Experimental studies. Sorption of radionuclides by colloids is affected by the same solution composition parameters discussed in the previous section on sorption processes. The important parameters include pH, redox conditions, the concentrations of competing cations such as Mg^{2+} and K^+, and the concentrations of organic ligands and carbonate. The high surface area of colloids leads to relatively high uptake of radionuclides compared to the rock matrix. This means that a substantial fraction of mobile radionuclides could be associated with carrier colloids in some systems. The association of radionuclides with naturally occurring colloids and studies of radionuclide uptake by colloids in laboratory systems give some indication of the potential importance of colloid-facilitated radionuclide transport in the environment as discussed below.

For example, Kim (1994) summarizes evidence for strong sorption of americium by silica and alumina colloids in simple $NaClO_4$ solutions ($K_d > 10^4$ at pH = 8 and is independent of ionic strength, temperature, and concentrations of both americium and colloids). He also reports that significant sorption of Np(V) by alumina colloids occurs at pH>7 at high colloid concentrations (>100 ppm) under the same conditions. Lieser *et al.* (1990) studied partitioning of strontium, caesium, thorium, and actinium between molecular (<0.002 μm) and particle-bound (0.002 μm to >0.45 μm) fractions in batch systems at low concentrations (several orders of magnitude below estimated solubility limits). Sediment–water systems comprised several particle-size fractions and natural waters of different salinities from the aquifers above the Gorleben salt dome were examined. In the groundwaters, colloids consisted primarily of clays, amorphous silica, and iron hydroxide. Appreciable fractions of the total concentrations of the cations in the groundwaters was associated with the large (>0.45 μm) and fine (0.002–0.45 μm) particles (in the order Ac = Th > Cs > Sr). When the groundwaters were passed through columns filled with sediments associated with the groundwaters, most of the caesium (99%) was retained (presumably by ion exchange) and the eluted caesium was primarily associated with fine-grained pseudocolloids. About 99% of the actinium and thorium were retained in the columns (presumably by chemisorption) and the colloidal fraction of the radionuclides dominated the effluent. In contrast, retention of strontium in the columns was less effective due to its lower potential for ion exchange.

Runde *et al.* (2002a) characterized plutonium precipitates and examined neptunium and plutonium uptake by inorganic colloidal particulates in J-13 water from the Yucca Mountain site. Plutonium solubilities determined experimentally at pH 6, 7, and 8.5 were about two orders of magnitude higher than those calculated using the existing thermodynamic database, indicating the influence of colloidal Pu(IV) species. Solid-phase characterization using X-ray diffraction revealed primarily Pu(IV) in all precipitates formed at pH 6, 7, and 8.5. As discussed previously, hematite, montmorillonite, and silica colloids were used for uptake experiments with ^{239}Pu(V) and ^{237}Np(V). The capacity of hematite to sorb plutonium significantly exceeded that of montmorillonite and silica. A low desorption rate was indicative of highly stable plutonium–hematite colloids, which may facilitate plutonium transport to the accessible environment. Plutonium(V) uptake on all mineral phases was far greater than Np(V) uptake, suggesting that a potential Pu(V)–Pu(IV) reductive sorption process was involved.

Microbial and humic colloids. The transport of radionuclides and metals adsorbed to microbes has been considered by a number of researchers including McCarthy and Zachara (1989), Han and Lee (1997), and Gillow *et al.* (2000). Because of their small size (<10 μm diameter), these colloids can be transported rapidly through fractured media and either filtered out or be transported through porous media. Microbes can sorb to geologic media, thereby retarding transport. Alternatively, under conditions of low nutrient concentrations, the microbes can reduce their size and adhesion capabilities and become more easily transported. Studies performed in support of WIPP compliance certification indicated that, under relevant redox conditions, microbially bound actinides contributed significantly to the concentration of mobile actinides in WIPP brines (Strietelmeyer *et al.*, 1999; Gillow *et al.*, 2000). For performance assessment calculations, the concentration of each actinide sorbed onto microbial colloids was estimated to be 3.1, 0.0021, 12.0, 0.3, and 3.6 times the dissolved concentration for thorium, uranium, neptunium, plutonium, and americium, respectively (US DOE, 1996).

Evidence for strong sorption of actinides and fission products by humic substances, both in dilute and high-ionic-strength media, is provided by experimental studies and thermodynamic calculations. Humic colloids are stable and occur in concentrations up to 0.4 g L^{-1} in dilute, shallow groundwaters overlying the Gorleben test site in Germany (Buckau *et al.*, 2000). Humic substances have experimentally been shown to strongly complex the trivalent actinides (Czerwinski *et al.*, 1996; Artinger *et al.*, 1998; Morgenstern *et al.*, 2000): Th(IV) (Nash and Choppin, 1980); U(VI) and probably U(IV) (Czerwinski *et al.*, 1994; Zeh *et al.*, 1997); and Np(V) (Kim and Sekine, 1991; Rao and Choppin, 1995; Marquardt *et al.*, 1996; Marquardt and Kim, 1998) at mildly acidic to neutral pH. Under basic conditions, actinide–humic substance complexation is strongly a function of the carbonate concentration, because carbonate competes effectively with the humic acid as a ligand (Zeh *et al.*, 1997; Unsworth *et al.*, 2002). Little data are available for tetravalent actinides, but Tipping (1993) suggests, based on thermodynamic modeling, that these should be even more strongly complexed by humic substances than other oxidation states.

Several studies have shown that actinide–humic acid complexes are thermodynamically stable in high-ionic-strength solutions (Czerwinski *et al.*, 1996; Marquardt *et al.*, 1996; Labonne-Wall *et al.*, 1999). However, destabilization of humic colloids at high ionic strength (Buckau *et al.*, 2000) and competition for humic acid sites by divalent metal cations (Tipping, 1993; Marquardt *et al.*, 1996) may limit the importance of colloidal transport of actinides in brines. In the WIPP performance assessment, the estimated contribution of actinide sorbed onto humic substances to the total mobile concentration was >0.01 times the dissolved fraction for An(V), 0.1–1.4 times the dissolved fraction for An(III) and An(VI), and 6.3 times the dissolved fraction for An(IV) (US DOE, 1996).

Transport of radionuclides by colloids. Several numerical models have been developed to assess the potential magnitude of colloidal-facilitated transport of radionuclides compared to the transport of dissolved species. Vilks *et al.* (1998) proposed a simple modification to the standard equation for the retardation factor. The equation applies to the ideal case where colloids are not trapped by the rock matrix and the composition of the colloids and the rock matrix are the same:

$$R_{F,\mathrm{eff}} = 1 + \frac{(1-\phi)\rho K_\mathrm{d}}{\phi(1+CFK_\mathrm{d})} \tag{8}$$

where $R_{F,\mathrm{eff}}$ is the effective retardation factor for radionuclides, including the effect of reversible sorption onto and transport by colloids. In the equation, ρ is the bulk density of the porous medium, ϕ is the porosity, K_d is the distribution coefficient for both the colloid and the rock matrix (mL g^{-1}), and C is the colloid concentration (mg L^{-1}). F is defined as

$$F = (A'_\mathrm{colloid}/A'_{IP}) \tag{9}$$

where A'_colloid and A'_IP are the specific surface areas of the colloid and rock matrix, respectively.

When the above conditions are not met, more complex models that account for colloid generation, irreversible sorption, differences between the sorptive capacity of colloids and rock matrix (i.e., different K_ds), and colloid filtration should be used. Avogadro and de Marsily (1984) developed a simple model involving colloid filtration under field conditions. Nuttall *et al.* (1991) developed a 2D population-balance model for radiocolloid transport that includes production and filtration of colloids under saturated and unsaturated conditions. van der Lee *et al.* (1992) and Smith and Degueldre (1993) developed numerical models for colloid-facilitated transport through fractured media that incorporate a finite number of sorption sites, Langmuir isotherms, and irreversible sorption. Finally, the incorporation of colloid transport in systems with slow desorption, contrasts between the sorptive capacity of colloids and the rock matrix, and variable K_ds as functions of solution composition are simulated by the LEHGC reactive transport model as described by Yeh *et al.* (1995) and Honeyman and Ranville (2002).

There is considerable debate concerning the potential importance of colloid-facilitated transport of radionuclides for the design and performance assessment of nuclear waste repositories and for risk assessments of radioactively contaminated sites. Honeyman and Ranville (2002) develop a framework to determine the conditions under which colloid-facilitated contaminant transport will be important compared to the transport of solution species. They conclude that such conditions will be relatively rare in the environment. In contrast, Penrose *et al.* (1990) and Nuttall *et al.* (1991) suggest that colloidal transport of radionuclides in the unsaturated zone can be important. They describe field evidence, laboratory results, and computer simulations that suggest that colloidal transport of strongly sorbing actinides such as plutonium and americium is potentially significant in the unsaturated zone and in shallow aquifers near Los Alamos, New Mexico. Similarly, Kersting *et al.* (1999) provide evidence that measurable amounts of plutonium and perhaps cobalt, europium, and caesium produced by nuclear weapons tests (1956–1992) at the Nevada Test site have been transported at least 1.3 km from the blast sites by colloids. They argue that models that do not include colloid-facilitated transport may significantly underestimate the extent of radionuclide migration. In contrast, Vilks (1994) proposes that colloids do not have to be considered in the safety assessment for the Canadian repository in granite. He argues that the clay-based buffer to be used in the repository will filter out any colloids produced by degradation of the waste package. In addition, the concentration of naturally occurring colloids is too low to provide a substantial transport vector for radionuclides that escape to the far field of the repository.

Four types of colloids were considered in the WIPP program: intrinsic actinide colloids, mineral colloids, microbes, and humic acid colloids (US DOE, 1996). Intrinsic actinide colloids, consisting of polymerized hydrated actinide hydroxides, are not stable in the neutral to moderately basic pH conditions expected in the WIPP, and were assumed not to contribute to the total actinide concentrations in solution. Mineral colloids are destabilized and tend to flocculate in the high-ionic-strength WIPP brines (Kelly *et al.*, 1999). In the performance assessment calculations for the WIPP, a highly conservative value of 2.6×10^{-8} mol actinide per liter, for each actinide, was assumed to be bound to mineral colloids and to contribute to the mobile fraction. Actinides sorbed onto microbes and humic acids were estimated to contribute significantly to the concentration of mobile actinides in WIPP brines as discussed above (Section 9.06.3.2.2).

Contardi *et al.* (2001) used an SCM to examine the potential effect of colloidal transport on the effective retardation factors for americium, thorium, uranium, neptunium, and plutonium in waters from the proposed repository site at Yucca Mountain. They found that colloidal transport reduced the effective retardation of strongly sorbed radionuclides such as americium and thorium by several orders of magnitude compared to simulations in which such transport was ignored. Uranium, neptunium, and Pu(V) are less strongly sorbed by colloids and therefore were relative unaffected by colloidal transport. They also described performance assessment calculations of the effect of colloid-facilitated radionuclide transport on the peak mean annual total effective dose equivalent (TEDE) from the proposed nuclear water repository. The colloid transport simulations showed no increase in the TEDE within a compliance period of 10^4 yr; however, at longer simulation times, as the waste container failures increased, the TEDE from the colloid models are up to 60 times that of the base (noncolloid) case. Such simulations are strongly dependent on scenario assumptions and can be used to provide conservative estimates of the potential importance of colloids for radionuclide transport. They are also useful in demonstrating the relative importance of processes included in the performance assessment models.

9.06.3.3.2 Microbe–actinide interactions

In addition to possible transport of radionuclides by microbial colloids, microbe–actinide chemical interactions are important for the genesis of uranium ore bodies, dissolution of radioactive waste, and remediation of contaminated sites. Chapelle (1993) provides a recent comprehensive treatment of microbial growth, metabolism, and ecology for geoscience applications. Suzuki and Banfield (1999) and Abdelouas *et al.* (1999) provide well-documented overviews of geomicrobiology of uranium with discussion of applications to environmental transport and remediation of sites contaminated with uranium and actinides. Microbial–uranium interactions have been studied for technological applications such as bioleaching, which is an important method for uranium extraction (Bertheolet, 1997). *In situ* stabilization of uranium plumes by microbial reduction is an important method for remediation (Barton *et al.*, 1996). Microbial ecology in high uranium environments has been studied in uranium mill tail wastes, and the effects of microbes on the stability of radioactive wastes buried in geological repositories are discussed by Francis (1994) and Pedersen (1996).

Microbes can control the local geochemical environment of actinides and affect their solubility and transport. Francis *et al.* (1991) report that oxidation is the predominant mechanism of dissolution of UO_2 from uranium ores. The dominant oxidant is not molecular oxygen but Fe(III) produced by oxidation of Fe(II) in pyrite in the ore by the bacteria *Thiobacillus ferroxidans*. The Fe(III) oxidizes the UO_2 to UO_2^{2+}. The rate of bacterial catalysis is a function of a number of environmental parameters including temperature, pH, TDS, f_{O2}, and other factors important to microbial ecology. The oxidation rate of pyrite may be increased by five to six orders of magnitude due to the catalytic activity of microbes such as *Thiobacillus ferroxidans* (Abdelouas *et al.*, 1999).

Suzuki and Banfield (1999) classify methods of microbial uranium accumulation as either *metabolism dependent* or *metabolism independent*. The former consists of precipitation or complexation with metabolically produced ligands, processes induced by active cellular pumping of metals, or enzyme-mediated changes in redox state. Examples include precipitation of uranyl phosphates due the activity of enzymes such as phosphatases, formation of chelating agents in response to metal stress, and precipitation of uraninite through enzymatic uranium reduction.

Metabolism-independent processes involve physicochemical interactions between ionic actinide species and charged sites in microorganisms; these can occur with whole living cells or cell fragments. Uranium can be accumulated in the cells by passive transport mechanisms across the cell membrane or by biosorption, a term that includes nondirected processes such as, absorption, ion exchange, or precipitation. Suzuki and Banfield (1999) describe the effects of pH and concentrations of other cations and anions on uranium uptake by a variety of organisms. Uptake of uranium by microbes can be described using the techniques and formalisms used to analyze the sorption of metals by metal hydroxide surfaces such as the Freundlich, Langmuir, and surface-complexation sorption models (Fein *et al.*, 1997; Fowel and Fein, 2000).

A large number of species of bacteria, algae, lichen, and fungi have been shown to accumulate high levels of uranium through these processes. Suzuki and Banfield provide examples of organisms, whose uranium uptake capacities range from approximately $50\ mg\ U\ g^{-1}$ to $500\ mg\ U\ g^{-1}$ dry cell weight. Maximum uptake occurs from pH 4–5. Microorganisms can develop resistance to the chemical and radioactive effects of actinides through genetic adaptation. In contaminated environments such as uranium mill tailings and mines, uranium accumulation levels exceed those observed in laboratory experiments with normal strains of bacteria. Suzuki and Banfield (1999) describe several mechanisms used by microbes to detoxify uranium.

Suzuki and Banfield (1999) discuss the similarities between the uranium–microbe interactions and transuranic–microbe interactions. Macaskie (1991) notes that it is possible to extrapolate the data for microbial uranium accumulation to other actinides. Hodge *et al.* (1973) observe that the biological behavior of uranium, thorium, and plutonium resemble that of ferric iron. Microbes can also affect the speciation and transport of multivalent fission products. For example, Fe^{3+}-reducing bacteria and sulfate-reducing bacteria can reduce soluble pertechnetate to insoluble Tc(IV), as discussed by Lloyd *et al.* (1997). For additional information about these topics, the reader is referred to the references cites above. Applications of these principles are described in the section on bioremediation later in this chapter.

9.06.4 FIELD STUDIES OF RADIONUCLIDE BEHAVIOR

9.06.4.1 Introduction

Studies of field sites establish the link between theoretical calculations, laboratory studies, and the behavior of radionuclides in the environment. The field observations may complement or in some cases conflict with information obtained in laboratory studies. Important field sites include the areas of anthropogenic radioactive contamination described in previous sections, such as the areas surrounding the Chernobyl reactor, the Hanford reservation, the Chleyabinsk-65 complex, the Hamr uranium mining district of the Czech Republic, the Nevada test site, releases from the Nagasaki atomic bomb detonation, and the Konegstein mine in Germany. Important sites of natural radioactivity (natural analogues) include the Pena Blanca deposit, the Alligator River Region, Cigar Lake, and the Oklo natural reactor. The Pena Blanca deposit in northern Mexico is an analogue for the proposed HLW repository in the unsaturated tuffs at Yucca Mountain (Pearcy *et al.*, 1994; Murphy, 2000). The Koongarra uranium deposit in the Alligator Rivers Region, Australia, has been widely studied in a coordinated international program (Payne *et al.*, 1992; Duerden *et al.*, 1992; van Maravic and Smellie, 1992; Davis, 2001). The Cigar Lake uranium deposit in altered sandstone in Saskatchewan, Canada is another analogue for nuclear waste repositories located below the groundwater table (Vilks *et al.*, 1993; Bruno *et al.*, 1997; Curtis, 1999). Natural analogue studies have also been carried out in the uranium, thorium, and REE ore bodies in altered volcanic rock at Pocos de Caldas, Brazil (Miekeley *et al.*, 1991) and in a uranium-bearing quartz dike/breccia complex in weathered fractured granite at the El Borrocal site near Madrid, Spain (Gomez *et al.*, 1992). The interested reader is referred to the sources cited above for detailed descriptions of how the geochemical and hydrologic characteristics of those sites are used to validate or calibrate hydrogeochemical models for radionuclide behavior.

The following section provides detailed information concerning the transport of radionuclides associated with two very different field analogues: the Chernobyl reactor accident and the Oklo Natural Reactor. These examples span wide temporal and spatial scales and include the rapid geochemical and physical processes important to nuclear reactor accidents or industrial discharges as well as the slower processes important to the geologic disposal of nuclear waste.

9.06.4.2 Short-term Behavior of Radionuclides in the Environment—Contamination from the Chernobyl Reactor Accident

On April 26, 1986, an explosion at the Chernobyl Nuclear Power Plant, and the subsequent fire in the graphite reactor core, released ~3% (6–8 Mt) of the total fuel inventory of the reactor core. During the initial explosion, most radionuclide release occurred as fragments of unoxidized uranium dioxide fuel, which were deposited mostly in a plume extending 100 km to the west of the plant. The core fire lasted 10 days and releases were again dominated by fuel particles. However, because the core was exposed, these particles were partially or completely oxidized. For the first four days, high temperatures in the reactor resulted in the release of volatile elements (xenon, krypton, iodine, tellurium, and caesium), much of which was deposited as condensed particles in plumes to the northwest, west, and northeast of the plant, extending as far as Scandinavia. Over the next six days, temperatures in the reactor decreased, and the release of volatile fission products decreased. The lower temperatures (600–1,200 K) favored oxidation of the nuclear fuel; therefore, fuel particles released during this phase were more heavily oxidized. Radionuclides released during this phase were deposited mostly in a southern plume, extending as far south as Greece.

A major part of the radionuclides released at Chernobyl were deposited as "hot particles," either fuel particles with an average median diameter of 2–3 μm or condensed particles. Within the exclusion zone, extending 30 km from the plant, more than 90% of the radioactive contamination was in the form of fuel particles (Kashparov *et al.*, 1999). This included ~80% of the ^{90}Sr and ~50–75% of the ^{137}Cs contamination. Particle size decreased with distance from the Chernobyl plant, and the proportion of condensed particles relative to the fuel particles increased. Other radionuclides released include the fission products ^{134}Cs, ^{144}Ce, ^{125}Sb, ^{106}Ru, ^{103}Ru, and ^{95}Zr, and the neutron activation products ^{110m}Ag and ^{54}Mn (Petropoulos *et al.*, 2001).

Considerable attention has been given to fuel particle behavior in the environment because of their importance to the total radionuclide release from Chernobyl. The transport properties of the radionuclides present in the fuel particles and type of hazard that they represented changed as the particles weathered. Initially, particle (and radionuclide) transport was governed by physical processes. Particles were transported both as aerosols in the atmosphere and as suspended load in runoff and rivers. The primary hazards represented by the particles were inhalation and dermal exposure. Particle α-activities were similar to that of the original nuclear fuel (2.5×10^8 Bq cm^{-3}; Boulyga *et al.* (1999)), while β- and γ-activities were somewhat lower due to loss of volatile fission products. The median fuel particle size released at Chernobyl (2–3μm) falls within the respirable fraction of aerosols (defined as <7 μm), and such particles are easily resuspended by anthropogenic or natural processes that disturb the soil. Fuel particles are not readily transported downward through the soil column, and soil sampling in the exclusion zone, carried out 10 yr after the accident, showed that the particles were still concentrated in the upper 5 cm (Kashparov *et al.*, 1999). Thirteen years after the accident, agricultural activities in the exclusion zone resulted in local airborne concentrations of hot particles of 50–200 particles m^{-3} (Boulyga *et al.*, 1999).

Radionuclides sequestered in the fuel particles are not immediately available to the biosphere, because the particles generally have a low solubility in water, simulated lung fluids, and HCl solutions (Chamberlain and Dunster, 1958; Oughton *et al.*, 1993; Salbu *et al.*, 1994). Release of biologically important radionuclides such as ^{90}Sr and ^{137}Cs into the biosphere requires weathering and dissolution of the fuel particles. Particle weathering rates vary with several source-related particle characteristics (particle size, oxidation state, and structure) and depend on environmental parameters such as soil pH and redox conditions. Field studies have shown that the fraction of exchangeable ^{90}Sr in soil increases as fuel particles dissolve (Petryaev *et al.*, 1991; Konoplev *et al.*, 1992; Baryakhtar, 1995; Konoplev and Bulgakov, 1999). Kashparov *et al.* (1999) and Konoplev and Bulgakov (1999) used these observations to estimate the fraction of undissolved fuel particles present in soils in the Chernobyl 30 km exclusion zone. They found that the unoxidized fuel particles, deposited from the initial explosion in a plume to the west of Chernobyl, are more resistant to leaching and dissolution than the oxidized fuel particles released during the subsequent fire and deposited to the north, northeast, and south of the plant. Dissolution rates for the unoxidized fuel particles were about one-third those of oxidized particles, with 27–79% remaining undissolved in 1995, while only 2–30% of the oxidized fuel remained. Fuel particle dissolution rates also increased with increasing soil acidity and with decreasing particle size (Kashparov *et al.*, 1999; Konoplev and Bulgakov, 1999).

These results are consistent with laboratory measurements of the effects of pH and oxidation state on the leachability of nuclear fuel particles (Kashparov *et al.*, 2000). The increased leachability of the oxidized fuel particles may be due to: (i) an increased solubility because of the change in oxidation state; (ii) the higher surface area of the highly fractured oxidized particles; or (iii) the diffusion of radionuclides (strontium, caesium) to grain boundaries and particle surfaces during the heating and oxidation process.

Once released from the fuel particles, strontium and caesium are available to the biosphere and can be taken up by plants or can be transferred down the soil profile. At present, the relative rates of each process are not well enough constrained to predict future levels of ^{90}Sr and ^{137}Cs in vegetation in the exclusion zone (Kashparov *et al.*, 1999).

In more distal areas, caesium and strontium concentrations in soils, lakes, and rivers were initially high from direct fallout but have progressively dropped as these elements move downward into the soil profile and are flushed or sedimented out of bodies of water. The estimated ecological half-life for ^{137}Cs in German forest soils is 2.8 ± 0.5 yr for the L horizon and 7.7 ± 4.9 yr for the Ah horizon (Rühm *et al.*, 1996). ^{90}Sr concentrations in the Black Sea had dropped to pre-Chernobyl levels by 1994, and ^{137}Cs is predicted to reach pre-accident levels by 2025–2030 (Kanivets *et al.*, 1999). The main causes of the decreases are radioactive decay and loss through the Bosporus Strait. However, the relative proportion of ^{90}Sr entering the Black Sea as river input is increasing as fuel particles in the major watersheds weather and release sequestered radionuclides. Smith *et al.* (1999, 2000) have shown that caesium removal from lakes and rivers is dominantly by lake outflow and by sedimentation. Caesium is strongly sorbed onto the frayed edge sites of illitic clay minerals, and caesium removal rates from lakes correlate with aqueous K^+ concentrations.

9.06.4.3 Natural Analogues for the Long-term Behavior of Radionuclides in the Environment—The Oklo Natural Reactor

Naturally occurring uranium deposits have been an important source of information on the long-term behavior of actinides and fission products in the environment. Of special interest are the Oklo and Bangombe deposits of Gabon, which hosted natural fission reactors ~2 Gyr ago (Pourcelot and Gauthier-Lafaye, 1999; Jensen and Ewing, 2001). The Gabon deposits occur in the Francevillian series, a 2.1 Ga sedimentary series consisting of sandstones, conglomerates, black shales, and volcaniclastic sediments. All uranium mineralization occurs in the basal sandstone formation near the upper contact with overlying black shales. The ore deposits have been interpreted as classic uranium roll-front deposits, which formed in oil traps where uranium-rich oxidizing fluids met reducing conditions in the hydrocarbon accumulations. The uranium is present primarily as uraninite, and uranium concentrations in the normal ore vary from 0.1 wt.% to 10 wt.%. In the reactor zones, the sandstone was highly fractured, and the mineralized stockwork ore initially contained up to 20 wt.% uranium.

As redox conditions continued to concentrate uranium in the ore zone, the conditions required to initiate and sustain criticality were achieved. With initiation of criticality, hydrothermal circulation cells formed in the sandstone, resulting in migration of silica and other components out of the reactor. Uraninite was concentrated in the core; uranium grades as high as 80 wt.% have been reported. In some reactors, the volume lost from the core was sufficient to cause slumping and collapse of the overlying beds. The reactor cores are generally 10–50 cm thick and are commonly overlain by a clay-rich hydrothermal gangue, consisting of magnesium chlorite and illite, known as the "reactor clays" or "argile de pile." During criticality and cooling, aluminum-rich chlorites formed in the reactor core (Pourcelot and Gauthier-Lafaye, 1999).

During operation, the reactor cores reached temperatures of 200–450 °C (Brookins, 1990; Pourcelot and Gauthier-Lafaye, 1999), perhaps as high as 1,000 °C (Holliger and Derillers, 1981). The reactors operated for $(1-8)\times 10^5$ yr (Gauthier-Lafaye *et al.*, 1996) and in the largest, consumed up to 1,800 kg of ^{235}U. Because of this, uranium from the reactor cores is depleted in ^{235}U, with this isotope constituting as little as 0.29% of the total (in "normal" ore, it would be 0.72%).

Analysis of the isotopic concentrations of uranium, thorium, fission products, and their daughters gives information on element mobility during and after criticality. There is evidence for some actinide and fission product migration during criticality, during dolerite dike intrusion, and due to recent supergene weathering. Element retentivity in the Oklo natural reactors has application to nuclear waste disposal, because the uranium oxide reactor core and the reactor clays have been described as analogous to spent nuclear fuel that is embedded in a clay-rich backfill (Menet *et al.*, 1992).

A summary of actinide and fission product behaviors at Oklo is given by Gauthier-Lafaye *et al.* (1996). In general, the core of the reactor consists mainly of uraninite with varying amounts of clay. Grains of metal/metal-oxide/sulfide are present in the uraninite, similar to metal/metal oxide grains found in depleted uranium fuel. As the reactors have seen little oxidation, actinides, and fission products that are compatible with the uraninite structure have largely been retained by the core. Evidence that uranium and plutonium did not migrate during criticality is given by the presence of large amounts of ^{232}Th in the reactor cores. ^{232}Th is formed by decay of ^{236}U and ^{240}Pu, both of which were produced by neutron capture processes during criticality. Thorium is of low abundance in the surrounding rocks, and its presence in the reactor zones is taken to indicate that uranium and plutonium were mostly retained in the reactor core. Plutonium was also retained by clays in the "argile de pile," as is evidenced by enrichments of ^{235}U, the long-lived daughter of ^{239}Pu, in some reactor clay samples (Pourcelot and Gauthier-Lafaye, 1999). Concentrations of ^{209}Bi, the stable daughter of ^{237}Np, suggest that neptunium was also mostly retained by the reactor core.

Many environmentally important fission products have short half-lives; hence their behavior in the reactors is determined by proxy, by examining the distribution of stable daughters. Such parent/daughter pairs include $^{90}Sr/^{90}Zr$, $^{137}Cs/^{137}Ba$, $^{135}Cs/^{135}Ba$, $^{129}I/^{129}Xe$, and $^{99}Tc/^{99}Ru$. The daughters commonly have a significantly different chemistry than the parents and may have migrated during criticality, the 800 Ma dike intrusion, or during supergene weathering. Despite this uncertainty, comparisons between measured concentrations and theoretical fission yields have provided valuable information on fission product mobility (Hidaka *et al.*, 1992). These studies suggest that retention of fission products by the reactors was variable. Gaseous and volatile fission products—xenon, krypton, iodine, cadmium, and caesium—were largely lost from the reactor. Highly mobile alkali metals and alkali earths—rubidium, strontium, and barium—were also lost.

Fission products that are compatible with the uraninite crystal structure—the REE, yttrium, neodymium, and zirconium—were largely retained in the uraninite core, the reactor clays, minor phosphate phases, and uranium and zirconium silicate phases (Gauthier-Lafaye *et al.*, 1996). Lighter REE—lanthanum, cerium, and praseodymium—were partially lost from the reactor. Finally, molybdenum, technetium, ruthenium, rhodium, and other metallic elements were retained in the metal/metal oxide inclusions and arsenide/sulfide inclusions in the core, and in the reactor clays (Hidaka *et al.*, 1993; Jensen and Ewing, 2001).

9.06.5 APPLICATIONS: DEALING WITH RADIONUCLIDE CONTAMINATION IN THE ENVIRONMENT

The previous sections of this chapter have briefly described the nature and locations of the most serious radioactive environmental contamination on the planet and have established the geochemical foundations for understanding the behavior of radionuclides in the environment. Applications of this information to remediating or assessing the risk posed by the contamination is the subject of this section of the review.

9.06.5.1 Remediation

In many remediation programs, simple excavation of contaminated soil and removal of contaminated groundwater by pumping are the preferred techniques. These techniques may be practical for removal of relatively small volumes of contaminated soils and water; however, after these source terms have been removed, large volumes of soil and water with low but potentially hazardous levels of contamination may remain. For poorly sorbing radionuclides, capture of contaminated water and removal of radionuclides may be possible using permeable reactive barriers and bioremediation. Alternatively, radionuclides could be immobilized in place by injecting agents that lead to reductive precipitation or irreversible sorption.

For strongly sorbing radionuclides, contaminant plumes will move very slowly and likely pose no potential hazards to current populations (Brady *et al.*, 2002). However, regulations may require cleanup of sites to protect present and future populations under a variety of future-use scenarios. In these cases, it may be necessary to use soil-flushing techniques to mobilize the radionuclides and then to collect them. Alternatively, it may be possible to demonstrate that contaminant plumes will not reach populations and that monitoring networks and contingency remedial plans are in place to protect populations if the plume moves more rapidly than predicted. This approach is called monitored natural attenuation (MNA) and is described in a later section.

9.06.5.1.1 Permeable reactive barriers

Permeable reactive barriers include reactive filter beds containing materials such as zero-valent iron (Fe^0), phosphate rock (apatite), silica sand, organic materials, or combinations of these materials (US EPA, 1999c). The barriers can be installed by digging a trench in the flow path of a contaminated groundwater plume and backfilling with reactive material or by injecting either a suspension of colloidal material or a solution containing a strong reductant (Cantrell *et al.*, 1995, 1997; Abdelouas *et al.*, 1999). The reactive filter material is used to reduce and precipitate the contaminant from solution while allowing the treated water to flow through the reactive bed. In some installations, the drain field can be designed to have removable cells should replacement and disposal of the reactive material be required. A classical funnel and gate arrangement can be used for this purpose. In this system, impermeable subsurface walls are used to direct the flow of the plume through a narrow opening where the reactive material is emplaced. A number of techniques applicable for remediation of radionuclide plumes are described in a collection of articles edited by Looney and Falta (2000) and in publications of the Federal Remediation Technologies Roundtable (e.g., US EPA, 2000a).

The use of Fe^0 to reduce and precipitate uranium out of solution has been shown to be effective by Gu *et al.* (1998) and Fiedor *et al.* (1998). The technique has been deployed in permeable reactive barriers at the Rocky Flats site in Colorado (Abdelaous *et al.*, 1999), the Y-12 Plant near Oak Ridge National Laboratories (Watson *et al.*, 1999), and other DOE sites. In this method, the Fe^0 reduces the U(VI) species to U(IV) aqueous species, which then precipitates as U(IV) solids (Gu *et al.*, 1998; Fiedor *et al.*, 1998). Reduction of U(VI) to U(IV) usually results in the precipitation of poorly crystalline U(IV) (e.g., uraninite, compositions ranging from UO_2 to $UO_{2.25}$) or mixed U(IV)/U(VI) solids (e.g., U_4O_9).

Uranium(VI) readily precipitates in the presence of phosphate to form a number of sparingly soluble U-phosphate phases (U phases, such as saleeite, meta-autunite, and autunite) and also is removed by sorption and co-precipitation in apatite. Several studies have shown that hydroxyapatite is extremely effective at removing heavy metals, uranium, and other radionuclides from solution (Gauglitz *et al.*, 1992; Arey and Seaman, 1999).

Apatite was shown to be effective at removing a number of metals including uranium at Fry Canyon, Utah (US EPA, 2000b). Krumhansl *et al.* (2002) reviews the sorptive properties of a number of other materials for backfills around nuclear waste repositories and permeable reactive barriers.

Injection of a reductant such as sodium dithionite creates a reducing zone that may be effective in immobilizing uranium and other redox-active radionuclides. The technique is known as *in situ* redox manipulation (ISRM). It has been shown to be moderately effective for chromium and is proposed for use at the Hanford site for remediation of a uranium groundwater plume (Fruchter *et al.*, 1996).

9.06.5.1.2 Bioremediation

A number of remediation techniques based on biological processes are in use at contaminated sites. Examples include use of microbes to sequester uranium and phytoremediation of a number of metals. The former method involves reductive reactions by bacteria, particularly those of sulfate reduction (Lovely and Phillips, 1992a) and direct reduction (Lovely and Phillips, 1992b; Truex *et al.*, 1997). Several techniques have been employed to generate high organic loading by growth of plant and algal biomass. Injection of nutrients into the subsurface and subsequent microbial bloom leads to the low redox conditions favorable for reductive reactions, a significant decrease in the solubility and consequently, removal of the metal onto the geomedia.

Abdelouas *et al.* (1999) provide a good review of remediation techniques for uranium mill tailings and groundwater plumes. Biological processes used in bioremediation include biosorption, bioaccumulation, and bioreduction. Biosorption includes uptake of uranium by ion-exchange or surface complexation by living microbes or the cell membranes of dead organisms. In bioaccumulation, the radionuclides are precipitated with enzymatic reactions (Macaskie *et al.* (1996) and Abdelouas *et al.* (1999) and references therein). Bioreduction includes both direct reduction of radionuclides by organisms and indirect reduction. The latter involves creation of reducing conditions by the activity of sulfate and iron-reducing microbes and the subsequent reduction of the radionuclides by produced reductants such as H_2 and H_2S. Abdelouas *et al.* (1999) provide an excellent review of laboratory and field studies of microbes such as various *Desulfovibrio* species that have been shown to be effective in reducing hexavalent uranium by both of these processes.

Phytoremediation has been used to remove uranium and strontium from groundwaters and surface waters. Studies have been conducted on the uptake of heavy metals, uranium, and other radionuclides (Cornish *et al.*, 1995; Abdelouas *et al.* (1999); both the uptake rates and the phytoconcentration of the radionuclides are high. The plants can be harvested and the volume of the residuals minimized by combusting the plant material.

9.06.5.1.3 Monitored natural attenuation

Natural attenuation encompasses processes that lead to reduction of the mass, toxicity, mobility, or volume of contaminants without human intervention. The US EPA has recently published guidelines for the use of MNA for a variety of contaminated sites (US EPA, 1997). For inorganic constituents, the most potentially important processes include dispersion and immobilization (reversible and irreversible sorption, co-precipitation, and precipitation) (Brady *et al.*, 1998). Studies of remediation options at UMTRA sites (Jove-Colon *et al.*, 2001) and the Hanford Site (Kelley *et al.*, 2002) have addressed the viability of adopting an MNA approach for uranium and strontium, respectively. As discussed below, different approaches are required to establish the viability of MNA for these radioelements.

Laboratory experiments, transport modeling, field data, and engineering cost analysis provide complementary information to be used in an assessment of the viability of an MNA approach for a site. Information from kinetic sorption/desorption experiments, selective extraction experiments, reactive transport modeling, and historical case analyses of plumes at several UMTRA sites can be used to establish a framework for evaluation of MNA for uranium contamination (Brady *et al.*, 1998, 2002; Bryan and Siegel, 1998; Jove-Colon *et al.*, 2001). The results of a recent project conducted at the Hanford 100-N site provided information for evaluation of MNA for a ^{90}Sr plume that has reached the Columbia River (Kelley *et al.*, 2002). The study included strontium sorption–desorption studies, strontium transport and hydrologic modeling of the near-river system, and evaluation of the comparative costs and predicted effectiveness of alternative remediation strategies.

It is likely that it will be easier to gain acceptance for an MNA approach for radionuclides such as ^{90}Sr compared to $^{235/238}U$. This is because ^{90}Sr has a short half-life and uniformly strong sorption, whereas uranium isotopes have very long half-lives and complex sorption behavior. Strontium transport merely needs to be slowed enough to allow radioactive decay to remove the strontium, whereas demonstrating sorption irreversibility might be a key component of an MNA remedy for uranium. MNA may be acceptable for uranium only if it can be shown that an appreciable fraction of the uranium is irreversibly sequestered on mineral sorption sites or physically occluded and cannot be leached out in the foreseeable future. Institutional controls are also important for MNA. Monitoring programs and contingency remediation plans are required as part of an overall MNA strategy (US EPA, 1997; Brady *et al.*, 1998).

9.06.5.2 Geochemical Models in Risk Assessment

9.06.5.2.1 Overview

In performance assessment models, simplified process models and sampling techniques are linked to provide a description of the release of radionuclides from idealized source terms, transport through engineered barriers and surrounding geomedia, and finally uptake by potentially exposed populations. The resulting doses are compared to environmental and health regulatory standards to estimate the risk posed by the releases. A basic overview of the process of risk assessment is presented by Fjeld and Compton (1999). Probabilistic performance assessment methods have been developed to provide a basis for evaluation of the risk associated with nuclear waste disposal in geological repositories (Cranwell *et al.*, 1987; Rechard, 1996, 2002; Wilson *et al.*, 2002). Similar approaches are used for LLW disposal and uranium mill tailings (Serne *et al.*, 1990). Development of risk assessment models by the European community is summarized in NEA (1991). The current status of risk assessment programs in several countries was reviewed in a session devoted to performance assessment at the *2001 Materials Research Society Symposium on the Scientific Basis for Nuclear Waste Management* (McGrail and Cragnolino, 2002).

Abstraction of the hydrogeochemical properties of real systems into simple models is required for risk assessment. Heterogeneities in geochemical properties along potential flow paths, uncertainties in or lack of thermodynamic and kinetic parameter values, and the lack of understanding of geochemical processes all necessitate the use of a probabilistic approach to risk assessment. System complexity and limitations in computer technology preclude precise representation of geochemical processes in risk assessment calculations. Uncertainties in properties of the engineered and natural barriers are incorporated into the risk assessment by using ranges and probability distributions for the parameter values (K_ds and maximum aqueous radionuclide concentrations) in Monte Carlo simulations, by regression equations to calculate sorption and solubility limits from sampled geochemical parameter ranges, and by the use of alternative conceptual models. Representation of the probabilistic aspects of geochemical processes in risk assessment is discussed in Siegel *et al.* (1983, 1992)), Chen *et al.* (2002), and Turner *et al.* (2002). Simplifications in solubility and sorption models used in performance assessment calculations for the WIPP and the proposed HLW repository at Yucca Mountain, respectively, are described below.

9.06.5.2.2 Solubility calculations for the waste isolation pilot plant

Performance assessment calculations of actinide speciation and solubility, and of the potential releases that could result if the repository is breached, were carried out as part of the CCA) for the waste isolation pilot plant (WIPP) (US DOE, 1996; US EPA, 1998a,b,c,d). The calculations modeled actinide behavior in a reference Salado brine and a less magnesium-rich brine from the Castile Formation as described previously (see Tables 6 and 8). The performance assessment calculations will be periodically repeated with updated parameter sets as part of site recertification.

Predicted repository conditions placed several constraints on the WIPP actinide speciation and solubility model. These conditions include: (i) high ionic strength, requiring the use of a Pitzer ion interaction model for calculating activity coefficients; (ii) the presence of magnesium oxide backfill, which will buffer P_{CO_2} and pH in the repository; and (iii) the presence of large amounts of iron and organics in the waste, establishing reducing conditions in the repository and constraining the actinides to their lower oxidation states.

The predicted waste inventory for the repository indicates that potentially significant quantities of the organic ligands—acetate, citrate, oxalate, and EDTA—will be present (US DOE, 1996). Actinide interactions with these compounds were not considered in the speciation and solubility modeling, as calculations suggested that they would be mostly complexed by transition metal ions (Fe^{2+}, Ni^{2+}, Cr^{2+}, V^{2+}, and Mn^{2+}) released by corrosion of the steel waste containers and waste components. A thermodynamic model of actinide–ligand interactions appropriate to brines will be included in solubility calculations for WIPP recertification.

Under many experimental conditions, it is difficult to maintain plutonium and some other actinides in a single oxidation state (Choppin, 1999; Neck and Kim, 2001). Depending on pH and solution composition, as many as four plutonium oxidation states may coexist, not necessarily in equilibrium. This leads to uncertainty in the oxidation state(s) present in experimental solutions. For this reason, there is little reliable speciation and solubility data available for Pu(III) and Pu(IV). WIPP solubility models were only developed for Am(III), Th(IV), and Np(V). Results of Am(III) calculations were used, through an oxidation state analogy, to predict the speciation of Pu(III) and to place an upper bound on its solubility. Similarly, results of the Th(IV) calculations were used to predict and bound the speciation and solubilities of Pu(IV), Np(IV), and U(IV). Recently, Wall *et al.* (2002) evaluated the appropriateness of the analogy and found that the predicted behavior of Am(III) was reasonably similar to that of Pu(III), while predicted solubilities for Th(IV) in Salado and Castile brines were 10–11 orders of magnitude higher than those for Pu(IV). Thus, Th(IV) is a highly conservative analogue for Pu(IV).

The An(V) model was developed using Np(V) but was not used for other actinides, because none are expected to be present in the +V oxidation state. Although U(VI) may be present, there were insufficient experimental data to develop an An(VI) model for the WIPP, and solubilities for this oxidation state were estimated from literature data rather than using a Pitzer model.

9.06.5.2.3 Models for radionuclide sorption at Yucca mountain

Since the 1970s, the US DOE has evaluated the suitability of Yucca Mountain, Nevada, as a potential site for a geologic repository for high-level nuclear waste. The proposed site is ~170 km northwest of Las Vegas, Nevada, and occupies a portion of the Nevada Test Site where nuclear weapons testing has been carried out since approximately 1945. The geochemical setting of the proposed repository site was described in Section 9.06.3.2.2 above. The results of performance assessment calculations for the proposed repository at Yucca Mountain suggest that the most significant contributors to risk are radionuclides that are highly soluble or poorly sorbing (^{99}Tc, ^{129}I, and ^{237}Np) in the oxidizing, sodium-bicarbonate-rich waters at the site. In addition, other radionuclides such as ^{239}Pu, ^{241}Am, ^{238}U, and ^{230}Th may be important due to colloidal transport or high dose-conversion factors.

Sorption of a radionuclide may vary drastically over postulated flow paths and over time at Yucca Mountain. Changes in mineralogy and transient concentrations of competing and complexing ligands may cause the sorption at any point along the flow path to change. One approach to represent this variability in performance assessment calculations is to sample K_ds for transport equations from a probability distribution based on experimental measurements as discussed above (Siegel *et al.*, 1983; Wilson *et al.*, 2002). Another approach is to calculate a range of K_ds from thermodynamic data for a range of groundwater compositions. Turner and Pabalon (1999) and Turner *et al.* (2002) outline two methods by which surface-complexation models can be used to obtain reasonable bounds on K_ds for stochastic performance assessment calculations for nuclear waste repositories. The authors represented the mineral surfaces of rocks at the Yucca Mountain site as two sorption sites (>SiOH and >AlOH). They used a DLM in the MINTEQA code (Allison *et al.*, 1991) to calculate ranges and distributions of K_ds for Am(III), Th(IV), Np(V), Pu(V), and U(VI) in a suite of groundwater compositions sampled at the site. The contours in calculated K_ds due to spatial variations in hydrochemistry were presented on maps of Yucca Mountain. In an alternate approach, Turner *et al.* (2002) used a simplified SCM to calculate neptunium sorption behavior over a wide range of pH and P_{CO_2}. K_ds can be sampled from such calculated response surfaces for use in simple transport codes in Monte Carlo simulations for performance assessment.

9.06.6 SUMMARY—CHALLENGES AND FUTURE RESEARCH NEEDS

Disposal of nuclear waste in geological repositories remains a topic of bitter controversy 30 yr after the nuclear waste program was initiated. Public opposition to geologic disposal occurs in all countries with active waste disposal programs. Many people have not accepted geologic disposal of nuclear waste and many environmental groups and some scientists call for monitored retrievable storage until additional information is gathered. The ability of the public to accept the risks associated with disposal and cleanup of nuclear contamination depends, in part, on our ability to predict the behavior of radionuclides in the environment. Although much data and model development have occurred since the 1970s, active research in a number of areas could potentially enhance our ability to predict nuclide migration accurately. These include: (i) better characterization of the chemical interactions between radionuclides and geomedia; (ii) better characterization of the sorptive media along potential flow paths between radionuclide sources and exposed populations; and (iii) cost-effective monitoring of potential radionuclides releases from waste sites. These areas are being pursued by advances in: (i) spectroscopic techniques and molecular simulation models; (ii) geostatistical models and computer simulations of radionuclide transport; (iii) improved geophysical and drilling techniques for characterizing the properties of geomedia; and (iv) improved monitoring technologies for potential radionuclide releases and exposures.

A large body of empirical sorption (K_d) data has been generated since the 1970s. One of the accomplishments of the 1990s has been the widespread awareness of the limitations of much of the data as discussed previously. It has been recognized that many of these data do not describe reversible equilibrium sorption. When precipitation or other mass transfer processes influence measurements of sorption, the resulting K_d will not provide accurate estimates of radionuclide velocities when used in transport equations. In addition, even when the K_d values represent only reversible sorption, they are valid only for the specific conditions of their experiment. Critically evaluated data sets that are useful for performance assessment of waste repositories have been assembled by several workers. It is hoped that when used with sampling schemes in Monte Carlo calculations, the data can be used to provide order-of-magnitude estimates and reasonable confidence intervals for radionuclide migration velocities that reflect uncertainties in radionuclide sorptive properties along flow paths.

Some workers argue that a more fundamental approach to sorption and solubility will lead to more accurate estimates of radionuclide transport behavior. Since the 1970s, much sorption data have been collected within the framework of SCMs. These data can be critically evaluated to determine if processes other than adsorption or ion exchange have occurred. They can also be applied over much wider ranges of solution compositions than empirical K_d measurements. Most of the available surface-complexation data have been collected in simple electrolyte solutions for single mineral phases. Application of SCMs to natural mineral assemblages is difficult due to the existence of multiple sorption sites. New experimental and theoretical methods for determination of the surface-complexation constants would improve our ability to apply the SCMs to natural systems. The resulting capacity to predict radionuclide sorption on natural materials based on fundamental properties could increase confidence in risk assessment and in the design of remediation of contaminated sites.

Cooperative international efforts such as those carried out by the NEA/OECD Thermodynamic Database Project allow sharing of the results of basic research, standardization of techniques for experiments, and establishment of reference values for thermodynamic and kinetic calculations. A comparable effort would be useful for sorption data modeling. A number of alternative SCMs are used in the literature. They are all based on conditional constants that are obtained by fitting equations or curves to experimental data obtained in solutions. The surface-complexation constants from different models cannot be combined. This is because the constants are dependent on the assumed stoichiometry of the surface species, the identity and properties of species present in solution, and properties of the electrical double layer. Recent advances in surface spectroscopy can remove some of the conditional nature of the sorption constants by providing direct information about the stoichiometry of the sorbing species. Molecular modeling could also be used to constrain the likely stoichiometry of surface species. These techniques would allow establishment of a set of reference surface species. High-quality existing sorption data sets could then be reinterpreted in light of the reference surface species and the chosen SCM. This effort would result in a much larger internally consistent set of thermodynamic data than currently exists.

Prediction of radionuclide transport in the environment is complicated by the heterogeneity of rocks along radionuclide migration paths. Estimates of expected values or ranges of radionuclide discharges at exposed populations can benefit from improvements in computer technology and applications of geostatistical methods to the geochemical properties of the flow field. Surface-complexation constants have been used to calculate sorption ratios for a suite of groundwater compositions and plotted on a map of the Yucca Mountain site. This method provides a framework to assess the range of sorption behavior at the proposed waste site. The approach could be combined with geostatistical simulation of the sorptive properties of the site (site density, surface areas, identity of sorbing sites) based on samples from boreholes. When used as input parameters to reactive transport codes, geostatistical simulation of the compositions of coexisting water and rock could be used to produce multiple realizations of radionuclide transport at a site. These multiple simulations of radionuclide transport will lead to greater confidence that reasonable upper limits for the release of radionuclides to the environment have been calculated.

Improved monitoring techniques that are cheaper and more robust with respect to the environment will allow networks of monitoring wells to be placed between sources of radionuclides such as repositories or disposal sites and potentially exposed populations. This will improve the acceptance of MNA. With improved modeling capabilities and better understanding of radionuclide interactions, public confidence in predictions of the risk associated with radioactive waste management will increase.

This overview of the geochemistry of radioactive contamination in the environment has included a summary of available data and conceptual models, descriptions of experimental and computational methods, and examples of applications of the information to address environmental problems. Historical improvements in our ability to predict the migration of radionuclides in the environment and the hazards they pose to humans and ecosystems have benefited from advances in the field of geochemistry in general. Geochemical conceptual models have progressed from early thermodynamic models through kinetic models, *ab initio* molecular models, and models incorporating the molecular biology of microbes. Future advances in these areas will lead to an improved understanding of radionuclide geochemistry and an improved ability to manage radioactive contamination in the environment.

ACKNOWLEDGMENTS

The authors would like to thank our colleagues who reviewed this document for technical content, including Pat Brady, Carlos Jove-Colon, Louise Criscenti, James Krumhansl, Yifeng Wang, Donald Wall, Lawrence Brush, and Natalie Wall. Amy Rein provided library support; Judy Campbell assisted in technical editing; Mona Aragon drafted several figures. James Leckie (Stanford University) did the calculations that are plotted in Figure 5. A special thanks to Arielle Siegel for helping to create the database used to manage the citation list. This work was supported in part by the US Department of Energy, Office of Science and Technology. Sandia National Laboratories is a multi-program laboratory operated by the Sandia Corporation, a Lockheed Martin Company, for the US Department of Energy Under Contract DE-AC04-94AL85000.

REFERENCES

Aarkrog A., Dahlgaard H., and Nielsen S. P. (1999) Marine radioactivity in the Arctic: a retrospect of environmental studies in Greenland waters with emphasis on transport of Sr-90 and Cs-137 with the East Greenland Current. *Sci. Tot. Environ.* **238**, 143–151.

Abdelouas A., Lutze W., and Nuttall E. (1999) Uranium contamination in the subsurface: characterization and remediation. In *Uranium, Mineralogy, Geochemistry, and the Environment* (eds. P. Burns and R. Finch). Mineralogical Society of America, vol. 38, pp. 433–473.

Allard B. (1983) Actinide chemistry in geologic systems. *Kemia* **10**(2), 97–102.

Allard B., Torstenfelt B., Andersson K., and Rydberg J. (1980) Possible retention of iodine in the ground. In *International Symposium on the Scientific Basis for Waste Management.* Plenum, New York, vol. 2, pp. 673–680.

Allison J., Brown D., and Novo-Gradac K. (1991) *MINTEQA2/PRODEFA2, a Geochemical Assessment Model for Environmental Systems.* Environmental Protection Agency.

AMAP (2002) *Arctic Monitoring and Assessment Program.* http://www.amap/no/.

Ames L. and Rai D. (1978) Radionuclide interactions with soil and rock media: Volume I. Processes influencing radionuclide mobility and retention, element chemistry and geochemistry, conclusions and evaluation. Pacific Northwest National Laboratory.

Andersson K. (1990) *Natural Variability in Deep Groundwater Chemistry and Influence on Transport Properties of Trace Radionuclides.* SKTR 90:17, Swedish Nuclear Power Inspectorate.

Andersson K., Evans S., and Albinsson Y. (1992) Diffusion of radionuclides in sediments—*in situ* studies. *Radiochim. Acta* **58/59**, 321–327.

Andrews J. N., Ford D. J., Hussain N., Trivedi D., and Youngman M. J. (1989) Natural radioelement solution by circulating groundwaters in the Stripa granite. *Geochim. Cosmochim. Acta* **53**, 1791–1802.

Arey J. and Seaman J. (1999) Immobilization of uranium in contaminated sediments by hydroxyapatite addition. *Environ. Sci. Technol.* **33**(2), 337–342.

Arnold T., Zorn T., Zanker H., Berhard G., and Nitsche H. (2001) Sorption behavior of U(VI) on phyllite: experiments and modeling. *J. Contamin. Hydrol.* **47**, 219–231.

Artinger R., Kienzler B., Schussler W., and Kim J. I. (1998) Effects of humic substances on the Am-241 migration in a sandy aquifer: column experiments with Gorleben groundwater/sediment systems. *J. Contamin. Hydrol.* **35**(1–3), 261–275.

ASTM (1987) 24-hour batch-type measurement of contaminant sorption by soils and sediments. In *Annual Book of ASTM Standards Water and Environmental Technology.* Philadelphia, Pennsylvania, vol. 11.04, pp. 163–167.

ATSDR (1990a) *Toxicological Profile for Thorium.* US Department of Heath and Human Services. Public Health Service. Agency for Toxic Substances and Disease Registry.

ATSDR (1990b) *Toxicological Profile for Radium.* US Department of Heath and Human Services. Public Health Service. Agency for Toxic Substances and Disease Registry.

ATSDR (1990c) *Toxicological Profile for Radon.* US Department of Heath and Human Services. Public Health Service. Agency for Toxic Substances and Disease Registry.

ATSDR (1999) *Toxicological Profile for Uranium.* US Department of Heath and Human Services. Public Health Service. Agency for Toxic Substances and Disease Registry.

ATSDR (2001) Appendix D. Overview of basic radiation physics, chemistry and biology. In *Draft Toxicological Profile for Americium—Draft for Public Comment—July 2001.* US Department of Health and Human Services, Public Health Service, Agency for Toxic Substances and Disease Registry, Atlanta, GA, pp. D.1–D.11.

Avogadro A. and de Marsily G. (1984) The role of colloids in nuclear waste disposal. In *Scientific Basis for Nuclear Waste Management VII,* Materials Research Society Proceedings (ed. G. L. McVay). North-Holland, New York, vol. 26, pp. 495–505.

Bargar J. R., Reitmeyer R., Lenhart J. J., and Davis J. A. (2000) Characterization of U(VI)-carbonato ternary complexes on hematite: EXAFS and electrophoretic mobility measurements. *Geochim. Cosmochim. Acta* **64**(16), 2737–2749.

Barnett M. O., Jardine P. M., and Brooks S. C. (2002) U(VI) adsorption to heterogeneous subsurface media: application of a surface complexation model. *Environ. Sci. Technol.* **36**, 937–942.

Barney G. S. (1981a) *Radionuclide Reactions with Groundwater and Basalts from Columbia River Basalt Formations.* Rockwell Hanford Operations.

Barney G. S. (1981b) *Evaluation of Methods for Measurement of Radionuclide Distribution in Groundwater/Rock Systems.* Rockwell Hanford Operations.

Barney G. S., Navratil J. D., and Schulz W. W. (1984) *Geochemical Behavior of Disposed Radioactive Waste*. American Chemical Society Symposium Ser. 246, Washington, DC.

Barton L. L., Choudhury K., Thompson B., Steenhoudt K., and Groffman A. R. (1996) Bacterial reduction of soluble uranium: the first step of *in situ* immobilization of uranium. *Radioact. Waste Manage. Environ. Restor.* **20**, 141–151.

Baryakhtar V. G. (1995) *Chernobyl Catastrophe*. Export Publishing House, Kiev, Russia.

Bates J. K., Bradley J. P., Teetsov A., Bradley C. R., and Buchholtz M. (1992) Colloid formation during waste form reaction: implications for nuclear wasted disposal. *Science* **256**, 649–651.

Bayley S. E., Siegel M. D., Moore M., and Faith S. (1990) *Sandia Sorption Data Management System Version 2 (SSDMS II) User's Manual*. Sandia National Laboratories.

Beck K. M. and Bennett B. G. (2002) Historical overview of atmospheric nuclear weapons testing and estimates of fall out in the continental United States. *Health Phys.* **82**(5), 591–608.

BEIR IV (1988) *Health Risks of Radon and Other Internally Deposited Alpha Emitters*. Committee on the Biological Effects of Ionizing Radiations, National Research Council, National Academy Press.

BEIR V (1988) *The Effects of Exposure to Low Levels of Ionizing Radiation*. Committee on the Biological Effects of Ionizing Radiations, National Research Council, National Academy Press.

Bergman R. (1994) The distribution of radioactive caesium in boreal forest ecosystems. In *Nordic Radioecology, the Transfer of Radionuclides through Nordic Ecosystems to Man: Studies in Environmental Science*. Studies in Environmental Science (ed. H. Dahlgaard). Elsevier, New York, vol. 62, pp. 335–379.

Bernhard G., Geipel G., Brendler V., and Nitsche H. (1998) Uranium speciation in waters of different uranium mining areas. *J. Alloy. Comp.* **271**, 201–205.

Bertetti F., Pabalan R. T., and Almendarez M. (1998) Studies of neptunium^{V} sorption on quartz, clinoptilolite, montmorillonite, and alumina. In *Adsorption of Metals by Geomedia* (ed. E. Jenne). Academic Press, San Diego, CA, pp. 131–148.

Berthelot D., Leudc L. G., and Ferroni G. D. (1997) Iron-oxidizing autotrophs and acidophilic heterotrophs from uranium mine environments. *Geomicrobiol. J.* **14**, 317–324.

Bertsch P. M. and Hunter D. B. (2001) Applications of synchrotron-based x-ray microprobes. *Chem. Rev.* **101**, 1809–1842.

Bertsch P. M., Hunter D. B., Sutton S. R., Bajt S., and Rivers M. L. (1994) Situ chemical speciation of uranium in soils and sediments by micro x-ray absorption spectroscopy. *Environ. Sci. Technol.* **28**, 980–984.

Bethke C. M. (1997) Modelling transport in reacting geochemical systems. *Comptes Rendus de l'Academie des Sciences: Sciences de la Terre et des Planetes* **324**, 513–528.

Bethke C. M. (1998) *The Geochemist's Workbench Release 3.0*. University of Illinois at Urbana-Champaign.

Bidoglio G., De Plano A., and Righetto L. (1989) Interactions and transport of plutonium-humic acid particles in groundwater environments. In *Scientific Basis for Nuclear Waste Management XII*, Materials Research Society Proceedings (eds. W. Lutze and R. Ewing). Materials Research Society, Pittsburgh, PA, vol. 127, pp. 823–830.

Bird G. W. and Lopato V. J. (1980) Solution interaction of nuclear waste anions with selected geological materials. *Int. Symp. Sci. Basis Waste Manage.*, 419–426.

Bock W. D., Bruhl H., Trapp T., and Winkler A. (1989) Sorption properties of natural sulfides with respect to technetium. *Int. Symp. Sci. Basis Waste Manage. VII*, 973–977.

Bors J., Martens H., and Kühn W. (1991) Sorption studies of radioiodine on soils with special references to soil microbial biomass. *Radiochim. Acta* **52/53**, 317–325.

Botov N. G. (1992) ALWP-67: A little-known big nuclear accident. In *High Level Radioactive Waste Management, Proceedings of the Third International Conference, April 12–16, 1992. Las Vegas, Nevada*. American Society of Civil Engineers, New York, pp. 2331–2338.

Boulyga S. F., Lomonosova E. M., Zhuk J. V., Yaroshevich O. I., Kudrjashov V. P., and Mironov V. P. (1999) Experimental study of radioactive aerosols in the vicinity of the Chernobyl nuclear power plant. *Radiat. Measure.* **30**, 703–707.

Brady P. V., Brady M. V., and Borns D. J. (1998) *Natural Attenuation, CERCLA, RBCA's and the Future of Environmental Remediation*. Lewis Publishers, New York.

Brady P. V., Papenguth H. W., and Kelly J. W. (1999) Metal sorption to dolomite surfaces. *Appl. Geochem.* **14**(5), 569–579.

Brady P. V., Jove-Colon C., Carr G., and Huang F. (2002) Soil Radionuclide Plumes. In *Geochemistry of Soil Radionuclides*, SSSA Special Publication Number 59 (eds. P. Zhang and P. Brady). Soil Science Society of America, Madison, Wisconsin, pp. 165–190.

Bronsted J. N. (1922) Studies on solubility: IV. The principals of the specific interactions of ions. *J. Am. Chem. Soc.* **44**, 877–898.

Brookins D. G. (1990) Radionuclide behaviour at the Oklo nuclear reactor, Gabon. *Waste Manage.* **10**, 285–296.

Brown G. E., Jr. and Sturchio N. C. (2002) An overview of synchrotron radiation applications to low temperature geochemistry and environmental science. In *Applications of Synchrotron Radiation in Low Temperature Geochemistry and Environmental Sciences*, Reviews in Mineralogy and Geochemistry (eds. P. A. Fenter, M. L. Rivers, N. C. Sturchio, and S. R. Sutton). Mineraological Society of America, Washington, DC, vol. 49, pp. 1–33.

Brown G. E., Jr., and Henrich V. E. (1999) Metal oxide surfaces and their interactions with aqueous solutions and microbial organisms. *Chem. Rev.* **99**, 77–174.

Brown G. E., Jr., Henrich Z. E., Casey W. H., Clark D. L., Eggleston C., Felmy A., Goodman D. W., Grätzel M., Maciel G., McCarthy M. I., Nealson K. H., Sverjensky D. A., Toney M. F., and Zachara J. M. (1999) Metal oxide surfaces and their interactions with aqueous solutions and microbial organisms. *Chem. Rev.* **99**, 77–174.

Browning L. and Murphy W. M. (2003) Reactive transport modeling in the geosciences. *Comput. Geosci.* **29**(3), 245–411.

Bruno J., Casas I., Cera E., and Duro L. (1997) Development and application of a model for the long-term alteration of UO_2 spent nuclear fuel: test of equilibrium and kinetic mass transfer models in the Cigar Lake ore deposit. *J. Contamin. Hydrol.* **26**, 19–26.

Bruno J., Cera E., Grive M., Pablo J. D., Sellin P., and Duro L. (2000) Determination and uncertainties of radioelement solubility limits to be used by SKB in the SR 97' performance assessment exercise. *Radiochim. Acta* **88**, 823–828.

Bryan C. R. and Siegel M. D. (1998) Irreversible adsorption of uranium onto iron oxides; a mechanism for natural attenuation at uranium contaminated sites. In *The Eighth Annual West Coast Conference on Contaminated Soils and Groundwater Abstracts and Supplemental Information*. Association for the Environmental Health of Soils, Oxnard, CA, 201pp.

Buckau G., Artinger R., Fritz P., Geyer S., Kim J. I., and Wolf M. (2000) Origin and mobility of humic colloids in the Gorleben aquifer system. *Appl. Geochem.* **15**(2), 171–179.

Bunzl K. and Schimmack W. (1988) Distribution coefficients of radionuclides in the soil: analysis of the field variability. *Radiochim. Acta* **44**(5), 355–360.

Burns P. and Finch R. (1999) In *Uranium: Mineralogy, Geochemistry, and the Environment*. Reviews in Mineralogy 38 (ed. P. H. Ribbe). Mineralogical Society of America, Washington, DC.

Byegård J., Albinsson Y., Skarnemark G., and Skålberg M. (1992) Field and laboratory studies of the reduction and sorption of technetium(VII). *Radiochim. Acta* **58/59**, 239–244.

Caceci M. S. and Choppin G. R. (1983) The first hydrolysis constant of uranium(VI). *Radiochim. Acta* **33**, 207–212.

Campbell J. E., Dillon R. T., Tierney M. S., Davis H. T., McGrath P. E., Pearson F. J. J., Shaw H. R., Helton J. C., and Donath F. A. (1978) *Risk Methodology for Geologic Disposal of Nuclear Waste: Interim Report, NUREG/CR-0458; SAND78-0029*. Sandia National Laboratories, Albuquerque, NM.

Cantrell K. J., Kaplan D. I., and Wietsma T. W. (1995) Zero-valent iron for the *in situ* remediation of selected metals in groundwater. *J. Hazard. Mater.* **42**, 201–212.

Cantrell K. J., Kaplan D. I., and Gilmore T. J. (1997) Injection of colloidal Fe^0 particles in sand with shear-thinning fluids. *J. Environ. Eng.* **123**, 786–791.

Capdevila H. and Vitorge P. (1998) Solubility product of $Pu(OH)_{4(am)}$. *Radiochim. Acta* **82**, 11–16.

Carroll S. A., Bruno J., Petit J. C., and Dran J. C. (1992) Interactions of U(VI), Nd and Th(IV) at the calcite-solution interface. *Radiochim. Acta* **58/59**, 245–252.

Cember H. (1996) *Introduction to Health Physics*. McGraw Hill, New York.

Chamberlain A. C. and Dunster J. (1958) Deposition of radioactivity in north-west England from the accident at Windscale. *Nature* **182**, 629–630.

Chapelle F. H. (1993) *Ground-Water Microbiology and Geochemistry*. Wiley, New York.

Chen Y., Loch A. R., Wolery T. J., Steinborn T. L., Brady P. V., and Stockman C. T. (2002) Solubility evaluation for Yucca Mountain TSPA-SR. In *Scientific Basis for Nuclear Waste Management XXV*, Materials Research Society Symposium Proceedings (eds. B. P. McGrail and G. A. Cragnolino). Materials Research Society, Pittsburgh, PA, vol. 713, pp. 775–782.

Chisholm-Brause C., Conradson S. D., Eller P. G., and Morris D. E. (1992) Changes in U(VI) speciation upon sorption onto montmorillonite from aqueous and organic solutions. In *Scientific Basis for Nuclear Waste Management XVI*, Materials Research Society Symposium Proceedings (eds. C. G. Interrante and R. T. Pabalan). Materials Research Society, Pittsburgh, PA, vol. 294, pp. 315–322.

Chisholm-Brause C., Conradson S. D., Buscher C. T., Eller P. G., and Morris D. E. (1994) Speciation of uranyl sorbed at multiple binding sites on montmorillonite. *Geochim. Cosmochim. Acta* **58**(17), 3625–3631.

Choppin G. R. (1999) Utility of oxidation state analogs in the study of plutonium behavior. *Radiochim. Acta* **85**, 89–95.

Choppin G. R. and Stout B. E. (1989) Actinide behavior in natural waters. *Sci. Tot. Environ.* **83**, 203–216.

Choppin G. R., Liljenzin J. O., and Rydberg J. (1995) *Radiochemistry and Nuclear Chemistry*. Butterworth-Heinemann Ltd., Oxford.

Ciavatta L. (1980) The specific interactions theory in evaluation ionic equilibria. *Ann. Chim.* **70**(11-1), 551–567.

Cleveland J. M. and Rees T. F. (1981) Characterization of plutonium in Maxey Flats radioactive trench leachates. *Science* **212**, 1506–1509.

Cleveland J. M., Rees T. F., and Nash K. L. (1983a) Plutonium speciation in water from Mono Lake, California. *Science* **222**, 1323–1325.

Cleveland J. M., Rees T. F., and Nash K. L. (1983b) Plutonium speciation in selected basalt, granite, shale, and tuff groundwaters. *Nuclear Technol.* **62**, 298–310.

Cochran T. B., Norris R. S., and Suokko K. L. (1993) Radioactive contamination at Chelyabinsk-65, Russia. *Ann. Rev. Energy Environ.* **18**, 507–528.

Comans R. N. J., Middleburg J. J., Zonderhuis J., Woittiez J. R. W., De Lange G. J., Das H. A., and Van der Weijden C. H. (1989) Mobilization of radiocaesium in pore water in lake sediments. *Nature* **339**(6223), 367–369.

Combes J., Chisholm-Brause D., Brown G. E. J., Parks G., Conradson S. D., Eller P. G., Triay I. R., Hobart D., and Meijer A. (1992) EXAFS spectroscopic study of neptunium (V) sorption at the a-FeOOH/water interface. *Environ. Sci. Technol.* **26**, 376–383.

Contardi J. S., Turner D. R., and Ahn T. M. (2001) Modeling colloid transport for performance assessment. *J. Contamin. Hydrol.* **47**(2–4), 323–333.

Cornish J. E., Goldberg W. C., Levine R. S., and Benemann J. R. (1995) Phytoremediation of soils contaminated with toxic elements and radionuclides. *Bioremediat. Inorg.* **3**, 55–62.

Couture R. A. and Seitz M. G. (1985) Sorption of anions and iodine by iron oxides and kaolinite. *Nuclear Chem. Waste Manage.* **4**, 301–306.

Cranwell R. M., Campbell J. E., Helton J. C., Iman R. L., Longsine D. E., Ortiz N. R., Runkle G. E., and Shortencarier M. J. (1987) *Risk Methodology for Geologic Disposal of Radioactive Waste: Final Report*. Sandia National Laboratories.

Cremers A., Elsen A., De Preter P., and Maes A. (1988) Quantitative analysis of radiocaesium retention in soils. *Nature* **335**, 247–249.

Criscenti L. J., Cygan R. T., Eliassi M., and Jove-Colon C. F. (2002) *Effects of Adsorption Constant Uncertainty on Contaminant Plume Migration*. US Nuclear Regulatory Commission.

Cui D. and Erikson T. (1996) Reactive transport of Sr, Cs, and Tc through a column packed with fracture-filling material. *Radiochim. Acta* **82**, 287–292.

Curtis D. (1999) Nature's uncommon elements: plutonium and technetium. *Geochim. Cosmochim. Acta* **63**(2), 275–285.

Czerwinski K., Buckau G., Scherbaum F., and Kim J. I. (1994) Complexation of the uranyl-ion with aquatic humic-acid. *Radiochim. Acta* **65**(2), 111–119.

Czerwinski K., Kim J. I., Rhee D. S., and Buckau G. (1996) Complexation of trivalent actinide ions (Am^{3+}, Cm^{3+}) with humic acid: the effect of ionic strength. *Radiochim. Acta* **72**(4), 179–187.

Cygan R. (2002) Molecular models of radionuclide interaction with soil minerals. In *Geochemistry of Soil Radionuclides* (eds. P. Zhang and P. Brady). Soil Science Society of America, Madison, Wisansin, pp. 87–110.

David F. (1986) Thermodynamic properties of lanthanide and actinide ions in aqueous solution. *J. Less-Common Metals* **121**, 27–42.

Davis J. and Kent D. (1990) Surface complexation modeling in aqueous geochemistry. In *Mineral-Water Interface Chemistry*, Reviews in Mineralogy 23 (eds. M. Hochella and A. White). Mineralogical Society of America, Washington, DC, pp. 177–260.

Davis J. A. (2001) *Surface Complexation Modeling of Uranium (VI) Adsorption on Natural Mineral Assemblages NUREG/CR-6708*. US Nuclear Regulatory Commission, Washington, DC.

Davis J. A. and Leckie J. O. (1978) Surface ionization and complexation at the oxide/water interface: II. Surface properties of amorphous iron oxyhydroxide and adsorption on metal ions. *J. Colloid Interface Sci.* **67**, 90–107.

Davis J. A., Coston J., Kent D., and Fuller C. (1998) Application of the surface complexation concept to complex mineral assemblages. *Environ. Sci. Technol.* **32**, 2820–2828.

Davis J. A., Kohler M., and Payne T. E. (2001) Uranium (VI) aqueous speciation and equilibrium chemistry. In *Surface Complexation Modeling of Uranium (VI) Adsorption on Natural Mineral Assemblages NUREG/CR-6708*. US Nuclear Regulatory Commission, Washington, DC, pp. 11–18.

Davis J. A., Payne T. E., and Waite T. D. (2002) Simulation of the pH and pCO_2 dependence of Uranium(VI) adsorption by a weathered schist with surface complexation models. In *Geochemistry of Soil Radionuclides* (eds. P. Zhang and P. Brady). Soil Science Society of America, pp. 61–86.

Dearlove J. P., Longworth G., Ivanovich M., Kim J. I., Delakowitz B., and Zeh P. (1991) A study of groundwater-colloids and their geochemical interactions with natural radionuclides in Gorleben aquifer systems. *Radiochim. Acta* **52/53**, 83–89.

Degueldre C. (1997) Groundwater colloid properties and their potential influence on radionuclide transport. *Mater. Res. Soc. Symp. Proc.* **465**, 835–846.

Degueldre C., Baeyens B., Goerlich W., Riga J., Verbist J., and Stadelmann P. (1989) Colloids in water from a subsurface fracture in granitic rock, Grimsel test site. *Geochim. Cosmochim. Acta* **53**, 603–610.

Degueldre C., Triay I. R., Kim J. I., Vilks P., Laaksoharju M., and Miekeley N. (2000) Groundwater colloid properties: a global approach. *Appl. Geochem.* **15**, 1043–1051.

Dent A. J., Ramsay J. D. F., and Swanton S. W. (1992) An EXAFS study of uranyl ion in solution and sorbed onto silica and montmorillonite clay colloids. *Coll. Interface Sci.* **150**, 45–60.

Douglas L. A. (1989) Vermiculites. In *Minerals in Soil Environments* (eds. J. B. Dixon and S. B. Week). Soil Science Society of America.

Duerden P., Lever D. A., Sverjensky D., and Townley L. R. (1992) *Alligator Rivers Analogue Project Final Report.* Sandia National Laboratories.

Dzombak D. and Morel F. (1990) *Surface Complexation Modeling: Hydrous Ferric Oxide.* Wiley, New York.

Efurd D. W., Runde W., Banar J. C., Janecky D. R., Kaszuba J. P., Palmer P. D., Roesnsch F. R., and Tait C. D. (1998) Neptunium and plutonium solubilities in a Yucca Mountain groundwater. *Environ. Sci. Technol.* **32**, 3893–3900.

Eisenbud M. (1987) *Environmental Radioactivity from Natural, Industrial, and Military Sources.* Academic Press, New York.

Erickson K. L. (1983) Approximations for adapting porous media radionuclide transport models to analysis of transport in jointed porous rocks. In *Scientific Basis for Nuclear Waster Management VI*, Materials Research Society Symposium Proceedings 15 (ed. D. Brookins). Elsevier, Amsterdam, pp. 473–480.

Erikson T. E., Ndalamba P., Bruno J., and Caceci M. (1992) The solubility of $TcO_2 \cdot nH_2O$ in neutral to alkaline solutions under constant P_{CO_2}. *Radiochim. Acta* **58/59**, 67–70.

Ewing R. (1999) Radioactivity and the 20th century. In *Uranium: Mineralogy, Geochemistry and the Environment*, Reviews in Mineralogy 38 (eds. P. C. Burns and R. J. Finch). Mineralogical Society of America, Washington, DC, pp. 1–22.

Fanghänel T. and Kim J. I. (1998) Spectroscopic Evaluation of Thermodynamics of Trivalent Actinides in Brines. *J. Alloy. Comp.* **271–273**, 728–737.

Fein J. B., Daughney J., Yee N., and Davis T. A. (1997) A chemical equilibrium model for the sorption onto bacterial surfaces. *Geochim. Cosmochim. Acta* **61**, 3319–3328.

Felmy A. R. and Rai D. (1999) Application of Pitzer's equations for modeling aqueous thermodynamics of actinide species in natural waters: a review. *J. Solut. Chem.* **28**(5), 533–553.

Felmy A. R. and Weare J. H. (1986) The prediction of borate mineral equilibria in natural waters: application to Searles Lake, California. *Geochim. Cosmochim. Acta* **50**(12), 2771–2783.

Felmy A. R., Dhanpat R., Schramke J. A., and Ryan J. L. (1989) The solubility of plutonium hydroxide in dilute solution and in high-ionic-strength chloride brines. *Radiochim. Acta* **43**, 29–35.

Felmy A. R., Rai D., and Fulton R. W. (1990) The solubility of $AmOHCO_{3(C)}$ and the aqueous thermodynamics of the system Na^+-Am^{3+}-HCO_3 − -CO_3^2 − -OH − -H_2O. *Radiochim. Acta* **50**(4), 193–204.

Felmy A. R., Rai D., and Mason M. J. (1991) The solubility of hydrous thorium(IV) oxide in chloride media: development of an aqueous ion-interaction model. *Radiochim. Acta* **55**(4), 177–185.

Fenter P. A., Rivers M. L., Sturchio N. C., and Sutton S. R. (2002) *Applications of Synchrotron Radiation in Low-Temperature Geochemistry and Environmental Science*, Reviews in Mineralogy and Geochemistry 49 (eds. J. J. Rosso and P. H. Ribbe). Mineralogical Society of America, Washington, DC.

Fiedor J. N., Bostic W. D., Jarabek R. J., and Farrell J. (1998) Understanding the mechanism of uranium removal from groundwater by zero-valent iron using x-ray photoelectron spectroscopy. *Environ. Sci. Technol.* **32**(10), 1466–1473.

Fjeld R. A. and Compton K. L. (1999) Risk assessment. In *Encyclopedia of Environmental Pollution and Cleanup* (eds. R. A. Meyers and D. K. Dittrick). Wiley, vol. 2, pp. 1450–1473.

Fowel D. A. and Fein J. B. (2000) Experimental measurements of the reversibility of metal adsorption reactions. *Chem. Geol.* **168**(1-2), 27–36.

Francis A. J. (1994) Microbiological treatment of radioactive wastes. In *Chemical Pretreatment of Nuclear Waste for Disposal* (eds. W. W. Schulz and E. P. Horwitz). Plenum, New York, pp. 115–131.

Francis A. J., Dodge C. J., Gillow J. B., and Cline J. E. (1991) Microbial transformations of uranium in wastes. *Radiochim. Acta* **52/53**, 311–316.

Fruchter J. S., Amonette J. E., Cole C. R., Gorby Y. A., Humphrey M. D., Istok J. D., Olsen K. B., Spane F. A., Szecsody J. E., Teel S. S., Vermeul V. R., Williams M. D., and Yabusaki S. B. (1996) *In situ* redox manipulation field injection test report—Hanford 100 H area. Pacific Northwest National Laboratory.

Fuger J. (1992) Thermodynamic properties of actinide aqueous species relevant to geochemical problems. *Radiochim. Acta* **58/59**, 81–91.

Fuger J., Khodakovsky I., Medvedev V., and Navratil J. (1990) *The Chemical Thermodynamics of Actinide Elements and Compounds: Part 12. The Actinide Aqueous Inorganic Complexes.* International Atomic Energy Agency, Vienna, Austria.

Gabriel U., Gaudet J. P., Spandini L., and Charlet L. (1998) Reactive transport of uranyl in a goethite column: an experimental and modeling study. *Chem. Geol.* **151**, 107–128.

Gauglitz R., Holterdorf M., Franke W., and Marx G. (1992) Immobilization of actinides by hydroxylapatite. In *Scientific Basis for Nuclear Waste, Management XV*, Materials Research Society Symposium Proceedings (ed. C. G. Sombret). Materials Research Society, Pittsburgh, PA, vol. 257, pp. 567–573.

Gauthier-Lafaye F., Holliger P., and Blanc P. L. (1996) Natural fission reactors in the Franceville basin, Gabon: a review of the conditions and results of a "critical event" in a geologic system. *Geochim. Cosmochim. Acta* **60**(23), 4831–4852.

Gillow J. B., Dunn M., Francis A. J., Lucero D. A., and Papenguth H. W. (2000) The potential for subterranean microbes in facilitating actinide migration at the Grimsel Test Site and Waste Isolation Pilot Plant. *Radiochim. Acta* **88**(9–11), 769–774.

Girvin D. C., Ames L. L., Schwab A. P., and McGarrah J. E. (1991) Neptunium adsorption on synthetic amorphous iron oxyhydroxide. *J. Colloid Interface Sci.* **141**(1), 67–78.

Glynn P. D. (2003) Modeling Np and Pu transport with a surface complexation model and spatially variant sorption capacities: implications for reactive transport modeling and performance assessments of nuclear waste disposal sites. *Comput. Geosci.* **29**(3), 331–349.

Goldsmith W. A. and Bolch W. E. (1970) Clay slurry sorption of carrier-free radiocations. *J. Sanitary Eng. Div. Am. Soc. Civil Eng.* **96**, 1115–1127.

Gomez P., Turrero M. J., Moulin V., and Magonthier M. C. (1992) Characterization of natural colloids in groundwaters of El Berrocal, Spain. In *Water-Rock Interaction* (eds. Y. K. Kharaka and A. S. Maest). Balkema, Rotterdam, pp. 797–800.

Goyer R. A. and Clarkson T. W. (2001) Toxic effects of metals. In *Casarett and Doull's Toxicology: The Basic Science of Poisons* (ed. C. D. Klaassen). McGraw-Hill, New York, pp. 811–867.

Grenthe I., Fuger J., Konings R. J. M., and Lemire R. J. (1992) *Chemical Thermodynamics of Uranium*. North-Holland, Amsterdam.

Grenthe I., Puigdomenech I., Sandino M. C., and Rand M. H. (1995) Chemical thermodynamics of uranium. In *Chemical Thermodynamics of Americium* (eds. R. J. Silva, G. Bidoglio, M. H. Rand, P. Robouch, H. Wanner, and I. Puigdomenech). Elsevier, Amsterdam, pp. 342–374.

Gu B. and Schultz R. K. (1991) *Anion Retention in Soil: Possible Application to Reduce Migration of Buried Technetium and Iodine: A Review*. US Nuclear Regulatory Commission, 32pp.

Gu B., Liang L., Dickey M. J., Yin X., and Dai S. (1998) Reductive precipitation of uranium(VI) by zero-valent iron. *Environ. Sci. Technol.* **32**, 3366–3373.

Guggenheim E. A. (1966) *Applications of Statistical Mechanics*. Claredon Press, Oxford.

Guillaumont R. and Adloff J. P. (1992) Behavior of environmental pollution at very low concentration. *Radiochim. Acta* **58**(59), 53–60.

Haines R. I., Owen D. G., and Vandergraaf T. T. (1987) Technetium-iron oxide reactions under anaerobic conditions: a Fourier transform infrared, FTIR study. *Nuclear J. Can.* **1**, 32–37.

Han B. S. and Lee K. J. (1997) The effect of bacterial generation on the transport of radionuclide in porous media. *Ann. Nuclear Energy* **24**(9), 721–734.

Harley N. H. (2001) Toxic effects of radiation and radioactive materials. In *Casarett and Doull's Toxicology*, 6th edn. (ed. C. D. Klaassen). McGraw-Hill, New York, pp. 917–944.

Harvie C. E. and Weare J. H. (1980) The prediction of mineral solubilities in natural waters: The Na–K–Mg–Ca–Cl–SO_4–H_2O system from zero to high concentration at 25 °C. *Geochim. Cosmochim. Acta* **44**(7), 981–997.

Harvie C. E., Möller N., and Weare J. H. (1984) The prediction of mineral solubilities in natural waters: the Na–K–Mg–Ca–H–Cl–SO_4–OH–HCO_3–CO_3–CO_2–H_2O system to high ionic strengths at 25 °C. *Geochim. Cosmochim. Acta* **48**, 723–751.

Helton J. C. and Davis F. J. (2002) *Latin Hypercube Sampling and the Propagation of Uncertainty in Analyses of Complex Systems*. Sandia National Laboratiories.

Hidaka H., Konishi T., and Masuda A. (1992) Reconstruction of cumulative fission yield curve and geochemical behaviors of fissiogenic nuclides in the Oklo natural reactors. *Geochem. J.* **26**, 227–239.

Hidaka H., Shinotsuka K., and Holliger P. (1993) Geochemical behaviour of ^{99}Tc in the Oklo natural fission reactors. *Radiochim. Acta* **63**, 19–22.

Hobart D. (1990) Actinides in the environment. In *Fifty Years with Transuranium Elements, Robert A. Welch Foundation Conference on Chemical Research 34*, Robert A. Welch Foundation, Houston, TX, pp. 379–436.

Hodge H. C., Stannard J. N., and Hursh J. B. (1973) Uranium, plutonium, and transuranic elements. In *Handbook of Experimental Pharmacology XXXVI*. Spriner, New York, pp. 980–995.

Higgo J. J. W. and Rees L. V. C. (1986) Adsorption of actinides by marine sediments: effect of the sediment/seawater ratio on the measured distribution ratio. *Environ. Sci. Technol.* **20**, 483–490.

Hölgye Z. and Malú M. (2000) Sources, vertical distribution, and migration rates of ^{239,240}Pu, ^{238}Pu, and ^{137}Cs in grassland soil in three localities of central Bohemia. *J. Environ. Radioact.* **47**, 135–147.

Holliger P. and Devillers C. (1981) Contribution to study of temperature in Oklo fossil reactors by measurement of lutetium isotopic ratio. *Earth Planet. Sci. Lett.* **52**(1), 76–84.

Honeyman B. D. and Ranville J. F. (2002) Colloid properties and their effects on radionuclide transport through soils and groundwater. In *Geochemistry of Soil Radionuclides*, SSSA Special Publication Number 59 (eds. P. Zhang and P. Brady). Soil Science Society of America, Madison, Wisconsin, pp. 131–164.

Hsi C.-K. and Langmuir D. (1985) Adsorption of uranyl onto ferric oxyhydroxides: application of the surface complexation site-binding model. *Geochim. Cosmochim. Acta* **49**, 1931–1941.

Hughes M. N. and Poole R. K. (1989) *Metals and Microorganisms*. Chapman and Hall, London.

Huie Z., Zishu Z., and Lanying Z. (1988) Sorption of radionuclides technetium and iodine on minerals. *Radiochim. Acta* **44/45**, 143–145.

Hunter K. A., Hawke D. J., and Choo L. K. (1988) Equilibrium adsorption of thorium by metal oxides in marine electrolytes. *Geochim. Cosmochim. Acta* **52**, 627–636.

Iman R. L. and Shortencarier M. J. (1984) *A Fortran 77 Program and User's Guide for the Generation of Latin Hypercube and Random Samples for Use with Computer Models, Nureg/Cr-3264, Sand83-2365*. Sandia National Laboratories.

Isaksson M., Erlandsson B., and Mattsson S. (2001) A 10-year study of the ^{137}Cs distribution in soil and a comparison of Cs soil inventory with precipitation-determined deposition. *J. Environ. Radioact.* **55**, 47–59.

Ivanovich M. (1992) *The Phenomenon of Radioactivity*. Oxford University Press.

Jackson R. E. and Inch K. J. (1989) The *in-situ* absorption of Sr-90 in a sand aquifer at the Chalk River nuclear laboratories. *J. Contamin. Hydrol.* **4**, 27–50.

Jacquier P., Meier P., and Ly J. (2001) Adsorption of radioelements on mixtures of minerals-experimental study. *Appl. Geochem.* **16**(1), 85–93.

Jenne E. A. (1998) *Adsorption of Metals by Geomedia: Variables, Mechanisms, and Model Applications*. Academic Press, Richland, Washington, 583pp.

Jensen K. and Ewing R. (2001) The Okelobondo natural fission reactor, southeast Gabon: geology, mineralogy, and retardation of nuclear-reaction products. *Geol. Soc. Am. Bull.* **113**(1), 32–62.

Jove-Colon C. F., Brady P. V., Siegel M. D., and Lindgren E. R. (2001) Historical case analysis of uranium plume attenuation. *Soil Sedim. Contamin.* **10**, 71–115.

Kanivets V. V., Voitsekhovitch O. V., Simov V. G., and Golubeva Z. A. (1999) The post-Chernobyl budget of ^{137}Cs and ^{90}Sr in the Black Sea. *J. Environ. Radioact.* **43**, 121–135.

Kaplan D., Serne R. J., Parker K. E., and Kutnyakov I. V. (2000) Iodide sorption to subsurface sediments and illitic minerals. *Environ. Sci. Technol.* **34**, 399–405.

Kashparov V. A., Lundin S. M., Khomutinin Y. V., Kaminsky S. P., Levchuk S. E., Protsak V. P., Kadygrib A. M., Zvarich S. I., Yoschenko V. I., and Tschiersch J. (2001) Soil contamination with ^{90}Sr in the near zone of the Chernobyl accident. *J. Environ. Radioact.* **56**, 285–298.

Kashparov V. A., Protsak V. P., Ahamdach N., Stammose D., Peres J. M., Yoschenko V. I., and Zvarich S. I. (2000) Dissolution kinetics of particles of irradiated Chernobyl nuclear fuel: influence of pH and oxidation state on the release of radionuclides in the contaminated soil of Chernobyl. *J. Nuclear Mater.* **279**, 225–233.

Kashparov V. A., Oughton D. H., Zvarich S. I., Protsak V. P., and Levchuk S. E. (1999) Kinetics of fuel particle weathering and ^{90}Sr mobility in the Chernobyl 30-km exclusion zone. *Health Phys.* **76**(3), 251–259.

Kaszuba J. P. and Runde W. (1999) The aqueous geochemistry of Np: dynamic control of soluble concentrations with applications to nuclear waste disposal. *Environ. Sci. Technol.* **33**, 4427–4433.

Katz J. J., Seaborg G. T., and Morse L. R. (1986) *The Chemistry of the Actinide Elements*. Chapman Hall, London.

Keeney-Kennicutt W. L. and Morse J. W. (1985) The redox chemistry of $Pu(V)O_2^+$ interaction with common mineral surfaces in dilute solutions and seawater. *Geochim. Cosmochim. Acta* **49**, 2577–2588.

Kelly J. W., Aguilar R., and Papenguth H. W. (1999) Contribution of mineral-fragment type pseudo-colloids to the mobile actinide source term of the Waste Isolation Pilot Plant (WIPP). In *Actinide Speciation in High Ionic Strength Media* (eds. D. T. Reed, S. B. Clark, and L. Rao). Kluwer/Plenum, New York, pp. 227–237.

Kelley M., Maffit L., McClellan Y., Siegel M. D., and Williams C. V. (2002) Hanford 100-N area remediation options evaluation summary report. Sandia National Laboratories.

Kent D., Tripathi V., Ball N., Leckie J., and Siegel M. (1988) Surface-complexation modeling of radionuclide adsorption in subsurface environments. US Nuclear Regulatory Commission.

Kersting A. B., Efurd D. W., Finnegan D. L., Rokop D. J., Smith D. K., and Thompson J. L. (1999) Migration of plutonium in ground water at the Nevada test site. *Nature* **397**, 56–59.

Khan S. A., Riaz-ur-Rehman, and Kahn M. (1994) Sorption of cesium on bentonite. *Waste Manage.* **14**(7), 629–642.

Kim J. I. (1993) The chemical behavior of transuranium elements and barrier functions in natural aquifer systems. In *Scientific Basis for Nuclear Waste Manage* (eds. C. G. Interrante and R. T. Pabalan). Materials Research Society, vol. 294, pp. 3–21.

Kim J. I. (1994) Actinide colloids in natural aquifer systems. *MRS Bull.* **19**(12), 47–53.

Kim J. I. and Sekine T. (1991) Complexation of neptunium(V) with humic-acid. *Radiochim. Acta* **55**(4), 187–192.

Kingston W. L. and Whitbeck M. (1991) Characterization of colloids found in various groundwater environments in central and southern Nevada. Water Resources Center Publication #45083. Desert Research Institute, University of Nevada System.

Knopp R., Neck V., and Kim J. I. (1999) Solubility, hydrolysis, and colloid formation of plutonium(IV). *Radiochim. Acta* **86**, 101–108.

Ko*β* V. (1988) Modelling of U(VI) sorption and speciation in a natural sediment-groundwater system. *Radiochim. Acta* **44/45**, 403–406.

Koch J. T., Rachar D. B., and Kay B. D. (1989) Microbial participation in iodide removal from solution by organic soils. *Can. J. Soil Sci.* **69**, 127–135.

Kohler M., Weiland E., and Leckie J. O. (1992) Metal-ligand-surface interactions during sorption of uranyl and neptunyl on oxides and silicates. In *Proceedings of 7th International Symposium Water–Rock Interaction*, pp. 51–54.

Kohler M., Curtis G., Kent D., and Davis J. A. (1996) Experimental investigation and modeling of uranium(VI) transport under variable chemical conditions. *Water Resour. Res.* **32**(12), 3539–3551.

Kohler M., Honeyman B. D., and Leckie J. O. (1999) Neptunium(V) sorption on hematite (a-Fe_2O_3) in aqueous suspension: the effect of CO_2. *Radiochim. Acta* **85**, 33–48.

Konoplev A., Bulgakov A., Popov V. E., and Bobovnikova T. I. (1992) Behaviour of long-lived radionuclides in a soil-water system. *Analyst* **117**, 1041–1047.

Konoplev A. V. and Bulgakov A. A. (1999) Kinetics of the leaching of Sr-90 from fuel particles in soil in the near zone of the Chernobyl power plant. *Atomic Energy* **86**(2), 136–141.

Krauskopf K. B. (1986) Aqueous geochemistry of radioactive waste disposal. *Appl. Geochem.* **1**, 15–23.

Kretzschmar R., Borovec M., Grollimund D., and Elimelech M. (1999) Mobile subsurface colloids and their role in contaminant transport. *Adv. Agronomy* **66**, 121–194.

Kruglov S. V., Vasil'eva N. A., Kurinov A. D., and Aleksakhin R. M. (1994) Leaching of radionuclides in Chernobyl fallout from soil by mineral acids. *Radiochemistry* **36**(6), 598–602.

Krumhansl J. L., Brady P. V., and Zhang P. (2002) Soil mineral backfills and radionuclide retention. In *Geochemistry of Soil Radionuclides*. SSSA Special Publication Number 59 (eds. P. Zhang and P. Brady). Soil Science Society of America, Madison, Wisconsin, pp. 191–210.

Krupka K. M. and Serne R. J. (2000) *Understanding Variation in Partition Coefficient, Kd, Values, Volume III: Review of Geochemistry and available Kd values for Americium, Arsenic, Curium, Iodine, Neptunium, Radium, and Technetium*. Pacific Northwest National Laboratory, Richland, WA.

Kudo A., Mahara Y., Santry D. C., Suzuki T., Miyahara S., Sugahara M., Zheng J., and Garrec J. (1995) Plutonium mass-balance released from the Nagasaki a-bomb and the applicability for future environmental-research. *Appl. Radiat. Isotopes* **46**(11), 1089–1098.

Labonne-Wall N., Choppin G. R., Lopez C., and Monsallier J. M. (1999) Interaction of uranyl with humic and fulvic acids at high ionic strength. In *Actinide Speciation in High Ionic Strength Media* (eds. D. T. Reed, S. B. Reed, and L. Rao). Kluwer/Plenum, New York, pp. 199–211.

Laflamme B. D. and Murray J. W. (1987) Solid/solution interaction: the effect of carbonate alkalinity on adsorbed thorium. *Geochim. Cosmochim. Acta* **51**, 243–250.

Langmuir D. (1997a) *Aqueous Environmental Chemistry*. Prentice Hall, Upper Saddle River, NJ.

Langmuir D. (1997b) The use of laboratory adsorption data and models to predict radionuclide releases from a geological repository: a brief history. *Mater. Res. Soc. Symp. Proc.* **465**, 769–780.

Langmuir D. and Herman J. S. (1980) The mobility of thorium in natural waters at low temperatures. *Geochim. Cosmochim. Acta* **44**, 1753–1766.

Laul J. C. (1992) Natural radionuclides in groundwaters. *J. Radioanalyt. Nuclear Chem.* **156**(2), 235–242.

Leckie J. O. (1994) Ternary complex formation at mineral/solution interfaces. In *Binding Models Concerning Natural Organic Substances in Performance Assessment: Proceedings of an NEA Workshop*. Nuclear Energy Agency, Bad Surzach, Switzerland, pp. 181–211.

Lefevre F., Sardin M., and Schweich D. (1993) Migration of Sr in clayey and calcareous sandy soil: precipitation and ion exchange. *J. Contamin. Hydrol.* **13**(4), 215–229.

Lemire R. J. (1984) *An Assessment of the Thermodynamic Behavior of Neptunium in Water and Model Groundwaters from 25 to 150°C*. AECL-7817, Atomic Energy of Canada Limited, Pinawa, Manitoba, Canada.

Lemire R. J. and Tremaine P. R. (1980) Uranium and plutonium equilibria in aqueous solutions to 200 °C. *J. Chem. Eng. Data* **25**, 361–370.

Lemire R. J., Fuger J., Nitsche H., and Potter P. (2001) *Chemical Thermodynamics of Neptunium and Plutonium*. Elsevier, Amsterdam.

Lenhart J. J. and Honeyman B. D. (1999) Uranium(VI) sorption to hematite in the presence of humic acid. *Geochim. Cosmochim. Acta* **63**(19/20), 2891–2901.

Lichtner P. (1996) Continuum formulation of multicomponent-multiphase reactive transport. In *Reactive Transport in Porous Media*, Reviews in Mineralogy (eds. P. Lichtner, C. I. Steefel, and E. H. Oelkers). Mineralogical Society of America, Washington, DC, vol. 34, pp. 1–81.

Lichtner P. C., Steefel C. I., and Oelkers E. H. (1996) *Reactive Transport in Porous Media*, Reviews in Mineralogy (ed. P. H. Ribbe). Mineralogical Society of America, Washington, DC.

Lieser K. H. and Bauscher C. (1988) Technetium in the hydrosphere and in the geosphere: II. Influence of pH, of complexing agents, and of some minerals on the sorption of technetium. *Radiochim. Acta* **44/45**, 125–128.

Lieser K. H. and Hill R. (1992) Hydrolysis and colloid formation of thorium in water and consequences for its migration behavior: comparison with uranium. *Radiochim. Acta* **56**(1), 37–45.

Lieser K. H. and Mohlenweg U. (1988) Neptunium in the hydrosphere and in the geosphere. *Radiochim. Acta* **43**, 27–35.

Lieser K. H. and Peschke S. (1982) The geochemistry of fission products (Cs, I, Tc, Sr, Zr, Sm). In *NEA/OECD Workshop on Geochemistry and Waster Disposal*. Geneva, Switzerland, pp. 67–88.

Lieser K. H., Ament A., Hill R. N., Singh U., and Thybusch B. (1990) Colloids in groundwater and their influence on migration of trace elements and radionuclides. *Radiochim. Acta* **49**, 83–100.

Lieser K. H., Hill R., Mühlenweg U., Singh R. N., Shu-De T., and Steinkopff T. (1991) Actinides in the environment. *J. Radioanalyt. Nuclear Chem.* **147**(1), 117–131.

Liu Y. and van Gunten H. R. (1988) Migration chemistry and behavior of iodine relevant to geological disposal of radioactive wastes: a literature review with a compilation of sorption data. Paul Scherrer Institute.

Lloyd J. R. and Macaskie L. E. (1996) A novel phosphorimager-based technique for monitoring the microbial reduction of technetium. *Appl. Environ. Microbiol.* **62**(2), 578–582.

Lloyd J. R., Cole J. A., and Macaskie L. E. (1997) Reduction of technetium from solution by *Escherichia coli. J. Bacteriol.* **179**, 2014–2021.

Lloyd J. R., Chenses J., Glasauer S., Bunker D. J., Livens F. R., and Lovely D. R. (2002) Reduction of actinides and fission products by Fe(III)-reducing bacteria. *Geomicrobiol. J.* **19**, 103–120.

Looney B. B. and Falta R. W. (2000). *Vadose Zone Science and Technology Solutions*. Battelle Press, Columbus, OH.

Lovely D. R. and Phillips E. J. P. (1992a) Reduction of uranium by *Desulfovibrio desulfuricans. Appl. Environ. Microbiol.* **58**, 850–856.

Lovely D. R. and Phillips E. J. P. (1992b) Bioremediation of uranium contamination with enzymatic uranium reduction. *Environ. Sci. Technol.* **26**, 2228–2234.

Luckscheiter B. and Kienzler B. (2001) Determination of sorption isotherms for Eu, Th, U, and Am on the gel layer of corroded HLW glass. *J. Nuclear Mater.* **298**, 155–162.

Lyalikova N. N. and Khizhnyak T. V. (1996) Reduction of heptavalent technetium by acidophilic bacteria on the genus *Thiobacillus. Microbiology* **65**(4), 468–473.

Macaskie L. E. (1991) The application of biotechnology to the treatment of water produced from the nuclear fuel cycle: biodegradation and the bioaccumulation as a means of treating radionuclide-containing streams. *Critical Rev. Biotechnol.* **11**, 41–112.

Macaskie L. E., Lloyd J. R., Thomas R. A. P., and Tolley M. R. (1996) The use of micro-organisms for the remediation of solutions contaminated with actinide elements, other radionuclides, and organic contaminants generated by nuclear fuel cycle activities. *Nuclear Technol.* **35**, 257–271.

Mahara Y. and Kudo A. (1995) Plutonium released by the Nagasaki a-bomb: mobility in the environment. *Appl. Radiat. Isotopes* **46**(11), 1191–1201.

Mahoney J. J. and Langmuir D. (1991) Adsorption of Sr on kaolinite, illite, and montmorillonite at high ionic strengths. *Radiochim. Acta* **54**, 139–144.

Mangold D. C. and Tsang C. F. (1991) A summary of subsurface hydrological and hydrochemical models. *Rev. Geophys.* **29**, 51–79.

Marquardt C. and Kim J. I. (1998) Complexation of Np(V) with fulvic acid. *Radiochim. Acta* **81**(3), 143–148.

Marquardt C., Herrmann G., and Trautmann N. (1996) Complexation of neptunium(V) with humic acids at very low metal concentrations. *Radiochim. Acta* **73**(3), 119–125.

McBride J. P., Moore R. E., Witherspoon J. P., and Blanco R. E. (1978) Radiological impact of airborne effluents of coal and nuclear power plants. *Science* **202**, 1045–1050.

McCarthy J. F. and Zachara J. M. (1989) Surface transport of contaminants. *Environ. Sci. Technol.* **23**, 496–502.

McCubbin D. and Leonard K. S. (1997) Laboratory studies to investigate short-term oxidation and sorption behavior of neptunium in artificial and natural seawater solutions. *Mar. Chem.* **56**, 107–121.

McGrail B. P. and Cragnolino G. A. (2002) *Scientific Basis for Nuclear Waste Management XXV*. Materials Research Society Symposium Proceedings 713, Materials Research Society, Pittsburgh, PA.

McKinley I. G. and Alexander J. L. (1993) Assessment of radionuclide retardation: uses and abuses of natural analogue studies. *J. Contamin. Hydrol.* **13**, 727–732.

McKinley I. and Scholtis A. (1992) A comparison of sorption databases used in recent performance assessments. In *Disposal of Radioactive Waste: Radionuclide Sorption from the Safety Evaluation Perspective, Proceedings of an NEA Workshop, Interlaken, Switzerland, 16–18 October 1991*, NEA-OECD, Paris, France, pp. 21–55.

McKinley J., Zachara J. M., Smith S. C., and Turner G. (1995) The influence of hydrolysis and multiple site-binding reactions on adsorption of U(VI) to montmorillonite. *Clays Clay Mineral.* **43**, 586–598.

Medvedev Z. A. (1979) *Nuclear Disaster in Urals*. Norton, New York.

Meece D. E. and Benninger L. K. (1993) The coprecipitation of Pu and other radionuclides with $CaCO_3$. *Geochim. Cosmochim. Acta* **57**, 1447–1458.

Menet C., Ménager M. T., and Petit J. C. (1992) Migration of radioelements around the new nuclear reactors at Oklo: analogies with a high-level waste repository. *Radiochim. Acta* **58–59**(2), 395–400.

Meyer R., Palmer D., Arnold W., and Case F. (1984) Adsorption of nuclides on hydrous oxides. Sorption isotherms on natural materials. *Geochem. Behavior Disp. Radioact. Waste*, 79–94.

Meyer R., Arnold W., Case F., and O'Kelley G. D. (1991) Solubilities of Tc(IV) Oxides. *Radiochim. Acta* **55**, 11–18.

Miekeley N., Coutinho de Jesus H, Porto da Siveira C. L., and Degueldre C. (1991) *Chemical and Physical Characterization of Suspended Particles and Colloids in Waters from the Osamu Utsumi and Morro de Ferro Analog Study Sites, Poco de Caldas, Brazil*. SKB Technical Report 90-18.

Morgenstern M., Lenze R., and Kim J. I. (2000) The formation of mixed-hydroxo complexes of Cm(III) and Am(III) with humic acid in the neutral pH range. *Radiochim. Acta* **88**(1), 7–16.

Moulin V. and Ouzounian G. (1992) Role of colloids and humic substances in the transport of radio-elements through the geosphere. *Appl. Geochem.* (1), 179–186.

Mucciardi A. N. (1978) Statistical investigation of the mechanics controlling radionuclide sorption: Part II. Task 4. In *Second Contractor Information Meeting*. Battelle Northwest Laboratory, Richland, WA, vol. II, pp. 333–425.

Mucciardi A. N. and Orr E. C. (1977) Statistical investigation of the mechanics controlling radionuclide sorption. In *Waste Isolation Safety Assessment Program, Task 4, Contractor Information Meeting Proceedings*. Battelle Northwest Laboratory, Richland, WA, pp. 151–188.

Muramatsu Y., Uchida S., Sriyotha P., and Sriyotha K. (1990) Some considerations on the sorption and desorption phenomena of iodide and iodate on soil. *Water Air Soil Pollut.* **49**, 125–138.

Murphy R. J., Lenhart J. L., and Honeyman B. D. (1999) The sorption of thorium(IV) and uranium(VI) to hematite in the presence of natural organic matter. *Physicochem. Eng. Aspects* **157**, 47–62.

Murphy W. M. (2000) Natural analogs and performance assessment for geologic disposal of nuclear water. In *Scientific Basis for Nuclear Waster Management XXII*,

Materials Research Society Symposium Proceedings (eds. R. W. Smith and D. W. Shoesmith). Materials Research Society, Warrendale, PA, vol. 608, pp. 533–544.

Nash K. L. and Choppin G. R. (1980) Interaction of humic and fulvic acids with Th(IV). *J. Inorg. Nuclear Chem.* **42**, 1045–1050.

National Academy of Sciences (NAS) (1996) *The Waste Isolation Pilot Plant: A Potential Solution for the Disposal of Transuranic Waste*. National Academy Press, Washington, DC.

National Academy of Sciences (1998) *Health Effects of Exposure to Radon, National Academy of Sciences Report BEIR VI*. National Academy Press.

National Institute of Health (1994) *Radon and Lung Cancer Risk: a Joint Analysis of 11 Underground Miner Studies*. US Department of Health and Human Services, National Institutes of Health.

National Research Council (1995) *Technical Bases for Yucca Mountain Standards*, Washington, DC, 205pp.

National Research Council (2001) *Science and Technology for Environmental Cleanup at Hanford*. National Academy Press.

NEA (1991) *Disposal of Radioactive Waste: Review of Safety Assessment Methods*. Nuclear Energy Agency, OECD.

NEA Nuclear Science Committee (1996) Survey of thermodynamic and kinetic databases. http://www.nea.fr/html/science/chemistry/tdbsurvey.Html.

Neck V. and Kim J. L. (2001) Solubility and hydrolysis of tetravalent actinides. *Radiochim. Acta* **89**(1), 1–16.

Neretnieks I. and Rasmuson A. (1984) An approach to modeling radionuclide migration in a medium with strongly varying velocity and block sizes along the flow path. *Water Resour. Res.* **20**(12), 1823–1836.

Nisbet A. F., Mocanu N., and Shaw S. (1994) Laboratory investigation into the potential effectiveness of soil-based countermeasures for soils contaminated with radiocaesium and radiostrontium. *Sci. Tot. Environ.* **149**, 145–154.

Nitsche H. (1991) Solubility studies of transuranium elements for nuclear waste-disposal: principles and overview. *Radiochim. Acta* **52–53**, 3–8.

Nitsche H., Muller A., Standifer E. M., Deinhammer R. S., Becraft K., Prussin T., and Gatti R. C. (1992) Dependence of actinide solubility and speciation on carbonate concentration and ionic strength in ground water. *Radiochim. Acta* **58/59**, 27–32.

Nitsche H., Gatti R. C., Standifer E. M., Lee S. C., Muller A., Prussin T., Deinhammer R. S., Maurer H., Becraft K., Leung S., and Carpenter S. A. (1993) Measured solubilities and speciations of neptunium, plutonium and americium in a typical groundwater (J-13) from the Yucca Mountain region. Los Alamos National Laboratory.

Nitsche H., Roberts K., Becraft K., Prussin T., Keeney D., Carpenter S. A., and Hobart D. E. (1995) *Solubility and Speciation Results from Over- and Undersaturation Experiments on Np, Pu, and Am in Water from Yucca Mountain Region Well UE 25p#1*. LA-13017-MS, Los Alamos National Laboratories, Los Alamos, NM.

Nordstrom D. K., Ball J. W., Donahoe R. J., and Whittemore D. (1989) Groundwater chemistry and water–rock interactions at Stripa. *Geochim. Cosmochim. Acta* **53**, 1727–1740.

Novak C. F. (1997) Calculation of actinide solubilities in WIPP SPC and ERDA-6 brines under MgO backfill scenarios containing either nesquehonite or hydromagnesite as the Mg-CO_3 solubility-limiting phase. Sandia National Labs, WIPP Records Center.

Novak C. F., Nitsche H., Silber H. B., Roberts K., Torretto P. C., Prussin T., Becraft K., Carpenter S. A., Hobart D. E., and AlMahamid I. (1996) Neptunium(V) and neptunium(VI) solubilities in synthetic brines of interest to the Waste Isolation Pilot Plant (WIPP). *Radiochim. Acta* **74**, 31–36.

Nuclear Energy Agency (2001) *Using Thermodynamic Sorption Models for Guiding Radioelement Distribution Coefficient (Kd) Investigations. Radioactive Waste Management*. Nuclear Energy Agency, Paris, France.

Nuttall H. E., Jain R., and Fertelli Y. (1991) Radiocolloid transport in saturated and unsaturated fractures. In *2nd Annual International Conference on High Level Radioactive Waste Management*, Las Vegas, NV, pp. 189–196.

Ogard A. E. and Kerrisk J. F. (1984) Groundwater chemistry along the flow path between a proposed repository site and the accessible environment. Los Alamos National Laboratory.

Ohnuki T. and Kozai N. (1994) Sorption characteristics of radioactive cesium and strontium. *Radiochim. Acta* **66/67**, 327–331.

Olofsson U. and Allard B. (1983) Complexes of actinides with naturally occurring organic substances. Literature Survey Technical. KBS Report 83–09.

Östhols E. (1995) Thorium sorption on amorphous silica. *Geochim. Cosmochim. Acta* **59**(7), 1235–1249.

Östhols E., Bruno J., and Grenthe I. (1994) On the influence of carbonate on mineral dissolution: III. The solubility of microcrystalline ThO_2 in CO_2–H_2O media. *Geochim. Cosmochim. Acta* **58**(2), 613–623.

Oughton D. H., Salbu B., Brand T. L., Day J. P., and Aarkrog A. (1993) Underdetermination of Strontium-90 in soils containing particles of irradiated uranium oxide fuel. *Analyst* **118**, 1101–1105.

Pabalan R. T., Turner D. R., Bertetti F. P., and Prikryl J. (1998) Uranium(VI) sorption onto selected mineral surfaces: key geochemical parameters. In *Adsorption of Metals by Geomedia* (ed. E. Jenne). Academic Press, San Diego, CA, pp. 99–130.

Panin A. V., Walling D. E., and Golosov V. N. (2001) The role of soil erosion and fluvial processes in the post-fallout redistribution of Chernobyl-derived caesium-137: a case study of the Lapki Catchment, central Russia. *Geomorphology* **40**, 185–204.

Papelis C., Hayes K. F., and Leckie J. O. (1988) *HYDRAQL: A Program for the Computation of Chemical Equilibrium Composition of Aqueous Batch Systems Including Surface-complexation Modeling of Ion Adsorption at the Solution Oxide/Solution Interface. 306*. Environmental Engineering and Science, Department of Civil Engineering, Stanford University, Stanford, CA, 131pp.

Park S.-W., Leckie J. O., and Siegel M. D. (1992) *Surface Complexation Modeling of Uranyl Adsorption on Corrensite from the Waste Isolation Pilot Plant site*. Sandia National Laboratories.

Parkhurst D. L. (1995) *User's Guide to PHREEQC, a Computer Model for Speciation, Reaction-path, Advective-transport and Inverse Geochemical Calculations*. US Geological Survey Water-resources Investigations Report 95-4227, US Geological Survey, 143pp.

Parrington J. R., Knox H. D., Breneman S. L., Baum E. M., and Feiner F. (1996) *Nuclides and Isotopes: Wall Chart Information Booklet*. General Electric, Schenectady, NY.

Payne T., Edis R., and Seo T. (1992) Radionuclide transport by groundwater colloids at the Koongarra Uranium Deposit. *Sci. Basis Nuclear Waste Manage.* **XV**, 481–488.

Payne T. E., Fenton B. R., and Waite T. D. (2001) Comparison of "in-situ distribution coefficients" with experimental Rd values for uranium (VI) in the Koongarra Weathered Zone. In *Surface Complexation Modeling of Uranium (VI) Adsorption on Natural Mineral Assemblages* (ed. J. A. Davis). Nuclear Regulatory Commission, Washington, DC, pp. 133–142.

Pearcy E., Prikryl J. D., Murphy W. M., and Leslie B. W. (1994) Alteration of uraninite from the Nopal I deposit, Pena Blanca District, Chihuahua, Mexico, compared to degradation of spent nuclear fuel in the proposed US high-level nuclear waste repository at Yucca Mountain, Nevada. *Appl. Geochem.* **9**, 713–732.

Pedersen K. (1996) Investigations of subterranean bacteria in deep crystalline bedrock and their importance for the disposal of nuclear waste. *Can. J. Microbiol.* **42**, 382–400.

Penrose W. R., Polzer W. L., Essington E. H., Nelson D. M., and Orlandini K. A. (1990) Mobility of plutonium and americium through a shallow aquifer in a semiarid region. *Environ. Sci. Technol.* **24**, 228–234.

Perfect D. L., Faunt C. C., Steinkampf W. C., and Turner A. K. (1995) *Hydrochemical Database for the Death Valley Region, Nevada and California.* US Geological Survey Open-File Report 94-305, US Geological Survey.

Petropoulos N. P., Anagnostakis M. J., Hinis E. P., and Simopoulos S. E. (2001) Geographical mapping and associated fractal analysis of the long-lived Chernobyl fallout radionuclides in Greece. *J. Environ. Radioact.* **53**, 59–66.

Petryaev E. P., Ovsyannikova S. V., Rubinchik S., Lubkina I. J., and Sokolik G. A. (1991) Condition of Chernobyl fallout radionuclides in the soils of Belorussia. Proceedings of AS of BSSR. *Physico-Energent. Sci.* **4**, 48–55.

Phillips S., Hale F., Silvester L., and Siegel M. (1988) *Thermodynamic Tables for Nuclear Waste Isolation: Volume 1. Aqueous Solutions Database.* Sandia National Laboratories.

Pitzer K. S. (1973) Thermodynamics of electrolytes: I. Theoretical basis and general equations. *J. Phys. Chem.* **77**, 268–277.

Pitzer K. S. (1975) Thermodynamics of electrolytes: V. Effects of higher-order electrostatic terms. *J. Solut. Chem.* **4**, 249–265.

Pitzer K. S. (1979) Theory: Ion interaction approach. In *Activity Coefficients in Electrolyte Solutions* (ed. R. M. Pytkowicz). CRC Press, Boca Raton, FL, pp. 157–208.

Pitzer K. S. (1987) A thermodynamic model for aqueous solutions of liquid-like density. In *Thermodynamic Modeling of Geological Materials: Minerals, Fluids, and Melts.* Reviews in Mineralogy (eds. I. S. E. Carmichael and H. P. Eugster). Mineralogical Society of America, Washington, DC, vol.17, pp. 97–142.

Pitzer K. S. (2000) *Activity Coefficients in Electrolyte Solutions*, 2nd edn. CRC Press, Boca Raton, FL.

Plummer L., Parkhurst D., Fleming G., and Dunkle S. (1988) *A Computer Program Incorporating Pitzer's Equations for Calculation of Geochemical Reactions in Brines.* US Geological Survey.

Pourcelot L. and Gauthier-Lafaye F. (1999) Hydrothermal and supergene clays of the Oklo natural reactors: conditions of radionuclide release, migration and retention. *Chem. Geol.* **157**, 155–174.

Prasad A., Redden G., and Leckie J. O. (1997) *Radionuclide Interactions at Mineral/Solution Interfaces in the Wipp Site Subsurface Environment.* Sandia National Laboratories.

Prikryl J. D., Jain A., Turner D. R., and Pabalan R. T. (2001) Uranium sorption behavior on silicate mineral mixtures. *J. Contamin. Hydrol.* **47**(2–4), 241–253.

Quigley M. S., Honeyman B. D., and Santschi P. H. (1996) Thorium sorption in the marine environment: equilibrium partitioning at the hematite/water interface, sorption/desorption kinetics and particle tracing. *Aquat. Geochem.* **1**, 277–301.

Rädlinger G. and Heumann K. G. (2000) Transformation of iodide in natural and wastewater systems by fixation on humic substances. *Environ. Sci. Technol.* **34**(18), 3932–3936.

Rai D., Serne R. J., and Swanson J. L. (1980) Solution species of plutonium in the environment. *J. Environ. Qual.* **9**, 417–420.

Rai D., Felmy A. R., and Ryan J. L. (1990) Uranium(VI) hydrolysis constants and solubility product of $UO_2 \times H_2O$. *Inorg. Chem.* **29**, 260–264.

Rai D., Felmy A. R., Sterner S. M., Moore D. A., Mason M. J., and Novak C. F. (1997) The solubility of Th(IV) and U(IV) hydrous oxides in concentrated NaCl and $MgCl_2$ solutions. *Radiochim. Acta* **79**(4), 239–247.

Rai D., Felmy A. R., Hess H. J., and Moore D. A. (1998) A thermodynamic model of solubility of $UO_{2(am)}$ in the aqueous $K^+-Na^+-HCO_3^--CO_3^{2-}-OH^--H_2O$. *Radiochim. Acta* **82**, 17–25.

Rao L. and Choppin G. R. (1995) Thermodynamic study of the complexation of neptunium(V) with humic acids. *Radiochim. Acta* **69**(2), 87–95.

Rard J. A. (1983) *Critical Review of the Chemistry and Thermodynamics of Technetium and some of its Inorganic Compounds and Aqueous Species.* Lawrence Livermore National Laboratory, University of California, 86pp.

Rard J. A., Rand M. H., Anderegg G., and Wanner H. (1999) *Chemical Thermodynamics of Technetium.* North-Holland, Amsterdam, Holland.

Read D., Hooker P. J., Ivanovich M., and Milodowski A. E. (1991) A natural analogue study of an abandoned uranium mine in Cornwall, England. *Radiochim. Acta* **52/53**, 349–356.

Rechard R. P. (1996) *An Introduction to the Mechanics of Performance Assessment Using Examples of Calculations Done for the Waste Isolation Pilot Plant between 1990 and 1992.* Sandia National Laboratories.

Rechard R. P. (2002) General approach used in the performance assessment for the Waste Isolation Pilot Plant. In *Scientific Basis for Nuclear Waste Management XXV.* Materials Research Society Symposium Proceedings (eds. B. P. McGrail and G. A. Cragnolino). Materials Research Society, Pittsburgh, PA, vol. 713, pp. 213–228.

Redden G., Li J., and Leckie J. O. (1998) Adsorption of U(VI) and citric acid on goethite, gibbsite, and kaolinite. In *Adsorption of Metals by Geomedia* (ed. E. Jenne). Academic Press, San Diego, CA, pp. 291–315.

Redden G. B. and Bencheikh-Latmar J. R. (2001) Citrate enhanced uranyl adsorption on goethite: an EXAFS analysis. *J. Colloid Interface Sci.* **244**(1), 211–219.

Reed D. T., Wygmans D. G., and Richman M. K. (1996) *Actinide Stability/Solubility in Simulated WIPP Brines: Interim Report under SNL WIPP Contract AP-2267.* Sandia National Laboratories.

Reeder R. J., Nugent M., Tait C. D., Morris D. E., Heald S. M., Beck K. M., Hess W. P., and Lanzirotti A. (2001) Coprecipitation of Uranium(VI) with calcite: XAFS, micro-XAS, and luminescence characterization. *Geochim. Cosmochim. Acta* **65**(20), 3491–3503.

Rees T. F., Cleveland J. M., and Nash K. L. (1983) The effect of composition of selected groundwaters from the basin and range province on plutonium, neptunium, and americium speciation. *Nuclear Technol.* **65**, 131–137.

Relyea J. F. (1982) Theoretical and experimental considerations for the use of the column method for determining retardation factors. *Radioact. Waste Manage. Nuclear Fuel Cycle* **3**(3), 151–166.

Robertson A. and Leckie J. O. (1997) Cation binding predictions of surface complexation models: effects of pH, ionic strength, cation loading, surface complex, and model fit. *J. Colloid Interface Sci.* **188**, 444–472.

Rosén K., Öborn I., and Lönsjö H. (1999) Migration of radiocaesium in Swedish soil profiles after the Chernobyl accident, 1987–1995. *J. Environ. Radioact.* **46**, 45–66.

Rühm W., Kammerer L., Hiersche L., and Wirth E. (1996) Migration of ^{137}Cs and ^{134}Cs in different forest soil layers. *J. Environ. Radioact.* **33**, 63–75.

Runde W. (2002) Geochemical interactions of actinides in the environment. In *Geochemistry of Soil Radionuclides* (eds. P. Zhang and P. Brady). Soil Science Society of America, pp. 21–44.

Runde W., Neu M. P., and Reilly S. D. (1999) Actinyl(VI) carbonates in concentrated sodium chloride solutions: characterization, solubility, and stability. In *Actinide Speciation in High Ionic Strength Media* (eds. D. T. Reed, S. B. Clark, and L. Rao). Kluwer/Plenum, New York, pp. 141–151.

Runde W., Conradson S. D., Efurd W., Lu N., VanPelt C. E., and Tait D. C. (2002a) Solubility and sorption of redox-sensitive radionuclides (Np, Pu) in J-13 water from Yucca Mountain site: comparison between experiment and theory. *Appl. Geochem.* **17**(6), 837–853.

Runde W., Neu M. P., Condrdson S. D., Li J., Lin M., Smith D. M., Van-Pelt C. E., and Xu Y. (2002b) Geochemical speciation of radionuclides in soil and solution. In *Geochemistry of Soil Radionuclides*. SSSA Special Publication Number 59 (eds. P. Zhang and P. Brady). Soil Science Society of America, Madison, Wisconsin, pp. 45–60.

Ryan J. N. and Elimelech M. (1996) Colloid mobilization and transport in groundwater. *Coll. Surf. A-Physicochem. Eng. Aspects* **107**, 1–56.

Salbu B., Krekling T., Oughton D. H., Ostby G., Kashparov V. A., Brand T. L., and Day J. P. (1994) Hot particles in accidental releases from Chernobyl and Windscale nuclear installations. *Analyst* **119**(1), 125–130.

Salbu B., Nikitin A. I., Strand P., Christensen G. C., Chumichev V. B., LInd B., Fjelldal H., Bergan T. D. S., Rudjord A. L., Sickel M., Valetova N. K., and Foyn L. (1997) Radioactive contamination from dumped nuclear waste in the Kara sea—results from the joint Russian–Norwegian expeditions in 1992–1994. *Sci. Tot. Environ.* **202**(1–3), 185–198.

Sanchez A. L., Murray J. W., and Sibley T. H. (1985) The adsorption of plutonium IV and V on goethite. *Geochim. Cosmochim. Acta* **49**, 2297–2307.

Santschi P. H. and Honeyman B. D. (1989) Radionuclides in aquatic environments. *Radiat. Phys. Chem.* **34**, 213–240.

Scatchard G. (1936) Concentrated solutions of strong electrolytes. *Phil. Mag.* **19**, 588–643.

Schramke J. A., Rai D., Fulton R. W., and Choppin G. R. (1989) Determination of aqueous plutonium oxidation states by solvent extractions. *J. Radioanalyt. Nuclear Chem.* **130**(2), 333–346.

Seaborg G. T. and Katz J. J. (1954) *The Actinide Elements*. McGraw-Hill, New York.

Serne R. J. and Gore V. L. (1996) Strontium-90 Adsorption-Desorption Properties and Sediment Characterization at the 100 N-area. Pacific Northwest National Laboratory.

Serne R. J. and Muller A. B. (1987) A perspective on adsorption of radionuclides onto geologic media. In *The Geological Disposal of High Level Radioactive Wastes* (ed. D. G. Brookins). Theophrastus Publications, pp. 407–443.

Serne R. J., Arthur R. C., and Krupka K. M. (1990) Review of Geochemical Processes and Codes for Assessment of Radionuclide Migration Potential at Commercial LLW Sites. 129pp., US Nuclear Regulatory Commission.

Shanbhag P. M. and Morse J. W. (1982) Americium interaction with calcite and aragonite sufaces in seawater. *Geochim. Cosmochim. Acta* **46**, 241–246.

Shutov V. N., Travnikova I. G., Bruk G. Y., Golikov V. Y., and Balanov M. I. (2002) Current contamination by Cs-137 and Sr-9- of the inhabited part of the Techa river basin in the Urals. *J. Environ. Radioact.* **61**, 91–109.

Siegel M. D. and Erickson K. L. (1984) Radionuclide releases from a hypothetical nuclear waste repository: potential violations of the proposed EPA standard by radionuclides with multiple aqueous species. In *Waste Manage. 84* (ed. R. G. Post). University of Arizona, Tuscon, AZ, vol. 1, pp. 541–546.

Siegel M. D. and Erickson K. L. (1986) Geochemical sensitivity analysis for performance assessment of HLW repositories: effects of speciation and matrix diffusion. *Proceedings of the Symposium on Groundwater Flow and Transport Modeling for Performance Assessment of Deep Geologic Disposal of Radioactive Waste: A Critical Evaluation of the State of the Art*. Sandia National Laboratories, Albuquerque, NM, pp. 465–488.

Siegel M. D., Chu M. S., and Pepping R. E. (1983) Compliance assessments of hypothetical geological nuclear waster isolation systems with the draft EPA standard. In *Scientific Basis for Nuclear Waste Management VI*. Materials Research Society Symposium Proceedings 15 (ed. D. G. Brookins). Elsevier, Amsterdam, Holland, pp. 497–506.

Siegel M. D., Leckie J. O., Phillips S. L., and Kelly W. R. (1989) Development of a methodology of geochemical sensitivity analysis for performance assessment. In *Proceedings of the Conference on Geostatistical, Sensitivity, and Uncertainty Methods for Ground-Water Flow and Radionuclide Transport Modeling*. San Francisco, CA, Battdle, pp. 189–211.

Siegel M. D., Holland H. D., and Feakes C. (1992) Geochemistry. In *Techniques for Determining Probabilities of Geologic Events and Processes, Studies in Mathematical Geology 4*. Int. Nat. Assoc. Math. Geology (eds. R. L. Hunter and C. J. Mann). Oxford University Press, New York, pp. 185–206.

Siegel M. D., Ward D. B., Bryan C. R., and Cheng W. C. (1995a) *Characterization of Materials for a Reactive Transport Model Validation Experiment*. Sandia National Laboratories.

Siegel M. D., Ward D. B., Bryan C. R., and Cheng W. C. (1995b) *Batch and Column Studies of Adsorption of Li, Ni, and Br by a Reference Sand for Contaminant Transport Experiments*. Sandia National Laboratories.

Silva R. J. and Nitsche H. (1995) Actinide environmental chemistry. *Radiochim. Acta* **70/71**, 377–396.

Silva R. J., Bidoglio G., Rand M. H., Rodouch P. B., Wanner H., and Puigdomenech I. (1995) *Chemical Thermodynamics of Americium*. Elsevier, New York.

Sims R., Lawless R., Alexander J., Bennett D., and Read D. (1996) Uranium migration through intact sandstone: effect of pollutant concentration and the reversibility of uptake. *J. Contamin. Hydrol.* **21**, 215–228.

SKI (1991) *SKI Project 90 Summary*. SKB Technical Report 91–23.

Slezak J. (1997) *National Experience on Groundwater Contamination Associated with Uranium Mining and Milling in the Czech Republic*. DIAMO s.p., Straz pod Ralskem, Czech Republic.

Slovic P. (1987) Perception of risk. *Science* **236**, 280–285.

Smith J. T. and Comans R. N. J. (1996) Modeling the diffusive transport and remobilization of ^{137}Cs in sediments: the effects of sorption kinetics and reversibility. *Geochim. Cosmochim. Acta* **60**, 995–1004.

Smith J. T., Comans R. N. J., and Elder D. G. (1999) Radiocaesium removal from European lakes and reservoirs: key processes determined from 16 Chernobyl-contaminated lakes. *Water Res.* **33**, 3762–3774.

Smith J. T., Comans R. N. J., Ireland D. G., Nolan L., and Hilton J. (2000) Experimental and *in situ* study of radiocaesium transfer across the sediment/water interface and mobility in lake sediments. *Appl. Geochem.* **15**(6), 833–848.

Smith P. A. and Degueldre C. (1993) Colloid-facilitated transport of radionuclides through fractured media. *J. Contamin. Hydrol.* **13**, 143–166.

Sokolik G. A., Ivanova T. G., Leinova S. L., Ovsiannikova S. V., and Kimlenko I. M. (2001) Migration ability of radionuclides in soil-vegetative cover of Belarus after Chernobyl accident. *Environ. Int.* **26**(3), 183–187.

Steefel C. I. and Van Cappellen P. (1998) Special Issue Reactive transport modeling of natural systems. *J. Hydrol.* **209**(1–4), 1–7.

Stout D. L. and Carrol S. A. (1993) *A Literature Review of Actinide-carbonate Mineral Interactions*. Sandia National Laboratories.

Strickert R., Friedman A. M., and Fried S. (1980) The sorption of technetium and iodine radioisotopes by various minerals. *Nuclear Technol.* **49**, 253–266.

Strietelmeyer B. A., Gillow J. B., Dodge C. J., and Pansoy-Hjelvik M. E. (1999) Toxicity of actinides to bacterial strains isolated from the Waste Isolation Pilot Plant (WIPP) Environment. In *Actinide Speciation in High Ionic Strength*

Media (eds. D. T. Reed, S. B. Clark, and L. Rao). Kluwer/ Plenum, New York, pp. 261–268.

Stumm W. (1992) *Chemistry of the Solid–Water Interface*. Wiley, New York.

Sumrall C. L. I. and Middlebrooks E. J. (1968) Removal of radioisotopes from water by slurrying with Yazoo and Zilpha clays. *J. Am. Water Works Assoc.* **60**(4), 485–494.

Suzuki Y. and Banfield J. F. (1999) Geomicrobiology of uranium. In *Reviews in Mineralogy*. Uranium: Mineralogy, Geochemistry and the Environment (eds. P. Burns and R. Finch). Mineralogical Society of America, Washington, DC, vol. 38, pp. 393–432.

Sverjensky D. A. (2003) Standard states for the activities of mineral surface sites and species. *Geochem. Cosmochim. Acta* **67**(1), 17–28.

Szecsody J. E., Zachara J. M., Chilakapati A., Jardine P. M., and Ferrency A. C. (1998) Importance of flow and particle-size heterogeneity in $Co^{II/III}$ EDTA reactive transport. *J. Hydrol.* **209**(1–4), 000.

Tamura T. (1972) Sorption phenomena significant in radioactive-waste disposal. In *Underground Waste Management and Environmental Implications*, American Association of Petroleum Geologists (ed. T. D. Cook). Tulsa, Oklahoma, pp. 318–330.

Tanaka S., Yamawaki M., Nagasaki S., and Moriyama H. (1992) Geochemical behavior of neptunium. *J. Nuclear Sci. Technol.* **29**(7), 706–718.

Tanaka T. and Muraoka S. (1999) Sorption characteristics of ^{237}Np, ^{238}Pu, ^{241}Am in sedimentary materials. *J. Radioanalyt. Nuclear Chem.* **240**(1), 177–182.

Thomas K. (1987) Summary of sorption measurements performed with Yucca Mountain, Nevada tuff samples and water from well J-13. Los Alamos National Laboratory.

Tien P.-L., Siegel M. D., Updegraff C. D., Wahi K. K., and Guzowski R. V. (1985) *Repository Site Data Report for Unsaturated Tuff, Yucca Mountain, Nevada*. US Nuclear Regulatory Commission.

Tipping E. (1993) Modeling the binding of europium and the actinides by humic substances. *Radiochim. Acta* **62**, 141–152.

Torstenfelt B. (1985a) Migration of the actinides thorium, protactinium, uranium, neptunium, plutonium, and americium in clay. *Radiochim. Acta* **39**, 105–112.

Torstenfelt B. (1985b) Migration of the fission products strontium, technetium, iodine and cesium in clay. *Radiochim. Acta* **39**, 97–104.

Trabalka J. R., Eyman L. D., and Auerbach S. I. (1980) Analysis of the 1957–1958 Soviet nuclear accident. *Science* **209**, 345–352.

Triay I. R., Mitchell A. J., and Ott M. A. (1992) *Radionuclide Migration Laboratory Studies for Validation of Batch-Sorption Data*. LA-12325-C, Los Alamos National Laboratory, Los Alamos.

Triay I. R., Furlano A. C., Weaver S. C., Chipera S. J., Bish D. L., Meijer A., and Canepa J. A. (1996) *Comparison of Neptunium Sorption Results Using Batch and Column Techniques*. Los Alamos National Laboratory.

Triay I. R., Meijer A., Conca J. L., Kung K. S., Rundberg R. S., Streitelmeier B. A., Tait C. D., Clark D. L., Neu M. P., and Hobart D. E. (1997) *Summary and Synthesis Report on Radionuclide Retardation for the Yucca Mountain Site Characterization Project LA-13262-MS*. Los Alamos National Laboratories.

Tripathi V. (1983) Uranium(VI) Transport modeling: geochemical data and sub-models. PhD Thesis, Stanford University.

Tripathi V. S., Siegel M. D., and Kooner Z. S. (1993) Measurements of metal adsorption in oxide-clay mixtures: "competitive-additivity" among mixture components. In *Scientific Basis for Nuclear Waste Management*. Materials Research Society, vol. XVI, pp. 791–796.

Truex M. J., Peyton B. M., Valentine N. B., and Gorby Y. A. (1997) Kinetics of U(VI) reduction by a dissimilatory Fe(III)-reducing bacterium under non-growth conditions. *Biotechnol. Bioeng.* **55**, 490–496.

Tsunashima A., Brindley W., and Bastovanov M. (1981) Adsorption of uranium from solutions by montmorillonite; compositions and properties of uranyl montmorillonites. *Clays Clay Min.* **29**(1), 10–16.

Turin H. J., Groffman A. R., Wolfsberg L. E., Roach J. L., and Strietelmeier B. A. (2002) Tracer and radionuclide sorption to vitric tuffs of Busted Butte. *Nevada. Appl. Geochem.* **17**(6), 825–836.

Turner D. and Sassman S. (1996) Approaches to sorption modeling for high-level waste performance assessment. *J. Contamin. Hydrol.* **21**, 311–332.

Turner D. R. (1991) *Sorption Modeling for High-level Waste Performance Assessment: a Literature Review*. Center for Nuclear Waste Regulatory Analyses.

Turner D. R. (1995) *A Uniform Approach to Surface Complexation Modeling of Radionuclide Sorption*. Center for Nuclear Waste Regulatory Analyses.

Turner D. R. and Pabalan R. T. (1999) Abstraction of mechanistic sorption model results for performance assessment calculations at Yucca Mountain. *Nevada. Waste Manage.* **19**, 375–388.

Turner D. R., Griffin T., and Dietrich T. (1993) Radionuclide sorption modeling using the MINTEQA2 speciation code. *Mater. Res. Soc. Symp. Proc.* 783–789.

Turner D. R., Pabalan R. T., and Bertetti F. P. (1998) Neptunium(V) sorption on montmorillonite: an experimental and surface complexation modeling study. *Clays Clay Min.* **46**, 256–269.

Turner G., Sachara J., McKinley J., and Smith S. C. (1996) Surface-charge properties and UO_2^{2+} adsorption on a surface smectite. *Geochim. Cosmochim. Acta* **60**, 3399–3414.

Turner D. R., Bertetti F. P., and Pabalan R. T. (2002) Role of radionuclide sorption in high-level waste performance assessment: approaches for the abstraction of detailed models. In *Geochemistry of Soil Radionuclides* (eds. P. Zhang and P. Brady). Soil Science Society of America, Madison, Wisconsin, pp. 211.

Uchida S., Tagami K., Rühm W., and Wirth E. (1999) Determination of ^{99}Tc deposited on the ground within the 30-km zone around the Chernobyl reactor and estimation of ^{99}Tc released into atmosphere by the accident. *Chemosphere* **39**(15), 2757–2766.

US Department of Energy (1980) *Project Review: Uranium Mill Tailings Remedial Action Project*. US Department of Energy, Washington, DC.

US Department of Energy (DOE) (1988) *Site Characterization Plan, Yucca Mountain Site, Nevada Research and Development Area, Nevada*. US Department of Energy, Office of Civilian Radioactive Waste Management.

US Department of Energy (DOE) (1996) *Title 40 CFR Part 191 Compliance Certification Application for the Waste Isolation Pilot Plant*. US Department of Energy, Carlsbad Area Office, vol. 1–21.

US Department of Energy (1997a) *Linking Legacies: Connecting the Cold War Nuclear Weapons Production Process to Their Environmental Consequences*. DOE/EM-0319, US Department of Energy, Washington, DC.

US Department of Energy (1997b) The Legacy Story. http://legacystory.apps.em.doe.gov/index.asp.

US Department of Energy (2001) *Status Report on Paths to Closure*. DOE/EM-0526, US Department of Energy, Office of Environmental Management, Washington, DC.

US Environmental Protection Agency (1996) 40 CFR Part 194: criteria for the Certification and Re-certification of the Waste Isolation Pilot Plant's Compliance With the 40 CFR Part 191 Disposal Regulations; Final Rule. Federal Register,. (No. 28), Office of the Federal Register, National Archives and Records Administration, Washington, DC, vol. 61, pp. 5224–5245.

US Environmental Protection Agency (EPA) (1997) Use of monitored natural attenuation at superfund, RCRA corrective action and underground storage tank sites, directive 9200.4-17. US Environmental Protection Agency, Office of Solid Waste and Emergency Response.

US Environmental Protection Agency (EPA) (1998a) *Compliance Application Review Documents for the Criteria for the Certification and Recertification of the Waste Isolation Pilot Plant's Compliance with the 40 CFR Part 191 Disposal Regulations: Final Certification Decision. CARD 23: Models and Computer Codes*. US Environmental Protection Agency, Office of Radiation and Indoor Air.

US Environmental Protection Agency (EPA) (1998b) *Technical Support Document for Section 194.23: Models and Computer Codes*. US Environmental Protection Agency, Office of Radiation and Indoor Air.

US Environmental Protection Agency (EPA) (1998c) *Technical Support Document for Section 194.23: Parameter Justification Report*. US Environmental Protection Agency, Office of Radiation and Indoor Air.

US Environmental Protection Agency (EPA) (1998d) *Technical Support Document for Section 194.24: EPA's Evaluation of DOE's Actinide Source Term*. US Environmental Protection Agency, Office of Radiation and Indoor Air.

US Environmental Protection Agency (1999a) *Understanding Variation in Paritition Coefficient, Kd, Values Volume 1: the Kd Model Methods of Measurement and Application of Chemical Reaction Codes*. EPA-402-R-99-044A, United States Environmental Protection Agency Office of Air and Radiation, Washington, DC.

US Environmental Protection Agency (EPA) (1999b) Understanding variation in partition coefficient, Kd, values: Volume II. Review of geochemistry and available Kd values for cadmium, cesium, chromium, lead, plutonium, radon, strontium, thorium, tritium (3H) and uranium. Prepared for the EPA by Pacific Northwest National Laboratory.

US Environmental Protection Agency (EPA) (1999c) *Field Applications of in situ Remediation Technologies: Permeable Reactive Barriers*. US Environmental Protection Agency.

US Environmental Protection Agency (EPA) (2000a). *Abstracts of Remediation Case Studies*. Federal Remediation Technologies Roundtable, vol. 4.

US Environmental Protection Agency (EPA) (2000b) *Field Demonstration of Permeable Reactive Barriers to Remove Dissolved Uranium from Groundwater, Fry Canyon, Utah*. US Environmental Protection Agency.

US Environmental Protection Agency (EPA) (2001) 40 CFR Part 197: public Health and environmental radiation protection standards for Yucca Mountain, NV; final rule. *Federal Register* **66**, 32074–32135.

US Nuclear Regulatory Commission (2001) 10 CFR Parts 2, 19, 20, 21, etc. disposal of high-level radioactive wastes in a proposed geological repository at Yucca Mountain, Nevada; final rule. *Federal Register* **66**(213), 55732–55816.

UNSCEAR (2000) *Sources and Effects of Ionizing Radiation*. Report of the United Nations Scientific Committee on the Effects of Atomic Radiation. United Nations.

Unsworth E. R., Jones P., and Hill S. J. (2002) The effect of thermodynamic data on computer model predictions of uranium speciation in natural water systems. *J. Environ. Moniter.* **4**, 528–532.

Vallet V., Schimmelpfennig B., Maron L., Teichteil C., Leininger T., Gropen O., Grenthe I., and Wahlgren U. (1999) Reduction of uranyl by hydrogen: an *ab initio* study. *Chem. Phys.* **244**(2–3), 185–193.

Van Cappellen P., Charlet L., Stumm W., and Wersin P. (1993) A surface complexation model of the carbonate mineral-aqueous solution interface. *Geochim. Cosmochim. Acta* **57**, 3505–3518.

van der Lee J., Ledoux E., and de Marsily G. (1992) Modeling of colloidal uranium transport in a fractured medium. *J. Hydrol.* **139**, 135–158.

Van Genuchten M. T. and Wierenga P. J. (1986) Solute dispersion coefficients and retardation factors. In *Methods of Soil Analysis, Part 1. Physical and Mineralogical Methods* (ed. A. Klute). American Society of Agronomy, Madison, WI, pp. 1025–1054.

van Maravic H. and Smellie J. (1992) *Fifth CEC Natural Analogue Working Group Meeting and Alligator Rivers Analogue Project (ARAP) Final Workshop*. Commission of the European Communities.

Vandergraaf T. T., Tichnor K. V., and George I. M. (1984) Reactions between technetium in solution and iron-containing minerals under oxic and anoxic conditions. In *Geochemical Behaviour of Disposed Radioactive Waste*. ACS Symposium Series 246 (eds. G. S. Barney, J. D. Navratil,, and W. W. Schultz). American Chemical Society, Washington, DC, pp. 25–44.

Vilks P. (1994) *The Role of Colloids and Suspended Particles in Radionuclide Transport in the Canadian Concept for Nuclear Fuel Waste Disposal*. AECL Research.

Vilks P., Miller H. G., and Doern D. C. (1991) Natural colloids and suspended particles in the Whiteshell Research area and their potential effect on radiocolloid formation. *Appl. Geochem.* **6**(5), 565–574.

Vilks P., Cramer J. J., Bachinski D. B., Doern D. C., and Miller H. G. (1993) Studies of colloids and suspended particles, Cigar Lake uranium deposit, Saskatchewan, Canada. *Appl. Geochem.* **8**, 605–616.

Vilks P., Caron F., and Haas M. (1998) Potential for the formation and migration of colloidal material from a near-surface waste disposal site. *Appl. Geochem.* **13**, 31–42.

Vodrias E. A. and Means J. L. (1993) Sorption of uranium by brine-saturated halite, mudstone, and carbonate minerals. *Chemosphere* **26**(10), 1753–1765.

Wagenpfeil F. and Tschiersch J. (2001) Resuspension of coarse fuel hot particles in the Chernobyl area. *J. Environ. Radioact.* **52**(1), 5–16.

Wagman D. D., Evans W. H., Parker V. B., Schumm R. H., and Halow I. (1982) The NBS tables of chemical thermodynamic properties. *J. Phys. Chem. Ref. Data* **11**, 2-1–2-34.

Wahlberg J. S. and Fishman M. J. (1962) *Adsorption of Cesium on Clay Minerals, Geological Survey Bulletin 1140-A*. US Government Printing Office.

Waite T. D., Davis J. A., Fenton B. R., and Payne T. E. (2000) Approaches to modeling uranium(VI) adsorption on natural mineral assemblages. *Radiochim. Acta* **88**, 687–693.

Wall N. A., Giambalvo E. R., Brush L. H., and Wall D. E. (2002) *The Use of Oxidation-state Analogs for WIPP Actinide Chemistry*. Unpublished presentation at the 223rd American Chemical Society National Meeting, April 7–11, 2002. Sandia National Laboratories.

Wang P., Andrzej A., and Turner D. R. (2001a) Thermodynamic modeling of the adsorption of radionuclides on selected minerals: I. Cations. Indust. Eng. Chem. Res. 40, 4428–4443.

Wang P., Andrzej A., and Turner D. R. (2001b) Thermodynamic modeling of the adsorption of radionuclides on selected minerals: II. Anions. *Indust. Eng. Chem. Res.* **40**, 4444–4455.

Ward D., Bryan C., and Siegel M. D. (1994) Detailed characterization and preliminary adsorption model for materials for an intermediate-scale reactive transport experiment. In *Proceedings of 1994 International Conference of High Level Radioactive Waste Management*, pp. 2048–2062.

Ward D. B., Brookins D. G., Siegel M. D., and Lambert S. J. (1990) Natural analog studies for partial validation of conceptual models of radionuclide retardation at the WIPP. In *Scientific Basis for Nuclear Waster Management XIV* (eds. T. A. Abrajano, Jr. and L. H. Johnson). Materials Research Society, Boston, MA, pp. 703–710.

Watson D., Gu B., Phillips D., and Lee S. Y. (1999) *Evaluation of Permeable Reactive Barriers for Removal of Uranium and other Inorganics at the Department of Energy Y-12 Plant, S-3 Disposal Ponds*. Oak Ridge National Laboratory, Environmental Sciences Division.

Westall J. and Hohl H. (1980) A comparison of electrostatic models for the oxide/solution interface. *Adv. Coll. Interface Sci.* **12**, 265–294.

Wildung R. E., Routson R. C., Serne R. J., and Garland T. R. (1974) Pertechnetate, iodide, and methyl iodide retention by surface soils. In *Pacific Northwest Laboratory Annual Report for 1974 to the USAEC Division of Biomedical and Environmental Research: Part 2. Ecological Sciences* (Manager, B. E. Vaughan), BNWL-1950 PT2, Pacific Northwest Laboratories, Richland, WA, pp. 37–40.

Wildung R. E., McFadden K. M., and Garland T. R. (1979) Technetium sources and behavior in the environment. *J. Environ. Qual.* **8**(2), 156–161.

Wilson M. L., Gauthier J. H., Barnard R. W., Barr G. E., Dockery H. A., Dunn E., Eaton R. R., Guerin D. C., Lu N., Martinez M. J., Nilson R., Rautman C. A., Robey T. H., Ross B., Ryder E. E., Schenker A. R., Shannon S. A., Skinner L. H., Halsey W. G., Gansemer J. D., Lewis L. C., Lamont A. D., Triay I. R. A. M., and Morris D. E. (1994) *Total-System Performance Assessment for Yucca Mountain-SNL. Second Iteration* vol. 2 *(TSPA-1993)*. SAND93-2675. Sandia National Laboratories, Albuquerque, NM.

Wilson M. L., Swift P. N., McNeish J. A., and Sevougian S. D. (2002) Total-system performance assessment for the Yucca Mountain Site. In *Scientific Basis for Nuclear Waste Manage. XXV* (eds. B. P. McGrail and G. A. Cragnolino). Materials Research Society, vol. 713, pp. 53–164.

Winkler A., Bruhl H., Trapp C., and Bock W. D. (1988) Mobility of technetium in various rock and defined combinations of natural minerals. *Radiochim. Acta* **44/45**, 183–186.

Wolery T. J. (1992a) *EQ3/EQ6, a Software Package for Geochemical Modeling of Aqueous Systems, Package Overview and Installation Guide (Version 7.0)*. Lawrence Livermore National Laboratory.

Wolery T. J. (1992b) *EQ3NR, A Computer Program for Geochemical Aqueous Speciation-Solubility Calculations: Theoretical Manual, User's Guide, and Related Documentation (Version 7.0)*. Lawrence Livermore National Laboratory.

WM Symposia (2001) WM 01 Proceedings, Feb. 24–28, 2001, Tucson, Arizona: HLW, LLW, mixed wastes and environmental restoration-working towards a cleaner environment. Waste Management, 2001.

Xia Y., Roa L., Rai D., and Felmy A. R. (2001) Determining the distribution of Pu, Np, and U oxidation states in dilute NaCl and synthetic brine solutions. *J. Radioanalyt. Nuclear Chem.* **250**(1), 27–37.

Yeh G. T., Carpenter S. L., Hopkins P. L., and Siegel M. D. (1995) *Users' Manual for LEHGC: A Lagrangian-Eulerian Finite-element Model of HydroGeoChemical Transport through Saturated-unsaturated Media-version 1.1*. Sandia National Laboratories.

Yeh G. T., Li M. H., and Siegel M. D. (2002) Fluid flow and reactive chemical transport in variably saturated subsurface media. In *Environmental Fluid Mechanics* (eds. H. Shen, A. Cheng, K. Wang, M. Teng, and C. Liu). American Society of Civil Engineers, pp. 207–256.

Yoshida S., Muramatsu Y., and Uchida S. (1998) Soil-solution distribution coefficients, Kds of I^- and IO_3^- for 68 Japanese soils. *Radiochim. Acta* **82**, 293–297.

Zeh P., Czerwinski K. R., and Kim J. I. (1997) Speciation of uranium in Gorleben groundwaters. *Radiochim. Acta* **76**, 37–44.

Zhang P. and Brady P. V. (2002) *Geochemistry of Soil Radionuclides*. Soil Science Society of America.

9.07
The Medical Geochemistry of Dusts, Soils, and Other Earth Materials

G. S. Plumlee and T. L. Ziegler
US Geological Survey, Denver, CO, USA

9.07.1 INTRODUCTION

"Town clenched in suffocating grip of asbestos"

USA Today, article on Libby, Montana, February, 2000

"Researchers find volcanoes are bad for your health... long after they finish erupting"

University of Warwick Press Release, 1999

"Toxic soils plague city—arsenic, lead in 5 neighborhoods could imperil 17,000 residents"

Denver Post, 2002

"Ill winds—dust storms ferry toxic agents between countries and even continents"

Science News, 2002

A quick scan of newspapers, television, science magazines, or the internet on any given day has a fairly high likelihood of encountering a story (usually accompanied by a creative headline such as those above) regarding human health concerns linked to dusts, soils, or other earth materials. Many such concerns have been recognized and studied for decades, but new concerns arise regularly.

Earth scientists have played significant roles in helping the medical community understand some important links between earth materials and human health, such as the role of asbestos mineralogy in disease (Skinner *et al.*, 1988; Ross, 1999; Holland and Smith, 2001), and the role of dusts generated by the 1994 Northridge, California, earthquake in an outbreak of Valley Fever (Jibson *et al.*, 1998; Schneider *et al.*, 1997).

Earth science activities tied to health issues are growing (Skinner and Berger, 2003), and are commonly classified under the emerging discipline of *medical geology* (Finkelman *et al.*, 2001; Selinus and Frank, 2000; Selinus, in press).

Medical geochemistry (also referred to as environmental geochemistry and health: Smith and Huyck (1999), Appleton *et al.* (1996)) can be considered as a diverse subdiscipline of medical geology that deals with human and animal health in the context of the Earth's geochemical cycle (Figure 1). Many medical geochemistry studies have focused on how chemical elements in rocks, soils, and sediments are transmitted via water or vegetation into the food chain, and how regional geochemical variations can result in disease clusters either through dietary deficiency of essential elements or dietary excess of toxic elements.

This chapter focuses on a somewhat narrower area of medical geochemistry: the study of mechanisms of uptake of earth materials by humans and animals and their reactions to these materials. In order for earth materials to affect health, they must first interact with the body across key interfaces such as the respiratory tract, gastrointestinal tract, skin, and eyes. In some way, all of these interfaces require the earth materials to interact chemically with water-based body fluids such as lung fluids, gastrointestinal fluids, saliva, or blood plasma.

The primary goal of this chapter, co-authored by a geochemist and a toxicologist, is to provide both geochemists and scientists from health disciplines with an overview of the potential geochemical mechanisms by which earth materials can influence human health. It is clear that significant opportunities for advancement in this arena will require continued and increased research collaborations between geochemists and their counterparts in the health disciplines.

9.07.2 EARTH MATERIALS LINKED TO HUMAN HEALTH

A wide variety of natural and anthropogenic materials and chemicals are recognized to influence human health. In this chapter, we consider earth materials to include a fairly broad range of solids and gases that: are produced by natural earth processes; are liberated from the earth as a result of human activities; or that are produced from the earth by humans and transformed for use in society. Examples of earth materials that have been the focus of health concerns (Tables 1 and 2) include:

- Mineral dusts of asbestos and some other asbestiform or fibrous minerals, silica, and coal. These include dusts generated by the natural weathering of rocks, and dusts generated

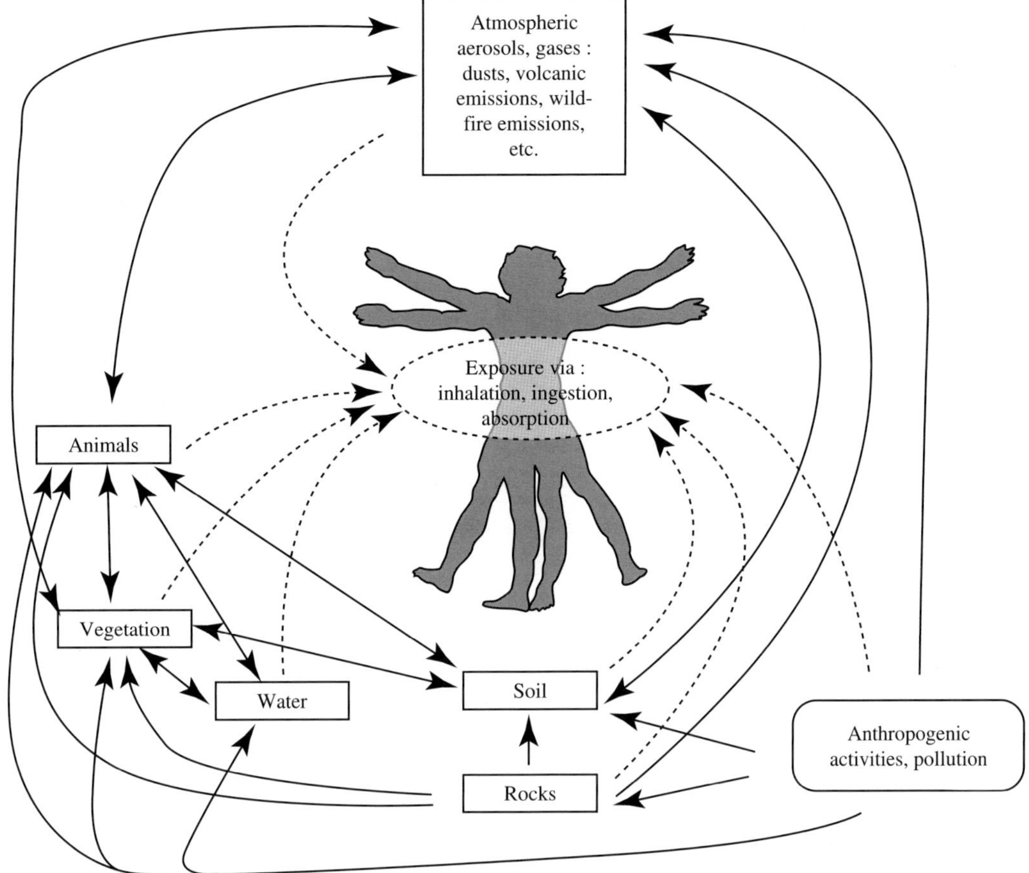

Figure 1 Potential human exposure routes within the earth's geochemical cycle can come from a wide variety of both natural and anthropogenic sources.

by anthropogenic activities such as mining, industrial processes, and construction.

- Volcanic ash and gases.
- Soils and dusts containing heavy metals, organic contaminants, or pathogens.
- Solid, gaseous, and aqueous wastes or by-products of mining, mineral processing, smelting, and energy production.
- Construction materials such as cement, concrete, aggregate, mineral and glass fiber insulation, and gypsum wallboard.
- Dusts released by building collapse or demolition.

A wide variety of elements found in earth materials can be associated with specific health problems (WHO, 1996; Taylor and Williams, 1998; Goyer and Clarkson, 2001). Essential/beneficial elements (Table 2) are those that are required for the proper physiologic function of the body, and so their associated health problems may result from either deficiencies or excesses. A variety of elements, primarily metals or metalloids, have no known natural physiological benefits but are considered toxic in excess exposure.

Environmental and health effects of radionuclides are summarized by Siegel and Bryan, (see Chapter 9.06). Potential environmental effects of hydrocarbons and organic chemicals are addressed in other chapters, and will not be discussed here. Environmental and health effects of arsenic, selenium, and mercury are addressed in greater detail elsewhere in this volume, including Chapters 9.02 and 9.04. Further information on the environmental geochemistry of metals is presented by Callender (see Chapter 9.03).

9.07.3 OVERVIEW OF THE HEALTH EFFECTS OF EARTH MATERIALS

There are a variety of diseases or health problems that have been tied to exposure to the earth materials discussed previously, or that result from industrial exposure to a wide variety of chemicals or materials. Because these are far too numerous to list in detail, the interested reader is referred to toxicological reviews such as those by Sullivan and Krieger (2001) and Klassen (2001). Examples of illnesses that are known, suspected or

Table 1 Examples of earth materials and their sources with known or postulated health effects.

Material	*Examples of potential sources*	*Primary exposure pathways,* health effects
Asbestos	Dusts from: industrial, commercial products (insulation, brake linings, building products, others) and activities; asbestos accessory minerals in other industrial/commercial products (some vermiculite, talc, other products); natural sources (via natural weathering, erosion of asbestos-bearing rocks) may provide low-level exposures.	*Inhalation*. Asbestosis; lung cancer; pleural effusion, thickening, plaques; mesothelioma cancer. Associated secondary illnesses include heart failure, lung infection. *Ingestion*. Has been proposed as a trigger of GI cancers, other health effects; however, links have not been demonstrated with certainty.
Crystalline silica	Dusts generated by mining, other industrial activities. Dusts produced by erosion of friable, silica-rich rocks (i.e., some ash flow tuffs, diatomaceous earth deposits).	*Inhalation*. Silicosis, industrial bronchitis with airflow limitation, progressive massive fibrosis. Associated illnesses include opportunistic infections, silica nethropathy, lung cancer.
Coal dust, coal fly ash	Dusts generated by coal mining, processing, and combustion activities.	*Inhalation*. "Black lung"— includes CWP, progressive massive fibrosis, chronic airway obstruction or bronchitis, emphysema; possible silicosis due to intermixed crystalline silica.
Other mineral dusts	Dusts of talc, kaolinite, other clays, micas, aluminosilicates. Sources include dusts generated from industrial, commercial activities and products.	*Inhalation*. Mineral-specific fibrosis, such as talcosis; also silicosis and asbestosis due to intermixed crystalline silica and asbestos.
Man-made mineral fibers	Glass wool (fiberglass); mineral wool (slag, rock wool).	*Inhalation*. Irritation of upper respiratory tract, skin. No ties to lung cancers, lung fibrosis.
Cement/concrete dust	Dusts from cement, concrete manufacturing. Concrete dusts generated by demolition, construction activities.	*Inhalation, contact with exposed mucous membranes or moist skin*. Irritation of eyes, throat, respiratory tract; ulceration of mucous surfaces. Effects largely tied to alkalinity of the dusts.
Volcanic ash	Atmospheric particulates generated by eruptions. Natural and anthropogenic disturbance of volcanic ash deposits, such as earthquakes, landslides, construction activities.	*Inhalation*. Irritation of respiratory tract; asthma; potential effects of crystalline silica and iron-rich particles within ash.
Volcanic gases, vog, laze	Sulfur dioxide, hydrogen fluoride, hydrogen chloride, other acid gases emanating from active volcanoes. Acidic aerosol droplets formed when hot lava contacts seawater and causes it to boil.	*Inhalation, contact with exposed mucous membranes or moist skin*. Irritation of eyes, throat, respiratory tract; ulceration of mucous surfaces. Effects largely tied to acidity of the gases and droplets. If gases are in sufficiently high concentration, toxic effects can also result.
Pathogens	Soils, and dusts generated from soils, that host pathogens such as bacteria and fungi.	*Inhalation, ingestion*. Asthma and pathogen-specific diseases such as Valley Fever (from the soil fungus *C. Immitis*) and anthrax (*B. anthracis*). *Percutaneous absorption*. Pathogen-related infections can develop through breaks in the skin.

References include: Holland and Smith (2001), Mossman and Gee (1989), Skinner *et al.* (1988), Selikoff and Lee (1978), SSDC (1988), Daroowalla (2001), Castranova and Vallyathan (2000), Castranova (2000), Daroowalla (2001), Hesterberg *et al.* (2001), CDC (1986), Baxter *et al.* (1999,1983), Wilson *et al.* (2000), Sutton *et al.* (1997), Bultman *et al.* (in press), and Griffin *et al.* (2002).

Table 2 Examples of heavy metals or metalloids, their potential sources (with an emphasis on earth materials), and health effects associated with their major exposure pathways.

Chemical element	*Examples of possible earth material sources enriched or depleted in element*	*Health effects associated with deficiency*	*Health effects associated with excess for the dominant exposure route(s)*
Essential/beneficial element			
Calcium (Ca)	Rocks and soils enriched in bioavailable calcium carbonate (limestone, caliche, etc.). Hard waters, waters affected by acid-rock drainage. Cement, concrete, fly ash, many other industrial/ commercial materials or by-products.	Bone deformities (rickets); tetany (spasms of extremities).	*Ingestion*. Atherosclosis, cataracts, gall stones.
Cobalt (Co)	Some types of nickel or silver ore deposits. Soils and waters affected by smelter emissions, mining wastes and by-products. Some rocks such as black shales, ultramafic rocks.	Anemia, anorexia.	*Ingestion*. Cardiomyopathy, hypothyroidism, polycythemia (excess of red blood cells), cancer. *Inhalation*. Respiratory irritation. "Hard metal" pneumoconiosis. *Percutaneous*. Allergic dermatitis.
Copper (Cu)	Cu-rich ore deposits (i.e., porphyry, massive sulfide, sediment-hosted). Soils and waters affected by smelter emissions, mining wastes and byproducts. Cu locally enriched in some rocks (continental redbed sediments, basalts).	Anemia, Menke's syndrome.	*Ingestion, Inhalation*. Wilson's disease (associated with Cu buildup in organs), intestinal and liver inflammation, hemolysis (destruction of red blood cells, with diffusion of hemoglobin into surrounding fluids), hyperglycemia.
Chromium (Cr)	Commonly enriched in ultramafic rocks and their associated ore deposits. Much naturally occurring Cr is relatively insoluble chromite. Soluble Cr may occur naturally in evaporative lake sediments or other evaporative environments, as a trace element within other soluble salts. Anthropogenic Cr can occur in soils, sediments, and waters affected by industrial wastes and byproducts (i.e., leather tanning, electroplating., cement use).	Cr(III) is essential— deficiencies result in defective glucose metabolism, hyperlipidemia, corneal opacity	*Inhalation, ingestion, percutaneous absorption*. Irritation of and generation of lesions in skin, respiratory tract, and gastric and intestinal mucosa; contact dermatitis; pulmonary edema. Acute kidney failure. Long-term risk for lung cancers. Pneumoconiosis from exposure to chromite ore dust.

(continued)

Table 2 (continued).

Chemical element	*Examples of possible earth material sources enriched or depleted in element*	*Health effects associated with deficiency*	*Health effects associated with excess for the dominant exposure route(s)*
Fluorine (F)	Commonly associated with limestones, fluorine-rich granites and associated ore deposits. May be present in coal combustion byproducts, fly ash.	Dental decay, possible growth retardation.	*Ingestion* > *inhalation*. At low levels: mottling of tooth enamel. At high levels: fluorosis—includes wide variety of health problems such as hyperparathyroidism, calcification of soft tissues, interference with collagen formation, severe skeletal deformity.
Iodine (I)	Brines. Some evaporite rocks, and sediments formed in evaporative environments.	Fetal problems, including abortion, stillbirths, congenital anomalies, neurological cretinism. Goiter, hypothyroidism, impaired mental function, increased susceptibility to radiation.	*Ingestion*. Hyperthyroidism
Iron (Fe)	Common rock-forming element. Enriched in many rocks, ores, soils, mine wastes, smelter emissions, etc. Pyrite (an iron sulfide) is a source of readily available iron and occurs in many different rocks and ores.	Anemias.	*Ingestion*. Most toxicity results from accidental ingestion of iron-containing medicines. Hemachromatosis, siderosis, cardiac failure, cancer.
Lithium (Li)	Salts from evaporative brines, evaporative lake sediments.	Manic depression.	*Ingestion*. Most toxicity results from excess medicinal use. Adverse neuromuscular, central nervous system, cardiovascular, gastrointestinal, and renal effects.
Magnesium (Mg)	Dolomite rock; cement and concrete; evaporative brines.	Convulsions.	*Ingestion*. Primarily from chronic use of Mg-containing drugs by people with renal dysfunction. Anesthesia, hypotension, electroardiograph abnormalities, secondary central nervous system effects. *Inhalation*. Mg-oxide can cause metal fume fever.
Molybdenum (Mo)	Mo-rich ore deposits (i.e., porphyry deposits); black shales.	Growth depression, keratinization effects, hyperurinemia.	*Ingestion*. High uric acid in serum and urine, loss of appetite, diarrhea, slow growth, anemia, "gout-like" lesions. Molybdenosis.

Manganese (Mn)	Present in limestones and other rocks formed as chemical precipitates. Common in many types of ore deposits.	Skeletal deformities, testicular dysfunction.	*Inhalation* > *ingestion* *Inhalation*. Mn-pneumonitis from acute exposure. Chronic exposure leads to manganism, other neurological and phychological disorders. *Ingestion*. Liver cirrhosis.
Selenium (Se)	Black shales, phosphatic sediments. Sometimes enriched in soluble evaporative salts formed in evaporative lake sediments, and in agricultural soils developed from Se-rich rocks. Smelter particulates and smelter-affected soils. *Some rock and sediment types, as well as their derived soils, are naturally depleted in Se, for example some dust-derived sediments (loess), some granites or gneissic metamorphic rocks.*	Liver necrosis, endemic cardio-myopathy (Kesham disease), osteoarthropathy (Kashin's Beck disease), membrane malfunction.	*Ingestion*. Teratogenesis (triggers birth defects), fetal toxicity, liver and kidney damage, cancer, brittle hair and nails, skin lesions, some effects on central, peripheral nervous systems. Selenosis.
Zinc (Zn)	Zn-rich ore deposits (i.e., massive sulfide, sediment-hosted). Soils and waters affected by smelter emissions, mining wastes and by-products. Enriched in some rocks, such as black shales, basalts.	Anorexia, dwarfism, anemia, hypogonadism, hyperkeratosis, acrodermatitis, enteropathica, depressed immune response, teratogenic effects.	*Ingestion*. Hyperchronic anemia. *Inhalation*. Metal fume fever at high doses.
Nonessential metals, toxic in excess			
Aluminum (Al)	Common rock-forming mineral, although most Al-bearing silicates quite insoluble. Potentially soluble Al-hydroxides, hydroxysulfates form in lateric ore deposits, tropical soils, and precipitate in streams affected by acid-rock drainage. Al-rich soluble salts can occur in evaporative lake sediments, and in mine wastes. Potentially reactive forms in cement, concrete, smelter emissions, coal fly ash.	None recognized.	*Ingestion* > *inhalation*. Osteomalacia (softening of bones). Neurotoxicity effects include: neurofibrillary tangles; behavioral changes; possible causative agent of Alzheimer's disease, some human dementia syndromes. Exposure to Al dusts can trigger lung fibrosis.

(continued)

Table 2 (continued).

Chemical element	*Examples of possible earth material sources enriched or depleted in element*	*Health effects associated with deficiency*	*Health effects associated with excess for the dominant exposure route(s)*
Arsenic (As)	Soils and waters affected by emissions from smelters, power plants. Soils and waters affected by mining wastes and by-products. Some playa lake sediments. Soils and dusts derived from naturally As-enriched rocks and sediments. Waters that have leached As from As-rich rocks, soils, and sediments. Pesticides, other industrial chemicals. By-products or wastes from chemical manufacturing or other industrial processes.	None recognized.	*Ingestion, inhalation*. Acute poisoning can lead to a wide variety of maladies, including: systemic hypotension; GI pain and bleeding; pulmonary edema; anemia, destruction of red blood cells; liver necrosis, kidney failure; encephalopathy and other central and peripheral nervous system disorders. Chronic toxicity can lead to: systemic hypotension; skin disorders such as eczema, hyperkeratosis, melanosis, ulceration, skin cancers; blood problems such as anemia, acute leukemia; kidney failure; delirium, encephalopathy, seizures, neuropathy.
Beryllium (Be)	Enriched in relatively insoluble form in silicates in pegmatites; also enriched in some coals and alkalic rocks and their associated mineral deposits. Soils and waters affected by emissions from coal-fired power plants.	None recognized.	*Inhalation*. Exposures primarily industrial. Chemical pneumonitis, berylliosis, cancer. *Percutaneous*. Contact dermatitis, formation of granulatomous lesions; lesions from exposure to soluble forms.
Cadmium (Cd)	Enriched in many zinc ores, black shales, phosphatic shales. Can be enriched in soils, sediments, and waters affected by: emissions from smelters, power plants; agricultural applications of sewage sludge; mining and industrial wastes and by-products; industrial wastes, by-products, and trash (i.e., battery production, leather tanning, electroplating, cement use).	None recognized.	*Inhalation, ingestion*. Acute exposure leads to: GI tract distress, gastroenteritis, liver and kidney damage, cardiomyopathy, metabolic acidosis, irritation of nasopharyngeal tract, pneumonitis. Chronic exposure can lead to: obstructive lung disease, bronchitis, emphysema, lung cancer, kidney damage, secondary skeletal system effects (osteoporosis, osteomalacia—brittleness and softening of bones; Itai-itai disease).

Lead (Pb)	Soils and waters affected by leaded gas, smelter emissions, mining wastes and by-products. Dusts, soils, and debris containing lead-bearing paint. Foods grown in lead-rich soils. Soils and dusts derived from naturally lead-enriched rocks. Waters that have leached lead from supply pipes.	None recognized.	*Inhalation, ingestion*. Acute poisoning leads to acute encphalopathy, renal failure and severe GI distress. Chronic poisoning leads to central nervous system problems, impaired neurobehavioral function, diminished gross and fine motor development in children, kidney disease, hypertension, anemia, and other hematologic effects.
Mercury (Hg)	Soils and waters affected by atmospheric deposition of Hg from volcanoes, smelters, power plants. Soils and waters affected by mining and industrial wastes and by-products. Environmental contributions from pesticides, paints, antiseptic agents, thermometers, vacuum pumps, and a wide variety of other commercial, industrial, and pharmaceutical chemicals and products.	None recognized.	*Inhalation, ingestion, percutaneous absorption*. Chronic exposure leads to: tremors and other central nervous system disorders; kidney damage, some pulmonary damage. Acute exposure produces: severe damage to the central nervous system peripheral nervous system, and GI system; kidney failure. Mercury is also a potent teratogen.
Nickel (Ni)	Enriched in ultramafic rocks and their associated mineral deposits; also enriched in many black shales, some phophatic shales. Soils, sediments, and waters affected by mining wastes, smelter emissions, power plants, industrial wastes and by-products.	None recognized.	*Inhalation*. Chronic bronchitis, emphysema, reduced lung capacity, and cancers of the lungs and nasal sinus. *Ingestion*. death (due to cardiac arrest), gastrointestinal effects (nausea, cramps, diarrhea, vomiting), effects on blood, liver, kidneys. Also neurological effects (giddiness, weariness).
Radionuclides (including uranium, radium, radon, thorium, plutonium)	A variety of rocks, soils, sediments, dusts, Tores, and other solids are enriched in uranium (U) and thorium, which can decay to radon, radium, and other daughter products. Ground waters that have traveled through U-rich rocks, soils, and sediments.	None recognized.	*Ingestion*. Common exposure route for uranium, which can triggeracute renal damage and failure. *Inhalation*. Primarily is exposure route for fine radionuclide particulates, radon gas, which can decay to other radioactive daughter products and lead to lung cancer.

This table is modeled after one in Taylor and Williams (1998) with additional information from Goyer and Clarkson (2001), ATSDR (1990; 1994; 1997; 1999a–d; 2000a,b), Geller (2001), WHO (1996), Gibly and Sullivan (2001), Derbyshire (in press), Fisher (2001), Yip *et al.* (2001), Waalkes *et al.* (2001), Cook *et al.* (1993), and Landrigan (1990). Information on element enrichments in earth materials is from Smith and Huyck (1999) and Plumlee (1999). For a discussion of elements not included in this table (e.g., thallium, antimony, bismuth, silver, tin, gallium, etc.) the reader is referred to summaries in Goyer and Clarkson (2001) and Sullivan and Krieger (2001).

postulated to result from excessive acute or chronic exposures to toxins that can be contained in earth materials, include (Tables 1 and 2):

- cancers of the lungs, skin, kidneys, mesothelium, bladder, and other organs;
- pneumoconioces such as asbestosis, silicosis, black lung disease, and other diseases tied to long-term exposure to mineral dusts. These may also trigger secondary diseases such as congestive heart failure;
- pathogen-triggered diseases where the pathogen is associated with dusts or soils (valley fever, naturally occurring anthrax);
- chronic lung illnesses, such as asthma and emphysema;
- chemical injury to, irritation of, or allergic reaction in the skin, eyes, throat, respiratory tract, and digestive tract. These include, for example, alkali or acid burns and ulceration, dermatitis, and general tissue inflammation. Membrane injuries can also result in exposure pathways for toxins and pathogens across the body's mucous membranes that result in secondary toxic effects or pathogenic infections;
- diseases of the kidneys (nethropathy triggered by a variety of toxins), liver, and other organs;
- diseases of the skeletal system (Itai-itai disease triggered by cadmium poisoning; fluorosis triggered by fluoride poisoning);
- unnamed combinations of asthma, coughing, shortness of breath, and tiredness;
- diseases of the central nervous system, such as: Parkinson's disease, Parkinson's dementia complex, and Lou Gehrig's disease (possibly tied to aluminum toxicity); encephalopathy (tied to poisoning from a number of metals); manganism; diminished mental capacity (lead, mercury poisoning); and encephalopathy; and
- diseases of the peripheral nervous system, including tremors, loss of motor skills, and others.

9.07.4 EXPOSURE PATHWAYS, ABSORPTION, BIODISTRIBUTION, METABOLISM, AND DETOXIFICATION OF POTENTIAL TOXINS

The types of health effects associated with exposure to a potentially toxic earth material depend upon:

- *dose response*—the intensity and duration of the exposure;
- *route of exposure*—the pathways by which exposure occurs;
- *solubility*—the degree to which the earth materials are solubilized in body fluids for a given exposure route;
- *toxicokinetics*—how potential toxins are processed in the body, including absorption, distribution, metabolism, and elimination (ADME).

Health effects of toxins can result from either local toxicity near the site of exposure or systemic toxicity in cells, tissues, and organs away from the site of exposure.

A detailed discussion of the many complex processes by which toxins are absorbed, modified, stored, and excreted by the body is far beyond the scope of this chapter. Interested readers are referred to toxicology overview volumes and texts such as Sullivan and Krieger (2001), and Klassen (2001).

9.07.4.1 Dose Response

> "All substances are poisonous, there are none which is not a poison; the right dose is what differentiates a poison from a remedy,"
>
> *Paracelsus*, 1538.

This frequently repeated quote underscores the long-recognized importance of understanding exposure intensity and duration (the dose) in assessing the health effects of potential toxins. The concept of *dose response* states that the greater amount of a toxin taken up by the organism, the greater the toxicological response (Rozman and Klaasen, 2001). However, the true determinant of toxicity is based on the concentration and form of the toxin at the site of action. The body has a remarkable capability for clearing or diminishing the toxic effects of a wide variety of toxins such as pesticides or asbestos fibers if the dose is small; toxicity results when the dose exceeds the body's inherent toxin mitigation mechanisms.

As shown by the examples in Table 2, many elements are essential to the effective functioning of the myriad biochemical processes in the body, yet can become toxic after intense exposures over short periods of time (acute toxicity) or moderately high exposures over longer periods of time (chronic toxicity). The threshold above which a material becomes toxic is a complex function of the substance, the exposure route, the chemical form of the substance as it is absorbed and presented at the site of action, and, to a lesser extent, the genetic makeup of the exposed individual (Sullivan *et al.*, 2001).

9.07.4.2 Routes of Exposure

The primary exposure pathways (Sipes and Badger, 2001) for toxins and pathogens include (see Figure 2):

- gastrointestinal tract (ingestion),
- respiratory tract (inhalation), and
- skin (percutaneous absorption).

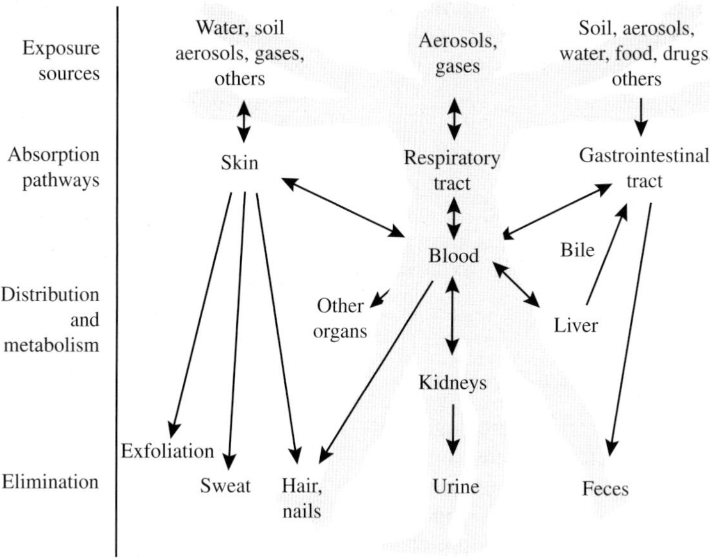

Figure 2 This schematic diagram shows the absorption pathways and systems of distribution, metabolism, and elimination for potential toxins. "Aerosols" include dusts, other solid particulates (such as smoke), and liquid droplets (such as fog, mists, etc.). Distribution may involve deposition of a toxin within a target organ and/or metabolism with or without excretion of the toxin by the target organ (after Goyer and Clarkson, 2001).

All exposure pathways can ultimately result in the absorption of soluble substances across the body's membranes (skin, eyes, respiratory, or digestive tracts), by passive or active diffusion, active transport, or cellular pinocytosis/phagocytosis (the engulfment of foreign particles by cells). The proportion of a substance in contact with a membrane that is absorbed is a complex function of many factors, including the concentration and chemical form of the substance, the relative chemical conditions ambient on either side of the membrane, and the surface area of the membrane with which the substance is in contact.

A key physiologic response when the body is exposed to foreign substances is the production of fluids at the site of exposure to help dilute, solubilize, or physically clear the substances. For example, inhalation of particulates leads to an increase in the amount of mucous and other fluids produced in the respiratory tract. Ingestion of food or other substances triggers the production of increased stomach acids to maintain the pH optimal for food digestion in the stomach.

A variety of terms have been established to help constrain the relative availability of toxins that can be readily released from administered substances into body fluids. *Bioavailability* is defined by toxicologists as the fraction of an administered dose of a substance that is absorbed via an exposure route and actually reaches the bloodstream (Ruby *et al.*, 1999; Lioy, 1990; Hamel, 1998). The *bioaccessibility* of a substance is the fraction that can be dissolved by body fluids (i.e., in the gastrointestinal tract or in the lungs), and that therefore is available for absorption (Ruby *et al.*, 1999; Hamel, 1998). Geochemists have defined *geoavailability* as the portion of a compound's total content in an earth material that can be liberated into the environment or biosphere through mechanical, chemical, or biological processes (Smith and Huyck, 1999); in a toxicological context, *geoavailability* is similar to *bioaccessibility*, but stresses the important role that the form of the toxin in an earth material plays in its overall toxicity.

For most potential toxins in earth materials, the following relationship generally holds:

$$\text{bioavailability} < \text{bioaccessibility} < \text{total concentration of a potential toxin in a substance}$$

A number of environmental regulations governing allowable toxin concentrations in materials such as soils are based on the total concentration of the toxin in the material, a parameter that is generally easiest to measure reproducibly. However, the above relationship shows that equating the total concentration of a toxin in an administered substance to its bioavailability is the worst-case scenario. This is truly valid only in the rare circumstances when the toxin is completely released from the administered substance and is in the appropriate chemical form to permit complete absorption.

As summarized by Jurinski and Rimstidt (2001), two terms have been proposed to characterize substances that can persist in the body for many years after exposure. *Biodurability* is a measure of a substance's resistance to clearance by dissolution in body fluids. In contrast, *biopersistence* is a measure of a substance's resistance to all chemical, physical, and biological clearance mechanisms. Although these terms were specifically defined for and are most commonly applied to inhaled substances, they are also potentially applicable for other exposure routes.

9.07.4.2.1 Gastrointestinal tract (ingestion)

In general, most earth materials are ingested inadvertently, such as particles cleared from the upper respiratory tract, particles ingested by infants or small children, or particles ingested with foodstuffs (such as soil on incompletely cleaned vegetables). In the past, inadvertent ingestion of soil particles on incompletely cleaned foodstuffs actually provided humans with a substantial source of some key mineral nutrients; this source has decreased substantially due to modern advances in food processing and cleaning technologies (Taylor and Williams, 1995; WHO, 1996, Oliver, 1997). In some societies, ingestion of soil (termed geophagia) is done purposely for nutritional or cultural reasons (Oliver, 1997).

Mastication breaks down ingested earth materials into small particles, less than ~500 μm to 1 mm in diameter. This increases the surface area available for chemical reaction with the saliva and digestive juices, and may help to increase the bioaccessibility of contaminants associated with earth materials.

The fate of ingested substances is a function of their particle makeup and size distribution, their chemical solubility in the digestive fluids, the presence of other material (such as food) in the digestive tract, and biologically mediated reactions (either with or without mediation of resident microbes) that may transform the earth materials and their degradation products in the gastrointestinal tract. The survival of ingested pathogens depends on their ability to survive in the chemical conditions of the gastrointestinal tract, particularly the acidic conditions in the stomach.

Most dissolution of ingested substances occurs in the stomach, and most absorption occurs in the intestinal tract (Sipes and Badger, 2001). However, materials that are dissolved in the stomach may not all be absorbed across the intestinal tract lining. For simple diffusion across the intestinal wall into the bloodstream, a toxin must be in the most lipid-soluble (nonionized) form, which is based on its acid/base characteristics (Rozman and Klaasen, 2001). For example, a weak acid is mainly in the nonionized (lipid soluble) form in the stomach and in the ionized form in the intestine, while bases are ionized in the stomach and nonionized in the intestine. It is also possible that materials dissolved in the acidic, oxidized conditions of the stomach tract may subsequently precipitate as solids or adsorb onto solids in the intestines, due to the more alkaline and less oxygenated conditions in the intestine.

Gastrointestinal absorption, and subsequent utilization and retention by the body, of essential trace elements such as zinc, copper, and selenium can also be enhanced or diminished by the presence or absence of other trace elements and chemicals in the diet (WHO, 1996). For example, cadmium and lead absorption is enhanced when dietetic intake of calcium, iron, and phosphate is low. Phytate, an organic phosphate that is abundant in diets high in unrefined grains, especially when accompanied by high dietetic calcium, helps suppress the uptake of potentially toxic elements such as lead and cadmium, but also inhibits the uptake of essential zinc (WHO, 1996).

9.07.4.2.2 Respiratory tract (inhalation)

The health effects of inhaled earth materials are a function of the type of material inhaled (solid, liquid, gas, or pathogen); the concentration of the material in the inhaled air, the chemical composition of the material, the solubility and reactivity of the material in respiratory tract fluids, and, for solid particulates, their shape and size distribution (Newman, 2001; Sipes and Badger, 2001).

The shape and size of solid particles influence both the depth to which they can be transported in the respiratory tract and the extent to which they can be cleared by various mechanisms (Newman, 2001; Schlesinger, 1995; Snipes, 1995). The largest inhaled particles (5 μm to somewhat greater than 10 μm) are deposited in the mucous linings of the nasopharyngeal tract (Figure 2). Progressively smaller particle sizes are deposited in successively deeper portions of the respiratory tract by entrapment in a layer of mucous lining the airways. The particle-laden mucous is cleared from the trachea and bronchi in part by coughing. In addition, the cells lining the trachea, bronchi, and bronchioles are ciliated. The cilia beat to transport the particle-laden mucous up and out of the respiratory tract. Particles cleared by coughing or mucociliary clearance are either expectorated or ingested.

Particles less than ~2 μm in size reach the alveoli, the deepest portions of the lungs, where the most active exchange of oxygen and carbon dioxide occurs. These very small particles are either trapped in the alveoli or are exhaled. In the alveoli, trapped particle clearance occurs either through dissolution in the fluid lining the alveoli or through phagocytosis (engulfing of the particles) by alveolar macrophage cells. The macrophages, whose purpose is to digest or clear respired particles, contain lysosomes with acidic pH and digestive enzymes such as acid hydrolases (Newman, 2001; Sipes and Badger, 2001; Brain, 1992). Macrophages that successfully engulf respired particulates are cleared upward in the airways, or into the lymph system or blood vessels. Another function of the macrophages, once they phagocytize particles, is to release chemicals into the surrounding epithelium that recruit other macrophages to the site to engulf other foreign particles (Lehnert, 1992).

The fate of inhaled gases in the respiratory tract is a function of the concentration of the gases in the inhaled air coupled with the solubility of the gases in the fluid lining the lungs. As summarized by Newman (2001), water-soluble gases such as sulfur dioxide tend to be absorbed at higher levels in the respiratory tract than less soluble gases such as nitrogen oxides. The greater the concentration of the gas, regardless of its solubility, the greater the likelihood that it will escape absorption in the upper respiratory tract and penetrate into the alveoli, where gas exchange with the blood is greatest.

Inhaled pathogens behave similarly to solid particulates in a physical sense. Many pathogens (such as soil fungus spores, viruses, bacteria, and bacterial spores) are in the appropriate size range to be able to penetrate to the alveoli, where they encounter warm, moist, nutrient-rich conditions that can promote pathogen development and absorption into the blood stream.

There has been increasing attention in recent years to the potential health effects of ultrafine particles less than 100 nm in diameter (see overview by Donaldson *et al.*, 2001). *In vivo* experiments with fine and ultrafine compounds of the same particles, such as metallic nickel, have shown that the ultrafine particles, for a given dose by mass, result in a greater inflammatory response than fine particles. This is probably due to the extremely small particle size and correspondingly high surface area, which are interpreted to inhibit phagocytosis, enhance oxidative stress, enhance inflammation in the lung epithelium, and permit the ultrafine particles to diffuse more readily into the lung interstitium.

9.07.4.2.3 Skin (percutaneous absorption)

Percutaneous exposures can occur either directly through the skin or through injuries to the skin. Some chemicals in gaseous or liquid form (such as methyl mercury or cyanide) can be absorbed directly through the skin. Materials that are soluble in skin perspiration can also be absorbed. Reactions of gases, solids, and liquids with the skin can lead to problems ranging from skin irritation to allergic reactions, and to chemical burns (Sullivan *et al.*, 2001). For example, contact of wet alkaline solids (such as cement or lye) with the skin can cause severe burns via a liquefaction necrosis process, which saponifies lipids, denatures proteins and collagen, and dehydrates cells. Inorganic acids (sulfuric, hydrochloric, nitric), which can form through the reaction of sulfur dioxide, hydrogen chloride, and nitric oxide gases with moisture in the respiratory tract, can cause tissue injury through dehydration and heat production that trigger denaturing of proteins and cell death. Hydrofluoric acid can trigger cell necrosis; in addition the fluoride ion can react with divalent cations in the tissues to precipitate fluoride salts that interrupt cell membrane function.

Exposure through breaks in the skin can result in the introduction of both toxins and pathogens. Toxins or toxic particles that are soluble in the blood plasma can be absorbed into the blood quite rapidly. For example, dermal exposure to chromic acid used in the electroplating industry can cause tissue damage, which then allows rapid uptake of hexavalent chromium ion and potential acute chromium intoxication.

9.07.4.3 Toxicokinetics (ADME) and Bioavailability

Toxicokinetics is the study of the time dependence of physiological processes on a toxin, and ultimately its response or effect (Boroujerdi, 2002a). The physiological processes, usually referred to as ADME, are absorption, distribution, metabolism, and elimination/excretion (Figure 3, Table 3). These are highly complex and toxin-specific. The chemical form of a potentially toxic substance, both as it is derived from an earth material and as it is transformed within the body, strongly influences its ADME as well as its bioavailability in the blood, organs, and tissues, thus dictating its toxicological effects.

Some substances are toxic at the site of exposure, because they can react chemically with the body fluids and tissues causing irritation, allergic reactions, or tissue damage. The example given previously for alkali- or acid-triggered cell necrosis via dermal exposure can also occur via respiration and ingestion exposures.

Other substances are toxic because they are not readily cleared by the body. Toxic effects of these types of substances in part result from the body's failed attempts to detoxify and/or excrete them. In addition, these substances may slowly react chemically with the body, leading to adverse effects.

The toxicity of substances that are readily soluble in the body fluids depend upon the exposure route, dose, chemical form of the substance at exposure, and processes that

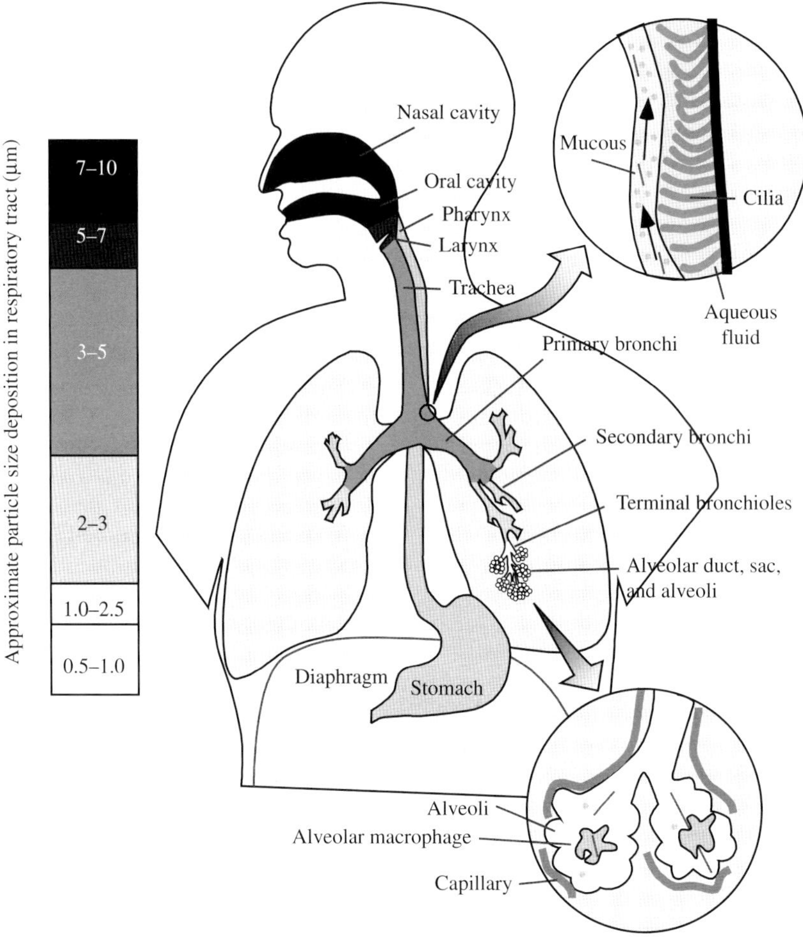

Figure 3 A schematic diagram of the respiratory system shows the fractionation of particle sizes that occurs with progressive depth in the system (after Newman, 2001).

chemically transform the substance during absorption, transport, and metabolism (Table 3). Toxicity that is dependent on the chemical form of a substance results from differences in the ways each form interacts chemically with the body.

For example, hexavalent chromium is more toxic than trivalent chromium, because it is more readily absorbed and transported across cell membranes; similar to sulfate and phosphate species, Cr(VI) can be transported via facilitated diffusion through nonspecific anion channels. Cr(VI) is readily reduced to Cr(III) in the acidic conditions of the stomach, but is less rapidly reduced in the pH 7.4 blood and tissues, through reactions with organic species such as ascorbate, glutathione, and amino acids (ATSDR, 2000b). Ingestion exposure to hexavalent chromium is therefore partly mitigated by conversion to trivalent chromium in the GI tract, where the Cr(III) compounds are less soluble, less readily absorbed, and are largely excreted in the feces. Chromium(VI) forms that are not rapidly reduced to Cr(III) are distributed by the blood to the organs and tissues, where continued reduction to Cr(III) occurs. This reduction process produces toxicity through the generation of reactive oxygen species (ROS) and reactive intermediates, which can react with and damage DNA, leading to cancer and other toxic effects. Inhalation exposure to Cr(VI) can produce toxicity in the lungs, because it is absorbed across cell membranes and is then reduced to Cr(III) within the cells. Chromium(III) that is absorbed or produced via Cr(VI) reduction is transported to the liver, where it is conjugated with glutathione and excreted in the bile. Chromium(III) species may also react to form insoluble chromium oxides.

Mercury, arsenic, and lead also provide examples of how differences in the form (valence, organic/inorganic, gas/aqueous, etc.) of a given element can result in strikingly different exposure pathways, absorption and transport mechanisms, toxicity mechanisms, metabolic processes, excretion/storage mechanisms, and the affected target organs (see Table 3).

Table 3 Biochemical mechanisms that influence the absorption, transport, storage, metabolism, excretion, and toxicity of selected potential toxins in the body.

Toxin, relative toxicities of forms, references	*Exposure pathways, absorption processes (for toxins soluble in body fluids)*	*Biochemical transport, storage, metabolism processes*	*Detoxification, excretion mechanisms*	*Source of toxicity*
Asbestos and asbestiform silicates *Amphibole asbestos, erionite > chrysotile asbestos*				
Holland and Smith (2001), van Oss *et al.* (1999), Werner *et al.* (1995), and Skinner *et al.* (1988)	*Inhalation, secondary ingestion.* Fibers are not readily cleared by macrophages, and are not completely dissolved in body fluids.	• Fibers persist in alveoli, leading to the buildup of scar tissue, and carcinogenic effects. • Fibers trigger immune response, with attack by macrophages. • Long-term chemical reactions between fibers and alveolar fluids, macrophages.	• Clearance of short fibers by mucociliary action, phagocytosis. • Some clearance by partial dissolution in lung, GI fluids.	• Incomplete fiber clearance, coupled with immune response, and macrophage death lead to scarring of lung tissue, decrease in blood oxygenation efficiency. • Release of reactive iron (and other redox-sensitive metals?) from fibers may trigger free radical generation, leading to DNA damage.
Arsenic (As) *Arsine gas > Inorganic As(III) compounds > organic As(III) compounds > inorganic As(V) compounds > organic As(V) compounds*				
Yip and Dart (2001) and ATSDR (2000a)	*Inhalation, ingestion (likely following mucociliary clearance of inhaled particles).* Easy absorption of organic forms, soluble inorganic forms. Rapid absorption of arsine gas	• As compounds bind to proteins in blood and are transported to liver, spleen, kidneys, GI tract, and other tissues or organs. • As compounds metabolized to less toxic methylated compounds in liver. • Interconversion of As(III), As(VI) forms.	• Conversion to methylated forms • Excreted in urine or stored in keratin-rich tissues such as nails, hair, skin	• As(III) reversibly binds to sulfhydryl groups, and inhibits critical sulfhydryl enzyme systems. • As(V) can also compete for phosphate in key biochemical reactions. • Chronic low-level exposure to As may stimulate growth of cells that produce keratin, leading to increased cellular division (and accompanying DNA replication) that creates allows greater opportunities for genetic damage.
Mercury (Hg) *Short-chain organic Hg (i.e., methyl, ethyl Hg) > aryl and long-chain organic Hg, inorganic Hg(II) > Hg^0*				
Yip *et al.* (2001)	*Inhalation.* Hg^0 vapor, organic Hg rapidly and efficiently absorbed. *Ingestion.* Minor Hg^0 absorption, except where mucous membranes are breached. Some absorption of soluble inorganic Hg(II). Rapid and easy absorption of organic Hg compounds due to high solubility in lipids. *Percutaneous.* Organic Hg compounds readily absorbed through skin.	• Hg^0 vapor rapidly diffuses into red blood cells and tissues, where it is oxidized to Hg(II). Some Hg^0 transported across blood-brain membrane, where it is oxidized to Hg(II). • Hg(II) distributed between plasma, red blood cells; only small amounts cross blood-brain barrier; most accumulates in renal cortex. • Aryl, long-chain organic Hg readily converted to Hg(II). • Short-chain organic Hg forms highly lipid soluble, readily cross cell membranes, placenta, and blood-brain barrier; only slowly transformed into Hg(II). Readily stored in a variety of tissues.	• Hg^0 excreted in urine, feces; some expired. • Hg(II) excreted in urine, feces; some reduction to Hg^0, which can be expired; some excretion in saliva, sweat. • Methyl Hg excreted in bile, feces, hair, and in toxic amounts in breast milk.	• Hg^0 acts as airway irritant, and cellular poison; can affect central nervous system due to ability to pass blood-brain barrier. • Hg(II) binds to sulfhydryl groups in proteins, which inhibits enzyme systems, and pathologically alters cell membranes. Strongest toxicologic effects are in the kidneys and central nervous system. • Methyl Hg toxicity primarily manifested in central nervous system, due to ability to cross blood-brain barrier.

(continued)

Table 3 (continued).

Toxin, relative toxicities of forms, references	*Exposure pathways, absorption processes (for toxins soluble in body fluids)*	*Biochemical transport, storage, metabolism processes*	*Detoxification, excretion mechanisms*	*Source of toxicity*
Lead (Pb) *Organic lead species[i.e., tetraethyl lead] > inorganic lead species*				
Keogh and Boyer (2001) and ATSDR (1999b)	*Inhalation*. Ready absorption of lead-rich fumes, volatile organic lead compounds, or soluble lead-rich particles. *Ingestion*. Absorption influenced by form, particle size, and iron and calcium absorption. More ingested lead absorbed in children than adults *Percutaneous*. Organic lead compounds readily absorbed.	• Most transport via binding to red blood cells, largely as complexes with hemoglobin, low molecular weight intracellular compounds, and membrane proteins • Distributed to bones, teeth, liver, lungs, kidneys, brain, and spleen. • Substitutes for calcium in bones; can be remobilized from bones, especially during pregnancy. • Crosses blood-brain barrier and concentrates in gray matter. • Mimics Ca in biochemical processes.	• Excretion through kidneys into urine (both dissolved and through shedding of renal tubular epithelial cells). • Possible excretion in bile.	• Affects many enzyme systems, including those critical to heme synthesis, those that maintain cell membranes, and those involved with steroid metabolism. • Also affects concentrations of neurotransmitters. Interferes with biochemical processes in the central nervous system involving calcium. • Accumulates in renal tubular cells, and interferes with other renal functions.
Chromium (Cr) Cr(VI) > Cr(III), Cr(II) > Cr^0 > $Cr_2O_{3\ (solid)}$				
ATSDR (2000), and Geller (2001)	*Inhalation*. Hexavalent salts readily solubized, absorbed; metallic Cr, many Cr(III) salts less readily solubilized and absorbed; Cr_2O_3 insoluble. *Ingestion*. Cr(VI) more readily absorbed than Cr(III), but Cr(VI) can be reduced by GI fluids to Cr(III). *Percutaneous*. Cr(VI) salts can be absorbed through intact skin. Skin irritation and ulceration by Cr(VI) compounds can greatly increase absorption.	• Cr(VI) can more easily pass through cell membranes than Cr(III) — in a process similar to sulfate and phosphate species, chromate enters cells via facilitated diffusion through nonspecific anion channels. • Cr(VI) is readily reduced to Cr(III) in the body by reactions with organic acids, amino acids (i.e., ascorbate and glutathione), and microsomal enzymes. • Cr(III) may precipitate as macromolecular oxides in body fluids.	• Excreted primarily in feces and urine; minor excretion in hair, nails. • Cr(III) is conjugated with glutathione in the liver, and excreted in the bile.	• Toxicity largely results from free radicals generated during the reduction of Cr(VI) to Cr(III). • Cr(VI) has greater toxicity than Cr(III) due to its greater ability to cross membranes and bind to intracellular proteins. • Some fibrosis, pulmonary effects result from biodurability of inhaled Cr_2O_3 in lungs. • Cr(VI) salts can irritate and cause ulcers in skin, other membranes.
Cadmium (Cd)				
ATSDR (1999c) and Waalkes *et al.* (2001)	*Ingestion*. Cd generally poorly absorbed. Metal–metal and metal–protein interactions influence extent of absorption; increased sorption with iron and calcium deficiency. *Inhalation*. Cd absorbed more readily, especially from fine soluble particulates	• Absorbed Cd bound to red blood cells and serum albumin (binding to sulfhydryl groups on proteins is especially strong). • Cd bound to serum albumin is metabolized in liver through binding reactions with metallothionein, which is then released by liver to the bloodstream. Cd-metallothionen is readily taken up and stored in the kidneys. • Cd can mimic Zn in metabolic processes.	• Ingested Cd mostly remains unabsorbed in GI tract and is excreted in feces. • Absorbed Cd is excreted very slowly; therefore Cd toxicity results due to its ability to accumulate in the body.	• Above a critical threshold concentration, Cd-metallothionen is toxic to kidneys. • Cd toxicity to kidneys can lead to degradation of vitamin D metabolism, which leads to osteoporosis; Cd-induced nephotoxicity can also lead to decreased calcium and phosphate retention, which can produce osteomalacia (weakening of bones).
Zinc (Zn)				
Goyer and Clarkson (2001), ATSDR (1994), and Fisher (2001)	*Ingestion*. Zn readily absorbed in intestine. *Inhalation*. Zn can be absorbed through alveolar epithelial cells	• Zn bound to albumin in plasma, from which it can be readily liberated to the tissues. • Major tissue storage sites include liver, pancreas, bone, kidney, and muscle. • Synthesis of Zn-metallo-thionein is stimulated in the liver, which facilitates the retention of zinc by hepatocytes. • Zinc in bone is relatively unavailable for use by other tissues.	• Excreted in feces via bile and pancreatic fluids. • Zinc excretion occurs slowly (1/2 life of 300 days).	• Gastrointestinal effects if ingested. • Zinc compounds linked to some contact dermatitis • Irritation of respiratory tract • Zn linked to degenerative diseases of nervous system; may contribute to formation of degenerative plaques in brains of Alzheimer's patients.

Nickel (Ni) *Soluble Ni compounds more acutely toxic; insoluble Ni compounds more carcinogenic*				
Sunderman (2001) and ATSDR (1997)	*Ingestion*. Some soluble Ni rapidly absorbed; amount absorbed decreases if food is present in the GI tract. *Inhalation*. Ni metal is retained in respiratory system for many years, and is slowly absorbed. More rapid absorption of soluble inorganic Ni. Ready absorption of Ni carbonyl gas. *Dermal absorption*. Ni absorbed into skin (leading to local inflammatory reactions), but not absorbed through skin into blood.	• Soluble inorganic Ni bound in plasma to amino acids, low molecular weight proteins (albumin, histidine, macroglobulin, nickeloplasmin). • Storage in a wide variety of tissue sites. • In rodent studies, Ni(II) induces formation of, but does not complex with, metallothionein.	• Urine is major excretion route for absorbed Ni. • Unabsorbed ingested Ni is excreted in feces. • Some excretion via bile, sweat, salivay, hair, fingernails, milk, etc.	• Increased risk of lung, nasal cancer through inhalation exposure of sparingly soluble Ni oxides and sulfides, and soluble Ni sulfates and chlorides. • Ni may mimic or substitute for essential elements (i.e., for Ca in the hypothalamic thermoregulatory center resulting in hypothermia; for Mg and Ca in enzyme processes).
Radon (Rn)				
ATSDR (1990)	*Inhalation*. Absorption of dissolved Rn gas through alveolar lining. Some Rn gas generated by decay of inhaled radioactive particles. Daughter particles produced by Rn decay (Pb^{210}, Bi^{210}, and Po^{210}) may be absorbed as well. *Ingestion*. Primarily of Rn dissolved in drinking water.	• Rn is inert noble gas that does not interact biochemically with the blood or tissues. • Rn daughter products behave biochemically in a manner similar to that of their nonradiogenic counterparts; for example, Pb^{210} can be stored in bones.	• Most inhaled Rn exhaled as gas. • $>90\%$ of ingested Rn is distributed to the lung, where it is rapidly exhaled.	• Toxicity primarily tied to effects of radioactive decay of Rn.

9.07.4.4 Toxins and Carcinogenesis: The Role of Earth Materials

Carcinogenesis is a term used to describe the process by which a cancer-causing agent, a carcinogen, induces a heritable altered, relatively autonomous cell growth commonly referred to as a malignant neoplasm (Pitot and Dragan, 2001). Carcinogens are or form reactive intermediates, which are electrophiles or free radicals that undergo covalent reactions with cellular macromolecules containing nucleophilic sites, such as DNA (Williams and Weisburger, 1991). Ultimately, a carcinogen induces a breakdown at the cellular level by altering the replication process of cells in the cellular DNA. The process of carcinogenesis is not a single event but a series of three biological stages termed *initiation*, *promotion*, and *progression* (Pitot and Dragan, 2001).

Initiation is the initial permanent and irreversible alteration made by a chemical carcinogen on the DNA of individual cells. The process of initiation can be altered by the high efficiency of DNA repair of the cell or detoxification of the carcinogen by the metabolic processes. Furthermore, not all initiated cells survive due to the normal process of apoptosis also referred to as programmed cell death (Wyllie, 1987). Those initiated cells that do survive may continue through the stages of carcinogenesis. *Promotion* is the proliferation of initiated cells, via the induction of changes in cell shape, growth rate, and other parameters. In contrast to initiation, promotion is a reversible process and involves multiple applications of a promoting agent. *Progression* is the process by which one or more proliferations of initiated and promoted cells undergo cellular evolution to a biologically malignant cell population. Carcinogenic agents can serve as initiating, promoting, or progression agents, or as various combinations of the three.

Elements commonly found in earth materials such as nickel, arsenic, chromium, cobalt, lead, manganese, beryllium, and some of their derivatives in specific valence states have been found to be carcinogenic (Williams and Weisburger, 1991). The particular stage(s) of carcinogenesis in which these elements participate is dependent on the form and mechanism of activity for the specific element, and each can play a role in more than one or all of the three stages. Uranium, polonium, radium, and radon gas are also classified as potent carcinogens; however, their activity is mainly attributed to their radioactive properties. The mechanisms by which asbestos minerals participate in carcinogenesis are currently under debate, but they may participate in both initiation and promotion (Wylie *et al.*, 1997).

9.07.5 THE CHEMICAL CONDITIONS OF THE HUMAN BODY FROM A GEOCHEMICAL PERSPECTIVE

The chemical compositions of various human body fluids play key roles in the stability and health effects of earth materials that are taken up by inhalation, ingestion, or dermal contact. Dissolution or precipitation of solids, uptake or evolution of gases, and viability of pathogens are all dependent upon the pH conditions, oxidation states, and types of complexing agents that are present in the different body fluids. The following discussion is condensed from information found in Scanlan *et al.* (1999), Letkeman (1996), Taylor and Williams (1998, 1995), Thomas (1997, 1989), Templeton (1995), Rhoades and Pflanzer (1992), Staub (1991), Cogan (1991), May and Williams (1980), Iyengar *et al.* (1978), May *et al.* (1977), Sahlin *et al.* (1977), Altman (1961), and Gamble (1942).

9.07.5.1 Blood Plasma

Plasma is the water-rich component of the blood within which the blood cells and platelets circulate. It plays an important role in material transport, pH regulation, and other metabolic processes in the human body. Plasma interactions with earth materials are restricted to breaks in the skin and vascular system.

Plasma can be highly dynamic in its composition, depending on its location within the body (i.e., in the arteries or veins), the activities of the individual (i.e., rest or exertion, eating or fasting, etc.), and many other factors. Nonetheless, some compositional generalizations can be made.

Inorganic electrolytes are important chemical components (Table 4, Figure 4), and from a geochemical perspective comprise a near-neutral pH, sodium-chloride-bicarbonate electrolyte solution with lesser amounts of calcium, potassium, magnesium, sulfate, and phosphate. A wide variety of trace metals are also present, generally complexed with many different inorganic and organic ligands; these trace metals include silicon, iron, manganese, cobalt, zinc, copper, chromium, and selenium.

A myriad of organic species are present in plasma (Table 4), many of which form strong complexes with major cations and trace metals. High-molecular-weight proteins such as albumin, globulins, and fibrinogen comprise ~7% of the plasma and help the plasma to maintain its high osmotic pressure. Also present are: organic acids such as lactate, citrate, and tartrate; amino acids such as glycine, alanine, histidine, cysteine, and cistine; and many other organic species such as

Table 4 The chemical composition of diverse body fluids. Blank entries indicate that no data have been found to date in the literature. For entries where an average value and ranges are available, the average value is listed first, followed by the range in parentheses.

Property or constituent	*Blood plasma*	*Interstitial fluid*	*Intracellular fluid*	*Gastric fluid*	*Intestinal fluid (duodenum)*	*Intestinal fluid (upper ileum)*	*Intestinal fluid (lower ileum)*	*Sweat*
pH	7.33–7.45	7.33–7.45	7	1.5–8.4	5.8–7.6	6.1	7.23	3.8–6.5
p_{O_2} (atm)	Arterial: 0.132 Venous: 0.02–0.053							Atmospheric
p_{CO_2} (atm)	Arterial: 0.053 Venous: 0.06							Atmospheric
Calcium (mg L^{-1})	100	100	40–60	72 (21–140)	124	53 (52–54)	74 (50–98)	10–80
Sodium (mg L^{-1})	3,265	3,333	160–230	1,126 (0–2,667)	1,950–3,300	2,974 (2,423–3,303)	2,975 (2,423–3,303)	240–3,120
Potassium (mg L^{-1})	156	156	6,060–6,256	454 (20–1,270)	39–430	438 (230–1,145)	438 (230–1,145)	210–1,260
Magnesium (mg L^{-1})	24	24	316–365	22–94		228 (184–279)	228 (184–279)	0.04–2.9
Chloride (mg L^{-1})	3,580	4,041	106–248	2,750–5,640	3,052 (1,800–4,700)	4,558	4,488 (4,392–4,545)	360–4,680
Sulfate (mg L^{-1})	48	48–192	961					7–74 (“S”)
Bicarbonate (mg L^{-1})	1,647	1,709–1,892	1,709	0–1,300	475 (245–1,287)	140	890 (635–1,037)	
Phosphate (inorg, as HPO_4^{2-}, mg L)$^{-1}$	63–123	63	3,200–5,800	11.7–42	51 (47–55)	58	63	
Silica (mg L^{-1})	9.2–16.9							
Iron (mg L^{-1})	1					170 (22-177)	170 (22–177)	
Manganese (mg L^{-1})	0.0006–0.08							
Copper (mg L^{-1})	0.6–1.4							
Cobalt (mg L^{-1})	0.001–0.01							
Aluminum (mg L^{-1})	0.05–0.3							
Zinc (mg L^{-1})	0.8–6							
Protein (total, mg L)$^{-1}$)	72,000			3,300				
Albumin (mg L^{-1})	48,000							
Globulin (mg L^{-1})	23,000							
Fibrinogen (mg L^{-1})	2,800							
Bilirubin (mg L^{-1})	4				55.6 (9–180)	84.5 (12–325)	84.5 (12–325)	
Mucoprotein (mg L^{-1})				0–460				
Amino acids								
Alanate (mg L^{-1})	33			18–27	31			
Arginate (mg L^{-1})	16			33–36	29			

(continued)

Table 4 (continued).

Property or constituent	*Blood plasma*	*Interstitial fluid*	*Intracellular fluid*	*Gastric fluid*	*Intestinal fluid (duodenum)*	*Intestinal fluid (upper ileum)*	*Intestinal fluid (lower ileum)*	*Sweat*
Asparaginate (mg L^{-1})	8							
Aspartate (mg L^{-1})	1			17–23	30			
Citrullinate (mg L^{-1})	5							
Cysteinate (mg L^{-1})	3							
Cistinate (mg L^{-1})	10			18–37	45			
Glycinate (mg L^{-1})	18			13–16	17			
Glutamic acid (mg L^{-1})	7			20–30	22			
Histidinate (mg L^{-1})	13			13–20	12			
Isoleucinate (mg L^{-1})	8			7–14	11			
Leucinate (mg L^{-1})	16			12–22	12			
Lysinate (mg L^{-1})	26			14–18	22			
Methionate (mg L^{-1})	4			8–15	20			
Phenylalanate (mg L^{-1})	11			8–18	17			
Prolinate (mg L^{-1})	24			17–32	30			
Serinate (mg L^{-1})	13			16–23	20			
Threoninate (mg L^{-1})	20			15–25	18			
Tryptophanate (mg L^{-1})	2			14–19	11			
Tyrosinate (mg L^{-1})	10			10–11	5			
Organic acids								
Citrate (mg L^{-1})	22							
Lactate (mg L^{-1})	164							
Malate (mg L^{-1})	4							
Oxalate (mg L^{-1})	1							
Pyruvate (mg L^{-1})	8							
Salicylate (mg L^{-1})	1							
Succinate (mg L^{-1})	5							
Ascorbate (mg L^{-1})	8			10				
Cholic acid (mg L^{-1})					1,300–4,600			
Glucuronic acid (mg L^{-1})				20				
Sialic acid (mg L^{-1})				73.1				
Histamine (mg L^{-1})				0.013–0.535				
Urea (mg L^{-1})	50.4			20				
Glucose (mg L^{-1})				3.5–12				
Fucose (mg L^{-1})				138				
Hexosamine (mg L^{-1})				327				

Sources: Altman (1961), Scanlan *et al.* (1999), Thomas (1997, 1989), May *et al.* (1977), Gamble (1942), Iyengar *et al.* (1978), Letkeman (1996), and Rhoades and Pflanzer (1992).

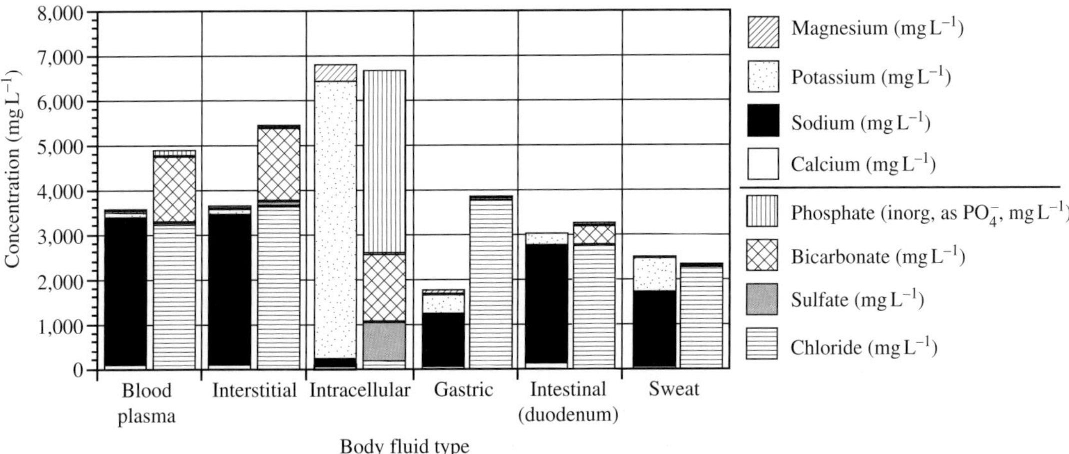

Figure 4 Concentrations of the major inorganic electrolyte species can vary substantially between the different body fluid types. These variations likely play an important role in the relative stability of a variety of minerals and earth material components in the body's different body fluid types. For a given fluid, cations are shown on the left, anions on the right.

peptides (i.e., glutathione), sugars (glucose and others), and fatty acids.

In order to sustain life and optimize the many biochemical reactions crucial for effective physiological function, the body strives to maintain its pH in the narrow range between 7.35 and 7.45 using a complex array of inorganic and organic buffers in the plasma and interstitial fluids. Excess hydrogen ions, such as lactic acid produced during exercise, are first buffered by reaction with bicarbonate ions in the plasma, producing dissolved CO_2 that is then removed as exhaled gas. The kidneys can also remove excess CO_2 through excretion of carbonic acid in the urine. If CO_2 is produced faster than it can be removed by ventilation or the kidneys, the body must turn to nonbicarbonate buffers in the blood, including hemoglobin, organic and inorganic phosphates, and plasma proteins.

Oxidation–reduction conditions (Figure 5) and processes in the plasma may be influenced by a wide variety of potential inorganic and organic redox buffers. Various organic redox couples that may be active, such as reduced/oxidized glutathione, ascorbic acid-dehydroascorbic acid, and cysteine-cystine, all have relatively reduced redox potentials in the vicinity of -100 mV to 0 mV; hence the overall Eh of the plasma is generally assumed to be in this range. Partial pressures of oxygen in arterial ($P_{O_2} = 0.132$ atm) and venous ($P_{O_2} = 0.02$–0.053 atm) plasma indicate that the plasma has, even in its most oxygen-depleted state in the veins, high dissolved oxygen concentrations that are well out of redox equilibrium with organic redox buffers and with all of the carbon-rich compounds found in the blood and tissues. A variety of redox-sensitive elements such as sulfur and iron can also be out of redox equilibrium, both internally (i.e., between reduced and oxidized forms of the same element) and externally with dissolved oxygen in the plasma. For example, various amino acids, glutathione, and proteins such as metallothioneins contain reduced sulfur in the form of sulfhydryl (HS) groups, but are common in plasma that is sulfate-rich. Hemoglobin, a protein, contains tightly bound ferrous iron, even though the iron atoms in hemoglobin are those with which transported oxygen associates or dissociates.

By exploiting a variety of complex chemical reactions involving proteins and other organic species, the body can selectively oxidize or reduce specific redox-sensitive elements in the plasma and the tissues as needed to meet its physiologic needs. For example, iron is routinely shifted between oxidation states by the action of specific enzymes and other organic compounds. When macrophages destroy aged blood cells and recycle hemoglobin, the ferrous iron in the hemoglobin is oxidized to ferric iron, which then combines with globulin proteins to produce transferrin. Transferrin carries the iron to storage sites such as the liver (where the iron is stored in oxidized form within the protein ferritin) or the bone marrow, where hemoglobin and the red blood cells are produced. Mobilization of iron from the ferritin stores, transport of iron across cell membranes, and production of heme (the iron-rich portion of hemoglobin), requires reduction of the ferric iron to ferrous iron, through the action of specific enzymes and other organic compounds such as flavins.

9.07.5.1.1 *Chemical complexing in the plasma*

The solubility of major cations, trace metals, and metalloids in the plasma is enhanced by

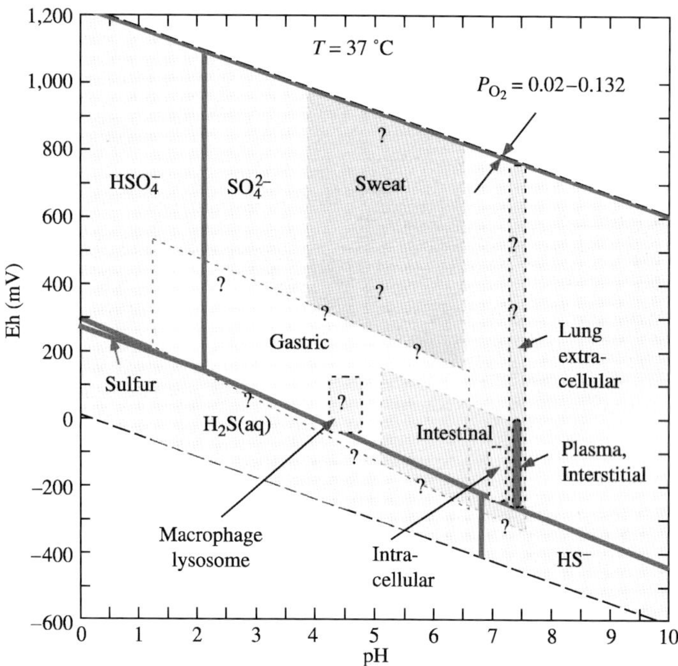

Figure 5 A plot showing variations in pH and speculated variations in Eh further illustrates the variability between different body fluid types. For comparison, the light gray area shows the Eh–pH stability field for water, the medium gray lines mark the stability fields of major inorganic sulfur species, and the darker gray line along the upper stability limit for water shows the range of Eh values in equilibrium with dissolved oxygen at arterial (upper edge of area) and venous (lower edge of area) oxygen pressures. The Eh ranges for the various body fluid types are highly speculative, and are postulated based on comparison to inferred Eh values for the plasma (black area) (see discussion in text). While there are indications that most body fluids have an overall Eh that is quite reduced and well out of equilibrium with dissolved oxygen in the plasma, there are substantial uncertainties in the extent to which the different potential redox couples in the body fluids reach equilibrium with each other and, especially in the case of sweat and lung fluids, dissolved oxygen.

complexation reactions with inorganic and, especially in the case of the trace metals, organic ligands in the plasma. Cation and metal species are distributed between several different states in the body:

- *Solids.* For example, calcium and chemically similar trace metals (such as lead, cadmium, and uranium) are incorporated into the phosphate phases of the teeth and bones.
- *Relatively inert or thermodynamically irreversible high-molecular-weight proteins.* These include, for example, the metal-binding proteins such as metallothioneins (including ceruloplasmin, a copper protein, and α_2 macroglobulin, a zinc protein), ferritin, and hemoglobin. Metals tend to be bound irreversibly to these proteins, and so are generally unavailable for participation in chemical reactions with other species.
- *Labile proteins.* These include proteins such as transferrin (which complexes iron, as discussed previously) and albumin (which complexes copper, zinc, and other metals).
- *Labile low-molecular-weight amino acids and organic acids.* These include, for example, complexes with amino acids such as cysteine and histidine, complexes with carboxylic acids such as lactate, and mixed ligand complexes involving more than one amino acid or carboxylic acid.
- *Labile inorganic complexes or aquated metal ions.* These include complexes with bicarbonate and chloride ions, and rarer hydroxy complexes.

In general, the latter three forms are viewed as those between which metals can be exchanged most readily in response to changing chemical conditions in the body. However, cations and metals can be mobilized under certain conditions from bones and other solids to the plasma, and can also be released from high-molecular-weight proteins if the proteins are degraded through disease or normal recycling mechanisms.

The chemical speciation of the different cations and metals strongly influences their chemical and physical behavior in the plasma and tissues. For example, chemical species with net neutral charges are most easily transferred across cell membranes and other membranes, and so are most readily absorbed by the body. Due to limitations in current chemical analysis techniques that preclude measurement of the actual concentrations of many complexes, thermodynamic-based speciation calculations have been used to estimate which complexes are predominant for a given metal, and whether or not their net charge will be amenable for their uptake across membranes into the body. Such calculations, although fraught with potentially substantial uncertainties in the thermodynamic data, etc., illustrate that each of the major cations and trace metals in plasma is probably complexed by a unique mix of ligands (Figures 6 and 7). The trace metals are most strongly complexed by amino acids, mixed ligand complexes with multiple amino or carboxylic acids (such as copper-cystinate-histidinate), and glutathione (Figures 6 and 7).

9.07.5.2 Interstitial Fluids

Interstitial fluids bathe the cells and tissues. With respect to the electrolytes and other constituents, physiology textbooks indicate that the interstitial fluids are generally similar in composition to the plasma, with the important

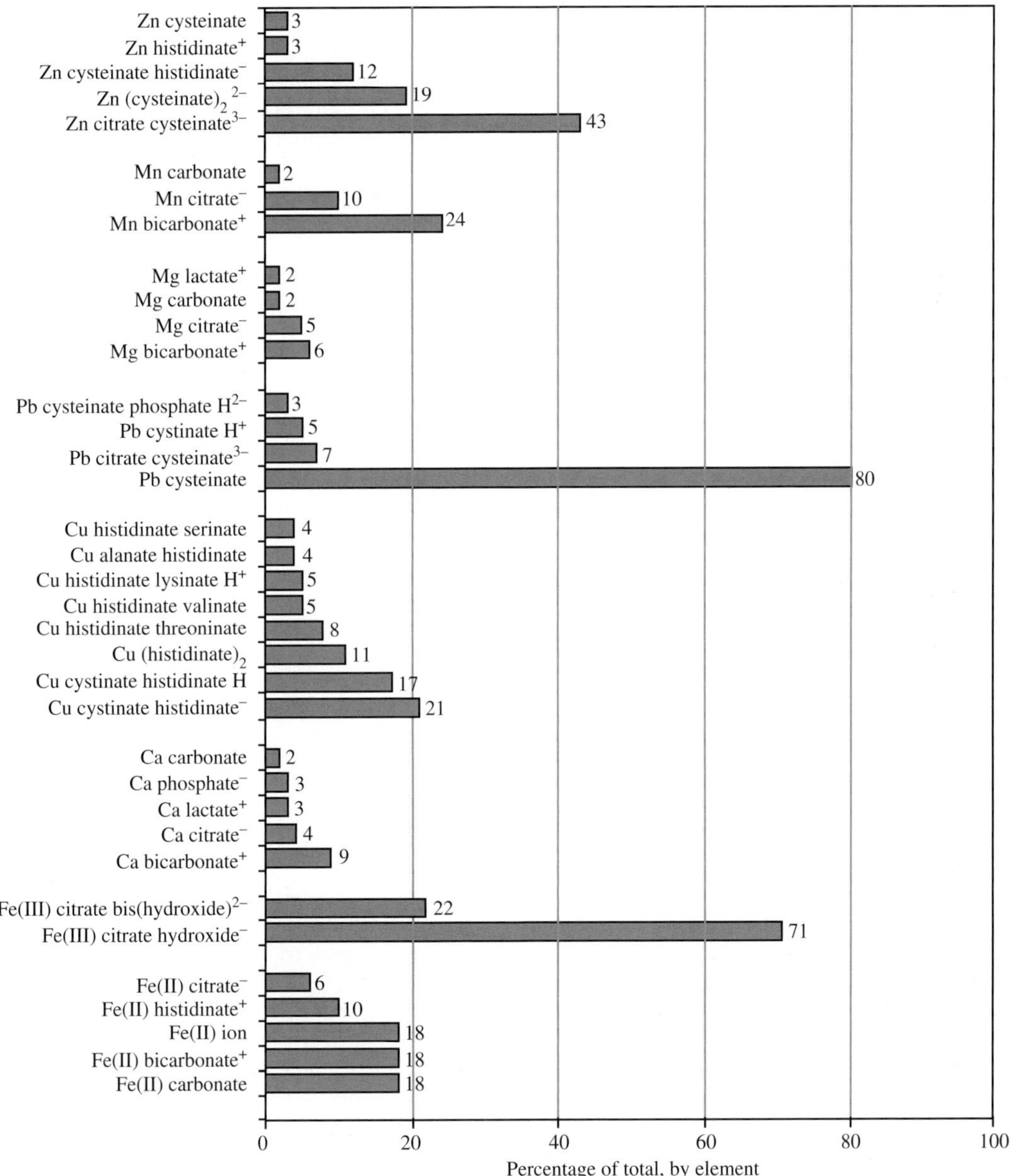

Figure 6 Chemical speciation calculations have been used (May *et al.*, 1977; May and Williams, 1980) to estimate the dominant low-molecular-weight organic and inorganic complexes for various metals in the plasma. The results show that each metal can be complexed by a unique set of ligands. Results of such speciation calculations can be used to infer the important chemical complexes and complexing ligands in other body fluids such as extracellular fluids.

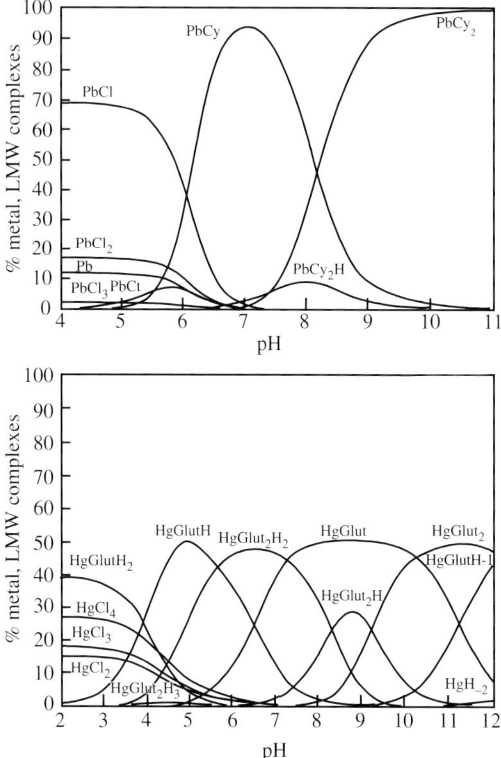

Figure 7 Letkeman (1996) used chemical speciation calculations to estimate the major complexes for lead and mercury as a function of pH in a fluid having overall plasma composition. The results show that most metal complexes with low-molecular-weight organic ligands such as amino acids diminish in importance with decreasing pH (such as in the gastric fluids) due to the increased protonation of the organic ligands. Cl—chloride; Cy—cysteinate; Ct—citrate; Glut—glutathione (reproduced by permission of the Division of Chemical Education Inc. from *J. Chem. Educat.* **1996**, *73(2)*, 165–170).

exception that they contain much lower concentrations of proteins (such as albumin and globulin) than the plasma (Table 4; Figures 4 and 5).

9.07.5.3 Intracellular Fluids

Intracellular fluids (also called the cytosol) are quite different compositionally from plasma and interstitial fluids (Table 4, Figures 4 and 5). The internal pH of many cells is maintained near 6.9–7.0. via various membrane transport mechanisms such as Na^+/H^+ and Cl^-/HCO_3^- exchangers, and various phosphate and protein buffers. In contrast to the plasma, the intracellular fluids have substantially lower concentrations of sodium, calcium, chloride, and bicarbonate and higher to substantially higher concentrations of potassium, magnesium, sulfate, inorganic and organic phosphates, proteins, and other organic species. The cells maintain their high K/Na via the sodium pump, also known as the Na^+/K^+ ATPase pump; the free energy derived from the hydrolysis of ATP to ADP is used to drive the transport of sodium out of the cell and potassium into the cell.

The redox potential within the cells is "substantially lower" than in the plasma (May and Williams, 1980), and may vary depending upon particular cellular biochemical activities and a myriad of potential redox couples such as reduced and oxidized glutathione species. In general, greater levels of reduced glutathione in the intracellular fluids than in the plasma may provide an indication of an overall lower oxidation condition within the cells. However, it is interesting to note that intracellular fluids have relatively high concentrations of dissolved sulfate (Table 4), in spite of the more reduced conditions inferred to be present.

The deleterious effects of toxins and processes that cause shifts in the redox balance within the cells are thought to lead to tissue damage and disease such as cancer. For example, generation of reactive superoxide and hydroxyl radicals by the intracellular reduction of metals such as Fe(III) (when present in excess of the amounts needed for proper cellular function) and Cr(VI) is thought to lead to cell membrane damage, destruction of enzymes and other proteins (through oxidation of sulfhydryl groups), and induction of breaks in DNA strands (Kawanishi, 1995; Rhoades and Pflanzer, 1992; Aust and Lund, 1990). Chemical species such as glutathione, ascorbic acid, and selenium, and enzyme systems such as catalase are examples of the chemicals that the body mobilizes to scavenge free radicals, but adverse effects occur when the free radicals are generated in excess and these defense mechanisms are overwhelmed (Kawanishi, 1995).

9.07.5.4 Gastrointestinal Fluids

The chemical composition of gastric fluids can vary substantially depending upon the presence or absence of food in the stomach. Under fasting conditions, the pH of gastric fluids is around 1.5, and their electrolyte composition is essentially that of a hydrogen-sodium-chloride solution with lesser amounts of potassium, calcium, and magnesium (Table 4; Figure 4). Organic species present in the gastric fluids include a wide range of amino acids, carboxylic acids, mucoproteins, other proteins, carbohydrates, and enzymes such as lipase, lysozyme, and pepsin that catalyze the breakdown of proteins and other components of food. As food is added to the stomach, the pH of the gastric fluids increases substantially due to neutralization by the food. In addition, the concentration of species such as bicarbonate, proteins, and amino acids increases due to the neutralization of the gastric acid and the chemical breakdown of the food. In addition, other components such as tannins and humic acids can be added to the gastric fluids from ingested food, drink, and other materials.

The redox conditions of the gastric fluids are probably quite variable, depending upon the dissolved oxygen content and the rates at which different organic redox couples reach equilibrium (Figure 5). Oxygen pressures may transiently reach near atmospheric (due to air swallowed with food or drink) but are probably shifted to substantially lower levels as food is digested and the oxygen in swallowed gases is consumed chemically. The fact that Cr(VI) is readily reduced to Cr(III) in the stomach (Table 3) provides indications that reduced conditions can be achieved in the stomach, and that redox couples active in the gastric fluids are probably not in redox equilibrium with dissolved oxygen.

Cation and metal speciation is likely to be substantially different in the gastric fluids than in the plasma, with decreased complexation with organic acids (due to increased acid protonation) and increased speciation as hydrated ions and chloride complexes (Figure 7).

Fluid secretions in the upper (duodenal) portion of the small intestine are alkaline, and help raise the pH of the partially digested food–gastric juice mixture called chyme (Table 4). Bile salts and pancreatic fluids are added to the chyme in the upper portions of the intestine. Bile aids in the digestion and absorption of lipids, whereas the pancreatic fluids contain enzymes and other organic species that assist with the continuing breakdown of proteins and carbohydrates. Redox conditions in the intestinal tract, although not specified in detail in the literature, are possibly somewhat more reduced than those generally present in the stomach, due to the increased separation from the periodic influx of atmospheric oxygen during swallowing. Metal complexes with amino acids and carboxylic acids probably increase in importance as the chyme moves into the intestine where its pH increases. Some metals may precipitate or sorb onto solids in the higher-pH environment of the intestine.

9.07.5.5 Extracellular Lung Fluids

Fluids lining the deeper portions of the lungs are a complex and dynamic mixture of several components (Goerke, 1998; de Meringo *et al.*, 1994; Scholze and Conradt, 1987; Reynolds and Chretien, 1984; Kanapilly, 1977; Gamble, 1942). *Surfactants* are composed predominantly of phospholipid-protein material such as dipalmitoyl phosphatidylcholine. They help the lungs lower their surface tension, maintain a wet surface for gas exchange, and reduce the amount of muscular effort needed to expand the lungs. *Mucus material* (composed primarily of glycomucoproteins) helps trap and clear foreign particles. An *aqueous serum transudate* phase helps with gas transport.

A variety of literature sources that discuss simulated lung fluids (SLFs) imply that the aqueous component of lung fluid is an electrolyte fluid generally similar in composition to that of interstitial fluid with pH near 7.4 (Table 4). Redox conditions in the lung fluids are not readily available from the literature. It is unclear which organic redox couples are active in the pulmonary fluids, and how these may drive redox equilibria to lower Eh values than those in equilibrium with the 0.132 atm PO_2 present in the alveoli.

9.07.5.6 Fluids in the Alveolar Macrophages

Once an alveolar macrophage engulfs a particle, the resulting phagocytic vacuole fuses with lysosomes to form a phagolysosome, within which the particle is digested. Respired foreign particles can therefore also come into contact with both macrophage cytoplasm and lysosomal fluids (Collier *et al.*, 1992). Some studies indicate that alveolar macrophages maintain a substantially lower pH in their intracellular fluids than most cells, from 3 to 4 (van Oss *et al.*, 1999; de Meringo *et al.*, 1994; Hume and Rimstidt, 1992; Jaurand *et al.*, 1984), whereas others report macrophage intracellular fluid pH values in the 6.5–6.9 range (Collier *et al.*, 1992). Electrolyte concentrations in the macrophage intracellular fluids, or cytosol, are inferred to be grossly similar to intracellular fluids of other types of cells (Table 4), with possible variations arising depending upon the cell function.

Compositions of lysosomal fluids in the alveolar macrophages have been characterized primarily in terms of their high concentrations of proteases, bactericidal enzymes such as lysozyme, and other cytotoxic chemicals (Burns-Naas *et al.*, 2001; Brain, 1992). The pH of lysosomal fluids is generally thought to be low, in the range from 4 to 5 (Collier *et al.*, 1992; Nyberg *et al.*, 1992; Johnson, 1994). These chemical features are designed to help the alveolar macrophages degrade foreign particles that they engulf.

9.07.5.7 Sweat

Sweat can vary substantially in composition depending upon activity level and other parameters (Table 4). For example, the pH can vary rather widely between 3.8 and 6.5. The dominant electrolytes in sweat are sodium, chloride, and phosphate. Urea, ammonium, and lactate, generated as metabolic wastes, are abundant, and a variety of amino acids are also present. Although sweat on the skin surface is most likely saturated with oxygen at atmospheric pressure, it is again unclear the extent to which various organic redox couples may shift redox equilibria to more reduced conditions.

9.07.5.8 Summary—Body Fluids from a Geochemical Perspective

As summarized in Table 4 and Figures 4–7, various human body fluids with which earth materials may come into contact exhibit a wide range of compositions, including their pH and Eh, electrolyte content, and concentrations of organic species such as amino acids, organic acids, and proteins. This compositional variability indicates that any given mineral or earth material, and its contained metals, may potentially behave quite differently from a geochemical perspective depending upon the exposure pathway, the resulting body fluid(s) it encounters, and how it may modify body fluid chemistry.

9.07.6 METHODS USED TO ASSESS THE INTERACTIONS OF EARTH MATERIALS WITH, AND THEIR TOXIC EFFECTS UPON, THE HUMAN BODY

There is a wide variety of chemical and toxicological methods that are used to assess the body's interactions with earth materials and their potential toxic effects. These can be grouped into several major categories, including: *in vitro* bioaccessibility and biodurability tests; *in vitro* toxicological tests; *in vivo* toxicological tests; *in vivo* bioaccessibility tests; and computer-based chemical modeling calculations. Due to length limitations, only a brief overview of the extensive literature available on these topics is possible in this chapter; a more detailed bibliography is available from the authors upon request.

Bioaccessibility tests (also called bioavailability or physiologically based extraction tests) have been used extensively to measure the short-term solubility of and metal extraction from earth materials and other substances in simulated gastric, intestinal, and lung fluids (Ruby *et al.*, 1992, 1996, 1999; Hamel, 1998; Battelle and Exponent, 2000; Oomen *et al.*, 2002; Mullins and Norman, 1994). *Biodurability* tests have been used extensively to measure the long-term solubility of earth materials and other substances such synthetic glass fibers in SLFs (Johnson and Mossman, 2001; Jurinski and Rimstidt, 2001; Werner *et al.*, 1995; Mattson, 1994a,b; Eastes *et al.*, 1996, 2000a,b). The goal of both types of tests is to understand the types and rates of chemical dissolution or alteration reactions that earth materials and other substances might undergo in the body via various exposure routes. The studies can vary considerably in their physical design; particle size, shape, and other characteristics of the minerals tested; fluid compositions; test duration; and other parameters. Key uncertainties that are inherent in both types of *in vitro* solubility-based tests are (1) how well they reproduce actual conditions in the body; and (2) how well the predicted results (such as particle dissolution rates, or types and relative abundances of trace elements solubilized from the particles) can be readily extrapolated to infer toxic responses *in vivo* (de Meringo *et al.*, 1994; Ruby *et al.*, 1999; Johnson and Mossman, 2001). Nonetheless, even the least sophisticated tests may provide at least some valuable information to the investigator regarding chemical reactions that may be occurring between the particles and fluids *in vivo*; this is especially true when the test results are interpreted in an appropriate mineralogical and geochemical context.

A variety of *in vitro* toxicity tests have been developed to model the effects of toxins on living cells or tissues. In these tests, a carrier medium (such as fetal bovine serum) containing given concentrations, or doses, of a particular toxin are added to cell cultures (cell lines). Various indicators of toxicity, cell morphology transformation, or cell proliferation are then measured after specified periods of time. The cell types used in a particular study can be chosen to approximate the types of cells that would be affected during actual exposure, such as respiratory cells or tissues. Toxicity indicators include, for example, measures of the percent of viable cells remaining at the end of the test (compared to a control line with no added toxin), and the concentrations various cytokines or other cytoplasmic enzymes induced from the cells by the toxin. Uncertainties with the *in vitro* toxicity tests include how comparable their results are to those of *in vivo* toxicity tests, and how well they reproduce actual physiological conditions and processes in the human body (Johnson and Mossman, 2001).

In vivo toxicity tests involve the direct exposure (via appropriate exposure routes) of living animals to variable doses of toxins over time, followed by measurement of toxic effects or exposure indicators. Inhalation tests either expose the subject animals to known concentrations of particles in an airstream, or utilize direct intratracheal implantation of the particles in the subject animals (e.g., studies summarized in Johnson and

Mossman, 2001; Buchet *et al.*, 1995). Ingestion tests usually provide the subject animals with a diet containing known concentrations of the target earth material being ingested (Ruby *et al.*, 1996; Schoof *et al.*, 1995). Examples of toxic effects that have been assessed include adverse changes in neurological or other physiological behavior in response to the exposure, the occurrence and severity of abnormal changes in pathologic tissue samples, such as fibrosis, tumor growth, or cell necrosis, measures of exposure and absorption such as metal concentrations in blood, urine, fur, or nails (Schoof *et al.*, 1995; Buchet *et al.*, 1995; Ruby *et al.*, 1996), and the burden and nature of foreign materials in lung tissues or other tissues (Churg *et al.*, 1989). An important uncertainty is how well the physiologic processes and endpoints of exposure, dose/response, toxin uptake, and toxicity measured in the *in vivo* tests on animals can be extrapolated to quantify similar processes and endpoints in the human body.

In vivo bioaccessibility (bioavailability) assessments are often conducted on individuals exposed to potential toxins to assess the extent to which the toxins have been absorbed, transported, and metabolized by the body. Bioavailability can be calculated as the fraction of the exposure level (dose) that reaches the systemic circulation (Boroujerdi, 2002b). The extent of absorption is determined by comparing the plasma concentration following a single intravenous administration to those following the primary exposure routes (Medinsky and Valentine, 2001). Such *in vivo* analyses can assess the whole body bioaccumulation of the material of concern as well as the particular tissues in which this deposition takes place. These models have been useful in toxin evaluation and are presently being used by some federal agencies in extrapolating experimental data to human risk situations (Pitot and Dragan, 2001).

Aqueous chemical speciation calculations (Alpers and Nordstrom, 1999; Bethke, 1996) have been used for some time to help understand the speciation of and trace metal chelation in diverse human body fluids such as plasma, wound fluids, saliva, sweat, fat emulsions, and gastrointestinal fluids (Taylor and Williams, 1995, 1998; Williams, 2000). In contrast, chemical speciation calculations have only been used infrequently in studies evaluating interactions between the human body and earth materials such as asbestos (Hume and Rimstidt, 1992; Gunter and Wood, 2000; Taunton *et al.*, 2002; Davis *et al.*, 1992, 1996). There are many potential uses of chemical speciation calculations in the interpretation of interactions between human body fluids and earth materials. For example, the interpretation of *in vitro* mineral solubility tests could be greatly improved by chemical speciation calculations on the resulting leach fluid. Comparison of the chemical compositions of many simulated body fluids used in *in vitro* extraction tests to results of chemical speciation calculations of the plasma and other body fluids (May *et al.*, 1977) indicates that the simplified simulated fluid compositions used in most studies may not include a broad enough array of organic ligands, especially those that are the most effective metal complexing agents (such as cysteine and glutathione). As a result, the *in vitro* tests may understate the true metal mobility from earth materials.

Chemical reaction path calculations (Alpers and Nordstrom, 1999; Bethke, 1996) have only seen limited use as applied to fluid-mineral interactions in the human body (e.g., Davis *et al.*, 1992). The potential further applications in this realm are intriguing, both for understanding chemical reactions between body fluids and earth materials, and in understanding potential changes in body fluid chemistry in response to physiological processes and therapeutic treatments such as toxic metal chelation therapy.

9.07.7 EARTH MATERIALS IN A BIOSOLUBILITY AND BIOREACTIVITY CONTEXT

Important geochemical factors that influence the health effects of earth materials (Figure 8) are: (i) their *biosolubility*, the extent to which they are soluble in the various body fluids (their) and (ii) their *bioreactivity*, the extent to which they can modify key body fluid parameters such as pH, concentrations of major electrolytes, and redox species. As discussed previously, *biodurable* minerals and earth materials are those that are generally bioinsoluble or sparingly biosoluble in body fluids, and so cannot be cleared rapidly by chemical dissolution. *Bioaccessible* earth materials are those that are readily biosoluble (and therefore can readily release toxins into the body fluids by their dissolution), or that may not be biosoluble but that contain readily bioaccessible toxins (e.g., heavy metals sorbed onto particle surfaces). A given earth material may vary in its biodurability or bioaccessibility depending upon the exposure route, due to differences in its solubility or toxin bioaccessibility in the particular body fluids present. Especially in the case of biodurable earth materials, geochemical processes controlled by particle surface chemistry and phenomena may be quite important in influencing dissolution rates (Guthrie, 1997; Hochella, 1993).

Earth materials with abundant soluble alkali components (such as cement or concrete) or acidic

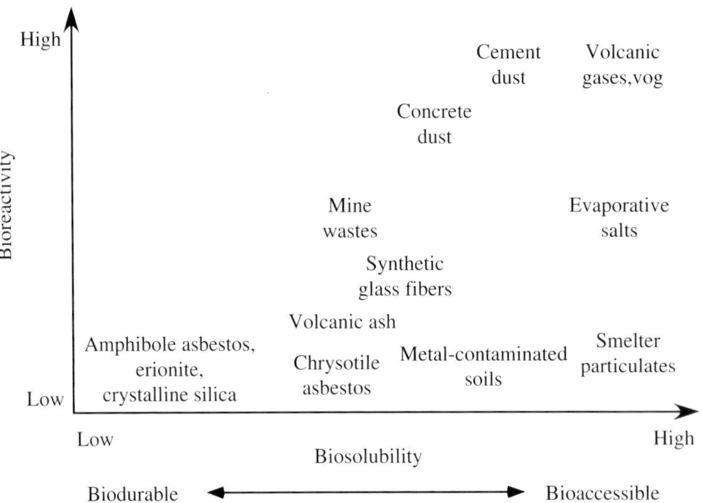

Figure 8 This schematic plot shows the inferred biosolubility and bioreactivity of general classes of earth materials. Many types of earth materials (i.e., mine wastes, volcanic ash, soils) can contain a complex variety of minerals having quite different biosolubilities and bioreactivities, so the particular location of a given earth material on the plot should be considered as an averaged approximation.

components (such as acid volcanic gases, or acid-generating salts in mine wastes) are the most bioreactive. However, even sparingly soluble, biodurable earth materials such as asbestos can be somewhat bioreactive by slowly reacting at their surfaces with, and releasing chemicals into, the surrounding fluids and tissues.

9.07.8 THE MEDICAL GEOCHEMISTRY OF BIODURABLE EARTH MATERIALS

Adverse health effects of biodurable earth materials result primarily from inhalation exposure to airborne particulates. The earth materials that are most commonly associated with adverse respiratory health effects include dusts of asbestos, erionite (a fibrous sodium-rich zeolite), crystalline silica, and coal. However, dusts from a wide variety of other biodurable earth materials (such as metal oxides, talc, kaolinite, feldspars, bentonite, fuller's earth, and micas) and non-earth-materials (such as wood dusts, wood fibers, glass fibers, and others) are also known to be associated with adverse health effects if exposures are of sufficient intensity and duration. Adverse health effects resulting from ingestion and percutaneous exposure to biodurable earth materials (either direct or indirect of particles cleared from the respiratory tract) have also been the focus of some investigation and speculation. After decades of research, there is a substantial and growing understanding of the important role that geochemical processes play in the health effects of biodurable minerals. However, there are still many areas in which conflicting information and interpretations reveal continuing uncertainties.

9.07.8.1 Asbestos, Erionite, and Other Fibrous Materials

Many different epidemiological and toxicological studies during the last several decades have shown that intense, generally prolonged inhalation exposure to asbestos and erionite is associated with elevated occurrences of diseases such as asbestosis, lung cancer, pleural effusions, pleural thickening, pleural plaques, and mesothelioma cancer. Although chrysotile asbestos, amphibole asbestos, and crionite fibers all can generate adverse health effects such as fibrosis, there is a recognition that the different forms are not equal in their pathogenicity, with amphibole asbestos and erionite considered as more pathogenic than chrysotile asbestos. See summaries in Nolan *et al.* (2001), Holland and Smith (2001), and Guthrie and Mossman (1993). No definitive links have been established between asbestos exposure and gastrointestinal health effects (Holland and Smith, 2001).

Most attention has focused on occupational exposures to chrysotile and amphibole asbestos as it has been mined, processed, and used commercially or industrially. However, there has been some renewed attention to potential occupational, residential, and environmental exposures to asbestos that occur as an accessory mineral in rocks such as serpentinite (Renner, 2000), and to deposits of other industrial minerals such as vermiculite (e.g., Libby, Montana: Lybarger *et al.*, 2001; Dearwent *et al.*, 2000; McDonald *et al.*, 1986; Wright *et al.*, 2002; Van Gosen *et al.*, 2002; Meeker *et al.*, in press; Wylie and Verkouteren, 2000). Environmental exposures to erionite have also been well documented as the cause for asbestos-related disease (studies summarized in Ross *et al.* (1993); Dumortier *et al.* (2001)).

9.07.8.2 Crystalline Silica

Crystalline silica includes the silica minerals quartz and its polymorphs such as cristobalite and tridymite, which have the same chemical formula but different crystal structures. Amorphous, non-crystalline silica can also occur in a wide variety of geologic environments (Ross, 1999).

Silicosis, a form of pulmonary fibrosis, is the primary health problem resulting from inhalation exposure to particles of crystalline silica (SSDC, 1988; NIOSH, 2002; Castranova, 2000; Castranova and Vallyathan, 2000). Other diseases associated with occupational inhalation exposure to crystalline silica include lung cancer, chronic obstructive pulmonary disease, nonmalignant respiratory disease, auto-immune related diseases (such as rheumatioid arthritis), renal diseases, and (as a complication of silicosis) increased risk of bacterial or fungal infections such as tuberculosis. Skin granulomas or obstructive lymphopathies may result from dermal exposure and uptake of silica particles (NIOSH, 2002).

9.07.8.3 Factors Influencing the Health Effects of Biodurable Minerals

9.07.8.3.1 Particle shape and size

During inhalation, fibrous particles with length generally less than 10–20 μm and diameters less than 0.5–1 μm are thought to flow aerodynamically into the deep alveoli, where they can lodge (Holland and Smith, 2001). Because the alveolar macrophages are not big enough to completely engulf these fibers, macrophage clearance of the fibers is limited. Various studies indicate that short fibers are less pathogenic than long fibers, in part because the long fibers are less readily cleared than the short fibers (Holland and Smith, 2001; van Oss *et al.*, 1999; Davis *et al.*, 1991), and possibly in part because of chemical differences between short and long fibers (Graham *et al.*, 1999). Fibers that are needle-like and rigid (van Oss *et al.*, 1999) may also physically penetrate the lung tissue, where they can then be transported by the lymph system or can physically migrate to the pleural or peritoneal spaces (Holland and Smith, 2001). Some reports conclude that amphibole fibers (which tend to be more straight and rigid than typically curved chrysotile fibers) are more likely to be able to penetrate deeper into the lung tissue than chrysotile fibers, and therefore are more likely to contribute to interstitial diseases such as mesothelioma (i.e., van Oss *et al.*, 1999). However, some recent studies indicate that short chrysotile fibers (less than ~5 μm) may also be able to penetrate and migrate to the pleural and peritoneal spaces, and can therefore also trigger mesothelioma and other diseases in these regions (Suzuki and Yuen, in press, 2001).

Crystalline silica particles (except for fibrous crystobalite and tridymite) tend to be more equant in shape rather than fibrous, and therefore small (<1.5–2 μm) particles that lodge deep in the respiratory system should be relatively amenable to macrophage clearance. Adverse health effects result when the clearance capacity via macrophage, chemical, and physical mechanisms, is exceeded by particle deposition rates. As with asbestos fibers (Snipes, 1995), small particles may be transported within phagocytic cells into the lymph system, and may also be able to penetrate the circulatory system, where they can then be transported to other organs and deposited. The fact that commonly nonfibrous crystalline silica can be pathogenic under high-dose exposures indicates that particle shape is not the sole factor in pathogenicity.

9.07.8.3.2 Particle solubility and dissolution rates

The rates at which inhaled particles dissolve *in vivo* are thought to play important roles in their biopersistence, and therefore their ability to trigger fibrosis, cancer, and other diseases. Dissolution rates are a complex function of the chemical solubility of particles in the body fluids, coupled with factors (such as crystal structure and mineral surface properties) that determine the rate at which particles dissolve. For example, several lines of evidence indicate that the amphibole asbestos minerals and erionite are less readily dissolved in lung, interstitial, and phagolysosomal fluids than chrysotile asbestos. This enables their fibers to persist for longer periods of time in the lungs and adjacent tissues, thereby imparting a greater potential to trigger fibrosis and cancer (Sébastien *et al.*, 1989; Churg *et al.*, 1993, 1989; Johnson and Mossman, 2001).

A variety of studies have used *in vitro* biodurability tests to the determine dissolution rate of various silicate minerals in simulated lung and lysosomal fluids (Hume and Rimstidt, 1992; van Oss *et al.*, 1999; Jurinski and Rimstidt, 2001; Werner *et al.*, 1995). Most *in vitro* studies of mineral solubility analyze changes over time in solution chemistry for a limited number of constituents such as aqueous silica, magnesium, and iron. With this information, coupled with analytical data on the composition of the fibers after being leached, dissolution rates can be inferred based on the total amount of these constituents released over time, and dissolution processes can be inferred based on element ratios in the leach fluid and the characteristics of the leached fibers. These studies have shown that the dissolution rate of chrysotile is quite high in SLFs, and are higher than for the amphibole asbestos minerals, erionite, talc, and quartz. A variety of studies have found that magnesium is preferentially leached from the surfaces of talc particles, chrysotile fibers, and crocidolite fibers, leaving behind a rind of leached material enriched in silica. Werner *et al.* (1995) also found that iron was leached preferentially from crocidolite in the presence of organic iron chelators. It has been proposed that this chemical leaching process weakens the fibers, thereby making them more susceptible to breakage and therefore less biopersistent. On the basis of these results, Jurinski and Rimstidt (2001) proposed that silica release from particle surfaces is the rate-limiting step in their dissolution.

9.07.8.3.3 Particle solubility: insights from chemical modeling calculations

Hume and Rimstidt (1992) used thermodynamic constraints to show that lung fluids should be substantially undersaturated with respect to chrysotile. Taunton *et al.* (2002) and Gunter and Wood (2000) used a thermodynamic phase diagram approach to model potential chemical reactions between asbestos and simplified lung fluids that might alter the asbestos to other minerals *in vivo*; their model results suggested that silica, and possibly talc, are predicted to precipitate as chrysotile dissolves; the precipitation of silica is consistent with the silica-rich leached rind produced by the *in vitro* solubility studies discussed above.

We have used chemical speciation calculations to predict the relative saturation indices of quartz and of different asbestos-forming minerals in equilibrium with solutions having inorganic electrolyte concentrations (Table 4) approximating those of interstitial fluids (approximating lung fluids), intracellular fluids, and inferred lysosomal fluids (Figure 9). The results illustrate both the challenges and benefits of applying chemical speciation calculations to understand mineral solubility in bioinorganic systems. For example, a large number of minerals are predicted to be quite highly supersaturated in the interstitial fluids, including hydroxyapatite, a key mineral component of bones (Skinner, 2000), and other minerals (such as calcite) that should from a kinetic standpoint, be able to precipitate *in vivo*. This suggests that the calculations may have used fluid compositions that were too simple (i.e., neglecting the organic acids and their calcium complexes may overestimate carbonate and hydroxyapatite saturation), or incorrect (i.e., perhaps aluminum concentrations are too high, and published phosphate concentrations may not differentiate inorganic from organic phosphate). Also, it may not be appropriate to extrapolate our knowledge of mineral precipitation kinetics in inorganic systems to conditions *in vivo*. However, even with these potential shortcomings, the simple speciation calculations depicted in Figure 9 do provide some potentially useful information. For example, quartz is calculated to be slightly supersaturated in both the interstitial and intracellular fluids, and several of its polymorphs, including chalcedony, are slightly undersaturated. This indicates that there may not be a strong thermodynamic driver for quartz dissolution *in vivo*, and helps explain the development of silica-rich rinds on dissolving silicate particles (Jurinski and Rimstidt, 2001) as in chrysotile dissolution:

$$\begin{aligned}&Mg_6(Si_4O_{10})(OH)_8 + 12H^+\\&\Rightarrow 6Mg^{2+} + 4SiO_{2\ aq} + 10H_2O4SiO_{2\ aq}\\&\Rightarrow 4SiO_{2\ \mathrm{Quartz}}\end{aligned}$$

However, dissolution studies indicate that quartz does dissolve, albeit slowly, *in vitro* (Jurinski and Rimstidt, 2001); this suggests that organic species or some other ligand not considered in these calculations may be helping to complex aqueous silica and therefore driving the dissolution. Further work is needed to evaluate the potential role of organic complexes with silica *in vivo*.

The various asbestos-forming minerals are predicted by the speciation calculations (Figure 9) to be moderately undersaturated in the pH 7.4 interstitial fluids and extremely undersaturated in the acidic lysosomal fluid compositions, and therefore should have a substantial tendency to dissolve. The amphiboles are predicted to be substantially more soluble than chrysotile, and so should have a greater tendency to dissolve than chrysotile, contrary to the observed relative dissolution rates. This provides support for the conclusion by Hume and Rimstidt (1992), and Jurinski and Rimstidt (2001) that kinetic factors play important roles in determining particle dissolution rates. Although erionite was not considered in the calculations, several sodic zeolites such as stilbite were considered. They are only slightly undersaturated in the lung fluid proxy at pH 7.4, suggesting that the biodurability of erionite may be due in part to its low solubility in the lung fluids.

The various asbestos-forming minerals (and silicates in general) are predicted by the speciation calculations (Figure 9) to become supersaturated

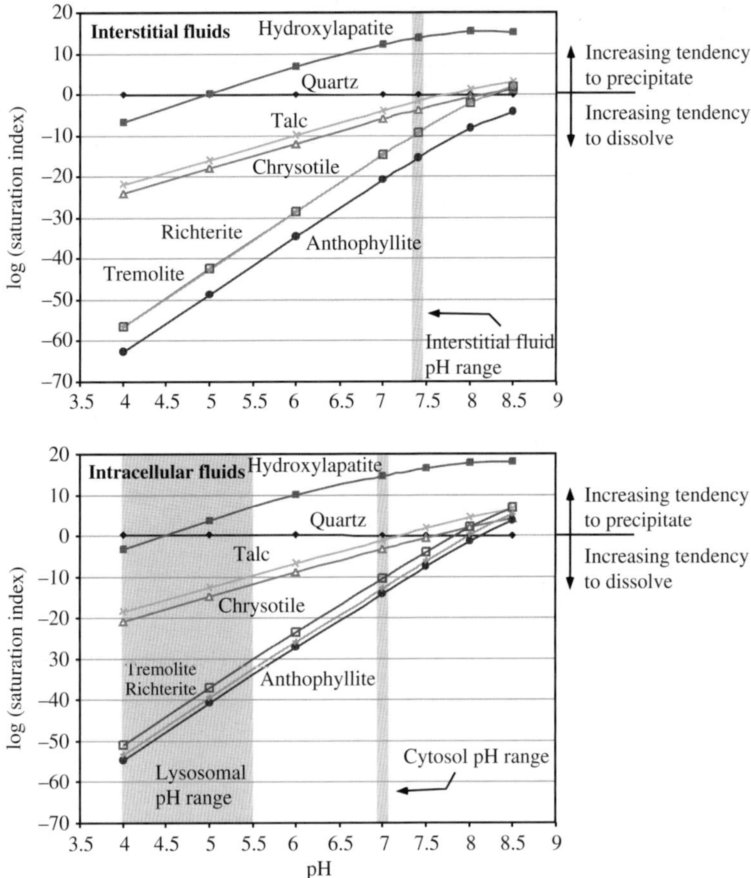

Figure 9 Plots showing the calculated mineral saturation indices as a function of pH for hydroxylapatite, quartz, and various asbestos-forming minerals in electrolyte solutions approximating the electrolyte compositions of lung fluids (approximated by interstitial fluids, upper plot) and intracellular fluids (lower plot). Electrolyte concentrations used as input were taken from Table 4. The CO_2 partial pressure was fixed at the value for venous plasma for each speciation at a different pH. Organic species such as amino acids and other organic acids were not included in the calculations, but likely would have the effect of decreasing the calculated saturation indices somewhat due to their complexation with cations.

in the interstitial fluids with shifts to higher pH (to near 8–8.5). Alkaline, reactive particles that are inhaled with the asbestos fibers, could increase the pH of the lung fluids. The solubility drivers for fiber dissolution might then be diminished. Silicate fiber dissolution should also consume acid according to the previous dissolution reaction for chrysotile, and the following dissolution reaction for tremolite:

$$Ca_2Mg_5(Si_8O_{22})(OH)_2 + 14H^+ \Rightarrow 2Ca^{2+} + 5Mg^{2+} + 8SiO_{2\,aq} + 8H_2O$$

It is interesting to speculate whether, in spite of the lung's attempts to produce more fluids at the site of foreign particle deposition, fiber dissolution could increase the pH of the local lung fluids so that they sufficiently become saturated with fibrous silicates, thereby limiting further dissolution.

9.07.8.3.4 Mitigating or exacerbating effects of trace elements and accessory minerals

Most toxicological studies focus on the toxicity or solubility behavior of a particular sample of a given mineral such as chrysotile, tremolite, or talc. They do not tend to examine the role that variations in morphology, trace element content, accessory minerals, and other characteristics between the same mineral from different samples in the same geologic locality, and between samples from different geologic localities, can play in *in* vitro and *in vivo* biodurability, and therefore toxicity. These parameters have been demonstrated to play important roles in the rate at which other minerals, such as sulfides, weather under environmental conditions (see summary in Plumlee, 1999), and so are also likely to be important for particle durability *in vivo*.

Hochella (1993) cites several studies, such as Nolan *et al.* (1991), in which the same mineral from different localities shows a variable range of carcinogenicity in laboratory animals. Johnson and Mossman (2001) summarize results of some studies that have found that the fiber length and biological activity can vary substantially between different chrysotile samples collected from different geological localities.

Ziegler *et al.* (2002) have systematically characterized and compared 5 sets of asbestos toxicological standards (5 amosites, 4 anthophyllites, 6 chrysotiles, 5 crocidolites, 4 tremolites). Substantial variability was found in a number of mineralogical parameters between different standards of the same asbestos minerals, such as: morphology (i.e., length, width, "curlyness," etc.), the types and abundances of accessory or contaminant minerals present with the fibers, and the major and trace element compositions. For example, a number of the toxicological standards contain accessory minerals such as quartz and others that, although present in generally low to moderate amounts, are themselves potentially toxic. In addition, a variety of heavy metals are present, some in readily leachable form, including redox-active and/or potentially toxic metals such as chromium and nickel. All of these additional characteristics could serve to confound the interpretation of toxicological studies. In fact, *in vitro* 24-hour biodurability and cell line toxicology tests of the different standards also produced highly variable results (Ziegler *et al.*, 2002). Although further work is needed (such as longer-term biodurability and *in vivo* toxicology tests on all the standards), our results to date confirm the results of previous studies that mineralogical and geochemical variations between samples of a given asbestos mineral may have important effects on their relative biodurability and toxicity.

Some accessory minerals that accompany the inhaled dose of particles may themselves be reactive (such as pyrite, an iron sulfide) and may be able to modify fluid chemistry sufficiently to enhance or diminish particle solubility, or to release redox-active species such as iron. For example, the well-documented decrease in crystalline silica toxicity when combined with other, nonsilica mineral particles (SSDC, 1988) implies that the other mineral particles are reacting chemically with the body fluids and the silica to modify the surface chemistry of the silica that induces ROS generation and cytotoxicity.

9.07.8.3.5 Particle crystal structure

Jurinski and Rimstidt (2001) concluded that particle dissolution rates are more strongly influenced by crystal structure than by bulk chemistry, citing the greater rates of chrysotile dissolution compared to those of talc *in vitro*. They concluded that chrysotile asbestos fibers dissolve more readily than talc, because their rather unique crystal structure (the fibers are constructed of sheets coiled around the fiber axis) allows material to be readily leached from layers as they unwrap along the entire fiber length, rather than just from the grain edges, as in the case of talc. This may also help to explain the greater dissolution rate of chrysotile asbestos fibers than amphibole asbestos fibers (which form by the separation of coarser crystals along irregular planes parallel to the long crystal axis; Ahn and Buseck (1991)), even though the body fluids are predicted to be more undersaturated with respect to the amphiboles than to chrysotile (Figure 9). Hochella (1993) and Crawford (1980) have also shown that mineral dissolution can occur preferentially along crystal structural defects.

9.07.8.3.6 Particle surface features

Additional features that have been cited as potential contributors to fiber pathogenicity include a variety of surface features such as surface- and near-surface composition; surface area, surface morphology (microtopography) and atomic structure, surface charge (coupled with its dependence on the pH and chemical composition of the surrounding fluids), and relative hydrophobicity versus hydrophilicity (Hochella, 1993; Guthrie, 1997; van Oss *et al.*, 1999; Johnson and Mossman, 2001; Giese and van Oss, 1993). The net surface charge of a particle can be either negative or positive, and may influence chemical interactions with surrounding fluids and tissues; however, van Oss *et al.* (1999) found no systematic correlation between surface charge and particle mineralogy. Although hydrophobic minerals can sorb biopolymers more strongly than hydrophilic minerals, and hydrophobic particles are more readily phagocytized than hydrophilic particles, van Oss *et al.* (1999) found no consistent correlation between particle hydrophobicity, hydrophilicity, and potential pathogenicity.

9.07.8.3.7 Fluid–mineral reactions that generate free radicals

A number of studies have proposed that chemical interactions between mineral particles and body fluids can lead to the production of free radicals and other ROS, which can in turn potentially trigger cell injury such as lipid peroxidation, DNA damage, and ultimately cell mutation and carcinogenesis (Hardy and Aust, 1995; Kamp *et al.*, 1992; Lund and Aust, 1992;

Aust and Lund, 1990; Werner *et al.*, 1995; Graham *et al.*, 1999; Shi *et al.*, 2001).

Much attention has focused on the potential for ferric iron released from asbestos fibers to react with organic reductants in the body fluids and subsequently to generate free radicals and ROS. Werner *et al.* (1995) used *in vitro* leach tests to show that organic chelators can effectively leach ferric iron from crocidolite fiber surfaces over prolonged periods of time. Lund and Aust (1992) used *in vitro* tests with crocidolite to show that increased mobilization of ferric iron from crocidolite surfaces correlated well with increased production of DNA strand breaks. Graham *et al.* (1999) concluded that grinding of iron-containing amosite fibers led to a decrease in the amount of ferric iron that could be leached from the fiber surface, thereby indicating another potential cause for long fibers to be more pathogenic than short fibers. In contrast to these studies, van Oss *et al.* (1999) were not convinced that asbestos iron content is an important factor in asbestos pathogenicity. They cited as an example the high pathogenicity of erionite, which is a sodium-rich zeolite that generally has a very low iron content. However, Hardy and Aust (1995) pointed out that the release of iron sorbed on erionite surfaces can trigger a short-term increase in DNA breaks. Werner *et al.* (1995) cited the low iron content of erionite to underscore the fact that the iron content of fibers is probably only one of several potential triggers for fiber pathogenicity.

Freshly ground particles of crystalline silica can develop surface Si—O bonds and other surface features that, when reacted a with oxygenated water-rich fluid can foster the generation of ROS (Shi *et al.*, 2001). Si—OH groups can form on the surface of silica particles as reactions with water hydrate surface Si—O bonds; the Si—OH groups are also tied to the formation of ROS. With increasing time after grinding, the generation of ROS diminishes. Such an effect may also occur with other silicates such as asbestos.

Mineral particles likely also indirectly trigger production of free radicals as a result of phagocyte activation and cytokine release (Kamp *et al.*, 1992; NIOSH, 2002).

9.07.8.3.8 Smoking—a confounding factor

Smoking is a well-known trigger of respiratory diseases such as emphysema and lung cancer. Workers who smoke and who are exposed to excessive levels of particulates such as silica or asbestos have a combined lung cancer risk that is greater than for smoking or particle exposure alone (Holland and Smith, 2001). As a result, smoking (and exposure to other environmental pollutants) can make it difficult, but not impossible, to interpret epidemiological data for silica- or asbestos-related disease.

9.07.8.3.9 Dose—a key factor

Based on a number of indications from the literature discussed above, it is clear that the dose (intensity and/or duration) of inhalation exposure to insoluble, sparingly soluble, and slowly dissolving particles is a key factor in particle pathogenesis. Particle composition, morphology, biodurability, surface effects, and other characteristics may influence the intensity and the duration threshold above which disease is triggered. For example, equant biodurable particles such as silica can cause disease, but the intensity and duration of exposure required are probably greater than for fibrous particles that are less easily cleared by the macrophages. Among fibrous particles, lower doses are seemingly required for those that are less readily dissolved, have key toxic elements such as iron, and have morphologies that enhance their ability to pierce and migrate into tissues. Clearly, a profitable area of future research will be the continued examination of the link between toxicity and dose *as a function of particle characteristics*.

9.07.9 THE MEDICAL GEOCHEMISTRY OF EARTH MATERIALS WITH READILY SOLUBLE, BIOACCESSIBLE, AND/OR BIOREACTIVE COMPONENTS

Most earth materials to which humans can be exposed are likely to contain a complex mixture of different minerals or materials, which may differ substantially in their relative biosolubility and bioreactivity, as well as the bioaccessibility and chemical form of potential toxins that they contain. In the following discussion, we will discuss various earth materials as examples of the role that mineralogy, chemical composition, and geochemical properties of complex materials can play in their bioaccessibility, biosolubility, bioreactivity, and therefore potential toxicity.

9.07.9.1 Mining Wastes, Tailings, Smelting By-products

A primary human health concern associated with mine wastes, tailings, and smelting by-products produced during the extraction of metals from metallic mineral deposits has been the incidental ingestion exposure, and resulting heavy metal uptake, especially for small children who play on waste piles, tailings, slag heaps, or soils containing or affected by these wastes or smelter emissions (Ruby *et al.*, 1993, 1996). There has also been some recognition for potential metal uptake via the inhalation of windblown particulates derived from mine wastes (Mullins and Norman, 1994). Silicosis has long been recognized as a potential health concern for miners (Daroowalla, 2001; NIOSH, 2002), especially in the past, when dust-control measures were not practiced during mining; however, it is not noted as a potential health concern resulting from environmental exposure to windblown dust from mine and processing wastes.

9.07.9.1.1 Mineralogy

Mine wastes are composed of unmineralized or mineralized, but low-grade, rock material that must be removed in order to mine high-grade mineralization. The material can range in size from coarse boulders to micrometer-size particles. Tailings are composed of mineralized material that has been ground to sand size or smaller, and from which most of the minerals of economic interest (such as gold, or sulfides of copper, lead, zinc, mercury, etc.) have been removed. Sulfide ore smelting results in the formation of waste slag and gaseous emissions containing both sulfur dioxide and particulates with high concentrations of the metals contained in the ores.

Mine wastes, tailings, and smelter slag and emissions can contain complex mixtures of minerals that are typically a predictable function of the geologic characteristics of the deposit type being mined, coupled with the ambient climate/environmental characteristics and the mining or mineral processing method used. See Chapter 9.05 and Plumlee (1999) for a detailed discussion.

Minerals present in mine wastes and tailings include:

- Primary minerals formed in the ore deposit prior to weathering and erosion, including a wide variety of metal sulfides and sulfosalts, metal oxides, metal- and alkaline-earth carbonates, sulfates, crystalline silica, clays, and other silicates. Many metal sulfides (especially iron sulfides such as pyrite), when exposed by erosion or mining to atmospheric oxygen and water, can form acid–rock drainage (ARD).
- Secondary metal oxides, carbonates, sulfates, and phosphates formed by weathering of the ore deposit prior to mining.
- Soluble metal sulfate salts formed by the evaporation of ARD.
- Iron and aluminum oxyhydroxides and hydroxysulfates with sorbed metals that form in ARD and in surface waters affected by ARD.

Smelter slag is typically composed primarily of an amorphous, silica- and calcium-rich material formed by the reactions of iron in the ores with silica and lime used as flux (Ruby *et al.*, 1996; Davis *et al.*, 1992, 1996; Plumlee, 1999). The slag also commonly contains remnant sulfides from ore that was not completely smelted, and a variety of oxides, silicates, chlorides, sulfates, and other minerals containing the metals originally present in the ores. Airborne particulates generated by smelting are a complex mixture of fine-grained (<100 μm) metal oxides, silica-rich phases, and sulfates.

Soils that have been contaminated by mine wastes, tailings, smelter slags, or smelter particulates can contain a complex mixture of: minerals present in the soils prior to contamination; minerals contributed by the contaminants; minerals formed by soil weathering, biological reactions, and chemical reactions with infiltrating waters and soil moisture; windblown dust, and other anthropogenic materials (Ruby *et al.*, 1999). For example, reactions of lead oxide with soil moisture in alkaline soils can precipitate lead carbonate, whereas reactions in acidic soils can precipitate lead sulfate.

Hence, mine wastes, tailings, and smelting by-products can contain a wide variety of minerals, including minerals that are bioreactive (such as acid-generating sulfides and evaporative sulfate salts), minerals that contain potentially bioaccessible heavy metals and metalloids (lead, cadmium, arsenic, mercury, zinc, copper, nickel, uranium, molybdenum, antimony, etc.), and minerals that are biodurable (such as quartz and, in some deposit types, asbestiform silicates).

9.07.9.1.2 Ingestion bioaccessibility

As summarized by Ruby *et al.* (1999), a variety of mineralogical characteristics control the oral bioaccessibility of metals such as lead and arsenic (Figure 10), including: the types and grain sizes of the minerals in which they occur, and their degree of encapsulation by other, less reactive or soluble minerals. In addition, based on observations made for the weathering of mine wastes (Plumlee, 1999), the reactivity (and hence bioaccessibility) of sulfides and other mineral groups can be strongly influenced by their crystal morphology (fibrous, framboidal, botryoidal, massive, blocky), and the concentrations and types of trace elements in their crystal structure.

Results of various *in vitro* and *in vivo* bioaccessibility studies (Ruby *et al.*, 1993, 1996, 1999; Davis *et al.*, 1992, 1996, 1993; Borch *et al.*, 1994; Dieter *et al.*, 1993) indicate that some aspects of the fate of metals or metalloids leached from mine wastes that enter the gastrointestinal system can be readily understood based on their mineralogical characteristics, coupled with a knowledge of how

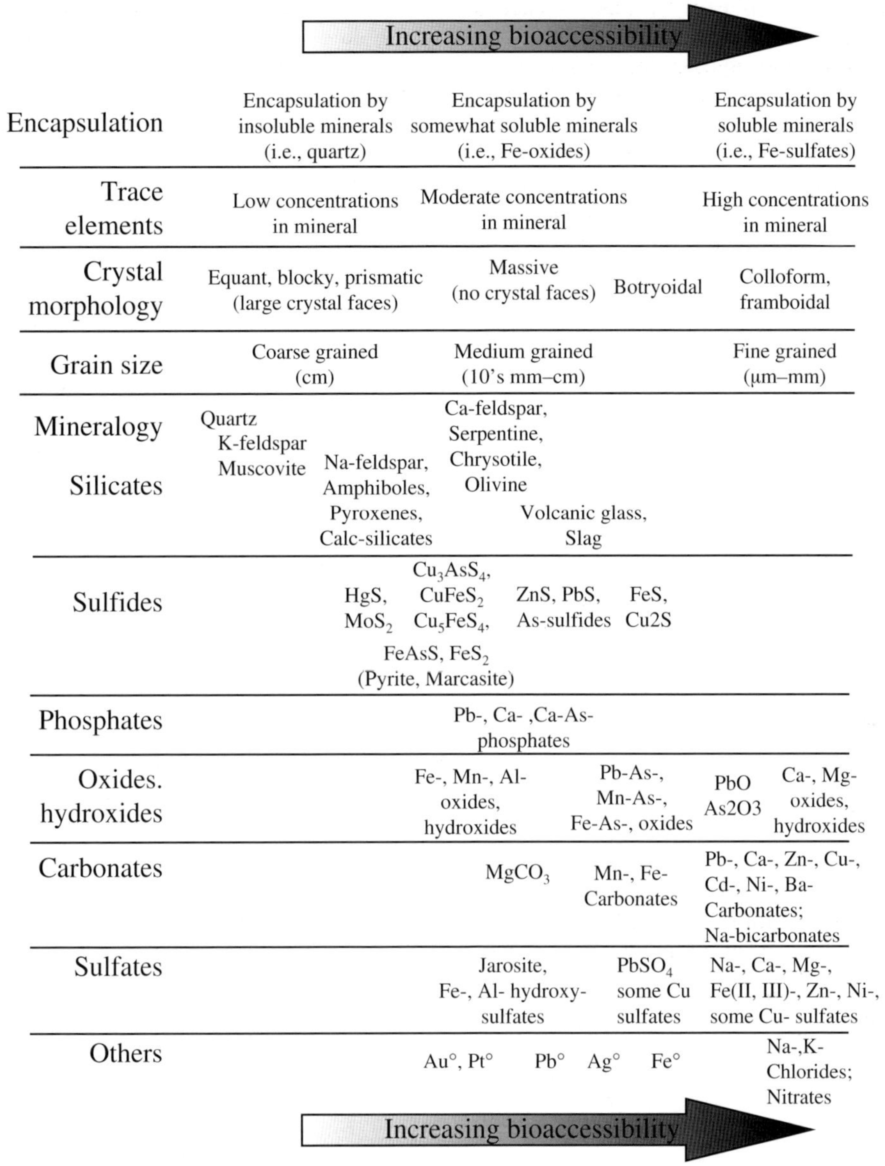

Increasing bioaccessibility →

Encapsulation	Encapsulation by insoluble minerals (i.e., quartz)	Encapsulation by somewhat soluble minerals (i.e., Fe-oxides)	Encapsulation by soluble minerals (i.e., Fe-sulfates)
Trace elements	Low concentrations in mineral	Moderate concentrations in mineral	High concentrations in mineral
Crystal morphology	Equant, blocky, prismatic (large crystal faces)	Massive (no crystal faces); Botryoidal	Colloform, framboidal
Grain size	Coarse grained (cm)	Medium grained (10's mm–cm)	Fine grained (μm–mm)
Mineralogy Silicates	Quartz; K-feldspar; Muscovite; Na-feldspar, Amphiboles, Pyroxenes, Calc-silicates	Ca-feldspar, Serpentine, Chrysotile, Olivine; Volcanic glass, Slag	
Sulfides	HgS, MoS_2	Cu_3AsS_4, $CuFeS_2$, Cu_5FeS_4; FeAsS, FeS_2 (Pyrite, Marcasite); ZnS, PbS, As-sulfides; FeS, Cu2S	
Phosphates		Pb-, Ca- ,Ca-As-phosphates	
Oxides. hydroxides		Fe-, Mn-, Al- oxides, hydroxides; Pb-As-, Mn-As-, Fe-As-, oxides	PbO As2O3; Ca-, Mg- oxides, hydroxides
Carbonates		$MgCO_3$; Mn-, Fe- Carbonates	Pb-, Ca-, Zn-, Cu-, Cd-, Ni-, Ba- Carbonates; Na-bicarbonates
Sulfates		Jarosite, Fe-, Al- hydroxy-sulfates; $PbSO_4$ some Cu sulfates	Na-, Ca-, Mg-, Fe(II, III)-, Zn-, Ni-, some Cu- sulfates
Others		Au°, Pt°; Pb°; Ag°; Fe°	Na-,K- Chlorides; Nitrates

Increasing bioaccessibility →

Figure 10 Influence of mineral type, crystal morphology, grain size, degree of encapsulation, and trace element content on bioaccessibility. Differences in bioaccessibility between different mineral types should be considered as qualitative, and may vary depending upon the chemistry of the surrounding fluids. For example, sulfides are all substantially much more bioaccessible under oxidizing conditions than reducing conditions. Similarly, carbonates are much more bioaccessible under acidic than alkaline conditions (sources Ruby *et al.*, 1999; Plumlee, 1999; Nordstrom and Alpers, 1999).

the metals or metalloids might respond (Smith and Huyck, 1999) to the redox, pH, and fluid compositions of the stomach and intestines. Although sulfides are likely to be highly undersaturated in the gastric fluids, particles broken from coarse sulfides should not dissolve substantially during the hour-scale residence in the stomach. Similarly, acid-stable, relatively insoluble sulfates and phosphates (such as jarosite, alunite, plumbojarosite, and others) should not dissolve appreciably while in the stomach. The most soluble, reactive carbonates may dissolve partly to completely, as may some fine-grained metal oxides. The fine-grained sulfides that weather most rapidly (especially those that have high trace element contents), as well as readily soluble metal sulfate and metal chloride salts, have the highest potential for the release of their contained metals into the stomach fluids. In the acidic conditions of the stomach, most base metals such as lead, iron, copper, zinc, cadmium, nickel, and cobalt are likely to be complexed primarily by chloride, less so by organic acids due to their degree of protonation. Elements such as arsenic, antimony, and molybdenum are likely to be present as oxyanion species, and so their concentration may not be enhanced by the complexing agents in the stomach. The redox response of species having varying redox state is uncertain, but may be quite variable in the stomach depending on the redox kinetics of the species, along with the redox kinetics of the multiple redox couples (such as dissolved oxygen or various organic couples) potentially active in the stomach fluids. For example, it is likely that aqueous ferrous iron liberated by pyrite oxidation will not be oxidized to ferric iron in the stomach by reaction with dissolved oxygen (if any has been introduced by swallowing); however, it is unclear to what extent ferric iron and other oxidized species (such as sulfate, arsenate, etc.) that are released from soluble salts may be reduced by reaction with organic compounds in the stomach fluids.

Once they have reached higher pH, reducing conditions of the intestinal tract (Davis *et al.*, 1992), sulfides should be more stable, and may actually precipitate if reduced sulfur is present. Other solids, such as hydroxides or hydroxysulfates of aluminum, and possibly iron, may also precipitate. The increased pH should also lead to the increased sorption onto particulates of various metals and metalloids such as lead and copper (Smith, 1999). However, *in vitro* tests (Ruby *et al.*, 1993) indicate that the increased complexing with unprotonated organic acids and enzymes helps offset the pH-driven precipitation and sorption of the base metals that were dominantly chloride-complexed in the stomach fluids. Arsenic and other oxyanionic species are likely to be sorbed as the stomach acids are neutralized, but may be partially desorbed once higher pH values are reached in the intestine (Ruby *et al.*, 1996).

9.07.9.1.3 Respiratory bioaccessibility

Only limited work has been done on the bioaccessibility of metals in windborne mine waste and tailings material, and so much must be inferred. Mullins and Norman (1994) analyzed the size distribution, metal content, and metal extraction by simulated biofluids (lung, gastric, intestinal) of surface materials (soils) collected from several mine waste piles in the Butte, Montana, district. They found that the concentrations of arsenic, cadmium, copper, manganese, and lead were commonly greatest in the smallest size fractions (<4.7 μm) of the waste dump material. The percentage of metals leached from the fine fraction was quite variable, but not in any consistent way, between different metals, different dumps, and different extraction fluids. The variability between metals extracted, dumps, and fluids results primarily from mineralogical variability within and between the dumps; however, the mineralogical characteristics of the dump materials were apparently not determined.

As with ingestion exposures, the geochemical aspects of inhalation exposures to mine waste and tailings particles are likely to be influenced strongly by the mineralogical characteristics in Figure 10. However, many aspects of the particle dissolution reactions and redox reactions are still quite speculative. Because the lung and macrophage fluids are likely to have Eh values favoring aqueous sulfate over sulfide (Figure 5), it is quite possible that particles broken from coarse, well-crystallized sulfides react and dissolve relatively slowly in the lung fluids and the macrophage lysosomal fluids, and may be somewhat biodurable. They might also serve as a long-term source of acid and metals that slowly leach into the surrounding fluids and tissues, possibly triggering inflammation or, in the case of variable oxidation state elements such as iron or arsenic, trigger production of free radicals and DNA breakage. Particles of fine-grained or trace-element-rich sulfides might be expected to dissolve more rapidly, possibly producing higher levels of acid and metals in their surroundings over a shorter period of time following exposure. The soluble metal sulfate salts probably dissolve quite quickly in the lungs, producing high levels of acid and metals over a relatively short period following exposure, as has been shown for coal dusts containing abundant pyrite and oxidation products such as ferrous sulfate (Huang *et al.*, 1993).

It is possible that a variety of secondary phases may precipitate in the lungs or in the macrophages, or may form reaction rinds on sulfide particles, as the result of sulfide oxidation or soluble salt dissolution *in vivo*. For example, phosphate (present in high concentrations in both the lung and macrophage fluids) may combine with calcium, aluminum, iron, lead, or other metals released from the salts to precipitate a variety of less soluble phosphate phases. Other secondary phases might include hydroxides or hydroxysulfates of aluminum or iron, or sulfates such as gypsum.

Because it is unclear which of many possible redox couples are active in the lung or macrophage fluids, it also is unclear what might happen to redox-sensitive species such as iron, arsenic, or manganese released from the sulfides and salts into the lungs or alveolar macrophages. Although dissolved oxygen concentrations in the lung fluids are likely high, and so might tend to shift reduced species to their oxidized forms, it is unclear whether organic ligands in the lung fluids could either inhibit this oxidation, and/or possibly shift oxidized species toward their reduced forms.

9.07.9.2 Coal and Coal Fly Ash

Coal is the collective term that refers to organic-rich sedimentary rock materials formed by the fossilization of plant matter. Coal can have different ranks, depending on the temperature and pressure to which the organic matter has been subjected following deposition. Anthracite has the highest rank and the highest carbon content (95%); progressively lower ranks of coal have a progressively lower carbon content. Coal can also have variable amounts of accessory materials interspersed with the organic components, including a variety of clays, carbonate minerals, crystalline silica or other silicate minerals (collectively termed ash), and iron sulfides such as pyrite and marcasite. In addition, coals can contain very high levels of a variety of potentially toxic metals and metalloids such as mercury, cadmium, and arsenic.

The primary health problems tied to coal are coal workers' pneumoconiosis (CWP, also called black lung disease), progressive massive fibrosis, silicosis, chronic bronchitis, and emphysema, which result from the prolonged inhalation exposure to coal dusts generated during mining or coal processing (Daroowalla, 2001; Castranova, 2000; Castranova and Vallyathan, 2000). These health effects have traditionally been thought to result from the biodurability of the organic-rich coal dusts (and accessory silica dusts) in the lungs, the resulting cytotoxicity of the particles, and the activation of macrophages with their production of oxidants and ROS (Castranova, 2000). Huang *et al.* (1998) proposed that the potential of a particular coal to generate CWP increases with increasing content of acid-soluble ferrous iron and with the decreasing acid-buffering capacity of coal (presumably reflecting decreasing carbonate mineral content). They proposed that the release of acid soluble iron is enhanced in coals with lower acid-buffering capacity, and that the release of ferrous iron and its subsequent oxidation in the pulmonary environment leads to oxidative stress, formation of free radicals, and resulting lung injury. In an earlier paper, Huang *et al.* (1993) proposed that ferrous iron is released during the solution of ferrous sulfates (such as melanterite) that formed as an oxidation product of pyrite under acidic conditions prior to inhalation; under alkaline conditions (e.g., if carbonates are present), pyrite oxidation would lead to the rapid oxidation of ferrous to ferric iron and preclude the buildup of ferrous sulfate on the oxidizing pyrite, thereby diminishing the potential for ferrous iron release in the pulmonary environment. It is possible, however, that the oxidation of sulfide particles themselves in the lungs could generate acid and release ferrous iron (and other potentially toxic elements) into the surrounding fluids and tissues.

Coal fly ash (CFA) is the particulate material, typically characterized by the presence of many silt- to clay-sized particles, which is produced by the combustion of ground coal (Jones, 1995). The particles are composed of quartz, mullite (an aluminum silicate), iron oxides, aluminosilicate glasses, and gypsum. SO_2 and a wide variety of trace elements can be sorbed on the particle surfaces. Depending on the proportions of carbonates to iron sulfides in the coal prior to combustion, CFA can generate highly alkaline (high carbonate/sulfide coal) to somewhat acidic (low carbonate/sulfide coal) water leach solutions. The types and concentrations of metals and metalloids released from CFA into water leach solutions depend on their initial concentration in the coal and on the alkalinity or acidity of the CFA. In general, metals such as cadmium, copper, manganese, iron, nickel, lead, and zinc are released in greatest quantities from acidic CFA, whereas elements that form oxyanionic species such as arsenic, boron, molybdenum, selenium, antimony, and vanadium are released in greatest quantities from alkaline CFA.

A variety of *in vitro* and *in vivo* toxicological studies have examined the potential health effects of CFA. Most of these studies have concluded that iron release from CFA can generate free radicals, and therefore can trigger DNA damage and toxicity (Smith *et al.*, 1998, 2000; Veranth *et al.*, 2000; van Maanen *et al.*, 1999; Chen *et al.*, 1990). Hence, the deleterious effects from inhalation exposure to CFA may be linked to its content of leachable iron, its alkali content, and the amounts of soluble salts that can dissolve to generate acid and thereby enhance iron release. However, the chemical reactions between CFA and lung and macrophage fluids, and the potential health effects of leachable metals and metalloids other than iron must be studied further.

9.07.9.3 Volcanic Ash, Gases, and Vog

Volcanic ash (VA) is composed of a mixture of pumice, glass shards, rock fragments, and variable proportions of crystals or crystal fragments of various silicate minerals such as (depending upon the composition of the magma being erupted) pyroxenes, feldspars, quartz, and cristobalite. In addition, small amounts of minerals such as clays may be present due to alteration of the silicates by acidic aerosols and gases. VA that is freshly erupted can have a wide variety of trace metals that are present in soluble halide salts or that are loosely sorbed onto particle surfaces; these metals and salts result from interactions with air of ash particles and the gaseous components (including acidic gases) of the volcanic plume (Smith *et al.*, 1982, 1983).

Health concerns regarding VA have focused primarily on respiratory effects in heavily exposed populations such as short-term respiratory irritation and longer-term development of pneumoconiosis. The potential toxicity of crystalline silica in the ash has been of particular concern (Baxter *et al.*, 1999; Wilson *et al.*, 2000; CDC, 1986; Vallyathan *et al.*, 1983a,b; Baxter *et al.*, 1983). These studies indicate that the potential toxicity of VA can vary between different ash eruptions from a given volcano and between different volcanoes. The variability is probably due to difference in the crystalline silica content and in the proportion of respirable particles.

Relatively little attention has focused on the potential health effects of metal release from respired ash. Chemical leach tests using water, dilute HCl, and carbonate–bicarbonate solutions of ash from Mount St. Helens (Smith *et al.*, 1983; Hinkley, 1987) and other active volcanoes (Smith *et al.*, 1982) have shown that a wide variety of cations, metals, and anions are readily leached from fresh ash, including, depending upon the volcano: Ca, Cl, SiO_2, SO_4, Mg, Na, Fe, Mn, in 1–100 mg L^{-1} concentrations (1 : 4 solid : liquid by weight); and Zn, Cd, and Pb in $\mu g\ L^{-1}$ concentrations. It is unclear whether sufficiently high concentrations of metals such as iron and manganese could be released from ash *in vivo* over either the short term or the long term to trigger free radical generation and DNA damage. Recent studies of ash from the Soufrière Hills volcano have found high surface reactivity and high levels of hydroxy radicals that were attributed to release of ferrous iron from the ash (Horwell *et al.*, in press).

Volcanic smog (known as vog) is a mixture of atmospheric gases and suspended liquid and solid particles. It forms by the reaction of sulfur dioxide and other volcanic gases with atmospheric moisture, gases, dust, and sunlight (Sutton *et al.*, 1997). Vog consists primarily of sulfuric acid and other sulfate compounds, and can contain a variety of heavy metals, including selenium, mercury, and arsenic (Sutton *et al.*, 1997). Laze, a volcanic haze, forms when molten lava flows into the sea and vaporizes seawater (Sutton *et al.*, 1997). It has many of the same characteristics as vog, with the exception that it probably contains higher levels of chloride and hydrochloric acid derived from seawater.

Adverse health effects that have been noted as the result of vog produced by active Hawaiian volcanoes include acid-triggered irritation of the mucous membranes (eyes, nose, and throat), increased asthma, respiratory distress, increased susceptibility to respiratory ailments, headaches, watery eyes, and lack of energy (Sutton *et al.*, 1997). These problems increased on the island of Hawaii during the eruption cycle of Kilauea volcano that began in 1986. Increased lead uptake by local residents has also been noted. This is thought to originate from water collected on roofs for domestic consumption, because vog-generated acid rain leaches lead from metal roofing, flashing, etc.

The respiratory health effects of vog and volcanic gases such as sulfur dioxide are tied to the generation of locally acidic environments in the lung and respiratory tract fluids by condensation of SO_2 and other acid gases, uptake of acid-sulfate aerosol droplets, and the dissolution of acid-bearing, sulfate- or chloride-rich salts from the vog particulates by the fluids lining the respiratory tract.

9.07.9.4 Dust from Owens Lake, California, and Other Dry Lake Beds

In its natural condition, Owens Lake was the terminal lake of the Owens River, which drains much of the eastern slope of the Sierra Nevada in central California. However, in the early 1900s, water withdrawal for municipal consumption in Los Angeles caused the lake to dry up, leaving behind a dry lake bed whose sediments have become the biggest single point source of dust in the United States (Raloff, 2001); during the passage of storm systems, the 24 h average concentrations of PM10 dust particles (those less than 10 μm in diameter) can exceed 150 $\mu g\ m^{-3}$, more than 10 times the maximum amount allowed under Federal air quality regulations. More recent studies indicate that Owens Lake dust is transported as far as 400 km to the east (Reheis *et al.*, 2002).

The Owens Lake dusts are derived from lake bed sediments containing abundant alkaline evaporative salts of sodium, chloride, carbonate/bicarbonate, and sulfate, including, e.g., halite, natron, thermonatrite, mirabilitie, and trona (Saint Amand *et al.*, 1986). In addition, the dusts contain a variety of silicates and other minerals derived from local alluvial material, and possibly some mine waste materials from the Cerro Gordo lead–zinc–silver mining district on the east end of Owens Lake.

Chemical analysis (Reheis *et al.*, 2002) of the <50 μm fraction of the dusts that originate from the dry Owens Lake bed indicate that they contain quite high levels of a variety of metals or metalloids including: iron (several percent); zinc (tens of thousands of ppm), manganese and lead (hundreds of ppm); arsenic, chromium, nickel, and lithium (tens of ppm), and uranium (several ppm). Given the oxidized conditions of the playa lake surface, it is likely that arsenic, chromium, and other redox-sensitive elements are present in their oxidized forms.

The Owens Lake dusts have been recognized as a potential human health hazard. Remediation efforts have recently been implemented to help mitigate dust generation from the dry lake beds. However, we are not aware of any systematic epidemiological or exposure studies that have been carried out to assess or document the health effects on people living near Owens Lake.

Chemical leach tests of the <50 μm size fraction of dust samples collected around Owens Lake, using water (Reheis *et al.*, 2001, and our unpublished data) and SLFs (our unpublished data), show that the dusts are sufficiently alkaline and reactive to shift the pH of water and SLF to values near 10.5 and 9.5, respectively. Arsenic, chromium, vanadium, molybdenum, lithium, zinc, and other trace metals or metalloids are readily solubilized from the dusts. The trace metals or metalloids leached in the greatest quantities are those that form oxyanion species or abundant carbonate complexes in solution, and that are therefore mobilized most effectively under the alkaline conditions generated by the alkaline dusts.

These results suggest that evaporative playa lake sediments and dusts generated from dry lake beds can be potentially significant sources of reactive, alkaline material with high levels of soluble, potentially toxic trace metals and metalloids. Further studies are needed to determine whether such dusts pose a substantial health hazard to those exposed to them on a regular basis.

9.07.9.5 Soils

Soils can exhibit a complex range of physical, mineralogical, and chemical characteristics that depend on many interrelated factors such as parental rock composition and mineralogy, climate, topography, vegetation amounts and types, water infiltration versus runoff, soil moisture, organic matter, amounts and types of anthropogenic contaminants, and many others. As a result, the bioreactivity, bioaccessibility, and biodurability of bulk soils can vary widely. Potential role of soils in human health has been discussed extensively (e.g., Oliver, 1997; Selinus, in press; see references in Appleton *et al.*, 1996; Ruby *et al.*, 1999). Two examples here illustrate the geochemical interactions between soils and body fluids.

9.07.9.5.1 Soils and neurodegenerative diseases

Although aluminum is typically poorly absorbed via either inhalation or ingestion routes (Goyer and Clarkson, 2001), uptake of bioaccessible aluminum has been speculated to be one of several possible causes for some degenerative neurological diseases such as Alzheimer's disease, Parkinsonism Dementia Complex (PDC), and amyotrophic lateral sclerosis (known as Lou Gehrig's disease, or ALS) (Goyer and Clarkson, 2001).

Epidemiological clusters of PDC–ALS on Guam were first recognized in the early 1940s. Epidemiological studies there have essentially ruled out a genetic cause for the diseases (Perl, 1997), and indicate a link to environmental factors. Incidence of PDC–ALS on Guam has diminished since the 1940s (Garruto *et al.*, 1985; Perl, 1997). The diseases are rare on the northern portions of the island that are underlain by limestones. They are much more prevalent on the southern volcanic portions of the island. A potential link to bedrock, soil, and water geochemistry has been proposed as a result of greater acid weathering on the volcanic-derived soils than on the more alkaline, limestone-derived soils, leading to a decrease in available calcium and magnesium in soils and waters, and an increase in available aluminum via either inhalation or ingestion uptake of soil particles (Garruto *et al.*, 1985, 1989; Crapper MacLachlan *et al.*, 1989).

Miller and Sanzolone (2002) subjected samples of both volcanic- and limestone-derived soils from Guam to detailed bulk chemical analysis and several types of chemical leach tests, including a leach test using SLFs. The leach tests indicated that, contrary to results of earlier studies, little aluminum is readily leached from the soils, but that calcium is readily leached. Based on these results, bioaccessible aluminum from soils does not appear to be a viable etiologic agent for the elevated PDC and ALS occurrences on Guam. However, manganese (a known neurotoxin), silica, barium, cobalt, nickel, uranium, and vanadium can be leached, depending upon the particular soil sample, in substantial quantities; these elements may warrant further investigation for their potential health effects.

9.07.9.5.2 Soil-borne pathogens

As summarized by Bultman *et al.* (in press), a number of soil-borne pathogens have been linked to a variety of diseases, including protozoa, bacteria, fungi, viruses, and prions. Pathogens considered to be soil-borne include those that complete all or some of their life cycles in soils, and those whose life cycle is spent primarily in other organisms but that can survive for some period of time if released into soils from their hosts. Depending upon the particular pathogen, exposure can result from inhalation of dusts generated from soil, direct or incidental (i.e., on foodstuffs) ingestion of soil, dermal contact with soil, and/or ingestion of water containing pathogens washed from soil. A few examples of soil-borne pathogens and their associated diseases include (Bultman *et al.*, in press; Griffin *et al.*, 2002): *Coccidioides immitis*, a soil fungus that causes Valley Fever, or coccidioidomycosis; *Aspergillus* spp., various species of a fungus found in soils, dusts, and waters that cause aspergillosis and increased asthma; *Bacillus anthracis*, a well known soil bacteria, which causes anthrax; *Hantavirus* spp., various species of viruses that cause Hantavirus pulmonary syndrome

As described by Bultman *et al.* (in press), the viability of soil microbes while in the soil can be significantly affected by a variety of soil characteristics, including: grain size; pH; water content; presence or absence of abundant organic matter, clays, or soluble salts. For example, *C. Immitis*, the etiological agent for valley fever, is endemic in much of the arid US southwest and in parts of South America, in hot, seasonally dry climates with short winters. It seems to be most abundant, and hence best competes against other microbes, in soils that: are alkaline; have abundant pore spaces; have <10% clay-sized material; have low amounts of organic matter; occur in seasonally wet/dry, hot climates; have elevated salinity, with abundant soluble evaporative salts; and, may have borate salts, which may act as antibiotics for bacteria that compete with the fungus.

It is interesting to speculate whether soil minerals that help enhance the viability of a pathogen in the soil can also help to enhance the pathogen's ability to infect a human. For example, alkaline soluble salts that favor *C. Immitis* in soils might, if inhaled, also be reactive enough to make the lung fluids more alkaline, thereby providing a more hospitable environment in which the spores could take hold to trigger valley fever. In addition, it is possible that elements or metals in the salts, such as boron, might also be toxic to the macrophages, thereby diminishing the body's ability to clear the spores.

9.07.9.6 Dusts Generated by the World Trade Center Collapse

The tragic attacks on and collapse of the World Trade Center (WTC) towers in New York City on September 11, 2001, generated a massive dust cloud that enveloped much of lower Manhattan. The dust cloud left behind deposits of dust, debris, and paper up to many inches thick, both outdoors and indoors in rooms where windows were open at the time of the collapse or (in the case of buildings close to Ground Zero) that were blown open by the force of the collapse.

Concerns immediately developed regarding the potential health risks associated with exposure to the dusts generated by the initial collapse and subsequent months of cleanup, as well as smoke from fires that smoldered within the debris at Ground Zero for some weeks after the attacks. A variety of health problems and toxicological effects have been documented to date (Stephenson, 2002; Gavett *et al.*, 2002; Prezant *et al.*, 2002; Scanlon, 2002; Rom *et al.*, 2002; Landrigan *et al.*, 2002; Landrigan, 2001). These include:

- short-term effects of dust exposure, such as intense burning of the eyes, mouth, and upper respiratory tract, nasal congestion, and gagging reflux of incidentally swallowed dust;
- sinusitis, laryngitis;
- development of the "WTC Cough," a persistent cough developed after prolonged exposure;
- other respiratory effects, such as shortness of breath, chronic chemically induced bronchitis, new-onset asthma, exacerbation of existing asthma;
- chemical pneumonitis (rare, three cases);
- acute eosinophilic pneumonia (rare);
- corrosive damage to and irritation of respiratory tract; and
- gastroesophageal reflux disease, resulting from corrosive damage to the gastrointestinal tract.

A number of excellent studies have been carried out to characterize the materials and chemical composition of settled dust deposits and airborne dust, smoke, and other aerosols generated by the WTC collapse (Lioy *et al.*, 2002; Lioy, 2002; Thurston *et al.*, 2002; Millette *et al.*, 2002; Chatfield and Kominsky, 2001; Clark *et al.*, 2001; USGS, 2002; Plumlee *et al.*, 2002). The following discussion is taken largely from results of USGS studies of settled dust deposits (Clark *et al.*, 2001; USGS, 2002; Plumlee *et al.*, 2002; Plumlee and Ziegler, unpubished data). Results of other studies are also cited where they provide information not obtained in the USGS study.

The settled dusts are quite heterogeneous in materials makeup, particle size, and chemical composition, both from sample to sample across lower Manhattan, and within a given sample down to the micrometer scale. The dusts are composed of particles of a wide variety of materials used in building construction and that are found within office buildings, including: glass fibers (mineral wool or slag wool used in ceiling tiles and insulation), gypsum (from wallboard), concrete, glass shards, paper, rock-forming minerals such as quartz (from aggregate in concrete, dimension stone and other sources), iron-rich particles (from steel beams and other sources), zinc-rich particles (presumably from metal ductwork), lead-rich particles (solder, lead oxide from paints), and others.

Chrysotile asbestos was found in most settled dust samples at levels generally less than 1%, but in levels as high as 20% by volume in material coating a steel beam in the debris at Ground Zero. Low levels of amphibole asbestos fibers were identified only in one settled dust sample collected north of the WTC complex (Chatfield and Kominsky, 2001).

Lioy *et al.* (2002) and Thurston *et al.* (2002) found that the dominant particle size of the settled dust samples is in the 2.5 μm to tens of μm range. These particles, if inhaled, would tend to be trapped in the uppermost portions of the respiratory tract. They found little asbestos in the <2.5 μm fraction of settled dust samples.

The major-element compositions (silicon, sulfur, magnesium, aluminum, iron, carbon) of the dusts represent the contributions of glass fibers, concrete, gypsum wallboard, metals, paper, and other materials within the office buildings (Clark *et al.*, 2001). Trace-element compositions enriched in a variety of metals (zinc, barium, lead, copper, chromium, molybdenum, antimony, titanium) reflect contributions from paints, lighting, electrical wires, pipes, computer equipment, electronics, and other diverse materials. Many of these metals are substantially enriched in the settled dust samples compared to soils of the eastern United States (USGS, 2002), and some, such as lead, are in some samples in excess of recommended residential soil standards set by some states (Sittig, 1994). Lioy *et al.* (2002) found a wide variety of organic chemicals in bulk settled dust samples (not separated by size), including polycyclic aromatic hydrocarbons, polychlorinated biphenyls, and many others.

Chemical leach tests on the bulk settled dust samples showed that the dusts are quite chemically reactive. Leach solutions have high alkalinities, due to the rapid partial dissolution of calcium hydroxide from concrete particles. Indoor dust samples produced higher pH levels (11.8–12.4) and alkalinities (~600 mg L^{-1} $CaCO_3$) than outdoor dusts (pH 8.2–10.4; alkalinity ~30 mg L^{-1} $CaCO_3$), indicating that outdoor dust samples had reacted with rainfall or other water prior to collection. Thurston *et al.* (2002) found that the leachate pH of the dusts decreased with decreasing particle size. Some metals or metalloids in the dusts (aluminum, chromium, antimony, molybdenum, barium, copper, zinc, cobalt, nickel) are readily leached by deionized water; many of these form oxyanion species or carbonate complexes that are most mobile at the alkaline pH's generated by the leachates.

Our recent chemical leach tests on settled dust samples using deionized water and SLF as the extracting fluids provide further insights into the potential chemical behavior of the dusts *in vivo* (Figures 11 and 12). The SLF produce smaller pH shifts than deionized water due to the buffering capacity of the SLF components (Figure 11). The concentration of phosphate in the SLF drops substantially, probably due to reactions with calcium released from the concrete particles and the resulting precipitation of insoluble calcium phosphate minerals. Metals such as copper and zinc are even more soluble in SLF than in water due to complexing with chloride, citrate, and glycine (Figure 11). The tests do not indicate that lead can be substantially dissolved from the dusts by either water or SLF, possibly because the lead occurs in relatively insoluble phases in the dusts and/or forms insoluble phosphate precipitates. Chemical speciation calculations of the leachate fluids produced by the SLF (Figure 12) indicate that the extraction fluids are highly supersaturated with a wide variety of silicates, including chrysotile and amphiboles.

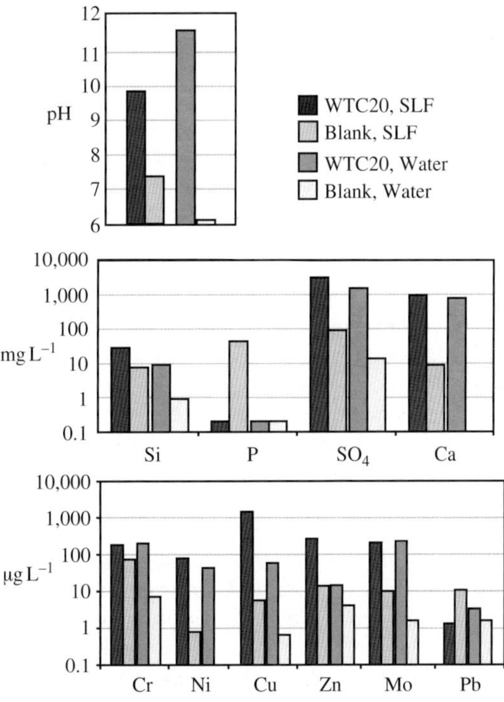

Figure 11 Plots comparing results of water and SLF leach tests performed on settled dusts generated by the WTC collapse. One part dust is added to 20 parts water or SLF at 37 °C and mixed for 24 h, with the leachate filtered (<0.45 μm) and analyzed. The composition of the SLF used in the extraction is a variation on the recipe provided by Bauer *et al.* (1997): pH 7.4; Na—150.7 mM; Ca—0.197 mM; NH_4—10 mM; Cl—126.4 mM; SO_4—0.5 mM; HCO_3—27 mM; HPO_4—1.2 mM; Glycine—5.99 mM; Citrate—0.2 mM. Other metals shown in the SLF blank were contributed as trace constituents of the various chemical reagents used to make up the fluids.

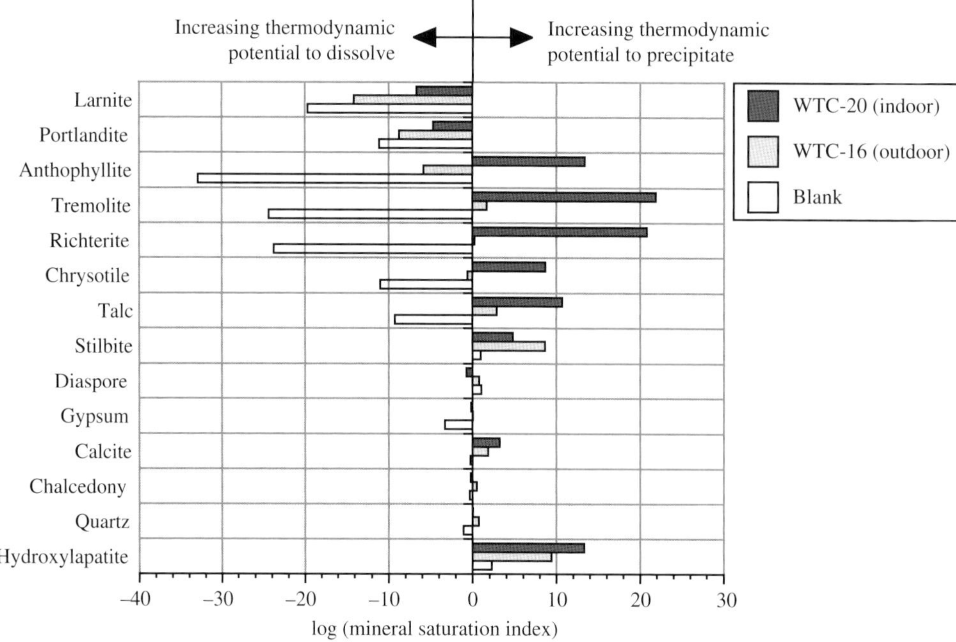

Figure 12 Plot comparing calculated saturation indices of various minerals of interest in SLF blank, and filtered (<0.45 μm) SLF leachates of an indoor WTC dust sample (WTC-20) and outdoor WTC dust sample (WTC-16). The extreme supersaturations in the WTC dust leachate samples may result in part (but not entirely) from inclusion of some colloidal material less than 0.45 μm in size in the analyzed filtrate.

In summary, the settled dusts generated by the WTC collapse are a complex and heterogeneous mixture of bioreactive, bioaccessible, and biodurable particles. Results of chemical leach tests using water and SLF indicate that a variety of chemical reactions may have occurred as inhaled dusts encountered fluids lining the respiratory tract. Although indications are that most particles were too coarse to reach the alveoli, highly alkaline concrete particles likely triggered the generation of caustic alkalinity where they came into contact with body fluids; this alkalinity is thought to have been the major cause of the chemically induced irritation of the eyes, mouth, throat, upper respiratory tract, and gastrointestinal tract (Stephenson, 2002). A variety of metals and metalloids, especially those mobile under relatively alkaline conditions, were probably also released from the dusts; it is unclear what, if any, toxicity such metal releases may have caused. Chemical reactions between the body fluids and the dust constituents may have triggered precipitation of secondary phases in the respiratory tract such as calcium and lead phosphates. Finally, the presence of readily soluble, alkaline material in the dusts may have enhanced the chemical stability of biodurable particles such as asbestos fibers in the respiratory system; however, it is unclear whether this enhanced stability would persist over the long term as the more soluble components of the dusts are removed by interaction with successive aliquots of lung fluids.

Perhaps most importantly, these results show that the potential health effects of complex earth materials cannot be assessed based solely on the toxicity of their together individual components—the integrated effects of the whole material must be considered, along with potentially complex chemical interactions between individual components and the body's fluids.

9.07.10 SUMMARY

In this chapter, we have provided an overview of the myriad potential geochemical and biochemical processes, coupled with their potential links to toxic responses, which can occur when earth materials come into contact with body fluids via inhalation, ingestion, or percutaneous exposure routes.

It is possible to group the health effects resulting from exposures to earth materials according to similarities in how the health effects originate:

- effects that result primarily when bioreactive earth materials substantially modify the chemical composition of body fluids and tissues (such as alkali or acid burns);
- effects that result from the solubilization of bioaccessible toxins from earth materials and the subsequent absorption, metabolism, and effects of the toxins in the body;

- effects that result from the exposure to earth materials that are insoluble (biodurable) in body fluids, and that trigger toxic responses as the body attempts to clear the materials;
- effects that result from pathogens associated with earth materials; and
- effects that result from the body's immune response to the earth materials or toxins contained within the earth materials.

Important unifying threads between most, if not all, of these different these types of health effects are that they:

- can be influenced strongly by the forms in which potentially toxic earth materials occur as they are delivered to the body (mineralogy; particle size and morphology; particle solubility, alkalinity, acidity; oxidation state of contained constituents; etc.) and
- ultimately require some level of chemical interactions between the earth materials and body fluids.

Both of these unifying threads implicitly require important roles for the earth scientist in helping to both characterize earth materials and understand geochemical processes in the context of the human body. In spite of the many *in vitro* and *in vivo* studies that have been carried out to address this complex but fascinating topic, there are many unresolved questions remaining, and therefore abundant opportunities for fruitful future collaborations between geochemists and their colleagues in the toxicology, chemical physiology, epidemiology, and other medical fields.

ACKNOWLEDGMENTS

The authors would like to acknowledge very helpful review comments from Blair Jones, Jim Crock, Barbara Sherwood-Lollar, and Dick Holland, as well as the comments made by a number of other scientists who informally reviewed portions of this manuscript.

REFERENCES

Ahn J. H. and Buseck P. R. (1991) Microstructures and fiber-formation mechanisms of crocidolite asbestos. *Am. Min.* **76**, 1467–1478.

Alpers C. N. and Nordstrom D. K. (1999) Geochemical modeling of water–rock interactions in mining environments. In *The Environmental Geochemistry of Mineral Deposits: Part A. Processes, Techniques and Health Issues*. (eds. G. S. Plumlee and M. J. Logsdon). *Soc. Econ. Geol. Rev. Econ. Geol.* **6A**, Soceity of Economic Geologists, Littleton, pp. 289–323.

Altman P. L. (1961) *Blood and Other Body Fluids*. (ed. D. S Ditmer). Committee on Biological Handbooks, Washington Federation of American Societies for Experimental Biology, Washington, DC, 540pp.

Appleton J. D., Fuge R., and McCall G. J. H. (eds.) (1996) *Environmental Geochemistry and Health, with Special Reference to Developing Countries*. Geological Society Special Publication No. 113. The Geological Society, London, 264pp.

ATSDR *Toxicological Profiles* ((1990)—*Radon*; (1994)—*Zinc*; (1997)—*Nickel*; (1999a)—*Mercury*; (1999b)—*Lead*; (1999c)—*Cadmium*; (1999d)—*Aluminum*; (2000a)—*Arsenic*; (2000b)—*Chromium*) US Department of Health and Human Services, Public Health Service, Agency for Toxic Substances and Disease Registry. Available from: http://www.atsdr.cdc.gov/toxpro2.html

Aust A. E. and Lund L. G. (1990) The role of iron in asbestos-catalyzed damage to lipids and DNA. *Biol. Oxidat. Syst.* **2**, 597–605.

Battelle and Exponent (2000) Guide for incorporating bioavailability adjustments into human health and ecological risk assessments at US Navy and Marine Corps facilities: Part 2. Technical background document for assessing metals bioavailability. Naval Facilities Engineering Command User's Guide UG-2041-ENV. Available from http://enviro.nfesc.navy.mil/erb/erb_a/support/wrk_grp/bio_a/bioa_guide_final2.pdf

Bauer J., Mattson S. M., and Eastes W. (1997) *In-vitro* acellular method for determining fiber durability in simulated lung fluid. Available from http://my.ohio.voyager.net/~eastes/fibsci/papers.html

Baxter P. J., Ing R., and Falk H. (1983) Mount St. Helens eruptions: the acute respiratory effects of volcanic ash in a North American community. *Arch. Environ. Health* **38**, 138–143.

Baxter P. J., Bonadonna C., Dupree R., Hards V. L., Kohn S. C., Murphy M. D., Nichols A., Nicholson R. A., Norton G., Searl A., Sparks R. S. J., and Vickers B. P. (1999) Cristobalite in volcanic ash of the Soufriere hills volcano, Montserrat, British West Indies. *Science* **283**, 1142–1145.

Bethke C. M. (1996) *Geochemical Reaction Modeling-concept and Applications*. Oxford University Press, New York, NY, 397pp.

Borch R. S., Hastings L. L., Tingle T. N., and Verosub K. L. (1994) Speciation and *in vitro* gastrointestinal extractability of arsenic in two California gold mine tailings. *EOS* **75**, 190.

Boroujerdi M. (2002a) Groundwork. In *Pharmacokinetics: Principles and Applications*, McGraw-Hill, New York, NY, pp. 1–38.

Boroujerdi M. (2002b) Bioavailability and Bioequivalence Evaluation. In *Pharmacokinetics: Principles and Applications*. McGraw-Hill, New York, NY, pp. 315–329.

Brain J. D. (1992) Mechanisms, measurement, and significance of lung macrophage function. *Environ. Health Perspect.* **97**, 5–10.

Buchet J. P., Lauwerys R. R., and Yager J. W. (1995) Lung retention and bioavailability of arsenic after single intratracheal administration of sodium arsenite, sodium arsenate, fly ash and copper smelter dust in the hamster. *Environ. Geochem. Health* **17**, 182–188.

Bultman M. W., Fisher F. S., and Pappagianis D. An overview of the ecology of soil-borne human pathogens. In *Medical Geology—Earth Science in Support of Public Health Protection* (ed. O. Selinus). Academic Press (in press).

Burns-Naas L. A., Meade B. J., and Munson A. E. (2001) Toxic responses of the immune system. In *Casarett and Doull's Toxicology: The Basic Science of Poisons*, 6th edn. (ed. C. D. Klassen). McGraw-Hill, New York, NY, pp. 419–471.

Castranova V. (2000) From coal mine dust to quartz: mechanisms of pulmonary pathogenicity. *Inhalat. Toxicol.* **12**, 7–14.

Castranova V. and Vallyathan V. (2000) Silicosis and coal workers pneumoconiosis. *Environ. Health Perspect.* **108**(suppl. 4), 675–684.

CDC (1986) Epidemiologic notes and reports, cytotoxicity of volcanic ash: assessing the risk for pneumoconiosis. CDC

MMWR Weekly, **35**, 265–267, http://www.cdc.gov/mmwr/preview/mmwrhtml/00000724.htm

Chatfield E. J. and Kominsky J. R. (2001) Summary report: characterization of particulate found in apartments after destruction of the World Trade Center, 42pp.

Chen L. C., Lam H. F., Kim E. J., Guty J., and Amdur M. O. (1990) Pulmonary effects of ultrafine coal fly ash inhaled by guinea pigs. *J. Toxicol. Environ. Health* **29**, 169–184.

Churg A., Wright J. L., Gilks B., and DePaoli L. (1989) Rapid short term clearance of chrysotile compared to amosite asbestos in the guinea pig. *Am. Rev. Resp. Disease* **139**, 885–890.

Churg A., Wright J. L., and Vedal S. (1993) Fiber burden and patterns of asbestos-related disease in chrysotile miners and millers. *Am. Rev. Resp. Disease* **148**, 25–31.

Clark R. N., Green R. O., Swayze G. A., Meeker G., Sutley S., Hoefen T. M., Livo K. E., Plumlee G., Pavri B., Sarture C., Wilson S., Hageman P., Lamothe P., Vance J. S., Boardman J., Brownfield I., Gent C., Morath L. C., Taggart J., Theodorakos P. M., and Adams M. (2001) *Environmental Studies of the World Trade Center Area after the September 11, 2001 Attack*. US Geological Survey, Open File Report OFR-01-0429, http://pubs.usgs.gov/of/2001/ofr-01-0429

Cogan M. G. (1991) *Fluid and Electrolytes, Physiology and Pathophysiology*. Appleton and Lange, Norwalk, 304pp.

Collier C. G., Pearce M. J., Hodgson A., and Ball A. (1992) Factors affecting the *in vitro* dissolution of cobalt oxide. *Environ. Health Perspect.* **97**, 109–113.

Cook M., Chappell W. R., Hoffman R. E., and Mangione E. J. (1993) Assessment of blood lead levels in children living in a historic mining and smelting community. *Am. J. Epidemiol.* **137**, 447–455.

Crapper MacLachlan D. R., MacLachlan C. D., Krishnan B., Krishnan S. S., Dalton A. J., and Steele J. C. (1989) Aluminum and calcium in soil and food from Guam, Palau, and Jamaica: implications for amyotrophic lateral sclerosis and parkinsonism-dementia syndromes of Guam. *Environ. Geochem. Health* **11**, 45–53.

Crawford D. (1980) Electron microscopy applied to studies of the biological significance of defects in crocidolite asbestos. *J. Micros.* **120**, 181–192.

Daroowalla F. M. (2001) Pneumoconioses. In *Clinical Environmental Health and Exposures*, 2nd edn. (ed. J. B. Sullivan, Jr. and G. Krieger). Lippincott Williams and Wilkins, Philadelphia, pp. 538–545.

Davis A., Ruby M. V., and Bergstrom P. D. (1992) Bioavailability of arsenic and lead in soils from the Butte, Montana, mining district. *Environ. Sci. Technol.* **26**, 461–468.

Davis A. M., Drexler J. W., Ruby M. V., and Nicholson A. (1993) Micromineralogy of mine wastes in relation to lead bioavailability, Butte, Montana. *Environ Sci. Technol.* **27**, 1415–1425.

Davis A., Ruby M. V., Bloom M., Schoof R., Freeman G., and Bergstrom P. D. (1996) Mineralogic constraints on the bioavailability of arsenic in smelter-impacted soils. *Environ. Sci. Technol.* **30**, 392–399.

Davis J. M. G., Addison J., McIntosh C., Miller B. G., and Niven K. (1991) Variations in the carcinogenicity of tremolite dust samples of differing morphology. *Ann. NY Acad. Sci.* 473–490.

Dearwent S., Imtiaz R., Metcalf S., and Lewin M. (2000) Health consultation—mortality from asbestosis in Libby, Montana (report dated December 12, 2000), Agency for Toxic Substances and Disease Registry, http://www.atsdr.cdc.gov/HAC/pha/libby/lib_toc.html

De Meringo A., Morscheidt C., Thélohan S., and Tiesler H. (1994) *In vitro* assessment of biodurability: acellular systems. *Environ. Health Perspect.* **102**(suppl. 5), 47–54.

Derbyshire E. Aerosolic mineral dust and human health. In *Medical Geology—Earth Science in Support of Public Health Protection* (ed. O. Selinus). Academic Press, London, (in press).

Dieter M. P., Matthews H. B., Jeffcoat R. A., and Moseman R. F. (1993) Comparison of lead bioavailability in F344 rats fed lead acetate, lead oxide, lead sulfide, or lead ore concentrate from Skagway, Alaska. *J. Toxicol. Environ. Health* **39**, 79–93.

Donaldson K., Stone V., Clouter A., Renwick L., and MacNee W. (2001) Ultrafine particles. *Occupat. Environ. Med.* **58**, 211–216.

Dumortier P., Çoplü L., Broucke I., Emri S., Selcuk T., De Maertelaer V., De Vuyst P., and Baris I. (2001) Erionite bodies and fibres in bronchoalveolar lavage (BALF) of residents from Tuzköy, Cappadocia, Turkey. *Occupat. Environ. Med.* **58**, 261–266.

Eastes W., Hadley J. G., and Bender J. (1996) Assessing the role of biological activity of fibres: insights into the role of fibre biodurability, Australia and New Zealand. *J. Occupat. Health Safety* **12**, 381–385.

Eastes W., Potter R. M., and Hadley J. G. (2000a) Estimating *in-vitro* glass fiber dissolution rate from composition. *Inhalat. Toxicol.* **12**, 269–280.

Eastes W., Potter R. M., and Hadley J. G. (2000b) Estimating rock and slag wool fiber dissolution rate from composition. *Inhalat. Toxicol.* **12**, 1127–1139.

Finkelman R. B., Skinner C. W., Plumlee G. S., and Bunnell J. E. (2001) Medical geology. *Geotimes* **46**, 20–23.

Fisher D. C. (2001) Zinc. In *Clinical Environmental Health and Exposures*, 2nd edn. (eds. J. B. Sullivan, Jr. and G. Krieger). Lippincott Williams and Wilkins, Philadelphia, pp. 902–905.

Gamble J. L. (1942) *Chemical Anatomy, Physiology, and Pathology of Extracellular Fluid: A Lecture Syllabus*, 6th edn. Harvard University Press, Cambridge, 164p.

Garruto, R. M., Yanagihara, R. and Gajdusek D. C. (1985) Disappearance of high-incidence amyotrophic lateral sclerosis and parkinsonism-dementia on Guam. *Neurology*. **35** 193–198.

Garruto R. M., Shankar S. K., Yanagihara R., Salazar A. M., Amyx H. L., and Gajdusek D. C. (1989) Low-calcium, high-aluminum diet-induced motor neuron pathology in cynomolgus monkeys. *Acta Neuropathol. (Berl)* **78**, 210–219.

Gavett S. H., Haykal-Coates N., McGee J. K., Highfill J. W., Ledbetter A. D., and Costa D. L. (2002) *Toxicological Effects of Fine Particulate Matter Derived from the Destruction of the World Trade Center.* US EPA Report EPA/600/R-02/028, 53pp.

Geller R. J. (2001) Chromium. In *Clinical Environmental Health and Exposures*, 2nd edn. (eds. J. B. Sullivan, Jr. and G. Krieger). Lippincott Williams and Wilkins, Philadelphia, pp. 926–929.

Gibly R. L. and Sullivan J. B., Jr. (2001) Manganese. In *Clinical Environmental Health and Exposures*, 2nd edn. (eds. J. B. Sullivan, Jr. and G. Krieger). Lippincott Williams and Wilkins, Philadelphia, pp. 930–937.

Giese R. F., Jr. and van Oss C. J. (1993) The surface thermodynamic properties of silicates and their interactions with biological materials. In *Health Effects of Mineral Dusts,* Rev. Min. 28 (eds. G. D. Guthrie and B. T. Mossman). Mineralogical Society of America, Washington, DC, pp. 327–346.

Goerke J. (1998) Pulmonary surfactant: functions and molecular composition. *Biochim. Biophys. Acta* **1408**, 79–89.

Goyer R. A. and Clarkson T. W. (2001) Toxic effects of metals. In *Casarett and Doull's Toxicology: the Basic Science of Poisons*, 6th edn. (ed. C. D. Klassen). McGraw-Hill, New York, NY, pp. 811–868.

Graham A., Higinbotham J., Allan D., Donaldson K., and Beswick P. H. (1999) Chemical differences between long and short amosite asbestos: differences in oxidation state and coordination sites of iron, detected by infrared spectroscopy. *Occupat. Environ. Med.* **56**, 606–611.

Griffin D. W., Kellogg C. A., Garrison V. A., and Shinn E. A. (2002) The global transport of dust. *Am. Sci.* **90**, 228–235.

Gunter M. E. and Wood S. A. (2000) Can chrysotile alter to tremolite in the human lung? Abstracts, American Geophysical Union Spring Meeting 2000, M51A-05, http://www.agu.org/dbasetop.html

Guthrie G. D. (1997) Mineral properties and their contributions to particle toxicity. *Environ. Health Perspect.* **105**(suppl. 5), 1003–1011.

Guthrie G. D. and Mossman B. T. (eds.) (1993). In *Health Effects of Mineral Dusts*, Rev. Mineral. 28 Mineralogical Society of America, 584pp.

Hamel S. L. C. (1998) The estimation of bioaccessibility of heavy metals in soils using artificial biofluids. PhD Dissertation, Graduate School, New Brunswick, Rutgers, and University of Medicine and Dentistry of New Jersey, Graduate School of Public Health, 153pp.

Hardy J. A. and Aust A. E. (1995) Iron in asbestos: chemistry and carcinogenicity. *Chem. Rev.* **95**, 97–118.

Hesterberg T. W., Anderson R., Bunn W. B., III, Chase G. R., and Hart G. A. (2001) Man-made mineral fibers. In *Clinical Environmental Health and Exposures,* 2nd edn. (eds. J. B. Sullivan Jr. and G. Krieger), Lippincott Williams and Wilkins, pp. 1227–1240.

Hinkley T. K. (ed.) (1987) Chemistry of ash and leachates from the May 18, 1980 eruption of Mount St. Helens, Washington. US Geological Survey Professional Paper 1397, 64pp.

Hochella M. F., Jr. (1993) Surface chemistry, structure, and reactivity of hazardous mineral dust. In *Health Effects of Mineral Dusts*. (eds. G. D. Guthrie and B. T. Mossman). *Rev. Min.* **28**, Mineralogical Society of America, Washington, DC, pp. 275–308.

Holland J. P. and Smith D. D. (2001) Asbestos. In *Clinical Environmental Health and Exposures*, 2nd edn. (eds. J. B. Sullivan, Jr. and G. Krieger). Lippincott Williams and Wilkins, Philadelphia, pp. 1214–1227.

Horwell C. J., Fenoglio I., Ragnarsdottir K. V., Sparks R. S. J., and Fubini B. Surface reactivity of volcanic ash from the Soufrière Hills volcano, Montserrat, West Indies with implications for health hazards. *Environ. Res.* (in press).

Huang X., Laurent P. A., Zalma R., and Pezerat H. (1993) Inactivation of 1-antitrypsin by aqueous coal solutions: possible relation to the emphysema of coal workers. *Chem. Res. Toxicol.* **6**, 452–458.

Huang X., Fournier J., Koenig K., and Chen L. C. (1998) Buffering capacity of coal and its acid-soluble Fe^{2+} content: possible role in coal workers' pneumoconiosis. *Chem. Res. Toxicol.* **11**, 722–729.

Hume L. A. and Rimstidt J. D. (1992) The biodurability of chrysotile asbestos. *Am. Min.* **77**, 1125–1128.

Iyengar G. V., Kollmer W. E., and Bowen H. M. J. (1978) *The Elemental Composition of Human Tissues and Body Fluids: A Compilation of Values for Adults.* Weinheim, Verlag Chemie, Weinheim, 151pp.

Jaurand M. C., Gaudichet A., Halpern S., and Bignon J. (1984) *In vitro* biodegradation of chrysotile fibres by alveolar macrophages and mesothelial cells in culture: comparison with a pH effect. *Br. J. Indust. Med.* **41**, 389–395.

Jibson R. W., Harp E. L., Schneider E., Hajjeh R. A., and Spiegel R. A. (1998) An outbreak of coccidioidomycosis (valley fever) caused by landslides triggered by the 1994 Northridge, California, earthquake. *Geol. Soc. Rev. Eng. Geol.* **12**, 53–61.

Johnson N. F. (1994) Phagosomal pH and glass fiber dissolution in cultured nasal epithelial cells and alveolar macrophages: a preliminary study. *Environ. Health Perspect.* **102**, 97–102.

Johnson N. F. and Mossman B. T. (2001) Dose, dimension, durability, and biopersistence of chrysotile asbestos. In *The Health Effects of Chrysotile Asbestos: Contribution of Science to Risk-management Decisions* (eds. R. P. Nolan, A. M. Langer, M. Ross, F. J. Wicks, and R. F. Martin). The Canadian Mineralogist Special Publication No. 5, Ottawa, ON, pp. 145–154.

Jones D. R. (1995) The leaching of major and trace elements from coal ash. In *Environmental Aspects of Trace Elements in Coal* (eds. D. J. Swaine and F. Goodarzi). Kluwer, Dordrecht, pp. 221–262.

Jurinski J. B. and Rimstidt J. D. (2001) Biodurability of talc. *Am. Min.* **86**, 392–399.

Kamp D. W., Graceffa P., Pryor W. A., and Wetzman S. A. (1992) The role of free radicals in asbestos-induced diseases. *Free Radical Biol. Med.* **12**, 293–315.

Kanapilly G. M. (1977) Alveolar microenvironment and its relationship to the retention and transport into the blood of aerosols deposited in the alveoli. *Health Phys.* **32**, 89–100.

Kawanishi S. (1995) Role of active oxygen species in metal-induced DNA damage. In *Toxicology of Metals—Biochemical Aspects* (eds. R. A. Goyer and M. G. Cherian). Handbook of Experimental Pharmacology, Springer, Berlin, vol. 115, pp. 349–372.

Keogh J. P. and Boyer L. V. (2001) Lead. In *Clinical Environmental Health and Exposures*, 2nd edn. (eds. J. B. Sullivan, Jr. and G. Krieger). Lippincott Williams and Wilkins, Philadelphia, pp. 879–888.

Klassen C. D. (ed.) (2001) *Casarett and Doull's Toxicology: The Basic Science of Poisons.* McGraw-Hill, New York, NY, 1236pp.

Landrigan P. (1990) Current issues in the epidemiology and toxicology of occupational exposure to lead. *Environ. Health Perspect.* **89**, 61–66.

Landrigan P. (2001) Health consequences of the 11 September 2001 attacks. *Environ. Health Perspect.* **109**, A514–A515.

Landrigan P., Levin S., Forman J., Berkowitz G., and Yehuda R. (2002) Confronting the health consequences of the World Trade Center attacks. *American Public Health Association 2002 Annual Meeting, Abstracts with Programs*, #51102. http://apha.confex.com/apha/130am/techprogram/meeting_130am.htm

Lehnert B. E. (1992) Pulmonary and thoracic macrophage subpopulations and clearance of particles from the lung. *Environ. Health Perspect.* **97**, 17–46.

Letkeman P. (1996) Computer-modelling of metal speciation in human blood serum. *J. Chem. Educat.* **73**, 165–170.

Lioy P. J. (1990) Assessing total human exposure to contaminants. *Environ. Sci. Technol.* **24**, 938–945.

Lioy P. J. (2002) Composition of dust/smoke from collapse of the World Trade Center and implications for human exposure. *American Public Health Association 2002 Annual Meeting, Abstracts with Programs*, #51098. http://apha.confex.com/apha/130am/techprogram/meeting_130am.htm

Lioy P., Weisel C. P., Millette J. R., Eisenreich S., Vallero D., Offenberg J., Buckley B., Turpin B., Zhong M., Cohen M. D., Prophete C., Yang I., Stiles R., Chee G., Johnson W., Porcja R., Alimokhtari S., Hale R. C., Weschler C., and Chen L. C. (2002) Characterization of the dust/smoke aerosol that settled east of the World Trade Center (WTC) in lower Manhattan after the collapse of the WTC 11 September 2001. *Environ. Health Perspect.* **110**, 703–714.

Lund L. G. and Aust A. E. (1992) Iron mobilization from crocidolite asbestos greatly enhances crocidolite-dependent formation of DNA single-strand breaks in ϕX174 RFI DNA. *Carcinogenesis* **13**, 637–642.

Lybarger J. A., Lewin M., Peipins L. A., Campolucci S. S., Kess S. E., Miller A., Spence M., Black B., and Weis C. (2001) *Year 2000 Medical Testing of Individuals Potentially Exposed to Asbestoform [sic] Minerals Associated with Vermiculite in Libby, Montana: a Report to the Community,* http://www.atsdr.cdc.gov/asbestos/doc_phl_testreport.html

Mattson S. M. (1994a) Glass fiber dissolution in simulated lung fluid and measures needed to improve consistency and correspondence to *in vivo* dissolution. *Environ. Health Perspect* **102**(suppl. 5), 87–90.

Mattson S. M. (1994b) Glass fibers in simulated lung fluid: dissolution behavior and analytical requirements. *Ann. Occupat. Hygiene* **38**, 857–877.

May P. M. and Williams D. R. (1980) The inorganic chemistry of iron metabolism. In *Iron in Biochemistry and Medicine* (eds. A. Jacobs and M. Worwood). Academic Press, London, vol. II, pp. 1–28.

May P. M., Linder P. W., and Williams D. R. (1977) Computer simulation of metal-ion equilibria in biofluids: models for the low-molecular-weight complex distribution of calcium(II), magnesium(II), manganese(II), iron(III), copper(II), zinc(II), and lead(II) ions in human blood plasma. *J. Chem. Soc. Dalton Trans.* 588–595.

McDonald J. C., McDonald A. D., Armstrong B., and Sebastien P. (1986) Cohort study of mortality of vermiculite miners exposed to tremolite. *Br. J. Indust. Med.* **43**, 436–444.

Medinsky M. A. and Valentine J. L. (2001) Toxicokinetics. In *Toxicology: The Basic Science of Poisons*, 6th edn. (ed. C. D. Klaasen). McGraw-Hill, pp. 225–237.

Meeker G. P., Bern A. M., Brownfield I. K., Sutley S. J., Hoefen T. M., Vance J. S., and Lowers H. A. The composition and morphology of prismatic, fibrous, and asbestiform amphibole from the Rainy Creek district, Libby, Montana. *Am. Min.* (in press).

Miller W. R. and Sanzolone R. F. (2002) Investigation of the possible connection of rock and soil geochemistry to the occurrence of high rates of neurodegenerative diseases on Guam and a hypothesis for the cause of the diseases. US Geological Survey Open-File Report 02-475, 42pp.

Millette J. R., Boltin R., Few P., and Turner W., Jr. (2002) Microscopical studies of world trade center disaster dust particles. *Microscope* **50**, 29–35.

Mossman B. T. and Gee J. B. (1989) Asbestos-related diseases. *New England J. Med.* **320**, 1721–1730.

Mullins M. J. P. and Norman J. B. (1994) Solubility of metals in windblown dust from mine waste dump sites. *Appl. Occupat. Environ. Hyg.* **9**, 218–223.

Newman L. S. (2001) Clinical pulmonary toxicology. In *Clinical Environmental Health and Exposures,* 2nd edn. (eds. J. B. Sullivan, Jr. and G. Krieger). Lippincott Williams and Wilkins, Philadelphia, pp. 206–223.

NIOSH (2002) Health effects of occupational exposure to respirable crystalline silica. NIOSH Hazard Review, DHHS (NIOSH) Publication 2002-129, National Institutes of Occupational Safety and Health, 126pp.

Nolan R. P., Langer A. M., and Herson G. B. (1991) Characterization of palygorskite specimens from different geological locales for health hazard evaluation. *J. Br. Indust. Med.* **48**, 463–475.

Nolan R. P., Langer A. M., Ross M., Wicks F. J., and Martin R. F. (eds.) (2001) *The Health Effects of Chrysotile Asbestos: Contribution of Science to Risk-Management Decisions*. The Canadian Mineralogist Special Publication No. 5, 304pp.

Nordstrom D. K. and Alpers C. N. (1999) Geochemistry of acid mine waters. In *The Environmental Geochemistry of Mineral Deposits: Part A. Processes, Techniques and Health Issues* (eds. G. S. Plumlee and M. J. Logsdon). *Soc. Econ. Geol. Rev. Econ. Geol.* **6A**, Society of Economic Geologists, Littleton, pp. 133–160.

Nyberg K., Johansson A. J., and Camner P. (1992) Phagolysosomal pH in alveolar macrophages. *Environ. Health Perspect.* **97**, 149–152.

Oliver M. A. (1997) Soil and human health: a review. *Euro. J. Soil Sci.* **48**, 573–592.

Oomen A. G., Hack A., Minekus M., Zeijdner E., Cornelis C., Schoeters G., Verstraete W., Van De Wiele T., Wragg J., Rompelberg C. J. M., Sips A. J. A. M., and Van Wijnen J. H. (2002) Comparison of five *in vitro* digestion models to study the bioaccessibility of soil contaminants. *Environ. Sci. Technol.* **36**, 3326–3334.

Perl D. P. (1997) Amyotrophic lateral sclerosis-parkinsonism-dementia complex of Guam. In *Neuropathology of Dementia* (eds. M. M. Esire and J. H. Morris). Cambridge University Press, Cambridge, pp. 268–292.

Pitot H. C., III and Dragan Y. P. (2001) Chemical carcinogenesis. In *Toxicology: The Basic Science of Poisons*, 6th edn. (ed. C. D. Klaasen). McGraw-Hill, New York, NY, pp. 241–319.

Plumlee G. S. (1999) The environmental geology of mineral deposits. In *The Environmental Geochemistry of Mineral Deposits: Part A. Processes, Techniques and Health Issues*, (eds. G. S. Plumlee and M. J. Logsdon). *Soc. Econ. Geol. Rev. Econ. Geol.* **6A**, Society of Economic Geologists, Littleton, 71–116.

Plumlee G. S., Hageman P., Ziegler T., Meeker G. P., Lamothe P. J., Theodorakos P., Sutley S. J., Clark R. N., Wilson S. A., Swayze G. A., Hoefen T. F., Taggart J., and Adams M. (2002) The geochemical composition and reactivity of dusts deposited by the September 11, 2001 World Trade Center collapse. *Geological Society of America Abstracts with Programs, 2002 Annual meeting.* http://gsa.confex.com/gsa/htsearch.cgi

Prezant D. J., Weiden M., Banauch G. I., McGuinness G., Rom W. N., Aldrich T. K., and Kelley K. J. (2002) Cough and bronchial responsiveness in firefighters at the World Trade Center site. *New England J. Med.* **347**, 806–815.

Raloff J. (2001) Ill winds—dust storms ferry toxic agents between countires and even continents. *Sci. News* **160**, 218–220.

Reheis M., Plumlee G., Lamothe P., Budahn J., Hageman P., Hinkley T., Meeker G., Gill T., Winn R., and Thames D. (2001) Potential health hazards of Owens Lake dust. *Geological Society of America 2001 Annual Meeting Abstracts*, Paper 100–100.

Reheis M. C., Budahn J. R., and Lamothe P. J. (2002) Geochemcial evidence for diversity of dust sources in the Southwestern US. *Geochim. Cosmochim. Acta* **66**, 1569–1587.

Renner R. (2000) Asbestos in the air. *Sci. Am.* February, 34.

Reynolds H. Y. and Chretien J. (1984) Respiratory tract fluids: analysis of content and contemporary use in understanding lung diseases. *Disease Monogr.* **30**, 1–103.

Rhoades R. and Pflanzer R. (1992) *Human Physiology,* 2nd edn. Sanders College Publishing, Harcourt Brace College Publishers, Fortworth, TX, 1058pp.

Rom W. M. N., Weiden M., Garcia R., Yie T. A., Vathesatogkit P., Tse D. B., McGuinness G., Roggli V., and Prezant D. (2002) Acute eosinophilic pneumonia in a New York City firefighter exposed to World Trade Center dust. *Am. Resp. Crit. Care Med.* **166**, 797–800.

Ross M. (1999) The health effects of mineral dusts. In *The Environmental Geochemistry of Mineral Deposits: Part A. Processes, Techniques, and Health Issues* (eds. G. S. Plumlee and M. J. Logsdon). *Soc. Econ. Geol. Rev. Econ. Geol.* **6A**, Society of Economic Geologists, Littleton, 339–356.

Ross M., Nolan R. P., Langer A. M., and Cooper W. C. (1993) Health effects of mineral dusts other than asbestos. In *Health Effects of Mineral Dusts*, Rev. Min. 28 (eds. G. D. Guthrie and B. T. Mossman). Mineralogical Society of America, Washington, DC, pp. 362–407.

Rozman K. K. and Klaasen C. D. (2001) Absorption, Distribution, and Excretion of Toxicants. In *Toxicology: The Basic Science of Poisons*. 6th edn, (ed. C. D. Klaasen). McGraw-Hill, New York, NY, pp. 107–132.

Ruby M. V., Davis A., Kempton J. H., Drexler J. W., and Bergstrom P. (1992) Lead bioavailability: dissolution kinetics under simulated gastric conditions. *Environ. Sci. Technol.* **26**, 1242–1248.

Ruby M. V., Davis A., Link T. E., Schoof R., Chaney R. L., Freeman G. B., and Bergstrom P. (1993) Development of an *in vitro* screening test to evaluate the *in vivo* bioaccessibility of ingested mine-waste lead. *Environ. Sci. Technol.* **27**, 2870–2877.

Ruby M. V., Davis A., Schoof R., Eberle S., and Sellstone C. M. (1996) Estimation of lead and arsenic bioavailbility using a physiologically based extraction test. *Environ. Sci. Technol.* **30**, 422–430.

Ruby M. V., Schoof R., Brattin W., Goldade M., Post G., Harnois M., Mosby D. E., Casteel S. W., Berti W., Carpenter M., Edwards D., Cragin D., and Chappell W. (1999) Advances in evaluating the oral bioavailability of inorganics in soil for use in human health risk assessment. *Environ. Sci. Technol.* **33**, 3697–3705.

Sahlin K., Alvestrand A., Bergstrom J., and Hultman E. (1977) Human intracellular pH and bicarbonate concentration as determined in biopsy samples from the quadriceps muscle of man at rest. *Clinic. Sci. Mol. Med.* **53**, 459–466.

Saint Amand P., Mathews L. A., Gaines C., and Reinking R. (1986) *Dust Storms from Owens and Mono Valleys, California*. Naval Weapons Center Technical Publication NWC TP 6731, Naval Weapons Center, China Lake, 79pp.

Scanlan C. L., Wilkins R. L., and Stoller J. K. (eds.) (1999) *Egan's Fundamentals of Respiratory Care,* 7th edn. Mosby, St. Louis, 1238pp.

Scanlon P. D. (2002) World trade center cough—a lingering legacy and a cautionary tale. *New England J. Med.* **347**, 840–842.

Schlesinger R. B. (1995) In Deposition and clearance of inhaled particles. In *Concepts in Inhalation Toxicology*, 2nd edn. (eds. R. O. McClellan and R. F. Henderson). Taylor and Francis, Washington, DC, pp. 191–224.

Schneider E., Spiegel R. A., Jibson R. W., Harp E. L., Marshall G. A., Gunn R. A., McNeil M. M., Pinner R. W., Baron R. C., Burger R. C., Hutwagner L. C., Crump C., Kaufman L., Reef S. E., Feldman G. M., Pappagianis D., and Werner B. (1997) A coccidioidomycosis outbreak following the Northridge, Calif, earthquake. *J. Am. Med. Assoc.* **277**, 904–908.

Scholze H. and Conradt R. (1987) An *in vitro* study of the chemical durability of siliceous fibres. *Ann. Occupat. Hyg.* **31**(4B), 683–692.

Schoof R., Butcher M. K., Sellstone C., Ball R. W., Fricke J. R., Keller V., and Keehn B. (1995) An assessment of lead absorption from soil affected by smelter emissions. *Environ. Geochem. Health* **17**, 189–199.

Sébastien P., McDonald J. C., McDonald A. D., Case B., and Harley R. (1989) Respiratory cancer in chrysotile textile and mining industries: exposure inferences from lung analysis. *Br. J. Indust. Med.* **46**, 180–187.

Selikoff I. J. and Lee D. H. (1978) *Asbestos and Disease*. Academic Press, London, pp. 34–50.

Selinus O. (ed.) *Medical Geology—Earth Science in Support of Public Health Protection*. Academic Press, London (in press).

Selinus O. and Frank A. (2000) Medical geology. In *Environmental Medicine* (ed. L. Moller). Joint Industrial Safety Council/Sweden, pp. 164–183.

Shi X., Ding M., Chen F., Wang L., Rojanasakul Y., Vallyathan V., and Castranova V. (2001) Reactive oxygen species and molecular mechanisms of silica-induced lung injury. *J. Environ. Pathol. Toxicol. Oncol.* **20**(suppl. 1), 85–93.

Sipes I. G. and Badger D. (2001) Principles of toxicology. In *Clinical Environmental Health and Exposures,* 2nd edn. (eds. J. B. Sullivan, Jr. and G. Krieger). Lippincott Williams and Wilkins, Philadelphia, pp. 49–67.

Sittig M. (1994) *World-wide Limits for Toxic and Hazardous Chemicals in Air, Water, and Soil*. Noyes Publications, Park Ridge, NJ, 792pp.

Skinner H. C. W. (2000) In praise of phosphates, or why vertebrates chose apatite to mineralize their skeletal elements. *Int. Geol. Rev.* **42**, 232–240.

Skinner H. C. W. and Berger A. R. (2003) *Geology and Health: Closing the Gap*. Oxford University Press, New York, 179pp.

Skinner H. C. W., Ross M., and Frondel C. (1988) *Asbestos and Other Fibrous Minerals*. Oxford University Press, Oxford, 204pp.

Smith D. B., Zielinski R. A., and Rose W. I., Jr. (1982) Leachability of uranium and other elements from freshly erupted volcanic ash. *J. Volcanol. Geotherm. Res.* **13**, 1–30.

Smith D. B., Zielinski R. A., Taylor H. E., and Sawyer M. B. (1983) Leaching characteristics of ash from the May 18, 1980, eruption of Mount St. Helens volcano, Washington. *Bull. Volcanol.* **46**, 103–124.

Smith K. R., Veranth J. M., Lighty J. S., and Aust A. E. (1998) Mobilization of iron from coal fly ash was dependent upon the particle size and the source of coal. *Chem. Res. Toxicol.* **11**, 1494–1500.

Smith K. R., Veranth J. M., Hu A. A., Lighty J. S., and Aust A. E. (2000) Interleukin-8 levels in human lung epithelial cells are increased in response to coal fly ash and vary with the bioavailability of iron, as a function of particle size and source of coal. *Chem. Res. Toxicol.* **13**, 118–125.

Smith K. S. (1999) Metal sorption on mineral surfaces: an overview with examples relating to mineral deposits. In *The Environmental Geochemistry of Mineral Deposits: Part A. Processes, Techniques and Health Issues*, (eds. G. S. Plumlee and M. J. Logsdon). *Soc. Econ. Geol. Rev. Econ. Geol.* **6A**, Society of Economic Geologists, Littleton, pp. 161–182.

Smith K. S. and Huyck H. L. O. (1999) An overview of the abundance, relative mobility, bioavailability, and human toxicity of metals. In *The Environmental Geochemistry of Mineral Deposits: Part A. Processes, Techniques and Health Issues* (eds. G. S. Plumlee and M. J. Logsdon). Soc. Econ. Geol. Rev. Econ. Geol. **6A**, Society of Economic Geologists, Littleton, pp. 29–70.

Snipes M. B. (1995) Pulmonary retention of particles and fibers: biokinetics and effects of exposure concentrations. In *Concepts in Inhalation Toxicology*, 2nd edn. (eds. R. O. McClellan and R. F. Henderson). Taylor and Francis, Washington, DC, pp. 225–256.

SSDC (1988) Diseases associated with exposure to silica and nonfibrous silicate minerals (Silicosis and Silicate Disease Committee). *Arch. Pathol. Lab. Med.* **112**, 673–720.

Staub N. C. (1991) *Basic Respiratory Physiology*. Churchill Livingstone, New York, NY, 242pp.

Stephenson J. (2002) Researchers probe health consequences following the World Trade Center attack. *J. Am. Med. Assoc.* **288**, 1219–1221.

Sullivan, J. B. Jr. and Krieger G. (eds.) (2001) *Clinical Environmental Health and Exposures*, 2nd edn. Lippincott Williams and Wilkins, Philadelphia, 1323pp.

Sullivan J. B., Jr., Levine R. J., Bangert J. L., Maibach H., and Hewitt P. (2001) Clinical dermatoxicology. In *Clinical Environmental Health and Exposures*, 2nd edn. (eds. J. B. Sullivan, Jr. and G. Krieger). Lippincott Williams and Wilkins, Philadelphia, pp. 182–206.

Sunderman F. W., Jr. (2001) Nickel. In *Clinical Environmental Health and Exposures*, 2nd edn. (eds. J. B. Sullivan, Jr. and G. Krieger). Lippincott Williams and Wilkins, Philadelphia, pp. 905–910.

Sutton J., Elias T., Hendley J. W., II, and Stauffer P. H. (1997) *Volcanic Air Pollution—A Hazard in Hawaii.* US Geological Survey Fact Sheet, http://wrgis.wr.usgs-gov/fact-sheet/fs169-97/

Suzuki Y. and Yuen S. R. (2001) Asbestos tissue burden study on human malignant mesothelioma. *Indust. Health* **39**, 150–160.

Suzuki Y. and Yuen S. R. Asbestos fibers contributing to the induction of human malignant mesothelioma. *Ann. New York Acad. Sci.* (in press).

Taylor D. M. and Williams D. R. (1995) *Trace Element Medicine and Chelation Therapy*. The Royal Society of Chemistry, London, 124pp.

Taylor D. M. and Williams D. R. (1998) Bio-inorganic chemistry and its pharmaceutical applications. In *Smith and Williams' Introduction to the Principles of Drug Design and Action* (ed. H. J. Smith). Harwood Academic, Amsterdam, pp. 509–538.

Taunton A., Wood S. A., and Gunter M. E. (2002) The thermodynamics of asbestos mineral dissolution and conversion in the human lung. In *Proceedings, 2002 Goldschmidt Conference, Davos, Switzerland*, A765. http://www.goldschmidt-conference.com/2002/gold2002/

Templeton D. M. (1995) Theraputic use of chelating agents in iron overload. In *Toxicology of Metals—Biochemical Aspects*, Handbook of Experimental Pharmacology (eds. R. A. Goyer and M. G. Cherian). Springer, Berlin, vol. 115, pp. 305–332.

Thomas C. L. (ed.) (1989) *Taber's Cyclopedic Medical Dictionary,* 16 edn. F. A. Davis Co., 2403pp.

Thomas C. L. (1997) *Taber's Cyclopedic Medical Dictionary* (ed.) 18 edn. F. A. Davis Co., Philadelphia, 2439pp.

Thurston G. D., Cohen M., Maciejczyk P., Cohen B., Kendall M., Heikkinen M., Lippmann M., Schuetz L., Costa M., and Chen L. (2002) Characterization of World Trade Center disaster airborne and settled particulate matter exposures. *American Public Health Association 2002 Annual Meeting, Abstracts with Programs*, #51104. http://apha.confex.com/apha/130am/techprogram/meeting_130am.htm

USGS (2002) *USGS Environmental Studies of the World Trade Center Area, New York City, after September 11, 2001.* US Geological Survey Fact Sheet FS-050-02.

Vallyathan V., Mentnech M. S., Stettler L. E., Dollberg D. D., and Green F. H. (1983a) Mount St. Helen's volcanic ash: hemolytic activity. *Environ. Res.* **30**, 349–360.

Vallyathan V., Mentnech M. S., Tucker J. H., and Green F. H. (1983b) Pulmonary response to Mount St. Helen's volcanic ash. *Environ. Res.* **30**, 361–371.

Van Gosen B. S., Lowers H. A., Bush A. L., Meeker G. P., Plumlee G. S., Brownfield I. K., and Sutley S. J. (2002) *Reconnaissance Study of the Geology of US Vermiculite Deposits—are Asbestos Minerals Common Constituents?* US Geological Survey Bulletin 2192, 8pp.

van Maanen J. M., Borm P. J., Knaapen A., van Herwijnen M., Schilderman P. A., Smith K. R., Aust A. E., Tomatis M., and Fubini B. (1999) *In vitro* effects of coal fly ashes: hydroxyl radical generation, iron release, and DNA damage and toxicity in rat lung epithelial cells. *Inhalat. Toxicol.* **11**, 1123–1141.

van Oss C. J., Naim J. O., Costanzo P. M., Giese R. F., Jr., Wu W., and Sorling A. F. (1999) Impact of different asbestos species and other mineral particles on pulmonary pathogenesis. *Clays Clay Min.* **47**, 697–707.

Veranth J. M., Smith K. R., Hugins F., Hu A. A., Lighty J. S., and Aust A. E. (2000) Mossbauer spectroscopy indicates that iron in an aluminosilicate glass phase is the source of the bioavailable iron from coal fly ash. *Chem. Res. Toxicol.* **13**, 161–164.

Waalkes M. P., Wahba Z. Z., and Rodriguez R. E. (2001) Cadnium. In *Clinical Environmental Health and Exposures*, 2nd edn. (eds. J. B. Sullivan, Jr. and G. Krieger). Lippincott Williams and Wilkins, Philadelphia, pp. 889–897.

Werner A. J., Hochella M. F., Jr., Guthrie G. D., Jr., Hardy J. A., Aust A. E., and Rimstidt J. D. (1995) Asbestiform riebeckite (crocidolite) dissolution in the presence of Fe chelators: implications for mineral-induced disease. *Am. Min.* **80**, 1093–1103.

WHO (1996) *Trace Elements in Human Nutrition and Health.* World Health Organization in collaboration with the Food and Agricultural Organization of the United nations and the International Atomic Energy Agency, 343pp.

Williams G. M. and Weisburger J. H. (1991) Chemical carcinogenesis. In *Toxicology: The Basic Science of Poisons*, 4th edn. (eds. M. O. Amdur, J. Doull, and C. D. Klaasen). Pergamon, London, pp. 127–200.

Williams D. R. (2000) Chemical speciation applied to bioinorganic chemistry. *J. Inorg. Biochem.* **79**, 275–283.

Wilson M. R., Stone V., Cullen R. T., Searl A., Maynard R. L., and Donaldson K. (2000) *In vitro* toxicology of respirable Montserrat volcanic ash. *Occupat. Environ. Med.* **57**, 727–733.

Wright R. S., Wright J. L., Abraham P., Harber B.R., Burnett P., Morris P. and West P. (2002) Fatal asbestosis 50 years after brief high intensity exposure in a vermiculite expansion plant. *Am. J. Resp. Crit. Care Med.* **165**, 1–5.

Wylie A. G. and Verkouteren J. R. (2000) Amphibole asbestos from Libby, Montana, aspects of nomenclature. *Am. Min.* **85**, 1540–1542.

Wylie A. G., Skinner H. C. W., Marsh J., Snyder H., Garzione C., Hodkinson D., Winters R., and Mossman B. T. (1997) Mineralogical features associated with cytotoxic and proliferative effects of fibrous talc and asbestos on rodent tracheal epithelial and pleural mesothelial cells. *Toxicol. Appl. Pharmacol.* **147**, 143–150.

Wyllie A. H. (1987) Apoptosis: cell death in tissue regulation. *J. Pathol.* **153**, 313–316.

Yip L. and Dart R. C. (2001) Arsenic. In *Clinical Environmental Health and Exposures*, 2nd edn. (eds. J. B. Sullivan, Jr. and G. Krieger). Lippincott Williams and Wilkins, Philadelphia, pp. 858–866.

Yip L., Dart R. C., and Sullivan J. B., Jr. (2001) Mercury. In *Clinical Environmental Health and Exposures*, 2nd edn. (eds. J. B. Sullivan, Jr. and G. Krieger). Lippincott Williams and Wilkins, Philadelphia, pp. 867–878.

Ziegler T. L., Plumlee G. S., Lamothe P. J., Meeker G. P., Witten M. L., Sutley S. J., Hinkley T. K., Wilson S. A., Hoefen T. F., Brownfield I. K., and Lowers H. (2002) Mineralogical, geochemical and toxicological variations of asbestos toxicological standards and amphibole samples from Libby, MT. In *Geological Society of America Abstracts with Programs, 2002 Annual Meeting*, http://gsa.confex.com/gsa/htsearch.cgi

9.08
Worldwide Eutrophication of Water Bodies: Causes, Concerns, Controls

E. E. Prepas
Lakehead University, Thunder Bay, ON, Canada

and

T. Charette
University of Alberta, Edmonton, AB, Canada

NOMENCLATURE

Glossary of Limnological Terms (from Ruttner, 1952; Crosby et al., 1990)

biomass Weight of living matter.

chlorophyll a Primary photosynthetic pigment. The concentration of chlorophyll *a* in water is an indicator of phytoplankton biomass. See *biomass*.

cyanobacteria Prokaryotic organisms in the phytoplankton community. Cyanobacteria are typified by cells without a nucleus or organelles and with photosensitive pigments dispersed throughout the cell. Some species are capable of obtaining nitrogen for metabolism from the atmosphere.

drainage basin The land area that contributes surface runoff to a water body. See "runoff."

epilimnion The warm uppermost layer in a thermally stratified water body that is subject to mixing by wind.

eutrophic Waters with a good supply of nutrients and hence a rich organic production.

freshwater Water with concentration of total dissolved solids below 500 mg L^{-1}.

hypolimnion The cool bottom layer in a thermally stratified water body that is separated from surface influences by a thermocline (see *thermocline*).

macrophytes Large aquatic plants.

mesotrophic Waters with a moderate supply of nutrients and organic production.

nutrient limitation Limitation of phytoplankton biomass by an insufficient nutrient supply compared to the demand.

oligotrophic Waters with a poor supply of nutrients and organic production.

phytoplankton The photosynthesizing portion of the plankton. See also *plankton*.

plankton The community of the free water.

primary production The production of organic matter from inorganic materials within a certain period of time by autotrophic organisms.

runoff The water reaching a lake, stream, or ocean after flow over land or through the surficial layers of the land.

Secchi disk depth The depth in water to which a Secchi disk (a 20 cm diameter disk with alternating black and white quadrants) can be seen from the surface. Secchi disk depth is an easy measurement of water transparency.

stratified Divided into layers. In stratified water bodies, there may be mixing within a layer but little mixing occurs between layers. Layers have different densities, which may be determined either by temperature and/or salinity.

thermocline The layer of water in a water body between the epilimnion and hypolimnion in which the temperature gradient is greatest and exceeds a change of 1 °C m^{-1} of depth.

trophic state The degree of fertility of a lake. See also *eutrophic*, *mesotrophic*, and *oligotrophic*.

zooplankton The animal portion of the plankton.

9.08.1 INTRODUCTION

9.08.1.1 Aspects of Worldwide Concern over Eutrophication

Eutrophication is the nutrient enrichment of waters that stimulates an array of symptomatic changes, that can include increased phytoplankton and rooted aquatic plant (macrophyte) production, fisheries and water quality deterioration, and other undesirable changes that interfere with water uses (Bartsch, 1972). The trophic state, or degree of fertility, of water bodies ranges from oligotrophic to mesotrophic to eutrophic with increasing supply of nutrients and organic matter (Table 1). Eutrophication is most often the result of an elevated supply of nutrients, particularly nitrogen and phosphorus, to surface waters that results in enhanced production of primary producers, particularly phytoplankton and aquatic plants.

Phytoplankton are unpleasant at high densities. The sight and smell of clots or masses of decaying phytoplankton decreases the recreational value of most waters and usually generates concerns among the public. Furthermore, blooms of toxin-producing phytoplankton can cause widespread illness. A bloom is a conspicuous concentration of phytoplankton, often concentrated at or near

Table 1 Mean annual values for the trophic classification system.

	Total phosphorus (μg L^{-1})	*Chlorophyll a* (μg L^{-1})	*Secchi disk depth* (m)
Ultra-oligotrophic	<4	<1	>12
Oligotrophic	<10	<2.5	>6
Mesotrophic	10–35	2.5–8	6–3
Eutrophic	35–100	8–25	3–1.5
Hypertrophic	>100	>25	<1.5

Source: OECD (1982).

the surface. It is difficult to quantify what constitutes a "bloom," but a rough estimate places it as a chlorophyll *a* concentration over 30 $\mu g\ L^{-1}$. Toxins produced by dinoflagellates such as *Pfiesteria* in marine environments of the northeastern US and red tides in tropical waters have caused massive fish kills, millions of dollars in losses to seafood-related industries, human memory loss, paralysis, and even death (Van den Hoeck *et al.*, 1995; Silbergeld *et al.*, 2000). Bloom-forming species of cyanobacteria can produce potent hepato-(liver) toxins termed microcystins that have been implicated in poisonings of domestic livestock, pets, wildlife, and susceptible humans (Codd, 1995; Dunn, 1996). In addition, an accumulation of dead phytoplankton in bottom waters of eutrophic systems can lead to high decomposition rates by bacteria. Dissolved oxygen consumption by decomposers, combined with a barrier to gas exchange (thermocline or ice cover), can reduce (hypoxia) or eliminate (anoxia) dissolved oxygen in bottom waters. (A *thermocline* is the junction between an upper layer of warm, less dense water (the epilimnion) and a deeper layer of cold water (the hypolimnion). When this stratification is in place, the typically oxygen-rich waters of the epilimnion do not mix with the waters of the hypolimnion.) Oxygen depletion is one of the most harmful side effects of eutrophication because it can cause catastrophic fish kills, devastating local fisheries.

The accumulation of plant biomass depends on the addition of factors that stimulate plant growth. On average, the macronutrients nitrogen and phosphorus are present in marine phytoplankton at an atomic ratio 16 : 1 (Redfield, 1958). The ratio of nitrogen to phosphorus in freshwaters tends to be greater than the ratio in phytoplankton; therefore, phosphorus most often limits the growth of phytoplankton. As a result, phosphorus enrichment of freshwater often causes its eutrophication (Schindler, 1977). In lakes, nitrogen is usually present in concentrations equal to or beyond what is required for aquatic plant growth because, unlike phosphorus, it has an atmospheric source. In marine systems, nitrogen concentrations are often limiting because bacterial nitrogen fixation, while a considerable source of nitrogen in lakes, not as important in marine waters. A wide variety of prokaryotic organisms (i.e., certain cyanobacteria, heterotrophic, and chemoautotrophic bacteria) can use nitrogen gas directly and incorporate it into organic compounds through a process called nitrogen fixation. Nitrogen fixation is an enzyme-catalyzed process that reduces nitrogen gas (N_2) to ammonia (NH_3). Nitrogen-fixing cyanobacteria make up less than 1% of the total biomass of phytoplankton in estuaries of the Atlantic coast of North America, whereas in lakes they often make up more than 50% of

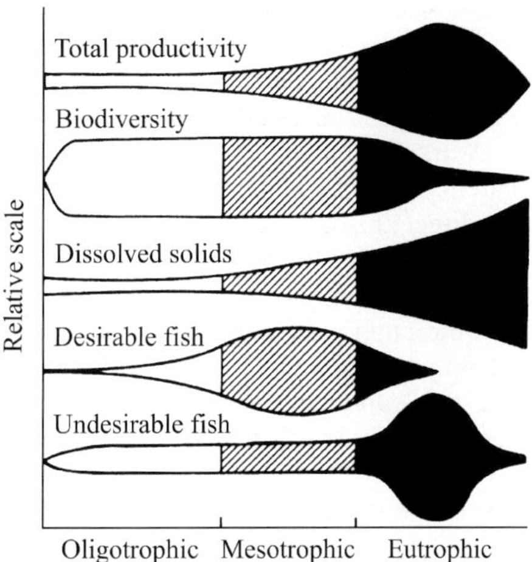

Figure 1 Suggested changes in various characteristics of lakes with eutrophication (reproduced by permission of Cambridge University Press from Welch, E. B., *Ecological Effects of Waste Water*, **1980**, p. 273).

phytoplankton biomass (reviewed in Howarth, 1988). An increase in water clarity can also spur the growth of aquatic vegetation in systems where the clarity of water is poor from high concentrations of suspended particles.

The biodiversity of most aquatic systems decreases with eutrophication (Figure 1). Phytoplankton species diversity is reduced in highly productive systems. Cyanobacteria are usually dominant in eutrophic systems because these organisms are better adapted to conditions of high nutrients (Smith, 1986; Trimbee and Prepas, 1987; Watson *et al.*, 1997). In addition, fish and macro-invertebrate species diversity can decrease with eutrophication. Depletion of dissolved oxygen in deep water is associated with eutrophication and can lead to a loss or displacement of species intolerant of such conditions (Ludsin *et al.*, 2001). In eutrophic lakes of North America, characteristic fish types are surface-dwelling, warm water fishes such as pike, perch, and bass, as compared to deep-dwelling, cold-water fishes like salmon, trout, and cisco (Ryding and Rast, 1989).

9.08.1.2 Indicators of Eutrophication and Sampling Methods

Trophic state can be measured directly via indicators of phytoplankton biomass and nutrient concentrations or indirectly by measuring water transparency (Table 1). Water transparency can be an accurate indicator of phytoplankton productivity, even though it is affected by nonbiological particles suspended in the water column. Secchi disk depth is the oldest, simplest, and quickest quantitative measure of water clarity. A Secchi disk is a black and white disk that is lowered down through the water column until it can no longer be seen. Secchi disk depth is the midpoint between the depth at which it disappears when lowered and reappears when it is pulled up again. A more objective method of measuring transparency is with a light meter, generally an upward-directed quantum sensor that determines the intensity of light (often measured in $\mu E\ m^{-2}\ s^{-1}$). A light meter can be used to measure the euphotic zone, or the portion of the water column receiving over 1% surface irradiation. In the euphotic zone, net photosynthesis takes place and phytoplankton and rooted aquatic plants (macrophytes) can grow. A third approach to determine water transparency involves measuring turbidity in water samples. Turbidity is a measure of the scattering of light caused by suspended particles (e.g., mud, silt, and phytoplankton) in the water column and is recorded in standard international Nephelometric Turbidity Units (NTU). Water transparency is a good indicator of eutrophication and can easily be measured by members of the general public.

Because eutrophication is associated with high phytoplankton numbers, the biomass of these organisms is a useful indicator of trophic state. A good descriptive measure of eutrophication and the general state of water is the presence of phytoplankton blooms. Phytoplankton blooms can cause visible coloration of surface waters or dense mats and clumps, usually accompanied by an unpleasant rotting smell. Cyanobacteria are the dominant type of phytoplankton in most freshwater blooms (Crosby *et al.*, 1990). Many species of cyanobacteria contain gas vacuoles in their cells that make them buoyant. As such, in highly eutrophic lakes, cyanobacterial mats and clumps can form and be transported by wind into foul smelling, decaying masses around the perimeter of a lake. In the sea, blooms are caused by species of nonbacterial forms of phytoplankton (e.g., dinoflagellates, haptophytes). An effective indicator of changes in phytoplankton biomass, and the most widely used, is the concentration of chlorophyll *a*, the primary photosynthetic pigment in all phytoplankton. Chlorophyll *a* is measured by filtering water, extracting chlorophyll from phytoplankton cells captured on a fine filter and measuring the color density of the extract with spectrophotometry.

Because an increased supply of nitrogen in marine waters and/or phosphorus in freshwaters is typically responsible for eutrophication, directly measuring the concentration of these nutrients can indicate trophic state. Nutrients in water are either dissolved or bound to particulates produced *in situ* and deposited into water bodies from groundwater, atmospheric deposition, and drainage basin runoff. To measure trophic state, nutrient concentrations are measured in samples taken from a water body when surface water nutrient concentrations are at their maximum after replenishment from bottom waters. Sampling is carried out during spring overturn (i.e., complete mixing of the water column) where nutrient concentrations in surface waters vary seasonally (i.e., systems with alternating thermal stratification and overturn with the seasons). In systems with less routine thermal dynamics (i.e., those with multiple breakdown of thermal stratification), additional sampling may be required to take account of changes in nutrient concentrations from multiple mixing events. Nutrient concentrations in surface water are relatively low (i.e., often few ppb) and are measured in raw water samples. The principal method of determining the total concentration of phosphorus and nitrogen in a water sample is by digestion or oxidation of organic matter to release bio-available inorganic nutrients (phosphate, nitrate, and ammonium). The addition of a mixed reagent creates color and its density is measured with spectrophotometry (Prepas and Rigler, 1982; Crumpton *et al.*, 1992).

9.08.2 NATURAL EUTROPHICATION

Because of the high retention of nutrients in most undisturbed land, very few aquatic systems are naturally eutrophic. However, nutrient loading, and thus trophic state, is influenced first by natural processes. Nutrient composition of soil and its exposure to draining water are important determinants of the chemistry of water exported from the drainage basin. Soil fertility is related to the chemistry of its parent rock material. Soils that lie on sedimentary rock tend to be more fertile than soils on hard igneous rock. For example, the classic prairie soils, famous for the production of corn, are of sedimentary origin (Jenny, 1980). Lakes on igneous geology can be naturally more nutrient-poor than lakes on sedimentary geology (Rawson, 1960). Further, there is more potential for leaching or biological decay of soils in drainage basins with longer exposure to water. Therefore, drainage basins that are relatively flat (i.e., low slopes), and those that are located in areas of low precipitation, are more likely to export water with high nutrient concentrations (D'Arcy and Carignan, 1997). Water bodies are closely linked to the characteristics of their drainage basins.

The drainage ratio ("surface area of drainage basin" to "water body") can represent the contribution of the drainage basin to the nutrient budget of receiving waters, relative to the atmosphere. The nutrient budget of a water body with a relatively small drainage ratio will be controlled primarily by atmospheric nutrient inputs because drainage basin contributions are relatively minor. The relative drainage basin size ("surface area of drainage basin" to "water body volume") can indicate how long it takes for the entire volume of water (and nutrients) in a water body to be flushed out of a system. Rapidly flushed systems with large relative drainage basins, such as streams, are more closely connected to their drainage basin than those that are flushed slowly, such as large lakes (Søballe and Kimmel, 1987). A large relative drainage basin, all else being equal, is typically associated with higher nutrient content in surface waters, because nutrient inputs per unit of time are greater. Within one geological setting, the relative drainage basin size of water bodies can account for most of the variability in phosphorus concentrations and phytoplankton biomass in aquatic systems (D'Arcy and Carignan, 1997; Prepas *et al.*, 2001b). The quantity of water and nutrients exported from a drainage basin also depends on evapotranspiration. Studies comparing drainage basins that vary in forest composition indicate that the quantity of water exported is a function of the effect of vegetation differences on transpiration rate (Bosch and Hewlett, 1982). Therefore, the amount (e.g., Hobbie and Likens, 1973) and type (e.g., Cronan and Aiken, 1985) of vegetation, in addition to relative drainage basin size, can influence water and nutrient export to water bodies.

Summer phosphorus concentrations in nutrient-rich water bodies can often depend on *in situ* rather than external loading. Internal phosphorus loading, or the recycling of phosphorus from bottom sediments, usually occurs and/or is enhanced after a long history of eutrophication and phosphorus enrichment of sediments due to high rates of organic sedimentation. In culturally eutrophic systems with a long history of external phosphorus enrichment, recovery can be delayed due to the release of biologically available phosphorus from bottom sediments. Sediment phosphorus release is regulated by the intensity of oxidizing or reducing conditions within an aquatic system as measured by the reduction–oxidation, or redox, potential. A reducing environment produced by the oxidation of sediment organic matter during bacterial decomposition causes dissolution of $FeOOHPO_4$ complexes and release of highly mobile and bio-available PO_4^{3-}. In productive systems, anoxic conditions can develop at the sediment–water interface after an extended period of thermal stratification that prevents atmospheric oxygen from reaching deeper waters. Oxygen depletion, and therefore internal loading, is most likely in relatively shallow lakes, because the hypolimnion (i.e., deep layer of cold water), no matter how transient, has a reduced water volume and thus dissolved oxygen pool. In shallow and fertile systems, external loading can be insignificant compared to internal loading on an annual basis (Riley and Prepas, 1984). All else being equal, most lakes become shallower over a long timescale through accumulation of bottom sediments from particles that have settled from the water column. Therefore, lakes tend to evolve towards eutrophy (Hutchinson, 1973).

9.08.3 CULTURAL EUTROPHICATION: CASE STUDIES

Water pollution shadows human population growth and development and is caused by diffuse (nonpoint) and concentrated (point) nutrient enrichment. Some of the most severe cases of anthropogenic eutrophication occurred soon after the Second World War, when rapid population growth and development followed. At that time, population growth and the concentration of nutrient-rich sewage enhanced the point-source nutrient loading. Deforestation, or clearing of native vegetation, is one of the most basic human alterations of the environment and can be an early and important nonpoint source of eutrophication. With the removal of actively transpiring vegetation, an excess of water in the ground is available for soil weathering (Roby and Azuma, 1995). Therefore, vegetation removal can increase the transfer of nutrients from the drainage basin to receiving waters. The largest improvements in water quality have often been associated with reduction or elimination of point-source inputs. Not surprisingly, the most important challenges for the future preservation of water quality lie in controlling nonpoint nutrient sources from intense activity such as agriculture and associated use of fertilizer, concentration of animal wastes, and vehicle emissions. Drainage basins with intensive agriculture can export five or more times the phosphorus and nitrogen of forested drainage basins (Dillon and Kirchner, 1975). Although eutrophication abatement has focused on enhanced nitrogen and phosphorus content of drainage basin runoff, concern over atmospheric nitrogen pollution has developed with increasing industrialization. Our case studies take us from the west coast of North America (Lake Tahoe and Lake Washington) across to Lake Erie and then to Chesapeake Bay on the east coast. Two further examples come from southern Europe (Lago Maggiore) and Africa (Lake Victoria; Figures 2 and 3). These six case studies highlight varying degrees of eutrophication and restoration of systems impacted by numerous sources of nitrogen and phosphorus pollution (Table 2).

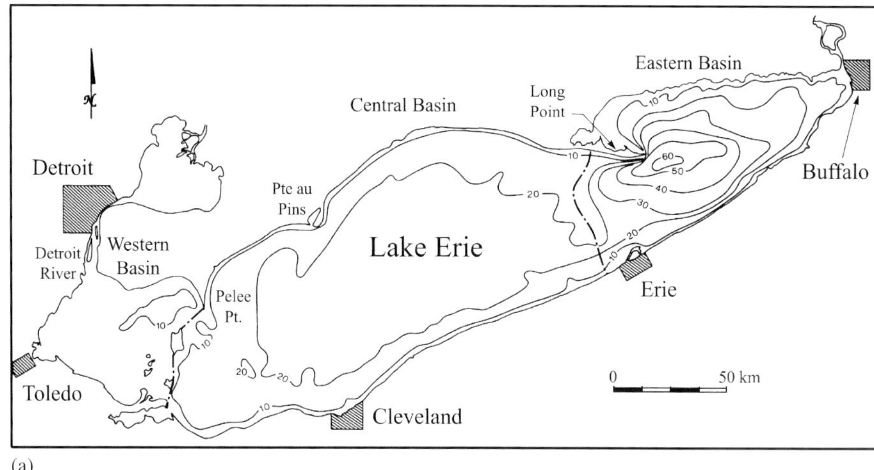

(a)

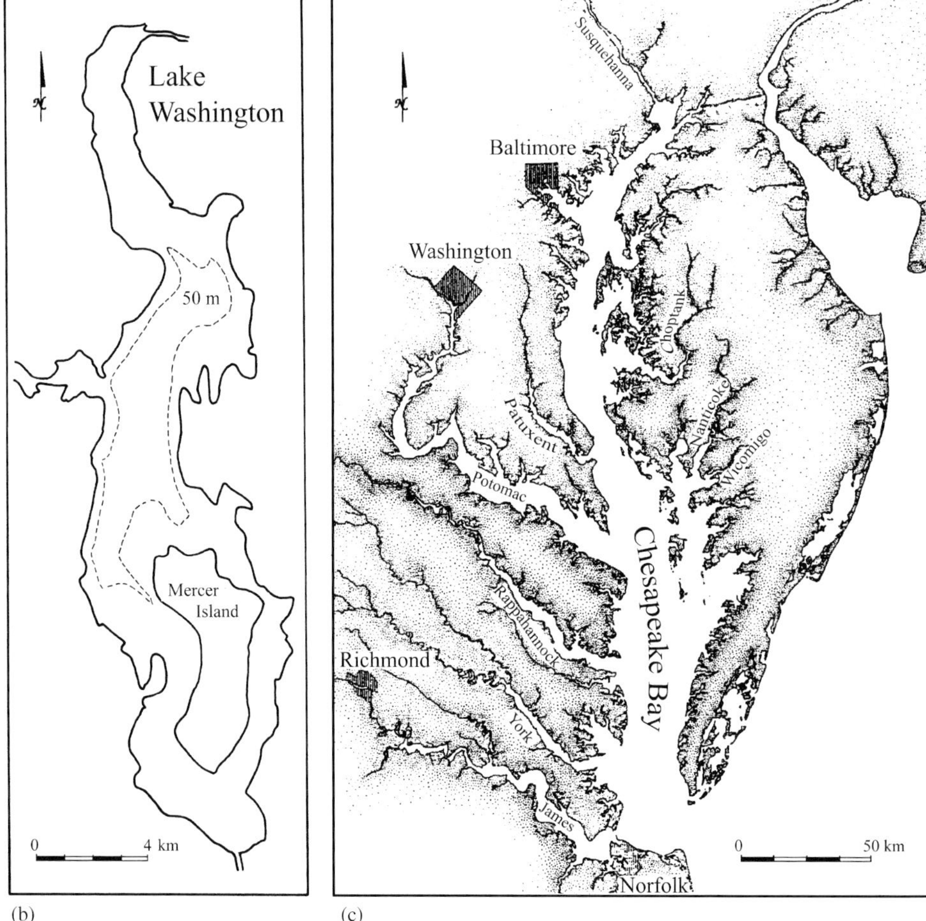

(b) (c)

Figure 2 (a) Lake Erie, (b) Lake Washington, and (c) Chesapeake Bay ((a) reproduced by permission of the Minister of Public Works and Government Services, 2002 from *J. Fish. Res. Board Can.*, **1975**, *33*, 355–370; (b) reproduced by permission of The American Society of Limnology and Oceanography, from *Limnol. Oceanogr.*, **1991**, *36*, 1031–1044; and (c) reproduced by permission of Inter-Research Science Publishers from *Mar. Ecol. Prog. Ser.*, **1994**, *104*, 267–291).

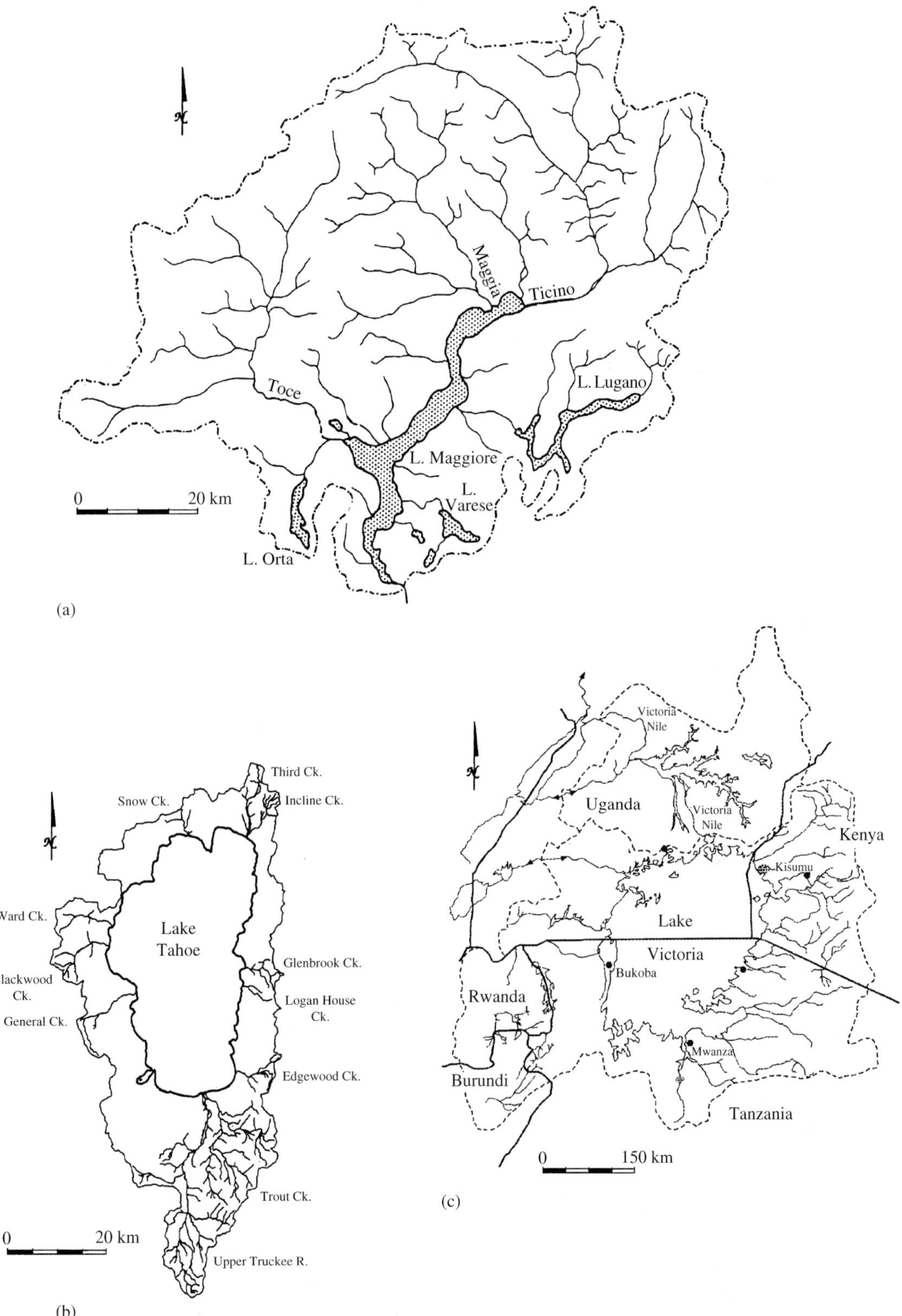

Figure 3 (a) Lago Maggiore, (b) Lake Tahoe, and (c) Lake Victoria and their drainage basins ((a) reproduced by permission of Istituto Italiano di Idrobiologia from *Mem. Ist. Ital. Idrobiol.*, **1995**, *53*, 53–73; (b) reproduced by permission of E. Schweizerbart Science Publishers from *Arch. Hydrobiol.*, **1995**, *135*, 1–21; and (c) reproduced by permission of Elsevier from *J. Environ. Manage.*, **2000**, *58*, 235–248).

Table 2 Characteristics of, and lessons from, case studies.

Water body	*Characteristics*
Lake Washington	Point-source (wastewater) nutrient stress. Mitigated via diversion of nutrient source downstream and chance biological control of phytoplankton. Rapid recovery.
Lake Erie	Point (wastewater) and nonpoint (agriculture) source stresses. Multiple basins illustrate effects of morphometry, stratification, and flushing rates on trophic status. Mitigated via legislated wastewater and laundry detergent nutrient targets. Partially recovered—depends on future nonpoint-source mitigation.
Lake Tahoe	Atmospheric pollution stress. Low drainage ratio = atmosphere important nitrogen source. Tertiary sewage treatment limited success because sensitivity to nitrogen rather than phosphorus. Mitigation via nonpoint-source controls but atmospheric sources difficult to control = further degradation likely.
Lago Maggiore	Point (wastewater + polluted upstream lakes) and nonpoint (agriculture) source stresses. Mitigated via wastewater treatment and recovery of upstream lakes. Partially recovered system.
Lake Victoria	Nonpoint-source stresses: atmospheric and land use + internal nutrient loading controlled by wet–dry seasons. Low drainage ratio and high basin evapotranspiration = atmosphere important nutrient source. Further degradation likely because little mitigation.
Chesapeake Bay	Mostly nonpoint (agriculture), some point (wastewater) source stress. Estuarine example where nitrogen-limitation important. Mitigation efforts via wastewater treatment (point source) + wetland restoration to serve as nutrient traps (nonpoint source). Early restoration phase.
Lessons	Point sources are easy to control. Nonpoint sources are difficult to control because of diffuse nature. Internal loading is important long-term feedback mechanism. P problems associated with land use changes. N problems associated with atmospheric pollution.

9.08.3.1 Lake Washington

Lake Washington (87.6 km^2; Figure 2) is a relatively deep lake (maximum depth 65.2 m) in Washington State, USA, with one of the finest histories of limnological observations due to the lifelong dedication of the late W. T. Edmondson. Classified as mesotrophic (water with moderate supply of nutrients and organic production) in recent history (Scheffer and Robinson, 1939), Lake Washington was used as a source of drinking water for some small communities until 1965. At that time, total and bio-available (soluble reactive) phosphorus concentrations were relatively low (16 $\mu g\ L^{-1}$ and 8 $\mu g\ L^{-1}$ (ppb), respectively), and there was no record of nuisance phytoplankton blooms.

Pollution was associated with water effluent from the city of Seattle. By 1963, water quality in Lake Washington was the poorest on record. At that point, sewage accounted for ~65% of all phosphorus inputs into Lake Washington (Edmondson, 1975). The average annual total phosphorus and winter phosphate concentrations peaked at 3 and 5 times, respectively, that in 1933 (Figure 4). Summer phytoplankton biomass (estimated as chlorophyll *a*) was over 10 times higher in 1963 than only 13 years earlier and Secchi disk depth dropped from 3.7 m to less than 1 m over the same time span (Figure 4). At the peak of water quality decline, as much as 98% of the volume of phytoplankton was made up of cyanobacteria, largely *Oscillatoria* species such as *O. rubescens* and *O. agardhii*, that are often present in eutrophic systems (Edmondson, 1969).

Public pressure to halt the severe deterioration of Lake Washington resulted in action. Between 1963 and 1968, sewage effluent from the treatment plants was progressively diverted to Puget Sound (Figure 4). The first diversion removed 28% of the effluent, stopping the eutrophication process. Recovery was quick with a sharp (65%) decrease in phytoplankton biomass two years after the beginning of diversion (Figure 4). By the end of the diversion process, water quality had returned to mesotrophic conditions

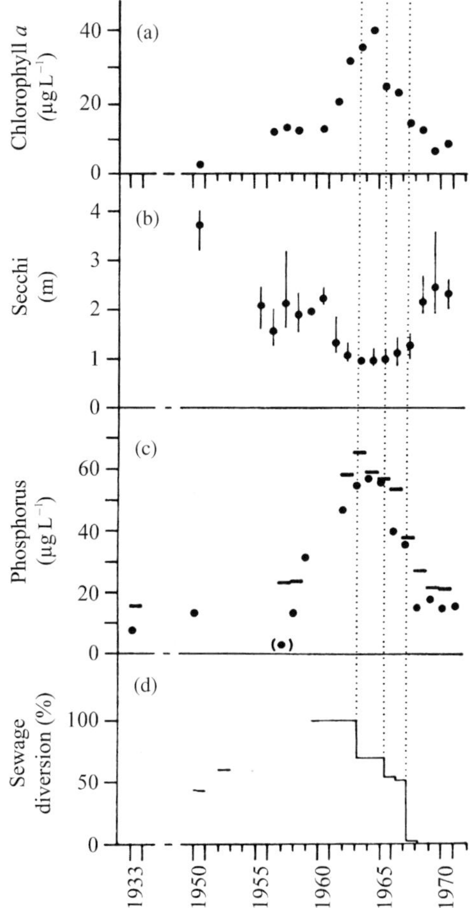

Figure 4 Changes in Lake Washington water quality in relation to sewage diversion. (a) Mean (July–August) lake surface (epilimnetic) chlorophyll *a* concentration. (b) Mean (July–August) Secchi disk depth. The vertical lines show the range. (c) Bars = annual mean of total phosphorus in the epilimnion. Circles = mean (January–March) bio-available phosphorus (phosphate) concentrations in the epilimnion. (d) Relative amount of sewage as indicated by the capacity of the treatment plants emptying into the lake, maximum taken as 100% (reproduced by permission of American Society of Limnology and Oceanography from *Nutrients and Eutrophication*, **1972**, pp. 172–193).

(chlorophyll *a*: 5–10 $\mu g\ L^{-1}$, Secchi disk depth: 2–2.5 m, total phosphorus: 20–30 $\mu g\ L^{-1}$). Recovery continued. By 1975, transparency increased to 4 m and total phosphorus and chlorophyll *a* concentrations decreased to 16 $\mu g\ L^{-1}$and 4 $\mu g\ L^{-1}$, respectively, similar to conditions in 1933. Thus, deterioration in the water quality of Lake Washington was reversible.

Biological regulation of phytoplankton biomass greatly improved water quality in Lake Washington. The appearance of the aggressive phytoplankton grazer, the microcrustacean *Daphnia* spp., coincided with an increase in water transparency to as much as 12.9 m by 1976. *Daphnia* appeared after the decline of inedible cyanobacteria (*Oscillatoria* spp.), and thrived due to a sustained 10% decrease in its voracious predator *Neomysis mercedis* in 1965. The decrease in *Neomysis* occurred at about the same time as the introduction of longfin smelt, a predator known to specialize in eating *Neomysis* (Edmondson, 1994). Helped by a rapid flushing rate (2.3 yr) and stable dissolved phosphorus inputs since the end of sewage diversion (Edmondson, 1994), Lake Washington remains mesotrophic.

9.08.3.2 Lake Erie

Lake Erie offers one of the best-publicized examples of the serious impacts of humans on water quality as well as a model of ecosystem rehabilitation, thanks to international cooperation. Of the five North American Great Lakes, Lake Erie was the most seriously impacted by cultural nutrient enrichment. Lake Erie is thought to be naturally more eutrophic than the other Great Lakes because of its shallower depth and the greater erodibility of nutrient-rich soils in its immediate drainage basin (i.e., excluding upstream Great Lakes). Lake Erie can be divided into three basins (Figure 2), distinguished by mean depth, that have distinct morphometric and trophic characteristics. The western basin (3,080 km^2) is the shallowest (mean depth 7.6 m), and most nutrient- and plankton-rich of the Lake Erie basins. The water column mixes completely throughout most of the year with brief and unpredictable periods of thermal stratification between May and September. The western basin is eutrophic, because it is shallower (see Section 9.08.2: Natural Eutrophication), and receives the largest external nutrient loads. The central basin, largest of the three in area (16,425 km^2) and volume has a mean depth of 18.6 m. Here, thermal stratification is established yearly between June and September, causing bottom waters to become hypoxic by late summer. The eastern basin (area 6,159 km^2) is by far the deepest of the three (mean depth 26 m). Thermal stratification is strongest in this basin and persists from June to late October or early November. The eastern basin is considered oligotrophic, and therefore the thick hypolimnion (bottom layer of cold water) experiences only small reductions in dissolved oxygen concentrations (Bartish, 1987). Thus, Lake Erie displays a longitudinal gradient in trophic status; from eutrophic waters in the western basin to progressively diluted and oligotrophic waters in the outflowing eastern basin.

Because most of the immediate drainage basin is composed of highly erodible sedimentary soils, it is believed that the eutrophication of Lake Erie began with land clearing associated with European settlement. Phosphorus loading increased slowly after forest removal in the drainage basin in the late nineteenth century, then increased exponentially over three decades up to the early 1970s (Snell, 1987). At that time, the western basin was highly eutrophic, the eastern basin was mesotrophic, and the central basin was meso- to eutrophic (M. Munawar and I. F. Munawar, 1976). When Lake Erie was in its worst state, total phosphorus load was estimated at 1.2×10^4 t yr^{-1}, four times higher than precolonization values (Vallentyne and Thomas, 1978). Parallel to the increasing load, phosphorus concentrations increased fivefold in the central basin, to ~25 μg L^{-1}. By the early 1960s, cyanobacterial blooms appeared in Lake Erie and phytoplankton abundance tripled from measurements taken 40 years earlier (Figure 5; Davis, 1964). The western basin and parts of the central basin developed dense mats of the cyanobacterium *Aphanizomenon flos-aquae* (Beeton, 1965). Furthermore, dissolved oxygen depletion rates in bottom waters increased (Figure 5) and extensive hypoxia developed: dissolved oxygen concentrations were low (3 mg L^{-1} (ppm) or less) in ~70% of the bottom waters of the central basin during late summer (Beeton, 1963). Major changes also occurred in the bottom fauna of the western basin; the benthic community dominated by the burrowing mayfly (*Hexagenia* spp.) was replaced

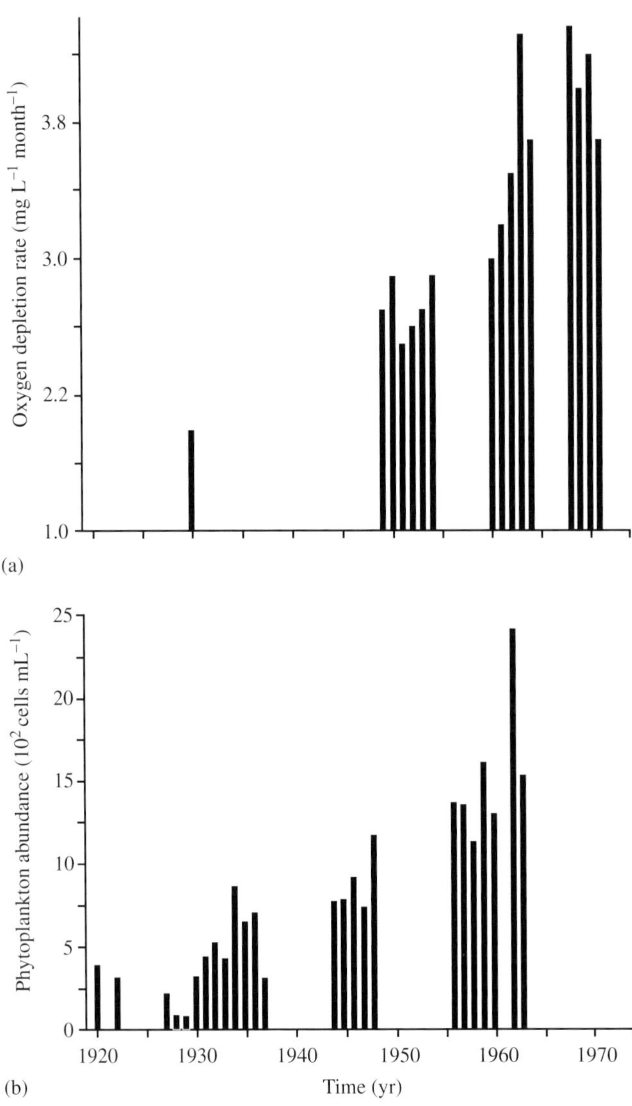

Figure 5 (a) Mean depletion rates for dissolved oxygen during summer in the bottom-water (hypolimnion) of central Lake Erie (reproduced by permission of the Minister of Public Works and Government Services, 2002 from *J. Fish. Res. Board Can.*, **1975**, *33*, 355–370). (b) Mean phytoplankton abundance in Lake Erie from 1920 to 1963 (reproduced by permission of American Society of Limnology and Oceanography from *Limnol. Oceanogr.*, **1964**, *9*, 275–283).

almost entirely by one dominated by midge larvae (chironomids) and oligochaetes tolerant of low-oxygen conditions. *Hexagenia* spp. disappeared shortly after two periods of anoxia in the early 1950s (Beeton, 1961). Furthermore, the fish fauna changed from natural cold-water fish populations (whitefish, sauger, cisco, blue-pike, and lake trout) to domination by warm-water species (yellow perch, smelt, fresh-water drum, white bass, carp, catfish, and walleye) characteristic of eutrophic lakes.

Due to the severity of eutrophication and extensive hypoxic conditions, programs were developed in the early 1970s to reduce phosphorus loads into the Great Lakes as part of the Great Lakes Water Quality agreement between Canada and the United States. The agreement required that phosphorus concentration of sewage effluent from all large municipal waste treatment plants be limited to 1 mg L^{-1}. A total of 7.5 billion dollars was spent on municipal waste facility enhancement. Total phosphorus loads from direct municipal discharges were reduced by more than 80% by the mid-1980s. Much of this reduction was a result of phosphorus restrictions in detergents; up to 50% of phosphorus in sewage originated from laundry detergents. The Lake Erie ecosystem responded quickly to these management efforts. Total phosphorus concentrations in the western basin had decreased by 50% only a decade after the Great Lakes Water Quality Agreement (Rosa, 1987). Over two decades after the Agreement, the rate of decline in total phosphorus concentrations was $\sim$0.44 μg L^{-1} yr^{-1} (Rosa, 1987; Bertram, 1993). During the same time period, total phytoplankton and cyanobacterial (i.e., *Aphanizomenon flos-aquae*) biomass decreased by about 65% and 89%, respectively (Makarewicz and Bertram, 1991), and mesotrophic phytoplankton species common before eutrophication (*A. formosa*) reappeared (Makarewicz, 1993). During that time, there was also a slight decrease in dissolved oxygen depletion rates in the deeper water (Bertram, 1993). Six fish species tolerant of hypoxic conditions declined in abundance and three species intolerant of degraded water quality recovered (Ludsin *et al.*, 2001).

Despite the dramatic recovery of Lake Erie through point-source reductions, the lake remains enriched in phosphorus. Over half of the external load of total phosphorus to Lake Erie is from tributaries draining cultivated land. Therefore, attainment of the desired total phosphorus load will depend on further nonpoint-source controls (Dolan, 1993). Lake Erie must be monitored regularly because of its dynamic nature due to heavy use as an international shipping route. Following its introduction to the Great Lakes about 20 years ago and prolific expansion, the zebra mussel (*Dreissena* spp.) has been partly responsible for tremendous improvements in water clarity in recent years (Makarewicz *et al.*, 1999). The Lake Erie restoration story continues to unfold.

9.08.3.3 Lake Tahoe

With its enchanting emerald green waters, Lake Tahoe lies on the border of the American states of Nevada and California, in the Sierra—Nevada mountain range (elevation 1,898 m; Figure 3). Because of its stunning mountainous setting and its naturally clear waters (Secchi disk depth 41 m in 1972), this large (500 km^2) and deep (maximum depth 505 m) lake has long been revered as one of the most beautiful water bodies in the world, thanks in part to its limnological guardian, C. R. Goldman. Naturally ultra-oligotrophic, Lake Tahoe receives low nutrient loads from the drainage basin. On an annual basis, phosphorus inputs to the lake are predominantly from terrestrial sources, similar to Lake Washington. In contrast, atmospheric sources of nitrogen, mostly nitrate, are more significant than drainage basin inputs because of Lake Tahoe's small drainage ratio (1.6; Figure 3). Due to the relative paucity of nitrogen, however, phytoplankton biomass in Lake Tahoe has been controlled, until recently, by the limited availability of ammonium and nitrate (Goldman *et al.*, 1993).

Because of human influences, nitrogen inputs to Lake Tahoe have steadily increased and have fueled a noticeable growth in phytoplankton biomass. Limnological changes started to occur by the middle of the nineteenth century, when the basin was logged to provide timber for mining operations (Jassby *et al.*, 2001). Lake Tahoe has also seen a tremendous population growth and drainage basin development since the 1950s (as of early 2000s: a resident population of 52,000; one million visitors yearly). Shoreline development proceeded with little consideration for runoff and erosion control. To complicate matters, Lake Tahoe's drainage basin has steep slopes that produce high-energy runoff. Therefore, nutrients leached from destabilized and eroding soil on the lakeshore have become a major source of nitrogen to the lake. However, atmospheric deposition provides most of the nitrogen load to Lake Tahoe and this has been increasing steadily due to local combustion of fossil fuels for transportation and heat, and upwind human development (cities of Los Angeles and San Francisco; Jassby *et al.*, 1994; Goldman, 1993). Phytoplankton are quite responsive to atmospheric loads of nitrogen, because they are deposited mostly in bio-available form (i.e., nitrate and ammonium). Conversely, total nitrogen from drainage basins contains a greater fraction of organic nitrogen, much of which is resistant to rapid bacterial degradation to bio-available forms (Kalff, 2002). As a result of increased deposition of atmospheric nitrogen, the rate of phytoplankton production (primary productivity) has more than quadrupled and Secchi disk depth has decreased ~30% since the late 1960s (Figure 6). With the increase in nitrogen relative to phosphorus loads, phytoplankton growth in Lake Tahoe has gone from limitation largely by nitrogen to strong phosphorus-limitation (Jassby *et al.*, 1995). Despite cultural eutrophication, Lake Tahoe remains ultra-oligotrophic (Secchi disk depth 25 m; Goldman *et al.*, 1993). However, because water and nutrient flushing times are long (650 yr), prevention of additional fertilization of Lake Tahoe is crucial. Reduction of within-basin nutrient sources (e.g., Best Management Practices (BMPs); see Section 9.08.4.2) has been the focus of water quality management. However, the fate of Lake Tahoe appears dim at this moment because atmospheric nutrient sources are variable; they depend on large-scale wind patterns and are difficult to control.

9.08.3.4 Lago Maggiore

Lago Maggiore is located in the foothills of the Alps, just north of the most industrialized part of Italy (Figure 3). With the CNR—Institute of Ecosystem Study (ISE) on its shore, Lago Maggiore has the privilege of being the most thoroughly investigated lake in Italy, both in terms of the number of key biogeochemical parameters measured and the duration of their monitoring. Lago Maggiore is deep (maximum depth 370 m) and only the top 100–150 m of its water column circulates once each year. However, the lake mixes completely over a varying number of years, the mixing being connected with very cold autumns and winters, and with windy periods during March and April, when thermal stratification is weaker (Ambrosetti and Barbanti, 1999). Before major eutrophication began, the lake was oligotrophic (spring total phosphorus concentrations below 10 $\mu g\ L^{-1}$, Secchi disk depth 10.7 m), and the growth of phytoplankton was limited by the availability of phosphorus (Mosello and Ruggiu, 1985).

Like many lakes around the world, Lago Maggiore was rapidly eutrophied during the 1960s causing a shift in trophic state from oligotrophic to mesotrophic a decade later. During this period, bio-available (soluble reactive) phosphorus concentrations tripled and nitrate concentrations doubled in surface waters and stabilized in subsequent years (Figure 7; Ambrosetti *et al.*, 1992; Ruggiu and Mosello, 1984). As agriculture

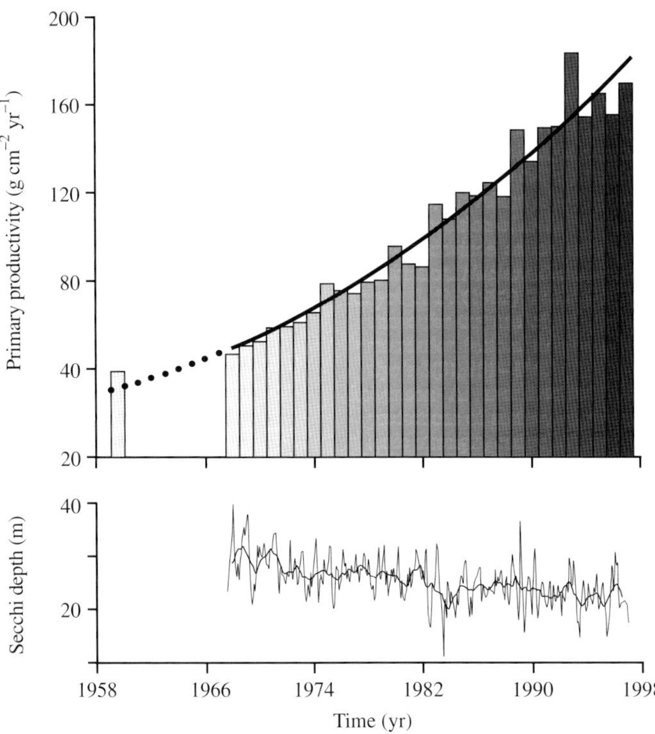

Figure 6 Mean annual primary productivity and mean monthly Secchi disk depth over time in Lake Tahoe (reproduced by permission of Backhuys Publishers from *The Great Lakes of the World (GLOW): Food-web, Health and Integrity*, **2001**, pp. 431–454).

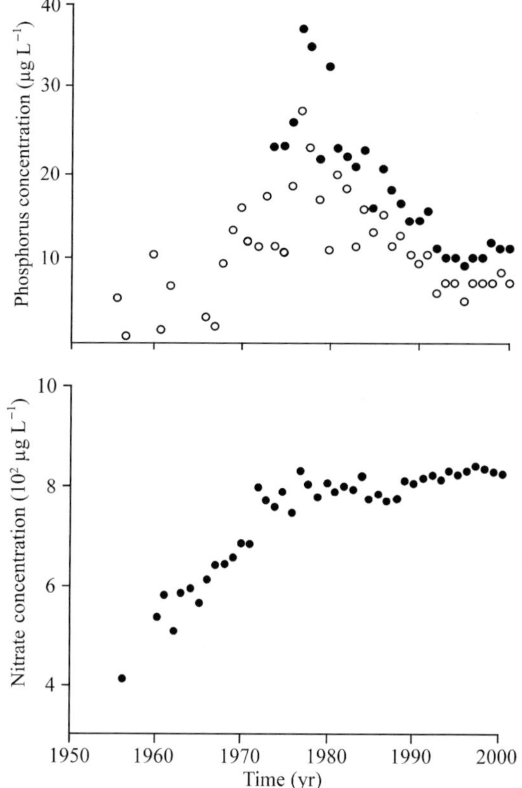

Figure 7 Bio-available (○) and total (•) phosphorus and nitrate concentrations in Lago Maggiore between 1956 and 1991. Mean volume weighted concentrations measured at overturn (reproduced by permission of Istituto Italiano di Idrobiologia from *Mem. Ist. Ital. Idrobiol.*, **1992**, *50*, 171–200; R. Mosello, CNR—Institute of Ecosystem Study, Pallanza, Italy, unpublished data from 1990 to 2000).

in Lago Maggiore's drainage basin is limited, urban sewage from a heavy population base (6.34×10^5 current inhabitants in the drainage basin) provides most of the phosphorus inputs to the lake. Lake nitrogen concentrations seem to be closely related to atmospheric inputs; just under 80% of the nitrogen reaching Lago Maggiore comes from the atmosphere through deposition (Mosello *et al.*, 2001), mostly from the highly industrialized Po Plain situated upwind. Furthermore, three of the biggest lakes draining into Lago Maggiore are highly eutrophic and are important nutrient sources (Mosello and Ruggiu, 1985). Total phytoplankton biomass in Lago Maggiore increased 2.5 times during the two decades after the 1950s (de Bernardi *et al.*, 1988). Over this time, cyanobacteria bloomed for the first time and became the dominant phytoplankton in Lago Maggiore (Manca *et al.*, 1992).

Lago Maggiore has been on the path of recovery since the early 1980s, mostly through the reduction of phosphorus loads. Construction of five treatment plants with a tertiary stage halved the phosphorus load from the shoreline. There has been a further reduction of phosphorus inputs from tributaries along which treatment plants were constructed. Restoration efforts of eutrophic upstream lakes, especially Lago di Lugano, also helped the recovery of Lago Maggiore. Phosphorus loads were also reduced by gradual reductions of the phosphorus content in detergents, from 8% before 1982 to 1% a decade later. (Provini *et al.*, 1992). As a result of these point-source reductions, phosphorus loads from the drainage basin decreased by half from the late 1970s to the late 1980s. Between 1979 and 2000, mean annual total phosphorus concentrations decreased fivefold to ~6.5 $\mu g\ L^{-1}$, while total- and nitrate-nitrogen concentrations have not changed much (Figure 7), probably because of substantial inputs from the atmosphere (Manca *et al.*, 1992; Istituto Italiano di Idrobiologia—CNR, 2001). In response to reduced phosphorus concentrations and because phytoplankton growth was limited by phosphorus, water quality improved and phytoplankton biodiversity increased (Manca *et al.*, 1992). Less than 10 years after the first bloom, cyanobacterial blooms were no longer seen and the dominance of this group had decreased.

9.08.3.5 Lake Victoria

Lake Victoria is located in eastern Africa's rift valley, shared by the countries of Kenya, Uganda, and Tanzania (Figure 3). This ancient lake is an important ecological resource because of its rich diversity of endemic fish species. Due to its impressive size (with a surface area of $6.88 \times 10^4\ km^2$, it is the second largest freshwater lake in the world after Lake Superior), Lake Victoria is also an important source of cheap protein for millions of people that reside on its shore. Furthermore, the lake regulates the activities of human populations that live downstream in water-scarce locations. Outflow occurs solely north into the White Nile, near Jinja, Uganda. The White Nile crosses into Sudan and links with the Blue Nile in the capital of Khartoum to form the main Nile flow through Egypt to the Mediterranean Sea. Lake Victoria supplies 14% of the water passing through the Nile and thus influences Nile baseflow. The 1964 floods in Cairo were caused by high lake levels due to record rainfalls (Yin and Nicholson, 1998). The water budget of Lake Victoria is controlled largely by evaporation and direct precipitation due to its small drainage ratio and because inputs from tributaries (20% of total water entering the lake) approximately equal outflow over the long term. The contribution of the atmosphere to the nutrient budget of semi-arid lakes is greater, in general, than in temperate systems because of the relatively high evapotranspiration rates in the former.

Nutrients and biota in Lake Victoria are highly dependent on the physical structure of the water column, which is in turn dependent on seasonality. Seasons in the African Great Lakes region are mediated by the annual cycle of Indian monsoon winds. There are two wet seasons: a longer and wetter rainy season occurs between March and May and the short rainy season falls between October and December. The dry or windy season extends from June to September and is characterized by a cooler climate with strong and persistent southerly winds. The thermocline is strongest during the wet seasons and weakens during the drier, windy season. During the windy season, heat loss through evaporation and turbulence caused by the wind enhances mixing in the water column. Thermocline depth increases substantially during the windy season and complete vertical mixing occurs (Ochumba, 1996). During the rainy season of September to April, runoff, wet deposition of nutrients, and thus, phytoplankton biomass are at their yearly maxima. During the windy season, the open offshore areas are mixed to the bottom and the phytoplankton is pushed to low-light depths, limiting their growth (Mugidde, 1993). Therefore, changes in the phytoplankton biomass of Lake Victoria are determined by seasonality.

Nutrients enter Lake Victoria mainly through two diffuse pathways, or nonpoint sources: land runoff and atmospheric deposition, together accounting for ~90% of phosphorus and ~94% of nitrogen input into the lake. About one-third of the nitrogen inputs to Lake Victoria are from atmospheric deposition and over half of the phosphorus comes from particles associated with runoff from agricultural lands (Scheren *et al.*, 2000). Nitrogen loading is dominated by biological fixation in Lake Victoria; the latter provides two-thirds of the total nitrogen supply (Mugidde, 2001). Water bodies in semi-arid climates like Lake Victoria are relatively more dependent on nutrient supply from the atmosphere than temperate systems. High evapotranspiration rates result in relatively less water and nutrients released from the drainage basin, given the relative size of the drainage basin. In addition, undisturbed tropical soils are high in iron and aluminum oxide concentrations that tightly bind phosphorus, making the latter unavailable for dissolution in runoff. Through a process called podzolization, soluble organic acids play an important role in removing iron and aluminum by chelation (Schlesinger, 1997). However, organic decomposition at the floor of tropical forests is so complete that nearly no soluble organic acids percolate through the soil profile, inhibiting podzolization. The growth of phytoplankton biomass in the East African lake appears to be nitrogen limited. Phytoplankton biomass in Lake Victoria increased only when nitrogen additions were made to the lake, either alone or in combination with phosphorus. However, phosphorus additions alone produced no detectable change in phytoplankton biomass (Lehman and Branstrator, 1994). The total nitrogen to total phosphorus ratio (average of 13 : 1 by weight for 1990 and 1992 to 1996) is slightly lower in Lake Victoria water than the ratio needed for optimum phytoplankton growth (around 16 : 1), providing more evidence for nitrogen limitation of the growth of phytoplankton biomass (Guildford and Hecky, 2000).

Thirty million inhabitants live in Lake Victoria's relatively small drainage basin (1.94×10^5 km^2), making it one of the most densely populated areas of Africa (~240 individuals km^{-2}). Rapid population growth (3% yr^{-1}) is adding to already intense environmental pressures. Approximately half of the drainage basin is cultivated and in some areas, cattle densities are as high as human densities. Most (71%) of the energy consumed in sub-Saharan Africa is in the form of fuel wood due to the high poverty level of the populace and because wood gathering imposes no personal financial burden (Davidson, 1992). Therefore, most of the drainage basin has been cleared for rangeland, cultivation, and wood for household burning.

Our knowledge of the response of Lake Victoria to local environmental degradation is based on data from scattered sources: water quality measurements from 1960 to 1961 (Talling and Talling, 1965) and the early 1990s (Mugidde, 1993), and sedimentary studies of past aquatic conditions (Hecky, 1993). Due to a lack of long-term field measurements, knowledge of nutrient dynamics in tropical lakes is severely limited. However, Lake Victoria is one of a handful of tropical lakes with seasonal water quality data. Nitrogen and phosphorus deposition to Lake Victoria's sediments increased at different times: nitrogen deposition increased early in the twentieth century until the 1960s when phosphorus deposition began a rapid increase. Bio-available phosphorus and the nitrogen concentrations of the lake water were consistently higher during the rainy season in 1990–1991, compared to 1960–1961 (Figure 8). Recent increases in lake nutrient concentrations may have been caused, in part, by greater atmospheric loading as a result of wood burning and increased soil erosion. Soil disturbances from deforestation and agriculture are likely to have reduced the soil retention capacity of the drainage basin. In response to increased nutrient inputs to Lake Victoria, phytoplankton biomass (estimated as chlorophyll *a*) was over 10-fold higher in 1990–1991, than in 1960–1961.

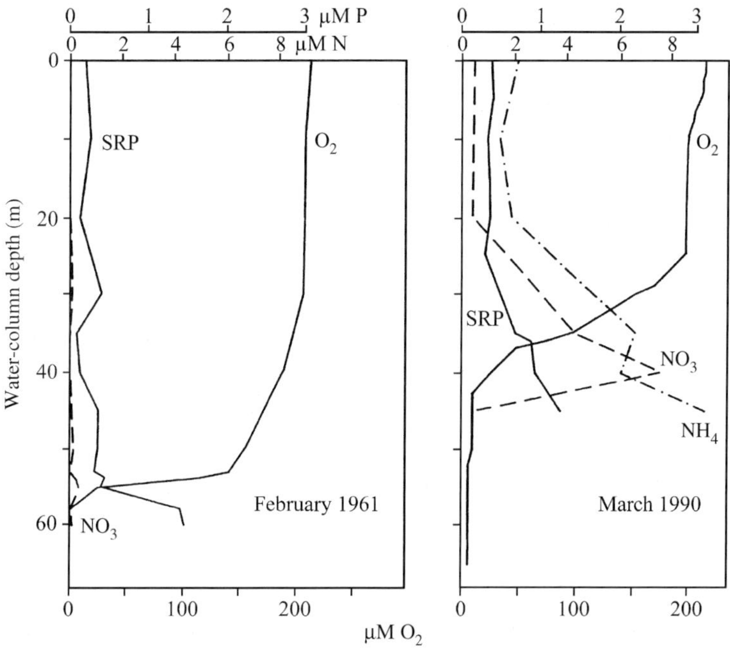

Figure 8 Depth profile of O_2, bio-available phosphorus (SRP), nitrate (NO_3), and ammonium (NH_4, 1990 only) concentrations in February 1961 and March 1990 during Lake Victoria stratification (reproduced by permission of E. Schweizerbart Science Publishers from *Verh. Int. Ver. Limnol.*, **1993**, *25*, 39–48).

Furthermore, the phytoplankton community structure has changed from a community dominated by diatoms, to one dominated by eutrophic species of cyanobacteria that produce substances toxic to humans (*Microcystis* sp., *Anabaena* sp.). Due to an increase in floating particles (e.g., phytoplankton) in the water column, water transparency and the size of the zone in which photosynthesis may occur (photic zone) have decreased. Over a period of 60 years, Secchi disk depth had dropped ~70% by 1990–1991 (Mugidde, 1993) and the area of the lake floor within the photic zone has been halved (Hecky, 1993). Decomposition of organic matter in the deep water has increased dissolved oxygen consumption. As a result, dissolved oxygen concentrations at lake depths of 40–60 m were consistently lower during the rainy season in 1990–1991 than 30 years earlier (Figure 8). Lake hypoxia is now more frequent, more persistent, and affects a greater fraction of Lake Victoria's bottom. Anoxia below a depth of 45 m now affects up to 50% of the lake's bottom area for prolonged periods of time (Hecky, 1993). Tropical water bodies are especially susceptible to dissolved oxygen depletion because of the inverse relationship between temperature and solubility of oxygen in natural waters. As temperature increases, the saturation concentration of oxygen in water decreases and the metabolic rate of decomposers (i.e., oxygen-users) increases (Lewis, 1987). Massive fish kills have been associated with cyanobacterial blooms and the mixing of anoxic waters from deep layers (Ochumba, 1987). Due to its large volume and relatively small outflow, the water of Lake Victoria has a long flushing time (140 yr; Bootsma and Hecky, 1993). Therefore, the recovery of the lake will be slow because pollutants are retained for a long time.

9.08.3.6 Chesapeake Bay

Chesapeake Bay (Figure 2) is the largest and, historically, the most productive estuary in the continental US, and it has been studied extensively for decades. Its drainage basin (165,760 km^2) spans six states and is highly populated (15 million inhabitants). However, nutrient contributions from urban effluent are thought to be minor compared to the nonpoint source of heavy fertilization for agriculture. Agriculture accounts for about half of the total nitrogen and phosphorus loads to the Bay (Magnien *et al.*, 1995). Because of its shallow depth (mean depth ~9 m), much of the estuary's bottom historically received enough light to support submerged plant biomass. Yet, Chesapeake Bay is deep enough to stratify strongly in the summer.

The growth of phytoplankton is typically nitrogen-limited in the ocean due to enhanced sediment release of phosphorus relative to nitrogen (Caraco *et al.*, 1990), slow rates of nitrogen fixation (Howarth, 1988), and high rates of denitrification (bacterial conversion of nitrate to nitrogen gas; Seitzinger, 1988). In the Chesapeake Bay estuary, the growth of phytoplankton is limited by nitrogen or phosphorus, depending on the season. Late winter/early spring is the period of greatest flow in Chesapeake Bay and its tributaries. Runoff from the drainage basin, rich in nitrate from fertilizer application, creates high nitrogen concentrations in Chesapeake Bay and its sub-estuaries. Therefore, the most severe phytoplankton blooms occur in the late winter/early spring season. Owing to the wealth of nitrogen, the ratio of dissolved inorganic nitrogen to phosphorus concentrations in late winter can be as high as 90 : 1 by weight and phytoplankton become phosphorus limited (D'Elia *et al.*, 1986). During the summer, when drainage basin runoff is at its lowest, nitrate concentrations in Chesapeake Bay are low and sometimes undetectable (Malone *et al.*, 1988). However, phosphorus concentrations increase due to internal release from bottom sediments through bacterial decomposition and chemical diffusion to top layers. As a result, during the summer the nitrogen to phosphorus ratio of Chesapeake Bay water can be as low as 5 : 1 by weight, and nitrogen limits the growth of phytoplankton (D'Elia *et al.*, 1986).

Eutrophication of Chesapeake Bay has been occurring since Europeans began aggressive deforestation of the drainage basin. By the end of the nineteenth century, 80% of the land was cleared, mostly for agriculture. Due to reduced transpiration potential, water export from the drainage basin increased during this time, resulting in a gradual increase in nitrogen concentrations in Chesapeake Bay. Total annual loading of nitrogen and phosphorus, mostly from terrestrial sources, have increased 6- and 17-fold, respectively, since pre-European times (Boynton *et al.*, 1995). Consequently, the eutrophication of Chesapeake waters has been increasing; the centric to pennate diatom ratio, a sedimentary indicator of eutrophication, increased from 1.3 before European settlement to 5.2 in recent sediments. Along with a doubling in human population and a tripling in inorganic fertilizer use, phytoplankton biomass (estimated as chlorophyll *a*) increased up to fivefold during the two decades prior to the 1970s. Phytoplankton biomass has remained similar or slightly higher since then and phytoplankton blooms are now more frequent and of longer duration (Harding, 1994). The neurotoxin-producing dinoflagellate *Pfiesteria piscicida* was first reported in the Bay in the early 1990s (Lewitus *et al.*, 1995) and was responsible for fish kills and impaired memory capacity in fishermen and field workers that sampled the estuary (Grattan *et al.*, 1998). Also, diatom biodiversity decreased from 100 to 68 species during the three decades before the early 1990s, a further indication of phytoplankton stress in the Bay (Cooper and Brush, 1991). Chesapeake Bay has a long history of eutrophication.

Eutrophication in Chesapeake Bay has reduced the habitats for fish, rooted aquatic plants, and bottom dwelling organisms. During blooms, phytoplankton not consumed by predators die and settle to the bottom where they is decomposed by bacteria. In response to increased phytoplankton production and sediment load from the drainage basin, sedimentation rates in Chesapeake Bay have increased from 1.5- to sevenfold since the late eighteenth century (Cooper and Brush, 1991). Anoxia associated with bacterial decomposition develops in the bottom layers of Chesapeake Bay, affecting many aquatic organisms (i.e., fish and bottom-dwellers such as oysters). The habitat of these organisms is reduced as they are forced to move to oxic parts of the estuary bottom. Seasonal anoxia has intensified between 1950 and 1980 time period and now occurs annually (Cooper and Brush, 1991; Malone *et al.*, 1988).

The distribution and abundance of submerged aquatic vegetation in Chesapeake Bay has decreased since the 1960s. Severe changes began when Tropical Storm Agnes (1972) caused a dramatic increase in sediment loading from the drainage basin to the estuary. Suspended sediment particles reduced water transparency and caused a drastic decrease in the distribution and abundance of submerged plants. A decade later, the major tributaries of the Bay were barren. Sedimentary analysis of seeds and pollen from aquatic plants of Chesapeake Bay indicates that seeds were deposited to sediments from the eighteenth century to the time of Tropical Storm Agnes; at that time they disappeared abruptly from the record (Orth and Moore, 1983). Such a major decline in aquatic plants is unprecedented in the Chesapeake Bay estuary.

Efforts to clean the Bay began when, in 1982, 1987, and 2000, the states covering most of the drainage basin (Virginia, Maryland, and Pennsylvania), the District of Columbia, the Chesapeake Bay Commission, and the US Environmental Protection Agency, signed agreements to protect and restore Chesapeake Bay's ecosystem. Part of the 1987 agreement was a commitment to achieve a 40% reduction in nitrogen and phosphorus loads to the Bay by 2000, in comparison with 1985. Bio-available phosphorus concentrations had decreased by at least half since the early 1970s due to tertiary phosphorus treatment of sewage effluent (Harding, 1994). The goals of the 2000 Agreement will require even greater reductions in point-source nitrogen loading from wastewater treatment facilities. Currently, 45% of the nitrogen flow from sewage treatment plants is treated and by 2010, 79% is to be treated. In an effort to reduce nonpoint-source nutrient loads,

2.5×10^4 acres of tidal and nontidal wetlands are to be restored by 2010. Also, by 2012, the rate of urban sprawl development of forest and agricultural land in the Chesapeake Bay drainage basin is to be reduced by 30% from the average over 1992–1997. It is too early to determine the level of success achieved by the efforts to clean up Chesapeake Bay.

9.08.4 EUTROPHICATION CONTROL

Eutrophication control methods are time- and watercourse-specific. To predict the success of methods under consideration, the predominant sources of nutrients must be identified and related to changes in water quality. Reduction of nutrient loads from point sources is often the simplest approach when nutrient enhancement stems from known sources, such as sewage. Where drainage basins were disturbed by forest harvesting, forest fires, road building, land development, agriculture, and urbanization, diffuse or nonpoint sources of nutrients may be the most important contributors to aquatic eutrophication. In water bodies, where conditions are such that anoxic water accumulates over the bottom sediments, *in situ* sources can provide most of the nutrients to the euphotic zone, a process termed internal loading. In this case, the management of internal sources may be necessary to complement external nutrient control.

9.08.4.1 Control of Concentrated (Point) Nutrient Sources

Historically, the primary method to control eutrophication has involved reducing or eliminating point sources of nutrients, mostly sewage from urban areas. Diversion of point sources, such as sewage effluent, downstream is a cost-effective method of nutrient load reduction into lakes. The outstanding improvements in water quality in Lake Washington encouraged other communities to consider this management solution. However, diversion is often not a practical management solution because it transfers the problem downstream, rather than eliminating it. Advanced wastewater treatment is a more popular point-source management solution. The addition of a tertiary stage of water treatment can precipitate phosphorus with aluminum sulfate (alum) or calcium hydroxide (slaked lime). However, tertiary treatment entails high initial capital and operational costs. It is therefore crucial to ensure that the aquatic system considered for receipt of tertiary treated water will benefit significantly from phosphorus removal. In Lake Tahoe, the initial strategy to improve water quality was to treat sewage through tertiary phosphorus removal and then to return the effluent to the lake. However, short-term bioassays indicated that phytoplankton biomass was stimulated by nitrogen, and not phosphorus. Thus, in Lake Tahoe, the nitrogen left in effluents after tertiary treatment would continue to enhance phytoplankton biomass (Goldman, 1993). In Shagawa Lake, Minnesota, phosphorus was the primary factor controlling phytoplankton biomass and 80% of the phosphorus was derived from sewage. An advanced wastewater treatment plant was constructed. This removed 99% of the phosphorus and within two years phytoplankton biomass (estimated as chlorophyll *a*) decreased and Secchi disk depth increased by 50% (Horne and Goldman, 1994). Another method of point-source nutrient removal that has proved effective is the construction of basins, or preimpoundments, that retain nutrient-rich water for a short period of time, allowing nitrogen- and phosphorus-bearing particles to settle out of the water column prior to entering downstream waters.

9.08.4.2 Control of Diffuse (Nonpoint) Nutrient Sources

Diffuse or nonpoint nutrient sources are often most important in drainage basins dominated by agricultural activity. In the USA, agriculture is the most important source of nutrients to lakes and rivers and the third largest source of nutrients after urban runoff and municipal sewage to estuaries (Parry, 1998). Unfortunately, the control of nutrients from nonpoint sources has proved most difficult. Control of runoff from nutrient-rich sources often relies on a suite of methods, together called "BMPs". One or a combination of BMPs may be needed for pollution reduction, these include:

(i) Soil stabilization to minimize the movement of soils and attached nutrients. The addition of chemical soil stabilizers such as high molecular weight anionic polyacrylamide can reduce soil loss ninefold and phosphorus loss five- to sevenfold (Lentz *et al.*, 1998). Grassed outlets can decrease sediment suspension by running water and reduce nutrient concentrations in solution. Revegetation can improve stabilization of soil surfaces through root networks and increased soil content of heavy organics. Conservation tillage reduces the erosive energy of rainfall by leaving crop residues at the land surface. Also, in some cases, buffer strips of vegetation left along the shoreline can absorb excess nutrients and water before they escape from the drainage basin. Livestock exclusion from watercourses prevents bank erosion and direct nutrient inputs from feces.

(ii) Interruption of overland flow. Water treatment techniques such as the creation of artificial wetlands that collect water and remove nutrients through aquatic plants. Basins can also be constructed to collect runoff water and allow settling of suspended sediment, often rich in phosphorus, before discharge downstream (Brown *et al.*, 1981). Other examples include flow regulators such as energy dissipators within culverts, reduced slope angles, and soil roughening.

(iii) Changes in chemical application techniques to minimize excess nutrient availability and reduce export. A decrease in road salt usage or a change to calcium chloride usage can help to preserve roadside ditch vegetation (Goldman and Lubnow, 1992). The use of pellet fertilizers, rather than granular and liquid fertilizers, can reduce the availability of nutrients to weathering by releasing nutrients more slowly. Soil and manure testing should be completed before application to determine the need for fertilization. Ploughing manure into the soil, and the proper timing of chemical applications, such as the application of fertilizer after spring thaw instead of in the fall, can reduce the vulnerability of added nutrients to removal in surface runoff. Additions of slaked lime or alum to manure can greatly reduce nitrate volatilization and phosphorus solubility (Moore and Miller, 1994).

(iv) Reduction of nutrients at their source. Genetic measures can be employed to increase phosphorus absorption capacity of livestock. Much of the phosphorus in corn grain is present in the form of phytic acid, a phosphorus source that is not digestible in monogastric animals and thus is often excreted in waste. Corn with reduced levels of phytic acid can be isolated with chemical mutant induction. Using this method, a 65% reduction in phytic acid has been recorded with no effect on total grain phosphorus content (Ertl *et al.*, 1998). Also, intensive grazing by livestock can reduce off-farm nutrient inputs because less feed is needed and imported. Nutrient inputs in feed and fertilizer exceed production outputs in crop and animal produce leaving the farm or drainage basin.

9.08.4.3 Control of Internal Nutrient Sources

Reduction of the visible symptoms of eutrophication (i.e., green scum) can be achieved by various methods. An indirect method that has received considerable attention, although it is still in the experimental phase, is biomanipulation or the manipulation of aquatic food chains to reduce phytoplankton biomass. The idea is to reduce predation on aquatic grazers such as large zooplankton, which feed on phytoplankton. This method has proven successful in increasing water transparency and dissolved oxygen concentrations and reducing phytoplankton biomass (Shapiro and Wright, 1984; Vighi *et al.*, 1995). However, stabilization of nonequilibrium populations is difficult and unpredictable over the long term (Kasprzak *et al.*, 1993). Another method of aquatic vegetation reduction involves mechanical harvesting of macrophytes or surface blooms. This method provides immediate relief from conditions that impair water recreation, but it requires repeated application, is costly, and often spreads the problem out over a larger area. Chemical control of phytoplankton blooms with algicides, such as copper sulfate, can also keep the biomass of phytoplankton at a minimum. However this method requires continuous applications; the suppression effects are only temporary because previously organic-bound phosphorus is released. Further, the use of herbicides such as copper sulfate raises concerns about their possible toxicity to other biota. Complete mechanical circulation of the water column, achieved by vigorous mechanical aeration of deep waters, can reduce phytoplankton biomass by pushing the phytoplankton to greater depths, where light is insufficient for their growth.

Sediments can be an important source of nutrients. The most popular method to prevent phosphorus release from sediments is inactivation. Phosphorus inactivation involves chemical treatment of the water column with aluminum sulfate/sodium aluminate (alum) or lime to precipitate phosphorus out of the water column. Once in the bottom sediments, the precipitate can continue to adsorb phosphorus and prevent its release to the overlying water (Prepas *et al.*, 2001a). Phosphorus inactivation can be highly effective: in a survey of treated American lakes, phosphorus loading from sediments was reduced by about two-thirds and treatments are expected to last 10 years and 15 years for shallow (polymictic) and deep (dimictic) lakes, respectively, before reapplication (Welch and Cooke, 1999). However, only relatively small lakes have potential for phosphorus inactivation because of the high costs of chemicals. Hypolimnetic oxygenation without destratification can eliminate the reducing environment at the sediment–water interface and diminish nutrient release from sediments. Hypolimnetic oxygenation in Amisk Lake, Alberta, from 1988 to 1993 increased dissolved oxygen concentrations in the hypolimnion fivefold to near an average of 5 mg L^{-1} in the treated basin. This change improved the habitat for fish and their food base. Also, surface water chlorophyll *a* and total phosphorus concentrations were reduced such that a lake previously classified as eutrophic was reclassified as mesotrophic (Prepas *et al.*, 1997). Withdrawal of nutrient-rich, hypoxic, water from the hypolimnion can reduce the transfer of nutrients from the deep water to surface waters, thereby reducing the potential for increased phytoplankton biomass. In 10 lakes, hypolimnetic withdrawal decreased epilimnetic total phosphorus concentration by 11% yr^{-1} on average. Operational costs are very low, especially when passive siphoning is employed (Nürnberg, 1987). Finally, mechanical removal of sediments by dredging is a very effective but expensive method to reduce nutrient inputs to systems with high internal loads. For example, in lake Trümmen, a shallow Swedish lake, cyanobacterial biomass decreased after dredging only 0.5 m from the lake bottom (Cronberg *et al*., 1975). The challenge of dredging, however, is to find a use, like crop fertilization, for waste sediment. Decreases in epilimnetic total phosphorus concentrations from internal loading control can often be detected within two to three years.

9.08.5 THE FUTURE—OPPORTUNITIES FOR EUTROPHICATION MANAGEMENT

The primary focus of any drainage basin management program should be to restrict external nutrient loads. With such action, lakes with small internal nutrient sources can be restored quickly, as in the case of phosphorus and in Lake Washington. However, where internal loading is a significant component of the annual nutrient load, restoration can be lengthy. For example, Shagawa Lake (Minnesota) is projected to take 80 years to achieve a 90% reduction in sediment phosphorus load (Chapra and Canale, 1991). Early action to prevent eutrophication, as was done in Lake Tahoe, can minimize accumulation of phosphorus-rich sediments. Otherwise, the removal of stockpiled phosphorus in bottom sediments can require a long and costly cleanup. There are too few Lake Washingtons and Lago Maggiores, where resources were relatively abundant to assess and to remedy the problem.

Worldwide, the greatest unresolved water quality concerns are in countries with the fewest resources, where demand for water is growing and research and restoration funding sources are scarce. The fate of aquatic resources in the world's poorest countries should be of concern to all countries. Because the toxicity of drinking water often increases with nutrient enrichment, excess nutrients should be treated as toxic substances and banned. There is an urgent need for the universal regulation of nutrients to protect drinking water supplies and aquatic biodiversity.

ACKNOWLEDGMENTS

We thank R. Mosello and L. Harding for supplying unpublished data, and R. Hecky, A. Litt, S. Abella, C. Goldman, J. Mackarewicz, J. Lehman, R. Mosello, D. Boesch, J. Burke, B. Sherwood-Lollar, and an anonymous reviewer for reviewing the manuscript, or parts thereof.

REFERENCES

Ambrosetti W. and Barbanti L. (1999) Deep water warming in lakes: an indicator of climatic change. *J. Limnol.* **58**, 1–9.

Ambrosetti W., Barbanti L., Mosello R., and Pugnetti A. (1992) Limnological studies on the deep southern Alpine lakes Maggiore, Lugano, Como, Iseo, and Garda. *Mem. Ist. Ital. Idrobiol.* **50**, 117–146.

Bartish T. (1987) A review of exchange processes among the three basins of Lake Erie. *J. Great Lakes Res.* **13**, 607–618.

Bartsch A. F. (1972) Nutrients and eutrophication—prospects and options for the future. In *Nutrients and Eutrophication: The Limiting Nutrient Controversy* (ed. G. E. Likens). The American Society of Limnology and Oceanography, Lawrence, Kansas, pp. 297–300.

Beeton A. M. (1961) Environmental changes in Lake Erie. *Trans. Am. Fish. Soc.* **90**, 153–159.

Beeton A. M. (1963) *Limnological Survey of Lake Erie 1959 and 1960. Great Lakes Fisheries Commission*, Ann Arbor, MI, Report no. 6.

Beeton A. M. (1965) Eutrophication of the St. Lawrence Great Lakes. *Limnol. Oceanogr.* **10**, 240–254.

Bertram P. E. (1993) Total phosphorus and dissolved oxygen trends in the central basin of Lake Erie, 1970–1991. *J. Great Lakes Res.* **19**, 224–236.

Bootsma H. A. and Hecky R. E. (1993) Conservation of the African great lakes: a limnological perspective. *Conserv. Biol.* **7**, 644–656.

Bosch J. M. and Hewlett J. D. (1982) A review of catchment experiments to determine the effect of vegetation changes on water yield and evapo-transpiration. *J. Hydrol.* **55**, 3–23.

Boynton W. R., Garber J. H., Summers R., and Kemp W. M. (1995) Inputs, transformations, and transport of nitrogen and phosphorus in Chesapeake Bay and selected tributaries. *Estuaries* **18**, 285–314.

Brown M. J., Bondurant C. E., and Brockway C. E. (1981) Ponding surface drainage water for sediment and phosphorus removal. *Trans. ASAE* **24**, 1478–1481.

Caraco N., Cole J., and Likens G. E. (1990) A comparison of phosphorus immobilization in sediments of freshwater and coastal marine systems. *Biogeochemistry* **9**, 277–290.

Chapra S. C. and Canale R. P. (1991) Long-term phenomenological model of phosphorus and oxygen for stratified lakes. *Water Res.* **25**, 707–715.

Codd G. A. (1995) Cyanobacterial toxins: occurrence, properties and biological significance. *Water Sci. Technol.* **32**, 149–156.

Cooper S. R. and Brush G. S. (1991) Long-term history of Chesapeake Bay Anoxia. *Science* **254**, 992–996.

Cronan C. S. and Aiken G. R. (1985) Chemistry and transport of soluble humic substances in forested watersheds of the Adirondack Park, New York, USA. *Geochim. Cosmochim. Acta* **49**, 1697–1706.

Cronberg G., Gelin C., and Larsson K. (1975) Lake Trummen restoration project: II. Bacteria, phytoplankton and phytoplankton productivity. *Verh. Int. Ver. Limnol.* **19**, 1088–1096.

Crosby J. M., Bradford M. E., Mitchell P. A., Prepas E. E., McIntyre L. G., Hart Buckland-Nicks L., and Hanson J. M. (1990) In *Atlas of Alberta Lakes* (eds. P. Mitchell and E. Prepas). University of Alberta Press, Edmonton, Alberta.

Crumpton W., Isehart G. M., and Mithell P. L. (1992) Nitrate and organic N analyses with second-derivative spectroscopy. *Limnol. Oceanogr.* **37**, 907–913.

D'Arcy P. and Carignan R. (1997) Influence of catchment topography on water chemistry in southeastern Québec Shield lakes. *Can. J. Fish. Aquat. Sci.* **54**, 2215–2227.

Davidson O. (1992) Energy issues in sub-Saharan Africa: future directions. *Ann. Rev. Energy Environ.* **17**, 359–404.

Davis C. C. (1964) Evidence for the eutrophication of Lake Erie from phytoplankton records. *Limnol. Oceanogr.* **9**, 275–283.

de Bernardi R., Giussani G., Manca M., and Ruggiu D. (1988) Long-term dynamics of plankton communities in Lago Maggiore (N. Italy). *Verh. Int. Ver. Limnol.* **23**, 729–733.

D'Elia C. F., Sanders J. G., and Boynton W. R. (1986) Nutrient enrichment studies in a coastal plain estuary: phytoplankton growth in large-scale, continuous cultures. *Can. J. Fish. Aquat. Sci.* **43**, 397–406.

Dillon P. J. and Kirchner W. B. (1975) The effects of geology and land use on the export of phosphorus from watersheds. *Water Res.* **9**, 135–148.

Dolan D. M. (1993) Point-source loadings of phosphorus to Lake Erie: 1986–1990. *J. Great Lakes Res.* **19**, 212–223.

Dunn J. (1996) Algae kills dialysis patients in Brazil. *Br. Med. J.* **312**, 1183–1184.

Edmondson W. T. (1969) Eutrophication in North America. In *Eutrophication: Causes, Consequences, Correctives*, Publ. 1700. National Academy of Science/National Research Council, Washington, DC, pp. 129–149.

Edmondson W. T. (1975) Recovery of Lake Washington from eutrophication. In *Recovery and Restoration of Damaged Ecosystems* (eds. J. Cairns, Jr., K. L. Dickson, and E. E. Henricks). University Press of Virginia, Charlottesville, VA, pp. 102–109.

Edmondson W. T. (1994) Sixty years of Lake Washington: a curriculum vitae. *Lake Reserv. Manage.* **10**, 75–84.

Ertl D. S., Young K. A., and Raboy V. (1998) Plant genetic approaches to phosphorus management in agricultural production. *J. Environ. Qual.* **27**, 299–304.

Goldman C. R. (1993) Failures, successes and problems in controlling eutrophication. *Mem. Ist. Ital. Idrobiol.* **52**, 79–87.

Goldman C. R. and Lubnow F. S. (1992) Seasonal influence of calcium magnesium acetate on microbioal processes in 10 northern Californian lakes. *Resour. Conserv. Recy.* **7**, 51–67.

Goldman C. R., Jassby A. D., and Hackley S. H. (1993) Decadal, interannual, and seasonal variability in enrichment bioassays at Lake Tahoe, California–Nevada, USA. *Can. J. Fish. Aquat. Sci.* **50**, 1489–1496.

Grattan L. M., Oldach D., Perl T. M., Lowitt M. H., Matuszak D. L., Dickson C., Parrott C., Shoemaker R. C., Kaufman C. L., Wasserman M. P., Hebel J. R., Charache P., and Morris J. G., Jr. (1998) Learning and memory difficulties after environmental exposure to waterways containing toxin-producing *Pfiesteria* or *Pfiesteria*-like dinoflagellates. *Lancet* **352**, 532–539.

Guildford S. J. and Hecky R. E. (2000) Total nitrogen, total phosphorus, and nutrient limitation in lakes and oceans: is there a common relationship? *Limnol. Oceanogr.* **45**, 1213–1223.

Harding L. W., Jr. (1994) Long-term trends in the distribution of phytoplankton in Chesapeake Bay: roles of light, nutrients, and streamflow. *Mar. Ecol. Prog. Ser.* **104**, 267–291.

Hecky R. E. (1993) The eutrophication of Lake Victoria, East Africa. *Verh. Int. Ver. Limnol.* **25**, 39–48.

Hobbie J. E. and Likens G. E. (1973) Output of phosphorus, dissolved organic carbon, and fine particulate carbon from Hubbard Brook watersheds. *Limnol. Oceanogr.* **18**, 734–742.

Horne A. J. and Goldman C. R. (1994) *Limnology*. McGraw-Hill, New York, NY.

Howarth R. W. (1988) Nutrient limitation of net primary production in marine ecosystems. *Ann. Rev. Ecol. Syst.* **19**, 89–110.

Hutchinson G. E. (1973) Eutrophication. *Am. Sci.* **61**, 269–279.

Istituto Italiano di Idrobiologia—CNR (2001) Ricerche sull'evoluzione del Lago Maggiore. Aspetti limnologici. Programma quinquennale 1998–2002. Campagna 2000. (ed. Commissione Internazionale per la protezione delle acque italo-svizzere). Ist. Ital. Idrobiol.—CNR, 77pp.

Jassby A. D., Reuter J. E., Axler R. P., Goldman C. R., and Hackley S. H. (1994) Atmospheric deposition of nitrogen and phosphorus in the annual nutrient load of Lake Tahoe (California–Nevada). *Water Resour. Res.* **30**, 2207–2216.

Jassby A. D., Goldman C. R., and Reuter J. E. (1995) Long-term change in Lake Tahoe (California–Nevada, USA) and its relation to atmospheric deposition of algal nutrients. *Arch. Hydrobiol.* **135**, 1–21.

Jassby A. D., Goldman C. R., Reuter J. E., Richards R. C., and Heyvaert A. C. (2001) Lake Tahoe: diagnosis and rehabilitation of a large mountain lake. In *The Great Lakes of the World (GLOW): Food-web, Health and Integrity* (eds. M. Munawar and R. E. Hecky). Backhuys Publishers, Leiden, The Netherlands, pp. 431–454.

Jenny H. (1980) *The Soil Resource*. Springer, New York.

Kalff J. (2002) *Limnology*. Prentice Hall, New Jersey.

Kasprzak P., Krienitz L., and Koschel R. (1993) Biomanipulation: a limnological in-lake ecotechnology of eutrophication management? *Mem. Ist. Ital. Idrobiol.* **52**, 151–169.

Lehman J. T. and Branstrator D. K. (1994) Nutrient dynamics and turnover rates of phosphate and sulfate in Lake Victoria, East Africa. *Limnol. Oceanogr.* **39**, 227–233.

Lentz R. D., Sojka R. E., and Robins C. W. (1998) Reducing phosphorus losses from surface-irrigated fields: emerging polyacrylamide technology. *J. Environ. Qual.* **27**, 305–312.

Lewis W. M., Jr. (1987) Tropical limnology. *Ann. Rev. Ecol. Syst.* **18**, 159–184.

Lewitus A. J., Hesian R. V., Kana T. M., Burkholder J. M., Glasgow H. B., and May E. (1995) Discovery of the "phantom" dinoflagellate in the Chesapeake Bay. *Estuaries* **18**, 373–378.

Ludsin S. A., Kershner M. W., Blocksom K. A., Knight R. L., and Stein R. A. (2001) Life after death in Lake Erie: Nutrient controls drive fish species richness, rehabilitation. *Ecol. Appl.* **11**, 731–746.

Magnien R., Boward D., and Bieber S. (1995) *The State of the Chesapeake 1995*. USEPA, Annapolis, MD.

Makarewicz J. C. (1993) Phytoplankton biomass and species composition in Lake Erie, 1970 to 1987. *J. Great Lakes Res.* **19**, 258–274.

Makarewicz J. C. and Bertram P. (1991) Evidence for the restoration of the Lake Erie ecosystem. *Bioscience* **41**, 216–223.

Makarewicz J. C., Lewis T. W., and Bertram P. (1999) Phytoplankton composition and biomass in the offshore waters of Lake Erie: pre- and post-*Dreissena* introduction (1983–1993). *J. Great Lakes Res.* **25**, 135–148.

Malone R. C., Crocker L. H., Pike S. E., and Wendler B. W. (1988) Influences of river flow on the dynamics of phytoplankton production in a partially stratified estuary. *Mar. Ecol. Prog. Ser.* **48**, 235–249.

Manca M., Calderoni A., and Mosello R. (1992) Limnological research in Lago Maggiore: studies on hydrochemistry and plankton. *Mem. Ist. Ital. Idrobiol.* **50**, 171–200.

Moore P. A., Jr. and Miller D. M. (1994) Decreasing phosphorus solubility in poultry litter with aluminum, calcium, and iron amendments. *J. Environ. Qual.* **23**, 325–330.

Mosello R. and Ruggiu D. (1985) Nutrient load, trophic conditions and restoration prospects of Lake Maggiore. *Int. Rev. ges. Hydrobiol.* **70**, 63–75.

Mosello R., Barbieri A., Brizzio M. C., Calderoni A., Marchetto A., Passera S., Rogora M., and Tartari G. (2001) Nitrogen budget of Lago Maggiore: the relative importance of atmospheric deposition and catchment sources. *J. Limnol.* **60**, 27–40.

Mugidde R. (1993) Changes in phytoplankton primary productivity and biomass in Lake Victoria (Uganda). *Int. Ver. Theor. Angew. Limnol. Verh.* **25**, 846–849.

Mugidde R. (2001) Nutrient status and planktonic nitrogen fixation in Lake Victoria, Africa. PhD Thesis, University of Waterloo.

Munawar M. and Munawar I. F. (1976) A lakewide study of phytoplankton biomass and its species composition in Lake Erie, April–December 1970. *J. Fish. Res. Board Can.* **33**, 581–600.

Nürnberg G. K. (1987) Hypolimnetic withdrawal as lake restoration technique. *J. Environ. Eng.* **113**, 1006–1017.

Ochumba P. B. O. (1987) Periodic massive fish kills in the Kenyan part of Lake Victoria. *Water Qual. Bull.* **12**, 119–122.

Ochumba P. B. O. (1996) Measurement of water currents, temperature, dissolved oxygen and winds on the Kenyan Lake Victoria. In *The Limnology, Climatology, and Paleoclimatology of the East African Lakes* (eds. T. C. Johnson and E. O. Odata). Gordon and Breach, Australia, pp. 155–167.

OECD (Organization for Economic Cooperation and Development) (1982) *Eutrophication of Waters: Monitoring, Assessment, and Control.* OECD, Paris.

Orth R. J. and Moore K. A. (1983) Chesapeake Bay: an unprecedented decline in submerged aquatic vegetation. *Science* **222**, 51–53.

Parry R. (1998) Agricultural phosphorus and water quality: a US Environmental Protection Agency perspective. *J. Environ. Qual.* **27**, 258–261.

Prepas E. E. and Rigler F. H. (1982) Improvements in quantifying the phosphorus concentration in lake water. *Can. J. Fish. Aquat. Sci.* **39**, 822–829.

Prepas E. E., Field K. M., Murphy T. P., Johnson W. L., Burke J. M., and Tonn W. M. (1997) Introduction to the Amisk Lake Project: oxygenation of a deep, eutrophic lake. *Can. J. Fish. Aquat. Sci.* **54**, 2105–2110.

Prepas E. E., Pinel-Alloul B., Chambers P. A., Murphy T. P., Reedyk S., Sandland G., and Serediak M. (2001a) Lime treatment and its effects on the chemistry and biota of hardwater eutrophic lakes. *Freshwater Biol.* **46**, 1049–1060.

Prepas E. E., Planas D., Gibson J. J., Vitt D. H., Prowse T. D., Dinsmore W. P., Halsey L. A., McEachern P. M., Paquet S., Scrimgeour G. J., Tonn W. M., Paszkowski C. A., and Wolfstein K. (2001b) Landscape variables influencing nutrients and phytoplankton communities in Boreal Plain lakes of northern Alberta: a comparison of wetland- and upland-dominated catchments. *Can. J. Fish. Aquat. Sci.* **58**, 1286–1299.

Provini A., Marchetti R., and Tartari G. (1992) The Italian lakes: trophic status and remedial measures. *Mem. Ist. Ital. Idrobiol.* **50**, 147–169.

Rawson D. S. (1960) A limnological comparison of twelve large lakes in Northern Saskatchewan. *Limnol. Oceanogr.* **5**, 195–211.

Redfield A. C. (1958) The biological control of chemical factors in the environment. *Am. Sci.* **46**, 205–221.

Riley E. T. and Prepas E. E. (1984) Role of internal phosphorus loading in two shallow, productive lakes in Alberta, Canada. *Can. J. Fish. Aquat. Sci.* **41**, 845–855.

Roby K. B. and Azuma D. L. (1995) Changes in a reach of a northern California stream following wildfire. *Environ. Manage.* **19**, 591–600.

Rosa F. (1987) Lake Erie central basin total phosphorus trend analysis from 1968 to 1982. *J. Great Lakes Res.* **13**, 684–696.

Ruggiu D. and Mosello R. (1984) Nutrient levels and phytoplankton characteristics in the deep southern alpine lakes. *Verh. Int. Ver. Limnol.* **22**, 1106–1112.

Ruttner F. (1952) *Fundamentals of Limnology*, 3rd edn. (Translated in 1953 by D. G. Frey and F. E. S. Fry). University of Toronto Press, Toronto, ON.

Ryding S.-O. and Rast W. (1989) *The Control of Eutrophication of Lakes and Reservoirs.* UNESCO, Paris.

Scheffer V. B. and Robinson R. J. (1939) A limnological study of Lake Washington. *Ecol. Monogr.* **9**, 95–143.

Scheren P. A. G. M., Zanting H. A., and Lemmens A. M. C. (2000) Estimation of water pollution sources in Lake Victoria, East Africa: application and elaboration of the rapid assessment methodology. *J. Environ. Manage.* **58**, 235–248.

Schindler D. W. (1977) Evolution of phosphorus limitation in lakes. *Science* **195**, 260–262.

Schlesinger W. H. (1997) *Biogeochemistry: An Analysis of Global Change.* Academic Press, San Diego, CA.

Seitzinger S. P. (1988) Denitrification in freshwater and coastal marine ecosystems: ecological and geochemical significance. *Limnol. Oceanogr.* **33**, 702–724.

Shapiro J. and Wright D. I. (1984) Lake restoration by biomanipulation. *Freshwater Biol.* **14**, 371–384.

Silbergeld E. K., Grattan L., Oldach D., and Morris J. G. (2000) *Pfiesteria*: harmful algal blooms as indicators of human: ecosystem interactions. *Environ. Res.* **82**, 97–105.

Smith V. H. (1986) Light and nutrient effects on the relative biomass of blue-green algae in lake phytoplankton. *Can. J. Fish. Aquat. Sci.* **43**, 148–153.

Snell E. A. (1987) Wetland distribution and conversion in southern Ontario. Working Paper No. 48. Inland Waters/Lands Directorate, Environment Canada.

Søballe D. M. and Kimmel S. T. (1987) Advection, phytoplankton, biomass, and nutrient transformations in a rapidly flushed impoundment. *Arch. Hydrobiol.* **105**, 187–203.

Talling J. F. and Talling I. B. (1965) The chemical composition of African lake waters. *Int. Rev. ges. Hydrobiol.* **50**, 421–463.

Trimbee A. M. and Prepas E. E. (1987) Evaluation of total phosphorus as a predictor of the relative biomass of blue-green algae with emphasis on Alberta lakes. *Can. J. Fish. Aquat. Sci.* **44**, 1337–1342.

Vallentyne J. E. and Thomas N. A. (1978) *Fifth Year Review of Canada–United States Great Lakes Water Quality Agreement.* International Joint Commission, Windsor, ON.

Van den Hoeck C., Mann D. G., and Jahns H. M. (1995) *Algae: An Introduction to Phycology.* Cambridge University Press, Cambridge, UK.

Vighi M., Sandroni D., and Ferri G. (1995) Biomanipulation of trophic chain in two small eutrophic lakes. *Mem. Ist. Ital. Idrobiol.* **53**, 157–175.

Watson S. B., McCauley E., and Downing J. A. (1997) Patterns in phytoplanktontaxonomic composition across temperate lakes of differing nutrient status. *Limnol. Oceanogr.* **42**, 487–495.

Welch E. B. and Cooke G. D. (1999) Effectiveness and longevity of phosphorus inactivation with Alum. *Lake Reserv. Manage.* **15**, 5–27.

Yin X. and Nicholson S. R. (1998) The water balance of Lake Victoria. *Hydrol. Sci. J.* **43**, 789–811.

9.09
Salinization and Saline Environments

A. Vengosh

Ben Gurion University of the Negev, Beer Sheva, Israel

9.09.1 INTRODUCTION

One of the most conspicuous phenomena of water-quality degradation, particularly in arid and semi-arid zones, is salinization of water and soil resources. Salinization is a long-term phenomenon, and during the last century many aquifers and river basins have become unsuitable for human consumption owing to high levels of salinity. Future exploitation of thousands of wells in the Middle East and in many other water-scarce regions in the world depends, to a large extent, on the degree and rate of salinization. Moreover, every year a large fraction of agricultural land is salinized and becomes unusable.

Salinization is a global environmental phenomenon that affects many different aspects of our life (Williams, 2001a,b): changing the chemical composition of natural water resources (lakes, rivers, and groundwater), degrading the quality of water supply to the domestic and agriculture sectors, contribution to loss of biodiversity, taxonomic replacement by halotolerant species (Williams, 2001a,b), loss of fertile soil, collapse of agricultural and fishery industries, changing of local climatic conditions, and creating severe health problems (e.g., the Aral Basin). The damage due to salinity in the Colorado River Basin alone, for example, ranges between \$500 and \$750 million per year and could exceed \$1 billion per year if the salinity in the Imperial Dam increases from 700 mg L^{-1} to 900 mg L^{-1} (Bureau of Reclamation, 2003, USA). In Australia, accelerating soil salinization has become a massive environmental and economic disaster. Western Australia is "losing an area equal to one football oval an hour" due to spreading salinity (Murphy, 1999). The annual cost for dryland salinity in Australia is estimated as AU\$700 million for lost land and AU\$130 million for lost production (Williams *et al.*, 2002). In short, the salinization process has become pervasive.

Salinity in water is usually defined by the chloride content (mg L^{-1}) or total dissolved solids content (TDS, mg L^{-1}or g L^{-1}), although the chloride comprises only a fraction of the total dissolved salts in water. The Cl/TDS ratio varies from 0.1 in nonmarine saline waters to ~0.5 in marine-associated saline waters. Water salinity is also defined by electrical conductivity (EC). In soil studies, the electrical conductivity and the ratio of $Na/\sqrt{(Ca + Mg)}$ (SAR) are often used as an indirect measure of soil salinity. In addition to chloride, high levels of other dissolved constituents may limit the use of water for domestic, agriculture, and industrial applications. In some parts of Africa, China, and India, for example, high fluoride content is associated with saline groundwater and causes severe dental and skeletal fluorosis (Shiklomanov, 1997). Hence, the "salinity" problem is only the "tip of the iceberg," as high levels of salinity are associated with high concentrations of other inorganic pollutants (e.g., sodium, sulfate, boron, fluoride), and bioaccumulated elements (e.g., selenium, and arsenic) (see Chapter 9.03).

The World Health Organization (WHO) recommends that the chloride concentration of the water supply for human consumption should not exceed 250 mg L^{-1}. Agriculture applications also depend upon the salinity level of the supplied water. Many crops, such as citrus, avocado, and mango, are sensitive to chloride concentration in irrigation water (an upper limit of 250 mg L^{-1}). In addition, long-term irrigation with water enriched with sodium results in a significant reduction in the hydraulic conductivity and hence the fertility of the irrigated soil. Similarly, the industrial sector demands water of high quality. For example, the high-tech industry requires a large amount of water with low levels of dissolved salts. Hence, the salinity level of groundwater is one of the limiting factors that determine the suitability of water for a variety of applications.

The salinity problem is a global phenomenon but it is more severe in water-scarce areas, such as arid and semi-arid zones. The increasing demand for water has created tremendous pressures on water resources that have resulted in lowering water level and increasing salinization. For example, in the Middle East salinity is the main factor that limits water utilization, and future prospects for water use in Israel, Palestinian Authority, and Jordan are overshadowed by the increasing salinization (Vengosh and Rosenthal, 1994; Salameh, 1996). The salinity problem has numerous grave economic, social, and political consequences, particularly in cross-boundary basins that are shared by different communities (e.g., Salinas Valley California; Vengosh *et al.*, 2002a), friendly states (e.g., salinization of the Colorado River along Mexico-US border; Stanton *et al.*, 2001), and hostile states (e.g., the Jordan River, Vengosh *et al.*, 2001; Aral Basin, Weinthal, 2002; Euphrates River, Beaumont, 1996; and the Nile River, Ohlsson, 1995).

Salinization of water resources also affects agricultural management. The type of irrigation water and its quality determine the salinity and fertility of the soil and eventually the quality of the underlying water resource. The use of treated wastewater or other marginal water (e.g., brackish water) depends on the salinity and the chemical composition of the water. Treated wastewater with high contents of chloride, sodium, and boron is suitable only for salt-tolerant crops and requires special treatment of the soil. Finally, high boron in irrigation water and consequently in soil water is also an important limiting factor for crops, as boron is an essential micronutrient for plants but becomes toxic at high levels (typically >0.75 mg L^{-1} in irrigation water).

This chapter investigates the different mechanisms and geochemistry of salinization in different parts of the world. The role of the unsaturated zone in shaping the chemical composition of dryland salinization is discussed. Special emphasis is on the anthropogenic effects and to man-made fluids and reused water, such as treated wastewater and agricultural drainage water. Two anthropogenic salinization cycles are introduced—the agricultural and the domestic cycles. Some useful geochemical fingerprinting tracers are also included for defining the sources of salinity. Finally, the chemical composition of future water resources is predicted, based on the chemical and isotopic fractionation associated with remediation and desalination.

9.09.2 RIVER SALINIZATION

More than one-half of the world's major rivers are being seriously depleted and polluted, degrading and poisoning the surrounding ecosystems, and threatening the health and livelihood of people, who depend on them for drinking water and irrigation. Rivers are being depleted because the global demand for water is rising sharply. The problem will be further aggravated by the need to supply food, drinking water, and irrigation waters for an additional two billion people on Earth by 2025 (Serageldin, 2000). The World Commission on Water for the twenty-first century (Serageldin, 2000) defines the most stressed rivers as: the Yellow River in China, Amu Darya and Syr Darya in Central Asia, Colorado River in the western USA, the Nile River in Egypt, Volga River in Russia, Ganges River in India, and the Jordan River in the Middle East.

Salinization of surface waters occurs mainly due to a combination of natural and anthropogenic processes. In the natural setting, particularly in the dryland environment, salts are deposited on the ground, stored in the subsurface, and transported to shallow groundwater that discharges into adjacent rivers. Some rivers flow through arid regions, although their source lies in wetter parts of upper basins (Colorado, Rio Grande, Orange, Nile, Ephrates, Tigris, Jordan, Indus, Murray; Figure 1). About 50% of arid land is located in "endorheic" regions whence there is no flow to the ocean. In these regions, rivers flow into lakes such as the Caspian, Aral, Chad, Great Salt, Eyre, Dead Sea, and Titicaca, which have no outlets.

Salinization of rivers also occurs due to human intervention, e.g., diversion of upstream natural flow, dam construction, and consequently significant reduction of natural flow discharge. For example, the Amu Darya and Syr Darya rivers in Central Asia were almost desiccated due to the diversion of water for cotton irrigation in the former Soviet Union (Weinthal, 2002). The estimated historical annual flow in these two rivers is $122 \times 10^9\ m^3$. By the mid-1980s, the Amu Darya and Syr Darya no longer flowed to the Aral Sea (Micklin, 1988, 1992).

In a classic paper, Pillsbury (1981) described how recycling of salts via irrigation and agricultural return flow controls the salinity of downstream rivers in arid zones. Once the natural salt balance is disturbed and salts begin to accumulate, either in the unsaturated zone or in drainage waters, the salinity increases. The salinity of the Colorado River is derived from a century of activity that includes upstream diversion of freshwater, massive irrigation, evapotranspiration and salt accumulation in the soil, and return of saline drainage back to the river (Pillsbury, 1981). Similarly, the rise of the salinity in the Nile Delta has been attributed to a disturbance of the natural salts balance after the construction of Assuan dam and the reduction of the natural outflow of salts from the Nile River to the Mediterranean Sea. About $50 \times 10^9\ m^3$ of the Nile water is used for irrigation, from which the recycling of drainage water increases the salinity of the northern part close to the outlets to the Mediterranean Sea (Kotb *et al.*, 2000). Saline drainage water is also the primary source of salinity of the Euphrates and Tigris rivers in Iraq (Robson *et al.*, 1983; Fattah and Abdul Baki, 1980), coupled with sewage pollution (Mutlake *et al.*, 1980; Al-Muhandis, 1977), and reduction of the discharge of the river. Since the mid-1960s, Turkey has embarked on a large-scale program for the development of southeastern Turkey using water as the main agent of change (Beaumont, 1996). The Southeastern Anatolian Project (GAP) with the giant Ataturk dam is one of the largest water projects in the world. It utilizes more than $13 \times 10^9\ m^3\ yr^{-1}$ from the Euphrates and Tigris rivers for irrigation. Consequently, the annual flow of the rivers reduced by 30–50% (of an annual natural discharge of $\sim 30 \times 10^9$; Beaumont, 1996). Massive irrigation in Syria and Iraq has resulted in the formation of saline agricultural return flows that are mixed with local shallow groundwater and then discharged back to the downstream sections of the river. As a result, the salinity of the Euphrates River, close to its confluence into the Persian Gulf, is higher than 3,000 mg L^{-1} (Robson *et al.*, 1983; Fattah and Abdul Baki, 1980).

In the dryland environment, river salinization occurs as a response to land clearing of deep-rooted natural vegetation, which accelerates recharge rates and causes groundwater tables to rise. In the Murray–Darling Basin in South Australia, soluble aerosols derived from the ocean are deposited in the drainage basin, concentrated by evapotranspiration, and discharged to the Murray River (Herczeg *et al.*, 1993). Large-scale clearing of natural vegetation and its replacement by annual

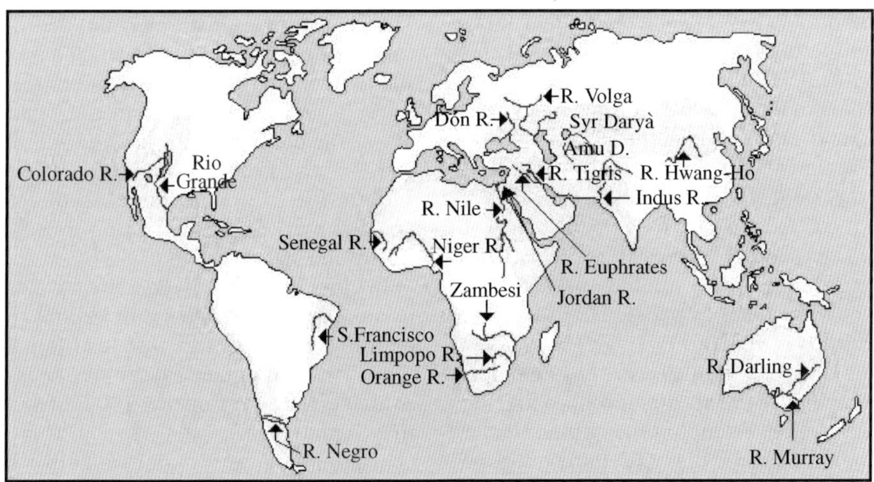

Figure 1 Major river systems and arid and semi-arid regions of the world.

crops and pastures has increased the amount of water leaking through the unsaturated zone. Excess irrigation has also increased groundwater levels (Williams *et al.*, 2002 and references there in). The rise of saline groundwater and mobilization of salts stored in the unsaturated zone has formed saline seepages that discharge to streams and the soil surface, which affects runoff salinity. In the Murray River in South Australia (Table 1) water salinity has been gradually increasing since the 1950s, and the 2000–2001 salinity level exceeded international water standards (defined as 800 μs cm^{-3}) for ~10% of the year (Williams *et al.*, 2002; Jolly *et al.*, 2000). In the Sandspruit River, a tributary of Berg River in the semi-arid western Cape Province of South Africa, salinization of the river (TDS range from 4,500 mg L^{-1} to 9,000 mg L^{-1}) is also attributed to accumulation and leaching of salts from the soil. Salt mass-balance estimations suggest that the salts are derived from both recycling of meteoric aerosols and soil leaching (Flugel, 1995).

A different type of river salinization in a dryland environment is represented by the Jordan River Basin along the border between Israel and Jordan. A 10-fold reduction of surface water flow in the Jordan River ((50–200) $\times 10^6$ m^3 today relative to ~1,400 $\times 10^6$ m^3 in historical times) and intensification of shallow groundwater discharge resulted in the salinization of the Jordan River. During August 2001, the salinity of the southern end of the Jordan River, just before its confluence into the Dead Sea, reached 11 g L^{-1}, a quarter of the Mediterranean seawater salinity. Based on Na/Cl, Br/Cl, $^{87}Sr/^{86}Sr$, $\delta^{11}B$, $\delta^{34}S_{sulfate}$, and $\delta^{18}O_{water}$. ($\delta^{11}B = [(^{11}B/ \ ^{10}B)_{sample}/(^{11}B/^{10}B)_{NBS\text{-}SRM\text{-}951} - 1] \times 10^3$), $\delta^{34}S_{sulfate} = [(^{34}S/ \ ^{32}B)_{sample}/(^{34}S/^{32}S)_{VCDT} - 1] \times 10^3$, and $\delta^{18}O_{water} = [(^{18}O/^{16}O)_{sample}/(^{18}O/^{16}O)_{SMOW} - 1] \times 10^3$) whereas NBS SRM 951, VCDT, and SMOW are international standards (Coplen *et al.*, 2002) variations, the salinity rise of the southern section of the Jordan River (Table 1; Figure 2) was attributed to the discharge of saline groundwater (Vengosh *et al.*, 2001). It was suggested that the shallow saline groundwater was formed by mixing of saline groundwater originated from leaching of local saline sediments and deep hypersaline brines. Likewise, the salinity of the Rio Grande River increases to ~2,000 mg L^{-1} along a distance of 1,200 km. The parallel decrease of the Br/Cl ratio of the salinized river implies discharge of saline groundwater that has interacted with halite deposits (Phillips *et al.*, 2002).

Salinization of rivers can also occur in temperate climatic zone owing to direct anthropogenic contamination. For example, the Rhine River suffers from the discharge of potash mine drainage brines since the opening of potash mines ~100 yr ago. Chloride levels and salt fluxes have increased by a factor of 15–20. The rise of the annual chloride load at the Rhine River mouth (recorded since 1880) reflects an increase from natural load of less than 5 kg s^{-1} to more than 300 kg s^{-1} in the 1960s (Meybeck and Helmer, 1989). Another example is the Arno River in northern Tuscany that is polluted by wastewaters, which has resulted in a downstream increase of Na^+, Cl^-, and SO_4^{2-} ions with a characteristic sulfur isotopic composition (Cortecci *et al.*, 2002).

In sum, there are four major sources of soluble salts in river basins: (i) meteoric salts; (ii) salts derived from water–rock interaction (e.g., dissolution of evaporitic rocks); (iii) salts derived from remnants of formation water entrapped in the basin; and (iv) anthropogenic salts (e.g., wastewater effluents). Meteoric salts are concentrated via in-stream net evaporation and evapotranspiration along the river flow. In addition, meteoric salts can be recycled through irrigation in the watershed and development of saline agricultural drainage water that flows to the river.

9.09.3 LAKE SALINIZATION

Naturally occurring salt lakes reflect hydrological equilibrium states that have been developed during geological time (see Chapter 5.13). The wide range of chemical composition of natural salt lakes reflects a variety of sources: salts that are derived from the evolution of freshwater inflows (e.g., Great Salt Lake, Utah; Spencer *et al.*, 1985), hydrothermal fluids (Qaidam Basin, China; Vengosh *et al.*, 1995), remnants of evaporated seawater (Dead Sea, Israel; Starinsky, 1974; Stein *et al.*, 1997), and accumulation of marine aerosols over an exposed continent (Lake Eyre, Australia; Chivas *et al.*, 1991; Vengosh *et al.*, 1991a; Herczeg *et al.*, 2001). Human-induced activities may modify the natural conditions, as in the case of the Dead Sea. The diversion of natural tributaries has significantly reduced the inflows, which has resulted in a sharp decrease in the water level of the Dead Sea. Every year the water level of the Dead Sea drops by a meter (Yechieli *et al.*, 1998; Salameh and Naser, 1999). In contrast, the term *lake salinization* used here refers to direct anthropogenic activity and the ongoing transformation of freshwater lakes into salt lakes.

In principle, the diversion of water from one basin to another or the use of natural river inflows are the basic processes that lead to lake salinization. The consequences can be devastating: for example, the diversion of ~100% of the water from the Amu Daryà and Syr Daryà rivers in central Asia to grow cotton and other crops led to the desiccation of the Aral Sea, a fivefold increase in its water salinity (59 g L^{-1} during 1991

Table 1 Chemical composition of saline water from various sources. The ion concentrations are reported in mg L^{-1}, whereas the ionic ratios are molar.

Site	*Source*	*TDS*	*Ca*	*Mg*	*Na*	*K*	*Cl*	SO_4	HCO_3	NO_3	*Br*	*B* (ppb)	*Na/Cl*	SO_4*/Cl*	*Br/Cl* $\times 10^{-3}$	*B/Cl*
Seawater (Red Sea)		41,390	418	1,442	12,396	516	23,290	3,077	161		75.6	5.3	0.86	0.05	1.5	0.8
Freshwater river																
Amzon River	1	39	5.2	1.0	1.5	0.8	1.1	1.7	20			6	2.1	0.6		18.2
Orinoco River	1	35	3.3	1.0	1.5	0.7	2.9	3.4	11			2	0.8	0.4		27.1
Mississippi (1905)	1	216	34	8.9	11	2.8	10.3	25.5	116				1.7	0.9		
Mackenzie	1	211	33	10.4	7	1.1	8.9	36.1	111			12	1.2	1.5		4.4
Danube	1	307	49	9	9	1	19.5	24	190				0.7	0.5		
Congo	1	34	2.4	1.4	2	1.4	1.4	1.2	13.4			3	2.2	0.3		7.2
Zambeze	1	58	9.7	2.2	4	1.2	1	3	25				6.7	1.1		
Saline river																
Murray River, South Australia	2	448	21	17	101	6	171	38	94		0.4		0.91	0.08	1.1	
Jordan River (south end)	3	1,109	545	705	23,00	170	5,370	1,650	254	20	81	2,800	0.66	0.11	6.7	1.7
Tigris River (Baghdad, 1977)	4	521	64	21.7	47.7		82.6	66.5	238	10		200	0.89	0.3		8.7
Saline Lakes																
Salton See (1989)	5	40,700	950	1,300	11,000	220	17,000	10,000	185		13	12,000	1.0	0.22	0.3	2.3
Aral Sea (1991)	6	59,120	1,020	3,600	14,600	640	22,650	16,180	430				0.99	0.26		
Caspian Sea	7	12,385	340	700	3,016	88	5,233	3,008					0.89	0.2		
Dead Sea, Rift valley	8	337,800	17,600	42,120	41,300	7,600	224,200	280	200		4,500	54,690	0.28	0.005	10	0.8
Seawater intrusion																
Coastal aquifer, Israel	9		980	245	2,830	22	6,304	470	206		21.3	950	0.69	0.03	1.5	5.0
Salinas Valley, California	10		410	126	450	12	1,670	212	62		5.4	245	0.42	0.05	1.5	4.8

(continued)

Table 1 (continued).

Site	*Source*	*TDS*	*Ca*	*Mg*	*Na*	*K*	*Cl*	SO_4	HCO_3	NO_3	*Br*	*B* (ppb)	*Na/Cl*	SO_4/Cl	*Br/Cl* $\times 10^{-3}$	*B/Cl*
Saline plumes and upcoming of brines																
Coastal aquifer, Israel (Beer Toviyya)	11	2,560	176	97	545	3.9	1,125	143	370	98	3.5	508	0.75	0.05	1.4	1.5
Ogallala aquifer, Texas	12	67,530	1,460	388	23,850	32	36,120	5,610	41		6.2		1.0	0.06	0.07	
Dammam aquifer, Kuwait	13	4,062	470	138	635	15.2	1,241	1,189	77	8		1,500	0.77	0.35		3.9
Jordan Valley, Jericho, Cenomanian aquifer	14	8,270	490	580	1,700	110	4,950	105	275	62	83		0.52	0.02	7.4	
Pleistocen aquifer	14	3,100	156	208	590	84	1,372	295	363	29	10		0.66	0.16	3.3	
Groundwater associated with freezing process																
Sweden	15	18,200	3,690	31	2,850	12	11,100	522	7		79		0.39	0.02	3.1	
Finland	15	14,400	3,900	13	1,500	7	8,900	1	21		77		0.25	<0.01	3.8	
Agricultural drainage																
San Joaquim Valley, California	16	4,580	192	242	952		499	2,650		44		5,940	2.9	1.9		3.9
Mendota, San Joaquim Valley, California	17	14,280	438	285	3,720	3.1	1,210	8,350	250	25	3.4		4.7	2.5	1.2	
Imperial Valley, California	5	6,715	310	330	1,420	19	1,200	3,000	383	53			1.9	0.9		
Imperial Valley, California	5	8,250	560	320	1,700	24	2,600	2,700	342	5	2.3	1,700	1.0	0.4	0.4	2.0
Wastewaters																
Dan Reclamation Project, Israel (1993)	18	1,300	80	30	264	34.2	361	112.5	420		0.4	500	1.1	0.11	0.5	6
Orleans, Cape Cod, Massachusetts	19	1,700	440	0.3	100	30	950	19	134	30		200	0.16	0.007		0.7

Sources: 1. E. K. Berner and R. A. Berner (1987); 2. Herczeg *et al.* (1993); 3. Vengosh *et al.* (2001); 4. Mutlak *et al.* (1980); 5. Schroedar and Rivera, (1993); 6. Linnikov and Podberezny (1996); 7. Peeters *et al.* (1999); 8. Starinsky (1974). 9. Vengosh *et al.* (1994); 10. Vengosh *et al.* (2002a); 11. Vengosh *et al.* (1999); 12. Mehta *et al.* (2000a); 13. Al-Ruwaith (1995); 14. Marie and Vengosh (2001); 15. Bein and Arad (1992); 16. Mitchel *et al.* (2000); 17. Kharaka *et al.* (1996); 18. Vengosh and Keren (1996); 19. DeSimone *et al.* (1997).

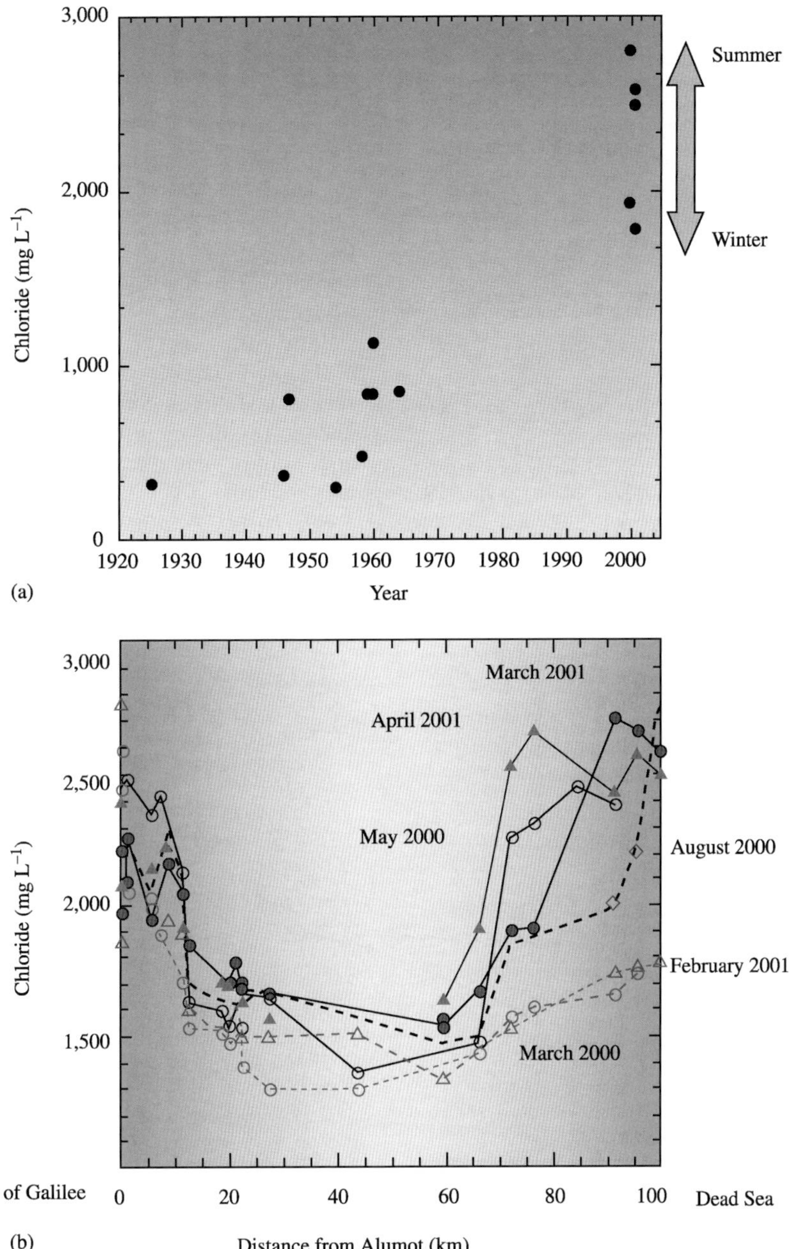

Figure 2 Salinization of the Jordan River. (a) Evolution of the chloride content during the twentieth century as recorded in Abdalla Bridge, the southern point of the Jordan River before its entry into the Dead Sea. (b) Chloride variation transects along the Jordan River. Distance in km refers to the beginning of the river flow (Alumot dam) below the Sea of Galilee (source Vengosh *et al.*, 2001).

(Linnikov and Podberezny, 1996), destruction of the fisheries industry, and accelerating health problems in the region (Micklin, 1988, 1992). Finally, draining saline agricultural return flow from the Imperial Valley in Southern California during the twentieth century into the perminal lake of the Salton Sea (Figure 3) resulted in salt accumulation and the creation of salt lake with a salinity of 43 g L^{-1} (Schroeder *et al.*, 1991, 2002; Schroeder and Rivera, 1993; Hely *et al.*, 1966; Amrhein *et al.*, 2001). The salinity rise of a lake may lead to a dramatic change in the ecological system and change in the species composition (e.g., rise of brine shrimp), which in turn will also modify the food chain mechanism in the lake, and consequently the bioaccumulation of toxic elements such as selenium (see Chapter 9.02; Glenn *et al.*, 1999).

Two types of lake salinization are considered. The first type is the imbalance between evaporation and water inflow, usually due to the diversion of freshwater discharge. This is demonstrated

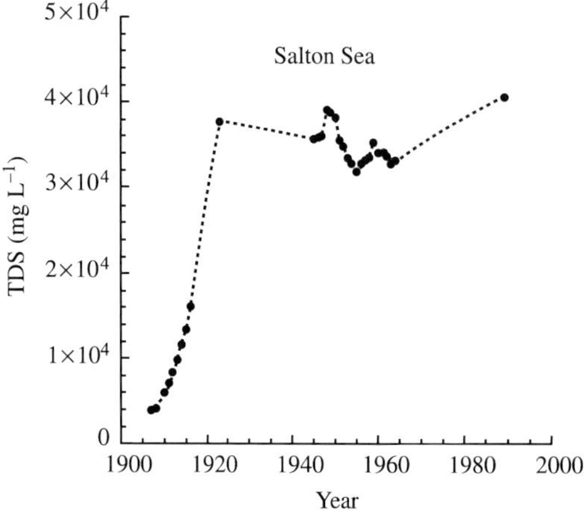

Figure 3 Historical salinization of the Salton Sea (California, USA) during the twentieth century (sources Schroeder *et al.*, 1991; Schroeder and Rivera, 1993; Amrhein *et al.*, 2001).

in the case of the Aral Sea (Table 1), where the chemical composition of the brackish water during early stages of salinization (salinity of 14 g L^{-1}in 1971) is identical to that of brines during later stages (salinity of 56 g L^{-1} during 1991; Figure 4; Linnikov and Podberezny, 1996). The second type of lake salinization is related to saline inflow coupled with evaporation within the lake. The source of the saline inflow can be anthropogenic, as in the case of the Imperial Valley in California, where agricultural return flow with a salinity of 1–3 g L^{-1} flows into the Salton Sea (Table 1; Figure 5). Alternatively, natural saline springs (salinity up to 40 g L^{-1}) flow to the Sea of Galilee (Figure 6) in the northern Jordan Valley and control the chemical composition of the lake (salinity of ~0.5 g L^{-1}; Kolodny *et al.*, 1999; Nishri *et al.*, 1999). In both cases, the chemical composition of the saline inflows dominates the composition of the salinized lake.

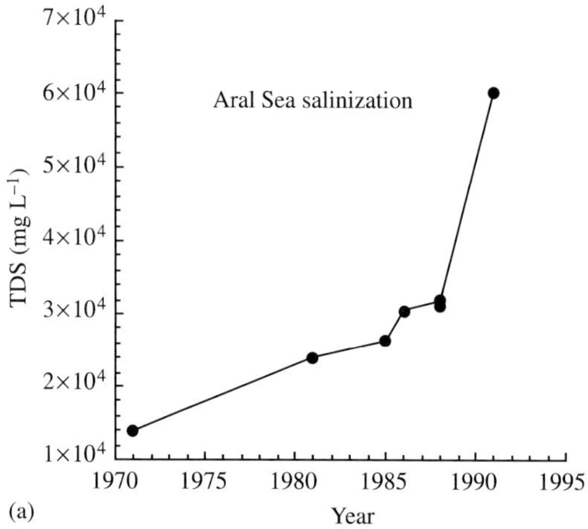

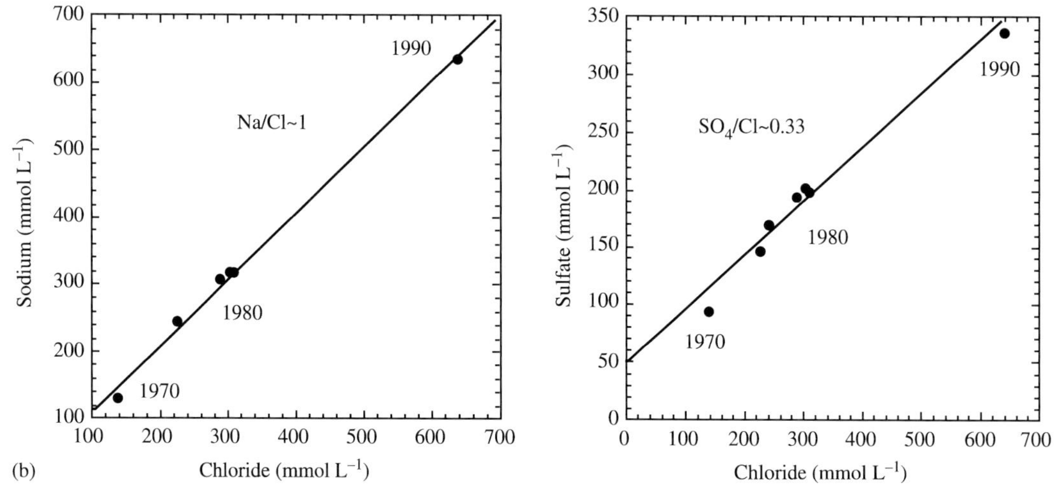

Figure 4 Salinization of Aral Sea. (a) The rise of the salinity during the 1970s and 1980s. (b) Sodium and sulfate versus chloride concentrations (in mmol L^{-1}) as measured in the lake during different times. Note the linear relationships between the different ions, indicating that the initial chemical composition of the saline lake was not modified during the salinization process (source Linnikov and Podberezny, 1996).

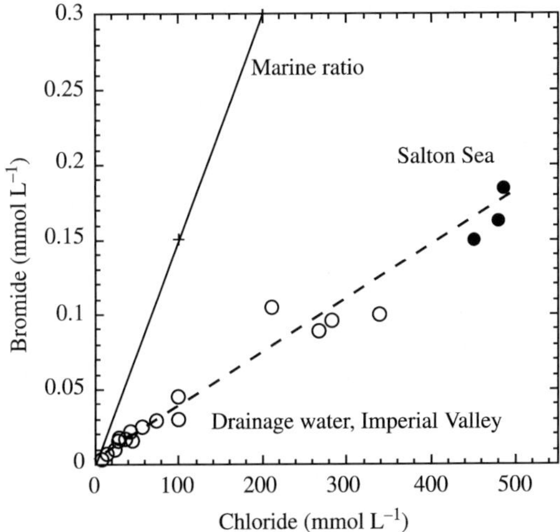

Figure 5 Bromide versus chloride concentrations (in mmol L^{-1}) of agricultural drainage waters (open circles) and the Salton Sea (closed circles) in Southern California, USA. Note the identical Br/Cl ratios measured in both the agricultural drainage waters and the saline lake, inferring that the lake was originated from agricultural drainage inflow. The Br/Cl ratios are lower than the marine ratio, indicating dissolution of halite mineral (sources Schroeder *et al.*, 1991; Schroeder and Rivera, 1993; Amrhein *et al.*, 2001).

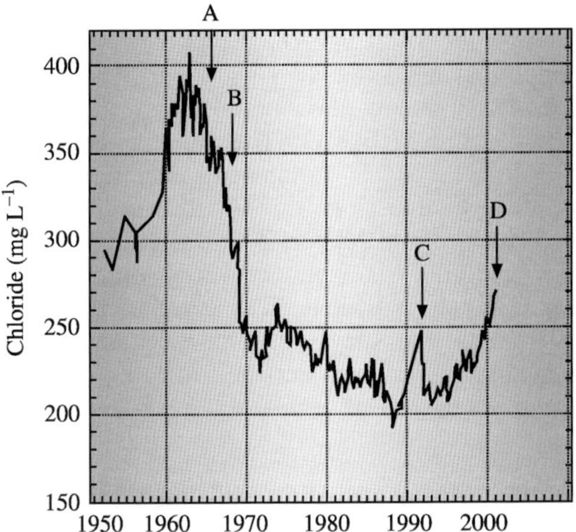

Figure 6 Historic evolution of the chloride content (in mg L^{-1}) of Sea of Galilee, Israel. The variations of the salinity of the lake reflect human operational stages: (i) during the mid-1960s the salinity increased due to the damming of the natural outflow of the lake; (ii) diversion of the saline springs that emerge to the lake (point A) initiated reduction of the salinity, followed by major flood events (point B) during the late 1960s; (iii) low level of the salinity marinated during the 1970s to 1990s as a result of a salt balance between freshwater input, saline water discharge, and pumping; and (iv) drought years (points C and D) resulted in significant reduction of the lake level and extensive evaporation and consequently salinization of the lake (data from the archive of the Hydrological Service, Israel).

In addition to the reduction of freshwater discharge, the fate and the rate of future salinization of lakes depends on the chemical composition of the lakes (see Chapter 5.13). Precipitation of minerals (Hardie and Eugster, 1970) and the activity of the saline water form important negative feedbacks that limit salt accumulation in the lakes. In this respect, the saturation level of calcite, gypsum, and halite in brines can determine the salinity and composition of the residual brines. In the Salton Sea, the lake water has a total salinity of 43 g L^{-1} and is oversaturated with respect to calcite, which controls the Ca^{2+} and HCO_3^- level of the lake. Biologic activity, sulfate reduction processes, and base-exchange reactions (Na^+ removal) add additional dissolved Ca^{2+}, which enhances calcite precipitation (Amrhein *et al.*, 2001). At higher salinity levels, such as in the Dead Sea (Table 1) with a salinity of 340 g L^{-1}, the low water activity and thus the low vapor pressure of the hypersaline brines imposes low evaporation rates. Yechieli *et al.* (1998) predicted that despite significant reduction in freshwater inflows to the Dead Sea, the Dead Sea will not totally desiccate but will reach a new equilibrium, given the low water activity. The geometry of the lake basin and the size of the surface area are also important factors that control the amount of evaporation (Yechieli *et al.*, 1998). In sum, while the rate of salinization is rapid during the early stages of salinization, it decreases at evolved stages due to salt precipitation and the low water activity of the brines. As the salinity increases, the rate of lake salinization is expected to decrease.

9.09.4 GROUNDWATER SALINIZATION

9.09.4.1 Seawater Intrusion

Salt-water intrusion is one of the most widespread and important processes that degrade water quality to levels exceeding acceptable drinking and irrigation water standards, and endanger future water exploitation in coastal aquifers. This problem is intensified by large population growth, and the fact that ~70% of the world population occupies the coastal plain zones. Human activities (e.g., pumping, agriculture, reuse of wastewater) increase the rate of salinization (Jones *et al.*, 1999). Salt-water intrusion into coastal aquifers has been identified in the eastern Atlantic (Meisler *et al.*, 1985; Wicks *et al.*, 1995; Wicks and Herman, 1996) and the western Pacific (Izbicki, 1991, 1996; Todd, 1989; Vengosh *et al.*, 2002a) coasts of the USA. Seawater intrusion occurs even under extremely humid climatic conditions, as in the Jeju volcanic island in South Korea, where the annual average rainfall is ~1,870 mm and the estimated groundwater recharge is 1.4×10^9 m^3 yr^{-1}.

Although the groundwater withdrawal in the volcanic island is only 5% of the estimated replenishment, Br/Cl and $\delta^{18}O$ data indicate that seawater has intruded 2.5 km inland due to the high conductivity and heterogeneity of the basaltic aquifers (Kim *et al.*, 2003). As a result of global warming, a rise of sea level could lead to seawater intrusion in similar volcanic islands in the Pacific and further salinization of their groundwater.

Fossil seawater that represents past invasions into coastal aquifers accompanying rise in sea level can be documented by the apparent old age inferred by the ^{14}C and ^{3}H data of saline waters as reported in Germany (Hahn, 1991), Belgium (De Breuck and De Moor, 1991), India (Sukhija, 1996), and Israel (Yechieli *et al.*, 2001).

In many coastal aquifers around the world, modern seawater intrusion commonly occurs owing to extensive freshwater withdrawals. Ocean water itself has a salinity of 35 g L^{-1} (TDS), while internal seas have higher (e.g., Mediterranean Sea, Red Sea; TDS ~40 gL^{-1}) or lower salinities. Nevertheless, seawater has a uniform chemistry due to the long residence time of the major constituents. There is a predominance of Na^+ and Cl^- with a molar ratio of 0.86, an excess of Cl^- over the alkali ions (Na + K), and Mg^{2+} greatly in excess of Ca^{2+} (Mg/Ca = 4.5–5.2; Table 1). Seawater has also a uniform Br/Cl (1.5×10^{-3}), $\delta^{18}O_{water}$ (0–1‰), $\delta^{34}S_{sulfate}$ (21‰), $\delta^{11}B$ (39‰), and $^{87}Sr/^{86}Sr$ (0.7092) values. In contrast, the chemical compositions of fresh groundwaters is highly variable, though typically they are composed of Ca–Mg–HCO_3. In most cases Ca^{2+} dominates over Mg^{2+}. The most striking phenomenon that characterizes seawater intrusion into coastal aquifers is the difference between the chemical composition of the saline water associated with salt-water intrusion and the theoretical mixture of seawater with groundwater (Jones *et al.*, 1999). In many cases, the saline water has a Ca-chloride composition (i.e., the ratio of $Ca/(SO_4 + HCO_3) > 1$) with low ratios of Na^+, SO_4^{2-}, K^+, and B to chloride relative to modern ocean water (Table 1).

The geochemical modification of seawater intrusion has been attributed to base-exchange reactions with the aquifer rocks (Sayles and Mangelsdorf, 1977; Appelo and Willemsen, 1987; Appelo and Geirnart, 1991; Appelo and Postma, 1993). Typically, cation exchangers in aquifers are clay minerals, organic matter, oxy-hydroxides, and fine-grained rock materials, which have mainly Ca^{2+} adsorbed on their surfaces. When seawater intrudes into a coastal freshwater aquifer, Na^+ replaces part of the Ca^{2+} on the solid surfaces. As a result, Na^+ is taken up by the solid phase, Ca^{2+} is released, the solute composition changes from a Mg-chloride into a Ca-chloride type water, the Na/Cl ratio decreases, and the (Ca + Mg)/Cl ratio increases (Custodio, 1987, 1997; Custodio *et al.*, 1993; Appelo and Postma, 1993; Jones *et al.*, 1999). Under such conditions, the relative enrichments in calcium and magnesium, normalized to chloride concentrations, should be balanced by the relative depletion of sodium (i.e., $\Delta(Ca + Mg) = -2\Delta Na$ molar units; Vengosh *et al.*, 2002a). Together with Ca^{2+}, exchangeable Sr^2 is released to the dissolved phase (Johnson and Depaolo, 1994). The $^{87}Sr/^{86}Sr$ ratio of adsorbed strontium can therefore affect the isotopic composition of strontium in the saline groundwater. In Salinas Valley, California, the $^{87}Sr/^{86}Sr$ ratio of the salinized water increases with the Ca^{2+} content (Figure 7), indicating a contribution of a high $^{87}Sr/^{86}Sr$ source from the clays (Vengosh *et al.*, 2002a). In contrast, in the basaltic aquifer of Jesu island in South Korea, the $^{87}Sr/^{86}Sr$ ratios show conservative relationships with salinity and reflect a mixing between seawater and local groundwater (Kim *et al.*, 2003). Flushing of the mixing zone by freshwater results in an opposite reaction: an uptake of Ca^{2+} and Mg^{2+} by the exchangers with concomitant release of Na^+. This is reflected in an increase of the Na/Cl ratio and a decrease of the (Ca + Mg)/Cl ratio, and formation of an Na–HCO_3 water type (Appelo, 1994).

Finally, seawater intrusion is also characterized by high $\delta^{34}S_{sulfate}$ (>20‰) and $\delta^{11}B$ (>39‰) values that are associated with low SO_4/Cl and B/Cl ratios below the marine ratios (0.05 and 8×10^{-4}, respectively). The relative depletion of sulfate and ^{34}S enrichment is attributed to sulfate reduction along the salt–freshwater interface (Krouse and Mayer, 2000). The removal of dissolved boron is explained via adsorption onto clay minerals in which ^{10}B is adsorbed preferentially onto the clays and the residual saline groundwater becomes enriched in ^{11}B (Vengosh *et al.*, 1994).

9.09.4.2 Mixing with External Saline Waters

Many studies of regional aquifer systems show the general sequence of major ion evolution from low-salinity Ca–Mg–HCO_3 type to saline Na–SO_4–Cl groundwater along a hydraulic gradient (Mazor, 1997; Herczeg *et al.*, 1991, 2001; Hendry and Schwartz, 1988). The ratio of Na^+ and Cl^- ions relative to the other dissolved salts (i.e., Na + Cl/TDS) increases with flow, as demonstrated in the saline groundwater from the Murray aquifer in South Australia (Herczeg *et al.*, 2001) and the Cretaceous sandstone aquifer of Milk River Formation, Alberta, western Canada (Hendry and Schwartz, 1988; Herczeg and Edmunds, 2000). The gradual increase of salinity and the chemical modification towards

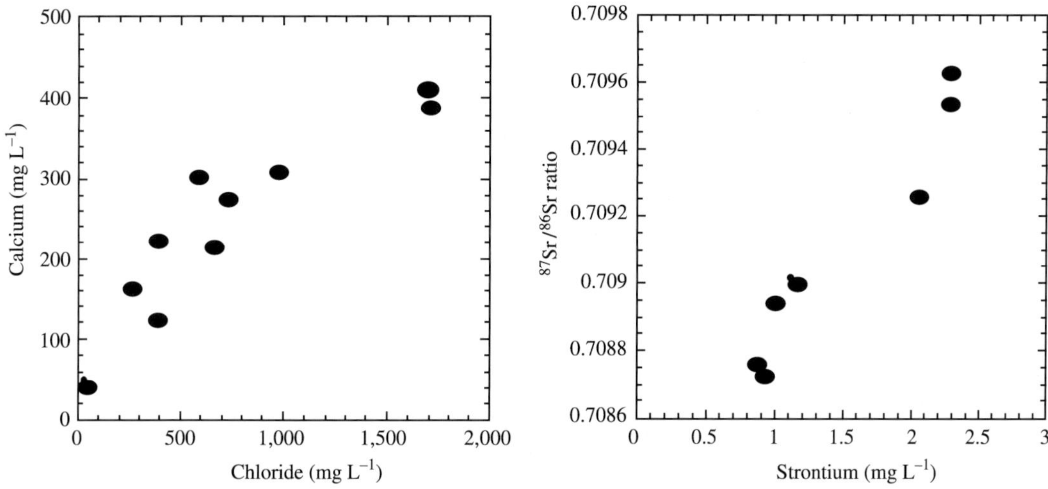

Figure 7 The variations of chloride, calcium, strontium (in mg L^{-1}), and ^{87}Sr/^{86}Sr ratios during intrusion of seawater into the coastal aquifer of Salinas Valley, California, USA. Note the linear relationships between all constituents, which indicates conservative mixing relationships between freshwater and modified seawater. These relationships suggest that base-exchange reactions occur at early stage of seawater intrusion and that Sr in exchange sites has a high ^{87}Sr/^{86}Sr ratio (source Vengosh *et al.*, 2002a).

predominance of chloride and sodium ions is a result of several possible processes: (i) advection and diffusion of saline fluids entrapped in an aquitard that is connected to an active aquifer and (ii) dissolution of soluble salts such as gypsum and halite minerals within the aquifer. For example, several models have been postulated to explain the increase of salinity along flow paths in the Cretaceous sandstone aquifer of the Milk River Formation, Alberta, western Canada. Hendry and Schwartz (1988) hypothesized that the salinity increase is derived from diffusion of solute from an underlying aquitard. The entrapment of saline fluids in geological units of low hydraulic conductivity that are connected to active aquifers may result in diffusion of solute and a gradual increase of salinity (Herczeg and Edmunds, 2000).

Likewise, salinization of many aquifers is induced by intrusion of underlying or adjacent saline groundwater or saline-water flow from adjacent or underlying aquifers (e.g., Magaritz *et al.*, 1984; Maslia and Prowell, 1990 , Vengosh *et al.*, 1999, 2002a,c; Sanchez-Martos and Pulido-Bosch, 1999; Sanchez-Martos *et al.*, 2001; Kloppman *et al.*, 2001; Hsissou *et al.*, 1999). For example, in the upper Floridan aquifer in Georgia, USA, faults breach the nearly impermeable units of the underlying confined aquifer and allow upward leaking of saline groundwater (Maslia and Prowell, 1990). Similarly, extensive saline plumes in the Ogallala aquifer in the southern High Plains, Texas, USA are attributed to cross-formational flow from underlying evaporite units (Mehta *et al.*, 2000a,b). Saline plumes formed from uprising of saline groundwater are also the major source of salinity in the Mediterranean coastal aquifer of Israel (Table 1 and Figure 8; Vengosh *et al.*, 1999). Salinization of groundwater is one of the critical water-quality problems in the Gaza Strip, where the salinity of groundwater exceeds the salinity level of international drinking water. The major source of the salinity is the natural flow of saline groundwater from the eastern part of the aquifer. The rate of flow and salinization have been accelerated due to over-exploitation in the Gaza Strip (Vengosh *et al.*, 2002c).

The impact of salinization is particularly conspicuous in aquifers where the freshwater is not renewable (i.e., fossil; see Chapter 5.15). Numerous studies have shown that groundwater resources across the Sahara and the Sahel region in northern Africa and in the arid zones of the Middle East are fossil and reflect paleorecharge during Late Pleistocene periods of higher humidity (e.g., Cook *et al.*, 1992; Cook and Herczeg, 2000; Phillips, 1994; Edmunds and Gaye, 1994; Gaye and Edmunds, 1996; Edmunds and Droubi, 1998; Edmunds *et al.*, 1999, 2001). Typically, fresh paleowaters overlie saline dense water bodies. With geological time, a fragile hydraulic equilibrium has been established between the two water bodies. Drillhole construction and groundwater abstraction affect this delicate hydrological balance. Exploitation of paleowaters, with no possibility of modern replenishment, can lead to rapid salinization of the groundwater resources. In the Middle East, salinization of fossil groundwater has devastating effects since in some cases they are the only source of potable water (Vengosh and Rosenthal, 1994).

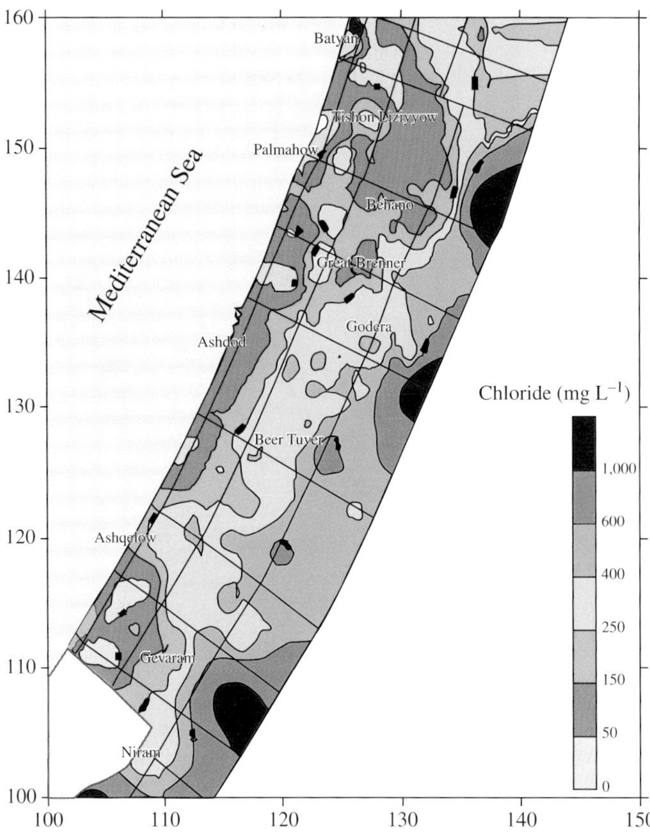

Figure 8 Chloride (in mg L^{-1}) distribution in groundwater from the central and southern part of the Mediterranean coastal aquifer of Israel. Note the saline plumes in the central and eastern parts of the aquifer.

For example, in the Damman carbonate aquifer in Bahrain, over-exploitation has resulted in a drop of the piezometric surface by 4 m and an increase in the salinity by over 3 g L^{-1} owing to a combination of leakage from a deeper aquifer, marine intrusion, sabkha water migration, and agricultural drainage. More than half of the area of the Damman aquifer has become saline as a result of over-exploitation (Edmunds and Droubi, 1998). In Israel, the abstraction of brackish paleowater from the Lower Cretaceous Nubian sandstone aquifer in areas along the Rift Dead Sea Valley has resulted in the intrusion of Ca-chloride brine and the salinization of the associated groundwater. In South Jordan the fossil (i.e., ^{14}C ages of $\sim 1.3 \times 10^4$ yr) freshwaters of the Disi–Mudawwara aquifer have been extracted since the early 1970s, which has resulted in a non-recoverable lowering of water levels by 20 m. The fresh groundwater body of Disi–Mudawwara aquifer is hydraulically interconnected with the overlying aquifer, in particular with the saline bearing of the Khrein confining unit. As such, early signs of salinization have already been recorded in the Disi–Mudawwara aquifer (Salameh, 1996). The future utilization of this valuable water resource, which is the only available freshwater resource in Jordan, thus primarily depends upon the rate and extent of the salinization process.

In this section, the chemical and isotopic compositions of different types of saline groundwater are examined. These include residual evaporated seawater entrapped as formation water, saline water generated by evaporite dissolution, and hydrothermal water. Residual evaporated seawater is typically characterized by a Ca-chloride composition ($Ca/(SO_4 + HCO_3) > 1$) with an Na/Cl ratio below seawater ratio value, Br/Cl $\geq$ seawater ratio (1.5×10^{-3}), relative depletion of sulfate ($SO_4/Cl < 0.05$), $\delta^{34}S_{sulfate} \geq 20‰$, and $\delta^{11}B \geq 39‰$ (Figure 9; Starinsky, 1974; Carpenter, 1978; Carpenter *et al.*, 1974; McCaffrey *et al.*, 1987; Wilson and Long, 1993; Vengosh *et al.*, 1992; Raab and Spiro, 1991). Marine brines can be derived from relics of evaporated seawater that was modified by water–rock interaction (e.g., dolomitization) and sulfate reduction (Starinsky, 1974; Stein *et al.*, 1997). Upon hydrological contact and mixing of brines with fresh groundwater, the chemical composition of the hypersaline brine dominates the composition of the mixture owing to the large difference in salinity between the brines and the freshwater. Consequently, the dilution factor has

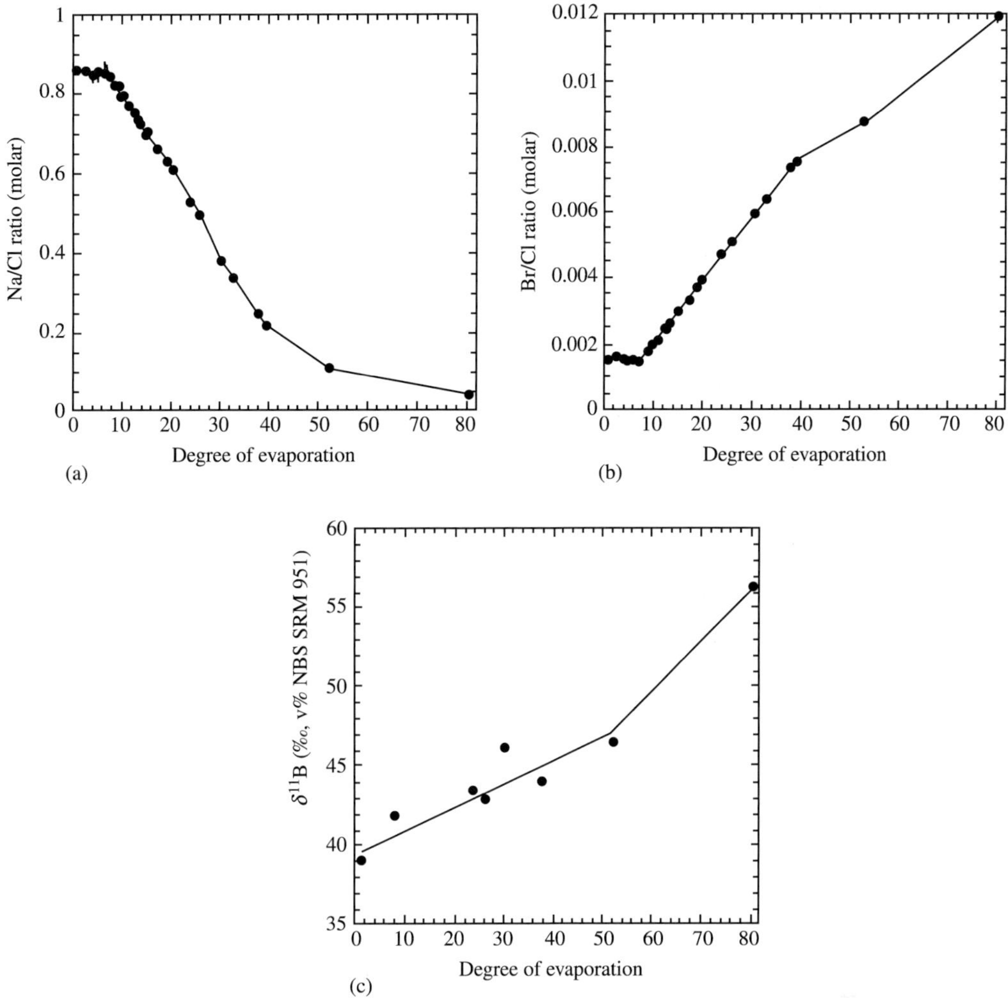

Figure 9 Variations of Na/Cl, Br/Cl, and $\delta^{11}B$ values during evaporation of seawater. Degree of evaporation is defined as the ratio of Br in the solution at different stages of evaporation to the initial Br in seawater (sources McCaffrey *et al.*, 1987; Raab and Spiro, 1991; Vengosh *et al.*, 1992).

only a minor effect on the chemical and isotopic compositions of saline groundwater derived from brine intrusion.

In the Dead Sea Rift Valley, deep pressurized brines that are residual evaporated seawater (>10-fold; Starinsky, 1974) are in hydrological contact with overlying fresh groundwater (Vengosh and Rosenthal, 1994; Marie and Vengosh, 2001). Once the natural hydrological balance is disturbed due to exploitation of the overlying fresh groundwater, the salinization process starts. In the Jericho area in the southern Jordan Valley, rapid rates of salinization due to rising underlying brines are found in wells located near the major Rift faults (Marie and Vengosh, 2001). The high salinity of the brines (>100 g L^{-1}; Starinsky, 1974) indicates that only a small fraction of brine is required to cause devastating salinization in overlying freshwater resources.

An additional source of salinity in polar areas is residual brines derived from the freezing of seawater (Nelson and Thompson, 1954; Herut *et al.*, 1990; Richardson, 1976; Marison *et al.*, 1999; Bottomley *et al.*, 1999; Yaqing *et al.*, 2000). Bein and Arad (1992) showed that deep saline groundwaters in Sweden and Finland have low Na/Cl and high Br/Cl ratios. Based on their chemical composition, it was argued that the saline waters are remnants of frozen seawater formed during the last glaciation, followed by dilution with meteoric water (Bein and Arad, 1992).

Dissolution of evaporite minerals in sedimentary basins is also a common cause of salinization. This type of salinization occurs in the Ogallala Formation in the Southern High Plains, Texas, USA (Mehta *et al.*, 2000a,b), the Dammam aquifer in Kuwait (Al-Ruwaih, 1995), the Nubian sandstone aquifer in the Sinai and Negev (Rosenthal *et al.*, 1998), and the Great Artesian Basin in

Australia (Love *et al.*, 2000; Herczeg *et al.*, 1991). In case of halite dissolution, salinity typically increases along flow lines with a predominance of Na^+ and Cl^- ions, Na/Cl ~ 1, and Br/Cl < seawater ratio (e.g., the Ogallala Formation; Table 1; Mehta *et al.*, 2000a,b). Salinization can also be due to the dissolution of marine sulfate as in Pinawa, Canada (Nesbitt and Cramer, 1993) and the Salinas Valley, California (Vengosh *et al.*, 2002a). Typically, gypsum dissolution produces saline water with $Ca/SO_4 \sim 1$. However, the dissolution of gypsum can also be associated with Ca^{2+} uptake on exchange substrates displacing Na^+ to the solution. The residual water has an $Na-SO_4$ composition (Nesbitt and Cramer, 1993). Dissolution of evaporitic sulfates would affect the sulfur isotopic composition of the water. The $\delta^{34}S_{sulfate}$ values of marine sulfate varied over geological time with a maximum near +35‰ in the Cambrian and less than +10‰ in the Permian. In contrast, oxidation of sedimentary sulfides, which are significantly depleted in ^{34}S relative to oceanic sulfates, would result in low $\delta^{34}S_{sulfate}$ values with typical range of −30‰ to +5‰ (Krouse and Mayer, 2000).

Finally, salinization of groundwater can result from mixing with hydrothermal saline fluids. Thermal fluids, both in marine and nonmarine settings, are often characterized by high salt content. In western Turkey, for example, the salinity of thermal water is in the range 2–66 g L^{-1}, with a high content of boron (>50 mg L^{-1}; Vengosh *et al.*, 2002b). In Mexico City, thermal water affects the quality of local groundwater (Edmunds *et al.*, 2002). In addition to chloride content, thermal waters are often enriched in sodium, boron, fluoride, arsenic, and other contaminants that present threat to the associated freshwater resources.

9.09.5 SALINIZATION OF DRYLAND ENVIRONMENT

Salinity in dryland environment is a natural phenomenon derived from a long-term accumulation of salts on the ground and a lack of adequate flushing in the unsaturated zone. Salt accumulation and efflorescent crusts have been documented in the upper unsaturated zone (e.g., Gee and Hillel, 1988; Nativ *et al.*, 1997; Leaney *et al.*, 2003) and in fracture surfaces (Weisbrod *et al.*, 2000) in many arid areas. The salt formation has been attributed to surface evaporation (Allison and Barnes, 1985), wetting and drying cycles (Drever and Smith, 1978), soil capillarity, and capillarity transport of water and salts from the bulk rock matrix towards fracture surfaces (Weisbrod *et al.*, 2000).

In the western United States, recharge typically occurs along surrounding highlands and groundwater flows towards the basin centers. Along the flow path the salinity of the groundwater increases by several orders of magnitude by both salt dissolution and extensive evaporation (Richter *et al.*, 1993). The chemistry of the residual saline groundwater is primarily controlled by the initial freshwater composition and the subsequent saturation with respect to typical minerals (calcite, gypsum, sepiolite, halite; Hardie and Eugster, 1970; Eugster and Jones, 1979). Evaporation and mineral precipitation control the salinity of groundwater in Deep Spring Lake (Jones, 1965), Death Valley (Hunt *et al.*, 1966), and the Sierra Nevada Basin (Garrels and MacKenzie, 1967).

While the source of the salts is natural, the process of salinization in dryland environment refers to human intervention. The most widespread phenomenon of dryland salinization is the response to land clearing and replacing the natural vegetation with annual crops and pastures. The natural vegetation in the arid and semi-arid zone uses any available water and, thus, the amount of water that leaks below the root zone is minimal, estimated to be 1–5 mm yr^{-1} in South Australia (Allison *et al.*, 1985). Over thousands of years, salts have accumulated in the unsaturated zone, and the salt discharge to the saturated zone was balanced by the slow recharge flux. Large-scale replacement of the natural vegetation with annual crops and pastures with short roots significantly increased the amount of water leaking beneath the root zone and increased the rate of salt discharge to the underlying groundwater (Williams *et al.*, 2002 and references therein). The salts that were stored in the root zone were flushed into the unsaturated and saturated zone, causing salinization of the underlying groundwater. Leaney *et al.* (2003) predicted that the salt flux from the unsaturated zone in the Murray–Darling Basin of South Australia (with soil salinity up to 1.5×10^4 mg L^{-1}) would increase the salinity of the shallow groundwater (~1,000 mg L^{-1}) by a factor of 2–6. The combination of increasing groundwater tables and salt fluxes in the unsaturated zone has caused devastating effects in the dryland environment. This was demonstrated in North America (Miller *et al.*, 1981), Argentina (Lavado and Taboada, 1987), India (Choudhari and Sharma, 1984), and South Africa (Flugel, 1995). However, the most dramatic large-scale salinization process occurs in the dryland environment of Australia (Allison *et al.*, 1985, 1990; Williams *et al.*, 2002; Peck and Hatton, 2003; Leaney *et al.*, 2003; Fitzpatrick *et al.*, 2000; Herczeg *et al.*, 1993, 2001).

The Australian case is used here to describe the chemical evolution of solutes in the dryland environment. The dryland salinization cycle (Figure 10) is a complex process that begins with salt accumulation on the ground,

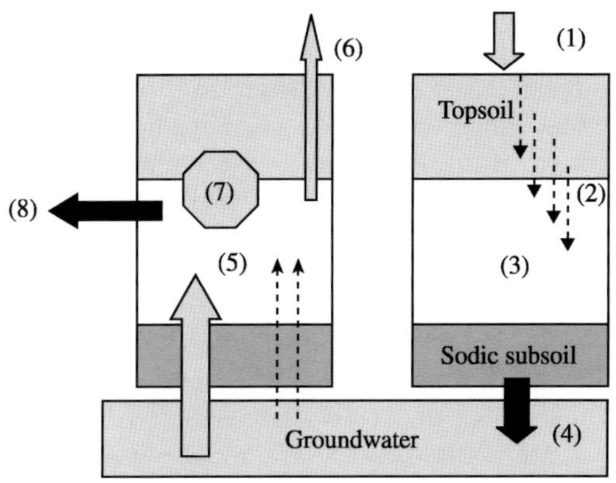

Figure 10 The dryland salinization cycle (the Australian model): (1) salt accumulation and precipitation of minerals; (2) selective dissolution and transport of soluble salts in the vadoze zone; (3) storage of salts influenced by soil permeability; (4) leaching and salinization of groundwater; (5) rise of saline groundwater; (6) capillarity evaporation of rising groundwater; (7) soil salinization; and (8) lateral solute transport and salinization of streams and rivers (after Fitzpatrick *et al.*, 2000).

evaporation, total desiccation, precipitation–dissolution of carbonate minerals, incongruent silicate mineral reactions, and precipitation–dissolution of gypsum and halite minerals (Hardie and Eugster, 1970; Eugster and Jones, 1979). The salts that accumulate in the soil are flushed into the vadose zone. Two factors control the flushing: the mineral solubility and the physical properties of soil. The differences in mineral solubilities cause chemical separation along the unsaturated zone: the higher the solubility of the mineral, the longer the travel of the salts that are derived from its dissolution. This process is also known as wetting and drying and involves the complete precipitation of dissolved salts during dry conditions and subsequent dissolution of soluble salts during wet conditions (Drever and Smith, 1978). This results in an uneven distribution of ions along the unsaturated zone: Ca^{2+}, Mg^{2+}, and HCO_3^- tend to accumulate at relatively shallow depths, SO_4^{2-} at intermediate depths, and Na^+ and Cl^- are highly mobilized to greater depths of the unsaturated zone. The Na/Ca ratio varies along the travel route of solutes in the vadose zone; Ca^{2+} is removed by the precipitation of carbonate minerals, whereas the concentration of Na^+ remains constant, or even increases by dissolution of halite. In addition, the mobilization of Na^+ triggers base-exchange reactions; Na^+ is adsorbed on clays and oxides, but the newly generated Ca^{2+} released from adsorbed sites is removed by the precipitation of soil carbonate.

The final chemical composition of the solutes generated in the dryland environment depends on the initial fluid composition. In Australia, the solutes are derived from marine aerosols that are deposited on the soil (Herczeg *et al.*, 2001 and references therein). Carbon-14 ages of soil solution reveal that most of the recharge occurred during wet climatic periods more than 2×10^4 yr ago (Leaney *et al.*, 2003). Consequently, with thousands of years of salt accumulation and numerous cycles through the unsaturated zone, the saline groundwater in the dryland environment has become "marine like" with Na/Cl and Br/Cl ratios identical to that of seawater (Herczeg *et al.*, 1993, 2001; Mazor and George, 1992). The Australian salts lakes, which represent groundwater discharge zones, are characterized by marine chemical and isotopic (sulfur and boron) compositions superimposed by internal lake processes (Chivas *et al.*, 1991; Vengosh *et al.*, 1991a). If the initial solutes are derived from water–rock interactions induced by the generation of acids from CO_2 accumulation and oxidation of organic matter in the soil, the final product is different. The "nonmarine" signature of the saline groundwater and salt lakes in the western USA reflects the role of water–rock interaction in shaping the chemical composition of both initial and evolved groundwater in the arid zone of the USA (Hardie and Eugster, 1970; Eugster and Jones, 1979).

The second factor that controls salinization of dryland environments is the physical characteristics (e.g., permeability) of the soil. In the arid zone of Australia, rainfall was not always sufficient to leach the salts, and the clay layers in deep sodic subsoil prevents the downward movement of water and salts, leading to a saline zone (Fitzpatrick *et al.*, 2000). The accumulation of salts in soil can therefore be natural, due to the decrease of soil permeability (referred to as "subsoil transient salinity"; Fitzpatrick *et al.*, 2000) or anthropogenic due to the rise of saline groundwater as evidenced in South Australia (Allison and Barnes, 1985; Allison *et al.*, 1990; Herczeg *et al.*, 1993). As much as the leaching process in the vadose zone controls the salinity of the underlying shallow groundwater, the soil chemistry is also influenced by the chemistry of the rising groundwater (Cox *et al.*, 2002). The selective leaching process affects the composition of underlying groundwater. In some cases, recharged groundwater is controlled by marine aerosols and/or halite dissolution and has a typical predominance of Na^+ and Cl^- (e.g., Cox *et al.*, 1996). In other cases, the solubility of gypsum produces saline water enriched in sulfate.

These two groundwater types exercise a direct control on the chemistry of secondary saline soils (Cox *et al.*, 1996; Fitzpatrick *et al.*, 2000). The rising of Na–Cl groundwater creates halite dominant soils, where chloride is the dominant anion. The rising of sulfate-enriched groundwater creates three types of soils: (i) gypsic soil—under aerobic conditions and saturation with calcium sulfate; (ii) sulfidic soil—under anaerobic conditions and sufficient quantities of bacteria that use the oxygen associated with sulfate and produces pyrite; and (3) sulfuric soil—exposure of pyrite to oxygen in the air causes oxidation of pyrite and formation of sulfuric acid, which consequently reduces the soil pH and enhances leaching of basic cations, anions, and trace elements into the soil solution (Fitzpatrick *et al.*, 2000).

The differentiation of soil permeability when soil becomes clogged with clay and mineral precipitation causes lateral flow of saline soil water and shallow groundwater towards low-lying areas. The final stage of the dryland cycle is salinization of adjacent streams and rivers. The chemical composition of the salinized river in the dryland environment reflects the net results of salt recycling between soil, subsoil, groundwater, secondary soil, soil solution, and surface water (Figure 10).

9.09.6 ANTHROPOGENIC SALINIZATION

9.09.6.1 Urban Environment and Sewage Salinization

Two major sources of salinity are important in the urban environment: sewage and road salt. The salinity of domestic wastewater is derived from both the salinity of the source water supply to the municipality and the salts added directly by humans (Figure 11). This includes the use of detergents, washing powders, and salts. In Israel, for example, the average net human contributions of chloride, sodium, and boron to domestic wastewater are 125 mg, 120 mg, and 0.6 mg, respectively, per liter of wastewater (Hoffman, 1993; Vengosh *et al.*, 1994). Chloride is derived from salts used in dishwashers and for refreshing ion-exchange columns, while sodium is also derived from detergents. First and secondary treatments of sewage, which form the most common procedure for sewage purification, do not remove the inorganic constituents, and consequently even treated sewage is relatively saline. In Israel the chloride content of domestic sewage effluent is between 300 mg L^{-1} and 400 mg L^{-1} (Vengosh *et al.*, 1994; Harussi *et al.*, 2001). In water-scare areas, treated sewage is an important water source that substitutes for freshwater irrigation. It is expected that future utilization of treated domestic sewage will become the major source for irrigation water in the western United States, Israel, and Jordan. The saline wastewater can affect soil salinity (Beltran, 1999) and the composition of underlying groundwater in phreatic aquifers (Ronen *et al.*, 1987; Vengosh and Keren, 1996; DeSimone *et al.*, 1997; Stigter *et al.*, 1998; Gavrieli *et al.*, 2001). In addition, direct pollution by sewage affects both the organic and inorganic composition of contaminated groundwater.

The typical chemical composition of saline domestic sewage is presented in Table 1. The use of household NaCl salt and detergents enriched in sodium and boron results in $Na/Cl > 1$, a low Br/Cl ratio (i.e., Br/Cl < seawater ratio due to halite dissolution; Vengosh and Pankratov, 1998), and a high B/Cl ratio (>seawater ratio of 8×10^{-4}). Detergents are enriched in boron due to the addition of Na-borate (derived from natural borate minerals) used as a bleaching agent (Waggott, 1969; Raymond and Butterwick,

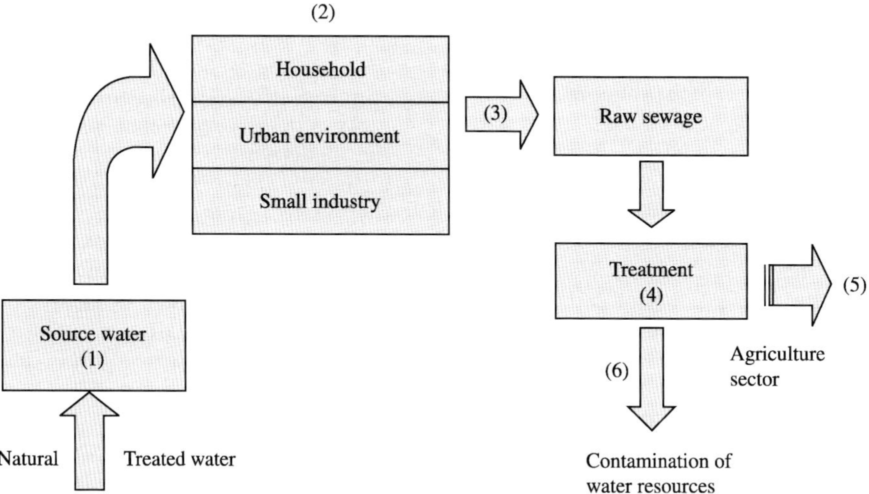

Figure 11 The domestic salinization cycle: (1) supply water to a municipality; (2) salts added within the urban environment; (3) sewage generation; (4) sewage treatment; (5) reuse of treated sewage for irrigation; and (6) contamination of water resources.

1992). Since Na-borates are mined from natural borate deposits (mainly in the western USA, Turkey, and China), they have typical boron isotopic compositions ($\delta^{11}B$ range of 0–10‰; Vengosh *et al.*, 1994). The anthropogenic sulfate is also isotopically distinguished, as indicated by the pollution of the Arno River in northern Tuscany with a distinctive sulfur isotopic ratio ($\delta^{34}S = 6$–8‰; Cortecci *et al.*, 2002).

Another source of urban salinization is the use of road de-icing salts. Salt has been used for road de-icing for several decades, particularly in the eastern and northeastern states in the United States and Canada. The use of road salt improves fuel efficiency and reduces accidents; at the same time, it causes salinization of associated groundwater. In 1990, more than 10.5 Mt of salt were used for road de-icing (Richter *et al.*, 1993). Brine generated in storage piles of salt (Wilmoth, 1972) and from the dissolution of salts that are applied directly to the roads (Howard and Beck, 1993; Williams *et al.*, 1999) can contaminate water resources. If salt is applied as a powder, its particles may become airborne and transported for considerable distances downwind (Jones and Hutchon, 1983; Richter *et al.*, 1993).

The most common road de-icing practice is to apply pure sodium chloride to main urban roads and highways. Calcium chloride is also used but not as frequently, since it is more expensive and is known to make road surfaces more slippery when wet (Richter *et al.*, 1993). Hence, the most important geochemical signal of road de-icing is typically an Na–Cl solution with an Na/Cl ratio close to unity (Table 1; Howard and Beck, 1993). The use of calcium chloride results in contamination with Ca-chloride solutions with an $Na/Cl \ll 1$.

9.09.6.2 Agricultural Drainage and the Unsaturated Zone

A survey of the quality of drainage water in the western United States shows that the drainage waters are typically saline, with elevated concentrations of selenium, boron, arsenic, and mercury (Hern and Feltz, 1998). In San Joaquin Valley, Salinas Valley, and the Imperial Valley in California, drainage waters are highly saline and contaminate the associated ground- and surface waters. The salinity of agriculture return flow is derived from two principal sources: added commercial chemicals (e.g. nitrogen fertilizers, gypsum, dolomite, and boron compounds), and the nature of the soil in which the irrigation water flows. In the western US, the occurrence of selenium in the drainage water is attributed to mobilization from oxidized alkaline soils derived predominantly from seleniferous marine sediments (Hern and Feltz, 1998; Grieve *et al.*, 2001; see Chapter 9.03). As a result, agricultural contaminants have caused significant changes in groundwater geochemistry and have induced water–rock interactions (Bölke, 2002).

The salinity of agricultural return flow depends on the balance between the amount of salt entering the soil and the amount of salt that is removed. A change from natural vegetation to agricultural crops and application of irrigation water adds salt to the system. Approximately 60% of the supplied irrigation water is consumed by growing crops, but the salts remain in the residual solution as they are not consumed by evaporation and transpiration. Consequently, adequate drainage is one of the key factors that determines soil salinization; nevertheless, the nonreactivity of the soluble salts (i.e., conservative and not consumed by the sediments) makes them a long-term hazard.

The salinity of agricultural return flow is also derived from the salinity of the soil. In arid and semi-arid areas the natural salinity of the soil is high, and thus the flushing with irrigation water enhances the dissolution of stored salts. Moreover, the selenium and boron anomalies associated with agricultural return flow in the San Joaquin Valley (Kharaka *et al.*, 1996) and the Imperial Valley (Glenn *et al.*, 1999) in California are directly derived from leaching the local soil. The soil reactivity is also enhanced by the composition of the irrigation water and added fertilizers. As high salinity is associated with high sodium content, exchangeable sodium replaces exchangeable calcium. As a result of this interaction, the soil becomes impermeable. The "sodium hazard" is expressed as "sodium adsorption ratio" (SAR), which represents the relative activity of sodium ions in exchange reactions with soil. Typically, irrigation water with SAR values higher than 10 are considered to be a sodium hazard (e.g., San Joaquin Valley; Mitchel *et al.*, 2000). In cases of high SAR values coupled with heavy soil, gypsum is applied to improve the physical properties of soil. The agricultural drainage salt generation during irrigation of crops in San Joaquin Valley in California exceeds 6×10^5 t annually and accumulates at a rate that is causing serious concern for the environment and the local agricultural industry (Jung and Sun, 2001).

The chemical composition of agricultural drainage (Figure 12) is influenced by the quality of the water source. In many water-scarce areas, the only available water for irrigation is the treated (usually only secondary treatment) domestic sewage. Hence, the end product of urban waste (Figure 11) can be the initial water for the agriculture cycle. Other types of marginal waters (e.g., brackish waters) may also affect the composition of the initial stage of the agricultural cycle. The second factor that determines the

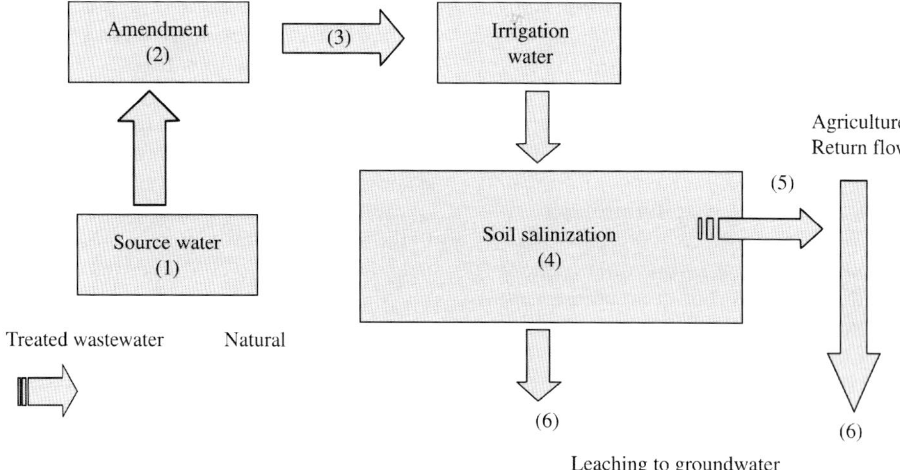

Figure 12 The agriculture salinization cycle: (1) supply water for irrigation; (2) salts added with fertilizers and amendment treatment; (3) generation of irrigation water; (4) soil salinization; (5) formation of agricultural return flow; and (6) contamination of water resources.

chemistry of the agriculture cycle is fertilizers and other types of ammandates that are added to the irrigation water. Nitrate is the dominant ion contributed by fertilizers. Bölke (2002) showed that artificial fertilizer is the largest nonpoint source of nitrogen in rural areas. This reflects the global increase in the use of nitrogen fertilizer since the middle of the twentieth century. Nitrate derived from fertilizers has relatively low $\delta^{15}N_{nitrate}$ values (−4‰ to +3‰; Kendall and Aravena, 2000 and references therein). Note that $\delta^{15}N_{nitrate} = [(^{15}N/^{14}N)_{sample}/(^{15}N/^{14}N)_{AIR} - 1] \times 10^3$; AIR refers to N_2 in air (Coplen *et al.*, 2002). Other important forms of nitrogen applied to crops are urea, $CO(NH_2)_2$, ammonia (NH_3), ammonium nitrate (NH_4NO_3), and animal manure (Bölke, 2002).

In addition to nitrogen application, other types of ammandates are added to the irrigation water. These include dolomite to provide calcium and magnesium for plant growth and to neutralize acid soils (Bölke, 2002), and gypsum, which is used in heavy soil with irrigation water having high SAR values in order to neutralize the exchangeable Na^+ in the soil (e.g., Salinas valley, California; Vengosh *et al.*, 2002a). In addition, application of KCl results in chloride contamination and use of Na-borate or Ca-borate fertilizers in boron-depleted soil contribute boron (Komor, 1997). The use of soil fumigants such as ethylene dibromide (EDB) (Bölke, 2002) can lead to degradation and release of bromide to groundwater, as demonstrated in shallow groundwater underlying strawberry cultivation in Salinas Valley, having conspicuously high Br/Cl ratios (Vengosh *et al.*, 2002a).

Another source of salinity that is associated with agricultural activity is animal waste, which includes cattle (Gooddy *et al.*, 2002; Hao and Chang, 2002, Harter *et al.*, 2002) and swine manure (Krapac *et al.*, 2002). Hao and Chang (2002) and Krapac *et al.* (2002) showed that, following long-term manure application, the salinity of soil and underlying groundwater increase significantly. Manure is characterized by high concentrations of chloride (1,700 mg L^{-1} in swine manure), sodium (840 mg L^{-1}), potassium (3,960 mg L^{-1}), and ammonium with $\delta^{15}N$ values >9‰ (Krapac *et al.*, 2002). The combination of a high K/Cl ratio, a high nitrogen concentration (ammonium that is transformed into nitrate in the unsaturated zone), and high $\delta^{15}N_{nitrate}$ make it possible to trace the impact of animal wastes on the quality of the underlying groundwater.

The third stage in the agriculture cycle (Figure 12) is transport and reactivity with the unsaturated zone. The unsaturated zone acts as a buffer and modifies the original chemical composition of the irrigation water (this is also true for contamination by wastewater and de-icing salts). Microbial oxidation of ammonium releases protons (H^+) that generate acidity along with nitrate in soils (Bölke, 2002) and induce dissolution of calcium carbonate minerals. Consequently, nitrate-rich waters in carbonate aquifers are associated with high calcium derived from dissolution of the aquifer matrix. For example, in the mid-Atlantic coastal plain in the US, groundwater recharged beneath fertilized fields has a unique $Ca-Mg-NO_3$ composition with positive correlations between nitrate and other inorganic constituents such as Mg^{2+}, Ca^{2+}, Sr^{2+}, Ba^{2+}, K^+, and Cl^- (Hamilton *et al.*, 1993; Hamilton and Helsel, 1995). In the Gaza Strip of the Mediterranean coastal aquifer, dissolution of the aquifer matrix is reflected by the $^{87}Sr/^{86}Sr$ ratios of nitrate-rich groundwater (Vengosh *et al.*, 2002c).

Using irrigation water enriched in sodium with high SAR values triggers ion-exchange reactions (Stigter *et al.*, 1998). The capacity of ion exchange on clay minerals is limited, however, and depends on various lithological (e.g., clay content) and environmental (e.g., pH, solute composition) factors. Nonetheless, the uptake of Na^+ and release of Ca^{2+} (and Mg^{2+}) is a major geochemical modifier associated with transport of irrigation water in the unsaturated zone. The Na/Cl ratio is a good indicator for the efficiency of the base-exchange reactions; low ratios (i.e., below the original value of the irrigation water) reflect continuation of exchange reactions, whereas an increase of Na/Cl ratio toward the original Na/Cl value of the irrigation water suggests exhaustion of the exchangeable sites and reduction in the clay capacity for exchange reactions (Vengosh and Keren, 1996). Gavrieli *et al.* (2001) showed that irrigation with calcium-rich water causes an opposite reaction, in which Na^+ is released and the residual groundwater has an Na/Cl ratio higher than that of the irrigation water. Similarly, the ability of the clay minerals to adsorb reactive elements (e.g., boron and potassium) is limited, and over long-term recharge or irrigation, the capacity of adsorption decreases (Vengosh and Keren, 1996; DeSimone *et al.*, 1997; Stigter *et al.*, 1998). In most cases, the potassium level in groundwater is low, and the K/Cl ratios in agricultural recharge are less than unity due to adsorption of potassium onto clay minerals (Bölke, 2002).

Leaching of the soil also contributes contaminants to the drainage waters (Johnson *et al.*, 1999; Sun, 2001; Tedeschi *et al.*, 2001; Gates *et al.*, 2002). The drainage waters of Imperial Valley and the Salton Sea in Southern California are characterized by Na/Cl > 1, high B/Cl and SO_4/Cl (ratios higher than in seawater) and typically low Br/Cl ratio (Table 1; data from Schroeder and Rivera, 1993). The Br/Cl ($(3-4)\times 10^{-4}$) in the drainage water is one order of magnitude lower than seawater (1.5×10^{-3}) or atmospheric ($>1.5\times 10^{-3}$) ratios (Figure 5; Davis *et al.*, 1998). This suggests that halite dissolution in the soil can be an important component of the salinity associated with agriculture return flow in the Imperial Valley.

The amount of organic load associated with anthropogenic contamination and redox conditions also controls the reactivity of the unsaturated zone, and consequently the salinity of associated water. In aerobic conditions, the degradation of organic matter results in accumulation of HCO_3^- and Ca^{2+} due to dissolution of the carbonate matrix of the host aquifer. In addition, the contents of sulfate and nitrate are expected to rise and to be conservatively transported in the unsaturated zone. Thus, the typically high SO_4/Cl and the high nitrate level, which represents the original composition of the irrigation water, are preserved. In contrast, under anaerobic conditions, sulfate and nitrate reduction enhances calcium carbonate precipitation, and hence sulfate (by sulfate reduction), nitrate (denitrification), and calcium (precipitation) are removed along the flow in the unsaturated zone (Gooddy *et al.*, 2002). In this case, the original chemical signature of the agricultural return flow is modified: the residual groundwater becomes relatively depleted in sulfate and nitrate concentrations with high $\delta^{34}S_{sulfate}$ and $\delta^{15}N_{nitrate}$ values relative to the irrigation water.

9.09.7 SOIL SALINIZATION

Soil salinization is a major process of land degradation that decreases soil fertility and is a significant component of desertification processes in the world's dryland (Thomas and Middleton, 1993). The World Bank states that soil salinization caused by inappropriate irrigation practices affects ~60 Mha, or 24% of all irrigated land worldwide. In Africa, salinization accounts for 50% of irrigated land (Thomas and Middleton, 1993; Ceuppens and Wopereis, 1999). Increasing soil salinization is also occurring in India, Pakistan, China, and Central Asia (Wichelns, 1999). In Egypt, ~35% of the agricultural land suffers from salinity (Kotb *et al.*, 2000; Kim and Sultan, 2002). Soil salinization is the first stage of environmental destruction caused by salinity, and is related to river and lake salinization. For example, the diversion of the Amu Darya and Syr Darya rivers not only caused a significant desiccation of the Aral Sea, but also caused salinization of associated agricultural land (Weinthal, 2002). In Australia, soil salinization is the most severe environmental problem of the continent, causing a dramatic change in landscape, industry, and the future of farmland (Dehaan and Taylor, 2002).

The accumulation of soluble salts in soil occurs when evaporation exceeds precipitation, and salts are not leached but remained in the upper soil layers in low-lying areas. Natural soil salinization, called "primary salinization," occurs in arid and semi-arid climatic zones. "Secondary salinization" is the term used to describe soil salinized as a consequence of direct human activities (Fitzpatrick *et al.*, 2000; Dehaan and Taylor, 2002).

Salinization of soil results from a combination of evaporation, salt precipitation and dissolution, salt transport, and ion exchange (Shimojima *et al.*, 1996). Excessive salinity in soil leads to toxicity in crops, reduction in soil fertility, reduction of availability of water to plants by reducing the osmostic potential of the soil solution, and a significant change in the hydraulic properties of soil (Hillel, 1980; Bresler *et al.*, 1982; Frenkel *et al.*, 1978; Ramoliya and Pandey, 2003). The quality of the irrigation water is also a major contributor to soil salinity. Irrigation with marginal waters (e.g., brackish water, wastewater) with a high content of soluble salts directly affects soil salinity (Sparks, 1995). For example, in the Nile Delta the highly saline drainage water is the principal source of soil salinization, and extensive irrigation without an adequate drainage system causes soil salinization (Kotb *et al.*, 2000).

In shallow groundwater conditions, water and dissolved salts move by capillary action to the soil surface. When the water evaporates at the surface, the salts are left behind. The excess salts cause many types of crops to wither and die. The salts prevent plant roots from making use of water in the soil. Plant roots absorb water from the soil through the process of osmosis. Osmosis moves water from an area of lower salt (higher water) concentration to an area of higher salt concentration. The salt concentration inside a normal plant cell is ~1.5%, so that water moves into root cells. In saline soils, the concentration of salt in the soil water can rise above 1.5% and prevent osmosis from moving water into the roots. This may cause water to move out of the roots, thereby dehydrating the plant (Poljakoff-Mayber and Gale, 1975).

Soil salinization is often associated with sodic soil. Natural or anthropogenic accumulation of sodium in the system leads to gradual replacement of divalent cations with Na^+ on the exchange complex of clay minerals. The increased level of adsorbed Na^+ causes the soil to become dispersed, which significantly reduces the soil porosity and permeability. This is a major problem for drainage of soil water and salt flush in the unsaturated zone. The predominance of Na^+ in the exchange phases occurs due to both high levels of Na^+ in the soil water and Ca^{2+} and Mg^{2+} precipitating as carbonate and sulfate minerals. In Australia, natural sodic soils, which are defined as soils in which the exchangeable sodium percentage is greater than 15%, are associated with saline soil ($EC > 8\ dS\ m^{-1}$; Fitzpatrick *et al.*, 2000). Sodic soils also develop under irrigation of water enriched in sodium (e.g., wastewater), which requires special reclamation measures such as using gypsum or $CaCl_2$ salts to remove the exchangeable Na^+ (Bresler *et al.*, 1982).

9.09.8 WETLAND SALINIZATION

More than one-fourth of the world's wetlands and more than 50% of the wetlands in the United States have been lost owing to their salinization. The destruction of wetlands has reduced the diverse assemblages of millions of waterfowl and shorebirds. Present and future water projects and salinization induced by agricultural and urban uses of water pose a threat to the bird populations in wetlands. Moreover, salinization is likely to harm the birds' reproductive capacity and place further stresses on their diversity (Williams, 2001a,b). Wetland destruction has been particularly devastating in arid zones. Half of Nevada's wetlands have been lost during the last 200 years (Titus and Richman, 2000). Inland wetland salinization occurs in the Tavlas de Daimiel National Park of Spain (Berzas *et al.*, 2000), which is a unique wetland area with a wide floral and faunal diversity, rich in many migratory birds. The wetland virtually dried up and salinized due to the use of its source waters (Berzas *et al.*, 2000).

Coastal wetland salinization is caused by a significant decrease of river discharge, diversion of freshwater for irrigation network, and upstream seawater intrusion under the influence of the tide (Zalidis, 1998). Indeed, salinization of low-lying coastal or deltaic plains is the most striking process that threatens the future of global wetlands. Coastal wetlands, those of southern United States, Southeast Asia, China, and northern Australia, are vulnerable to saltwater intrusion. Their vulnerability is likely to increase with the expected sea-level rise during the next century due to global warming. Even in near-pristine environments, such as the mouth of the Yukon and Kuskokwim rivers in northern Alaska, salinization due to sea-level rise threatens to destroy the diverse freshwater habitat. In northern Australia, saltwater intrusion has been identified in the extensive low-lying coastal plains, particularly in the Lower Mary River area, changing the system from a predominantly freshwater wetland environment to one dominanted by saltwater. The dramatic environmental changes that have taken place in the Lower Mary River since the early 1950s could be a possible analogue to the more widespread salinization that will accompany future sea-level rise (Mulrennan and Woodroffe, 1998). EPA studies (Titus and Richman, 2000) estimated that sea level is likely to rise by 90 cm along the US coast by the middle of next century. Consequently, low-lying areas 1.5 m below sea level would be flooded by spring high tides. A survey of the coastal levels along the coastal plain of the US reveals that the areas most vulnerable to sea-level rise are those of North Carolina, Louisiana, the Florida Everglades, and virtually all barrier islands and coastal wetlands. The most striking finding is that in North Carolina, an area equivalent to that of the Netherlands is expected to be flooded by seawater (Titus and Richman, 2000).

9.09.9 ELUCIDATING THE SOURCES OF SALINITY

Identifying the origin of salinity in water is crucial for water management, model prediction, and remediation. Yet the variety of salinization sources makes this task difficult. Ground- and surface-water salinization can result from point sources (e.g., leakage or recharge of domestic wastewater) or nonpoint sources (e.g., agriculture return flows, irrigation with sewage effluent). Salinization can be derived from natural (geo-genetic) processes such as seawater intrusion or saline-water flow from adjacent or underlying aquifers (e.g., Maslia and Prowell, 1990; Vengosh *et al.*, 1999, 2002a; Sanchez-Martos and Pulido-Bosch, 1999; Sanchez-Martos *et al.*, 2001; Kloppman *et al.*, 2001; Hsissou *et al.*, 1999). Alternatively, salinization can also be induced by direct anthropogenic contamination. The multiple salinity sources therefore present a real challenge to water agencies and regulatory bodies.

The key for tracing salinity sources is the assumption that the chemical composition of the original saline source is preserved during the salinization process. Due to the large differences between the solute content of saline and freshwaters, the chemical composition of the contaminated water mimics the composition of the saline source. However, the original composition of the saline source can be modified once it is masked by water–rock interactions. For example, the composition of seawater is significantly modified as it intrudes into coastal aquifers. Consequently, diagnostic tracers, in order to be useful, must be conservative. The overlap and similarity of the chemical composition of different saline sources makes the tracing task even more difficult. For example, the Na/Cl ratio can be a good tracer for distinguishing marine (e.g., seawater intrusion with $Na/Cl < 0.86$) from nonmarine or anthropogenic sources ($Na/Cl \geq 1$). However, the reactivity of Na^+ in the unsaturated zone can reduce the Na/Cl ratio even in nonmarine settings (e.g., Vengosh and Keren, 1996; DeSimone *et al.*, 1997; Stigter *et al.*, 1998; Gavrieli *et al.*, 2001). It is therefore essential to use assemblages of diagnostic chemical and isotopic tracers for accurate delineation of the salinity sources. Mazor (1997) and Herczeg and Edmunds (2000) reviewed most of the geochemical tools that are useful for discriminating between different sources of solutes in groundwater systems. The source of solutes in the salinized environment can be derived from (i) mixing of meteoric water with saline water such as seawater, connate fluids, and hydrothermal waters trapped within or outside the aquifer; (ii) dissolution of evaporites left behind after the last seawater or brine retreat; (iii) weathering of the aquifer minerals; (iv) accumulation of salts derived from long-term deposition of atmospheric fallout; (v) sewage (domestic or industrial) contamination; and (vi) salinization by agricultural return flows. Each of these sources has a unique and distinctive chemical and isotopic composition. The integration of geochemical and isotopic tracers can be used to help resolve these multiple sources. Summary of the geochemical characteristics of these sources is given in Table 2.

Reactive species are important tools to evaluate the level of interaction with the host sediments/rocks of the investigated aquifer. Strontium isotopes are an excellent tracer to delineate the type of rock source and weathering process (see Chapter 5.12), the flow path of groundwater in a watershed of different lithological compositions (e.g., Starinsky *et al.*, 1980, 1983; Banner and Hanson, 1990; Banner *et al.*, 1989; Bullen *et al.*, 1996), the composition of exchangeable strontium in clay minerals (Johnson and Depaolo, 1994; Amstrong *et al.*, 1998; Vengosh *et al.*, 2002a), the origin of dissolved strontium in rivers (Singh *et al.*, 1998; Vengosh *et al.*, 2001), and the interaction between groundwater, lake water, and aquifer minerals (Katz and Bullen, 1996; Lyons *et al.*, 1995).

In carbonate or calcareous sand aquifers, the strontium isotopic composition of saline groundwater should mimic that of the host aquifer rocks. In cases of lack of similarity, the groundwater must interact with external rock sources. For example, saline groundwaters from central Missouri, USA have high $^{87}Sr/^{86}Sr$ ratios that are considerably more radiogenic than the host Mississippian carbonates. The high $^{87}Sr/^{86}Sr$ ratios reflect deep subsurface migration and water–rock interaction with Paleozoic and Precambrian strata (Banner *et al.*, 1989). Similarly, the $^{87}Sr/^{86}Sr$ ratios of saline groundwaters in the Mediterranean coastal aquifer of Israel are considerably lower than that of the aquifer carbonate matrix, indicating an external source for the saline groundwaters (Vengosh *et al.*, 1999).

Strontium isotopes can also be used to trace salinization by agricultural return flow. Bölke and Horan (2000) showed that some fertilizers, and hence agricultural recharge, have a high radiogenic $^{87}Sr/^{86}Sr$ ratio, which differs significantly from that of strontium acquired by water–rock interactions in the aquifer. Although denitrification and carbonate dissolution may alter the strontium isotopic ratio, the association of distinctive $^{87}Sr/^{86}Sr$ ratios and nitrate concentrations may indicate the impact of fertilizers on groundwaters (Bölke and Horan, 2000).

In contrast, other tracers can be considered as universal—i.e., their use is not related to the specific local geology or hydrological system.

Table 2 Typical chemical and isotopic characteristics of major saline sources.

Source	*TDS* ($g\ L^{-1}$)	*Na/Cl* (molar ratio)	SO_4/Cl ($\times 10^{-3}$)	*Br/Cl molar ratio* ($\times 10^{-3}$)	*B/Cl molar ratio* ($\times 10^{-3}$)	$\delta^{11}B$ (per mil)	$\delta^{34}S$ (per mil)	$^{36}Cl/Cl$ (10^{-15})
Seawater	35	0.86	0.05	1.5	0.8	39	21	<5
Replies of evaporated seawater (brines)	>35	<0.86	>0.05	>1.5	<0.8	>39	>21	<5–100
Evaportie dissolution	>1	1	≫0.05	<1.5	<0.8	20–30	<21	<5
Hydrothermal water	0.2 to >1	>1	≫0.05	<1.5	>5	0 ± 5	≪21	<5
Domestic wastewater	~1	>1	>0.05	<1.5	5	0–10	6–10	50 to >100
Agricultural return flow	0.5–5	>1	≫0.05	<1.5	>0.8	20–30		50 to >100

In this respect, the Br/Cl (Edmunds, 1996; Davis *et al.*, 1998; Vengosh and Pankratov, 1998), B/Cl, $\delta^{11}B$ (Bassett, 1990; Bassett *et al.*, 1995; Palmer and Swihart, 1996; Vengosh *et al.*, 1991a,b, 1992, 1994; Vengosh and Spivack, 2000), $\delta^{34}S_{sulfate}$ (Krouse and Mayer, 2000), $^{36}Cl/Cl$ (Phillips *et al.*, 1986; Phillips, 2000; see Chapter 5.15), and $\delta^{18}O_{water}$ (see Chapter 5.11) variations can be used to discriminate between multiple salinity sources. The variations of some of these tracers are presented in Figures 13 and 14. A clear distinction is shown between marine sources such as seawater intrusion or marine-derived brines (e.g., Dead Sea), nonmarine (e.g., evaporite dissolution), and anthropogenic (e.g., sewage effluents, agricultural return flows) sources.

The use of boron and sulfur isotopes is constrained by their reactivity in the hydrological system. Boron tends to be adsorbed on clay minerals and oxides, particularly under highly saline conditions. During the adsorption process $^{10}B(OH)_4^-$ is incorporated preferentially into adsorbed sites, whereas the residual dissolved boron in the form of $B(OH)_3$ is enriched in ^{11}B. The magnitude of the boron isotope fractionation varies between 20‰ and 30‰, and as a result seawater and evaporated brines have $\delta^{11}B$ values >40‰ (60‰ in the Dead Sea; Figure 15; Vengosh *et al.*, 1991a,b, 1992, 1994; Vengosh and Spivack, 2000).

The sulfur ($\delta^{34}S_{sulfate}$) and oxygen in sulfate ($\delta^{18}O_{sulfate}$) isotope variations provides a distinction between seawater ($\delta^{34}S_{sulfate}$ = 21‰; $\delta^{18}O_{sulfate}$ = 9.5‰), evaporated seawater (>21‰), dissolution of marine sulfate minerals

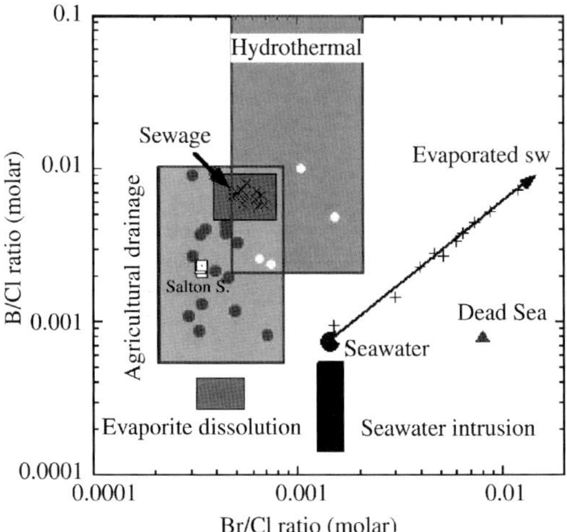

Figure 13 Elucidation of saline sources by using the variations of Br/Cl and B/Cl ratios. Note the expected geochemical distinction between seawater, evaporated seawater, brines (e.g., Dead Sea), hydrothermal fluids, sewage effluents, agricultural drainage (e.g., Salton Sea), and evaporite dissolution.

(depends on time of deposition with a range of +10–35‰ and +9–20‰), fertilizers (+6–11‰), domestic sewage (6–8‰), and sulfate derived from the oxidation of reduced inorganic sulfur components (−30 to +5‰; −10‰ to +5‰; Krouse and Mayer, 2000, and references therein). However, the $\delta^{34}S_{sulfate}$ values are modified by sulfate reduction and the residual sulfate becomes enriched in ^{34}S and ^{18}O.

The use of the stable isotopes of oxygen and hydrogen for tracing the origin of salinity is straightforward in the case of river salinization,

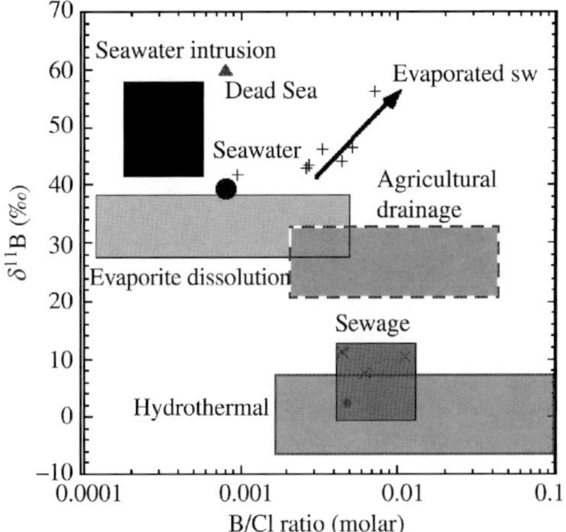

Figure 14 Elucidation of saline sources by using the variations of B/Cl and $\delta^{11}B$ values. Note the expected geochemical distinction between seawater, evaporated seawater, brines (e.g., Dead Sea), hydrothermal fluids, sewage effluents, agricultural drainage, and evaporite dissolution.

but it is more problematic in groundwater studies. In many cases, the original $\delta^{18}O_{water}$ and $\delta^{2}H$ values of the saline sources are completely modified by mixing (dilution) with meteoric water (e.g., Banner *et al.*, 1989; Hanor, 1994; Bergelson *et al.*, 1999). Nevertheless, the $\delta^{18}O_{water}$ and $\delta^{2}H$ values can be used to calculate mixing relationships in the case of seawater intrusion (Jones *et al.*, 1999; Yechieli *et al.*, 2001) or reflect agricultural return flows (Davisson and Criss, 1993). It is also possible to use the stable isotopes for detecting the origin of saline groundwater: remnant of diluted evaporated seawater and formation waters are characterized by low $\delta^{18}O_{water}$ and $\delta^{2}H$ values relative to direct seawater intrusion. For example, Herczeg *et al.* (2001) showed that groundwater in the Murray Basin is depleted in ^{18}O and ^{2}H relative to possible mixing between meteoric water and seawater. This indicates that the salinity of the groundwater in the Murray Basin is derived from atmospheric fallout, not from direct seawater intrusion. Similarly, Bergelson *et al.* (1999) showed that the saline springs that emerge at the Sea of Galilee in the Jordan Rift valley have low $\delta^{18}O_{water}$ and $\delta^{2}H$ values to account for direct seawater intrusion and must reflect dilution of hypersaline brines. Finally, anthropogenic saline sources, such as domestic wastewater that is treated or stored in open reservoirs, are enriched in ^{18}O and ^{2}H, and thus can be used to trace their impact on contaminated groundwater (Vengosh *et al.*, 1999).

The radioactive isotopes ^{36}Cl and ^{129}I behave conservatively in the hydrological system and are important tools for studying salinization processes. Since 1952, atmospheric thermonuclear testing

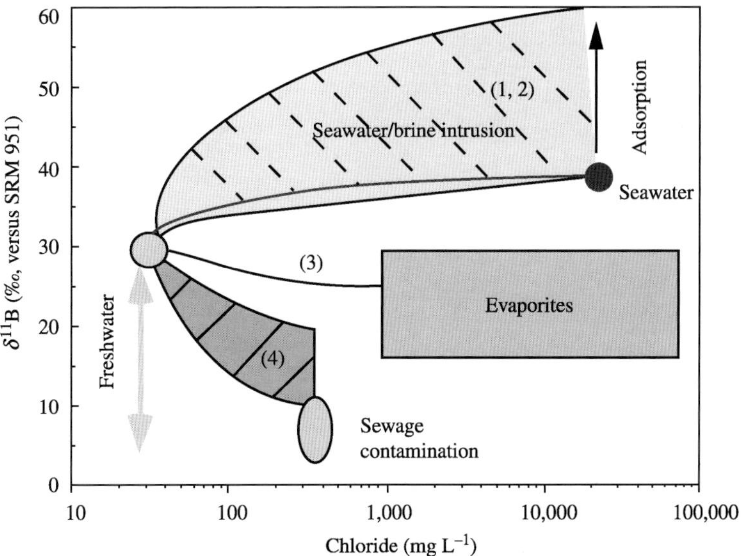

Figure 15 Schematic illustration of the expected boron isotope variations upon salinization by (1) seawater intrusion; (2) mixing with brines; (3) dissolution of marine evaporites; and (4) mixing with sewage effluents. Note the enrichment in ^{11}B associated with adsorption process (sources Vengosh *et al.*, 1992, 1994).

has significantly changed the ^{36}Cl budget of the atmosphere. The atmospheric ^{36}Cl fallout was several orders of magnitudes higher than normal flux, particularly between 1952 and 1970. Hence, the bomb ^{36}Cl-pulse is an important hydrological tracer that is preserved in the hydrological system. Its signal is modified by dispersion and natural decay of ^{36}Cl (half-life ~3.01×10^5 yr; Phillips, 2000, and references therein). Meteoric chloride is therefore characterized by a high ^{36}Cl/Cl ratio that is preserved during evaporative concentration and recycling of salts via precipitation–dissolution (Phillips, 2000). In contrast, other saline sources such as road salts and oilfield brines have significantly lower ^{36}Cl/Cl ratios. The distinction between meteoric and groundwater chloride (Figure 16) has been used in several studies to estimate the relative proportion of meteoric and subsurface chlorine sources in the Jordan Valley and Dead Sea (Paul *et al.*, 1986; Magaritz *et al.*, 1990; Yechieli *et al.*, 1996), closed basins in Antarctica (Lyons *et al.*, 1998), and Lake Magadi of the East African Rift (Phillips, 2000).

Similarly, ^{129}I is a naturally occurring, cosmogenic, and fissiogenic isotope ($T_{1/2} = 15.7$ Myr; Fabryka-Martin *et al.*, 1991; FabrykaMartin, 2000). Like ^{3}H, ^{14}C, and ^{36}Cl, ^{129}I was produced in bomb tests, but in greater abundance above the natural level. While some of the other anthropogenic radionuclides have returned to near pre-bomb levels in the surface environment, the ^{129}I level continues to be elevated due to emissions from nuclear fuel reprocessing facilities, which are transported via the atmosphere on a hemispheric scale. Modern meteoric waters are expected to have large ^{129}I/^{127}I ratios ($>1{,}000 \times 10^{-12}$) relative to that in old groundwater ($<10^{-12}$; Moran *et al.*, 1999, 2002). By combining the values of the isotopic ratios ^{36}Cl/Cl and ^{129}I/I, Ekwurzel *et al.* (2001) were able to discriminate between different saline sources in the Souss-Massa Basin in Morocco, particularly between modern saline recharge and saline groundwater (Figure 17).

Geochemical tracers are useful for delineating the sources of salinity in groundwater systems; they can also be used for studying river salinization (Herczeg *et al.*, 1993; Vengosh *et al.*, 2001; Cortecci *et al.*, 2002; Phillips *et al.*, 2002; Mills *et al.*, 2002). River salinization can be derived from surface evaporation along the river flow, recycling of meteoric salts via irrigation and formation of agricultural return flows, discharge of saline groundwater derived from the dissolution of evaporites or from mixing with remnants of formation waters, and from anthropogenic contamination like discharge of sewage effluents.

Each of these sources is characterized by identifiably different chemical and isotopic compositions. Hence, the different salinization processes induce different chemical and isotopic changes in the salinized river. By using conservative tracers of Br/Cl, $\delta^{18}O_{water}$, and ^{36}Cl/Cl ratios, it is possible to discriminate between these processes (Figure 18). Recycling of meteoric salts by in-stream river salinization results in

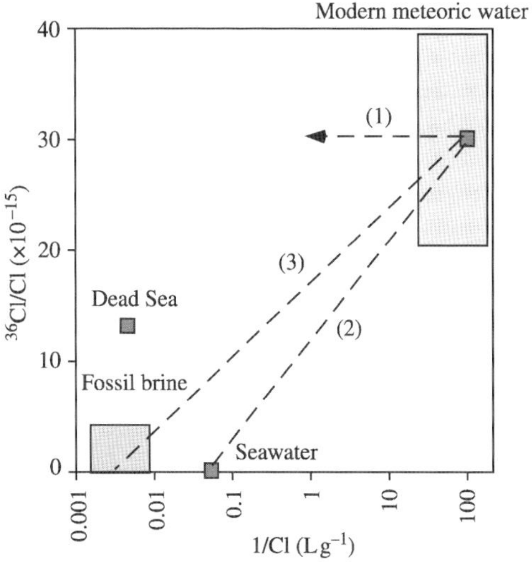

Figure 16 Schematic illustration of the expected ^{36}Cl/Cl variations versus reciprocal of chloride (L g^{-1}) upon salinization by: (1) surface evaporation; (2) seawater intrusion; and (3) mixing with brines and/or dissolution of evaporates. The relatively high ^{36}Cl/Cl ratio in the Dead Sea suggests mixing with meteoric chloride (Yechieli *et al.*, 1996) (sources Magaritz *et al.*, 1990; Yechieli *et al.*,1996; Phillips, 2000).

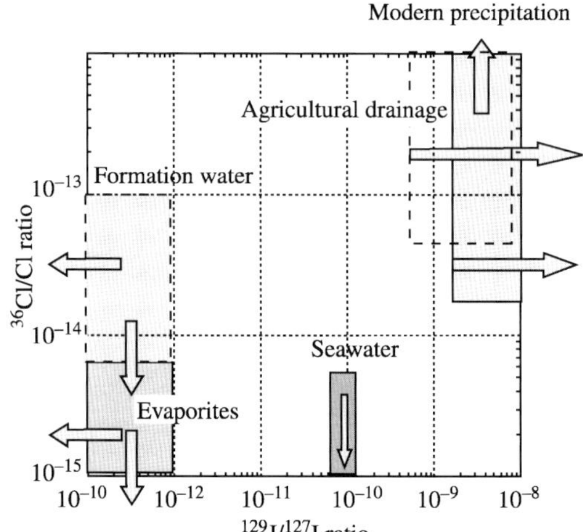

Figure 17 Elucidation of saline sources by using the variations of ^{36}Cl/Cl and ^{129}I/^{127}I. Note the isotopic distinction between modern meteoric water and agricultural return relative to formation water, evaporite dissolution, and seawater. Arrows represent possible extended ranges (sources Ekwurzel *et al.*, 2001; Moran *et al.*, 2002).

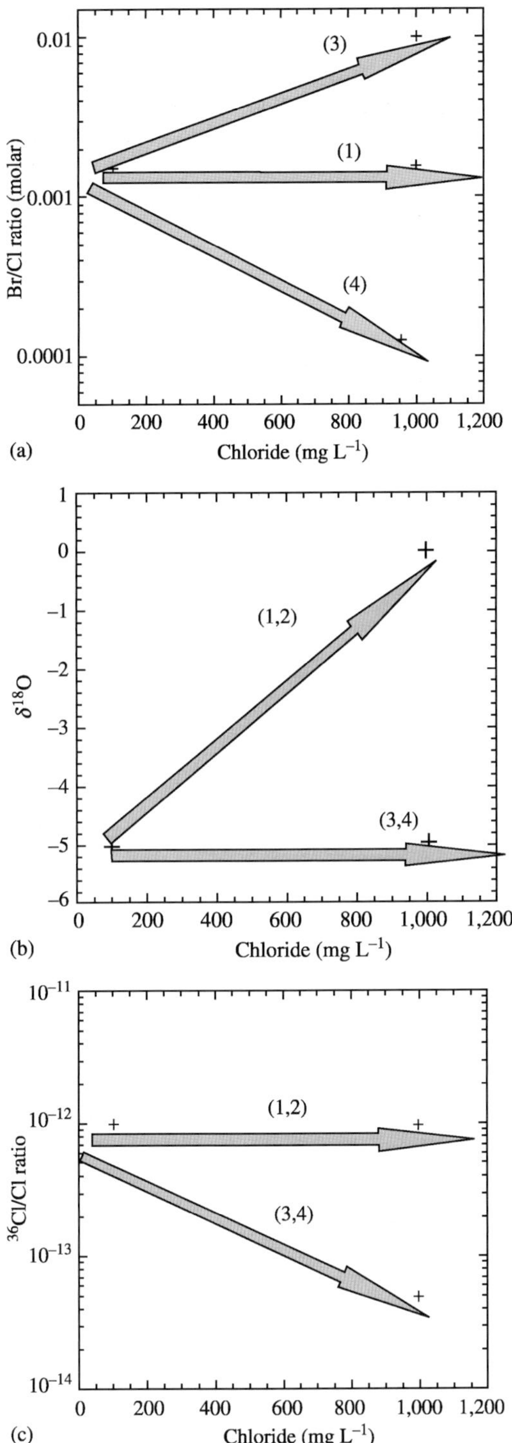

Figure 18 Schematic illustration of possible Br/Cl (a), $\delta^{18}O_{water}$ (b), and $^{36}Cl/Cl$ (c) evolution scenarios upon river salinization: (1) surface evaporation (and/or evapotranspiration) of the river water; (2) recycling of salts via formation of agricultural return flow and discharge to the river; (3) discharge of external groundwater to the river with brine or old formation water components (i.e., high Br/Cl); and (4) discharge of external groundwater salinized by dissolution of evaporite deposits.

increasing $\delta^{18}O_{water}$ and δ^2H values with flow distance due to evaporation, but the Br/Cl and $^{36}Cl/Cl$ ratios in the river are not expected to change with increasing salinity (Figure 18). Salinization via agricultural return flow is likely to increase the $\delta^{18}O_{water}$ values and may change the Br/Cl ratio, but the $^{36}Cl/Cl$ is not changed (i.e., short-term recycling of modern meteoric chlorine with a high $^{36}Cl/Cl$ ratio). However, in dryland environment the long-term storage of chlorine in the subsurface buffers the original $^{36}Cl/Cl$ ratio. Dissolution of halite deposits and the formation of saline groundwater are expected to form saline groundwater with a low Br/Cl ratio (e.g., Rio Grande Basin; Phillips *et al.*, 2002; Mills *et al.*, 2002) and a low $^{36}Cl/Cl$ ratio. In contrast, an increase of the Br/Cl ratio with salinity reflects discharge of saline groundwater that originated from mixing with relics of evaporated seawater that is entrapped in the basin (e.g., Jordan River; Vengosh *et al.*, 2001). In such a scenario, the $^{36}Cl/Cl$ ratio is also expected to be low. Finally, sewage pollution increases the salinity, nitrate (with $\delta^{15}N_{nitrate} > 10‰$), boron ($\delta^{11}B < 10‰$ in domestic sewage; Vengosh *et al.*, 1994), and sulfate ($\delta^{34}S_{sulfate} < 0‰$ in industrial areas; Cortecci *et al.*, 2002) levels in the contaminated river.

9.09.10 REMEDIATION AND THE CHEMICAL COMPOSITION OF TREATED WATER RESOURCES

As the world experiences an explosive population growth of 50%, demands for food production will result in the conversion of prime agricultural land to urban uses and the conversion of more marginal lands to agricultural uses. At the turn of the twenty-first century, 70% of the world's freshwater is used for irrigated agriculture. In order to meet global food demands, agricultural production will have to increase by two- to threefold during the next few decades, and the exploitation of freshwater resources will increase further (Hern and Feltz, 1998). The expected increase of the human population during the next century (8–12 billion by the year 2050, relative to the population of over 6 billion at the end of the twentieth century; Rouch, 1994) will increase the demands for freshwater and cultivation of marginal land, and is likely to cause severe adverse environmental impacts such as salinization of the over-exploited water resources and direct contamination by agricultural drainage water (Bölke, 2002).

In principle, two solutions can be adopted by water agencies facing long-tem salinization of water resources—dilution and desalination. Dilution is the cheapest solution for pollution, given the availability of freshwater resources. In many depleted or salinized aquifers, artificial recharge of imported water compensates for overexploitation or reduces salinization rate. Artificial recharge modifies the oxygen and deuterium isotopic composition of the natural groundwater. In some cases where the recharge water is derived from upstream and higher elevations, the recharge causes lower $\delta^{18}O_{water}$ and δ^2H values (e.g., the recharge of the Colorado River aqueduct in Orange County, California; Davisson *et al.*, 1999; Williams, 1997). In contrast, in cases where the recharge water is derived from open reservoirs (e.g., recharge of the Salinas River, California; Vengosh *et al.*, 2002a) or lakes (recharge of imported water from the Sea of Galilee into the Mediterranean coastal aquifer in Israel; Vengosh *et al.*, 1999), their $\delta^{18}O_{water}$ and δ^2H values are significantly higher than that of the natural groundwater. Interbasin transfers are, however, no longer a possible solution for most water-scarce countries.

Desalination, in contrast, is the ultimate solution for the long-term sustainability of water-poor countries. The water crisis in the Middle East, for example, can only be resolved by large-scale desalination of seawater and brackish water (Glueckstern, 1992). By 1998, 1.25×10^4 desalination units around the word produced about $23 \times 10^6\ m^3\ d^{-1}$ (Glueckstern and Priel, 1998). In Saudi Arabia, 35 desalination plants were built to produce potable water from seawater and brackish water using multistage flush system (MS) and reverse osmosis (RO). Presently, Saudi Arabia is the largest producer of desalinated water in the world with a water production of $\sim 1{,}000 \times 10^6\ m^3\ yr^{-1}$ (Abderrahmann, 2000). In Israel, desalinated seawater is expected to produce about one-fifth ($500 \times 10^6\ m^3\ yr^{-1}$) of the country's annual water use during the 2000s. In California, desalinization of wastewater is one component of the overall water management scheme for continued exploitation of a stressed coastal aquifer as demonstrated in Orange County south of Los Angeles. Desalination of secondary treated wastewater and recharge into the aquifer is used simultaneously with the recharging of the external source (Colorado River) and continued pumping of the aquifer supply water to local inhabitants and agriculture section (Orange County Water District, 1994).

The chemical composition of water resources in the twenty-first century is likely to reflect these human interventions: the creation of new water and the contamination of the depleted water resources. RO has become the cheapest and hence most frequently used technique for desalination (Glueckstern and Priel, 1998). The generation of desalinated water is associated with chemical transformations that are controlled by the preferential selectivity of RO membrane. The ion selectivity through the membrane depends on their masses, size, charge, and reactivity. Rejection by RO membrane is stronger for heavier ions, thus the Na/Cl ratio is modified from 0.86 in seawater to 0.94 in desalinated water (Table 1; data from Vengosh *et al.*, in preparation). The preferential selection by reverse osmosis membranes of doubly charged ions (Ca^{2+}, Mg^{2+}, SO_4^{2-}) over singly charged ions (Cl^-, Na^+) also modifies the original chemical composition and ionic ratios of the desalted water. Thus, desalted water is modified into an Na–Cl water type (Cl + Na/TDI ~ 0.93, in equivalent units), and the Ca/Cl, Mg/Cl, SO_4/Cl, and Ca/Na ratios of RO desalted water are fivefold lower relative to those in the original seawater (Table 1). Similarly, nanofiltration membranes, with a negatively charged hydrophobic rejection layer, tend to selectively reject multivalent ions but have low rejection efficiency for monovalent ions (Kharaka *et al.*, 1996). Finally, the RO rejection of uncharged boric acid is low, which results in enrichment of boron in desalinated water (B/Cl ratio of 2×10^{-2} relative to 8×10^{-4} in the original seawater) with a seawater isotopic signature ($\delta^{11}B \sim 39‰$; Vengosh *et al.*, in preparation).

The $\delta^{18}O_{water}$ and δ^2H values of desalinated seawater are identical to those in seawater (Vengosh *et al.*, in preparation). Consequently, recharge of freshwater derived from desalination either by direct leakage in the urban environment or via wastewater irrigation would have $\delta^{18}O_{water}$ and ^2H values that lie along a mixing line between the natural and seawater values (Figure 19). The slope of the *anthropogenic line* depends on the $\delta^{18}O_{water}$ and ^2H values in the local area and is always lower than that of the natural meteoric line (Craig, 1961, Gat, 1974).

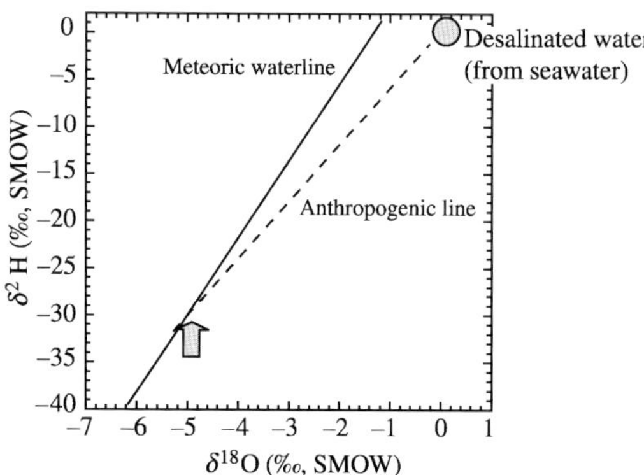

Figure 19 δ^2H versus $\delta^{18}O_{water}$ values of reverse osmosis desalted seawater and the global meteoric waterline (Craig, 1961). The anthropogenic line lies on a possible mixing relationship between natural waters along the meteoric waterline (arrow) and the desalted seawater. Future formation of a large volume of desalted seawater expects to cause infiltration (e.g., leakage, reuse) of freshwater with high δ^2H and $\delta^{18}O$ values relative to natural replenishment.

Post-treatment of desalinated water usually increases the Ca^{2+} and HCO_3^- (interaction with lime) contents of the water supply to the municipality. In some cases, the addition of sulfuric acid increases the sulfate content in the post-treated water. Nevertheless, the extremely high $\delta^{18}O_{water}$, B/Cl, and $\delta^{11}B$ values are not affected, so that desalinated water can be distinguished from natural water sources (Vengosh *et al.*, in preparation).

In addition to creating new water, anthropogically induced salinization modifies the chemical composition of natural water resources. Meybeck and Helmer (1989) showed that polluted rivers are systematically enriched in Cl^-, Na^+, SO_4^{2-}, and NO_3^-, relative to pristine rivers. The salinization of future water resources is expected to be derived primarily from contamination by domestic wastewater and agricultural return flow. First and secondary sewage treatments (the common sewage treatment) do not affect the inorganic chemical composition of treated wastewater (Vengosh *et al.*, 1994). The reuse of treated sewage in agriculture and the release of domestic wastewater into natural water resources will change the chemical composition of the environment. As of early 2000s, $\sim 450 \times 10^9$ m^3 of wastewater are globally discharged into rivers, streams, and lakes every year (Hinrichsen and Tacio, 2002). In the Middle East, reuse of domestic treated sewage is becoming one of the important water sources, given the depletion of natural water resources. In Israel, about 300×10^6 m^3 of treated sewage is used annually for irrigation (Blitz *et al.*, 2000). The impact of effluents enriched in sodium, sulfate, and boron are expected to modify the soil salinity (e.g., increase of sodicity), the composition of the unsaturated zone, and consequently the underlying groundwater resources. Long-term application and numerous cycles of reused wastewater are likely to affect the underlying water resources.

The second major source of salinization that is expected to shape environmental chemistry is agricultural return flow. Based on studies of irrigation drainage in the western United States, Hern and Feltz (1998) predicted that the future use of marginal land for agriculture is expected to form saline drainage effluents enriched in selenium, boron, and arsenic. Hence, the composition of drainage water (e.g., Imperial Valley, California; Table 1) is expected to affect the soil, the unsaturated zone, and the associated water resources. The bioaccumulation of selenium is turning out to be one of the most severe environmental hazards in the western United States (Hern and Feltz, 1998). Boron enrichment in soil and contaminated water will reduce soil productivity for certain crops and will increase boron levels in contaminated water. Consequently, future salinization of water resources due to the increasing contribution of agricultural return flows will have much wider environmental implications.

In conclusion, the human impact on the salinity of the environment has caused dramatic changes in the chemical composition of soil and water resources. The continued stress on land and water resources has shifted the natural balance and has accelerated the natural salinization process, particularly in arid and semi-arid climatic zones. In addition, direct anthropogenic salinization is expected to modify the environment. The inorganic chemical composition of the pristine environment is now diminishing.

REFERENCES

Abderrahmann W. A. (2000) Application of Islamic legal principles for advanced water management. *Water Inter.* **25**, 513–518.

Abramovitz J. (1996) Imperiled waters, impoverished futute: the decline of freshwater ecosystem. *Worldwatch Paper 128*, Worldwatch Institute, Washington, DC.

Allison G. B. and Barnes C. J. (1985) Estimation of evaporation from normally 'dry' lake in south Australia. *J. Hydrol.* **78**, 229–242.

Allison G. B., Stone W. J., and Hughes M. W. (1985) Recharge in karst and dune elements of a semi-arid landscape as indicated by natural isotopes and chloride. *J. Hydrol.* **76**, 1–25.

Allison G. B., Cook P. G., Barnett S. R., Walker G. R., Jolly I. D., and Hughes M. W. (1990) Land clearance and river salinization in the western Murray basin, Australia. *J. Hydrol.* **119**, 1–20.

Al-Muhandis M. H. (1977) Pollution of river water in Iraq. *Hydro. Sci. Publ.* **123**, 467–470.

Al-Ruwaih F. M. (1995) Chemistry of groundwater in the Damman aquifer, Kuwait. *Hydrogeol. J.* **3**, 42–55.

Amrhein C., Crowley D., Holdren G. C., Kharaka J. K., Parkhust D. L., Pyles J., Schroeder R. A., and Wenghorst P. A. (2001) *Effect of Salt Precipitation on Historical and Projected Salinities of the Salton Sea.* Summary comments from Workshop at the University of California, Riverside, January 2001.

Amstrong S. C., Sturchio N. C., and Hendry M. J. (1998) Strontium isotopic evidence on the chemical evolution of pore waters in the Milk River aquifer, Alberta, Canada. *Appl. Geochem.* **13**, 463–475.

Appelo C. A. J. (1994) Cation and proton exchange, pH variations, and carbonate reaction in freshening aquifer. *Water Resour. Res.* **30**, 2793–2805.

Appelo C. A. J. and Geirnart W. (1991) Processes accompanying the intrusion of salt water. In *Hydrogeology of Salt Water Intrusion—a Selection of SWIM Papers II*, 291–304.

Appelo C. A. J. and Postma D. (1993) *Geochemistry, Groundwater and Pollution.* A. A. Balkema Publishers, Rotterdam, 536p.

Appelo C. A. J. and Willemsen A. (1987) Geochemical calculations and observations on salt water intrusions, a combine geochemical/mixing cell model. *J. Hydrol.* **94**, 313–330.

Banner J. L. and Hanson G. N. (1990) Calculation of simultaneous isotopic and trace element variations during water–rock interaction with applications to carbonate diagenesis. *Geochim. Cosmochim. Acta* **54**, 3123–3137.

Banner J. L., Wasserburg G. J., Dobson P. F., Carpenter A. B., and Moore C. H. (1989) Isotopic and trace element constraints on the origin and evolution of saline groundwaters from centural Missouri. *Geochim. Cosmochim. Acta* **53**, 383–398.

Bassett R. L. (1990) A critical evaluation of the available measurements for stable isotopes of boron. *Appl. Geochem.* **85**, 541–554.

Bassett R. L., Buszka P. M., Davidson G. R., and Damaris C.-D. (1995) Identification of groundwater solute sources using boron isotopic composition. *Environ. Sci. Technol.* **29**, 2915–2922.

Beaumont P. (1996) Agricultural and environmental changes in the upper Euphrates catchment of Turkey and Syria and their political and economic implications. *Appl. Geogr.* **16**, 137–157.

Bein A. and Arad A. (1992) Formation of saline groundwaters in the Baltic region through freezing of seawater during glacial periods. *J. Hydrol.* **140**, 75–87.

Beltran J. M. (1999) Irrigation with saline water: benefits and environmental impact. *Agri. Water Manage.* **40**, 183–194.

Bergelson G., Nativ R., and Bein A. (1999) Salinization and dilution history of groundwater discharge into the Sea of Galilee, the Dead Sea Transform, Israel. *Appl. Geochem.* **14**, 91–118.

Berner E. K. and Berner R. A. (1987) *The Global Water Cycle: Geochemistry and Environment*. Prentice-Hall, Englewood Cliffs, NJ, 398pp.

Berzas J. J., Garcia L. F., Rodrigues R. C., and Martin-Alvarez P. J. (2000) Evolution of the water quality of a managed natural wetland: Tables de Daimiel National Park (Spain). *Water Res.* **34**, 3161–3170.

Blitz N., Shwartz D., and Nosbaum S. (2000) *The Water in Israel 1998: Consumption and Production*. Israel Water Commission, Tel Aviv, 42pp.

Bölke J. K. (2002) Groundwater recharge and agricultural contamination. *Hydrogeol. J.* **10**, 153–179.

Bölke J. K. and Horan M. (2000) Strontium isotope geochemistry of groundwaters and streams affected by agriculture, Locust Grove, MD. *Appl. Geochem.* **15**, 599–609.

Bottomley D. J., katz A., Chan L. H., Starinsky A., Douglas M., Clark I. D., and Raven K. G. (1999) The origin and evolution of Canadian Shield brines: evaporation or freezing of seawater? New lithium isotope and geochemical evidence from Slave Craton. *Chem. Geol.* **155**, 295–320.

Bresler E., McNeal B. L., and Carter D. L. (1982) *Saline and Sodic Soils*. Advanced series in agricultural sciences 10, Springer, Berlin, 236pp.

Bullen T. D., Krabbenhoft D. P., and Kendall C. (1996) Kinetic and mineralogic controls on the evolution of groundwater chemistry and $^{87}Sr/^{86}Sr$ in a sandy silicate aquifer, northern Wisconsin, USA. *Geochim. Cosmochim. Acta* **60**, 1807–1821.

Bureau of Reclamation (2003) *Colorado River Basin Salinity Control Program*. US Department of the Interior (http://www.uc/usbr/gov/progact/salinity/).

Carpenter A. B. (1978) Origin and chemical evolution of brines in sedimentary basins. *Okl. Geol. Surv. Circular* **79**, 60–77.

Carpenter A. B., Trout M. L., and Pickett E. E. (1974) Preliminary report on the origin and chemical evolution of lead and zinc reach oil field brines in Central Mississippi. *Econ. Geol.* **69**, 1191–1206.

Ceuppens J. and Wopereis M. C. S. (1999) Impact of non-drained irrigated rice cropping on soil salinization in the Senegal River Delta. *Geoderma* **92**, 125–140.

Chivas A. R., Andrew A. S., Lyons W. B., Bird M. I., and Donnelly T. H. (1991) Isotopic constraints on the origin of salt in the Australian playa: 1. Sulfur. *Paleogeogr. Paleoclimatol. Paleoecol.* **84**, 309–331.

Choudhari J. S. and Sharma K. D. (1984) Stream salinity in the Indian arid zone. *J. Hydrol.* **71**, 149–163.

Cook P. and Herczeg A. (2000) *Environmental Tracers in Groundwater Hydrology*. Kluwer, Boston, Dorderecht, London.

Cook P., Edmunds W. M., and Gaye C. B. (1992) Estimate paleorecharge and paleoclimate from the unsaturated zone profiles. *Water Resour. Res.* **28**, 2721–2731.

Coplen T. B., Hopple J. A., Bohlke J. K., Perser H. S., Rieder S. E., Krouse H. R., Rosman K. J. R., Ding T., Vocke R. D., Jr., Revesz K. M., Lamberty A., Taylor P., and De Bievre P. (2002) *Compilation of Minimum and Maximum Isotope Ratios of Selected Elements in Naturally Occurring Terrestrial Materials and Reagents*. Water-Resource Investt. Report 01-4222, 98pp.

Cortecci G., Dinelli E., Bencini A., Adorni-Braccesi A., and La Ruffa G. (2002) Natural and anthropogenic SO_4 sources in the Arno river catchment, northern Tuscany, Italy: a chemical and isotopic reconnaissance. *Appl. Geochem.* **17**, 79–92.

Cox J. W., Fritsch E., and Fitzpatrick R. W. (1996) Interpretation of soil features produced by modern and ancient processes in degraded landscapes: VII. Water duration. *Austral. J. Soil Res.* **34**, 803–824.

Cox J. W., Chittleborough D. J., Brown H. J., Pitman A., and Varcoe J. (2002) Seasonal changes in hydrochemistry along a toposequence of texture-contrast soils. *Austral. J. Soil Res.* **40**, 1–24.

Craig H. (1961) Isotopic variations in natural waters. *Science* **133**, 1702–1703.

Custodio E. (1987) Salt–fresh water interrelationship under natural conditions (chap. 3) and Effect of human activities on salt–freshwater relationships in coastal aquifers (chap. 4). In *Groundwater Problems in Coastal Areas*, UNESCO-IHP, pp. 14–112.

Custodio E. (1997) Studying, monitoring and controlling seawater intrusion in coastal aquifers. In *Guidelines for Study, Monitoring and Control*. FAO Water Reports, 11, pp. 7–23.

Custodio E., Bayo A., Pascual M., and Boch X. (1993) Results from studies in several karst formations in southern Catalonia (Spain). In *Hydrogeological Processes in Karst Terrains*. Proc. Int. Symp. and Field Seminar (eds. G. Gunay, I. Johnson, and W. Back). IAHS, Antalya, Turkey, vol. 207, pp. 295–326.

Davis S. N., Wittemore D. O., and Febryka-Martin J. (1998) Uses of chloride/bromide ratios in studies of potable water. *Ground Water* **36**, 338–350.

Davisson M. L. and Criss R. E. (1993) Stable isotope imaging of a dynamic groundwater system in the southwestern Sacramento Valley, CA. *J. Hydrol.* **144**, 213–246.

Davisson M. L., Hudson G. B., Esser B. K., and Ekwurzel B. (1999) Tracing and age-dating recycled wastewater recharged for potable reuse in a seawater injection barrier, southern California. In *Int. Symp. Isotope Techniques in Water Resources Development and Management*. IAEA, Vienna, Austria, May 1999 (IAEA-SM-361/36).

De Breuck W. and De Moor G. (1991) The evaluation of coastal aquifer of Belgium. In *Hydrogeology of Salt Water Intrusion—a Selection of SWIM Papers II*, 35–48.

Dehaan R. L. and Taylor G. R. (2002) Field-derived spectra of salinized soils and vegetation as indicators of irrigation-induced soil salinization. *Remote Sen. Environ.* **80**, 406–417.

DeSimone L. A., Howes B. L., and Barlow P. M. (1997) Mass-balance analysis of reactive transport and cation exchange in a plume of wastewater-contaminated groundwater. *J. Hydrol.* **203**, 228–249.

Drever J. I. and Smith C. I. (1978) Cyclic wetting and drying of the soil zone as an influence on the chemistry of groundwater in arid terrains. *Am. J. Sci.* **278**, 1448–1454.

Edmunds W. M. (1996) Bromine geochemistry of British groundwaters. *Mineral. Mag.* **60**, 275–284.

Edmunds W. M. and Droubi A. (1998) Groundwater salinity and environmental change. In *Isotope Techniques in the Study of Environmental Change. Proc. Int. Symp. Isotope Technol. Study of the Past and Current Environment. Changes in the Hydrosphere and Atmosphere*. IAEA, Vienna, pp. 503–518.

Edmunds W. M. and Gaye C. B. (1994) Estimating the spatial variability of groundwater recharge in the Sahel using chloride. *J. Hydrol.* **156**, 47–59.

Edmunds W. M., Fellman E., and Goni I. B. (1999) Lakes, groundwater and paleohydrology in the Sahel of NE Nigeria: evidence from hydrogeochemistry. *J. Geol. Soc. London* **156**, 345–355.

Edmunds W. M., Fellman E., Goni I. B., and Prudhomme C. (2001) Spatial and temporal distribution of groundwater recharge in northern Nigeria. *Hydrogeol. J.* **10**, 205–215.

Edmunds W. M., Carrillo-Rivera J. J., and Cardona A. (2002) Geochemical evolution of groundwater beneath Mexico City. *J. Hydrol.* **258**, 1–24.

Ekwurzel B., Moran J. E., Hudson G. B., Bissani M., Blake R., Krimissa M., Mosleh N., Marah H., Safsaf N., Hsissou Y., and Bouchaou L. (2001) An isotopic investigation of salinity and water sources in the Souss-Massa basin, Morocco. SWICA M3. In *Int. Conf. Salt Water Intrusion and Coastal Aquifers, Monitoring, Modeling and Management*, abstracts, Essaoouira, Morocco, April 2001.

Eugster H. P. and Jones B. F. (1979) Behaviour of major solutes during closed-basin brine evolution. *Am. J. Sci.* **279**, 609–631.

Fabryka-Martin J. (2000) Iodine-129 as a groundwater tracer. In *Environmental Tracers in Subsurface Hydrology* (eds. P. Cook and A. L. Herczeg). Kluwer Academic, Boston.

Fabryka-Martin J., Whittemore D. O., Davis S. N., Kubik P. W., and Sharma P. (1991) Geochemistry of halogens in the Milk River aquifer, Alberta, Canada. *Appl. Geochem.* **6**, 447–464.

Fattah Q. N. and Abdul Baki S. J. (1980) Effects of drainage systems on water quality of major Iraqi rivers. *Int. Assoc. Hydrol. Sci. Publ.* **130**, 265–270.

Fitzpatrick R. W., Merry R., and Cox J. (2000) What are saline soils and what happens when they are drained? *J. Austral. Assoc. Nat. Res. Manage. (AANRM)* **6**, 26–30.

Flugel W.-A. (1995) River salinization due to dryland agriculture in the western cape Province, Republic of South Africa. *Envron. Int.* **21**, 679–686.

Frenkel H., Goertzen J. O., and Rhoades J. D. (1978) Effects of clay type and content, exchangeable sodium percentage, and electrolyte concentration on clay dispersion and soil hydraulic conductivity. *Soil Sci. Soc. Am. J.* **42**, 32–39.

Garrels R. M. and MacKenzie F. T. (1967) Origin of the chemical composition of some springs and lakes. In *Equilibrium Concepts in Natural Water Systems*. Advances in Chemistry Series 67 (ed. R. F. Gould). American Chemical Society, pp. 222–242.

Gat J. R. (1974) Local variability of the isotope composition of groundwater. In *Isotope Techniques in Groundwater Hydrology*. IAEA, Vienna. vol. II, pp. 51–60.

Gates T. K., Burkhalter J. P., Labadie J. W., Vallian T. J. C., and Broner I. (2002) Monitoring and modeling flow and salt transport in a salinity-threatened irrigated valley. *J. Irrig. Drainage Eng. ASCE* **128**, 87–99.

Gavrieli I., Yechieli Y., Kass A., Vengosh A., and Starinsky A. (2001) *The Impact of Wastewater Irrigation on the Chemistry of Groundwater in the Israeli Coastal Aquifer: a Case Study from Kibbutz Nitzanim Area*. Geological Survey of Israel, Report GSI/35/01, Jerusalem, Israel.

Gaye C. B. and Edmunds W. M. (1996) Groundwater recharge estimation using chloride, stable isotopes and tritium profiles in the sands of northwest Senegal. *Environ. Geol.* **27**, 246–251.

Gee G. W. and Hillel D. (1988) Groundwater recharge in arid regions: review and critique of estimation methods. *Hydrol. Proc.* **2**, 255–266.

Glenn E. P., Cohen M. J., Morrison J. I., Valdes-Casillas C., and Fitzsimmons K. (1999) Science and policy dilemmas in the management of agricultural wastewaters: the case of the Salton Sea, CA. *Environ. Sci. Policy* **2**, 413–423.

Glueckstern P. (1992) Desalination—present and future. *Water Irrig. Rev.* **12**, 4–11.

Glueckstern P. and Priel M. (1998) Advanced concept of large desalination system for Israel. International Water Services Association Conference, Membranes in drinking and Industrial water production, Amsterdam, Holland, September, 1998.

Gooddy D. C., Clay J. W., and Bottrell S. H. (2002) Redox-driven changes in porewater chemistry in the unsaturated zone of the chalk aquifer beneath unlined cattle slurry lagoons. *Appl. Geoch.* **17**, 903–921.

Grieve C. M., Poss J. A., Suarez D. L., and Dierig D. A. (2001) Lesquerella growth and selenium uptake affected by saline irrigation water composition. *Indust. Crop Product.* **13**, 57–65.

Hahn J. (1991) Aspects of groundwater salinization in the Wittmund (East Freisland) coastal area. In *Hydrogeology of Salt Water Intrusion—a Selection of SWIM Papers, II*, 251–270.

Hamilton P. A. and Helsel D. R. (1995) Effects of agriculture on groundwater quality in five regions of the United States. *Ground Water* **33**, 217–226.

Hamilton P. A., Denver J. M., Phillips P. J., and Shedlock R. J. (1993) *Water Quality Assessment of the Delmarva Penisula, Delaware, Maryland, and Virginia: Effects of Agricultural Activities on the Distribution of Nitrate and other Inorganic Constituents in the Surficial Aquifers*. US Geol. Surv. Open-File Rep. 93-40, 149pp.

Hanor J. S. (1994) Origin of saline fluids in sedimentary basins. In *Geofluids: Origin, Migration, and Evolution of Fluids in Sedimentary Basins* (ed. J. Parnell). *Geol. Soc. Spec. Publ.* **78**, 151–174.

Hao X. and Chang C. (2002) Does long-term heavy cattle manure application increase salinity of a clay loam soil in semi-arid southern Alberta? *Agr. Ecosys. Environ.* **1934**, 1–16.

Hardie L. A. and Eugster H. P. (1970) The evolution of closed-basin brines. *Mineral Soc. Am. Spec. Publ.* **3**, 273–290.

Harter T., Davis H., Mathews M. C., and Meyer R. D. (2002) Shallow groundwater quality on dairy farms with irrigated forage crops. *J. Contamin. Hydrol.* **55**, 287–315.

Harussi Y., Rom D., Galil N., and Semiat R. (2001) Evaluation of membrane process to reduce the salinity of reclaimed wastewater. *Desalination* **137**, 71–89.

Hely A. G., Hughes G. H., and Irelan B. (1966) *Hydrologic Regimen of Salton Sea, CA*. Geological Survey Professional Paper 486-C,. US Government Printing Office, Washington.

Hendry M. J. and Schwartz F. W. (1988) An alternative view on the origin of chemical and isotopic patterns in groundwater from the Milk River aquifer, Canada. *Water Resour. Res.* **24**, 1747–1763.

Herut B., Starinsky A., Katz A., and Bein A. (1990) The role of seawater freezing in formation of subsurface brines. *Geochim. Cosmochim. Acta* **54**, 13–21.

Herczeg A. L. and Edmunds W. M. (2000) Inorganic ions as tracers. In *Environmental Tracers in Subsurface Hydrology* (eds. P. Cook and A. L. Herczeg). Kluwer Academic, Boston.

Herczeg A. L., Torgersen T., Chivas A. R., and Habermehl M. A. (1991) Geochemistry of Groundwaters from the Great Artesian basin, Australia. *J. Hydrol.* **126**, 225–245.

Herczeg A. L., Simpson H. J., and Mazor E. (1993) Transport of soluble salts in a large semi-arid basin: River Murray, Australia. *J. Hydrol.* **144**, 59–84.

Herczeg A. L., Dogramaci S. S., and Leaney W. J. (2001) Origin of dissolved salts in a large, semi-arid groundwater system: Murray Basin, Australia. *Mar. Freshwater Res.* **52**, 41–52.

Hern J. and Feltz H. R. (1998) Effects of irrigation on the environment of selected areas of the western United States and implications to world population growth and food production. *J. Environ. Manage.* **52**, 353–360.

Hillel D. (1980) *Fundamentals of Soil Physics*. Academic Press, New York, pp. 233–244.

Hinrichsen D. and Tacio H. (2002) The coming freshwater crisis is already here. In *Finding the Source: the Linkages between Population and Water*. Woodrow Wilson International Center for Scholars, Environmental Change and Security Project, Washington, DC, pp. 1–26.

Hoffman D. (1993) *Use of Potassium and Nanofiltration Plants for Reducing the Salinity of Urban Wastewater*. Adam Technical and Economic Services, Agriculture Ministry of Israel, Water Department 084/91.

Howard K. W. F. and Beck P. J. (1993) Hydrochemical implications of groundwater contamination by road de-icing chemicals. *J. Contamin. Hydrol.* **12**, 245–268.

Hsissou Y., Mudry J., Mania J., Bouchaou L., and Chauve P. (1999) Use of the Br/Cl ratio to determine the origin of the salinity of groundwater: an example from the Souss Plain, Morocco. *Surf. Geosci.* **328**, 381–386.

Hunt C. B., Robinson T. W., Bowles W. A., and Washburn A. L. (1966) *Hydrologic Basin, Death Valley, CA*. US Geological Survey Professional paper, 400-B, B-456-B457.

Izbicki J. A. (1991) Chloride sources in a California coastal aquifer. In *Ground Water in the Pacific Rim Countries*. Conference Proceedings, Irrigation Division, American Society of Consulting Engineers, Honolulu, Hi, July 23–35, 1991.

Izbicki J. A. (1996) *Seawater Intrusion in Coastal California Aquifer*. US Geological Survey Fact Sheet FS-125-96.

Johnson T. M. and Depaolo D. J. (1994) Interpretation of isotopic data in groundwater—rock systems: model development and application to Sr isotope data from Yucca Mountain. *Water Resour. Res.* **30**, 1571–1587.

Johnson J. S., Baker L. A., and Fox P. (1999) Geochemical transformations during artificial groundwater recharge: soil-water interactions of inorganic constituents. *Water Res.* **33**, 196–206.

Jolly I. D., Williamson D. R., Gilfedder M., Walker G. R., Morton R., Zhang L., Dowling T. I., Dyce P., Nathan R. J., Nandakumar N., Clarke R., Mcneil V., Robinson G., and Jones H. (2000) Historical stream salinity trends and catchment salt balance in the Murray–Darling basin, Australia. *Mar. Freshwater Res.* **52**, 53–65.

Jones B. F. (1965) Geochemical evolution of closed basin water in the western Great Basin. In *Second Symposium on Salt* (ed. J. L. Rau). *Northern Ohio Geol. Soc.* vol. 1, pp. 181–200.

Jones B. F., Vengosh A., Rosenthal E., and Yechieli Y. (1999) *Chapter 3*: Geochemical investigations. In *Seawater Intrusion in Coastal Aquifers—Concepts, Methods and Practises* (eds. J. Bear, A. H. D. Cheng, S. Sorek, D. Ouazar, and I. Herrera). Kluwer Academic, Dordrecht, The Netherlands, pp. 51–72.

Jones P. H. and Hutchon H. (1983) Road salt in the environment. In *6th Int. Symp. Salt, Salt Institute, Alexandria, Virginia* (eds. B. C. Schreiber and H. L. Harner). vol. 2, pp. 615–619.

Jung J. and Sun G. (2001) Recovery of sodium sulfate from farm drainage salt for use in reactive dyeing of cotton. *Environ. Sci. Technol.* **35**, 3391–3395.

Katz B. G. and Bullen T. D. (1996) The combined use of $^{87}Sr/^{86}Sr$ and carbon and water isotopes to study the hydrochemical interaction between groundwater and lake-water in mantled karst. *Geochim. Cosmochim. Acta* **60**, 5075–5087.

Kendall C. and Aravena R. (2000) Nitrate isotopes in groundwater systems. In *Environmental Tracers in Subsurface Hydrology* (eds. P. Cook and A. L. Herczeg). Kluwer Academic, Boston, pp. 261–297.

Kharaka Y. K., Ambates G., and Presser T. S. (1996) Removal of selenium from contaminated agricultural drainage water by nanofiltration membranes. *Appl. Geochem.* **11**, 797–802.

Kolodny Y., Katz A., Starinsky A., Moise T., and Simon E. (1999) Chemical tracing of salinity sources in lake Kinneret (Sea of Galilee), Israel. *Limnol. Oceanogr.* **44**, 1035–1044.

Kloppman W., Negel P., Casanova J., Klinge H., Schelkes K., and Guerrot C. (2001) Halite dissolution derived brines in the vicinity of a Permian salt dome (N German basin). Evidence from boron, strontium, oxygen, and hydrogen isotopes. *Geochim. Cosmochim. Acta* **65**, 4087–4101.

Komor S. C. (1997) Boron contents and isotopic compositions of hog manure, selected fertilizers, and water in Minnessota. *J. Environ. Qual.* **26**, 1212–1222.

Kotb T. H. S., Watanabe T., Ogino Y., and Tanji K. T. (2000) Soil salinization in the Nile Delta and related policy issues in Egypt. *Agri. Water Mange.* **43**, 239–261.

Kim J. and Sultan M. (2002) Assesment of long-term hydrologic impacts of lake Nasser and related irrigation projects in southwestern Egypt. *J. Hydrol.* **262**, 68–83.

Kim Y., Lee K.-S., Koh D.-C., Lee D.-H., Lee S.-G., Park W.-B., Koh G.-W., and Woo N.-C. (2003) Hydrogeochemical and isotopic evidence of groundwater salinization in a coastal aquifer a case study in Jeju volcanic island, Korea. *J. Hydrol.* **270**, 282–294.

Krapac I. G., Dey W. S., Roy W. R., Smyth C. A., Storment E., Sargent S. L., and Steele J. D. (2002) Impacts of swine manure pits on groundwater quality. *Environ. Pollut.* **120**, 475–492.

Krouse H. R. and Mayer B. (2000) Sulphur and oxygen isotopes in sulphate. In *Environmental Tracers in Subsurface Hydrology* (eds. P. Cook and A. L. Herczeg). Kluwer Academic, Boston, pp. 195–231.

Lavado R. S. and Taboada M. A. (1987) Soil salinization as an effect of grazing in native grassland soil in the Flooding Pampa of Argentina. *Soil Use Manage.* **3**, 143–148.

Leaney F. W., Herczeg A. L., and Walker G. R. (2003) Salinization of a fresh paaeo-groundwater resources by enhanced recharge. *Ground Water* **41**, 84–92.

Linnikov O. D. and Podberezny V. L. (1996) Prevention of sulfate scale formation in desalination of Aral Sea water. *Desalination* **105**, 143–150.

Love A. J., Herczeg A. L., Sampson L., Cresswell R. G., and Fifield L. K. (2000) Sources of chloride and implications for 36Cl dating of old groundwater, southwestern Great Artesian Basin, Australia. *Water Resour. Res.* **36**, 1561–1574.

Lyons W. B., Tyler S. W., Gaudette H. E., and Long D. T. (1995) The use of strontium isotopes in determining groundwater mixing and brine fingering in a playa spring zone, Lake Tyrrel, Australia. *J. Hydrol.* **167**, 225–239.

Lyons W. B., Welch K. A., and Sharma P. (1998) Chlorine-36 in the waters of McMurdo Dry valley lakes, southern Victoria land Antartica: revisited. *Geochim. Cosmochim. Acta* **62**, 185–192.

Mazor E. (1997) *Chemical and Isotopic Groundwater Hydrology, the Applied Approach*. Dekker, London.

Mazor E. and George R. (1992) Marine airbone salts applied to trace evaporation, local recharge, and lateral groundwater flow in Western Australia. *J. Hydrol.* **139**, 63–77.

Magaritz M., Nadler A., Kafri U., and Arad A. (1984) Hydrogeochemistry of continental brackish waters in the southern coastal plain, Israel. *Chem. Geol.* **42**, 159–176.

Magaritz M., Kaufman A., Paul M., Boaretto E., and Hollos G. (1990) A new method to determine regional evapotranspiration. *Water Resour. Res.* **26**, 1759–1762.

Maslia M. L. and Prowell D. C. (1990) Effect of faults on fluid flow and chloride contamination in a carbonate aquifer system. *J. Hydrol.* **115**, 1–49.

Marie A. and Vengosh A. (2001) Sources of salinity in groundwater from Jericho area, Jordan valley. *Ground Water* **39**, 240–248.

Marison G. M., Farren R. E., and Komrowski A. J. (1999) Alternative pathways for seawater freezing. *Cold Reg. Sci. Technol.* **29**, 259–266.

McCaffrey M. A., Lazar B., and Holland H. D. (1987) The evaporation path of seawater and the co-precipitation of Br and K with halite. *J. Sedimen. Petrol.* **57**, 928–937.

Mehta S., Fryar A. E., and Banner J. L. (2000a) Controls on the regional-scale salinization of the Ogallala aquifer, Southern High Plains, Texas. *Appl. Geochem.* **15**, 849–864.

Mehta S., Fryar A. E., Brady R. M., and Morin R. H. (2000b) Modeling regional salinization of the Ogallala aquifer, Southern High Plains, Texas. *J. Hydrol.* **238**, 44–64.

Meisler H., Leahy P. P., and Knober L. L. (1985) *Effect of Eustatic Sea Level Changes on Saltwater–Freshwater Relations in the Northern Atlantic Coastal Plain.* USGS Water Supply Paper, 2255.

Meybeck M. and Helmer R. (1989) The quality of rivers: from pristine state to global pollution. *Palaeogeogr. Palaeoclimatol. Palaeoecol. (Global Planet Change Section)*, **75**, 283–309.

Micklin P. P. (1988) Desiccation of the Aral Sea: a water management disaster in the Soviet Union. *Science* **241**, 1170–1176.

Micklin P. P. (1992) The Aral crisis: introduction to the special issue. *Post-Soviet Geogr.* **33**, 269–282.

Miller M. R., Brown P. L., and Donovan J. J. (1981) Saline deep development and control in North America Great Plains—hydrogeological aspects. *Agri. Water Manage.* **4**, 115–141.

Mills S. K., Phillips F. M., Hogan J., and Hendrickx J. M. H. (2002) Deep thoughts: what is the influence of deep groundwater discharge on salinization of the Rio Grande? *Am. Geol. Soc. Ann. Meeting*, Denver, October, 2002.

Mitchel J. P., Shennan C., Singer M. J., Peters D. W., Miller R. O., Prichard T., Grattan S. R., Rhoades J. D., May D. M., and Munk D. S. (2000) Impacts of gypsum and winter cover crops on soil physical properties and crop productivity when irrigated with saline water. *Agri. Water Manage.* **45**, 55–71.

Moran J. E., Oktay S., Santschi P. H., and Schink D. R. (1999) Atmospheric dispersal of 1^{29}I from nuclear fuel reprocessing facilities. *Environ. Sci. Technol.* **33**, 2536–2542.

Moran J. E., Oktay S., and Schink D. R. (2002) Sources of iodine and iodine 129 in rivers. *Water Resour. Res.* **38** No. 8, 10.1029/2001WR000622.

Mulrennan M. E. and Woodroffe C. D. (1998) Saltwater intrusion into coastal plains of the Lower Mary River, Northern Territory, Australia. *J. Environ. Manage.* **54**, 169–188.

Murphy J. (1999) *Salinity our Silent Disaster.* Australian Broadcasting Corporation (http://abc.net.au/science/slab/salinity/default.htm).

Mutlake S. M., Salih B. M., and Tawfiq S. J. (1980) Quality of the Tigris river passing through Baghdad for irrigation. *Water Air Soil Pollut.* **13**, 9–16.

Nativ R., Adar E., Dahan O., and Nissim I. (1997) Water salinization in arid regions—observation from the Negev desert, Israel. *J. Hydrol.* **196**, 271–296.

Nelson K. H. and Thompson T. G. (1954) Deposition of salts from seawater by figid concentration. *J. Mar. Res.* **13**, 166–182.

Nesbitt H. W. and Cramer J. J. (1993) Genesis and evolution of HCO_3-rich and SO_4-rich groundwaters of Quaternary sediments, Pinawa, Canada. *Geochim. Cosmochim. Acta* **57**, 4933–4946.

Nishri A., Stiller M., Rimmer A., Geifman Y., and Krom M. (1999) Lake Kinneret (The Sea of Galilee): the effects of diversion of external salinity sources and the probable chemical composition of the internal salinity sources. *Chem. Geol.* **158**, 37–52.

Ohlsson L. (1995) *Hydropolitics.* Zed Books, London, 230pp.

Orange County Water District (1994) *The Groundwater Management Plan.* Orange County, Fountain Valley, CA.

Palmer M. R. and Swihart G. H. (1996) Boron Isotope Geochemistry: an overview. In *Boron: Mineralogy, Petrology, and Geochemistry.* Rev. Mineral. (eds. E. S. Grew and L. M. Anovitz). Mineralogical Society of America, vol. 33, pp. 709–744.

Paul M., Kaufman A., Magaritz M., Fink D., Henning W., Kaim R., Kutschera W., and Meirav (1986) A new 36Cl hydrological model and 36Cl systematics in the Jordan River/Dead Sea system. *Nature* **321**, 511–515.

Peck A. J. and Hatton T. (2003) Salinity and the discharge of salts from catchments in Australia. *J. Hydrol.* **272**, 191–202.

Peeters F., Kipfer R., Achermann D., Hofer M., Aeschbach-Hertig W., Beyorle U., Imboden D. M., Rozanski K., and Fröhlich K. (1999) Analysis of deep-water exchange in the Caspian Sea based on environmental tracers. *Deep-Sea Res.* **47**, 621–654.

Phillips F. M. (1994) Environmental tracers for water movement in desert soils of the American Southwest. *Soil Sci. Soc. Am. J.* **58**, 14–24.

Phillips F. M. (2000) Chlorine-36. In *Environmental Tracers in Groundwater Hydrology* (eds. P. Cook and A. Herczeg). Kluwer Publisher, Boston, Dorderecht, London, pp. 479–485.

Phillips F. M., Bentley H. W., Davis S. N., Elmore D., and Swannick G. B. (1986) Chlorine-36 dating of very old groundwater: II. Milk River aquifer, Alberta. *Water Resour. Res.* **22**, 2003–2016.

Phillips F. M., Hogan J. F., Mills S. K., and Hendrickx J. M. H. (2002) Environmental tracers applied to quantifying causes of salinity in arid-region rivers: preliminary results from the Rio Grande, southwestern USA. In *Water Resources Perspectives: Evaluation, Management and Policy* (eds. A. S. Alsharhan and W. W. Wood). Elsevier, Amsterdam.

Pillsbury A. F. (1981) The salinity of rivers. *Sci. Am.* **245**, 54–65.

Poljakoff-Mayber A. and Gale J. (1975) *Plants in Saline Environment.* Ecol. Studies, 15, Springer, Berlin, 213pp.

Raab M. and Spiro B. (1991) Sulfur isotopic variations during seawater evaporation with fractional crystallization. *Chem. Geol.* **86**, 323–333.

Ramoliya P. J. and Pandey N. A. (2003) Effect of salinization of soil on emergence, growth and survival of seedings of Codia rothii. *Forest Ecol. Manage.* **176**, 185–194.

Raymond K. and Butterwick L. (1992) Perborate. In *Detergents* (ed. N. T. de Qude). Springer, New York, pp. 288–318.

Richardson C. (1976) Phase relationships in sea ice as a function of temperature. *J. Glaciol.* **17**, 507–519.

Richter B. C., Kreitler C. W., and Bledsoe B. E. (1993) *Geochemical Techniques for Identifying Sources of Groundwater Salinization.* C. K. Smoley, 258pp.

Robson J. F., Stoner R. F., and Perry J. H. (1983) Disposal of drainage water from irrigated alluvial plains. *Water Sci. Technol.* **16**, 41–55.

Ronen D., Magaritz M., Almon E., and Amiel A. (1987) Anthropogenic anoxification ("Eutrophication") of the water table region of a deep phreatic aquifer. *Water Resour. Res.* **23**, 1554–1560.

Rouch W. (1994) Population: the view from Cairo. *Science* **265**, 1164–1167.

Rosenthal E., Jones B. F., and Weinberger G. (1998) The chemical evolution of Kurnob Group paleowater in the Sinai–Negev province—a mass-balance approach. *Appl. Geochem.* **13**, 553–569.

Salameh E. (1996) *Water Quality Degradation in Jordan.* The Higher Council of Science and Technoloy, Royal Society for the Conservation of Nature, Jordan, 179pp.

Salameh E. and Naser H. (1999) Does the actual drop in the Dead Sea level reflect the development of water resources within its drainage basin. *Acta Hydrochem. Hydrobiol.* **27**, 5–11.

Sanchez-Martos F. and Pulido-Bosch A. (1999) Boron and the origin of salinization in an aquifer in Southeast Spain. *Surface Geosci.* **328**, 751–757.

Sanchez-Martos F., Pulido-Bosch A., Molina-Sanchez L., and Vallejos-Izquierdo A. (2001) Identification of the origin of salinization in groundwater using minor ions (Lower Andarax, Southeast Spain). *Sci. Total Environ.* **297**, 43–58.

Sayles F. L. and Mangelsdorf P. C. (1977) The equilibrium of clay minerals with seawater, exchange reactions. *Geochim. Cosmochim. Acta* **41**, 951–960.

Schroeder R. A. and Rivera M. (1993) *Physical, Chemical, and Biological Data for Detailed Study of Irrigation Drainage in the Salton Sea Area, California*. US Geological Survey Open-file Report 93-83, Sacramento, CA.

Schroeder R. A., Setmire J. G., and Densmore J. N. (1991) Use of stable isotopes, tritium, soluble salts, and redox-sensitive elements to distinguish groundwater from irrigation water in the Salton Sea basin. In *Irrigation and Drainage*, Proc. 1991 national conference, Honolulu, Hawaii, July 22-26, 1991 (ed. W. F. Ritter). American Society of Civil Engineers.

Schroeder R. A., Orem W. H., and Kharaka Y. K. (2002) Chemical evolution of the Salton Sea, California: nutrient and selenium dynamics. *Hydrobiology*, vol. 473, pp. 23–45.

Serageldin I. (2000) *World's Rivers in Crisis some are Dying; others could Die*. The World Commission on Water for the 21st Century (http://www.worldwatercouncil.org/Vision/6902B03438178538C125683A004BE974.htm).

Shimojima E., Yoshioka R., and Tamagawa I. (1996) Salinization owing to evaporation from bare-soil surfaces and its influences on the evaporation. *J. Hydrol.* **178**, 109–136.

Singh S. K., Trivedi J. R., Pande K., Ramesh R., and Krishnawami S. (1998) Chemical and strontium, oxygen, and carbon isotopic compositions of carbonates from the Lesser Himalaya: implications to the strontium isotope composition of the source waters of the Ganga, Ghaghara, and the Indus rivers. *Geochim. Cosmochim. Acta* **62**, 743–755.

Sparks D. L. (1995) *Environmental Soil Chemistry*. Academic Press, 267pp.

Spencer R. J., Eugster H. P., Jones B. F., and Retting S. L. (1985) Geochemistry of Great Salt lake, Utah: I. Hydrochemistry since 1850. *Geochim. Cosmochim. Acta* **49**, 727–737.

Shiklomanov I. A. (1997) *Comprehensive Assessment of the Freshwater Resources of the World*. United Nation Commission for Sustainable Development, World Meteorological Organization and Stockholm Environment Institute, Stockholm, 88pp.

Stanton J., Olson D. K., Brook J. H., and Gordon R. S. (2001) The environmental and economic feasibility of alternative crops in arid areas: considering mesquite in Baja California, Mexico. *J. Arid Environ.* **48**, 9–22.

Starinsky A. (1974) Relationship between calcium–chloride brines and sedimentary rocks in Israel. PhD Thesis, The Hebrew University, Jerusalem, 176pp.

Starinsky A., Bielsky M., Lazar B., Wakshal E., and Steinitz G. (1980) Marine $^{87}Sr/^{86}Sr$ ratios from the Jurassic to Pleistocene: evidence from groundwaters in Israel. *Earth Planet. Sci. Lett.* **47**, 75–80.

Starinsky A., Bielski M., Ecker A., and Steinitz G. (1983) Tracing the origin of salts in groundwater by Sr isotopic composition (the crystalline complex of the southern Sinai, Egypt). *Isotope Geosci.* **1**, 257–267.

Stein M., Starinsky A., Katz A., Goldstein S. L., Machlus M., and Schramm A. (1997) Strontium isotopic, chemical, and sedimentological evidence for the evolution of Lake Lisan and the Dead Sea. *Geochim. Cosmochim. Acta* **61**, 3975–3992.

Stigter T. Y., van Ooijen S. P. J., Post V. E. A., Appelo C. A. J., and Carvalho Dill M. M. (1998) A hydrogeological and hydrochemical explanation of the groundwater composition under irrigated land in a Mediterranean environment, Algarve, Portugal. *J. Hydrol.* **208**, 262–279.

Sukhija B. S. (1996) Differential of paleomarine and modern seawater intruded salinities in coastal groundwaters (of Karaikal and Tanjavur, India) based on inorganic chemistry, organic biomarker fingerprints and radiocarbon dating. *J. Hydrol.* **174**, 173–201.

Sun J. J. (2001) Recovery of sodium sufate from farm drainage salt for use in reactive dyeing of cotton. *Environ. Sci. Technol.* **35**, 3391–3395.

Tedeschi A., Beltran A., and Aragues R. (2001) Irrigation management and hydrosalinity balance in a semi-arid area of the middle Ebro river basin (Spain). *Agr. Water Manage.* **49**, 31–50.

Titus J. G. and Richman C. (2000) Maps of lands vulnerable to sea level rise: Modeled elevation along US Atlantic and Gulf coasts. *Climate Res.* **18**, 205–228.

Thomas D. S. G. and Middleton N. J. (1993) Salinization: new perspectives on a major desertification issue. *J. Arid Environ.* **24**, 95–105.

Todd D. K. (1989) *Sources of Saline Intrusion in the 400-foot Aquifer, Castroville Area, California*. Report for Monterey County Flood Control and Water Conservation District Salinas, CA, 41p.

Vengosh A. and Rosenthal A. (1994) Saline groundwater in Israel: its bearing on the water crisis in the country. *J. Hydrol.* **156**, 389–430.

Vengosh A. and Keren R. (1996) Chemical modifications of groundwater contaminated by recharge of sewage effluent. *J. Contamin. Hydrol.* **23**, 347–360.

Vengosh A. and Pankratov I. (1998) Chloride/bromide and chloride fluoride ratios of domestic sewage effluents and associated contaminated groundwater. *Ground Water* **36**, 815–824.

Vengosh A. and Spivack A. J. (2000) Boron in Ground Water. In *Environmental Tracers in Groundwater Hydrology* (eds. P. Cook and A. Herczeg). Kluwer, Boston, Dorderecht, London, pp. 479–485.

Vengosh A., Chivas A. R., McCulloch M. T., Starinsky A., and Kolodny Y. (1991a) Boron-isotope geochemistry of Australian salt lakes. *Geochim. Cosmochim. Acta* **55**, 2591–2606.

Vengosh A., Starinsky A., Kolodny Y., and Chivas A. R. (1991b) Boron-isotope geochemistry as a tracer for the evolution of brines and associated hot springs from the Dead Sea, Israel. *Geochim. Cosmochim. Acta* **55**, 1689–1695.

Vengosh A., Starinsky A., Kolodny Y., Chivas A. R., and Raab M. (1992) Boron isotope variations during fractional evaporation of sea water: new constraints on the marine vs. nonmarine debate. *Geology* **20**, 799–802.

Vengosh A., Heumann K. G., Juraske S., and Kasher R. (1994) Boron isotope application for tracing sources of contamination in groundwater. *Environ. Sci. Technol.* **28**, 1968–1974.

Vengosh A., Chivas A. R., Starinsky A., and Kolodny Y. (1995) Boron isotope geochemistry of nonmarine brines from the Qaidam Basin (China). *Chem. Geol.* **120**, 135–154.

Vengosh A., Spivack A. J., Artzi Y., and Ayalon A. (1999) Boron, strontium, and oxygen isotopic and geochemical constraints for the origin of salinity in groundwater from the Mediterranean coast of Israel. *Water Resour. Res.* **35**, 1877–1894.

Vengosh A., Farber E., Shavit U., Holtzman R., Segal M., Gavrieli I., and Bullen T. M. (2001) Exploring the sources of salinity in the Middle East: a hydrologic, geochemical and isotopic study of the Jordan River. In *Proc. 10th Int. Symp. Water–Rock Interaction, WRI-10, Villasimuis, Sardinia, Italy, June 2001* (ed. R. Cidu). A. A. Balkema, Sardinia, Italy, vol. 1, pp. 71–79.

Vengosh A., Gill J., Davisson M. L., and Huddon G. B. (2002a) A multi isotope (B, Sr, O, H, C) and age dating (^{3}H-^{3}He, ^{14}C) study of groundwater from Salinas Valley, California: hydrochemistry, dynamics, and contamination processes. *Water Resour. Res.* **38**(9), 1–17.

Vengosh A., Helvaci C., and Karamanderesi I. H. (2002b) Geochemical constraints for the origin of thermal waters from western Turkey. *Appl. Geochem.* **17**, 163–183.

Vengosh A., Marei A., Guerrot C., and Pankratov I. (2002c) An enigmatic salinity source in the Mediterranean coastal aquifer and Gaza Strip: utilization of isotopic (B, Sr, O) constraints for searching the sources of groundwater contamination. Goldschmidt Conference, Davos, Switzerland, August 18–23, 2002.

Vengosh A., Kloppmann W., and Pankratov I. (in preparation) The geochemical imprint of anthropogenic freshwater: chemical and isotopic fractionation of reverse osmosis desalination (in preparation).

Waggott A. (1969) An investigation of the potential problem of increasing boron concentrations in rivers and water courses. *Water Res.* **3**, 749–765.

Weinthal E. (2002) *State-making and Environmental Cooperation*. MIT Press, Cambridge, MA.

Weisbrod N., Nativ R., Adar E., and Ronen D. (2000) Salt accumulation and flushing in unsaturated fractures in an arid environment. *Ground Water* **38**, 452–461.

Wichelns D. (1999) An economic model of waterlogging and salinization in arid regions. *Ecol. Econ.* **30**, 475–491.

Wicks C. M. and Herman J. S. (1996) Regional hydrogeochemistry of a modern coastal mixing zone. *Water Resour. Res.* **32**, 401–407.

Wicks C. M., Herman J. S., Randazzo A. F., and Jee J. L. (1995) Water–rock interactions in a modern coastal mixing zone. *Geol. Soc. Am. Bull.* **107**, 1023–1032.

Williams A. E. (1997) Stable isotope tracers: natural and anthropogenic recharge, Orange County, California. *J. Hydrol.* **201**, 230–248.

Williams D. D., Williams N. E., and Cao Y. (1999) Road salt contamination of groundwater in a major metropolitan area and development of biological index to monitor its impact. *Water Res.* **34**, 127–138.

Williams J., Walker G. R., and Hatton T. J. (2002) Dryland salinization: a challenge for land and water management in the Australian landscape. In *CAB International 2002, Agriculture, Hydrology and Water Quality* (eds. P. M. Haygarth and S. C. Jarvis), pp. 457–475.

Williams W. D. (2001a) Anthropogenic salinization of inland waters. *Hydrobiology* **466**, 329–337.

Williams W. D. (2001b) Salinization: unplumbed salt in a parched landscape. *Water Sci. Technol.* **43**, 85–91.

Wilmoth B. A. (1972) Salty groundwater and meteoric flushing of contaminated aquifers in West Virginia. *Ground Water* **10**, 99–105.

Wilson T. P. and Long D. T. (1993) Geochemistry and isotope chemistry of Michigan Basin brines: Devonian formations. *Appl. Geochem.* **8**, 81–100.

Yaqing W., Xiaobai C., Guanglan M., Shaoqing W., and Zhenyan W. (2000) On changing trends of dD during seawater freezing and evaporation. *Cold Reg. Sci. Technol.* **31**, 27–31.

Yechieli Y., Ronen D., and Kaufman A. (1996) The source and age of groundwater brines in the Dead Sea area, as deduced from ^{36}Cl and ^{14}C. *Geochim. Cosmochim. Acta* **60**, 1909–1916.

Yechieli Y., Gavrieli I., Berkowitz B., and Ronen D. (1998) Will the Dead Sea die? *Geology* **26**, 755–758.

Yechieli Y., Sivan O., Lazar B., Vengosh A., Ronen D., and Herut B. (2001) ^{14}C in seawater intruding into the Israeli Mediterranean coastal aquifer. *Radiocarbon* **43**(2B), 773–781.

Zalidis G. (1998) Management of river water for irrigation to mitigate soil salinization on a coastal wetland. *J. Environ. Manage.* **54**, 161–167.

9.10
Acidification and Acid Rain

S. A. Norton
University of Maine, Orono, ME, USA

and

J. Veselý
Czech Geologic Survey, Prague, Czech Republic

9.10.1 INTRODUCTION

Air pollution by acids has been known as a problem for centuries (Ducros, 1845; Smith, 1872; Camuffo, 1992; Brimblecombe, 1992). Only in the mid-1900s did it become clear that it was a problem for more than just industrially developed areas, and that precipitation quality can affect aquatic resources (Gorham, 1955). The last three decades of the twentieth century saw tremendous progress in the documentation of the chemistry of the atmosphere, precipitation, and the systems impacted by acid atmospheric deposition. Chronic acidification of ecosystems results in chemical changes to soil and to surface waters and groundwater as a result of reduction of base cation supply or an increase in acid (H^+) supply, or both. The most fundamental changes during chronic acidification are an increase in exchangeable H^+ or Al^{3+} (aluminum) in soils, an increase in H^+ activity ($\simeq$concentration) in water in contact with soil, and a decrease in alkalinity in waters draining watersheds. Water draining from the soil is acidified and has a lower pH ($= -\log [H^+]$). As systems acidify, their biotic community changes.

Acidic surface waters occur in many parts of the world as a consequence of natural processes and also due to atmospheric deposition of strong acid (e.g., Canada, Jeffries *et al.* (1986); the United Kingdom, Evans and Monteith (2001); Sweden, Swedish Environmental Protection Board (1986); Finland, Forsius *et al.* (1990); Norway, Henriksen *et al.* (1988a); and the United States (USA), Brakke *et al.* (1988)). Concern over acidification in the temperate regions of the northern hemisphere has been driven by the potential for accelerating natural acidification by pollution of the atmosphere with acidic or acidifying compounds. Atmospheric pollution (Figure 1) has resulted in an increased flux of acid to and through ecosystems. Depending on the ability of an ecosystem to neutralize the increased flux of acidity, acidification may increase only imperceptibly or be accelerated at a rate that endangers the existing biota. Concerns about acid (or acidic) rain in its modern sense were publicized by the Swedish soil scientist Svante Odén (1968). He argued, initially in the Swedish press, that long-term increases in the atmospheric deposition of acid could lower the pH of surface waters, cause a decline in fish stocks, deplete soils of nutrients, and accelerate damage to materials. By the 1970s, acidification of surface waters was reported in many countries in Europe as well as in North America. The late twentieth-century rush to understand the impact of acid rain was driven by: (i) reports of damaged or threatened freshwater fisheries and (ii) damaged forests. Perhaps the earliest linkage between acidic surface water and damage to fish was made by Dahl (1921) in southern Norway. There, spring runoff was sufficiently acidic to kill trout. It was not until the 1970s that a strong link was established between depressed pH, mobilization of aluminum from soil, and fish status (Schofield and Trojnar, 1980). The relationship between acidification of soils and forest health started with hypotheses in the 1960s and has slowly developed. Acid rain enhances the availability of some nutrients (e.g., nitrogen), and may either enhance or diminish the availability of others (e.g., calcium, magnesium, potassium, and phosphorus). Damage to anthropogenic structures, human health, and visibility have also raised concerns. The history of these early developments was summarized by Cowling (1982). Since the 1970s, sulfur and nitrogen emissions to the atmosphere have been reduced by 50–85% and 0–30%, respectively, both in North America and Europe. The emission reductions have occurred as a consequence of knowledge gained and economic factors. While recovery of water quality is underway in some areas, problems of acidification persist, and are now complicated by the effects of climate change (Schindler, 1997).

9.10.2 HOW DO WE DESCRIBE ACIDIFICATION?

Acidity of waters is typically expressed by the pH ($= -\log[H^+]$) as an intensity factor and by acid-neutralizing capacity (ANC), or alkalinity (ALK), as a capacity factor. The latter is commonly expressed in $\mu eq\ L^{-1}$. An *acidic* water has a pH below 7.0. Many papers suggest that acidic water should be defined as a water that

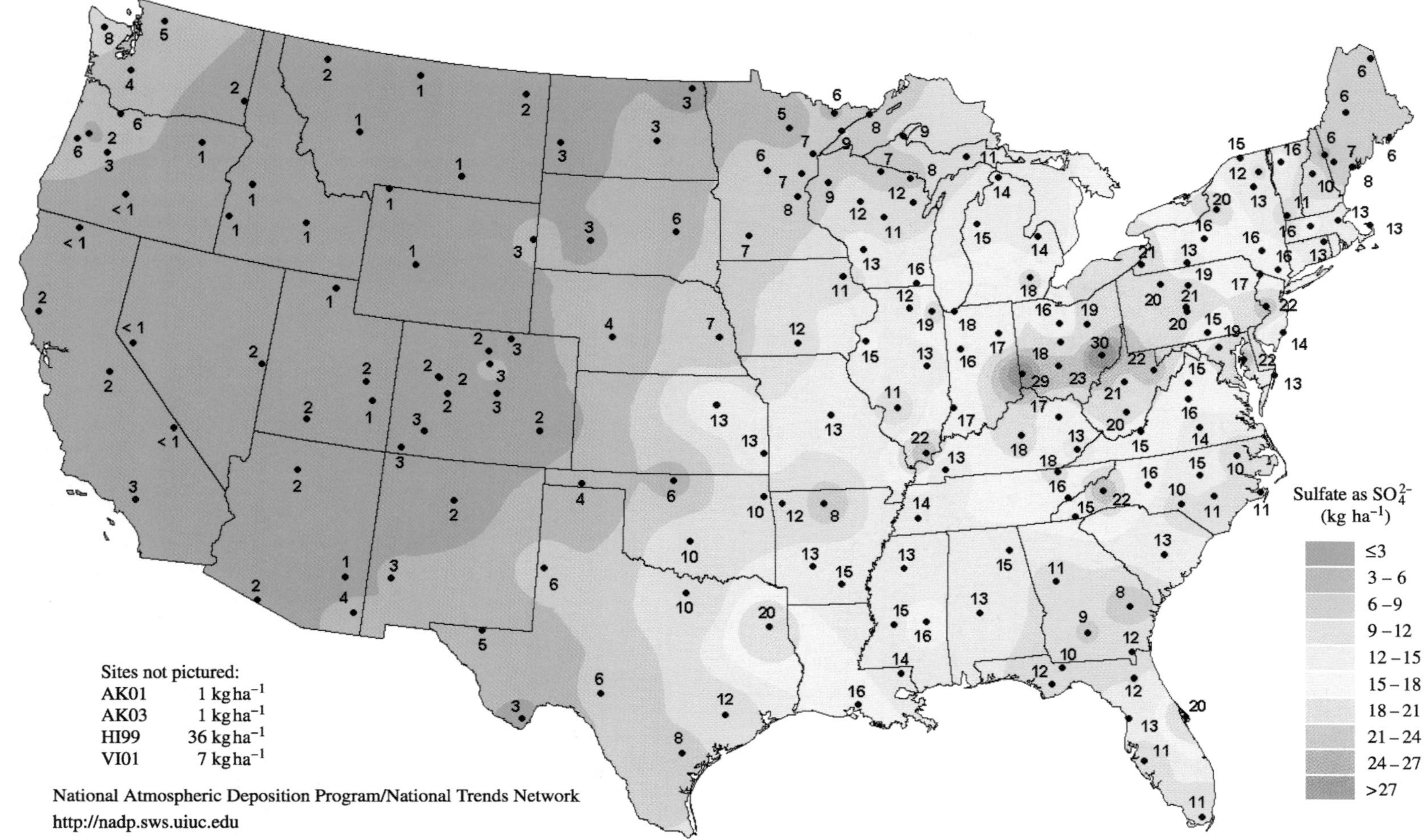

(a)

Figure 1 Deposition of (a) SO_4 and (b) NO_3 in the USA for 2001 (source National Atmospheric Deposition Program, 2002), (c) SO_4 and (d) NO_3 in Europe for 1999 (source EMEP, 2002).

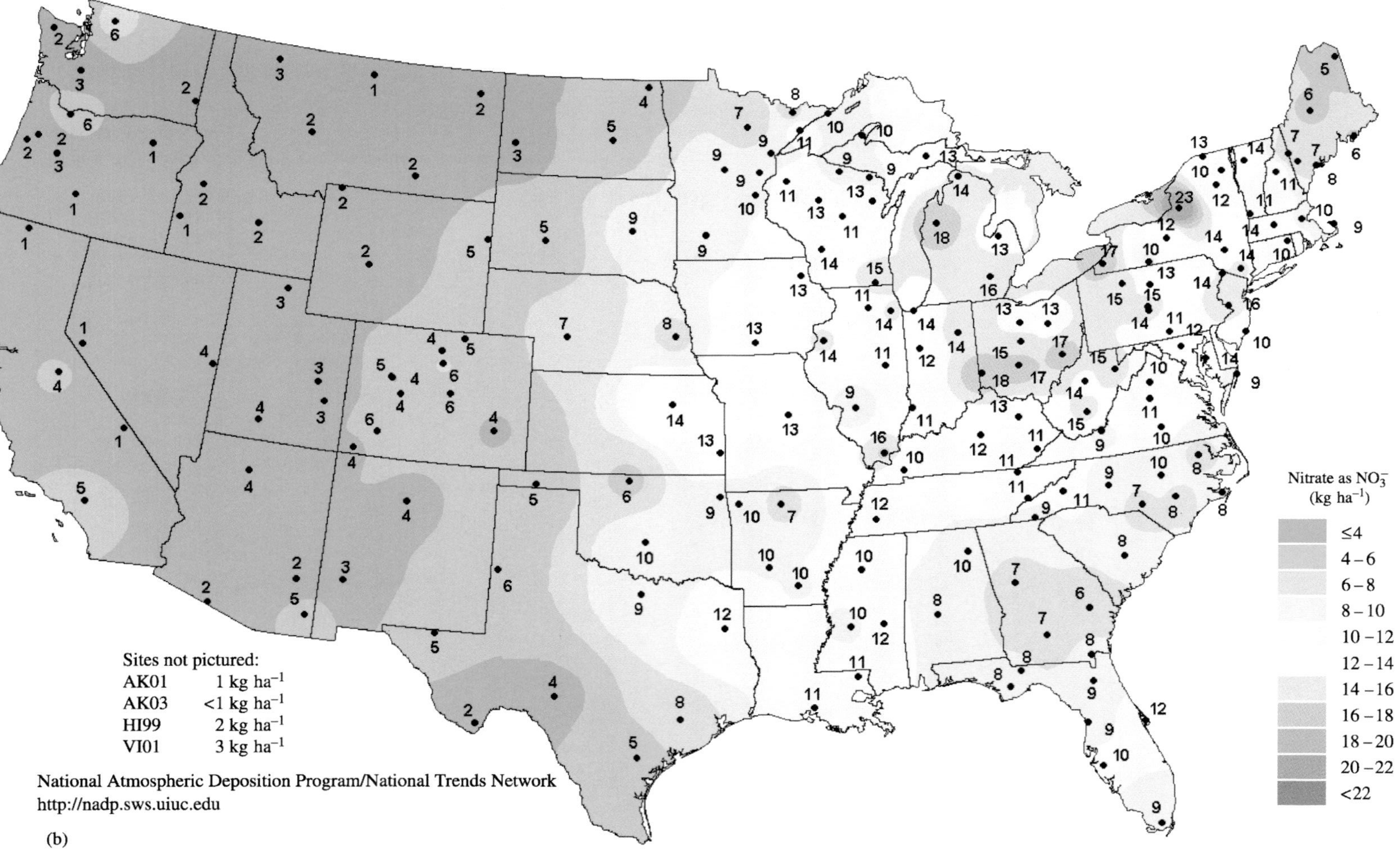

Figure 1 (continued).

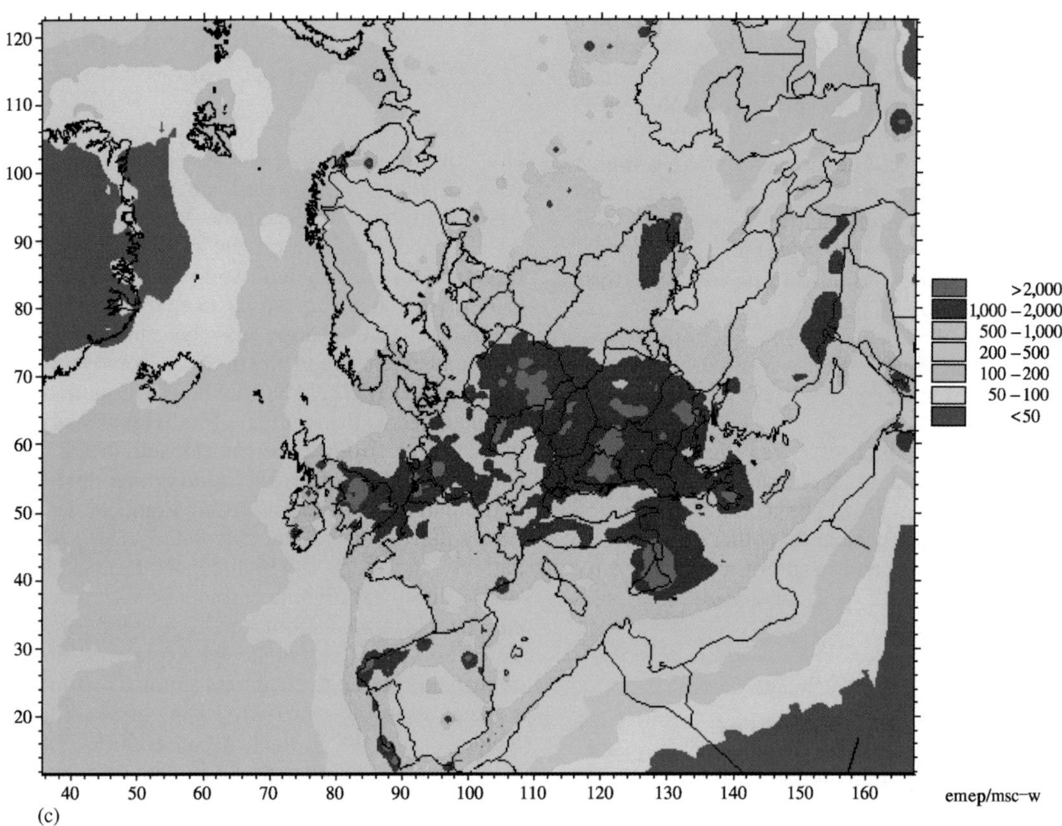

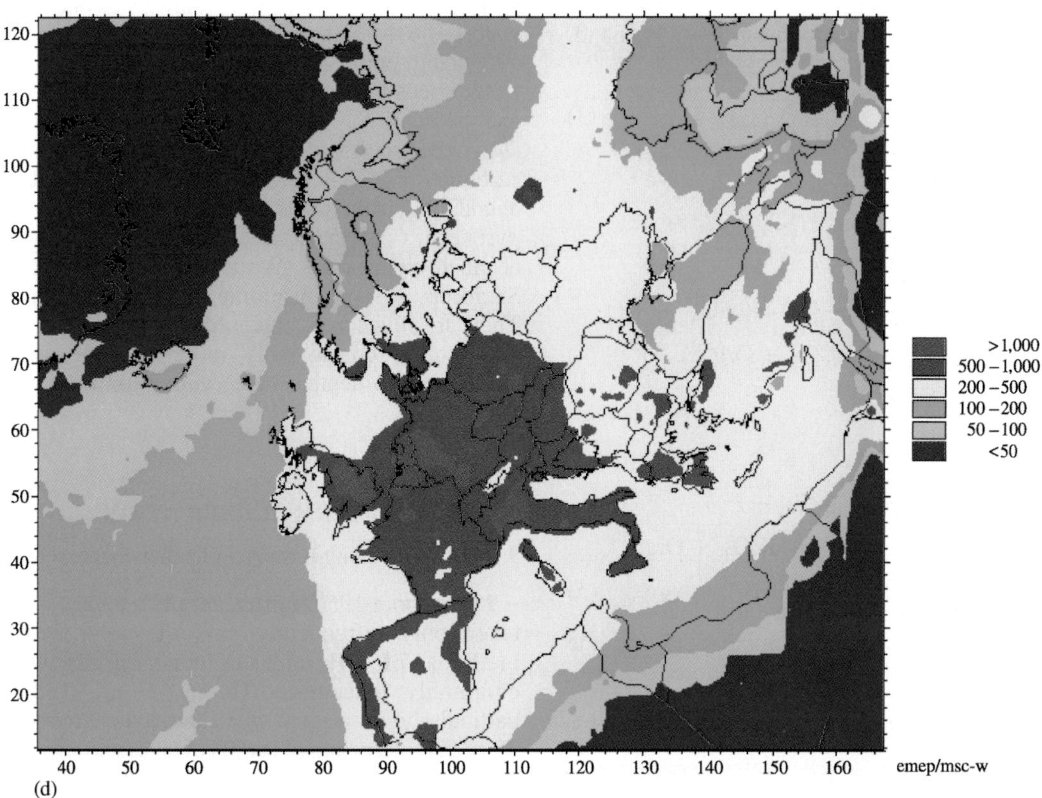

Figure 1 (continued).

has a pH below that of pure water in equilibrium with atmospheric CO_2, 5.65. Whichever definition is adopted, the process of water *acidification* involves a decrease in pH and ANC, with accompanying secondary changes. Adverse biological effects of acidification start at ANC higher than ~0, or pH ~ 5.6. Sites termed "insensitive" to acidification from acidic deposition have surface waters with ANC > 200 μeq L^{-1}.

Early definitions of alkalinity took the form (in equivalents of charge):

$$\text{Carbonate alkalinity} = (HCO_3^- + CO_3^{2-} + OH^-) - (H^+) \quad (1)$$

This carbonate alkalinity was determined by titration with acid to a defined pH end point. As understanding grew about water chemistry, it became abundantly clear that other ions played a role in alkalinity and a more comprehensive definition of alkalinity was advanced:

$$\text{Alkalinity} = (HCO_3^- + CO_3^{2-} + OH^- + Org^{x-} + \text{other weak acid anions}) - (H^+ + Al^{+3-n}(OH^-)_n) \quad (2)$$

where Org^{x-} is the organic anions from the dissociation of dissolved organic acids and $Al^{+3-n}(OH^-)_n$ represents variously charged (hydroxylated) species of aluminum in solution. The ANC of a solution can be defined as charge balance ANC (Reuss and Johnson, 1986; Hemond, 1990). In equivalents

$$\sum(+) = \sum(-) \quad (3)$$

or again, in equivalents,

$$(Ca^{2+} + Mg^{2+} + Na^+ + K^+ + (Al^{+3-n}(OH^-)_n) + H^+) = (OH^- + F^- + NO_3^- + SO_4^{2-} + Cl^- + CO_3^{2-} + HCO_3^- + Org^{x-}) \quad (4)$$

Rearranging we get

$$(Ca^{2+} + Mg^{2+} + Na^+ + K^+) - (SO_4^{2-} + NO_3^- + Cl^- + F^-) = (OH^- + CO_3^{2-} + HCO_3^- + Org^{x-}) - (H^+ + Al^{+3-n}(OH^-)_n) \quad (5)$$

The right-hand side of Equation (5) is the expanded definition of ALK. Therefore,

$$\sum(\text{Strong base cations}) - \sum(\text{Strong acid anions}) = \text{alkalinity} = \text{acid neutralizing capacity} = \sum(\text{Weak acid anions}) - \sum(\text{Weak acid cations}) \quad (6)$$

or

$$\text{ANC} = \text{ALK} = (\text{SBC}) - (\text{SAA}) = (\text{WAA}) - (\text{WAC}) \quad (7)$$

The ANC or ALK are usually measured by Gran titration. In this process, some Org^{x-} and $Al^{+3-n}(OH^-)_n$ are titrated, contributing to the ANC. In this chapter we equate ALK with the term ANC. Concentrations on the right-hand side of Equation (2) all vary with soil partial pressure of CO_2 (P_{CO_2}), but vary so as to maintain electroneutrality. ANC change may be caused by any independent change in any of the summative terms in Equation (5). The ANC defined by Equation (7) is calculated as the residual of individual analyses of water for strong base cations and strong acid anions. The sum of errors in individual analyses, particularly if concentrations are high, can lead to substantial errors in the calculated ANC. Alternatively, in waters with high concentrations of sea salt or dissolved organic carbon (DOC), the ANC can be calculated from the right-hand side of Equation (2) as (carbonate ALK plus (estimated contribution of DOC to anions)) minus (estimated concentration of ionic Al) (Evans *et al.*, 2001a; Köhler *et al.*, 1999). Each mg of DOC L^{-1} adds 3–6 μeq L^{-1} to carbonate ALK. The discrepancy between ANC and carbonate ALK due to DOC and/or Al species for low-ANC waters commonly exceeds 50 μeq L^{-1}. Commonly, measured cationic (+) charge exceeds anionic (−) charge in water. The difference, anionic deficit, is typically assigned to unmeasured organic anions.

We focus here on surface waters draining soils and bedrock for which chemical weathering is slow. Bedrock lithologies that contain free carbonate minerals (e.g., calcite, $CaCO_3$) and/or abundant ferromagnesian silicate minerals (e.g., pyroxene, $(Ca,Mg,Ca)SiO_3$) release base cations at much higher rates (White and Brantley, 1995; Sverdrup, 1990), consuming H^+ in the process. They are much less susceptible to acidification, and require considerably more time or more strong acid to deplete base cations from the soil and bedrock.

9.10.3 LONG-TERM ACIDIFICATION

9.10.3.1 Has Long-term Acidification Occurred?

Up to about 1960, freshwater pH readings were taken only infrequently, and the colorimetric methods of pH measurements used were commonly inaccurate (Haines *et al.*, 1983), particularly in waters with low ionic strength. Thus, reconstruction of the pH of lakes and streams from the literature, even for the mid-1950s, is problematic. However, dated lake sediment cores have now been analyzed for fossil diatom and chrysophyte assemblages. Statistical interpretation of these fossil assemblages enables us to make inferences about longer-term trends in past environmental lake conditions, especially their pH (Battarbee *et al.*, 1990; Charles and Smol, 1988; Dixit *et al.*, 1992), but also DOC, alkalinity, and dissolved aluminum (Davis, 1987). Decreases in atmospheric deposition of acidic compounds since 1990 have coincided with increases in measured lake-water ANC and pH, as well as the pH inferred from fossil remains. These changes have been observed directly, and this lends confidence in the reconstructions of earlier natural acidification.

Reconstruction of the pH history of several northern hemisphere lakes using fossil diatom assemblages showed that the lakes were alkaline shortly after the last deglaciation but had become markedly acidified by the Early Holocene (the last 10^4 yr of Earth history) (Whitehead *et al.*, 1986; Ryan and Kahler, 1987; Renberg, 1990). Neutral to alkaline pH in soil and surface water, shortly after deglaciation, was caused by weathering of an abundance of finely divided "rock flour" and highly soluble minerals such as calcite ($CaCO_3$) to produce positive ANC in watersheds. In Sweden, the initial decrease in diatom-inferred pH from as high as 8 to as low as 6 after deglaciation was followed by a long-term decrease from ~6 to 5 caused by the development of vegetation and soils, and by the release of organic acids to some lakes (Renberg, 1990). These natural soil-forming processes acidified soil and surface waters over thousands of years in glaciated terrain before the onset of modern acidic precipitation ("acid rain"). Through time, weathering rates slowed as small particles were weathered rapidly, soluble minerals were dissolved and depleted in upper soil horizons, and weathering reactions became diffusion controlled on larger grains. As a result, natural acid inputs were neutralized less completely as the weathering rate decreased, and the pH of solutions draining from the upper soil layers to surface waters decreased.

For most streams and lakes in glaciated terrain, the pH was rarely <5 until the post-1800 period. Exceptions include lakes rich in organic acidity (high DOC). Historic pH values inferred from the species composition of diatoms in sediment of soft water lakes in the Sierra Nevada Mountains (California) and the Alps showed fluctuations that were attributed to climate changes during the nineteenth century (Whiting *et al.*, 1989; Psenner and Schmidt, 1992). Cold periods were associated with lower pH values. During warm periods with increased evaporation, longer water retention times, stronger lake stratification, and enhanced assimilation of organic sulfur and nitrogen the pH was higher. These temperature-driven fluctuations of pH were disrupted by the onset of significant acidic deposition shortly before 1900. In acid-sensitive systems, anthropogenic acidification has been much faster than natural acidification and the pH and ANC decreased below the natural prehistoric minima for many lakes, as inferred from diatom and chrysophyte species.

Neutralization of acidity (commonly incorrectly termed acid buffering) is caused primarily by the weathering of calcium from bedrock and soil (see Volume 5). Magnesium is most commonly the second most important cation released during weathering. Rarely, there are watersheds with unusual bedrock (e.g., serpentinite or serpentine ($Mg_3Si_2O_5(OH)_4$), or unmetamorphosed ultramafic rocks (e.g., dunite or olivine (Mg_2SiO_4) that produce $Mg-HCO_3$ surface waters; Krám *et al.* (1997)). Acidification is greatest in regions where bedrock and soils are more chemically resistant to weathering, where soils and glacial deposits are thinner, rainfall is greater, and production of organic acids is higher. For example, much of Fennoscandinavia, Scotland, much of Wales, the Adirondack Mountains of New York, USA, the Muskoka Region of Ontario, and eastern Nova Scotia, Canada have acidified lake districts as a consequence of atmospheric inputs of SO_4 and NO_3, in combination with granitic or quartzite/shale bedrock and relatively high naturally occurring DOC. Surface waters draining mafic and calcareous bedrock are generally not affected by acidification because of their high weathering rates. Waters draining noncalcareous sandstone, granite, and schist are more likely to be impacted by acid rain (Kuylenstierna and Chadwick, 1989). The low content of easily weathered minerals (calcite, biotite, amphibole, pyroxene, and calcic feldspar), coupled with short hydrologic flow paths, restrict the amount of reaction between precipitation and mineral phases resulting in high susceptibility to acidification. Few localities have sulfide mineralization sufficient to supply H_2SO_4 in significant amounts. Where this occurs, it is most commonly caused by land disturbance associated with construction (e.g., Hindar and Lydersen, 1994) (see Chapter 5.07), mining or quarrying activities (i.e., acid mine drainage, recent volcanism, or recent deglaciation (Engstrom and Wright, 1984; Engstrom *et al.*, 2000).

Soils play an important role in providing defenses against acidification, either through slowing of sulfur transport through forest ecosystems, retention of SO_4, or by the release of base cations from the exchangeable pool. Such soils have a long-term ability to retard acidification during increasing SO_4 deposition and to retard recovery as SO_4 deposition declines (David *et al.*, 1991a; Kopáček *et al.*, 2001a). Maps of soil type, base saturation (BS) (see Section 9.10.4.5), and cation- and anion-exchange capacity have been particularly useful in predicting regions where short-term acidification is likely. In recently glaciated regions, soils may develop on till or other glacial deposits that are mineralogically different from the bedrock (Johansson and Tarvainen, 1997). Well-drained forest soils such as those in southern Europe or central and southern USA that predate the Quaternary continental glaciations are commonly rich in sesquioxides (aluminum and iron oxides and hydroxides) that have a high SO_4 adsorption capacity in unglaciated terrains (e.g., Cosby *et al.*, 1986). Even young postglacial soils have a substantial ability to adsorb SO_4 (see Chapter 5.07 and Kahl *et al.* (1999)). Organic cycling of deposited sulfate-S plays an equally important role in polluted coniferous forest in the Czech Republic (and probably elsewhere; Houle and Carignan (1995)) where over 80% of the total sulfur content in soils is organically bound (Novák *et al.*, 2003).

Changes in the hydrology of soils may mobilize substantial quantities of acidic compounds. For example, runoff from marine sediments in Finland eustatically uplifted since deglaciation has high concentrations of SO_4 from the oxidation of sulfide minerals contained within these postglacial sediments (Forsius *et al.*, 1990). Runoff from recently drained peatlands may contain substantial concentrations of SO_4 derived from the oxidation of sulfide minerals and organically bound sulfur, as well as elevated DOC. Studies of the chemistry of wetlands (Gorham *et al.*, 1985; Bayley *et al.*, 1988) show empirically and experimentally that SO_4 from atmospheric sources or added as a treatment is removed from bog water by some combination of precipitation as sulfide minerals, transformation into reduced organic sulfur, or reduction to H_2S and emission to the atmosphere. Reduced sulfur may be reoxidized during lower groundwater levels and may then be leached from the system as the water table rises. DOC from wetlands may contribute substantial acidity to runoff, particularly if the level of water tables varies.

The base cation status of surface water is also controlled partly by hydrology. Areas of steep topography generally have thinner soils, and a shorter contact time between soil solutions and mineral soil; these conditions decrease the extent of chemical weathering and the rate of alkalinity production. Seepage lakes containing neither inlets nor outlets may recharge the local groundwater system. In these areas the lake-water chemistry may be similar to atmospheric deposition, modified by evaporation and in-lake processes. If groundwater flow paths change so that more water flows through a seepage lake, ANC of the lake typically increases, because the entering groundwater has been in contact with soil minerals. The residence time of water in lakes is important in determining how much the in-lake processes may alter the acid–base status in the short term. Important processes include: (i) biotic reduction of SO_4 and NO_3 with net storage in sediment, producing alkalinity (Schindler, 1986); (ii) sediment–water interaction, releasing base cations to the water column and generating alkalinity (Schiff and Anderson, 1986); (iii) evaporative concentration, increasing strong acid acidity (or basicity) (Webster and Brezonik, 1995); and (iv) breakdown of DOC, either biological or by ultraviolet (UV) radiation, with a reduction in organic acidity (Kortelainen *et al.*, 1992).

9.10.3.2 What Controls Long-term Acidification?

The long-term average pH of surface waters draining temperate humid watersheds is determined by the rate of supply of acidity to the watersheds and the rate of release of base cations by chemical weathering. In the absence of acid rain, rainwater unaffected by humans is influenced by several factors, including equilibration with atmospheric CO_2, soil and volcanic dust and gases, marine aerosols, and volatile organic hydrocarbons. Alkaline dust, such as occurs in the mid-western United States, may raise the pH of precipitation by several pH units (NADP, 2002). In forested areas with low dust input, and in the absence of air pollution, the dominant factor is atmospheric CO_2. The dissolving of atmospheric CO_2 ($10^{-3.5}$ bar) in liquid precipitation produces weakly acidic precipitation (~pH 5.6). The CO_2 dissolves to form carbonic acid, H_2CO_3; H^+, HCO_3^-, and CO_3^{2-} are produced by the dissociation of the H_2CO_3. For pH < 6.5, H_2CO_3 is the dominant carbon-bearing species; HCO_3^- dominates in the pH range from 6.5 to 10.3. Respiration in soils produces substantially enhanced P_{CO_2} (commonly higher than 10^{-2} bar or 1% by mass), tending to depress the pH of soil solutions (Norton *et al.*, 2001). Degassing of the excess CO_2 from soil water as it enters surface water causes an increase in pH (H^+ decrease). The ANC and ALK do not change as a result of CO_2 degassing if the solution is not in contact with solids.

Natural acidity is contributed from emissions of acidic or acidifying compounds from volcanoes (Pyle *et al.*, 1996; Camuffo, 1992), including compounds of S, N, Cl (chloride), and NH_3 (ammonia), from the ocean (e.g., methyl sulfonate) (Charlson *et al.*, 1987), and from wetlands (e.g., H_2S) (Gorham *et al.*, 1987).

Soil water may contain substantial concentrations of DOC produced by the metabolism of organisms. The DOC includes a spectrum of weak to fairly strong organic acids that dissociate, yielding additional H^+ and organic anions (Org^{x-}). The concentration of (Org^{x-}) can be estimated from chemical analyses of the major base cations, acid anions, and ALK as an "anion deficit," measured by titration with acid, or calculated from DOC concentrations in conjunction with models of organic acidity (Oliver *et al.*, 1983; Driscoll *et al.*, 1994; Köhler *et al.*, 1999).

The supply of natural acidity for weathering depends mostly on respiration in soils, producing elevated P_{CO_2} and DOC. These natural sources of acidity are supplemented by anthropogenic emissions to the atmosphere (Reuss *et al.*, 1987), dominated by the burning of fossil fuels, nonferrous smelting, high-temperature and high-pressure combustion, and agriculture. These sources are responsible for much of the strong acid anions found in eastern Asian, European, and North American precipitation. Only SO_4 and NO_3 are regionally significant and important in short-term (recent) acidification caused by atmospheric deposition. Land application of fertilizers such as $(NH_4)_2SO_4$ is an additional source of nitrogen and sulfur atmospheric deposition of NH_4 and NH_3, derived from agricultural products, is an important source of audity to much of Europe and Central North America (EMEP, 2003; NADP, 2002). Industrial emissions of Cl (as HCl) may also contribute to acidification (Veselý *et al.*, 2002a).

In watersheds where mechanical erosion is low (low topographic relief), release of base cations (which neutralizes acidity) diminishes with time as finer-grained and more-soluble minerals are selectively dissolved or leached from the upper parts of soil profiles (Taylor and Blum, 1995) and secondary minerals develop, including clay minerals and amorphous hydroxide phases rich in aluminum and iron. Anthropogenic acidification is superimposed on natural acidification processes. Concentrations of base metals (e.g., calcium and magnesium) in ^{210}Pb-dated lake sediment cores (less than 200 yr history) in regions with strong acidification stress commonly decrease as acidification inferred from diatom changes (Norton *et al.*, 1992a). These data are consistent with accelerated loss of base cations from soils, stream sediments, and shallow water lake sediments (prior to erosion and deposition in the deep lake environment) as well as progressive loss of base cations from already deposited lake sediment in a progressively more acidic water column environment.

One of the consequences of long-term and recent soil acidification is increased export of aluminum (Cronan and Schofield, 1979; Driscoll and Postek, 1996), iron (Borg, 1986; Norton *et al.*, 2003), and manganese (Borg, 1986) from soils into streams. Lake sediment records show highly variable consequences for these three metals. Iron typically is enriched in lake sediments deposited during acidification of soils. This is a natural consequence of the higher pH and Eh of lakes than of groundwater. The iron enrichment is commonly accompanied by enrichment of zinc and manganese (Norton *et al.*, 1992a). However, manganese may bypass the lake and be depleted in sediments deposited in a more acidic environment (Norton *et al.*, 1990). Most studies of aluminum budgets indicate that lakes serve as a trap for most of the aluminum leached from soils.

Surface water is a mixture of precipitation and water that has penetrated to some depth into soil or bedrock. During high discharge, flow paths in soil are dominated by shallow routing. During base flow, during relatively dry periods, discharge to streams is dominated by deeper groundwater. This groundwater has higher concentrations of base cations, lower DOC, commonly somewhat higher SO_4 and F (fluoride), and almost invariably a higher ANC and pH. Typically, concentrations of major base cations in water are in the order:

$$\text{precipitation} < \text{surface water} < \text{deep soil water} < \text{groundwater}$$

Apparently, deeper soils and bedrock flow paths have been subjected to throughputs of less water and with a higher pH, and are thus not as highly weathered and not as depleted of their more soluble minerals.

The timescale of the acidification process is dictated by weathering rates, the abundance of the constituent minerals, temperature, and moisture supply (Figure 2). The pH of surface waters in recently glaciated terrane may be as high as 8.5, decreasing several pH units through time if the soils become depleted of labile base cations (Figure 3(a)) or as surface waters become more dystrophic (DOC rich) due to paludification and vegetation changes. The timescale of the rapid lowering of pH typically ranges from a few hundred to a few thousand years in terrane containing no appreciable carbonate or ferromagnesian minerals in the bedrock (Figure 2). Whitehead *et al.* (1986, 1989) used fossil diatoms in lake sediment cores to demonstrate that diatom communities in lake-water columns change from alkalophilic to acidophilic, as chemical weathering depletes soil of labile base cations. Veselý *et al.* (2002b) demonstrated that the concentration of arsenic, copper, and zinc decreased relative to that of the more insoluble metals in progressively younger sediments ($\sim 1.15 \times 10^4$ yr ago to present) in Plešné Lake in the Czech Republic. They interpreted the changes to indicate the progressive depletion of relatively easily mobilized metals from soil developing on till during soil acidification (Figure 3(b)). Substituting space for time, Engstrom *et al.* (2000) demonstrated for a region with relatively homogeneous bedrock and undergoing deglaciation that the pH of the most

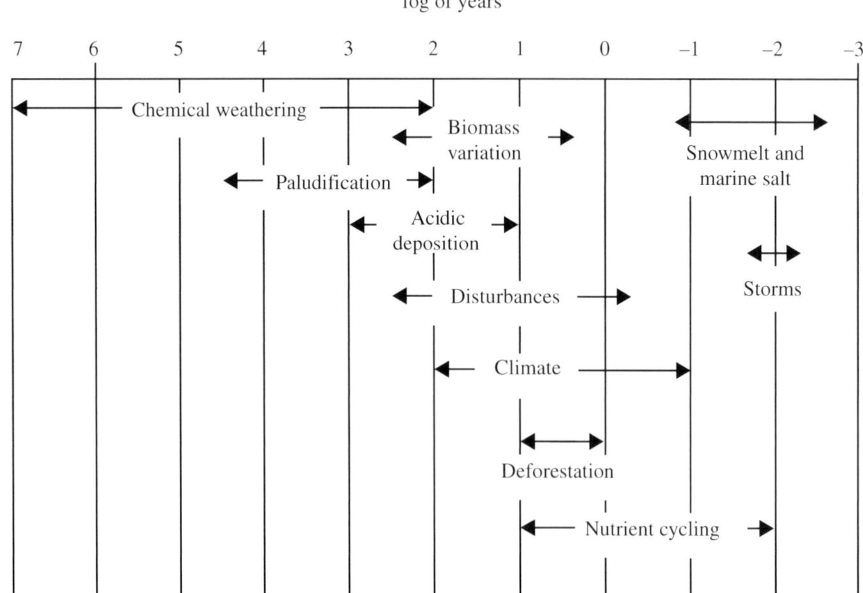

Figure 2 Timescale of process leading to increased acidification.

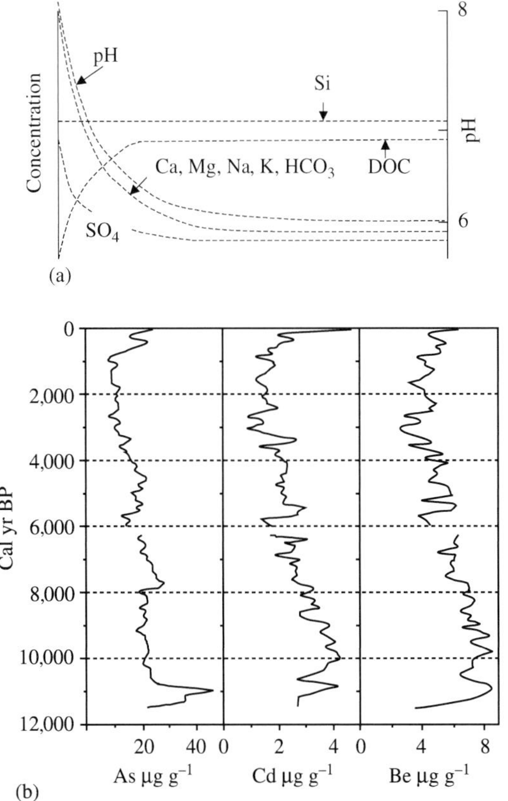

Figure 3 (a) Schematic chemistry of runoff during long-term acidification. (b) Decrease of relatively soluble trace elements (As, Be, and Cd) in progressively younger sediments in Plešné Lake, Czech Republic. The trend reversed during the last ca. 2,000 yr because of atmospheric pollution. High concentrations of As just after deglaciation (at the beginning of the Holocene) are likely caused by rapid weathering.

recently deglaciated lakes is higher and the chemistry changes systematically and rapidly with time (Figure 4). Numerous chronosequence studies of freely drained soils developing on glaciated surfaces show progressive soil horizon development (e.g., Crocker and Major, 1955), depletion of rapidly weathering minerals in upper soils (Frogner, 1990; Swoboda-Colberg and Drever, 1993), accumulation of organic matter in and on top of mineral soil, accumulation of aluminum and iron in lower mineral soil horizons, and decreasing pH. The pH in organic-rich surface layers is dominated by the production and dissociation of organic acids, whereas the pH of lower mineral soil horizons is dominated by silicate weathering driven by CO_2 acidity (see Volume 5).

In unglaciated regions with low topography, soil development in well-drained sites may produce base cation depletion extending to a depth of many meters, accompanied by concurrent and spatially coincident enrichment of aluminum and iron secondary phases. These hydroxide phases, together with organic forest floor, are particularly important in controlling the effects of acid rain through SO_4^{2-} adsorption and binding with organics (Cosby *et al.*, 1986).

9.10.3.3 The Chemistry of Weathering—Long-term Control of Acidification

Although weathering (see Volume 5) plays an important role in the neutralization of acids, the rates at which base cations are released in natural systems from weathering are not well known.

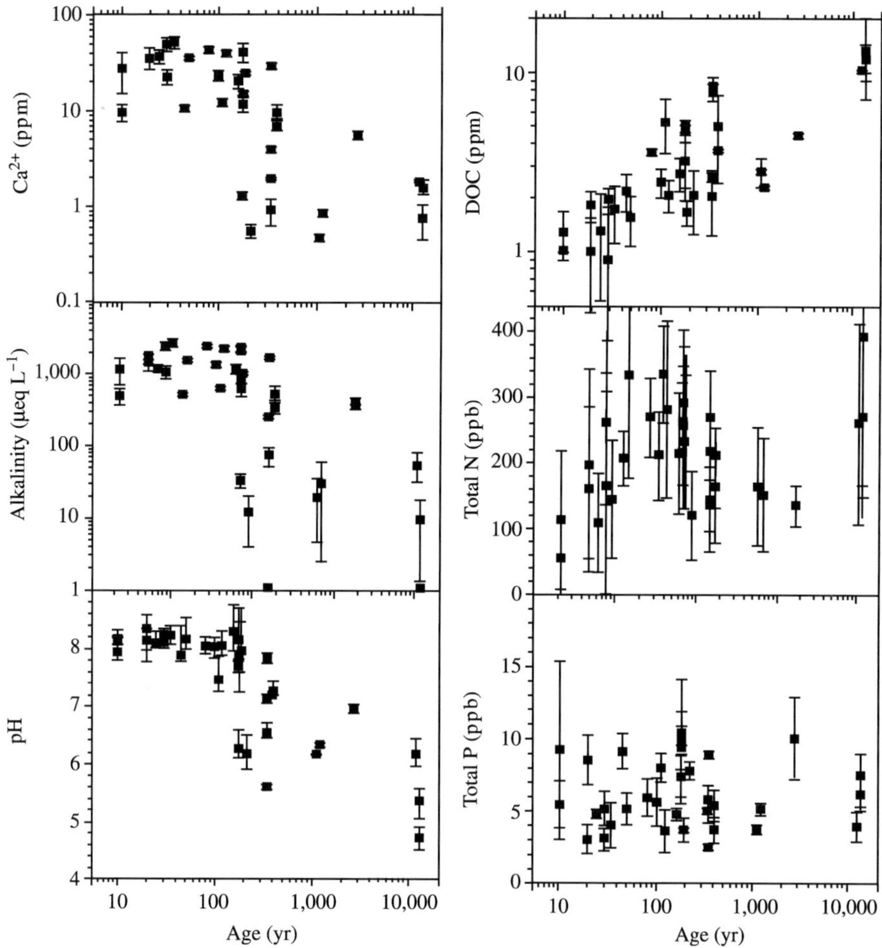

Figure 4 Chronosequence lake chemistry from a deglaciated region in Alaska, USA, showing age of the lake versus various chemical parameters (Engstrom *et al.*, 2000) (reproduced by permission of Nature Publishing Group from *Nature*, **408**, 161–166).

Weathering rates vary with bedrock and soil composition, concentration of organic ligands, temperature, soil moisture and its pH, precipitation amounts and their pH, redox (reduction/oxidation) conditions, and vegetation (e.g., Sverdrup, 1990; White and Brantley, 1995). Weathering rates generally increase with decreasing concentration of silica in the bedrock or soil parent material, increasing flow rate of water through soils, and increasing temperature. Under less acidic conditions, weathering of aluminosilicate minerals may lead to the formation of insoluble clay minerals (incongruent weathering) plus release of base cations to solution:

$$\text{primary (Ca, Mg, Na, K)} - \text{(alumino-silicate)} + H^+ \\ = \text{secondary minerals} + \text{(base cations)}_{aq} \\ + H_4SiO_{4(aq)} \quad (8)$$

These secondary minerals (e.g., kaolinite = $Al_2Si_2O_5(OH)_4$, or "gibbsite" = $Al(OH)_3$) may dissolve if the pH of the soil solution decreases as a consequence of acidification. Weathering can lead directly to the complete (congruent) dissolution of primary minerals if water is sufficiently acidic and plentiful. It is widely accepted that the dissolution of primary minerals is governed by surface reactions. Bacteria, algae, fungi, and lichens are abundant on rock and detritus surfaces. These organisms excrete H^+, organic acids, and strongly complexing acids (e.g., oxalic and citric) that increase the rates of chemical weathering (Sverdrup, 1990; Volume 5). Forest trees may even access base cations from inside mineral grains by means of mycorrhizal fungi, thus partly bypassing the soil solution (Jongmans *et al.*, 1997; Blum *et al.*, 2002).

Weathering is stimulated by the presence of carbonic (H_2CO_3) and organic acids (DOC). Organic acidity is produced primarily in the organic-rich layer of forested ecosystems (the "forest floor"). The pH of soil solutions in organic-rich forest soils may be depressed below 4 because of the dissociation of organic acids, caused largely by dissociation of carboxylic groups (–COOH). Much of the DOC acidity is lost to aerobic respiration as soil solutions descend through the mineral soil, producing substantial amounts of CO_2. The carbonic acid acidity is produced throughout the soil column but typically reaches a maximum below the forest floor. Aerobic respiration can increase the P_{CO_2} as much as two orders of magnitude above the atmospheric value, to produce an equilibrium pH with water as low as 4.5. Chemical weathering rates of alumino-silicate minerals increase at lower pH in laboratory experiments, whether it is caused by DOC or CO_2. Complexation of aluminum and iron by organic ligands enhances the mobilization of these metals, but at the same time reduces the toxicity of aluminum.

Laboratory- and field-based experiments have yielded a better understanding of the rate at which base cations can be supplied to offset the input of strong mineral acids (Bain and Langan, 1995; Swoboda-Colberg and Drever, 1993). For many elements, chemical budgets for watersheds have been constructed by the simplistic relationship:

$$\text{Weathering rate} = \text{stream output} - \text{atmospheric input} \quad (9)$$

Equation (9) ignores numerous quantitatively important processes, including changes in biomass (living and dead), changes in exchangeable soil pools for cations and anions, changes in dry deposition, and even loss to the atmosphere (e.g., for nitrogen). Thus, a more realistic equation for an element might be

$$\begin{aligned}\text{Weathering rate} &= (\text{stream water output} + \text{atmospheric re-emission}) - (\text{atmospheric input (wet)} \\ &\quad + \text{atmospheric input (dry)} \\ &\quad \pm \Delta\text{biomass storage} \\ &\quad \pm \Delta\text{exchangeable soil pools}) \quad (10)\end{aligned}$$

The weathering rate is calculated by adding and subtracting many fluxes, each of which is difficult to assess (Pačes, 1983; Velbel, 1985; White and Blum, 1995). No long-term calibrated watershed studies measure all these variables well, and thus the weathering rates are only approximate. For example, Bormann and Likens (1979) calculated that calcium weathering from 1963 to 1974 was $\sim$11.5 kg ha^{-1} yr^{-1} at the Hubbard Brook Experimental Forest (HBEF), New Hampshire, USA, assuming that all the excess calcium was from weathering of primary minerals. At HBEF, surface waters have very low alkalinity, have likely been acidified, and are susceptible to episodic acidification. Bailey *et al.* (2003), based on a more complete analysis, were not able to resolve the primary weathering from depletion of exchangeable base cations from the soil. At an ecologically and geologically similar site at Bear Brook in Maine USA, calcium weathering rates were calculated on a similar basis. Assuming steady state for many of the unknown variables, the rate ranged between 11 kg ha^{-1} yr^{-1} and 15 kg ha^{-1} yr^{-1} from 1988 to 1992 in a reference watershed (Norton *et al.*, 1999).

Numerous studies suggest that the relative proportion of calcium derived from the watershed compared to the calcium derived from atmospheric inputs can be inferred by using strontium isotope data. The strontium isotope ratios for the bedrock, atmospheric input, and output are combined in a linear mixing model to infer the ultimate sources of calcium (bedrock/soil complex versus atmosphere). The explicit assumption in this technique is that calcium and strontium behave similarly during all biogeochemical processes. The assumption has been challenged by Bullen *et al.* (2002).

Several types of evidence from soils have been used to estimate long-term weathering rates. The historical approach uses the reduction in the base cation concentrations (Johansson and Tarvainen, 1997) or labile minerals (e.g., biotite and hornblende (Frogner, 1990)) in the soil profile with respect to chemically unaltered C-horizon soil. If the age of the soil is known (e.g., post-Wisconsinan in North America, Weichselian in Europe), this method provides long-term average weathering rates that are generally greater than present-day rates. Weathering rate decreases with soil age following a power-law equation (Taylor and Blum, 1995). Modern weathering rates may be only 10–20% of the rates immediately after deglaciation (Figure 3(a) and (b)).

Several studies have suggested that acid deposition has accelerated weathering rates (e.g., Miller *et al.*, 1993). However, Norton *et al.* (1999) concluded on the basis of runoff chemistry at the Bear Brook Watershed, Maine, USA that chemical weathering was not affected during a 10 yr artificial acidification of an entire watershed. There, Swoboda-Colberg and Drever (1993) acidified *in situ* soil columns, after removal of the forest floor (organic horizon). They stripped exchangeable cation pools from the mineral soil with strong acid leaching until the rate of cation leaching reached steady state. This steady-state loss was attributed to primary chemical weathering. The values (kg ha^{-1} yr^{-1}) were 200–400 times higher than those based on the whole watershed (Equation (10), simplified). Dahlgren *et al.* (1990) used reconstructed soils from the watershed, and determined that experimental acidification of the columns with H_2SO_4 was accompanied by increased leaching of base cations from the exchangeable pools and dissolution of aluminum from a solid phase. Silica release was not enhanced by their acid treatment.

Generally, silica is relatively unchanged in all the acidification experiments, suggesting no substantial change in congruent weathering rates. In summary, most field experiments and watershed studies suggest that variations in short-term base cation supply are governed by ion-exchange equilibria, not changes in weathering rates. As exchangeable base cation supplies become depleted and the pH declines, mobilization of toxic aluminum becomes important.

Weathering rates of isolated minerals have been determined for most important rock-forming minerals (Sverdrup, 1990). Typically, sized and cleaned mineral fragments are subjected to a pH-stated solution, a closed (batch) system solution, or recirculated solutions (with or without controlled pH). "Weathering" is tracked through time by changes in the concentration of liberated elements. Temperature, pH, and the concentration of various species of DOC are important variables. Experimental rates of weathering (typically expressed as mol m^{-2} s^{-1}) are generally two to three orders of magnitude higher than field rates. These differences are partly an artifact of differing experimental methods, non-steady-state processes (Holdren and Adams, 1982), differences in hydrological conditions between the field and laboratory, and effective mineral surface area in contact with reacting water. Chemical weathering in soil may be retarded and virtually stopped in dry periods (Zilberbrand, 1999). Alternatively, as acidic soil solutions dry, the pH drops, and ionic strength increases, weathering rates should increase, but in a restricted volume of soil solution. Clearly, extrapolating experimental weathering rates determined in the laboratory to the field and regionalization of the results is problematic. Unfortunately, such data are important to realistically calibrate some dynamic and static models of soil and water acidification (e.g., PROFILE, Sverdrup and De Vries, 1994).

9.10.4 SHORT-TERM AND EPISODIC ACIDIFICATION

Atmospheric deposition of excess sulfur and nitrogen compounds ("acid rain") may substantially accelerate acidification, because H^+ typically accompanies the sulfur and nitrogen in deposition (as H_2SO_4 and HNO_3). The sulfur and nitrogen may then be temporarily stored in a variety of ways, or leached directly to surface water, largely in oxidized form (SO_4^{2-} and NO_3^-). These excess leached anions must be instantaneously accompanied by cations to maintain electrical neutrality. Consequently, soils must provide sufficient cationic charge to balance the anion flux. Soils neutralize the deposited acids by adsorbing some SO_4^{2-}, NO_3^-, and H^+. They compensate for excess SO_4^{2-} and NO_3^- flux by desorbing exchangeable base cations or aluminum, or dissolving solid-phase aluminum. Soils with abundant labile base cations, occurring either in soluble primary minerals or as exchangeable ions on the soil complex, can neutralize all the deposited acids, yielding comparable increases in both SAA and SBC (Equation (7)), thereby increasing ionic strength (Galloway *et al.*, 1983). Silicate weathering reactions are relatively slow. Thus, if exchangeable base cations in soil are desorbed more rapidly than they are provided by chemical weathering and atmospheric deposition, soil acidification occurs and surface-water ANC and pH decline, as more of the charge balance is maintained by the export of H^+ and Al^{3+}. The soil- and surface water thus become more susceptible to short-term episodic acidification events (Wigington *et al.*, 1996), lasting from hours to months. The events are caused by a variety of mechanisms including pulsed inputs of water (high discharge from snowmelt or rain), oxidized sulfur and nitrogen from organic or inorganic pools (Dillon *et al.*, 1997), marine aerosols (the salt effect; Wright *et al.* (1988)), and pulsed release of DOC (Hruška *et al.*, 2001). These mechanisms are controlled by climate. Depending on watershed characteristics, episodic acidification may be manifested by and caused by increased concentrations of SO_4, NO_3, DOC, or Cl. Episodic acidification is a naturally occurring phenomenon that may progress to chronic acidification if the rate of base cation removal continues to exceed cation supply by primary weathering and atmospheric input. This state is commonly characterized by permanent ANC < 0 and pH < 5. Conversely, chronic acidification by strong acids makes systems more susceptible to episodic acidification. Many aspects of episodic acidification have been thoroughly reviewed by Wigington *et al.* (1990).

9.10.4.1 High Discharge from Snowmelt and Rain

During snow melt, pollutants are preferentially eluted early. Consequently, acidic pulses are released and may enter streams and lakes with little contact with soils (Johannessen and Henriksen, 1978; Jeffries, 1990). Base cation concentrations become diluted concurrently with elevated concentrations of strong acid anions. The associated pH and alkalinity depressions may have severe biological impacts on fish and other biota during their sensitive early life stages. The temperature of meltwater may be close to 0 °C and it tends not to mix downward with more dense lake water if the lake is covered with ice. The result is a shallow layer of relatively low pH water directly beneath the ice. Inorganic aluminum and DOC interactions cause variability in fish mortality together with duration of exposure in an acidic episode (Baldigo and Murdoch, 1997). Acidic episodes kill fish long before the water is chronically acidic; recovery from acidic episodes is a key to biotic recovery. The pH may also be depressed in the epilimnion of ice-covered lakes, due to the accumulation of CO_2 from aerobic respiration. Normally, such a pH depression is not accompanied by an increased concentration of potentially toxic aluminum. Conversely, anaerobic respiration causes SO_4 and NO_3 reduction in hypolimnia, which may substantially raise the pH. Changing flow paths during hydrological events are of overhelming importance in controlling the chemical character of episodes in streams (Davies *et al.*, 1992). For example, acid-sensitive fish species were absent in streams of the northeastern USA that had a median pH $<5.0-5.2$ and inorganic aluminum $>100-200\ \mu g\ L^{-1}$ during high flow (Baker *et al.*, 1996). In circumneutral streams, dilution of calcium has an important negative effect (Tranter *et al.*, 1994).

9.10.4.2 Pulsed Release of SO_4 and NO_3 from Soils

Episodic release of SO_4 and/or NO_3 from soils may depress pH and ANC, on a timescale of individual high discharge events, or seasonally. Episodically elevated concentrations of SO_4 in runoff may be caused by prolonged drought, lowering of the groundwater table, oxidation of sulfur stored in organic matter, and leaching during the periods of subsequent higher discharge (Dillon *et al.*, 1997). Increases in DOC may accompany the elevated SO_4, enhancing the depression of pH. Normal fluctuations of hydrology are typically unaccompanied by substantial variations in stream SO_4, because of anion-exchange equilibria in soils and stream sediment. The flux of NO_3 is dominated more by biological processes, being strongly diminished in many streams during the growing season versus the dormant season for vegetation. Consequently, many watersheds have a strong annual cyclicity in the release of NO_3. Superimposed on this is the short-term release of NO_3 caused by flushing of mineralized nitrogen (as NO_3) from shallow soils during periods of higher flow. It is common for NO_3 to vary more than SO_4 in runoff (on both a percentage and absolute basis)(Figure 5), apparently because most soils have a low capacity for NO_3 exchange.

9.10.4.3 Marine Aerosols

Deposition of marine salt aerosol causes episodic acidification of runoff near the coast by alteration of cation-exchange equilibria within strongly acidic soil. During the "sea-salt effect," marine aerosol Na^+ and Mg^{2+} displace primarily H^+, Al^{n+}, and Ca^{2+} from soil exchange sites. The Na/Cl and Mg/Cl equivalent ratios in runoff may decline well below those of marine aerosols, 0.86 and 0.2, respectively, as Na^+ and Mg^{2+} cations are adsorbed by the soil. The ANC of runoff is reduced while base saturation and soil pH are increased very slightly. The process has been demonstrated experimentally at the laboratory scale (Skartveit, 1981) and at the watershed scale (Wright *et al.*, 1988). Individual high salt inputs may be reflected in surface-water chemistry for months to a few years (Kirchner *et al.*, 2000; Norton and Kahl, 2000; Evans *et al.*, 2001c).

Wiklander (1975) suggested that the following reaction (11) is important in acidification of leachate by the sea-salt effect:

$$R{-}H + Na^+ + Cl^- = R{-}Na + H^+ + Cl^- \quad (11)$$

where R indicates a negatively charged functional group in the soil exchange complex. The maximum effect from sea-salt input occurs in thin soils

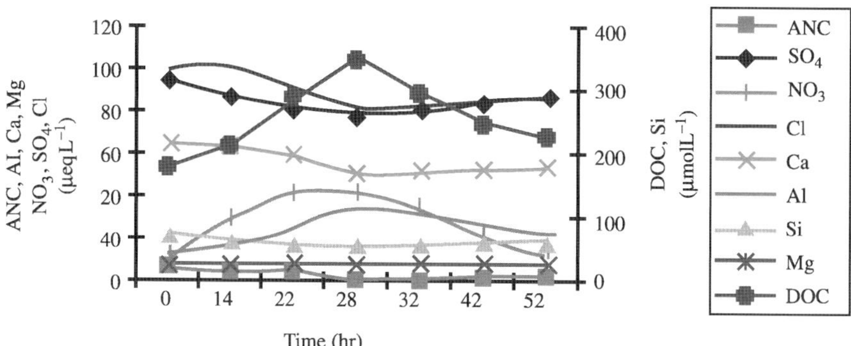

Figure 5 High-frequency chemistry at West Bear Brook, Maine, USA for a snowmelt event starting on February 24, 2002. Maximum discharge was at 28 h. Acidification was marked by decreased pH (not shown; from 5.5 to 5.0) and ANC, increased NO_3 (shown 10X), DOC, Al, and P (not shown), and slight dilution of Si, Ca, Mg, Cl, and SO_4 (R. L. Reinhardt and S. A. Norton, unpublished).

that have low BS and low-cation-exchange capacity (CEC). Most soils have a relatively low chloride (chloride) exchange capacity. Thus, Cl behaves conservatively. The sites most responsive to sea-salt inputs are those at an intermediate distance from the coast, where occasional major sea-salt inputs can generate large proportional changes in the chloride and other marine ions (Harriman *et al.*, 1995). Major regional events with pH depressions sufficient to kill fish have been documented by Hindar *et al.* (1994). These events may occur in areas even where acidification from strong acids is absent. Salt-driven acidification episodes may be relatively common in acidic bogs, although they have not been well documented. Pugh *et al.* (1996) demonstrated a salt acidification effect in an ombrotrophic/poor fen site. The salt originated from road runoff, but the chemical changes were otherwise the same.

9.10.4.4 Organic Acidity

The most acidic stream flows in polluted regions commonly have disproportionately higher concentrations of H^+, NO_3^-, and inorganic aluminum. In unpolluted regions, as in northern Sweden (Hruška *et al.*, 2001) or in North Shore rivers, Quebec, Canada episodic acidification may be primarily caused by dilution of base cations and increased organic anions during high discharge (Campbell *et al.*, 1992). For example, the "anion deficits" increased there from around 35 $\mu eq\,L^{-1}$ to 70–100 $\mu eq\,L^{-1}$ representing up to 20 mg DOC L^{-1} during high flow at snowmelt, causing pH to drop from ~7 to ~5. The episodic pH depression associated with DOC operates independently of anthropogenic acidity and likely is responsible for substantial episodic acidification. Fortunately, as DOC and H^+ increase, much of any potentially toxic metals that are mobilized (especially aluminum) becomes complexed with the DOC and is thus biologically much less reactive.

9.10.4.5 Neutralization of Episodic and Longer-term Acidity

Chemical weathering in soils provides base cations relatively slowly, but the total long-term neutralization capacity of soils is typically large. In contrast, ion-exchange reactions (cation and anion exchange) in soil are relatively rapid, and tend to buffer H^+, SO_4^{2-}, and ANC concentrations. Solids in soils and sediments have surface charges that permit exchange of ions. For example

$$\text{Soil-Ca}^{2+} + 2H^+ = \text{Soil-2H}^+ + Ca^{2+} \quad (12)$$

$$\text{Soil}\begin{smallmatrix}\diagup \text{Al-OH}\\ \diagdown \text{Al-OH}\end{smallmatrix} + SO_4^{2-} = \text{Soil}\begin{smallmatrix}\diagup \text{Al} \diagdown\\ \diagdown \text{Al} \diagup\end{smallmatrix} SO_4^{2-} + 2OH^- \quad (13)$$

Laboratory and *in situ* experiments have demonstrated that stream and lake sediments in contact with the water column are effective ion exchangers for base cations (e.g., Cook *et al.*, 1986; Oliver and Kelso, 1983), anions (Norton *et al.*, 2000), and aluminum (Tipping and Hopwood, 1988; Henriksen *et al.*, 1988b). Consequently, pH depression is lessened during episodic acidification, and recovery of ANC is retarded during de-acidification (recovery). The sum of the exchange sites for cations (expressed in centimoles of charge per 100 g soil) is termed the cation exchange capacity. The percentage of the cation-exchange sites occupied by the base cations (calcium, magnesium, sodium, and potassium) is termed the base saturation. The rest of the cation-exchange sites are occupied primarily by aluminum and hydrogen. If desorption of base cations exceeds the weathering rate releasing those base cations, the BS of soil decreases, site occupancy by aluminum and hydrogen increases, and acidification is underway. As the pH of soil water decreases below 5.5–5.0, and soil cation-exchange sites become depleted in base cations, desorption of exchangeable aluminum and dissolution of aluminum solid secondary phases may become important. In very acidic forest soils, dissolution of iron may also contribute to the neutralization (Ulrich, 1983; Borg, 1986; Matschullat *et al.*, 1992; Norton *et al.*, 2003). Dissolution of solid manganese phases may be important, but rarely. The depletion of base cations from exchangeable positions in the soil is a reversible phenomenon. As the acid stress decreases during reduced pollution, soils adsorb base cations, retarding the recovery of alkalinity, thereby prolonging the period of depressed pH (see Section 9.10.7).

Several less important mechanisms, with varying timescales, also contribute to acid neutralization. (i) In soils rich in iron and aluminum secondary phases, excess SO_4 from the atmosphere may be reversibly adsorbed or desorbed, thereby retarding acidification and recovery from acidification, respectively (David *et al.*, 1991a,b). Such soils are common in unglaciated areas. (ii) As pH declines, dissociated inorganic carbon species protonate toward H_2CO_3 and aqueous CO_2. (iii) Protonation, condensation, and precipitation of organic acids increases. (iv) Hydroxylated aluminum species shift toward Al^{3+}. The buffering capacity provided by organic acids and aqueous aluminum species reduces the sensitivity to episodic changes in pH due to the addition of strong mineral acid or marine aerosols. In combination with the cleavage of higher molecular organic matter (Steinberg and Kühnel, 1987), acidification decreases DOC and water color, and increases transparency. The altered light regime in water (including UV permeability) has consequences for the chemistry, the biota, and the thermal structure of lakes (Schindler *et al.*, 1996).

The substantial decrease of the atmospheric input of sulfur in Europe and North America since the early 1980s has caused a general decline in surface-water SO_4 (Stoddard *et al.*, 1998; Evans *et al.*, 2001b). Many soils have switched from being a sink to a source of sulfur (Driscoll *et al.*, 1998; Prechtel *et al.*, 2001). Just as sorption of SO_4 delays acidification (Equation (13)), desorption delays chemical recovery from acidification and will continue to do so for decades. As atmospheric deposition of SO_4 decreased, fluxes of SO_4 in runoff from watersheds with thin (e.g., alpine) soil and low SO_4 adsorption capacity have decreased rapidly (Kopàček *et al.*, 2001a) in comparison to watersheds with deeply weathered and thick soils (Alewell, 2001). Stable isotope ($\delta^{34}S$) studies and budget calculations suggest that the pool of organic sulfur in forest floor and biological sulfur turnover are important contributors to SO_4 export. Sulfur isotope studies indicate that ~30% of sulfate in stream water in polluted areas of the Czech Republic was organically cycled (Novák *et al.*, 2000).

Regardless of long- or short-term acidification, the rate of acidification is determined by changes in the rate of supply of SBC and SAA. More specifically, the difference (SBC–SAA) is crucial. As acid anion loading is reduced (particularly for SO_4), the equilibrium between soils and infiltrating soil water should shift. SBC should decline, but less than the reduction in SAA, leading to recovery (increase) of ANC (e.g., Evans *et al.*, 2001b). If acid loading is reduced but is still high enough to leach base cations faster than they are re-supplied from mineral weathering, long-term acidification of runoff will still occur. Permanently restoring pre-acidification runoff ANC requires reducing acid loading and base cation leaching enough so that mineral weathering can replace exchangeable bases depleted from soils. Despite recent, comparable rates of SAA decrease in sites in northern Europe and North America, many American sites show no changes in ANC or pH recovery. Some North American sites have SBC declining faster than SAA; the reverse is observed for most sites in Europe. The behavior of base cations plays an important role in creating these continental-scale differences (Skjelkvåle *et al.*, 2001a). There is no generally accepted explanation for the steep decline of base cations at many North American sites (Driscoll *et al.*, 1989; Likens *et al.*, 1996; Lawrence *et al.*, 1999; Fernandez *et al.*, in press).

9.10.5 OTHER EFFECTS OF ACIDIFICATION

9.10.5.1 Release of Aluminum and Other Metals

Acidification of surface waters to a pH ~ 5 causes sharp increases in concentrations of aluminum (Schecher and Driscoll, 1987; Driscoll and Postek, 1996) and trace metals (e.g., beryllium (Veselý *et al.*, 2002c), cadmium, manganese, and zinc) (Figure 6) (Veselý and Majer, 1996). Increased concentration of dissolved inorganic aluminum during soil and water acidification is the primary cause of fish mortality in acidic waters (Baker *et al.*, 1996). The controls on aluminum concentration as pH declines include ion exchange, dissolution of solid-phase aluminum (typically referred to as "gibbsite" = $Al(OH)_3$) with variable solubility constants (Mulder and Stein, 1994), and equilibrium with solid aluminum-organic complexes (Cronan *et al.*, 1986). Concurrently, the aluminum aqueous speciation changes toward more of the uncomplexed free ion. Free Al^{3+} and H^+ outcompete trace metal cations on soil exchange sites, the concentrations of OH^- and HCO_3^- ligands are lower, and the importance of organo-complexes declines as pH declines. Potentially toxic substances such as Al^{3+} become more biologically available. Increases of aluminum in a low DOC stream, caused by artificial acidification of the watershed, were entirely inorganic aluminum and apparently related to solubility of an unidentified aluminum phase (Postek *et al.*, 1996). Mobilization of aluminum from acid soils is perhaps the most detrimental result of the acidification of surface waters. Knowledge of the mechanisms regulating the release of aluminum from amorphous inorganic and organic compounds in soil is uncertain (LaZerte and Findeis, 1995; Mulder and Stein, 1994). Laboratory experiments suggest a combination of kinetically limited aluminum release from primary and secondary minerals and organic compounds together with the complexation of aluminum with DOC (Berggren and Mulder, 1995). During acid episodes aluminum can exceed 1 mg L^{-1}; such high concentrations can be long lasting in heavily polluted regions (Veselý *et al.*, 1998b). The dissolution of aluminum is further enhanced by the formation of soluble complexes, especially with fluoride. Mixing of these acidic aluminum-rich waters with higher pH waters commonly causes the precipitation, which also present problems for fish (Weatherley *et al.*, 1991; Rosseland *et al.*, 1992; Reinhardt *et al.*, 2003).

Many pollutants other than sulfur and nitrogen are found in modern atmospheric deposition. The history of deposition of these pollutants

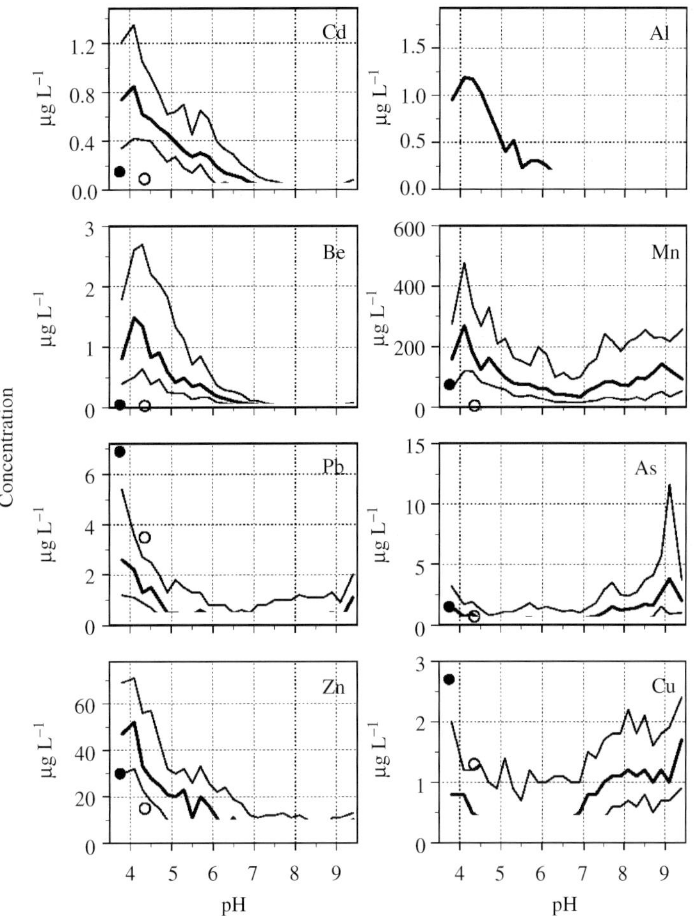

Figure 6 Relationship between median, upper, and lower quartiles of Al, As, Be, Cd, Cu, Mn, Pb, and Zn, and pH in Czech Republic brooks in the late 1980s. Values were calculated after sorting the water samples ($n = 12{,}988$) into groups with 0.2 pH unit ranges. Volume-weighted average concentrations in bulk precipitation (o) and throughfall (●) in 1991 samples from the Bohemian Forest are shown (after Veselý and Majer, 1996, 1998). The declines of Be, Cd, Mn, and Zn at very low pH are likely caused by depletion of exchangeable trace metals from the watershed soils.

has been determined by chemical and isotopic analyses of ice cores (e.g., Boutron *et al.*, 1995), lake sediment cores (e.g., Renberg *et al.*, 2000), peat cores (e.g., Shotyk *et al.*, 1996), soils (Bindler *et al.*, 1999), and vegetation (Steinnes, 1995). Most metals have enhanced mobility as a consequence of acidification. Mercury and lead are relatively conservative, in large part because they are fixed by solid organic matter. Atmospheric deposition of lead increased in the northern hemisphere to more than 50 times the pre-industrial rates. Lead deposition peaked in the 1970s. It has declined since then to values that are generally less than 5% of their peak values. The biological impact of atmospherically deposited lead, except in grossly polluted regions, has not been demonstrated to be significant. Mercury (Chapter 9.04) has increased in atmospheric deposition by more than 100% since the mid-1800s, peaking in the 1970s, and declining since then. Mercury becomes methylated, enters the food chain through phytoplankton, is magnified as much as 10^6 in the food chain, and reaches maximum concentration in fish. The increase of SO_4 in the environment from acid rain may have stimulated methylation of mercury because of enhanced biotic reduction of SO_4, thus increasing methyl-Hg in fish and humans who consume the fish. Enhanced methylation may be linked with biotic reduction of SO_4 in wetlands, groundwater, and lakes (see Chapter 9.04). The trend for the deposition of cadmium from the atmosphere is generally parallel to that of mercury for eastern Canada and the USA (Alfaro-De la Torre and Tessier, 2002; S. A. Norton, unpublished).

9.10.5.2 Nutrient Availability

The initial response to acid rain is an increase in the export of base cations and, commonly, NO_3. This essentially corresponds to a period of fertilization of plants in forests and in surface waters. However, prolonged increased leaching of exchangeable base cations and subsequent decline in exchangeable pools of base cations in soils (Kirchner, 1992; Likens *et al.*, 1996; Fernandez *et al.*, 2003) may have a long-term impact on terrestrial ecosystem health. As the molar ratio (Ca + Mg + K)/Al in soil solution declines during acidification, nutrient uptake by roots may be impaired. Limited calcium or magnesium uptake, associated with elevated aluminum concentrations in the rooting zone, slows growth and decreases the stress tolerance of trees (Cronan and Grigal, 1995).

Unpolluted forest ecosystems export mostly organically bound N, instead of inorganic N ($NO_3 + NH_4$) (e.g., Perakis and Hedin, 2002). Increased input of N may first be favorable in N limited ecosystems. Turnover of mineralized N in forest floor is generally an order of magnitude higher than atmospheric input of inorganic N, which creates only a small addition to a large N soil pool. Nitrogen demands on the part of vegetation must be satisfied first and a certain amount of excessive N is probably immobilized in forest organic matter before N saturation and NO_3 leaching occur (Aber *et al.*, 1989, 1998; Stoddard *et al.*, 2001). Ecosystems vary widely in their capacity to retain N inputs. Excess N is exported mostly as NO_3, increasing the concentration of strong anions in water, contributing to acidification. Nitrate leakage is greatest from high-elevation, steep sites, and from mature forest with high soil N stores and low soil C/N ratio (Fenn *et al.*, 1998), and lowest from watersheds containing extensive wetlands. Concentrations and seasonality of NO_3 in stream water are used as indices of N saturation (Stoddard, 1994). Mosello *et al.* (2000) and Kopàček *et al.* (2001a) indicated that retention of N in watersheds decreased with time under acidification stress. However, reduction of acidic deposition may be connected to slightly elevated N retention (Lorz *et al.*, 2002; Veselý *et al.*, 1998a, 2002a).

Export of NO_3 in surface waters is thought to be linked with microbial activity and the soil C/N ratio (Yoh, 2001). Empirical data showed that a C/N ratio of the forest floor of <25 and a throughfall of 9–10 kg N ha^{-1} yr^{-1} is the threshold for leaching NO_3 in Europe (Gundersen *et al.*, 1998; Dise and Wright, 1995). The slope of the relationship between nitrogen input and NO_3 leached was twice as high for sites where C/N < 25. Higher rates of NO_3 leaching also occurred at sites with pH < 4.5 and high nitrogen input (MacDonald *et al.*, 2002). However, nitrogen leakage at relatively low deposition occurred at high-elevation alpine sites in the Rocky Mountains of Colorado, USA (Williams *et al.*, 1996). The export of NO_3 from nitrogen-saturated forest corresponded also with the soil's potential net nitrification and land-use history. While net nitrogen mineralization did not vary with land-use history, nitrification rates doubled at old-growth sites in comparison with hardwood forest disturbed by fire and harvesting about a century ago (Goodale and Aber, 2001). Enhanced nitrification at old-growth sites may have resulted from excess nitrogen accumulation relative to accumulation of carbon in soils. Carbon mineralization rates and C/N ratios were comparable in spruce forest soils of two neighboring watersheds in the Bohemian Forest, while potential net nitrogen mineralization and nitrification differed by 50–70%; higher potentials were associated with higher leaching of NO_3 (Kopàček *et al.*, 2002a).

The productivity of temperate lakes and freshwaters is generally limited by the availability of phosphorus. Phosphorus occurs in many rocks, primarily in minerals with relatively high weathering rates. Consequently, soil profiles may become depleted in phosphorus minerals. Much phosphorus is concentrated in organic-rich soils and is strongly recycled or sequestered by adsorption on aluminum- and iron-rich soil layers (Kuo and Lotse, 1974). Lakes predisposed to acidification, thus, have low concentrations of base cations and phosphorus. Acidification of catchments may result in a slightly increased export of dissolved phosphorus from soils (Roy *et al.*, 1999). They and Reinhardt *et al.* (2003) found that two acidifying streams contained high concentrations of particulate acid-soluble aluminum and iron hydroxides and acid-soluble particulate, phosphorus during acidic episodes. Particulate phosphorus was 10–50 times higher than dissolved phosphorus. Ionic aluminum species hydrolyze downstream or in lakes at higher pH. Polymeric aluminum species are formed with large specific surfaces and with a strong affinity for PO_4^{3-}. The phosphorus in acidified streams and lakes (typically with a pH in the 5.5–6.5 range) may be scavenged by aluminum- or iron-rich particles. If the aluminum hydroxide is deposited as a sediment, the flux of phosphorus into sediment may be irreversible (Kopàček *et al.*, 2001b), even during periods of hypolimnetic anoxia when pH typically increases, iron hydroxide dissolves, and adsorbed phosphorus is normally released to the water column (Einsele, 1936). Thus, stream acidification may lead to downstream oligotrophication as suggested by Dickson (1978).

9.10.6 EFFECTS OF CLIMATE ON ACIDIFICATION

Variations in climate may change the concentration and fluxes of solutes and compounds, thereby altering the acid–base status of runoff.

Year-to-year changes need to be distinguished from biologically driven seasonal cycles that control the concentration of a number of components in runoff to varying degrees. These components include NO_3, SO_4, K, and Mg (e.g., Likens *et al.*, 1994). These changes are difficult to distinguish from the effects of climate change that occur over decades to centuries because the possible ecosystem responses are very complex. Much of the variance is related to the amount and temporal distribution of precipitation and variations in temperature. Even increases in precipitation may not result in increasing discharge if accompanied by higher temperature (Clair and Ehrman, 1996). As major atmospheric circulation patterns change, the input of marine aerosols may also be substantially altered. Empirical evidence has emerged suggesting that temperature variability may partially control important abiotic reactions involving aluminum (Lydersen *et al.*, 1990; Veselý *et al.*, in press).

9.10.6.1 NO_3

Current mean annual concentrations of NO_3 in stream water are generally related to the magnitude of nitrogen deposition and the nitrogen-saturation status of the forest (Aber *et al.*, 1989; Stoddard *et al.*, 2001; Wright *et al.*, 2001). During the forest growing season, NO_3 export is typically suppressed because of biological uptake, both above and below ground. Drought leads to the oxidation of organic nitrogen in dried soils. Nitrate concentrations in many watersheds are commonly highest during the spring snowmelt period and lowest during summer base flow. This pattern implies that nitrification occurs during the winter months, and that NO_3 accumulates in the soil until it is flushed by snowmelt. However, peak spring and winter NO_3 concentrations vary markedly from year to year. Cyclic inter-annual variations in these peak NO_3 concentrations have been ascribed to summer drought (Ulrich, 1983; Reynolds *et al.*, 1992; Harriman *et al.*, 2001), cold dry winters (Mitchell *et al.*, 1996), and variable mean annual temperature (Murdoch *et al.*, 1998). Synchronous variation in NO_3 concentrations among lakes and strong negative correlation with the winter North Atlantic Oscillation Index (NAO) and mean winter temperature occurred in the UK. Low NAO winters increased NO_3 in the monitored waters of the UK (Monteith *et al.*, 2000).

Snow acts as an insulator to prevent freezing of forest soils. Physical disruption during soil freezing can increase fine root mortality and reduce plant nitrogen uptake, allowing soil NO_3 levels to increase even without an increase in net mineralization or nitrification (Groffman *et al.*, 2001). The annual mean soil temperature influences mineralization, and may shift the C/N ratio of forest floor as much as 0.5 per 1 °C (Yoh, 2001). Thus, climate may be a primary factor controlling the soil C/N level. The C/N ratio probably regulates the NO_3 production due to a metabolic balance between carbon and nitrogen. In soils with a higher C/N ratio, nitrification is probably insignificant and decomposition rates are high. Surprisingly, artificially warmed soils in a coniferous forest in Maine USA had lower rates of nitrogen cycling, and forest floor nitrogen concentration was a better predictor of potential net nitrogen mineralization than was total carbon or the C/N ratio (Fernandez *et al.*, 2000). Generally, the C/N ratio alone is a poor predictor of nitrogen leaching or retention.

9.10.6.2 SO_4

In oligotrophic boreal lakes of Ontario, Canada, a drought in the 1980s decreased the water table and lake levels, exposing littoral sediments containing reduced sulfur. Sulfur was reoxidized to SO_4^{2-} and mobilized to the lakes during subsequent wet periods (Jeffries *et al.*, 1995; Dillon *et al.*, 1997). During the drought the length of the ice-free season, duration of stratification, depth of the photic zone, and light extinction increased, while precipitation and nutrient inputs to the lakes decreased. Drought occurred in Ontario in years following strong El Niño/ENSO events (Dillon *et al.*, 1997), another major circulation feature in addition to the NAO (Jones *et al.*, 2001). When the ENSO Index was strongly negative, the frequency of drought during the following summer was high. If long-term changes in global or regional climate alter the frequency or magnitude of El Niño/ENSO-related droughts (Dai and Wigley, 2000), the recovery of acidified lakes will be longer and more complex.

9.10.6.3 CO_2

Carbon dioxide was the first atmospheric gas shown to be increasing because of human activities, and it is often implicated in global warming. Small seasonal variations reflect the net respiration/photosynthesis of the hemisphere and possible forcing by El Niño (Bacastow, 1976). The recent increase in atmospheric CO_2 of ~0.5% yr^{-1} is well documented with direct measurements since 1958 and indirectly by ice core analyses (Schneider, 1989). It is largely due to burning of fossil fuels, deforestation, and land-use changes. The atmospheric CO_2 increase has no direct impact on acidification but may be linked indirectly through climate changes caused by greenhouse effects due to increased atmospheric CO_2. Wright (1998) experimentally increased ambient air CO_2 by 100% in a mini-watershed in Risdalsheia, southern Norway.

This increase, in combination with warming of the soil by 3–5 °C, produced an increased concentration of NO_3 in the runoff. The ecosystem switched from being a net sink to a net source of inorganic nitrogen, probably due to an acceleration of the decomposition of soil organic matter at the higher temperatures.

The increase of atmospheric CO_2 may have decreased the pH of precipitation very slightly, but P_{CO_2} in soils is far more important for the acid–base status of surface waters. Variation in forest soil P_{CO_2} is related to the temperature and moisture content of soils as well as the release of excess soil CO_2 to the atmosphere. Warmer conditions increase the rate of microbial and root respiration in the soil, thereby increasing soil P_{CO_2} above the long-term average value and producing short-term increases in runoff ANC, and vice versa. Norton *et al.* (2001) found that intra-seasonal variations in P_{CO_2} caused by variable snowpack thickness could induce variation in ANC in runoff of 10–15 $\mu eq\ L^{-1}$. Such variability is comparable to variability in ANC caused by a 15–20 $\mu eq\ L^{-1}$ change in SO_4 in runoff. Decline in soil P_{CO_2}, despite increased temperature and possibly increased soil respiration could result from a lower soil moisture content and a greater efflux of soil CO_2.

9.10.6.4 Organic Acids

The export of DOC from forests is controlled by production, decomposition, sorption, and flushing, which are all connected to local climate (Kalbitz *et al.*, 2000). For example, organic carbon discharge from sub-watersheds in the Rhode River watershed, Maryland, USA varied eightfold (Correll *et al.*, 2001). Temperature effects on DOC concentrations were weak and fluxes were not correlated with temperature in a Norway spruce forest in Germany (Michalzik and Matzner, 1999). Drought in western Ontario, Canada caused a decline in DOC export from watersheds, or more removal from lake-water columns because of the longer water residence times (Schindler *et al.*, 1996). In a whole watershed acidification experiment, Gjessing (1992) studied the response of Lake Skjervatjern, western Norway, to additions of NH_4NO_3 and H_2SO_4 and also cycles of drought. DOC became lower as a result of drought-related decreased input of DOC from the watershed and in-lake processes that consumed DOC. It is unclear why DOC concentrations have generally increased recently in European and North American soft waters (Bouchard, 1997; Evans and Monteith, 2001). Anesio and Granéli (2003) suggest that DOC is more photoreactive in acidified waters. Conversely, with recovery underway as a consequence of reduced SO_4 deposition, photoreactivity may be decreasing, resulting in an increase in DOC. If the quality of the DOC remains constant while the concentration increases, recovery of ANC in a regime of decreasing atmospheric pollution (sulfur and nitrogen) will be retarded but aluminum toxicity to fish will decrease as a consequence of increased complexation of aluminum and DOC.

9.10.6.5 Evaporation/Hydrology

In many lakes of the upper Midwest in the USA, the widespread decrease in lake SO_4 observed farther east was absent during a 4 yr drought that caused evaporative concentration of the already acidic seepage lakes (Webster and Brezonik, 1995). Lower than normal precipitation reduced seepage lake-water levels and groundwater elevations. A decrease and eventual cessation of groundwater inflow caused by the drought led to losses of ANC, calcium, and magnesium in the lakes. In groundwater-dominated (seepage) lakes in Wisconsin, USA there was no relationship between the concentration of solutes and precipitation (Webster *et al.*, 2000). Landscape position defined by the spatial position of a lake within a hydrologic flow system accounted for differences in chemical response to drought (Webster *et al.*, 1996). In surface-water-dominated lakes of Ontario, Canada the response of conservative solutes such as calcium and chlorine in low-ANC lakes was directly related to changes in precipitation. Concentrations were negatively correlated with the amount of precipitation.

9.10.6.6 Marine Aerosols

The sea-salt effect, which has been considered only as an episodic process, may also operate over decadal periods related to the NAO. The NAO is derived from the atmospheric pressure difference between the Azores and Iceland, and strongly affects winter temperatures and precipitation in regions bordering the North Atlantic (Hurrell, 1995; Jones *et al.*, 2001). High winter NAO index values are associated with wet and warm, frontally dominated winter weather in northwestern Europe. Such periods coincide with more marine salt input and increasing chloride concentrations (Evans *et al.*, 2001c).

9.10.6.7 Biological Feedbacks

Climate variability and atmospheric emissions alter vegetation and microbial activity. Elevated concentrations of atmospheric CO_2 can increase forest nutrient uptake, as does increased nitrogen deposition. Warming lengthens the growing season, increases net primary production (uptake) and decomposition, thereby accelerating nutrient cycling. Evolution of the vegetation community structure (or alteration by humans) may produce substantial changes in dry deposition of acidic compounds, evapotranspiration, hydrology, and base cation sequestration in biomass. Forested ecosystems may respond to climate warming by a temporal increase in inorganic nitrogen leaching caused by enhanced mineralization (Wright, 1998; Mol-Dijkstra and Kros, 2001). Simulations by the NuCM model at six USA sites suggested that increasing temperature caused nitrogen release from forest floors. At nitrogen-saturated sites nitrogen leaching increased. At the nitrogen-limited sites, increased growth (uptake) occurred (Johnson *et al.*, 2000). All these complex processes impact the acid–base balance of runoff water.

9.10.7 ACIDIFICATION TRAJECTORIES THROUGH TIME

A long-term natural acidification trajectory, as described by, for example, calcium concentration in runoff, is shown schematically in Figure 3(a). Such a curve incorporates the long-term average chemistry as it responds to depletion of calcium-rich minerals in the soil profile. Variations in the trend may be induced by variations in hydrology, temperature, biomass accumulation, and the concentration of weak acids (carbonic and organic). With the onset of acidification induced by atmospheric deposition of strong acid, the export of calcium should increase, either because of desorption of base cations from the soil complex or because of an increase in the rate of mineral weathering. Although variations in mineral weathering as a consequence of acidification were studied intensively in the 1980s (e.g., Swoboda-Colberg and Drever, 1993; Schnoor, 1990), experimental studies at both the laboratory scale and watershed scale (e.g., Dahlgren *et al.*, 1990; Fernandez *et al.*, 2003) indicate that the increased export of base cations is largely attributable to desorption.

Within a yearly cycle in watersheds that are nearly at steady state with respect to acid–base status, the concentration of calcium and other base cations in runoff commonly varies inversely with discharge (e.g., Feller and Kimmins, 1979). This relationship is caused primarily by the dilution of runoff with precipitation. Although concentrations may vary considerably, base cation ratios commonly remain relatively constant (Norton *et al.*, 1995) (Figure 7), indicating that short-term variability is controlled by ion-exchange equilibria among aluminum, hydrogen, calcium, magnesium, potassium, and sodium, with sodium and

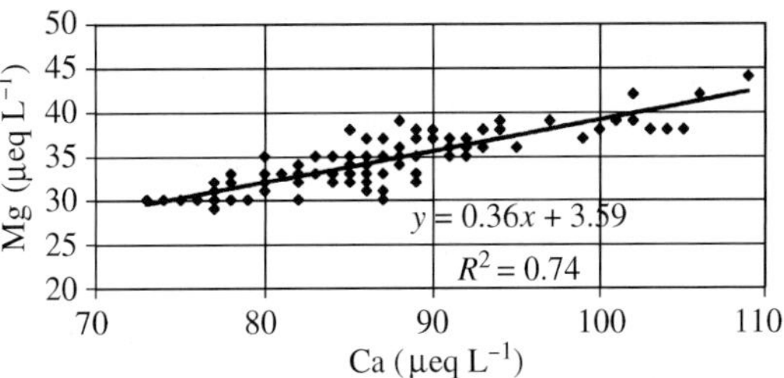

Figure 7 High frequency sample variability for Ca and Mg in a stream at Bear Brook Watershed, Maine, USA (S. A. Norton, unpublished).

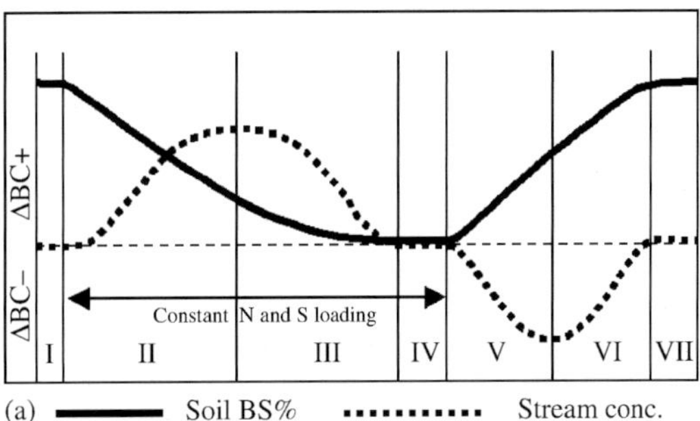

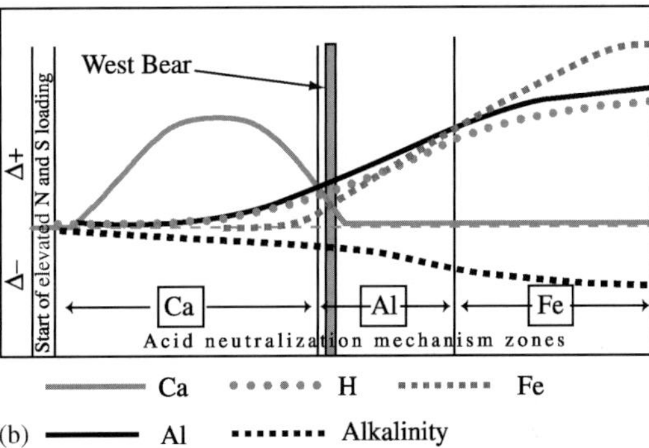

Figure 8 (a) Changes in base cations in stream runoff as a consequence of acidification and deacidification. (b) Acidification trajectories during acidic deposition for Ca, Al, and Fe (after Fernandez *et al.*, 2003).

potassium playing only a minor role. If acid loading is substantially increased in a system, the exchangeable base cation soil pool resists acidification by releasing base cations, and the soils adsorb SO_4. The trajectory of changes in the calcium concentration in runoff during acidification related to anthropogenic acidity should follow the path shown on Figure 8(a). This evolution (Galloway *et al.*, 1983) divides the history of anthropogenic acidification and recovery into seven stages.

Stage 1 corresponds to a "steady state" where base cation concentrations in runoff are relatively constant, averaged over periods longer than an annual cycle. BS of soils is also relatively constant. The rate of export of base cations is equal to the rate of chemical weathering of minerals containing these elements. The end of stage 1 is a step increase in acid loading to a constant value.

Stage 2 corresponds to the period of increasing export of base cations as a consequence of the increased acid loading from the atmosphere. Concurrently, ANC of the stream and BS of the soils decrease. Ultimately, as acidification progresses and base cations become depleted in the exchangeable pool in the soil, base cation concentrations peak and then start to decrease, as stage 3 commences.

Stage 3 is a period of decreasing base cation concentrations in runoff (and continued decrease in ANC and increase in dissolved aluminum (Figures 8(a) and (b)) and hydrogen and decreasing BS in the soils.

Stage 4 marks the runoff chemistry returning to stage 1 values and once again the rate of export equals the weathering rate. However, BS of the soil is now lower and episodic acidification is easily induced.

Stage 5 starts when the excess acid loading from the atmosphere stops. With continued weathering, part of the supply of base cations is adsorbed onto the impoverished soil exchange complex, lowering the runoff cation concentrations to values below pre-acidification values. The concentration of base cations decreases to a minimum value.

Stage 6 is an increase in the concentration of base cations as the BS and stream concentrations approach their pre-acidification steady-state values.

Stage 7 is the recovered steady state, equivalent to stage 1.

This scenario suggests that anthropogenic acidification, unlike natural acidification, is reversible. The oversimplified acidification trajectory is linked with a timescale that is highly site specific and the details of the processes will depend on many factors. Nonetheless, there is strong evidence for such a scenario, including laboratory studies of soils (Dahlgren *et al.*, 1990), ecosystem-level experiments (Norton *et al.*, 2003), and empirical data (Jenkins *et al.*, 2001). Regrettably, there are not sufficiently long time series to document all stages of this process, but studies of watersheds not in steady state can be placed within the evolutionary scheme. In watersheds with low base cation reservoirs, anthropogenic acidification may deplete base cations rather quickly or already has done so. In these watersheds, aluminum (Neal *et al.*, 1997; Veselý *et al.*, 1998b) or iron (e.g., Borg, 1986) may be the dominant cation exported. In a long-term acidification experiment in Maine, USA, acidification of a watershed has been accelerated from stage 1 through stages 2 and 3 in a decade (Norton *et al.*, 2003). As base cations became depleted, aluminum replaced base cations as the principal neutralizer of incoming acid, and iron started to be mobilized as the more easily mobilized forms of soil aluminum began to be depleted (Figure 8(b)).

9.10.8 LONGITUDINAL ACIDIFICATION

In general, soil water pH increases downward from the forest floor in well-drained forest soils. Thicker mineral soils typically have higher pH lower in the soil profile. Consequently, headwater streams generally increase in pH downstream (Driscoll *et al.*, 1988) because

(i) Flow paths are typically shallow at higher elevations.

(ii) Soils are typically thicker at lower elevation.

(iii) Higher elevations more commonly support nondeciduous forest, whereas deciduous vegetation is typically more common at lower elevations.

(iv) Larger watersheds typically have more heterogeneous geology, with the possibility of higher ANC-producing bedrock/soil.

(v) Lower elevation terrain is typically underlain by bedrock with higher ANC-producing chemistry.

Consequently, watersheds are more acid-sensitive at the top and acidify from the top down (Hauhs, 1989; Norton, 1989; Matschullat *et al.*, 1992). They recover from the bottom up. The first systems to acidify are the last to recover.

Base cations removed from the upper part of watersheds are generally lost from the system; mostly remaining in solution. Some of the base cations are temporarily stored in stream sediments or in wetlands. Mobilized aluminum (dominantly) (Roy *et al.*, 1999), iron, and manganese (Borg, 1986) move downstream where they may be precipitated in the higher pH regions of streams (Veselý *et al.*, 1985) or lakes (Kopàček *et al.*, 2001b), scavenging trace metals and phosphorus. These aluminum- and iron-rich precipitates represent translocated ANC. The neutralizing capacity, represented by exchangeable base cations and aluminum- and iron-rich precipitates must be stripped by progressive acidification before an entire stream can become chronically acidic (Norton *et al.*, 1990).

Flat topography and a cool, moist climate provide favorable conditions for organic matter accumulation. Bogs and fens along the surface-water flow path may alter water chemistry substantially, commonly by adding DOC. All other chemical factors being equal, humic waters (those rich in DOC) are more acidic. These waters, commonly naturally acidic, are more common at low elevations on flat terrane with relatively low precipitation such as in Finland (Kortelainen and Mannio, 1990). Streams that are recharged primarily by springs have a relatively stable chemistry (Lange *et al.*, 1995).

9.10.9 SOME AREAS WITH RECENTLY OR POTENTIALLY ACIDIFIED SOFT WATERS

9.10.9.1 Eastern Canada

Much of eastern Canada is underlain by slowly weathering rocks with few carbonate minerals. This area includes large parts of Ontario, Quebec, Newfoundland, Labrador, and Nova Scotia. The regional climate ranges from relatively dry continental in western Ontario, the site of the classic Experimental Lakes Area (e.g., Schindler *et al.*, 1990), to relatively wet cool maritime (see Jeffries (1997) for an overview). Sudbury, Ontario, once the site of the world's largest point source of SO_2 emissions, has been the focus of chemical and biological studies describing the trajectory of acidification and recovery (Gunn, 1995) after the point source emissions were dramatically reduced. This west-to-east transect also corresponds to a strong gradient of atmospheric deposition of SO_4 and NO_3. The highest pollution is in eastern Ontario and western Quebec. Consequently, the various lake districts have distinctly different water qualities. The Ontario and Quebec lakes are dominated by relatively low DOC drainage lakes, whose modified acid–base chemistry is dominated by SO_4 input. Nova Scotian lakes and streams, alternatively, are more easily impacted by atmospheric deposition because of their relatively low calcium and high DOC. The time series of lake chemistries are long, high quality, and commonly of high spatial and temporal resolution so that both the processes of acidification and recovery can be studied.

9.10.9.2 Eastern United States

The Adirondack Mountain Region of New York was the first area in the United States identified to be under stress from acidic deposition. It is in the region of highest sulfur and nitrogen deposition in the USA (Figure 1), annually receives between 1 m and 1.5 m of precipitation, and is underlain by large areas of low calcium bedrock and thin soils developed from till. In 1985, the USEPA conducted a national survey to characterize terrane at risk from anthropogenic acidification (Brakke *et al.*, 1988). They identified additional areas that were currently receiving significant quantities of acid rain and had bedrock and soils that placed them at risk of acidification. Since then, there have been few comprehensive federally sponsored surveys of lake or stream chemistry. Fewer sites have been studied, but these have been investigated more intensively (e.g., Stoddard *et al.*, 1998). In the last half of the 1990s, SO_4 in deposition declined, but recovery of ANC in acidified systems has been slow or nonexistent (see Section 9.10.13). At some localities, base cations are decreasing faster than strong acid anions (NO_3 and SO_4) in surface water, implying that important contemporary processes other than anthropogenic acidification and recovery may be occurring (Yanai *et al.*, 1999).

South of the glaciated region in eastern USA, there are few lakes but many streams draining old, thin soils developed on low calcium bedrock. Webb *et al.* (1989) surveyed 344 trout streams draining the Blue Ridge Mountains in Virginia and found many low-alkalinity streams with low calcium and magnesium. Sulfate concentrations, while the highest of all ions, were maintained at less than one third of equilibrium values, based on soil characteristics and atmospheric deposition of SO_4. The soils, dominantly ultisols, strongly adsorb SO_4 and had not yet equilibrated with the elevated concentrations in deposition, thus delaying acidification. The desorption of the SO_4 in the future may delay recovery.

Some of the longest time series of environmental measurements of soils and soil solutions, precipitation quality, and nutrient cycling for the USA come from the Walker Branch Watershed, Tennessee, and Coweeta Watershed, North Carolina. In the Hubbard Brook Experimental Forest, NH, much research there was originally designed to evaluate the impact of forest practices on soil fertility. A common characteristic of the older soils is the relatively low concentration of calcium and magnesium because of long-term acidification. The accumulation of abundant sesquihydroxide secondary phases in the soils results in adsorption of excess SO_4 from the atmosphere, even as base cations are being depleted by acidic deposition (Johnson *et al.*, 1982, 1988).

9.10.9.3 British Isles

The United Kingdom Acid Waters Monitoring Network (UKAWMN) was established in 1988 to assess the effect of emission reduction on selected acid-sensitive sites in the UK (Evans and Monteith, 2001). Widespread declines in non-marine SO_4 concentration did not occur between 1988 and mid-1990s. Sulfate concentrations have declined substantially since the mid-1990s (Harriman *et al.*, 2001; Ferrier *et al.*, 2001). The most strikingly consistent observation from UKAWMN is the gradual regional increase in DOC, which has slowed recovery. The increase in organic acidity has been of a similar magnitude to the decrease in mineral acidity (Evans and Monteith, 2001). There has also been a general reduction in winter storms connected with sea-salt effect from the early 1990s with a reduction of chlorine and a slight reduction in acidity. Since the early 1950s, patterns of storminess over the UK connected with marine salt input have oscillated on an ~10 yr cycle (Evans *et al.*, 2001c). Long-term studies at several research sites have led to the detection and understanding of processes in addition to acidification due to air pollution (e.g., Plynlimon Wales; Neal, 1997).

A stratified, pseudorandom survey of 200 lakes in the Republic of Ireland showed that lake chemistry was dominated by marine aerosol deposition (Aherne *et al.*, 2002). The acidity of lakes with a pH < 6 was dominated by organic acidity, followed by SO_4 (primarily in the east) and NO_3.

9.10.9.4 Scandinavia

Acidification of freshwaters has been and remains a major environmental problem for the three Nordic countries (Finland, Norway, and Sweden) where freshwaters have low ionic strength and low concentrations of nutrients. This is mainly due to low bedrock weathering rates and thin soils (Henriksen *et al.*, 1998; Skjelkvåle *et al.*, 2001b). From western Norway to eastern Finland, there is a gradient from high (~3 m) to low (~0.5 m) precipitation rates and thin and patchy soils in mountainous areas to forested areas with relatively thick soils. Clear-water lakes with low base cation concentrations and low ANC dominate in western Norway and brown, high DOC lakes with higher concentrations of solutes dominate in Finland (Skjelkvåle *et al.*, 2001a; Mannio, 2001). In the northern half of Sweden, differences in water chemistry can be attributed largely to differences in climate (Fölster and Wilander, 2002). Surveys of 485 lakes in Norway conducted in 1986 and again in 1995 reveal widespread chemical recovery from acidification (Henriksen *et al.*, 1998). At first, most of the decrease in nonmarine SO_4 was compensated by a decrease in base cations; ANC, therefore, remained unchanged. Then, as SO_4 continued to decrease, the concentration of nonmarine calcium and magnesium stabilized and ANC increased (Skjelkvåle *et al.*, 1998). No further increases of NO_3 occurred in the 1990s. This suggests that nitrogen saturation is a very long term process in Nordic countries. In the late 1990s, DOC or TOC (total organic carbon) increased significantly in southeastern Norway and southern Sweden, whereas there was no trend in TOC in most of the humic lakes of Finland.

9.10.9.5 Continental Europe

Geology, weathering rates, and soil composition in much of continental Europe provide a greater acid neutralizing ability than in most of Scandinavia. The highly variable bedrock type and soil thickness create a spatial heterogeneity with regard to acidification and the chemical recovery of surface waters. Acidification of higher elevation watersheds was rapid and recovery quicker in sensitive areas of central Europe (Veselý *et al.*, 2002a; Kopàček *et al.*, 2001a), in response to the largest changes in atmospheric deposition. However, streams draining watersheds on deeply weathered preglacial soils have had no or only a slight decrease in SO_4 concentrations and no chemical recovery, even with a substantial decline of acidic deposition (Alewell, 2001; Alewell *et al.*, 2001). Variable soils together with regional variations in deposition produced isolated, severely acidified regions at higher elevations of the Czech Republic (Veselý and Majer, 1996, 1998). At these sites there has been a regional decline in NO_3 concentrations (up to 60%) since the mid-1980s (Veselý *et al.*, 2002a). However, in northwestern Italy, NO_3 concentrations increased by ~25% from the 1970s to the 1990s (Mosello *et al.*, 2000). Sites in the central and southern parts of The Netherlands experience high levels of NH_3/NH_4 deposition, which yields a high potential for nitrification and acidification (De Vries *et al.*, 1995).

While the impacts of acidification on fish are the main concern in Scandinavia and North America, damage to planted coniferous forests at high elevation sites in central Europe is critical. For example, in the Czech Republic ~10^5 ha of *Picea abies* have died, and more than 50% of the forest has suffered irreversible damage (Moldan and Schnoor, 1992). Acid deposition and elevated ozone levels are predisposing factors for damage to tree crowns.

9.10.9.6 South America

Much of South America is relatively free of significant air pollution. Nitrate concentrations in remote Chilean and Argentinian streams are low. Dissolved organic nitrogen (DON) is responsible for most of the high nitrogen losses, 0.2–3.5 kg N ha^{-1} yr^{-1}, from these forests (Perakis and Hedin, 2002). In the Amazonian areas there is strong internal recycling of nitrogen with little export from the nondisturbed watersheds. The rain is slightly acidic (pH 5.2) and there is soil input of 3.7–8.7 kg SO_4–S ha^{-1} yr^{-1} and ~0.8 kg NO_3–N ha^{-1} yr^{-1} (Forti *et al.*, 2001). Abundant DOC depresses the pH of some larger tributary streams of the Amazon (Williams, 1968). The MAGIC acidification model has been used to assess the effects of the conversion of tropical Amazonian rain forest watersheds to pasture (Neal *et al.*, 1992). The modeling demonstrated the sensitivity of tropical rainforest runoff to deforestation, even without climate changes (Forsius *et al.*, 1995).

9.10.9.7 Eastern Asia

In contrast to North America and Europe, emissions of SO_2, NO_x, and NH_3 are rising markedly in many developing countries. In East Asia (Japan, Korea, China, Mongolia, and Taiwan), emission of SO_2, NO_x, and NH_3 are projected to increase by about 46%, 95%, and 100%, respectively, by 2030 (Klimont *et al.*, 2001). By 2020, Asian emissions of SO_2, NO_x, and NH_3 may equal or exceed the combined emissions of Europe and North America (Galloway, 1995). Annual mean concentrations of SO_2 in air as high as 1,300 ppb have been observed in industrial areas of Thailand (Wangwongwatava, 2001). Fujian and Guizhou provinces in south China received significantly acidic rain. Although total sulfur deposition in China is relatively high (Larssen *et al.*, 1998), alkaline dust and NH_3 emissions neutralize much of the SO_4 acidity in precipitation. Total deposition of sulfur compounds over Japan was more than twice the human emissions because of volcanic activity. The emission of SO_2 from Japan is less than 5% of the total emissions in East Asia.

9.10.10 EXPERIMENTAL ACIDIFICATION AND DEACIDIFICATION OF SOFT WATER SYSTEMS

Numerous experiments have been conducted to understand the chemical linkages between atmospheric deposition of acidic compounds and acidification of soils, lakes, and streams. Experiments have included additions of acid, exclusion of acids, and the application of limestone ($CaCO_3$) or other acid neutralizing compounds to add ANC directly to surface water or soils. Many of the studies are discussed in Dise and Wright (1992), Rasmussen *et al.* (1993), and Jenkins *et al.* (1995). We highlight a few studies here.

9.10.10.1 Experimental Acidification of Lakes

Lake 223 in the Experimental Lakes Area (ELA) in northwestern Ontario, Canada, and Little Rock Lake (LRL) in northern Wisconsin, USA were progressively acidified by in-lake addition of H_2SO_4. The pH was lowered from original values of 6.1–6.8 to 4.7–5.1 (Schindler *et al.*, 1990). Both lakes generated an important part of their ANC internally by reduction of SO_4, and to a lesser extent by reduction of NO_3. ANC production increased in LRL as ~50% of the H_2SO_4 added was neutralized. Acidification disrupted nitrogen cycling. Nitrification was inhibited in Lake 223. In LRL, nitrogen fixation was greatly decreased at the lower pH.

Lake 302S in the ELA was experimentally acidified from an original pH of 6.0–6.7 to a final pH of 4.5. The DOC drastically declined from 7.2 to 1.4–1.6 mg L^{-1}, comparable to clear arctic and alpine lakes (Schindler *et al.*, 1996). The increased clarity allowed greater penetration of solar radiation, including UV. The thermal structure of the water column was altered, impacting circulation and nutrient regeneration. Results from both Lake 223 and lakes near Sudbury, Canada, suggest a rapid recovery of lacustrine communities when acidification is reversed. Fish reproduction resumed at pH values similar to those at which reproduction ceased when the lake was being acidified.

Dystrophic (humic-rich) Lake Skjervatjern and its watershed, western Norway, were acidified with a combination of H_2SO_4 and NH_4NO_3 as part of the HUMEX experiment (Humic Experiment) (Gjessing, 1992). The experiment studied the role of humic substances during acidification of surface waters, and the impacts of acidic deposition on chemical and biological properties of humic water (Kortelainen *et al.*, 1992). The watershed and lake were divided for the treatment. The DOC of Lake Skjervatjern was partly controlled by a variable retention time in the lake basin. DOC decreased with increasing retention time. DOC and water color decreased during acidification, followed by an increase perhaps caused by fertilization of the watershed with nitrogen. Periods of high precipitation and discharge coincided with increases in the concentration and quality of DOC (Gjessing *et al.*, 1998).

9.10.10.2 Experimental Acidification of Wetlands

Bayley *et al.* (1988) added H_2SO_4 to a bog in western Ontario, Canada to evaluate the chemical and biological response of wetlands to atmospheric inputs of SO_4. Sulfate is substantially retained in true bogs, or reduced to H_2S and re-emitted to the atmosphere. The fate of NO_3 in true bogs seems quite clear; bogs always have lower concentrations of NO_3 than precipitation. The fate of SO_4 is less clear because they are very strong intra-annual variations in concentration due to oscillating oxidation and reduction processes within the bogs. Both sulfur and nitrogen may be stored in reduced form as organic matter, released to the atmosphere as gaseous compounds, or leached to surface water as organically bound sulfur and nitrogen or as oxidized anions. It seems unlikely that true bogs with a pH near 4 can be significantly acidified by any reasonable loading of acid from the atmosphere (Gorham *et al.*, 1987).

9.10.10.3 Experimental Acidification of Terrestrial Ecosystems

At the HBEF, New Hampshire USA, Likens *et al.* (1977) pioneered the paired watershed approach to biogeochemical experiments, including many factors related to the impact of acidification. Since the mid-1970s, they have examined long- and short-term data for precipitation chemistry, stream chemistry, cation supply from various sources, variable forestry practices, and smaller-scale experiments with moisture, soil CO_2, salt additions, and calcium amendments to the forest floor. The long-term, high-resolution data derived there have brought much insight into the variability of our chemical climate and have emphasized the necessity of long-term data to sort out important long-term process from short-term variability (Driscoll *et al.*, 1989).

Watersheds in the Fernow Experimental Forest, West Virginia, USA have been artificially acidified with $(NH_4)_2SO_4$ (Adams *et al.*, 1997; Edwards *et al.*, 2002). They have observed appreciable retention of SO_4 in the older, unglaciated soils, and increased export of calcium and especially magnesium as a consequence of the treatment.

Risdalsheia, Norway (Wright *et al.*, 1993) is located in southern Norway in a region substantially impacted by acidification during the twentieth century. It has been the site for many paired-watershed chemical manipulations. A mini-watershed, including the canopy, was covered by a transparent roof to exclude ambient acid precipitation. "Clean" reconstituted rain with natural concentrations of sea salts was applied underneath the roof. Loading of SO_4, NO_3, and NH_4 was reduced by ~80%; the remaining 20% occurred as dry deposition of gases and particles. Later, the same mini-watershed was subjected to 3–5 °C warming and elevated CO_2 (560 ppmv). The flux of nitrogen in runoff increased by ~5–12 mmol m^{-2} yr^{-1} (Wright, 1998), probably due to increased mineralization and nitrification rates in the soils because of higher temperature. The pH did not increase substantially, because it was buffered by high concentrations of DOC. Long-term simulations by the Simulation Model for Acidification's Regional Trends (SMART 2) model (Mol-Dijkstra and Kros, 2001) predicted a long-term increase of nitrogen in runoff.

Wright also conducted a series of experiments at Sogndal, western Norway, a pollution-free site which receives more than one meter of rain. The paired watershed design was used. H_2SO_4 and a combination of H_2SO_4 and HNO_3 were added to the terrestrial part of several watersheds over a long period of time. The acidification trajectory involved an increased export of base cations and aluminum, and decreased ANC and pH (Frogner, 1990). Episodic acidification due to atmospheric deposition of marine aerosols was also demonstrated experimentally. Diluted seawater (~600 mg Cl L^{-1}) simulating a salty rain event was added to a mini-watershed using irrigation equipment. The runoff response included depressed pH and ANC, and increased export of aluminum, calcium, and magnesium desorbed from the soil (Wright *et al.*, 1988).

Gårdsjøn, Sweden is the home of the famous roof experiment (Andersson and Olsson, 1985; Hultberg and Skeffington, 1999). The watershed complex includes terrestrial and aquatic experiments, which include acid additions, terrestrial liming, and acid exclusion. Located in southwestern Sweden, this terrestrial and aquatic system had been acidified by anthropogenic acidity from the atmosphere. In a somewhat analogous experiment to that at Risdalsheia, Norway, a roof was installed below the canopy of a small watershed to catch the throughfall. The precipitation plus chemicals leached from the canopy were reconstituted to contain only the normal marine salts and were then distributed below the roof. This was essentially a step function reduction in acid rain, designed to assess the recovery of this ecosystem. A sub-watershed was part of a European network of manipulations where excess nitrogen (as NH_4 or NO_3), labeled with ^{15}N, was added (Wright and Tietema, 1995; Wright and Rasmussen, 1998). This tracer enabled scientists to track the processing, sequestration, and release of nitrogen in the ecosystems. The NITREX (Nitrogen saturation Experiments) (Dise and Wright, 1992) project involved chemical manipulations with nitrogen and nitrogen isotopes in Norway, Sweden, Wales, Switzerland, The Netherlands, and Denmark (Wright and Rasmussen, 1998).

Bear Brook Watershed in Maine (BBWM), USA is a paired watershed (~10 ha each). One forested watershed is treated with $(NH_4)_2SO_4$, the other is a reference (Norton and Fernandez, 1999). The major changes in stream chemistry observed during the 1989–2001 manipulation period included a decrease in pH and ANC, and an increase in the export of base cations, SO_4, NO_3, Al, P, Fe, and Mn (Norton *et al.*, 2003). DOC and silica remained essentially constant, suggesting that acidification from a pH of ~5.2 to 4.6 was largely controlled by increased nitrification, increased SO_4 flux, desorption of base cations, and dissolution of secondary aluminum, iron, and manganese phases from the soil. Fernandez *et al.* (2003) indicate that the excess base cation export in the runoff was matched by loss of base cations from soil in the experimentally acidified watershed, in agreement with MAGIC model predictions.

9.10.10.4 Experimental Acidification of Streams

Chemical acidification experiments in streams have the advantage of a chemical and biological reference (upstream of any chemical additions), easy sampling, and repeatability. Understanding the role of sediments during episodic acidification also yields information about the behavior of the soils from which much of the sediment is derived. The classic experiment for stream acidification was conducted by Hall *et al.* (1980) at the HBEF, New Hampshire, USA. They demonstrated that stream sediments yield base cations and aluminum, thereby resisting the depression of the pH by acid addition (HCl). While much of this acid neutralizing ability was exhausted, stream sediment clearly played a role in ameliorating episodic acidification and recovery. Similar experiments have been conducted in the UK (Tipping and Hopwood, 1988), Norway (Henriksen *et al.*, 1988b), and Maine, USA (Norton *et al.*, 1992b, 2000). In summary, the studies indicate that stream sediments (inorganic and organic) are a pool of reversibly exchangeable base cations, aluminum (probably precipitated plus exchangeable), and other trace metals such as beryllium, cadmium, iron, and manganese. Stream sediments also have SO_4 adsorption capacity; this contributes to the delay of and diminishes episodic acidification, and delays recovery. Stream water has the ability to buffer excursions of pH if there is sufficient DOC weak acid acidity, as demonstrated by Hruška *et al.* (1999).

9.10.11 REMEDIATION OF ACIDITY

9.10.11.1 Liming

Limestone ($CaCO_3$ or something chemically similar that yields alkalinity during dissolution) has been used as an antidote for acidification in terrestrial and aquatic systems. There is no general agreement regarding the usefulness of liming as a countermeasure to anthropogenic acidification. Nonetheless, all countries with significant acidification problems have studied this method of remediation. Many of these studies were reviewed by Porcella *et al.* (1989). Liming generates ANC but not in a way consistent with the behavior of the natural system, and generally not at the same place as natural processes. Thus, while pH may be restored to pre-acidification levels in soil or surface waters, the chemistry of the system may not resemble pre-acidification conditions, and the biota may not recover along the same trajectory that they followed during acidification.

For example, Dillon *et al.* (1979) followed the response of several lakes in the vicinity of Sudbury, Ontario, Canada. Wright (1985) conducted a three-lake study at the Hovvatn site in southern Norway. There, two lake basins were limed in 1981 by several methods, including dispersal on ice and along the shore. A contiguous lake with similar initial chemistry served as a reference lake. A major finding was the rapid rate of loss of the alkalinity, generated by the dissolution of $CaCO_3$, due to flushing of the lake with acidic water from the watershed. Similar findings appear in most studies, although the details of the response are complicated by the hydrology of individual lakes. It is clear that direct addition of the $CaCO_3$ to a lake has to be done on a more-or-less continuing basis to avoid large excursions in pH or a return to acidic conditions. The cost of areal applications of lime is enormous, commonly on the order of \$100 (US) per ton. Nonetheless, some nations have adopted liming as a strategy to protect or restore surface-water pH. For example, Sweden has been expending substantial funds in a program that limes hundred of lakes per year, on a rotating basis. Other experiments have included liming low pH, high aluminum streams during episodic acidification. The net effect is to raise pH, ANC, and calcium, but the accompanying precipitation of aluminum presents a serious stress, at least for anadromous fish (Rosseland *et al.*, 1992). Liming directly on the land (e.g., at Gårdsjøn, Sweden; Hultberg and Skeffington (1999)) is also partially effective in restoring exchangeable calcium reservoirs in soil, but a considerable dose is necessary to improve ANC in draining waters. Reacidification of limed lakes may result in elevated trace metal concentrations as sediments desorb metals and adsorption of metals in the water column is diminished.

9.10.11.2 Nutrient Additions to Eliminate Excess NO_3

Forests have a large capacity to retain nitrogen and to increase growth even after years of large anthropogenic inputs of nitrogen (Prescott *et al.*, 1995). Similarly, adding phosphorus increased nitrogen uptake by trees and significantly reduced the NO_3 concentration in the soil solution (Stevens *et al.*, 1993). This technique would presumably lead to reduced export of NO_3 in surface water. Addition of phosphorus to lakes (Davison *et al.*, 1995) and a stream (Hessen *et al.*, 1997) reduced the NO_3 concentration, increased ANC, and increased the pH of the water. Nitrate is assimilated as organic matter by growing phytoplankton (Equation (14)) which is removed by sedimentation. Modest phosphorus addition

$$106\,CO_2 + 138\,H_2O + 16\,NO_3^- \rightarrow (CH_2O)_{106}(NH_3)_{16} + 16\,OH^- + 138\,O_2 \quad (14)$$

($<15\ \mu g\ P\ L^{-1}$) instead of liming may avoid eutrophication in a cost-effective way, increase pH, and reduce NO_3. However, in lakes where aluminum is being precipitated (see Section 9.10.5.2), the addition of phosphorus may have a very transitory effect as it is irreversibly scavenged from the water column.

9.10.11.3 Land Use

9.10.11.3.1 Deforestation

Understanding rates and processes of immobilization and leaching of nitrogen from soil organic matter with different levels of nitrogen deposition is crucial for assessing future acidification. The tight internal nitrogen cycle is broken as summertime soil temperature, moisture, stream discharge, and storm-peak discharge increase, and when the vegetation is disturbed by harvesting, forest decline, fire, wind-throw, insect defoliation, and canopy destruction by ice storms. Harvesting effects have been studied extensively in North America, particularly at Hubbard Brook (Bormann and Likens, 1979; Lawrence *et al.*, 1987) and in Wales (Reynold *et al.*, 1995). Deforestation in Wales resulted in a 5–7 yr long NO_3 pulse in stream water related to increased mineralization and nitrification in the soil, and to an increase in inorganic nitrogen available for leaching due to decreased biological uptake. Temporarily enhanced production of NO_3^- and H^+ increases adsorption of SO_4 to soil (Gbondo-Tugbawa *et al.*, 2002), and H^+ may exchange for aluminum and calcium in the soil (Henriksen and Kirkhusmo, 2000). The removal of biomass (harvesting) involves a removal of base cations, contributing to base cation decrease in streams. This increases the susceptibility of streams to other acidifying stresses. Forests disturbed by harvesting, wind, or fire during the past several centuries have long-term impacts on their vulnerability to nitrogen saturation and their capacity to store carbon (Goodale and Aber, 2001). Decreasing the canopy causes a decrease in the flux of pollutants, because dry deposition is decreased (Hultberg and Grennfelt, 1992).

9.10.11.3.2 Afforestation

Aggrading forests contribute to surface-water acidification in a number of ways. Water discharge decreases because of enhanced evapotranspiration, causing evaporative concentration of pollutants. Hydrological pathways become modified (Waters and Jenkins, 1992), and coniferous afforestation commonly raises DOC. Dry deposition of acidifying pollutants to a forest canopy increases as the canopy develops. Total deposition of sulfur and other pollutants in throughfall in forests is typically several times higher than bulk deposition outside the forest (Beier *et al.*, 1993; Hansen *et al.*, 1994; Rustad *et al.*, 1994). Episodic acidification caused by marine aerosols is probably enhanced in degree and frequency in forested areas, particularly in polluted areas (Jenkins *et al.*, 1990). Inputs of pollutants in fog and rime ice (e.g., Ferrier *et al.*, 1994) increase with the development of a canopy, increasing the interception cross area or leaf area index.

The Plynlimon, Wales studies (Neal, 1997) have been built on a program of conifer afforestation, largely since World War II. Forest plots of various ages and their surface-water discharge and chemistry have been studied for several decades to help define the role of forests in acidification processes. Base cations and NH_4 are sequestered in biomass (Nilsson *et al.*, 1982) with a concurrent release of H^+. The accumulation rates of nitrogen and base cations in biomass decrease as the forest matures and consequently the acidifying affect of an aggrading forest decreases as standing biomass and dead biomass reach steady state. The time to steady state is species specific (Emmett *et al.*, 1993).

9.10.12 CHEMICAL MODELING OF ACIDIFICATION OF SOFT WATER SYSTEMS

9.10.12.1 Steady-state Models

The response of the chemistry of soil, soil solutions, and surface waters to acid deposition has been simulated using steady-state and dynamic models since the early 1980s. Models were developed to: (i) interpret the past, (ii) guide future research, (iii) support policy decision (Forsius *et al.*, 1997; Henriksen and Posch, 2001), and (iv) explain observed ecosystem behavior (Gbondo-Tugbawa *et al.*, 2002). Initial steady-state models used empirical data, either in space or time to understand regional lake chemistry (e.g., using lake populations; Henriksen (1980)) and the chemical behavior of streams (e.g., Christophersen and Neal, 1990). Kirchner (1992) developed a watershed-based static model based on runoff chemistry for assessing acidification vulnerability. Empirical models remain powerful tools even as dynamic models have grown in complexity in parallel with computer development.

"Critical loads" are quantitative estimates of the exposure to one or more pollutants below which significant harmful effects on specified sensitive components of the environment do not occur, at least according to present knowledge (Nilsson and Greenfelt, 1988). The exceedance of critical loads for impacts on soils and surface waters have been the basis of negotiations for emission reductions in Europe. The concept of critical loads has spread to Canada but not to the United States. The steady-state model for critical loads implies that only the final results of a certain deposition level are considered. The time required to reach the final state is not considered. Critical loads of acidity for surface waters are set so that the input of acids to a watershed does not exceed the weathering rate less a stated amount of ANC (0–50 μeq L^{-1}) in the long term. Early studies of critical loads focused mainly on sulfur. As sulfur in atmospheric deposition has declined, nitrogen has become a more important focus. The steady-state water chemistry (SSWC) model (Henriksen *et al.*, 1995) considered only the extant nitrogen leaching level. The first-order acidity balance (FAB) model (Henriksen and Posch, 2001) for lakes assumed the nitrogen-immobilization rate to be equal to the long-term annual amount of nitrogen that is used for alteration of the C/N ratio in the soil plus nitrogen lost to denitrification and retained within lakes. This is the worst case of NO_3 leaching to surface waters. The FAB model, for example, predicts that more of the Norwegian lakes (46% in comparison to 37% by the SSWC model) may experience exceedance of critical loads for acidifying deposition in the future (Kaste *et al.*, 2002). Emission reductions adopted in Gothenburg in 1999 may reduce the area of exceedance of critical load for acidity from 93 Mha in Europe in 1990 to 15 Mha in 2010 (UN-ECE, 1999).

Critical loads for pollutant metals (especially cadmium, mercury, and lead) are being considered in Europe (Skjelkvåle and Ulstein, 2002).

9.10.12.2 Dynamic Models

The static models for critical loads for sulfur and nitrogen neglect the time component of acidification. Static models exclude long-term processes, including acid neutralization by soil adsorption of SO_4 and desorption of cations. Both processes delay acidification and recovery. The numerical models Integrated Lake Water Acidification Study (ILWAS) (Gherini *et al.*, 1985), MAGIC (Cosby *et al.*, 2001), SMART (De Vries *et al.*, 1989), soil acidification in forest ecosystems (SAFE) (Warfvinge *et al.*, 1993), PnET-BGC/CHESS (Krám *et al.*, 1999), AHM (Alpine hydrochemical model) (Meixner *et al.*, 2000), PROFILE (Sverdrup and Warfringe, 1993), and nutrient cycling model (NuCM) (Johnson *et al.*, 2000) are based on mathematical formulations of hydrological and biogeochemical processes in soils and waters. The models have been used both for forecasting and hindcasting water quality in watersheds based on data on topography, meteorology, soil chemistry, weathering, and acidic deposition. The reliability of model predictions increases with the length of time covered by the observation data used for calibration of the models. The models, although varying in detail, are based on similar principles: charge balance of ions in the soil solution and mass balance of the elements considered. They describe the changes in the element pools over time.

The MAGIC and SMART models have been developed for regional scale applications. These models have a high degree of process aggregation to minimize the data requirements for application at large scales and multiple sites. The opposite is true for models having relatively complex process formulations, which are developed for application on a site scale, for example, ILWAS. ILWAS is perhaps the most mechanistic and synthetic model, providing detailed descriptions of watershed acidification.

MAGIC is perhaps the model most widely applied in acidification and recovery studies. It is a process-oriented model, which lumps key soil processes at the watershed scale in monthly or yearly time steps (Cosby *et al.*, 1985). MAGIC has a soil–soil solution equilibrium section in which the chemical composition of soil solutions is governed by SO_4 adsorption, CO_2 equilibria, cation exchange, and leaching of aluminum. The mass balance section assumes that the flux of major ions is governed by atmospheric inputs, chemical weathering inputs, net uptake in biomass, and loss to runoff. MAGIC version 7 incorporated the major controls on nitrogen fluxes through time (Cosby *et al.*, 2001). Nitrogen leaching to surface waters is a function of total inorganic nitrogen deposition, plant uptake of nitrogen, soil uptake or immobilization of nitrogen, and nitrification of reduced nitrogen. The MAGIC 7 model assumes that the net retention or release of incoming inorganic nitrogen in the soil is determined by the C/N ratio of soil organic matter. MAGIC 7 emphasizes the central importance of future nitrogen emission controls.

Modeling by MAGIC is fairly successful in predicting sulfur and cation dynamics (Figure 9) in freshwater, although stable sulfur isotopes indicate that biological sulfur turnover (not modeled) is also an important process (Alewell, 2001; Gbondo-Tugbawa *et al.*, 2002; Novák *et al.*, 2000). Similarly, nitrogen-cycle modeling is still under development and further studies are needed to verify nitrogen immobilization processes under varying nitrogen and sulfur deposition levels. The size, kinetics, and uptake capacity of soils are critical factors determining their response to increased nitrogen loading and may be related also to land-use history (Magill *et al.*, 1997; Goodale and Aber, 2001).

The SMART model (De Vries *et al.*, 1989) estimates long-term chemical changes in soil and soil water in response to changes in atmospheric deposition. The model structure is based on the mobile anion concept, incorporating the charge balance principle. SMART 2 adds forest growth and biocycling processes, which allow soil

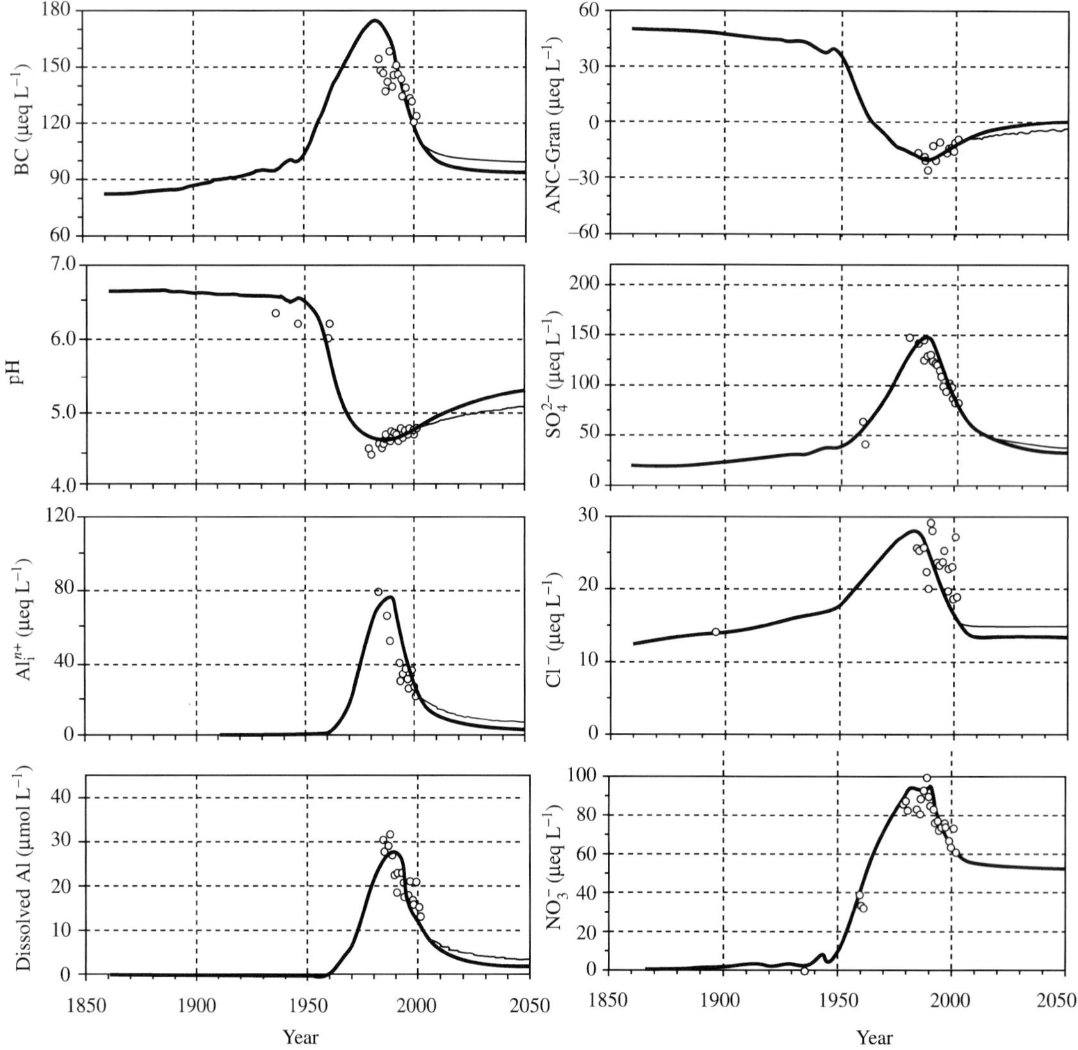

Figure 9 Measured (open circles) and MAGIC 7 modeled (lines) trends in the Černé Lake, Bohemian Forest, Czech Republic. The model was calibrated on long-term (up to 70 yr) data. The best agreement was attached with a three-times higher SO_4 adsorption capacity than determined experimentally, and slightly decreasing N retention during the acidic deposition peak. The forecast is based on the 2000 deposition value (thin line) and the projected decreased S and N emission/deposition rate in central Europe according to Gothenburg Protocol (UN–ECE, 1999 (thick line)) (source V. Majer *et al.*, in review).

nitrogen availability and forest growth to be modeled (Mol-Dijkstra and Kros, 2001). In SMART 2, total nutrient uptake is described as a demand function, which consists of the maintenance uptake in leaves and net growth uptake in stems. Immobilization of nitrogen is dependent on the soil C/N ratio.

The assumption employed in models is that equilibrium chemistry is applicable in all relevant situations. This implies that the reaction of soil pH and other parameters to a change in input is virtually instantaneous and that processes such as diffusion can be neglected. Long-term, large-scale acidification models are difficult to calibrate and validate because of the paucity of sufficient long-term (>50 yr) observations.

9.10.13 CHEMICAL RECOVERY FROM ANTHROPOGENIC ACIDIFICATION

The many spatial and temporal lake and stream surveys conducted in many countries plus experiments have demonstrated the linkages between the emissions of acidic compounds to the atmosphere and terrestrial and aquatic acidification. Mandated and implemented reductions in air pollution in North America and Europe have occurred since the 1980s. The long-term experiment of ecosystem acidification from air pollution is seeing a reduction of the dose.

Long-term acidification related to soil depletion of soluble alkalinity producing minerals is irreversible without soil scarification or substantial

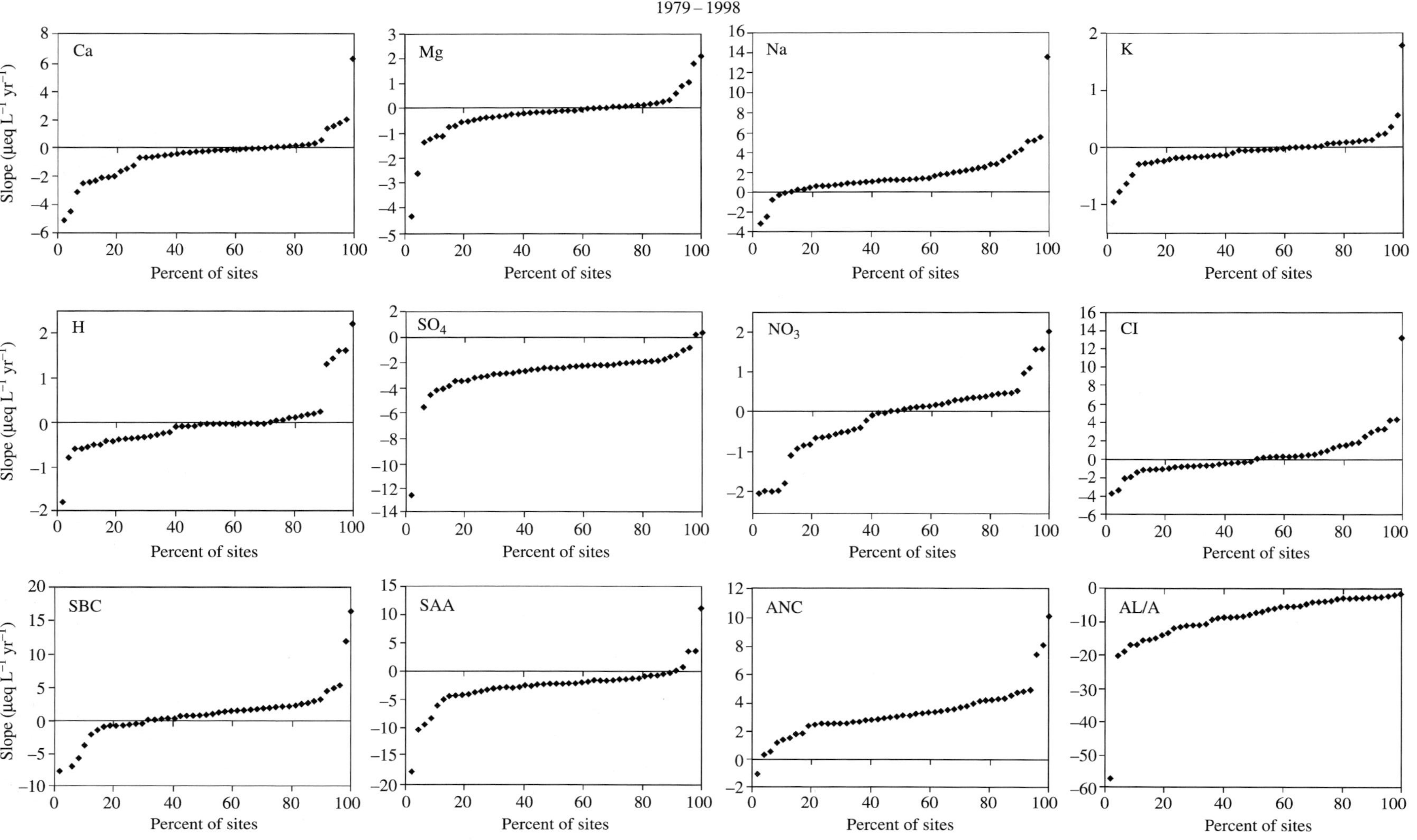

Figure 10 Recovery of regional alkalinity (ANC) and associated chemical changes in a suite of lakes in Scotland, UK (Ferrier *et al.*, 2001) (reproduced by permission of European Geoscience Union from *Hydrol. Earth Syst. Sci.*, **2001**, *5*, 421–431).

soil amendments that, as for liming lakes, must be an ongoing process. Shorter-term acidification due to atmospheric deposition is reversible if the flux of mobile acidic anions (primarily SO_4 and NO_3) can be reduced to the point where chemical weathering can provide sufficient base cations to allow soils to recharge their BS (stages 5 and 6, Figure 8(a)). Recovery from acidification caused by acidic deposition has been demonstrated in whole ecosystem experiments (e.g., in Sweden by Hultberg and Skeffington (1999), and in Norway by Wright *et al.* (1993)) and has probably occurred in some sensitive central European lakes impacted by local smelting in pre-industrial times. Partial recovery of ANC has occurred in many areas impacted by acid rain. The general pattern of recovery involves a reduction of SO_4 (and in some localities NO_3) in surface water that exceeds reductions in base cations associated with the recovery of soil BS. ANC increases because of a decline of SO_4 relative to base cations in runoff. Repeated surveys in the UK, Scandinavia, the Czech Republic and elsewhere indicate that since 1990 many, but not all, individual lakes and streams have recovered (Jenkins *et al.*, 2001) (Figure 10). In the northeastern USA, recovery of ANC has accompanied the reduction of atmospheric SO_4, although atmospheric NO_3 has remained relatively constant (Stoddard *et al.*, 1998, 1999, 2001). Considerable recovery in sensitive areas of central Europe has included not only reduction in SO_4, but also in Al, NO_3, and Cl concentrations (Veselý *et al.*, 2002a). The chemical path of recovery may follow a hysteresis loop (Kopàček *et al.*, 2002b), that is also suggested by the studies of artificial acidification and recovery of stream sediments (Figure 11). A simple linear recovery from reduced SO_4 and NO_3 atmospheric deposition is unlikely due to concurrent variations in climate (Section 9.10.6). For example, an increase in temperature (~1.3 °C) over 17 yr accelerated the decrease of inorganic aluminum during recovery by ~13% in strongly acidified lakes in the Czech Republic; this temperature increase was the second-most important cause of aluminum reduction after the SAA decrease (Figure 12) (Veselý *et al.*, in press).

The importance of these secondary effects on acid–base status, metal concentration, and toxicity are still being studied. The recovery from acidification has been simulated by MAGIC for many watersheds, but sufficient time has not elapsed since acid inputs declined to assess the accuracy of the model predictions (Majer *et al.*, in review).

In general, systems that had been acidified most, recovered ANC most rapidly. However, many aquatic systems in North America and Europe show little or no recovery, or even continue to acidify as base cations decline faster than the reduction in SO_4 plus NO_3. The causes of continued acidification during reduced air pollution may be the complex interaction of many secondary processes including changes in climate-driven hydrology and temperature, biomass uptake, net respiration in soils, nitrogen-processing, and long-term fluxes of marine aerosols. These secondary effects may mask the

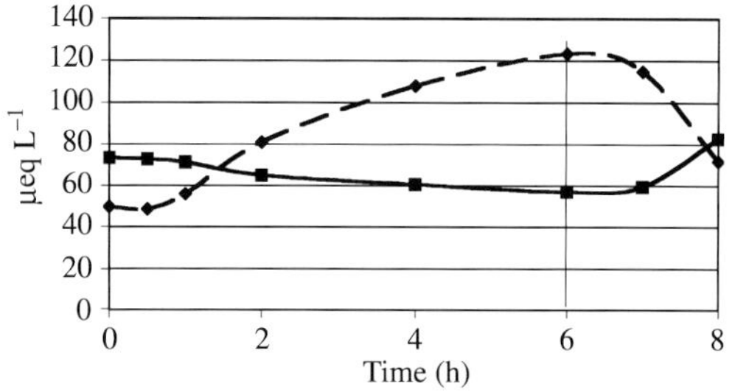

Figure 11 Experimental episodic acidification and recovery of a low-ANC stream in Maine, USA showing Ca and SO_4 dynamics. Change in concentration of SO_4 (solid line) and Ca (dashed line), 50 m downstream from an HCl addition that ended at 6 h (vertical line). Note adsorption and desorption of SO_4 and desorption and adsorption (after 8 h) of Ca (S. A. Norton, unpublished).

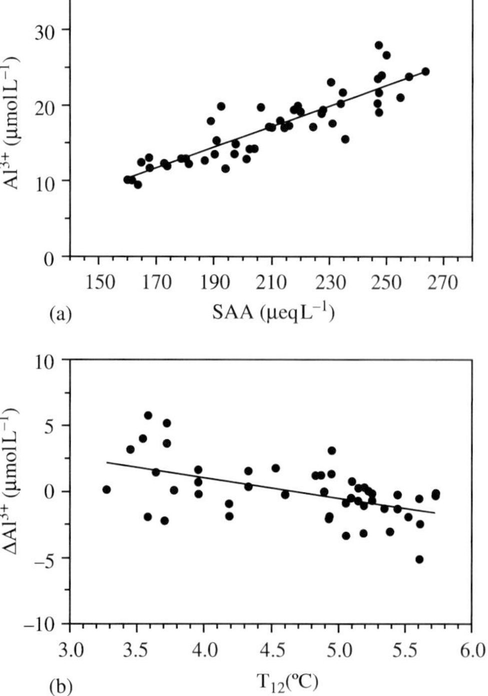

Figure 12 (a) The linear relationship between Al^{3+} and the sum of strong acid anions (SAA) for Cerné Lake, Czech Republic, 1984–2001. (b) The relationship between the residuals of the Al^{3+}–SAA relationship (a) and the mean temperature for the 12 months preceding the sample (T_{12}) from Cerné Lake, Czech Republic, 1984–2001.

recovery of systems from reduced sulfur input for some time.

ACKNOWLEDGMENTS

The authors are very appreciative of many thoughtful interactions with colleagues, too numerous to identify, in Europe, North America, and Asia concerning acid rain and its impacts. They have been generous with ideas and presented the authors opportunities to understand the breadth and depth of research related to anthropogenic acidification. Anthropogenic acidification was perhaps the most important and heavily researched environmental problem in large areas of the northern hemisphere for three decades. The attention paid to the terrestrial and aquatic resources at risk has provided the world with an understanding of how they function. This is critical for preparing us to assess the effects of climate change on these resources.

The body of research on ecosystem acidification has been built substantially since the 1970s by hundreds of scientists in several dozen countries. Several billion dollars (US equivalent) have been spent achieving a significantly better understanding of ecosystems prior to acid rain, their present status, and their probable future in the face of various scenarios of atmospheric pollution. More than 100 books and thousands of articles have been published on the topic since the 1970s. The literature we cite identifies many of the individuals and organizations that are responsible for this progress. Not everybody could be recognized, of course. One amazing characteristic of the research effort has been the international cooperation that evolved as a consequence of the realization that atmospheric pollution knows no political boundries. Its consequences can be understood and remediation implemented only with international political cooperation, driven by excellent science. That cooperation is one of the important secondary results of the effort. The authors thank the people who, early on, had the vision to recognize that human activity can degrade our chemical climate through atmospheric pollution and who stimulated so many people to look, experiment, and understand. These people include, but are not limited to, Svante Odén in Sweden, David Schindler and Peter Dillon in Canada, Ellis Cowling and Gene Likens in the USA, Arne Henriksen and Richard Wright in Norway, and Bernard Ulrich in the former West Germany.

REFERENCES

Aber J. D., Nadelhoffer K. J., Steudler P., and Mellillo J. M. (1989) Nitrogen saturation in northern forest ecosystems. *Bioscience* **39**, 378–386.

Aber J., McDowell W., Nade lhoffer K., Magill A., Berntson G., Kamakea M., McNulty S., Currie W., Rustad L., and Fernandez I. (1998) Nitrogen saturation in temperate forests ecosystems. *Bioscience* **48**, 921–934.

Adams M. B., Angrandi T. R., and Kochenderfer J. N. (1997) Stream water and soil solution responses to 5 years of nitrogen and sulfur additions at the fernow experimental forest, West Virginia. *Forest Ecol. Manage.* **95**, 79–91.

Aherne J., Kelly-Quinn M., and Farrell E. P. (2002) A survey of lakes in the Republic of Ireland: hydrochemical characteristics and acid sensitivity. *Ambio* **31**, 452–459.

Alewell C. (2001) Predicting reversibility of acidification: the european sulfur story. *Water Air Soil Pollut.* **130**, 1271–1276.

Alewell C., Armbruster M., Bittersohl J., Evans C. D., Meesenberg H., Moritz K., and Prechtel A. (2001) Are there signs of acidification reversal in freshwaters of the low mountain ranges in Germany? *Hydrol. Earth Syst. Sci.* **5**, 367–378.

Alfaro-De la Torre M. C. and Tessier A. (2002) Cadmium deposition and mobility in the sediments of an acidic oligotrophic lake. *Geochim. Cosmochim. Acta* **66**, 3549–3562.

Andersson F. and Olsson B. (eds.) (1985) Lake Gårdsjön—an acid forest lake and its catchment. *Swedish Nat. Sci. Res. Council: Ecol. Bull.* **37**, 336p.

Anesio A. M. and Granéli W. (2003) Increased photoreactivity of DOC by acidification: implications for the carbon cycle in humic lakes. *Limnol. Oceanogr.* **48**, 735–744.

Bacastow R. B. (1976) Modulation of atmospheric carbon dioxide by the southern oscillation. *Nature* **261**, 116–118.

Bailey S. W., Buso D. C., and Likens G. E. (2003) Implications of sodium mass balance for interpreting the calcium cycle of a northern hardwood ecosystem. *Ecology* (in press).

Bain D. C. and Langan S. J. (1995) Weathering rates in catchments calculated by different methods and their relationship to acidic inputs. *Water Air Soil Pollut.* **85**, 1051–1056.

Baker J. P., VanSickle J., Gagen C. J., DeWalle D. R., Sharpe W. E., Carline R. F., Baldigo B. P., Murdoch P. S., Bath D. W., Kretser W. A., Simonin H. A., and Wigington P. J. (1996) Episodic acidification of small streams in the Northeastern US: effects on fish populations. *Ecol. Appl.* **6**, 422–437.

Baldigo B. P. and Murdoch P. S. (1997) Effect of stream acidification and inorganic aluminum on mortality of brook trout (*Salvelinus fontinalis*) in the Catskill mountains, New York. *Can. J. Fish Aquat. Sci.* **54**, 603–615.

Battarbee R. W., Mason J., Renberg I., and Talling J. F. (eds.) (1990) *Palaeolimnology and Lake Acidification.* The Royal Society London, 327.

Bayley S. E., Parker B., Vitt D., and Rosenburg D. (1988) *Experimental Acidification of a Freshwater Wetland.* Wildlife Toxicology Fund, Final Report.

Beier C., Hansen K., and Gundersen P. (1993) Spatial variability of throughfall fluxes in a spruce forest. *Environ. Pollut.* **81**, 257–267.

Berggren D. and Mulder J. (1995) The role of organic matter in controlling aluminum solubility in acid mineral soils. *Geochim. Cosmochim. Acta* **59**, 4167–4180.

Bindler R., Brännvall M.-L., and Renberg I. (1999) Natural lead concentrations in pristine boreal forest soils and past pollution trends: a reference for critical load models. *Environ. Sci. Technol.* **33**, 3362–3367.

Blum J. D., Klaue A., Nezat C. A., Driscoll C. T., Johnson C. E., Siccama T. G., Eagars C., Fahey T., and Likens G. E. (2002) Mycorrhizal weathering of apatite as an important calcium source in base-poor forest ecosystems. *Nature* **417**, 729–731.

Borg H. (1986) Metal speciation in acidified mountain streams in central Sweden. *Water Air Soil Pollut.* **30**, 1007–1014.

Bormann F. H. and Likens G. E. (1979) *Pattern and Process in a Forested Ecosystem.* Springer, New York.

Bouchard A. (1997) Recent lake acidification and recovery trends in southern Quebec, Canada. *Water Air Soil Pollut.* **94**, 225–245.

Boutron C. F., Candelone J.-P., and Hong S. (1995) Greenland snow and ice cores: unique archives of large-scale pollution of the troposphere of the northern hemisphere by lead and other heavy metals. *Sci. Tot. Environ.* **160/161**, 233–241.

Brakke D. F., Landers D. H., and Eilers J. M. (1988) Chemical and physical characteristics of lakes in the Northeastern US. *Environ. Sci. Technol.* **22**, 155–163.

Brimblecombe P. (1992) History of atmospheric acidity. In *Atmospheric Acidity* (eds. M. Radojevic and R. M. Harrison). Elsevier, London, pp. 267–304.

Bullen T., Wiegand B., Chadwick O., Vitousek P., Bailey S., and Creed I. (2002) New approaches to understanding calcium budgets in watersheds: Sr isotopes meet Ca isotopes. In *BIOGEOMON Int. Conf. Reading, England*. University of Reading, England, 18p.

Campbell P. G. C., Hansen H. J., Dubreuil B., and Nelson W. O. (1992) Geochemistry of Quebec north shore salmon rivers during snowmelt: organic acid pulse and aluminum mobilization. *Can. J. Fish Aquat. Sci.* **49**, 1938–1952.

Camuffo D. (1992) Acid rain and the deterioration of monuments: how old is the problem? *Atmos. Environ.* **26B**, 241–247.

Charles D. F. and Smol J. P. (1988) New methods for using diatoms and chrysophytes to infer past pH of low alkalinity lakes. *Limnol. Oceanogr.* **33**, 1451–1462.

Charlson R. J., Lovelock J. E., Andreae M. O., and Warren S. G. (1987) Oceanic phytoplankton, atmospheric sulfur, cloud albedo and climate. *Nature* **326**, 655–661.

Christophersen N. and Neal C. (1990) Linking hydrochemical, geochemical and soil chemical processes on the catchment scale: an interplay between modeling and field work. *Water Resour. Res.* **26**, 3077–3086.

Clair T. A. and Ehrman J. M. (1996) Variations in discharge and dissolved organic carbon and nitrogen export from terrestrial basins with changes in climate: a neural network approach. *Limnol. Oceanogr.* **41**, 921–927.

Cook R. B., Kelly C. A., Schindler D. W., and Turner M. A. (1986) Mechanisms of hydrogen ion neutralization in an experimentally acidified lake. *Limnol. Oceanogr.* **31**, 134–148.

Correll D. L., Jordan T. E., and Weller D. E. (2001) Effects of precipitation, air temperature, and land use on organic carbon discharges from Rhode river watersheds. *Water Air Soil Pollut.* **128**, 139–159.

Cosby B. J., Hornberger G. M., Galloway J. N., and Wright R. F. (1985) Modelling the effects of acid deposition: assessment of a lumped-parameter model of soil water and stream water chemistry. *Water Resour. Res.* **21**, 51–63.

Cosby B. J., Hornberger G. M., Wright R. F., and Galloway J. N. (1986) Modeling the effects of acid deposition: controls of long-term sulfate dynamics by soil sulfate adsorption. *Water Resour. Res.* **22**, 1283–1302.

Cosby B. J., Ferrier R. C., Jenkins A., and Wright R. F. (2001) Modelling the effects of acid deposition: refinements, adjustments and inclusion of nitrogen dynamics in the MAGIC model. *Hydrol. Earth Syst. Sci.* **5**, 499–517.

Cowling E. B. (1982) Acid precipitation in historical perspective. *Environ. Sci. Technol.* **16**, 110A–123A.

Crocker R. L. and Major J. (1955) Soil development in relation to vegetation and surface age at Glacier bay, Alaska. *J. Ecol.* **43**, 427–448.

Cronan C. S. and Grigal D. F. (1995) Use of calcium/aluminum ratios as indicators of stress in forest ecosystems. *J. Environ. Qual.* **24**, 209–226.

Cronan C. S. and Schofield C. L. (1979) Aluminum leaching response to acidic precipitation effects on high-elevation watersheds in the Northeast US. *Science* **204**, 304–306.

Cronan C. S., Walker W. J., and Bloom P. R. (1986) Predicting aqueous aluminum concentrations in natural waters. *Nature* **324**, 140–143.

Dahl K. (1921) Research on the die-off of brown trout in mountain lakes in southwestern Norway. *Norsk Jæger-og Fiskeforenings Tidsskrift* **49**, 249–267 (in Norwegian).

Dahlgren R. A., McAvoy D. C., and Driscoll C. T. (1990) Acidification and recovery of a spodisol Bs horizon from acidic deposition. *Environ. Sci. Technol.* **24**, 531–537.

Dai A. G. and Wigley T. M. L. (2000) Global patterns of ENSO-induced precipitation. *Geophys. Res. Lett.* **27**, 1283–1286.

David M. B., Fasth W. J., and Vance G. F. (1991a) Forest soil response to acid and salt additions of sulfate: I. Sulfur constitutents and net retention. *Soil Sci.* **151**, 136–145.

David M. B., Vance G. F., and Fasth W. J. (1991b) Forest soil response to acid and salt additions of sulfate: II. Aluminum and base cations. *Soil Sci.* **151**, 208–219.

Davies T. D., Tranter M., Wigington P. J., and Eshleman K. N. (1992) Acidic episodes in surface waters in Europe. *J. Hydrol.* **132**, 25–69.

Davis R. B. (1987) Paleolimnological diatom studies of acidification of lakes by acid rain: an application of Quaternary science. *Quat. Sci. Rev.* **6**, 147–163.

Davison W., George D. G., and Edwards N. J. A. (1995) Controlled reversal of lake acidification by treatment with phosphate fertilizer. *Nature* **377**, 504–507.

De Vries W., Leeters E. E. J. M., and Hendriks C. M. A. (1995) Effects of acid deposition on Dutch forest ecosystems. *Water Air Soil Pollut.* **85**, 1063–1068.

De Vries W., Posch M., and Kämäri J. (1989) Simulation of the long-term soil response to acid deposition in various buffer ranges. *Water Air Soil Pollut.* **48**, 349–390.

Dickson W. (1978) Some effects of the acidification of Swedish lakes. *Verh. Int. Ver. Limnol.* **20**, 851–856.

Dillon P. J., Yan N. D., Scheider W. A., and Conroy N. (1979) Acidic lakes in Ontario, Canada: characterization, extent and responses to base and nutrient additions. *Arch. Hydrobiol. Beih.* **13**, 317–336.

Dillon P. J., Molot L. A., and Futter M. (1997) The effect of El Niño-related drought on the recovery of acidified lakes. *Environ. Monitor Assess.* **46**, 105–111.

Dise N. B. and Wright R. F. (1992) The NITREX Project. Ecosystem Research Report #2, Commission of the European Communities, Brussels, 101p.

Dise N. B. and Wright R. F. (1995) Nitrogen leaching from European forests in relation to nitrogen deposition. *Forest Ecol. Manage.* **71**, 153–162.

Dixit S. S., Smol J. P., Kingston J. C., and Charles D. F. (1992) Diatoms: powerful indicators of environmental change. *Environ. Sci. Technol.* **26**, 23–33.

Driscoll C. T. and Postek K. M. (1996) The chemistry of Aluminum in Surface Waters. In *The Environmental Chemistry of Aluminum* (ed. G. Sposito). CRC Press, Boca Raton, pp. 363–418.

Driscoll C. T., Yatsko C. P., and Unangst F. J. (1988) Longitudinal and temporal trends in the water chemistry of the north branch of the Moose river. *Biogeochemistry* **3**, 37–61.

Driscoll C. T., Likens G. E., Hedin L. O., Eaton J. S., and Bormann F. H. (1989) Changes in the chemistry of surface waters: 25-year results at the hubbard brook experimental forest. *Environ. Sci. Technol.* **23**, 137–143.

Driscoll C. T., Lehtinen M. D., and Sullivan T. J. (1994) Modeling the acid-base chemistry of organic solutes in Adirondack, New York, lakes. *Water Resour. Res.* **30**, 297–306.

Driscoll C. T., Likens G. E., and Church M. R. (1998) Recovery of surface waters in the Northeastern US from decreases in atmospheric deposition of sulfur. *Water Air Soil Pollut.* **105**, 319–329.

Ducros M. (1845) Observation d' une pluie acide. *J. Pharm. Chim.* **3**, 273–277.

Edwards P. J., Wood F., and Kochenderfer J. N. (2002) Baseflow and peakflow chemical responses to experimental applications of ammonium sulfate to a forested watershed in north-central West Virginia, USA. *Hydrol. Proc.* **16**, 2287–2310.

Einsele W. (1936) Über die Beziehungen des Eisenkreislaufs zum Phosphatkreislauf im eutrophen See. *Arch. Hydrobiol.* **29**, 664–686.

Emmett B. A., Reynolds B., Stevens P. A., Norris D. A., Hughes S., Gorres J., and Lubrecht I. (1993) Nitrate leaching from afforested Welsh catchment—interactions between stand age and nitrogen deposition. *Ambio* **22**, 386–394.

Engstrom D. R. and Wright H. E., Jr. (1984) Chemical stratigraphy of lake sediments as a record of environmental change. In *Lake Sediments and Environmental History* (eds. E. Y. Haworth and J. W. G. Lund). University of Minnesota Press, Minneapolis Minn., pp. 11–67.

Engstrom D. R., Fritz S. C., Almendinger J. E., and Juggins S. (2000) Chemical and biological trends during lake evolution in recently deglaciated terrain. *Nature* **408**, 161–166.

Evans C. D. and Monteith D. T. (2001) Chemical trends at lakes and streams in the UK acid waters monitoring network, 1988–2000: evidence for recent recovery at a national scale. *Hydrol. Earth Syst. Sci.* **5**, 351–366.

Evans C. D., Harriman R., Monteith D. T., and Jenkins A. (2001a) Assessing the suitability of acid neutralising capacity as a measure of long-term trends in acid waters based on two parallel datasets. *Water Air Soil Pollut.* **130**, 1541–1546.

Evans C. D., Cullen J. M., Alewell C., Kopàček J., Marchetto A., Moldan F., Prechtel A., Rogora M., Veselý J., and Wright R. (2001b) Recovery from acidification in European surface waters. *Hydrol. Earth Syst. Sci.* **5**, 283–297.

Evans C. D., Monteith D. T., and Harriman R. (2001c) Long-term variability in the deposition of marine ions at west coast sites in the UK acid waters monitoring network: impacts on surface water chemistry and significance for trend determination. *Sci. Tot. Environ.* **265**, 115–129.

Feller M. C. and Kimmins J. P. (1979) Chemical characteristics of small streams near haney in southwestern British Columbia. *Water Resour. Res.* **15**, 247–258.

Fenn M. E., Poth M. A., Aber J. D., Baron J. S., Bormann B. T., Johnson D. W., Lemly A. D., McNulty S. G., Ryan D. E., and Stottlemyer R. (1998) Nitrogen excess in north American ecosystems: predisposing factors, ecosystem responses, and management strategies. *Ecol. Appl.* **8**, 706–733.

Fernandez I. J., Simmons J. A., and Briggs R. D. (2000) Indices of forest floor nitrogen status along a climate gradient in maine, USA. *Forest Ecol. Manage.* **134**, 77–187.

Fernandez I. J., Rustad L. E., Norton S. A., Kahl J. S., and Cosby B. J. (2003) Experimental acidification causes soil base cation depletion at the Bear Brook Watershed in Maine. *Soil Sci. Soc. J.* (in press).

Ferrier R. C., Jenkins A., and Elston D. A. (1994) The composition of rime ice as an indicator of the quality of winter deposition. *Environ. Pollut.* **87**, 259–266.

Ferrier R. C., Helliwell R. C., Cosby B. J., Jenkins A., and Wright R. F. (2001) Recovery from acidification of lochs in Galloway, Southwest Scotland, UK: 1979–1998. *Hydrol. Earth Syst. Sci.* **5**, 421–432.

Fölster J. and Wilander A. (2002) Recovery from acidification in Swedish forest streams. *Environ. Pollut.* **117**, 379–389.

Forsius M., Kämäri J., Kortelainen P., Mannio J., Verta M., and Kinnunen K. (1990) Statistical lake survey in Finland: regional estimates of lake acidification. In *Acidification in Finland* (eds. P. Kauppi, P. Anttila, and K. Kenttämies). Springer, Berlin, pp. 759–780.

Forsius M. C., Neal C., and Jenkins A. (1995) Modeling perspective of the deforestation impact in stream water quality of small preserved forested areas in the Amazonian rainforest. *Water Air Soil Pollut.* **79**, 325–337.

Forsius M., Alveteg M., Jenkins A., Johanson M., Kleemola S., Lükewille A., Posch M., Sverdrup H., and Walse C. (1997) MAGIC, SAFE, and SMART model applications at integrated monitoring sites: effects of emission reduction scenarios. *Water Air Soil Pollut.* **105**, 21–30.

Forti M. C., Carvalho A., Melfi A. J., and Montes C. R. (2001) Deposition patterns of SO_4, NO_3 and H^+ in the Brazilian territory. *Water Air Soil Pollut.* **130**, 1121–1126.

Frogner T. (1990) The effect of acid deposition on cation fluxes in artificially acidified catchments in Western Norway. *Geochim. Cosmochim. Acta* **54**, 769–780.

Galloway J. N. (1995) Acid deposition: perspectives in time and space. *Water Air Soil Pollut.* **85**, 15–24.

Galloway J. N., Norton S. A., and Church M. R. (1983) Freshwater acidification from atmospheric deposition of sulfuric acid: a conceptual model. *Environ. Sci. Technol.* **17**, 541–545.

Gbondo-Tugbawa S. S., Driscoll C. T., Mitchell M. J., Aber J. D., and Likens G. E. (2002) A model to simulate the response of a northern hardwood forest ecosystem to change in S deposition. *Ecol. Appl.* **12**, 8–23.

Gherini S. A., Mok L., Hudson R. J. M., Davis G. F., Chen C. W., and Goldstein R. A. (1985) The ILWAS model, formulation and application. *Water Air Soil Pollut.* **26**, 425–459.

Gjessing E. T. (1992) The HUMEX project: experimental acidification of a catchment and its humic lake. *Environ. Int.* **18**, 535–543.

Gjessing E. T., Riise G., and Lydersen E. (1998) Acid rain and natural organic matter (NOM). *Acta Hydrochim. Hydrobiol.* **26**, 131–136.

Goodale C. L. and Aber J. D. (2001) The long-term effects of land-use history on nitrogen cycling in northern hardwood forests. *Ecol. Appl.* **11**, 253–267.

Gorham E. (1955) On the acidity and salinity of rain. *Geochim Cosmochim. Acta* **7**, 231–239.

Gorham E., Eisenriech S. J., Ford J., and Santelman M. V. (1985) The chemistry of bog waters. In *Chemical Processes in Lakes* (ed. W. Stumm). Wiley, New York, pp. 339–362.

Gorham E., Janssens J. A., Wheeler G. A., and Glaser P. H. (1987) The natural and anthropogenic acidification of peatlands. In *Effects of Atmospheric Pollutants on Forests, Wetlands, and Agricultural Ecosystems* (eds. T. C. Hutchinson and K. M. Memma). Springer, Berlin, pp. 493–512.

Groffman P. M., Driscoll C. T., Fahey T. J., Hardy J. P., Fitzhugh R. D., and Tierney G. L. (2001) Effects of mild winter freezing on soil nitrogen and carbon dynamics in a northern hardwood forest. *Biogeochemistry* **56**, 191–213.

Gundersen P., Callesen I., and De Vries W. (1998) Nitrate leaching in forest ecosystems is related to forest floor C/N ratio. *Environ. Pollut.* **102**, 403–407.

Gunn J. M. (ed.) (1995) *Restoration and Recovery of an Industrial Region*. Springer, NY, 358p.

Haines T. A., Akielaszek J. J., Norton S. A., and Davis R. B. (1983) Errors in pH measurement with colorimetric indicators in low alkalinity waters. *Hydrobiologia* **107**, 57–61.

Hall R., Likens G. E., Fiance S. B., and Hendrey G. R. (1980) Experimental acidification of a stream in the hubbard brook experimental forest, New Hampshire. *Ecology* **61**, 976–989.

Hansen K., Draaijers G. P. J., Ivens W. P. M. F., Gundersen P., and Leeuwen N. F. M. (1994) Concentration variations in rain and throughfall collected sequentially during individual rain events. *Atmos. Environ.* **28**, 3195–3205.

Harriman R., Anderson H., and Miller J. D. (1995) The role of sea-salts in enhancing and mitigating surface water acidity. *Water Air Soil Pollut.* **85**, 553–558.

Harriman R., Watt A. W., Christie A. E. G., Collen P., Moore D. W., McCartney A. G., Taylor E. M., and Watson J. (2001)

Interpretation of trends in acidic deposition and surface water chemistry in Scotland during the past three decades. *Hydrol. Earth Syst. Sci.* **5**, 407–420.

Hauhs M. (1989) Lange Bramke: an ecosystem study of a forested catchment. In *Acid Precipitation: 1. Case Studies* (eds. D. C. Adriano and M. Havas). Springer, New York, pp. 275–304.

Hemond H. F. (1990) Acid neutralizing capacity, alkalinity, and acid-base status of natural waters containing organic acids. *Environ. Sci. Technol.* **24**, 1486–1489.

Henriksen A. (1980) Acidification of freshwaters—a large scale titration. *Ecological Impact of Acid Precipitation.* SNSF Project, Oslo, 68–74.

Henriksen A. and Kirkhusmo L. A. (2000) Effects of clear-cutting of forest on the chemistry of a shallow groundwater aquifer in southern Norway. *Hydrol. Earth Syst. Sci.* **4**, 323–331.

Henriksen A. and Posch M. (2001) Steady-state models for calculating critical loads of acidity for surface waters. *Water Air Soil Pollut. Focus* **1**, 375–398.

Henriksen A., Lien L., Traaen T. S., Sevalrud I. S., and Brakke D. F. (1988a) Lake acidification in Norway—present and predicted chemical status. *Ambio* **17**, 259–266.

Henriksen A., Wathne B. M., Røgeberg E. J. S., Norton S. A., and Brakke D. F. (1988b) The role of stream substrates in aluminum mobility and acid neutralization. *Water Res.* **22**, 1069–1073.

Henriksen A., Posch M., Hultberg H., and Lien L. (1995) Critical loads of acidity for surface waters—can the ANC_{limit} be considered variable? *Water Air Soil Pollut.* **85**, 2419–2424.

Henriksen A., Skjelkvåle B. L., Mannio J., Wilander A., Harriman R., Curtis C., Jensen J. P., Fjeld E., and Moiseenko T. (1998) Northern european lake survey—1995. Finland, Norway, Sweden, Denmark, Russian Kola, Russian Karelia, Scotland and Wales. *Ambio* **27**, 80–91.

Hessen D. O., Henriksen A., and Smelhus A. M. (1997) Seasonal fluctuations and diurnal oscillations in nitrate of a heathland brook. *Water Res.* **31**, 1813–1817.

Hindar A. and Lydersen E. (1994) Extreme acidification of a lake in southern Norway caused by weathering of sulfide-containing bedrock. *Water Air Soil Pollut.* **77**, 17–25.

Hindar A., Henriksen A., Torseth K., and Semb A. (1994) Acid water and fish death. *Nature* **372**, 327–328.

Holdren G. R. and Adams J. E. (1982) Parabolic dissolution kinetics of silicate minerals: an artifact of nonequlibrium precipitation processes? *Geology* **10**, 186–190.

Houle D. and Carignan R. (1995) Role of SO_4 adsorption and desorption in the long-term S-budget of a coniferous catchment on the Canadian shield. *Biogeochemistry* **28**, 161–182.

Hruška J., Köhler S., and Bishop K. (1999) Buffering processes in a boreal dissolved organic carbon-rich stream during experimental acidification. *Environ. Pollut.* **106**, 55–65.

Hruška J., Laudon H., Johnson C. E., Köhler S., and Bishop K. (2001) Acid/base character of organic acids in boreal streams during snowmelt. *Water Resour. Res.* **37**, 1013–1026.

Hultberg H. and Grennfelt P. (1992) Sulfur and sea-salt deposition as reflected by throughfall and runoff chemistry in forested catchments. *Environ. Pollut.* **75**, 215–222.

Hultberg H. and Skeffington R. (1999) Experimental reversal of Acid rain effects. In *The Gårdsjøn Roof Project*. Wiley, New York, 484p.

Hurrell J. W. (1995) Decadal trend in the North Atlantic Oscillation: regional temperatures and precipitation. *Science* **269**, 676–679.

Jeffries D. S. (1990) Buffering of pH by sediments in streams and lakes. In *Soils, Aquatic Processes, and Lake Acidification, 4, Advances in Environment Science, Acidic Precipitation* (eds. S. A. Norton, S. E. Lindberg, and A. L. Page). Springer, New York, pp. 107–132.

Jeffries D. S. (1997) *Canadian Acid Rain Assessment: 3. Aquatic Effects*. Ministry of the Environment, Burlington, Canada.

Jeffries D. S., Wales D. L., Kelso J. R. M., and Linthurst R. A. (1986) Regional chemical characteristics of lakes in North America. Part 1. Eastern Canada. *Water Air Soil Pollut.* **31**, 551–569.

Jeffries D., Clair T. A., Dillon P. J., Papineau M., and Stainton M. P. (1995) Trends in surface water acidification at ecological monitoring sites in Southeastern Canada. *Water Air Soil Pollut.* **85**, 577–582.

Jenkins A., Cosby B. J., Ferrier R. C., Walker T. A. B., and Miller J. D. (1990) Modelling stream acidification in afforested catchments—an assessment of the relative effects of acid deposition and afforestation. *J. Hydrol.* **120**, 163–181.

Jenkins A., Ferrier R. C., and Kirby C. (eds.) (1995) *Ecosystem Manipulation Experiments: Scientific Approaches, Experimental Design and Relevant Results*. Ecosystems Research Report 20, Commission of the European Communities, Brussels.

Jenkins A., Ferrier R. C., and Wright R. F. (eds.) (2001) Assessment of recovery of European surface waters from acidification 1970–2000. *Hydrol. Earth Syst. Sci.* **5**, 273–542.

Johannessen M. and Henriksen A. (1978) Chemistry of snowmelt water: changes in concentration during melting. *Water Resour. Res.* **14**, 615–619.

Johansson M. and Tarvainen T. (1997) Estimation of weathering rates for critical load calculations in Finland. *Environ. Geol.* **29**, 158–164.

Johnson D. W., Henderson G. S., Huff D. D., Lindberg S. E., Richter D. D., Shriner D. S., Todd D. E., and Turner J. (1982) Cycling of organic and inorganic sulfur in a chestnut oak forest. *Oecologia* **54**, 141–148.

Johnson D. W., Henderson G. S., and Todd D. E. (1988) Changes in nutrient distribution in forests and soils of walker branch watershed, Tennessee, over an eleven-year period. *Biogeochemistry* **5**, 275–293.

Johnson D. W., Susfalk R. B., Gholz H. L., and Hanson P. J. (2000) Simulated effects of temperature and precipitation change in several forest ecosystems. *J. Hydrol.* **235**, 183–204.

Jongmans A. G., Van Breemen N., Lundstrøm U., Van Hees P. A. W., Finlay R. D., Srinivasan M., Unestam T., Giesler R., Melkerud P. A., and Olsson M. (1997) Rock-eating fungi. *Nature* **389**, 682–683.

Jones P. D., Osborn T., and Briffa K. R. (2001) The evolution of climate over the last millenium. *Science* **292**, 662–667.

Kahl J., Norton S., Fernandez I., Haines T., Rustad L., Nodvin S., Scofield J., Strickland T., Erickson H., Wigington P., and Lee J. (1999) Nitrogen and sulfur input-output budgets in the experimental and reference watersheds, bear brook watershed in Maine (BBWM). *Environ. Monitor Assess.* **55**, 113–131.

Kalbitz K., Solinger J.-H., Park B., Michalzik B., and Matzner E. (2000) Control on the dynamics of dissolved organic matter: a review. *Soil Sci.* **165**, 277–304.

Kaste O., Henriksen A., and Posch A. (2002) Present and potential nitrogen outputs from Norwegian soft water lakes—an assessment made by applying steady-state first-order acidity balance (FAB) model. *Hydrol. Earth Syst. Sci.* **6**, 101–112.

Kirchner J. W. (1992) Heterogeneous geochemistry of catchment acidification. *Geochim. Cosmochim. Acta* **56**, 2311–2327.

Kirchner J. W., Feng X., and Neal C. (2000) Fractal stream chemistry and its implications for contaminant transport in catchments. *Nature* **403**, 524–527.

Klimont Z., Cofala J., Schopp W., Amann M., Streets D. G., Ichikawa Y., and Fujita S. (2001) Projections of SO_2, NO_x,

NH_3 and VOC emissions in East Asia. *Water Air Soil Pollut.* **130**, 193–198.

Köhler S. J., Hruška J., and Bishop K. (1999) Influence of organic acids site density on pH modelling of Swedish lakes. *Can. J. Fish Aquat. Sci.* **56**, 1461–1470.

Kopàček J., Veselý J., and Stuchlík E. (2001a) Sulphur and nitrogen fluxes and budgets in the bohemian forest and tatra mountains during the industrial revolution (1850–2000). *Hydrol. Earth Syst. Sci.* **5**, 391–405.

Kopàček J., Ulrich K., Hejzlar J., Borovec J., and Stuchlík E. (2001b) Natural inactivation of phosphorus by aluminum in atmospherically acidified water bodies. *Water Res.* **35**, 3783–3790.

Kopàček J., Kana J., Šantrucková H., Porcal P., Hejzlar J., Picek T., Šimek M., and Veselý J. (2002a) Physical, chemical, and biochemical characteristics of soils in watersheds of the Bohemian Forest Lakes: II. Certovo and Cerné Lake. *Silva Gabreta* **8**, 67–93.

Kopàček J., Stuchlík E., Veselý J., Schaumburg J., Anderson I. C., Fott J., Hejzlar J., and Vrba J. (2002b) Hysteresis in reversal of central European mountain lakes from atmospheric acidification. *Water Air Soil Pollut. Focus* **2**, 91–114.

Kortelainen P. and Mannio J. (1990) Organic acidity in Finnish lakes. In *Acidification in Finland* (eds. P. Kauppi, P. Anttilla, and K. Kenttämies). Springer, Berlin, pp. 849–864.

Kortelainen P., David M. B., Roila T., and Makinen I. (1992) Acid–base characteristics of organic-carbon in the HUMEX Lake Skjervatjern. *Environ. Int.* **18**, 621–629.

Krám P., Hruška J., Wenner B. S., Driscoll C. T., and Johnson C. E. (1997) The biogeochemistry of base cations in two forest catchments with contrasting lithology in the Czech Republic. *Biogeochemistry* **37**, 173–202.

Krám P., Santore R. C., Driscoll C. T., Aber J. D., and HruŤka J. (1999) Application of the forest-soil-water model (PnET-BGC/CHESS) to the Lysina catchment, Czech Republic. *Ecol. Model.* **120**, 9–30.

Kuo S. and Lotse E.G. (1974) Kinetics of phosphate adsorption and desorption by hematite and gibbsite. *Soil. Sci.* **116**, 400–406.

Kuylenstierna J. C. I. and Chadwick M. J. (1989) The relative sensitivity of ecosystems in Europe to the indirect effects of acidic depositions. In *Regional Acidification Models* (eds. J. Kämäri, D. F. Brakke, A. Jenkins, and R. F. Wright). Springer, Berlin, pp. 3–22.

Lange H., Hauhs M., and Schmidt S. (1995) Long-term sulfate dynamics at Lange Bramke (Harz) used for testing two acidification models. *Water Air Soil Pollut.* **79**, 339–351.

Larssen T., Xiong J. L., Vogt R. D., Seip H. M., Liao B. H., and Yhao D. W. (1998) Studies of soils, soil water and stream water at a small catchment near Guiyang, China. *Water Air Soil Pollut.* **101**, 137–162.

Lawrence G. B., Fuller R. D., and Driscoll C. T. (1987) Release of aluminum following whole-tree harvesting at the hubbard brook experimental forest, New Hampshire, USA. *J. Environ. Qual.* **16**, 383–390.

Lawrence G. B., David M. B., Lovett G. M., Murdoch P. S., Burns D. A., Stoddard J. L., Baldigo B. P., Porter J. H., and Thompson A. W. (1999) Soil calcium status and the response of stream chemistry to changing acidic deposition rates. *Ecol. Appl.* **9**, 1059–1072.

LaZerte B. D. and Findeis J. (1995) The relative importance of oxalate and pyrophosphate extractable aluminum to the acidic leaching of aluminum in podzol B horizons from the Precambrium Shield, Ontario, Canada. *Can. J. Soil. Sci.* **75**, 43–54.

Likens G. E., Bormann F. H., Pierce R. S., Eaton J. S., and Johnson N. M. (1977) *Biogeochemistry of a Forested Ecosystem.* Springer, New York.

Likens G. E., Driscoll C. T., Buso D. C., Siccama T. G., Johnson C. E., Lovett G. M., Ryan D. F., Fahey T., and Reiners W. A. (1994) The biogeochemistry of potassium at Hubbard Brook. *Biogeochemistry* **25**, 61–125.

Likens G. E., Driscoll C. T., and Buso D. C. (1996) Long-term effects of acid rain: response and recovery of a forest ecosystem. *Science* **272**, 244–246.

Lorz C., Armbruster M., and Feger K. H. (2002) Temporal development of water chemistry in three forested watersheds in Germany as influenced by contrasting deposition regimes—results from long-term monitoring and model applications. In *BIOGEOMON Book of Abstracts*, University of Reading, UK, 141p.

Lydersen E., Salbu B., Polèo A. B. S., and Muniz I. P. (1990) The influence of temperature on aqueous aluminum chemistry. *Water Air Soil Pollut.* **51**, 203–215.

MacDonald J. A., Dise N. B., Matzner E., Armbruster M., Gundersen P., and Forsius M. (2002) Nitrogen input together with ecosystem nitrogen enrichment predict nitrate leaching from European forests. *Global Change Biol.* **8**, 1028–1033.

Magill A. H., Aber J. D., Hendricks J. J., Bowden R. D., Melillo J. M., and Steudler P. A. (1997) Biogeochemical response of forest ecosystems to simulated chronic nitrogen deposition. *Ecol. Appl.* **7**, 402–415.

Majer V., Cosby B. J., Kopáček J., and Veselý J. (xxxx) Predicting reversibility of central european mountain ecosystems from acidification with the MAGIC model: Part I. The Bohemian Forest. *Hydrol. Earth Syst. Sci.* (in review).

Mannio J. (2001) *Responses of Headwater Lakes to Air Pollution Changes in Finland.* PhD Thesis, University of Helsinki.

Meixner T., Bales R. C., Williams M. W., Campbell D. H., and Baron J. S. (2000) Stream chemistry modeling of two watersheds in the front range, Colorado. *Water Resour. Res.* **36**, 77–87.

Matschullat J., Andreae H., Lessman D., Malessa V., and Siewers U. (1992) Catchment acidification—from the top down. *Environ. Pollut.* **77**, 143–150.

Michalzik B. and Matzner E. (1999) Dynamics of dissolved organic nitrogen and carbon in a central european Norway spruce ecosystems. *Euro. J. Soil Sci.* **50**, 579–590.

Miller E. K., Blum J. D., and Friedland A. J. (1993) Determination of soil exchangeable-cation loss and weathering rates using Sr isotopes. *Nature* **362**, 438–441.

Mitchell M. J., Driscoll C. T., Kahl J. S., Likens G. E., Murdoch P. S., and Pardo L. H. (1996) Climatic control of nitrate loss from forested watersheds in Northeast US. *Environ. Sci. Technol.* **30**, 2609–2612.

Moldan B. and Schnoor J. L. (1992) Czechoslovakia-examining a critically ill environment. *Environ. Sci. Technol.* **26**, 14–21.

Mol-Dijkstra J. P. and Kros H. (2001) Modelling effects of acid deposition and climate change on soil and run-off chemistry at Risdalsheia, Norway. *Hydrol. Earth Syst. Sci.* **5**, 499–517.

Monteith D. T., Evans C. D., and Reynolds B. (2000) Are temporal variations in the nitrate content of UK upland freshwaters linked to the North Atlantic Oscillation? *Hydrol. Process.* **14**, 1745–1749.

Mosello R., Marchetto A., Brizzio M. C., Rogora M., and Tartari G. A. (2000) Results from the Italian participation in the international co-operative programme on assessment and monitoring of acidification of rivers and lakes (ICP Waters). *J. Limnol.* **59**, 47–54.

Mulder J. and Stein A. (1994) The solubility of aluminum in acidic forest soils: long-term changes due to acid deposition. *Geochim. Cosmochim. Acta* **58**, 85–94.

Murdoch P. S., Burns D. A., and Lawrence G. B. (1998) Relation of climate change to the acidification of surface waters by nitrogen deposition. *Environ. Sci. Technol.* **32**, 1642–1647.

National Atmospheric Deposition Program (NRSP-3)/National Trends Network (2002) NADP Program Office, Illinois State Water Survey, 2204 Griffith Dr., Champaign, IL 61820, www.nadp.sws.uuc.edu

Neal C. (1997) Water quality of the plynlimon catchments. *Hydrol. Earth Syst. Sci.* **1**, 381–763.

Neal C., Fisher R., Smith C. J., Hill S., Neal M., Conway T., Ryland G. P., and Jefferey H. A. (1992) The effects of tree harvesting on stream-water quality at an acidic and acid-sensitive spruce forested area: plynlimon, mid-wales. *J. Hydrol.* **135**, 305–319.

Neal C., Wilkinson J., Neal M., Harrow M., Wickham H., Hill L., and Morfitt C. (1997) The hydrochemistry of the headwaters of the River Severn, Plynlimon. *Hydrol. Earth Syst. Sci.* **1**, 583–617.

Nilsson J. and Greenfelt P. (1988) *Critical Loads for Sulphur and Nitrogen*. Nordic Council of Ministers and the United Nations Economic Commission for Europe (ECE) Nord, 15, Gotab Pub., Stockholm, 418p.

Nilsson S. I., Miller H. G., and Miller J. D. (1982) Forest growth as a possible cause of soil and water acidification: an examination of the concepts. *Oikos* **39**, 40–49.

Norton S. A. (1989) Watershed acidification—a chromatographic process. In *Regional Acidification Models* (eds. J. Kamari, D. F. Brakke, A. Jenkins, S. A. Norton, and R. F. Wright). Springer, Berlin, pp. 89–101.

Norton S. A. and Fernandez I. J. (eds.) (1999) *The Bear Brook Watershed in Maine: A Paired Watershed Experiment—the First Decade (1987–1997)*. Kluwer, Dordrecht, 250p.

Norton S. A. and Kahl J. S. (2000) Impacts of marine aerosols on surface water chemistry at bear brook watershed, Maine. *Verh. In. Ver. Limnol.* **27**, 1280–1284.

Norton S. A., Kahl J. S., Henriksen A., and Wright R. F. (1990) Buffering of pH depressions by sediments in streams and lakes. In *Soils, Aquatic Processes, and Lake Acidification, Acidic Precipitation* (eds. S. A. Norton, S. E. Lindberg, and A. L. Page). Springer, New York, vol. 4, pp. 133–157.

Norton S. A., Bienert R. W., Binford M. W., and Kahl J. S. (1992a) Stratigraphy of total metals in PIRLA sediment cores. *J. Paleolimnol.* **7**, 191–214.

Norton S. A., Brownlee J. C., and Kahl J. S. (1992b) Artificial acidification of a non-acidic and an acidic headwater stream in Maine, USA. *Environ. Pollut.* **77**, 123–128.

Norton S. A., Kahl J. S., Scofield J. P., and Fernandez I. J. (1995) Altered soil-soil water interactions at an artificially acidified watershed at bear brook watershedat bear brook watershed, Maine. In *Ecosystem Manipulation Experiments: Scientific Approaches, Experimental Design, and Relevant Results*, Ecosystem Research Report No. 20 (eds. A. Jenkins, R. C. Ferrier, and C. Kirby). Commission of the European Communities, Brussels, pp. 227–235.

Norton S., Kahl J., Fernandez I., Haines T., Rustad L., Nodvin S., Scofield J., Strickland T., Erickson H., Wigington P., and Lee J. (1999) The bear brook watershed, Maine (BBWM) USA. *Environ. Monitor Assess.* **55**, 7–51.

Norton S. A., Wagai R., Navrátil T., Kaste J. M., and Rissberger F. A. (2000) Response of a first-order stream in Maine to short-term in-stream acidification. *Hydrol. Earth Syst. Sci.* **4**, 383–391.

Norton S. A., Cosby B. J., Fernandez I. J., Kahl J. S., and Church M. R. (2001) Long-term and seasonal variations in CO_2: linkages to catchment alkalinity generation. *Hydrol. Earth Syst. Sci.* **5**, 83–91.

Norton S. A., Fernandez I. J., Kahl J. S., and Reinhardt R. L. (2003) Acidification trends and the evolution of neutralization mechanisms through time at the bear brook watershed in Maine (BBWM), USA. *Water Air Soil Pollut.* (in press).

Novák M., Kirchner J. W., Groscheová H., Havel M., Cernú J., Krejcí R., and Buzek F. (2000) Sulphur isotope dynamics in two central european watersheds affected by high atmospheric deposition of SO_x. *Geochim. Cosmochim. Acta* **64**, 367–383.

Novák M., Buzek F., Harrison A. F., Prechová E., Jacková I., and Fottová D. (2003) Similarity between C, N and S stable isotope profiles in european spruce forest soils: implications for the use of $\delta^{34}S$ as a tracer. *Appl. Geochem.* **18**, 765–779.

Odén S. (1968) *The Acidification of the Atmosphere and Precipitation and its Consequences in the Natural Environment*. SNSR, Stockholm.

Oliver B. G. and Kelso J. R. M. (1983) A role for sediments in retarding the acidification of headwater lakes. *Water Air Soil Pollut.* **20**, 379–389.

Oliver B. G., Thurman E. M., and Malcolm R. L. (1983) The contribution of humic substances to the acidity of colored natural waters. *Geochim. Cosmochim. Acta* **47**, 2031–2035.

Pačes T. (1983) Rate constants of dissolution derived from the measurements of mass balance in hydrochemical catchments. *Geochim. Cosmochim. Acta* **47**, 1855–1863.

Perakis S. S. and Hedin L. O. (2002) Nitrogen loss from unpolluted South American forests mainly via dissolved organic compounds. *Nature* **415**, 416–419.

Porcella D. B., Schofield C. L., DePinto J. V., Driscoll C. T., Bukaveckas P. A., Gloss S. P., and Young T. C. (1989) Mitigation of acidic conditions in lakes and streams. In *Acidic Precipitation, Soils, Aquatic Processes, and Lake Acidification* (eds. S. A. Norton, S. E. Lindberg, and A. L. Page). Springer, New York, vol. 4, pp. 159–186.

Postek K. M., Driscoll C. T., Kahl J. S., and Norton S. A. (1996) Changes in the concentrations and speciation of aluminum in response to an experimental addition of ammonium sulfate to the bear brook watershed, Maine USA. *Water Air Soil Pollut.* **85**, 1733–1738.

Prechtel A., Alewell C., Armbruster M., Bittersohl J., Cullen J. M., Evans C. D., Helliwell R. C., Kopáček J., Marchetto A., Matzner E., Messenburg H., Moldan F., Moritz K., Veselý J., and Wright R. F. (2001) Response of sulphur dynamics in european catchments to decreasing sulphate deposition. *Hydrol. Earth Syst. Sci.* **5**, 311–325.

Prescott C. E., Kishchuk B. E., and Weetman G. F. (1995) Long term effects of repeated N fertilization and straw application in a jack pine forest: 3. Nitrogen availability in the forest floor. *Can. J. Forest Res.* **25**, 1991–1996.

Psenner R. and Schmidt R. (1992) Climate driven pH control of remote alpine lakes and effects of acid deposition. *Nature* **356**, 781–783.

Pyle D. M., Battie P. D., and Bluth G. J. S. (1996) Sulphur emissions to the stratosphere from explosive volcanic eruptions. *Bull. Volcanol.* **57**, 663–671.

Pugh A. L., Norton S. A., Schauffler M., Jacobson G. L., Jr, Kahl J. S., Brutsaert W. F., and Mason C. F. (1996) Interactions between peat and salt-contaminated runoff in alton bog, Maine USA. *J. Hydrol.* **182**, 83–104.

Rasmussen L., Brydges T., and Mathy P. (1993) *Experimental Manipulations of Biota and Biogeochemical Cycling in Ecosystems: Approach, Methodologies, Findings*. Ecosystem Research Report #4, Commission of the European Communities, Brussels.

Reinhardt R. L., Norton S. A., Handley M., and Amirbahman A. (2003) Mobilization of and linkages among P, Al, and Fe during high discharge episodic acidification at the Bear Brook Watershed in Maine, USA. *Water Air Soil Pollut.* (in press).

Renberg I. (1990) A 12,600 year perspective of the acidification of Lilla Öresjön, southwest Sweden. *Phil. Trans. Roy. Soc. London* **237B**, 357–361.

Renberg I., Brännvall M.-L., Bindler R., and Emteryd O. (2000) Atmospheric lead pollution history during four millenia (2000 BC to 2000 AD) in Sweden. *Ambio* **29**, 150–156.

Reuss J. O. and Johnson D. W. (1986) *Acid Deposition and Acidification of Soils and Waters*. Ecological Studies. Springer, NY, vol. 59.

Reuss J. O., Cosby B. J., and Wright R. F. (1987) Chemical processes governing soil and water acidification. *Nature* **329**, 27–32.

Reynolds B., Emmett B. A., and Woods C. (1992) Variations in streamwater nitrate concentrations and nitrogen budgets over 10 years in a headwater catchments in Mid-Wales. *J. Hydrol.* **136**, 150–155.

Reynolds B., Stevens P. A., Hughes S., Parkinson J. A., and Weatherley N. S. (1995) Stream chemistry impacts of conifer harvesting in Welsh catchments. *Water Air Soil Pollut.* **79**, 147–170.

Rosseland B. O., Blakar I. A., Bulger A., Kroglund F., Kvellestad A., Lydersen E., Oughton D. H., Salbu B., Staurnes M., and Vogt R. D. (1992) The mixing zone between limed and acidic river waters: complex aluminium chemistry and extreme toxicity for salmonids. *Environ. Pollut.* **78**, 3–8.

Roy S. J., Norton S. A., and Kahl J. S. (1999) Phosphorous dynamics at bear brooks, Maine USA. *Environ. Monitor Assess.* **55**, 133–147.

Rustad L. E., Kahl J. S., Norton S. A., and Fernandez I. J. (1994) Multi-year estimates of dry deposition at the bear brook watershed in Eastern Maine, USA. *J. Hydrol.* **162**, 319–336.

Ryan D. F. and Kahler D. M. (1987) Geochemical and mineralogical indications of pH in lakes and soils in central new hamphire in the early Holocene. *Limnol. Oceanogr.* **32**, 751–757.

Schecher W. D. and Driscoll C. T. (1987) An evaluation of uncertainty associated with the aluminum equilibrium calculations. *Water Resour. Res.* **23**, 525–534.

Schiff S. L. and Anderson R. F. (1986) Alkalinity production in epilimnetic sediments: acidic and non-acidic lakes. *Water Air Soil Pollut.* **31**, 941–948.

Schindler D. W. (1986) The significance of in-lake production of alkalinity. *Water Air Soil Pollut.* **239**, 149–157.

Schindler D. W. (1997) Widespread effects of climatic warming on freshwater ecosystems in North America. *Hydrol. Process.* **11**, 1043–1067.

Schindler D. W., Frost T. M., Mills K. H., Chang P. S. S., Davies I. J., Findlay L., Malley D. F., Shearer J. A., Turner M. A., Garrison P. J., Watras C. J., Webster K., Gunn J. M., Brezonik P. L., and Swenson W. A. (1990) Comparisons between experimentally-acidified and atmospherically-acidified lakes during stress and recovery. *Proc. Roy. Soc. Edinburgh, Sec. B-Biol. Sci.* **97**, 193–226.

Schindler D. W., Curtis P. J., Parker B. R., and Stainton M. P. (1996) Consequences of climate warming and lake acidification for UV-B penetration in North American boreal lakes. *Nature* **379**, 705–708.

Schneider S. (1989) The changing climate. *Sci. Am.* **261**, 274.

Schnoor J. L. (1990) Kinetics of chemical weathering: a comparison of laboratory and field weathering rates. In *Aquatic Chemical Kinetics* (ed. W. Stumm). Wiley, New York, pp. 475–504.

Schofield C. L. and Trojnar J. R. (1980) *Polluted Rain*. Plenum, NY, 347p.

Shotyk W., Cheburkin A. K., Appleby P. G., Fankhauser A., and Kramers J. D. (1996) Two thousand years of atmospheric arsenic, antimony, and lead deposition recorded in an ombrotrophic peat bog profile, Jura Mountains, Switzerland. *Earth Planet. Sci. Lett.* **145**, E1–E7.

Skartveit A. (1981) Relationships between precipitation chemistry, hydrology, and runoff acidity. *Nordic Hydrol.* **12**, 55–60.

Skjelkvåle B. L. and Ulstein M. (2002) Proceedings from the workshop on heavy metals (Pb, Cd, and Hg) in surface waters: monitoring and biological impact. *Norwegian Inst. Water Research Report*, pp. 21–22.

Skjelkvåle B. L., Wright R. F., and Henriksen A. (1998) Norwegian lakes show widespread recovery from acidification: results of national surveys of lakewater chemistry 1986–1997. *Hydrol. Earth Syst. Sci.* **2**, 555–562.

Skjelkvåle B. L., Stoddard J. L., and Anderson T. (2001a) Trends in surface water acidification in Europe and North America. *Water Air Soil Pollut.* **130**, 787–792.

Skjelkvåle B. L., Mannio J., Wilander A., and Andersen T. (2001b) Recovery from acidification of lakes in Finland, Norway and Sweden 1990–1999. *Hydrol. Earth Syst. Sci.* **5**, 327–337.

Smith R. A. (1872) *Air and Rain: The Beginnings of Chemical Climatolgy*. Longmans, Green, London.

Steinberg C. and Kühnel W. (1987) Influence of cation acids on dissolved humic substances under acidified conditions. *Water Res.* **21**, 95–98.

Steinnes E. (1995) A critical evaluation of the use of naturally growing moss to monitor the deposition of atmospheric metals. *Sci. Tot. Environ.* **160/161**, 243–249.

Stevens P. A., Harrison A. F., Jones H. E., Williams T. G., and Hughes S. (1993) Nitrate leaching from a Sitka spruce plantation and the effect of fertilisation with phosphorus and potassium. *Forest Ecol. Manage.* **58**, 233–247.

Stoddard J. L. (1994) Long-term changes in watershed retention of nitrogen: its causes and aquatic consequences. In *Environmental Chemistry of Lakes and Reservoirs*, ACS Advances in Chemistry No. 237 (ed. L. A. Baker). American Chemical Society, Washington, pp. 223–284.

Stoddard J. L., Driscoll C. T., Kahl J. S., and Kellogg J. H. (1998) A regional analysis of lake acidification trends for the Northeastern US, 1982–1994. *Environ. Monitor Assess.* **51**, 399–413.

Stoddard J. L., Jeffries D. S., Luke wille A., Clair T. A., Dillon P. J., Driscoll C. T., Forsius M., Johannessen M., Kahl J. S., Kellogg J. H., Kemp A., Mannio J., Montieth D. T., Murdoch P. S., Patrick S., Rebsdorf A., Skjelkvåle B. L., Stainton M. P., Traem T., van Dam H., Webster K. E., Wietine J., and Wilander A. (1999) Regional trends in aquatic recovery from acidification in North America and Europe. *Nature* **401**, 575–578.

Stoddard J. L., Traaen T. S., and Skjelkvåle B. L. (2001) Assessment of nitrogen leaching ICP-Waters sites (Europe and North America). *Water Air Soil Pollut* **130**, 781–786.

Sverdrup H. (1990) *The Kinetics of Base Cation Release Due to Chemical Weathering*. Lund University Press, Lund, Sweden, 246p.

Sverdrup H. and Warfringe P. (1993) Calculating field weathering rates using a mechanistic geochemical model PROFILE. *Appl. Geochem.* **8**, 273–283.

Sverdrup H. and De Vries W. (1994) Calculating critical loads for acidity with the simple mass balance method. *Water Air Soil Pollut.* **72**, 143–162.

Swedish Environmental Protection Board (1986) *Acid and Acidified Waters*. Swedish Environ. Prot. Board, Solna.

Swoboda-Colberg N. G. and Drever J. I. (1993) Mineral dissolution rates in plot scale field and laboratory experiments. *Chem. Geol.* **105**, 51–69.

Taylor A. and Blum J. D. (1995) Relation between soil age and silicate weathering rates determined from the chemical evolution of a glacial chronosequence. *Geology* **23**, 979–982.

Tipping E. and Hopwood J. (1988) Estimating streamwater concentrations of aluminum released from streambeds during 'acid episodes'. *Environ. Tech. Lett.* **9**, 703–712.

Tranter M., Davies T. D., Wigington P. J., and Eshleman K. N. (1994) Episodic acidification of fresh-water systems in Canada—physical and geochemical processes. *Water Air Soil Pollut.* **72**, 19–39.

Ulrich B. (1983) Soil acidity and its relation to acid deposition. In *Effects of Accumulation of Air Pollutants in Forest Ecosystems* (eds. B. Ulrich and J. Pankrath). D. Reidel, Dordrecht, pp. 127–146.

UN-ECE. (1999) The 1999 Gothenburg Protocol to Abate Acidification, Eutrophication and Ground-level Ozone. http://www.unece.org/env/lrtap.

Velbel M. A. (1985) Geochemical mass balance and weathering rates in forested watersheds of the southern blue ridge. *Am. J. Sci.* **285**, 904–930.

Veselý J. and Majer V. (1996) The effect of pH and atmospheric deposition on concentrations of trace elements in acidified freshwaters: a statistical approach. *Water Air Soil Pollut.* **88**, 227–246.

Veselý J. and Majer V. (1998) Hydrogeochemical mapping of Czech freshwaters. *Bull. Czech Geol. Surv.* **73**, 183–192.

Veselý J., Šulcek Z., and Majer V. (1985) Acid–base changes in streams and their effect on the contents of some heavy metals in stream sediment. *Bull. Czech Geol. Surv.* **60**, 9–23.

Veselý J., Hruška J., Norton S. A., and Johnson C. E. (1998a) Trends in the chemistry of acidified Bohemian lakes from 1984–1995: I. Major solutes. *Water Air Soil Pollut.* **108**, 107–127.

Veselý J., Hruška J., and Norton S. A. (1998b) Trends in water chemistry of acidified bohemian lakes from 1984–1995: II. Trace elements and aluminum. *Water Air Soil Pollut.* **108**, 425–443.

Veselý J., Majer V., and Norton S. A. (2002a) Heterogeneous response of central European streams to decreased acidic atmospheric deposition. *Environ. Pollut.* **120**, 275–281.

Veselý J., Norton S. A., Majer V., and Kopàček J. (2002b) Lake sediment evidence of air pollution from pre-historic copper and bronze production. In *BIOGEOMON Book of Abstracts*. University of Reading, UK, 245p.

Veselý J., Norton S. A., Skrivan P., Majer V., Krám P., Navrátil T., and Kaste J. M. (2002c) Environmental Chemistry of Beryllium. In *Beryllium—Mineralogy, Petrology and Geochemistry*, Rev. Min. Geochem. (ed. E. S. Grew). Mineralogical Society of America, Washington, vol. 50, pp. 291–317.

Veselý J., Majer V., Kopàček J., and Norton, S.A. (2003) Increased temperature decreases aluminum concentrations in Central European lakes recovering from acidification. *Limnol.Oceanogr.* (in press).

Wangwongwatava S. (2001) Step-by-step approach to establish acid deposition monitoring network in East Asia (EANET): Thailand's experiences. *Water Air Soil Pollut.* **130**, 151–162.

Warfvinge P., Falkengren-Grerup U., and Sverdrup H. (1993) Modelling long-term base cation supply in acidified forest stands. *Environ. Pollut.* **80**, 1–14.

Waters D. and Jenkins A. (1992) Impacts of afforestation on water quality trends in two catchments in Mid-Wales. *Environ. Pollut.* **77**, 167–172.

Weatherley N. S., Rutt G. P., Thomas S. P., and Ormerod S. J. (1991) Liming acid stream: aluminium toxicity to fish in mixing zones. *Water Air Soil Pollut.* **55**, 345–353.

Webb J. R., Cosby B. J., Galloway J. N., and Hornberger G. M. (1989) Acidification of native brook trout streams in Virginia. *Water Resour. Res.* **25**, 1367–1377.

Webster K. E. and Brezonik P. L. (1995) Climate confounds detection of chemical trends to acid deposition in upper midwest lakes in the USA. *Water Air Soil Pollut.* **85**, 1575–1580.

Webster K. E., Kratz T. K., Bowser C. J., Magnuson J. J., and Rose W. J. (1996) The influence of landscape position on lake chemical responses to drought in Northern Wisconsin. *Limnol. Oceanogr.* **41**, 977–984.

Webster K. E., Soranno P. A., Baines S. B., Kratz T. K., Bowser C. J., Dillon P. J., Campbell P., Fee E. J., and Hecky R. E. (2000) Structuring features of lake districts: landscape controls on lake chemical responses to drought. *Freshwater Biol.* **43**, 499–515.

White A. F. and Blum A. E. (1995) Effects of climate on chemical weathering in watersheds. *Geochim. Cosmochim. Acta* **59**, 1729–1747.

White A. F. and Brantley S. L. (eds.) (1995) In *Chemical Weathering Rates of Silicate Minerals*. Reviews in Mineralogy, Volume 31, Mineralogical Society of America, Washington, DC, 583p.

Whiting M. C., Whitehead D. R., Holmes R. W., and Norton S. A. (1989) Paleolimnological reconstruction of recent acidity changes in four Sierra Nevada lakes. *J. Paleolimnol.* **2**, 284–304.

Whitehead D. R., Charles D. F., Jackson S. T., Reed S. E., and Sheehan M. C. (1986) Late glacial and Holocene acidity changes in Adirondack (NY) lakes. In *Diatoms and Lake Acidity* (ed. J. P. Smol, R. W. Battarbee, R. B. Davis, and J. Meriläinen). Junk, Dordrecht, pp. 251–274.

Whitehead D. R., Charles D. F., Jackson S. T., Smol J. P., and Engstrom D. R. (1989) The developmental history of Adirondack (NY) lakes. *J. Paleolimnol.* **2**, 185–206.

Wigington P. J., Davies T. D., Tranter M., and Eshleman K. N. (1990) Episodic acidification of surface waters due to acidic deposition. US Natl. Acid Precip. Assess. Prog. State of Science Report 12, Washington DC, 200p.

Wigington P. J., DeWalle D. R., Nurdoch P. S., Kretser W. A., Simonin H. A., VanSickle J., and Baker J. P. (1996) Episodic acidification of small streams in the Northeastern US: ionic controls of episodes. *Ecol. Appl.* **6**, 389–407.

Wiklander L. (1975) The role of neutral salts in the ion exchange between acid precipitation and soil. *Geoderma* **14**, 93–105.

Williams P. M. (1968) Organic and inorganic constituents of the Amazon River. *Nature* **216**, 937–938.

Williams M. W., Baron J. S., Caine N., Sommerfeld R., and Sanford R. (1996) Nitrogen saturation in the rocky mountains. *Environ. Sci. Technol.* **30**, 640–646.

Wright R. F. (1985) Chemistry of lake hovvatn, Norway, following liming and reacidification. *Can. J. Fish Aquat. Sci.* **42**, 1103–1113.

Wright R. F. (1998) Effect of increased carbon dioxide and temperature on runoff chemistry at a forested catchment in southern Norway. *Ecosystems* **1**(2), 216–225.

Wright R. F. and Rasmussen L. (eds.) (1998) The whole ecosystem experiments of the NITREX and EXMAN projects. *Forest Ecol. Manage.* **101**, 1–363.

Wright R. F. and Tietema A. (eds.) (1995) Nitrex. *Forest Ecol. Manage.* **71**, 1–169.

Wright R. F., Norton S. A., Brakke D. F., and Frogner T. (1988) Acidification of stream water by whole-catchment experimental addition of dilute seawater. *Nature* **334**, 422–424.

Wright R. F., Lotse E., and Semb A. (1993) RAIN project: results after 8 years of experimentally reduced acid deposition to a whole catchment. *Can. J. Fish Aquat. Sci.* **50**, 258–268.

Wright R. F., Alewell C., Cullen J. M., Evans C. D., Marchetto A., Moldan F., Prechtel A., and Rogora M. (2001) Trends in nitrogen deposition and leaching in acid-sensitive streams in Europe. *Hydrol. Earth Syst. Sci.* **5**, 299–310.

Yanai R. D., Siccama T. G., Arthur M. A., Federer C. A., and Friedland A. J. (1999) Accumulation and depletion of base cations in forest floors in the Northeastern US. *Ecology* **80**, 2774–2787.

Yoh M. (2001) Soil C/N ratio as affected by climate: an ecological factor of forest NO_3 leaching. *Water Air Soil Pollut.* **130**, 661–666.

Zilberbrand M. (1999) On equlibrium constants for aqueous geochemical reactions in water unsaturated soils and sediments. *Aquat. Geochem.* **5**, 195–206.

9.11
Tropospheric Ozone and Photochemical Smog

S. Sillman
University of Michigan, Ann Arbor, MI, USA

9.11.1 INTRODUCTION

The question of air quality in polluted regions represents one of the issues of geochemistry with direct implications for human well-being. Human health and well-being, along with the well-being of plants, animals, and agricultural crops, are dependent on the quality of air we breathe. Since the start of the industrial era, air quality has become a matter of major importance, especially in large cities or urbanized regions with heavy automobile traffic and industrial activity.

Concern over air quality existed as far back as the 1600s. Originally, polluted air in cities resulted from the burning of wood or coal, largely as a source of heat. The industrial revolution in England saw a great increase in the use of coal in rapidly growing cities, both for industrial use and domestic heating. London suffered from devastating pollution events during the late 1800s and early 1900s, with thousands of excess deaths attributed to air pollution (Brimblecombe, 1987). With increasing use of coal, other instances also occurred in continental Europe and the USA. These events were caused by directly emitted pollutants (primary pollutants), including sulfur dioxide (SO_2), carbon monoxide (CO), and particulates. They were especially acute in cities with northerly locations during fall and winter when sunlight is at a minimum. These original pollution events gave rise to the term "smog" (a combination of smoke and fog). Events of this type have become much less severe since the 1950s in Western Europe and the US, as natural gas replaced coal as the primary source of home heating, industrial smokestacks were designed to emit at higher altitudes (where dispersion is more rapid), and industries were required to install pollution control equipment.

Beginning in the 1950s, a new type of pollution, photochemical smog, became a major concern. Photochemical smog consists of ozone (O_3) and other closely related species ("secondary pollutants") that are produced photochemically from directly emitted species, in a process that is driven by sunlight and is accelerated by warm temperatures. This smog is largely the product of gasoline-powered engines (especially automobiles), although coal-fired industry can also generate photochemical smog. The process of photochemical smog formation was first identified by Haagen-Smit and Fox (1954) in association with Los Angeles, a city whose geography makes it particularly susceptible to this type of smog formation. Sulfate aerosols and organic particulates are often produced concurrently with ozone, giving rise to a characteristic milky-white haze associated with this type of air pollution.

Today ozone and particulates are recognized as the air pollutants that are most likely to affect human health adversely. In the United States, most major metropolitan areas have periodic air pollution events with ozone in excess of government health standards. Violations of local health standards also occur in major cities in Canada and in much of Europe. Other cities around the world (especially Mexico City) also experience very high ozone levels. In addition to urban-scale events, elevated ozone occurs in region-wide events in the eastern USA and in Western Europe, with excess ozone extending over areas of 1,000 km^2 or more. Ozone plumes of similar extent are found in the tropics (especially in Central Africa) at times of high biomass burning (e.g., Jenkins *et al.*, 1997; Chatfield *et al.*, 1998). In some cases ozone associated with biomass burning has been identified at distances up to 10^4 km from its sources (Schultz *et al.*, 1999).

Ozone also has a significant impact on the global troposphere, and ozone chemistry is a major component of global tropospheric chemistry. Global background ozone concentrations are much lower than urban or regional concentrations during pollution events, but there is evidence that the global background has increased as a result of human activities (e.g., Wang and Jacob, 1998; Volz and Kley, 1988). A rise in global background ozone can make the effects of local pollution events everywhere more acute, and can also cause ecological damage in remote locations that are otherwise unaffected by urban pollution. Ozone at the global scale is also related to greenhouse warming.

This chapter provides an overview of photochemical smog at the urban and regional scale, focused primarily on ozone and including a summary of information about particulates. It includes the following topics: dynamics and extent of pollution events; health and ecological impacts; relation between ozone and precursor emissions, including hydrocarbons and nitrogen oxides (NO_x); sources, composition, and fundamental properties of particulates; chemistry of ozone and related species; methods of interpretation based on ambient measurements; and the connection between air pollution events and the chemistry of the global troposphere. Because there are many similarities between the photochemistry of ozone during pollution events and the chemistry of the troposphere in general, this chapter will include some information about global tropospheric chemistry and the links between urban-scale and global-scale events. Additional treatment of the global troposphere is found in Volume 4 of this work. The chemistry of ozone formation discussed here is also related to topics discussed in greater detail elsewhere in this volume (see Chapters 9.10 and 9.12) and in Volume 4.

9.11.2 GENERAL DESCRIPTION OF PHOTOCHEMICAL SMOG

9.11.2.1 Primary and Secondary Pollutants

The term "primary pollutants" refers to species whose main source in the atmosphere is direct emissions or introduction from outside (e.g., from soils). These species are contrasted with secondary pollutants, whose main source is photochemical production within the atmosphere.

The distinction between primary and secondary pollutants is conceptually useful, because primary and secondary species usually show distinctly different patterns of diurnal and seasonal variation in polluted regions of the atmosphere. The ambient concentrations of primary species are controlled largely by proximity to emission sources and rates of dispersion. The highest concentrations of these species tend to occur at nighttime or early morning and in winter in northerly locations, because atmospheric dispersion rates are slowest at these times. By contrast, high concentrations of secondary species such as ozone are often associated with atmospheric conditions that favor photochemical production. The highest concentrations of ozone usually occur during the afternoons and in summer (in mid-latitudes) or during the dry season (in the tropics). The highest concentrations of ozone and other secondary species also occur at locations significantly downwind of emission sources, rather than in immediate proximity to precursor emissions. These diurnal and seasonal cycles are discussed in detail in Section 9.11.4.

9.11.2.2 Ozone

Ozone occurs naturally in the troposphere, largely as a result of downward mixing from the stratosphere. This downward mixing includes both direct transport of ozone and transport of NO_x, which leads to photochemical formation of ozone in the troposphere. Ozone mixing ratios in the stratosphere (from ~20 km to ~60 km above ground level) are as high as 1.5×10^4 ppb. This is higher, by a factor of 100, than ozone concentrations at ground level, even in the most polluted regions. Approximately 95% of the Earth's ozone is located in the stratosphere. Ozone concentrations in the troposphere are much lower, and generally decrease from the top of the troposphere to ground level. The ozone that is transported downward from the stratosphere is removed through photochemical processes in the troposphere (which include both production and removal of ozone, but with removal rates exceeding production rates). Ozone is also removed through dry deposition at the Earth's surface. Removal of ozone in the troposphere happens on a timescale of approximately three months. In the absence of human activities, ozone concentrations would vary from 200 ppb in the upper troposphere to 10–20 ppb at ground level.

9.11.2.2.1 Urban ozone

Ozone is formed in polluted urban areas by phototochemical reactions involving two classes of precursors: hydrocarbons (or, more generally, volatile organic compounds or VOCs) and oxides of nitrogen (NO and NO_2). During a typical urban air pollution event peak ozone reaches a value of 120–180 ppb. Until recently, hour average ozone concentrations in excess of 125 ppb were counted as an air quality violation in the USA. During the 1970s and 1980s, such violations occurred as often as 180 days per year in the Los Angeles metropolitan area. Stringent control measures have succeeded in lowering the frequency and severity of these air quality violations, but Los Angeles still records violations on ~40 days per year. Houston also has frequent air quality violations. Mexico City's early 2000s ozone levels are comparable to Los Angeles' levels in the 1970s with ozone in excess of 125 ppb on ~180 days per year (e.g., see Sosa *et al.*, 2000). Events with ozone in excess of 200 ppb are quite rare and generally occur only in cities with the most severe ozone problems. Ozone as high as 490 ppb has been recorded in Los Angeles (NRC, 1991) and in Mexico City.

Most other major cities in the USA and in Europe also record events with ozone in excess of 125 ppb, but these occur only a few times per year. Severe air pollution events occur less frequently in these cities, because the meteorological conditions that favor rapid formation of ozone (high sunlight, warm temperatures, and low rates of dispersion) occur less frequently. Significant excess ozone is formed only when temperatures are above 20 °C, and smog events are usually associated with temperatures of 30 °C or higher. In the major cities of northeastern USA and northern Europe, ozone levels exceed 80 ppb on ~30–60 days per year. At other times, a combination of cool temperatures and/or clouds prevents ozone formation, regardless of the level of precursor emissions.

The most severe pollution events occur when a combination of light winds and suppressed vertical mixing prevents the dispersion of pollutants from an urban center. The process of ozone formation typically requires several hours and occurs only at times of bright sunlight and warm temperatures. For this reason, peak ozone values typically are found downwind of major cities rather than in the urban center. During severe events with light winds, high ozone concentrations are more likely to occur closer to the city center.

9.11.2.2.2 Regional pollution events and long-distance transport

Peak ozone in urban plumes is found most commonly 50–100 km downwind of the city center. Once formed, ozone in urban plumes has an effective lifetime of approximately three days. For this reason, urban plumes with high ozone concentrations can travel great distances. Transport of ozone can be even longer in the middle and upper troposphere, where the lifetime of ozone extends to three months.

Although peak ozone most commonly occurs 50–100 km downwind from urban centers, plumes with high ozone have frequently been observed at distances of 300 km or more from their source regions. Ozone in excess of 150 ppb has frequently been observed on Cape Cod (Massachusetts), attributed to emissions in the New York metropolitan area 400 km away. Similar excess ozone has been observed along the shores of Lake Michigan, apparently due to emissions in the Chicago area that have traveled 300 km across the lake. Ozone as high as 200 ppb has been observed in Acadia National Park, Maine, attributed to transport from Boston (300 km away) and New York (700 km away) (see Figure 1). The plume from the New York–Boston corridor has also been observed by aircraft over the North Atlantic Ocean at a distance of several hundred kilometers from the source region (Daum *et al.*, 1996). In Europe, plumes from cities in Spain have been predicted to travel over the Mediterranean Sea, again for distances several hundred kilometers from the source region (Millan *et al.*, 1997).

Well-defined plumes with excess ozone are also associated with large coal-fired power plants in the USA. The largest power plants can have rates of NO_x emissions that are comparable to the summed emission rate from a city as large as Washington DC. Because these power plants have relatively low emissions of CO or volatile organics, the rate of ozone formation is slower and peak ozone occurs further downwind (100–200 km). Ozone as high as 140 ppb has been observed in plumes from individual power plants (Miller *et al.*, 1978; White *et al.*, 1983; Gillani and Pleim, 1996; Ryerson *et al.*, 1998, 2001). These plumes have been observed by aircraft for distances up to 12 h downwind from the plume sources. It is more difficult to observe plumes at greater downwind

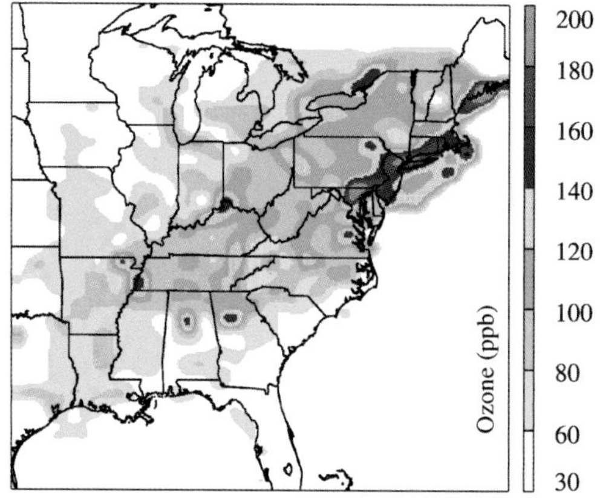

Figure 1 Peak ozone concentrations in the eastern United States during a severe air pollution event (June 15, 1988) based on surface observations at 350 EPA monitoring sites. The shadings represent values of 30–60 ppb (lightest shading) to 180–210 ppb (darkest shading) with 30 ppb intervals in between. Values reported for Canada and the Atlantic Ocean are inaccurate, since no observations were available for these locations (first printed in Sillman, 1993) (reproduced by permission of Annual Reviews from *Annual Reviews of Energy and the Environment*, **1993**, *18*, 31–56).

distances, because well-defined plumes are usually dispersed following overnight transport (Clarke and Ching, 1983).

In addition to transport in well-defined plumes, ozone in excess of 80 ppb is found to extend over broad regions (500 × 500 km or larger) during region-wide events. These events are associated with stagnant high-pressure systems that bring several consecutive days of high temperatures, sunlight, and suppressed atmospheric mixing to a polluted area. Region-wide events of this type have been observed frequently in the eastern USA and less frequently in Western Europe. Elevated ozone during these events affects rural areas as well as urban and suburban locations. These events are caused by the combined effect of emissions from large and small cities, industries and power plants rather than emissions from a single urban center, and can include ozone that has formed and accumulated over a period of several days. Emissions from cities within the affected region create plumes with additional excess ozone added to (and subsequently contributing to) the regional background.

An event of this sort is illustrated in Figure 1. During this event ozone in excess of 90 ppb was observed at every surface monitoring site (including both urban and rural sites) over an area extending from Ohio to Virginia and Maine, an area ~1,000 × 1,000 km. Higher ozone (150–200 ppb) was found throughout the cities of the northeast corridor (Washington, Philadelphia, New York, and Boston). The plume of high ozone in the northeast corridor extended to Maine. Locally high ozone was also found near several other cities.

There has been considerable speculation that the rapid worldwide growth of cities and industries will cause the scale of these region-wide events to increase and eventually to involve intercontinental transport. Ozone concentrations of 80 ppb have been observed at Sable Island, Nova Scotia, transported from source regions 2,000 km distant in the USA (e.g., Parrish *et al.*, 1993). Layers of elevated ozone aloft over both the South Atlantic Ocean and the western Pacific have been attributed to biomass burning in Africa (Jenkins *et al.*, 1997; Chatfield *et al.*, 1998; Schultz *et al.*, 1999). Model calculations have estimated that emissions from East Asia can cause a significant increase in background ozone levels in the western USA (Horowitz and Jacob, 1999; see also Lelieveld and Dentener, 2000). Measurements have identified possible transport from North America in air over Europe (Stohl and Trickl, 1999). Long-range transport of sulfate aerosols is also possible (Barth and Church, 1999). While such long-range transport of ozone is possible, it might best be viewed as part of the global tropospheric balance.

9.11.2.2.3 Ozone and the global troposphere

Surface ozone concentrations in the remote northern hemisphere range from 20 ppb to 40 ppb, with a seasonal maximum in May. Background ozone in the southern hemisphere is significantly lower (20–25 ppb). These background ozone concentrations are affected by global-scale photochemistry, which includes both production and destruction of ozone. It is believed that emissions resulting from human activities have increased the global background ozone, especially in the northern hemisphere (e.g., see Wang and Jacob, 1998; Lelieveld and Dentener, 2000; and Chapters 4.01 and 4.03).

Although the transport of ozone plumes from source regions provides dramatic evidence of the global impact of human activities, the chemical content of the global troposphere is more likely to be affected by photochemistry during average conditions rather than by episodic transport events. The rates of photochemical production and destruction of ozone in the free troposphere (defined as the region extending from the top of the planetary boundary layer, ~2–3 km above ground, to the top of the troposphere) is much larger in terms of total molecules produced than the rate of production in polluted source regions. Production rates in polluted regions are much higher on a per-volume basis, but the volume of the free troposphere is large enough, so that photochemical production there greatly exceeds the amount of ozone molecules produced in source regions.

NO_x, the critical precursor for ozone formation, typically has daytime concentrations of 5–20 ppb in urban areas, 0.5–1 ppb in polluted rural areas during region-wide events, and 10–100 ppt in the remote troposphere. An NO_x concentration of 1 ppb is associated with ozone formation at rates of 2–5 ppb h^{-1}, which is fast enough to allow ozone concentrations to increase to 90 ppb when air stagnates in a polluted region for two days or more. Ozone production rates as high as 100 ppb h^{-1} have been observed in urban locations (e.g., in the recently completed Texas Air Quality Study in Houston (Kleinman *et al.*, 2002)).

Production rates are much slower in the free troposphere, and loss usually exceeds production. However, NO_x concentrations of 100 ppt, which are much too small to allow the formation of episodic high ozone levels, would still allow ozone to remain at a steady-state concentration of ~80 ppb. As of early 2000s, the level of background ozone in the lower troposphere (20–40 ppb) is closely related to the photochemical steady state, achieved over several months, based on concentrations of NO_x and organics in the remote troposphere.

There is an obvious close relation between smog events in polluted regions and conditions in the global troposphere, because the global troposphere is strongly impacted by pollutants that are emitted primarily in cities or polluted regions. However, the relation between polluted regions and the global troposphere can often be counter-intuitive. In general, the rate of ozone formation per NO_x ("ozone production efficiency," discussed in more detail below) is higher when NO_x concentrations are lowest. Consequently, the global impact of emissions is actually higher when there is rapid dispersion of pollutants from a polluted region. The exported pollutants produce more total ozone (though with lower peak concentration) when they undergo photochemical processing in downwind rural areas or in the remote troposphere rather than in a polluted region during a stagnation event. Ozone precursors lead to ozone formation in the remote troposphere even when local conditions (e.g., clouds, low temperatures) prevent the formation of ozone in the polluted source region.

9.11.2.2.4 Ozone precursors: NO_x, CO, and volatile organics

Ozone in urban areas is produced from two major classes of precursors: NO_x, consisting of nitric oxide (NO) and nitrogen dioxide (NO_2), and VOCs). The ozone formation process is also closely associated with the OH radical (see below for a complete description). The process of ozone formation is initiated by the reaction of organics (usually primary hydrocarbons) with OH. The subsequent reaction sequence involves NO_x and results simultaneously in the production of ozone, oxidation of organics to CO_2, and oxidation of NO_x to nitric acid (HNO_3). In urban areas, the ozone formation process is also accompanied by the conversion of NO_x to organic nitrates such as peroxyacetyl nitrate ($CH_3CO_3NO_2$, often abbreviated as PAN), which have the effect of transporting NO_x to the remote troposphere.

In addition to their impact on ozone, NO_x and VOCs are associated with various other pollutants which impact human health and activities. Nitrogen dioxide (NO_2) causes impairment of lung functions and generally has the same level of toxicity as ozone, although ambient concentrations are usually much lower. NO_2 is produced in the atmosphere by chemical conversion from directly emitted nitric oxide (NO), although some NO_2 is also emitted directly into the atmosphere. NO_2 is usually grouped together with NO as NO_x because conversion from NO to NO_2 is rapid (with timescales of 5 min or less), and because the ambient mixing ratios of NO_2 show a pattern of behavior that resembles primary rather than secondary pollutants. Nitric acid contributes to acid rain and to the formation of particulates (see Section 9.11.2.3). Both primary and secondary VOCs include species that are directly toxic (see Chapters 9.12 and 9.13), and secondary organics produced from directly emitted VOCs are major components of particulates. Carbon monoxide is also a toxic gas, although ambient concentrations are rarely high enough to raise health concerns.

In the remote troposphere the ozone formation process is initiated primarily by the oxidation of CO and methane (CH_4) rather than volatile organics. Carbon monoxide and CH_4 are both long-lived species (two-month lifetime for CO, nine years for CH_4) and are widely distributed in the remote troposphere. They have less impact on urban photochemistry, because most of the CO emitted in urban areas is exported to the remote troposphere. Shorter-lived volatile organics (with lifetimes ranging from one hour to three days) are more important in terms of urban photochemistry, because they undergo reactions rapidly enough to contribute to ozone formation during local air pollution events. Alkenes, aromatics, and oxygenated organic species such as formaldehyde (HCHO) are especially important in terms of urban photochemistry, because they initiate reaction sequences that generate additional OH radicals (which catalyze further ozone production) in addition to producing ozone directly. This is discussed further in Chapter 9.12.

The relative impact of NO_x and VOCs on ozone formation during pollution events represents a major source of uncertainty. The chemistry of ozone formation is highly nonlinear, so that the exact relation between ozone and precursor emissions depends on the photochemical state of the system. Under some conditions, ozone is found to increase with increasing NO_x emissions and to remain virtually unaffected by changes in VOCs. Under other conditions, ozone increases rapidly with increased emission of VOCs and decreases with increasing NO_x. This split into NO_x-sensitive and NO_x-saturated (or VOC-sensitive) photochemical regimes is a central feature of ozone chemistry and a major source of uncertainty in formulating pollution control policy.

An analogous split between NO_x-sensitive and NO_x-saturated chemistry occurs in the remote troposphere, but the implications are somewhat different. Increased CO and VOCs always contribute to increased ozone in the remote troposphere, even under NO_x-sensitive conditions (Jaegle *et al.*, 1998, 2001), whereas ozone in polluted regions with NO_x-sensitive chemistry is largely insensitive to CO and VOCs.

NO_x emissions in polluted regions originate from two major sources: gasoline- and diesel-powered vehicles (primarily automobiles) and coal-fired power plants. Volatile organics are also generated largely by gasoline- and diesel-powered vehicles and by a variety of miscellaneous sources, all involving petroleum fuel or petroleum products. Coal-fired industry does not generate significant amounts of organics.

Although the question of NO_x- versus VOC-sensitivity and the related policy issue of NO_x versus VOC controls are complex and uncertain, there are a few concepts and trends that are useful for gaining a good general understanding of ozone generation. NO_x-sensitive conditions occur when there are excess VOCs and a high ratio of VOCs to NO_x, while VOC-sensitive conditions occur when there is excess NO_x and low VOCs/NO_x. The ratio of summed VOCs/NO_x ratio, weighted by the reactivity rate of each individual VOC, provides a good indicator for NO_x- versus VOC-sensitive chemistry.

Among freshly emitted pollutants, the initial rate of ozone formation is often controlled by the amount and chemical composition of VOCs. For this reason, ozone formation in urban centers is often (but not always) controlled by VOCs. As air moves downwind, ozone formation is increasingly controlled by NO_x rather than VOCs (Milford *et al.*, 1989). Ozone in far downwind and rural locations is often (but not always) controlled by upwind NO_x emissions (Roselle and Schere, 1995). Rural areas also tend to have NO_x-sensitive conditions due to the impact of biogenic VOCs (see Section 9.11.2.2.5). However, this description represents a general trend only and is not universally valid. NO_x-sensitive conditions can occur even in large urban centers, and VOC-sensitive conditions can occur even in aged plumes. For a more complete discussion, see NARSTO (2000) and Sillman (1999).

9.11.2.2.5 Impact of biogenics

In addition to anthropogenic sources, there are significant biogenic sources of organics. Isoprene (C_5H_8) is emitted by a variety of deciduous trees (especially oaks), and these emissions have a significant impact on ozone formation. Terpenes (e.g., α-pinene, $C_{10}H_{10}$) are emitted primarily by conifers, and are precursors of particulates (see Section 9.11.2.3). Emission of biogenic VOCs often equals or exceeds the rate of emission of anthropogenic VOCs at the regional scale, and even within urban areas biogenic VOCs can account for a significant percentage of total VOC reactivity. Biogenic VOCs are especially important because they have a relatively short lifetime (one hour or less) and consequently contribute to local ozone formation during pollution events.

The major significance of biogenic VOCs with regard to ozone lies in their implications for the effectiveness of control strategies and their impact on ozone–NO_x–VOC chemistry. The presence of biogenic VOCs in polluted regions effectively increases the VOCs/NO_x ratio, especially when ratios are weighted by the rate of reactivity. Consequently, regions with biogenic VOCs are more likely to have ozone formation that is sensitive to NO_x rather than to VOCs (Chameides *et al.*, 1988; Pierce *et al.*, 1998; Simpson, 1995).

Biogenic sources of NO_x are generally too small to contribute significantly to pollution events. Biogenic emissions represent ~5% of total NO_x emissions in the USA (compared to 50% of total VOCs) (Williams *et al.*, 1992). Biogenic NO_x emissions can be important in intensively farmed regions, where soil emission of NO_x is enhanced by heavy use of fertilizers.

9.11.2.3 Particulates

Technically, the term "aerosol" refers to a mixture of solid and liquid particles suspended in a gaseous medium, whereas the term "suspended particulates" refers to the suspended particles themselves. In practice, the terms "aerosols" and "particulates" are often used interchangeably. Particulates, or aerosols, have wide-ranging impacts on both human activities and environmental quality. First, aerosols have been identified as one of the major health hazards, affecting the respiratory system, associated with air pollution (along with ozone). Second, because degradation of visibility in polluted air is due almost entirely to aerosols (Seinfeld and Pandis, 1998), particulates are the most noticeable form of air pollution. Third, removal of acidic aerosols from the atmosphere, through deposition on soils and water surfaces or through rainout, can cause ecological damage. Acid rain (see Chapter 9.10) is the best-known example of this type of damage. Fourth, sulfate aerosols affect the global climate directly (by enhancing atmospheric reflectivity) and indirectly (by affecting the growth and reflectivity of clouds). This is believed to have a significant cooling effect on the atmosphere, although there is large uncertainty for assessing the impact of human activities on climate (Houghton *et al.*, 2001). Other aerosols (e.g., black carbon) rapidly absorb solar radiation and can possibly increase global warming (Jacobson, 2002; Chung and Seinfeld, 2002).

As was the case with ozone, aerosols also occur naturally in the atmosphere. Aerosols play an important role in the atmosphere's hydrologic cycle. Formation of cloud droplets occurs on hygroscopic aerosols, and nucleation of ice also needs a particle to initiate ice formation. Precipitation, which is enhanced by the presence of large aerosols or ice, strongly depends on these ice and cloud condensation nuclei. Most (though not all) of the damaging effects are due to anthropogenic rather than naturally occurring aerosols.

Aerosols are composed of a large variety of species, from both natural and anthropogenic material. Naturally occurring aerosols include sea salt, mineral dust, pollens and spores, organic aerosols derived from biogenic VOCs, and sulfate aerosols derived from reduced sulfur gases. Anthropogenic aerosols consist of soot (also known as black carbon), sulfate derived from sulfur dioxide emitted from coal-burning, organics derived from anthropogenic VOCs, and fly ash. Biomass burning (either naturally occurring or anthropogenic) also creates aerosols. Aerosols include both primary species and species produced by photochemical reactions.

Aerosols are generally divided into two groups, fine particles (with size below 2.5 μm) and larger coarse particles, because of their distinct impacts on human activities and environmental quality.

The fine and coarse particles differ from each other in terms of their origin, chemical composition, and their atmospheric effects. Coarse particles are usually the result of mechanical wear on preexisting solid substances: mineral dust (i.e., soil particles), sea salt, solid organic matter from plants, flakes from automobile tires and from buildings, etc. By contrast, fine particles are largely the result of chemistry—either combustion chemistry in fires, smokestacks, and internal combustion engines or photochemistry in the atmosphere, cloud droplets and water aerosols. Chemical and photochemical production followed by condensation (in combustion processes) or nucleation (in the atmosphere) results in the formation of solid particles of very small size (0.005–0.1 μm), a size range that is referred to as the *nuclei mode*. After formation, these particles lead to formation of larger particle sizes (0.1–1 μm) through coagulation, or through deposition of chemically formed particulate material on existing nuclei. These *accumulation mode* particles are the major cause of health and visibility effects associated with particulates. The process of coagulation does not lead to significant numbers of particles of a size greater than 2.5 μm, so that the fine particles (nuclei and accumulation mode) have distinctly different sources than the coarse particles.

Observed distributions of particle sizes (Figure 2) often show two separate peaks in particle mass, at 0.1–1 μm and at 3–20 μm, reflecting the different origins of coarse and fine particles.

Apart from their origin, there are other important differences between coarse and fine particulates. Fine particulates are usually acidic in nature and rapidly dissolve in water (and, in some cases, are formed through aqueous chemistry in cloud droplets or water aerosols).

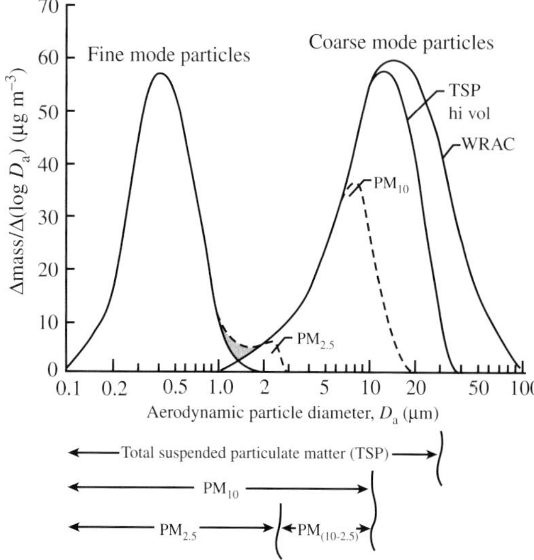

Figure 2 Representative example of a mass distribution of ambient particulate matter as a function of particle diameter. Mass distribution per particle size interval is shown as Δmass/Δ(log D_a) (in μg m^{-3}) plotted against particle size (D_a) in micrometers. The figure also shows the range of aerosol sizes included in various methods of aerosol measurement: wide range aerosol classifiers (WRAC), total suspended particulate (TSP) samplers, PM_{10} and $PM_{2.5}$ samplers (source Lippman and Schlesinger, 2000) (reproduced by permission of Annual Reviews from *Annual Review of Public Health* **2000**, *21*, 309–333).

Coarse particulates are usually nonacidic and hydrophobic. Coarse particulates are removed from the atmosphere through gravitational settling, on timescales of one day or less, depending on the size of the individual particles. Fine particulates are too small to be removed by gravitational settling. They are removed either by rainout (wet deposition) or by direct deposition to ground surfaces (dry deposition). The dry deposition rate for fine particulates is typically 0.1 cm s^{-1} (Seinfeld and Pandis, 1998), and allows these particulates to remain in an atmospheric boundary layer of typical depth for 10 days or more. Removal by wet deposition is often more rapid (2–5 days, Rasch *et al.*, 2000), although this depends on local climatology. The slow removal rate allows the concentration of fine particulates to build up during multiday pollution events. Fine particulates are often transported for distances of 300 km or more in the atmosphere, more readily than coarse particulates (although coarse mineral dust from desert regions can be transported over continental distances, e.g., Prospero, 1999). Fine particulates have a much larger impact on atmospheric visibility than coarse particulates. Most importantly, adverse health impacts are associated primarily with fine particulates (see Section 9.11.2.4).

The major sources of fine particles in the ambient atmosphere are:

- sulfates, which are produced photochemically from sulfur dioxide;
- organics, which are produced chemically from both anthropogenic and biogenic VOCs;
- black carbon (soot), which is emitted directly from anthropogenic industry and transportation; and
- nitrates, which are formed from NO_x.

Sulfates and organic compounds are typically the largest components of fine particulates in urban and industrialized regions. Sulfate aerosols have historically been the largest aerosol component, especially in regions with coal-fired industry. They are currently the dominant aerosol in the eastern and midwestern US, although their concentrations may be significantly reduced in the future by planned reductions of sulfur emissions. Nitrates form aerosols in combination with ammonia (NH_3), usually from agricultural sources. The resulting ammonium nitrate aerosol (NH_4NO_3) is often a significant component of particulates in locations that include both intensive agriculture and urban NO_x sources in close proximity (e.g., Los Angeles, Milan). Organic particulates from biomass burning (largely in the tropics) also contribute significantly to the total amount of particulates at the global scale. Trace metals (iron, lead, zinc, mercury, etc.) are also present in small quantities as aerosol components. These contribute little to the total particulate mass but are often of concern because they may individually be toxic (see Chapter 9.03).

Particulate concentrations in the atmosphere are frequently expressed in terms of total suspended particulates (TSP, as mass per unit volume) or as the summed mass of particulates with individual size below a given diameter (typically $PM_{2.5}$ for particulates smaller than 2.5 μm, or PM_{10} for particulates smaller than 10 μm). In urban centers in the USA and Europe, typical particulate concentrations are 15–30 μg m^{-3} for $PM_{2.5}$, 30–50 μg m^{-3} for PM_{10}, and 50–100 μg m^{-3} for TSP, with peak concentrations approximately 3 times higher (e.g., Baltensberger *et al.*, 2002; Blanchard *et al.*, 2002; Jacobson, 1997; Seinfeld and Pandis, 1998). Rural and remote PM_{10} concentrations are typically below 10 μg m^{-3}. Several rapidly growing mega cities in industrializing regions (Delhi, Mumbai, Cairo, Mexico City, Bangkok) have TSP in excess of 400 μg m^{-3} (Mage *et al.*, 1996). Particulate concentrations as high as 5,000 μg m^{-3} (TSP) have been observed during historically severe episodes, e.g., in the Ruhr Valley, Germany in 1966, and in London in 1952 (Anderson, 1999; Brimblecombe, 1987). Particulate concentrations (as TSP, PM_{10}, and $PM_{2.5}$)

are most frequently measured, but often the chemical composition of particulates is equally important, in terms of atmospheric analysis and environmental impacts.

The distribution of aerosols (shown in Figure 2—per unit mass) can also be expressed in terms of the total number of particles, which places greater emphasis on the smaller particles and provides information about the nuclei mode and the process of accumulation. Aerosol concentrations are sometimes expressed in terms of aerosol surface area, which is closely related to visibility impacts.

The role of chemistry in producing sulfate particulates is especially noteworthy. Sulfates are emitted directly from coal-fired industries, but most atmospheric sulfate is produced photochemically from sulfur dioxide (SO_2, also emitted from coal-fired industries). Sulfates are produced in two ways: by gas-phase oxidation (a process which is often linked to ozone formation); and by aqueous-phase reactions. The aqueous reactions can proceed on timescales of a few minutes or less in cloud droplets, so that atmospheric sulfur is rapidly converted to sulfates in the presence of clouds. This cloud-formed sulfate is often removed from the atmosphere by rain (see Chapter 9.10) or dispersed vertically by dynamics associated with clouds. Sulfates can also be produced in hygroscopic or wetted aerosols, which are present in the atmosphere at times of high relative humidity, and in fog. Dangerously high levels of sulfates were generated during the classic London fogs of the early 1900s, which combined stagnant meteorology, fog, and high sulfur concentrations (Brimblecombe, 1987).

Ozone air pollution events generally have lower concentrations of sulfates than the winter fog events, because the dynamics that lead to ozone production usually have much more rapid vertical dilution than the fog events. However, ozone events can lead to a significant enhancement of sulfates, especially at the regional scale. The same photochemical processes that lead to ozone formation also cause rapid photochemical conversion of sulfur dioxide to sulfates. Conversion of sulfur dioxide to sulfates during air pollution events occurs on a timescale of 1–2 days. This allows for significant accumulation of sulfates during regional air pollution events.

A thorough review of issues associated with aerosol science is currently in preparation (NARSTO, 2003).

9.11.2.4 Environmental and Health Impacts

The impact of ozone, acid aerosols and other related pollutants on human health has been the subject of intense scrutiny (Lippman, 2000; Lippman and Schlesinger, 2000; Hoening, 2000; Anderson, 1999; Wilson and Spengler, 1996; Dockery *et al.*, 1993). There is evidence that ambient levels of both ozone and acid aerosols have significant health impacts. In addition, ozone has been linked with both damage to agricultural crops (Mauzerall and Wang, 2001) and forests (US Congress, 1989). Particulates are responsible for most of the visibility degradation associated with air pollution (Seinfeld and Pandis, 1998).

The most direct and striking evidence for health effects from air pollution is found for particulates. A series of studies in the USA have shown that mortality rates correlate with exposure to particulates (Dockery *et al.*, 1993; Lippman and Schlesinger, 2000). These studies found that high exposure to particulates was correlated with increased mortality from respiratory or cardiopulmonary diseases, but not with increased death rates from other causes. The increase in mortality rates was significant (10–26%) and the associated premature deaths were believed to represent a 2–3 year shortening of life spans. The associated range of particulate concentrations was 30–80 $\mu g\ m^{-3}$ for total suspended particulates (TSPs) and 10–30 $\mu g\ m^{-3}$ for fine particulates ($PM_{2.5}$), approximately half of which consisted of sulfates. The studies were able to rule out statistically alternate causes of increased mortality, including incidence of smoking, presence of air-borne allergens, temperature, humidity, or the presence of other air pollutants. Increased mortality was specifically associated with fine particulates rather than coarse particulates, and was linked to both sulfate and nonsulfate fine particulates. There have also been numerous episodic events in the USA and in Europe (including the well-known London fogs) in which elevated particulates, SO_2, and other primary pollutants have been correlated with excess deaths and admissions to hospitals (Brimblecombe, 1987; Anderson, 1999). Although particulates have been correlated with excess mortality, the physiological cause of damage from particulates is less clear. It is also uncertain whether the impact of particulates is due to their chemical composition or only to the particulate size.

Ambient ozone has not been clearly linked to excess mortality, possibly because it is difficult to separate ambient ozone statistically from other possible causative factors. Ambient ozone is strongly correlated with temperature (see Section 9.11.4), so that excess deaths associated with ozone are hard to separate from deaths associated with heat. However, ambient levels of ozone have been linked to impairment of respiratory functions both in laboratory studies and in studies of individuals under ambient conditions. A 10–20% reduction in forced expiratory volume (FEV) was found to result from exposure to ozone mixing ratios of 80–100 ppb for 6 h, or for exposure to 180–200 ppb for 2 h. Studies with laboratory animals suggest that this type of impairment can lead to permanent lung damage. Studies have also identified increased inflammation of the lungs, coughing, and other asthmatic symptoms following exposure to ambient ozone as low as 80 ppb (Brauer and Brook, 1997). Autopsies of auto accident victims in Los Angeles also showed evidence of long-term lung damage. See Lippman and Schlesinger (2000) and Lippman (2000) for a complete summary.

Based on the above evidence, the US Environmental Protection Agency (EPA) proposed in 1997 to strengthen and change the format of the National Ambient Air Quality Standards for both particulates and ozone. The previous standards (dating from the 1970s) were a 1 h maximum mixing ratio of 125 ppb for ozone and an annual average concentration of 75 $\mu g\ m^{-3}$ (PM_{10}) for particulates. The proposed new standard for ozone is an 8 h average mixing ratio of 85 ppb (Brauer and Brook, 1997). The change to a standard based on 8 h averages rather than 1 h peak concentrations was based on the studies described above that showed damage resulting from prolonged exposure. The proposed health standard for particulates would be based on $PM_{2.5}$ rather than PM_{10}, because $PM_{2.5}$ reflects the concentration of fine particulates more closely than PM_{10} (Wilson and Suh, 1997; see Figure 2), and would be 15 $\mu g\ m^{-3}$ for annual average concentrations and 50 $\mu g\ m^{-3}$ for 24 h average peak concentrations. Implementation of the new standards was delayed by court challenges and the change of administration in the USA in 2000, and it is currently uncertain whether the new standards will be implemented.

Acute health effects have also been identified as a result of exposure to NO_2 and CO, but these effects were found only for exposure to NO_2 above 250 ppb and CO above 10^4 ppb, amounts that were 10 times higher than ambient concentrations (Bascomb *et al.*, 1996; Anderson, 1999). These species are primarily of concern as possible indoor air pollutants, because ambient mixing ratios are often significantly higher indoors (Jones *et al.*, 1999).

Visibility degradation associated with air pollution is almost entirely due to fine particulates, although coarse particulates and a few gaseous species (e.g., NO_2) also may contribute (Seinfeld and Pandis, 1998). Visibility degradation associated with fine particulates occurs through the process of Mie scattering, which is most efficient for particulate sizes close to the wavelength of visible light (0.4–0.7 μg).

Damage to agricultural crops from air pollution is primarily associated with ozone, while ecological damage is associated with both ozone and the deposition of acid particulates (see Chapter 9.10). Ozone enters plants through the plant stomata and can interfere with various cell functions. Many plants respond to elevated ozone by closing the stomata, which limits internal damage but slows growth rates. Negative impacts include reduced rate of plant photosynthesis, increased senescence and reduced rates of reproduction. These impacts can result from ozone levels as low as 60–80 ppb, a level frequently surpassed in rural and agricultural areas subject to regional air pollution events. Reductions in crop yields have been found for ozone as low as 40 ppb, especially for soyabeans, which are particularly vulnerable to ozone damage (Mauzerall and Wang, 2002). It has been estimated that crop damage from ozone in the USA causes monetary losses of \$1–2 billion per year (US Congress, 1989). Similar monetary losses have been estimated for China, and forecasts suggest that a 20% reduction of crop yields due to ozone could occur in China by 2020 (Aunon *et al.*, 2000). For a review on agricultural impacts, see Mauzerall and Wang, 2002.

9.11.2.5 Long-term Trends in Ozone and Particulates

Attempts to identify changes in air pollutant concentrations over time have been the subject of great interest, both as a basis for evaluating the effectiveness of existing controls and as a way of identifying the sources of possible problems for the future. Evaluating trends for ozone is difficult, because day-to-day and seasonal variations in ozone depend largely on meteorology. This is especially true in the northern USA and Western Europe, where annual changes in ozone depend on the frequency of occurrence of hot dry conditions that promote ozone formation. In addition, ozone concentrations are of interest in terms of extreme events rather than climatic averages. Trends in the occurrence of these extreme events are difficult to evaluate in a way that is statistically robust. Trends in primary pollutants are easier to evaluate, because these pollutant concentrations are less dependent on meteorology.

Fiore *et al.* (1998) evaluated 10-year trends in ozone in the USA by statistically filtering out changes in meteorology. They found a statistically significant downward trend in Southern California and in New York, but not elsewhere. The downward trend was somewhat more pronounced when trends were evaluated based on more extreme events. Estimates for changes in NO_x in the USA during the 1990s suggest either a modest (10–15%) reduction (Fenger, 1999) or no reduction (Butler *et al.*, 2001; Bowen *et al.*, 2001).

Significant downward trends were found for SO_2 (Fenger *et al.*, 1999). Kuebler *et al.* (2001) reported a significant decrease in precursor emissions in Switzerland but little change in ozone, possibly due to transport from elsewhere in Europe or to an increase in global background ozone.

Ozone in Los Angeles in particular has decreased significantly since the 1970s and early 1980s, when the US air quality standard (125 ppb) was exceeded on ~180 days per year.

As mentioned above, there have been sharp reductions in both sulfate particulates and soot since the 1950s, largely due to the conversion from coal to natural gas as the primary source of home heating (Fenger, 1999).

Concentrations of both ozone and particulates have increased since the early 1980s in many parts of the developing world, including cities in Asia and Latin America (e.g., Mage *et al.*, 1996). It is possible that continued growth will lead to increases in global background concentrations in the future (e.g., Lelieveld and Dentener, 2000).

9.11.3 PHOTOCHEMISTRY OF OZONE AND PARTICULATES

9.11.3.1 Ozone

The central concepts of ozone photochemistry are: the split into NO_x-sensitive and NO_x-saturated photochemical production regimes, the role of the OH radical, and the concept of ozone production efficiency per NO_x.

9.11.3.1.1 Ozone formation

The ozone formation process is almost always initiated by a reaction involving a primary hydrocarbon (abbreviated here as RH), other organic or CO with the OH radical. The reaction with OH (Equation (1)) removes a hydrogen atom from the hydrocarbon chain, which then acquires O_2 from the atmosphere to form a radical with the form RO_2. For example, methane (CH_4) reacts with OH to form the RO_2 radical CH_3O_2 propane (C_3H_8) reacts to form $C_3H_7O_2$, etc. The equivalent reaction for CO (Equation (2)) forms HO_2, a radical with many chemical similarities to the various RO_2 radicals:

$$RH + OH \xrightarrow{[O_2]} RO_2 + H_2O \qquad (1)$$

$$CO + OH \xrightarrow{[O_2]} HO_2 + CO_2 \qquad (2)$$

This is followed by reactions of RO_2 and HO_2 with NO (Equations (3) and (4)), resulting in the conversion of NO to NO_2:

$$RO_2 + NO \xrightarrow{[O_2]} R'CHO + HO_2 + NO_2 \qquad (3)$$

$$HO_2 + NO \rightarrow OH + NO_2 \qquad (4)$$

Here, R′CHO represents intermediate organic species, typically including aldehydes and ketones. Photolysis of NO_2 results in the formation of atomic oxygen (O), which reacts with atmospheric O_2 to form ozone:

$$NO_2 + h\nu \rightarrow NO + O \qquad (5)$$

The conversion of NO into NO_2 is the characteristic step that leads to ozone formation, and the rate of conversion of NO to NO_2 is often used to represent the ozone formation rate. The process does not remove either OH or NO from the atmosphere, so that the OH and NO may initiate additional ozone-forming reactions.

For $NO_x > 0.5$ ppb (typical of urban and polluted rural sites in the eastern USA and Europe) Equations (3) and (4) represent the dominant reaction pathways for HO_2 and RO_2 radicals. In this case the rate of ozone formation is controlled largely by the rate of the initial reaction with hydrocarbons or CO (Equations (1) and (2)). Analogous reaction sequences lead to the formation of various other gas-phase components of photochemical smog (e.g., formaldehyde (HCHO) and PAN) and to the formation of organic aerosols.

9.11.3.1.2 Odd hydrogen radicals

The rate of ozone production is critically dependent on the availability of odd hydrogen radicals (defined by Kleinman (1986) as the sum of OH, HO_2, and RO_2) and in particular by the OH radical. The OH radical is important because reaction sequences that lead to either the production or removal of many tropospheric pollutants are also initiated by reactions involving OH. In particular, the ozone production sequence is initiated by the reaction of OH with CO (reaction (1)) and hydrocarbons (reaction (2)). The split into NO_x-sensitive and VOC-sensitive regimes, discussed below, is also closely associated with sources and sinks of radicals.

Odd hydrogen radicals are produced by photolysis of ozone, formaldehyde, and other intermediate organics:

$$O_3 + h\nu \xrightarrow{[H_2O]} 2OH + O_2 \qquad (6)$$

$$HCHO + h\nu \xrightarrow{2O_2} 2HO_2 + CO \qquad (7)$$

They are removed by reactions that produce peroxides and nitric acid:

$$HO_2 + HO_2 \rightarrow H_2O_2 + O_2 \quad (8)$$

$$RO_2 + HO_2 \rightarrow ROOH + O_2 \quad (9)$$

$$OH + NO_2 \rightarrow HNO_3 \quad (10)$$

Formation of PAN is also a significant sink for odd hydrogen. Removal of OH occurs on timescales of 1 s or less, and removal of odd hydrogen radicals as a group usually occurs on timescales of 5 min or less.

The supply of radicals is dependent on photolytic reactions, so that most significant gas-phase chemistry only occurs during the daytime. The supply of radicals is also linked to the availability of water vapor (H_2O) through Equation (6). Both the photochemical production and loss of pollutants are slower in winter, due to lack of sunlight and lower H_2O. Photochemical loss rates are also slower in the upper troposphere, where temperatures are lower and mixing ratios of H_2O are much smaller.

The role of NO_x as a sink for odd hydrogen radicals (primarily through Equation (10)) is especially noteworthy. When ambient NO_x mixing ratios are large, reaction (10) drives the radical mixing ratios to very low levels, with the result that photochemical production and loss rates are much slower. Under such NO_x-saturated conditions, rates of production of ozone and other secondary species and rates of photochemical removal of pollutants are significantly slowed.

9.11.3.1.3 O_3, NO, and NO_2

Rapid interconversion among the species O_3, NO, and NO_2 occurs through reactions (5) and (11),

$$NO + O_3 \rightarrow NO_2 + O_2 \quad (11)$$

followed by reaction of atomic oxygen with O_2 to form O_3. Taken together, reactions (1) and (2) produce no net change in ozone. Each of these reactions occurs rapidly, on a timescale of 200 s or less. Typically, the two major components of NO_x (NO and NO_2) adjust to establish a near-steady state between Equation (1) and Equation (2). However, these reactions can lead to a significant decrease in ozone concentrations in the vicinity of large NO_x sources. More than 90% of NO_x emissions consist of NO rather than NO_2, so that the process of approaching a steady state among reactions (5) and (11) (which usually has NO_2 mixing ratios equal to or greater than those of NO) involves conversion of O_3 to NO_2. This process (sometimes referred to as NO_x titration) is important mainly in plumes extending from large point sources. NO_x mixing ratios in these plumes can be 100 ppb or higher, and O_3 is often depressed to very low levels. These plumes typically lead to a net increase in ozone, but only following dispersion and travel downwind (e.g., Gillani *et al.*, 1996; Ryerson *et al.*, 1998, 2001). The process of NO_x titration also can lead to very low O_3 in urban centers at night, when the reverse reaction (Equation (5)) is zero. Ambient O_3 is usually low at night (<30 ppb) due to removal through deposition in the shallow nocturnal boundary layer. Nighttime NO_x emissions followed by reaction (11) often drive these concentrations to 1 ppb or lower.

The process of NO_x titration has little impact on O_3 during conditions that favor photochemical production, because production rates (via reactions (1)–(5)) greatly exceed losses associated with NO_x titration.

Analyses of ozone chemistry often use the concept of odd oxygen, $O_x = O_3 + O + NO_2$ (Logan *et al.*, 1981) as a way to separate the process of NO_x titration from the processes of ozone formation and removal that occur on longer timescales. Odd oxygen is unaffected by reactions (5) and (11) and remains constant in situations dominated by NO_x titration, such as the early states of a power plant plume. Production of odd oxygen occurs only through NO_x–VOC–CO chemistry, and loss of odd oxygen occurs through conversion of NO_2 to PAN and HNO_3 or through slower ozone loss reactions (e.g., Equation (6)), rather than through the more rapid back-and-forth reactions (1) and (2). The chemical lifetime of odd oxygen relative to these losses is typically 2–3 days in the lower troposphere. This lifetime is often more useful for describing atmospheric processes associated with ozone than the chemical lifetime of ozone relative to reaction (11).

The lifetime of odd oxygen in the middle and upper troposphere is much longer (one month or more) because the rate of removal via reaction (6) ($O_3 + h\nu \rightarrow 2OH$) depends on the availability of H_2O. Typically, H_2O mixing ratios in the middle and upper troposphere are lower than at ground level by a factor of 10 or more.

The ambient ratio NO_2/NO is controlled by a combination of the interconversion reactions (Equations (11) and (5)) and the ozone-producing reactions (3) and (4). Because the ozone-producing reactions involve conversion of NO to NO_2 and affect the ratio NO_2/NO, measured values of this ratio can be used (especially in the remote troposphere) to identify the process of ozone formation. When the ratio NO_2/NO is higher than it would be if determined solely by reactions (5) and (11), it provides evidence for ozone formation (e.g., Ridley *et al.*, 1992). Reactions (3)–(5), and (11) can be combined to derive the summed concentration of HO_2 and RO_2 radicals from measured O_3,

NO, NO_2, and solar radiation (e.g., Duderstadt *et al.*, 1998; see also Trainer *et al.* (2000) for complete citations).

9.11.3.1.4 O_3–NO_x–VOC sensitivity and OH

The split into NO_x-sensitive and NO_x-saturated regimes is often illustrated by an isopleth plot, which shows ozone as a function of its two main precursors, NO_x and VOCs. As shown in Figure 3, the rate of ozone production is a nonlinear function of the NO_x and VOC concentrations. When VOCs/NO_x ratios are high, the rate of ozone production increases with increasing NO_x. Eventually, the ozone production rate reaches a local maximum and subsequently decreases as NO_x concentrations are increased further. This local maximum (the "ridge line") defines the split into NO_x-sensitive and NO_x-saturated regimes. At high VOCs/NO_x ratios the rate of ozone production increases with increasing NO_x and is largely insensitive to VOCs. At low VOCs/NO_x ratios (above the "ridge line" in Figure 3) the rate of ozone production increases with increasing VOCs and decreases with increasing NO_x. This split into NO_x-sensitive and NO_x-saturated regimes has immediate implications for the design of effective control policies for ozone. Ozone in a NO_x-sensitive region can be reduced only by reducing NO_x. Ozone in a VOC-sensitive region can be reduced either by reducing VOCs or by reducing NO_x to extremely low levels.

Isopleth plots such as Figure 3 are commonly used to show ozone concentrations as a function of NO_x and VOC emission rates. Representations of ozone concentrations versus emission rates are always critically dependent on assumptions about rates of vertical mixing and the time allowed for photochemical production of ozone. The format shown here, relating ozone production rates to NO_x and VOCs, is also sensitive to various assumptions (ozone and water vapor concentrations, solar radiation, and the specific composition of VOCs) but shows less variation than ozone concentrations.

The split into NO_x-sensitive and NO_x-saturated regimes is closely related to the cycle of odd hydrogen radicals, which controls the rate of photochemical production of ozone. The NO_x-saturated regime occurs when the formation of nitric acid via (10) represents the dominant sink for odd hydrogen, while the NO_x-sensitive regime occurs when the peroxide-forming reactions (Equations (8) and (9)) represent the dominant sinks (Sillman *et al.*, 1990; Kleinman, 1991). When nitric acid dominates, the ambient level of OH decreases with increasing NO_x and is largely unaffected (or increases slightly) with increasing VOCs. The rate of ozone formation is determined by the rate of the reaction of hydrocarbons and CO with OH (Equations (1) and (2)), and consequently increases with increasing VOC and decreases with increasing NO_x. This is the VOC-sensitive regime. When peroxides represent the dominant sink for odd hydrogen, the sum $HO_2 + RO_2$ is relatively insensitive to changes in NO_x or VOCs. The rate of ozone formation, approximately equal to the rate of reactions (5) and (6), increases with increasing NO_x and is largely unaffected by VOC. This is the NO_x-sensitive regime. These patterns can be seen in Figures 4 and 5, which show OH and $HO_2 + RO_2$ as a function of NO_x and VOC for conditions corresponding to the isopleths in Figure 3.

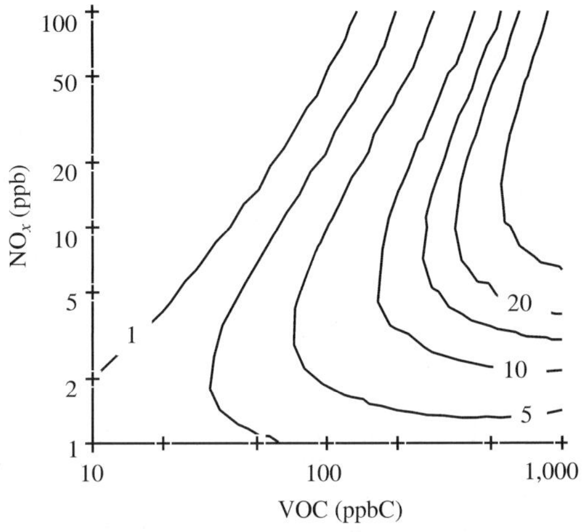

Figure 3 Isopleths giving net rate of ozone production (ppb per hour, daytime average, solid line) as a function of VOC (ppbC) and NO_x (ppb) (source Sillman, 1999) (reproduced by permission of Elsevier from *Atmos. Environ.* **1999**, *33*, 1821–1845).

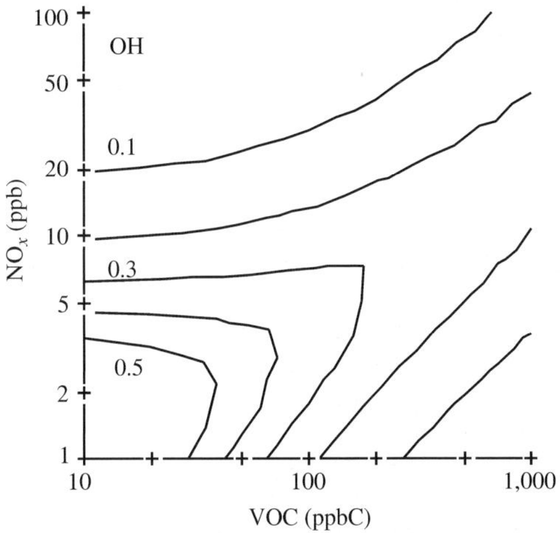

Figure 4 Isopleths showing the concentration of OH (ppt) as a function of VOC (ppbC) and NO_x (ppb) for mean summer daytime meteorology and clear skies, based on zero-dimensional calculations shown in Milford *et al.* (1994) (reproduced by permission of American Geophysical Union from *J. Geophys. Res.* **1994**, *99*, 3533–3542) and in Figure 3. The isopleths represent 0.1, 0.2, 0.3, 0.4, and 0.5 ppt.

The "ridge line" in Figure 3 that separates NO_x-sensitive and VOC-sensitive chemistry corresponds to high OH, while $HO_2 + RO_2$ is highest in the region corresponding to NO_x-sensitive chemistry. OH is lowest for conditions with either very high NO_x (due to removal of OH through formation of nitric acid, reaction (10)) or very low NO_x (due to the slow rate of conversion from HO_2 to OH through reaction (4)).

This somewhat mechanistic description can be summarized in broadly conceptual terms as follows: when the source of odd hydrogen radicals (equal to Equations (6) and (7)) exceeds the source of NO_x (related to Equation (10)), then chemistry follows a NO_x-sensitive pattern. When the source of NO_x approaches or exceeds the source of radicals, then chemistry shows an NO_x-saturated pattern and O_3, OH, and the rate of photochemistry in general decrease with further increases in NO_x (Kleinman, 1991, 1994).

The split into NO_x-sensitive and NO_x-saturated regimes affects formation rates of many other secondary species, including gas-phase species and particulates. There are significant variations for individual species: organics such as PAN and organic particulates show greater sensitivity to VOCs, while nitrate formation shows greater sensitivity to NO_x and almost never decreases with increasing NO_x. However, many of the features of the O_3–NO_x–VOC sensitivity shown in Figure 3 also appear for these other species.

Jaegle *et al.* (1998, 2001) has described an analogous split between NO_x-sensitive and NO_x-saturated photochemical regimes in the remote troposphere. However, most analyses show that ozone in the remote troposphere would increase in response to increases in either NO_x, CO, methane (CH_4), or various VOCs. Ozone at the global scale is also affected by the complex coupling between these species and OH (e.g., Wild and Prather, 2000) and shows a very different sensitivity to precursors than in polluted regions.

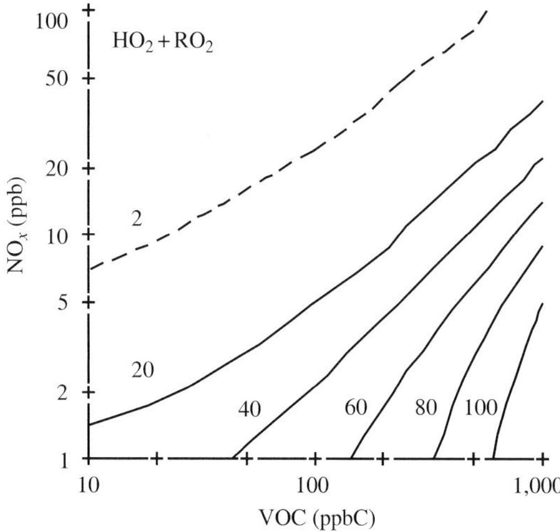

Figure 5 Isopleths showing the concentration of $HO_2 + RO_2$ (ppt) as a function of VOC (ppbC) and NO_x (ppb) for mean summer daytime meteorology and clear skies, based on zero-dimensional calculations shown in Milford *et al.* (1994) (reproduced by permission of American Geophysical Union from *J. Geophys. Res.* **1994**, *99*, 3533–3542) and in Figure 3. The isopleths represent 2 ppt (dashed line) and 20, 40, 60, 80, and 100 ppt (solid lines).

9.11.3.1.5 Ozone production efficiency

The concept of ozone production efficiency per NO_x was developed by Liu *et al.* (1987), Lin *et al.* (1988), and Trainer *et al.* (1993) as a method for evaluating ozone production, especially in the global troposphere. Ozone production efficiency represents the ratio of net production of ozone to removal of NO_x ($= P(O_3 + NO_2)/L(NO_x)$). Liu *et al.* and Lin *et al.* found that production efficiencies are highest at low NO_x concentrations, even when VOC concentration is assumed to increase with increasing NO_x. Lin *et al.* also found that production efficiencies increase with VOCs. In theory, ozone production efficiencies are given by the ratio between reactions (1) + (2) and Equation (10), i.e., by the ratio of the sum of reactivity-weighted VOCs and CO to NO_x, although they are also influenced by the rate of formation of organic nitrates. An updated analysis (Figure 6) showed the same pattern but with lower

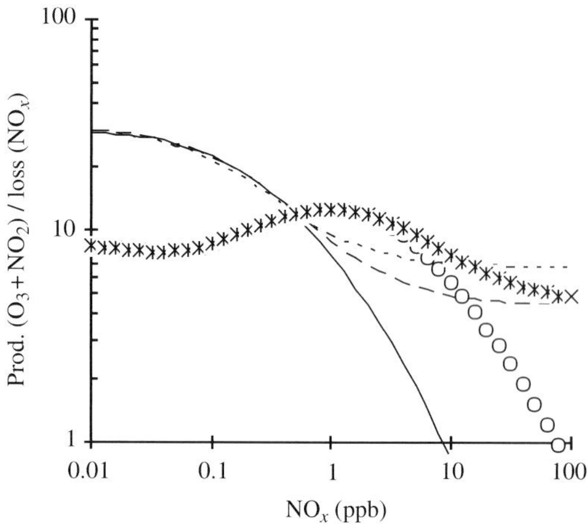

Figure 6 Ozone production efficiency, expressed as the rate of production of odd oxygen ($O_3 + NO_2$) divided by the loss of NO_x, from steady state calculations. The calculations assume: (i) CO and CH_4 only (solid line); (ii) anthropogenic VOC with VOC/NO_x = 10 (dashed line); (iii) anthropogenic VOC with VOC/NO_x = 20 (short dashed line); (iv) CH, CH_4 and 1 ppb isoprene (circles); and (v) anthropogenic VOC/NO_x = 10and 1 ppb isoprene (asterisks) (source Sillman, 1999 (reproduced by permission of Elsevier from *Atmos. Environ.* **1999**, *33*, 1821–1845); based on similar unpublished analyses by Greg Frost, NOAA Aeronomy lab).

values than initially reported by Liu and Lin. Ozone production efficiencies in polluted regions are likely to be even lower than shown in Figure 6, because these calculations typically do not include removal of ozone (even though removal of NO_x is directly linked to removal of ozone through the reaction sequence (11) followed by (10)), and also do not count the net formation of PAN or the nighttime formation of HNO_3 in the sum of NO_x losses. Recent studies (e.g., Sillman *et al.*, 1998; Ryerson *et al.*, 1998, 2001; Nunnermacker *et al.*, 1998; Trainer *et al.*, 2000) estimated an ozone production efficiency of 3–5 during pollution events.

9.11.3.2 Chemistry of Aerosols

Formation of aerosols in the gas phase is often initiated by reaction of precursors with the OH radical, and is therefore sensitive to many of the same factors (sunlight, NO_x, and VOCs) that affect ozone formation. This is especially true for secondary organic aerosols. Formation of organic aerosols is often initiated by the reactions between VOCs and OH (equivalent to Equation (1)), followed by reaction with NO (equivalent to Equation (2)) to produce secondary organics. These secondary organics often also react with OH, leading to additional reaction sequences analogous to Equations (1) and (2). Secondary organics are also formed from reactions between VOC and O_3. These processes are closely linked with the ozone formation process and have the same complex dependence on OH, NO_x, and VOCs as ozone. They differ from the ozone formation process in that the secondary organics are associated only with specific organic precursors, whereas virtually all reactive organics can initiate reaction sequences that lead to ozone formation.

Formation of organic aerosols from photochemically produced VOCs occurs when the mixing ratios of individual VOCs exceeds the saturation mixing ratio (usually expressed in terms of vapor pressure). Organic aerosols can also be formed by condensation and formation of a solution with preexisting particles, a process that can occur at vapor pressures below the saturation mixing ratio. Generally, only organics with seven or more carbon atoms lead to the formation of particulates, and yields tend to be higher for species with higher carbon number. The most common secondary organics are alkyl dicarboxylic acids with the form $HOOC(CH_2)_nCOOH$, for $n = 1–8$.

Production of sulfate and nitrate aerosols is also initiated by gas-phase reactions involving OH. Nitrate aerosols are produced from NO_x through reaction (10), followed by the combination of HNO_3 with atmospheric NH_3 to produce ammonium nitrate aerosol (NH_4NO_3). Production of nitrates can also occur at night through a reaction on aerosol surfaces:

$$NO_2 + NO_3 \rightarrow N_2O_5 \tag{12}$$

$$N_2O_5 + H_2O \xrightarrow{\text{aerosol}} 2HNO_3 \tag{13}$$

where NO_3 is produced by the nighttime gas-phase reaction of NO_2 with O_3. The resulting nitric acid again forms an aerosol by combining with ammonia. This reaction sequence is significant only at night, because during the daytime NO_3 rapidly photolyzes. The nighttime reaction produces significant amounts of nitric acid and has a significant impact on both urban smog chemistry and global tropospheric chemistry (Dentener and Crutzen, 1993), although the reaction of NO_2 with OH (Equation (11)) is the largest source of nitric acid in the atmosphere. The rate of nitrate formation shows a complex dependence on NO_x and VOCs, in a manner similar to O_3 (Meng *et al.*, 1997).

The amount of nitrate aerosol is controlled partly by the production of HNO_3 (via Equations (11)–(13)) and partly by the availability of NH_3 to form ammonium nitrate. Ammonium nitrate aerosol forms in equilibrium with gas-phase ammonia and nitric acid (Equation (14)):

$$NH_3(g) + HNO_3(g) \leftrightarrow NH_4NO_3(s) \tag{14}$$

The availability of ammonia in the atmosphere is affected by the amount of other acid aerosols, such as sulfates, which also form compounds with ammonia. As a result, formation of nitrate aerosol can show complex dependence on emission rates of NH_3 and SO_2, as well as on NO_x and VOCs (Pandis, 2003).

Production of sulfates occurs through the gas-phase reaction of SO_2 with OH. Sulfates are also produced in the aqueous phase, in both cloud droplets and liquid water aerosols. The aqueous reaction sequences that lead to formation of sulfates are initiated by the reaction of dissolved SO_2 (in the forms of H_2SO_3, HSO_3^-, or SO_3^{2-}) with one of three species: aqueous hydrogen peroxide (H_2O_2), O_3, or OH. Trace metals can also catalyze the formation of sulfates. Aqueous hydrogen peroxide and ozone are both formed by dissolution of gas-phase species, and their abundance is determined by gas-phase photochemistry (Equation (9) for production of hydrogen peroxide). The rate of sulfate formation during episodes may be limited by the availability of these oxidants, especially during winter in the mid-latitudes. The seasonal cycle of SO_2 is affected by seasonally varying emission rates, abundance of clouds and meteorological dispersion, in addition to the availability of oxidants (e.g., Rasch *et al.*, 2000).

Once formed, sulfates readily combine with atmospheric ammonia to form ammonium sulfate ($(NH_4)_2SO_4$) or ammonium bisulfate (NH_4HSO_4). Formation of these species can be a significant sink of atmospheric ammonia. As mentioned above, the interaction between sulfates, nitrates, and ammonia is a cause of nonlinearity in aerosol source–receptor relationships.

Trace metals also undergo complex chemical transformations in the aqueous phase, including reactions that form sulfates. These metals enter the aqueous phase through dissolved particulates. Trace metals are especially important for the chemical transformation of mercury. These reactions are the source of considerable uncertainty at present, due in part to large variations in the trace metal composition of particulates.

9.11.3.3 Ozone–Aerosol Interactions

In recent years there has been considerable interest in the possibility that heterogeneous reactions on aerosol surfaces affect gas-phase photochemistry, including both urban smog chemistry and the chemistry of the global troposphere (e.g., Jacob, 2000). Several types of possible interaction have been identified. Photolysis rates can be significantly lower during pollution events with high particulate concentrations, and the reduced photolysis rates can slow the rate of formation of ozone and other secondary reaction products (Dickerson *et al.*, 1997). Significant conversion of NO_x to nitric acid occurs in nighttime reactions on aerosol surfaces (Dentener and Crutzen, 1993). This reaction can affect the formation of both ozone (by removing NO_x) and aerosol nitrate during pollution events. It also can represent a significant sink for NO_x in the global troposphere. Heterogeneous conversion of NO_2 to HONO, leading to the formation of OH radicals, may also impact ozone formation rates in large cities.

Formation of ammonium nitrate aerosols also affects the global troposphere by transporting NO_x from polluted regions to remote locations (Horowitz *et al.*, 1998). Gas-phase organic nitrates such as PAN, formed in polluted regions and exported to the remote troposphere, are often a significant source of NO_x in remote locations. Because ammonium nitrate is relatively long-lived (with a lifetime of days to weeks, similar to other fine particulates) it can also transport NO_x to the remote troposphere.

As discussed above, the formation of ozone and aerosols in polluted regions are frequently coupled, because they involve many of the same precursors and have closely related precursor sensitivity (Meng *et al.*, 1997).

9.11.4 METEOROLOGICAL ASPECTS OF PHOTOCHEMICAL SMOG

Meteorology affects the development of photochemical smog in two ways: through dynamics (especially vertical mixing) and through temperature.

9.11.4.1 Dynamics

Dynamics has a direct effect on smog formation because it affects the rate of dilution of emitted pollutants and dispersion out of the polluted region. During stagnant conditions, polluted air is exported from an urban area very slowly, and may remain within a polluted region (or recirculate through the region) for several days. The most important aspect of dynamics is vertical mixing.

Vertical mixing over land follows a diurnal cycle that is strongly influenced by sunlight. At nighttime the land surface cools, producing a stable atmospheric structure near the surface (with cooler air below) that suppresses vertical mixing. Under these conditions, vertical mixing near the surface is driven solely by friction between the surface and the synoptic wind, and is opposed by the thermally stable vertical structure. During the night and early morning, pollutants emitted near the surface are often mixed only to a height of 100 m. Elevated emissions from smokestacks (which often rise to an altitude of 500–1,000 m due to the heat content of the emitted species) remain in a narrow atmospheric layer aloft and are not mixed to the surface. During the daytime, solar radiation warms the surface, and this heat causes convective mixing of the atmosphere. On sunny afternoons, this convective mixing typically reaches a height of 2,000 m. This mixing has the effect of diluting primary emissions, but it also causes elevated plumes from smokestacks to mix down to the surface. It also causes pollutants that were formed in the convective mixed layer on the previous day to mix down to the surface.

The region of the atmosphere that is in direct contact with the surface (on a timescale of 1 h or less) is commonly referred to as the boundary layer or mixed layer. Technically, the boundary layer refers to the region of the atmosphere that is dynamically influenced by the surface (through friction or convection driven by surface heating). Less formally, the boundary layer is used to represent the layer of high pollutant concentrations in source regions. The top of the boundary layer in urban areas is characterized by a sudden decrease in pollutant concentrations and usually by changes in other atmospheric features (water vapor content, thermal structure, and wind speeds).

Concentrations of primary pollutants show a diurnal cycle that is a direct result of the pattern of vertical mixing in the atmosphere. Concentrations of NO_x, CO, and SO_2 and elemental carbon particulates are typically higher at night than at midday, even though emission rates are ~3 times higher during the day. Peak concentrations of primary pollutants usually occur during early and mid-morning, when emission rates are high and rates of vertical mixing are still relatively low. Pollutant concentrations decrease during the day as convective mixing increases and reach the diurnal minimum from noon to 6 p.m. (see Figure 7).

Ozone and other secondary pollutants show a very different diurnal pattern. Because ozone is only produced photochemically during the day, it typically reaches a diurnal maximum during the afternoon. Ozone is removed near the surface at night, both through surface deposition and through nighttime reactions with primary pollutants, and consequently reaches its minimum at night. Nighttime ozone concentrations are typically lower in urban centers than in the surrounding rural area because primary pollutants react to remove ozone in the nighttime surface layer. During large-scale pollution episodes ozone may decrease to very low values at ground level at

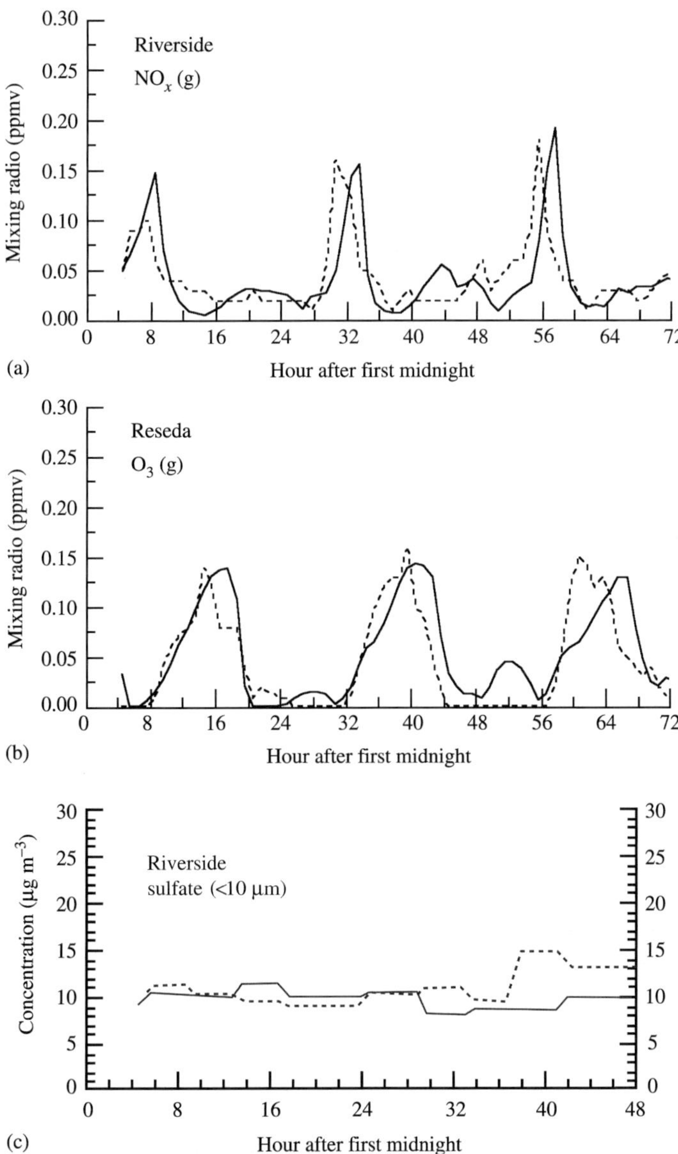

Figure 7 Time series for (a) NO_x and (b) O_3 (in parts per million, ppmv), and (c) sulfate (PM_{10}, $\mu g\ m^{-3}$) versus hour at Riverside, CA, August 26–28, 1988. Dashed lines represent measurements, and solid lines represent model predictions (source Jacobson *et al.*, 1996; (reproduced by permission of Elsevier from *Atmos. Environ.* **1996**, *30*, 1939–1963) Jacobson, 1997 (reproduced by permission of Elsevier from *Atmos. Environ.* **1997**, *31*, 587–608)).

night, but high ozone concentrations remain in a layer (typically 200–2,000 m above the surface). Ozone concentrations increase rapidly during the morning as convective mixing begins and ozone from above the ground is mixed down to the surface. Subsequently, ozone concentrations continue to increase as photochemistry results in additional ozone formation. A similar diurnal pattern is shown by HNO_3, PAN, and many other secondary pollutants. Other secondary species such as sulfate particulates and many secondary hydrocarbons are not removed from the atmosphere at night. These species show much smaller diurnal variations than either ozone or the primary species (see Figure 7).

Due to the contrasting diurnal patterns of primary and secondary pollutants, ozone and other secondary pollutants appear to anticorrelate with their precursor species (NO_x and VOCs) if measured concentrations are plotted for a full diurnal cycle. Peak ozone occurs during and after the time of maximum sunlight, which is also the time of maximum vertical dilution. Concentrations of primary pollutants reach their diurnal minimum at this time.

Severe pollution episodes that involve primary pollutants are associated with meteorology that suppresses vertical mixing during the daytime. This is most common in winter in northerly locations (when solar radiation is minimal) and is often accompanied by fog near the surface. These conditions do not produce high levels of ozone and other secondary pollutants, because low temperatures and lack of sunlight prevent ozone formation. The warm temperatures and high sunlight that lead to rapid ozone formation also tend to create a relatively deep convective mixed layer. Severe ozone events are often associated with persistent high-pressure systems and subsidence inversions, which limit vertical dispersion and prevent the formation of clouds (which would vent the polluted boundary layer). However, the height of the daytime mixed layer is usually 1,500–2,000 m even during these events.

The city of Los Angeles is uniquely vulnerable to ozone because dynamics in Los Angeles can cause air to be trapped in a relatively shallow layer even at times of warm temperatures and high solar radiation. This occurs when relatively cool marine air is trapped in the Los Angeles basin under much warmer air from the inland desert. Convective mixing heights during these events are often just 500 m.

9.11.4.2 Ozone and Temperature

As stated above, ozone in polluted regions shows a strong dependence on temperature. This dependence on temperature is important as a basis for understanding variations in ozone concentrations from year to year or between cities. As shown in Figure 8, elevated ozone is always associated with temperatures in excess of 20 °C and often with temperatures above 30 °C. In the eastern USA and Europe, year-to-year variations in ozone concentrations are often the result of variations in temperature and cloud cover, rather than changes in emission of pollutants.

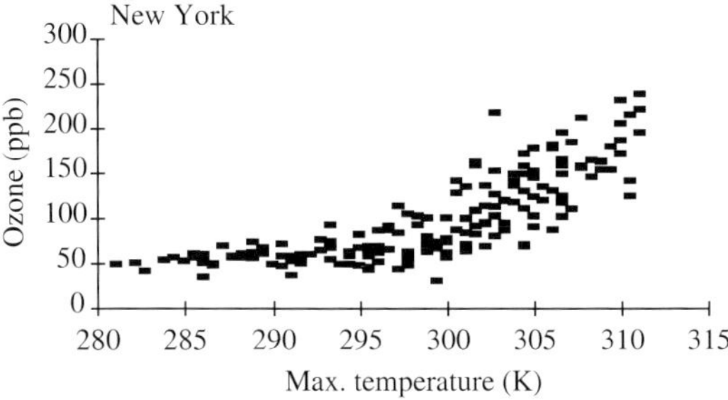

Figure 8 Diurnal peak O_3 (ppb) versus maximum surface temperature observed in the New York–New Jersey–Connecticut metropolitan area for April 1 through September 30, 1988 (source Sillman and Samson, 1995) (reproduced by permission of American Geophysical Union from *J. Geophys. Res.* **1995**, *100*, 11497–11508).

The reason for the dependence on temperature is due largely to the chemistry of ozone formation. Cardellino and Chameides (1990) and Sillman and Samson (1995) found that the temperature dependence was associated with the temperature-dependent decomposition rate of PAN. PAN becomes longer lived at lower temperatures, and formation of PAN results in the removal of NO_x, hydrocarbons, and odd hydrogen radicals (described below), all of which suppress ozone formation. PAN, also a component of photochemical smog, tends to reach maximum values at intermediate temperatures (5–10 °C). Jacob *et al.* (1993) proposed that ozone correlates with temperature partly because the meteorological conditions that favor ozone formation (high solar radiation and light winds) tend to be associated with warm temperatures. In addition, the emission rate of biogenic hydrocarbons (a major ozone precursor, discussed above) increases sharply with increasing temperature.

The dependence of ozone on temperature is an example of a phenomenon that only affects ozone in polluted regions and does not have a similar impact on the global troposphere. Warm temperatures and stagnant conditions in polluted regions lead to ozone formation, which is subsequently exported to the free troposphere. In the absence of these conditions, ozone precursors are exported to the free troposphere, and eventually lead to ozone formation at a slightly higher rate (Sillman and Samson, 1995). PAN formed in polluted regions eventually decomposes in the free troposphere, producing NO_x and leading to ozone formation.

9.11.5 NEW DIRECTIONS: EVALUATION BASED ON AMBIENT MEASUREMENTS

Some of the most important new results in recent years involve the use of high-quality field measurements as a basis for deriving information about photochemical smog.

Since the original analysis of smog formation by Haagen-Smit and Fox (1954), analysis of photochemical smog has relied mainly on two methods: calculations with photochemical models to identify the chemical relation between ozone and its precursors, and experiments involving smog chambers that duplicate the smog formation process in a laboratory setting. Since 1954, the model calculations have expanded from zero-dimensional models (representing only photochemical transformations) to three-dimensional models with sophisticated treatment of emissions and horizontal and vertical transport. The photochemical transformations associated with urban smog involve thousands of individual species (mostly VOCs), which cannot be conveniently included in model calculations. Instead, models use condensed photochemical mechanisms that approximate the chemistry of the thousands of individual VOCs by a smaller number of either lumped species (which represent the composite of several individual species) or surrogate species (which use the chemistry of selected individual species to represent a group of related species). Condensed mechanisms are also necessary, because the reaction rates and products have not been measured for all individual VOC species and are often inferred by analogy with species for which experimental data are available. The accuracy of condensed mechanisms is tested by comparing results of these mechanisms to smog chamber experiments (e.g., Stockwell *et al.*, 1997; Zaveri and Peters, 1999; Dodge, 2000). An extensive summary of contemporary research in the USA is available as part of a recent National Academy of Science review. This includes reviews of gas-phase and aerosol measurements in polluted regions (Parrish and Fehsenfeld, 2000; Trainer *et al.*, 2000; Kleinman, 2000; Blanchard, 2000; McMurry *et al.*, 2000), monitoring networks in the USA (Demerjian *et al.*, 2000), gas-phase and heterogeneous chemistry (Atkinson, 2000; Dodge, 2000; Jacob, 2000), anthropogenic and biogenic emission inventories (Sawyer *et al.*, 2000; Placet *et al.*, 2000; Guenther *et al.*, 2000), meteorological and photochemical models (Seaman, 2000; Russell and Dennis, 2000).

Despite new advances, models for photochemical smog formation are subject to a number of uncertainties. Estimates for emission rates represent the largest uncertainty. In 1995 it was discovered that biogenic VOC emissions in the USA had been underestimated by a factor of 4 (Geron *et al.*, 1994). This discovery led to a major change in model predictions for the impact of NO_x and VOCs on ozone formation in the eastern USA (Pierce *et al.*, 1998; see also Chameides *et al.*, 1988) and led to greater emphasis on the control of NO_x emission sources. Recent results from the Texas Air Quality study also suggest that emissions of anthropogenic VOC from industrial sources are much larger than represented in emission inventories. The accuracy of emission inventories effectively limits our ability to understand the process of smog formation.

Representation of meteorology during smog events has been greatly improved in recent years by the development of mesoscale meteorological models using sophisticated assimilations of ambient data (Seaman, 2000). Uncertainties related to dynamics are likely to remain, because the smog events are usually associated with light and variable wind speeds. These stagnant conditions make it difficult to estimate accurately rates of dispersion of air pollutants through either measurements or dynamical models.

The accuracy of our understanding of photochemical mechanisms is an additional source of uncertainty. Known uncertainties in photochemical reaction rates and stoichiometries cause an uncertainty of 20% in calculated ozone formation rates (Gao *et al.*, 1996).

Perhaps the biggest challenge is the need to evaluate model predictions and assumptions based on ambient measurements. Until recently, research results and regulations involving photochemical smog were based largely on model simulations, which were evaluated only in terms of their ability to reproduce the observed ozone concentration during specific smog events (NRC, 1991; NARSTO, 2000). This type of evaluation is problematic, because model calculations can yield reasonably accurate ozone concentrations in comparison with measurements even when individual assumptions (e.g., about emission rates) are incorrect. In particular, comparison with ambient ozone levels provides no basis for evaluating the accuracy of model predictions for the response of ozone to changes in emission rates of NO_x and VOCs. It frequently happens that different model scenarios for the same event give opposite predictions for the impact of NO_x versus VOC controls, without showing a difference in predicted ozone concentrations. (Sillman *et al.*, 1995; Reynolds *et al.*, 1996).

During the decade prior to 2003 there was a major expansion in the range of ambient measurements of species related to smog formation, especially in polluted rural environments. These measurements included primary NO_x and VOCs, a range of long-lived secondary reaction products and, in latter years, direct measurement of the OH and HO_2 radicals that control the photochemistry of smog formation. These measurements should permit a more extensive evaluation of the current understanding of the smog formation process, and either validate or correct current assumptions. Critical issues that can be addressed by existing field measurements include the accuracy of emission inventories; accuracy of predictions for the impact of NO_x versus VOCs on ozone formation; rates of ozone formation per NO_x, and the general accuracy of our understanding of ozone chemistry.

One of the most useful analyses based on ambient measurements is that of the correlation between ozone and the sum of reactive nitrogen species (NO_y, defined as the sum of NO_x, HNO_3, aqueous NO_3^-, organic nitrates, etc., but not including NH_3) and the correlation between ozone and the sum of NO_x reaction products (NO_z, defined as $NO_y - NO_x$). Trainer *et al.* (1993) used these correlations to estimate the ozone production efficiency per NO_x. The ozone production efficiency is related to the observed slope between ozone and NO_z, although the actual ozone production efficiency is lower than the observed slope due to the more rapid removal of NO_z (Chin *et al.*, 1994; Sillman *et al.*, 1998). The correlation between O_3 and NO_z, along with other similar correlations involving ozone, reactive nitrogen, and hydrogen peroxide, is also related to the split between NO_x-sensitive and NO_x-saturated photochemistry. Ozone production efficiency per NO_x in models is consistently higher for NO_x-sensitive conditions than for NO_x-saturated conditions, so that ratios such as O_3/NO_z are expected to be higher for NO_x-sensitive locations relative to NO_x-saturated locations (Sillman, 1995; Sillman and He, 2002).

Measured correlations between O_3, NO_y, and NO_z (Figure 9) illustrate the distinctly different chemistry and meteorology for different types of environments: polluted rural sites in the eastern USA; the moderately polluted cities of Nashville (USA) and Paris; and the extreme case of Los Angeles. Nashville, Paris, and the polluted rural sites all have features that are typical of the majority of polluted events and locations. All are continental sites and (in contrast to Los Angeles) do not have unusual mesoscale meteorology. Elevated ozone occurs during events characterized by warm temperatures and sunshine, and is formed in a well-mixed daytime convective layer extending to 1,500–2,000 m or higher. Chemistry is active, and by the afternoon most of the directly emitted NO_x has been converted to PAN, HNO_3, and other reaction products. The sum NO_y is dominated by these reaction products rather than by primary NO_x. The ozone mixing ratio increases with increasing NO_y, and the O_3-NO_y slope is surprisingly similar at all three locations. By

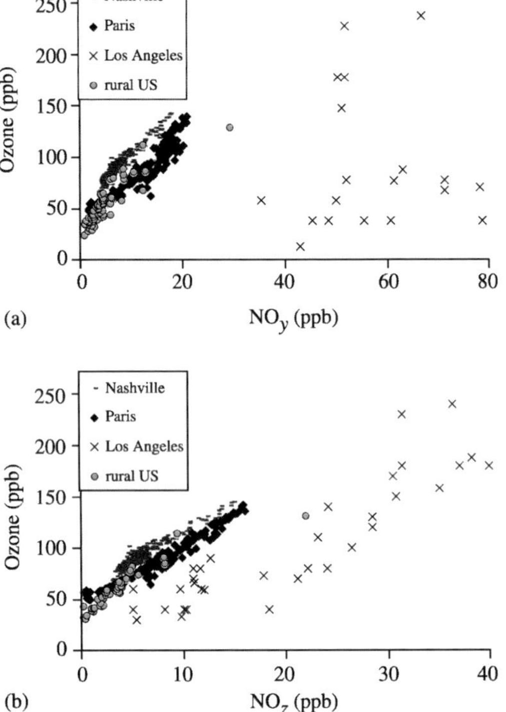

Figure 9 Measured correlations between (a) O_3 and NO_y, and (b) O_3 and ($NO_y - NO_x$) (also known as NO_z), all in ppb. Measurements are shown from field campaigns in Nashville (pink dashes), Paris (blue diamonds), Los Angeles (Xs) and from four rural sites in the eastern US (green circles) (sources measurements reported by Sillman *et al.*, 2002 (reproduced by permission of American Geophysical Union from *J. Geophys. Res.* **2002**, in press); Trainer *et al.*, 1993 (reproduced by permission of American Geophysical Union from *J. Geophys. Res.* **1993**, *98*, 2917–2926)).

contrast, Los Angeles has much higher mixing ratios of NO_y, NO_x, and other primary pollutants. The density of emissions in Los Angeles is not significantly higher than in Paris or in other large cities, but in Los Angeles the daytime convective mixed layer is frequently capped by an inversion and extends only to no more than 500 m. The high NO_x suppresses photochemical activity, and most reactive nitrogen remains in the form of NO_x. In contrast to the other sites, O_3, and NO_y are poorly correlated. Measurements also show that O_3 in Los Angeles is strongly correlated with the sum of NO_x reaction products (NO_z), because the same photochemical processes that lead to the formation of O_3 also lead to the conversion of NO_x to reaction products. However, the O_3-NO_z slope is significantly lower for Los Angeles than for the other sites. The low O_3-NO_z slope is evidence that photochemistry in Los Angeles differs significantly from photochemistry elsewhere (probably reflecting strongly NO_x-saturated conditions in Los Angeles).

Similar measured correlations have identified distinctly different photochemical conditions (probably reflecting strongly NO_x-saturated conditions) in plumes from power plants (e.g., Gillani and Pleim, 1996; Ryerson *et al.*, 1998, 2001) and at rural sites in the USA during autumn (Jacob *et al.*, 1995; Hirsch *et al.*, 1996).

Correlations among related primary species are useful as a basis for evaluating emission inventories (Trainer *et al.*, 2000; Parrish *et al.*, 1998; McKeen *et al.*, 1996). Because the emission sources of individual VOCs and NO_x are often co-located (even when the specific sources differ), these species show a strong correlation in the ambient atmosphere (see Figure 10). Isoprene is an obvious exception because its source is biogenic and not collocated with anthropogenic sources. Measured correlations between individual VOCs and NO_y provide a basis for evaluating the VOC/NO_x emissions ratio, which is a critical parameter for determining O_3–NO_x–VOC sensitivity (see Figure 3).

Based on existing model capabilities and field measurements, it should be possible to evaluate and modify current models for ozone photochemistry with the following goals:

(i) show individual VOC mixing ratios and VOC reactivity (defined as the sum of individual primary VOC weighted by their respective rate constants with OH) that is consistent with ambient measurements;

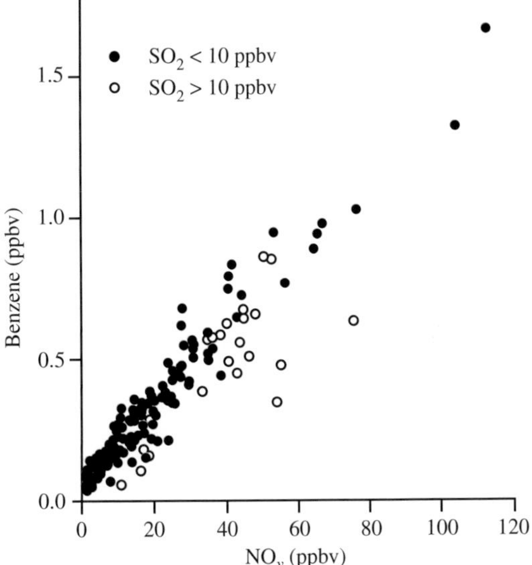

Figure 10 Measured correlations between Benzene and NO_x during winter at an urban site (Boulder, CO) and at a rural site (Idaho Hill, CO) (source Trainer *et al.*, 2000 (reproduced by permission of Elsevier from *Atmos. Environ.* **2000**, *34*, 2045–2062); based on measurements reported by Goldan *et al.*, 1995, 1997 (reproduced by permission of American Geophysical Union from *J. Geophys. Res.* **1995**, *100*, 22771–22785 and **1997**, 102, 6315–6324)).

(ii) show reasonable agreement with measured isoprene and HCHO, if available; and

(iii) show reasonable agreement with measured correlation between ozone and reactive nitrogen species and/or sums.

The first two goals would have the effect of evaluating the accuracy of emission inventories, while the third goal would evaluate the accuracy of photochemical production rates. Model applications that met these criteria would be very likely to predict the relation between ozone and precursor emissions correctly. Conversely, major errors in photochemistry or in emission inventories would be quickly apparent in this type of analysis. Analyses of these species should also facilitate the identification of long-term trends in both ozone and its precursors.

The relation between ambient levels and precursor emissions is somewhat easier to identify for particulates than for ozone, because the chemical composition of individual particulates provides evidence for their origin: sulfate particulates are associated with sulfur dioxide emissions; organic particulates with specific VOCs, and so on. A variety of statistical methods have been used to identify source types for particulates based on chemical composition, especially in terms of trace metal components (e.g., Henry, 1992; Seinfeld and Pandis, 1998). This type of analysis requires sophisticated measurements of the chemical composition of individual particulates, rather than the more common measurement of summed concentrations. Statistical methods have also been used to gain information about ozone and ozone precursors (e.g., Buhr *et al.*, 1995; Stehr *et al.*, 2000).

The source and composition of organic aerosols are still a major area of uncertainty. The composition of organic particulates consists of hundreds of individual compounds, and much of their chemical content has not been identified. The relative amounts of primary versus secondary organics contained in these aerosols, and the relative contribution of anthropogenic versus biogenic emissions and of different anthropogenic emission sources are all uncertain. In addition, the formations of secondary organics, sulfate, and nitrate aerosols are all affected by nonlinear photochemical processes, which lead to uncertainty concerning source–receptor relationships. It is possible, e.g., that reduction of sulfur emissions can free up ammonia and lead to higher concentrations of nitrate aerosols, thus counteracting the benefits of reduced sulfates (Pandis, 2003).

The most challenging research area pertaining to photochemical smog is the attempt to measure or infer concentrations of primary radical species (OH, HO_2 and/or $HO_2 + RO_2$) and to compare these measurements with calculations based on the current understanding of photochemical mechanisms. Measured or inferred radical species also would allow the instantaneous rate of ozone production to be calculated and compared to expected values. Summed $HO_2 + RO_2$ can be inferred indirectly through measurements of NO, NO_2, O_3, and the photolysis rate of NO_2 as described above (Cantrell *et al.*, 1993; see Trainer *et al.*, 2000 for complete references). The most recent analyses have combined measured OH and/or HO_2 with NO and NO_2 as a test of consistency (Thornton *et al.*, 2002). Results of these attempts have been somewhat ambiguous. Measured OH can be compared with expected values by performing a constrained steady-state calculation in which OH is calculated based on measured concentrations of long-lived species. This is possible only if a complete set of measured VOC is available, since OH concentrations depend on the removal rate through OH–VOC reactions. Recent evaluations have often shown discrepancies between measured and model OH concentrations by up to a factor of 2 (Poppe *et al.*, 1994; Eisele *et al.*, 1996; McKeen *et al.*, 1997; George *et al.*, 1999; Frost *et al.*, 1999; Carslaw *et al.*, 1999; Volz-Thomas and Kolahgar, 2000; Tan *et al.*, 2001). In some cases where measured OH was lower than model values, it appeared likely that OH was removed by additional unmeasured VOCs (McKeen *et al.*, 1997). However, the accuracy of OH and HO_x predictions in models has not been adequately demonstrated. Thornton *et al.* (2002) also reported apparent inconsistencies between measured radical concentrations and broader ozone photochemistry.

ACKNOWLEDGMENTS

I want to thank Mary Barth and Spyros Pandis for their thorough review of this document and many helpful suggestions, especially with regard to aerosols. Support for this work was provided by the National Science Foundation (NSF) under grants #ATM-9713567 and #ATM-0207841. Although support was by NSF, it does not necessarily reflect the views of the agency and no official endorsement should be inferred.

REFERENCES

Anderson H. R. (1999) Health effects of air pollution episodes. In *Air Pollution and Health* (eds. S. T. Holgate, J. M. Samet, H. S. Koren, and R. L. Maynard). Academic Press, London, pp. 460–482.

Atkinson R. (2000) Atmospheric chemistry of VOCs and NO_x. *Atmos. Environ.* **34**, 2063–2102.

Aunon K., Bernston T. K., and Seip H. M. (2000) Surface ozone in China and its possible impact on crop yields. *Ambio* **29**, 294–301.

Baltensberger U., Streit N., Weingartner E., Nyeki S., Prevot A. S. H., Van Dingenen R., Virkkula A., Putaud J.-P., Even A., ten Brink H., Blatter A., Neftel A., and Gaggeler H. W. (2002) Urban and rural aerosol characterization of summer smog events during the PIPAPO field campaign in Milan, Italy. *J. Geophys. Res.* **107**, 10.1029/2001JD001292.

Barth M. C. and Church A. T. (1999) Regional and global distributions and lifetimes of sulfate aerosols from Mexico City and southeast China. *J. Geophys. Res.* **104**, 30231–30239.

Bascomb R., Bromberg P. A., Costa D. L., Devlin R., Dockery D. W., Frampton M. W., Lambert W., Samet J. M., Speizer F. E., and Utell M. (1996) Health effects of outdoor air pollution. *Am. J. Resp. Crit. Care Med.* **153**, 477–498.

Blanchard C. L. (2000) Ozone process insights from field experiments: Part III. Extent of reaction and ozone formation. *Atmos. Environ.* **34**, 2035–2043.

Blanchard P., Brook J. R., and Brazal P. (2002) Chemical characterization of the organic fraction of atmospheric aerosol at two sites in Ontario, Canada. *J. Geophys. Res.* **107** NO, doi: 10.1029/2001JD000627.

Bowen J. L. and Valiela I. (2001) Historical changes to atmospheric nitrogen deposition to Cape Cod, Massachusetts, USA. *Atmos. Environ.* **35**, 1039–1051.

Brauer M. and Brook J. R. (1997) Ozone personal exposures and health effects for selected groups residing in the Fraser Valley. *Atmos. Environ.* **31**, 2113–2121.

Brimblecombe P. (1987) *The Big Smoke: A History of Air Pollution in London Since Medieval Times.* Methuen, London.

Buhr M., Parrish D., Elliot J., Holloway J., Carpenter J., Goldan P., Kuster W., Trainer M., Montzka S., McKeen S., and Fehsenfeld F. C. (1995) Evaluation of ozone precursor source types using principal component analysis of ambient air measurements in rural Alabama. *J. Geophys. Res.* **100**, 22853–22860.

Butler T. J., Likens G. E., and Stunder B. J. B. (2001) Regional-scale impacts of Phase I of the Clean Air Act Amendments in the USA: the relation between emissions and concentrations, both wet and dry. *Atmos. Environ.* **35**, 1015–1028.

Cantrell C. A., Shelter R. E., Calvert J. G., Parrish D. D., Fehsenfeld F. C., Goldan P. D., Kuster W., Williams E. J., Westberg H. H., Allwine G., and Martin R. (1993) Peroxy radicals as measured in ROSE and estimated from photostationary state deviations. *J. Geophys. Res.* **98**, 18355–18366.

Cardellino C. A. and Chameides W. L. (1990) Natural hydrocarbons, urbanization, and urban ozone. *J. Geophys. Res.* **95**, 13971–13979.

Carslaw N., Creasey D. L., Heard D. E., Lewis A. C., McQuaid J. B., Pilling M. J., Monks P. S., Bandy B. J., and Penkett S. A. (1999) Modeling OH, HO_2 and RO_2 radicals in the marine boundary layer: 1. Model construction and comparison with field measurements. *J. Geophys. Res.* **104**, 30241–30256.

Chameides W. L., Lindsay R. W., Richardson J., and Kiang C. S. (1988) The role of biogenic hydrocarbons in urban photochemical smog: Atlanta as a case study. *Science* **241**, 1473–1474.

Chatfield R. B., Vastano J. A., Li L., Sachse G. W., and Conners V. S. (1998) The great African plume from biomass burning: generalizations from a three-dimensional study of TRACE A carbon monoxide. *J. Geophys. Res.* **103**, 28059–28078.

Chin M., Jacob D. J., Munger J. W., Parrish D. D., and Doddridge B. G. (1994) Relationship of ozone and carbon monoxide over North America. *J. Geophys. Res.* **99**, 14565–14573.

Chung S. H. and Seinfeld J. H. (2002) Global distribution and climate forcing of carbonaceous aerosols. *J. Geophys. Res.* **107**, doi: 10.1029/2001JD001397.

Clarke J. F. and Ching J. K. S. (1983) Aircraft observations of regional transport of ozone in the northeastern United States. *Atmos. Environ.* **17**, 1703–1712.

Daum P. H., Kleinman L. I., Newman L., Luke W. T., Weinstein-Lloyd J., Berkowitz C. M., and Busness K. M.

(1996) Chemical and physical properties of anthropogenic pollutants transported over the North Atlantic during NARE. *J. Geophys. Res.* **101**, 29029–29042.

Demerjian K. L. (2000) A review of national monitoring networks in North America. *Atmos. Environ.* **34**, 1861–1884.

Dentener F. J. and Crutzen P. J. (1993) Reaction of N_2O_5 on tropospheric aerosols: impact on the global distribution of NO_x, O_3, and OH. *J. Geophys. Res.* **98**, 7149–7162.

Dickerson R. R., Kondragunta S., Stenchicov G., Civeraolo K. L., Doddridge B. G., and Holben B. N. (1997) The impact of aerosols on solar ultraviolet radiation and photochemical smog. *Science* **278**, 827–830.

Dockery D. W., Pope C. A., Xu X., Spengler J. D., Ware J. H., Fay M. E., Ferris B. G., and Speizer F. E. (1993) An association between air pollution and mortality in six US cities. *New England J. Med.* **329**, 1753–1759.

Dodge M. C. (2000) Chemical oxidant mechanisms for air quality modeling: critical review. *Atmos. Environ.* **34**, 2103–2130.

Duderstadt K. A., Carroll M. A., Sillman S., Wang T., Albercook G. M., Feng L., Parrish D. D., Holloway J. S., Fehsenfeld F., Blake D. R., Blake N. J., and Forbes G. (1998) Photochemical production and loss rates at Sable Island, Nova Scotia during the North Atlantic Regional Experiment 1993 Summer Intensive. *J. Geophys. Res.* **103**, 13531–13555.

Eisele F. L., Tanner D. J., Cantrell C. A., and Calvert J. G. (1996) Measurements and steady-state calculations of OH concentrations at Mauna Loa Observatory. *J. Geophys. Res.* **101**, 14665–14679.

Fenger J. (1999) Urban air quality. *Atmos. Environ.* **33**, 4877–4900.

Fiore A. M., Jacob D. J., Logan J. A., and Yin J. H. (1998) Long-term trends in ground level ozone over the contiguous United States, 1980–1995. *J. Geophys. Res.* **103**, 1471–1480.

Frost G. J., Trainer M., Mauldin R. L., III, Eisele F. L., Prevot A. S. H., Flocke S. J., Madronich S., Kok G., Schillawski D., Baumgardner D., and Bradshaw J. (1999) Photochemical modeling of OH levels during the First Aerosol Characterization Experiment (ACE 1). *J. Geophys. Res.* **104**, 16041–16052.

Gao D., Stockwell W. R., and Milford J. B. (1996) Global uncertainty analysis of a regional-scale gas-phase chemical mechanism. *J. Geophys. Res.* **101**, 9071–9078.

George L. A., Hard T. M., and O'Brian R. J. (1999) Measurement of free radicals OH and HO_2 in Los Angeles smog. *J. Geophys. Res.* **104**, 11643–11655.

Geron C. D., Guenther A. B., and Pierce T. E. (1994) An improved model for estimating emissions of volatile organic compounds from forests in the eastern United States. *J. Geophys. Res.* **99**, 12773–12791.

Gillani N. V. and Pleim J. E. (1996) Sub-grid-scale features of anthropogenic emissions of NO_x and VOC in the context of regional Eulerian models. *Atmos. Environ.* **30**, 2043–2059.

Goldan P. D., Trainer M., Kuster W. C., Parrish D. D., Carpenter J., Roberts J. M., Yee J. E., and Fehsenfeld F. C. (1995) Measurements of hydrocarbons, oxygenated hydrocarbons, carbon monoxide, and nitrogen oxides in an urban basin in Colorado: implications for emissions inventories. *J. Geophys. Res.* **100**, 22771–22785.

Goldan P. D., Kuster W. C., and Fehsenfeld F. C. (1997) Nonmethane hydrocarbon measurements during the tropospheric OH photochemistry experiment. *J. Geophys. Res.* **102**, 6315–6324.

Guenther A., Geron C., Pierce T., Lamb B., Harley P., and Fall R. (2000) Natural emissions of nonethane volatile organic compounds, carbon monoxide, and oxides of nitrogen from North America. *Atmos. Environ.* **34**, 2205–2230.

Haagen-Smit A. J. and Fox M. M. (1954) Photochemical ozone formation with hydrocarbons and automobile exhaust. *J. Air Pollut. Control Assoc.* **4**, 105–109.

Henry R. C. (1992) Dealing with near collinearity in chemical mass balance receptor models. *Atmos. Environ.* **26A**(5), 933–938.

Hirsch A. I., Munger J. W., Jacob D. J., Horowitz L. W., and Goldstein A. H. (1996) Seasonal variation of the ozone production efficiency per unit NO_x at Harvard Forest, Massachusetts. *J. Geophys. Res.* **101**, 12659–12666.

Hoening J. Q. (2000) *Health Effects of Ambient Air Pollution: How Safe is the Air We Breathe?* Kluwer Academic, Norwell, MA.

Horowitz L. W. and Jacob D. J. (1999) Global impact of fossil fuel combustion on atmospheric NO_x. *J. Geophys. Res.* **104**, 23823–23840.

Horowitz L. W., Liang J., Gardner G. M., and Jacob D. J. (1998) Export of reactive nitrogen from North America during summertime: sensitivity to hydrocarbon chemistry. *J. Geophys. Res.* **103**, 13451–13476.

Houghton J. T., Ding Y., Griggs D. J., Noguer M., van der Linden P. J., Dai X., Maskell K., and Johnson C. A. (eds.) (2001) *Climate Change 2001: The Scientific Basis. Contribution of Working Group I to the Third Assessment Report of the Intergovernmental Panel on Climate Change.* Cambridge University Press, Cambridge.

Jacob D. J. (2000) Heterogeneous chemistry and tropospheric ozone. *Atmos. Environ.* **34**, 2131–2160.

Jacob D. J., Logan J. A., Gardner G. M., Yevich R. M., Spivakowsky C. M., Wofsy S. C., Sillman S., and Prather M. J. (1993) Factors regulating ozone over the United States and its export to the global atmosphere. *J. Geophys. Res.* **98**, 14817–14827.

Jacob D. J., Heikes B. G., Dickerson R. R., Artz R. S., and Keene W. C. (1995) Evidence for a seasonal transition from NO_x- to hydrocarbon-limited ozone production at Shenandoah National Park, Virginia. *J. Geophys. Res.* **100**, 9315–9324.

Jacobson M. Z. (1997) Development and application of a new air pollution modeling system: Part III. Aerosol-phase simulations. *Atmos. Environ.* **31**, 587–608.

Jacobson M. Z. (2002) Control of fossil-fuel particulate black carbon and organic matter, possibly the most effective method of slowing global warming. *J. Geophys. Res.* **107**, doi: 10.1029/2001JD001376.

Jacobson M. Z., Lu R., Turco R. P., and Toon O. P. (1996) Development and application of a new air pollution modeling system: Part I. Gas-phase simulations. *Atmos. Environ.* **30**, 1939–1963.

Jaegle L., Jacob D. J., Brune W. H., Tan D., Faloona I., Weinheimer A. J., Ridley B. A., Campos T. L., and Sachse G. W. (1998) Sources of HO_x and production of ozone in the upper troposphere over the United States. *Geophys. Res. Lett.* **25**, 1705–1708.

Jaegle L., Jacob D. J., Brune W. H., and Wennberg P. O. (2001) Chemistry of HO_x radicals in the upper troposphere. *Atmos. Environ.* **35**, 469–490.

Jenkins G. S., Mohr K., Morris V. R., and Arino O. (1997) The role of convective processes over the Zaire–Congo Basin to the southern hemispheric ozone maximum. *J. Geophys. Res.* **102**, 18963–18980.

Jones A. P. (1999) Indoor air quality and health. *Atmos. Environ.* **33**, 4535–4564.

Kleinman L. I. (1986) Photochemical formation of peroxides in the boundary layer. *J. Geophys. Res.* **91**, 10889–10904.

Kleinman L. I. (1991) Seasonal dependence of boundary layer peroxide concentration: the low and high NO_x regimes. *J. Geophys. Res.* **96**, 20721–20734.

Kleinman L. I. (1994) Low- and high-NO_x tropospheric photochemistry. *J. Geophys. Res.* **99**, 16831–16838.

Kleinman L. I. (2000) Ozone process insights from field experiments: Part II. Observation-based analysis for ozone production. *Atmos. Environ.* **34**, 2023–2034.

Kleinman L. I., Daum P. H., Imre D., Lee Y.-N., Nunnermacker L. J., Springston S. R., Weinstein-Lloyd J., and Rudolph J. (2002) Ozone production rate and hydrocarbon reactivity in five urban areas: a cause of high ozone concentration in Houston. *Geophys. Res. Lett.* **29**, 10.1029/2001GL014569.

Kuebler J., van den Bergh H., and Russell A. G. (2001) Long-term trends of primary and secondary pollutant concentrations in Switzerland and their response to emission controls and economic changes. *Atmos. Environ.* **35**, 1351–1363.

Lelieveld J. and Dentener F. J. (2000) What controls tropospheric ozone? *J. Geophys. Res.* **105**, 3531–3552.

Lin X., Trainer M., and Liu S. C. (1988) On the nonlinearity of tropospheric ozone. *J. Geophys Res.* **93**, 15879–15888.

Lippman M. (ed.) (2000) *Environmental Toxicants: Human Exposures and Their Health Effects.* Wiley-Interscience, New York.

Lippman M. and Schlesinger R. B. (2000) Toxicological bases for the setting of health-related air pollution standards. *Annu. Rev. Public Health* **21**, 309–333.

Liu S. C., Trainer M., Fehsenfeld F. C., Parrish D. D., Williams E. J., Fahey D. W., Hubler G., and Murphy P. C. (1987) Ozone production in the rural troposphere and the implications for regional and global ozone distributions. *J. Geophys. Res.* **92**, 4191–4207.

Logan J. A., Prather M. J., Wofsy S. C., and McElroy M. B. (1981) Tropospheric chemistry: a global perspective. *J. Geophys. Res.* **86**, 7210–7254.

Mage D., Ozolins G., Peterson P., Webster A., Orthofer R., Vandeweerd V., and Gwynne M. (1996) Urban air pollution in the megacities of the world. *Atmos. Environ.* **30**, 681–686.

Mauzerall D. L. and Wang X. (2001) Protecting agricultural crops from the effects of tropospheric ozone exposure: reconciling science and standard setting in the United States, Europe, and Asia. *Ann. Rev. Energy Environ.* **26**, 237–268.

McKeen S. A., Liu S. C., Hsie E.-Y., Lin X., Bradhaw J. D., Smyth S., Gregory G. L., and Blake D. R. (1996) Hydrocarbon ratios during PEM-WEST A: a model perspective. *J. Geophys. Res.* **101**, 2087–2109.

McKeen S. A., Mount G., Eisele F., Williams E., Harder J., Goldan P., Kuster W., Liu S. C., Baumann K., Tanner D., Fried A., Sewell S., Cantrell C., and Shetter R. (1997) Photochemical modeling of hydroxyl and its relationship to other species during the Tropospheric OH Photochemistry Experiment. *J. Geophys. Res.* **102**, 6467–6493.

McMurry P. H. (2000) A review of atmospheric aerosol measurements. *Atmos. Environ.* **34**, 1959–2000.

Meng Z., Dabdub D., and Seinfeld J. H. (1997) Chemical coupling between atmospheric ozone and particulate matter. *Science* **277**, 116–119.

Milford J., Russell A. G., and McRae G. J. (1989) A new approach to photochemical pollution control: implications of spatial patterns in pollutant responses to reductions in nitrogen oxides and reactive organic gas emissions. *Environ. Sci. Technol.* **23**, 1290–1301.

Milford J., Gao D., Sillman S., Blossey P., and Russell A. G. (1994) Total reactive nitrogen (NO_y) as an indicator for the sensitivity of ozone to NO_x and hydrocarbons. *J. Geophys. Res.* **99**, 3533–3542.

Millan M. M., Salvador R., Mantilla E., and Kallos G. (1997) Photooxidant dynamics in the Mediterranean basin during summer: results from European research projects. *J. Geophys. Res.* **102**, 8811–8823.

Miller D. F., Alkezweeny A. J., Hales J. M., and Lee R. N. (1978) Ozone formation related to power plant emissions. *Science* **202**, 1186–1188.

NARSTO (2000) *An Assessment of Tropospheric Ozone Pollution: A North American Perspective*. The NARSTO Synthesis Team (available at http://camraq.owt.com/Narsto).

NARSTO (2003) NARSTO Particulate Matter Science for Policy Makers: A NARSTO Assessment Parts 1 and 2, EPRI 1007735, February, 2003 (available at http://www.cgenv.com/Narsto).

National Research Council (NRC) (1991) Committee on Tropospheric Ozone Formation and Measurement. *Rethinking the Ozone Problem in Urban and Regional Air Pollution.* National Academy Press, Washington, DC.

Nunnermacker L. J., Imre D., Daum P. H., Kleinman L., Lee Y. N., Lee J. H., Springston S. R., Newman L., Weinstein-Lloyd J., Luke W. T., Banta R., Alvarez R., Senff C., Sillman S., Holdren M., Keigley G. W., and Zhou X. (1998) Characterization of the Nashville urban plume on July 3 and July 18, 1995. *J. Geophys. Res.* **103**, 28129–28148.

Pandis S. (2003) Atmospheric aerosol processes: how particles change when suspended in air. In *NARSTO Fine Particle Assessment*, chap. 2 (available at http://www.cgenv.com/Narsto).

Parrish D. D. and Fehsenfeld F. C. (2000) Methods for gas-phase measurements of ozone, ozone precursors, and aerosol precursors. *Atmos. Environ.* **34**, 1921–1957.

Parrish D. D., Holloway J. S., Trainer M., Murphy P. C., Forbes G. L., and Fehsenfeld F. C. (1993) Export of North American ozone pollution to the North Atlantic Ocean. *Science* **259**, 1436–1439.

Parrish D. D., Trainer M., Young V., Goldan P. D., Kuster W. C., Jobson B. T., Fehsenfeld F. C., Lonneman W. A., Zika R. D., Farmer C. T., Riemer D. D., and Rodgers M. O. (1998) Internal consistency tests for evaluation of measurements of anthropogenic hydrocarbons in the troposphere. *J. Geophys. Res.* **103**, 22339–22359.

Pierce T., Geron C., Bender L., Dennis R., Tonnesen G., and Guenther A. (1998) Influence of increased isoprene emissions on regional ozone modeling. *J. Geophys. Res.* **103**, 25611–25630.

Placet M., Mann C. O., Gilbert R. O., and Niefer M. J. (2000) Emissions of ozone precursors from stationary sources: a critical review. *Atmos. Environ.* **34**, 2183–2204.

Poppe D., Zimmermann J., Bauer R., Brauers T., Bruning D., Callies J., Dorn H.-P., Hofzumahaus A., Johnen F.-J., Khedim A., Koch H., Koppman R., London H., Muller K.-P., Neuroth R., Plass-Dulmer C., Platt U., Rohrer F., Roth E.-P., Rudolph J., Schmidt U., Wallasch M., and Ehhalt D. (1994) Comparison of measured OH concentrations with model calculations. *J. Geophys. Res.* **99**, 16633–16642.

Prospero J. M. (1999) Long-term measurements of the transport of African mineral dust to the southeastern United States: implications for regional air quality. *J. Geophys. Res.* **104**, 15917–15927.

Rasch P. J., Barth M. C., Kiehl J. T., Schwartz S. E., and Benkovitz C. M. (2000) A description of the global sulfur cycle and its controlling processes in the National Center for Atmospheric Research Community Climate Model, Version 3. *J. Geophys. Res.* **105**, 1367–1385.

Reynolds S., Michaels H., Roth P., Tesche T. W., McNally D., Gardner L., and Yarwood G. (1996) Alternative base cases in photochemical modeling: their construction, role, and value. *Atmos. Environ.* **30**, 1977–1988.

Ridley B. A., Madronich S., Chatfield R. B., Walega J. G., Shetter R. E., Carroll M. A., and Montzka D. D. (1992) Measurements and model simulations of the photostationary state during the Mauna Loa Observatory Photochemistry Experiment: implications for radical concentrations and ozone production and loss rates. *J. Geophys. Res.* **97**, 10375–10388.

Roselle S. J. and Schere K. L. (1995) Modeled response of photochemical oxidants to systematic reductions in anthropogenic volatile organic compound and NO_x emissions. *J. Geophys. Res.* **100**, 22929–22941.

Russell A. and Dennis R. (2000) NARSTO critical review of photochemical models and modeling. *Atmos. Environ.* **34**, 2283–2324.

Ryerson T. B., Buhr M. P., Frost G. J., Goldan P. D., Holloway J. S., Hubler G., Jobson B. T., Kuster W. C., McKeen S. A., Parrish D. D., Roberts J. M., Sueper D. T., Trainer M., Williams J., and Fehsenfeld F. C. (1998) Emissions lifetimes and ozone formation in power plant plumes. *J. Geophys. Res.* **103**, 22569–22584.

Ryerson T. B., Trainer M., Holloway J. S., Parrish D. D., Huey L. G., Sueper D. T., Frost G. J., Donnelly S. G., Schauffler S., Atlas E. L., Kustler W. C., Goldan P. D., Hubler G., Meagher J. F., and Fehsenfeld F. C. (2001) Observations of ozone formation in power plant plumes and implications for ozone control strategies. *Science* **292**, 719–723.

Sawyer R. F., Harley R. A., Cadle S. H., Norbeck J. M., Slott R., and Bravo H. A. (2000) Mobile sources critical review: 1998 NARSTO assessment. *Atmos. Environ.* **34**, 2161–2182.

Schultz M. G., Jacob D. J., Wang Y., Logan J. A., Atlas E. L., Blake D. R., Blake N. J., Bradshaw J. D., Browell E. V., Fenn M. A., Flocke F., Gregory G. L., Heikes B. G., Sachse G. W., Sandholm S. T., Shetter R. E., Singh H. B., and Talbot R. W. (1999) On the origin of tropospheric ozone and NO_x over the tropical South Pacific. *J. Geophys. Res.* **104**, 5829–5843.

Seaman N. L. (2000) Meteorological modeling for air quality assessments. *Atmos. Environ.* **34**, 2231–2260.

Seinfeld J. H. and Pandis S. N. (1998) *Atmospheric Chemistry and Physics: From Air Pollution to Climate Change*. Wiley, New York.

Sillman S. (1993) Tropospheric ozone: the debate over control strategies. *Ann. Rev. Energy Environ.* **18**, 31–56.

Sillman S. (1995) The use of NO_y, H_2O_2 and HNO_3 as indicators for O_3–NO_x–ROG sensitivity in urban locations. *J. Geophys. Res.* **100**, 14175–14188.

Sillman S. (1999) The relation between ozone, NO_x and hydrocarbons in urban and polluted rural environments. Millennial review series. *Atmos. Environ.* **33**, 1821–1845.

Sillman S. and He D. (2002) Some theoretical results concerning O_3–NO_x–VOC chemistry and NO_x–VOC indicators. *J. Geophys. Res.* **107**, D22,4659, doi: 10.1029/2001JD001123.

Sillman S. and Samson P. J. (1995) The impact of temperature on oxidant formation in urban, polluted rural, and remote environments. *J. Geophys. Res.* **100**, 11497–11508.

Sillman S., Logan J. A., and Wofsy S. C. (1990) The sensitivity of ozone to nitrogen oxides and hydrocarbons in regional ozone episodes. *J. Geophys. Res.* **95**, 1837–1851.

Sillman S., Al-Wali K., Marsik F. J., Nowatski P., Samson P. J., Rodgers M. O., Garland L. J., Martinez J. E., Stoneking C., Imhoff R. E., Lee J.-H., Weinstein-Lloyd J. B., Newman L., and Aneja V. (1995) Photochemistry of ozone formation in Atlanta, GA: models and measurements. *Atmos. Environ.* **29**, 3055–3066.

Sillman S., He D., Pippin M., Daum P., Kleinman L., Lee J.-H., and Weinstein-Lloyd J. (1998) Model correlations for ozone, reactive nitrogen, and peroxides for Nashville in comparison with measurements: implications for VOC–NO_x sensitivity. *J. Geophys. Res.* **103**, 22629–22644.

Sillman S., Vautard R., Menut L., and Kley D. (2002) O_3–NO_x–VOC sensitivity and NO_x–VOC indicators in Paris: results from models and ESQUIF measurements. *J. Geophys. Res.* (in press).

Simpson D. (1995) Biogenic emissions in Europe: 2. Implications for ozone control strategies. *J. Geophys. Res.* **100**, 22891–22906.

Sosa G., West J., San Martini F., Molina L. T., and Molina M. J. (2000) *Air Quality Modeling and Data Analysis for Ozone and Particulates in Mexico City*. MIT Integrated Program on Urban, Regional and Global Air Pollution Report No. 15 (available from http://eaps.mit.edu/megacities/index.html).

Stockwell W. R., Kirchner F., and Kuhn M. (1997) A new mechanism for regional atmospheric chemistry modeling. *J. Geophys. Res.* **102**, 25847–25879.

Stehr J. W., Dickerson R. R., Hallock-Waters K. A., Doddridge B. G., and Kirk D. (2000) Observations of NO_y, CO, and SO_2 and the origin of reactive nitrogen in the eastern US. *J. Geophys. Res.* **105**, 3553–3563.

Stohl A. and Trickl T. (1999) A textbook example of long-range transport: simultaneous observation of ozone maxima from stratospheric and North American origin in the free troposphere over Europe. *J. Geophys. Res.* **104**, 30445–30462.

Tan D., Faloona I., Brune W. H., Shepson P., Couch T. L., Sumner A. L., Thornberry T., Carroll M. A., Apel E., Riemer D., and Stockwell W. (2001) HO_x budgets in a deciduous forest: results from the PROPHET summer 1998 campaign. *J. Geophys. Res.* **106**, 24407–24428.

Thornton J. A., Wooldridge P. J., Cohen R. C., Martinez M., Harder H., Brune W. H., Williams E. J., Fehsenfeld F. C., Hall S. R., Shetter R. E., Wert B. P., and Fried A. (2002) Observations of ozone production rates as a function of NO_x abundances and HO_x production rates in the Nashville urban plume. *J. Geophys. Res.* **107**, doi: 10.1029/2001JD000932.

Trainer M., Parrish D. D., Buhr M. P., Norton R. B., Fehsenfeld F. C., Anlauf K. G., Bottenheim J. W., Tang Y. Z., Wiebe H. A., Roberts J. M., Tanner R. L., Newman L., Bowersox V. C., Maugher J. M., Olszyna K. J., Rodgers M. O., Wang T., Berresheim H., and Demerjian K. (1993) Correlation of ozone with NO_y in photochemically aged air. *J. Geophys. Res.* **98**, 2917–2926.

Trainer M., Parrish D. D., Goldan P. D., Roberts J., and Fehsenfeld F. C. (2000) Review of observation-based analysis of the regional factors influencing ozone formation. *Atmos. Environ.* **34**, 2045–2062.

US Congress Office of Technology Assessment (1989) *Catching Our Breath: Next Steps for Reducing Urban Ozone*. OTA-O-412, US Government Printing Office, Washington, DC.

Volz A. and Kley D. (1988) Evaluation of the Montsouris series of ozone measurements made in the nineteenth century. *Nature* **332**, 240–242.

Volz-Thomas A. and Kolahgar B. (2000) On the budget of hydroxyl radicals at Schauinsland during the Schauinsland Ozone Precursor Experiment (SLOPE96). *J. Geophys. Res.* **105**, 1611–1622.

Wang Y. and Jacob D. J. (1998) Anthropogenic forcing on ozone and OH since preindustrial times. *J. Geophys. Res.* **103**, 31123–31135.

White W. H., Patterson D. E., and Wilson W. E., Jr. (1983) Urban exports to the nonurban troposphere: results from project MISTT. *J. Geophys. Res.* **88**, 10745–10752.

Wild O. and Prather M. J. (2000) Excitation of the primary tropospheric chemical mode in a global three-dimensional model. *J. Geophys. Res.* **105**, 24647–24660.

Williams E. J., Guenther A., and Fehsenfeld F. C. (1992) An inventory of nitric oxide emissions from soils in the United States. *J. Geophys. Res.* **97**, 7511–7519.

Wilson R. and Spengler J. D. (eds.) (1996) *Particles in Our Air: Concentrations and Health Effects*. Harvard School of Public Health, distributed by Harvard University Press, Cambridge, MA.

Wilson W. E. and Suh H. H. (1997) Fine particles and coarse particles: concentration relationships relevant to epidemiologic studies. *J. Air Waste Manage. Assoc.* **47**, 1238–1249.

Zaveri R. A. and Peters L. K. (1999) A new lumped structure photochemical mechanism for large-scale applications. *J. Geophys. Res.* **104**, 30387–30415.

9.12
Volatile Fuel Hydrocarbons and MTBE in the Environment

I. M. Cozzarelli

US Geological Survey, Reston, VA, USA

and

A. L. Baehr

US Geological Survey, W. Trenton, NJ, USA

9.12.1 INTRODUCTION

9.12.1.1 Scope of the Problem

Petroleum hydrocarbons (hydrocarbons that result from petroleum products such as oil, gasoline, or diesel fuel) are among the most commonly occurring and widely distributed contaminants in the environment. Volatile hydrocarbons are the lighter fraction of the petroleum hydrocarbons and, together with fuel oxygenates, are most often released from crude oil and liquid petroleum products produced from crude oil. The demand for crude oil stems from the world's ever-growing energy need. From 1970 to 1999, primary energy production of the world grew by 76% (Energy Information Administration, 2001), with fossil

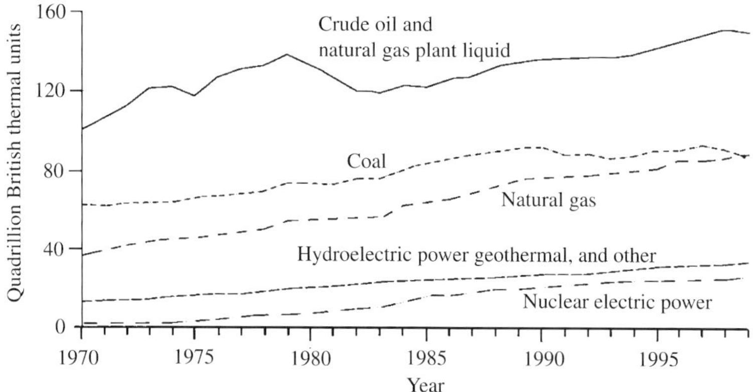

Figure 1 World primary energy production by source from 1970 to 1999 (Energy Information Administration, 2001).

fuels (crude oil, natural gas, and coal) accounting for ~85% of all energy produced worldwide (Figure 1). World crude oil production reached a record 68 million barrels (bbl) per day (1.08×10^{10} L d^{-1}) in 2000. The world's dependence on oil as an energy source clearly is identified as contributing to global warming and worsening air and water quality.

Petroleum products are present in Earth's subsurface as solids, liquids, or gases. This chapter presents a summary of the environmental problems and issues related to the use of liquid petroleum, or oil. The focus is on the sources of volatile hydrocarbons and fuel oxygenates and the geochemical behavior of these compounds when they are released into the environment. Although oxygenates currently in commercial use include compounds other than methyl *t*-butyl ether (MTBE), such as ethanol (ETOH), most of the information presented here focuses on MTBE because of its widespread occurrence. The environmental impact of higher molecular weight hydrocarbons that also originate from petroleum products is described in (Chapter 9.13, Abrajano *et al.*).

Crude oil occurs within the Earth and is a complex mixture of natural compounds composed largely of hydrocarbons containing only hydrogen and carbon atoms. The minor elements of sulfur, nitrogen, and oxygen constitute less than 3% of most petroleum (Hunt, 1996). Releases to the environment occur during the production, transport, processing, storage, use and disposal of these hydrocarbons. The petroleum industry classifies oil naturally occurring in a liquid phase as either conventional oil or nonconventional oil. Conventional oils can be explored and produced by conventional primary and secondary recovery techniques, whereas nonconventional oils, such as heavy oils, tar sands, and synthetic oils, are more difficult to extract from the host rock (Tissot and Welte, 1984). The history of liquid petroleum usage dates back centuries. Marco Polo described its use as a fuel for lamps in 1291 (Testa and Winegardner, 2000). In the mid-nineteenth century, Native Americans used crude oil skimmed from creeks and rivers as a medicinal ointment (Energy Information Administration, 1999). Commercial production in the United States began in 1859, with kerosene production for use in lamps. At the turn of the century, emphasis shifted to gasoline with the invention of the automobile, and in the 1930s and 1940s a substantial market for heating oil developed.

Table 1 United States and world energy consumption by source (January 2000).

Energy Source	*United States*		*World*	
	BBOE	*Percent*	*BBOE*	*Percent*
Petroleum	6.8	41	27.1	40
Natural gas	4.2	24.2	14.4	21.2
Coal	3.9	23.2	16.7	24.7
Nuclear	1.3	7.9	4.5	6.7
Hydroelectric	0.73	3.7	1.6	2.4
Biomass			3.4	5
Total	7.6	100	69.3	100

Source: Edwards (2001). BBOE = billion barrels of oil equivalent.

In 2000, over 85% of the world's energy came from fossil fuels (Table 1), with oil production alone supplying 40% of that energy (Edwards, 2001). Oil, with its high British Thermal Unit (BTU) density and ease of transport, is the most valuable fuel in the world today, and the demand for world crude oil is ever increasing. Annual world crude-oil demand increased from 22 billion to 28 billion bbl of oil (BBO) ($(3.5-4.5)\times10^{12}$ L) from 1990 to 2000. Increasing population growth and industrialization in developing countries is expected to drive this annual oil demand up 1.5% annually to ~38 BBO (6.0×10^{12} L) by 2030. As demand increases and known oil reserves are depleted, more emphasis will be placed on new discoveries and improved recovery technologies. Sources such as tar sands and oil shales are likely to become more important, bringing environmental problems inherent to the production, transport, and processing of these resources to new sectors of the environment.

Reserves of crude oil, the raw material used to make petroleum products, are not evenly distributed around the world. The production levels of the major oil-producing nations in the world, shown in Figure 2, are based on data collected by the US Department of Energy's (DOEs) Energy Information Administration (EIA). Nations in the Organization of Petroleum Exporting Countries (OPEC) produce ~43% of the world's total of nearly 68 million bbl per day (bpd) (1.08×10^{10} L d^{-1}), with Saudi Arabia as the world's largest producer at 8.3 million bpd (12% of total) (1.33×10^9 L d^{-1}), followed by Russia and the former Soviet Union at 6.4 million bpd (9%) (1.02×10^9 L d^{-1}). The United States produced ~5.7 million bpd (8%) (9.06×10^8 L d^{-1}). Oil and gas are being produced in virtually all geographic areas intersecting a wide range of ecological habitats, including tropical rain forests, Middle Eastern deserts, Arctic regions, and deep offshore marine environments.

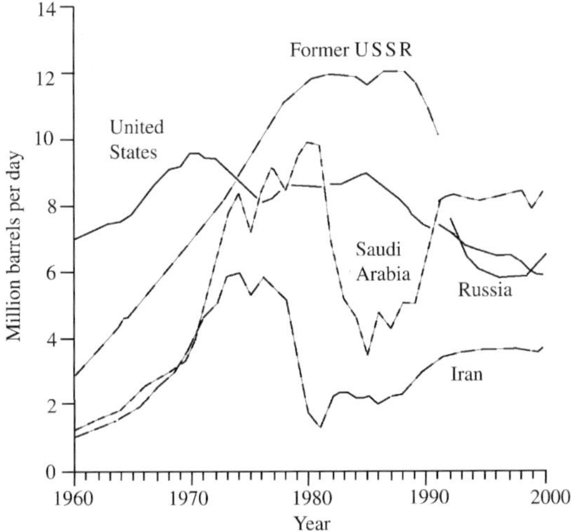

Figure 2 World crude-oil production from leading producers, 1960–2000, in million barrels per day (Energy Information Administration, 2001).

The petroleum industry has two main components, exploration/recovery and refinement/delivery. Crude oil is delivered to refineries and refined products are delivered to consumers through a large transportation network that includes tankers, barges, pipelines, and railroads. This transportation network links suppliers and producers across the world. Above-ground tanks, below-ground tanks and caverns, and offshore facilities store petroleum at various stages in this distribution system. The largest underground storage facilities in the US are part of the US. Strategic Petroleum Reserve maintained by the US government for times of supply shortage (Energy Information Administration, 1999). The global and complex nature of this industry means that every environmental compartment (air, land, water, biota) and all nations are affected by petroleum spills and waste products.

Petroleum products are released into the environment inadvertently and intentionally at all stages of petroleum use, from exploration, production, transportation, storage, use and disposal (Figure 3). The more volatile components of oil partition preferentially into the atmosphere and hydrosphere, and to a lesser extent, the geosphere (e.g., sorption onto sediments), and biosphere (i.e., bioaccumulation), whereas the higher molecular

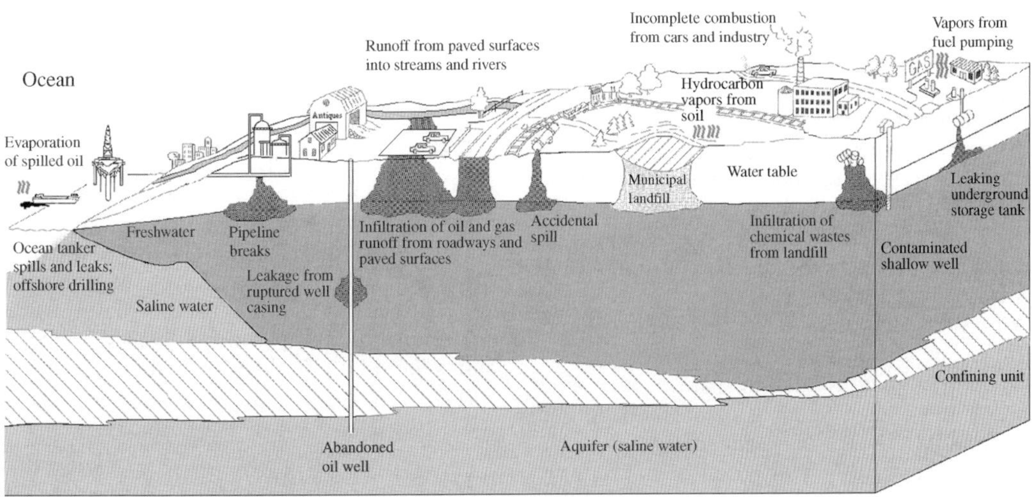

Figure 3 Modes of contamination by petroleum products during production, transport, use, and disposal (after Fetter, 1993).

weight fraction of oil is less soluble and less volatile and impacts the biosphere and geosphere more directly as described by Abrajano *et al.* (see Chapter 9.13). Usage causes compounds from petroleum products to separate and enter the environment. For example, incomplete fuel combustion in an automobile engine or a home heating furnace results in the release of compounds to the atmosphere, where they can be widely dispersed. Compounds in the exhaust gas of a motor boat partition in river or lake water, allowing for mixing and transport in the water body. These examples are illustrative of chronic low-level and widespread release of compounds referred to as nonpoint sources.

Concentrated releases of compounds to the environment generally are associated with product mishandling and accidents during storage and transportation. For example, gasoline leaking from an underground storage tank (UST) releases compounds to the air, water, and solids of the subsurface. An oil tanker spill at sea can result in the rapid dispersal of large amounts of product. These two examples of more concentrated release are referred to as point sources. However, in the case of a tanker spill, the point release can result in a serious regional problem such as the spill of the *Exxon Valdez* that spilled oil that spread 750 km from the original spill site and impacted 1,750 km of shoreline (Wolfe *et al.*, 1994).

Nonpoint sources can cause compounds to be dispersed throughout the hydrosphere, which can complicate efforts to evaluate the effect of spills. Current sampling and analytical techniques allow for the detection of volatile organic compounds (VOCs) at low concentrations (0.1 $\mu g\,L^{-1}$ and less). As a result, VOCs have been detected frequently in ambient shallow groundwater in urban areas across the country in studies conducted as part of the US Geological Survey (USGS) National Water-Quality Assessment (NAWQA) program (Squillace *et al.*, 1996). The importance of VOCs at low concentrations in ambient water is unclear. If these low concentrations are the result of plumes emanating from point sources, then concentrations could possibly increase with time. If the source were diffuse (nonpoint), as in the case of atmospheric sources, then changes in VOC concentrations in groundwater over time would be constrained by atmospheric concentrations.

Widely distributed and frequent point source spills can be considered nonpoint sources when the scale of concern is large in space or time. For example, the United States generates 1.3 billion gallons (4.9×10^9 L) of used motor oil each year, of which ~13% is disposed of improperly (Motor Oil Facts, http://www.epa.gov/seahome/housewaste/src/oilfact.htm). This large quantity of improperly disposed motor oil could be a regional source of MTBE and benzene, toluene, ethylbenzene, and xylenes (BTEX) as used motor oil contains BTEX and MTBE, ~500–2,000 $mg\,L^{-1}$ for BTEX and 100 $mg\,L^{-1}$ for MTBE (Baker *et al.*, 2002; Chen *et al.*, 1994). In a survey of 946 domestic wells designed to assess ambient groundwater quality in Maine, MTBE was detected in 15.8% of the wells, mostly at concentrations less than 0.1 $\mu g\,L^{-1}$ (State of Maine Bureau of Health, 1998). Small and widely distributed spills of fuel-related products not necessarily associated with leaking storage tanks were identified as likely sources.

9.12.1.2 Petroleum Chemical Composition

Petroleum in its natural form is of limited use. Therefore, further processing, or refining, is needed to create the wide array of petroleum products in use today (Energy Information Administration, 1999). The refining process results in products from the original crude oil with different hydrocarbon compositions. In industrialized societies, the most notable products are motor gasoline and distillate fuels such as home heating fuel (Figure 4). The United States leads the world in the generation of refined petroleum products, with gasoline accounting for just under half of the entire product.

A refinery converts crude oil into useful products by multistep processes that may include distillation, extraction, catalytic or thermal cracking, and final hydrotreating to remove unwanted compounds. The chemical components of the crude oil are separated by their different physical and chemical properties, thus creating products that have different compositions. The kinds of refined products generated from crude oil and their boiling point ranges are illustrated in Figure 5 (NRC, 1985). For example, gasoline is one of the lightest fractions and contains the lighter hydrocarbons (C_4–C_8 range, where n in C_n refers to the number of carbon atoms in the hydrocarbon chain), whereas bunker fuel (fuel for ships that may be a mixture of two or more of the distillate cuts shown in Figure 5 or it may be a residual oil from a distillation run (NRC, 1985)) has a higher boiling point and contains mostly the heavier components ($>C_{14}$). Thus, the effect of release of these refined products into the environment affects environmental compartments such as soils, the atmosphere, or freshwater bodies in different ways.

9.12.1.2.1 Crude oil

Crude oil within the Earth is of biological origin, formed when the remains of once-living

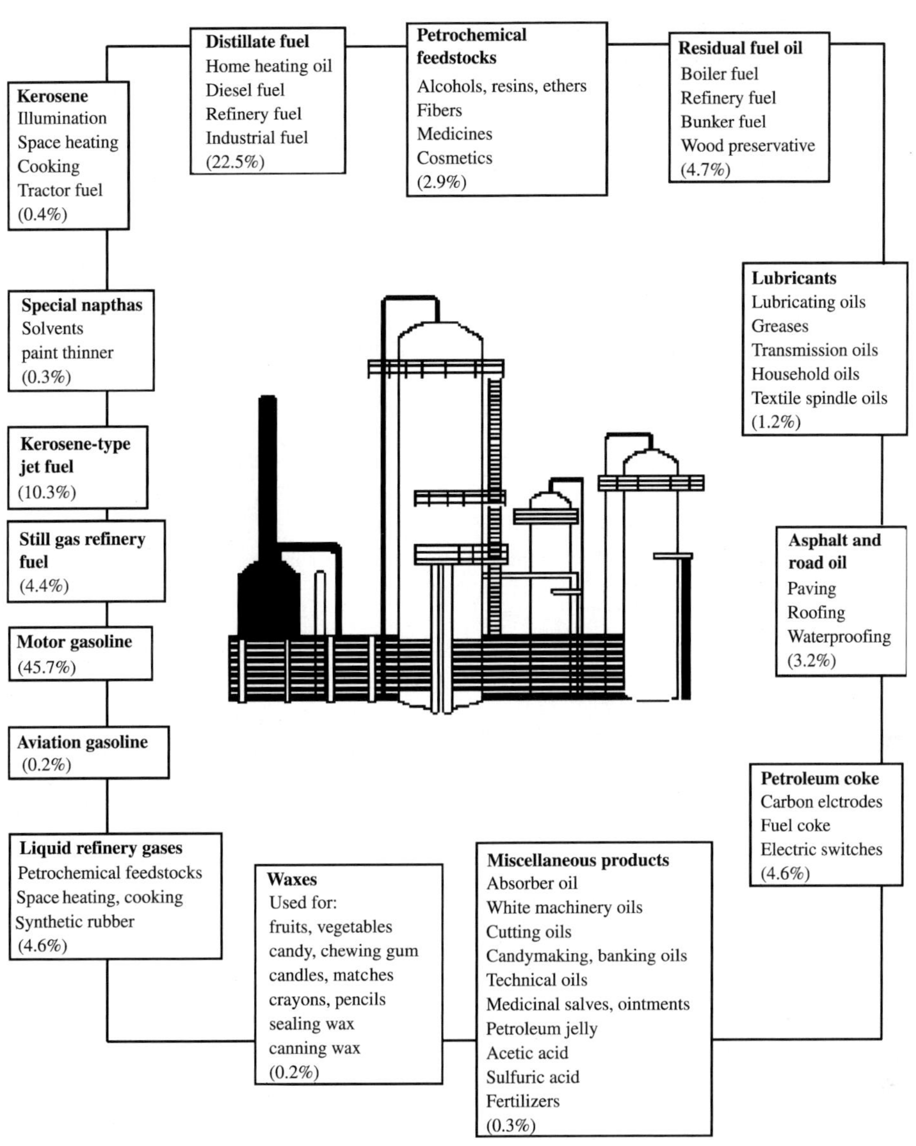

Figure 4 Petroleum products and uses (Energy Information Administration, 1999).

organisms were buried and subjected to heat and pressure over millions of years. The composition of the petroleum formed in a given geologic deposit depends chiefly on the nature of the original organic material and the basin's thermal history (Tissot and Welte, 1984). Petroleum hydrocarbons are a complex mixture of over 1,000 compounds with different physical and chemical properties. The fate of these compounds in the environment is dependent on their chemical structure.

The main groups of compounds in crude oils are saturated hydrocarbons (such as normal and branched alkanes and cycloalkanes that contain no double bonds), aromatic hydrocarbons, resins and asphaltenes (higher molecular weight polycyclic compounds containing nitrogen, sulfur, and oxygen (NSO)), and organometallic compounds (Tissot and Welte, 1984). Unsaturated hydrocarbons (nonaromatic hydrocarbons containing one or more double or triple bonds), such as olefins, essentially are absent. Most producible crude oils contain, on average, 30–35% each of *n*- and isoalkanes, cycloalkanes, and aromatics (Table 2) although there can be a large variation in the actual composition for individual crudes. The most abundant hydrocarbons in crude oil are in the light fraction ($<$ or $\approx C_{15}$). For example, in most crude oils the abundance of *n*-alkanes decreases beyond *n*-decane. In the aromatic fraction, benzene, toluene, and other low molecular weight alkylbenzenes are abundant (Tissot and Welte, 1984). Because of the process by which oil is

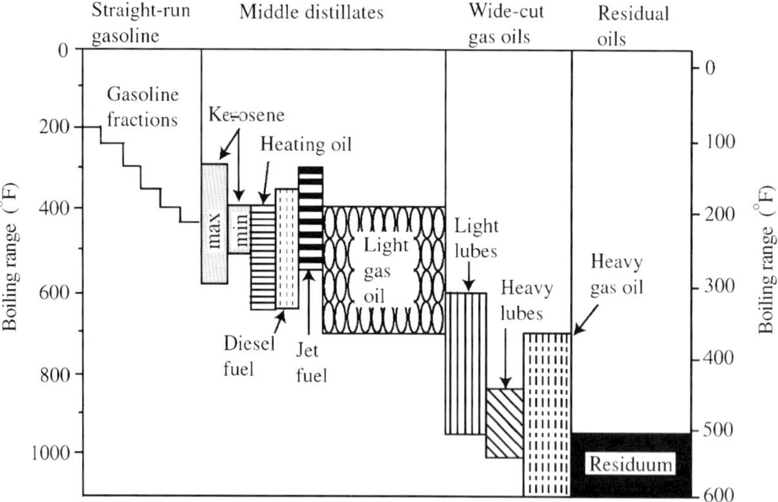

Figure 5 Common products obtained from crude-oil distillation and cracking, and their boiling point range (after National Research Council, 1985).

Table 2 Average composition of hydrocarbons in normal producible crude oils (% by weight of hydrocarbon) (total samples—517).

Compound group	*%*
n- + isoalkanes	33.3
Cycloalkanes	31.9
Aromatics	34.5
Saturated : aromatics	2.8
Alkanes : saturates	0.49

After Tissot and Welte (1984).

made, no two oils have the exact same history, and, therefore, no two crude oils have the same chemical composition.

Among the classes of compounds in crude oil, the alkanes, cycloalkanes, and aromatic fractions contain compounds that are volatile and are described in more detail in this chapter. The NSO compounds, resins and asphaltenes, and organometallic compounds are nonvolatile and, in general, are less mobile in the environment. A discussion of these compounds is beyond the scope of this chapter. Examples of some of the volatile hydrocarbons present in crude oil for the major compound classes are shown in Figure 6. Benzene, toluene, xylenes, ethylbenzene, isopropylbenzene, naphthalene, methyl- and dimethylnaphthalenes are among the most important volatile aromatic hydrocarbons, because they are widely used and, once released into the environment, their mobility and toxicity make them a significant environmental threat (Merian and Zander, 1982). The occurrence of these compounds in the environment is almost exclusively anthropogenic, with the exception of natural hydrocarbon seeps.

9.12.1.2.2 Fuels

Fuel products account for ~90% of all the petroleum used in the United States. The leading fuel produced in the world is gasoline (Energy Information Administration, 1999). Because of its widespread use and the fact that it is composed of that fraction of crude oil with lower boiling points, gasoline is the single largest source of volatile hydrocarbons to the environment (Table 3). Motor gasoline comes in various blends with properties that affect engine performance. All motor gasolines are made of the relatively volatile components of crude oil.

Other fuels include distillate fuel oil (diesel fuel and heating oil), jet fuel, residual fuel oil, kerosene, aviation gasoline, and petroleum coke (Figure 4). In the petroleum refining process, heat distillation is first used to separate different hydrocarbon components. The lighter products are liquefied petroleum gases and gasoline, whereas the heavier products include heavy gas oils. Liquefied petroleum gases include ethane, ethylene, propane, propylene, *n*-butane, butylenes, and isobutane. Jet fuels are kerosene-based fuels that fall into the lighter distillate range of refinery output (Figure 5).

9.12.1.2.3 Fuel additives

In the United States, passage of the 1990 Clean Air Act Amendments (CAAAs) led to the development of Federal programs that required oxygen-containing compounds be blended in gasoline to reduce carbon monoxide emissions. Examples of compounds used as oxygenates are methyl *t*-butyl ether (MTBE), ethanol (EtOH), ethyl *t*-butyl ether (ETBE), *t*-amyl methyl ether

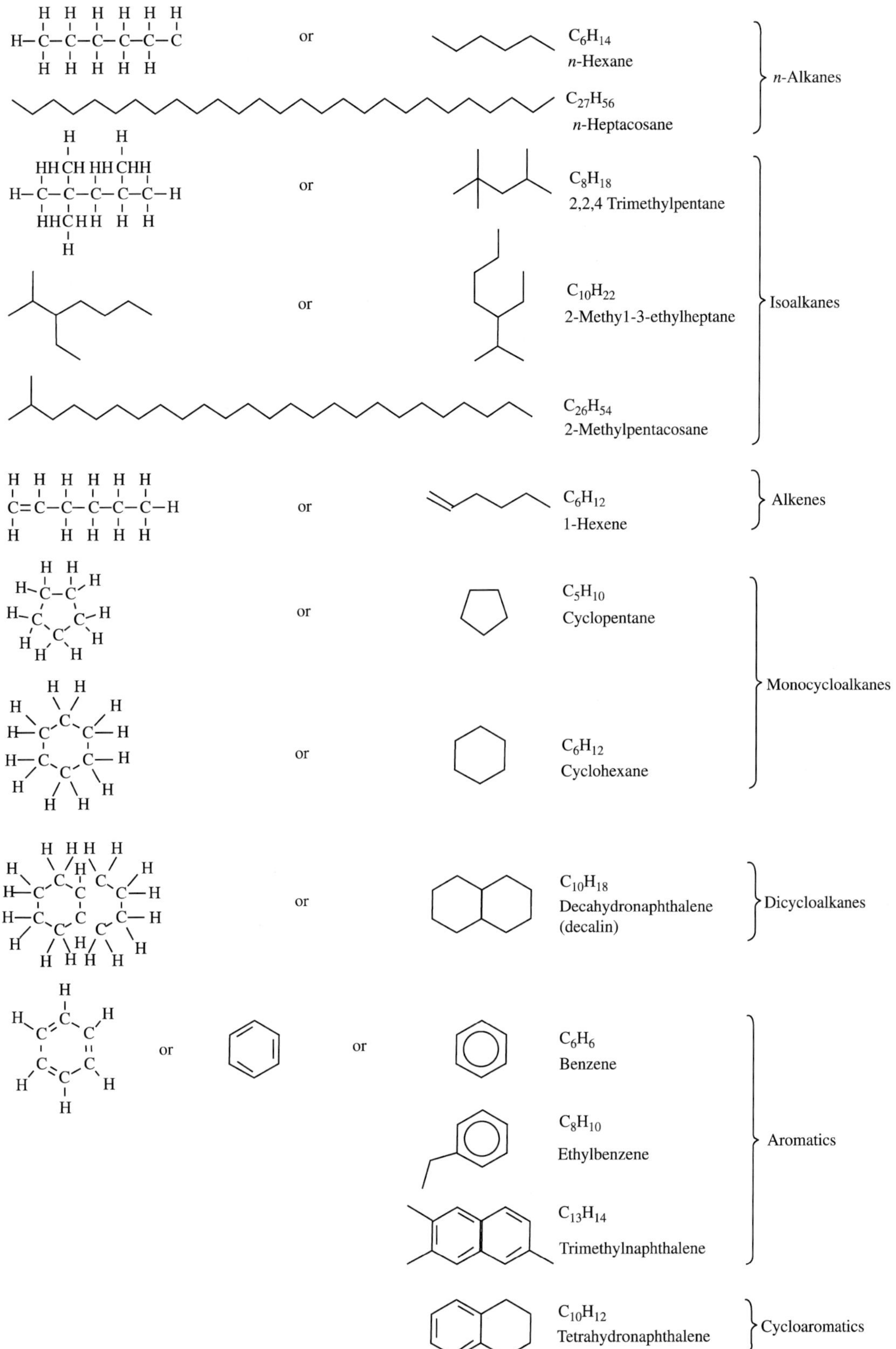

Figure 6 A sample of volatile hydrocarbons present in crude oil for the major compound classes, *n*-alkanes, isoalkanes, *n*-alkenes, cycloalkanes, aromatics, and cycloaromatics (after Tissot and Welte, 1984).

Table 3 Estimated composition of VOC emissions from gasoline evaporation at ambient air temperature.

Compound group	%
Alkanes	
Propane	1.0
n-Butane	17.0
Isobutane	7.0
n-Pentane	8.0
Isopentane	28.0
Hexane	20.0
Other alkanes $>C_6$	5.0
Alkenes	
Butene	3.0
Pentene	5.0
Other alkenes $>C_5$	3.0
Aromatic hydrocarbons	
Benzene	1.0
Toluene	1.5
Xylene	0.5

Source: Friedrich and Obermeier (1999).

(TAME), and diisopropyl ether (DIPE) (Zogorski *et al.*, 1997). MTBE (chemical formula $C_5H_{12}O$), is the most common gasoline oxygenate. MTBE use began in the 1970s as an octane booster to replace lead additives (Hogue, 2000), and, at that time, it was used in concentrations of 2–7% by volume in gasoline (Gullick and LeChevallier, 2000).

As mandated by CAAA, gasoline used in metropolitan areas where carbon-monoxide standards are not attained must contain no less than 2.7% oxygen by weight. By volume, this requirement corresponds to 14.8% and 7.3% for MTBE and EtOH, respectively. Gasoline oxygenated to this extent contains ~110 g of MTBE per liter. As a result of its adoption as the primary oxygenate to satisfy the needs of the CAAA, MTBE increased in rank of use among organic chemicals produced from 39th in 1970 to 4th in 1998 (Thompson *et al.*, 2000). The predominant use of MTBE as an oxygenate is because it is inexpensive to produce and has favorable blending characteristics with gasoline (Gullick and LeChevallier, 2000).

The US Environmental Protection Agency (USEPA) established a drinking water health advisory of 20–40 $\mu g\ L^{-1}$ MTBE in December 1997 (USEPA, 1987). MTBE is the only oxygenate whose occurrence in water supplies and the ambient environment has been widely documented. It has been found in 5–10% of community drinking-water supplies in areas where gasoline with high MTBE concentrations are used (Zogorski *et al.*, 1997). Approximately 3% of these supplies are above the 20 $\mu g\ L^{-1}$ limit of the advisory range (Gullick and LeChevallier, 2000). MTBE in community drinking water supplies has been reported in concentrations of 610 $\mu g\ L^{-1}$ in supply wells for the city of Santa Monica, Calif. (Johnson *et al.*, 2000) and over 10 $mg\ L^{-1}$ elsewhere in the US private wells (Davis and Powers, 2000) primarily because of releases from leaking USTs (LUSTs). The USEPA included MTBE on the Contaminant Candidate List in 1999.

9.12.1.2.4 Solvents, lubricants and petrochemical feedstocks

The nonfuel use of petroleum is small compared with fuel use. Nonfuel uses include solvents such as those used in paints, lacquers and printing inks, lubricating oils and greases, waxes used in candy making, packaging, and candles, petroleum jelly, asphalt, petroleum coke, and petrochemical feedstocks. Lubricating oils are complex mixtures containing linear and branched paraffins, cyclic alkanes and aromatic hydrocarbons ($>C_{15}$ with boiling points between 300 °C and 600 °C) (Vazquez-Duhalt, 1989). As with all petroleum products, the exact nature of the chemical composition varies with the starting crude-oil composition and the refinery process. In general, lubricating oils are 73–80% aliphatic hydrocarbons, 11–15% monoaromatic hydrocarbons, and 4–8% polyaromatic and polar compounds. As previously noted, motor oil accumulates additional compounds, such as MTBE and BTEX, as it circulates through the engine (Chen *et al.*, 1994).

Petrochemical feedstocks are used for the manufacture of chemicals, synthetic rubber, and a variety of plastics, and form the basis of synthetic fibers and drugs. Petrochemical feedstocks include solvents such as ethylene, propylene, butylenes, benzene, toluene, and xylenes. Large amounts of solvents are used in the chemical industry as reaction media, as a base for paints and thinners, and as propellants for sprays. The VOCs emitted as a result of solvent use are composed of a larger variety of species than those released from the use of fossil fuels (Friedrich and Obermeier, 1999). Depending on the product, the emissions of hydrocarbons during solvent use vary appreciably. The highest emissions of pure hydrocarbons come from adhesives, which release vapors containing typically ~20% by weight of pure toluene. With the exception of solvents, most products in this category contain largely the nonvolatile fraction of petroleum and therefore do not represent significant sources of volatile hydrocarbons to the environment.

9.12.1.3 Ecological Concerns and Human Exposure Pathways

The human exposure routes from hydrocarbons in the environment include inhalation, ingestion,

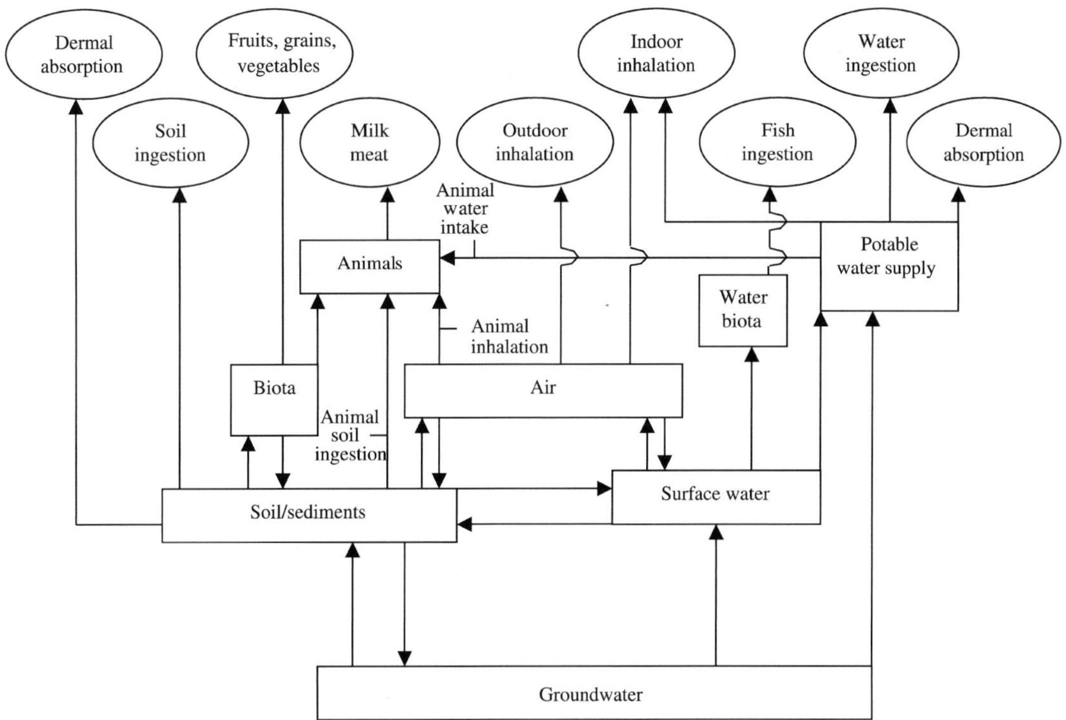

Figure 7 Modes of exposure to hydrocarbon contaminants (after Kostecki and Calabrese, 1991).

and dermal adsorption. The pathways of exposure can include direct contact with the hydrocarbon phase itself, inhalation of hydrocarbon vapors in indoor or outdoor air, and ingestion of contaminated food or water (Figure 7). Among children, ingestion of contaminated soils also is a potential concern (Calabrese *et al.*, 1991). Ingestion of unintentional hydrocarbon contamination of food is rare and direct contact with hydrocarbons typically is limited to workers in the petroleum industry. Therefore, it does not appear to be a major risk for the general population (Daugherty, 1991).

Exposure to indoor hydrocarbon vapors can occur as a consequence of changes of land use from commercial or industrial to residential. Residual hydrocarbons in soils or groundwater may result in a chronic vapor exposure pathway. Analysis of risk associated with exposure to hydrocarbons typically is undertaken in a multistep approach known as risk assessment. A thorough discussion of the use of risk assessment at contaminated sites is provided in Chapter 9.01. Assessing risks posed by hydrocarbon spills or wastes is complex and involves estimates of chemical concentrations at each potential exposure point, identification of the potential populations that may be exposed, and assessment of exposure pathways, intake rates, and the toxicity of the chemicals of concern.

Determining a safe level of exposure to petroleum hydrocarbons is a difficult process because petroleum is a complex mixture of thousands of compounds with varying degrees of toxicity to human health and the environment (Miller *et al.*, 2000). No toxicity data are available for the vast majority of these compounds. Evaluation of the potential impact of petroleum released into the environment is complicated further by the fact that the composition of the released hydrocarbons changes over time because of natural degradation processes. One approach to assess the potential health effects of hydrocarbon exposure is to first use the entire complex mixture in a battery of short- and long-term toxicological tests to determine the combined toxicity of the complex mixture. Unfortunately, these data are not available for most petroleum products. Toxicity data on whole mixtures of petroleum products are only available for gasoline, jet fuel, and mineral oil. No whole mixture toxicity data are available for crude petroleum or drilling fluids or cuttings (Miller *et al.*, 2000). A major shortcoming with the whole mixture approach is that once in the environment, the parent product separates into fractions based on differences in individual environmental transport characteristics (Miller *et al.*, 2000). Because of these differences in transport characteristics, the mixture to which a receptor is exposed will vary in space and time. Thus, toxicity data on the original product will not necessarily reflect the toxicity of the product once it is released into the environment.

Specific toxicity data are available for only ~95 individual hydrocarbons of the over 1,000 components of petroleum, and only 25 have sufficient toxicological data to develop exposure criteria (Miller *et al*., 2000). One approach to estimating exposure effects for a complex petroleum mixture is to base the estimates on the 25 compounds for which toxicity data exist. Individual toxicity data are largely available for the more toxic compounds and, thus, this approach may be overly conservative for more complex mixtures of petroleum, although the importance of simultaneous exposure to the multiple components of petroleum largely is unknown. Recent work has been based on dividing petroleum into fractions based on structure (such as aromatic or aliphatic) and molecular weight. Compounds greater than C_{38} are not considered bioavailable by dermal or oral routes (Miller *et al*., 2000). Indicator compounds are used to represent the toxicity of each fraction and different cleanup criteria are developed for each petroleum fraction. Most evaluations of the health effects of contamination consider single point sources and there is little understanding of the effect of multiple points of exposure over time. Assessment of cumulative exposure risks is especially important for nonpoint sources of contamination that may result in longer-term exposure.

Among the volatile fuel hydrocarbons, benzene, a known carcinogen is of utmost concern from a human-health perspective. Alkenes and alkylbenzenes are considered more toxic than alkanes (Oestermark and Petersson, 1993). All alkenes are potentially genotoxic because of metabolic formation of epoxides (Oestermark and Petersson, 1993). Of particular concern are ethene and 1,3-butadiene. Although alkanes are considered less toxic, in general, than alkenes or alkylbenzenes, hexane is a known neurotoxin (Perbellini *et al*., 1980). Calabrese and Kenyon (1991) report on the toxicological assessment of numerous volatile hydrocarbons including benzene, toluene, ethylbenzene, xylenes, and *n*-hexane.

9.12.2 THE PETROLEUM INDUSTRY

9.12.2.1 Petroleum Exploration, Production, and Processing

During petroleum exploration and production operations, hydrocarbons can be released to the environment by spillage of the crude oil itself or by the release or disposal of waste generated during the exploration and production process. The main categories in the petroleum refining process include separation (e.g., distillation), conversion (e.g., thermal cracking, alkylation), treatment (e.g., hydrosulfurization), and blending (Friedrich and Obermeier, 1999). Many of these processes release volatile hydrocarbons directly to the atmosphere and additional releases to the atmosphere as well as to the geosphere occur because of leaks from equipment and spills.

The techniques needed to remove or extract crude oil from the earth vary depending on the type of rock the oil is found in (such as sandstone or limestone) and its porosity and permeability. Oil typically is removed from the subsurface by wells drilled into the formation. Oil wells are onshore and offshore. The petroleum underground usually is under great pressure and will, therefore, flow into the lower pressure environment created by the open well. With high enough pressure, the oil will flow up through the well. Many wells, however, especially those that have been in production a long time (tens of years), require artificial lift methods, or pumping, because pressure within the reservoir is insufficient to lift the oil to the surface. When oil no longer flows sufficiently using these methods, secondary and tertiary, or enhanced, recovery methods are used. These methods may use water, gases, or other chemicals to flush the oil out of the rock, or use heat to thin the oil and increase its flow rate. These methods produce not only crude oil but also water or other chemical wastes that contain hydrocarbons.

In terms of volume and toxicity, drilling fluids and drill cuttings are among the most significant waste streams from exploration and development activities in the oil-and-gas industry (USEPA, 1987). The composition of drilling fluids is complex and varies depending upon the specific down-hole conditions such as the geology of the formation. A major part of the undiscovered oil and natural-gas resources outside the Middle East lie in offshore areas (USGS, 2000). In 1999 over 150 countries had some type of drilling or production operations offshore (Jones *et al*., 2000). Cuttings, the geological core materials brought to the surface by the drilling fluids, primarily are formation rock along with drilling fluids that adhere to their surface. Because there presently is no known use for the cuttings offshore, disposal is the same as with the used fluid—either discharge, hauled to shore for disposal, or injection back into the formation (Jones *et al*., 2000). The volume of cuttings generated can be 4–5 times more than the borehole volume because of excessive caving of formations into the borehole.

Produced water is the aqueous phase separated from recovered hydrocarbons produced during oil well production, and injection water used to maintain pressure in the reservoir. Over the life of an oil field, the volumes and composition of produced water varies greatly. Produced water discharges are continuous and represent the largest volume waste stream in production operations (Stephenson, 1992). The quantity discharged at any particular treatment facility depends on the number of wells handled by that facility. The discharge rates range from less than 500 bbl d^{-1} (7.9×10^4 L d^{-1}) (most of the platforms in the Gulf of Mexico) to over 6×10^5 bbl d^{-1} (9.5×10^7 L d^{-1}) (three production processing platforms in the South Java Sea).

Drilling fluids can be water-based fluids (WBFs), oil-based fluids (OBFs), and synthetic-based fluids (SBFs). An OBF has diesel, mineral oil, or some other oil as its continuous phase. Although wells drilled with OBF technology produce less waste volumes than those drilled with WBFs, they contain priority pollutants and have far greater toxicity. In many cases offshore, this waste is barged to shore for land disposal or recycled because of restrictions on offshore discharge. Despite their unique and valuable properties, the use of OBF is declining because of the added cost of barging waste fluids to shore and the development of more attractive alternatives such as SBFs (Jones *et al.*, 2000).

Drilling and production discharges to the marine environment present environmental concerns different from those in onshore areas. Discharge of OBF cuttings poses the most severe environmental impact on the seafloor of the three fluid types. On the seafloor, OBF cuttings can increase oil in the sediments and decrease biological abundance and diversity of immobile bottom-dwelling organisms (Hartley and Watson, 1993). The overall effect of an OBF cuttings discharge is, in part, determined by the concentration of the oil remaining on the cuttings and its toxicity. In zones severely affected by OBF cuttings discharge, the rate of recovery depends upon the energy level of the environment and the seafloor ecology. At many North Sea drilling sites, cuttings are found in large piles 100–150 feet (30–46 m) high and 200 feet (61 m) across on the seafloor (Jones *et al.*, 2000). Studies on the impact of oil wet cutting discharges in the oceans have been conducted for over 16 years. Benthic studies around offshore platforms using oil-based drilling muds have shown that the greatest disturbances are near the platform. Davies *et al.* (1984, 1989) found that within 500 m of the platform hydrocarbon levels reached 10^1–10^5 times background levels. Addy *et al.* (1984) discovered that the effect of drilling discharges were most severe with diesel-based muds followed by mineral oils, and the lowest effect were with water-based muds. Hydrocarbon inputs into the world's oceans have been the subject of various National Research Council (NRC) reports focusing on the sources, fate and effects of these hydrocarbons (NRC, 1985, 2003).

A substantial number of investigations of the surface-water and groundwater impacts of onshore produced water releases have been conducted in the United States since the 1960s. Much of the focus has centered on the effect of the high concentrations of salts in oil-field brines on soil fertility and erosion, vegetation health, and groundwater quality. It is estimated that there are nearly 683 mi^2 (1,769 km^2) of brine-impacted soil in Texas alone (McFarland *et al.*, 1987). Evaluating the effect of nearly 3.5 million oil and gas wells drilled in the conterminous US since the early 1800s is complicated in that there is no systematic collection of information at the state or national scale describing past produced water releases either as to location or volumes of water released. Evaluation of aquifer or watershed susceptibility to this source is being investigated by the USGS using geographic information system (GIS) techniques (Otton and Mercier, 2000). Recently, the USGS (Breit *et al.*, 2000) began reviewing data from 77,650 archived records of water samples, with the goal of developing a database with internal consistency and one that would be useful for assessing the general patterns related to the chemical compositions of formation waters from different geologic settings. This database provides useful information on inorganic chemical species present in the oil-field brines, but no similar effort yet has been undertaken to compile a comprehensive database of the dissolved hydrocarbon concentrations in these waters, so the potential cumulative effect of this source on the environment largely is unknown.

Other sources of pollution during oil exploration and production are evaporative losses of hydrocarbons, estimated at 45–68 Mt yr^{-1} (Merian and Zander, 1982) and the catastrophic release of crude oil because of accidents such as well-site explosions. Evaporative losses during production of crude oil largely result in the release of *n*-alkanes up to C_8 (octane) into the atmosphere (Friedrich and Obermeier, 1999). During petroleum processing, the major sources of VOC emissions to the atmosphere are vacuum distillation (0.05 kg of VOC emissions per cubic meter of refinery feed), catalytic cracking (0.25–0.6 kg of VOC per cubic meter of feed), coking (0.04 kg VOC per cubic meter of feed), chemical sweetening and asphalt blowing (27 kg VOC per metric ton of asphalt).

Analysis of spill data at the Beykan oil field in Turkey indicated that over a 6 yr period (1989–1995) that 252 spills were recorded resulting in a net release of 395 t of oil (Ünlü and Demirekler, 2000). One of the more famous cases of petroleum contamination of the environment included a huge oil spill in the Santa Barbara Channel in 1969 due to a blow out of a Union Oil Co. of California development well that leaked more than 5×10^4 bbl (7.9×10^6 L) of oil into the channel. Highly publicized contamination of coastal waters by petroleum spills, such as the Santa Barbara spill, are credited as the catalyst that thrust the modern environmental movement to the forefront of American conscience and politics in the early 1970s (Williams, 1991).

9.12.2.2 Petroleum Transportation and Storage

Transportation of crude oil and petroleum products occurs predominantly by tanker, pipeline, railway tank cars, and tank trucks, whereas storage occurs largely in surface or underground tanks and in underground caverns. The distribution and storage of crude oil and refined products result in releases of significant amounts of hydrocarbons to the atmosphere, surface waters, soils, and groundwater. Groundwater contamination by crude oil, and other petroleum-based liquids, is a particularly widespread problem. An estimated average of 83 crude-oil spills occurred per year during 1994–1996 in the United States, each spilling $\sim 5 \times 10^4$ bbl (7.9×10^6 L) of crude oil (Delin *et al.*, 1998). The USEPA estimates that there are 698,000 Federally regulated USTs buried at over 269,000 sites nationwide (as of March 31, 2002, http://www.epa.gov/swerust1/overview.htm). LUSTs are the most common source of groundwater contamination (http://www.epa.gov/swerust1/pubs/ustprogramfacts.pdf) and have resulted in a significant impact on groundwater quality.

Atmospheric emissions of volatile hydrocarbons from this sector of the petroleum industry have been summarized by Friedrich and Obermeier (1999). Gasoline, because it contains components with a high volatility, is the most important source of VOCs to the atmosphere during petroleum handling and storage. VOCs released into the atmosphere can undergo photochemical reactions with nitrous oxides in the atmosphere, producing ground-level ozone, or photochemical smog, as described in more detail in (Chapter 9.11, Sillman). The estimated composition of VOC emissions because of evaporation of gasoline is given in Table 3.

Seaports handle large amounts of petroleum products and have facilities on site to store these products. At the Port of Los Angeles alone, for example, there are ~500 above-ground storage tanks with a capacity to hold 500 million gallons (1.9×10^9 L) of product (Rice, 1991). Despite this, the lasting effects of oil spills on the open seas may be less than the effect of the many subsurface leaks and spills of petroleum products or the atmospheric pollution caused by the use of petroleum products, oil spills on the open seas create a major public policy impact because they spread rapidly and provide powerful visual images. According to Williams (1991) and Wolfe *et al.* (1994), after being relatively quiet in the 1980s, the American environmental movement was revitalized by the massive and well-publicized oil spill, the *Exxon Valdez*. The publicized images of environmental devastation had a galvanizing effect on the causes of environmental organizations.

Pipelines usually are used to transport crude oil from the exploration and discovery site to the refinery location, as well as to transport petroleum products to end users. Crude oil travels through pipelines under pressure at speeds up to 6 mi h^{-1} (1 mi = 1.609344 km). Pipeline breaches can result in explosive releases of crude oil or petroleum products. One of the better-studied pipeline breaks occurred near Bemidji, Minnesota in 1979, when the land surface and shallow subsurface were contaminated when a crude-oil pipeline burst, spilling $\sim 1.7 \times 10^6$ L (~10,700 bbl) of crude oil onto a glacial outwash deposit (Baedecker *et al.*, 1993; Delin *et al.*, 1998; Eganhouse *et al.*, 1993). Crude oil sprayed on the land surface covering an area estimated at 7,500 m^2 (Delin *et al.*, 1998) and percolated through the unsaturated zone to the water table near the rupture site. Some of the sprayed oil also flowed over the surface toward a small wetland. Aromatic hydrocarbons from this spill contaminated the groundwater, as well as the sediments and unsaturated zone soil gas. The lighter aliphatic fractions were found predominantly in the unsaturated zone whereas the aromatic hydrocarbons from C_6 to C_{10}, which are soluble in water and have lower Henry's law constants, were transported downgradient in groundwater, farther than other organic compounds (Baedecker and Eganhouse, 1991; Eganhouse *et al.*, 1993). The major hydrocarbon types found downgradient from the oil were volatile aromatic hydrocarbons in groundwater and nonvolatile aliphatic hydrocarbons in sediment. The higher molecular weight hydrocarbons, predominantly normal alkanes in the C_{11} to C_{33} range and the isoprenoid hydrocarbons, pristane and phytane, have low water solubilities and occurred in groundwater only near the oil source. In a study conducted eight years later, Furlong *et al.* (1997) found that the *n*-alkanes largely were absent from the groundwater at the Bemidji site, even close to the oil, whereas PAH (polycyclic aromatic hydrocarbons) persisted in water downgradient from the oil. Additional information about the fate of PAH in the environment has been provided by Abrajano *et al.* (Chapter 9.13).

Among the hundreds of thousands of USTs that are subject to federal regulation, releases have been confirmed at 422,573 of these sites (http://www.epa.gov/swerust1/pubs/ustprogram facts.pdf). It has been estimated that ~95% of USTs store petroleum fuels (Lund, 1995) and that gasoline usage represents ~80% of the volume of motor fuel consumption (Zogorski *et al.*, 1997). The UST problem, therefore, is largely one of gasoline released to the subsurface. The impact of petroleum releases in the subsurface on groundwater quality and restoration of contaminated aquifers has been the subject of numerous symposia, books and reviews (e.g., Testa and Winegardner, 2000; Ward *et al.*, 1997; Wiedemeier *et al.*, 1999).

MTBE is much more water-soluble (5×10^4 mg L^{-1}) than the BTEX compounds (ranging in solubility from 1,780 mg L^{-1} for benzene to 152 mg L^{-1} for ethylbenzene) and is present in oxygenated gasoline in larger amounts than the BTEX compounds. Therefore, when oxygenated gasoline is spilled, MTBE is the principal contaminant in groundwater on a mass basis. Although MTBE has been used as a gasoline additive since the late 1970s, its usage dramatically increased in 1992 in response to the CAAA requirement to oxygenate gasoline. Many sites of UST investigation prior to 1992 may have had MTBE in groundwater that was not detected because MTBE was not included in routine schedules of sample analysis at spill sites prior to that time.

Perhaps the most publicized case of MTBE contamination of groundwater is that involving public water supply wells in Santa Monica, CA. In August 1995, the city of Santa Monica discovered MTBE in drinking-water supply wells through routine analytical testing of well water. In 1996, levels of MTBE at the city's Charnock Wellfield rose to more than 600 ppb and, by June 13, 1996, all of the five supply wells at the city's Charnock Wellfield were shut down because of persistent and increasing concentrations of MTBE contamination (USEPA Fact Sheet available at the website: http://www.epa.gov/region09/cross_pr/mtbe/charnock/

9.12.2.3 Petroleum Usage

Transportation, or mobile sources, account for an estimated release of 34 Mt of hydrocarbons into the atmosphere each year (Merian and Zander, 1982) and urban runoff as a source of petroleum to the marine environment has been well documented (Eganhouse and Kaplan, 1981; Latimer *et al.*, 1990). In urban atmospheres, the composition of volatile aromatic hydrocarbons has been found by numerous investigators to be similar to that of gasoline, indicating gasoline as a source (Merian and Zander, 1982). The air in rural areas contains benzene at concentrations of 0.001–0.003 mg m^{-3}, whereas in urban areas concentrations typically average 0.04 mg m^{-3} of benzene. Worldwide, the estimated contribution to the atmosphere from evaporation of motor fuels alone is ~10^5 t benzene, 0.5×10^5 t toluene and 0.15×10^5 t higher aromatics per year (Merian and Zander, 1982). Incomplete combustion of fossil fuels also can lead to the emission of hydrocarbons and other gases into the atmosphere. Emissions from internal combustion engines include significant amounts of alkanes such as methane, butane, and isobutane, alkenes such as ethene, propene, and isobutene, and aromatics such as benzene, toluene, xylenes, and ethylbenzene (Friedrich and Obermeier, 1999). In the transportation sector, two-stroke engines, such as in mopeds, release the highest amounts of VOCs to the atmosphere, whereas gasoline engines equipped with three-way catalysts have the lowest emissions of VOCs.

Worldwide petroleum consumption rose from ~61 million bpd (9.7×10^9 L d^{-1}) in 1981 to ~76 million bpd (1.2×10^{10} L d^{-1}) in 2001. Over these two decades the US has accounted for ~26% of this consumption—19.65 million bpd in 2001 (Energy Information Administration, 2001). Of these 19.7 million bpd (3.1×10^9 L d^{-1}), the equivalent of 8.5 million bpd (1.4×10^9 L d^{-1}) are consumed as gasoline, which resulted in the usage of 13 million gallons per day (4.9×10^7 L d^{-1}) of MTBE (US Department of Energy, Energy Information Administration, Alternatives to Traditional Transportation Fuels, 1999, Washington, 2000, website http://www.eia.doe.gov). Therefore, it is not surprising that MTBE is the second most frequently detected VOC in ambient groundwater (exceeded only by chloroform) as determined by sampling conducted by the USGS (Squillace *et al.*, 1999). Although spilled oxygenated gasoline is the most apparent point source of MTBE in groundwater, the source of MTBE is not always known, especially when it is detected at low concentrations. In a survey of 946 domestic wells designed to assess ambient groundwater quality in Maine, MTBE was detected in 15.8% of the wells, mostly at concentrations less than 0.1 g L^{-1} (State of Maine Bureau of Health, 1998). Small and widely distributed spills of fuel-related products not necessarily associated with LUSTs were identified as likely sources.

Another possible nonpoint source of MTBE is the atmosphere; concentrations of MTBE in the atmosphere above urban areas can be sufficiently high to cause its detection in groundwater at concentrations of 0.1 μg L^{-1} or less (Baehr *et al.*, 1999). In their study of southern New Jersey groundwater quality, USGS researchers (Baehr *et al.*, 1999; Stackelberg *et al.*, 1997) compared atmospheric concentration data of MTBE and other VOCs over a 2 yr period. Among the VOCs most frequently detected in groundwater (trichloromethane (chloroform), MTBE, 1,1,1-trichloroethane (TCA), and tetrachloroethene (PCE)), only atmospheric concentrations of MTBE were sufficiently high to cause detection in water samples. It was concluded that the presence of VOCs (other than MTBE) resulted from nonatmospheric sources, whereas the atmospheric source of MTBE potentially could cause detection at the 0.1 μg L^{-1} concentration, provided MTBE was not degraded in the unsaturated zone (Baehr *et al.*, 2001).

9.12.2.4 Disposal of Petroleum Wastes

Fuels and heating oils are intended to be consumed during usage with little or no waste produced aside from gases released into the atmosphere. One exception is used motor oil that is disposed of in landfills or down-storm drains and can be a source of groundwater contamination. Bjerg *et al.* (see Chapter 9.16) evaluate the landfill as a source of volatile hydrocarbons to groundwater and present the concentrations typically found in leachate-impacted aquifers. Used motor oil (as discussed in Section 9.12.1.1) which contains BTEX and MTBE from interaction with gasoline during use, also presents a major environmental problem from improper disposal, partly because it is so widely used and handled. Almost 60% of the motor oil used in the US is consumed by vehicle owners changing their own oil (Motor Oil Facts, http://www.epa.gov/seahome/housewaste/src/oilfact.htm).

9.12.3 TRANSPORT PROCESSES

The presence of petroleum compounds through the environment is initiated by product usage and the partitioning of compounds from the initial product to multiple phases in the environment. Exposure pathways then are established because the compounds are transported within the bulk fluid phases of the atmosphere and hydrosphere or as solid phases in the case of suspended sediment or biota. The phases relevant to the transfer of compounds from petroleum product sources and the processes by which these transfers occur are listed in Table 4. The product phase usually is an organic liquid; however, compressed gases and waste slurries are examples of gaseous and aqueous phase sources, respectively. When the product is released, the primary factor governing the rate of mass transfer of product constituents to the environment is the concentration of the constituents at phase interfaces. Models of chemical equilibrium are used to estimate these concentrations. For example, if the product is liquid and immiscible in water, then the equilibrium concentrations at the product and aqueous phase interface, by the process of solubilization (Table 4), is expressed as

$$C_k = \gamma_k \chi_k C_{\text{sol}} \quad (1)$$

where C_k is the aqueous phase concentration of the kth constituent, C_{sol} is the solubility of the constituent (g cm^{-3}), γ_k is the activity coefficient, and χ_k is the mole fraction of the compound in the product (organic liquid) phase. The index k identifies the constituent in a product consisting of N compounds ($k = 1, 2, 3, \ldots, N$). If the product consists of nearly a pure compound (e.g., reagent grade benzene), then $\gamma_k \chi_k \rightarrow 1$.

Equation (1) is used to determine equilibrium between free products and dilute aqueous phases in terms of pure compound properties (solubility) adjusted for the composition of the product (mole fraction). The activity coefficient reflects the effect of phase composition on the equilibrium relation (nonideal behavior). If $\gamma_k \approx 1$, then Equation (1) reduces to Raoult's law, which states idealized

Table 4 Pathways relevant to the phase transfer of compounds from petroleum product sources to environmental compartments by partition processes.

Transfer from	*Examples*	*Transfer to*	*Process*
Free product	Gasoline, jet fuel, diesel fuel, crude oil, heating oil, motor oil, lubricants	Water Air Solid phases	Solubilization Volatilization Sorption
Water	Ocean, estuary, river, lake, precipitation, unsaturated-zone moisture, groundwater	Air Solid phases	Volatilization Sorption
Air	Atmosphere, unsaturated zone gas	Water Solid phases	Dissolution Sorption
Solid phases	Atmospheric particles, soil organic matter, suspended sediment, aquifer solids, biota	Water Air	Desorption Volatilization

behavior in both phases. Raoult's law provides a working approximation for predicting the level of water contamination. Solubilities and other properties for select compounds are listed in Table 5. The list provides a comparison between the different classes of compounds. More comprehensive lists of properties are provided in the cited references. The CAS number reported is useful when using online databases (e.g., Syracuse Research Corporation, website: http://esc.syrres.com/ accessed February 8, 2003). The temperature dependence of solubility generally is neglected at Earth surface and near-surface temperatures.

The co-solvent effect is the term commonly used to identify nonideal behavior involving aqueous/immiscible phase equilibrium. The effect can result in higher aqueous concentrations ($\gamma_k > 1$) than those predicted by Raoult's law. For traditional fuels and other petroleum products discussed here, the co-solvent effect is negligible because the total concentration in the aqueous phase is on the order of 100 mg L^{-1} (~0.01% hydrocarbons). This mixture is dilute enough so that the hydrocarbon solutes do not appreciably interact. Activity coefficients can be calculated using the UNIFAC method (Fredenslund *et al.*, 1975). The high concentrations of fuel oxygenates like MTBE in groundwater at spill sites have raised the question as to whether or not they could increase aqueous concentrations of the BTEX group (Zogorski *et al.*, 1997) via the co-solvent effect. It has been shown that the co-solvency effect arises only when the co-solvent is present in water at 1% (10^4 mg L^{-1}) or higher (Pinal *et al.*, 1990, 1991). These concentrations are much higher than typically would be encountered in the environment. For example, gasoline that contains 15% MTBE by volume, when equilibrated with water, results in no more than $7{,}500 \text{ mg L}^{-1}$ (~0.75%) of MTBE in water (Barker *et al.*, 1991). No co-solvency effect for BTEX was noted for gasolines containing any of the following compounds and volume percentages: 15% MTBE, 10% EtOH, 10% TAME, and 10% isopropyl alcohol by volume (Barker *et al.*, 1991; Poulsen *et al.*, 1992). In experiments with MeOH, no co-solvency effect was noted until the MeOH concentrations in water exceeded 8% by volume at which point aqueous BTEX concentrations increased in proportion to increasing methanol content of the aqueous phase (Barker *et al.*, 1991; Poulsen *et al.*, 1992).

Crude oil and its derived products are mixtures of many types of compounds. Although the exact composition of a product is desired to make equilibrium calculations, this information generally is unavailable because product quality is based on engineering specifications like octane number and not on a specific composition. Furthermore, many compounds of interest are present in products in small quantities: therefore, estimates for χ_k can be off by multiples, resulting in inaccurate equilibrium concentration prediction. Approximations to χ_k, however, can be made given some knowledge of product composition for the purpose of estimating the magnitude of equilibrium concentrations. The mole fraction is defined in terms of concentration in the product phase as

$$\chi_k = \frac{I_k/w_k}{\sum_{j=1}^{N}(I_j/w_j)} \tag{2}$$

where χ_k is the mole fraction, I_k is the concentration of the compound in the product phase (in g cm^{-3}, e.g., grams of benzene per cubic centimeter of crude oil) and w_k is the molecular weight (g mol^{-1}). The volumetric fraction is defined as

$$\chi_k^v = \frac{I_k/v_k}{\sum_{j=1}^{N} I_j/v_j} \tag{3}$$

where v_k (Table 5) is the specific volume of a compound ($\text{cm}^3 \text{ g}^{-1}$) in the mixture that can be approximated by the reciprocal of the compound's density.

The mass fraction is defined as

$$\chi_k^m = \frac{I_k}{\sum_{j=1}^{N} I_j} = \frac{I_k}{\rho} \tag{4}$$

where $\rho = \sum_{j=1}^{N} I_j$ is the density of the product.

To meet requirements that wintertime fuel contain 2.7% oxygen by weight, MTBE is added to gasoline so that the final product consists of 14.7% MTBE by volume ($\chi_k^v = 0.147$). MTBE and the other compounds in gasoline have similar specific volumes, therefore, $\chi_k^v \approx \chi_k^m$. Finally, the molecular weight of MTBE is on the order of the average molecular weight of gasoline compounds, therefore, $\chi_k \approx \chi_k^m \approx \chi_k^v$. A constituent breakdown for a gasoline oxygenated with MTBE (Lahvis, private communication) is given in Table 6. The compositional breakdown shows the similarity between the definitions of volume, mass, and mole fraction. The gasoline, on a mass basis, consists of ~15.5% MTBE, 32% aromatics, and 39% alkanes; however, MTBE is the primary contaminant in the aqueous phase on a mass basis (~$8{,}000 \text{ mg L}^{-1}$) followed by the aromatic constituents (~100 mg L^{-1}). Although alkanes constitute the majority of mass in the gasoline, their contribution to water contamination (~11 mg L^{-1}) is relatively small compared to other gasoline compounds. The contribution of alkenes and cycloalkanes (~8.0 mg L^{-1} and 1 mg L^{-1}, respectively) to water contamination also is relatively small. If MTBE was not blended into the gasoline, then the total hydrocarbon solubility would be approximately $1.15 (94.6 + 11.4 + 7.7 + 1.3) = 132.3 \text{ mg L}^{-1}$.

Table 5 Solubilities and other properties of select petroleum hydrocarbons and fuel additives.

	Formula	*CAS number*	*Molecular weight w_k* (g mol^{-1})	*Specific volume, v_k at 25 °C* (cm^3 g^{-1})	*Solubility, C_{sol} at 25 °C* (mg L^{-1})	*Vapor pressure p at 25 °C* (atm)	*Henry H_k at 25 °C*	*Octanol/water* $\log_{10} K_{OW}$
Aromatic hydrocarbons[a]								
Benzene	C_6H_6	71-43-2	78.1	1.14	1,780	0.125	0.225	2.13
Methylbenzene (toluene)	C_7H_8	108-88-3	92.1	1.15	515	0.038	0.274	2.69
1,4-Dimethylbenzene (*p*-xylene)	C_8H_{10}	106-42-3	106.2	1.16	185	0.012	0.271	3.18
1,2-Dimethylbenzene (*o*-xylene)	C_8H_{10}	95-47-6	106.2	1.14	175	0.012	0.286	3.15
1,3-Dimethylbenzene (*m*-xylene)	C_8H_{10}	108-38-3	106.2	1.13	159	0.011	0.296	3.20
Ethylbenzene	C_8H_{10}	100-41-4	106.2	1.15	152	0.013	0.358	3.13
Isopropylbenzene	C_9H_{12}	98-82-8	120.2	1.16	61	0.006	0.485	3.63
1,2,4-Trimethylbenzene	C_9H_{12}	95-63-6	120.2	1.14	57	0.003	0.230	3.60
n-Propylbenzene	C_9H_{12}	103-65-1	120.2	1.16	52	0.004	0.420	3.69
n-Butylbenzene	$C_{10}H_{14}$	104-51-8	134.2	1.16	15	0.001	0.494	4.26
Alkanes								
n-Butane	C_4H_{10}	106-97-8	58.1	1.73	61.4	2.398	92.8	
n-Pentane	C_5H_{12}	109-66-0	72.2	1.6	38.5	0.675	51.7	3.45
n-Hexane	C_6H_{14}	110-54-3	86.3	1.52	9.5	0.199	74.0	4.11
n-Heptane	C_7H_{16}	142-82-5	100.2	1.46	3	0.060	82.3	5.00
n-Octane	C_8H_{18}	111-65-9	114.2	1.42	0.7	0.018	118.5	5.15
Alkenes								
1-Butene	C_4H_8	106-98-9	56.1	1.68	222	0.293	3.0	
1-Pentene	C_5H_{10}	109-67-1	70.1	1.56	148	0.839	16.2	2.2
1-Hexene	C_6H_{12}	592-41-6	84.2	1.49	50	0.245	16.8	3.39
2-Heptene	C_7H_{14}	592-77-8	98.2	1.43	15	0.064	17.0	
1-Octene	C_8H_{16}	111-66-0	112.2	1.4	2.7	0.023	38.9	4.57
Cycloalkanes								
Cyclopentane	C_5H_{10}	287-92-3	70.1	1.34	156	0.418	7.7	
Cyclohexane	C_6H_{12}	110-82-7	84.2	1.29	55	0.125	7.8	2.86
Methylcyclohexane	C_7H_{14}	108-87-2	98.2	1.3	14	0.061	17.5	
Propylcyclopentane	C_8H_{16}	2040-96-2	112.2	1.3	2	0.016	37.1	
Ethers (oxygenates additives)[b]								
Methyl *t*-butyl ether (MTBE)	$C_5H_{12}O$	1634-04-4	88.2	1.34	50,000	0.329	0.024	1.20
Ethyl *t*-butyl ether (ETBE)	$C_6H_{14}O$	637-92-3	102.2	1.37	26,000	0.200	0.032	1.74
t-Amyl methyl ether (TAME)	$C_6H_{14}O$	994-05-8	102.2	1.3	20,000	0.090	0.019	
Diisopropyl ether (DIPE)	$C_6H_{14}O$	108-20-3	102.2	1.35	9,000	0.197	0.092	1.52

The Henry coefficient is obtained from the solubility and vapor pressure data from this table as follows: $H_k = (w_k p/RT)/C_{sol}$. Other values are provided in Table 6.
[a] Mackay *et al.* (1992a,b, 1993). [b] Zogorski *et al.* (1997).

Table 6 Composition of a gasoline oxygenated with MTBE and associated equilibrium concentrations in water.

Compound	*Formula*	*Molecular weight* w_k, (g mol^{-1})	*Volumetric fraction*, χ_k^v	*Mass fraction*, χ_k^m	*Mole fraction*, χ_k	*Solubility*, C_{sol} (mg L^{-1}) *(from Table 5)*	*Concentration* $C_k = \chi_k C_{sol}$ (mg L^{1})
MTBE	$C_5H_{12}O$	88.1	0.15	0.1544	0.1589	50,000	7945
Aromatics							
Benzene	C_6H_6	78.1	0.0136	0.0165	0.0192	1780	34.2
Toluene	C_7H_8	92.1	0.0699	0.0834	0.0822	515	42.3
Xylenes	C_8H_{10}	106.2	0.0912	0.0963	0.0823	175	14.4
C_9	C_9H_{12}	120.2	0.063	0.0751	0.0567	57	3.2
C_{10}	$C_{10}H_{14}$	134.2	0.0306	0.0364	0.0246	15	0.4
C_{11}	$C_{11}H_{16}$	148.2	0.0134	0.016	0.0098	10.5	0.1
C_{12}	$C_{12}H_{18}$	162.3	0.0007	0.0008	0.0005	5	0
Subtotal			0.28	0.32	0.28		94.6
Alkanes							
C_4	C_4H_{10}	58.1	0.053	0.0442	0.069	61.4	4.2
C_5	C_5H_{12}	72.2	0.137	0.1184	0.1488	40	6
C_6	C_6H_{14}	86.2	0.114	0.1045	0.11	9.5	1
C_7	C_7H_{16}	100.2	0.0687	0.0646	0.0585	3	0.2
C_8	C_8H_{18}	114.2	0.0617	0.0597	0.0474	0.5	0
Subtotal			0.43	0.39	0.43		11.4
Alkenes							
C_4	C_4H_8	56.1	0.0011	0.0009	0.0015	222	0.3
C_5	C_5H_{10}	70.1	0.0333	0.0295	0.0382	148	5.7
C_6	C_6H_{12}	84.2	0.029	0.0269	0.029	50	1.5
C_7	C_7H_{14}	98.2	0.0135	0.0132	0.0122	15	0.2
C_8	C_8H_{16}	112.2	0.0017	0.0017	0.0014	2.7	0
Subtotal			0.08	0.07	0.08		7.7
Cycloalkanes							
C_5	C_5H_{10}	70.1	0.002	0.002	0.0027	160	0.4
C_6	C_6H_{12}	84.2	0.0109	0.0117	0.0127	55	0.7
C_7	C_7H_{14}	98.2	0.0137	0.0146	0.0135	14	0.2
C_8	C_8H_{16}	112.2	0.0092	0.01	0.0081	2	0
Subtotal			0.04	0.04	0.04		1.3

Equilibrium between product and gaseous phases is expressed in a way analogous to the product and aqueous phase equilibrium as

$$G_k = \gamma_k \chi_k G_{\text{vap}} \tag{5}$$

where G_k is the concentration of the compound in the gaseous phase (g L^{-1}), G_{vap} is the concentration in the gaseous phase in equilibrium with the pure compound, and, as for the aqueous/product phase equilibrium, γ_k and χ_k are the activity coefficient and mole fraction in the product phase, respectively. Again, this is a form of Raoult's law, and generally it is assumed that $\gamma_k = 1$ (ideal behavior for equilibrium involving gaseous phase).

Gaseous phase data often are reported in terms of partial pressure. The ideal gas law is used to convert to concentration units as:

$$G_{\text{vap}} = \frac{\omega_k}{RT} p \tag{6}$$

where p is the partial pressure of the pure compound in atmospheres, T is temperature in K, R is the gas constant, $R = 82.05783$ cm^3 atm mol^{-1} K^{-1}. Values for p at 25 °C are provided in Table 5.

The temperature dependence of p is modeled by the Clausius–Clapeyron equation as

$$\frac{1}{p}\frac{\mathrm{d}p}{\mathrm{d}T} = \frac{\Delta H}{RT^2} \tag{7}$$

where ΔH is the latent heat of vaporization or change in enthalpy associated with the phase change (cm^3 atm mol^{-1}). Integration of Equation (7) yields

$$\ln(p/p_0) = A - B/T \tag{8}$$

where

$$A = \frac{\Delta H}{RT_0}, \qquad B = \frac{\Delta H}{R}$$

and p_0 is the value for p at temperature T_0.

If ΔH is given in units of calorie per mole and temperature in °C, then

$$A = \frac{(0.5035)\Delta H}{(T_0 + 273.15)} \quad \text{and} \quad B = (0.5035)\Delta H$$

because 1 cal mol^{-1} = 41.32 cm^3 atm mol^{-1}. This equation results from the following: 1 atm = 101,325 Pa, 1 Pa = 1 kg m^{-1} s^{-2}, 1 J = 1 kg m^2 s^{-2}, 1 cal = 4.187 J, and R = 82.05783 cm^3 atm mol^{-1} K^{-1}. The calculation of the temperature dependence of vapor pressure for select compounds is presented in Table 7. For example, the concentration of benzene in air at a filling station would be ~4 times higher on a warm day (35 °C) compared to a cold day (5 °C).

Equilibrium between aqueous and gaseous phases is modeled according to Henry's law as

$$G_k = H_k C_k \tag{9}$$

where H_k is the Henry's law coefficient (dimensionless) for the compound. Because both the aqueous and gaseous phases are assumed to be dilute with respect to the organic compounds, there is no activity coefficient or nonideal behavior associated with this interface. The temperature dependence of H_k is calculated similarly to vapor pressure and is referred to as vant Hoff equation:

$$\frac{\mathrm{d}\ln(H_k)}{\mathrm{d}T} = \frac{\Delta H_{\text{Henry}}}{RT^2} \tag{10}$$

where ΔH_{Henry} is the change in enthalpy associated with the aqueous/gaseous phase transfer (in cm^3 atm mol^{-1}). Integration of Equation (10) yields

$$\ln(H_k/H_{k_0}) = A - B/T \tag{11}$$

where

$$A = \frac{\Delta H_{\text{Henry}}}{RT_0}, \qquad B = \frac{\Delta H_{\text{Henry}}}{R}$$

and H_{k_0} is the value for H_k at temperature T_0. If ΔH_{Henry} is given in units of cal mol^{-1} and temperature in °C, then

$$A = \frac{(0.5035)\Delta H_{\text{Henry}}}{(T_0 + 273.15)} \quad \text{and}$$

$$B = (0.5035)\Delta H_{\text{Henry}}$$

The temperature dependence for MTBE and select aromatics are given in Table 8. The parameter values were selected from references cited in the review by Rathbun (1998). Henry's law coefficients at 25 °C for other compounds are provided in Table 5.

The temperature dependence of H_k for MTBE can result in appreciable changes in MTBE concentration in precipitation. For example, given the same gaseous concentrations in the atmosphere, the concentration of MTBE at equilibrium in rain at 10 °C compared to that at 30 °C is ~6 times greater. Therefore, MTBE detection would be expected to be higher in precipitation and runoff during cooler months (Baehr *et al.*, 1999).

Sorption involves mass transfer of VOCs to a solid phase such as particulates in the atmosphere, bed sediments of a stream or lake, or the solid matrix of porous media. Solid surfaces attract water electrostatically, except for cases involving extremely dry sediment when an aqueous phase surrounding the solid surfaces is involved in the process. Sorption, therefore, has been defined as

Table 7 Vapor pressure of petroleum compounds as a function of temperature $\ln(p/p_0) = A - B/(T + 273.15)$ where $A = [(0.5035)\Delta H]/(T_0 + 273.15)$ and $B = (0.5035)\Delta H$.

Compound	ΔH[a] (cal mol^{-1})	$p(T)$ (in atm)			
		5 °C	*15 °C*	*25 °C*[b]	*35 °C*
Aromatic hydrocarbons					
Benzene	8079.8	0.047	0.078	0.125	0.195
Methylbenzene (toluene)	9078.1	0.012	0.022	0.038	0.062
1,4-Dimethylbenzene (*p*-xylene)	10126.6	0.003	0.006	0.012	0.020
1,2-Dimethylbenzene (*o*-xylene)	10372.6	0.003	0.006	0.012	0.020
1,3-Dimethylbenzene (*m*-xylene)	10186.3	0.003	0.006	0.011	0.019
Ethylbenzene	10088.4	0.004	0.007	0.013	0.022
Isopropylbenzene	10778.6	0.002	0.003	0.006	0.011
1,2,4-Trimethylbenzene	11447.3	0.001	0.001	0.003	0.005
n-Propylbenzene	11038.9	0.001	0.002	0.004	0.008
n-Butylbenzene	12266.5	0.000	0.001	0.001	0.003
Alkanes					
n-Butane	5020.3	1.304	1.787	2.398	3.158
n-Pentane	6312.4	0.314	0.466	0.675	0.954
n-Hexane	7537.6	0.080	0.128	0.199	0.301
n-Heptane	8734.2	0.021	0.036	0.060	0.097
n-Octane	9909.2	0.005	0.010	0.018	0.031
Alkenes					
1-Butene	4829.2	0.163	0.221	0.293	0.382
1-Pentene	6083.1	0.401	0.587	0.839	1.171
1-Hexene	7310.7	0.101	0.159	0.245	0.365
2-Heptene	8672.1	0.022	0.038	0.064	0.102
1-Octene	9634.6	0.007	0.013	0.023	0.039
Cycloalkanes					
Cyclopentane	6811.6	0.183	0.281	0.418	0.608
Cyclohexane	7883.9	0.048	0.079	0.125	0.193
Methyl-cyclohexane	8445.2	0.022	0.037	0.061	0.097
Propyl-cyclopentane	9811.3	0.005	0.009	0.016	0.028
Ethers (oxygenates additives)					
Methyl*t*-butyl ether (MTBE)	7122.0	0.139	0.217	0.329	0.486

[a] Values from Lide (2002). [b] Values for $T_0 = 25$ °C from Table 6.

Table 8 Henry's Coefficient of MTBE and monoaromatic hydrocarbons as a function of temperature $\ln(H_k/H_{k_0}) = A - B/T$ where $A = [(0.5035)\Delta H_{\text{Henry}}]/(T_0 + 273.15)$ and $B = (0.5035)\Delta H_{\text{Henry}}$.

Compound	ΔH_{Henry} (cal mol^{-1})	ΔH_{k_0} at $T_0 = 25$ °C dimension-less	H_k			
			$T = 0$ °C	$T = 10$ °C	$T = 20$ °C	$T = 30$ °C
MTBE[a]	15,334.6	0.026	0.002	0.007	0.017	0.04
Aromatics[b]						
Benzene	7,720	0.195	0.059	0.098	0.156	0.242
Toluene	8,574	0.227	0.06	0.105	0.177	0.288
1,4-Dimethylbenzene (*p*-xylene)	9,755.7	0.244	0.054	0.102	0.184	0.32
1,2-Dimethylbenzene (*o*-xylene)	9,676.3	0.173	0.039	0.073	0.131	0.226
1,3-Dimethylbenzene (*m*-xylene)	8,570	0.247	0.066	0.115	0.193	0.314
Ethylbenzene	10,113.2	0.274	0.057	0.111	0.205	0.363

[a] From Robbins *et al.* (1993). [b] From Dewulf *et al.* (1995).

any accumulation of a dissolved organic compound by solid particles (Voice and Weber, 1983). Sorption of VOCs is conceptually different than partitioning into the bulk fluid phases of the environment because the composition of the solid must be taken into consideration to obtain an equilibrium relation. The natural organic content of sediment is fundamental to sorption because VOCs are hydrophobic. Sediments in lake beds that are rich in natural organic matter, such as humin or humic acid, would hold more VOCs than a sand largely devoid of natural organic matter.

Sorption of VOCs involves the processes of adsorption and partitioning. Partitioning is the incorporation of the VOC into the natural organic matter associated with the solid and is analogous to the dissolution of an organic compound into an organic solvent. Adsorption is the formation of a chemical or physical bond between the VOC and the mineral surface of a solid particle (Rathbun, 1998). The equilibrium relation between aqueous and solid phase concentrations then is expressed as

$$S = [f_{oc}K_{oc} + (1 - f_{oc})K_m]C \quad (12)$$

where S is the mass of the VOC sorbed per unit mass of solid (g kg^{-1}), C is the concentration in the aqueous phase (g L^{-1}), f_{oc} is the weight fraction of organic carbon in the solid (dimensionless), K_{oc} is the normalized sorption coefficient because of organic matter partitioning (liter per kilogram of organic carbon), and K_m is the sorption coefficient because of mineral adsorption (liter per kilogram of mineral sediment). Karickhoff (1984) presented guidelines for estimating the relative contribution of adsorption and partitioning. For small ($<C_{10}$) nonpolar organics, which describes essentially all of the compounds listed in Table 5 (except for ether oxygenates), adsorption is important compared to partitioning only if $f_{oc} <$ 0.02. The threshold for ethers such as MTBE, which are in the category of neutral organics with polar functional groups, is $f_{oc} < 0.04$. Rathbun (1998) provides a review of K_{oc} values for aromatic hydrocarbons and various sediments.

The octanol–water partition coefficient, K_{ow}, for an organic compound is the ratio of the compound concentration in octanol saturated with water to that in water saturated in octanol. This property of the VOC can be used to estimate K_{oc} as it is directly related to the tendency of a compound to partition in natural organic matter from aqueous solution (Karickhoff, 1981). The octanol–water partition coefficient also provides a measure of the propensity of a compound to bioaccumulate (Rathbun, 1998) in fatty tissue of aquatic biota. Values of K_{ow} are provided in Table 5.

Equation (12) is referred to as a linear isotherm. Linearity assumes that sorption sites on the solid surfaces are unlimited; therefore, the model may be applicable only for lower concentrations. An alternative formulation is

$$S = K_f C^{1/n} \quad (13)$$

where K_f and n are constants of the Freundlich-type isotherm (Voice and Weber, 1983). Values for n usually range from 0.7 to 1.1 (Lyman *et al.*, 1990).

Whereas phase partitioning describes how molecular constituents of petroleum products interact at interfaces, such as between a liquid petroleum and aqueous phase, once transferred to an environmental compartment, the compounds can undergo physical mixing processes, and chemical and biologically mediated transformations. The phase transfers described above can be viewed as the beginning of the overall process of transport through the atmosphere, hydrosphere, geosphere, or biosphere. Methods of predicting transport are discipline-specific, dependent on the scale of investigation, and which part of the environment (e.g., atmosphere, ocean, streams, groundwater) is most affected. Mathematical models of transport quantify the movement, dilution, and chemical and biochemical attenuation of contaminants through the relevant environmental compartments. The bases of transport models rely on conservation principles (conservation of mass, momentum, and energy) that are represented as a series of equations such as those described by Domenico and Schwartz (1998).

Although a detailed discussion of physical transport processes is beyond the scope of this chapter, it is important to note that quantification of the bulk movement of air in the atmosphere or water in the hydrosphere can provide a qualitative interpretation of chemical transport. A simplistic example is that stagnant air masses induce conditions favorable for smog formation because chemicals can increase in concentration over an urban area. Therefore, basic weather predictions can be used to anticipate air quality. In general, models of fluid movement define the flow field in which contaminants can be transported via advection. For example, the area over which water recharging a surficial aquifer eventually will travel to the public supply well (well contributing area, Figure 8). The delineation of the well contributing area was achieved with a groundwater flow model simulation. In aggregate, land use within the contributing area caused low-level detection of MTBE and other compounds in samples taken from the well. The accumulation of point sources within the contributing area can be interpreted as an urban nonpoint source. Estimated travel times of groundwater from point of recharge to the well can be calculated from a groundwater flow model simulation and are useful in determining the effect of land use throughout the contributing area (Kauffman *et al.*, 2001).

Accumulation of point sources from land use of petroleum products also can affect surface waters. For example, urban riverine inputs of volatile hydrocarbons to the marine environment have been studied in the coastal waters of Spain by Gomez-Belinchon *et al.* (1991). Volatile petroleum hydrocarbon inputs from two rivers were found to account for a mass flux of 47 t yr^{-1} and 96 t yr^{-1} of alkylbenzenes and 38 t yr^{-1} and 66 t yr^{-1} of n-alkanes each. Although this was a significant mass flux, the

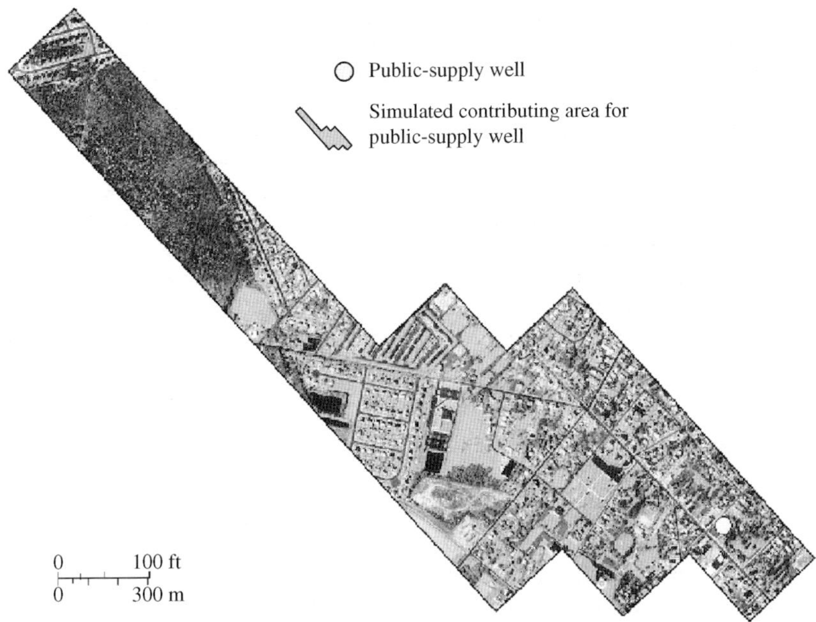

Figure 8 Simulated contributing areas to a shallow and moderate-depth monitoring and public-supply well (after Kauffman *et al.*, 2001).

concentrations of alkylbenzenes and *n*-alkanes in this coastal area originating from marine traffic actually outnumbered the contribution from the rivers.

9.12.4 TRANSFORMATION PROCESSES

Chemical and biological transformation processes, in addition to the physical transport processes, control the ultimate fate of hydrocarbons released into the environment. The transformation reactions differ depending on the environmental compartment within which the hydrocarbons reside and vary with chemical structure. When hydrocarbons are released to the atmosphere or surface waters, photochemical oxidation, an abiotic process, can occur. In soils and groundwater and surface water, biologically mediated degradation of hydrocarbons is the most important transformation process. In the absence of light, chemical degradation reactions at Earth surface temperature and pressure are relatively unimportant compared to biologically mediated degradation reactions.

9.12.4.1 Abiotic Transformations

Approximately 25% of the average oil spill on the open ocean evaporates. In the gaseous state, hydrocarbons are readily photooxidized (NRC, 1985). The dissolved fraction of petroleum also is subject to photo-oxidation. Mill *et al.* (1980) documented the photo-oxidation of isopropylbenzene in the presence of humic substances. A review of the photochemical oxidation of petroleum in water reported that oxidation of alkanes and alkylbenzenes, such as *s*-butylbenzene and *t*-butylbenzene, has been observed in photo-oxidation studies and the products produced have included acids, carbonyl compounds, alcohols, peroxides and ethers (Payne and Phillips, 1985). Photo-oxidation in the atmosphere of volatile aromatic hydrocarbons (benzene, toluene, and xylenes) has been studied in the presence of catalysts, such as nitrous oxide and carbon monoxide, as reviewed by Merian and Zander (1982). The products formed from these reactions include nitro-substituted phenols and aldehydes. A review of gas-phase reactions in the troposphere for alkanes, alkenes, alkynes, oxygenates, and aromatic hydrocarbons is presented by Atkinson (1990).

The largest sink for alkanes in the atmosphere is reaction with OH and NO_3 radicals. The formation of photochemical smog is described in detail in (Chapter 9.11, Sillman). Mono-aromatic hydrocarbons react only slowly with O_3 and NO_3 radicals in the troposphere. The only important atmospheric processes for mono-aromatic hydrocarbons, and naphthalene and dinaphthalenes are reactions with OH radicals (Atkinson, 1990). The products of these reactions include aldehydes, cresols, and, in the presence of NO, benzylnitrates. Methane can be an important contributor to ozone formation, especially in the remote troposphere, as described in (Chapter 9.11, Sillman).

Photo-oxidation has been reported at crude-oil spills. It results in depletion of *n*-alkanes below nC_{15} and alkylaromatics such as C_1- and C_2-substituted naphthalenes relative to unoxidized oil (Payne and Phillips, 1985). In terms of a material balance, photo-oxidation has been found to be a minor process (NRC, 1985) but does result in changes in the residual oil composition and can affect the subsequent behavior of an oil spill on the open ocean (Payne and Phillips, 1985). Auto-oxidation reactions of hydrocarbons in the absence of light have not been well studied.

9.12.4.2 Biotic Transformations

Whereas chemical oxidation processes may occur largely in the atmosphere, and on land surface or open waters where sunlight is a catalyst, biologically mediated processes dominate in soils and groundwater. The more water-soluble components of crude oil and petroleum products discussed in Section 9.12.1.2, benzene and the lower molecular weight alkylbenzenes are the hydrocarbons most frequently reported in groundwater downgradient from spills and leaks. These hydrocarbons are biologically reactive and their fate in the subsurface is controlled by microbiological as well as physical and chemical processes.

It has been known for almost 90 years that certain microorganisms are able to degrade petroleum hydrocarbons and use them as a sole source of carbon and energy for growth. The early studies of bacterial degradation of hydrocarbons are summarized by Gibson and Subramanian (1984), while reviews of the microbial metabolism of hydrocarbons were presented by Davis (1967), Atlas (1984), Chapelle (1993), and Rosenberg and Ron (1996). There are four types of microbial metabolism that release energy: photometabolism, fermentation, aerobic respiration, and anaerobic respiration (Reineke, 2001). Fermentation does not require oxygen or other electron acceptors such as NO_3^-, or $Fe(III)_s$, and depends on the capability of microorganisms to use part of the organic molecule as an electron acceptor. The respiration of organic material, however, does require an electron acceptor and proceeds as a series of coupled oxidation and reduction steps. In almost all shallow subsurface environments, petroleum hydrocarbons can serve as electron donors in microbial metabolism (Wiedemeier *et al.*, 1999). The oxidation of hydrocarbons results in the release of electrons that are transferred to oxygen, or, in the case of anoxic environments, alternate electron acceptors such as sulfate. Respiration reactions release more energy for cell growth than fermentation reactions and respiration reactions have been studied much more extensively. Aerobic and anaerobic respiration reactions are described in more detail in the following sections for aromatic and aliphatic hydrocarbons as well as for fuel oxygenates.

There are several factors that have been shown to affect the biodegradability of petroleum hydrocarbons, including hydrocarbon structure (i.e., aliphatic or aromatic), degree of branching and saturation, and chain length (Baker and Herson, 1994). In comparison to hydrocarbons, far less is known about the biodegradability of MTBE and other fuel oxygenates. Under aerobic conditions, MTBE degradation has been documented (Wilson *et al.*, 2002), but bacterial growth is slow (see Section 9.12.4.2). There is recent evidence that anaerobic MTBE degradation occurs (e.g., Kolhatkar *et al.*, 2002; Wilson *et al.*, 2000), yet these processes presently remain largely untested.

9.12.4.2.1 Aerobic processes

Oxygen is the preferred electron acceptor by microorganisms because of the high-energy yield of these processes. Aerobic degradation of hydrocarbons can occur when indigenous populations of bacteria capable of aerobic degradation of hydrocarbons are supplied with molecular oxygen and nutrients required for cell growth. The aerobic pathway of hydrocarbon degradation has been studied extensively in petroleum-contaminated soils. The capability to degrade hydrocarbons in aerobic environments is present in a wide variety of bacteria and fungi that contain the genetic capability of incorporating molecular oxygen into the hydrocarbon structure (Gibson and Subramanian, 1984; Rosenberg and Ron, 1996), thus, creating the widespread potential for hydrocarbon oxidation wherever oxygen and nutrients are available. In fact, studies involving complex mixtures of hydrocarbons, such as gasoline, have demonstrated that microorganisms can degrade most of the hydrocarbons present in gasoline (Jamison *et al.*, 1975).

The biodegradation of aromatic hydrocarbons under aerobic conditions was reviewed by Gibson and Subramanian (1984), Smith (1994), Rosenberg and Ron (1996), and Bosma *et al.* (2001). Both prokaryotic and eukaryotic microorganisms have the enzymatic potential to oxidize aromatic hydrocarbons (Rosenberg and Ron, 1996). Bacteria and fungi degrade aromatic hydrocarbons in different ways, however. Bacteria are able to utilize the compounds as a sole source of carbon and energy, whereas fungi appear to cometabolize aromatic hydrocarbons to hydroxylated products (Reineke, 2001). The first step in fungal metabolism is the formation of an epoxide (Reineke, 2001), whereas bacteria initiate the oxidation of unsubstituted aromatic compounds by incorporating both atoms of molecular oxygen into the aromatic ring to form a *cis*-dihydrodiol as reviewed by Gibson and Subramanian (1984). Further oxidation of the *cis*-dihydrodiol leads to catechol formation (Figure 9). The aromatic ring then is cleaved by the *ortho*- or *meta*-cleavage pathways, ultimately producing low molecular weight compounds such as pyruvate and acetaldehyde, which can be further oxidized via the Krebs cycle (e.g., Baker and Herson, 1994). In the case of toluene, for example, the presence of an alkyl substituent group on the benzene ring presents microorganisms with an additional site of attack and oxidation of this side chain results in the formation of benzyl alcohols, benzaldehydes, and alkylbenzoic acids (Figure 9).

The oxidation of straight-chain alkanes proceeds, in general, by oxidation of the terminal methyl group (Figure 10) to form intermediates including the corresponding alcohols, aldehydes, or carboxylic acids (Bouwer and Zehnder, 1993). In the presence of oxygen, a class of enzymes called oxygenases mediates this reaction. Fatty acids derived from alkanes then are oxidized further to acetate and propionate by β-oxidation (Rosenberg and Ron, 1996). Unsaturated hydrocarbons, such as alkenes and alkynes, are degraded by similar mechanisms (Bouwer and Zehnder, 1993). The most degradable *n*-alkanes are the C_{10} to C_{18} compounds, whereas longer-chain alkanes are less biodegradable because their lower aqueous solubilities limit uptake by microorganisms (Bosma *et al.*, 2001). Although *n*-alkanes are the predominant hydrocarbons by volume in many petroleum products, such as crude oils, those with chain lengths larger than C_{15} also are typically less mobile in subsurface and aquatic environments, again because of their relatively low solubilities (Eganhouse *et al.*, 1993).

Aerobic degradation reactions require available moisture, nitrogen, and phosphorus, in addition to molecular oxygen (Rosenberg and Ron, 1996). In oil spills in aquatic environments, where oxygen and moisture are plentiful, the limitation to biodegradation, therefore, tends to be nutrient availability. A review of the basic requirements for aerobic biodegradation and the quantification of this process at sites contaminated with hydrocarbons is presented by Rifai (1997).

The degradation processes and rates of transformation of fuel oxygenates, such as MTBE, and potential metabolites of degradation are not well understood. Although early studies indicated MTBE was resistant to microbial attack, more recent laboratory studies have shown that MTBE can biodegrade under oxygen-rich conditions. These studies have been reviewed by Deeb *et al.* (2000). Mixed and pure bacterial cultures have been used in the study of MTBE biodegradation under carefully controlled conditions. One of the first studies of a bacterial culture capable of MTBE biodegradation was of a microbial consortium enriched from activated sludge (Salanitro *et al.*, 1994). In this study, MTBE was utilized as a primary-growth substrate with the production of *t*-butyl alcohol (TBA) as a transient intermediate, although growth rates were slow. At MTBE concentrations above 5 mg L^{-1}, removal rates decreased. Studies with related cultures showed that MTBE was biodegraded at concentrations as high as 80 mg L^{-1} in laboratory microcosm experiments, where these cultures were added to soils and groundwater (Salanitro *et al.*, 1999). Other mixed and pure culture studies

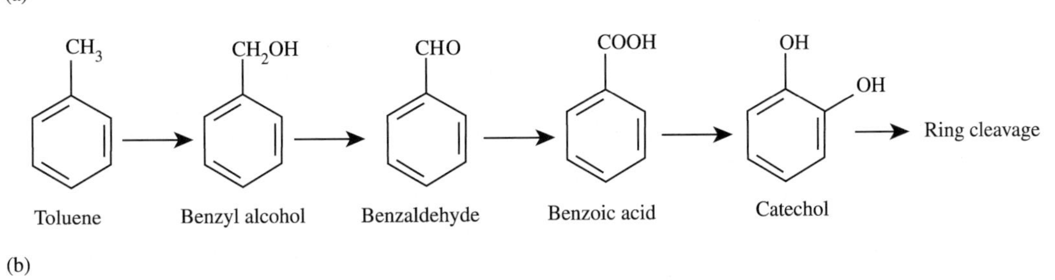

Figure 9 Aerobic toluene degradation pathway by (a) aromatic ring attack by dioxygenation and (b) side-chain attack by stepwise oxidation (after Smith, 1990).

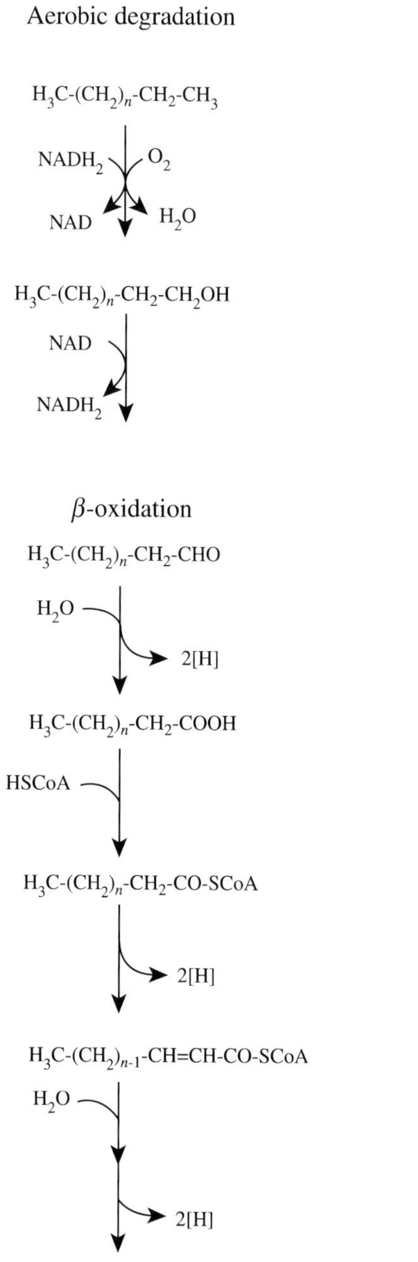

Figure 10 Alkane degradation under aerobic conditions, showing incorporation of oxygen from molecular oxygen into the aliphatic compound, producing a fatty acid. Fatty acids are oxidized further by β-oxidation. [H] indicates reducing equivalents that are either required or formed in the reaction step (after Bouwer and Zehnder (1993).

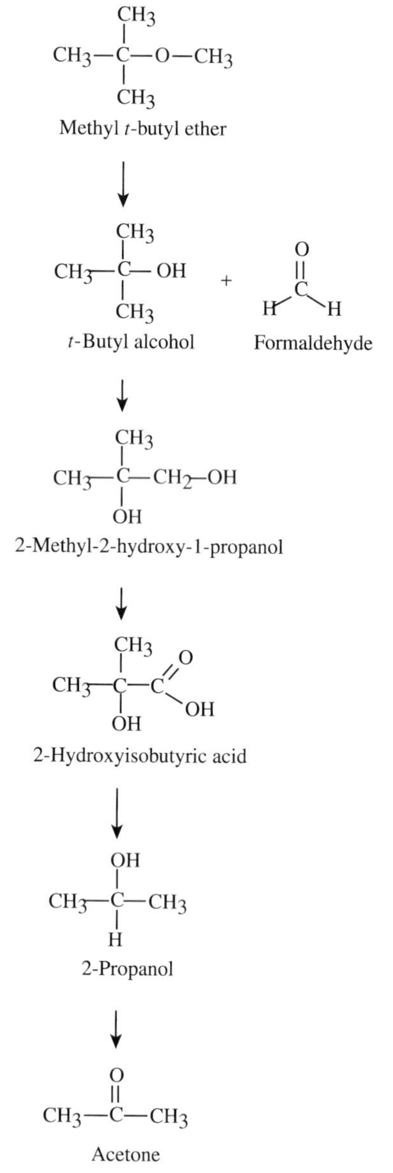

Figure 11 A demonstrated pathway for MTBE biodegradation under aerobic conditions (after Deeb *et al.*, 2000).

demonstrated the metabolism of MTBE in the presence of humic acids (Fortin and Deshusses, 1999) or within a restricted pH range between 6.5 and 7.8 (Eweis *et al.*, 1997).

The pathways of MTBE degradation have not been fully described but results from several studies support the suggestion of the cleavage of the ether bond of MTBE yielding TBA and formaldehyde as shown in Figure 11 (Deeb *et al.*, 2000). Aerobic degradation of MTBE also has been reported by co-metabolism in the presence of alkanes or aromatics. Whereas some bacterial cultures grown on alkanes either can utilize MTBE as a sole carbon and energy source, or degrade it by co-metabolism, bacteria grown on aromatics appear to degrade MTBE by co-metabolism only (Deeb *et al.*, 2000).

Although research on the biodegradation of MTBE under natural conditions is limited, the potential for significant *in situ* aerobic transformation of MTBE in surface water was demonstrated by Bradley *et al.* (2001), who identified MTBE-degrading capabilities in microorganisms indigenous to stream and lake-bed sediments at 11 different sites. MTBE biodegradation under aerobic conditions in aquifers contaminated with petroleum products has been demonstrated at the Canadian Forces Base (CFB) Borden (Schirmer and Barker, 1998; Schirmer *et al.*, 1999), although at slow rates, much slower than typically observed for BTEX compounds. Field experiments involving diffusive addition of oxygen into an aquifer with an existing MTBE plume (Vandenberg Air Force Base, CA) demonstrated that native aerobic MTBE-degrading microorganisms were stimulated to degrade MTBE (Wilson *et al.*, 2002).

Because MTBE usually is associated with other organic compounds, such as those in gasoline, the effect of co-contaminants on MTBE degradation is of interest. Studies have shown that the biodegradation rates of MTBE can either be enhanced or reduced by the presence of other organic contaminants in groundwater plumes, and although most studies do not indicate a negative effect on BTEX biodegradation rates from the presence of MTBE (Deeb *et al.*, 2000), this result is not always the case. In a study of the effect of MTBE addition to microcosms containing toluene and benzene, McMahon (1995) found that toluene degradation rates decreased by 10% and benzene degradation decreased by up to 5 times compared to controls containing no MTBE. Laboratory studies by Deeb and Alvarez-Cohen (2000) demonstrated no effect on aerobic BTEX biodegradation rates in a toluene-enriched mixed culture from a gasoline-contaminated aquifer that contained 2–100 mg L^{-1} MTBE. When added alone, MTBE was not biodegraded, indicating that non-MTBE degrading cultures were apparently not inhibited by the presence of MTBE. In studies by Schirmer *et al.* (1999) MTBE biodegradation only occurred in the absence of BTEX in laboratory columns containing aquifer material from four different field sites. Koenigsberg *et al.* (1999) demonstrated the inhibition of benzene biodegradation in the presence of MTBE, as well as inhibition of MTBE degradation in the presence of benzene, in a benzene-grown culture that was able to cometabolize MTBE, presumable because of competition for the same enzymes.

9.12.4.2.2 *Anaerobic processes*

Anoxic conditions frequently develop in subsurface environments affected by high concentrations of dissolved hydrocarbons because of rapid aerobic biodegradation rates and the limited supply of oxygen. In the absence of oxygen, the oxidized forms of other inorganic species, and some organic species such as humic substances, are used by microorganisms as electron acceptors. Because of the prevalence of petroleum product leaks and spills in oxygen-restricted subsurface environments, these reactions are the hydrocarbon degradation reactions studied most often. Anaerobic degradation reactions are described in more detail in (Chapter 9.16, Bjerg *et al.*) for another common contaminant, landfill leachate that likewise occurs in oxygen-restricted environments (Christensen *et al.*, 2000). There are many studies in which the nonconservative behavior of dissolved aromatic hydrocarbons from anoxic groundwater has been attributed to anaerobic degradation reactions (e.g., Barker *et al.*, 1986; Barker and Wilson, 1997; Cozzarelli *et al.*, 1990; Cozzarelli *et al.*, 1999; Eganhouse *et al.*, 1996; Reinhard *et al.*, 1984; Schwarzenbach *et al.*, 1983; Wilson *et al.*, 1990).

The most commonly available electron acceptors in subsurface environments include both solid and dissolved phase species. Bjerg *et al.* (Chapter 9.16, table 4) review these processes in landfill-impacted aquifers and provide quantitative values for the oxidative capacity for two aquifers. As shown in Table 9, the reactions involving mineralization of aromatic and aliphatic hydrocarbons under a range of redox conditions, are exergonic, and, therefore, considered useful to microorganisms. The biggest factor in determining the energy produced for the reaction is the type of available electron acceptor, not the type of hydrocarbon (i.e., aliphatic versus aromatic). Reactions proceed based largely on the free-energy yield, which is greatest for the denitrification pathway and least for the methanogenic pathway (Stumm and Morgan, 1996). In aquifers, as geochemical conditions change, a sequence of reactions occurs, reflecting the ecological succession of progressively less efficient modes of metabolism. The energy yields for common oxidation–reduction reactions have been reviewed by Wiedemeier *et al.* (1999) and Spormann and Widdel (2000).

Thus, natural attenuation studies of hydrocarbons often focus on the availability of electron acceptors (e.g., Barker and Wilson, 1997; Cozzarelli *et al.*, 1995; Gieg *et al.*, 1999; McGuire *et al.*, 2002; Skubal *et al.*, 2001). Numerous studies have focused on the availability of electron acceptors in both the sediment (e.g., Baedecker *et al.*, 1993; Bekins *et al.*, 2001a; Chapelle *et al.*, 2002; Cozzarelli *et al.*, 2001a; Heron and Christensen, 1994; Heron and Christensen, 1995; Heron and Christensen, 1995) and the aqueous phase (e.g., Ball and Reinhard, 1996; Cozzarelli *et al.*, 1999) as a key control of the fate of hydrocarbons in subsurface environments. In most sediment, $Fe(III)_s$, as iron oxides, is abundant

Table 9 Theoretical stoichiometry and energetics of mineralization of aromatic and aliphatic hydrocarbons under a range of redox conditions.

Chemical equation	*Change in free energy* ΔG^0
Denitrification	
$C_6H_5(CH_3) + 7.2NO_3^- + 0.2H^+ \rightarrow 7HCO_3^- + 3.6N_2 + 0.6H_2O$	-493.6 kJ mol^{-1} NO_3^-
$C_6H_{14} + 7.6NO_3^- + 1.6H^+ \rightarrow 6HCO_3^- + 3.8N_2 + 4.8H_2O$	-492.8 kJ mol^{-1} NO_3^-
$C_8H_{18} + 10NO_3^- + 2H^+ \rightarrow 8HCO_3^- + 5N_2 + 6H_2O$	-493.1 kJ mol^{-1} NO_3^-
Nitrate ammonification (example of toluene)	
$C_6H_5(CH_3) + 4.5NO_3^- + 2H^+ + 7.5H_2O \rightarrow 7HCO_3^- + 4.5NH_4^+$	-493.1 kJ mol^{-1} NO_3^-
Iron(III) reduction (selected examples)	
$C_6H_5(CH_3) + 36Fe(OH)_3 + 29HCO_3^- + 29H^+ \rightarrow 36FeCO_3 + 87H_2O$	-39.1 kJ mol^{-1} Fe
$C_6H_5(CH_3) + 36\alpha-FeO(OH) + 29HCO_3^- + 29H^+ \rightarrow 36FeCO_3 + 51H_2O$	-12.3 kJ mol^{-1} Fe
Sulfate reduction	
$C_6H_5(CH_3) + 4.5SO_4^{2-} + 2H^+ + 3H_2O \rightarrow 7HCO_3^- + 4.5H_2S$	-45.6 kJ mol^{-1} SO_4^{2-}
$C_6H_{14} + 4.75SO_4^{2-} + 3.5H^+ \rightarrow 6HCO_3^- + 4.75H_2S + H_2O$	-44.2 kJ mol^{-1} SO_4^{2-}
$C_8H_{18} + 6.25SO_4^{2-} + 4.5H^+ \rightarrow 8HCO_3^- + 6.25H_2S + H_2O$	-44.6 kJ mol^{-1} SO_4^{2-}
Methanogenesis (selected example)	
$C_6H_5(CH_3) + 7.5H_2O \rightarrow 2.5HCO_3^- + 4.5CH_4 + 2.5H^+$	-28.5 kJ mol^{-1} CH_4

After Spormann and Widdel (2000).
Notes on compounds: Free energy formation (ΔG_f^0, in kJ mol^{-1}) of hydrocarbons used in the presented equations: $C_6H_5(CH_3)$, toluene (lq): +114.2; C_6H_{14}, hexane (lq): −3.8; C_8H_{18}, octane (lq): +6.41. Free energy formation (ΔG_f^0, in kJ mol^{-1}) of iron oxides used in the presented equations: $Fe(OH)_3$, ferric hydroxide (amorphous), −699; α-FeO(OH), goethite (crystalline): −488.6. For adequate comparison of the free energetics of the indicated reactions, ΔG^0-values have to be related to the same number of electrons transferred. Hence, the values for denitrification and iron(III) reduction must be multiplied by 1.6 and 8, respectively, resulting in free energy changes per 8 mol electrons (viz. per 1.6 mol NO_3^- and 8 mol Fe, respectively).

and readily reduced by microorganisms. Although less abundant, manganese oxides (Mn $(IV)_s$) are easily reducible by microorganisms. Anaerobic bacteria have been isolated that can also use other metals as electron acceptors, such as the oxyanions of arsenate or selenate (Stolz and Oremland, 1999) and uranium (Abdelouas *et al.*, 2000; Francis *et al.*, 1994; Lovley *et al.*, 1991; Lovley and Anderson, 2000; Lovley and Phillips, 1992; Tebo and Obraztsova, 1998). Anaerobic microbial oxidation of toluene coupled to humus respiration was demonstrated by use of enriched anaerobic sediments from the Amsterdam petroleum harbor (APH) and the Rhine River (Cervantes *et al.*, 2001); the humic acids were utilized as terminal electron acceptors.

In most aquifers, nitrate and sulfate are supplied during groundwater recharge by precipitation. In coastal areas, mixing with seawater and contamination of groundwater with fertilizers can significantly increase the concentrations of these constituents. A number of studies have demonstrated a heterogeneous distribution of hydraulic conductivity within a given sedimentary deposit (e.g., Barber *et al.*, 1992; Bekins *et al.*, 2001a; Cozzarelli *et al.*, 1999; Davis *et al.*, 1993; Hess *et al.*, 1992; Robin *et al.*, 1991) that may control the introduction of these electron acceptors to the contaminated zone. *In situ* microorganisms are poised to take advantage of these changes in electron acceptor availability and the dominant microbial degradation reactions shift as a result of these changes. These shifting reactions result in the development of redox zones, as determined by geochemical and microbial signatures that are variable in space and time. These concepts are illustrated in (Chapter 9.16), Figure 3, and are not reproduced here.

Among the hydrocarbons, the volatile aromatic hydrocarbons, exemplified by the BTEX compounds, have been the most studied in terms of their anaerobic biodegradation potential. This results because of their toxicity and much greater water solubility and anaerobic degradability relative to the aliphatic, alicyclic, and polycyclic hydrocarbons. Many approaches have been used to study the fate of hydrocarbons under anoxic conditions and these have included field studies, microcosm experiments using sediments and water from contaminated sites, and mixed-culture and pure-culture laboratory studies. The first evidence of anaerobic degradation of monoaromatic hydrocarbons in the environment was observed in methanogenic microcosm experiments with petroleum-contaminated sediments (Ward *et al.*, 1980). Shortly thereafter, Schwarzenbach *et al.* (1983) and Reinhard *et al.* (1984) reported the selective removal of alkylbenzenes in anoxic zones of contaminated groundwater, providing evidence for the anaerobic degradation of these compounds by aquifer microorganisms. There were numerous studies in the 1980s focusing on anaerobic degradation of aromatic hydrocarbons, including studies by Vogel and Grbić-Galić (1986), who used an anaerobic methanogenic consortium derived from sewage sludge to verify that oxidation of these substrates occurred in the absence of molecular oxygen. In addition, Wilson *et al.* (1986) reported the degradation of benzene and alkylbenzenes in methanogenic microcosms derived from a landfill-leachate-contaminated aquifer. Both of these laboratory studies, involving mixed cultures of bacteria, highlighted the importance of microorganisms in the degradation of alkylbenzenes under methanogenic conditions. At the same time, evidence that these processes might be important in determining the fate of alkylbenzene in the environment was reported by Eganhouse *et al.* (1987), who documented the apparent *in situ* microbial degradation of selected alkylbenzenes in the iron-reducing and methanogenic zones of an aquifer contaminated with crude oil near Bemidji, Minnesota.

Many subsequent laboratory biodegradation studies, conducted using soils or sediment from contaminated sites, have confirmed the transformation of benzene and alkylbenzenes under nitrate-reducing (Hutchins *et al.*, 1991; Kuhn *et al.*, 1988; Major *et al.*, 1988; Zeyer *et al.*, 1986), iron-reducing (Baedecker *et al.*, 1993; Lovley *et al.*, 1989) and sulfate-reducing (Beller *et al.*, 1992a; Edwards *et al.*, 1992; Haag *et al.*, 1991) conditions. Pure-culture studies of nitrate-reducing, (e.g., Dolfing *et al.*, 1990; Evans *et al.*, 1991), iron-reducing (e.g., Lovley and Lonergan, 1990), and sulfate-reducing (e.g., Beller *et al.*, 1996; Rabus *et al.*, 1993) bacteria have provided insight into the biochemical mechanisms involved in the anaerobic degradation of alkylbenzenes and have identified specific microorganisms that can use these compounds as sole carbon sources. Spormann and Widdel (2000) provide a recent summary of the pure-culture studies of alkylbenzene metabolism.

While anaerobic alkylbenzene degradation was well documented by the early 1990s, conclusive evidence for the anaerobic degradation of benzene remained controversial until the mid-1990s (Krumholz *et al.*, 1996). It has since been well-documented. Kazumi *et al.* (1997) and Lovley (2000) reviewed the evidence for anaerobic degradation of benzene. Microbial degradation of benzene has been noted in slurries constructed with sediments from various geographical locations, ranging in composition from aquifer sands to fine-grained estuarine muds, under methanogenic, sulfate-reducing, and iron-reducing conditions (Kazumi *et al.*, 1997), demonstrating the likely widespread occurrence of this process in the environment. Most elusive has been the identification of a microorganism capable of growth with benzene as a sole substrate. Anaerobic degradation of benzene has also been linked to nitrate reduction in both enrichment cultures (Burland and Edwards, 1999) and pure cultures (Coates *et al.*, 2001).

The microbial metabolism of aromatic hydrocarbons has been reviewed extensively since the mid-1990s (Berry *et al.*, 1987; Colberg and Young, 1995; Heider and Fuchs, 1997; Holliger *et al.*, 1997; Krumholz *et al.*, 1996; Phelps and Young, 2001; Spormann and Widdel, 2000), and an examination of these reviews reveals that our understanding of the biodegradation pathways is constantly changing. Although a detailed discussion of the specific metabolic pathways is beyond the scope of this chapter, it is important to briefly discuss the types of transformations and intermediate products that result. The metabolic intermediates of anaerobic microbial oxidation of petroleum hydrocarbons include aromatic, aliphatic, and alicyclic organic acids, phenols, and aldehydes (e.g., Beller *et al.*, 1992b; Cozzarelli *et al.*, 1990, 1994; Grbić-Galić and Vogel, 1987; Heider *et al.*, 1999; Kuhn *et al.*, 1988; Wilkes *et al.*, 2000). Benzoic acid has been reported in laboratory studies of anaerobic toluene and benzene degradation by both mixed cultures (e.g., Grbić-Galić and Vogel, 1987) and pure cultures (for a review of these studies see Beller (2000)). *m*-Toluic acid has been identified as an intermediate by pure cultures of *m*-xylene degrading denitrifying bacteria (Seyfried *et al.*, 1994), whereas co-metabolism of *o*- and *p*-xylenes by toluene-degrading cultures has been found to result in the accumulation of *o*- and *p*-toluic acids (e.g., Beller *et al.*, 1996; Rabus and Widdel, 1995).

Although a pathway involving the oxidation of the methyl substituent group of alkylbenzenes has been proposed in studies of the anaerobic degradation of toluene under nitrate-reducing (Kuhn *et al.*, 1988), iron-reducing (Lovley and Lonergan, 1990), sulfate-reducing (Haag *et al.*, 1991) methanogenic (Grbić-Galić and Vogel, 1987) conditions, the mechanism of biochemical activation of alkylbenzenes has been elusive. Toluene and xylene oxidation by fumarate addition has also been documented, and this mechanism results in the production of a new group of organic acid intermediates, benzylsuccinic acids, first identified by Evans *et al.* (1992) and Beller *et al.* (1992b). These investigators reported the accumulation of benzylsuccinic acid and benzylfumaric acid (later shown to be E-phenyliataconate (Beller, 2000)) in pure cultures of toluene-degrading nitrate reducers and in enrichment cultures of toluene- and *ortho*-xylene-degrading sulfate reducers, respectively. This pathway produces benzylsuccinates or methylbenzylsuccinates (Beller and Spormann, 1997a,b; Biegert *et al.*, 1996; Evans *et al.*, 1992) by a novel mechanism in which the methyl group of the parent hydrocarbon is added to the double bond of fumarate to form benzylsuccinic acid (Figure 12). Benzylsuccinic acid appears to be the first intermediate produced during anaerobic toluene metabolism by pure cultures of nitrate-, iron-, and sulfate-reducing bacteria and in mixed

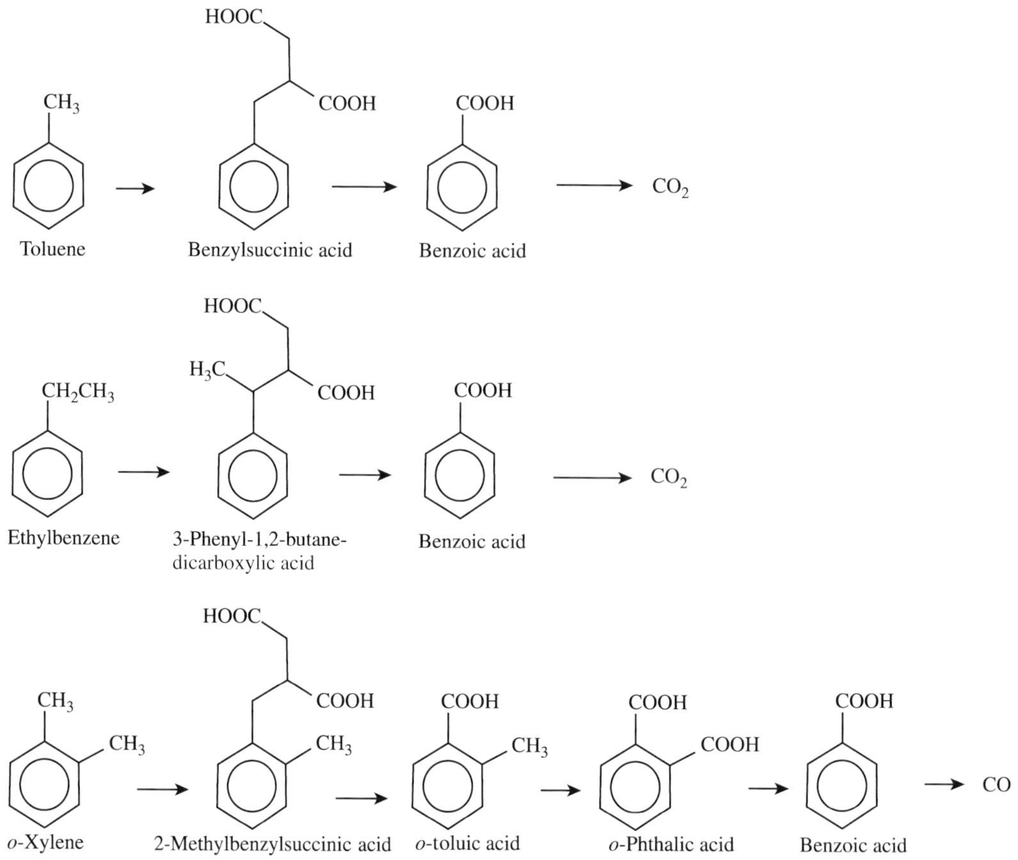

Figure 12 Proposed pathways of anaerobic degradation of toluene, ethylbenzene, and *o*-xylene by the fumarate addition pathway (after Elshahed *et al.*, 2001).

cultures of methanogens (these studies are reviewed by Beller (2000, 2002) with subsequent formation of benzoic acid. The enzyme responsible for catalyzing this step of the reaction is benzylsuccinate synthase (BSS) and the identification of bacteria harboring genes for BSS provides a new tool for quantifying hydrocarbon-degrading bacteria in environmental samples (Beller *et al.*, 2002). Recent evidence indicates that this same mechanism may extend to other alkylbenzenes as well. Wilkes *et al.* (2000) reported the selective removal of alkylbenzenes from crude oil by sulfate-reducing enrichment culture with the production of methylbenzylsuccinates and alkylbenzoic acids. In laboratory incubations under sulfate reducing and methanogenic conditions, sediment-associated microorganisms from a gas condensate-contaminated aquifer anaerobically biodegraded toluene, ethylbenzene, xylene, and toluic acid isomers to their corresponding benzylsuccinic acid derivatives with stoichiometric amounts of sulfate consumed or methane produced (Elshahed *et al.*, 2001). The benzylsuccinates were biodegraded further to toluates, phthalates, and benzoate. Some of the metabolites were also detected in groundwater samples from an aquifer where alkylbenzene concentrations decreased over time, suggesting that anaerobic microbial metabolism of these contaminants also occurred *in situ*.

Although oxidized aromatic compounds clearly are important metabolites in the anaerobic degradation of aromatic hydrocarbons, the conditions under which these intermediates accumulate in the environment are poorly understood. Although benzylsuccinic acid has been documented as a metabolic intermediate of toluene degradation under diverse electron-accepting conditions, the extracellular yield of this metabolite under methanogenic conditions can be low (Beller and Edwards, 2000) and sometimes it is not detected in subsurface environments where toluene is actively degrading under anaerobic conditions (Beller, 2002). Other investigators have found that the degradation of aromatic hydrocarbons, such as toluene, can occur in enrichment cultures without the accumulation of any extracellular intermediates under both denitrifying and iron-reducing conditions (Baedecker *et al.*, 1993; Kuhn *et al.*, 1988; Lovley *et al.*, 1989; Lovley and Lonergan, 1990), whereas transient formation of intermediates of toluene degradation, including *p*-cresol, *o*-cresol, and benzoic acid, occurred under methanogenic conditions (Grbić-Galić and Vogel, 1987). In the *in situ* environment, as well as in laboratory experiments containing mixed microbial communities, the intermediates themselves are subject to degradation. In subsurface environments contaminated with petroleum products, the availability of electron acceptors has been shown to be an important control on the apparent rates of microbial degradation of aromatic compounds *in situ* and the accumulation of metabolic intermediates, such as trimethylbenzoates (Cozzarelli *et al.*, 1994). Identification of the metabolites of biodegradation thus can be used as indicators of the degradation of hydrocarbons in petroleum-contaminated environments (see Section 9.12.5.1).

Because of its abundance in anoxic aquatic environments and its importance as a greenhouse gas, methane transformation by anaerobic oxidation has been the subject of numerous studies. The rates of anaerobic methane oxidation and the environments where it has been found were reviewed by Spormann and Widdel (2000). In marine systems, sulfate reduction has been shown to be an important part of the methane oxidation process. Landfills, however, not hydrocarbon contaminations *per se*, are the main source of anthropogenic methane emissions in the US and, therefore, methane degradation processes are not discussed further in this chapter (see Chapter 9.16 for a discussion of methane generation from landfills).

Among the low molecular weight *n*-alkanes, hexane and octane have been the most frequently studied by microbiologists in their attempt to elucidate the metabolic pathways of aliphatic hydrocarbon biodegradation. Evidence for the utilization of *n*-alkanes by anaerobic bacteria was first indicated in the 1940s by Novelli and ZoBell (1944) and Rosenfeld (1947). Depletion of long-chain *n*-alkanes ($C_{15}-C_{34}$) from diesel fuel and crude oil under sulfate-reducing conditions has been reported by Coates *et al.* (1997) and Caldwell *et al.* (1998). Degradation of *n*-alkanes has also been reported for *n*-hexadecane (Bregnard *et al.*, 1996; Rabus *et al.*, 2001) and *n*-hexane under nitrate-reducing conditions (Bregnard *et al.*, 1996; Rabus *et al.*, 2001) and for *n*-hexadecane under methanogenic conditions (Anderson and Lovley, 2000). Spormann and Widdel (2000) provide a review of the anaerobic bacteria identified in pure culture that are able to metabolize these compounds. There have been few detailed studies of the mechanism of alkane degradation under anaerobic conditions. Rabus *et al.* (2001) has proposed a fumarate addition pathway for the initial reaction of *n*-hexane by a dentitrifying bacterium, producing 1-methylpentylsuccinate. Another pathway of degradation under sulfate-reducing conditions appears to involve oxidation of the terminal carbon to form the corresponding carboxylic acid (So and Young, 1999).

The anaerobic degradation of alicyclic hydrocarbons has only recently been documented by (Rios-Hernandez *et al.*, 2003), who investigated the metabolism of ethylcyclopentane (ECP) by sulfate-reducing enrichment cultures obtained from a gas condensate-contaminated aquifer. During biodegradation of ECP, intermediates produced included ethylcyclopentylsuccinic acids, ethylcylcopentylpropionic acid, ethylcyclopentylcarboxylic acid, and ethylsuccinic acid. Based on the identification of these intermediates, Rios-Hernandez *et al.* (2003) proposed that alicyclic hydrocarbons such as ECP can be activated anaerobically under sulfate-reducing conditions by addition to the double bond of fumarate to form the first intermediates, the benzylsuccinate derivatives. This result suggests that the anaerobic metabolites of alicyclic hydrocarbons could be used as indicators of their anaerobic degradation *in situ*, an approach that has been used in the study of alkanes and aromatic hydrocarbon degradation (Rios-Hernandez *et al.*, 2003).

Although a few recent studies indicate anaerobic degradation MTBE can occur, other studies has been demonstrated that MTBE to be relatively recalcitrance in anaerobic aquifers (e.g., Bradley *et al.*, 1999; Suflita and Mormile, 1993) and little is known about the microbial metabolism of this compound under anoxic conditions. One of the first indications of anaerobic MTBE transformation came when Yeh and Novak (1994) found that MTBE could be degraded in microcosms constructed of clay soils and incubated under methanogenic conditions. Mormile *et al.* (1994) studied aquifer material collected from the Ohio River that had been impacted by oil storage and barge loading facilities. MTBE degradation was associated with the appearance of TBA. In a subsequent study with aquifer material that had been contaminated with gasoline, Landmeyer *et al.* (1998) documented slow but measurable rates of MTBE biodegradation, although only ~3% of the added MTBE was degraded. Follow-up studies by Finneran and Lovley (2001) demonstrated that MTBE degradation rates increased after an adaptation period when Fe(III) and humic substances were added to sediments from this same site. In addition, Finneran and Lovley (2001) demonstrated the complete degradation of MTBE with the simultaneous production of CH_4 in bottom sediment from the Potomac River. Bacteria in the aquatic sediments also degraded TBA under anaerobic conditions. A recent EPA study (Wilson *et al.*, 2000) provides perhaps the best evidence that degradation of MTBE occurs in the subsurface under methanogenic conditions. In this investigation, removal of MTBE in microcosms constructed with sediment from the former fuel-farm site at the US Coast Guard Support Center in Elizabeth City, North Carolina, occurred in both the presence and absence of BTEX.

9.12.5 ENVIRONMENTAL RESTORATION

9.12.5.1 Natural Attenuation Processes

Natural attenuation of petroleum hydrocarbons in the environment has been the focus of many studies over the past decade and has become widely applied as a remedial option (National Research Council (US) Committee on Intrinsic Remediation, 2000). Recognition of the broad scope of the problem of groundwater contamination and the high cost of monitoring engineered groundwater clean-up systems has contributed to the skyrocketing interest in monitored natural attenuation as a remediation strategy (Bekins *et al.*, 2001b). The conditions under which natural attenuation is likely to be an effective means of aquifer restoration were discussed by Barker and Wilson (1997) for BTEX compounds and by Chapelle (1999) for petroleum hydrocarbons. The theory, mechanisms, and applications of natural attenuation for fuels, including numerous case studies, recently were presented by Wiedemeier *et al.* (1999) and, therefore, do not require reiteration here. However, limits on the applicability of natural attenuation and some of the problems faced when attempting to document natural attenuation warrant discussion.

Heterogeneities in physical and chemical properties of aquifer sediments can constrain the possibilities for hydrologic transport and biogeochemical reactions and complicate the study of the fate of petroleum contaminants *in situ*. The potential difficulties in monitoring contaminant migration in heterogeneous systems were first illustrated by Patrick and Barker (1985) in a natural-gradient tracer test involving addition of dissolved benzene, toluene, and xylenes in groundwater. In the unconfined shallow glacio-fluvial sand aquifer, the vertical scale of heterogeneity was on the order of 0.01 m, and hydraulic conductivity contrasts of up to an order of magnitude were demonstrated among sediment layers. Resulting retardation of hydrocarbons along individual layers and concentration–depth profiles at individual locations appeared erratic. In a study of a coastal plain aquifer, the rates of biodegradation of toluene and benzene were observed to vary as a function of sediment type due to different levels of microbial activity (Aelion, 1996). Natural attenuation of petroleum hydrocarbon in a heterogeneous aquifer in Denmark showed that strong seasonal variations in the flow field affected the availability of electron acceptors and the shape and extent of the hydrocarbon plume (Mossing *et al.*, 2001).

Because of the complexities of the subsurface environment, documenting the natural attenuation of hydrocarbons requires multiple lines of evidence (National Research Council (US) Committee on Intrinsic Remediation, 2000; Wiedemeier *et al.*, 1995) and a multiscale approach. It has been demonstrated that closely spaced sampling is required to identify microbiological processes in contaminated aquifers (Bekins *et al.*, 1999; Smith *et al.*, 1991). In fact, detailed studies of microbial processes may require a microscale approach to account for the microbe–mineral interactions that have been documented by Bennett *et al.* (2000). Coupling the study of microbiological and hydrogeologic processes is a particular challenge because the scales used in these investigations usually are quite different. This difference has resulted in limited knowledge of the interaction between hydrological and microbiological features of subsurface environments (Haack and Bekins, 2000).

Scientists conducting research as part of the USGS Toxic Substances Hydrology Program (for more information see http://toxics.usgs.gov) have been characterizing chemical heterogeneity in an effort to better understand the natural attenuation of hydrocarbons in surficial aquifers. In one example of this approach, the scale of biogeochemical reactions was investigated in a physically and chemically heterogeneous surficial coastal plain aquifer contaminated by a gasoline spill (Cozzarelli *et al.*, 1995, 1999). Hydrocarbons were degraded by microbially mediated reactions that varied over short vertical distances (tens of centimeters) and time (months) (Figure 13). Anaerobic processes, iron reduction and sulfate reduction, dominated within a low-permeability clay unit present between 4 and 4.5 m bls (below land surface), whereas in the more permeable sandy layers (less than 4 m bls) nitrate reduction and aerobic degradation occurred. Mixing with oxygenated recharge water resulted in increased degradation of hydrocarbons near the water table (at 3.26 m bls, Figure 13) whereas mixing with less contaminated groundwater limited the spread of hydrocarbons at the base of the plume. Assessment of aquifer heterogeneities and groundwater contamination was possible because of sampling at the submeter scale, a finer resolution than is attempted in many remedial investigations of contaminated aquifers. The information obtained in this type of study is essential to the development of models capable of simulating the fate of hydrocarbons at the scale of a contaminated site.

Steep gradients in the concentrations of organic and inorganic carbon, and the reactants and products associated with hydrocarbon degradation

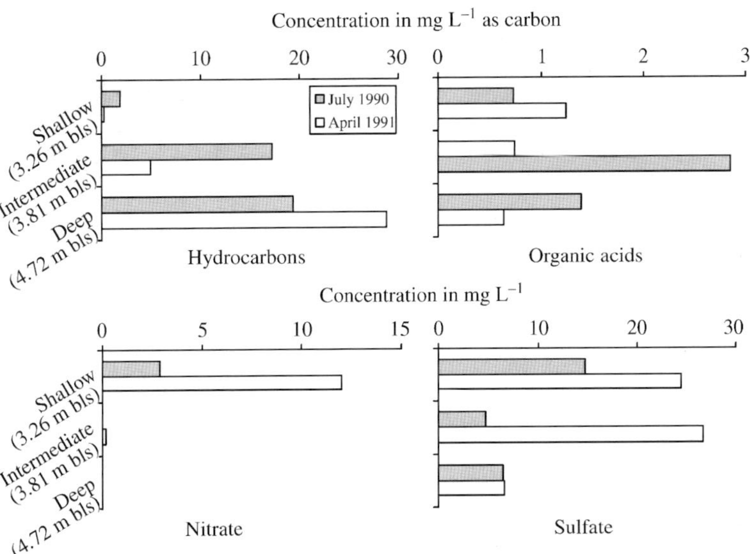

Figure 13 Concentration profiles for groundwater at the Galloway Township gasoline study site in July 1990 and April 1991(after Cozzarelli *et al.*, 1999). The two shallow samples are within the sandy sediments (3.26 m and 3.81 m bls) and one deeper sample (at 4.72 m bls) is within the low-permeability clay-rich layer.

also have been observed by investigators at a site contaminated with crude oil at Bemidji, MN (Bekins *et al.*, 2001a; Cozzarelli *et al.*, 2001b). The 16 yr study of the hydrocarbon plume documented that the extent of contaminant migration and compound-specific behavior has changed as redox reactions, most notably iron reduction, have evolved over time. Previous work at the Bemidji site documented the loss of benzene and alkylbenzenes in groundwater downgradient from the crude-oil source (Baedecker *et al.*, 1993; Cozzarelli *et al.*, 1990; Eganhouse *et al.*, 1993, 1996) and established that microorganisms degrade these compounds in the anoxic aquifer (Baedecker *et al.*, 1993; Cozzarelli *et al.*, 1994; Lovley *et al.*, 1989). Zones of maximum hydrocarbon contamination at this site were found within the anoxic plume and were on the order of only 1–2 m in thickness (Bekins *et al.*, 2001a; Cozzarelli *et al.*, 2001a). Within the narrow anoxic zones, anaerobic degradation reactions resulted in significant loss of hydrocarbons as groundwater moved downgradient. Over time, on the order of years, concentration changes at a small scale, determined from analysis of pore-water samples drained from aquifer cores, indicated that the hydrocarbon plume was growing slowly as sediment iron oxides were depleted. Depletion of the unstable Fe(III) oxides near the subsurface crude-oil source caused the zone of maximum concentrations of BTEX to spread within the anoxic plume (Cozzarelli *et al.*, 2001a). Microbial data from the same profiles through the anaerobic portion of the contaminated aquifer (Figure 14) clearly show areas where the microbial populations have progressed from iron-reducers to methanogens (Bekins *et al.*, 2001a). In monitoring the remediation of hydrocarbon-contaminated groundwater by natural attenuation, these investigators found that subtle concentration changes in observation-well data from the anoxic zone were diagnostic of depletion of the intrinsic electron-accepting capacity of the aquifer and may allow early prediction of growth of a hydrocarbon plume.

New analytical tools and approaches are being used to document the biodegradation of contaminants *in situ*. These include use of compound-specific stable isotope ratios (Dempster *et al.*, 1997; Wilkes *et al.*, 2000), identification of metabolism products (e.g., Beller, 2000), comprehensive two-dimensional gas chromatography (e.g., Gaines *et al.*, 1999), and molecular techniques to identify microbial diversity and function (Kleikemper *et al.*, 2002). Because disappearance of parent compounds is not always attributable to destructive attenuation mechanisms, the presence of oxidized metabolites of hydrocarbons sometimes is used as conclusive evidence of *in situ* microbial transformation. Beller (2000) reviews the use of metabolic indicators for detecting anaerobic degradation of alkylbenzenes; examples of these metabolites are shown in Figure 15. The metabolites of alkylbenzenes most frequently reported in aquifers contaminated with petroleum products are benzoic and alkylbenzoic acids. These compounds were first identified in a petroleum-contaminated aquifer by Cozzarelli *et al.* (1990), who found methyl-, dimethyl-, and trimethylbenzoic acids in anoxic groundwater downgradient from a crude-oil spill in Bemidji,

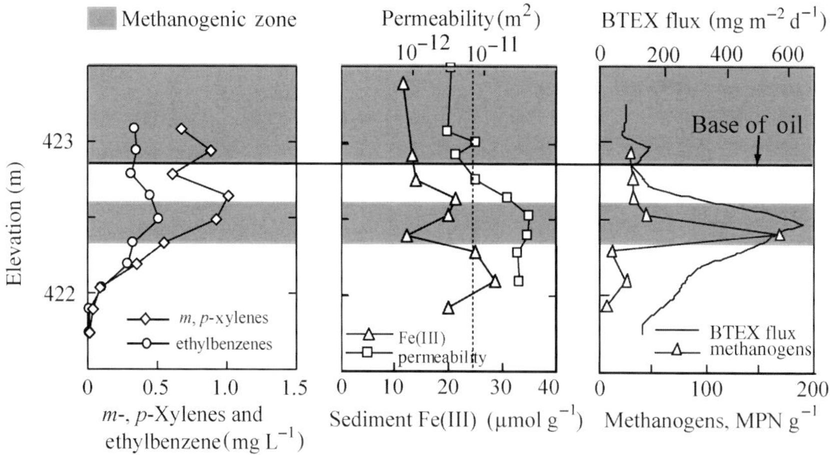

Figure 14 Microbial and chemical profiles from the crude-oil-contaminated aquifer near Bemidji, Minnesota. Methanogenic zones are shaded grey (after Bekins *et al.*, 2002).

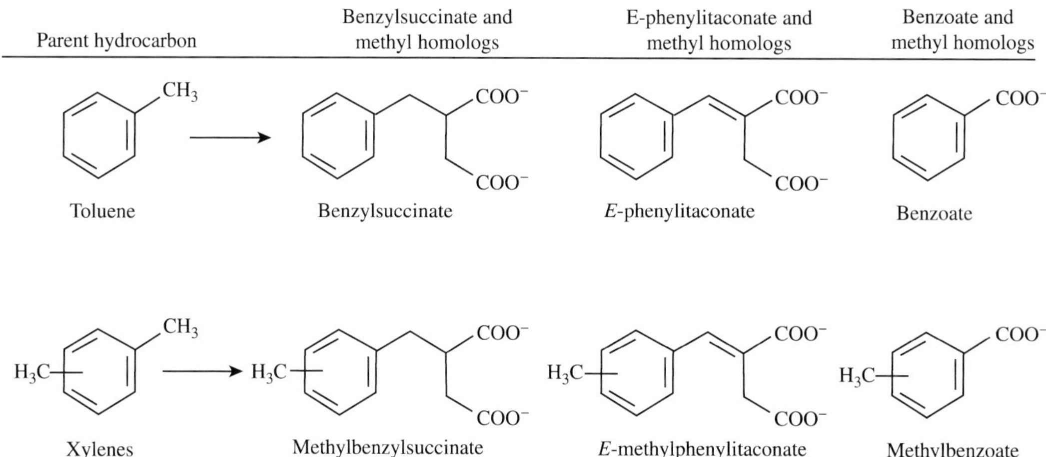

Figure 15 Potential indicators of *in situ* metabolism of toluene and xylene under anaerobic conditions (after Beller, 2000).

Minnesota, where microbial degradation of alkylbenzenes was occurring. Others have observed these compounds in association with hydrocarbon contamination including Fang and Barcelona (1999), who identified dimethyl- and trimethylbenzoic acids in an anoxic aquifer contaminated with JP-4 jet fuel at Wurtsmith Air Force Base, Oscoda, Michigan.

Beller *et al.* (1995) first proposed the use of benzylsuccinic acid and its methylated analogs as useful indicators of anaerobic toluene and xylene degradation in a controlled-release experiment at Seal Beach (California). During the two-month injection experiment, toluene, *o*-xylene, and *m*-xylene were depleted in the groundwater and the benzylsuccinate and *E*-phenylitaconate homologs corresponding to each of these hydrocarbons appeared in the groundwater. Concentration trends provided conclusive evidence that these metabolites were produced during anaerobic metabolism of the precursor hydrocarbons. Similar alkylbenzene transformation products were detected in a fuel-contaminated aquifer in Fallon (Nevada) (Beller, 2000). In a highly reduced gas-condensate contaminated aquifer, Elshahed *et al.* (2001) found benzylsuccinic acid analogs of ethylbenzene, *o*-, *m*-, and *p*-xylene metabolism in sulfate-reducing zones of the aquifer. Anaerobic metabolism of alkanes *in situ* also has been indicated by the presence of alkylsuccinates in six aquifers contaminated with petroleum products (Gieg and Suflita, 2002). The widespread occurrence of these metabolites at sites with both low and high hydrocarbon concentrations suggests that anaerobic degradation reactions are common in subsurface environments and that detection of metabolites may be a useful indicator of this process.

Numerous studies over the past two decades have established that hydrocarbon-degrading microorganisms are ubiquitous in soils, aquatic sediments, and aquifers. However, one aspect of documenting natural attenuation of hydrocarbons is demonstrating that the microbial populations necessary to carry out the specific transformations, outlined in Section 9.12.4.2, are active. This activity presents a practical constraint for many studies of natural attenuation because of the difficulty in accurately defining microbial activity in environmental samples (Bekins *et al.*, 2002). Culture-based and molecular analyses of microbial populations in subsurface contaminant plumes have revealed important adaptation of microbial populations to plume environmental conditions (Haack and Bekins, 2000). Results of recent studies at the Bemidji crude-oil site demonstrate that the subsurface geochemical and hydrological conditions appreciably affect subsurface microbial-community structure (Bekins *et al.*, 2001a, 2002).

Microbial population dynamics in petroleum-contaminated environments sometimes are characterized based on the analysis of lipids, such as phospholipid ester-linked fatty acids (PLFA). These techniques can provide information on a variety of microbial characteristics and overall community composition (Green and Scow, 2000). For example, Fang and Barcelona (1998) examined microbial biomass and community structure in a glaciofluvial aquifer located at Wurtsmith Air Force Base, Michigan, which had been contaminated with JP-4 fuel hydrocarbons released after the crash of a tanker aircraft. The fatty acids identified in the aquifer sediment ranged from C_{12}–C_{20}, including saturated, monounsaturated, branched and cyclopropyl fatty acids, whereas the polyunsaturated fatty acids were virtually absent. The results of the PLFA analysis indicated considerable microbial heterogeneity of bacteria in the subsurface. The hydrocarbon-contaminated anaerobic zones had higher microbial biomass and metabolically more diverse microbial communities than those in aerobic zones.

Nucleic-acid approaches, which involve the extraction of microbial DNA and RNA from environmental samples, have been applied to the study of natural attenuation of petroleum products since the 1990s. With these techniques, the specific genes that are responsible for hydrocarbon degrading capabilities in bacteria can be measured. These tools are especially useful for studying microbial diversity at impacted sites (Madsen, 2000). Various recent applications of these techniques in petroleum-contaminated environments are reviewed by Haack and Bekins (2000). For example, Stapleton and Sayler (1999) monitored changes in the molecular microbial ecology at the Natural Attenuation Test Site (NATS) at Columbus Air Force Base, Mississippi by identifying hydrocarbon-degrading genes *in situ*. Exposure of indigenous microorganisms to BTEX and naphthalene was evaluated using an array of gene probes targeting common genotypes associated with the aerobic biodegradation of these compounds. Each of the targeted genotypes showed appreciable responses to hydrocarbon exposure and, combined with increased aerobic degradation potentials of the added hydrocarbons, provided conclusive evidence that an aerobic contaminant-degrading community successfully developed within the plume. The Natural Attenuation Study at Columbus AFB represents a successful application of molecular techniques in linking adaptations of indigenous microorganisms to hydrocarbon exposure. Currently, however, the number of known genes responsible for hydrocarbon degradation is small and broader application of this technology awaits the identification of the nucleic-acid sequences of additional degradative genes (Haack and Bekins, 2000).

The natural attenuation potential of many compounds is not yet well understood (National Research Council (US) Committee on Intrinsic Remediation, 2000). Although natural attenuation of the aromatic hydrocarbon components of gasoline has been well established, much less is known about the efficacy of natural attenuation of MTBE. Natural attenuation of MTBE recently has been observed in oxygenated environments, even at low concentrations. Baehr *et al.* (2001) showed that unsaturated zone gas concentrations of MTBE were less than atmospheric concentrations because of aerobic degradation of very low MTBE concentrations in the soil zone. As described in the anaerobic processes section of this chapter, the study of the microbial metabolism of MTBE under anaerobic conditions (frequently encountered in the subsurface) is still in its infancy. Documenting the degradation of this compound at the field scale has been problematic because of the slow rates of biodegradation of MTBE. One recent study by Wilson *et al.* (2000), conducted at the former Fuel Farm Site at the US Coast Guard Support Center at Elizabeth City, North Carolina demonstrated MTBE removal in a methanogenic aquifer in which BTEX compounds undergo extensive anaerobic oxidation. The apparent attenuation of MTBE in the anoxic plume (average first-order attenuation rate of MTBE in the groundwater was calculated at 2.7 per year (Wilson *et al.*, 2000)) was consistent with the biodegradation rates measured in laboratory experiments.

Tools, such as compound-specific isotope analyses (CSIA), can be used to look for evidence of biodegradation at contaminated sites. In the laboratory (e.g., Ahad *et al.*, 2000; Harrington *et al.*, 1999; Hunkeler *et al.*, 2001a;

Mancini *et al.*, 2003) and field (Kelley *et al.*, 1997; Mancini *et al.*, 2002; Stehmeier *et al.*, 1999) there have been important applications of CSIA to aromatic hydrocarbons to investigate both aerobic and anaerobic biodegradation. For compounds such as MTBE, where the intermediates of biodegradation (such as TBA) are transitory and difficult to quantify, additional lines of evidence such as CSIA may be particularly important. For example, the biodegradation of MTBE to TBA proceeds, by cleavage of the ether bond, at different rates for the two stable isotopes of carbon ^{12}C and ^{13}C (Gray *et al.*, 2002; Zwank *et al.*, 2002). Degradation of MTBE leads to enrichment in the heavier ^{13}C isotope in the remaining MTBE pool. This isotopic enrichment effect has been observed in aerobic microcosm experiments conducted with sediment from the CFB Borden site (Ontario, Canada) during biodegradation of both MTBE and TBA (Hunkeler *et al.*, 2001b). Similar carbon enrichment factors (on the order of -1.5 per mil to -1.8 per mil) and even larger hydrogen isotope enrichment factors were observed in MTBE-degrading microcosms using sediment from Vandenberg Air Force Base, CA (Gray *et al.*, 2002). Kolhatkar *et al.* (2002) have demonstrated similar large isotope fractionation effects during anaerobic biodegradation of MTBE at a gasoline release site. These studies indicate that carbon isotopic fractionation offers the potential to aid in the assessment of MTBE biodegradation at field sites.

9.12.5.2 Engineered or Enhanced Remediation

Engineered remediation of hydrocarbon-contaminated environments focuses on the destruction, dilution, or physical removal of the hydrocarbons to reduce their concentrations to levels that are environmentally acceptable. There are *in situ* technologies, involving treatment of the contaminants in place, as well as *ex situ* technologies, which involve physical removal of the contaminated material to another location. The environmental compartment within which the hydrocarbons reside largely dictates the remediation approach. Sometimes the approach involves the complete or partial transfer of the contaminant from one environmental compartment to another, thereby transferring the risk. For example, in the treatment of contaminated soil, one approach is aeration, which increases the inhalation exposure risk of workers and nearby residents (NRC, 1993). Remedial technologies used in the restoration of petroleum-contaminated aquifers have been reviewed by Testa and Winegardner (2000). The restoration of contaminated aquifers requires that several objectives be accomplished, including both contaminant plume and source containment as well as plume and source removal. Different technological approaches often are used to address each of these objectives and may include combinations of physical, chemical, and biological processes.

In beach sediments or in the terrestrial environment, where physical removal of the contaminant is difficult and expensive and *ex situ* clean-up technologies have appreciable limitations (NRC, 1993), a common clean-up approach is enhanced bioremediation. Enhanced bioremediation is an intensive area of research and numerous books have been written on this topic (Baker and Herson, 1994; R. L. Crawford and D. L. Crawford, 1996; Riser- Roberts, 1992). Enhanced bioremediation focuses on aiding the destruction of contaminants by microorganisms by providing electron acceptors, donors, nutrients, or microorganisms to the environmental system. The presence of hydrocarbons in the environment usually brings about a selective enrichment of indigenous hydrocarbon-degrading microorganisms *in situ* and, thus, seeding a contaminated site with microorganisms usually is not necessary (Rosenberg and Ron, 1996). Addition of nutrients or electron acceptors, however, is a common approach, and has been effective in increasing biodegradation of hydrocarbons following surface water petroleum spills. As discussed in Section 9.12.4.2.1, in aquatic environments, microbial requirements for oxygen and moisture are not limiting, but available nutrients, such as nitrogen and phosphorus, usually are limiting (Rosenberg and Ron, 1996). For example, commercial nitrogen- and phosphorus-containing fertilizers that have an affinity for hydrocarbons have been used to treat oil contaminated shorelines.

Following the *Exxon Valdez* oil tanker spill, ~100 mi (161 km) of shoreline were treated with fertilizers, in the largest marine bioremediation project ever undertaken. Despite numerous studies of the remediation project, no conclusive evidence for the long-term impact of fertilizer treatment on the removal of oil from the beaches was obtained (Rosenberg and Ron, 1996). More success has been achieved using time-release fertilizers to treat oil-contaminated sand. Rosenberg *et al.* (1992) demonstrated significant biodegradation of hydrocarbons at Zvulon Beach in Israel that was heavily contaminated with heavy crude oil, after addition of controlled-release, polymeric, hydrophobic fertilizer (because bacteria that can hydrolyze this fertilizer are rare, these were added in this case).

Natural attenuation studies of MTBE are limited and the physicochemical characteristics of MTBE, including relatively high aqueous solubility and low Henry's law constant, are such that removal by traditional treatment technologies, such as pump-and-treat, is difficult.

For example, a common bioremediation approach in the unsaturated subsurface is known as "bioventing," which involves the forced air movement of oxygen through the petroleum-contaminated soil (Hinchee, 1995). The presence of MTBE in addition to petroleum hydrocarbon compounds complicates the use of these techniques because the volatility and aerobic biodegradation potential of MTBE and BTEX, for example, are different and the effects of their co-occurrence are largely unknown. The presence of MTBE in many cases of petroleum product releases to the environment has, therefore, appreciably changed the restoration options. Because of the slow rate of anaerobic degradation of MTBE, enhancement of aerobic degradation by enhanced oxygenation of an aquifer has been the focus of recent *in situ* bioremediation approaches (Koenigsberg *et al.*, 1999).

9.12.6 CHALLENGES

The use of petroleum products is integral to the functioning of industrialized societies. Increased worldwide energy consumption results as economies expand to meet the instinctive need to improve living standards. Increased energy demand translates directly to increased use and release of petroleum products because use of alternative energy sources (e.g., nuclear, wind, hydroelectric, and solar power) remains a secondary strategy. This release of petroleum compounds to the environment results in human and ecological exposure. Knowledge of the fate of the compounds throughout the environment, their distribution in the hydrosphere, biosphere, and geoshpere, at a scale relevant to human consumption, and the toxicological consequences associated with chronic exposure to mixtures of petroleum and other compounds (e.g., pesticides) has not advanced to the point where environmental costs associated with product usage can be measured.

Much research has focused on the transport of BTEX contaminants in groundwater that results from releases from USTs. An evolving consensus within the research community is that natural attenuation from microbial breakdown and volatilization limits the movement of BTEX in ground and surface waters so that only wells or streams in close proximity of a spill are affected. This consensus has been supported by results of the USGS in its assessment of ambient shallow groundwater throughout urban areas, as BTEX compounds are not detected frequently. Although it has been established that detailed field investigations are essential for a thorough evaluation of natural attenuation, allowing the documentation of the biogeochemical processes responsible for contaminant destruction (Bekins *et al.*, 2001b; National Research Council (US) Committee on Intrinsic Remediation, 2000), few of these studies have been undertaken. New tools are needed to allow the study of the fate of petroleum hydrocarbons and fuel oxygenates in diverse environments. Long-term detailed monitoring programs are essential to develop conceptual models of natural attenuation and studies need to allow the recognition that our understanding of microbial transformation pathways is constantly changing.

Field investigations of spatial and temporal variability of biogeochemical processes and resulting changes in groundwater and aquifer composition provide insight into how naturally occurring microorganisms degrade hydrocarbons in contaminated aquifers such as those underlying the Galloway and Bemidji sites. It is clear that understanding the environmental conditions that favor biodegradation depends on our ability to measure these processes at an appropriate scale. This knowledge forms the basis for applying either intrinsic or engineered *in situ* bioremediation as a groundwater cleanup strategy. Yet, long-term studies of the natural degradation of contaminants in the field are rare and studies of the impact of changes in important factors, such as the source term, are lacking. In addition, the fate of hydrocarbons in geologically heterogeneous systems is largely unknown and determining the appropriate spatial scale at which to sample in heterogeneous systems presents a major challenge to geochemists and microbiologists and will differ from site to site. New analytical tools will aid in the verification and quantification of contaminant fate processes in the environment. Understanding the environmental conditions that favor intrinsic bioremediation will depend largely on the ability to characterize and quantify chemical and microbiological heterogeneity at a scale appropriate to the reactions being measured.

MTBE usage in gasoline increased dramatically in the 1990s in response to programs in the United States (CAAAs) and Europe to improve air quality. As a result, when oxygenated gasoline is spilled, MTBE is the principal contaminant in groundwater on a mass basis. MTBE degrades more slowly and is less volatile than BTEX. As a result, MTBE has been observed by many investigators to migrate farther in aquifers than BTEX at gasoline-spill sites. Further, MTBE frequently is detected in ambient groundwater. The implication of the detection of MTBE at low concentrations in ambient groundwater is unclear. If the low concentrations of VOCs and MTBE found in regional water-quality studies (multi-state) are the result of evolving plumes emanating from USTs, then concentrations possibly could increase with time. If the source were the atmosphere, changes in VOC concentrations in groundwater over time would be constrained by atmospheric concentrations. In addition, the toxicological implication of increased exposure to MTBE is not known.

The effect of contaminants entering the environment in mixtures still is largely unknown. Biodegradation by indigenous microorganisms is now recognized as the major attenuation mechanism for BTEX compounds in both oxic and anoxic groundwater environments. The potential negative consequence of MTBE co-occurrence with hydrocarbons on the biodegradation rates of BTEX in contaminated aquifers is, therefore, of major concern. Thus far, studies investigating this effect have yielded inconsistent results. At the field scale, definitive evidence is lacking that MTBE appreciably inhibits degradation of BTEX compounds. In contrast, the degradation of MTBE itself does appear to be retarded by the presence of other more degradable organic compounds, including BTEX. Concerns over MTBE in drinking-water supplies have prompted elimination of MTBE from gasoline in some areas; however the effect of replacement oxygenates, such as ethanol, methanol, or *t*-amyl methyl ether, on BTEX transport and groundwater quality is unknown. In the case of ethanol, a highly degradable compound, it seems likely that the presence of this compound with hydrocarbons will alter the microbial degradation rates of the hydrocarbons. Church *et al.* (2000) have demonstrated the effects of ethanol on benzene plume lengths. They concluded that in cases with low dissolved oxygen concentrations ($2\ \mathrm{mg\ L^{-1}}$), benzene plume lengths increased over 100% when ethanol was added compared to the benzene plume length simulated with gasoline only (containing no fuel oxygenates).

Whereas MTBE usage has resulted in improved air quality in urban areas, it has challenged the UST spill natural attenuation paradigm because of its widespread occurrence. The classic environmental problem of improving atmosphere quality in exchange for degradation of the hydrosphere, or vice versa, has been demonstrated. The potential impact on water quality of other fuel oxygenates remains largely untested. With the exception of cases of well contamination by MTBE at high concentration, the detrimental effects of MTBE in aquifers at the regional scale are perceived, because the cumulative effect of MTBE extent and distribution resulting from the vast network of USTs has not been evaluated. The decision to modify oxygenated fuel usage, therefore, poses a major challenge to regulatory agencies as a balance between perception and scientific information is sought.

ACKNOWLEDGMENTS

The authors thank Jeanne Jaeschke and Robin Krest of the USGS for assistance with the reference database, tables, and figures. This manuscript was improved by thoughtful review comments of Robert Eganhouse of the USGS and an anonymous reviewer. Sections of the manuscript were improved by comments from Barbara Bekins of the USGS and Harry Beller of the Lawrence Livermore National Laboratory. This work was supported by the USGS National Research Program and Toxic Substances Hydrology Program.

REFERENCES

Abdelouas A., Lutze W., Nuttall E. H., Strietelmeier B. A., and Travis B. J. (2000) Biological reduction of uranium in groundwater and subsurface soil. *Sci. Tot. Environ.* **250**(1), 21–37.

Addy J. M., Hartley J. P., and Tibbetts P. J. C. (1984) Ecological effects of low toxicity oil based mud drilling in the Beatrice oilfield. *Mar. Pollut. Bull.* **15**(12), 429–436.

Aelion C. M. (1996) Impact of aquifer sediment grain size on petroleum hydrocarbon distribution and biodegradation. *J. Contamin. Hydrol.* **22**(1–2), 21–109.

Ahad J. M. E., Lollar B. S., Edwards E. A., Slater G. F., and Sleep B. E. (2000) Carbon isotope fractionation during anaerobic biodegradation of toluene: implications for intrinsic bioremediation. *Environ. Sci. Technol.* **34**(5), 892–896.

Anderson R. T. and Lovley D. R. (2000) Hexadecane decay by methanogenesis. *Nature* **404**, 722–723.

Atkinson R. (1990) Gas-phase tropospheric chemistry of organic compounds: a review. *Atmos. Environ.* **24A**(1), 1–41.

Atlas R. M. (1984) *Petroleum Microbiology*. Macmillan, New York, 692pp.

Baedecker M. J. and Eganhouse R. P. (1991) Partitioning and transport of hydrocarbons from crude oil in a sand and gravel aquifer. In *201st National Meeting of the American Chemical Society*, Washington, DC, pp. 463–466.

Baedecker M. J., Cozzarelli I. M., Eganhouse R. P., Siegel D. I., and Bennett P. C. (1993) Crude oil in a shallow sand and gravel aquifer: III. Biogeochemical reactions and mass balance modeling in anoxic groundwater. *Appl. Geochem.* **8**, 569–586.

Baehr A. L., Stackelberg P. E., and Baker R. J. (1999) Evaluation of the atmosphere as a source of volatile organic compounds in shallow ground water. *Water Resour. Res.* **35**(1), 127–136.

Baehr A. L., Charles E. G., and Baker R. J. (2001) Methyl *tert*-butyl ether degradation in the unsaturated zone and the relation between MTBE in the atmosphere and shallow groundwater. *Water Resour. Res.* **37**(2), 223.

Baker K. H. and Herson D. S. (1994) *Bioremediation*. McGraw-Hill, New York, 376pp.

Baker R. J., Best E. W., and Baehr A. L. (2002) Used motor oil as a source of MTBE, TAME, and BTEX to ground water. *Ground Water Monitor. Remediat.* **22**(4), 46–51.

Ball H. A. and Reinhard M. (1996) Monoaromatic hydrocarbon transformation under anaerobic conditions at seal beach, California: laboratory studies. *Environ. Toxicol. Chem.* **15**(2), 114–122.

Barber L. B., II, Thurman E. M., and Runnells D. D. (1992) Geochemical heterogeneity in a sand and gravel aquifer: effect of sediment mineralogy and particle size on the sorption of chlorobenzenes. *J. Contamin. Hydrol.* **9**, 35–54.

Barker J. F. and Wilson J. T. (1997) Natural biological attenuation of aromatic hydrocarbons under anaerobic conditions. In *Subsurface Restoration* (eds. C. H. Ward, J. A. Cherry, and M. R. Scalf). Ann Arbor Press, Chelsea, MI, pp. 289–300.

Barker J. F., Tessmann J. S., Plotz P. E., and Reinhard M. (1986) The organic geochemistry of a sanitary landfill leachate plume. *J. Contamin. Hydrol.* **1**, 171–189.

Barker J. F., Gillham R. W., Lemon L., Mayfield C. I., Poulsen M., and Sudicky E. A. (1991) Chemical fate and impact of oxygenates in ground water: solubility of BTEX from gasoline-oxygenate mixtures. American Petroleum Institute Publication 4531, American Petroleum Institute, Washington, DC, 90pp.

Bekins B. A., Baehr A. L., Cozzarelli I. M., Essaid H. I., Haack S. K., Harvey R. W., Shapiro A. M., Smith J. A., and Smith R. L. (1999) Capabilities and challenges of natural attenuation in the subsurface: lessons from the US Geological Survey Toxic Substances Hydrology Program. In *US Geological Survey Toxic Substances Hydrology Program: Proceedings of the Technical Meeting*, USGS, Denver, CO, WRI Report 99-4018C, pp. 37–56.

Bekins B. A., Cozzarelli I. M., Godsy E. M., Warren E., Essaid H. I., and Tuccillo M. E. (2001a) Progression of natural attenuation processes at a crude-oil spill site: II. Controls on spatial distribution of microbial populations. *J. Contamin. Hydrol.* **53**, 387–406.

Bekins B. A., Rittmann B. E., and MacDonald J. A. (2001b) Natural attenuation strategy for groundwater cleanup focuses on demonstrating cause and effect. *EOS Trans.* **53**, 57–58.

Bekins B. A., Cozzarelli I. M., Warren E., and Godsy E. M. (2002) Microbial ecology of a crude oil contaminated aquifer. In *Proceedings of the Groundwater Quality 2001 Conference held at Sheffield UK, June 2001*, IAHS Publication 275-2002, pp. 57–63.

Beller H. and Edwards E. (2000) Anaerobic toluene activation by benzylsuccinate synthase in a highly enriched methanogenic culture. *Appl. Environ. Microbiol.* **66**(12), 5503–5505.

Beller H. R. (2000) Metabolic indicators for detecting *in situ* anaerobic alkylbenzene degradation. *Biodegradation* **11**(2–3), 125–139.

Beller H. R. (2002) Analysis of benzylsuccinates in groundwater by liquid chromatography/tandem mass spectrometry and its use for monitoring *in situ* BTEX biodegradation. *Environ. Sci. Technol.* **36**(12), 2724–2728.

Beller H. R. and Spormann A. M. (1997a) Anaerobic activation of toluene and *o*-xylene by addition to fumarate in denitrifying strain T. *J. Bacteriol.* **170**(3), 670–676.

Beller H. R. and Spormann A. M. (1997b) Benzylsuccinate formation as a means of anaerobic toluene activation by sulfate-reducing strain PRTOL1. *Appl. Environ. Microbiol.* **63**(9), 3729–3731.

Beller H. R., Grbić-Galić D., and Reinhard M. (1992a) Microbial degradation of toluene under sulfate-reducing conditions and the influence of iron on the process. *Appl. Environ. Microbiol.* **58**(3), 786–793.

Beller H. R., Reinhard M., and Grbić-Galić D. (1992b) Metabolic by-products of anaerobic toluene degradation by sulfate-reducing enrichment cultures. *Appl. Environ. Microbiol.* **58**(9), 3192–3195.

Beller H. R., Ding W.-H., and Reinhard M. (1995) Byproducts of anaerobic alkylbenzene metabolism useful as indicators of *in situ* bioremediation. *Environ. Sci. Technol.* **29**(11), 2864–2870.

Beller H. R., Spormann A. M., Sharma P. K., Cole J. R., and Reinhard M. (1996) Isolation and characterization of a novel toluene-degrading, sulfate-reducing bacterium. *Appl. Environ. Microbiol.* **62**(4), 1188–1196.

Beller H. R., Kane S. R., Legler T. C., and Alvarez P. J. J. (2002) A real-time polymerase chain reaction method for monitoring anaerobic, hydrocarbon-degrading bacteria based on a catabolic gene. *Environ. Sci. Technol.* **36**(18), 3977–3984.

Bennett P. C., Hiebert F. K., and Rogers J. R. (2000) Microbial control of mineral-groundwater equilibria: macroscale to microscale. *Hydrogeol. J.* **8**(1), 47–62.

Berry D. F., Francis A. J., and Bollag J.-M. (1987) Microbial metabolism of homocyclic and heterocyclic aromatic compounds under anaerobic conditions. *Microbiol. Rev.* **51**(1), 43–59.

Biegert T., Fuchs G., and Heider J. (1996) Evidence that anaerobic oxidation of toluene in the denitrifying bacterium thauera aromatica is initiated by formation of benzylsuccinate from toluene and fumarate. *Euro. J. Biochem.* **238**(3), 661–668.

Bosma T. N. P., Harms H., and Zehnder A. J. B. (2001) Biodegradation of xenobiotics in environment and technosphere. In *The Handbook of Environmental Chemistry*, Volume 2, part K-Biodegradation and Persistence (ed. B. Beek). Springer, Berlin, pp. 163–202.

Bouwer E. J. and Zehnder A. (1993) Bioremediation of organic compounds-putting microbial metabolism to work. *Bioremediation* **11**(August), 360–367.

Bradley P. M., Landmeyer J. E., and Chapelle F. H. (1999) Aerobic mineralization of MTBE and tert-butyl alcohol by stream-bed sediment microorganisms. *Environ. Sci. Technol.* **33**(11), 1877–1879.

Bradley P. M., Landmeyer J. E., and Chapelle F. H. (2001) Widespread potential for microbial MTBE degradation in surface-water sediments. *Environ. Sci. Technol.* **35**(4), 658–662.

Bregnard T. P.-A., Hohener P., Haner A., and Zeyer J. (1996) Degradation of weathered diesel fuel by microorganisms from a contaminated aquifer in aerobic and anaerobic microcosms. *Environ. Toxicol. Chem.* **15**(3), 299–307.

Breit G. N., Kharaka Y., and Rice C. A. (2000) Database of the composition of water produced with oil and gas. In *7th International Petroleum Environmental Conference*, pp. 903–942.

Burland S. and Edwards E. A. (1999) Anaerobic benzene biodegradation linked to nitrate reduction. *Appl. Environ. Microbiol.* **65**(2), 529–533.

Calabrese E. J. and Kenyon E. M. (1991) *Air Toxics and Risk Assessment.* Lewis Publishers, Chelsea, MI.

Calabrese E. J., Stanek E. S., and Gilbert C. E. (1991) A preliminary decision framework for deriving soil ingestion rates. In *Hydrocarbon Contaminated Soils and Groundwater: Analysis, Fate, Environmental and Public Health Effects, Remediation* (eds. P. T. Kostecki and E. J. Calabrese). Lewis Publishers, Chelsea, MI, pp. 301–311.

Caldwell M. E., Garrett R. M., Prince R. C., and Suflita J. M. (1998) Anaerobic degradation of long-chain n-alkanes under sulfate-reducing conditions. *Environ. Sci. Technol.* **32**, 2191–2195.

Cervantes F., Dijksma W., and Duong-Dac T. (2001) Anaerobic mineralization of toluene by enriched sediments with quinoes and humus as terminal electron acceptors. *Appl. Environ. Microbiol.* **67**(10), 4471–4478.

Chapelle F. H. (1993) *Groundwater Microbiology and Geochemistry*. Wiley, New York.

Chapelle F. H. (1999) Bioremediation of petroleum hydrocarbon-contaminated groundwater: the perspectives of history and hydrology. *Ground Water* **37**(1), 122–132.

Chapelle F. H., Bradley P. M., and Lovely D. R. (2002) Rapid evolution of redox processes in a petroleum hydrocarbon-contaminated aquifer. *Ground Water* **40**(4) 353–360.

Chen C. S.-H., Delfino J. J., and Rao P. S. C. (1994) Partitioning of organic and inorganic components from motor oil into water. *Chemosphere* **28**(7), 1385–1400.

Christensen T. H., Bjerg P. L., Banwart S. A., Jakobsen R., Heron G., and Albrechtsen H.-J. (2000) Characterization of redox conditions in groundwater contaminant plumes. *J. Contamin. Hydrol.* **45**, 165–241.

Church C. D., Trartnyek P. G., Pankow J., Landmeyer J. E., Baehr A. L., Thomas M. A., and Schirmer M. (2000) Effects of Environmental conditions on MTBE degradation in model column aquifers. In *Proceedings of the Technical Meeting Of USGS Toxic Substances Hydrology Program, Charleston, South Carolina, 8–12 March, 1999*, pp. 93–101.

Coates J. D., Woodward J., Allen J., Philp P., and Lovely D. R. (1997) Anaerobic degradation of polycyclic aromatic hydrocarbons and alkanes in petroleum contaminated marine harbor sediments. *Appl. Environ. Microbiol.* **63**(9), 3589–3593.

Coates J. D., Chakraborty R., Lack J. G., O'Connor S. M., Cole K. A., Bender K. S., and Achenbach L. A. (2001) Anaerobic benzene oxidation coupled to nitrate reduction in pure culture by two strains of dechloromonas. *Nature* **411**(6841), 1039–1043.

Colberg P. J. S. and Young L. Y. (1995) Anaerobic degradation of nonhalogenated homocyclic aromatic compounds coupled with nitrate, iron, or sulfate reduction. In *Microbial Transformation and Degradation of Toxic Organic Chemicals* (eds. L. Y. Young and C. E. Cerniglia). Wiley-Liss, New York, pp. 307–330.

Cozzarelli I. M., Eganhouse R. P., and Baedecker M. J. (1990) Transformation of monoaromatic hydrocarbons to organic acids in anoxic groundwater environment. *Environ. Geol. Water Sci.* **16**(2), 135–141.

Cozzarelli I. M., Baedecker M. J., Eganhouse R. P., and Goerlitz D. F. (1994) The geochemical evolution of low-molecular-weight organic acids derived from the degradation of petroleum contaminants in groundwater. *Geochim. Cosmochim. Acta* **58**(2), 863–877.

Cozzarelli I. M., Herman J. S., and Baedecker M. J. (1995) Fate of microbial metabolites of hydrocarbons in a coastal plain aquifer: the role of electron acceptors. *Environ. Sci. Technol.* **29**(2), 458–469.

Cozzarelli I. M., Herman J. S., Baedecker M. J., and Fischer J. M. (1999) Geochemical heterogeneity of a gasoline-contaminated aquifer. *J. Contamin. Hydrol.* **40**(3), 261–284.

Cozzarelli I. M., Bekins B. A., Baedecker M. J., Aiken G. R., Eganhouse R. P., and Tuccillo M. E. (2001a) Progression of natural attenuation processes at a crude-oil spill site: I. Geochemical evolution of the plume. *J. Contamin. Hydrol.* **53**, 369–385.

Cozzarelli I. M., Eganhouse R. P., Godsy E. M., Warren E., and Bekins B. A. (2001b) Measurement of biodegradation of hydrocarbons *in situ* under different redox conditions. In *20th Int. Mtg of Organic Geochemistry, 10–14 Sept. 2001*, 2pp.

Crawford R. L. and Crawford D. L. (1996) *Bioremediation Principles and Applications*. University Press, Cambridge.

Daugherty S. (1991) Regulatory approaches to hydrocarbon contamination from underground storage tanks. In *Hydrocarbon Contaminated Soils and Groundwater: Analysis, Fate, Environmental and Public Health Effects, Remediation* (eds. P. T. Kostecki and E. J. Calabrese). Lewis Publishers, Chelsea, MI, pp. 24–63.

Davies J. M., Addy J. M., Blackman R. A. A., Blanchard J. R., Ferbrache J. E., Moore D. C., Somerville H. J., Whitehead A., and Wilkinson T. (1984) Environmental effects of the use of oil-based drilling muds in the north sea. *Mar. Pollut. Bull.* **15**(10), 363–370.

Davies J. M., Bedborough D. R., Blackman R. A. A., Addy J. M., Appelbee J. F., Grogan W. C., Parker J. G., and Whitehead A. (1989) The environmental effect of oil-based mud drilling in the North Sea. In *Drilling Wastes: Proceedings of the 1988 Conference on Drilling Wastes* (eds. F. R. Engelhardt, J. P. Ray, and A. H. Gillam). Elsevier Applied Science London and New York, pp. 59–90.

Davis J. B. (1967) *Petroleum Microbiology*. Elsevier, New York.

Davis J. M., Lohmann R. C., Phillips F. M., Wilson J. L., and Love D. W. (1993) Architecture of the sierra ladrones formation, central new Mexico: depositional controls on the permeability correlation structure. *Geol. Soc. Am. Bull.* **105**, 998–1007.

Davis S. W. and Powers S. E. (2000) Alternative sorbents for removing MTBE from gasoline-contaminated ground water. *J. Environ. Eng.* **126**, 354–360.

Deeb R. A. and Alvarez-Cohen L. (2000) Aerobic biotransformation of gasoline aromatics in multicomponent mixtures. *Bioremed. J.* **4**(2), 171–179.

Deeb R. A., Scow K. M., and Alvarez-Cohen L. (2000) Aerobic MTBE biodegradation: an examination of past studies, current challenges and future research directions. *Biodegradation* **11**, 171–186.

Delin G. N., Essaid H. I., Cozzarelli I. M., Lahvis M. H., and Bekins B. A. (1998) Ground water contamination by crude oil near Bemidji, Minnesota. *US Geol. Surv.* 4.

Dempster H. S., Lollar B. S., and Feenstra S. (1997) Tracing organic contaminants in groundwater: a new methodology using compound-specific isotopic analysis. *Environ. Sci. Technol.* **31**(11), 3193–3197.

Dewulf J., Drijvers D., and Van Langenhove H. (1995) Measurement of Henry's Law constant as a function of temperature and salinity for the low temperature range. *Atmos. Environ.* **29**(3), 323–331.

Dolfing J., Zeyer J., Binder-Eicher P., and Schwarzenbach R. P. (1990) Isolation and characterization of a bacterium that mineralizes toluene in the absence of molecular oxygen. *Arch. Microbiol.* **154**, 336–341.

Domenico P. A. and Schwartz F. W. (1998) *Physical and Chemical Hydrogeology*, 2nd edn. Wiley, New York.

Edwards E. A., Wills L. E., Reinhard M., and Grbić-Galić D. (1992) Anaerobic degradation of toluene and xylene by aquifer microorganisms under sulfate-reducing conditions. *Appl. Environ. Microbiol.* **58**(3), 794–800.

Edwards J. D. (2001) Twenty-first-century energy: decline of fossil fuel, increase of renewable nonpolluting energy sources. In *Petroleum Providences of the Twenty-First Century* (eds. M. W. Downey, J. C. Threet, and W. A. Morgan). AAPG Memoir, American Association of Petroleum Geologists, Tulsa, OK, vol. 74, pp. 21–34.

Eganhouse R. P. and Kaplan I. R. (1981) Extractable organic matter in urban stormwater runoff: 1. Transport dynamics and mass emission rates. *Environ. Sci. Technol.* **15**(3), 300–315.

Eganhouse R. P., Dorsey T. F., and Phinney C. S. (1987) Transport and fate of monoaromatic hydrocarbons in the subsurface, Bemidji, Minnesota, research site. In *US Geological Survey Program on Toxic Waste Ground-water Contamination: Proceedings of the Third Technical Meeting*, C-29, USGS Open File Report 87-109.

Eganhouse R. P., Baedecker M. J., Cozzarelli I. M., Aiken G. R., Thorn K. A., and Dorsey T. F. (1993) Crude oil in a shallow sand and gravel aquifer: II. Organic geochemistry. *Appl. Geochem.* **8**, 551–567.

Eganhouse R. P., Dorsey T. F., Phinney C. S., and Westcott A. M. (1996) Processes affecting the fate of monoaromatic hydrocarbons in an aquifer contaminated by crude oil. *Environ. Sci. Technol.* **30**(11), 3304–3312.

Elshahed M. S., Gieg L. M., McInerney M. J., and Suflita J. M. (2001) Signature metabolites attesting to the *in situ* attenuation of alkylbenzenes in anaerobic environments. *Environ. Sci. Technol.* **35**(4), 682–689.

Energy Information Administration (1999) *Petroleum: An Energy Profile*. Department of Energy, DOE/EIA-0545, Washington, DC, pp. 1–79.

Energy Information Administration (2001) *Annual Energy Review 2000*. Energy Information Administration, DOE/EIA-0384(00), Washington, DC, 414pp.

Evans P. J., Mang D. T., and Young L. Y. (1991) Degradation of toluene and *m*-xylene and transformation of *o*-xylene by denitrifying enrichment cultures. *Appl. Environ. Microbiol.* **57**(2), 450–454.

Evans P. J., Ling W., Goldschmidt B., Ritter E. R., and Young L. Y. (1992) Metabolites formed during anaerobic transformation of toluene and *o*-xylene and their proposed relationship to the initial steps of toluene mineralization. *Appl. Environ. Microbiol.* **58**(2), 496–501.

Eweis J. B., Chang D. P. Y., Schroeder E. D., Scow K., Morton R. L., and Caballero R. C. (1997) Meeting the challenge of MTBE biodegradation. In *Proceedings of the 90th Annual Meeting and Exhibition of the Air and Waste Management Association 8–13 June, 1997*, Toronto, Ontario, Canada.

Fang J. and Barcelona M. J. (1998) Biogeochemical evidence for microbial community change in a jet fuel hydrocarbons-contaminated aquifer. *Org. Geochem.* **29**(4), 899–907.

Fang J. and Barcelona M. J. (1999) Evolution of aromatic hydrocarbon and metabolic intermediate plumes in a shallow sand aquifer contaminated with jet fuel. In *Petroleum Hydrocarbons and Organic Chemicals in Ground Water: Prevention, Detection, and Remediation Conference*, (ed. Anita Stanley), National Groundwater Association, pp. 51–56.

Fetter C. W. (1993) *Contaminant Hydrology*. Macmillan, New York.

Finneran K. T. and Lovley D. R. (2001) Anaerobic Degradation of Methyl *tert*-Butyl Ether (MTBE) and *tert*-Butyl Alcohol (TBA). *Environ. Sci. Technol.* **35**(9), 1785–1790.

Fortin N. Y. and Deshusses M. A. (1999) Treatment of Methyl *tert*-Butyl ether vapors in biotrickling filters: 2. Analysis of the rate-limiting step and behavior under transient conditions. *Environ. Sci. Technol.* **33**(17), 2987–2991.

Francis A. J., Dodge C. J., Lu F., Halada G. P., and Clayton C. R. (1994) XPS and XANES studies of uranium reduction by *Clostridium* sp. *Environ. Sci. Technol.* **28**, 636–639.

Fredenslund A., Jones R., and Prausnitz J. M. (1975) Group-contribution estimation of activity coefficients in nonideal liquid mixtures'. *J. Am. Inst. Chem. Eng.* **21**(6), 1086–1099.

Friedrich R. and Obermeier A. (1999) Anthropogenic emissions of volatile organic compounds. In *Reactive Hydrocarbons in the Atmosphere* (ed. C. N. Hewitt). Academic Press, San Diego, CA, pp. 1–39.

Furlong E. T., Koleis J. C., and Aiken G. R. (1997) Transport and degradation of semivolatile hydrocarbons in a petroleum-contaminated aquifer, Bemidji, Minnesota. In *Molecular Markers in Environmental Geochemistry* (ed. R. P. Eganhouse). American Chemical Society, Washington, DC, vol. 671, pp. 398–412.

Gaines R. B., Frysinger G. S., Hendrick-Smith M. S., and Stuart J. D. (1999) Oil spill source identification by comprehensive two-dimensional gas chromatography. *Environ. Sci. Technol.* **33**(12), 2106–2112.

Gibson D. T. and Subramanian V. (1984) Microbial degradation of aromatic hydrocarbons. In *Microbial Degradation of Organic Compounds* (ed. D. T. Gibson). Dekker, New York, vol. 13, pp. 181–252.

Gieg L. M., Kolhatkar R. V., McInerney M. J., Tanner R. S., Harris S. H., Sublette K. L., and Suflita J. M. (1999) Intrinsic bioremediation of petroleum hydrocarbons in a gas condensate-contaminated aquifer. *Environ. Sci. Technol.* **33**(15), 2550–2560.

Gieg L. M. and Suflita J. M. (2002) Detection of anaerobic metabolites of saturated and aromatic hydrocarbons in petroleum-contaminated aquifers. *Environ. Sci. Technol.* **36**(17), 3755–3762.

Gomez-Belinchon J. I., Grimalt J. O., and Albaiges J. (1991) Volatile organic compounds in two polluted rivers in Barcelona (Catalonia, Spain). *Water Res.* **25**(5), 577–589.

Gray J., Couloume-Lacrampe G., Gandhi D., Scow K., Wilson R. D., MacKay D. M., and Lollar S. B. (2002) Carbon and hydrogen isotopic fractionation during biodegradation of methyl *tert*-butyl ether. *Environ. Sci. Technol.* **36**(9), 1931–1938.

Grbić-Galić D. and Vogel T. M. (1987) Transformation of toluene and benzene by mixed methanogenic cultures. *Appl. Environ. Microbiol.* **53**(2), 254–260.

Green C. T. and Scow K. M. (2000) Analysis of phospholipid fatty acids (PLFA) to characterize microbial communities in aquifers. *Hydrogeol. J.* **8**(1), 126–141.

Gullick R. W. and LeChevallier M. W. (2000) Occurrence of MTBE in drinking water sources. *J. Am. Water Works Assoc.* **92**, 100–113.

Haack S. K. and Bekins B. A. (2000) Microbial populations in contaminant plumes. *Hydrogeol. J.* **8**(1), 63–76.

Haag F., Reinhard M., and McCarty P. L. (1991) Degradation of toluene and *p*-xylene in anaerobic microcosms: evidence for sulfate as a terminal electron acceptor. *Environ. Toxicol. Chem.* **10**, 1379–1389.

Harrington R. R., Poulson S. R., Drever J. I., Colberg P. J. S., and Kelly E. F. (1999) Carbon isotope systematics of monoaromatic hydrocarbons: vaporization and adsorption experiments. *Org. Geochem.* **30**(8A), 765–775.

Hartley J. P. and Watson T. N. (1993) Investigation of a North Sea oil platform drill cuttings pile. In *25th Annual SPE et al. Offshore Technology Conference*, pp. 749–756.

Heider J. and Fuchs G. (1997) Anaerobic metabolism of aromatic compounds. *Euro. J. Biochem.* **243**(3), 577–596.

Heider J., Spormann A. M., Beller H. R., and Widdel F. (1999) Anaerobic bacterial metabolism of hydrocarbons. *FEMS Microbiol. Rev.* **22**, 459–473.

Heron G. and Christensen T. H. (1994) The role of aquifer sediment in controlling redox conditions in polluted groundwater. In *Transport and Reactive Processes in Aquifers/ Proceedings of the IAHR/AIRH Symposium on Transport and Reactive Processes in Aquifers*, A. A. Balkema.

Heron G. and Christensen T. H. (1995) Impact of sediment-bound iron on redox buffering in a landfill leachate polluted aquifer (Vejen, Denmark). *Environ. Sci. Technol.* **29**, 187–192.

Hess K. M., Herkelrath W. N., and Essaid H. I. (1992) Determination of subsurface fluid contents at a crude-oil spill site. *J. Contamin. Hydrol.* **10**, 75–96.

Hinchee R. E. (1995) Applied bioremediation of petroleum hydrocarbons. In *3rd International in situ and On-Site Bioremediation Symposium*, Battelle Press, 550pp.

Hogue C. (2000) Getting the MTBE out. *Chem. Eng. News* **78**(13), 6.

Holliger C., Gaspard S., Glod G., Heijman C., Schumacher W., Schwarzenbach R. P., and Vazquez F. (1997) Contaminated environments in the subsurface and bioremediation: organic contaminants. *FEMS Microbiol. Rev.* **20**, 517–523.

Hunkeler D., Anderson N., Aravena R., Bernasconi S. M., and Butler B. J. (2001a) Hydrogen and carbon isotope fractionation during aerobic biodegradation of benzene. *Environ. Sci. Technol.* **35**(17), 3462–3467.

Hunkeler D., Butler B. J., Aravena R., and Barker J. F. (2001b) Monitoring biodegradation of Methyl *tert*-Butyl ether (MTBE) using compound-specific carbon isotope analysis. *Environ. Sci. Technol.* **35**(4), 676–681.

Hunt J. M. (1996) *Petroleum Geochemistry and Geology*. W. H. Freeman, New York.

Hutchins S. R., Sewell G. W., Kovacs D. A., and Smith G. A. (1991) Biodegradation of aromatic hydrocarbons by aquifer microorganisms under denitrifying conditions. *Environ. Sci. Technol.* **25**(1), 68–76.

Jamison V. W., Raymond R. L., and Hudson J. O. J. (1975) Biodegradation of high-octane gasoline in groundwater. *Developm. Indust. Microbiol.* **16**, 305–312.

Johnson R., Pankow J., Bender D., Price C., and Zogorski J. (2000) MTBE: to what extent will past releases contaminate community water supply wells. *Environ. Sci. Technol.* **34**(9), 210a–217a.

Jones F. V., Leuterman A. J. J., and Still I. (2000) Discharge practices and standards for offshore operations around the world. In *7th International Petroleum Environmental Conference*, Albuquerque, NM, November 2000, National Energy Technology Laboratory, pp. 903–942.

Karickhoff S. W. (1981) Semi-empirical estimation of sorption of hydrophobic pollutants on natural sediments. *Chemosphere* **10**(8).

Karickhoff S. W. (1984) Organic pollutant sorption in aquatic systems. *J. Hydraulic Eng.* **110**(6), 707–735.

Kauffman L. J., Baehr A. L., Ayers M. A., and Stackelberg P. E. (2001) Effects of land use and travel time on the distribution of nitrate in the Kirkwood–Cohansey aquifer system in southern New Jersey. US Geological Survey, Water Resource Investigation Report 01-4117, 49pp.

Kazumi J., Caldwell M. E., Suflita J. M., Lovley D. R., and Young L. Y. (1997) Anaerobic degradation of benzene in diverse anoxic environments. *Environ. Sci. Technol.* **31**(3), 813–818.

Kelley C. A., Hammer B. T., and Coffin R. B. (1997) Concentrations and stable isotope values of BTEX in gasoline-contaminated groundwater. *Environ. Sci. Technol.* **31**(9), 2469–2472.

Kleikemper J., Schroth M. H., Sigler W. V., Schmucki M., Bernasconi S. M., and Zeyer J. (2002) Activity and diversity of sulfate-reducing bacteria in a petroleum hydrocarbon-contaminated aquifer. *Appl. Environ. Microbiol.* **68**(4), 1516–1523.

Koenigsberg S., Sandefur C., Mahaffey W., Deshusses M. A., and Fortin N. (1999) Peroxygen mediated bioremediation of MTBE. *In Situ Bioremediation of Petroleum Hydrocarbon and Other Organic Compounds: Proceedings from the Fifth International In Situ and On-Site Bioremediation Symposium*, Bettelle Press, pp. 13–18.

Kolhatkar R., Kuder T., Philp P., Allen J., and Wilson J. T. (2002) Use of compound-specific stable carbon isotope analyses to demonstrate anaerobic biodegradation of MTBE in groundwater at a gasoline release site. *Environ. Sci. Technol.* **36**(23), 5139–5146.

Kostecki P. T. and Calabrese E. J. (1991) *Hydrocarbon Contaminated Soils and Groundwater: Analysis, Fate, Environmental and Public Health Effects Remediation.* Lewis Publishers, Chelsea, MI.

Krumholz L. R., Caldwell M. E., and Suflita J. M. (1996) Biodegradation of 'BTEX' hydrocarbons under anaerobic conditions. In *Bioremediation: Principles and Applications* (eds. R. Crawford and D. Crawford). Cambridge University Press, Cambridge, pp. 61–99.

Kuhn E. P., Zeyer J., Eicher P., and Schwarzenbach R. P. (1988) Anaerobic degradation of alkylated benzenes in denitrifying laboratory aquifer columns. *Appl. Environ. Microbiol.* **54**(2), 490–496.

Landmeyer J. E., Chapelle F. H., Bradley P. M., Pankow J. F., Church C. D., and Tratnyek P. G. (1998) Fate of MTBE relative to benzene in a gasoline-contaminated aquifer (1993–98). *Ground Water Monitor. Remediat.* **18**(4), 93–102.

Latimer J. S., Hoffman E. J., Hoffman G., Fasching J. L., and Quinn J. G. (1990) Sources of petroleum hydrocarbons in urban runoff. *Water Air Soil Pollut.* **52**, 1–21.

Lide D. R. (2002) *CRC Handbook of Chemistry and Physics.* CRC Press, Boca Raton, FL.

Lovley D. R. (2000) Anaerobic benzene degradation. *Biodegradation* **11**(2–3), 107–116.

Lovley D. R. and Anderson R. T. (2000) Influence of dissimilatory metal reduction on fate of organic and metal contaminants in the subsurface. *Hydrogeol. J.* **8**(1), 77–88.

Lovley D. R. and Lonergan D. J. (1990) Anaerobic oxidation of toluene, phenol, and *p*-Cresol by the dissimilatory iron-reducing organism, GS-15. *Appl. Environ. Microbiol.* **56**(6), 1858–1864.

Lovley D. R. and Phillips E. J. (1992) Reduction of uranium by Desulfovibrio desulfuricans. *Appl. Environ. Microbiol.* **58**(3), 850–856.

Lovley D. R., Baedecker M. J., Lonergan D. J., Cozzarelli I. M., Phillips E. J. P., and Siegel D. I. (1989) Oxidation of aromatic contaminants coupled to microbial iron reduction. *Nature* **339**, 297–299.

Lovley D. R., Phillips E. J. P., Gorby Y. A., and Landa E. R. (1991) Microbial reduction of uranium. *Nature* **350**, 413–416.

Lund L. (1995) Changes in UST and LUST: the federal perspective. *Tank Talk* **10**(2–3), 7.

Lyman W. J., Reehl W. J., and Rosenblatt D. H. (1990) *Handbook of Chemical Property Estimation Methods: Environmental Behavior of Organic Compounds.* American Chemical Society, Washington, DC.

Mackay D., Shiu W. Y., and Ma K. C. (1992a) *Illustrated Handbook of Physical–Chemical Properties and Environmental Fate for Organic Chemicals Volume I: Monoaromatic Hydrocarbons, Chlorobenzenes, and PCBs.* Lewis Publishers, Chelsea, MI.

Mackay D., Shiu W. Y., and Ma K. C. (1992b) *Illustrated Handbook of Physical–Chemical Properties and Environmental Fate for Organic Chemicals Volume II: Polynuclear Aromatic Hydrocarbons, Polychlorinated Dioxins, and Dibenzofurans.* Lewis Publishers, Chelsea, MI.

Mackay D., Shiu W. Y., and Ma K. C. (1993) *Illustrated Handbook of Physical–Chemical Properties and Environmental Fate for Organic Chemicals Volume III: Volatile Organic Chemicals.* Lewis Publishers, Chelsea, MI.

Madsen E. L. (2000) Nucleic-acid characterization of the identity and activity of subsurface microorganisms. *Hydrogeol. J.* **8**(1), 112–125.

Major D. W., Mayfield C. I., and Barker J. F. (1988) Biotransformation of benzene by denitrification in aquifer sand. *Ground Water* **26**(1), 8–14.

Mancini S. A., Couloume G. L., Jonker H., van Breukelen B. M., Groen J., Volkering F., and Lollar B. S. (2002) Hydrogen isotopic enrichment: an indicator of biodegradation at a petroleum hydrocarbon contaminated field site. *Environ. Sci. Technol.* **36**(11), 2464–2470.

Mancini S. A., Ulrich A. C., Couloume G. L., Sleep B., Edwards E. A., and Lollar B. S. (2003) Carbon and hydrogen isotopic fractionation during anaerobic biodegradation of benzene. *Appl. Environ. Microbiol.* **69**(1), 191–198.

McFarland M. L., Ueckert D. N., and Hartmann S. (1987) Revegetation of oil well reserve pits in west Texas. *J. Range Manage.* **40**(2), 122.

McGuire J. T., Long D. T., Klug M. J., Haack S. K., and Hyndman D. W. (2002) Evaluating behavior of oxygen, nitrate, and sulfate during recharge and quantifying reduction rates in a contaminated aquifer. *Environ. Sci. Technol.* **36**(12), 2693–2700.

McMahon P. B. (1995) Effect of fuel oxidants on the degradation of gasoline components in aquifer sediments. In *3rd International Symposium on In Situ and On-Site Bioreclamation*, San Diego, April 1995.

Merian E. and Zander M. (1982) Volatile aromatics. In *The Handbook of Environmental Chemistry: Anthropogenic Compounds*, Volume 3, Part B (eds. O. Hutzinger, K. J. Bock, K. A. Daum, E. Merian, L. W. Newland, C. R. Pearson, H. Stache, and M. Zander). Springer, Berlin, pp. 117–161.

Mill T., Hendry D. G., and Richardson H. (1980) Free radical oxidants in natural waters. *Science* **207**, 886–887.

Miller D., Rivera R. G., Travis C. C., Solis R., and Calva L. (2000) An approach to soil restoration of hydrocarbon contamination using a carbon fractionation approach. In *7th International Petroleum Environmental Conference*, Albuquerque, NM, November 2000, National Energy Technology Laboratory, pp. 1264–1286.

Mormile M. R., Liu S., and Suflita J. M. (1994) Anaerobic biodegradation of gasoline oxygenates: extrapolation od information to multiple sites and redox conditions. *Environ. Sci. Technol.* **28**(9), 1727–1732.

Mossing C., Larsen L. C., Hansen H. C. L., Seifert D., and Bjerg P. L. (2001) Monitored natural attenuation (MNA) of petroleum hydrocarbons in a heterogenous aquifer affected by transient flow. In *The Sixth International in situ and on-site Bioremediation Symposium*, Battelle Press, vol. 2, pp. 11–18.

National Research Council (NRC) (1985) *Oil in the Sea: Inputs, Fates, and Effects.* National Academy Press, Washingon, DC.

National Research Council (NRC) (1993) In Situ *Bioremediation When does it Work?* National Academy Press, Washington, DC.

National Research Council (US) Committee on Intrinsic Remediation (2000). *Natural Attenuation for Groundwater Remediation*. National Academy Press, Washington, DC.

National Research Council (2003). *Oil in the Sea III: Inputs, Fates, and Effects*. National Academies Press, Washington, DC.

Novelli G. D. and ZoBell C. E. (1944) Assimilation of petroleum hydrocarbons by sulfate-reducing bacteria. *J. Bacteriol.* **47**, 447–448.

Oestermark U. and Petersson G. (1993) Volatile hydrocarbons in exhaust from alkylate-based petrol. *Chemosphere* **27**(9), 1719–1728.

Otton J. K. and Mercier T. J. (2000) Nationals watershed and aquifer susceptibility-evaluation of the impact of oil and gas production operation. In *7th International Petroleum Environmental Conference*, Albuquerque, NM, November 2000, National Energy Technology Laboratory, pp. 961–978.

Patrick G. C. and Barker J. F. (1985) A natural-gradient tracer study of dissolved benzene, toluene and xylenes in ground water. In *Proceedings of the Second Canadian/American Conference on Hydrogeology*, National Water Well Association, pp. 141–147.

Payne J. R. and Phillips C. R. (1985) Photochemistry of petroleum in water. *Environ. Sci. Technol.* **19**(7), 569–579.

Perbellini L., Brugnone F., and Pavan I. (1980) Identification of the metabolites on *n*-hexane, cyclohexane, and their isomers in men's urine. *Toxicol. Appl. Pharmacol.* **53**, 220–229.

Phelps C. D. and Young L. Y. (2001) Biodegradation of BTEX under anaerobic conditions: a review. *Adv. Agron.* **70**, 329–357.

Pinal R. P., Rao S. C., Lee L. S., and Cline P. V. (1990) Cosolvency of partially miscible organic solvents on the solubility of hydrophobic organic chemicals. *Environ. Sci. Technol.* **24**(5), 639–647.

Pinal R. P., Lee L. S., and Rao S. C. (1991) Prediction of the solubility of hydrophobic compounds in nonideal solvents. *Chemosphere* **22**(9–10), 939–951.

Poulsen M., Lemon L., and Barker J. F. (1992) Dissolution of monoaromatic hydrocarbons into groundwater from gasoline-oxygenate mixtures. *Environ. Sci. Technol.* **26**(12), 2483–2489.

Rabus R. and Widdel F. (1995) Anaerobic degradation of ethylbenzene and aromatic hydrocarbons by new denitrifying bacteria. *Arch. Microbiol.* **163**, 96–103.

Rabus R., Nordhaus R., Ludwig W., and Widdel F. (1993) Complete oxidation of toluene under strictly anoxic conditions by a new sulfate-reducing bacterium. *Appl. Environ. Microbiol.* **59**(5), 1444–1451.

Rabus R., Wilkes H., Behrends A., Armstroff A., Fischer T., Pierik A. J., and Widdel F. (2001) Anaerobic initial reaction of *n*-alkanes in a denitrifying bacterium: evidence for (1-methylpentyl) succinate as initial product and for involvement of an organic radical in *n*-hexane metabolism. *J. Bacteriol.* **183**(5), 1707–1715.

Rathbun R. E. (1998) Transport, behavior, and fate of volatile organic compounds in streams. US Geological Survey Professional Paper 1589, 151pp.

Reineke W. (2001) Aerobic and Anaerobic Biodegradation Pontentials of Microorganisms. In *The Handbook of Environmental Chemistry*, Volume 2 Part K-Biodegradation and Persistence (ed. B. Beek). Springer, Berlin.

Reinhard M., Goodman N. L., and Barker J. F. (1984) Occurrence and distribution of organic chemicals in two landfill leachate plumes. *Environ. Sci. Technol.* **18**(12), 953–961.

Rice D. W. (1991) Unique problems of hydrocarbon contamination for ports. In *Hydrocarbon Contaminated Soils and Groundwater: Analysis, Fate, Environmental and Public Health Effects, Remediation* (eds. P. T. Kostecki and E. J. Calabrese). Lewis Publishers, Chelsea, MI, pp. 71–75.

Rifai H. S. (1997) Natural aerobic biological attenuation. In *Subsurface Restoration* (eds. C. H. Ward, J. A. Cherry, and M. R. Scalf). Ann Arbor Press, Chelsea, MI, 411pp.

Rios-Hernandez L. A., Gieg L. M., and Suflita J. M. (2003) Biodegradation of an alicyclic hydrocarbon by a sulfate-reducing enrichment from a gas condensate-contaminated aquifer. *Appl. Environ. Microbiol.* **69**(1), 434–443.

Riser-Roberts E. (1992) *Bioremediation of Petroleum Contaminated Sites*, CRC Press, Boca Raton.

Robbins G. A., Wang S., and Stuart J. D. (1993) Using the static headspace method to determine Henry's Law constants. *Anal. Chem.* **65**(21), 3113–3118.

Robin M. J. L., Sudicky E. A., Gillham R. W., and Kachanoski R. G. (1991) Spatial variability of strontium distribution coefficients and their correlation with hydraulic conductivity in the canadian forces base bordon aquifer. *Water Resour. Res.* **27**(10), 2619–2632.

Rosenberg E., Legmann R., Kushmaro A., Taube R., Adler E., and Ron E. Z. (1992) Petroleum bioremediation: a multiphase problem. *Biodegradation* **3**, 337–350.

Rosenberg E. and Ron E. Z. (1996) Bioremediation of petroleum contamination. In *Bioremediation: Principles and Applications* (eds. R. L. Crawford and D. L. Crawford). Cambridge University Press, New York.

Rosenfeld W. D. (1947) Anaerobic oxidation of hydrocarbons by sulfate-reducing bacteria. *J. Bacteriol.* **54**, 664–665.

Salanitro J. P., Diaz L. A., Williams M. P., and Wisniewski H. L. (1994) Isolation of a bacterial culture that degrades methyl *t*-butyl ether. *Appl. Environ. Microbiol.* **60**(7), 2593–2596.

Salanitro J. P., Spinnler G. E., Neaville C. C., Maner P. M., Stearns S. M., and Johnson P. C. (1999) Demonstration of the enhanced MTBE bioremediation (EMB) *in situ* process. In *In Situ Bioremediation of Petroleum Hydrocarbon and Other Organic Compounds: Proceedings from the Fifth International In Situ and On-site Bioremediation Symposium*, Bettelle Press, vol. 3, pp. 37–46.

Schirmer M. and Barker J. F. (1998) A study of long-term MTBE attenuation in the borden aquifer, ontario, Canada. *Ground Water Monitor. Remediat.* **18**(2), 113–122.

Schirmer M., Butler B. J., Barker J. F., Church C. D., and Schirmer K. (1999) Evaluation of biodegradation and dispersion as natural attenuation processes of MTBE and benzene at the borden field site. *Phys. Chem. Earth Part B-Hydrol. Oceans Atmos.* **24**(6), 557–560.

Schwarzenbach R. P., Giger W., Hoehn E., and Schneider J. K. (1983) Behavior of organic compounds during infiltration of river water to groundwater: field studies. *Environ. Sci. Technol.* **17**(8), 472–479.

Seyfried B., Glod G., Schocher R. J., Tschech A., and Zeyer J. (1994) Initial reactions in the anaerobic oxidation of toluene and *m*-xylene by denitrifying bacteria. *Appl. Environ. Microbiol.* **60**(11), 4047–4052.

Skubal K. L., Barcelona M. J., and Adriaens P. (2001) An assessment of natural biotransformation of petroleum hydrocarbons and chlorinated solvents at an aquifer plume transect. *J. Contamin. Hydrol.* **49**(1–2), 151–169.

Smith M. R. (1990) The biodegradation of aromatic hydrocarbons by bacteria. *Biodegradation* **1**, 191–206.

Smith M. R. (1994) The physiology of aromatic hydrocarbon degrading bacteria. In *Biochemistry of Microbial Degradation* (ed. C. Ratledge). Kluwer, Dordrecht, pp. 347–378.

Smith R. L., Harvey R. W., and LeBlanc D. R. (1991) Importance of closely spaced vertical sampling in delineating chemical and microbiological gradients in groundwater studies. *J. Contamin. Hydrol.* **7**, 285–300.

So C. M. and Young L. Y. (1999) Initial reaction in anaerobic alkane degradation by a sulfate reducer, strain AK-01. *Appl. Environ. Microbiol.* **65**(12), 5532–5540.

Spormann A. M. and Widdel F. (2000) Metabolism of alkylbenzenes, alkanes, and other hydrocarbons in anaerobic bacteria. *Biodegradation* **11**, 85–105.

Squillace P. J., Zogorski J. S., Wilber W. G., and Price C. V. (1996) Preliminary assessment of the occurrence and possible sources of MTBE in groundwater in the US, 1993–1994. *Environ. Sci. Technol.* **30**(5), 1721–1730.

Squillace P. J., Moran M. J., Lapham W. W., Price C. V., Clawges R. M., and Zogorski J. S. (1999) Volatile organic compounds in untreated ambient groundwater of the US, 1985–1995. *Environ. Sci. Technol.* **33**(23), 4176–4187.

Stackelberg P. E., Hopple J. A., and Kauffman L. J. (1997) Occurrence of nitrate, pesticides, and volatile organic compounds in the Kirkwood–Cohansey aquifer system, southern New Jersey. US Geological Survey, Water Resources Investigation Report, 97-4241, 8pp.

Stapleton R. and Sayler G. (1999) Changes in subsurfacing catabolic gene frequencies during natural attenuation of petroleum hydrocarbons. *Environ. Sci. Technol.* **34**(10), 1991–1999.

State of Maine Bureau of Health (1998) *The Presence of MTBE and other Gasoline Compounds in Maine's Drinking Water—a Preliminary Report.* Bureau of Health, Department of Human Services, Bureau of Waste Management and Remediation, Department of Environmental Protection, Maine Geological Survey, Department of Conservation.

Stehmeier L. G., Francis M. M., Jack T. R., Diegorb E., Winsorb L., and Abrajano T. A. J. (1999) Field and *in vitro* evidence for *in-situ* bioremediation using compound-specific 13C/12C ratio monitoring. *Org. Geochem.* **30**, 821–833.

Stephenson M. T. (1992) A survey of produced water studies. In *Produced Water: Technological/Environmental Issues and Solutions* (eds. J. P. Ray and F. R. Englehardt). Plenum, New York, pp. 1–11.

Stolz J. F. and Oremland R. S. (1999) Bacterial respiration of arsenic and selenium. *FEMS Microbiol. Rev.* **23**(5), 615–627.

Stumm W. and Morgan J. J. (1996) *Aquatic Chemistry, Chemical Equilibria and Rates in Natural Waters*, 3rd edn. Wiley, New York.

Suflita J. M. and Mormile M. R. (1993) Anaerobic biodegradation of known and potential gasoline oxygenates in the terrestrial subsurface. *Environ. Sci. Technol.* **27**(5), 976–978.

Tebo B. M. and Obraztsova A. Y. (1998) Sulfate-reducing bacterium grows with Cr(VI), U(VI), Mn(IV), and Fe(III) as electron acceptors. *FEMS Microbiol. Lett.* **162**, 193–198.

Testa S. M. and Winegardner D. L. (2000) *Restoration of Contaminated Aquifer: Petroleum Hydrocarbons and Organic Compounds.* Lewis Publishers, Chelsea, MI, 446pp.

Thompson G. W., Jollett M. R., Cadena F., and Weisman C. (2000) Removal of MTBE using organozeolites. In *7th International Petroleum Environmental Conference*, Allouquerque, NM, Novermber 2000, National Energy Technology Laboratory, pp. 813–844.

Tissot B. P. and Welte D. H. (1984) *Petroleum Formation and Occurrence*, Springer, Berlin.

Ünlü K. and Demirekler E. (2000) Modeling water quality impacts of petroleum contaminated soils in a reservoir catchment. *Water Air Soil Pollut.* **120**(1–2), 169–193.

US Environmental Protection Agency (USEPA) (1987) Report to Congress: Management of Wastes from Exploration, Development, and Production of Crude Oil, Natural Gas, and Geothermal Energy. US Environmental Protection Agency.

US Geological Survey (USGS) (2000) USGS World Petroleum Assessment 2000, New estimates of undiscovered oil and natural gas, including reserve growth, outside the United Sates. USGS.

Vazquez-Duhalt R. (1989) Environmental impact of used motor oil. *Sci. Tot. Environ.* **79**, 1–23.

Vogel T. M. and Grbić-Galić D. (1986) Incorporation of oxygen from water into toluene and benzene during anaerobic fermentative transformation. *Appl. Environ. Microbiol.* **52**(1), 200–202.

Voice T. C. and Weber W. J. (1983) Sorption of hydrophobic compounds by sediments, soils, and suspended solids: I. Theory and background. *Water Res.* **17**(10), 1433–1441.

Ward C. H., Cherry J. A., and Scalf M. R. (1997) *Subsurface Restoration.* Ann Arbor Press, Chelsea, MI.

Ward D. M., Atlas R. M., Boehm P. D., and Calder J. A. (1980) Microbial biodegradation and chemical evolution of oil from the Amoco spill. *Ambio* **9**, 277–283.

Wiedemeier T. H., Wilson J. T., Kampbell D. H., Miller R. N., and Hansen J. E. (1995) Technical protocol for implementing intrinsic remediation with long-term monitoring for natural attenuation of fuel contamination dissolved in groundwater. US Air Force Center for Environmental Excellence.

Wiedemeier T. H., Rifai H., Newell C., and Wilson J. T. (1999) *Natural Attenuation of Fuels and Chlorinated Solvents in the Subsurface*, Wiley, New York.

Wilkes H., Boreham C., Harms G., Zengler K., and Rabus R. (2000) Anaerobic degradation and carbon isotopic fractionation of alkylbenzenes in crude oil by sulphate-reducing bacteria. *Org. Geochem.* **31**(1), 101–115.

Williams B. (1991) *US Petroleum Strategies in the Decade of The Environment.* Pennwell Publishing Company.

Wilson B. H., Smith G. B., and Rees J. F. (1986) Biotransformations of selected alkylbenzenes and halogenated aliphatic hydrocarbons in methanogenic aquifer material: a microcosm study. *Environ. Sci. Technol.* **20**(10), 997–1002.

Wilson B. H., Wilson J. T., Kampbell D. H., Bledsoe B. E., and Armstrong J. M. (1990) Biotransformation of monoaromatic and chlorinated hydrocarbons at an aviation gasoline spill site. *Geomicrobiol. J.* **8**, 225–240.

Wilson J. T., Cho J. S., Wilson B. H., and Vardy J. A. (2000) *Natural Attenuation of MTBE in the Subsurface Under Methanogenic Conditions.* US Environmental Protection Agency, pp. 1–59.

Wilson R. D., Mackay D. M., and Scow K. M. (2002) *In situ* MTBE biodegradation supported by diffusive oxygen release. *Environ. Sci. Technol.* **36**, 190–199.

Wolfe D. A., Galt J. A., Short J., Watabayashi G., O'Claire C., Rice S., Michel J., Payne J. R., Braddock J., Hanna S., and Sale D. (1994) The fate of the oil spilled from the exxon valdez. *Environ. Sci. Technol.* **28**(13), 561–567.

Yeh C. K. and Novak J. T. (1994) Anaerobic biodegradation of gasoline oxygenates in soils. *Water Environ. Res.* **66**(5), 744–752.

Zeyer J., Kuhn E. P., and Schwarzenbach R. P. (1986) Rapid microbial mineralization of toluene and 1,3-dimethylbenzene in the absence of molecular oxygen. *Appl. Environ. Microbiol.* **52**(4), 944–947.

Zogorski J. S., Morduchowitz A., Baehr A. L., Bauman B. J., Conrad D. L., Drew R. T., Korte N. E., Lapham W. W., Pankow J. F., and Washington E. R. (1997) Fuel oxygenates and water quality: Chap. 2. In Interagency Assessment of Oxygenated Fuels, National Science and Technology Council, Washington DC, Office of Science and Technology Policy, The Executive Office of the President, pp. 2(1)–2(80).

Zwank L., Schmidt T., Haderlein S., and Berg M. (2002) Simultaneous determination of fuel oxygenates and BTEX using direct aqueous injection gas chromatography mass spectrometry (DAI/GC/MS). *Environ. Sci. Technol.* **36**(9), 2054–2059.

9.13 High Molecular Weight Petrogenic and Pyrogenic Hydrocarbons in Aquatic Environments

T. A. Abrajano Jr. and B. Yan

Rensselaer Polytechnic Institute, Troy, NY, USA

and

V. O'Malley

Enterprise Ireland, Glasnevin, Republic of Ireland

9.13.1 INTRODUCTION

Geochemistry is ultimately the study of sources, movement, and fate of chemicals in the geosphere at various spatial and temporal scales. Environmental organic geochemistry focuses such studies on organic compounds of toxicological and ecological concern (e.g., Schwarzenbach *et al.*, 1993, 1998; Eganhouse, 1997). This field emphasizes not only those compounds with potential toxicological properties, but also the geological systems accessible to the biological receptors of those hazards. Hence, the examples presented in this chapter focus on hydrocarbons with known health and ecological concern in accessible shallow, primarily aquatic, environments.

Modern society depends on oil for energy and a variety of other daily needs, with present mineral oil consumption throughout the 1990s exceeding 3×10^9 t yr^{-1} (NRC, 2002). In the USA, e.g., ~40% of energy consumed and 97% of transportation fuels are derived from oil. In the process of extraction, refinement, transport, use, and waste production, a small but environmentally significant fraction of raw oil materials, processed products, and waste are released inadvertently or purposefully into the environment. Because their presence and concentration in the shallow environments are often the result of human activities, these organic materials are generally referred to as "environmental contaminants." Although such reference connotes some form of toxicological or ecological hazard, specific health or ecological effects of many organic "environmental contaminants" remain to be demonstrated. Some are, in fact, likely innocuous at the levels that they are found in many systems, and simply adds to the milieu of biogenic organic compounds that naturally cycle through the shallow environment. Indeed, virtually all compounds in crude oil and processed petroleum products have been introduced naturally to the shallow environments as oil and gas seepage for millions of years (NRC, 2002). Even high molecular weight (HMW) polyaromatic compounds were introduced to shallow environments through forest fires and natural coking of crude oil (Ballentine *et al.*, 1996; O'Malley *et al.*, 1997). The full development of natural microbial enzymatic systems that can utilize HMW hydrocarbons as carbon or energy source attests to the antiquity of hydrocarbon dispersal processes in the environment. The environmental concern is, therefore, primarily due to the rate and spatial scale by which petroleum products are released in modern times, particularly with respect to the environmental sensitivity of some ecosystems to these releases (Schwarzenbach *et al.*, 1993; Eganhouse, 1997; NRC, 2002).

Crude oil is produced by diagenetic and thermal maturation of terrestrial and marine plant and animal materials in source rocks and petroleum reservoirs. Most of the petroleum in use today is produced by thermal and bacterial decomposition of phytoplankton material that once lived near the surface of the world's ocean, lake, and river waters (Tissot and Welte, 1984). Terrestrially derived organic matter can be regionally significant, and is the second major contributor to the worldwide oil inventory (Tissot and Welte, 1984; Peters and Moldowan, 1993; Engel and Macko, 1993). The existing theories hold that the organic matter present in crude oil consists of unconverted original biopolymers and new compounds polymerized by reactions promoted by time and increasing temperature in deep geologic formations. The resulting oil can migrate from source to reservoir rocks where the new geochemical conditions may again lead to further transformation of the petrogenic compounds. Any subsequent changes in reservoir conditions brought about by uplift, interaction with aqueous fluids, or even direct human intervention (e.g., drilling, water washing) likewise could alter the geochemical makeup of the petrogenic compounds. Much of our understanding of environmental sources and fate of hydrocarbon compounds in shallow environments indeed borrowed from the extensive geochemical and analytical framework that was meticulously built by petroleum geochemists over the years (e.g., Tissot and Welte, 1984; Peters *et al.*, 1992; Peters and Moldowan, 1993; Engel and Macko, 1993; Moldowan *et al.*, 1995; Wang *et al.*, 1999; Faksness *et al.*, 2002).

Hydrocarbon compounds present in petroleum or pyrolysis by-products can be classified based on their composition, molecular weight, organic structure, or some combination of these criteria. For example, a report of the Committee on Intrinsic Remediation of the US NRC classified organic contaminants into HMW hydrocarbons, low molecular weight (LMW) hydrocarbons, oxygenated hydrocarbons, halogenated aliphatics, halogenated aromatics, and nitroaromatics (NRC, 2000). Hydrocarbons are compounds comprised exclusively of carbon and hydrogen and they are by far the dominant components of crude oil, processed petroleum hydrocarbons (gasoline, diesel, kerosene, fuel oil, and lubricating oil), coal tar, creosote, dyestuff, and pyrolysis waste products. These hydrocarbons often occur as mixtures of a diverse group of compounds whose behavior in near-surface environments is governed by their chemical structure and composition, the geochemical conditions and media of their release, and biological factors, primarily microbial metabolism, controlling their transformation and degradation.

Hydrocarbons comprise from 50% to 99% of compounds present in refined and unrefined oil, and compounds containing other elements such as oxygen, nitrogen, and sulfur are present in relatively smaller proportions. Hydrocarbon compounds have carbons joined together as single C—C bonds (i.e., alkanes), double or triple C=C bonds (i.e., alkenes or olefins), or via an aromatic ring system with resonating electronic structure (i.e., aromatics). Alkanes, also called paraffins, are the dominant component of crude oil, with the carbon chain forming either straight (*n*-alkanes), branched (iso-alkanes), or cyclic (naphthenes) arrangement of up to 60 carbons (Figure 1). Aromatic compounds are the second major component of crude oil, with asphalthenes, consisting of stacks of highly polymerized aromatic structures

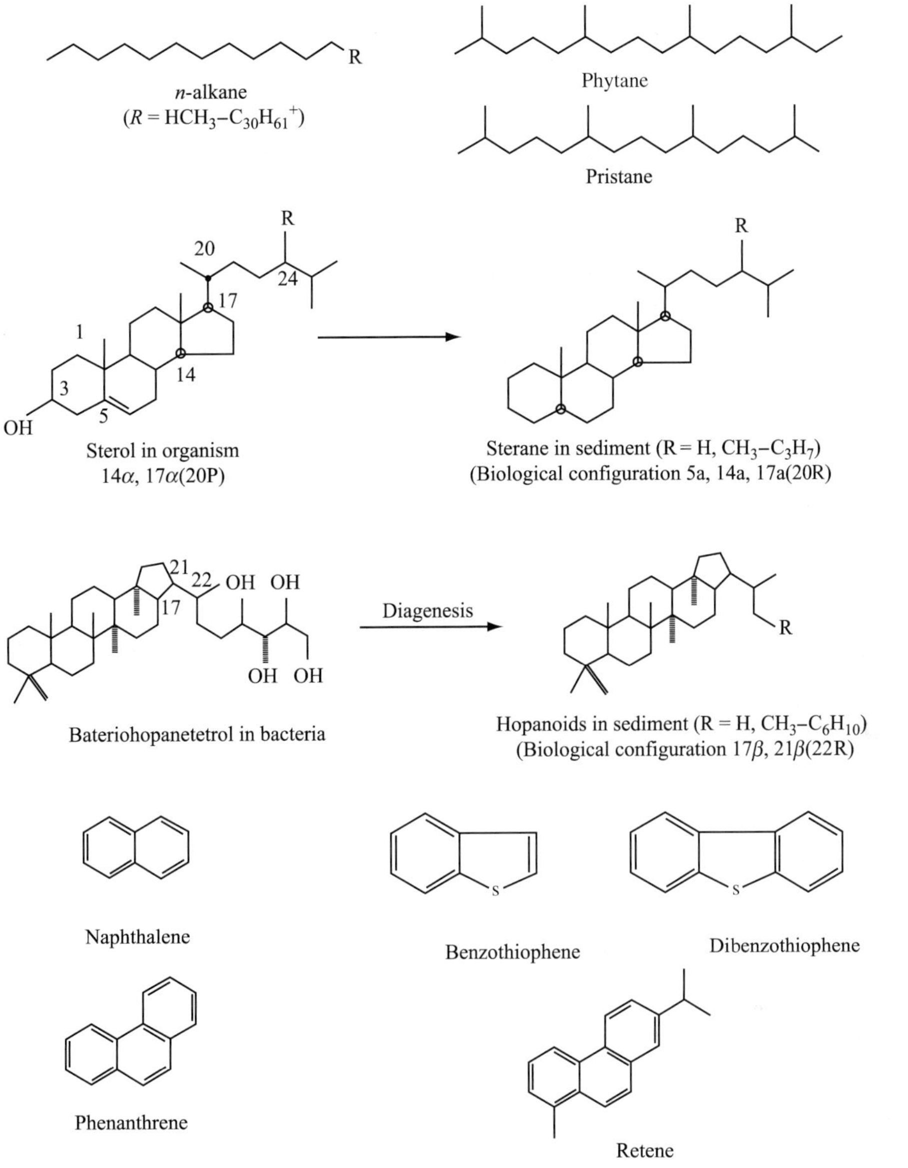

Figure 1 Examples of types of hydrocarbons found in crude oil and mentioned in the text.

(average of 16 rings), completing the list of major oil hydrocarbon components. Also shown in Figure 1 are several important classes of compounds that are extensively used in "fingerprinting" crude oil or petroleum sources: sterols derived from steroid, hopanol derived from bacteriohopanetetrols, and pristane and phytane derived from phytol (from chlorophyll) during diagenesis.

Polycyclic aromatic hydrocarbons (PAHs) that are made up of two or more fused benzene rings are minor components of crude oil (Figures 1 and 2), but they are by far the most important HMW compounds in terms of chronic environmental impact. Indeed, total PAH loading is used as the surrogate for the overall estimation of petroleum toxicity effects in environmental assessments (e.g., Meador *et al.*, 1995; NRC, 2002). PAHs are characterized by two or more fused benzene rings (Figure 2), and many have toxic properties including an association with mutagenesis and carcinogenesis (e.g., Cerniglia, 1991; Neilson, 1998). The World Health Organization (WHO) and the US Environmental Protection Agency (USEPA) have recommended 16 parental (unsubstituted rings) PAHs as priority pollutants (Figure 2). Although petroleum-sourced PAHs are major contributors in many surface and subsurface aquatic environments, another major contributor of PAHs to the environment is pyrolysis of fuel and other biomass. The latter are referred to as pyrogenic

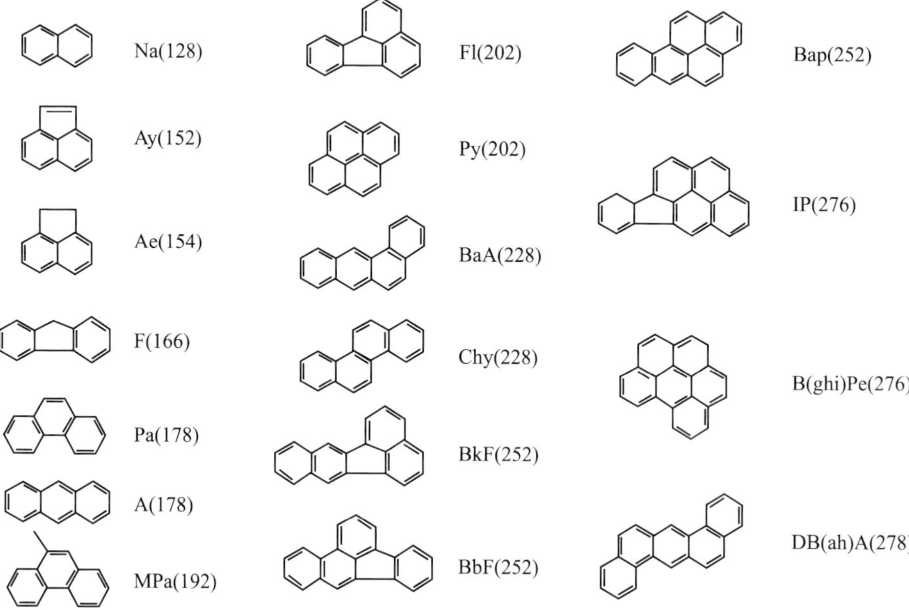

Figure 2 Structures of polycyclic aromatic hydrocarbons. Symbols used in this figure and text: Na (naphthalene), Ay (acetonaphthylene), Ae (acenaphthene), Fl (fluorene), Pa (phenanthrene), A (anthracene), MPa (methyl phenanthrene), F (fluoranthene), Py (pyrene), BaA (benz(a)anthracene), Chy (chrysene), BkF (benzo(k)fluoranthene), BbF (benzo(b)fluoranthene), BaP (benzo(a)pyrene), IP (indenopyrene), B(ghi)Pe (benzo(ghi)perylene), and Db(ah)A (dibenzo(ah)enthracene).

PAHs to distinguish them from the petrogenic PAHs derived directly from uncombusted petroleum, coal, and their by-products. Natural sources such as forest fires could be important in less inhabited and remote watersheds, but anthropogenic combustion of fossil fuel (e.g., petroleum, coal) and wood is the dominant source of pyrogenic PAHs (Neff, 1979; Bjorseth and Ramdahl, 1983; Ballentine *et al.*, 1996; O'Malley *et al.*, 1997).

9.13.2 SCOPE OF REVIEW

A number of previous reviews and textbooks on organic contaminant behavior in geochemical environments have already considered the general physical, chemical, and biological behavior of organic contaminants on the basis of their structure and composition. This review will focus on the geochemical behavior of a group of organic compounds referred to as HMW hydrocarbons, but will emphasize a class of compounds known as PAHs. Focus on PAHs is justified by their known toxicity and carcinogecity, hence the environmental concern already noted above. Monocyclic aromatic hydrocarbons comprise the LMW end of the aromatic hydrocarbon spectrum, which are discussed in more detail with other hydrocarbon fuels in see Chapter 9.12 of this volume.

Other chapters in this volume examine the LMW hydrocarbons and oxygenated hydrocarbons (see Chapter 9.12), halogenated compounds (see Chapter 14), and pesticides (see Chapter 9.15). Other general reviews and textbooks that summarize the sources and geochemical fate and transport of hydrocarbons in a variety of geological media are also available (e.g., Moore and Ramamoorthy, 1984; Schwarzenbach *et al.*, 1993, 1998; Eganhouse, 1997; Volkman *et al.*, 1997; NRC, 2000, 2002; Neilson, 1998; Abdul-Kassim and Simoneit, 2001; Beek, 2001). The following discussions focus on shallow aquatic environments, especially sediments, given the highly hydrophobic (mix poorly with water) and lipophilic (mix well with oil/fat) nature of most HMW hydrocarbon compounds. Nevertheless, the readers should recognize that in spite of our focus on a specific group of hydrocarbon compounds, the behavior of these compounds in aquatic systems will have broad similarity with the behavior of many other hydrophobic and lipophilic compounds discussed in other chapters in this volume. Indeed, most of these compounds are studied simultaneously in shallow aquatic environments, sharing not only common geochemical behavior but also analytical procedures for extraction, isolation, and characterization (e.g., Peters and Moldowan, 1993; Eganhouse, 1997; Abdul-Kassim and Simoneit, 2001).

9.13.3 SOURCES

9.13.3.1 Petrogenic Hydrocarbons

Worldwide use of petroleum outpaced coal utilization by the 1960s, and accidental oil discharge and release of waste products CO_2, soot (black carbon), and PAHs also increased concomitantly. The more recent report of the NRC (2002) shows that oil production crept up from 7 Mt d^{-1} in the 1970s to 11 Mt d^{-1} by the end of 2000. The total discharge of petroleum into the world's ocean was estimated to be between 0.5 Mt and 8.4 Mt of petroleum hydrocarbons annually (NRC, 2002), which roughly constitutes 0.1% of the annual oil consumption rate. Of this total, 47% are derived from natural seeps, 38% from consumption of petroleum (e.g., land-based runoff, operational discharges, and atmospheric deposition), 12% from petroleum transport, and 3% from petroleum extraction. Used crankcase oils or engine lubricating oils are an important specific source of PAHs in urban environments. The world production of crankcase oil is estimated to be ~40 Mt yr^{-1} and 4.4% of that is estimated to eventually reach aquatic environments (NRC, 1985).

Transportation-related release, in order of decreasing annual input, includes accidental releases during tank vessel spill, intentional ballast discharge, pipeline spills, and coastal facility spills. Spectacular oil releases recorded by massive oil spills from grounded tankers tend to capture the public attention, although chronic releases from operational discharges and land releases are quantitatively more important (NRC, 2002). Tank vessel spills account for less than 8% of worldwide petroleum releases in the 1990s (NRC, 2002). For example, oil slick formed from ballast discharges in the Arabian Sea was estimated to exceed 5.4×10^4 m^3 for 1978 (Oostdam, 1980). For comparison, the volume released by the 2002 Prestige oil spill off the Northwest coast of Spain is 1.3×10^4 m^3. Other major oil spills with larger releases include the Torrey Canyon (1.17×10^5 m^3), Amoco Cadiz (2.13×10^5 m^3), Ixtoc blowout (5.3×10^5 m^3), Exxon Valdez (5.8×10^4 m^3), and 1991 Gulf War ($>10^6$ m^3). Studies by Kvenvolden *et al.* (1993a, 1993b, 1995) also showed the substantially greater input of long-term chronic releases of California oil in the Prince Williams Sound sediments compared to oil released from the Exxon Valdez. A report by the NRC (2002) points to the dramatic decline in oil spilled in North American waters during the 1990s compared to the previous decades (Figure 3), with vessel spills accounting for only 2% of total petroleum release to US waters. The 1980s recorded the largest number (391) and volume (2.55×10^5 m^3) of oil spilled to North American waters.

Land releases into groundwater aquifers, lakes, and rivers are dominated by urban runoff and municipal/industrial discharges, but the actual amounts are difficult to quantify (NRC, 2002). Petroleum discharges from underground storage tanks are by far the dominant source of hydrocarbons in groundwater (see Chapter 9.12). However, surface releases from bulk supply depot, truck stops, industrial refueling facilities, and oil storage terminals could also be locally significant sources. It is noteworthy that substantial releases of hydrocarbons may also emanate from natural deposits of oil, heavy oil, shale oil,

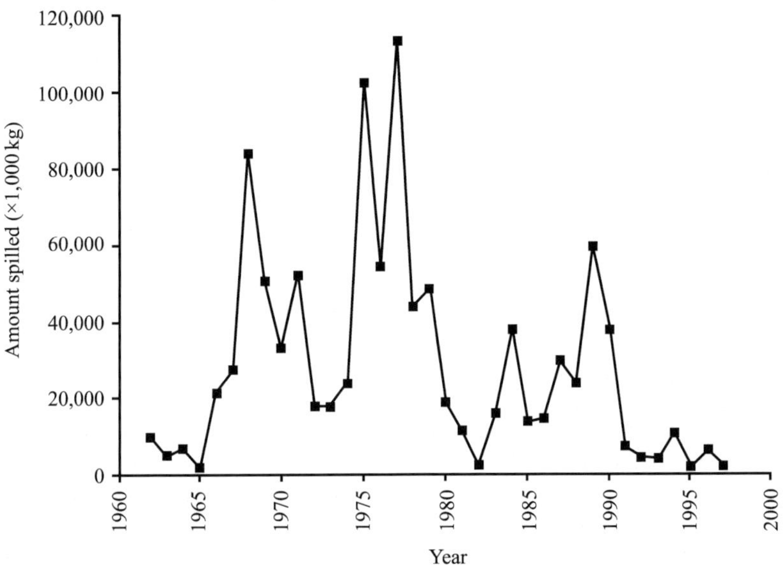

Figure 3 Oil spill trend in North American waters during the 1990s compared to the previous decades (data source NRC, 2002).

and bitumen. For example, the Athabasca River in northern Canada shows concentration levels of oil and grease of 850 $\mu g\ L^{-1}$ and >3,000 $\mu g\ L^{-1}$, respectively, from natural bitumen deposits (Moore and Ramamoorthy, 1984). Likewise, Yunker *et al.* (1993) and Yunker (1995) showed dominant input of natural hydrocarbons in the Mackenzie River. Indeed, natural seeps account for over 60% of petroleum releases to North American waters (NRC, 2002).

Petroleum hydrocarbon sources to North American and worldwide waters were summarized in a report by NRC (2002). In many cases of large petroleum spills, the specific source of petroleum spill is evident, and no geochemical fingerprinting is required to establish the source. Nevertheless, the inventory of petroleum compounds and biomarkers that are eventually sequestered in bottom sediments need not reflect sole derivation from a single source, even in cases of massive oil spills in the area (e.g., Kvenvolden *et al.*, 1995; Wang *et al.*, 1999). Where a mass balance of petroleum sources is required to properly design remediation or identify a point source, molecular methods for distinguishing sources of hydrocarbons have come to the fore.

Several geochemical methods for allocating sources of petrogenic hydrocarbons that are released to aquatic systems have been successfully applied (e.g., O'Malley *et al.*, 1994; Whittaker *et al.*, 1995; Abdoul-Kassim and Simoneit, 1995; Kvenvolden *et al.*, 1995; Wang and Fingas, 1995; Dowling *et al.*, 1995; Bieger *et al.*, 1996; Kaplan *et al.*, 1997; Eganhouse, 1997; Volkman *et al.*, 1997; Mansuy *et al.*, 1997; Hammer *et al.*, 1998; Wang *et al.*, 1999; McRae *et al.*, 1999, 2000; Mazeas and Budzinski, 2001; Hellou *et al.*, 2002; Faksness *et al.*, 2002; NRC, 2002; Lima *et al.*, 2003). The fingerprints used are either molecular or isotopic, and are variably affected by weathering processes. They include overall molecular distribution of hydrocarbons (e.g., range of carbon numbers and odd–even predominance), source-specific biomarkers (e.g., terpanes and steranes), so-called "diagnostic molecular ratios," and stable isotope compositions. The level of specificity by which a source can be pinpointed is dependent on the fingerprint used to tag the specific source and the multiplicity of hydrocarbon sources involved. Faksness *et al.* (2002) presented a flow chart for oil spill identification using a tiered molecular discrimination scheme based on the overall hydrocarbon distribution, source-specific markers, and diagnostic ratios (Figure 4) target compounds have been used for source identification of spilled oil including: (i) saturated hydrocarbons + pristane and phytane; (ii) volatile alkylated aromatics including benzene, toluene, ethyl benzene, and xylene (BTEX); (iii) alkylated and nonalkylated PAHs and heterocyclics; and (iv) terpanes and steranes (Volkman *et al.*, 1997; Wang *et al.*, 1999). For example, Wang and Fingas (1995) and Douglas *et al.* (1996) suggested that alkylated dibenzothiophenes are sufficiently resilient to a wide range of weathering reactions to be useful fingerprints for sources of crude oil. Kvenvolden *et al.* (1995) used the abundances of sterane and hopane biomarkers to differentiate specific crude oil sources in Prince William Sound, Alaska. The use of resilient biomarker signatures has matured to the point that they are widely used for specific litigation cases for assigning liability for oil releases (e.g., Kaplan *et al.*, 1997; Wang *et al.*, 1999).

The molecular distribution of PAHs in petroleum and crankcase oils is quite distinct from pyrogenic sources that will be discussed in the following section (Figure 5). This contrast provides an excellent basis for source apportionment of HMW hydrocarbons in the environment. Petrogenic PAHs consist primarily of two- and three-ring parental and methylated compounds with lower concentrations of HMW PAHs (Figure 5) (Bjorseth and Ramdahl, 1983; Pruell and Quinn, 1988; Vazquez-Duhalt, 1989; Latimer *et al.*, 1990; O'Malley, 1994). PAH formation during oil generation is attributed both to the aromatization of multi-ring biological compounds (e.g., sterols) and to the fusion of smaller hydrocarbon fragments into new aromatic structures (e.g., Radke, 1987; Neilson and Hynning, 1998; Simoneit, 1998). Steroids are probably the most well understood in terms of biological origin and geological fate, and a simplification of the proposed pathways of sterol diagenesis and catagenesis is summarized in Figure 6 (after Mackenzie, 1984). Crude oils usually formed at temperatures below 150 °C have a predominance of alkylated (i.e., possessing alkyl side chains) over parental PAHs. O'Malley (1994) characterized hundreds of variably used crankcase oils, and of these samples, 25.5% comprised of four- and five-ring parental compounds, and virgin crankcase oil samples were found to contain no resolvable PAHs (cf. Pruell and Quinn, 1988; Latimer *et al.*, 1990). Since virgin crankcase oils are devoid of measurable PAHs, the most probable source of PAHs in used crankcase oil are the thermal alteration reactions (i.e., aromatization) of oil components such as terpenoids in the car engine (Pruell and Quinn, 1988; Latimer *et al.*, 1990). As with true diagenetic PAHs (Figure 6), the distribution of PAHs in used crankcase oils apparently depends on several factors including temperature of reaction, engine design, and general operating conditions of the engine. Since the engine operating temperatures are generally high, the likely products resulting from these reactions include three-, four-, and five-ring

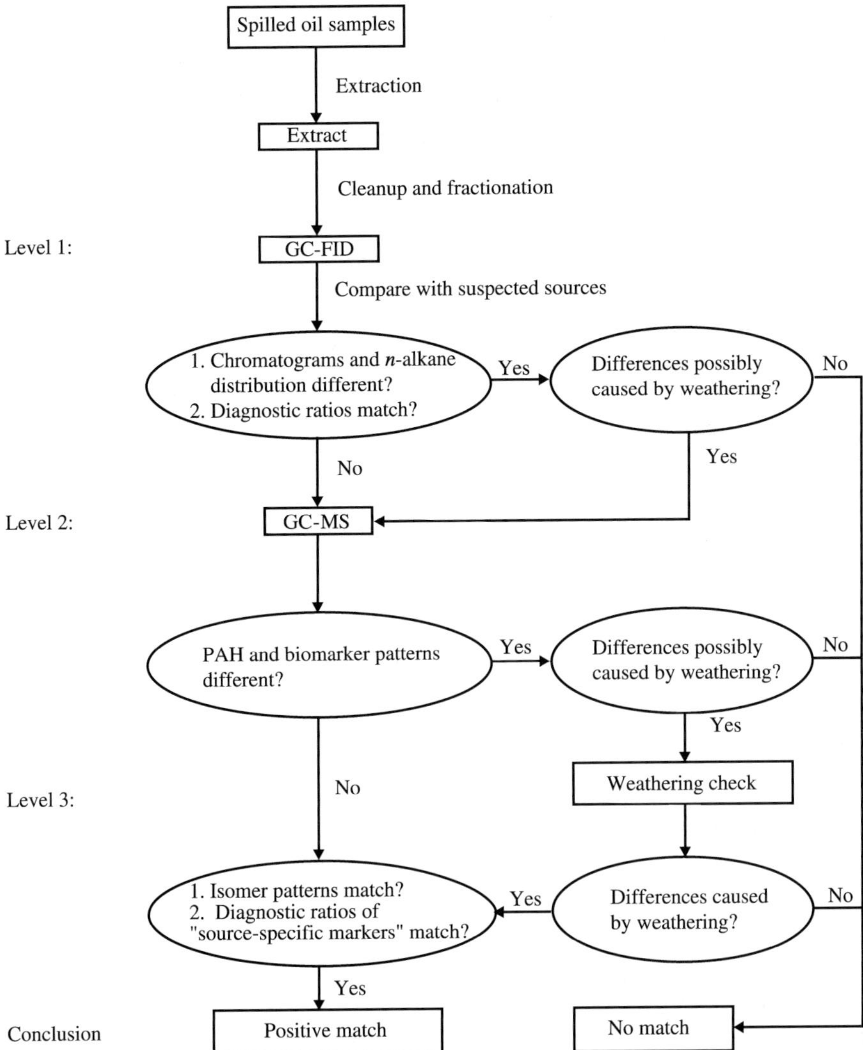

Figure 4 Tiered oil spill source identification scheme using molecular chemistry. The increasing level of source specificity required (down the diagram) is provided by global distributions of *n*-alkanes (level 1), PAH and biomarker distribution patterns (level 2), and isomeric and other diagnostic marker ratios (level 3), respectively (reproduced by permission of Nordtest from *Revision of the Nordtest Methodology for Oil Spill Identification*, **2002**, 110).

unsubstituted compounds such as Pa, Fl, Py, BaA, Chy, BeP, and BaP (see Figure 2 caption for abbreviations). Pruell and Quinn (1988) also suggested that LMW compounds (one- and two-ring) might be accumulated in crankcase oils from admixed gasoline. The relative stability of pyrogenic PAHs to weathering makes them attractive markers for discrimination of oil sources (Figure 5) (Pancirov and Brown, 1975; O'Malley, 1994; Wang *et al.*, 1999). However, Volkman *et al.* (1997) advocate caution in the use of aromatic compounds because of the relatively small variations between different oils and the differential aqueous solubility of these compounds (see below). Some HMW PAHs may be present in crude oil, albeit, at very low concentrations. Finally, crude oils are also rich in heterocyclic species particularly thiophenes (e.g., dibenzothiophenes). Significant molecular variations of these compounds have been reported for individual crude oil samples, and these are mainly attributed to oil origin and maturity (Neff, 1979). Other important potential sources of petrogenic PAHs identified in sedimentary environments are asphalt and tire and brake wear (Wakeham *et al.*, 1980a; Broman *et al.*, 1988; Takada *et al.*, 1990; Latimer *et al.*, 1990; Reddy and Quinn, 1997).

Although previous workers have suggested bacterial synthesis as a possible source of biogenic PAHs in modern sediments, Hase and Hites (1976) have shown that bacteria more likely only bioaccumulate them from the growth medium. Anaerobic aromatization of tetracyclic triterpenes appeared to have been demonstrated by Lohmann *et al.* (1990) by incubating radiolabeled β-amyrin,

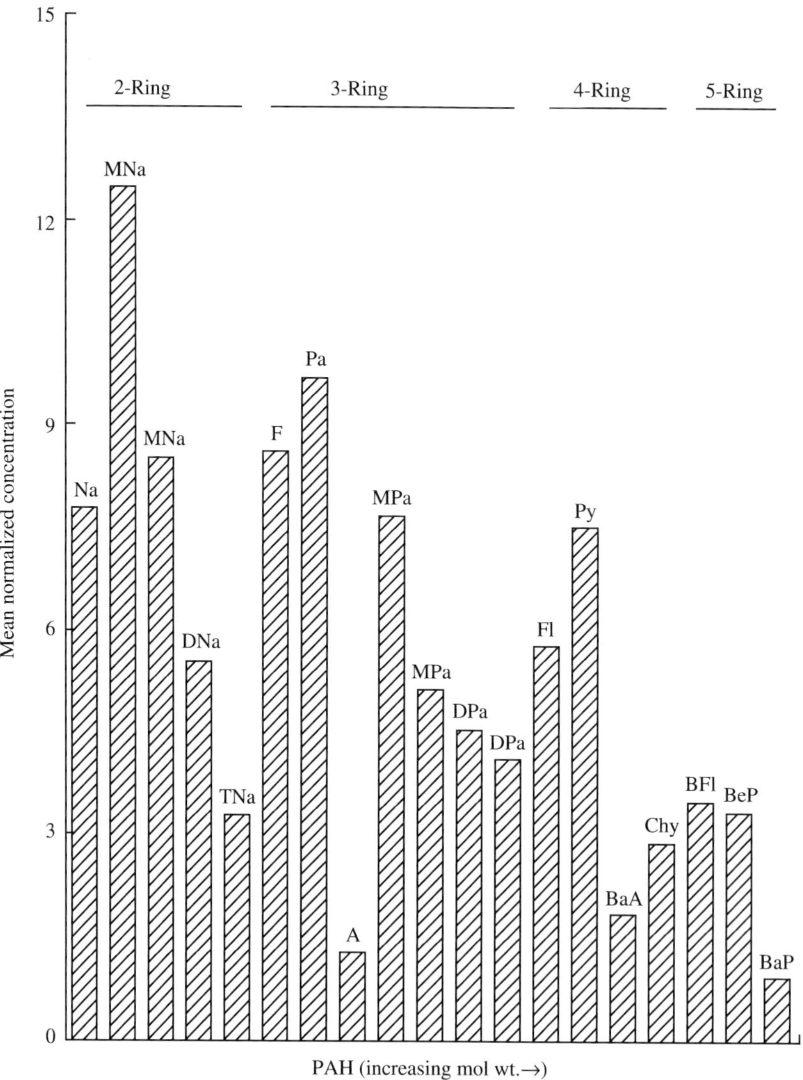

Figure 5 PAHs distribution in petrogenic sources. Compound symbols as in Figure 2 with the modifier "M" (methylated), "D" (dimethylated), and "T" (trimethylated) added for alkylated PAH.

but the quantitative importance of these synthesis pathways remains unresolved (e.g., Neilson and Hynning, 1998). The early diagenesis of sedimentary organic matter is certain to lead to the formation of a number of PAHs from alicyclic precursors (e.g., Neilson and Hynning, 1998; Simoneit, 1998). Examples of reactions are the diagenetic production of phenanthrene and chrysene derivatives from aromatization of pentacyclic triterpenoids originating from terrestrial plants, early diagenesis of abietic acids to produce retene (Figure 7), and *in situ* generation of perylene from perylene quinones (Youngblood and Blumer, 1975; Laflamme and Hites, 1978; Wakeham *et al.*, 1980b; Venkatesan, 1988; Lipiatou and Saliot, 1991; Neilson and Hynning, 1998; Simoneit, 1998; Wang *et al.*, 1999). Perylene and retene are the two most prominent PAHs found in recently deposited sediments (Lipiatou and Saliot, 1991). The origin of perylene has been linked to terrestrial precursors (4,9-dihydroxyperylene-3,10-quinone, the possible candidate), marine precursors, and anthropogenic inputs (Blumer *et al.*, 1977; Laflamme and Hites, 1978; Prahl and Carpenter, 1979; Venkatesan, 1988; Lipiatou and Saliot, 1991). The diagenetic pathway for retene is well constrained (Figure 7), but it can also be produced from wood combustion (Wakeham *et al.*, 1980b; Lipiatou and Saliot, 1991; Neilson and Hynning, 1998).

9.13.3.2 Pyrogenic Sources of HMW Hydrocarbons

The burial maturation of sedimentary organic matter leading to oil and coal formation and possible biosynthesis are only two of three possible

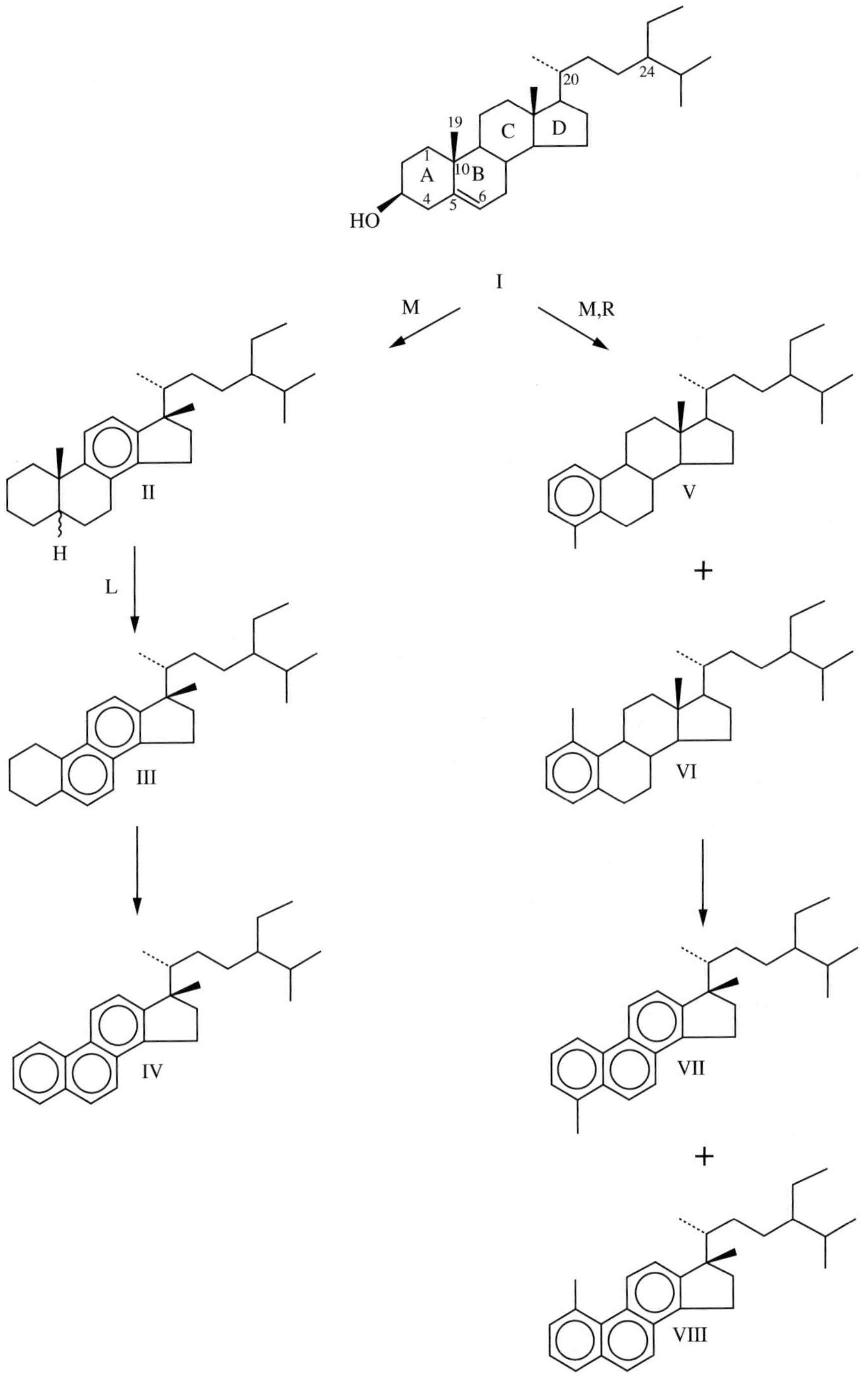

Figure 6 Diagenetic conversion of sterol (I) to various aromatic hydrocarbons during diagenesis. Abbreviation "M" implies that the reaction involved multiple steps, "R" represents aromatization of the A ring, and "L" represents the aromatization of the B ring. Compounds II, V, and VI are intermediates for the formation of triaromatic steroids (after Mackenzie, 1984).

pathways for generating HMW hydrocarbons. Pyrolysis or incomplete combustion at high temperatures also generates a wide variety of LMW and HMW hydrocarbons depending on the starting materials, environmental condition of pyrolysis, and kinetic factors (e.g., gas circulation). Whereas lower molecular compounds generated by pyrolysis have generated significant interest (e.g., butadiene, formaldehyde), the condensed structures from naphthalene to "black carbon" or soot has been the focus of interest amongst the HMW hydrocarbons.

Indeed, pyrolysis of petroleum and fossil fuel comprise quantitatively the most important

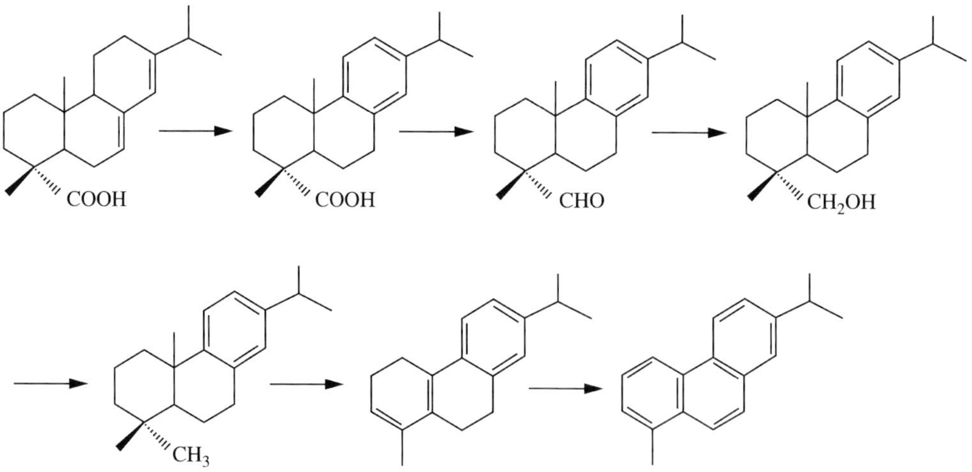

Figure 7 Diagenetic pathway for formation of retene from abietic acid (after Wakeham *et al.*, 1980a).

source of PAHs in modern sediments (Youngblood and Blumer, 1975; Laflamme and Hites, 1978; Neff, 1979; Wakeham *et al.*, 1980a,b; Sporstol *et al.*, 1983; Bjorseth and Ramdahl, 1983; Kennicutt II *et al.*, 1991; Lipiatou and Saliot, 1991; Canton and Grimalt, 1992; Brown and Maher, 1992; Steinhauer and Boehm, 1992; Yunker *et al.*, 1993, 1995; O'Malley *et al.*, 1996; Lima *et al.*, 2003). For example, sedimentary PAH distribution shows some common molecular features including the dominance of four- and five-ring PAHs (Fl, Py, BaA, Chy, BeP, and BaP), Pa/A ratio between 2 and 6, high Pa/MPa ratio and Fl/Py ratio close to unity. As noted above, the common characteristics related to direct petrogenic-related sources are a series of two- and three-ring parental and alkylated compounds (Na, MNa, Pa, and MPa), low Pa/MPa and Fl/Py ratios, and an unresolved complex mixture (UCM) (Kennicutt II *et al.*, 1991; Volkman *et al.*, 1992; Wang *et al.*, 1999). Although these sedimentary hydrocarbons may bear the imprint of petrogenic sources, the overall molecular attributes are signatures of high-temperature pyrolysis of fossil fuels and natural sources (e.g., forest fires) (Youngblood and Blumer, 1975; Laflamme and Hites, 1978; Lake *et al.*, 1979; Killops and Howell, 1988; Ballentine *et al.*, 1996). Individual markers such as perylene and retene, which are thought to be formed by the diagenetic alteration of biogenic compounds, are also typical of recently deposited sediments (Wakeham *et al.*, 1980a; Venkatesan, 1988; Lipiatou and Saliot, 1991; Yunker *et al.*, 1993, 1995), although combustion sources of perylene are also known (e.g., Wang *et al.*, 1999).

The mechanisms by which pyrolytic production of PAHs occur are complex, and have been widely studied since the 1950s (Badger *et al.*, 1958; Howsam and Jones, 1998). Pyrolytic production of PAHs is generally believed to occur through a free radical pathway, wherein radicals of various molecular weights can combine to yield a series of different hydrocarbon products. Therefore, the formation of PAHs is thought to occur in two distinct reaction steps: pyrolysis and pyrosynthesis (Lee *et al.*, 1981). In pyrolysis, organic compounds are partially cracked to smaller unstable molecules at high temperatures. This is followed by pyrosynthesis or fusion of fragments into larger and relatively more stable aromatic structures. Badger *et al.* (1958) was the first to propose this stepwise synthesis using BaP from free radical recombination reactions (Figure 8). Compounds identified in subsequent studies suggest that the C_2 species react to form C_4, C_6, and C_8 species, and confirm the mechanisms proposed by Badger *et al.* (1958) (Howsam and Jones, 1998; Figure 8). Despite the large quantities of different PAHs formed during primary reactions, only a limited number enter the environment. This is because initially formed PAHs themselves can be destroyed during combustion as a result of secondary reactions that lead to the formation of either higher condensed structures or oxidized carbon

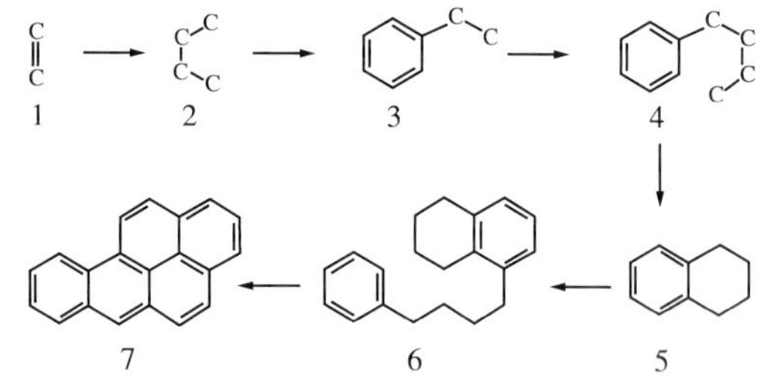

Figure 8 Stepwise pyrosynthesis of benzo(a)pyrene through radical recombination involving acetylene (1), a four carbon unit such as vinylacetylene or 1,3-butadiene, and styrene or ethylbenzene (3) (reproduced by permission of Royal Society of Chemistry from *J. Chem. Soc.*, **1958**, *1958*, 2449–2461).

(Howsam and Jones, 1998). For example, the pyrolysis of naphthalene can yield a range of HMW species such as perylene and benzofluoranthenes, possibly as a result of cyclodehydrogenation of the binaphthyls (Howsam and Jones, 1998). This may be particularly important for compounds that are deposited on the walls of open fireplaces along with soot particulates close to the hot zone of the flame.

PAHs isolated from important pyrogenic sources vary widely in composition, and they are quite distinct from the PAH distribution of petrogenic PAHs (Figure 9) (Alsberg *et al.*, 1985; Westerholm *et al.*, 1988; Broman *et al.*, 1988; Freeman and Cattell, 1990; Takada *et al.*, 1990; Rogge *et al.*, 1993; O'Malley, 1994; Howsam and Jones, 1998). For example, individual fireplace soot samples, from hard- and softwood-burning open fireplaces, were consistently dominated by three-, four-, and five-ring parental PAHs with generally lower concentrations of methylated compounds (Figure 9) (cf. Freeman and Cattell, 1990; Howsam and Jones, 1998). Vehicular emission and soot samples are also generally characterized by the presence of pyrolysis-derived three-, four-, and five-ring parental PAHs (Wakeham *et al.*, 1980a; Stenberg, 1983; Alsberg *et al.*, 1985; Westerholm *et al.*, 1988; Broman *et al.*, 1988; Takada *et al.*, 1990). The range and concentration of PAHs that accumulate in car mufflers are dependent on car age, engine operating conditions, catalytic converter efficiency, and general driving conditions (Pedersen *et al.*, 1980; Stenberg, 1983; Rogge *et al.*, 1993). Rogge *et al.* (1993) also reported greater PAH emissions occurred from noncatalytic automobiles compared to vehicles with catalytic systems. Perylene, a pentacyclic hydrocarbon, reported to be predominantly of diagenic origin (Laflamme and Hites, 1978; Prahl and Carpenter, 1979; Wakeham *et al.*, 1980b; Simoneit, 1998; Wang *et al.*, 1999) was also identified in some of the investigated car soots samples, although in very low concentrations (Blumer *et al.*, 1977). Lipiatou and Saliot (1992) reported that perylene may also be derived from coal pyrolysis. In contrast to the pyrogenic sources, Figure 9 also shows the enhanced concentration of two-ring compounds, especially a whole series of methylated phenanthrenes and naphthalenes in petrogenic sources.

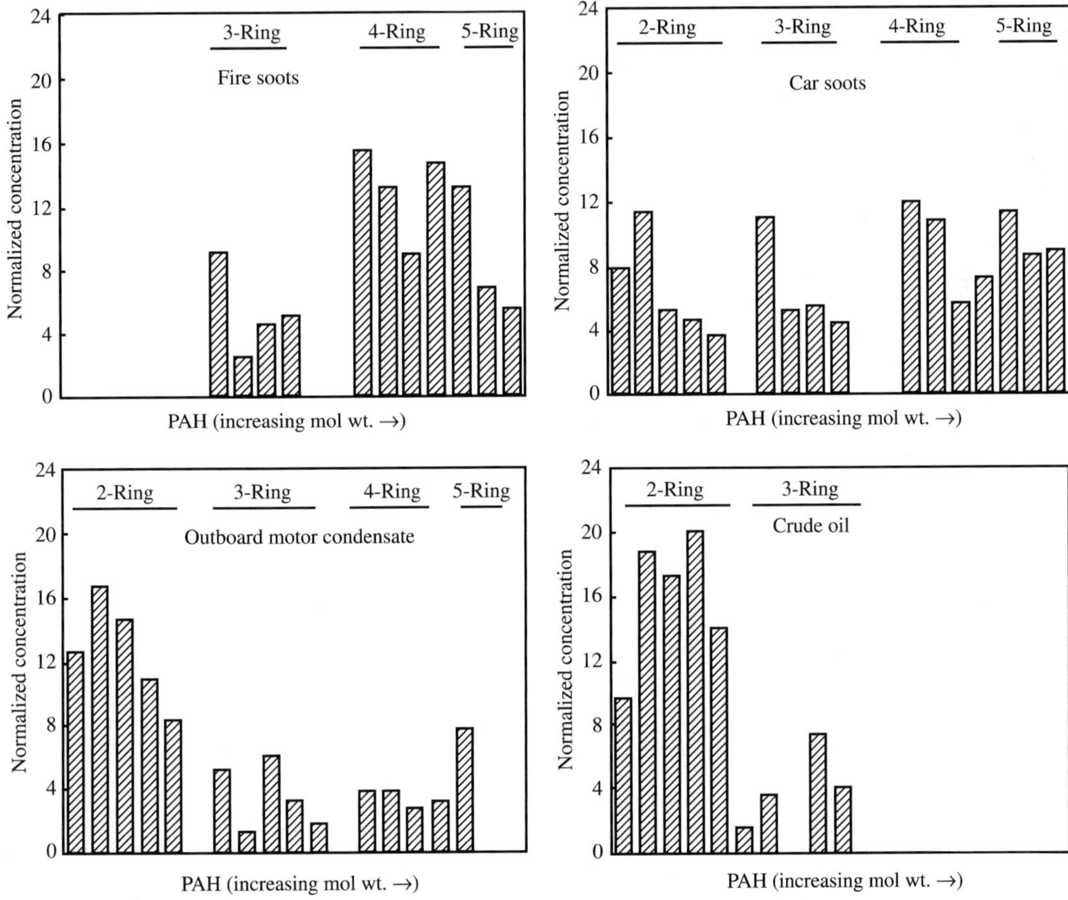

Figure 9 Molecular distribution/signatures of fire soot, car soot, outboard motor concentrate, and crude-oil PAH sources (after O'Malley, 1994).

9.13.4 PATHWAYS

HMW hydrocarbons can enter surface aquatic systems directly by spillage, accidental release, and natural oil seeps, or indirectly through sewers, urban, and highway runoff. Once hydrocarbons enter aquatic environments, they become rapidly associated with particulate matter, and are deposited in bottom sediments of surface waters or sorbed onto aquifer materials in groundwater systems (Radding *et al.*, 1976; Schwarzenbach *et al.*, 1993; Luthy *et al.*, 1997) (Figure 10). In surface-water systems, physical factors, such as turbulence, stability and composition of colloidal particles, deep-water currents, surface waves, and upwelling influence the length of time they remain suspended in the water column. The rate at which hydrocarbons are incorporated into bottom sediments is controlled by sedimentation rate, bioturbation, and bottom sediment–water column exchanges. In groundwater systems, the transport of hydrocarbons depends on whether or not they comprise a separate nonaqueous phase liquid (NAPL), groundwater flow velocities, and geochemical partitioning that are discussed below. HMW hydrocarbons likewise enter groundwater environments directly or indirectly from domestic and industrial effluents and urban runoff, and direct spillage of petroleum and petroleum products (e.g., ballast discharge, underground storage tanks).

An important alternate pathway of hydrocarbons to aquatic systems, however, is deposition of airborne particulates including anthropogenic (e.g., soot particles) and natural biogenic (e.g., monoterpenes, difunctional carboxylic acids) aerosols (Simoneit, 1984, 1986; Strachan and Eisenreich, 1988; Baker and Eisenreich, 1990; Eisenreich and Strachan, 1992; Currie *et al.*, 1999). Indeed, the importance of "organic aerosols" from anthropogenic and natural sources is now extensively recognized (e.g., Simoneit, 1984, 1986; Ellison *et al.*, 1999). A major part of the extractable and elutable organic matter in urban aerosols consists of an UCM, mainly the branched and cyclic hydrocarbons that originate from car exhaust (e.g., Simoneit, 1984; Rogge *et al.*, 1993). The resolved organic compounds in aerosol extracts consist of *n*-alkanes and fatty acids and PAHs. Whereas health, environmental, and climatic concerns have targeted the reaction products and intermediates formed from tropospheric reactions of labile hydrocarbons and carboxylic acids in these aerosols, the focus of toxicological concern has been on the PAHs.

In the case of PAHs, it is generally recognized that virtually all emissions to the atmosphere are indeed associated with airborne aerosols (Suess, 1976; Simoneit, 1986; McVeety and Hites, 1988; Strachan and Eisenreich, 1988; Baek *et al.*, 1991; Eisenreich and Strachan, 1992; Currie *et al.*, 1999; Offenberg and Baker, 1999). For example, Eisenreich and Strachan (1992) and Strachan and Eisenreich (1988) showed that upwards of 50% of the PAH inventory of the Great Lakes is deposited via atmospheric fallout. PAHs are initially generated in the gas phase, and then as the vapor cools; they are adsorbed onto soot particulates (Howsam and Jones, 1998). The highest concentrations of HMW PAHs in airborne particulates occur in the $<5\ \mu m$ particle size range (Pierce and Katz, 1975; Offenberg and Baker, 1999). PAH distribution between the gas and particulate phase is generally influenced by the following factors: vapor pressure as a function of ambient temperature, availability of fine particulate material, and the affinity of individual PAHs for the particulate organic matrix (Goldberg, 1985; Baek *et al.*, 1991; Schwarzenbach *et al.*, 1993). Atmospheric concentrations of PAHs are normally high in winter

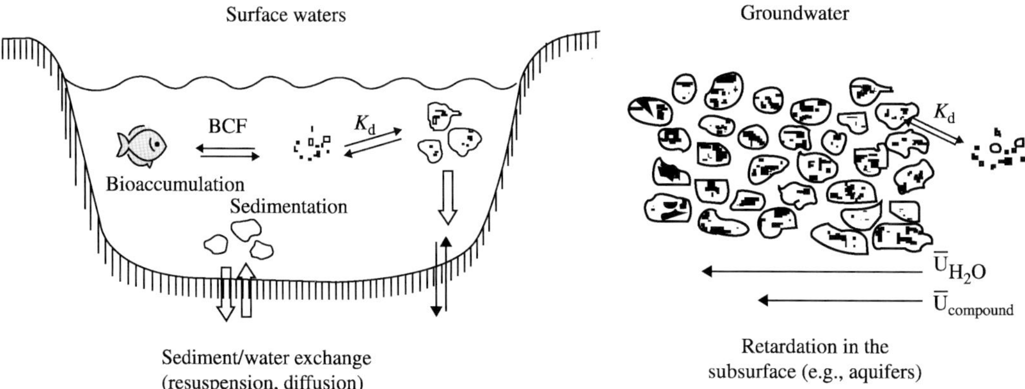

Figure 10 Solid–water exchange processes in groundwater and surface-water environments illustrating sorption to particulates (K_d) and biological concentration factor (BCF). Retardation in groundwater aquifer signified by the difference between water and compound velocities (U) (after Schwarzenbach *et al.*, 1998).

and low during the summer months (Pierce and Katz, 1975; Gordon, 1976; Howsam and Jones, 1998), an observation attributed to increased rates of photochemical activity during the summer and increased consumption of fossil fuels during the winter period.

Residence times of particulate PAHs in the atmosphere and their dispersal by wind are determined predominantly by particle size, atmospheric physics, and meteorological conditions (Howsam and Jones, 1998; Offenberg and Baker, 1999). The main processes governing the deposition of airborne PAHs include wet and dry deposition and, to a smaller extent, vapor phase deposition onto surfaces. Particles between 5 μm and 10 μm are generally removed rapidly by sedimentation and by wet and dry deposition (Baek *et al.*, 1991). However, PAHs associated with fine particulates (<1–3 μm) can remain suspended in the atmosphere for a sufficiently long time to allow dispersal over hundreds or thousands of kilometers (McVeety and Hites, 1988; Baek *et al.*, 1991).

PAHs in domestic sewage are predominantly a mixture of aerially deposited compounds produced from domestic fuel combustion and industrial and vehicle emissions, combined with PAHs from road surfaces that have been flushed into sewage. Road surface PAHs are derived primarily from crankcase oil, asphalt, and tire and brake wear (Wakeham *et al.*, 1980a; Broman *et al.*, 1988; Takada *et al.*, 1990; Rogge *et al.*, 1993; Reddy and Quinn, 1997). Some PAHs are removed from sewage during primary treatment (sedimentation). Not all urban runoff enters the sewer system, and some sewer designs allow runoff to be independently discharged to aquatic systems without primary treatment. The quantity of runoff from an urban environment is generally governed by the fraction of paved area within a catchment and the annual precipitation. During periods of continued rainfall, road surfaces are continually washed and the contributions of PAHs to watersheds are generally low. However, significant episodic contributions to aquatic systems can occur after prolonged dry periods or during spring snow melt (Hoffman *et al.*, 1984; Smirnova *et al.*, 1998). The distribution and quantity of PAHs in industrial effluents depends on the nature of the operation and on the degree of treatment prior to discharge. In some urban areas, industrial effluents are combined with domestic effluents prior to treatment, or they are independently treated before being discharged to sewer systems. Once particle-associated hydrocarbons are deposited in sediments or sorbed onto aquifer material, the subsequent fate is determined by the physics of sediment and water transport, and the biogeochemical reactions described below.

9.13.5 FATE

Shallow geochemical environments consist of solid, aqueous, and air reservoirs and their interfaces. Hydrocarbon compounds partition into these various reservoirs in a manner determined by the structure and physical properties of the compounds and the media, and the mechanism of hydrocarbon release. The structure and physical properties of the compounds and media understandably impact their sorption, solubility, volatility, and decomposition behavior (e.g., Schwarzenbach *et al.*, 1993). In addition, hydrocarbon partitioning in real systems is holistically a disequilibrium process; hence, the distribution of hydrocarbons depends as much on the pathway taken as on the final physical state of the system (e.g., Schwarzenbach *et al.*, 1993; Luthy *et al.*, 1997). Shallow aquatic systems may tend towards some equilibrium distribution (Figure 10), but this is seldom, if ever, truly attained.

Processes affecting hydrocarbon distribution and fate in multiphase aquatic systems can be characterized either as inter-media exchange or as transformational reactions (Mackay, 1998). The latter pertains to processes that involve molecular transformation of the compound, whereas the former is concerned with the movement of the molecularly intact compound from one medium to another. For the purpose of this review, we focus on the three most important exchange processes dictating the geochemical fate of hydrocarbons in aquatic environments: sorption, volatilization, and water dissolution. Sorption characterizes exchange between water and particulate phases (sediments or aquifer media), volatility characterizes exchange between air and water or air and solid, and dissolution pertains to the ability of contaminants to be present as true solutes in the aqueous media. A process of potential importance in some oil spills, emulsification, is not covered here but the readers are referred to a recent document published by NRC (2002), and references cited therein. Similarly, discussion of molecular reaction will focus on the two dominant transformation processes of hydrocarbons in geologic media: photolytic and biological transformation. Photolysis pertains to light-assisted chemical reactions that can affect compounds in the atmosphere and in the photic zone of water columns. Biological transformation is often accomplished by microorganisms, and could take place in aerobic or anaerobic environments. Purely chemical transformations, including hydrolysis, redox, and elimination reactions, are not examined here, because they are unlikely to be the dominant reaction pathways in shallow aquatic systems (NRC, 2002). For example, the theoretical pK_a values of hydrocarbons are exceedingly high; hence, they tend not to participate in acid–base reactions. Redox reactions involving alkenes can take place at surface geochemical conditions, but this compound group is not present in major amounts in petroleum hydrocarbon spills or pyrogenic products.

Hydrocarbons are hydrophobic and lipophilic compounds. As free liquid, gas, or solid phases, they are quite immiscible in water primarily because of their low polarity. In a competition between aqueous solution, air and solids, HMW hydrocarbons tend to partition heavily into the solid phases, and hence they are also sometimes referred to as "particle-associated compounds." This is true in both the atmospheric reservoir, where they associate with atmospheric particulates, and surface and groundwater systems, where they exhibit affinity for suspended particles and aquifer solids. In the case of nonhalogenated NAPL, the separate liquid phase is generally lighter than the aqueous phase, and hence they tend to physically "float" on the aqueous surface. Such is the case for oil spills either in surface water (e.g., ocean, streams, and lake) or in groundwater (e.g., underground storage tanks).

The hydrophobicity can be expressed by the dimensionless octanol/water partition coefficient (K_{OW}):

$$K_{OW} = C_{i,O}/C_{i,W}$$

where $C_{i,O}$ is the concentration of i in octanol and $C_{i,W}$ is the concentration of i in water. Many of the thermodynamic properties describing the partitioning of hydrocarbons in air, water, biota, or solid have been successfully, albeit empirically, related to K_{OW}. The value of K_{OW} among petroleum hydrocarbons, especially PAHs, tends to be very high (10^3 to $>10^6$), and this preference for the organic phase, either as a free phase or in organic particulates, is a major control on the fate and distribution of hydrocarbons in aquatic systems. The lipophilic affinity of PAHs is also a major contributor to enrichment of these compounds in organisms and organic-rich sediments.

9.13.5.1 Sorption

Solid–water exchange of hydrocarbons is a primary control on their aqueous concentration in groundwater and surface aquatic environments (Figures 10 and 11). The specific partitioning of hydrocarbons between solid and aqueous phase could be described by the distribution coefficient (K_d):

$$K_d = C_{i,\mathrm{solid}}/C_{i,W}$$

where $C_{i,\mathrm{solid}}$ is the concentration of i in the solid (mol kg^{-1}) and $C_{i,W}$ is the concentration of i in water (mol L^{-1}). For hydrophobic compounds, K_d tends to be constant only within a limited

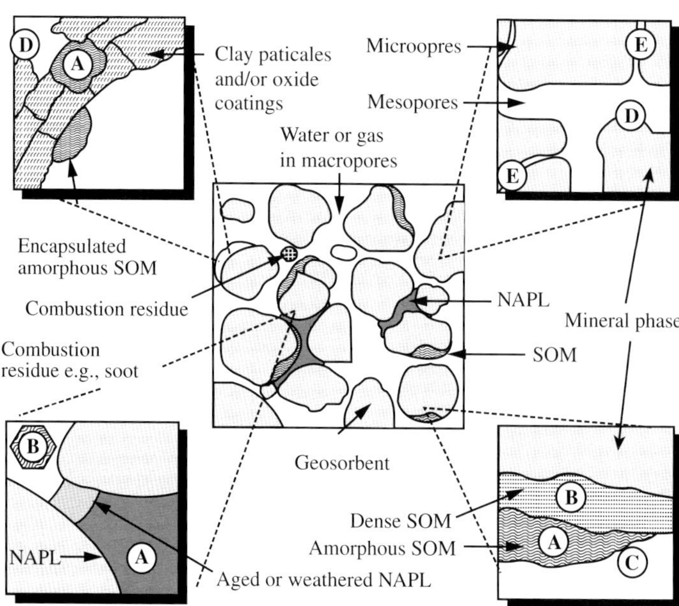

Figure 11 Heterogenous solid compartments in soils and sediments. NAPL signifies nonaqueous phase liquid and SOM represents soil organic matter. **A**—absorption into amorphous organic matter or NAPL, **B**—absorption into soot or black carbon, **C**—adsorption onto water-wet organic surfaces, **D**—adsorption to non-porous mineral surfaces, and **E**—adsorption into microvoids or microporous minerals (reproduced by permission of American Chemical Society from *Environ. Sci. Technol.*, **1997**, *31*, 3341–3347).

concentration range, and the more general Freundlich isotherm is commonly used:

$$C_{i,\text{solid}} = K_\text{d} C_{i,\text{W}}^n$$

where n is a parameter relating to the nonlinearity of the sorption process (e.g., Schwarzenbach *et al.*, 1993; Luthy *et al.*, 1997). Hydrocarbons tend to partition preferentially to organic matter; hence, the K_d is often written to take explicit account of this preference:

$$K_\text{d} = C_{i,\text{OM}} f_\text{OC} / C_{i,\text{W}}^n = K_{i,\text{OC}} f_\text{OC}$$

where $K_{i,\text{OC}}$ is the distribution coefficient between organic matter and aqueous phase and f_OC is the fraction of organic carbon (~0.5 fraction of organic matter, f_OM) in the solid substrate. Note that the above formulation assumes insignificant sorption on mineral matter, which is generally true for "threshold" $f_\text{OC} > 0.0005$ (Schwarzenbach *et al.*, 1993). Typical soils have a f_OC range of 0.005–0.05, whereas coarse aquifers have a f_OC range of 0–0.025.

"Linear free energy relationships" are quite frequently used for estimating K_d or $K_{i,\text{OC}}$. This linear correlation comes in the form

$$\log K_{i,\text{OC}} = a \log K_\text{OW} - b$$

where a and b are linear fitting parameters, and K_OW is the aforementioned octanol/water partition coefficient (e.g., Chiou *et al.*, 1979, 1998; Karickhoff, 1981; Schwarzenbach *et al.*, 1993). For example, $K_{i,\text{OC}}$ of PAHs could be estimated from their K_OW by assuming that $K_{i,\text{OC}} = 0.41 K_\text{OW}$ (Karickhoff's, 1981).

Hydrocarbon sorption on sediments or atmospheric particulates can involve either absorption or adsorption, which connotes surface attachment or subsurface dissolution, respectively. No distinction between these two processes is made here, except to note that sorption in geochemical systems often involves both (Kleineidam *et al.*, 2002). Laboratory experiments have shown that HMW petroleum hydrocarbons rapidly associate with sediment surfaces, with sorption uptake of up to 99% in 48 h at hydrocarbon concentrations of 1–5 mg L^{-1} (e.g., Knap and Williams, 1982). The rate and extent of sorption depends on the competitive affinity of the compounds between the aqueous or gaseous solution and the solid surface. The latter, in turn, depends on the nature of the crystallographic or amorphous substrate, especially the presence, type, and amount of organic matter in the solid and colloidal phase (e.g., Witjayaratne and Means, 1984; Luthy *et al.*, 1997; Ramaswami and Luthy, 1997; Gustafsson and Gschwend, 1997; Villholth, 1999; MacKay and Gschwend, 2001). The extent of preference for the solid surface is often described in terms of the sorption coefficient, K_d, as defined above.

The use of single distribution coefficient for organic matter would imply a constancy of sorption behavior on all organic substrates, and that the nature of organic matter substrate is not critical for sorption assessment. The impact of the heterogeneous nature of organic matter substrates on geosorption has been explored, and it has become evident that the wide variety of organic matter properties requires accounting of substrate-specific K_OM values (e.g., Luthy *et al.*, 1997; Kleineidam *et al.*, 1999, 2002; Chiou *et al.*, 1998; Bucheli and Gustafsson, 2000; Karapanagioti *et al.*, 2000) (Figure 11). For example, Karapanagioti *et al.* (2000) and Kleineidam *et al.* (1999) demonstrated the distinct Freundlich K_d (and $K_{i,\text{OC}}$) for phenanthrene sorption on different organic substrates as well as mineral matter substrates (Figure 12) (Karickhoff *et al.*, 1979). Bulk $K_{i,\text{OC}}$ characteristics of naphthalene, phenanthrene, and pyrene were examined by Chiou *et al.* (1998), and they concluded that significant differences exist in $K_{i,\text{OC}}$ of pristine terrestrial soil organic matter and sediment organic matter, and that sorption is enhanced in sediments with organic contaminants present. Furthermore, there is now increasing recognition of the explicit role of black carbon (soot) sorption particularly in regards to explaining much higher field measured K_d values compared to previously suggested "overall" K_d values (e.g., Chiou *et al.*, 1979; Karickhoff *et al.*, 1979; Karickhoff, 1981; Gustafsson and Gschwend, 1997; Accardi-Dey and Gschwend, 2003). It appears that the cohesive compatibility between PAHs and aromatic components of the organic substrate is a major factor

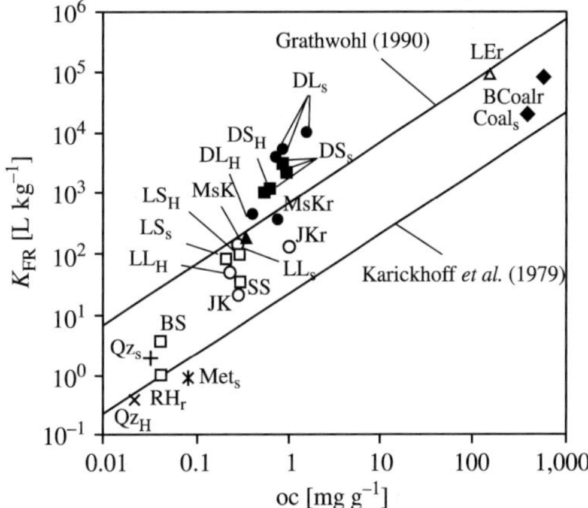

Figure 12 Phenanthrene sorption coefficients on different organic substrates. K_FR is the Freundlich coefficient and oc is organic carbon content. Remaining symbols are per table 2 of Kleineidam *et al.* (1999): D/L—dark/light, L—limestone, S—sandstone, Met—igneous and metamorphic rocks, Qz—quartz, and BS—bituminous shale (reproduced by permission of American Chemical Society from *Environ. Sci. Technol.*, **1999**, *33*, 1637–1644).

in enhancement of sorption (e.g., Gustafsson and Gschwend, 1997; Chiou *et al.*, 1998; Bucheli and Gustafsson, 2000; MacKay and Gschwend, 2001; Accardi-Dey and Gschwend, 2003). Accardi-Dey and Gschwend (2003) proposed treating organic sorption as a composite of organic carbon absorption and black carbon adsorption:

$$K_d = K_{i,OC} f_{OC} + f_{BC} K_{BC} C_{i,W}^{n-1}$$

where f_{BC} is the fraction of black carbon in the sample, K_{BC} is the black carbon distribution coefficient, and n is the Freundlich exponent as before. Note also that f_{OC} is now the fraction of organic carbon that excludes the black carbon component. Whereas this approach apparently explained discrepancies in modeled and field-calculated K_d and the nonlinear K_d observed in the laboratory, it remains to be seen if non-BC organic carbon can indeed be represented by a single K_d (i.e., $=K_{i,OC}$) in the general case.

9.13.5.2 Volatilization

When petroleum products enter surface-water systems, the lighter aliphatic and aromatic hydrocarbons spread out along the surface of the water and evaporate. The volatilization half-life of naphthalene, e.g., is 0.5–3.2 h (Mackay *et al.*, 1992). The naphthalene that does not evaporate sorbs to the particulate matter as noted above or is transformed into water in oil emulsion. In general, evaporation is the primary mechanism of loss of volatile and semivolatile components of spilled oil (e.g., Fingas, 1995; Volkman *et al.*, 1997; NRC, 2002). After a few days of oil spill at sea, ~75 % of light crude, 40% of medium crude, and 10 % of heavy crude oil will be lost by evaporation (NRC, 2002). The specific evaporation rates are influenced by a number of factors including meteorological factors, stratification of the water column, flow and mixing of water, and sequestration by mineral and natural organic sorbent matter.

The partitioning between air and water is characterized by the Henry's law constant (K_H):

$$K_H = P_{i,AIR,}/C_{i,W}$$

where $P_{i,AIR,}$ is the partial pressure of i in and $C_{i,W}$ is the molarity of i in the aqueous phase. K_H is often estimated from $P_{0,}/C_{i,sat}$, where P_0 is the vapor pressure of the compound and $C_{i,sat}$ is the solubility (see below). High K_H implies higher volatility, and results from a combination of high vapor pressure and low aqueous solubility. The vapor pressure of hydrocarbon compounds, along with their affinity for the sorption surfaces, dictates the partitioning between the atmosphere and atmospheric particulate phases. Hence, P_0 is a primary control on the atmospheric fate and transport of hydrocarbons. For example, a hydrocarbon that is tightly bound to the particulate phase is less likely to be altered during transportation and deposition (Baek *et al.*, 1991). The subsequent fate of hydrocarbons deposited in surface waters is further influenced volatilization behavior because of possible surface losses to the atmosphere (Strachan and Eisenreich, 1988; Baker and Eisenreich, 1990; Eisenreich and Strachan, 1992; Lun *et al.*, 1998; Gustafson and Dickhut, 1997).

The dependence of vapor pressure on temperature follows from a simplified solution to the Clapeyron equation (Schwarzenbach *et al.*, 1993):

$$\ln P_0 = -B/T + A$$

where $B = \Delta H_{vap}/R$ and $A = \Delta S_{vap}/R$. ΔH_{vap} and ΔS_{vap} are, respectively, the molar enthalpy and molar entropy of vaporization and R is the gas constant (8.314 Pa m^3 mol^{-1} K^{-1}). ΔH_{vap} is principally the energy required to break van der Waals and hydrogen bonds in going from the condensed phase to vapor. ΔS_{vap} is a compound-specific constant, which has been shown to correlate with compound boiling point (Schwarzenbach *et al.*, 1993). Increase in temperature results in more condensed phase hydrocarbons moving into the vapor phase, while a temperature decrease results in more vapor phase hydrocarbons appearing in the condensed phase (e.g., Lane, 1989; Baker and Eisenreich, 1990; Mackay, 1998). Thus, particulate hydrocarbon deposition is likely greater during colder periods, but the effect is highest on the lower molecular weight compounds (e.g., naphthalene) with significant vapor pressure. Likewise, the relative behavior of hydrocarbons between water surface and air is dictated by their relative vapor pressures such that a net flux of PAHs into the atmosphere can take place in warm summer months and the reverse taking place during the colder months (e.g., Baker and Eisenreich, 1990; Gustafson and Dickhut, 1997). Mackay (1998) and Mackay and Callcott (1998) provided more detailed analysis of this partitioning behavior using the so-called "fugacity approach." To mass balance PAH partitioning between the atmosphere and water, they used the dimensionless parameter K_{AW}, the air–water partition coefficient, as $K_{AW} = K_H/RT$ (cf. Baker and Eisenreich, 1990; Schwarzenbach *et al.*, 1993). Along with solid–water partitioning discussed above and solubility values discussed below, they showed mass balance partitioning models for PAHs in a model "seven-phase geomedia" (Mackay and Callcott, 1998). For a detailed discussion of the "fugacity approach," the reader is referred to Mackay (1998).

As noted above, volatilization losses from the aqueous phase to the atmosphere are also influenced by the aqueous solubility of the compound (i.e., $P_0/C_{i,\text{sat}}$). Higher solubility results in lower K_H for compounds of identical vapor pressures. Although the impact of solubility on volatilization loss is critical for LMW hydrocarbons, the effect of solubility on the behavior of HMW compounds has broader significance for understanding their transport and fate (see below).

9.13.5.3 Water Dissolution and Solubility

HMW hydrocarbons have a wide range of solubility, reported as S or $C_{i,\text{W}}^{\text{sat}}$ (i.e., aqueous concentration at saturation), but these solubilities are generally very low (e.g., Readman *et al.*, 1982; Means and Wijayaratne, 1982; Whitehouse, 1984; Mackay and Callcott, 1998). For example, water solubility of PAHs tends to decrease with increasing molecular weight from 4×10^{-4} M for naphthalene to 2×10^{-8} M for benzo(a)pyrene (Schwarzenbach *et al.*, 1993). Linear fused PAHs (e.g., naphthalene and anthracene) also tend to be less soluble than angular or pericondensed structures (e.g., phenanthrene and pyrene). Furthermore, alkyl substitution decreases water solubility of the parental PAHs. In spite of their low solubility, dissolution is one of the most important media exchange processes leading to the primary destruction pathway of hydrocarbons in aquatic systems—biodegradation. Metabolic utilization of hydrocarbons requires that they be transported as dissolved components into the cells, a process often assisted by extracellular enzymes released by the microorganisms after a quorum has been reached (Ramaswami and Luthy, 1997; Neilson and Allard, 1998). The presence of dissolved humic acids in solution also enhances solubility (Chiou *et al.*, 1979, 1983; Fukushima *et al.*, 1997; Chiou and Kyle, 1998). Solubility enhancement may also be achieved using surfactants, a method employed in many previous PAH and oil cleanup efforts.

Temperature affects aqueous solubility in a manner dictated by the enthalpy of solution (ΔH_s^0) according to (Schwarzenbach *et al.*, 1993)

$$\ln C_{i,\text{w}}^{\text{sat}} = -\Delta H_s^0/RT + \Delta S_s^0/R$$

The temperature effects on solubility for liquid and solid hydrocarbons vary greatly because of the large melting enthalpy component to ΔH_s^0 for solids (for which all PAHs are at room temperature). In general, PAH solubility increases with temperature, prompting present interest in thermophilic degradation of HMW PAHs. Hydrocarbons, including PAHs, also exhibit "salting-out" effects in saline solutions. This effect is exemplified by the so-called Setschenow formulation (Schwarzenbachs *et al.*, 1993): $\log C_{i,\text{salt}}^{\text{sat}} = \log C_{i,\text{W}}^{\text{sat}} - K_s S$ where $C_{i,\text{salt}}^{\text{sat}}$ is the saturation concentration or solubility in the saline solution, $C_{i,\text{W}}^{\text{sat}}$ is the saturation concentration in pure water defined previously, K_s is the Setschenow constant, and S is the total molar salt concentration. For example, the Setschenow constant for pyrene at 25 °C is 0.31–0.32 (Schwarzenbach *et al.*, 1993) so that at the salinity of seawater, $C_{i,\text{W}}^{\text{sat}}$, is 54% greater than $C_{i,\text{salt}}^{\text{sat}}$. This salting-out effect could be large enough to manifest in the localization of PAH contamination in estuarine environments (e.g., Whitehouse, 1984).

HMW hydrocarbons and especially PAHs with no polar substituents are very hydrophobic, which along with their lipophilic characteristics and low vapor pressure explain why they are not efficiently transported in aqueous form.

9.13.5.4 Photochemical Reactions

Photochemical reactions involving electromagnetic radiation in the UV-visible light range can induce structural changes in organic compounds. Direct photochemical reactions occur when the energy of electronic transition in the compounds corresponds to that of the incident radiation, with the compound acting as the light-absorbing molecule (i.e., chromophore). Hence, the structure of hydrocarbons determines the extent by which they are prone to photodecomposition, but photolytic half-lives are also significantly dependent on compound concentration and substrate properties (e.g., Behymer and Hites, 1985, 1988; Paalme *et al.*, 1990; Reyes *et al.*, 2000; NRC, 2002). In general, aromatic and unsaturated hydrocarbons are more prone to UV absorption and decomposition, with increasing numbers of conjugated bonds resulting in lower energy required for electronic transition. Photodissociation is not likely an important weathering mechanism for HMW straight-chain hydrocarbons, because these compounds do not absorb light efficiently (e.g., Payne and Phillips, 1985). Nevertheless, these hydrocarbons may be transformed through the process of indirect photodissociation wherein another molecule (e.g., humic and fulvic acids) or substrate (mineral or organic) acts as the chromopore (NRC, 2002). Aromatic structures are prone to direct photochemical reaction in a manner that depends on molecular weight and degree of alkylation.

Laboratory experiments have shown that PAHs are photoreactive under atmospheric conditions (Zafiriou, 1977; Behymer and Hites, 1988; Schwarzenbach *et al.*, 1993; Reyes *et al.*, 2000) and in the photic zone of the water column (Zepp and Schlotzhauer, 1979; Payne and Phillips, 1985; Paalme *et al.*, 1990). The existence of PAH oxidation products in atmospheric particulate matter indicates that PAHs react with oxygen or ozone in the atmosphere (Schwarzenbach *et al.*, 1993; Howsam and Jones, 1998), but the reaction with hydroxyl (OH) radicals during daylight conditions is considered to be the major reaction sink of these compounds in the atmosphere. There is great variation in the reported half-lives of various PAHs due to photolysis, largely because of the differences in the nature of the substrate on which the PAHs are adsorbed and the degree to which they are bound (Behymer and Hites, 1985; Paalme *et al.*, 1990; Schwarzenbach *et al.*, 1993). Carbon content and the color of substrates are important factors in controlling PAH reactivity (Behymer and Hites, 1988; Reyes *et al.*, 2000), and the suppression of photochemical degradation of PAHs adsorbed on soot and fly-ash has been attributed to particle size and substrate color. Darker substrates absorb more light and thereby protect PAHs from photolytic degradation reactions (Behymer and Hites, 1988).

Photo-oxidation by singlet oxygen appears to be the dominant chemical degradation process of PAHs in aquatic systems (Lee *et al.*, 1978; Hinga, 1984; Payne and Phillips, 1985). The degree to which PAHs are oxidized in an aqueous system depends on the PAH type and structure, water column characteristics (such as oxygen availability), temperature and depth of light penetration, and residence time in the photic zone (Payne and Phillips, 1985; Paalme *et al.*, 1990; Schwarzenbach *et al.*, 1993). Paalme *et al.* (1990) have shown that the rate of photochemical degradation of different PAHs in aqueous solutions can differ by a factor of >140 depending on their chemical structure, with perylene, benzo(b)fluoranthene, and coronene showing the greatest stability. Alkyl PAHs are more sensitive to photo-oxidation reactions than parental PAHs, probably due to benzyl hydrogen activation (e.g., Radding *et al.*, 1976). Payne and Phillips (1985) reported that benz(a)-anthracene and benzo(a)pyrene are photolytically degraded 2.7 times faster in summer than in winter. Once deposited in sediments, photoxidation reactions are generally significantly reduced due to anoxia and limited light penetration. PAH residence times in sediments depend on the extent of physical, chemical, and biological reactions occurring at the sediment–water interface, as well as the intensity of bottom currents (Hinga, 1984).

9.13.5.5 Biodegradation

Biodegradation is perhaps the most important reaction mechanism for the degradation of hydrocarbons in aquatic environments. Aliphatic compounds in crude oil and petroleum products are readily degraded, with a prominent initial microbial preference for straight chain compounds (e.g., Atlas and Bartha, 1992; Prince, 1993; Volkman *et al.*, 1997; Wang *et al.*, 1999; Heider *et al.*, 1999; Bosma *et al.*, 2001; NRC, 2002). The aerobic pathway shows conversion of alkane chains to fatty acids, fatty alcohols and aldehyde, and carboxylic acids that are then channeled into the central metabolism for subsequent β-oxidation (Figure 13; cf. see Chapter 9.12). Anaerobic degradation proceeds with nitrate, Fe^{3+}, or sulfate as the terminal electron acceptor, with no intermediate alcohols in the alkane degradation (Figure 13). The degradation pathway involves an O_2-independent oxidation to fatty acids, followed by β-oxidation. Sulfate reducers apparently show specificity towards utilization of short chain alkanes (C_6–C_{13}) (Bosma *et al.*, 2001). Laboratory experiments on complex oil-blends further showed composition changes accompanying biodegradation. For example, Wang *et al.* (1998) examined the compositional evolution of Alberta Sweet Mixed Blend oil upon exposure to defined microbial inoculum, with total petroleum hydrocarbons (9–41%), and specifically the total saturates (5–47%) and *n*-alkanes (>90%), showing varying degrees of degradation. They further showed that susceptibility to *n*-alkane degradation is an inverse function of chain length, the branched alkanes are less susceptible than straight-chain *n*-alkanes, and that the most resilient saturate components are the isoprenoids, pristane and phytane. Even amongst isoprenoids, however, there is still a notable inverse dependence between chain length and degradation susceptibility. The relative susceptibility of *n*-alkanes to biodegradation compared to the isoprenoids is the basis for using C_{17}/pristane and *n*-C_{18}/phytane ratios for distinguishing biodegradation from volatilization effects because the latter discriminates primarily on the basis of molecular weight. Additionally, Prince (1993) noted the relative resilience of polar compounds compared to corresponding hydrocarbons. Interesting deviations from this generalized biodegradation pattern were nevertheless noted by others, including the observation that some Exxon Valdez spill site microbial communities preferentially degraded naphthalene over hexadecane at the earliest stages of biodegradation (e.g., Sugai *et al.*, 1997).

Numerous cases of crude oil and refined petroleum spill into surface environments have provided natural laboratories for examination of the biodegradation of petrogenic compounds in a variety of environmental conditions (e.g., Kaplan *et al.*, 1997; Prince, 1993; Wang *et al.*, 1998; NRC, 2002). The observations on compositional patterns of biodegradation noted above are generally replicated in these natural spills. For example, a long-term evaluation of compositional variation of Arabian light crude in a peaty mangrove environment was conducted by Munoz *et al.* (1997), who showed the same pattern of initial preferential loss

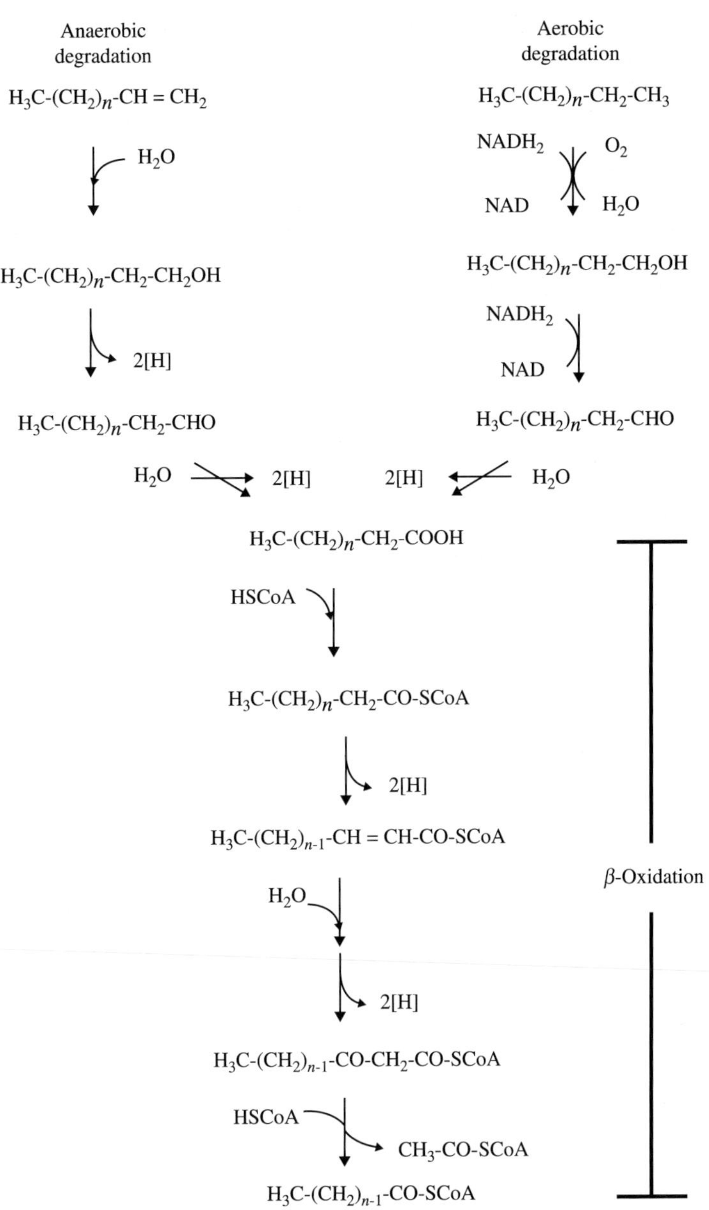

Figure 13 Generalized aerobic and anaerobic biodegradation pathways for *n*-alkanes (reproduced by permission of Springer-Verlag from *The Handbook of Environmental Chemistry*, **2001**, pp. 163–202).

of *n*-alkanes followed by isoprenoids, and ultimately the biomarkers in the order steranes, hopanes, bicyclic terpanes, tri- and tetracyclic terpanes, diasteranes, and the aromatic biomarkers. However, an interesting report on the biodegradation patterns in the crude-oil spill site in Bemidji, Minnesota, showed an apparent reversal in the biodegradation preference of *n*-alkanes, wherein HMW homologues show faster degradation rates (Hostetler and Kvenvolden, 2002). The overall resilience of terpane and sterane compounds to biodegradation has been well recognized; hence, biomarker ratios are widely used as indicators of oil spill sources even in highly weathered oils (e.g., Volkman *et al.*, 1997).

Aromatic hydrocarbons in the atmosphere and open waters may undergo volatilization and photodecomposition, but microbial degradation is the dominant sink below the photic zone (Gibson and Subramanian, 1984). As with aliphatic hydrocarbons, biodegradation of aromatic compounds involves the introduction of oxygen into the molecule forming catechol (cf. Chapter 9.12). Microbial degradation has also been recognized as the most prominent mechanism for removing PAHs from contaminated environments (Neilson, 1998; NRC, 2000, 2002). Microbial adaptations may result from chronic exposure to elevated concentrations as shown by the higher biodegradation rates in

PAH-contaminated sediments than in pristine environments (Neilson, 1998; NRC, 2000). Nevertheless, it is also known that preferential degradation of PAHs will not occur in contaminated environments where there are more accessible forms of carbon (NRC, 2000). A summary of the turnover times for naphthalene, phenanthrene, and BaP in water and sediment is shown in Table 1. In general, PAH biodegradation rates are a factor of 2–5 slower than degradation of monoaromatic hydrocarbons (cf. Chapter 9.12), and of a similar magnitude as HMW *n*-alkanes (C_{15}–C_{36}) under similar aerobic conditions. The ability of microorganisms to degrade fused aromatic rings is determined by their combined enzymatic capability which can be affected by several environmental factors (McElroy *et al.*, 1985; Cerniglia, 1991; Neilson and Allard, 1998; Bosma *et al.*, 2001). The most rapid biodegradation of PAHs occurs at the water–sediment interface (Cerniglia, 1991). Prokaryotic microorganisms primarily metabolize PAHs by an initial dioxygenase attack to yield *cis*-dihydrodiols and finally catechol (e.g., Figure 14). Biodegradation and utilization of lower molecular weight PAHs by a diverse group of bacteria, fungi, and algae has been demonstrated (Table 2; Neilson and Allard, 1998; Bosma *et al.*, 2001). For example, many different strains of microorganisms have the ability to degrade napththalene including *Pseudomonas*, *Flavobacterium*, *Alcaligenes*, *Arthrobacter*, *Micrococcus*, and *Bacillus*.

The degradation pathways of higher molecular weight PAHs—such as pyrene, benzo(e)pyrene, and benzo(a)pyrene—are less well understood (Neilson and Allard, 1998). Because these compounds are more resistant to microbial degradation processes, they tend to persist longer in contaminated environments (Van Brummelen *et al.*, 1998; Neilson and Allard, 1998; Bosma *et al.*, 2001). However, the degradation of fluoranthene, pyrene, benz(a)anthracene, benzo(a)-pyrene, benzo(b)fluorene, chrysene, and benzo(b)fluoranthene has been reported in laboratory conditions (Barnsley, 1975; Mueller *et al.*, 1988, 1990; Schneider *et al.*, 1996; Ye *et al.*, 1996; Neilson and Allard, 1998). Elevated temperatures increase the rate of biotransformation reactions *in vitro*. For example, laboratory studies have shown a 50% loss of phenanthrene after 180 d at 8 °C in water compared to 75% loss in 28 d at 25 °C (Sherrill and Sayler, 1981; Lee *et al.*, 1981). It is widely believed that PAHs with three or more condensed rings tend not to act as sole

Table 1 Turnover times for naphthalene, phenanthrene, and benzo(a)pyrene in water and sediment.

PAH and environment	*Temp.* (°C)	*Times* (d)
Naphthalene		
Estuarine water	13	500
Estuarine water	24	30–79
Estuarine water	10	1–30
Seawater	24	330
Seawater	12	15–800
Estuarine sediment	25	21
Estuarine sediment		287
Estuarine sediment	22	34
Estuarine sediment	30	15–20
Estuarine sediment	2–22	13–20
Stream sediment	12	>42
Stream sediment	12	0.3
Reservoir sediment	22	62
Reservoir sediment	22	45
Phenanthrene		
Estuarine sediment	25	56
Estuarine sediment	2–22	8–20
Reservoir sediment	22	252
Reservoir sediment	22	112
Sludge-treated soil	20	282
Benzo(a)pyrene		
Estuarine water	10	2–9,000
Estuarine sediment	22	>2,800
Estuarine sediment	2–22	54–82
Stream sediment	12	>20,800
Stream sediment	12	>1,250
Reservoir sediment	22	>4,200
Sludge-treated soil	20	>2,900

Source: O'Malley (1994).

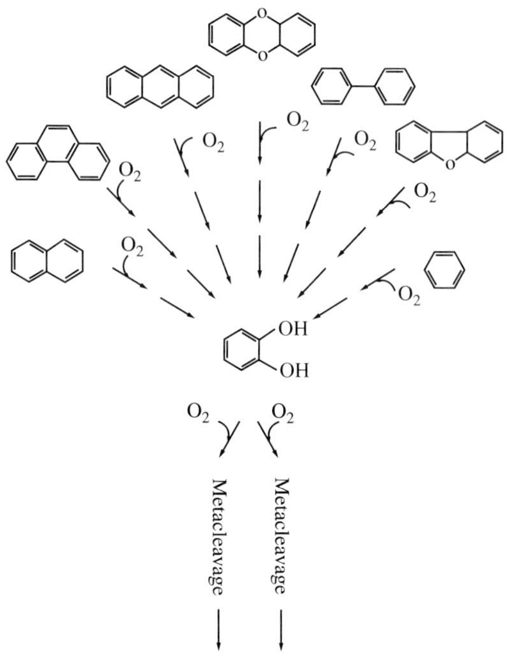

Figure 14 Generalized aerobic biodegradation pathways for aromatic hydrocarbons (after Bosma *et al.*, 2001).

Table 2 Biodegradation and utilization of lower molecular weight PAHs by a diverse group of bacteria, fungi, and algae.

Organism	*Substrate*
Pseudomonas sp.	Naphthalene
	Phenanthrene
	Anthracene
	Fluoranthene
	Pyrene
Flavobacteria sp.	Phenanthrene
	Anthracene
Alcaligenes sp.	Phenanthrene
Aeromonas sp.	Naphthalene
	Phenanthrene
Beijerenckia sp.	Phenanthrene
	Anthracene
	Benz(a)anthracene
	Benzo(a)pyrene
Bacillus sp.	Naphthalene
Cunninghamella sp.	Naphthalene
	Phenanthrene
	Benzo(a)pyrene
Micrococcus sp.	Phenanthrene
Mycobacterium sp.	Phenanthrene
	Phenanthrene
	Fluorene
	Fluoranthene
	Pyrene

Source: O'Malley, 1994.

substrates for microbial growth, but may be the subject of co-metabolic transformations. For example, co-metabolic reactions of pyrene, 1,2-benzanthracene, 3,4-benzopyrene, and phenanthrene can be stimulated in the presence of either naphthalene or phenanthrene (Neilson and Allard, 1998). Nevertheless, the degradation of PAHs even by co-metabolic reactions is expected to be very slow in natural ecosystems (e.g., Neilson and Allard, 1998).

Biodegradation effects on aromatic hydrocarbons is a subject of much interest both from the standpoint of characterizing oil spill evolution and engineered bioremediation (Wang *et al.*, 1998; Neilson and Allard, 1998). Wang *et al.* (1998) noted that the susceptibility to biodegradation increases with decreasing molecular weight and degree of alkylation. For example, the most easily degradable PAHs examined are the alkyl homologues of naphthalene, followed by the alkyl homologues of dibenzothiophene, fluorene, phenanthrene, and chrysene. Also noteworthy is the observation that microbial degradation is isomer specific, leading to the suggestion that isomer distribution of methyl dibenzothiophenes are excellent indicators of degree of biodegradation (Wang *et al.*, 1998). Finally, in spite of the focus of biodegradation studies on aerobic degradation, recent work has demonstrated the capacity of sulfate-reducing bacteria in degrading HMW PAH (e.g., BaP) (Rothermich *et al.*, 2002).

9.13.6 CARBON ISOTOPE GEOCHEMISTRY

The present review of hydrocarbon sources, pathways, and fate in aquatic environments highlights the current state of understanding of HMW hydrocarbon geochemistry, but it also provides a useful starting point for exploring additional approaches to unraveling the sources and fate of hydrocarbons in aquatic environments. It is clear that the degree and type of hydrocarbons ultimately sequestered in particulates, sediments, or aquifer materials depend not only on the nature and magnitude of various source contributions, but also on the susceptibility of the hydrocarbons to various physical, chemical, and microbial degradation reactions. As we have already shown, the latter alters the overall molecular signatures of the original hydrocarbon sources, complicating efforts to apportion sources of hydrocarbons in environmental samples. Nevertheless, resilient molecular signatures that either largely preserve the original source signatures or alter in a predictable way have been employed successfully to examine oil sources as already discussed (Wang *et al.*, 1999). Additionally, carbon-isotopic composition can be used to help clarify source or "weathering reactions" that altered the hydrocarbons of interest. In what follows, we will examine the emerging application of carbon-isotopic measurements in unravelling the sources and fate of PAHs in shallow aquatic systems. Similar approaches to studying aliphatic compounds have been employed in a number of hydrocarbon apportionment studies (e.g., O'Malley, 1994; Mansuy *et al.*, 1997; Dowling *et al.*, 1995), and the approach has indeed been a principal method used for oil–oil and oil source rock correlation for decades (e.g., Sofer, 1984; Schoell, 1984; Peters *et al.*, 1986; Faksness *et al.*, 2002). The key to using carbon isotopes for understanding the geochemistry of HMW hydrocarbons in shallow aquatic systems is to distinguish two reasons why the abundance of stable isotopes in these compounds might vary: (i) differences in carbon sources and (ii) isotope discrimination introduced after or during formation. In what follows, we will first describe and compare the carbon-isotope systematics in the pyrogenic and petrogenic PAH sources. Then we will examine possible changes in the carbon-isotope compositions in these compounds as a result of one or more weathering reactions. Finally, we will use the PAH inventory of an estuarine environment in eastern Canada as an example of how the molecular characteristics discussed earlier in this chapter can be blended with compound-specific carbon-isotope signatures to distinguish PAH sources and pathways.

9.13.6.1 Carbon Isotope Variations in PAH Sources

9.13.6.1.1 Pyrogenesis

The isotopic signature imparted on individual PAHs during formation is determined by both the isotopic composition of the precursor compounds and the formation conditions. Since these two factors can vary widely during pyrolysis or diagenesis, the potential exists for PAHs produced from different sources and from a variety of processes to have equally variable isotopic signatures. Isotopic characterization of bulk organic and aromatic fractions has been performed for many years (e.g., Sofer, 1984; Schoell, 1984), but the advent of compound-specific isotopic characterization methods for individual organic compounds has immensely increased the database for assessing the range of PAH isotopic compositions.

The specific mechanisms controlling the isotopic signatures of individual pyrogenic PAHs are only partly understood (O'Malley, 1994; Currie *et al.*, 1999). It is expected that the pyrolysis of isotopically distinct precursor materials would result in PAHs with different isotopic signatures. Due to the nature of PAH formation pathways discussed earlier in this chapter, the $\delta^{13}C$ of pyrolysis-derived compounds may be dictated by a series of primary and secondary reactions that precursor and intermediate compounds undergo prior to the formation of the final PAHs. The isotopic effects associated with ring cleavage reactions of intermediate precursors (LMW compounds) may result in ^{13}C enriched higher molecular weight species formed by the fusion of these reduced species. Variations in the $\delta^{13}C$ of the higher molecular weight condensed compounds may, therefore, depend on the $\delta^{13}C$ of the precursor and intermediate species. In the absence of pyrosynthetic recombination reactions, produced PAHs will have isotopic compositions that are largely dictated by those of the original precursor compounds. Alkylation or dealkylation reactions at specific sites of a parent molecule will alter the isotopic composition of that compound only to the extent that the alkyl branch is isotopically different from the substrate PAHs.

O'Malley (1994) performed the first characterization of compound-specific carbon-isotope characterization of primary and secondary PAH sources. Irrespective of the range observed in the $\delta^{13}C$ of the wood-burning soot PAHs, the overall trend in the mean $\delta^{13}C$ values indicates that lower molecular weight compounds (three-ring) are isotopically more depleted than four-ring PAHs, whereas five-ring compounds have $\delta^{13}C$ values similar to the three-ring species (Figure 15). Since PAHs are the products of incomplete combustion, the range of isotopic values generated during pyrolysis is related to the isotopic signature of their initial precursors and the fractionations that are associated with primary and secondary reactions noted above. The trend observed in the $\delta^{13}C$ of the three-, four-, and five-ring PAHs in this

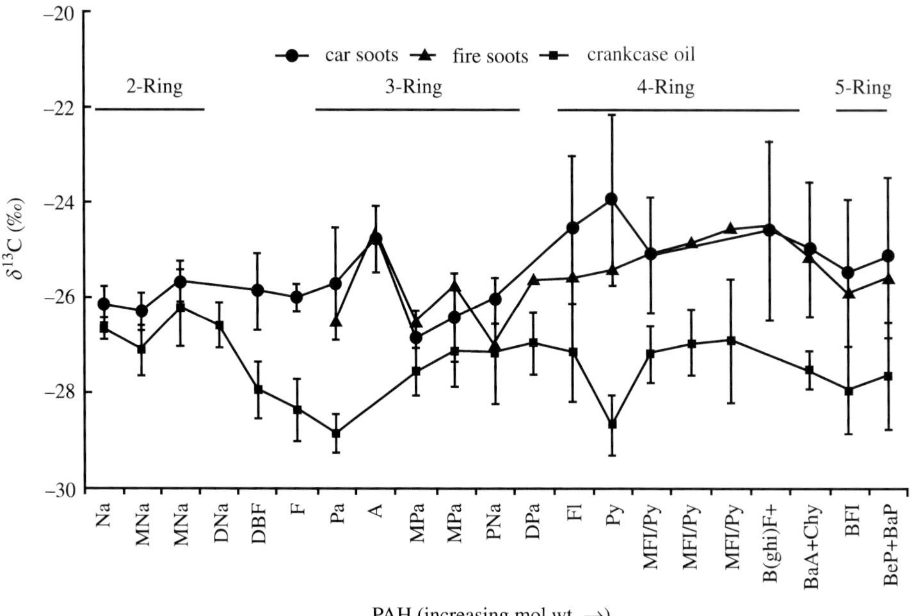

Figure 15 $\delta^{13}C$ of PAHs from car soot (automobile emission; 10 samples), fire soot (wood burning; 11 samples), and crankcase oil (12 samples). Error bars are 2σ variation within each source type (after O'Malley, 1994).

pyrolysis process indicates the following possibilities: (i) the individual compounds were derived from isotopically distinct precursors in the original pyrolysis source that underwent pyrolysis and pyrosynthesis possibly at different temperature ranges; (ii) the overall isotopic variation was primarily dictated by the formation and carbon branching pathways that occurred during pyrosynthesis; or (iii) some combination of (i) and (ii).

Benner *et al.* (1987) analyzed the $\delta^{13}C$ of various wood components and reported that cellulose, hemicellulose, and an uncharacterized fraction (solvent extractable) were, respectively, 1.3‰, 0.3‰, and 1.0‰ more enriched in ^{13}C than the total wood tissue, whereas lignins were 2.6‰ more depleted than the whole plant. Wood combustion in open wood fireplaces could resemble a stepped combustion process where hemicellulose normally degrades first, followed by cellulose and then the lignins. If the carbon-isotopic compositions of hemicellulose, cellulose, and lignin are sole factors in determining the overall isotopic signature of the wood-burning soot, then the $\delta^{13}C$-depleted PAHs may be primarily derived from lignin, while the more enriched compounds may originate from the heavier cellulose or uncharacterized fractions. The uncharacterized fraction of wood also contains certain compounds that are good PAH precursors (e.g., terpenes, fatty acids, and other aromatic compounds including acids, aldehydes, and alcohols).

As an alternative, the trend in the isotopic signature of wood-burning soot PAHs may be controlled by secondary condensation and cyclodehydrogenation reactions during pyrolysis to produce more condensed HMW compounds. Therefore, during the pyrolysis process, LMW compounds, generally formed at lower temperatures, may be actual precursors to the higher molecular weight species that are formed subsequently. Holt and Abrajano (1991) concluded from kerogen partial combustion studies that ^{13}C-depleted carbon is preferentially oxidized during partial combustion leaving a ^{13}C-enriched residue (cf. Currie *et al.*, 1999). Therefore, the PAH-forming radicals produced at lower temperatures are expected to be depleted in ^{13}C compared to radicals formed at higher temperatures. Despite the difference in the combustion/pyrolysis processes, materials, and conditions, the $\delta^{13}C$ of the parental PAHs isolated from car soots follow a similar trend to the wood-burning soot PAHs (Figure 15). This similarity argues for the prevailing role of pyrolytic secondary reactions in imparting the resulting PAH $\delta^{13}C$ values. Unlike open wood burning, the combustion process in the gasoline engine may be described as spontaneous or shock combustion (Howsam and Jones, 1998), whereby gasoline is initially vaporized and mixed with air prior to combustion in the combustion chamber. The PAHs produced during this process are reported to be primarily dependent on the fuel : air ratio and the initial aromaticity of the fuel (Begeman and Burgan, 1970; Jensen and Hites, 1983).

The complex controls on the carbon isotopic composition of produced PAHs during wood burning is also apparent in PAH produced by coal pyrolysis (McRae *et al.*, 1999, 2000; Reddy *et al.*, 2002). PAHs that have been isotopically characterized by McRae *et al.* (1999) showed progressive ^{12}C enrichment in successively higher temperatures of formation (−24‰ to −31‰). One analysis of a coal tar standard reference material (SRM) by Reddy *et al.* (2002) yielded compound-specific $\delta^{13}C$ values in the ^{13}C-enriched end of this range (−24.3‰ to −24.9‰). MacRae *et al.* (1999) reasoned that the isotopic changes observed at increasing temperature reflect the competition between original PAHs in the coal (mostly two and three rings), and those produced by high-temperature condensation and aromatization. Interestingly, the polycondensed higher molecular weight PAHs showed enrichment in ^{12}C, perhaps indicating the incorporation of the early ^{12}C-enriched fragments produced by during pyrolysis (cf. Holt and Abrajano, 1991). Creosote produced from coal tar likewise exhibits a wide range of $\delta^{13}C$ composition (e.g., Hammer *et al.*, 1998), mimicking the observation made by McRae *et al.* (1999).

9.13.6.1.2 Petrogenesis

Carbon isotopic compositions of PAHs in petroleum and petroleum source rocks are among the earliest targets of compound-specific carbon-isotopic analysis. Freeman (1991) analyzed individual aromatic compounds in the Eocene Messel Shale and demonstrated the general tendency of aromatic compounds to be consistently enriched in ^{13}C compared to the aliphatic fraction. This pattern is consistent with diagenetic effects associated with aromatization of geolipids (Sofer, 1984; Simoneit, 1998). As already noted, several examples of diagenetically produced PAHs (e.g., substituted Pa, Chy, retene, and perylene) result from the aromatization of precursors such as pentacyclic triterpenoids (Wakeham *et al.*, 1980b; Mackenzie, 1984; Peters and Moldowan, 1993; Neilson and Hynning, 1998; Simoneit, 1998). The aromatization of natural compounds to produce PAHs is not expected to result in significant isotopic alterations, a contention supported by the successful use of compound-specific carbon-isotope signatures to trace precursor–product relationships in diagenetic systems (e.g., Hayes *et al.*, 1989; Freeman, 1991). It is notable, however, that the range of

$\delta^{13}C$ values to be expected from potential precursors (e.g., triterpenoids) remains to be fully defined. O'Malley (1994) reported depletion in ^{13}C (up to 2‰) methylated naphthalene homologues compared to their corresponding parental compound. The $\delta^{13}C$-depleted isotopic values observed in these petrogenic-associated PAHs may be dictated by the similarly depleted nature of the precursor compounds. Published $\delta^{13}C$ values of petrogenic-associated PAHs are limited, and it is more than likely that further analysis will reveal more variations that would reflect the isotopic composition of precursor compounds and the nature and degree of diagenetic reactions.

In contrast to the wood fireplace and car soot signatures, the $\delta^{13}C$ of the PAHs isolated from crankcase oil were generally depleted in ^{12}C (Figure 15). Also, the range of $\delta^{13}C$ values measured was only marginally enriched compared to the range of bulk isotopic values (−29.6‰ and −26.8‰) reported for virgin crankcase oil. Unlike fuel pyrolysis, where a significant portion of the precursor source materials is initially fragmented into smaller radicals prior to PAH formation, the PAHs in used crankcase oil are largely derived from a series of thermally induced aromatization reactions of natural precursors that exist in the oil. Crankcase oils derived from petroleum consist primarily of naphthenes with minor *n*-alkane content ($<C_{25}$) and traces of steranes and triterpanes (Simoneit, 1998). Of the parental compounds isolated from the crankcase oils, Pa and Py were generally more depleted in ^{13}C than Fl, BaA/Chy, BFl, and BeP/BaP (Figure 15). The difference in the isotopic values of these two groups of parental compounds suggests that they may be derived from different precursors in the oil. The similarity in the isotopic values of Pa and Py indicates that they may be derived from precursor materials, which have a common origin or a common synthesis pathway that is distinct from those followed by other HMW PAHs. The assignment of aromatic structures to their biological precursors in crude oils has proved difficult (Radke, 1987). However, it is now well established that the saturated or partially unsaturated six-membered rings in steroids and polycyclic terpenoids undergo a process of stepwise aromatization (MacKenzie, 1984; Simoneit, 1998).

Our measurements of crude and processed oil products other than crankcase oil are shown in Figure 16, but these measurements are largely restricted to naphthalene and methylated naphthalene (unpublished data). The PAHs isolated from the bulk outboard motor condensate are isotopically very similar to the mean isotopic values of the PAHs in used crankcase oil, except that the outboard motor condensates were generally more enriched in ^{13}C (Figure 16). This similarity to

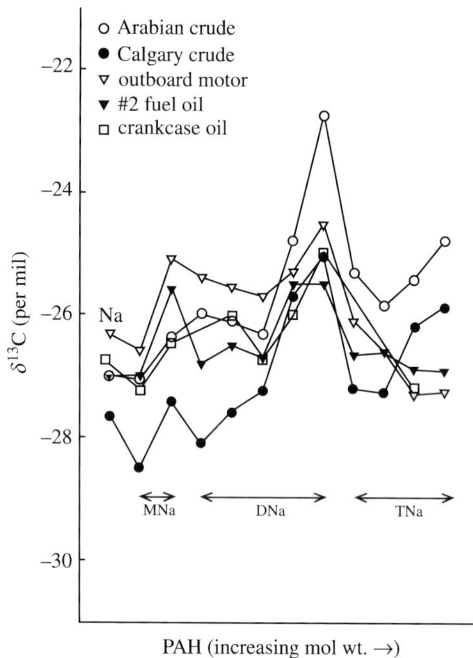

Figure 16 $\delta^{13}C$ of crude oil, outboard motors, and fuel oil. Also shown is the crankcase oil $\delta^{13}C$ signature.

crankcase oils was also observed in their PAH molecular signatures. The apparent enrichment in ^{13}C of naphthalene and its methylated derivatives in the outboard motor samples may be due to the partial combustion of gasoline that is characteristic of two-stroke outboard motor engines. The $\delta^{13}C$ of the PAHs isolated from the crude petroleum samples have similar trends (compound-to-compound variations), except that the Calgary crude PAHs are generally lighter (−32.2‰ to −25.1‰) than the Arabian sample (−30.0‰ to −22.7‰; Figure 16). This variation in isotopic values is possibly related to the origin or maturity of these samples. Differences in the $\delta^{13}C$ values of the isomeric compounds suggest that these compounds may originate from different precursor products or by different formation pathways. Naphthalene and methylated naphthalenes in #2 fuel oil had isotopic values within the range observed for the crude oils and other petroleum products investigated ($\delta^{13}C$). Mazeas and Budzinski (2001) also reported compound-specific $\delta^{13}C$ measurement of an oil sample, and their values fell in the range of − 24‰ to − 29‰. However, their measurements of PAHs from a crude oil SRM showed more enriched $\delta^{13}C$ values (− 22‰ to − 26‰). The latter values are the most ^{12}C-enriched values we have seen in petroleum-related compounds to date.

9.13.6.2 Weathering and Isotopic Composition

Once PAHs are emitted to the environment, the extent of alteration of the isotopic signature during transformation reactions depends on the nature of the carbon branching pathways involved in the reaction. Isotopic fractionation as a result of kinetic mass-dependent reactions (e.g., volatilization or diffusion) is unlikely to be significant, because the reduced mass differences between the light and heavy carbon in these HMW compounds are small. Substitution reactions, resulting in the maintenance of ring aromaticity, are also not likely to significantly alter the isotopic signature of the substrate molecule, although it is possible that the extent of substitution could result in kinetic isotope effects (KIEs). Experiments conducted to examine possible KIEs on biotic and abiotic transformation (e.g., volatilization, sorption, photodegradation, and microbial degradation) have shown the overall resilience of the compound-specific carbon-isotope signatures inherited from PAH sources (O'Malley, 1994; O'Malley *et al.*, 1994; Stehmeir *et al.*, 1999; Mazeas *et al.*, 2002). The HMW of these compounds apparently preempts significant carbon-isotopic discrimination during abiotic transformation processes (e.g., volatilization and photodegradation). For example, no fractionation from photolytic degradation was observed on the $\delta^{13}C$ of standard PAHs experimentally exposed to natural sunlight. The $\delta^{13}C$ of F, Pa, Fl, Py, and BbF during the various exposure periods were particularly stable despite up to a 40% reduction in the initial concentration of pyrene (Figure 17). The absence of significant KIE in the systems studied here may be related to the relatively HMW of the compounds and low reduced mass differences involved.

Microbial degradation experiments likewise showed negligible shifts in the $\delta^{13}C$ values of Na and Fl even after >90% biodegradation (Figure 18) (O'Malley *et al.*, 1994). Bacteria degrade PAHs by enzymatic attack which can be influenced by a number of environmental factors (Cerniglia, 1991). Alteration of the original isotopic values of Na and Fl may have been expected, since most PAH degradation pathways involve ring cleavage (loss of carbon) (Mueller *et al.*, 1990; Cerniglia, 1991). For bacterial degradation to occur, the PAHs are normally incorporated into the bacterial cells where enzymes are produced (van Brummelen *et al.*, 1998; Neilson and Allard, 1998). The pathways by which PAHs enter bacterial cells may dictate whether isotopic alterations will occur as a result of microbial degradation. This pathway is not well understood, but it is postulated that bacteria either ingests PAHs that are associated with organic matter, or because of their lipophilic properties, PAHs may diffuse in solution across the cell wall membrane. Hayes *et al.* (1989) suggested that organisms do not isotopically discriminate between heavy and light carbon when ingesting organic material from homogenous mixtures. In this case, the isotopic ratio of the unconsumed residue will not be different from the starting material. Even if cell wall diffusion was the predominant pathway, isotopic fractionation will again be limited by

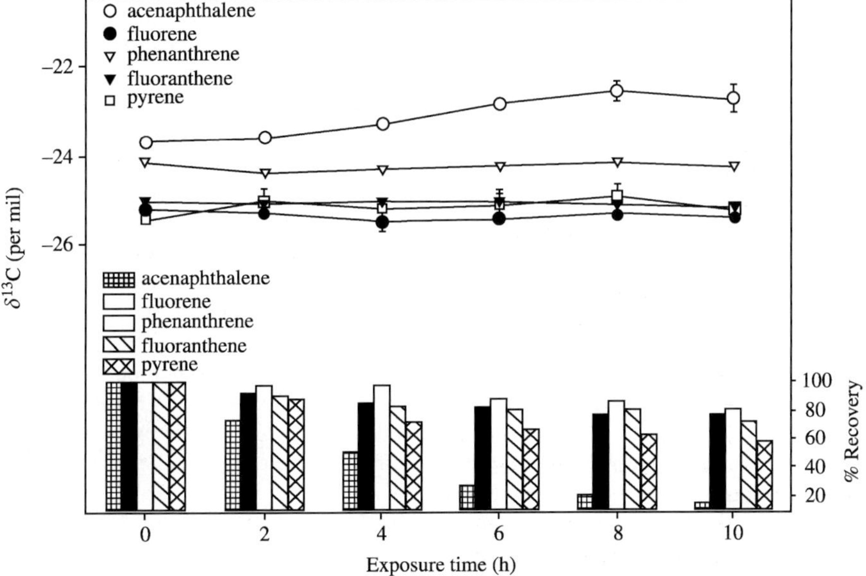

Figure 17 Carbon-isotope effects of PAH photolysis. Top line diagram shows negligible shifts in $\delta^{13}C$ for F, Pa, Fl, and P. Only Ae shows a shift in $\delta^{13}C$ (0.5‰) beyond differences in replicate experiments (1σ = error bar or size of data point). Lower histogram shows percentage of PAHs that were not photodegraded.

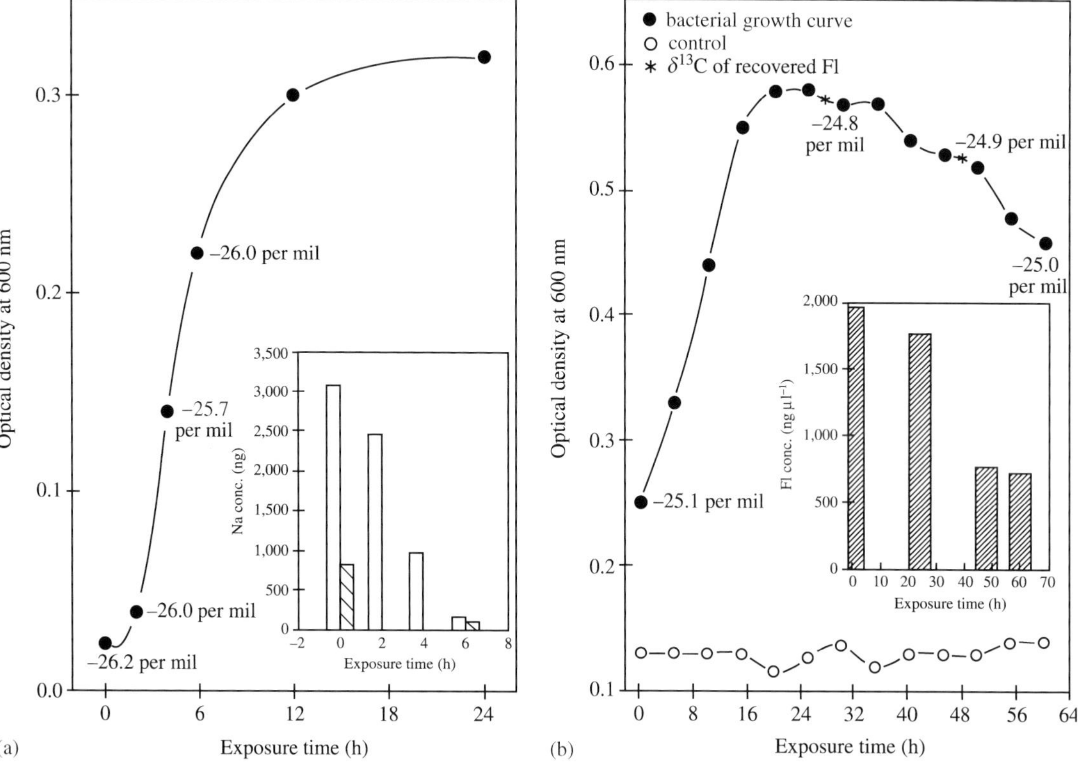

Figure 18 Carbon-isotope effects of PAH biodegradation: (a) biodegradation of Na with *Pseudomonas putida*, Biotype B ATCC 17484 and (b) biodegradation of Fl with *Pseudomonas paucimobilis*, strain EPA 505. Inset shows decline in PAH concentration over the same time period.

the small relative mass difference between ^{12}C and ^{13}C containing PAHs. It is for this reason that most isotopic fractionations in biological systems can be traced back to the initial C_1 (i.e., CO_2 and CH_4) fixation steps. Since no enrichments were observed in the biodegradation residue of naphthalene and fluoranthene, no observable mass discrimination can be attributed to biodegradation for these compounds. If the controlling steps for the biodegradation of higher molecular weight PAHs are similar, isotopic alterations resulting from microbial degradation would likely be insignificant in many natural PAH occurrences. It is notable that recent experiments on naphthalene and toluene show that some carbon-isotopic discrimination can occur under certain conditions (Meckenstock *et al.*, 1999; Diegor *et al.*, 2000; Ahad *et al.*, 2000; Morasch *et al.*, 2001; Stehmeir *et al.*, 1999). Yanik *et al.* (2003) also showed some carbon-isotope fractionation, interpreted as biodegradation isotope effect, in field weathered petroleum PAHs. If the presence of isotopic discrimination during biodegradation is shown for some systems, it is likely contingent on the specific pathways utilized during microbial metabolism. This may be a fruitful area of further inquiry.

9.13.6.3 Isotopic Source Apportionment of PAHs in St. John's Harbor: An Example

The earliest example of using compound-specific carbon-isotope compositions for apportioning PAH sources in aquatic systems is the study of St. John's Harbor, Newfoundland by O'Malley *et al.* (1994, 1996). This is an ideal setting for utilizing carbon isotopes because the molecular evidence is equivocal. The surface sediments in St. John's Harbor were characterized by high but variable concentrations of three-, four-, and five-ring parental PAHs suggestive of pyrogenic sources, the presence of methylated PAHs and a UCM indicative of petrogenic inputs, and occasional presence of retene and perylene, implicating diagenetic input. The relatively low Pa/A ratio (1.9) in the surface sediment, which is indicative of significant pyrolysis contributions, is similar to those calculated in previous studies (Killops and Howell, 1988; Brown and Maher, 1992). The Fl/Py ratio (1.2), indicative of pyrolysis, is comparable to the Fl/Py ratios reported for sediments from other similar systems (e.g., Penobscot Bay, Boston Harbor, Severn Estuary, Bedford Harbor, and Halifax Harbor) (Killops and Howell, 1988; Pruell and Quinn, 1988;

Gearing *et al.*, 1981; Hellou *et al.*, 2002). Similarly, the BaA/Chy ratio (0.80) also indicates that BaA and Chy are of pyrogenic origin. However, unlike Fl and Py, the generally higher BaA/Chy ratio suggests that the primary source of these particular compounds may be related to car emissions. The BaP/BeP and the Pa/MPa ratios of the sediment cannot be clearly reconciled with any of the primary source signatures. This could indicate that these ratios have limited value for source apportionment because of the reactivity of BaP and methylated Pa in the environment (e.g., Canton and Grimalt, 1992).

The reactivity of MPa and BaP seems to have resulted in the alteration of their source compositional ratios during transportation/deposition, and thus may not be a reliable source indicator. In addition to the compositional attributes of the parental PAHs, the presence of methylated compounds and a UCM in the surface sediments indicates possible petrogenic inputs of PAHs to the Harbor. The UCM is generally indicative of petroleum and petroleum products, and is a widely used indicator of petrogenic contamination in sediments (Prahl and Carpenter, 1979; Volkman *et al.*, 1992, 1997; Simoneit, 1998). It is commonly assumed that a UCM consists primarily of an accumulation of multibranched structures that are formed as a result of biodegradation reactions of petroleum (Volkman *et al.*, 1992). Since no clear indication of petroleum-derived inputs can be discerned from the compositional ratios of the prominent PAHs in the sediments, it is apparent that the isomeric ratios of the prominent petrogenic PAHs are masked by pyrogenic-derived components.

There is no characteristic trend in the $\delta^{13}C$ between the LMW and HMW PAHs in the surface sediments of St. John's Harbor. However, the isotopically enriched A and the trend between Pa and MPa are indicative of pyrogenic PAH sources, whereas the marginally depleted Py compared to Fl suggests inputs of petrogenic sources. The isotopic signatures of the PAHs from all the surface samples are broadly similar, and can be quantitatively related to the prominent primary sources (Figure 19). Significant enrichments of Pa, MPa, and A relative to the four- and five-ring compounds are consistently observed in all the individually analyzed samples. This may be attributed to the input of road sweep material via the sewer system. The consistency in the $\delta^{13}C$ of Pa (−25.9‰ to −25.2‰) in the surface sediments suggests that it is possibly derived from an isotopically enriched source of pyrogenic origin. Evidence of some petrogenic input is, nevertheless, indicated by the relatively ^{13}C-depleted pyrene, a trend that was consistently observed in most of the samples investigated. The results of mass balance mixing calculation

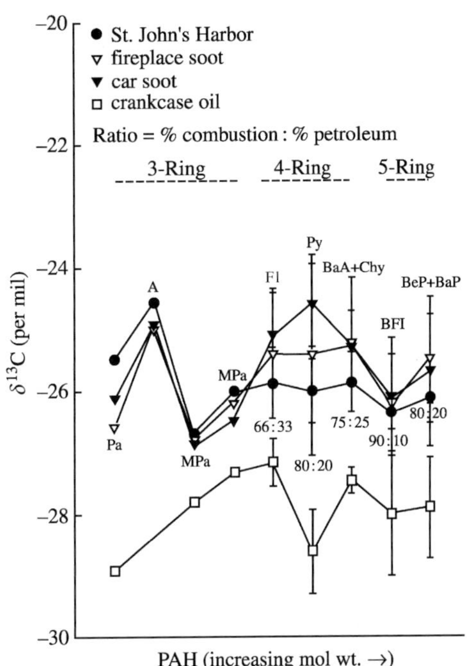

Figure 19 $\delta^{13}C$ of St. John's Harbor sediments compared to the three primary PAH sources in the Harbor. Error bars show 2σ variability in the source $\delta^{13}C$. Ratios shown (e.g., 80 : 20) are the approximate balance of crankcase oil and fireplace soot (assuming two-component mixing) in the Harbor sediments.

are diagrammatically shown in Figure 20. The $\delta^{13}C$ of the four- and five-ring PAHs isolated from the various sediments generally form positive mixing arrays between the wood-burning and car soot and crankcase oil end-members, quantitatively substantiating molecular data that mixtures of these prominent primary sources could explain the range of $\delta^{13}C$ values observed in these particular sediments. The validity of these mixing arrays is further supported by unchanging relative positions of the individual sites (e.g., E, F, and J; Figure 20) in the mixing array (accounting for experimental error) regardless of the projection used. In general, as indicated by the dominance of the three-, four-, and five-ring parental PAHs in the molecular signatures, PAHs of pyrolysis origin seem to be dominant contributors to the Harbor sediments. This is shown by the bulk of the data points, which are concentrated close to the two pyrolysis/combustion end-members (Figure 20). From the position of the mean values on the mixing curves, the average pyrolysis input is estimated to be ~70% while the remaining 30% of the PAHs seem to be derived from crankcase oil contributions. Sites H and J are particularly enriched in pyrolysis-derived PAHs, while the maximum input of crankcase oil (≈50%) consistently occurred at site E indicating the heterogeneous nature of these sediments which was not

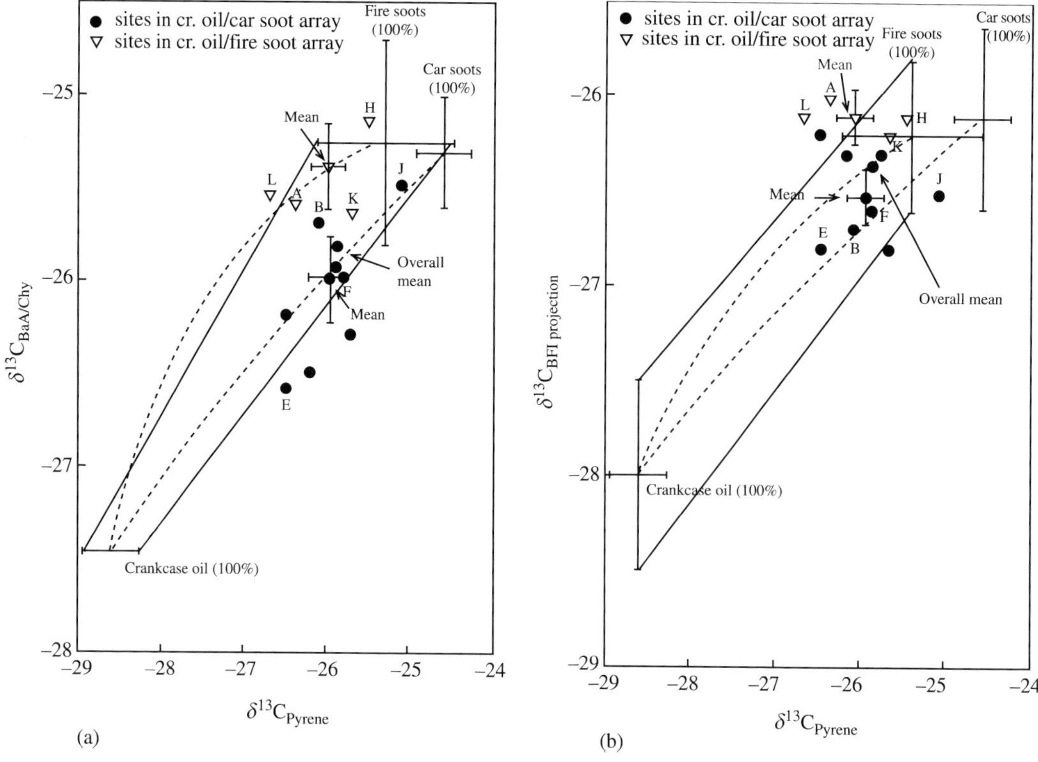

Figure 20 Carbon-isotopic mixing diagram for St. John's Harbor sediments using identified PAH source end-members: (a) $\delta^{13}C_{Pyrene}$ versus $\delta^{13}C_{BaA+Chy}$ and (b) $\delta^{13}C_{Pyrene}$ versus $\delta^{13}C_{BFI}$ projection. Data points are subdivided into retene- (▽) and non-retene-containing samples (●).

evident from the molecular signatures. The percentage input of crankcase oil is similar to the range (20–33%) estimated from Figure 19.

The relative importance of the two primary pyrolysis sources (wood-burning soot versus car soot) as PAH contributors to the sediments is also evident in these mixing curves. With the exception of sites A, H, K, and L, which plot closer to the wood-burning soot/crankcase oil mixing array, all the remaining sites tend to be concentrated close to the car soot/crankcase oil mixing curve (Figure 20). This indicates that a significant portion of the pyrolysis-derived PAHs in the Harbor is of car emission origin. Despite the lack of distinct evidence in the molecular signature data to suggest the importance of car soot emissions in the pyrolysis signatures, inputs of car soot PAHs to the sediments are indicated by the low Pa/A and elevated BaA/Chy compositional ratios. The dominance of car soot PAHs compared to wood-burning soots suggests that a significant part of wood-burning emissions is being transported away from the sampling area due to wind dispersion, and that car soot inputs mainly occur as a result of road runoff. The presence of retene, identified in most of the surface sediments samples, is mainly related to early diagenesis rather than from wood-burning sources, despite having a similar $\delta^{13}C$ value to these sources. This is supported by the apparent absence of retene in the open-road and sewage sample molecular signatures, which are identified to be the predominant pathways for PAH input to the Harbor. There was also a notable absence of retene in the atmospheric air samples collected in the St. John's area. The input of petroleum or petroleum products to the sediments is not surprising since most of the individual samples are characterized by the presence of a prominent UCM.

The prominence of car soot and crankcase oil PAHs in sediments supports the suggestion that road runoff and runoff via the sewer system, and not atmospheric deposition, is the main pathway for PAH input to the Harbor sediments. This is also reflected by the similarity in the isotopic values of the PAHs in the sewage and sediment samples (Figure 21). Input of road sweep materials to the sediments was also supported by the presence of asphalt-like particulates similar to those observed in the open-road sweeps. Due to the absence of major industries in the St. John's area, sewage is dominated by domestic sources. Sewage PAH content is, therefore, predominantly derived from aerially deposited pyrolysis products, car emissions, crankcase oil by direct

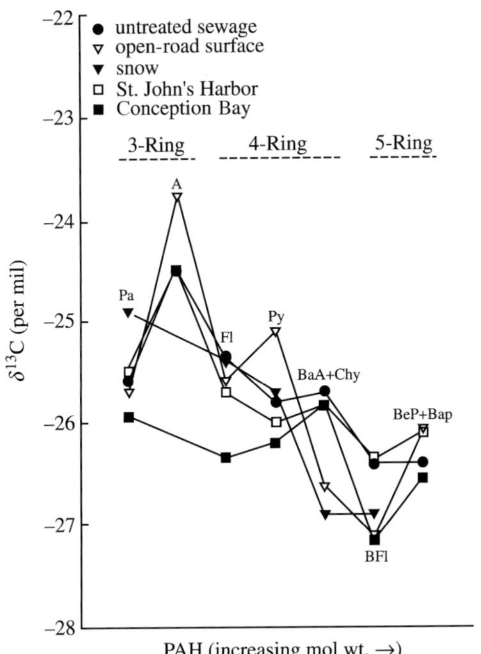

Figure 21 Comparison of PAH $\delta^{13}C$ from the St. John's Harbor sediment, road surface sweep, road-side snow pile, and sewer effluent. Also shown for comparison is the average (18 samples) PAH $\delta^{13}C$ of sediments from a separate harbor outside of the St. John's Harbor watershed (Conception Bay).

spillage and engine loss (NRC, 2002), and road sweep products such as weathered asphalt, tire, and brake wear. The occurrence of early diagenesis in these sediments is supported by the presence of perylene in the deeper samples which is suggested to be derived from diagenetic alteration reactions based on the Py/Pery ratio. Reliable isotopic measurements could not be performed on perylene due to coelution with unidentified HMW biological compounds.

In conclusion, using both the molecular signatures and the $\delta^{13}C$ of the PAHs isolated from St. John's Harbor, sources of pyrogenic and petrogenic PAHs along with diagenetic sources are positively identified. From a combination of the molecular abundance data and the $\delta^{13}C$ of the individual PAHs, car emissions are identified as important contributors of pyrolysis-derived compounds, whereas an average of 30% was derived from crankcase oil contributions. Direct road runoff, runoff via the storm sewers, and snow dumping are identified to be the most likely pathways for PAH input to the Harbor using both the molecular signature and carbon-isotope data. In this particular site, the combined use of molecular and carbon isotopic data elucidated the prominent sources of PAHs in the sediments. However, the use of isotopic data has the added advantage that some quantification of sources can be undertaken with more confidence (O'Malley *et al.*, 1996). It should be noted that these mixing equations used only the more stable higher molecular weight four- and five-ring compounds to avoid uncertainties related to possible weathering fractionations that could affect lower molecular weight compounds.

9.13.7 SYNTHESIS

The unwanted release of petrogenic and pyrogenic hydrocarbon contaminants in the environment represents a major global challenge for pollution prevention, monitoring, and cleanup of aquatic systems. Understanding the geochemical behavior of these compounds in aqueous environments is a prerequisite to identifying point and nonpoint sources, understanding their release history or predicting their transport and fate in the environment. This review examined the physical and chemical properties of HMW hydrocarbons and environmental conditions and processes that are relevant to understanding of their geochemical behavior, with an explicit focus on polycyclic (or polynuclear) aromatic hydrocarbons (PAHs). PAHs represent the components of greatest biological concern amongst pyrogenic and petrogenic hydrocarbon contaminants. The molecular diversity of PAHs is also useful for holistically characterizing crude oil and petrogenic sources and their weathering history in aquatic systems. The majority of PAHs resident in surface aquatic systems, including lake, river, and ocean sediments, and groundwater aquifers are derived from pyrolysis of petroleum, petroleum products, wood, and other biomass materials. Finally, HMW hydrocarbons share the hydrophobic and lipophilic behavior of PAH; hence, all the discussions of the geochemical behavior of PAHs are indeed relevant to HMW hydrocarbons in general.

The molecular distribution and compound-specific carbon-isotopic composition of hydrocarbons can be used to qualify and quantify their sources and pathways in the environment. Molecular source apportionment borrows from molecular methods that were developed and applied extensively for fundamental oil biomarker studies, oil–oil and oil source rock correlation analysis. Additionally, petroleum refinement produces well-defined mass and volatility ranges that are used as indicators of specific petroleum product sources in the environment. Compound-specific carbon-isotopic measurement is a more recent addition to the arsenal of methods for hydrocarbon source apportionment. Carbon isotopic discrimination of *n*-alkanes, biomarkers, and PAHs has shown that the technique is highly complementary to molecular apportionment methods.

HMW hydrocarbons are largely hydrophobic and lipophilic; hence, they become rapidly associated with solid phases in aquatic environments. This solid partitioning is particularly pronounced in organic-matter-rich sediments, with the result that the bulk solid-water partition coefficient is generally a linear function of the fraction of organic carbon or organic matter present in the sediments. Under favorable conditions, organic-rich sediments preserve potential records of the sources and flux of hydrocarbons in a given depositional system. The bulk molecular compositions of hydrocarbons are altered in aquatic environments as a result of selective dissolution, volatilization, sorption, photolysis, and biodegradation. This review has summarized our understanding of the effect of these "weathering reactions" on molecular and isotopic composition. Understanding hydrocarbon weathering is important for understanding sources of past deposition, and they may also provide insights on the timing of hydrocarbon release. Carbon isotope compositions of HMW hydrocarbons are not significantly altered by most known weathering and diagenetic reactions, enabling the use of isotopic signatures as source indicators of highly weathered hydrocarbons.

Several additional areas of research will likely yield major additional advances in understanding of petrogenic and pyrogenic hydrocarbon geochemistry. For example, developments in compound-specific isotope analysis of D/H ratios in individual hydrocarbon compounds will provide additional constraints on the origin, mixing, and transformation of hydrocarbons in shallow aquatic systems. Bacterial degradation experiments similar to those presented in this review have been completed, and no D/H fractionations in naphthalene and fluoranthene were observed (Abrajano, unpublished). This contrasts with large, but bacterial species-dependent, D/H fractionation already observed for toluene (e.g., Ward *et al.*, 2000; Morasch *et al.*, 2001). The dependence of D/H fractionation on the degrading microorganisms and the environment of degradation for mono- and PAHs are clearly fertile areas of further investigation. Similarly, the characterization of D/H signatures of PAH sources is a topic of ongoing interest. Compound-specific measurement of δD of fire soot (12 samples) reveals a wide range from $-40‰$ to $-225‰$, which overlaps with the observed range ($-50‰$ to $-175‰$) for crankcase oil (18 samples) (Abrajano, unpublished). Whereas source discrimination of the type shown for carbon isotopes does not appear promising for hydrogen, the factors governing the acquisition of D/H signatures of PAHs will likely yield unique insight into their pathways and transformation in the environment (e.g., Morasch *et al.*, 2001).

Likewise, the continued development of compound-specific radiocarbon dating techniques ushers an exciting dimension to the study of PAHs and other hydrocarbons in modern sediments (e.g., Eglinton *et al.*, 1997; Eglinton and Pearson, 2001; Pearson *et al.*, 2001; Reddy *et al.*, 2002). In particular, compound-specific radiocarbon determination offers a clear distinction to be made between "new carbon" (e.g., wood burning) and "old carbon" (petroleum and petroleum combustion).

Finally, there is now increasing recognition of a critical need for detailed characterization of the mechanisms of solid phase association of hydrocarbons in sediments or aquifer solid phase. The likelihood that the sorption/desorption behavior, bioavailability, and overall cycling of hydrocarbons depend critically on the nature of solid phase association provides a strong impetus for future work.

ACKNOWLEDGMENTS

Financial support for our continuing studies of hydrocarbon sources and pathways in surface and groundwater systems from the US National Science Foundation (Hydrologic Sciences, EAR-0073912) is gratefully acknowledged. The RPI Isotope Hydrology and Biogeochemistry Facility was also partially supported by the US National Science Foundation (BES-9871241). The St. John's Harbor study was initiated through generous support from the Natural Science and Engineering Research Council and the Department of Fisheries and Oceans (Canada).

REFERENCES

Abdul-Kassim T. A. T. and Simoneit B. R. T. (1995) Petroleum hydrocarbon fingerprinting and sediment transport assessed by molecular biomarker and multivariate statistical analysis in the eastern harbour of Alexandria, Egypt. *Mar. Pollut. Bull.* **30**, 63–73.

Abdul-Kassim T. A. T. and Simoneit B. R. T. (2001) Pollutant-Solid phase Interactions: mechanism, chemistry and Modeling. In *The Handbook of Environmental Chemistry*. Springer, Berlin, vol. 5, part E, 314pp.

Accardi-Dey A. M., and Gschwend P. (2003) Reinterpreting literature sorption data considering both absorption into organic carbon and adsorption onto black carbon. *Environ. Sci. Technol.* **37**, 99–106.

Ahad J. M. E., Lollar B. S., Edwards E. A., Slater G. F., and Sleep B. E. (2000) Carbon isotope fractionation during anaerobic biodegradation of toluene: implications for intrinsic bioremediation. *Environ. Sci. Technol.* **34**, 892–896.

Alsberg T., Stenberg U., Westerholm R., Strandell M., Rannug U., Sundvall A., Romert L., Bernson V., Pettersson B., Toftgard R., Franzen B., Jansson M., Gustafsson J. A., Egeback K. E., and Tejle G. (1985) Chemical and biological characterization of organic material from gasoline exhaust particles. *Environ. Sci. Technol.* **19**, 43–49.

Atlas R. M. and Bartha R. (1992) Hydrocarbon biodegradation and oil spill bioremediation. *Adv. Microb. Ecol.* **12**, 287–338.

Badger G. M., Buttery R. G., Kimber R. W., Lewis G. E., Moritz A. G., and Napier I. M. (1958) The formation of aromatic hydrocarbons at high temperatures: Part 1. Introduction. *J. Chem. Soc.* **1958**, 2449–2461.

Baek S. O., Field R. A., Goldstone M. E., Kirk P. W., Lester J. N., and Perry R. (1991) A review of atmospheric polycyclic aromatic hydrocarbons: sources, fate and behaviour. *Water Air Soil Pollut.* **60**, 279–300.

Baker J. E. and Eisenreich S. J. (1990) Concentrations and fluxes of polycyclic aromatic hydrocarbons and polychlorinated biphenyls across the air—water interface of lake superior. *Environ. Sci. Technol.* **24**, 342–352.

Ballentine D. C., Macko S. A., Turekian V. C., Gilhooly W. P., and Martincigh B. (1996) Tracing combustion-derived atmospheric pollutants using compound-specific carbon isotope analysis. *Org. Geochem.* **25**, 97–108.

Barnsley E. A. (1975) The bacterial degradation of fluoranthene and benzo(a)pyrene. *Can. J. Microbiol.* **21**, 1004–1008.

Beek B. (2001) The Handbook of Environmental Chemistry. In *Park K: Biodegradation and Persistence* Springer, Berlin, vol. 2.

Begeman C. R. and Burgan J. C. (1970) Polynuclear hydrocarbon emission from automotive engines. SAE Paper 700469, Society of Automotive Engineers, Detroit, MI.

Behymer T. D. and Hites R. A. (1985) Photolysis of polycyclic aromatic hydrocarbons adsorbed on simulated atmospheric particulates. *Environ. Sci. Technol.* **19**, 1004–1006.

Behymer T. D. and Hites R. A. (1988) Photolysis of polycyclic aromatic hydrocarbons adsorbed on fly ash. *Environ. Sci. Technol.* **22**, 1311–1319.

Benner R., Fogel M. L., Sprague E. K., and Hodson R. E. (1987) Depletion of ^{13}C in lignin and its implications for stable carbon isotope studies. *Nature* **329**, 708–710.

Bieger T., Hellou J., and Abrajano T. A. (1996) Petroleum biomarkers as tracers of lubricating oil contamination. *Mar. Pollut. Bull.* **32**, 270–274.

Bjorseth A. and Ramdahl T. (1983) Sources and emissions of PAHs. In *Handbook of Polycyclic Aromatic Hydrocarbons Emission Sources and Recent Progress in Analytical Chemistry*. Dekker, 432pp.

Blumer M., Blumer W., and Reich T. (1977) Polycyclic aromatic hydrocarbons in soils of a mountain valley: correlation with highway traffic and cancer incidence. *Environ. Sci. Technol.* **11**(12), 1082–1084.

Bosma T. N. P., Harms H., and Zehnder J. B. (2001) Biodegradation of xenobiotics in the environment and technosphere. In *The Handbook of Environmental Chemistry, Park K: Biodegradation and Persistence* (ed. B. Beek). Springer, Berlin, vol. 63, 202p.

Broman D., Coimsjo A., Naf C., and Zebuhr Y. (1988) A multi-sediment trap study on the temporal and spatial variability of polycyclic aromatic hydrocarbons and lead in an anthropogenic influenced archipelago. *Environ. Sci. Technol.* **22**(10), 1219–1228.

Brown G. and Maher W. (1992) The occurrence, distribution and sources of polycyclic aromatic hydrocarbons in the sediments of the georges river estuary, Australia. *Org. Geochem.* **18**(5), 657–658.

Bucheli T. D. and Gustafsson O. (2000) Quantification of the soot-water distribution coefficient of PAHs provides mechanistic basis for enhanced sorption observations. *Environ. Sci. Technol.* **34**, 5144–5151.

Canton L. and Grimalt J. O. (1992) Gas chromatographic-mass spectrometric characterization of polycyclic aromatic hydrocarbon mixtures in polluted coastal sediments. *J. Chromatogr.* **607**, 279–286.

Cerniglia C. E. (1991) Biodegradation of Organic Contaminants in Sediments: overview and examples with polycyclic aromatic hydrocarbons. In *Organic Substances and Sediments in Water* (ed. R. A. Baker). Lewis Publishers, Chelsea, MI, vol. 3, 267–281.

Chiou C. T. and Kyle D. E. (1998) Deviations from sorption linearity on soils of polar and nonpolar organic compounds at low relative concentrations. *Environ. Sci. Technol.* **32**, 338–343.

Chiou C. T., Peters L. J., and Freed V. H. (1979) Physical concept of soil-water equilibria for nonionic organic compounds. *Science* **206**, 831–832.

Chiou C. T., Porter P. E., and Schmeddling D. W. (1983) Partition equilibria of nonionic organic compounds between soil organic matter and water. *Environ. Sci. Technol.* **17**, 227–231.

Chiou C. T., McGroddy S. E., and Kile D. E. (1998) Partition charateristics of polycyclic aromatic hydrocarbons on soils and sediments. *Environ. Sci. Technol.* **32**, 264–269.

Currie L. A., Klouda G. A., Benner B. A., Garrity K., and Eglinton T. I. (1999) Isotopic and molecular marker validation, including direct molecular 'dating' (GC/AMS). *Atmos. Environ.* **33**, 2789–2806.

Diegor E., Abrajano T., Patel T., Stehmeir L., Gow J., and Winsor L. (2000) Biodegradation of aromatic hydrocarbons: microbial and isotopic studies, goldschmidt conference (Cambridge publications). *J. Conf. Abstr.* **5**, 350.

Douglas G. S., Bence E. A., Prince R. C., McMillen S. J., and Butler E. L. (1996) Environmental stability of selected petroleum hydrocarbon source and weathering ratios. *Environ. Sci. Technol.* **30**, 2332–2339.

Dowling L. M., Boreham C. J., Hope J. M., Marry A. M., and Summons R. E. (1995) Carbon isotopic composition of ocean transported bitumens from the coastline of Australia. *Org. Geochem.* **23**, 729–737.

Eganhouse R. P. (ed.) (1997) Molecular markers in environmental geochemistry. In *ACS Symposium Series*, American Chemical Society, Washington, DC, 426p.

Eglinton T. I. and Pearson A. (2001) Ocean Process tracers: single compound radiocarbon measurements. In *Encyclopedia of Ocean Sciences*. Academic Press, London.

Eglinton T. I., Benitez-Nelson B. C., Pearson A., McNichol A. P., Bauer J. E., and Druffel E. R. M. (1997) Variability in radiocarbon ages of individual organic compounds from marine sediments. *Science* **277**, 796–799.

Eisenreich S. J. and Strachan W. M. J. (1992) Estimating atmospheric deposition of toxic substances to the Great Lakes—an update. In *A Workshop Held at Canadian Centre for Inland Waters, Burlington, Ontario, Jan. 31–Feb. 2, (1992)*, 29pp.

Ellison G. B., Tuck A. F., and Vaida V. (1999) Atmospheric processing of organic aerosols. *J. Geophys. Res.* **104**, 11633–11641.

Engel M. H. and Macko S. A. (1993) Organic Geochemistry. In *Principles and Applications*. Plenum, New York, 861p.

Faksness L. G., Weiss H. M., and Daling P. S. (2002) *Revision of the Nordtest Methodology for Oil Spill Identification*. Sintef Applied Chemistry Report STF66 A02028, N-7465 Trondheim, Norway, 110p.

Fingas M. F. (1995) A literature review of the physics and predictive modeling of oil spill evaporation. *J. Hazards Mater.* **42**, 157–175.

Freeman K. H. (1991) The carbon isotopic composition of individual compounds from ancient and modern depositional environments. PhD Thesis, Indiana University, pp. 42–92.

Freeman D. J. and Cattell F. C. (1990) Wood burning as a source of polycylic aromatic hydrocarbons. *Environ. Sci. Technol.* **24**, 1581–1585.

Fukushima M., Oba K., Tanaka S., Kayasu K., Nakamura H., and Hasebe K. (1997) Elution of pyrene from activated carbon into an aqueous system containing humic acid. *Environ. Sci. Technol.* **31**, 2218–2222.

Gearing P. J., Gearing J. N., Pruell R. J., Wade T. L., and Quinn J. G. (1980) Partitioning of no. 2 fuel oil in controlled estuarine ecosystems, sediments and suspended particulate matter. *Environ. Sci. Technol.* **14**, 1129.

Gibson D. T. and Subramanian V. (1984) Microbial degradation of aromatic hydrocarbons. In *Microbial Degradation of Organic Compounds* (ed. D. T. Gibson). Dekker, New York, pp. 181–252.

Goldberg E. D. (1985) *Black Carbon in the Environment: Properties and Distribution.* Wiley, New York, 198p.

Gordon R. J. (1976) Distribution of airborne polycyclic aromatic hydrocarbons throughout Los Angeles. *Environ. Sci. Technol.* **10**, 370–373.

Gustafsson O. and Gscwend P. M. (1997) Soot as a strong partition medium for polycyclic aromatic hydrocarbons in aquatic systems. In *1997, Molecular Markers in Environmental Geochemistry, ACS Symposium Series* (ed. R. P. Eganhouse). American Chemical Society, Washington, DC, pp. 365–381.

Gustafson K. E. and Dickhut R. M. (1997) Gaseous exchange of polycyclic aromatic hydrocarbons across the air-water interface of southern chesapeake bay. *Environ. Sci. Technol.* **31**, 1623–1629.

Hammer B. T., Kelley C. A., Coffin R. B., Cifuentes L. A., and Mueller J. G. (1998) δ^{13}C values of polycyclic aromatic hydrocarbons collected from two creosote-contaminated sutes. *Chem. Geol.* **152**, 43–58.

Hase A. and Hites R. A. (1976) On the origins of polycyclic aromatic hydrocarbons in recent sediments: biosynthesis by anaerobic bacteria. *Geochim. Cosmochim. Acta* **40**, 1141–1143.

Hayes J. M., Freeman K. H., Popp B. N., and Hoham C. H. (1989) Compound-specific isotopic analysis: a novel tool for reconstruction of ancient biogeochemical processes. *Adv. Org. Geochem.* **16**(4–6), 1115–1128.

Heider J., Spormann A. M., Beller H. R., and Widdel F. (1999) Anaerobic bacterial metabolism of hydrocarbons. *FEMS Microbiol. Rev.* **22**, 459–473.

Hellou J., Steller S., Leonard J., and Albaiges J. (2002) Alkanes, terpanes, and aromatic hydrocarbons in surficial sediments of Halifax Harbor. *Polycyc. Aromat. Comp.* **22**, 631–641.

Hinga K. R. (1984) The fate of polycyclic aromatic hydrocarbons in enclosed marine ecosystems. PhD Thesis, University of Rhode Island.

Hoffman E. J., Mills G. L., Latimer J. S., and Quinn J. G. (1984) Urban runoff as a source of polycyclic hydrocarbons to coastal environments. *Environ. Sci. Technol.* **18**, 580.

Holt B. and Abrajano T. A. (1991) Chemical and isotopic alteration of organic matter during stepped combustion. *Anal. Chem.* **63**, 2973–2978.

Hostetler F. and Kvenvolden K. (2002) Forensic implications of spilled oil biodegradation in anaerobic environments. *Geol. Soc. Am. Abstr. Prog.*, Denver, Colorado, October, 2002.

Howsam M. and Jones K. C. (1998) Sources of PAHs in the environment. In *PAHs and Related Compounds the Handbook of Environmental Chemistry* (ed. A. H. Neilson). Springer, Berlin, vol. 3, part I, pp. 138–174.

Jensen T. E. and Hites R. A. (1983) Aromatic diesel emissions as function of engine conditions. *Anal. Chem.* **55**, 594–599.

Kaplan I. R., Galperin Y., Lu S. T., and Lee R. P. (1997) Forensic environmental geochemistry: differentiation of fuel types, their sources and release time. *Org. Geochem.* **27**, 289–317.

Karapanagioti H., Kleineidam S., Sabatini D., Grathwohl P., and Ligouis B. (2000) Impacts of heterogeneous organic matter on phenanthrene sorption: equilibrium and kinetic studies with aquifer material. *Environ. Sci. Technol.* **34**, 406–414.

Karickhoff S. W. (1981) Semi-empirical estimation of sorption of hydrophobic pollutants on natural sediments and soils. *Chemosphere* **10**, 833–846.

Karickhoff S. W., Brown D. S., and Scott T. A. (1979) Sorption of hydrophobic pollutants on natural sediments. *Water Res.* **13**, 241–248.

Kennicutt M. C., II, Brooks J. M., and McDonald T. J. (1991) Origins of hydrocarbons in bering sea sediments: I. Aliphatic hydrocarbons and fluorescence. *Org. Geochem.* **17**(1), 75–83.

Killops S. D. and Howell V. J. (1988) Sources and distribution of hydrocarbons in Bridgewater Bay (Severn Estuary, UK) intertidal surface sediments. *Estuar. Coast. Shelf Sci.* **27**, 237–261.

Kleineidam S., Rugner H., Ligouis B., and Grathwol P. (1999) Organic matter facie and equilibrium sorption of phenanthrene. *Environ. Sci. Technol.* **33**, 1637–1644.

Kleineidam S., Schuth C. H., and Grathwol P. (2002) Solubility-normalized combined adsorption-partitioning sorption isotherms for organic pollutants. *Environ. Sci. Technol.* **36**, 4689–4697.

Knap A. H. and Williams P. J. L. (1982) Experimental studies to determine the fate of petroleum hydrocarbons in refinery effluent on an estuarine system. *Environ. Sci. Technol.* **16**, 1–4.

Kvenvolden K. A., Hostettler F. D., Rapp J. B., and Carlson P. R. (1993a) Hydrocarbons in oil residue on beaches of islands of Prince William Sound, Alaska. *Mar. Pollut. Bull.* **26**, 24–29.

Kvenvolden K. A., Carlson P. R., Threlkeld C. N., and Warden A. (1993b) Possible connection between two Alaskan catastrophies occurring 25 yr apart (1964 and 1989). *Geology* **21**, 813–816.

Kvenvolden K. A., Hostettler F. D., Carlson P. R., Rapp J. B., Threlkeld C., and Warden A. (1995) Ubiquitous tar balls with a California source signature on the shorelines of Prince William Sound, Alaska. *Environ. Sci. Technol.* **29**(10), 2684–2694.

Laflamme R. E. and Hites R. A. (1978) The global distribution of polycyclic aromatic hydrocarbons in recent sediments. *Geochim. Cosmochim. Acta* **42**, 289–303.

Lake J. L., Norwood C., Dimock D., and Bowen R. (1979) Origins of polycyclic aromatic hydrocarbons in estuarine sediments. *Geochim. Cosmochim. Acta* **43**, 1847–1854.

Lane D. A. (1989) The fate of polycyclic aromatic compounds in the atmosphere during sampling. In *Chemical Analysis of Polycyclic Aromatic Compounds, Chemical Analysis Series.* (ed. Vo-Dinh). Wiley-Interscience, vol. 101, pp 30–58.

Latimer J. S., Hoffman E. J., Hoffman G., Fasching J. L., and Quinn J. G. (1990) Sources of petroleum hydrocarbons in urban runoff. *Water Air Soil Pollut.* **52**, 1–21.

Lee M. L., Novotny M. V., and Bartle K. D. (1981) *Analytical Chemistry of Polycyclic Aromatic Compounds.* Academic Press, New York.

Lee R. F., Gardner W. S., Anderson J. W., Blaylock J. W., and Barwell C. J. (1978) Fate of polycyclic aromatic hydrocarbons in controlled ecosystems enclosures. *Environ. Sci. Technol.* **12**, 832–838.

Lima A. C., Eglinton T. I., and Reddy C. M. (2003) High resolution record of pyrogenic polycyclic aromatic hydrocarbon deposition during the 20th Century. *Environ. Sci. Technol.* **37**, 53–61.

Lipiatou E. and Saliot A. (1991) Fluxes and transport of anthropogenic and natural polycyclic aromatic hydrocarbons in the western Mediterranean sea. *Mar. Chem.* **32**, 51–71.

Lohmann F., Trendel J. M., Hetru C., and Albrecht P. (1990) C-29 trtiated β amyrin: chemical synthesis aiming at the study of aromatization processes in sediments. *J. Label. Comp. Radiopharm.* **28**, 377–386.

Lun R., Lee K., De Marco L., Nalewajko C., and Mackay D. (1998) A model of the fate of polycyclic aromatic hydrocarbons in the saguenay ford. *Environ. Toxicol. Chem.* **17**, 333–341.

Luthy R. G., Aiken G. R., Brusseau M. L., Cunningham S. D., Gschwend P. M., Pignatello J. J., Reinhard M., Traina S. J., Weber W. J., and Westall J. (1997) Sequestration of hydrophobic organic contaminants by geosorbents. *Environ. Sci. Technol.* **31**, 3341–3347.

MacKay A. A. and Gschwend P. (2001) Enhanced concentration of PAHs in a coal tar site. *Environ. Sci. Technol.* **31**(35), 1320–1328.

Mackay D. (1998) Multimedia mass balance models of chemical distribution and fate. In *Chapter 8 of Ecotoxicology* (eds. G. Schuurmann and B. Markert). Wiley, NY and Spektrum, Berlin, pp. 237–257.

Mackay D. and Callcott D. (1998) Partitioning and physical chemical properties of PAHs. PAHs and Related Compounds. In *The Handbook of Environmental Chemistry, Volume 3, Part I,* (ed. A.H. Neilson). Springer, Berlin, Heidelberg, pp. 325–346.

Mackay D., Shiu W. Y., and Ma K. C. (1992). *Illustrated Handbook of Physical–Chemical Properties and Environmental Fate for Organic Chemicals, volume II, 1992.* Lewis, Boca Raton.

MacKenzie A. S. (1984) Application of biological markers in petroleum geochemistry. In *Advances in Petroleum Geochemistry* (eds. J. Brooks and D. H. Welte). Academic Press, London, vol. 1, pp. 115–214.

Mansuy L., Philp R. P., and Allen J. (1997) Source identification of oil spills based on the isotopic composition of individual components in weathered oil samples. *Environ. Sci. Technol.* **31**, 3417–3425.

Mazeas L. and Budzinski H. (2001) Polycyclic aromatic hydrocarbon $^{13}C/^{12}C$ ratio measurement in petroleum and marine sediments: application to standard reference material and a sediment suspected of contamination from Erika oil spill. *J. Chromatogr. A* **923**, 165–176.

Mazeas L., Budzinski H., and Raymond N. (2002) Absence of stable carbon isotope fractionation of saturated and polycyclic aromatic hydrocarbons during aerobic bacterial degradation. *Org. Geochem.* **33**, 1259–1272.

McElroy A. E., Farrington J. W., and Teal J. M. (1985) Bioavailability of polycyclic aromatic hydrocarbons in the aquatic environment. In *Metabolism of Polycyclic Aromatic Hydrocarbons in the Aquatic Environment* (ed. U. Varanasi). CRC Press, Boca Raton, pp. 1–39.

McRae C., Sun C. G., Snape C. E., Fallick A., and Taylor D. (1999) $\delta^{13}C$ values of coal-derived PAHs from different processes and their application to source apportionment. *Org. Geochem.* **30**, 881–889.

McRae C., Snape C. E., Sun C. G., Fabbri D., Tartari D., Trombini C., and Fallick A. (2000) Use of compound-specific stable isotope analysis to source anthropogenic natural gas-derived polycyclic aromatic hydrocarbons in lagoon sediments. *Environ. Sci. Technol.* **34**, 4684–4686.

McVeety B. D. and Hites R. A. (1988) Atmospheric deposition of polycyclic aromatic hydrocarbons to water surfaces: a mass balance approach. *Atmos. Environ.* **22**(3), 511–536.

Meador J., Stein J., Reichert W., and Varanasi U. (1995) Bioaccumulation of polycyclic aromatic hydrocarbons by marine organisms. *Rev. Environ. Contamin. Toxicol.* **143**, 80–164.

Means J. C. and Wijayaratne R. (1982) Role of natural colloids in the transport of hydrophobic pollutants. *Science* **215**, 968.

Meckenstock R. U., Morasch B., Wartmann R., Schinck B., Annweller E., Michaelis W., and Richnow H. H. (1999) $^{13}C/^{12}C$ isotope fractionation of aromatic hydrocarbons during microbial degradation. *Environ. Microbiol.* **1**, 409–414.

Moldowan J. M., Dahl J. E., McCaffrey M. A., Smith W. J., and Fetzer J. C. (1995) Application of biological marker technology to bioremediation of refinery by-products. *Energy Fuels* **9**(1), 155–162.

Moore J. W. and Ramamoorthy S. (1984) *Organic Chemicals in Natural Waters Applied Monitoring and Impact Assessment.* Springer, NY, 289pp.

Morasch B., Schink B., Richnow H., and Meckenstock R. U. (2001) Hydrogen and carbon isotope fractionation upon bacterial toluene degradation: mechanistic and environmental aspects. *Appl. Environ. Microbiol.* **67**, 4842–4849.

Mueller J. G., Chapman P. J., and Pritchard P. H. (1988) Action of a fluoranthene-utilizing bacterial community on polycyclic aromatic hydrocarbons components of Creosote. *Appl. Environ. Microbiol.* **55**(12), 3085–3090.

Mueller J. G., Chapman P. J., Blattmann B. O., and Pritchard P. H. (1990) Isolation and characterization of a flouranthene-utilizing strain of *Pseudomonas paucimobilis. Appl. Environ. Microbiol.* **56**(4), 1079–1086.

Munoz D., Guiliano M., Doumenq P., Jacquot F.,‘ Scherrer P., and Mille G. (1997) Long term evolution of petroleum biomarkers in mangrove soil (Guadaloupe). *Mar. Pollut. Bull.* **34**, 868–874.

Neff J. M. (1979) *Polycyclic Aromatic Hydrocarbons in the Aquatic Environment. Sources, Fates and Biological Effects.* Applied Science Publishers, London.

Neilson A. H. (1998) *PAHs and Related Compounds Biology.* Springer, NY, 386p.

Neilson A. H. and Allard A. S. (1998) Microbial metabolism of PAHs and heteroarenes. In *The Handbook of Environmental Chemistry.* (ed. A. H. Neilson) Springer, Berlin, Heidelberg, vol. 3, part I, pp. 2–80.

Neilson A. and Hynning P. (1998) PAHs: Products of chemical and biochemical transformation of alicyclic precursors. In *The Handbook of Environmental Chemistry* (ed. A. H. Neilson). Springer, Berlin, Heidelberg, vol. 3, part 1, pp. 224–273.

NRC (1985) Oil in the Sea: Inputs, Fates, and Effects. Committee on Oil in the Sea: Inputs, Fates, and Effects, National Research Council, 601p.

NRC (2000) Natural Attenuation for Groundwater Remediation, Committee on Intrinsic Remediation, Water Science and Technology Board,. Board on Radioactive Waste Management, National Research Council, 292pp.

US National Research Council (NRC) (2002) Oil in the Sea III: Inputs, Fates, and Effects. Committee on Oil in the Sea: Inputs, Fates, and Effects, National Research Council, 446p.

Offenberg J. H. and Baker J. E. (1999) Aerosol size distributions of polycyclic aromatic hydrocarbons in urban and over-water atmospheres. *Environ. Sci. Technol.* **33**, 3324–3331.

O'Malley V. P. (1994) Compound-specific carbon isotope geochemistry of polycyclic aromatic hydrocarbons in eastern newfoundland estuaries. PhD Thesis, Memorial University of Newfoundland.

O'Malley V. P., Burke R. A., and Schlotzhauer W. S. (1997) Using GC-MS/Combustion/IRMS to determine the $^{13}C/^{12}C$ ratios of individual hydrocarbons produced from the combustion of biomass materials—application to biomass burning. *Org. Geochem.* **27**, 567–581.

O'Malley V., Abrajano T. A., and Hellou J. (1994) Determination of $^{13}C/^{12}C$ Ratios of individual PAHs from environmental samples: can PAHs sources be source apportioned? *Org. Geochem.* **21**, 809–822.

O'Malley V., Abrajano T. A., and Hellou J. (1996) Isotopic apportionment of individual polycyclic aromatic hydrocarbon sources in St. Johns Harbour. *Environ. Sci. Technol.* **30**, 634–638.

Oostdam B. L. (1980) Oil pollution in the persian gulf and approaches 1978. *Mar. Pollut. Bull.* **11**, 138–144.

Paalme L., Irha N., Urbas E., Tsyban A., and Kirso U. (1990) Model studies of photochemical oxidation of carcinogenic polyaromatic hydrocarbons. *Mar. Chem.* **30**, 105–111.

Pancirov R. J. and Brown R. A. (1975). Analytical methods for PAHs in crude oils, heating oils and marine tissues. In *Conference Proceedings on Prevention and Control of Oil Pollution.* Washington, DC, API, 103–113.

Payne J. R. and Phillips C. R. (1985) Photochemistry of petroleum in water. *Environ. Sci. Technol.* **19**, 569.

Pearson A., McNichol A. P., Benitez-Nelson B. C., Hayes J. M., and Eglinton T. I. (2001) Origins of lipid biomarkers in santa monica basin surface sediment: a case study using compound-specific Δ14C analysis. *Geochim. Cosmochim. Acta* **65**, 3123–3137.

Pedersen P. S., Ingwersen J., Nielsen T., and Larsen E. (1980) Effects of fuel, lubricant, and engine operating parameters on the emission of polycyclic aromatic hydrocarbons. *Environ. Sci. Technol.* **14**(1), 71–79.

Peters K. E. and Moldowan J. M. (1993) The Biomarker Guide, Interpreting molecular fossils in petroleum and ancient sediments. Prentice-Hall, Englewood Cliffs, NJ.

Peters K. E., Moldowan J. M., Schoell M., and Hempkins W. B. (1986) Petroleum isotopic and biomarker composition related to source rock organic matter and depositional environment. *Org. Geochem.* **10**, 17–27.

Peters K. E., Scheuerman G. L., Lee C. Y., Moldowan J. M., Reynolds R. N., and Pena M. M. (1992) Effects of refinery processes on biological markers. *Energy Fuels* **6**, 560–577.

Pierce R. C. and Katz M. (1975) Dependency of polynuclear aromatic hydrocarbon content on size distribution of atmospheric aerosols. *Environ. Sci. Technol.* **9**, 347–353.

Prahl F. G. and Carpenter R. (1979) The role of zooplankton fecal pellets in the sedimentation of polycyclic aromatic hydrocarbons in dabob bay, Washington. *Geochim. Cosmochim. Acta* **43**, 1959–1972.

Prince R. C. (1993) Petroleum oil spill bioremediation in marine environments. *Crit. Rev. Microbiol.* **19**, 217–239.

Pruell J. R. and Quinn J. G. (1988) Accumulation of polycyclic aromatic hydrocarbons in crankcase oil. *Environ. Pollut.* **49**, 89–97.

Radding S. B., Mill T., Gould C. W., Liu D. H., Johnson H. L., Bomberger D. C., and Fojo C. V. (1976) The environmental fate of selected polynuclear aromatic hydrocarbons. US Environmental Protection Agency, EPA560/5-75-009.

Radke M. (1987) Organic Geochemistry of aromatic hydrocarbons. In *Advances in Petroleum Geocheistry* (eds. J. Brooks and D. H. Welte). Academic Press, New York, vol. 2, pp. 141–207.

Ramaswami A. and Luthy R. G. (1997) Mass transfer and bioavailability of PAHs compounds in coal tar NAPL-slurry systems: 1. Model development. *Environ. Sci. Technol.* **31**, 2260–2267.

Readman J. W., Mantoura R. F. C., Rhead M. M., and Brown L. (1982) Aquatic distribution and heterotrophic degradation of polycyclic aromatic hydrocarbons (PAHs) in the tamer estuary. *Estuar. Coast. Shelf Sci.* **14**, 369–389.

Reddy C. and Quinn J. G. (1997) Environmental chemistry of benzothiazoles derived from rubber. *Environ. Sci. Technol.* **31**, 2847–2853.

Reddy C., Reddy C. M., Pearson A., Xu L., McNichol A., Benner B. A., Wise S. A., Klouda G. A., Currie L. A., and Eglinton T. I. (2002) Radiocarbon as a tool to apportion the sources of polycyclic aromatic hydrocarbons and black carbon in environmental samples. *Environ. Sci. Technol.* **36**, 1774–1787.

Reyes C. S., Medina M., Crespo-Hernandez C., Cedeno M. Z., Arce R., Rosario O., Steffenson D. M., Ivanov I. N., Sigman M. E., and Dabestani R. (2000) Photochemistry of pyrene on unactivated and activated silica surfaces. *Environ. Sci. Technol.* **34**, 415–421.

Rogge W. F., Hildemann L. M., Mazurek M. A., Cass G. R., and Simoneit B. R. (1993) *Environ. Sci. Technol.* **27**, 1892–1904.

Rothermich M. M., Hayes L. A., and Lovley D. (2002) Anaerobic, sulfate-dependent degradation of polycyclic aromatic hydrocarbons in petroleum-contaminated harbor sediment. *Environ. Sci. Technol.* **36**, 4811–4817.

Schneider J., Grosser R., Jayasimhulu K., Xue W., and Warshawsky D. (1996) Degradation of pyrene, benz(a)-anthracene, and benz(a)pyrene by *Mycobacterium* sp. Strain RGHII-135, isolated from a former coal gasification site. *Appl. Environ. Microbiol.* **157**, 7–12.

Schoell M. (1984) Stable isotopes in petroleum research. In *Advances in Petroleum Geochemistry* (eds. J. Brooks and D. H. Welte). Academic Press, London, vol. 1, pp. 215–245.

Schwarzenbach R. P., Gschwend P. M., and Imboden D. M. (1993) *Environmental Organic Chemistry*. Wiley, NY, 681p.

Schwarzenbach R. P., Haderlein S. B., Muller S. R., and Ulrich M. M. (1998) Assessing the dynamic behavior of organic contaminants in natural waters. In *Perspectives in Environmental Chemistry* (ed. D. L. Macalady). Oxford University Press, NY, pp. 138–166.

Sherrill T. W. and Sayler G. S. (1981) Phenanthrene biodegradation in freshwater environments. *Appl. Environ. Microbiol.* **39**, 172–178.

Simoneit B. R. T. (1984) Organic matter of the troposphere: III. Characterization and sources of petroleum and pyrogenic residues in aerosols over the Western United States. *Atmos. Environ.* **18**, 51–67.

Simoneit B. R. T. (1986) Characterization of organic constituents in aerosols in relation to their origin and transport: a review. *Int. J. Environ. Analyt. Chem.* **23**, 207–237.

Simoneit B. R. T. (1998) Biomarker PAHs in the environment. In *PAHs and Related Compounds Chemistry* (ed. A. H. Neilson). Springer, NY, pp. 175–215.

Smirnova A., Abrajano T., Smirnov A., and Stark A. (1998) Distribution and sources of polycyclic aromatic hydrocarbons in the sediments of lake Erie. *Org. Geochem.* **29**, 1813–1828.

Sofer Z. (1984) Stable carbon isotope compositions of crude oils: application to source depositional environments and petroleum alteration. *Am. Assoc. Petrol. Geol. Bull.* **68**(1), 31–48.

Sporstol S., Gjos N., Lichtenthaler R. G., Gustavsen K. O., Urdal K., Oreld F., and Skel J. (1983) Source identification of aromatic hydrocarbons in sediments using GC-MS. *Environ. Sci. Technol.* **17**(5), 282–286.

Stehmeir L. G., Francis M. McD., Jack T. R., Diegor E., Winsor L., and Abrajano T. A. (1999) Field and *in vitro* evidence for *in-situ* bioremediation using compound-specific $^{13}C/^{12}C$ ratio monitoring. *Org. Geochem.* **30**, 821–834.

Steinhauer M. S. and Boehm P. D. (1992) The composition and distribution of saturated and aromatic hydrocarbons in nearshore sediments, river sediments, and coastal peat in the Alaskan beaufort sea: implications for detecting anthropogenic hydrocarbon inputs. *Mar. Environ. Res.* **33**, 223–253.

Stenberg, U. R. (1983) PAHs Emissions from automobiles. In *Handbook of Polycyclic Aromatic Hydrocarbons*, Volume 2. Emission Sources and Recent Progress in Analytical Chemistry, Marcel Dekker, New York, pp. 87–111.

Strachan W. M. J. and Eisenreich S. J. (1988) Mass balancing of toxic chemicals in the Great Lakes: the role of atmospheric deposition. International Joint Commission, Windsor, Ontario.

Suess M. J. (1976) The environmental load and cycle of polycyclic aromatic hydrocarbons. *Sci. Tot. Environ.* **6**, 239–250.

Sugai S. F., Lindstrom J. E., and Braddock J. F. (1997) Environmental influences on the microbial degradation of Exxon Valdez oil spill on the shorelines of Prince William Sound, Alaska. *Environ. Sci. Technol.* **31**, 1564–1572.

Takada H., Onda T., and Ogura N. (1990) Determination of polycyclic aromatic hydrocarbons in urban street dusts and their source materials by capillary gas chromatography. *Environ. Sci. Technol.* **24**(8), 1179–1186.

Tissot B. P. and Welte D. H. (1984) *Petroleum Formation and Occurrence*, 2nd edn. Springer, Berlin, 699p.

Van Brummelen T. C., van Hattum B., Crommentuijn T. and Kalf D. E. (1998) Bioavailability and ecotoxicology of PAHs. In *PAHs and Related Compounds Chemistry* (ed. A. H. Neilson). Springer, NY, pp. 175–215.

Vazquez-Duhalt R. (1989) Environmental impact of used motor oil. *Sci. Tot. Environ.* **79**, 1–23.

Venkatesan M. I. (1988) Occurrence and possible sources of perylene in marine sediments-a review. *Mar. Chem.* **25**, 1–27.

Villholth K. G. (1999) Colloid characterization and colloidal phase partitioning of polycyclic aromatic hydrocarbons in two creosote-contaminated aquifers in Denmark. *Environ. Sci. Technol.* **33**, 691–699.

Volkman J. K., Holdsworth D. G., Neill G. P., and Bavor H. J. (1992) Identification of natural, anthropogenic and petroleum hydrocarbons in aquatic sediments. *Sci. Tot. Environ.* **112**, 203–219.

Volkman J. K., Revill A. T., and Murray A. P. (1997) Applications of biomarkers for identifying sources of natural and pollutant hydrocarbons in aquatic environments. In *Molecular Markers in Environmental Geochemistry*, ACS Symposium Series (ed. R. P. Eganhouse). American Chemical Society, Washington, DC, pp. 110–132.

Wakeham S. G., Schaffner C., and Giger W. (1980a) Polycyclic aromatic hydrocarbons in recent lake sediments: II. Compounds derived from biogenic precursors during early diagenesis. *Geochim. Cosmochim. Acta* **44**, 415–429.

Wakeham S. G., Schaffner C., and Giger W. (1980b) Polycyclic aromatic hydrocarbons in recent lake sediments: I. Compounds having anthropogenic origins. *Geochim. Cosmochim. Acta.* **44**, 403–413.

Wang Z. and Fingas M. (1995) Use of methyldibenzothiophenes as markers for differentiation and source identification of crude and weathered oils. *Environ. Sci. Technol.* **29**, 2842–2849.

Wang Z., Fingas M., Blenkinsopp S., Sergey G., Landriault M., Sigouin L., Foght J., Semple K., and Westlake S. W. S. (1998) Comparison of oil composition changes due to biodegradation and physical weathering in different oils. *J. Chromatogr.* **A809**, 89–107.

Wang Z., Fingas M., and Page D. S. (1999) Oil spill identification. *J. Chromatogr.* **A843**, 369–411.

Ward J. A. M., Ahad J. M. E., Lacrampe-Couloume G., Slater G. F., Edwards E. A., and Lollar B. S. (2000) Hydrogen isotope fractionation during methanogenic degradation of toluene: potential for direct verification of bioremediation. *Environ. Sci. Technol.* **34**, 4577–4581.

Westerholm R., Stenbcrg U., and Alsberg T. (1988) Some aspect of the distribution of polycyclic aromatic hydrocarbons (PAHs) between particles and gas phase from diluted gasoline exhausts generated with the use of a dilution tunnel, and its validity for measurement in air. *Atmos. Environ.* **22**(5), 1005–1010.

Whitehouse B. G. (1984) The effects of temperature and salinity on the aqueous solubility of polynuclear aromatic hydrocarbons. *Mar. Chem.* 14.

Whittaker M., Pollard S. J. T., and Fallick T. E. (1995) Characterisation of refractory wastes at heavy oil-contaminated sites: a review of conventional and novel analytical methods. *Environ. Technol.* **16**, 1009–1033.

Wijayaratne R. D. and Means J. C. (1984) Adsorption of polycyclic aromatic hydrocarbons by natural estuarine colloids. *Mar. Environ. Res.* **11**, 77–89.

Yanik P., O'Donnell T. H., Macko S. A., Qian Y., and Kennicutt M. C. (2003) The isotopic compositions of selected crude oil PAHs during biodegradation. *Org. Geochem.* **34**, 291–304.

Ye D., Siddiqui M. A., Maccubbin A. E., Kumar S., and Sikka H. C. (1996) Degradation of polynuclear aromatic hydrocarbons by *Sphingomonas paucimbilis*. *Environ. Sci. Technol.* **30**, 136–142.

Youngblood W. W. and Blumer M. (1975) Polycyclic aromatic hydrocarbons in the environment: homologous series in soils and recent marine sediments. *Geochim. Cosmochim. Acta* **39**, 1303–1314.

Yunker M. B., MacDonald R. W., Cretney W. J., Fowler B. R., and McLaughlin F. A. (1993) Alkane, terpene, and polycyclic aromatic hydrocarbon geochemistry of the mackenzie river and mackenzie shelf: riverine contributions to beaufort sea coastal sediment. *Geochim. Cosmochim. Acta.* **57**, 3041–3061.

Yunker M. B., MacDonald R. W., Veltkamp D. J., and Cretney W. J. (1995) Terrestrial and marine biomarkers in a seasonally ice-covered arctic estuary-integration of multivariate and biomarkers approaches. *Mar. Chem.* **49**, 1–50.

Zafiriou O. C. (1977) Marine organic photochemistry previewed. *Mar. Chem.* **5**, 497–522.

Zepp R. G. and Schlotzhauer P. F. (1979) Photoreactivity of selected aromatic hydrocarbons in water. In *Polynuclear Aromatic Hydrocarbons* (eds. P. W. Jones and P. Leber). Ann. Arbor Sci. Publ., Ann. Arbor MI, pp. 141–158.

9.14
Biogeochemistry of Halogenated Hydrocarbons

P. Adriaens and C. Gruden

The University of Michigan, Ann Arbor, MI, USA

and

M. L. McCormick

Hamilton College, Clinton, NY, USA

9.14.1 INTRODUCTION

Halogenated hydrocarbons originate from both natural and industrial sources. Whereas direct anthropogenic emissions to the atmosphere and biosphere are often easy to assess, particularly when they are tied to major industrial activities, the attribution of emissions to other human activities (e.g., biomass burning), diffuse sources (e.g., atmospheric discharge, run off), and natural production (e.g., soils, fungi, algae, microorganisms) are difficult to quantify. The widespread occurrence of both alkyl and aryl halides in groundwater, surface water, soils, and various trophic food chains, even those not affected by known point sources, suggests a substantial biogeochemical cycling of these compounds (Wania and Mackay, 1996; Adriaens *et al.*, 1999; Gruden *et al.*, 2003). The transport and reactive fate mechanisms controlling their reactivity are compounded by the differences in sources of alkyl-, aryl-, and complex organic halides, and the largely unknown impact of biogenic processes, such as enzymatically mediated halogenation of organic matter, fungal production of halogenated hydrocarbons, and microbial or abiotic transformation reactions (e.g., Asplund and Grimvall, 1991; Gribble, 1996; Watling and Harper, 1998; Oberg, 2002). The largest source may be the natural halogenation processes in the terrestrial environment, as the quantities detected often exceed the amount that can be explained by human activities in the surrounding areas (Oberg, 1998). Since biogeochemical processes result in the distribution of a wide range of halogenated hydrocarbon profiles, altered chemical structures, and isomer distributions in natural systems, source apportionment (or environmental forensics) can often only be resolved using multivariate statistical methods (e.g., Goovaerts, 1998; Barabas *et al.*, 2003; Murphy and Morrison, 2002).

This chapter will describe the widespread occurrence of halogenated hydrocarbons, interpret their distribution and biogeochemical cycling in light of natural and anthropogenic sources, biotic and abiotic reactivity, and prevailing cycling mechanisms. Specific emphasis will be placed on the potential role of biotic and abiotic transformation reactions in soil, water, and sediment environments resulting in environmental sequestration and phase transfer.

9.14.2 GLOBAL TRANSPORT AND DISTRIBUTION OF HALOGENATED ORGANIC COMPOUNDS

The biogeochemistry of halogenated pollutants has to be reviewed within the context of transport and phase partitioning, and the impact of these processes on their distribution and reactivity in various environmental compartments. Whereas the authors recognize that these atmospheric processes dominate the global biogeochemistry of halogenated hydrocarbons, there are substantial differences in the types of biogeochemical controls that impact biogenic, anthropogenic, and complex halogenated organic matter at local (contaminant hot spots), regional (e.g., Great Lakes), and global (e.g., temperate and polar latitudes) scales. Since this chapter aims to discuss persistent organic pollutants (POPs), biogenic pollutants (e.g., the halomethanes), and anthropogenic non-POPs (e.g., chloroethenes, chlorofluoro-hydrocarbons), an operational separation between POPs and non-POPs based on their distribution mechanisms and physical–chemical properties which render them as belonging to either class is useful to provide context for the remainder of this chapter.

9.14.2.1 Persistent Organic Pollutants

POPs are generally considered as long-lived organic compounds that become concentrated as they move through the food chain, exhibiting toxic effects on animal reproduction, development, and immunological function. These compounds are dominated by chlorinated hydrocarbons, including many pesticides: DDT, hexachlorobenzene (HCB), chlordane, heptachlor, toxaphene, aldrin, dieldrin, endrin, mirex, polychlorinated biphenyls (PCBs), and chlorinated dibenzo-*p*-dioxins and furans (PCDD/F) (Bidleman, 1999). The chemistry of these compounds is very complex and spans orders of magnitude variability in their solubility, vapor pressure, and partitioning behavior as a function of chlorine content and substitution pattern. For example, there are 210 congeners of PCBs, 135 congeners of furans, and 75 congeners of dioxins. In addition, many compounds are chiral in nature (e.g., the hexachlorocyclohexane-HCH group), a property which also confers differential reactivity on the molecule. Based on observations dating back to mid-1960s, it has become apparent that POPs have the capability of being transported over thousands of kilometers (e.g., Sladen *et al.*, 1966; Peterle, 1969), during which these compounds have the opportunity to fractionate, react, and bioaccumulate in global food chains.

Several global maps have been generated that document the distribution of these compounds based on the analysis of tree bark samples (Simonich and Hites, 1995), pine needles and lichens (Ockenden *et al.*, 1998a), rural soils and isolated water bodies (Baker and Hites, 2000a,b), and butter samples (Kalantzi *et al.*, 2001). Aside from global latitudinal fractionation (e.g., Ockenden *et al.*, 1998b), analysis for organochlorine insecticides and PCBs over the Great Lakes indicate the existence of local and regional volatilization pathways, rather than long-range transport (e.g., McConnell *et al.*, 1998). These regional and global distribution phenomena and their implications have been recognized in a number of multilateral venues (Montreal, Stockholm, UN, etc.), and the need for identifying and classifying chemicals for their propensity to long-range transport, and characterization of POP distribution mechanisms is an urgent topic of research.

9.14.2.1.1 Global distribution mechanisms

Even though substantial experimentation and monitoring of POPs in the atmosphere, the world's soils, vegetation, and oceans indicate a widespread (mainly latitudinal) distribution, the present knowledge on the global dynamics of POPs is incomplete. The processes controlling their distribution can only be assessed by integration of models with the available datasets (Dachs *et al.*, 2002), or by constructing hypothetical environments to explore how a given chemical is likely to partition, react, and be transported (MacKay and Wania, 1995). The advantage of the latter approach is that chemical behavior is the only variable under consideration, and problems associated with environmental characterization and limiting chemical analyses are avoided. The former allows for a validation of this approach with multivariate data sets. Based on these approaches, two main mechanisms controlling air–water fluxes and enrichment of contaminants in polar regions and food webs have been advanced: temperature-controlled global fractionation and condensation, and oceanic controlled phytoplankton uptake with particle sinking.

For a quarter of a century, the concept of global distillation has been developed to describe the tendency of certain contaminants to evaporate from temperate and tropical regions and condense in cold climates (Goldberg, 1975). More recently, a "grasshopper effect" analogy was used to describe the tendency of POPs to undergo cycles of deposition and re-emission during transport (Wania and MacKay, 1996). Accordingly, properties such as vapor pressure and partition coefficient will impact this tendency for any given POP compound, resulting in distinctive condensation temperatures and degree of fractionation with latitude. These cycles are thought to be strongly influenced by diurnal and seasonal variability of environmental conditions, such as temperature, due to its influence on air–surface and air–water partitioning. Indeed, the influence of air–water exchange has been shown to dominate depositional processes in many aquatic systems for a wide range of POPs such as PCBs and HCHs (Totten *et al.*, 2001; Wania and MacKay, 1996). Large-scale surveys of organochlorine pesticides and PCBs in ocean air and water show features which support these effects (Iwata *et al.*, 1993), and many aryl halides in atmospheric air masses show strong temperature-dependent diurnal cycling (e.g., Lee *et al.*, 2000), providing support for rapid air–surface exchange of semi-volatile organic compounds. On a regional scale, significant temperature-dependent air–water exchange of toxaphene (polychlorinated bornanes and bornenes) in the Great Lakes has been demonstrated, whereby the colder temperatures and lower sedimentation rates in Lake Superior are responsible for its higher aqueous concentrations (Swackhamer *et al.*, 1999).

Large differences in exchange fluxes were calculated depending on mechanistic assumptions. Consider the following two limiting cases: (i) all of the POPs in water are truly dissolved and available to participate in gas exchange, and (ii) all of the POPs are bound to particles or colloidal material and unable to revolatilize. When the fugacity in surface water was taken into account, net deposition occurred only in colder regions; in the case of (i), air-to-sea deposition occurred in all ocean regions (Iwata *et al.*, 1993). Further, the controlling influence of temperature in determining the transport and sinks of POPs via cold condensation, global distillation, and latitudinal fractionation has been supported by climatic models (Wania and MacKay, 1996). Experimental observations have validated this model for terrestrial and limnic systems, but are scarce in the marine environment (e.g., Grimalt *et al.*, 2001; Meijer *et al.*, 2002), indicating that other mechanisms may play a role in the global distribution of POPs.

Vertical scavenging of POPs in oceans following sorption to particulate organic matter, and subsequent removal from participation in dynamic air–water exchange as the result of particulate sinking to deep waters and sediments has been demonstrated, resulting in vertical profiles in the water column (e.g., Dachs *et al.*, 1999a,b). Hence, deposition fluxes of POPs in the water column may at least in part be considered to represent a significant control for revolatilization as they are impacted by sinking particles and biogeochemical processes such as phytoplankton uptake (Dachs *et al.*, 2002). Support for the role of phytoplankton was derived from findings that air–water exchange and phytoplankton uptake behave as coupled processes in aquatic environments (Dachs *et al.*, 1999a,b). These mutual interferences in atmospherically driven aquatic systems result from (i) the impact of the magnitude of air–water exchange on phytoplankton trophic status (biomass and growth rate), and (ii) the impact of air–water exchange on POPs concentrations in phytoplankton (Dachs *et al.*, 2000). These combined processes result in lower PCB concentrations in zooplankton at increased biomass concentrations, due to a dilution effect (particle dilution). Similar effects are expected for phytoplankton. Further, higher growth rates lead to lower PCB concentrations in the phytoplankton due to dilution by the new organic matter introduced in the ecosystem (growth dilution) (Dachs *et al.*, 2000). Since biomass productivity at low latitudes is limited, it would be expected that POPs uptake and settling processes would not be very important at these latitudes.

Indeed, Dachs *et al.* (2002) demonstrated, using a combination of field measurements of atmospheric PCBs, PCDDs, and PCDFs, and remote sensing data of ocean temperature, wind speed and chlorophyll, that deposition in mid-high latitudes is driven by sinking marine particulate matter, rather than by cold condensation. However, the relative contribution of this process is highly dependent on the physical chemical properties of the POPs under consideration.

9.14.2.1.2 Contaminant classification

Hence, it appears that, independent of the distribution mechanism, physical–chemical properties govern the impact of the various mechanisms on the POPs, resulting in differential latitudinal fractionation (Bidleman, 1999; Wania and MacKay, 1996). To capture partitioning behavior, Wania and MacKay (1996) proposed a classification of POPs as a function of octanol–air partition coefficient (K_{OA}), vapor pressure of the subcooled liquid (PL), and condensation temperature (T_c). As shown in Table 1, four categories of global transport behavior are proposed, ranging from low to high mobility, for a range of POPs.

It has been argued that the K_{OA} values can be used as a unifying property for describing volatilization of POPs from soils and sorption to aerosols. The limited experimentally obtained values typically are supplemented by estimates from octanol–water and air–water partition coefficients. The value of condensation temperature lies in its ability to estimate sorption of atmospheric contaminants to aerosols (Bidleman, 1988). At T_c, the chemical is equally partitioned between the gas phase and aerosols. Since POPs exist in the atmosphere both as gases (vapor phase), and in condensed form adsorbed to aerosol particles, the characteristic temperature of condensation provides a measure of depositional preference (Wania and MacKay, 1996). Hence, based on the integration of this information, it is argued that PCBs, PCDDs, and most organochlorine pesticides will preferentially accumulate in mid-latitude to polar regions. Since these regions also exhibit the highest phytoplankton biomass (Dachs *et al.*, 2002) as a source for POPs uptake and particle sinking, they may represent a sink for POPs.

9.14.2.2 Biogenic Pollutants and Anthropogenic Non-POPs

In accordance with the rationale presented earlier, biogenic organohalides such as the halomethanes, and anthropogenic sources such as the chlorofluorocarbons (CFC), methylchloroform, and carbon tetrachloride are highly mobile and are atmospherically dispersed without significant depositional impacts (Table 1), as they tend to exhibit log K_{OA} values of less than 6.5 (Wania, 2003). Their global environmental impacts are predominantly derived from their long residence times which allow for substantial migration into the stratosphere. Here, solar UV photolysis results in dechlorination and ozone depletion reactions. This is particularly the case for the chlorofluorocarbons (Sherwood Rowland, 1991). Despite their global distribution, some deposition as a result of atmospheric gas exchange with natural waters has resulted in the presence of CFCs in young groundwaters (recharged within the last 50 years), resulting in diurnal, weekly and seasonal patterns (Ho *et al.*, 1998), which has allowed for groundwater dating provided information on the local or regional emissions is available. Similarly, evidence has been presented suggesting that anaerobic sediments and soils may represent a sink for CFCs, as a result of microbial uptake

Table 1 Global transport behavior of POPs.

Classification	*Low mobility*	*Relatively low mobility*	*Relatively high mobility*	*High mobility*
Global Transport Behavior	Rapid deposition and retention close to source	Preferential deposition and accumulation in mid-latitudes	Preferential deposition and accumulation in polar latitudes	Worldwide atmospheric dispersion, no deposition
Chlorobenzenes			5–6 chlorines	0–4 chlorines
PCBs	8–9 chlorines	4–8 chlorines	1–4 chlorines	
PCDD/Fs	4–8 chlorines	2–4 chlorines	0–1 chlorine	
Organochlorine pesticides	mirex	Polychlorinated camphenes, DDTs, chlordanes	HCB, HCHs dieldrin	
←				→
log K_{OA}	10	8	6	
log P_L	−4	−2	0	
T_c (°C)	+30	−10	−50	
←				→

After Wania and MacKay (1996).

(Lovley and Woodward, 1992). Similarly, the distribution of biogenic halomethanes in the atmosphere as the result of emissions from tropical forests and marine biological activity exhibits environmental impact derived from atmospheric oxidation reactions, and wet deposition of the resulting acid intermediates. The air–water exchange for the halomethane compounds is thought to be limited, and the net fluxes are in the direction of (re-)volatilization. The presence of other alkyl halides of anthropogenic origin such as the chloroethenes and tetrachloromethane in groundwaters and surface waters are usually considered to be the result of local discharges (e.g., Lendvay *et al.*, 1998a,b), rather than from atmospheric sources as described for the CFCs.

The impact of deposition on global distribution has been noted for the CFC replacements hydrochlorofluorocarbons (HCFCs), the chlorinated solvents tetrachloroethene (PCE), and trichloroethene (TCE), as these compounds undergo gas phase oxidation and photochemical degradation, resulting in the formation of carbonyl halides (e.g., CCl_2O) and haloacetyl halides (e.g., bromo-, chloro-, and fluoroacetates). As these compounds are polar and water soluble, they are transported via aerosols, rain, and fog, which impacts their tropospheric lifetime and depositional fluxes (Rompp *et al.*, 2001; de Bruyn *et al.*, 1995). It is not clear whether and to what extent there is evidence of latitudinal fractionation of these compounds.

9.14.3 SOURCES AND ENVIRONMENTAL FLUXES

Halogenated hydrocarbons in the atmosphere and biosphere are derived from a large number of natural and anthropogenic sources. The ultimate environmental sinks for most halogenated compounds, whether they are released to the atmosphere or directly discharged in waterways, are the Earth's soils, sediments, and waterways (Figure 1). Little is known about the relative contributions of point sources and nonpoint sources of contamination. The global distribution and fluxes of aryl (e.g., polychlorinated biphenyls, dibenzo-*p*-dioxins and dibenzofurans, and halogenated pesticides) and alkyl (e.g., C_1–C_4 chlorinated, fluorinated, iodinated, and brominated alkanes) halides are better established. Through a comparative assessment of areas affected by urbanization and relatively pristine areas such as remote lakes, anthropogenic sources (e.g., industrial effluents) may be distinguished from natural halogenated hydrocarbon production mechanisms and pathways. Since the 1990s, it has become clear that most halogens, particularly chlorine, bromine, and iodine, participate in a complex biogeochemical cycle. Whereas environmental phase partitioning and transport are expected to dominate the environmental behavior of halogenated hydrocarbons, their reactivity in soils and sediments impact the natural and anthropogenic sources and global mass balances of these chemicals. The focus of this chapter will be on quantifying the impacts of reactivity on environmental behavior, while recognizing the multiplicity of controls on their biogeochemistry.

9.14.3.1 Adsorbable Organic Halogens

Frequently, halogenated hydrocarbons in environmental matrices are quantified, in bulk, by a standard analytical procedure for adsorbable organic halogens (AOX). Originally conceived to monitor the formation of chlorinated organic compounds in drinking water, the data derived from environmental characterization have been interpreted as an indicator of anthropogenic

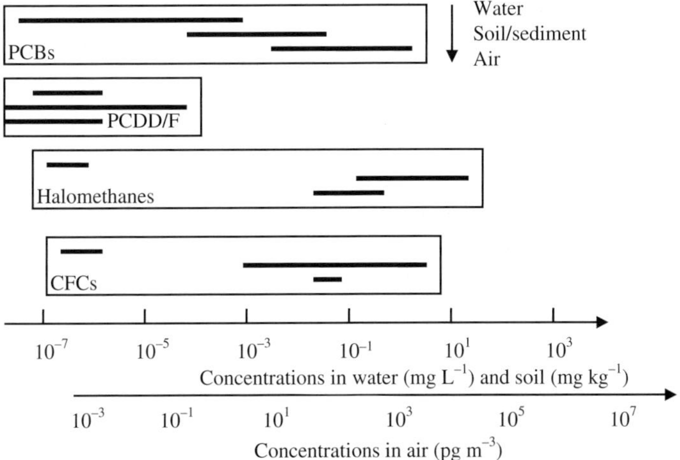

Figure 1 Distribution of selected aliphatic and aromatic organohalides in various environmental compartments.

activity (Kuhn *et al.*, 1977). For example, in Germany, AOX concentrations of 5 μg L^{-1} were used to discriminate between negligible and moderate influence of industrial activity (Hoffman, 1986). Later studies indicated that the apparent relationship between AOX and industrial activity was affected by another environmental variable, the amount of organic matter present (Asplund *et al.*, 1989). In a subsequent review on the natural occurrence of halogenated hydrocarbons in 135 lakes in southern Sweden (Asplund and Grimvall, 1991), AOX concentrations of 11–185 μg Cl L^{-1} were observed, with correlations to both color (pigmentation, Pt) and total organic carbon (TOC) (Figure 2). The ratio of halogenated hydrocarbons to organic carbon for soil samples was very stable (0.2–2.8 mg Cl per g C). In surface water samples of diverse origin, the ratio was higher (0.8–14.5 mg Cl per g C) and exhibited more variation. Mass balances for AOX, based on long-range atmospheric deposition of industrial compounds predicted an average concentration of only 15 μg Cl L^{-1} in bulk precipitation, indicating that natural production may be a substantial source of halogenated hydrocarbons. This hypothesis was further supported by the observation that AOX concentrations in runoff were, on average, 10-fold higher than in precipitation.

Indeed, in recent years evidence has emerged for the existence of a natural chlorine cycle (Figure 3) involving production and mineralization of halogenated hydrocarbons at the air–soil interface affecting deposition, volatilization, and leaching processes (Oberg, 2002). This is not surprising considering that chlorine, one of the most abundant elements, represents the sixth most common constituent of soil organic matter (0.01–0.5% DW) and is deposited globally through the action of sea spray (range: <1 kg ha^{-1} inland to 100 kg ha^{-1} in coastal areas). Hence, up to tons of AOX can be found per km^2 in some environments (Oberg, 1998). Despite published reviews on the subject (e.g., Gribble, 1996; Oberg, 2002), discussion on the chlorine cycle is not yet fully integrated in the biogeochemistry literature on the microbial ecology of biodegradation and biotransformation (Adriaens *et al.*, 1999). The number of naturally produced chlorinated compounds is on the order of 1,500, and covers alkenes, alkanes, terpenes, steroids, fatty acids, and glycopeptides, including compounds previously assumed to emanate exclusively from anthropogenic activity (Adriaens *et al.*, 1999; Oberg, 2002). Most of these compounds have been shown to undergo microbial and abiotic transformation reactions resulting in the liberation of the halogen ions in the environment (Section 9.14.5). As opposed to mineralization, halogenated hydrocarbons are formed in many micro- and macro-ecosystems via the following

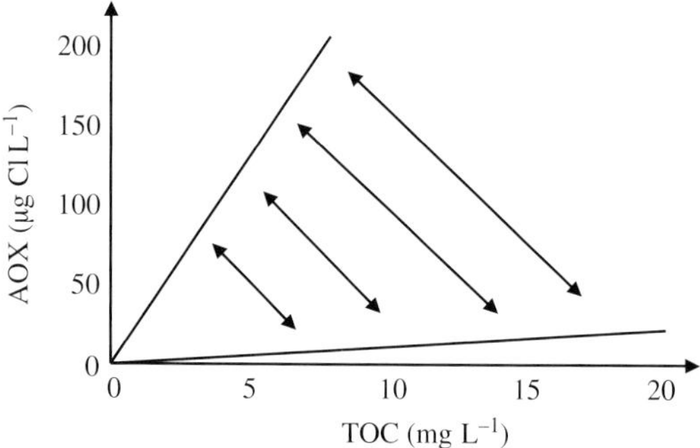

Figure 2 Concentration ranges (in delineated area) of AOX as a function of color and TOC in Swedish lakes and rivers (after Asplund and Grimvall, 1991).

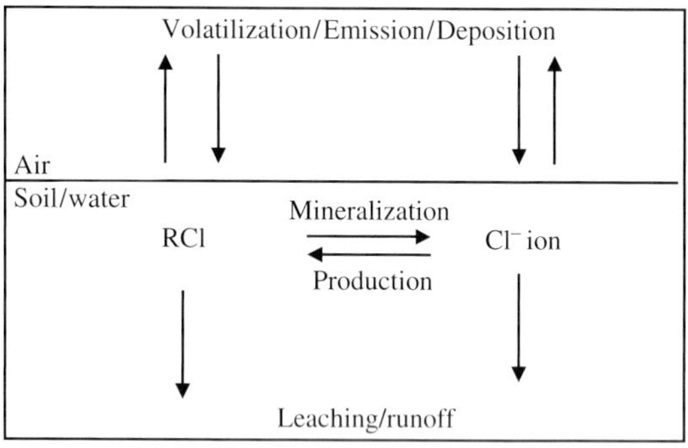

Figure 3 A conceptual model of the chlorine cycle in soils, with dominant processes indicated (after Oberg, 2002).

processes: intra- and extracellular defense mechanisms (e.g., Neidleman and Geigert, 1986), biosynthesis (Harper *et al.*, 1990), and production of reactive (e.g., HOCl) catalysts to oxidize organic substrates (Oberg *et al.*, 1997; Johansson *et al.*, 2000). From a biotransformation perspective, the halogenated (chlorinated) substrates have been shown to serve as a source of carbon and/or energy generation to (mostly) bacteria (Adriaens and Hickey, 1994; Adriaens and Vogel, 1995; Adriaens *et al.*, 1999; Adriaens and Barkovskii, 2002). The structural similarity between anthropogenic and natural compounds may explain why some microorganisms have evolved the capability to respire or otherwise degrade selected halogenated hydrocarbons (van der Meer *et al.*, 1992; Holliger and Schraa, 1994; Copley, 1998).

These observations point towards the shortcomings of using AOX as the sole source and mass balance indicator for halogenated hydrocarbons, and this operational designation would only be useful when known point sources of contamination (e.g., pulp mills and chemical manufacturing) or natural production (e.g., forest soils) are present. Sufficient chemical resolution of the composition of AOX will be required to further dissect the source attribution of selected aliphatic and aromatic compounds (Dahlman *et al.*, 1993, 1994; Johansson *et al.*, 1994; Laniewski *et al.*, 1995).

9.14.3.2 Alkyl Halides

The natural and anthropogenic production of alkyl halides has recently been studied (Laturnus *et al.*, 1995, 2002; Watling and Harper, 1998; Carpenter *et al.*, 1999; Giese *et al.*, 1999). This group of compounds is composed of methanes (CFC, tetrachloro-, trichloro-, chloro-, bromo- and iodo-, diiodo-, and chloroiodo-), propanes (1- and 2-iodo-, 1-iodo-2-methyl-), butanes (1- and 2-iodo), acetates (trifluoro-), and ethenes (tetrachloro-, trichloro-). Whereas naturally produced alkyl halides tend to be substituted with one or two halogens, those of industrial origin are generally poly-substituted (Key *et al.*, 1997). Chloromethane and bromomethane are the most abundant halogenated hydrocarbons in the atmosphere (Butler, 2000; Harper *et al.*, 2001), with the former catalyzing an estimated 17% of ozone destruction. Only CFC11 ($CFCl_3$) and CFC12 (CF_2Cl_2) exhibit greater ozone-depleting effects (Harper, 2000). In contrast to the CFCs, chloro- and bromomethane are mainly of nonindustrial origin. For example, as of early 2000s, 50% of the global annual input (4×10^{12} t) of chloromethane cannot be accounted for (Butler, 2000). Important sources that have been previously omitted from mass balance calculations include halogenated hydrocarbons resulting from natural oxidation processes during degradation of organic matter, biomass burning, oceanic emissions, wood rotting fungi, and release by higher plant species (Keppler *et al.*, 2000; Harper, 2000; Watling and Harper, 1998; Rhew *et al.*, 2000; Laturnus *et al.*, 2002). Even though a significant contributor of halomethanes was believed to be of marine origin, the ocean is, for the most part, undersaturated in bromomethane and insufficiently supersaturated in methyl chloride to explain more than a small percentage of the total atmospheric flux (Figure 1; Butler, 2000).

Evidence has indicated that terrestrial–coastal ecosystems, and abiotic formation mechanisms under acidic, iron-reducing conditions may help explain the flux deficit (Rhew *et al.*, 2000; Yokouchi *et al.*, 2000; Keppler *et al.*, 2000; Dimmer *et al.*, 2001). For example, the median biotic fluxes of halomethanes ($CHCl_3$, CH_3Br, CH_3Cl, CH_3I) from peat land ecosystems were estimated to be on the order of $(0.9–4.5) \times 10^9$ g yr^{-1}, with the highest values incurred for chloroform and the lowest for bromomethane (Dimmer *et al.*, 2001). Marine algae and vegetation in coastal marshes have been identified as potential sources of bromo- and chloromethane (Rhew *et al.*, 2000), as well as of volatile iodinated $C_1–C_4$ hydrocarbon production (Giese *et al.*, 1999). Fluxes from coastal marshes may account for up to 10% of atmospheric halomethanes, and those of microalgae ranging from 0.005% to 3%. Wood-rotting fungi (particularly *Phellinius* and *Inonotus* Basidiomycetes) have long been recognized as significant sources of chloromethane emissions, resulting from biosynthesis pathways (Watling and Harper, 1998). Estimations of the global fungal contributions are dependent on assumptions for the amount of woody tissue decomposed, chloride content of wood, and the global abundance of chloromethane-releasing species, and are on the order of 1.6×10^5 t yr^{-1} with 75% emanating from tropical forest. Those same characteristics and enzyme systems which allow Basidiomycetes to decompose complex organic matter also allow these fungi to degrade complex halogenated hydrocarbons such as polychlorinated biphenyls (PCBs) and chlorophenols to low molecular weight compounds which are released in the environment and often incorporated in soil organic matter (de Jong and Field, 1997). However, the relative trends of formation as compared to degradation have not been assessed, and thus the impact of environmental transformation processes to halogenated hydrocarbon flux reduction is not known.

Lastly, a relatively recently identified source is the abiotic formation of halocarbons in acidic soils amended with the halide ion in the presence of elevated concentrations of iron(III). Considering the importance of insoluble iron(III) oxide and oxyhydroxides in the sedimentary environment, elevated salinities in soils, and the availability of organic carbon, the potential of this mechanism to release halomethanes is substantial (Keppler *et al.*, 2000). Interestingly, this reaction is independent of microbial activity and sunlight. One important means to help discriminate between the contributions of the various biotic and abiotic sources to haloalkane production may be the application of carbon isotope ratios (Harper *et al.*, 2001). Isotopic fractionation values have now become available for biogenic sources, allowing investigators to discriminate between fungal and plant contributions to chloromethane production.

9.14.3.3 Aryl Halides

Substantial insights have been gleaned into the sources and biogeochemistry of target aryl halides (mostly POPs), such as fluorinated organics (Key *et al.*, 1997), polychlorinated dibenzo-*p*-dioxins (PCDD) and furans (Baker and Hites, 2000a,b; Cole *et al.*, 1999; Gruden *et al.*, 2003; Gaus *et al.*, 2002; Lohmann *et al.*, 2000), polychlorinated biphenyls (Ockenden *et al.*, 1998a,b; Breivik *et al.*, 2002a,b), toxaphene (Swackhamer *et al.*, 1999; James *et al.*, 2001), and organochlorine pesticides (McConnel *et al.*, 1998; Lee *et al.*, 2000). Based on a comprehensive inventory of the types and sources of chemical contamination in the environment, Swoboda-Colberg (1995) identified the petrochemical and pesticide industry as the main sources of aryl halides. Particularly, the herbicide and wood treatment industry have used (i) simple aromatic compounds (e.g., chlorobenzenes, chlorophenols) and their derivatives (e.g., chlorophenoxyacetates), (ii) cyclodienes (e.g., heptachlor, dieldrin, heptasulfan), and (iii) organonitrogen pesticides (e.g., alachlor, linuron). The natural production of aryl halides discharged into the environment by plants, microorganisms, marine organisms and other processes is rapidly being recognized as a fully integrated component of the biosphere (Gribble, 1996). This fraction is comprised of chlorinated pyrroles and indoles, as well as phenols, phenolic ethers, benzenes, hydroquinone, and orcinol methyl ethers (Figure 4(a)). Combustion and biotic formation processes have been identified as sources for the natural (non-anthropogenic) production of aryl halides, such as PCDD/F (Bumb *et al.*, 1980) and other aromatically bound halogens (Dahlman *et al.*, 1993; Gribble, 1996). Particularly Basidiomycetes have a widespread capacity for halogenated hydrocarbon biosynthesis and degradation (de Jong and Field, 1997). Many aryl halides have biotic origin, provided the right precursor molecule is present. For example, aerobic wastewater treatment processes and extracellular soil enzymes such as peroxidases and laccases have been demonstrated to produce PCDD and PCDF from chlorophenol precursors (Svenson *et al.*, 1989). Anaerobic processes result in natural dechlorination processes and the formation of lesser chlorinated, and thus more soluble and volatile, compounds (Häggblom, 1992; Adriaens *et al.*, 1999; Gruden *et al.*, 2003). Even including these processes, it has been suggested that environmental levels exceed known natural and anthropogenic sources by a significant margin, resulting in the conundrum that measurements of depositional fluxes exceed known emissions.

In a review, Baker and Hites (2000a) conducted an extensive mass balance investigation on PCDD/F, and estimated that annual emissions are on the order of 3,000 kg yr^{-1} with depositional fluxes totaling $(0.3-1) \times 10^4$ kg yr^{-1}. Most of the mass balance discrepancy was found to be due to the octachlorinated dioxin congener (OCDD), which is dominant in most terrestrial environments. Based on time trends of dioxin profiles in Siskiwit Lake (Lake Superior) sediment cores, it was argued that photochemical synthesis of OCDD from pentachlorophenol (PCP) in atmospheric condensed water may be a more substantial source of OCDD than combustion (Baker and Hites, 2000b). With sediments identified as the ultimate sinks for aryl halides, there is substantial interest in ongoing attenuation processes relevant to these systems. Gevao *et al.* (1997), using PCB analysis in a dated sediment core (Estwaithe Water, UK), demonstrated a predominance of lesser-chlorinated congeners in more recent sediments. This trend may indicate a postdepositional mobility of PCBs

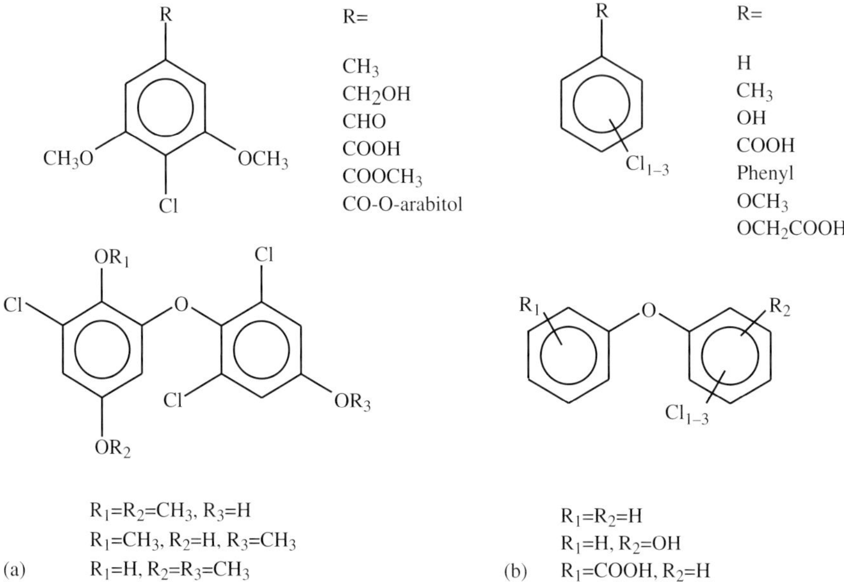

Figure 4 Selected structural features of aromatic organohalides produced by Basisiomycetes in soils (a), and degraded by microorganisms (b).

favoring the diffusive transport of the more soluble and volatile congeners of the overall pool of PCBs in these sediments.

Further evidence for re-emission of the PCB sediment burden via the water column to the atmosphere was obtained by Jeremiasson *et al.* (1994) in a mass balance study in Lake Superior. A similar mechanism may be operative in the Hudson–Raritan estuary, where dioxin profiles at the air–water interface were dominated by dichlorinated PCDD (diCDD) (Lohmann *et al.*, 2000). Fugacity calculations indicated the water column as the source, however, dioxin profiles in the sediment column were only analyzed for tetra- to octaCDD. It is unclear whether the diCDD constitute original source material or are the result of environmental transformation (dechlorination) processes (Gruden *et al.*, 2003). Gaus *et al.* (2002) provided supporting evidence for the potential contribution of natural dechlorination reactions to the alteration of dioxin profiles in sediments. Based on analysis of a global range of sediments, the authors observed the existence of a "sediment pattern" rich in 1,4-chlorines which increased with time (depth), indicating that the chlorines in 2, 3, 7, and 8 positions may have been selectively removed. The nomenclature is shown below:

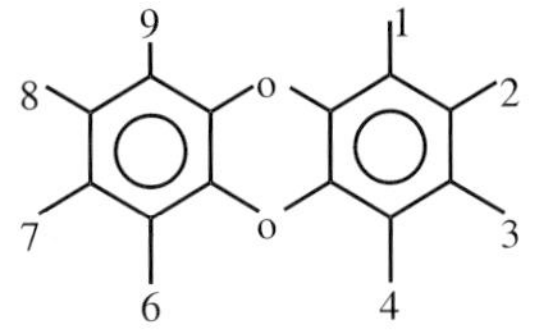

9 1
8 O 2
7 O 3
6 4

Nomenclature for PCDDs

To demonstrate the relevance of dioxin formation processes in sediments, Barabas *et al.* (2003) applied environmental forensics tools (polytopic vector analysis) to deconvolute sediment (Passaic River, NJ) dioxin patterns in a number of source and reactivity constituents. In addition to the validation of a half dozen source patterns, the authors were able to extract a dechlorination pattern which was depleted in heptaCDD and enhanced in 2,3,7,8-tetraCDD. This pattern was responsible for up to 7% (mean: 3%) of the total sample variance in a 12-mile stretch of the river, and the loading of the profile generally increased with depth, indicating a correlation between time and pattern occurrence. Further analysis of structured (undisturbed) sediment cores indicated that the dechlorination contribution to the presence of TCDD was ~90–100% in samples with low total dioxin loadings. Considering the similarity in structural features between biogenic aryl halides and aryl halides that have been biologically degraded, these versatile microbial degradation processes likely contribute significantly to the chlorine cycle (Figure 4(b)).

9.14.4 CHEMICAL CONTROLS ON REACTIVITY

The existence of a chlorine cycle and the scattered evidence of biogeochemical cycles for halogenated hydrocarbons involve a wide range of environmentally relevant reaction mechanisms and pathways leading to their widespread distribution and matrix-dependent profiles. The extent to which biotic and abiotic reactions influence the chlorine (halogen) cycle depends on complex interactions between the intrinsic molecular properties of these compounds and characteristics of the environment.

9.14.4.1 Phase Partitioning

The elemental composition, structure, and substituents found in natural and anthropogenic halogenated hydrocarbons influence the physical–chemical properties of these compounds, and thus their distribution between the various environmental phases as well as their intrinsic molecular reactivity. Vapor pressure, aqueous solubility, octanol–water partition coefficient, and octanol–air partition coefficient are the predominant properties that dictate the partitioning of the compounds between the various environmental phases (e.g., Cole *et al.*, 1999; Harner *et al.*, 2000). Fugacity calculations are the primary means for interpreting, correlating, and predicting the multimedia concentrations of halogenated hydrocarbons (MacKay, 1991). Since the three "solubilities" (air, water, octanol) generally demonstrate a linear but inverse correlation to LeBas molar volume, it would follow that bulkier molecules are less soluble in all three matrices. For example, the aqueous solubilities of halogenated C_1 and C_2 alkanes and alkenes are 2–8 times higher than those of the most soluble (lesser chlorinated) PCBs; the solubility range of PCB congeners themselves spans five orders of magnitude between mono- and octaCDD (Schwarzenbach *et al.*, 1993). The presence of polar functional groups such as hydroxyl, carboxyl, or methyl substituents tends to increase the solubility. Conversely, the less water-soluble compounds have the tendency to partition into solvent, lipid, or organic phases.

These properties and principles govern the transfer of organic chemicals between different environmental compartments. Of particular interest to halogenated hydrocarbon reactivity discussed in this treatise is the air–water interface (emissions, water column outgassing), and the solid–water interface (sorptive processes, sequestration, mineral-mediated reactions). For example, the role of air–water diffusive exchange in large aquatic systems may provide a source or sink for volatile compounds such as the halomethanes

(Butler, 2000), PCBs (Achman *et al.*, 1993; McConnell *et al.*, 1998; Zhang *et al.*, 1999), toxaphene (Swackhamer *et al.*, 1999), organochlorine insecticides (McConnell *et al.*, 1998), and PCDD/F (Lohmann *et al.*, 2000). The solid–water interface is perhaps the most dominant phase transfer affecting reactivity in the terrestrial environment, as halogenated hydrocarbons undergo a wide range of matrix interactions, including nondestructive processes such as sorption and sequestration (Luthy *et al.*, 1997), and destructive processes such as microbial, organic- and mineral-mediated transformations (Adriaens *et al.*, 1999; Adriaens and Barkovskii, 2002). Sorption and sequestration processes are strongly influenced by the structural features (backbone and functional groups), and the characteristics of the naturally occurring humic substances (e.g., Dahlman *et al.*, 1993). It is widely believed that these sequestration processes limit the availability of halogenated hydrocarbons to biotic and abiotic reactions or catalysts. These complex biogeochemical interactions will be discussed in more detail in Section 9.14.5.

9.14.4.2 Reaction Energetics

The propensity of halogenated hydrocarbons to react in the various relevant environmental compartments is often explained in terms of the energetic properties associated with their chemical structure (Dolfing and Harrison, 1992; Lynam *et al.*, 1998; Tratnyek and Macalady, 2000). These properties include, but are not limited to, redox potential ($E^{0\prime}$, mV), Gibbs free energy of formation ($\Delta G^{0\prime}$, kJ/reaction), carbon–halogen bond strength, and ionization potential. Figure 5(a) illustrates that the redox potential of most halogenated hydrocarbons corresponds to the range where microbial iron- and nitrate-reduction prevail. Hence, these compounds are likely susceptible to biological reduction (resulting in dechlorination/dehalogenation) as they reside in a range where the addition of electrons is energetically favorable.

By similar reasoning, we can further conclude that oxidation reactions will be most favorable for the lesser-halogenated isomers. This concept is illustrated in Figure 5(b) where relative rates of oxidation and reduction are plotted as a function of level of chlorination. In practice, the literature has revealed that there may be a (chemical-dependent) zone where oxidation and reduction reactions are equally likely. This zone captures aryl halides with four to six chlorines, and alkyl halides with one to three chlorines per molecule. Figure 5(c) shows a trend between the Gibbs free energy for reductive dechlorination and the HOMO–LUMO gap (HOMO = highest occupied molecular orbital, LUMO = lowest unoccupied molecular orbital). It has been argued that the size of the gap is directly correlated to compound reactivity, as it reflects the energy barrier associated with electron transfer. Hence, higher chlorinated (halogenated) compounds would, according to this plot, be chemically more reactive than lesser-halogenated isomers. It should be noted that there has been very limited biochemical validation of this concept, though correlations are apparent during reductive dechlorination and OH radical oxidation. Other molecular descriptors of chemical reactivity of halogenated hydrocarbons, such as electronic, steric, and hydrophobic parameters, have widely been used to predict reaction kinetics as a function of substituent effects, and to probe reaction mechanisms (Peijnenburg *et al.*, 1992; Lynam *et al.*, 1998; Zhao *et al.*, 2001; Lindner *et al.*, 2003).

9.14.5 MICROBIAL BIOGEOCHEMISTRY AND BIOAVAILABILITY

Microbiota, reactive surfaces, and natural organic matter, represent the main causative agents affecting halogenated hydrocarbon transformation and mineralization reactions in the terrestrial environment. This is accomplished by a complex web of interactions between the ecophysiology (structure and function) of microbial communities and the inorganic geochemical characteristics (dissolved and solid phases), which is mainly controlled by the energetics of the prevailing oxidation and reduction reactions involved in the carbon cycle (Adriaens *et al.*, 1999; Adriaens and Barkovskii, 2002). The extent to which these reactions affect the fate of halogenated hydrocarbons is determined by the governing matrix interactions.

9.14.5.1 Ecological Considerations

A fundamental issue in the biogeochemistry of natural environmental systems is the means by which organic carbon transport, geochemical processes and microbial activity combine to produce spatial and temporal variations in redox zonation. The development of redox zones is based on the availability of labile (i.e., biodegradable) organic carbon which can be oxidized, and solid or soluble inorganic compounds which can be used as electron acceptors (reduced) by microbial activities (Chapelle *et al.*, 1996; Christensen *et al.*, 2000). The succession of terminal electron accepting processes (TEAPs) proceeds in order of decreasing redox potential and free-energy yield, and is generally in the order of: oxygen reduction, nitrate reduction, manganese reduction, iron

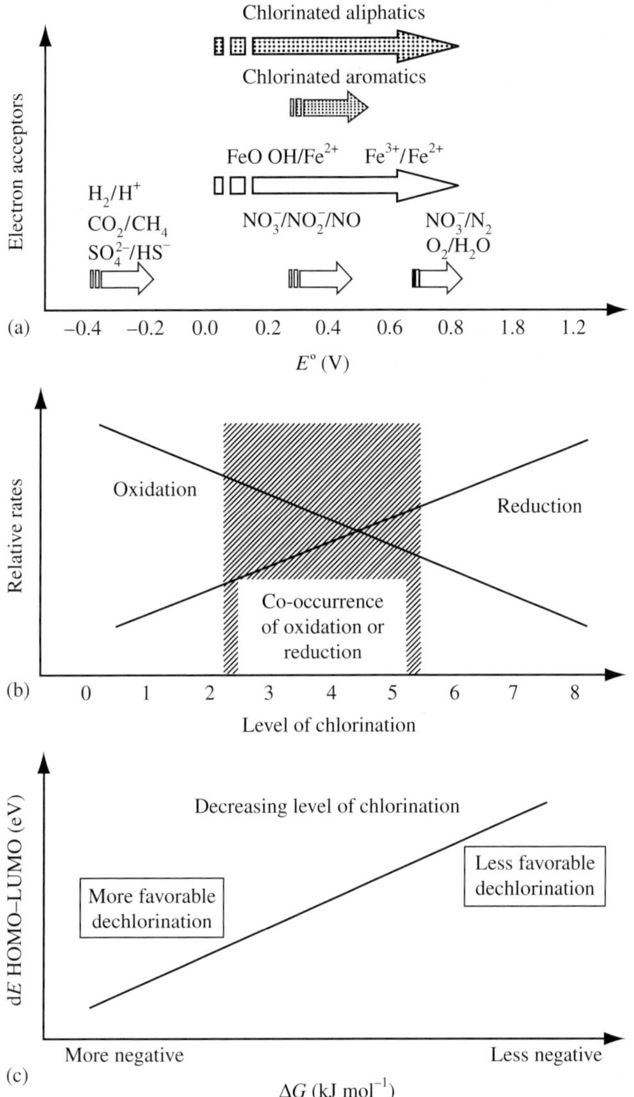

Figure 5 Thermodynamic parameters associated with organohalide reactivity: (a) electon accepting species versus redox potential; (b) relative importance of oxidation versus dechlorination rates as a function of number of chlorines [1–8]; and (c) trend between energy difference of the HOMO and LUMO electron layers and Gibbs free energy of reaction.

reduction, sulfate reduction, and methanogenesis. For more in-depth discussion on this topic, please refer to Bjerg *et al.* (Chapter 9.16). Since the reduction potential of a given redox couple depends on both the abundance and speciation (free, complexed, solid) of the electron donors and acceptors, the prevailing TEAPs in anaerobic environments are often dictated by the available mineral forms of iron and sulfur (Heron and Christensen, 1995; Postma and Jacobson, 1996; Jacobson *et al.*, 1998). The extent of reduction of these species is then controlled by the availability of labile organic matter. Since the development of redox zones is a nonequilibrium process it has been argued that fermentation reactions, which provide soluble labile organic compounds (acetate, hydrogen), are the rate limiting processes in establishing the prevailing TEAP (Postma and Jacobson, 1996).

Generally, in the presence of labile organics or contamination, the TEAP sequence results in increasingly oxidized conditions at distance from the source of organic. Whereas the redox clines (distances for redox gradients) are on the order of millimeters within soil aggregates and vertical sediment cores (Henrichs and Reeburgh, 1987; Fenchel *et al.*, 1998), they can range from meters to tens of meters in groundwater systems (Chapelle, 1993; Christensen *et al.*, 2000). Largely controlled by solute transport in the matrix under consideration (refer to Section 9.14.5.2), the occurrence and extent of each TEAP is dependent on the concentrations of both the electron donor and electron acceptor, the presence and activities of microbial populations, and the temporal changes in site hydrology (Christensen *et al.*, 1997; Hunter *et al.*, 1999). Hence, to interpret TEAP conditions in natural systems, the relative depletion of electron acceptors (or oxidation capacity, OXC) or accumulation of reduced respiration products (total reduction capacity, TRC) is usually complemented with redox and dissolved hydrogen gas measurements, and verified with microbial community analysis (Chapelle *et al.*, 1996; Ludvigsen *et al.*, 1997; Lendvay *et al.*, 1998a,b; Skubal *et al.*, 2001; McGuire *et al.*, 2000). TEAP conditions and redox zonation affect the chemical form, mobility, and persistence of many anthropogenic contaminants and natural halogenated hydrocarbons alike, due to the direct or indirect activities (e.g., the accumulation of reactive iron oxides or sulfides) of the prevailing microbial communities (e.g., Adriaens *et al.*, 1999; McCormick *et al.*, 2002a). The various catabolic contributions of microbiota to environmental transformation processes and pathways will be discussed in Section 9.14.6.

Limited quantitative information is available on the fluxes and reactivity of organic matter in soils and subsurface environments (Figure 6). With concentrations of low molecular weight (labile) organic matter (whether dissolved or solid-associated) in uncontaminated aquifers on the order of ng L^{-1} to μg L^{-1}, the rates of

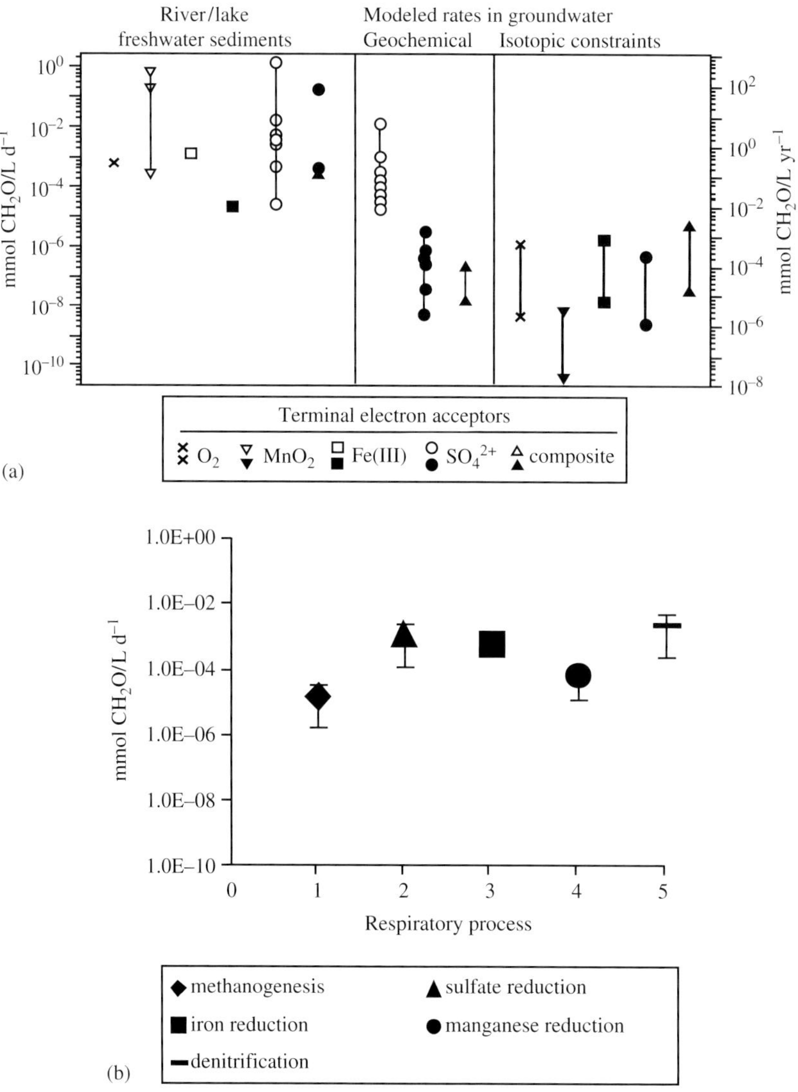

Figure 6 Distribution of reported organic matter turnover under the various TEAPs in pristine surface water sediments and groundwater aquifers (a) and landfill-leachate contaminated groundwater (b). The respiration rates in pristine groundwater ((a) middle and right panels) were modeled using total electron acceptor species concentration (geochemical) and carbon isotope fractionation (isotopic constraints) (after Murphy and Schramke, 1998 and Ludvigsen *et al.*, 1998).

biogeochemical transformations in these oligotrophic environments tend to be orders of magnitude lower than those found in surface soils or lake and marine sediments (e.g., Capone and Kiene, 1988; Murphy and Schramke, 1998; Hunter *et al.*, 1999). Estimates for carbon oxidation rates in pristine groundwater systems have been measured and calculated based on geochemical modeling to be $\sim 10^{-3}-10^{-8}$ mmol $CH_2O\ L^{-1}\ d^{-1}$ (Murphy and Schramke, 1998), and approximately one to two orders of magnitude higher in contaminated environments (Ludvigsen *et al.*, 1998). These rates are not constant, but are dependent on the fluxes of dissolved organic carbon (DOC) and geochemical species (e.g., Skubal *et al.*, 2001).

In sediments, the rates of redox zone development in sediment depth profiles are strongly correlated with sedimentation rates (Henrichs and Reeburgh, 1987). Indeed, spatial and temporal variations in microbial processes have been observed in terrestrial, estuarine, and coastal aquatic sediments: aerobic and denitrifying activity is confined to the top few centimeters of sediments, sulfate-reduction has been observed over 50–60 cm of sediment thickness, which is underlain by a zone of methanogenic activity (Skyring, 1987; Carlton and Klug, 1990). Thus, sulfate-reduction is expected to be the dominant process responsible for organic matter oxidation to carbon dioxide in estuarine and coastal sediments, unless high sedimentation ratios prevail and methanogenic conditions develop in the deeper layers (Skyring, 1987). Sulfate-reducing processes significantly affect porewater and sediment chemistry, including a decrease in pH with a concurrent rise in alkalinity and sulfides, carbonate precipitation, ammonia production, and reduction of iron(III) hydroxide minerals by sulfides. Estimated carbon oxidation fluxes from sediments of varying geochemistry range from 3 to 458 mmol $C\ m^{-2}\ d^{-1}$ (Capone and Kiene, 1988). The predominant global contributions to anaerobic carbon metabolism occur in shallow sediments (7.5–1,300 g $C\ m^{-2}\ yr^{-1}$) and estuaries and bays (2–150 g $C\ m^{-2}\ yr^{-1}$). Of the various electron acceptors, sulfate ((370–1,300) $\times 10^{12}$ g $C\ yr^{-1}$) is the predominant contributor to global carbon production as compared to oxygen ((150–1,900) $\times 10^{12}$ g $C\ yr^{-1}$), nitrate ((20–84) $\times$ 10^{12} g $C\ yr^{-1}$), and methane ((30–52) $\times 10^{12}$ g $C\ yr^{-1}$) (Henrichs and Reeburgh, 1987).

In topsoils, the development of redox zones is directly related to the soil volumetric water content (cm^3 of water per cm^3), as this determines the composition and activity of soil biota (Fenchel *et al.*, 1998). Water potential is affected by both solute and matrix characteristics, which subsequently affect the ecophysiology of microorganisms. The soil biota which are of interest to this chapter include bacteria, and low-water potential-tolerant fungi. The impact of water and oxygen limitations on the terrestrial nitrogen cycle have been well documented and affect the relative rates of ammonification, respiration, nitrification, and denitrification (Tiedje, 1988). The carbon equivalent oxidation fluxes due to denitrification globally amount to (13–233) $\times 10^{12}$ g $C\ yr^{-1}$ (including wetlands). Taking into account all soil respiration processes on the order of 6×10^{16} g $C\ yr^{-1}$ are liberated into the atmosphere (Schlesinger, 1991). Hence, the flux of oxidation equivalents in topsoils and sediments govern the geochemical cycling of the subsurface and, by governing relevant biogenic production and biomineralization activity, has potentially the greatest impact on the chlorine cycle.

9.14.5.2 Matrix Interactions

The physical–chemical properties of halogenated hydrocarbons in general, and of aryl halides (mainly POPs) in particular, include relatively low vapor pressures (aryl halides), high octanol–water partition coefficients, very high octanol–air partition coefficients, and generally low solubility. These features, in combination with the complex chemical characteristics of natural soil and sediment organic matter, affect the availability of these compounds to biochemical or abiotic reactions (Luthy *et al.*, 1997). Additionally, due to the multitude of reactive phenolic and carboxylic functional groups in sediment organic matter and many aryl halides (or their transformation products), ample opportunity exists for cross-linking and other sequestration reactions (Luthy *et al.*, 1997; Adriaens *et al.*, 1999; Adriaens and Barkovskii, 2002) which remove these contaminants effectively out of the biogeochemical cycle controlled by transport and phase partitioning processes.

The incorporation of parent halogenated hydrocarbons or their transformation products in soil organic matter has been indicated by the formation of operationally defined "polar material" and polymeric humic substances by addition or condensation reactions (Scheunert *et al.*, 1992; Adriaens and Barkovskii, 2002). Particularly under aerobic conditions, phenolic dimers or polymers, humic addition products, and condensation products have been observed (Andreux *et al.*, 1993). Apparently a significant fraction of anthropogenic halogenated hydrocarbon molecules (and their transformation products) can ultimately be incorporated in natural soil constituents. A factor which complicates the evaluation of soil-bound residues is that all organic matter recovered from natural sites, including that which is not contaminated, is chlorinated to some extent due to the incorporation of inorganic salts in low and high molecular weight humic-like substances (e.g., Oberg, 1998). Numerous attempts have been made to identify the types of structures in organic matter found in the environment (e.g., unpolluted waters, marine sediments, and coniferous forest soils) that are chlorinated. Using an oxidative degradation technique, several mono- and dichlorinated aromatic structures, exhibiting hydroxy-, methoxy-, and ethoxy-substituents have been detected (e.g., Dahlman *et al.*, 1993); however, the chlorine bound to such structures corresponded to a small fraction of the total organic chlorine (AOX). Considering that the natural AOX can often exceed the anthropogenic AOX in concentration, the absolute determination of matrix-bound material as the result of microbial degradation reactions as compared to microbial formation reactions may be difficult to estimate.

Transport in porous matrices also affects the availability of halogenated hydrocarbons to microbial or abiotic reactions. In soils or sediments where advective fluid flow is minimal, transport is governed by diffusion, whereas in groundwater systems both advective and diffusive transport processes are operative. Field measurements for sorbed halogenated hydrocarbons in sediments indicate that transport of even the most soluble components is on the order of decade-long time frames for centimeters distance (e.g., Gevao *et al.*, 1997; Zhang *et al.*, 1999). Halogenated hydrocarbons in topsoils (and water bodies), which are exposed to significant diurnal or seasonal temperature changes, undergo advection-driven (heat) air–surface exchange processes in timeframes on the order of hours to days (e.g., Lee *et al.*, 2000). Similarly, in low organic sandy/silty groundwater aquifers, advective transport, in combination with substantial longitudinal dispersion and sorption mechanisms, controls the movement of halogenated hydrocarbons on the order of centimeter per day (Wiedemeier *et al.*, 1999).

9.14.6 ENVIRONMENTAL REACTIVITY

Halogenated compounds have been distributed in the ambient environment in one form or another for millions of years. Even compounds such as dioxins appear to have been formed in million-year old ball clays (Ferrario and Byrne, 2000) by as of yet unknown mechanisms, and are by-products during natural incineration processes such as forest fires. In addition, plants, algae, and microorganisms produce and release these compounds for various physiological and ecological reasons, indicating that some latent activity for biosynthesis and biodegradation may have evolutionary roots. Thus, it is not surprising that our knowledge of microbial degradation pathways for stable environmental contaminants keeps expanding, as new enrichment techniques aimed at harnessing and exploring the microbial ecological potential are developed (Palleroni, 1995; Fathepure and Tiedje, 1999).

Aside from microbial mediation, halogenated hydrocarbons and inorganic halide ions may be transformed in the environment by a variety of abiotic species including reduced or oxidized organic matter, reactive surfaces, and other soluble species (Adriaens *et al.*, 1996; Fu *et al.*, 1999; Butler and Hayes, 2000; McCormick *et al.*, 2002a). By definition, abiotic reaction implies that no direct microbial mediation is required. Thus, these reactions are not necessarily limited to the terrestrial environment (e.g., photolysis, hydroxide radical-mediated oxidation). Indirectly, anaerobic microbial populations such as iron-, manganese-, and sulfate-reducers are responsible for the formation of reactive biogenic minerals and surfaces (Fredrickson and Gorby, 1996; Rugge *et al.*, 1998; Thamdrup, 2000; McCormick *et al.*, 2002a). Considered collectively, the abundance of photoreactive particles in atmospheric aerosols (Andreae and Crutzen, 1997), as well as the reducing power of the estimated $(1{,}500-2{,}000) \times 10^{12}$ g organic carbon stored in soil layers (Post *et al.*, 1982), indicate that terrestrial and atmospheric abiotic reactivity has the potential to exceed biotic transformation in the global biogeochemical chlorine (halogen) cycle. While limited mass balances have been developed on the global distribution and fluxes of halogenated hydrocarbon formation, estimates of degradation fluxes in the global environment have not been developed, and the information available is at best limited to local contamination burdens. To limit the extent of the review, only those processes that have applicability to natural terrestrial environmental systems will be described.

9.14.6.1 Microbial Reactivity

The microbiology of halogenated hydrocarbon transformation has been discussed in a number of treatises over the years (Vogel *et al.*, 1987; Häggblom, 1992; Mohn and Tiedje, 1992; Adriaens and Hickey, 1994; Adriaens and Vogel, 1995; Adriaens *et al.*, 1999; Adriaens and Barkovskii, 2002). Since the present chapter includes alkyl and aryl halides, this section will review the most salient features of the enzymes and degradation processes relevant to these compounds, under aerobic and anaerobic conditions. Among all microbiota that may participate in the halogen cycle, fungi will be featured separately, as they have been found to substantially contribute to halogenated hydrocarbon formation, as well as biodegradation (de Jong and Field, 1997).

The evolution of regulators and catabolic genes has resulted in an extraordinary metabolic versatility of microorganisms to transform or mineralize halogenated hydrocarbons, either via growth (energy-coupled) or nongrowth (cometabolic; nonenergetic) processes (Table 2). Whereas the latter are often operationally defined based on specific laboratory conditions, increasing evidence suggests that many aryl and alkyl halides are involved as fortuitous inducers controlling pathway-specific regulation, even though they do not serve as substrates for catabolic enzymes (Fathepure and Tiedje, 1999; Copley, 1998). This gratuitous induction may then result in a partial transformation of the halogenated substrate, often resulting in enzyme inhibition (van Hylckema-Vlieg *et al.*, 2000). The cometabolic nature of dehalogenation has, however, undergone a substantial revision during the 1990s, due to the discovery of a novel microbial respiration process, dehalorespiration. This process allows many anaerobic microorganisms to generate energy through the promotion of a proton-motive force resulting from the dechlorination step (Holliger and Schraa, 1994; Fathepure and Tiedje, 1999). Hence, for these organisms to thrive in contaminated environmental systems, halogenated compounds (e.g., chloroethenes, chlorobenzoates, chlorophenols, PCDD) need to be present as the electron acceptors, and simple organic acids (e.g., acetate, formate, ...) as the electron donors. The presence of dehalorespiring microorganisms has been demonstrated globally, in halogenated hydrocarbon-impacted environments. Considering the redox potentials of aryl and alkyl halides (Figure 5(a)), these processes are energetically very favorable as they are at par with iron and nitrate reduction. Indeed, environmental transformation of halogenated hydrocarbons has been observed across the entire spectrum of respiratory TEAP zones (Adriaens *et al.*, 1999).

Acclimation of microorganisms to halogenated hydrocarbons in the environment has resulted in a wide range of aerobic and anaerobic mechanisms to initiate aryl halide degradation, including oxidative, hydrolytic, and reductive dehalogenation, via novel recruitment and adaptation of proteins (Copley, 1998; van der Meer *et al.*, 1992). Whereas the oxidative and reductive processes occur predominantly under strictly aerobic and anaerobic conditions, respectively, hydrolytic processes prevail in aerobic and micro-aerophilic conditions. Some studies have indicated, e.g., the importance and ubiquity of haloalkane dehalogenase, dichloromethane dehalogenase, tetrachloro-hydroquinone dehalogenase, and perchloroethylene and trichloroethylene reductive dehalogenases (reviewed in Copley, 1998). It appears that complete or partial halogen removal is the predominant first step during microbial transformation of many important halogenated pollutants, resulting in the accumulation of hydroxylated and/or lesser halogenated intermediates (Table 3). Hydroxylated aromatics are considered "activated" as the functional group destabilizes the aromatic Π-electrons. In contrast, hydroxyalkanes are precursors for microbial dehydrogenase and hydroxylase activity. The compounds are then rendered more susceptible to other chemical transformations, such as ring cleavage and the formation of carboxylated acids, resulting in the production of catabolic intermediates to enter the tricarboxylic acid (TCA) or derivative cycles. The extensive removal of halogens through reductive dehalogenation under anaerobic conditions allows either for aerobic processes to commence, or for hydrogenase- and hydratase-type of anaerobic ring reduction and hydrolysis to produce catabolic intermediates

Table 2 Examples of halogenated and nonhalogenated compounds degraded via microbial growth and cometabolic processes.

Process	*Halogenated*	*Nonhalogenated*
Mineralization		
Aerobic	Mono-, dichlorobenzenes/benzoates/phenols; 2,4-D; 2,4,5-T; chlorinated aliphatic acids	BTEX, saturated/unsaturated hydrocarbons, naphthalene, creosote compounds
Anaerobic	Chlorobenzoates/phenols, dichloromethane	BTEX, aliphatic acids
Co-metabolism		
Aerobic	Di-pentachlorobenzenes/benzoates/ phenols/biphenyls; mono-dichlorodioxins/furans; chloroethenes;	PAH; long chain/branched hydrocarbons
Anaerobic	Di-hexachlorobenzenes, di-decachloro biphenyls; di-octachlorodioxins; PCE, trichloromethane; DDT; Lindane	PAH, aliphatics

Abbreviations: 2,4,-D: 2,4-dichlorophenoxyacetic acid; 2,4,5-T: trichlorophenoxyacetic acid; BTEX: benzene, toluene, ethylbenzene, xylenes; PAH: polycyclic aromatic hydrocarbons.

Table 3 Enzyme classes capable of transforming and dehalogenating organohalides under aerobic and anaerobic conditions.

Enzyme class	*Substrates*	*Reaction*
Monooxygenases (e.g., methane, peroxidases...)	Alkanes (C_1–C_{18}) Alkenes (C_2) Aromatics (phenol, toluene,...)	$RX + O_2 \rightarrow ROH + H_2O + X^-$ $ArX + O_2 \rightarrow ArOH + H_2O + X^-$
Dioxygenases	Alkenes (C_2) Aromatics (benzene, naphthalene,...)	$RX = RX + O_2 \rightarrow$ unstable oxirane + halogenated by-products $ArX + O_2 \rightarrow Ar(OH)_2$
Ring-cleavage dioxygenases	(Halogenated) 1,2-dihydroxy-aromatics	$Ar(OH)_2 + O_2 \rightarrow$ (halogenated) *ortho-* or *meta*-ring fission products
Dehalogenases (haloalkane, DCM, PCE/TCE reductive, aromatic hydrolytic)	Aliphatic acids, Aromatic acids	$RX + H_2O \rightarrow ROH + HX$ (aerobic) $RX + H_2 \rightarrow RH + HX$ (anaerobic) $ArX + H_2O \rightarrow ArOH + HX$ (aerobic) $ArX + H_2 \rightarrow RH + HX$ (anaerobic)

Abbreviations: DCM: dichloromethane; PCE: perchloroethene; TCE: trichloroethene.

(Adriaens and Hickey, 1994; Adriaens *et al.*, 1999).

Observations of these reactions in the natural environment indicate that they take place on the order of years per chlorine removed for aryl halides in sediments (Brown *et al.*, 1987; Adriaens *et al.*, 1999; Fu *et al.*, 2001; Gaus *et al.*, 2002), and months to years for alkyl halides or pesticides in groundwater (Potter and Carpenter, 1995; Wiedemeier *et al.*, 1999; Skubal *et al.*, 2001). Accurate calculations account for the metabolic rates measured in sediments or soils (Henrichs and Reeburgh, 1987; Schlesinger, 1991; Murphy and Schramke, 1998). For example, Fu *et al.* (2001) estimated the rate of cometabolic dioxin dechlorination in estuarine Passaic River sediments at 3–8 pg of diCDD per gram sediment per year, based on sediment respiration rates, and the ratios of dioxin dechlorination as compared to methane production, which represented the dominant TEAP under which these reactions occurred.

Whereas all previously described degradation reactions are catalyzed by archea and bacteria or by unspecified populations under various TEAPs, Basidiomycetes or wood-rotting fungi appear to be responsible for both biotic halogenated hydrocarbon production (e.g., Figure 4), and their transformation or mineralization (de Jong and Field, 1997). The source of organic chlorine in soil, and its subsequent volatilization has been ascribed to haloperoxidases, a group of enzymes that catalyzes the halogenation of humic substances in the presence of hydrogen peroxide and halide ions (Neidleman and Geigert, 1986). Chloroperoxidases oxidize chloride, bromide, and iodide and have been extracted from the fungus *Caldariomyces fumago*, in soil extracts and in spruce forest soils (Laturnus *et al.*, 1995; Asplund and Grimvall, 1991; Asplund *et al.*, 1993). The enzyme activity is pH-dependent (2.5–4.0), and its specific activity decreases with depth. The main reactive chlorine formed is HOCl, which then reacts with any other organic substrate for biodegradation purposes, while forming organic chlorine as a by-product (Johansson *et al.*, 2000). Aside from this haloperoxidase-like activity in (mainly) forest soils, halogenated hydrocarbons (>80 identified to date), and halomethanes are produced by 68 genera from 20 different Basidiomycete families (de Jong and Field, 1997; Watling and Harper, 1998). The aryl halides include anisyl (methoxylated chlorobenzyl) compounds (15–75 mg kg^{-1} wood or litter), monomeric and polymeric hydroquinone methyl ethers (74–2,400 mg kg^{-1}), orcinol (dimethoxy) methyl ethers, anthraquinones, and a wide range of other compounds (de Jong and Field, 1997). Many of these compounds are produced for physiological functions such as antibiotic activity and lignin degradation. The haloalkanes are dominated by chloro- and bromomethane, as indicated earlier.

Besides halogenated hydrocarbon production, the Basidiomycetes are also capable of aryl halide mineralization, and hence part of the *de novo* biosynthesized halogenated hydrocarbons may be fully mineralized. The remainder becomes incorporated into humus (de Jong and Field, 1997), and may produce PCDD/F as a by-product (de Jong *et al.*, 1994). The white-rot fungi, which are capable of degrading hemicellulose, lignin, and cellulose through an extracellular, oxidative enzyme system, also have the capability to biotransform a wide range of chlorinated pollutants via a nonspecific free-radical mechanism (e.g., Bumpus *et al.*, 1985; Barr and Aust, 1994). These pollutants include aliphatic halogenated hydrocarbons such as chloroform, trichloroethylene, and tetrachloromethane (Table 3). Aromatic halogenated hydrocarbons degraded by white rot fungi include chlorolignins (bleach kraft mill effluents), pentachlorophenol, PCDD/F, PCBs, and pesticides such as DDT (1,1,1-trichloro-2,2-bis(4-chlorophenyl)ethane) and atrazine with mineralization rates ~10–70% (reviewed in de Jong and Field, 1997). The balance between production and mineralization of the more complex aromatic by-products is unclear, but all evidence points towards a net accumulation (Oberg, 2002). For example, Watling and Harper (1998) describe the accumulation of chloroanisyl compounds at 15–75 mg kg^{-1} in wood or litter colonized by fungi in forest environments, and the presence of chlorinated hydroquinone metabolites at levels ranging from 74 mg kg^{-1} to 2,400 mg kg^{-1} in fungal tissue.

9.14.6.2 Surface-mediated Reactivity

Since the 1990s, there has been a trend toward using zero-valent metals to dehalogenate groundwater contaminants via surface-mediated reactions (e.g., Gillham and O'Hannesin, 1994; Matheson and Tratnyek, 1994). Since these reactive metals are not indigenous to natural environments, this section will consider the analogous manner in which biogenic ferrous and sulfidic minerals may contribute to the chlorine cycle. Iron is the fourth most abundant element in the Earth's crust, accounting for 3.6% by mass (on average) of Earth's surface rocks (Martin and Meybeck, 1979). In contrast, sulfur is the fourteenth most abundant element in the Earth's crust with the content of surface rocks varying from 0.027% to 0.240% by mass (Bowen, 1979). This abundance in many natural environments (soils, groundwater, surface water) sustains substantial dissimilative iron reduction and sulfate reduction in the presence of labile organic matter. It is then not surprising that these respiratory processes, and their associated microbial communities (DIRB: dissimilatory iron reducing bacteria; SRB: sulfate reducing bacteria) are dominant in marine, estuarine, and many groundwater environments. More importantly with respect to the present chapter, DIRB and SRB activity results in the accumulation of a variety of reactive iron-oxides, -hydroxides, and -sulfides as their respiratory end-products (Fredrickson *et al.*, 1998; Lovley, 1991; Rickard, 1969; Rickard *et al.*, 1995; Vaughan and Lennie, 1991; Zachara *et al.*, 2002).

The variety of common oxidation states encountered for iron(II, III) and sulfur(−II, 0, II, IV, VI) results in complex and varied speciation for these elements in natural environments. Limiting the scope to only those oxidation states and species that commonly support anaerobic respiration, narrows the discussion to primarily the oxides of iron(III) and sulfur(II, IV, VI). Oxides of Fe(III) are primarily found as insoluble oxide or hydroxide particle coatings and aggregates (Thamdrup, 2000), and can comprise several mass percent of freshwater or marine sediments (Coey *et al.*, 1974; Raiswell and Canfield, 1998; Thamdrup, 2000). The ferrous iron, Fe(II), that is produced as a consequence of iron respiration may accumulate in solution, adsorb to the surfaces of surrounding minerals or become incorporated into new biogenic minerals through a variety of chemical or microbially mediated pathways (schematically represented in Figure 7). In contrast to the oxides of Fe(III), the oxides of sulfur form soluble oxyanions that may be found in a variety of mixed oxidation-state polynuclear complexes (Erlich, 1996). Those most commonly known to support anaerobic respiration include sulfate (SO_4^{2-}), sulfite (SO_3^{2-}), and thiosulfate ($S_2O_3^{2-}$). However, elemental sulfur can also serve as a terminal electron acceptor for some species of eubacteria and archea (Schauder and Kroeger, 1993). The ultimate and most common product of sulfur respiration is sulfide, S(−II). In the presence of iron, biogenic S(−II) can precipitate as numerous insoluble sulfides, including pyrite (FeS_2), mackinawite (FeS_{1-x}), pyrrhotite ($Fe_{1-x}S$), and greigite (Fe_3S_4) (Rickard, 1969; Rickard *et al.*, 1995). As indicated by Figure 7, collectively DIRB and SRB can significantly alter the chemical and mineralogical speciation in their environment.

Despite the importance of biogeochemical iron- and sulfur-cycling (Rickard *et al.*, 1995; Schlesinger, 1991; Lovley, 1997; Erlich, 1996), the ability of biogenic ferrous and sulfidic minerals to mediate organohalide transformations has received limited attention. The thermodynamic feasibility of these reactions to occur under iron-reducing conditions is illustrated in Figure 8, which incorporates some of the dominant redox couples (mineral and biological) relevant to this TEAP. The range of substrates presently known to undergo dechlorination in the presence of reduced iron oxides and iron sulfides includes, among others: tetrachloromethane, hexachloroethane,

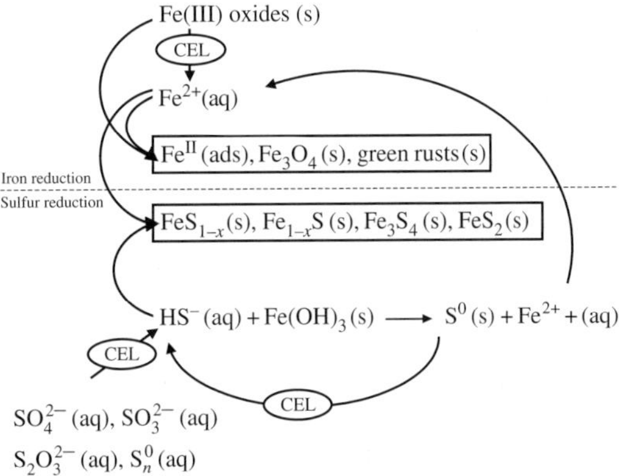

Figure 7 Formation of mineral and sorbed Fe(II) species under iron-reducing conditions (source McCormick *et al.*, 2002b).

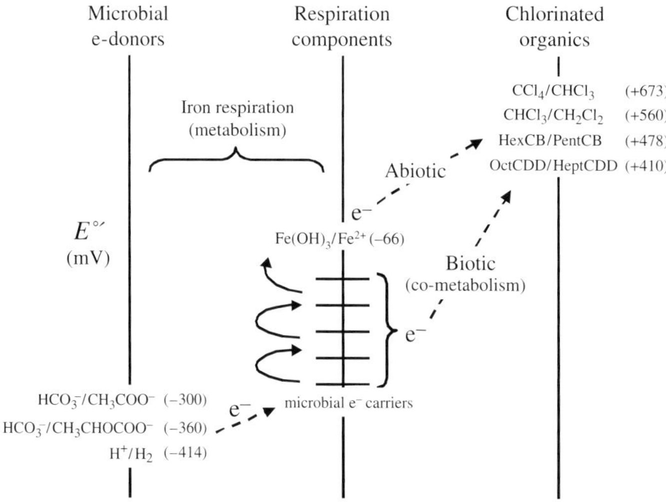

Figure 8 Comparison of microbial, mineral, and chloromethane redox pairs under iron-reducing conditions (after McCormick *et al.*, 2002a).

tetra- and trichloroethylene, bromoform, chloropicrine, hexachlorobenzene, and possibly PCBs (Kriegman-King and Reinhard, 1992; Svenson *et al.*, 1989; Butler and Hayes, 2000; Cervini-Silva *et al.*, 2000). Since most of these studies were confined to abiotic laboratory systems, it is difficult to assess the relative impact of surface-mediated reactions on organohalide fluxes in the natural environment. However, some insights may be gleaned from a recent study by McCormick *et al.* (2002a), who compared the reaction kinetics of tetrachloromethane transformation in the presence of the dissimilative iron-reducer *Geobacter metallireducens* and biogenic magnetite generated by this same strain. Due to the abundance of reduced mineral surfaces, the abiotic contribution to CCl_4 transformation was found to be at least two orders of magnitude greater than the cell mediated (co-metabolic) reaction (McCormick *et al.*, 2002a). Considering that the reported dechlorination rates observed with iron sulfides are orders of magnitude faster than those with reduced iron oxides (Butler and Hayes, 2000), it is likely that the contribution of these solids will exceed the microbial rates as well.

9.14.6.3 Organic-matter-mediated Reactivity

Soil and sediment organic matter affect environmental transformation reactions through direct participation in reduction (e.g., dechlorination) and oxidation (halogenation) reactions (e.g., Svenson *et al.*, 1989; Barkovskii and Adriaens, 1998; Fu *et al.*, 1999; Keppler *et al.*, 2000), or by serving as an electron acceptor for microbial respiration (Lovley *et al.*, 1996). The latter reaction has been coupled to the microbial capability to oxidize vinyl chloride and *cis*-dichloroethene (Bradley *et al.*, 1998). With quinonic and phenolic groups constituting from 13% to 56% (molar concentration) of all oxygen functional groups in natural organic matter (Schlesinger, 1991), environmental transformation reactions mediated by natural organic matter may potentially contribute significantly to the fate of halogenated hydrocarbons (Oberg, 1998, 2002).

Couples such as hydroquinone/quinone have been hypothesized to dominate the redox properties of humic and fulvic acids, and to act either as electron transfer mediators or as the direct donors of electrons for dechlorination reactions (Schwarzenbach *et al.*, 1990; Dunnivant *et al.*, 1992). For example, it has been shown in sediment–water systems that the rates of alkyl halide reduction increase with organic matter content (Peijnenburg *et al.*, 1992). Further support for this hypothesis was obtained by Svenson *et al.* (1989), who reported a first-order dependence between rates of hexachloroethane reduction and hydroxyl concentrations. Aside from alkyl halides, structural features of organic matter have been shown to catalyze (Fu *et al.*, 1999) or accelerate (Barkovskii and Adriaens, 1998) the dechlorination of dioxins (Figure 9).

A correlation between reaction rates, molecular structure of the humic or fulvic acid, and content of reactive sites is more difficult to demonstrate. It has been hypothesized that the hydroquinone or quinone is the main reactive site for electron transfer during dechlorination reactions. Phenolic acidity, as based on the inflection point during titration of organic matter, is indicative of the hydroquinone content within humic materials. Published information indicates that the quinone content of humic acids is generally higher than for fulvic acid (Stevenson, 1994).

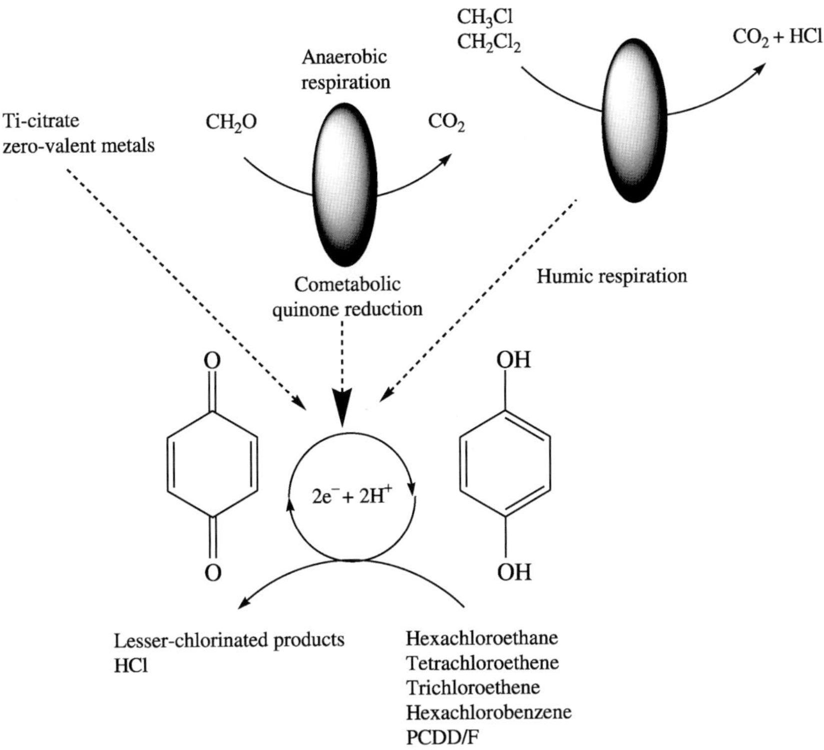

Figure 9 Model representation of organic-mediated dehalogenation reactions in the presence (oval) and absence of microorganisms.

In their study on the reductive dehalogenation of hexachloroethane, carbon tetrachloride, and bromoform, Svenson *et al.* (1989) found that the addition of humic acid or organic matter from aquifer material to aqueous solutions containing bulk electron donors increased the reduction rate by up to 10-fold. Although they did not determine the phenolic acidity, Peijnenburg *et al.* (1992) established a positive correlation between the reductive rate constants of a range of aryl halides and the organic carbon content of sediment systems. According to the rationale outlined earlier, a humic material with high hydroquinone/quinone content should facilitate a more effective electron transfer to electron acceptor molecules. Just as organic matter-mediated formation has been argued to contribute substantially to the global production of halomethanes, the involvement of humics in environmental dehalogenation reactions of trace halogenated hydrocarbons may be more substantial than currently assumed. Support for this hypothesis may be derived from observations that dioxin dechlorination patterns in sediment systems (Albrecht *et al.*, 1999) indicate the influence of a humic-dechlorination signature in model systems (Fu *et al.*, 1999). The predominant congener endpoints observed in the model systems of Fu *et al.* (1999) were also observed at the air–water interface in the estuary from which the sediments were obtained (Lohmann *et al.*, 2000), pointing towards the possible occurrence of a dioxin cycling mechanism influenced by organic catalysis.

9.14.6.4 Predictive Models: Structure–Reactivity Relationships

Structure–activity relationships (SARs) between chemical reactivity and molecular descriptors (Figure 10) for both alkyl and aryl halides are valuable tools to explore mechanistic determinants influencing activity, and to predict rate constants for environmental transformation within classes of compounds (Wolfe *et al.*, 1980; Moore *et al.*, 1990; Hermens *et al.*, 1995). An abundance of these statistically significant correlations based on appropriate predictor variables is available, yet very few of these yield useful quantitative structure–activity relationship expressions (QSARs). One of the limitations is the scarcity of relationships between readily available molecular descriptors and important microbial transformation reactions. Another relates to the lack of understanding of the contaminant–matrix interactions that govern the degradation process. Perhaps not surprisingly, the most robust SARs and QSARs have been generated for strictly abiotic transformation

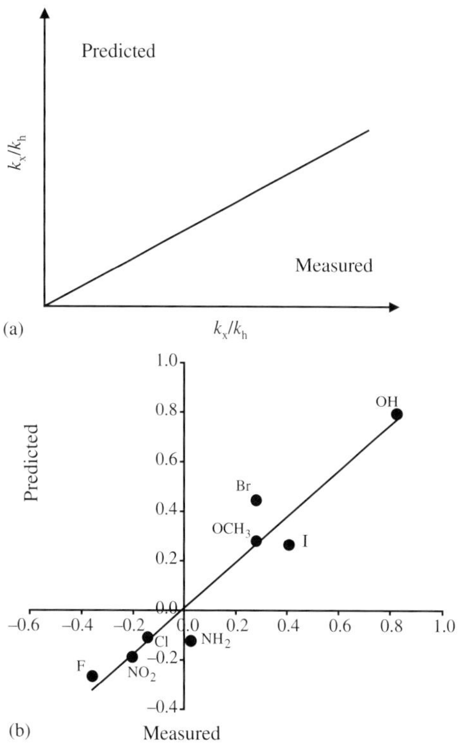

$k_x/k_h = 3.36 - 7.58(2.68)\,^{12}C + 0.59(0.25)\,E_S - 0.56(0.23)\log K_{ow}$; $R^2 = 0.92$

Figure 10 Theoretical (a) and practical (b) representation of QSARs. Panel b describes a QSAR for the methanotrophic oxidation (activity of methane monooxygenase) of *ortho*(C_{12})-substituted biphenyls. The structural backbone was biphenyl, and the substituents considered included all halogens, methyl-, methoxy-, hydroxyl-, nitro-, and amino-moieties (Lindner *et al.*, 2003). The molecular descriptors used in (b) are ^{12}C (charge on the *ortho*-carbon), E_s (Taft's steric parameter), and log K_{ow}.

processes such as hydrolysis, direct photolysis, and oxidation. More recently, correlations have been generated describing fortuitous degradation processes (e.g., reductive dechlorination) or for enzymes exhibiting broad substrate specificities (e.g., soluble methane monooxygenase) (Roberts *et al.*, 1993; Cozza and Woods, 1992; Rorije *et al.*, 1995; Lynam *et al.*, 1998; Lindner *et al.*, 2003).

Most QSARs are based on the Hansch method, where biological response is expressed as a linear function of hydrophobic, electronic, and steric properties (refer to equation below and Figure 10; k_x = rate constant for substituted compound; k_h = rate constant for unsubstituted compound):

$$\frac{k_x}{k_h} = \underbrace{aX + bY + cZ}_{\text{Electronic, Polar, Structural}} + \underbrace{dX' + eY' + fZ'}_{\text{Quantum-chemical}} + \underbrace{gE}_{\text{Environmental}} + \text{error}$$

Descriptors: Electronic Polar Structural; Quantum-chemical; Environmental

Ab initio and semi-empirical molecular orbital calculations have been used to obtain molecular descriptors such as electron density distributions of the highest occupied Π orbital, carbon–chlorine bond charges, heats of formation and ionization potentials, or HOMO–LUMO gaps (Okey and Stensel, 1996; Lynam *et al.*, 1998; Zhao *et al.*, 2001). One or more of these parameters have been found to be strongly correlated to the observed dechlorination pathways of dioxins and substituted benzenes (Cozza and Woods, 1992; Lynam *et al.*, 1998), to the preferred methane monooxygenase oxidation pathways of *ortho*-substituted biphenyls (Lindner *et al.*, 2003), or dioxygenase-mediated oxidation of dioxins and furans (Damborsky *et al.*, 1998). For example, Figure 10 shows a SAR between the measured and predicted oxygen uptake rates exhibited by a methanotrophic bacterium, *Methylosinus trichosporium* OB3b, against *ortho*(designated as C_{12})-substituted biphenyls. This particular SAR includes electronic (carbon charge), structural (compound width), and polarity (octanol–water partitioning coefficient) descriptors; no environmental or quantum-chemical descriptors were considered (Lindner *et al.*, 2003).

Robust QSARs were developed to predict the reductive transformation constants of 45 halogenated monoaromatic hydrocarbons and 13 halogenated aliphatic hydrocarbons in anoxic sediments using either Hansch-type descriptors (the carbon–halogen bond strength, the summation of the Hammett sigma constants and inductive constants for the additional substituents, and the steric factors for these substituents) (Peijnenburg *et al.*, 1992) or quantum-chemical descriptors (Rorije *et al.*, 1995; Zhao *et al.*, 2001). Whereas the electronic properties of the substituents exhibited a positive effect (increased the reaction rate), steric and bond strength factors decreased the predicted reaction rate. When the sediment organic matter content was included in the prediction, the electronic properties and quantum-chemical descriptors became less influential in predicting reaction rates. This indicates that the predicted reaction rate in more complex environmental samples is controlled by mass transfer limitations.

9.14.7 IMPLICATIONS FOR ENVIRONMENTAL CYCLING OF HALOGENATED HYDROCARBONS

The goal of this chapter was to assess the influence of environmental phase partitioning and transport on the biogeochemical cycling and terrestrial reactivity of halogenated hydrocarbons. Whereas the global environmental behavior with respect to POPs and non-POPs distribution can be described within regional or latitudinal contexts, the impact of terrestrial environmental transformations on contaminant biogeochemistry is of major importance at contaminant "hot spots." The biogeochemistry of volatile and semi-volatile halogenated hydrocarbons is very diverse and scale-dependent, and differs in terms of major environmental reservoirs, transport pathways and transformation reactions. Particularly pertaining to the latter, the environmental degradation or mineralization fluxes and their contribution to the global halogenated hydrocarbon biogeochemical cycles are difficult to quantify, because of: (i) the local nature of the available databases of reactive surfaces, reactive organic matter, and microbial reactivity; (ii) the differences in source characteristics of volatile, semi-volatile, and complex halogenated organic compounds; and (iii) the slow natural reaction rates (especially degradation) relative to the overall environmental cycling of these contaminants.

Despite these limitations, the literature on the natural attenuation of anthropogenic substrates in terrestrial environments has provided some insights into environmental transformations affecting halogenated substrates under a wide range of environmental conditions, and may allow some extrapolations to regional or global impact (e.g., Wiedemeier *et al.*, 1999; Lendvay *et al.*, 1998a,b; Skubal *et al.*, 2001). First, microbially mediated natural attenuation processes are abundant in contaminated environments, and affect the fate of hydrocarbons (e.g., Christensen *et al.*, 1997; Bekins *et al.*, 2001; Cozzarelli *et al.*, 2001), chlorinated solvents (e.g., Lendvay *et al.*, 1998a,b; Skubal *et al.*, 2001), chlorinated pesticides (e.g., Potter and Carpenter, 1995; Zipper *et al.*, 1998), and PCDD/F (Lohmann *et al.*, 2000; Fu *et al.*, 2001; Gaus *et al.*, 2002) or PCBs (Brown *et al.*, 1987; Flanagan and May, 1993) in a wide range of natural environments (soils, surface water sediments, groundwater). Second, there has been a bias towards the impact of biological processes on contaminant fate, and hence, the potential influence of abiotic reaction pathways is rarely considered. Finally, there is a dearth of quantitative contaminant mass balance estimates and fluxes associated with natural contaminated groundwater, sediment or soils, and those that are available are very local in scale.

In an attempt to assess the quantitative contribution of reactive processes to the overall cycling of anthropogenic source contamination, two examples will be considered: (i) vinyl chloride oxidation at groundwater–surface water interface (GSI) St. Joseph, Michigan (e.g., Lendvay *et al.*, 1998a,b), and (ii) dioxin dechlorination in estuarine sediment cores collected from the Passaic River, New Jersey (Albrecht *et al.*, 1999; Fu *et al.*, 2001). Case study 1 is relevant to the Great Lakes region, as 28% of the Michigan shoreline is affected by anthropogenic pollutants, with 600 sites contaminated by chlorinated solvents. The total estimated fluxes of all chlorinated solvents into Lakes Michigan and Huron, based on two sites recently investigated which were $\sim 10-30\ \mathrm{kg\ yr^{-1}}$ (Lendvay *et al.*, 1998a,b, 2002), would total $(6-18)\times 10^3$ kg annually. Anaerobic dechlorination processes upgradient from the shoreline form lesser-chlorinated ethenes such as vinyl chloride and ethene, only 1–8% of which are then aerobically degraded at the GSI. Considering that this site and the Lake Michigan shoreline in general represent a "high energy" GSI (in terms of seiche activity and reoxygenation of the aquifer) and may represent a "best-case scenario" for natural attenuation at GSIs, the total contribution of microbial processes to minimizing alkyl halide fluxes to the Great Lakes is minimal, since >90% of the dissolved dechlorinated contaminant burden is discharged to the Lakes. Considering the elevated solubility of (lesser chlorinated) solvents (Schwarzenbach *et al.*, 1993) limited transfer to the air–water boundary layer should be expected. However, the global fluxes of vinyl chloride and perhaps dichloroethenes discharged to surface water bodies as the result of anaerobic dechlorination processes in contaminant hot spots may be substantial. As indicated earlier, chloroethenes and particularly their atmospheric degradation products (carbonyl halides and acetyl halides) are subject to atmospheric transport and deposition and their source areas may be terrestrial or estuarine in origin.

The situation is somewhat different for POPs, such as chlorodioxins. Some estimates have indicated that up to 7% of the dioxin profile variance of the Passaic River sediments may be due to natural dechlorination (Barabas *et al.*, 2003). Laboratory studies have indicated that mainly mono- and dichlorinated congeners can be expected to accumulate, at dechlorination rates on the order of one chlorine removed every seven years (Fu *et al.*, 2001). One scenario suggests the lesser-chlorinated compounds would selectively diffuse up from buried sediments to the shallow layers (Gevao *et al.*, 1997), where they then dissolve or are transported via particulate material (since lesser-chlorinated dioxins are sparingly soluble) and accumulate at the air–water interface (Lohmann *et al.*, 2000). Estimated rates for diCDD production are in the picogram per gram sediment per year range, which could result in a global biotic production of tens of kilograms per year of diCDD alone, considering the volume of dioxin-contaminated sediments worldwide (~100 Mt; 3–15 Mt in the US). Similar rates, but higher concentrations can be expected for PCBs as the volume and concentrations of PCB-contaminated sediments are higher (Gevao *et al.*, 1997; Breivik *et al.*, 2002a,b). Average sediment and soil PCB concentrations are two to three orders of magnitude higher than PCDD, while the total impacted soil/sediment volume is probably similar to that of PCDD.

Overall, a meaningful synthesis of the processes influencing the distribution and cycling of aliphatic and aryl halides (Figure 11) within the context of biogeochemical controls requires a thorough

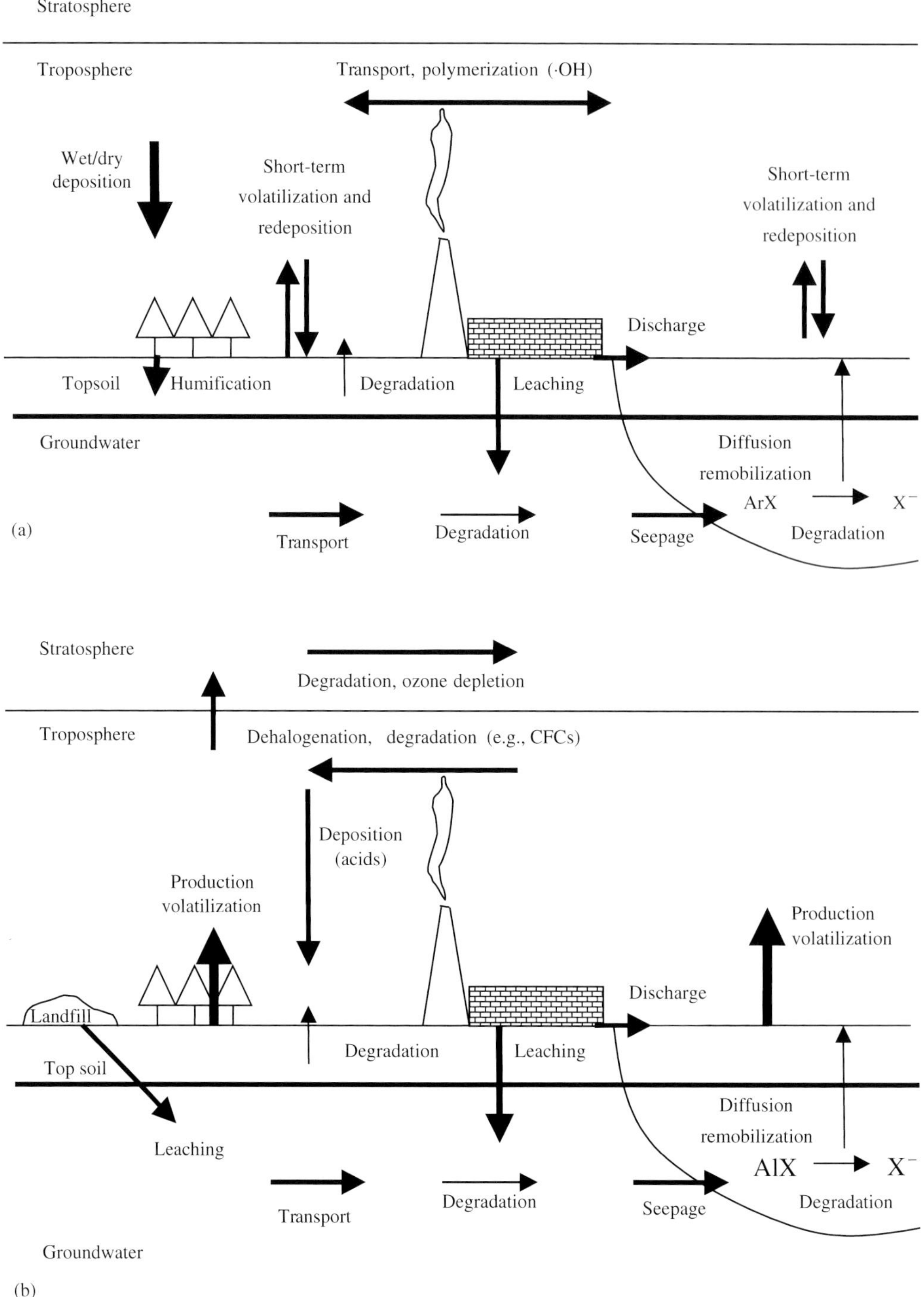

Figure 11 Schematic representation of biogeochemical cycling of aryl (a) and alkyl (b) organohalides in environmental systems (the thickness of the arrows indicate the relative importance of each process or pathway; note that the net flux of alkyl halides is towards volatilization, that of aryl halides towards deposition).

understanding of the sources, reservoirs, transport pathways, transformation reactions, and scale dependence of each. Semi-volatile POPs, which include aromatic and alicyclic compounds, are mainly anthropogenic in nature. They exhibit a tendency to fractionate latitudinally, and undergo rapid air–water exchange processes resulting in frequent regional and latitudinal deposition and revolatilization. Based on mass flux estimates, the net flux is in the direction of deposition, which may eliminate a substantial fraction of the total POP load from the cycling process due to carbon sinking, soil or sediment biotransformation processes, POP sequestration in natural organic matter, and uptake by vegetation. Point releases of POPs in waterways sediments limit their cycling to the local or regional scale, through sequestration in sediment organic matter, re-release due to sediment disturbance, biogenic gas ebullition and upward diffusion of the more soluble compounds, ultimately impacting the air–water interface in estuaries. Preliminary results indicate that these processes impacting point sources may result in a net local or regional efflux to the atmosphere. Non-POP alkyl halides, which represent signature compounds derived from both biogenic and anthropogenic sources, are globally characterized by a net volatilization flux to the atmosphere and troposphere. Deposition fluxes are thought to be less significant for the neutral parent compounds, but substantial for the atmospheric (ionic) degradation products. Considering the fairly rapid atmospheric reactions in clouds, the deposition fluxes are regional in nature, and no latitudinal fractionation has been observed. The multitude of point source releases of these compounds have resulted in locally impacted groundwaters and surface waters, where natural degradation reactions substantially contribute to their fate and transport, but are unlikely to impact the global biogeochemistry. Hence, Figure 11 is characterized by a difference in the relative importance of the fate pathways, sources, and sinks between POPs and non-POP halogenated hydrocarbons. Overall, the impact of biogenic production pathways, and degradation reactions on the scale of biogeochemical cycling is not clear at this time. Moreover, the impact of terrestrial environmental contaminant formation, transformation, and sequestration reactions on explaining the global halocarbon deficit remains an open question.

9.14.8 KNOWLEDGE GAPS AND FERTILE AREAS FOR FUTURE RESEARCH

This chapter discussed the biogeochemistry of halogenated hydrocarbons in the environment with emphasis on environmental flux and reactivity, to develop a better understanding of the chlorine (halogen) cycle. A secondary goal was to integrate information on biotic and abiotic organochlorine production with their degradation processes in light of chemical and environmental controls on reactivity.

As has been indicated by many experts in the area of global cycling of POPs, analytical methods, and environmental characterization present a challenge to closing the mass balance of the chlorine cycle. AOX and aryl halides, as well as their transformation products tend to be associated with soil, sediment, and planktonic organic matter, either due to sorption or covalent bonding, which complicates the development of accurate mass balances and flux calculations, and quantitative analysis of natural transformation processes. As of early 2000s, several approaches are being explored to enable interpretation of the environmental fate of halogenated hydrocarbons. On a global and regional scale, the evaluative environments often used for modeling partitioning and transport behavior are being validated using increasingly comprehensive datasets on contaminants and environmental controlling variables such as wind speed, sediment resuspension, and remote sensing data on plankton blooms. Despite the availability of data and integrated models, the mechanistic understanding of global transport processes is highly dependent on adequate system description and characterization, which adds substantial spatial and temporal uncertainty to the interpretation of the data.

To address the issue of contaminant specificity relative to environmental reactivity and global transport, QSARs are becoming increasingly robust and will advance the modeling approaches to a new level (Grammatica *et al.*, 2002). A fundamental assumption for this approach is that a change in compound structure results in a similar change in its behavior in the system. The parameters influencing reactivity should be chosen carefully, and relationships should be obtained using unbiased statistical analysis. Next-generation QSARs are currently being developed which include information on the genetic and physiological diversity of microorganisms within a given ecosystem. Another approach is based on contaminant fingerprint recognition in the environment, as a function of known sources and probable abiotic or biotic transformation processes. These fingerprints include congener, isomer, and enantiomer profiles, as well as isotopic fractionation. Evidence is emerging that these fingerprints may be able to discriminate between biotic and abiotic reactivity affecting halogenated hydrocarbons, as well as distinguish various source and sink contributions.

As testified by the multitude of reviews over the years, the fate of halogenated organic compounds as affected by microbial activity is extremely difficult to summarize, both with respect to pathways of transformation and reaction kinetics. This is due both to the great diversity in chemical structures (multiple and variable substituents) and the range of acclimation mechanisms and enzyme specificities. This review demonstrates that transition of mechanistic observations in the laboratory to fate prediction in environmental matrices is complicated by multiple attenuation processes and an overall lack of mass balance (due to either irreversible sequestration mechanisms or incomplete analysis for transformation products). Considering that these poorly understood issues have not even been resolved for PCBs and halomethanes, arguably the best-studied subsets of aromatic and aliphatic compounds; the state of the art with respect to their role and place in the chlorine (halogen) cycle has to be considered maturing at best.

Despite these limitations, great strides have been made during the 1990s with respect to understanding halogenated hydrocarbon biogeochemistry in terrestrial environments. Consider, e.g., the progress made in elucidating the genetic basis for microbial adaptation to structurally similar chemicals, and the requirements for transcriptional activation of key-metabolic enzymes in aryl and alkyl halide pathways. The breakthrough discovery that halorespiration and halogenation of natural organic matter may be more widespread than was previously assumed, opens up a vast array of new possibilities to understand the reactive component affecting the biogeochemistry of organohalides. Developments in molecular ecological characterization have afforded substantial advances in this area. For example, we can now monitor population shifts within complex communities as the result of aryl and alkyl halide induced stresses. Complementary to microbial research, the thermodynamic considerations of which transformations are likely to occur have aided in directing fate mechanism studies. First-principle (*ab initio*) model predictions have been shown to describe radical-catalyzed oxidations and reductions, and field observations have proven to be particularly useful to "calibrate" fundamental research investigations on environmental fate predictions. Finally, the quantitative extrapolation of global POP uptake and the impact of sinking carbon has advanced a complementary mechanism for contaminant cycling to the temperature-controlled global fractionation deposition and revolatilization processes.

Ultimately, research in the halogen cycles and the interface between natural and anthropogenic fluxes is directed towards risk quantification and reduction, through research questions such as: "Do environmental transformation reactions reduce the exposure risk from organohalides?," "What is the role for organohalide transformation processes in controlling biogeochemical cycling?," "Do environmental sequestration and carbon sinking permanently remove POPs and other halogenated hydrocarbons from the global biogeochemical cycle?," or "What characteristics differentiate a POP from other environmental contaminants, and help explain their global distribution?" As of early 2000s, the incomplete information presents an enormous challenge to the regulatory community, as it is unclear how the available data should be incorporated into risk-based decision models. Hence, the field of biogeochemical contaminant cycling would significantly benefit from the weighted integration of phase partitioning, transport, and reactivity processes, by informing uncertainty-based geostatistical approaches across spatial (local impact on regional and global processes) and temperal (diurnal temperature controls on long-term reactivity) scales of interrogation.

ACKNOWLEDGMENTS

The authors thank the Office of Naval Research, the National Science Foundation, the US Environmental Protection Agency (STAR Fellowship to MLM), the National Oceanic and Atmospheric Administration, and the Michigan Department for Environmental Quality for financial support. Since this document has not been submitted for review to either agency, no endorsement should be inferred. The lead author expresses his gratitude to all his students, postdocs, and technicians who have contributed to research providing insights in this important area during the last decade.

REFERENCES

Achman D. R., Hornbuckle K. C., and Eiseneich S. E. (1993) Volatilization of polychlorinated biphenyls from Green Bay, Lake Michigan. *Environ. Sci. Technol.* **27**, 75–87.

Adriaens P. and Barkovskii A. L. (2002) Fate and microbial degradation of organohalides. In *Encyclopedia of Environmental Microbiology* (ed. G. Britton). Wiley, New York, NY, pp. 1238–1252.

Adriaens P. and Hickey W. J. (1994) Physiology of biodegradative microorganisms. In *Biotechnology for the Treatment of Hazardous Waste* (ed. D. L. Stoner). Lewis Publishers, Chelsea, MI, pp. 45–71.

Adriaens P. and Vogel T. M. (1995) Treatment processes for chlorinated organics. In *Microbiological Transformation and Degradation of Toxic Organic Chemicals* (eds. L. Y. Young and C. Cerniglia). Wiley, New York, NY, pp. 427–476.

Adriaens P., Chang P. L., and Barkovskii A. L. (1996) Dechlorination of chlorinated PCDD/F by organic and inorganic electron transfer molecules in reduced environments. *Chemosphere* **32**, 433–441.

Adriaens P., Barkovskii A. L., and Albrecht I. D. (1999) Fate of chlorinated aromatic compounds in soils and sediments.

In *Bioremediation of Contaminated Soils*, Soil Science Society of America/American Society of Agronomy Monograph (eds. D. C. Adriano, J.-M. Bollag, W. T. Frankenberger, and R. Sims). Soil Science Society of America Press, Madison, WI, pp. 175–212.

Albrecht I. D., Barkovskii A. L., and Adriaens P. (1999) Production and dechlorination of 2,3,7,8-tetrachlorodibenzo-*p*-dioxin (TCDD) in historically-contaminated estuarine sediments. *Environ. Sci. Technol.* **33**, 737–744.

Andreae M. O. and Crutzen P. J. (1997) Atmospheric aerosols: biogeochemical sources and role in atmospheric chemistry. *Science* **276**, 1052–1058.

Andreux F., Scheunert I., Adrian P., and Schiavon M. (1993) The binding of pesticide residues to natural organic matter, their movement and their bioavailability. In *Fate and Prediction of Environmental Chemicals in Soils, Plants, and Aquatic Systems* (ed. M. Mansour). Lewis Publishers, Boca Raton, FL, pp. 133–149.

Asplund G. and Grimvall A. (1991) Organohalogens in nature. *Environ. Sci. Technol.* **25**, 1346–1350.

Asplund G., Grimvall A., and Petterson C. (1989) Naturally produced organic halogens (AOX) in humic substances from soil and water. *Sci. Tot. Environ.* **81/82**, 239–248.

Asplund G., Christiansen J. V., and Grimvall A. (1993) A chloroperoxidase-like catalyst in soil: detection and characterization of some properties. *Soil Biol. Biochem.* **25**, 41–46.

Baker J. I. and Hites R. A. (2000a) Is combustion the major source of polychlorinated dibenzo-*p*-dioxins and dibenzofurans? A mass balance investigation. *Environ. Sci. Technol.* **34**, 2879–2886.

Baker J. I. and Hites R. A. (2000b) Siskiwit Lake revisited: time trends of polychlorinated dibenzo-*p*-dioxin and dibenzofuran deposition at Isle Royale, Michigan. *Environ. Sci. Technol.* **34**, 2887–2891.

Barabas N., Goovaerts P., and Adriaens P. (2003) Modified polytopic vector analysis to identify and quantify a dioxin dechlorination signature in sediments: 2. Application to the Passaic River superfund site. *Environ. Sci. Technol.* (accepted).

Barkovskii A. L. and Adriaens P. (1998) Affect of model humic constituents on microbial reductive dechlorination of polychlorinated dibenzo-*p*-dioxins. *Environ. Toxicol. Chem.* **17**, 1013–1021.

Barr D. P. and Aust S. D. (1994) Mechanisms white rot fungi use to degrade pollutants. *Environ. Sci. Technol.* **28**, A78–A87.

Bekins B. A., Cozzarelli I. M., Godsy E. M., Warren E., Essaid H. I., and Tuccillo M. E. (2001) Progression of natural attenuation processes at a crude oil spill site: 2. Control on spatial distribution of microbial populations. *J. Contamin. Hydrol.* **53**, 387–406.

Bidleman T. F. (1988) Atmospheric processes. *Environ. Sci. Technol.* **22**, 361–367.

Bidleman T. F. (1999) Atmospheric transport and air–surface exchange of pesticides. *Water Air Soil. Pollut.* **115**, 115–166.

Bowen H. J. M. (1979) *Environmental Chemistry of the Elements*. Academic Press, London.

Bradley P. M., Chapelle F. H., and Lovley D. R. (1998) Humic acids as electron acceptors for anaerobic microbial oxidation of vinyl chloride and dichloroethene. *Appl. Environ. Microbiol.* **64**, 3102–3105.

Breivik K., Sweetman A., Pacyna J. M., and Jones K. C. (2002a) Towards a global historical emission inventory for selected PCB congeners—a mass balance approach: 1. Global production and consumption. *Sci. Tot. Environ.* **290**, 181–198.

Breivik K., Sweetman A., Pacyna J. M., and Jones K. C. (2002b) Towards a global historical emission inventory for selected PCB congeners—a mass balance approach: 2. Emissions. *Sci. Tot. Environ.* **290**, 199–224.

Brown J. F., Jr., Bedard D. L., Brennan M. J., Carnahan J. C., Feng H., and Wagner R. E. (1987) Polychlorinated biphenyl dechlorination in aquatic sediments. *Science* **236**, 709–712.

Bumb R. R., Crummett W. B., Cutie S. S., Gledhill J. R., Hummel R. H., Kagel R. O., Lamparski L. L., Luoma E. V., Miller D. L., Nestrick T. J., Shadoff L. A., Stehl R. H., and Woods J. S. (1980) Trace chemistries of fire: a source of chlorinated dioxins. *Science* **210** 385–386.

Bumpus J. A., Tien M., Wright D., and Aust S. D. (1985) Oxidation of persistent environmental pollutants by a white rot fungus. *Science* **228**, 1434–1436.

Butler E. C. and Hayes K. F. (2000) Kinetics of the transformation of halogenated aliphatic compounds by iron sulfide. *Environ. Sci. Technol.* **34**, 422–429.

Butler J. H. (2000) Better budgets for methyl halides? *Nature* **403**, 260–261.

Capone D. G. and Kiene R. P. (1988) Comparison of microbial dynamics in marine and freshwater sediments: contrasts in anaerobic carbon catabolism. *Limnol. Oceanogr.* **33**, 725–749.

Carlton R. G. and Klug M. J. (1990) Spatial and temporal variations in microbial processes in aquatic sediments: implications for the nutrient status of lakes. In *Sediments: Chemistry and Toxicity of In-place Pollutants* (eds. R. Baudo, J. Giesy, and H. Muntau). Lewis Publishers, Ann Arbor, MI, pp. 107–127.

Carpenter L. J., Sturges W. T., Penkett S. A., Liss P. S., Alicke B., Hebestreit K., and Platt U. (1999) Short-lived alkyl iodides and bromides at Mace Head, Ireland: links to biogenic sources and halogen oxide production. *J. Geophys. Res.-Atm.* **104**, 1679–1689.

Cervini-Silva J., Wu J., Larson R. A., and Stucki J. W. (2000) Transformation of chloropicrin in the presence of iron-bearing minerals. *Environ. Sci. Technol.* **34**, 915–917.

Chapelle F. H. (1993) *Groundwater Microbiology and Geochemistry*. Wiley-Liss, New York.

Chapelle F. H., Haack S. K., Adriaens P., Henry M. A., and Bradley P. M. (1996) Comparison of Eh and H_2 measurements for delineating redox processes in a contaminated aquifer. *Environ. Sci. Technol.* **30**, 3565–3569.

Christensen T. H., Kjeldsen P., Albrechtsen H. J., Heron G., Nielsen P. H., Bjerg P. L., and Holm P. E. (1997) Attenuation off landfill leachate pollutants in aquifers. *Crit. Rev. Environ. Sci. Technol.* **24**, 119–202.

Christensen T. H., Bjerg P. L., Banwart S. A., Jacobson R., Heron G., and Albrechtsen H.-J. (2000) Characterization of redox conditions in groundwater contaminant plumes. *J. Contamin. Hydrol.* **45**, 165–241.

Coey J. M., Schindler D. W., and Weber F. (1974) Iron compounds in lake sediments. *Can. J. Earth Sci.* **11**, 1489–1493.

Cole J. G., MacKay D., Jones K. C., and Alcock R. E. (1999) Interpreting, correlating, and predicting the multimedia concentrations of PCCDD/Fs in the United Kingdom. *Environ. Sci. Technol.* **33**, 399–405.

Copley S. D. (1998) Microbial dehalogenases: enzymes recruited to convert xenobiotic substrates. *Curr. Op. Chem. Biol.* **2**, 613–617.

Cozza C. L. and Woods S. L. (1992) Reductive dehclorination pathways for substituted benzenes: a correlation with electronic properties. *Biodegradation* **2**, 265–278.

Cozzarelli I. M., Bekins B. A., Baedecker M. J., Aiken G. R., Eganhouse R. P., and Tuccillo M. E. (2001) Progression of natural attenuation processes at a crude oil spill site: I. Geochemical evolution of the plume. *J. Contamin. Hydrol.* **53**, 369–385.

Dachs J., Eisenreich S. J., Baker J. E., Ko F. X., and Jeremiason J. (1999a) Coupling of phytoplankton uptake and air-water exchange of persistent organic pollutants. *Environ. Sci. Technol.* **33**, 3653–3660.

Dachs J., Bayona J. M., Ittekkot V., and Albaiges J. (1999b) Monsoon-driven vertical fluxes of organic pollutants in the Western Arabian Sea. *Environ. Sci. Technol.* **33**, 3949–3956.

Dachs J., Eisenreich S. J., and Hoff R. M. (2000) Influence of eutrophication on air–water exchange, vertical fluxes and phytoplankton concentrations of persistent organic pollutants. *Environ. Sci. Technol.* **34**, 1095–1102.

Dachs J., Lohmann R., Ockenden W. A., Mejanelle L., Eisenrich S. J., and Jones K. C. (2002) Oceanic biogeochemical controls on global dynamics of persistent organic pollutants. *Environ. Sci. Technol.* **36**, 4229–4237.

Dahlman O., Morck R., Ljungquist P., Reimann A., Johansson C., Boren H., and Grimvall A. (1993) Chlorinated structural elements in high molecular weight organic matter from unpolluted waters and bleached kraft effluents. *Environ. Sci. Technol.* **27**, 1616–1620.

Dahlman O., Reimann A., Ljungquist P., Morck R., Johansson C., Boren H., and Grimvall A. (1994) Characterization of chlorinated aromatic structures in high molecular weight BMKE materials and in fulvic acids from industrially unpolluted waters. *Water Sci. Technol.* **29**, 81–91.

Damborsky J., Lynam M., and Kuty M. (1998) Structure-biodegradability relationships for chlorinated dibenzo-*p*-dioxins and dibenzofurans. In *Biodegradation of Dioxins and Furans* (ed. R. Wittich). Landes, TX, pp. 165–228.

deJong E. and Field J. A. (1997) Sulfur tuft and turkey tail: biosynthesis and biodegradation of organohalogens by Basidiomycetes. *Ann. Rev. Microbiol.* **51**, 375–414.

deJong E., Field J. A., Spinnler H. E., Wijnberg J. B. P. A., and de Bont J. A. M. (1994) Significant biogenesis of chlorinated aromatics by fungi in natural environments. *Appl. Environ. Microbiol.* **60**, 264–270.

deBruyn W. J., Shorter J. A., Davidovits P., Worsnop D. R., Zahniser M. S., and Kolb C. E. (1995) Uptake of haloacetyl and carbonyl halides by water surfaces. *Environ. Sci. Technol.* **29**, 1179–1185.

Dimmer C. H., Simmonds P. G., Nickless G., and Bassford M. R. (2001) Biogenic fluxes of halomethanes from Irish peatland ecosystems. *Atmos. Environ.* **35**, 321–330.

Dolfing J. and Harrison B. K. (1992) The Gibbs free energy of halogenated aromatic compounds and their potential role as electron acceptors in anaerobic environments. *Environ. Sci. Technol.* **26**, 2213–2218.

Dunnivant F. M., Schwarzenbach R. P., and Macalady D. L. (1992) Reduction of substituted nitrobenzenes in aqueous solutions containing natural organic matter. *Environ. Sci. Technol.* **26**, 2133–2141.

Erlich H. L. (1996) *Geomicrobiology*. Dekker, New York, NY.

Fathepure B. Z. and Tiedje J. M. (1999) Anaerobic bioremediation: microbiology, principles, and applications. In *Bioremediation of Contaminated Soils*, Soil Science Society of America/American Society of Agronomy Monograph (eds. D. C. Adriano, J.-M. Bollag, and W. T. Frankenberger, Jr., and R. C. Sims). Soil Science Society of America Press, Madison, WI, pp. 339–386.

Fenchel T., King G. M., and Blackburn T. H. (1998) *Bacterial Biogeochemistry: The Ecophysiology of Mineral Cycling*. Academic Press, New york.

Ferrario J. and Byrne C. (2000) The concentration and distribution of 2,3,7,8-dibenzo-*p*-dioxins/-furans in chickens. *Chemosphere* **40**, 221–224.

Flanagan W. P. and May R. J. (1993) Metabolite detection as evidence for naturally occurring aerobic PCB biodegradation in Hudson River sediments. *Environ. Sci. Technol.* **27**, 2207–2212.

Fredrickson J. K. and Gorby Y. A. (1996) Environmental processes mediated by iron-reducing bacteria. *Curr. Opin. Biotech.* **7**, 287–294.

Fredrickson J. K., Zachara J. M., Kennedy D. W., Dong H., Onstott T. C., Hinman N. W., and Li S. W. (1998) Biogenic iron mineralization accompanying the dissimilatory reduction of hydrous ferric oxide by a groundwater bacterium. *Geochim. Cosmochim. Acta* **62**, 3239–3257.

Fu Q., Barkovskii A. L., and Adriaens P. (1999) Reductive transformation of dioxins: an assessment of the contribution of dissolved organic matter to dechlorination reactions. *Environ. Sci. Technol.* **33**, 3837–3842.

Fu Q. S., Barkovskii A. L., and Adriaens P. (2001) Dioxin cycling in aquatic sediments: the Passaic River estuary. *Chemosphere* **43**, 643–648.

Gaus C., Brunskill G. J., Connell D. W., Prange J., Muller J. F., Papke O., and Weber R. (2002) Transformation processes, pathways, and possible sources of distinctive polychlorinated dibeno-*p*-dioxin signatures in sink environments. *Environ. Sci. Technol.* **36**, 3542–3549.

Gevao B., Hamilton-Taylor J., Murdoch C., Jones K. C., Kelly M., and Tabner B. J. (1997) Depositional time trends and remobilization of PCBs in lake sediments. *Environ. Sci. Technol.* **31**, 3274–3280.

Giese B., Laturnus F., Adams F. C., and Wiencke C. (1999) Release of volatile iodinated C1–C4 hydrocarbons by marine microalgae from various climate zones. *Environ. Sci. Technol.* **33**, 2432–2439.

Gillham R. W. and O'Hannesin S. F. (1994) Enhanced degradation of halogenated aliphatics by zero-valent iron. *Ground Water* **32**(6), 958–967.

Goldberg E. D. (1975) Synthetic organohalides in the sea. *Proc. Roy. Soc. London Ser. B* **189**, 277–289.

Goovaerts P. (1998) Geostatistical tools for characterizing the spatial variability of microbiological and physico-chemical soil properties. *Biol. Fertil. Soils* **27**, 315–334.

Grammatica P., Pozzi S., Consonni V., and DiGuardo A. (2002) Classification of environmental pollutants for global mobility potential. *SAR QSAR Environ. Res.* **13**, 205–217.

Gribble G. W. (1996) Naturally occurring organohalogen compounds—a comprehensive survey. *Progr. Chem. Org. Nat. Prod.* **68**, 1–423.

Grimalt J. O., Fernandez P., Berdie L., Vilanova R. M., Catalan J., Psenner R., Hofer R., Appleby P. G., Rosseland B. O., Lien L., Massabuau J. C., and Battarbee R. W. (2001) Selective trapping of organochlorine compounds in mountain lakes of temperate areas. *Environ. Sci. Technol.* **35**, 2690–2697.

Gruden C., Fu Q. S., Barkovskii A. L., Albrecht I. D., Lynam M. M., and Adriaens P. (2003) Dechlorination of dioxins in sediments: catalysts, mechanisms, and implications for remedial strategies and dioxin cycling. In *Dehalogenation: Microbial Processes and Environmental Applications* (eds. M. M. Häggblom and I. D. Bossert). Wiley, pp. 347–372.

Häggblom M. M. (1992) Microbial breakdown of halogenated aromatic pesticides and related compounds. *FEMS Microbiol. Rev.* **103**, 29–72.

Harner T., Green N. J. L., and Jones K. C. (2000) Measurements of octanol-air partition coefficients for PCDD/Fs: a tool in assessing air–soil equilibrium status. *Environ. Sci. Technol.* **34**, 3109–3114.

Harper D. B. (2000) The global chloromethane cycle: biosynthesis, biodegradation and metabolic role. *Nat. Prod. Rep.* **17**, 337–348.

Harper D. B., Buswell J. A., Kennedy J. T., and Hamilton J. T. G. (1990) Chloromethane, methyl donor in veratryl alcohol biosynthesis in Phanerochaete chrysosporium and other lignin-degrading fungi. *Appl. Environ. Microbiol.* **56**, 3450–3457.

Harper D. B., Kalin R. M., Hamilton J. T. G., and Lamb C. (2001) Carbon isotope ratios for chloromethane of biological origin: potential tool in determining biological emissions. *Environ. Sci. Technol.* **35**, 3616–3619.

Henrichs S. M. and Reeburgh W. S. (1987) Anaerobic mineralization of marine sediment organic matter: rates and the role of anaerobic processes in the ocean carbon economy. *Geomicrobiol. J.* **5**, 191–237.

Hermens J., Balaz S., Damborsky J., Karcher W., Muller M., Peijnenburg W., Sabljic A., and Sjostrom M. (1995) Assessment of QSARs for predicting fate and effects of chemicals in the environment: an international European project. *SAR QSAR Environ. Res.* **3**, 223–236.

Heron G. and Christensen T. H. (1995) Affect of sediment-bound iron on redox buffering in a landfill leachate polluted aquifer (Vejen, Denmark). *Environ. Sci. Technol.* **29**, 187–192.

Ho D. T., Schlosser P., Smethie W. M., Jr., and Simpson H. J. (1998) Variability in atmospheric chlorofluorocarbons (CCl_3F and CCl_2F_2) near a large urban area: implications for groundwater dating. *Environ. Sci. Technol.* **32**, 2377–2382.

Hoffman H.-J. (1986) Untersuchung der AOX-gehalte von Bayerischen fluessen. *Muench. Beitr. Abwasser Fish Flussbiol.* **40**, 445–459.

Holliger C. and Schraa G. (1994) Physiological meaning and potential for application of reductive dechlorination by anaerobic bacteria. *FEMS Microbiol. Rev.* **15**, 297–305.

Hunter K. S., Wang Y., and Van Cappellen P. (1999) Kinetic modeling of microbially-driven redox chemistry of subsurface environments: coupling transport, microbial metabolism and geochemistry. *J. Hydrol.* **209**, 53–80.

Iwata H., Tanabe S., Sokal N., and Tatsukawa R. (1993) Distribution of persistent organochlorines in the oceanic air and surface seawater and the role of ocean on their global transport and fate. *Environ. Sci. Technol.* **27**, 1080–1098.

Jacobson R., Albrechtsen H.-J., Rasmussen M., Bay H., Bjerg P. L., and Christensen T. H. (1998) H_2 concentrations in a landfill leachate plume (Grindsted, Denmark): in site energetics of terminal electron acceptor processes. *Environ. Sci. Technol.* **32**, 2142–2148.

James R. R., McDonald J. G., Symonik D. M., Swackhamer D. L., and Hites R. A. (2001) Volatilization of toxaphene from Lakes Michigan and Superior. *Environ. Sci. Technol.* **35**, 3653–3660.

Jeremiasson J. D., Hornbuckle K. C., and Eisenreich S. J. (1994) PCBs in Lake Superior, 1978–1992: Decreases in water concentrations reflect loss by volatilisation. *Environ. Sci. Technol.* **28**, 903–914.

Johansson C., Pavasars I., Boren H., and Grimvall A. (1994) A degradation procedure for determination of halogenated structural elements in organic matter from marine sediments. *Environ. Int.* **20**, 103–111.

Johansson E., Kranz-Rulcker C., Zhang B. X., and Oberg G. (2000) Chlorination and biodegradation of lignin. *Soil Biol. Biochem.* **32**, 102–132.

Kalantzi O. I., Alcock R. E., Johnson P. A., Santillo D., Stringer R. L., Thomas G. O., and Jones K. C. (2001) The global distribution of PCBs and organochlorine pesticides in butter. *Environ. Sci. Technol.* **35**, 1013–1018.

Keppler F., Eiden R., Niedan V., Pracht J., and Scholer H. F. (2000) Halocarbons produced by natural oxidation processes during degradation of organic matter. *Nature* **403**, 298–301.

Key B. D., Howell R. D., and Criddle C. S. (1997) Fluorinated organics in the biosphere. *Environ. Sci. Technol.* **31**, 2445–2454.

Kriegman-King M. R. and Reinhard M. (1992) Transformation of carbon tetrachloride in the presence of sulfide, biotite, and vermiculite. *Environ. Sci. Technol.* **26**, 2198–2206.

Kuhn W., Fuchs F., and Sontheimer H. (1977) Untersuchungen zur Bestimmung des organisch gebundenen Chlors mit Hilfe eines neuartigen Anreicherungsverfahrens. *Z. Wasser Abwasser Forsch.* **6**, 192–194.

Laniewski K., Boren H., Grimvall A., Jonsson S., and von Sydow L. (1995) Chemical characterization of adsorbable organic halogens (AOX) in precipitation. In *Naturally Produced Organohalogens* (eds. A. Grimvall and E. W. B. de Leer). Kluwer, Dordrecht, The Netherlands, pp. 113–130.

Laturnus F., Mehrtens G., and Gron C. (1995) Haloperoxidase-like activity in spruce forest soil—A source of volatile halogenated organic compounds. *Chemosphere* **31**, 3709–3719.

Laturnus F., Haselmann K. F., Borch T., and Gron C. (2002) Terrestrial natural sources of trichloromethane (chloroform, CHCl3)—an overview. *Biogeochemistry* **60**, 121–139.

Lee R. G. M., Burnett V., Harner T., and Jones K. C. (2000) Short-term temperature dependent air–surface exchange and atmospheric concentrations of polychlorinated napthalenes and organochlorine pesticides. *Environ. Sci. Technol.* **34**, 393–398.

Lendvay J. M., Dean S. M., and Adriaens P. (1998a) Temporal and spatial trends in biogeochemical conditions at a groundwater-surface water interface: implications for natural bioattenuation. *Environ. Sci. Technol.* **32**, 3472–3478.

Lendvay J. M., Sauck W. A., McCormick M. L., Barcelona M. J., Kampbell D. H., Wilson J. T., and Adriaens P. (1998b) Geophysical characterization, redox zonation, and contaminant distribution at a groundwater-surface water interface. *Water Resour. Res.* **34**, 3545–3559.

Lendvay J. M., Barcelona M. J., Daniels G., Dollhopf M., Fathepure B. Z., Gebhard M., Heine R., Hickey R., Krajmalnik-Brown R., Löffler F. E., Major Jr., C L., Petrovskis E., Shi J., Tiedje J. M., and Adriaens P. (2002) Bioreactive barriers: bioaugmentation and biostimulation for chlorinated solvent remediation. *Environ. Sci. Technol.* **37**, 1422–1431.

Lindner A. S., Whitfield C., Chen N., Semrau J. D., and Adriaens P. (2003) Quantitative structure-biodegradation relationships for *ortho*-substituted biphenyl compounds with whole-cell methanotroph, *Methylosinus trichosporium* OB3b. *Environ. Toxicol. Chem.* (in press).

Lohmann R., Nelson E., Eisenreich S. J., and Jones K. C. (2000) Evidence for dynamic air–water exchange of PCDD/Fs: a study in the Raritan Bay/Hudson River estuary. *Environ. Sci. Technol.* **34**, 3086–3093.

Lovley D. R. (1991) Dissimilatory Fe(III) and Mn(IV) reduction. *Microbiol. Rev.* **55**, 259–287.

Lovley D. R. (1997) Microbial Fe(III) reduction in subsurface environments. *FEMS Microbiol. Rev.* **20**, 305–313.

Lovley D. R. and Woodward J. C. (1992) Consumption of freons CFC-11 and CFC-12 by anaerobic sediments and soils. *Environ. Sci. Technol.* **26**, 925–929.

Lovley D. R., Coates J. D., Blunt-Harris E. L., Phillips E. J. P., and Woodward J. C. (1996) Humic substances as electron acceptors for microbial respiration. *Nature* **382**, 445–448.

Ludvigsen L., Albrechtsen H.-J., Holst H., and Christensen T. H. (1997) Correlating phospholipid fatty acids (PLFA) in a landfill leachate polluted aquifer with biogeochemical factors by multivariate statistical methods. *FEMS Microbiol. Rev.* **20**, 447–460.

Ludvigsen L., Albrechtsen H. J., Heron G., Bjerg P. L., and Christensen T. H. (1998) Anaerobic microbial redox processes in a landfill leachate contaminated aquifer (Grindsted, Denmark). *J. Contamin. Hydrol.* **33**, 273–291.

Luthy R. G., Aiken G. R., Brusseau M. L., Cunningham S. D., Gschwend P. M., Pignatello J. J., Reinhard M., Traina S. J., Weber W. W., and Westall J. C. (1997) Sequestration of hydrophobic organic contaminants by geosorbents. *Environ. Sci. Technol.* **31**, 3341–3347.

Lynam M., Damborsky J., Kuty M., and Adriaens P. (1998) Molecular orbital calculations to probe mechanism of reductive dechlorination of polychlorinated dioxins. *Environ. Toxicol. Chem.* **17**, 998–1005.

MacKay D. (1991) *Multimedia Environmental Models: The Fugacity Approach.* CRC Press/Lewis Publishers, Boca Raton, FL.

MacKay D. and Wania F. (1995) Transport of contaminants to the Arctic: partitioning, processes and models. *Sci. Tot. Environ.* **160/161**, 25–38.

Martin J. M. and Meybeck M. (1979) Elemental mass-balance of material carried by major world rivers. *Mar. Chem.* **7**, 173–206.

Matheson L. J. and Tratnyek P. G. (1994) Reductive dehalogenation of chlorinated methanes by iron metal. *Environ. Sci. Technol.* **28**, 2045–2053.

McConnell L. L., Bidleman T. F., Cotham W. E., and Walla M. D. (1998) Air concentrations of organochlorine insecticides and polychlorinated biphenyls over Green Bay, WI, and the four lower Great Lakes. *Environ. Pollut.* **101**, 391–399.

McCormick M. L. (1999) Reductive dechlorination of chlorinated solvents under iron reducing conditions. PhD Disssertation, The University of Michigan.

McCormick M. L., Bouwer E. J., and Adriaens P. (2002a) Carbon tetrachloride transformation in a defined iron-reducing culture: relative kinetics of biotic and abiotic reactions. *Environ. Sci. Technol.* **36**, 403–410.

McCormick M. L., Jung P. T., Koster van Groos P. G., Hayes K. F., and Adriaens P. (2002b) Assessing biotic and abiotic contributions to chlorinated solvent transformation in iron-reducing and sulfidogenic environments. In *Groundwater Quality: Natural and Enhanced Restoration of Groundwater Pollution* (eds. S. F. Thornton and S. E. Oswald). International Association of Hydrological Sciences, Oxfordshire, UK, pp. 119–125.

McGuire J. T., Smith E. W., Long D. T., Hyndman D. W., Haack S. K., Klug M. J., and Velbel M. A. (2000) Temporal variations in parameters reflecting terminal electron accepting processes in an aquifer contaminated with waste fuel and chlorinated solvents. *Chem. Geol.* **169**, 471–485.

Meijer S. N., Steinnes E., Ockenden W. A., and Jones K. C. (2002) Influence of environmental variables on the spatial distribution of PCBs in Norwegian and UK soils: implications for global cycling. *Environ. Sci. Technol.* **36**, 2146–2153.

Mohn W. W. and Tiedje J. M. (1992) Microbial reductive dehalogenation. *Microbiol. Rev.* **56**, 482–507.

Moore S. A., Pope J. D., Barnett J. T. Jr., and Suarez L. A. (1990) Structure-activity relationships and estimation techniques for biodegradation of xenobiotics. EPA 600/S3-89/080, Research and Development Report, US Environmental Protection Agency, Athens, GA.

Murphy E. M. and Schramke J. A. (1998) Estimation of microbial respiration rates in groundwater by geochemical modeling constrained with stable isotopes. *Geochim. Cosmochim. Acta* **62**(21/22), 3395–3406.

Murphy L. and Morrison R. D. (2002) *Introduction to Environmental Forensics*. Academic Press, San Diego, 560pp.

Neidleman S. L. and Geigert J. (1986) *Biohalogenation: Principles, Basic Roles and Applications*. Ellis Horwood, Chichester.

Oberg G. (1998) Chloride and organic chlorine in soil. *Acta Hydrochim. Hydrobiol.* **26**, 137–144.

Oberg G. (2002) The natural chlorine cycle—fitting the scattered pieces. *Appl. Microbiol. Biotechnol.* **58**, 565–581.

Oberg G., Brunberg H., and Hjelm O. (1997) Production of organically-bound halogens during degradation of birch wood by common white rot fungi. *Soil Biol. Biochem.* **29**, 191–197.

Ockenden W. A., Steinnes E., Parker C., and Jones K. C. (1998a) Observations on persistent organic pollutants in plants: implications for their use as passive air samplers and for POP cycling. *Environ. Sci. Technol.* **32**, 2721–2726.

Ockenden W. A., Sweetman A. J., Prest H. F., Steinnes E., and Jones K. C. (1998b) Toward an understanding of the global atmospheric distribution of persistent organic pollutants: the use of semi-permeable membrane devices as time-integrated passive samplers. *Environ. Sci. Technol.* **32**, 2703–2795.

Okey R. W. and Stensel H. D. (1996) A QSBR development procedure for aromatic xenobiotic degradation by unacclimated bacteria. *Water Environ. Res.* **65**, 772–780.

Palleroni N. J. (1995) Microbial versatility. In *Microbial Transformation and Degradation of Toxic Organic Chemicals* (eds. L. Y. Young and C. E. Cerniglia). Wiley, New York, NY, pp. 3–27.

Peijnenburg W. J. G. M., t'sHart M. J., den Hollander H. A., van de Meent D., Verboom H. H., and Wolfe N. L. (1992) QSARs for predicting reductive transformation rate constants of halogenated aromatic hydrocarbons in anoxic sediment systems. *Environ. Toxicol. Chem.* **11**, 310–314.

Peterle T. J. (1969) DDT in Antarctic snow. *Nature* **224**, 620–623.

Post W. M., Emmanuel W. R., Zinke P. J., and Stangenberger A. L. (1982) Soil carbon pools and world life zones. *Nature* **298**, 156–159.

Postma D. and Jacobson R. (1996) Redox zonation: equilibrium constraints on the Fe(III)/SO_4-reduction interface. *Geochim. Cosmochim. Acta* **60**, 3169–3175.

Potter T. L. and Carpenter T. L. (1995) Occurrence of Alachlor environmental degradation products in groundwater. *Environ. Sci. Technol.* **29**, 1557–1564.

Raiswell R. and Canfield D. E. (1998) Sources of iron for pyrite formation in marine sediments. *Am. J. Sci.* **298**, 219–245.

Rhew R. C., Miller B. R., and Weiss R. F. (2000) Natural methyl bromide and methyl chloride emissions from coastal salt marshes. *Nature* **403**, 292–295.

Rickard D., Martin A. A., and Luther G. W., III (1995) Chemistry of iron sulfides in sedimentary environments. *Geochemical Transformations of Sedimentary Sulfur*. ACS Symposium Series 612, ACS Press, Washington, DC, pp. 168–193.

Rickard D. T. (1969) The microbiological formation of iron sulphides. *Stockholm Contr. Geol.* **20**, 49–66.

Roberts A. L., Jeffers P. M., Wolfe N. L., and Gschwend P. M. (1993) Structure-reactivity relationships in dehydrohalogenation reactions of polychlorinated and polybrominated alkanes. *Crit. Rev. Environ. Sci. Technol.* **23**, 1–39.

Rompp A., Klemm O., Fricke W., and Frank H. (2001) Haloacetates in fog and rain. *Environ. Sci. Technol.* **35**, 1294–1298.

Rorije E., Langenberg J. H., Richter J., and Peijnenburg W. J. G. M. (1995) Modeling reductive dehalogenation with quantum chemically derived descriptors. *SAR QSAR Environ. Res.* **4**, 237–252.

Rugge K., Hofstetter T. B., Haderlein S. B., Bjerg P. L., Knudsen S., Zraunig C., Mosbæk H., and Christensen T. H. (1998) Characterization of predominant reductants in an anaerobic leachate-contaminated aquifer by nitroaromatic probe compounds. *Environ. Sci. Technol.* **32**, 23–31.

Schauder R. and Kroeger A. (1993) Bacterial sulfur respiration. *Arch. Microbiol.* **159**, 491–497.

Scheunert I., Mansour M., and Andreux F. (1992) Binding of organic pollutants to soil organic matter. *Int. J. Environ. Anal. Chem.* **46**, 189–222.

Schlesinger W. H. (1991) *Biogeochemistry: an Analysis of Global Change*. Academic Press, San Diego, CA.

Schwarzenbach P. R., Stierly R., Lanz K., and Zeyer J. (1990) Quinone and iron porphyrin mediated reduction of nitroaromatic compounds in homogenous aqueous solutions. *Environ. Sci. Technol.* **24**, 1566–1574.

Schwarzenbach R. P., Gschwendt P. M., and Imboden D. M. (1993) *Environmental Organic Chemistry*. Wiley, New York, NY.

Sherwood Rowland F. (1991) Stratospheric ozone in the 21st century: the chlorofluorocarbon problem. *Environ. Sci. Technol.* **25**, 622–628.

Simonich S. L. and Hites R. A. (1995) Global distribution of persistent organochlorine compounds. *Nature* (London) **269**, 1851–1854.

Skubal K. L., Barcelona M. J., and Adriaens P. (2001) A field and laboratory assessment of natural bioattenuation in an aquifer contaminated by mixed organic waste. *J. Contamin. Hydrol.* **49**, 151–171.

Skyring G. W. (1987) Sulfate reduction in coastal ecosystems. *Geomicrobiol. J.* **5**, 295–374.

Sladen W. J. L., Menzie C. M., and Reichel W. L. (1966) DDT residues in Adelie penguins and a crabeater seal from Antarctica. *Nature* **210**, 670–673.

Svenson A., Kjeller L. O., and Rappe C. (1989) Enzyme-mediated formation of 2,3,7,8-tetrasubstituted chlorinated dibenzodioxins and dibenzofurans. *Environ. Sci. Technol.* **23**, 900–202.

Stevenson F. J. (1994) *Humus Chemistry: Genesis, Composition, Reactions*. Wiley, New York, NY.

Swackhamer D. L., Schottler S., and Pearson R. F. (1999) Air–water exchange and mass balance of toxaphene in the Great Lakes. *Environ. Sci. Technol.* **33**, 3864–3872.

Swoboda-Colberg N. G. (1995) Chemical contamination of the environment: sources, types, and fate of synthetic organic chemicals. In *Microbial Transformation and Degradation of Toxic Organic Chemicals* (eds. L. Y. Young and C. E. Cerniglia). Wiley, New York, NY, pp. 27–77.

Thamdrup B. (2000) Bacterial manganese and iron reduction in aquatic sediments. *Adv. Microb. Ecol.* **16**, 41–84.

Tiedje J. M. (1988) Ecology of denitrification and dissimilatory nitrate reduction to ammonium. In *Biology of Anaerobic Microorganisms* (ed. A. J. B. Zehnder). Wiley, New York, NY, pp. 179–244.

Totten L. A., Brunciak P. A., Gigliotti C. L., Dachs J., Glenn T. R., Nelson E. D., and Eisenreich S. J. (2001) Dynamic air-water exchange of polychlorinated biphenyls in the New York–New Jersey Harbor estuary. *Environ. Sci. Technol.* **35**, 3834–3840.

Tratnyek P. G. and Macalady D. L. (2000) Oxidation–reduction reactions in the aquatic environment. In *Handbook of Property Estimation Methods for Chemicals* (eds. R. S. Boethling and D. Mackay). Lewis Publishers, Boca Raton, FL, pp. 383–415.

van der Meer J. R., de Vos W. M., Harayama S., and Zehnder A. J. B. (1992) Molecular mechanisms of genetic adaptation to xenobiotic chemicals. *Microbiol. Rev.* **56**, 677–694.

van Hylckema-Vlieg J. E. T., Poelarends G. J., Mars A. E., and Jansen D. B. (2000) Detoxification of reactive intermediates during microbial metabolism of halogenated compounds. *Curr. Op. Microbiol.* **3**, 257–262.

Vaughan D. J. and Lennie A. R. (1991) The iron sulphide minerals: their chemistry and role in nature. *Sci. Progr. Edinburgh* **75**, 371–388.

Vogel T. M., Criddle C. S., and McCarty P. L. (1987) Transformations of halogenated aliphatic compounds. *Environ. Sci. Technol.* **21**, 722–736.

Wania F. (2003) Assessing the potential of persistent organic chemicals for long-range transport and accumulation in polar regions. *Environ. Sci. Technol.* **37**, 1344–1351.

Wania F. and MacKay D. (1996) Tracking the distribution of persistent organic pollutants. *Environ. Sci. Technol.* **30**, 390A–396A.

Watling R. and Harper D. B. (1998) Chloromethane production by wood-rotting fungi and an estimate of the global flux to the atmosphere. *Mycol. Res.* **102**, 769–787.

Wiedemeier T. H., Rifai H. S., Newell C. N., and Wilson J. T. (1999) Natural attenuation of fuels and chlorinated solvents in the subsurface. Wiley, New York, NY.

Wolfe N. L., Paris D. F., Steen W. C., and Baughman G. L. (1980) Correlation of microbial degradation rates with chemical structure. *Environ. Sci. Technol.* **14**, 1143–1144.

Yokouchi Y., Noijiri Y., Barrie L. A., Toom-Sauntry D., Machida T., Inuzuka Y., Akimoto H., Li H.-J., Fujinuma Y., and Aoki S. (2000) A strong source of methyl chloride to the atmosphere from tropical coastal land. *Nature* **403**, 295–298.

Zachara J. M., Kukkadapu R. K., Fredrickson J. K., Gorby Y. A., and Smith S. C. (2002) Biomineralization of poorly crystalline Fe(III) oxides by dissimilatory metal reducing bacteria (DMRB). *Geomicrobiol. J.* **19**, 179–207.

Zhang H., Eisenreich S. J., Franz T., Baker J. E., and Offenberg J. H. (1999) Evidence for increased gaseous PCB fluxes to Lake Michigan from Chicago. *Environ. Sci. Technol.* **33**, 2129–2137.

Zhao H., Chen J., Quan X., Yang F., and Peijnenburg W. J. G. M. (2001) Quantitative structure-property relationship study on reductive dehalogenation of selected halogenated aliphatic hydrocarbons in sediment slurries. *Chemosphere* **44**, 1557–1563.

Zipper C., Suter M. J.-F., Haderlein S. B., Gruhl M., and Kohler H.-P. E. (1998) Changes in the enantiomeric ratio of (R) to (S) mecoprop indicate *in situ* biodegradation of this chiral herbicide in a polluted aquifer. *Environ. Sci. Technol.* **32**, 2070–2076.

9.15
The Geochemistry of Pesticides

J. E. Barbash

United States Geological Survey, Tacoma, WA, USA

NOMENCLATURE

a.i.	active ingredient
f_{oc}	mass fraction of organic carbon in soil (m/m)
k	rate constant for the rate-limiting step of a chemical reaction (units depend on reaction order)
$1/n$	Freundlich exponent (dimensionless)
pK_a	the negative log of the equilibrium constant for the interconversion between a Brønsted acid and its conjugate base (dimensionless)
A	Arrhenius pre-exponential factor (same units as k)
C_{aq}	concentration of solute dissolved in water (m/v)
C_{oc}	amount of solute sorbed to soil organic carbon (m/m)
C_α	carbon atom representing the site of a particular reaction on a molecule (the "*alpha* carbon")
C_β	carbon atom (the "*beta* carbon") immediately adjacent to C_α
E_a	activation energy (energy mol^{-1})
H	Henry's Law constant (various forms)

K_f Freundlich partition coefficient (v/m)
K_{oc} soil organic carbon-water partition coefficient (v/m)
K_{ow} octanol–water partition coefficient (dimensionless)
K_p soil–water partition coefficient (v/m)
K_{SA} soil–air partition coefficient (form depends upon units for H)
R universal gas constant (form depends on units for E_a)
S_{N^2} bimolecular nucleophilic substitution reaction
T temperature
T_{max} temperature above which substrate decomposition or enzyme inactivation occurs
T_{min} temperature below which biological functions are inhibited
T_{opt} temperature of maximum biotransformation rate
μ_{max} maximum rate of biotransformation (m/v/t)

9.15.1 INTRODUCTION

The mid-1970s marked a major turning point in human history, for it was at that moment that the ability of the Earth's ecosystems to absorb most of the biological impacts of human activities appears to have been exceeded by the magnitude of those impacts. This conclusion is based partly upon estimates of the rate of carbon dioxide emission during the combustion of fossil fuels, relative to the rate of its uptake by terrestrial ecosystems (Loh, 2002). A very different threshold, however, had already been crossed several decades earlier with the birth of the modern chemical industry, which produced novel substances for which no such natural assimilative capacity existed. Among these new chemical compounds, none has posed a greater challenge to the planet's ecosystems than synthetic pesticides, compounds that have been intentionally released into the hydrologic system in vast quantities—several hundred million pounds of active ingredient per year in the United States alone (Donaldson *et al.*, 2002)—for many decades. To gauge the extent to which we are currently able to assess the environmental implications of this new development in the Earth's history, this chapter presents an overview of current understanding regarding the sources, transport, fate and biological effects of pesticides, their transformation products, and selected adjuvants in the hydrologic system. (Adjuvants are the so-called "inert ingredients" included in commercial pesticide formulations to enhance the effectiveness of the active ingredients.)

9.15.1.1 Previous Reviews of Pesticide Geochemistry

Pesticides have been in widespread use since the Second World War, and their environmental effects have been of concern for at least four decades (Carson, 1962). As a result, numerous reviews have been published summarizing the results from field and laboratory studies of their distribution, transport, fate and biological effects in the hydrologic system. The principal features of many of these reviews have been summarized elsewhere for the atmosphere (Majewski and Capel, 1995), vadose zone and groundwater (Barbash and Resek, 1996), surface waters (Larson *et al.*, 1997), stream sediments and aquatic biota (Nowell *et al.*, 1999). Other compilations of existing information on these topics have focused either on particular pesticide classes (e.g., Erickson and Lee, 1989; Weber, 1990; Pehkonen and Zhang, 2002), or on individual pesticides (e.g., Moye and Miles, 1988). Information on the biological effects of pesticides on both target and non-target organisms has been extensively reviewed (e.g., Matsumura, 1985; Murphy, 1986; Kamrin, 1997). Valuable overviews of environmental organic chemistry are also available (e.g., Capel, 1993; Schwarzenbach *et al.*, 1993).

9.15.1.2 Scope of This Review

This chapter focuses on the sources, transport, fate and biological effects of synthetic organic pesticides, their transformation products and volatile pesticide adjuvants—collectively referred to herein as *pesticide compounds*—in the hydrologic system. Although there are thousands of substances that are currently registered for use as pesticide adjuvants in the United States (USEPA, 2002), discussion of these chemicals will be limited to those that are volatile organic compounds because they are the adjuvant group whose geochemical behavior has been most thoroughly documented. (Volatile organic compounds are commonly used as solvents in commercial pesticide formulations (e.g., Wang *et al.*, 1995).) Most of the pesticide compounds examined in this chapter are registered for use in the United States. Similarly, discussion of the use of these compounds focuses primarily on their applications within the United States. For space considerations, pesticides that are wholly inorganic (e.g., sulfur, chromated copper arsenate, copper oxychloride), or contain metals (e.g., tributyltin) or metalloids (e.g., arsenicals) are not included. Topics for which comprehensive, up-to-date reviews have already been published will receive less attention than those for which fewer reviews are available. Consequently, greater emphasis will be placed on the factors that influence the rates and mechanisms of transformation of pesticide compounds than on the biological effects of these substances, their patterns of use and occurrence, or their partitioning among environmental media.

9.15.1.3 Biological Effects of Pesticide Compounds

All pesticides are designed to kill or otherwise control specific animals or plants, so a great deal is known about the acute biological effects of these chemicals on their target organisms. Insecticides (including most fumigants) act as either physical poisons, protoplasmic poisons, stomach poisons, metabolic inhibitors, neurotoxins or hormone mimics (Matsumura, 1985). Herbicides control or kill plants through a variety of mechanisms, including the inhibition of biological processes such as photosynthesis, mitosis, cell division, enzyme function, root growth, or leaf formation; interference with the synthesis of pigments, proteins or DNA; destruction of cell membranes; or the promotion of uncontrolled growth (William *et al.*, 1995). Fungicides act as metabolic inhibitors (Matheron, 2001). Most rodenticides are either anticoagulants, stomach poisons or neurotoxins (Meister, 2000). Because most pesticides are poisons, considerable knowledge has also developed regarding the acute effects of these compounds on humans (e.g., Murphy, 1986).

Far more elusive, however, are the myriad sublethal effects on non-target organisms (including humans) of chronic exposure to pesticide compounds. Of considerable concern in this regard is endocrine disruption, a phenomenon whose effects were first discovered five decades ago, but whose widespread impacts across a broad range of organisms and ecosystems have become known only since the 1990s (Colborn *et al.*, 1993, 1996). Other effects of chronic pesticide exposure in wildlife—some of which may themselves be related to endocrine disruption—include impaired homing abilities in fish (Scholz *et al.*, 2000), eggshell thinning in birds, reduced immune function, liver and kidney damage, teratogenicity, neurotoxicity and reduced tolerance to cold stress (Murphy, 1986; Smith, 1987; Colborn *et al.*, 1993). Recent evidence also suggests that many of the malformations and population declines that have been observed in amphibians over the past several decades may have been facilitated by exposure to pesticides at concentrations commonly encountered in the hydrologic system (e.g., Sparling *et al.*, 2001; Hayes *et al.*, 2002). Documented or suspected effects in humans include cancer (Patlak, 1996; Schreinemachers, 2000), immune system suppression (Repetto and Baliga, 1996), learning disorders (Guillette *et al.*, 1998), Parkinson's disease (Betarbet *et al.*, 2000), fetal death (Bell *et al.*, 2001), birth defects and shifts in sex ratios (Garry *et al.*, 2002).

When, as is often the case, individual pesticides are present in the hydrologic system in combination with other pesticides (e.g., Larson *et al.*, 1999; Kolpin *et al.*, 2000; Squillace *et al.*, 2002), the combined toxicity of the different chemicals may be either antagonistic, additive, or synergistic relative to the effects and toxicities of the individual pesticides alone, depending upon the compounds and conditions (e.g., Thompson, 1996; Pape-Lindstrom and Lydy, 1997). In addition, pesticide adjuvants may be responsible for substantial increases in the toxicity of commercial formulations, relative to that of the active ingredient alone (e.g., Bolognesi *et al.*, 1997; Oakes and Pollak, 1999; Lin and Garry, 2000). Water-quality criteria that have been established for the protection of aquatic life or human health in relation to individual pesticides, however, do not currently account for the potential effects of transformation products, adjuvants or other pesticides that may also be present.

9.15.1.4 Variations in Pesticide Use over Time and Space

In 1964, an estimated 617 million pounds of pesticide active ingredients (a.i.) were sold for either agricultural or nonagricultural use in the United States (Figure 1). This amount reached a maximum of 1.14 billion pounds a.i. in 1979 and decreased to 906 million pounds a.i. by 1987 (Donaldson *et al.*, 2002). However, despite overall decreases in total cropland area over the past decade, as well as extensive efforts to reduce pesticide use through the introduction of integrated pest management practices and genetically engineered crops (US General Accounting Office, 2001), Figure 1 indicates that the total mass of pesticides applied in the United States in 1999 (912 million pounds a.i.) was essentially the same as it was in 1987. Worldwide use in 1999 totalled 5.7 billion pounds of active ingredient (Donaldson *et al.*, 2002). Relatively detailed estimates of agricultural pesticide use during the 1990s are available for individual counties across the United States (US Geological Survey, 1998; Thelin and Gianessi, 2000). Estimates of pesticide use in nonagricultural settings, however, are available only on a national scale, despite the fact that such use represented ~23% of the total mass of a.i. sold in the United States in 1999 (Donaldson *et al.*, 2002). Summaries of the agricultural use of specific pesticides in individual countries around the world are available from the Database on Pesticides Consumption, maintained by the United Nations Food and Agriculture Organization (2003).

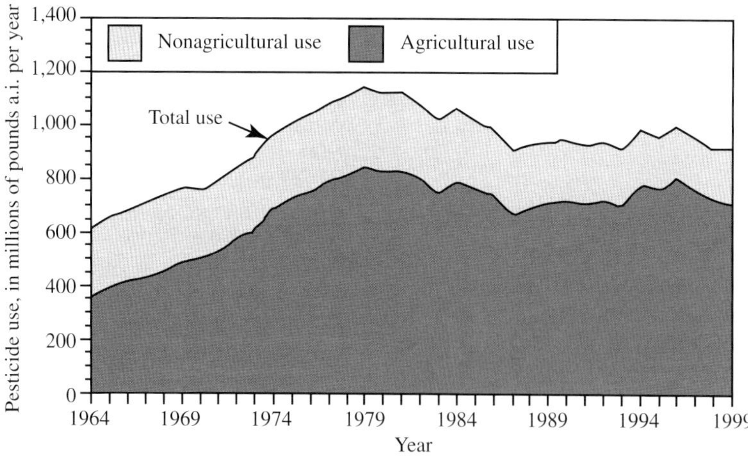

Figure 1 Estimated amounts of pesticides use in the United States for agricultural and nonagricultural purposes from 1964 to 1999 (source Donaldson *et al.*, 2002).

9.15.1.5 Environmental Distributions in Relation to Use

Figure 2 illustrates the variety of routes by which pesticides are dispersed within the hydrologic system after they are released, either intentionally through application or unintentionally through spills or other accidents. As might be expected, spatial patterns of pesticide detection in air (Majewski *et al.*, 1998), surface waters (Capel *et al.*, 2001), groundwaters (Barbash *et al.*, 1999), bed sediments and aquatic biota (Wong *et al.*, 2001) have generally been found to correspond with their patterns of application.

One of the best measures of our ability to predict the spatial distributions of any anthropogenic contaminant in the hydrologic system, however, is the degree to which its total mass can be accounted for after its release into the environment. Mass-balance studies of applied pesticides since the mid-1970s have had only limited success in this regard, even under highly controlled conditions. Of the initial amounts applied during numerous investigations—nearly all of which were conducted in agricultural settings—the proportions of applied active ingredient detected in the hydrologic system have typically been 3% or less in surface waters (Wauchope, 1978; Capel *et al.*, 2001), 5% or less in vadose zone waters or tile drainage (Flury, 1996) and 5% or less in groundwater (Barbash and Resek, 1996). Estimates of the proportion of applied pesticides that move offsite in spray drift range from 1% to 75% (with considerable variation among different compounds, depending upon their volatility), while the amounts lost to the atmosphere through volatilization from the soil following application have been estimated at between 0.2% and 90%. Offsite losses through both spray drift and volatilization from soil depend upon a variety of factors such as pesticide properties, application method, formulation and weather conditions (Majewski and Capel, 1995). High proportions—and often the majority—of the applied mass of pesticide active ingredients have been measured in association with plant tissues and surface soils within the first few hours to days after application, but the percentages of applied mass detected in these media typically drop below 30% within a few months (Barbash and Resek, 1996). Much of the pesticide mass that has remained unaccounted for during these studies—but did not move offsite in the air, groundwater or surface waters—may have formed covalent bonds with plant tissues or soil organic matter to form *bound residues* (e.g., Harris, 1967; Nicollier and Donzel, 1994; Xu *et al.*, 2003) or undergone transformation to CO_2 or products for which chemical analyses are rarely conducted (Barbash and Resek, 1996). The fate of applied adjuvants–which usually constitute the majority of the mass of commercial pesticide formulations (Tominack, 2000)—is almost entirely unknown.

9.15.1.6 Overview of Persistence in the Hydrologic System

Given their broad diversity of chemical structures, it is not surprising that pesticide compounds exhibit a wide range of persistence in the hydrologic system. Such variability may be observed among different compounds within a given environmental medium (i.e., the atmosphere, water, soil, aquatic sediment or biological tissues) or for the same compound in different environmental media. Since there are many thousands of compounds that are, or have been used as pesticides, space considerations preclude a characterization of the environmental reactivity of all of them in this chapter. However, Table 1

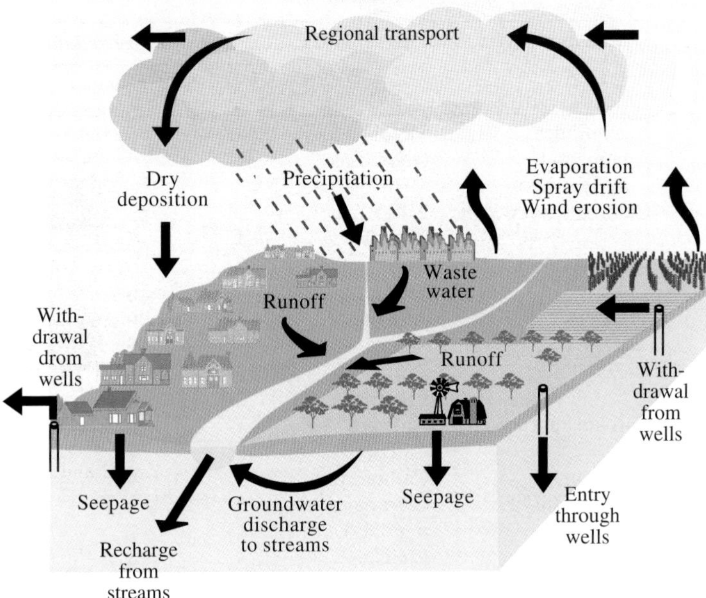

Figure 2 Pesticide movement in the hydrologic system (Barbash and Resek (1996); reproduced by permission of CRC Press, from *Pesticides in Groundwater—Distribution Trends, and Governing Factors*, **1996**).

provides a brief overview of the persistence of several common herbicides, insecticides, fumigants, and fungicides (representing some of the principal chemical classes of each) in the atmosphere, surface water, soil and aqueous sediments. These data were taken from Mackay *et al.* (1997), who drew on an extensive body of published data to determine a "half-life class" for 42 pesticide compounds in each of these four environmental media. As will be seen from the discussion in this chapter, the persistence of a given compound in a specific environmental setting is influenced by many different physical, chemical and biological factors. Each estimate of the "half-life class" in Table 1 thus represents a generalized mean value selected from what was often a wide range available from previous studies. Consequently, the reader is referred to Mackay *et al.* (1997) for a summary of the methods by which these data were generated.

The data in Table 1 indicate that the persistence of pesticide compounds generally increases among the four environmental media in the following order: atmosphere < surface waters < soils < aquatic sediments. According to Mackay *et al.* (1997), this pattern is a reflection of several circumstances, including the following: (i) chemical reaction rates in water are generally slower than those in air; (ii) pesticides in aquatic sediments and soils are exposed to less sunlight than those in the atmosphere or surface waters, and are therefore less subject to photochemical reactions; and (iii) pesticides sorbed to aquatic sediments and soils are often less accessible for biotransformation than those in the aqueous phase. The data in Table 1 also suggest that pesticide compounds within the same chemical class may show similar patterns of persistence in a given environmental setting. However, variations in structure among compounds within the same chemical class may also result in substantial variations in reactivity in the same medium (cf. diazinon versus malathion).

9.15.2 PARTITIONING AMONG ENVIRONMENTAL MATRICES

The large-scale movement of persistent pesticide compounds within the hydrologic system is controlled primarily by their rates of advection in water and air masses, as well as by the movement of biota in which they might bioaccumulate. However, the partitioning of these compounds among different environmental media occurs in response to differences in their chemical potential, or *fugacity*, among these media (Mackay, 1979). Given the broad diversity of chemical structures that pesticide compounds encompass, it is not surprising that their affinities for different environmental media also span a wide range. For example, if they are sufficiently persistent, hydrophobic compounds such as DDT, dieldrin, chlordane, and other organochlorine insecticides (OCs) will, over time, preferentially accumulate in organic soils, lipid-rich biological tissues, and other media with

Table 1 Persistence of some commonly used pesticides in the atmosphere, surface water, soil, and aquatic sediments.

Use class	*Chemical class*	*Example(s)*	*Suggested half-life class in*			
			Atm.	*Surface water*	*Soil*	*Aquatic sediment*
Herbicides	Acetanilides	Metolachlor	4	6	6	7
	Amino acid derivatives	Glyphosate	4	6	6	7
	Chlorophenoxy acids	2,4-D	2	3	5	6
		2,4,5-T	3	5	5	6
	Dinitroanilines	Isopropalin	2	5	6	7
		Trifluralin	4	6	6	7
	Triazines	Atrazine	1	8	6	6
		Simazine	3	5	6	7
	Ureas	Diuron	2	5	6	7
		Linuron	2	5	6	7
Insecticides	Carbamates	Aldicarb	1	5	6	8
		Carbaryl	3	4	5	6
		Carbofuran	1	4	5	6
	Organochlorines	Chlordane	3	8	8	9
		p,p'-DDT	4	7	8	9
		Lindane	4	8	8	9
	Organophosphates	Chlorpyrifos	2	4	4	6
		Diazinon	5	6	6	7
		Malathion	2	3	3	5
Fumigants	Organochlorines	Chloropicrin	4	3	3	4
Fungicides	Imides	Captan	2	2	5	5
	Organochlorines	Chlorothalonil	4	4	5	6
Half-life class definitions						
Class	Mean half-life	Range (h)				
1	5 h	<10				
2	~1 d	10–30				
3	~2 d	30–100				
4	~1 week	100–300				
5	~3 weeks	300–1,000				
6	~2 months	1,000–3,000				
7	~8 months	3,000–10,000				
8	~2 yr	10,000–30,000				
9	~6 yr	>30,000				

Source: Mackay *et al.* (1997).

high levels of organic carbon. By contrast, more volatile compounds, such as chlorofluorocarbons (several of which are used as adjuvants (USEPA, 2002)) and fumigants, will tend to reside primarily in soil gases and the atmosphere until they degrade. (The effects of transformation on the partitioning of pesticide compounds will be discussed in a later section.)

Mackay *et al.* (1997) provide detailed examples of fugacity calculations to illustrate how variations in the physical and chemical properties of pesticides affect their partitioning among environmental media. Figure 3 displays the results from some of these calculations for three of the pesticides listed in Table 1. Consistent with the expectations described above, these computations predict that following their release into the hydrologic system, the relatively water-soluble herbicide atrazine will come to reside mostly in the aqueous phase, the more hydrophobic insecticide chlorpyrifos will tend to concentrate in soil, and the volatile fumigant chloropicrin will be present primarily in the vapor phase. Results from such calculations for a given compound help to focus attention on the media where it is likely to be present in the highest concentrations, and thus where a detailed understanding of its persistence and biological effects may be most critical.

9.15.2.1 Partitioning between Soils, Sediments, and Natural Waters

The movement of pesticide compounds between the solid and aqueous phases exerts considerable influence over the transport, persistence, and bioavailability of these compounds in natural waters. (The *bioavailability* of a compound is the extent to which it is accessible for uptake by living organisms. The term *natural waters* is used herein

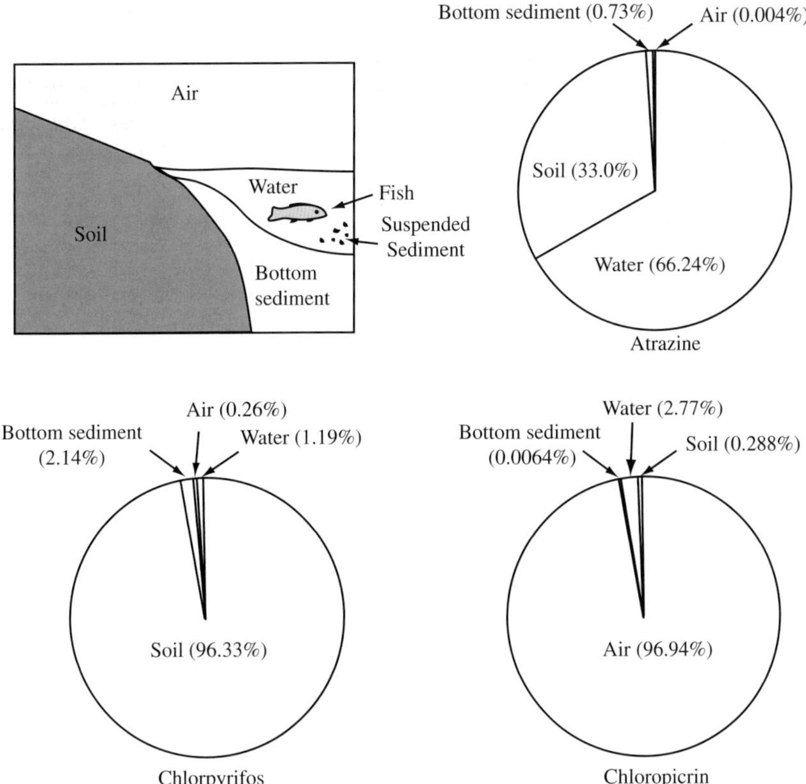

Figure 3 Distributions of atrazine, chlorpyrifos, and chloropicrin among air, surface water, soils, and aqueous sediments, based on fugacity calculations. Percentages sum to less than 100% because partitioning into fish and suspended sediment was not accounted for (Mackay *et al.* (1997); reproduced by permission of CRC Press, Lewis Publishers from *Illustrated Handbook of Physical–Chemical Properties and Environmental Fate for Organic Chemicals, Volume V. Pesticide Chemicals*, **1997**).

to refer to water occurring anywhere within the hydrologic system—including precipitation, surface waters, vadose-zone water, and groundwater—regardless of whether or not it has been affected by human activities.) The principal phases among which this partitioning takes place are the aqueous solution itself, natural organic matter (NOM), mineral surfaces, and biological tissues. Because water in the vadose and saturated zones is in such intimate contact with natural solids, such partitioning is presumed to exert a more substantial influence over the movement and persistence of pesticide compounds below the land surface than in surface waters or the atmosphere.

Sorptive interactions with NOM are particularly important for neutral pesticide compounds, including those that are *Brønsted acids* (i.e., chemical species that can donate a hydrogen atom to another species, known as a *Brønsted base* (Stumm and Morgan, 1981)). This is especially true in soils with mass fractions of organic carbon (f_{oc}) of 0.001 or more (Schwarzenbach and Westall, 1981). This threshold, however, is likely to vary inversely with the octanol–water partition coefficient (K_{ow}) among different compounds (McCarty *et al.*, 1981). The discovery that soil–water partition coefficients (K_p) for neutral organic compounds often vary among different soils in direct relation to f_{oc} led to the development of the organic carbon–water partition coefficient, or K_{oc} ($K_{oc} = K_p/f_{oc}$; Hamaker and Thompson, 1972). The association of these compounds with NOM is commonly viewed as being analogous to their dissolution in an organic solvent, especially since K_{oc} values are known to be inversely related to water solubility, directly related to K_{ow} and largely independent of competitive effects among solutes (Chiou, 1998).

Despite their normalization to f_{oc}, K_{oc} values for individual pesticide compounds still vary among different soils and sediments, though to a much lesser extent than K_p values (Curtis *et al.*, 1986). These variations in K_{oc}, which typically span a factor of 10 or less for individual pesticide compounds (e.g., Mackay *et al.*, 1997), are presumed to arise from variations in the sorption properties of the biogenic materials of which NOM is comprised (Shin *et al.*, 1970), changes in the chemical properties of NOM caused by weathering (Chiou, 1998) or, for ionic compounds or Brønsted acids, variations in solution properties such as pH and salinity (Schwarzenbach *et al.*, 1993).

The exchange of pesticide compounds between aqueous solution and the sorbed phase in soils is not instantaneous. Indeed, the more hydrophobic the compound, the longer the time required to reach sorption equilibrium. This phenomenon has been attributed to the effect of hydrophobicity on the rate at which an organic molecule diffuses through the polymeric structure of NOM within soil particles or aggregates (Curtis *et al.*, 1986; Brusseau and Rao, 1989). Support for this explanation is provided by the fact that the amount of time required for pesticides to reach sorption equilibrium has been observed to be longer for soils containing higher amounts of NOM (e.g., Moreau and Mouvet, 1997).

A K_p value (and its corresponding K_{oc}) is most commonly determined for a given solute in a specific soil–water system by computing the slope of a *sorption isotherm* (i.e., a graph of sorbed concentration versus dissolved concentration at equilibrium over a range of solute loadings). The widespread use of K_{oc} values assumes that the sorption isotherm is linear, i.e., that the quantitative relation between the amount of solute sorbed to the soil organic carbon (C_{oc}) and the dissolved concentration (C_{aq}) is of the following form (Hamaker and Thompson, 1972):

$$C_{oc} = K_{oc} C_{aq}$$

However, the sorption of a number of pesticide compounds to some soils has been found to be more accurately described by the nonlinear Freundlich isotherm (e.g., Hamaker and Thompson, 1972; Widmer and Spalding, 1996), i.e.,

$$C_{oc} = K_f C_{aq}^{1/n}$$

Published values of Freundlich parameters (K_f and $1/n$) are relatively sparse, but are currently available for at least 60 pesticide compounds (Barbash, unpublished compilation).

Sorption to mineral surfaces (as opposed to NOM) is generally viewed as more of a displacement than a dissolution phenomenon. Because mineral surfaces tend to be more polar than NOM, sorption to the former is more substantial for polar and ionic compounds than for those that are more hydrophobic (Curtis *et al.*, 1986; Chiou, 1998). Furthermore, since most NOM and mineral surfaces exhibit either a neutral or negative charge, sorption to soils and sediments is considerably stronger for pesticide compounds that are positively charged in solution—such as paraquat or diquat—than for neutral species, and weaker still for anions. As a consequence, measured K_p values in soils exhibit little dependence upon pH for pesticide compounds that are not Brønsted acids or bases (Macalady and Wolfe, 1985; Haderlein and Schwarzenbach, 1993). However, for those that are Brønsted acids or bases, K_p values increase dramatically as the pH is reduced below the pK_a value(s) for the compound (Haderlein and Schwarzenbach, 1993; Broholm *et al.*, 2001). (The pK_a of a Brønsted acid/base pair is the negative log of the equilibrium constant for their interconversion through the gain or loss of a proton. As such, the pK_a also represents the pH value below which the concentration of the acid exceeds that of the base, and above which the base dominates.) Schellenberg *et al.* (1984) introduced equations that may be used to quantify the effects of pH variations on K_p for Brønsted acids and bases.

Increases in temperature lead to a decrease in K_p values for most pesticide compounds (e.g., Katz, 1993), but some display the reverse trend (Hamaker and Thompson, 1972; Padilla *et al.*, 1988; Haderlein and Schwarzenbach, 1993; Chiou, 1998). In most environmental settings, however, the effect of temperature on sorption is expected to be relatively minor—and, indeed, is nearly always neglected. Precipitation from solution may be significant for some pesticides, as is the case for glyphosate, which has been shown to form relatively insoluble metal complexes with Fe^{III}, Cu^{II}, Ca^{II}, and Mg^{II} at circumneutral pH (Subramaniam and Hoggard, 1988). (Roman numerals are used as superscripts in this chapter to denote chemical species that may be present either as dissolved ions or as part of the solid phase.)

While much of the preceding discussion has focused on the effect of water–solid partitioning on the movement of pesticide compounds below the land surface, such partitioning may also influence the transport of these chemicals in surface waters (Figure 4). Compounds exhibiting a pronounced affinity for natural sediments—either because of hydrophobicity, low water solubility, or other chemical characteristics—are transported primarily with suspended sediments in surface waters, rather than in the aqueous phase (Wauchope, 1978). The deposition of sediments to which persistent pesticide compounds are sorbed leads to substantial increases in the residence time of these compounds in aquatic ecosystems. For this reason, detailed analyses of sediment cores obtained from reservoirs around the country have proved to be useful for observing long-term trends in the concentrations of OCs in aquatic environments over several decades (Van Metre *et al.*, 1998).

In addition to sorption and precipitation, the diffusive exchange of pesticide compounds between *mobile* and *immobile waters* also influences the rates at which these solutes move through the hydrologic system—or, more specifically, through the vadose and saturated zones. (The term *mobile water* refers to subsurface

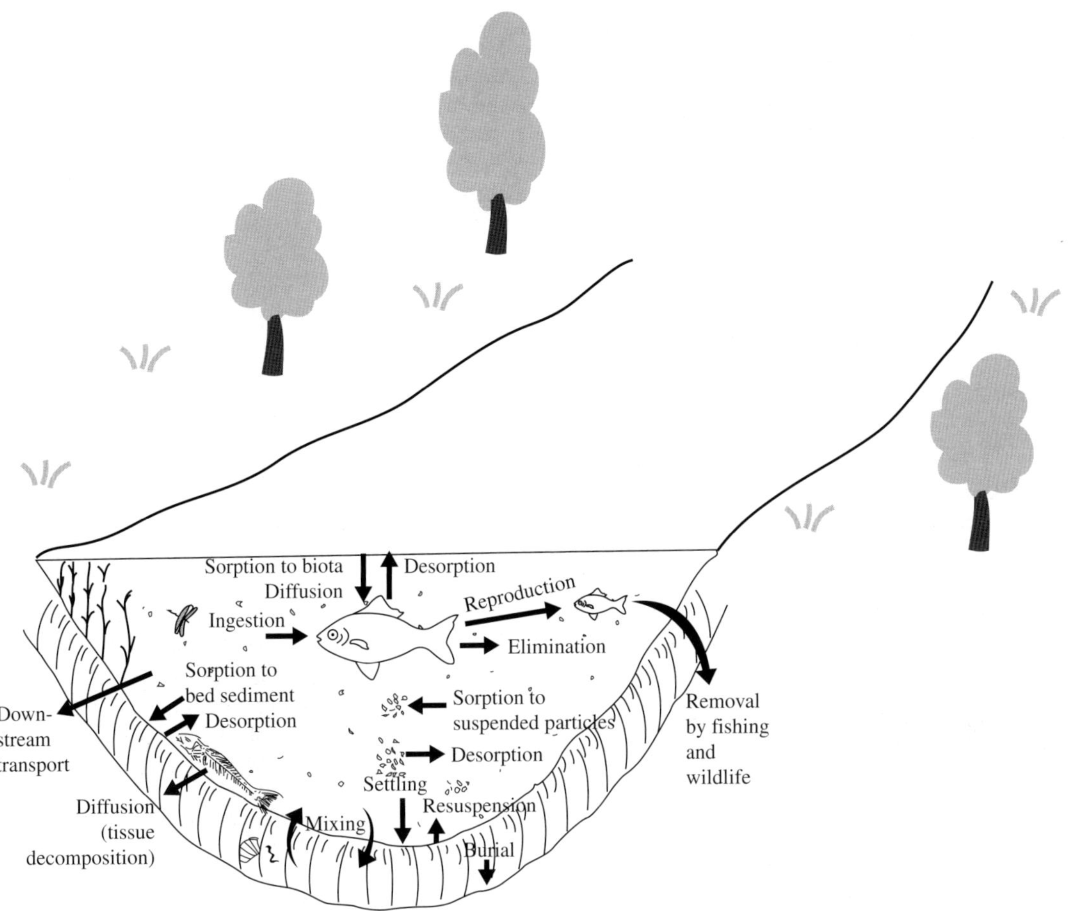

Figure 4 Pesticide movement to, from, and within sediment and aquatic biota in surface waters (Nowell *et al.* (1999); reproduced by permission of CRC Press from *Pesticides in Stream Sediment and Aquatic Biota—Distribution, Trends and Governing Factors*, **1999**).

water that moves by comparatively rapid, advective flow along preferred flow paths within soils and other geologic media. By contrast, *immobile water* resides within the interior pores of soil particles and aggregates, and therefore migrates much more slowly, if at all.) This exchange is associated with the macroscopic phenomenon known as *preferential transport*, and can affect solute transport in ways that are similar to the effects of sorption, including the tailing of solute breakthrough curves, long-term uptake and release of solutes over time, and much of what is often erroneously attributed to the formation of bound residues (Barbash and Resek, 1996). The diffusion of pesticide compounds from mobile waters into zones of immobile water is also believed to exert substantial effects on bioavailability, since pesticide molecules may diffuse into the interior pore spaces of soil particles or aggregates and become inaccessible to organisms that might otherwise be able to degrade them (Zhang *et al.*, 1998).

9.15.2.2 Partitioning between Aquatic Biota and Natural Waters

The movement and persistence of pesticide compounds in the hydrologic system are also affected by partitioning and transformation in the tissues of aquatic biota. Pesticide compounds accumulate in aquatic biota as a result of either passive partitioning from the water column or the ingestion of sediment or other organisms already containing the chemicals. The distribution of pesticides and other organic compounds between water and biological tissues has been most commonly described using a *bioconcentration factor* (BCF). (Compilations of BCF values for pesticides include those assembled by Kenaga (1980) and Mackay *et al.* (1997).) Since both the biota and sediments in aquatic ecosystems are in nearly constant contact with the water itself, the concentrations of pesticide compounds in aquatic sediments have been used as indicators of the anticipated levels of these substances in aquatic biota. This approach, most commonly implemented through the use of a *biota-sediment accumulation factor* (BSAF), has been shown to produce remarkably consistent results for a wide range of aquatic environments and organisms across the United States (Wong *et al.*, 2001). Partitioning-based approaches for predicting pesticide concentrations in biota, however, do not account for the metabolism of these compounds *in vivo*. Nowell *et al.* (1999) provide a detailed examination of the history, theoretical basis, assumptions, and limitations of the BSAF model, as well as a comprehensive summary of existing data on the occurrence of pesticide compounds in aquatic fauna and flora across the United States.

9.15.2.3 Partitioning between the Earth's Surface and the Atmosphere

Several types of observations suggest that atmospheric transport is principally responsible for the fact that pesticide residues are now likely to be detected in every terrestrial and marine ecosystem on the surface of our planet (e.g., Iwata *et al.*, 1993; Majewski and Capel, 1995). First, as noted earlier, a large proportion of applied pesticide compounds may be transported in the air away from the original application sites as a result of either spray drift, wind erosion of soil, or volatilization from the plant, soil, and water surfaces to which the chemicals were applied (Majewski and Capel, 1995). Second, while terrestrial and aquatic ecosystems are spatially discontinuous, the atmosphere is a single, comparatively well-mixed medium that is in direct contact with the entire surface of the Earth. Finally, the atmosphere exhibits mixing times that, by comparison with the transformation rates of most of the compounds of interest (e.g. Table 1), are comparatively rapid—on the order of weeks to months for mixing within each of the northern and southern hemispheres, and 1–2 years for exchange between the hemispheres (Majewski and Capel, 1995). However, an understanding of the processes and factors that control the dissemination of pesticide compounds across the globe depends upon a knowledge not only of large-scale spatial patterns of pesticide use (e.g., US Geological Survey, 1998) and atmospheric transport (Majewski and Capel, 1995; Bidleman, 1999), but also of the mechanisms and rates of partitioning of these compounds between air and the materials at the Earth's surface.

9.15.2.3.1 Movement between air and natural waters

The parameter used most often to quantify partitioning between air and water, the Henry's Law constant (H), takes several forms, but is frequently calculated as the ratio between the vapor pressure and the aqueous solubility of the subcooled liquid. Suntio *et al.* (1988) listed the assumptions upon which the use of H to describe this partitioning is based, described methods for converting between the different forms in which H is expressed, and assembled a compilation of H values for 96 pesticides. The rate of transfer of a pesticide compound between air and water is a function of the contrast between its fugacities in the two phases (Bidleman, 1999). This exchange rate is also controlled by the rates of diffusion of the compound through the thin, adjacent films of air and water that comprise the interface between the two phases, as well as by wind speed, current velocity, turbulence, and the other factors that influence the thickness of these films (Thomas, 1990a).

While it is obvious that pesticide compounds will migrate away from their application sites immediately following their release, the directions of large-scale movement of persistent compounds over longer timescales may be more difficult to anticipate. Fugacity calculations and *enantiomer ratios*, however, have proved to be valuable tools for revealing the large-scale patterns of atmospheric transport of persistent pesticides. (Enantiomers are pairs of compounds that have the same chemical composition, but spatial arrangements of atoms that differ in such a way that the two molecules are mirror-images of one another.) The first of these two methods involves estimating the fugacities of a given pesticide compound in pairs of environmental media that are in intimate physical contact (e.g., air/water or water/soil) in order to predict its future movement. If the fugacities of the compound in the two media are not equal, their ratio provides an indication of the likely direction of future exchange of the compound between the media of interest.

Enantiomer-based methods exploit the fact that some pesticide compounds are applied in known ratios of enantiomers, most commonly as *racemic* mixtures, i.e., 1:1 ratios (Buser *et al.*, 2000; Monkiedje *et al.*, 2003). Although most abiotic transformation and partitioning processes are not affected by the structural differences between enantiomers (Bidleman, 1999), the biotransformation of some pesticide compounds has been found to be an *enantioselective* process, i.e., one that exhibits a preference for one enantiomer over the other (e.g., Harner *et al.*, 1999; Monkiedje *et al.*, 2003). The measurement of enantiomer concentration ratios for a pesticide compound that is applied as a racemic mixture but may undergo enantioselective biotransformation in the environment can thus provide an indication of whether or not the compound has undergone biotransformation since it was applied—and thus a rough indication of its residence time in the hydrologic system (Bidleman, 1999).

The use of both fugacity calculations and enantiomer ratios in the Great Lakes, for example, has revealed alternating, seasonal cycles of net deposition and net volatilization of OCs to and from the lake surfaces. However, these methods have also indicated that the concentrations of OCs in the lakes are in approximate long-term equilibrium with the overlying air. Thus, significant reductions in the levels of OCs in the lakes are unlikely to occur until their concentrations in the atmosphere decline. Similar observations have been made in other parts of the world, including Lake Baikal and Chesapeake Bay. Other investigations have indicated that global-scale patterns of deposition for some OCs may be more strongly controlled by temperature than by spatial patterns of application, with net volatilization occurring in more temperate regions and net deposition to soil and other solid surfaces—including snow and ice—in polar regions (Bidleman, 1999). This pattern of movement, sometimes referred to as *global distillation*, may help explain why such high levels of OCs are commonly detected in polar ecosystems (e.g., Nowell *et al.*, 1999; Wania, 2003).

9.15.2.3.2 Movement between air, soil, and plant surfaces

The volatilization of a pesticide compound from soil is controlled by three general processes (Thomas, 1990b): (i) upward advection in the soil from capillary action caused by water evaporating at the surface (the *wick effect*), (ii) partitioning between the solid, liquid, and gas phases within the soil, and (iii) transport away from the soil surface into the atmosphere. If a compound is less volatile than water, the wick effect may cause it to concentrate at the soil surface, resulting in its precipitation from solution, an increase in its volatilization rate, and/or a suppression of the water evaporation rate. Thomas (1990b) summarizes a variety of methods that have been devised for estimating the rates of pesticide volatilization from soil following either surface- or depth-incorporated applications. Woodrow *et al.* (1997) found the rates of pesticide volatilization from recently treated soil, water bodies, and "noninteractive" surfaces (including freshly treated plants, glass, and plastic) to be highly correlated with simple combinations of the vapor pressure, water solubility, K_{oc}, and application rates of the compounds of interest.

Pesticides in the vapor phase show considerable affinity for dry mineral surfaces, especially those of clays. However, since mineral surfaces generally exhibit a greater affinity for water than for neutral organic compounds, the vapor-phase sorption of pesticides to dry, low-f_{oc} materials diminishes markedly with increasing relative humidity (RH) as the pesticide molecules are displaced from the mineral surfaces by the adsorbed water (Chiou, 1998). One important consequence of this phenomenon is that the wetting of dry soils to which a pesticide has previously been applied (and thus sorbed) may lead to a sudden increase in the concentration of the compound in the overlying air (e.g., Majewski *et al.*, 1993). At or above the RH required to cover all of the mineral surfaces with adsorbed water (a water content that some studies suggest is commonly at or below the wilting point), the sorption of pesticide vapor becomes controlled by interactions with the soil organic matter—rather than by competition with water—and the amount of pesticide taken up by the soil depends upon the soil f_{oc}. Organic matter has a much lower affinity for water than that exhibited by mineral surfaces. As a result, in organic rich soils, f_{oc} exerts considerably more influence over pesticide uptake than does RH, even under dry conditions (Chiou, 1998). Soil–air partition coefficients (K_{SA}) have typically been computed as the ratio between soil/water and air/water partition coefficients for the compounds of interest (e.g., $K_{SA} = K_p/H$), although methods for the direct measurement of K_{SA} have also been devised (Hippelein and McLachlan, 1998; Meijer *et al.*, 2003).

Consistent with the observations for soil/air partitioning, plant/air partition coefficients for pesticides have been found to be correlated with K_{ow}/H (Bacci *et al.*, 1990). (K_{ow} is commonly used as a measure of partitioning between biological tissues and water.) A direct correlation observed by Woodrow *et al.* (1997) between vapor pressure and the rates of volatilization from the surfaces of recently treated plants suggests that chemical interactions with plant surfaces exert only a minor influence on pesticide volatilization from plants during the first few hours following application. For persistent OCs that were banned in previous decades but are still widely detected in the environment, enantiomer ratios and other methods have been employed to distinguish between inputs from "old" sources—i.e., the ongoing atmospheric exchange of these compounds with soils, plants, and surface waters many years after their application—and fresh inputs of the compounds in countries where their use continues, either legally or illegally (Bidleman, 1999).

9.15.3 TRANSFORMATIONS

All transformations of pesticide compounds in the hydrologic system are initiated by either *photochemical* or *thermal* processes, depending upon whether or not the reactions are driven by solar energy. Following its introduction into the environment, the persistence of a pesticide molecule is determined by its chemical structure, as well as by the physical, chemical, and biological characteristics of the medium in which it is located. If it is exposed to sunlight of sufficient intensity within the appropriate wavelength range, the molecule may either be promoted to a higher-energy state and react via *direct photolysis*, or undergo *indirect photolysis* by reacting with another species that has itself been promoted (either directly or indirectly) to a higher-energy level by sunlight.

At the same time, however, the pesticide molecule may also be susceptible to thermal reactions. At a given temperature, the distribution of kinetic energy among all molecules of a particular chemical species exhibits a specific statistical form known as the *Boltzmann distribution*. For any given reaction, only those molecules possessing a kinetic energy exceeding a specific threshold, the *activation energy*, are likely to undergo that transformation (Atkins, 1982). Thermal reactions are those that occur as a result of collisions between molecules exceeding this energy barrier, either with or without biochemical assistance. Thus, if a pesticide molecule does not react by photolysis (either directly or indirectly), its persistence will be determined by the distribution of kinetic energy among the other chemical species with which it may undergo thermal reactions within the medium of interest.

Because pesticide compounds exhibit such a broad range of chemical structures, the variety of pathways by which they are transformed in the hydrologic system is also extensive, but all may be classified according to the manner in which the overall oxidation state of the molecule is altered, if at all. *Neutral* reactions leave the oxidation state of the original, or *parent* compound unchanged, while *electron-transfer* reactions (also referred to as *oxidation–reduction* or *redox* reactions) involve either an increase (*oxidation*) or a decrease (*reduction*) in oxidation state. Both neutral and electron-transfer mechanisms have been identified for thermal and photochemical transformations of pesticides. Most of the known transformation pathways involve reactions with other chemical species, but some are unimolecular processes—most commonly those that occur through direct photolysis. In many cases, pesticides may react via combinations of different types of reactions occurring simultaneously, sequentially or both (e.g., Figure 5). Several previous reviews have provided comprehensive summaries of the pathways that have been observed for the transformation of pesticide compounds in natural systems by either

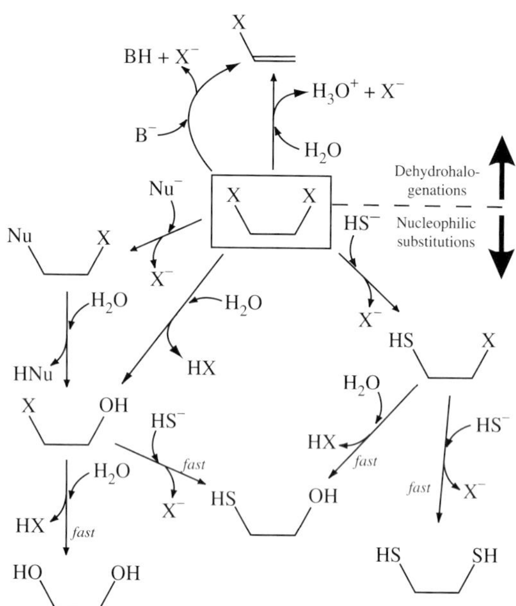

Figure 5 Chemical reactions leading to the dehalogenation of a 1,2-dihaloethane (X = Br for 1,2-dibromoethane (EDB), X = Cl for 1,2-dichloroethane) in an aqueous solution containing the bisulfide anion (HS^-) and one or more additional nucleophiles (Nu^-) that react with the compound via nucleophilic substitution (e.g., nitrate, hydroxide or a buffer conjugate base), or bases (B^-) that react via dehydrohalogenation (e.g., hydroxide or a buffer conjugate base) (after Barbash, 1993).

photochemical (e.g., Mill and Mabey, 1985; Harris, 1990b) or thermal mechanisms (e.g., Kearney and Kaufman, 1972; Castro, 1977; Alexander, 1981; Bollag, 1982; Vogel *et al.*, 1987; Kuhn and Suflita, 1989; Scow, 1990; Coats, 1991; Barbash and Resek, 1996).

Pesticide compounds may be transformed with or without the assistance of living organisms, depending upon compound structure and the biogeochemical environment. Distinguishing between biological and abiotic mechanisms of transformation, however, is not always straightforward. For several pesticide compounds, both abiotic and microbially mediated transformations may occur simultaneously or sequentially (e.g., Skipper *et al.*, 1967; Graetz *et al.*, 1970; Lightfoot *et al.*, 1987; Vogel and McCarty, 1987; Oremland *et al.*, 1994; and Gan *et al.*, 1999). Some transformations may occur either with or without biological assistance (e.g., Wolfe *et al.*, 1986; Jafvert and Wolfe, 1987; Mandelbaum *et al.*, 1993; Loch *et al.*, 2002). Other reactions appear to be primarily, if not exclusively abiotic (Konrad *et al.*, 1967; Haag and Mill, 1988a). Under the conditions of most natural waters, the *mineralization* of pesticides containing carbon–carbon bonds—i.e., their complete conversion to simple products such as CO_2, H_2O, and halide ions—does not take place abiotically (Alexander, 1981), although the abiotic conversion of CCl_4 to a variety of single-carbon products in the presence of dissolved sulfide and clay surfaces has been documented (Kriegman-King and Reinhard, 1992). The cleavage of aromatic rings also does not appear to occur readily through abiotic means in natural waters. With the exception of direct photolysis, however, one or more organisms have been found to be capable of facilitating all of the major types of pesticide transformation reactions listed above (e.g., Kearney and Kaufman, 1972; Castro, 1977; Alexander, 1981; Zepp and Wolfe, 1987; Kuhn and Suflita, 1989; Scow, 1990; Coats, 1991; Barbash and Resek, 1996).

As noted earlier, transformation reactions of pesticide compounds are typically faster in air than in water (Mackay *et al.*, 1997). However, general trends among the rates of different reactions *per se* are elusive, since chemical structure and the physical, chemical, and biological characteristics of the reaction medium may be as important in determining reaction rate as the nature of the reaction in question. The following discussion describes the circumstances under which pesticide compounds undergo each major type of reaction, and examines the various ways in which the rates of these transformations are controlled by chemical structure and the biogeochemical environment.

9.15.3.1 Photochemical Transformations

Pesticide compounds that may undergo direct photolysis in the hydrologic system are those for which the wavelengths required for bond breakage fall within the range of the solar spectrum (e.g., Zepp *et al.*, 1975). Because of this constraint, relatively few pesticide compounds undergo direct photolysis, but several chlorophenoxy acids (and their esters), nitroaromatics, triazines, organophosphates (OPs), OCs, carbamates, polychlorophenols, and fumigants have been shown to do so (e.g., Crosby and Leitis, 1973; Dilling *et al.*, 1984; Zepp *et al.*, 1984; Mill and Mabey, 1985; Harris, 1990b; Chu and Jafvert, 1994; Mansour and Feicht, 1994). Most phototransformations of pesticide compounds occur through indirect photolysis, as a result of reaction with another species, known as a *sensitizer*, or a sensitizer-produced *oxidant*. The most common sensitizers for the phototransformation of pesticide compounds in natural waters are the humic and fulvic acids derived from NOM (e.g., Mansour and Feicht, 1994). Sensitizer-produced oxidants include hydrogen peroxide, ozone, singlet oxygen, hydroxyl, peroxy, and nitrate radicals, and photo-excited triplet diradicals (Cooper and Zika, 1983; Mill and Mabey, 1985; Mackay *et al.*, 1997). The photoproduction of hydrogen peroxide has also been shown to be catalyzed by algae (Zepp, 1988).

For some pesticide compounds, such as dinitroaniline herbicides (Weber, 1990), phototransformation occurs primarily in the vapor phase, rather than in the dissolved or sorbed phases. Perhaps the most environmentally significant pesticide phototransformation in the atmosphere, however, is the photolysis of the fumigant methyl bromide, since the bromine radicals created by this reaction are 50 times more efficient than chlorine radicals in destroying stratospheric ozone (Jeffers and Wolfe, 1996). Detailed summaries of the rates and pathways of phototransformation of pesticides and other organic compounds in natural systems, and discussions of the physical and chemical factors that influence these reactions, have been presented elsewhere (e.g., Zepp *et al.*, 1984; Mill and Mabey, 1985; Harris, 1990b).

9.15.3.2 Neutral Reactions

The neutral reactions responsible for transforming pesticide compounds in the hydrologic system—listed in roughly decreasing order of the number of compounds known to be affected—include nucleophilic substitution, dehydrohalogenation, rearrangement, and addition. Figure 6 displays some examples of these reactions for pesticides. Nucleophilic substitution involves the replacement of a substituent on the molecule (the *leaving group*) by an attacking species (the *nucleophile*). (A leaving group is any part of a molecule that is removed during a chemical reaction (March, 1985).) Electrophilic substitution reactions are also well known in organic chemistry (March, 1985). However, since nucleophiles are substantially more abundant than electrophiles in most natural waters, reactions with electrophiles are of relatively minor importance in the hydrologic system, and confined primarily to photolytic and biologically mediated transformations (Schwarzenbach *et al.*, 1993).

Not surprisingly, the nucleophilic substitution reactions that have been studied most extensively for pesticide compounds in natural waters (e.g., Mabey and Mill, 1978; Washington, 1995) are those involving the three solutes that are present in all aqueous systems, i.e., H_2O, its conjugate base (OH^-), and its conjugate acid (H_3O^+). These reactions are referred to as *neutral*, *base-catalyzed*, and *acid-catalyzed hydrolysis*, respectively. The first two of these reactions involve the direct displacement of the leaving group by the nucleophile (H_2O or OH^-), while in the third case, protonation near the *alpha* carbon (C_α, i.e., the one from which the leaving group is displaced) decreases the electron density on C_α,

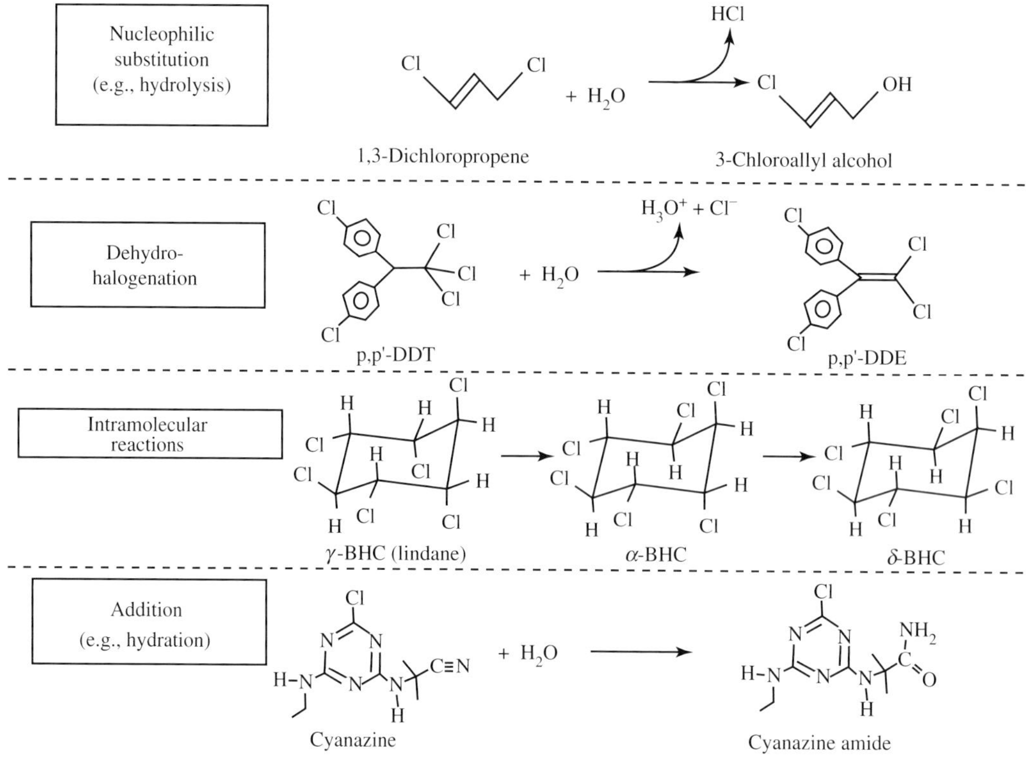

Figure 6 Some examples of neutral reactions of pesticide compounds: nucleophilic substitution (e.g., hydrolysis; Roberts and Stoydin, 1976); dehydrohalogenation (Kuhn and Suflita, 1989); intramolecular reactions (e.g., rearrangement; Newland *et al.*, 1969); and addition (e.g., hydration, Sirons *et al.* 1973).

rendering it more susceptible to nucleophilic attack by H_2O (March, 1985). Some plants employ catalysts to hydrolyze pesticides in their tissues as a detoxification mechanism (Beynon *et al.*, 1972). Extracellular biochemicals that may catalyze the hydrolysis of pesticide compounds include protease, esterase, and phosphatase enzymes (Huang and Stone, 2000). Other microbial enzymes that facilitate pesticide hydrolysis have been summarized by Bollag (1982). As discussed later, metals may also catalyze hydrolysis reactions.

Among the other nucleophiles whose reactions with pesticide compounds may be significant in the hydrologic system, perhaps the most important are reduced sulfur anions (Barbash and Reinhard, 1989a), particularly bisulfide, polysulfides (e.g., Barbash and Reinhard, 1989b; Miah and Jans, 2001; Lippa and Roberts, 2002; Loch *et al.*, 2002) and thiosulfate (e.g., Ehrenberg *et al.*, 1974; Wang *et al.*, 2000). Reactions with these nucleophiles are often an indirect mechanism of biotransformation, as these and other reduced sulfur anions are derived primarily from biological activity. However, living organisms also actively employ a variety of nucleophilic enzymes to detoxify halogenated compounds by displacing halide. In addition to the enzyme-catalyzed hydrolyses mentioned earlier, another environmentally important example of these reactions is the displacement of chloride from chloroacetanilide herbicides by glutathione and glutathione-*S*-transferase (Figure 7) to form the corresponding ethanesulfonic and oxanilic acids (Field and Thurman, 1996). (The initial step in this example is an illustration of a *synthetic reaction*, or *conjugation*, in which a part of the original molecule is replaced with a substantially larger moiety (Bollag, 1982; Coats, 1991).) Work by Loch *et al.* (2002), however, suggests that the ethanesulfonic acid products might also be generated through abiotic reactions of chloroacetanilide herbicides with reduced sulfur species.

Other nucleophiles of potential importance include pH buffer anions (Figure 5) since, as noted later, buffers are commonly used to stabilize pH during laboratory studies of pesticide transformations. Although it is only a weak nucleophile, the nitrate anion has also been found (Barbash and Reinhard, 1992b; Barbash, 1993) to displace bromide from the fumigant 1,2-dibromoethane (ethylene dibromide, or EDB) in aqueous solution (Figure 5), an observation worth noting because nitrate is probably the most widespread groundwater contaminant in the world.

Another neutral mechanism by which some pesticide compounds may be transformed is

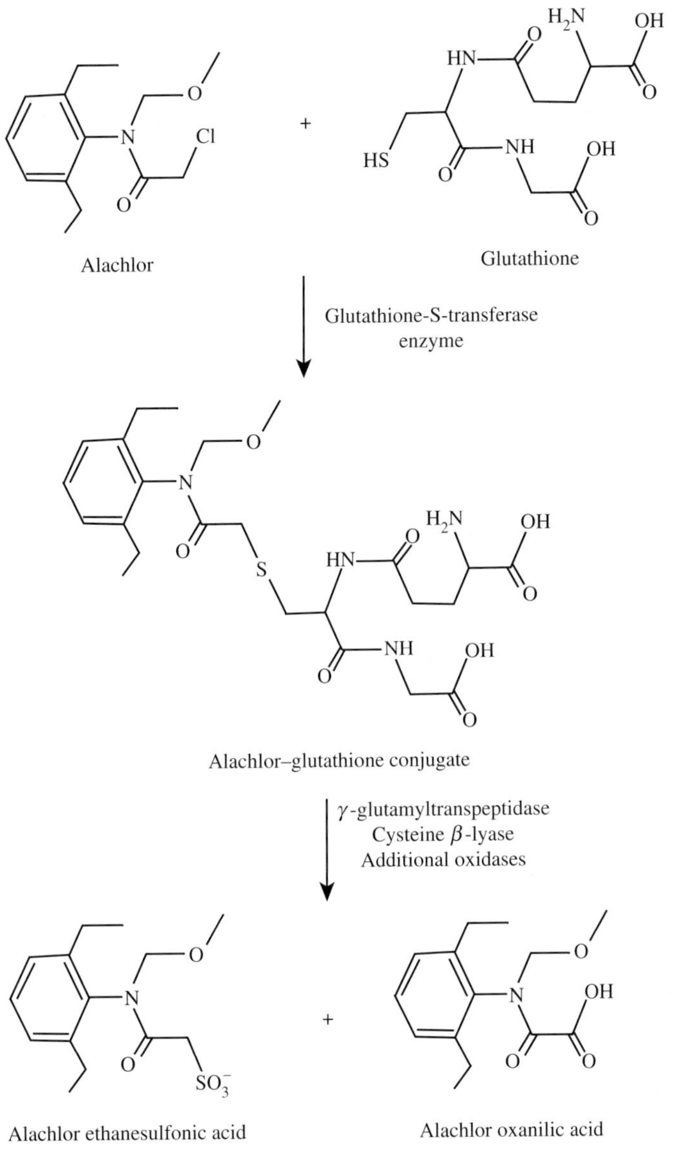

Alachlor + Glutathione

Glutathione-S-transferase enzyme

Alachlor–glutathione conjugate

γ-glutamyltranspeptidase
Cysteine β-lyase
Additional oxidases

Alachlor ethanesulfonic acid + Alachlor oxanilic acid

Figure 7 Formation of ethanesulfonic acid and oxanilic acid metabolites from the reaction of a chloroacetanilide herbicide (alachlor) with glutathione and other enzymes (Field and Thurman (1996); reproduced by permission of American Chemical Society from *Environ. Sci. Technol.*, **1996**, *30*, 1413–1417).

dehydrohalogenation, which involves the removal of a proton and a halide ion (HX) from a pair of adjacent carbon atoms (e.g., Figures 5 and 6). Under the comparatively mild conditions of most natural waters, only singly bonded carbons are likely to undergo this reaction, leading to the formation of the corresponding alkene. Pesticide compounds that have been observed to undergo dehydrohalogenation under environmentally relevant conditions include a number of fumigants, insecticides, and volatile adjuvants (Burlinson *et al.*, 1982; Vogel and Reinhard, 1986; Vogel and McCarty, 1987; Haag and Mill, 1988a; Cline and Delfino, 1989; Jeffers *et al.*, 1989; Kuhn and Suflita, 1989; Deeley *et al.*, 1991; Barbash and Reinhard, 1992a,b; Ngabe *et al.*, 1993).

Two other neutral mechanisms of pesticide transformation are intramolecular reactions and additions (Figure 6). Intramolecular reactions of pesticide compounds may either leave the overall chemical composition of the parent unchanged (*rearrangements* (Russell *et al.*, 1968; Newland *et al.*, 1969)), slightly altered (Coats, 1991), or substantially modified (Wei *et al.*, 2000). Addition reactions involve the coupling of a compound containing a double or triple bond with another molecule; examples involving pesticide compounds include the hydration of cyanazine to cyanazine amide (Sirons *et al.*, 1973) and the hypothesized addition of H_2S to polychloroethenes to form the corresponding polychloroethanethiols (Barbash and Reinhard, 1992a; Barbash, 1993).

9.15.3.3 Electron-transfer Reactions

Thermodynamic considerations dictate that the likelihood with which a pesticide compound in a particular geochemical setting will undergo oxidation or reduction is governed in large part by the tendency of the other chemical species present to accept or donate electrons, respectively (McCarty, 1972). In addition to the more transient oxidants mentioned earlier in relation to photochemical reactions, those that are most commonly present in natural waters—listed in roughly decreasing order of their tendency to accept electrons—include O_2, NO_3^-, Mn^{IV}, Fe^{III}, SO_4^{2-}, and CO_2. Conversely, the most common reductants in natural waters include CH_4 and other forms of reduced organic carbon, S^{II}, Fe^{II}, Mn^{II}, NH_4^+, and H_2O (Christensen *et al.*, 2000). Extensive discussion of the chemistry of these and other naturally occurring oxidants and reductants in the hydrologic system has been provided elsewhere in this volume (see Chapters 9.12 and 9.16).

Many redox transformations of pesticide compounds that may take place in the hydrologic system have been observed to occur abiotically (e.g., Kray and Castro, 1964; Castro and Kray, 1966; Wade and Castro, 1973; Mochida *et al.*, 1977; Klecka and Gonsior, 1984; Jafvert and Wolfe, 1987; Tratnyek and Macalady, 1989; Baxter, 1990; Kriegman-King and Reinhard, 1992; Curtis and Reinhard, 1994; Strathmann and Stone, 2001; Carlson *et al.*, 2002). Among the most common are reactions with both organic and inorganic forms of reduced iron, as well as reactions with reduced forms of nonferrous metals, such as Cr^{II}, Cu^{I}, Cu^{II}, and Sn^{II}. Reactions with these other metals may be important in natural waters affected by the use of copper-based fungicides or algicides for agriculture, aquaculture, or aquatic weed control; the discharge of mine drainage or tanning wastes; or the use of paints containing organotin fungicides on boats.

While many redox transformations of pesticide compounds can occur abiotically, virtually all such reactions in natural systems are facilitated, either directly or indirectly, by biological processes (Wolfe and Macalady, 1992). Some pesticide compounds may be taken up by living organisms and directly oxidized or reduced through the involvement of a variety of redox-active biomolecules (Bollag, 1982). Enzymes that have been found to be responsible for the biological oxidation of pesticide compounds include methane mono-oxygenase (Little *et al.*, 1988) and mixed-function oxidases (Ahmed *et al.*, 1980); those involved in reduction reactions consist primarily of transition-metal complexes centered on iron, cobalt, or nickel (e.g., Gantzer and Wackett, 1991). In addition to their ability to oxidize or reduce pesticide compounds directly, living organisms also exert indirect control over these reactions by regulating the predominant terminal electron-accepting processes (TEAPs) in natural waters (Lovley *et al.*, 1994).

Although a large number of studies have been conducted to examine the persistence of pesticide compounds under the geochemical conditions encountered in soils and natural waters (Barbash and Resek, 1996), relatively few have reported these observations in terms of the dominant TEAPs under which these transformations are likely to occur. Figure 8 summarizes the results from some of the studies from which such information may be extracted. Most of the examples shown involve reductive dehalogenation, reflecting the fact that, as noted by Wolfe and Macalady (1992), most investigations of redox reactions involving pesticides have focused on this type of transformation. For each reaction shown, the shorter the arrow, the more precisely the results from the cited studies could be used to determine the boundary separating the regions of redox stability for parent and product(s). Thus, for example, the specific TEAP under which bromacil undergoes reductive debromination appears to have been more precisely determined than that under which trichloromethane undergoes dechlorination. (Data from other studies not used for the construction of Figure 8, however, may have helped to define these boundaries more precisely.) The figure also shows that the same type of reaction may require different TEAPs in order for the transformation of different compounds to take place. For example, while the dechlorination of tetrachloromethane to trichloromethane can occur under denitrifying conditions, the dechlorination of 2,4,5-T to 2,4-D and 2,5-D requires sulfate-reducing conditions. Figure 8 also shows that transformations of the same parent compound may yield different products under the influence of different TEAPs (tetrachloromethane, dicamba).

For neutral reactions, the nature of the attacking species can often be inferred from the composition and structure of the reaction products. By contrast, the specific oxidants or reductants with which pesticide compounds undergo electron-transfer reactions in the hydrologic system are usually unclear (Wolfe and Macalady, 1992). This uncertainty arises both from the variety of different species that can serve as electron donors or acceptors in natural systems (many of which were listed earlier), and from the potential involvement of electron-transfer agents that may

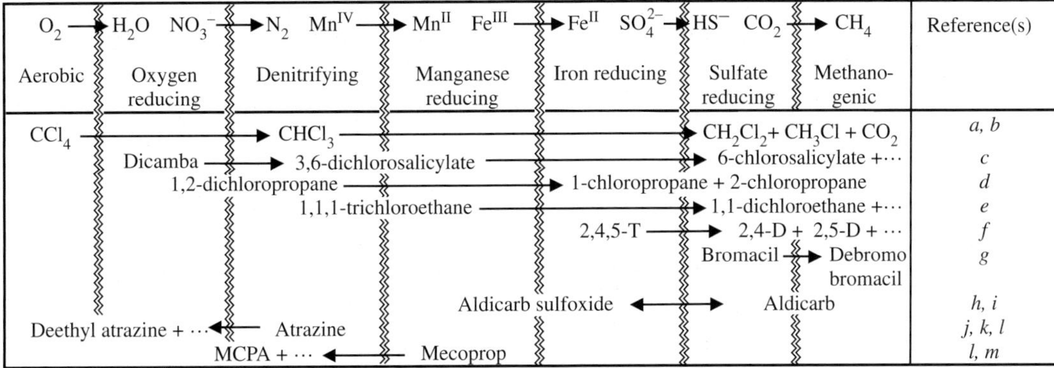

Figure 8 Some examples of electron-transfer reactions of pesticide compounds in relation to dominant terminal electron accepting processes (TEAPs) in natural waters. TEAP sequence based on Christensen *et al.* (2000). Manganese, iron, sulfur, or carbon may be present in either the dissolved or solid phase. Location of each compound indicates the condition(s) under which it has been found to be relatively stable in natural waters. Incomplete lists of transformation products denoted by ellipsis (...); compounds shown are those inferred to have been derived directly from the parent compound. One-way arrows denote essentially irreversible reactions; two-way arrow denotes a reversible reaction. The various references cited are: a, Egli *et al.* 1988; b, Picardal *et al.*, 1995; c, Milligan and Häggblom, 1999; d, Tesoriero *et al.*, 2001; e, Klecka *et al.*, 1990; f, Gibson and Suflita, 1986; g, Adrian and Suflita, 1990; h, Miles and Delfino, 1985; i, Lightfoot *et al.*, 1987; j, Nair and Schnoor, 1992; k, Papiernik and Spalding, 1998; l, Rügge *et al.* 1999; and m, Agertved *et al.*, 1992.

act as intermediaries in these reactions (Figure 9). The most common electron-transfer agents (or *electron carriers*) in natural waters appear to be redox-active moieties associated with NOM, since the rates of reduction of pesticides have been found to vary directly with the NOM content of soils (Glass, 1972) and sediments (Wolfe and Macalady, 1992). The specific components of NOM that are responsible for promoting the reduction of pesticide compounds in natural systems are elusive, but laboratory studies have demonstrated that these reactions may be facilitated by several different redox-active moieties likely to be present in NOM (Klecka and Gonsior, 1984; Tratnyek and Macalady, 1989; Schwarzenbach *et al.*, 1990; Gantzer and Wackett, 1991; Curtis and Reinhard, 1994; Chiu and Reinhard, 1995; Garrison *et al.*, 2000). These moieties are commonly thought to serve primarily as electron carriers in such reactions by cycling back and forth between electron acceptance from a "bulk" electron donor (e.g., solid-phase iron sulfides (Kenneke and Weber, 2003)) and electron donation to the pesticide compound (Figure 9). However, Glass (1972) proposed an alternate system in which iron serves as the electron carrier, and NOM as the bulk electron donor.

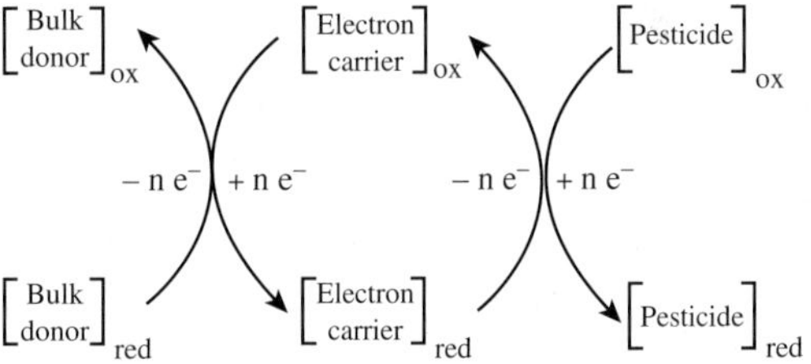

Figure 9 Hypothesized mechanism for the reductive transformation of a pesticide compound through the transfer of electrons from a bulk electron donor (e.g., FeS (Kenneke and Weber, 2003)) to a pesticide molecule (e.g., methyl parathion (Tratnyek and Macalady, 1989)) by an electron carrier (e.g., hydroquinone (Schwarzenbach *et al.*, 1990; reproduced by permission of American Chemical Society from *Environ. Sci. Technol.*, **1990**, *24*, 1566–1574).

9.15.3.4 Governing Factors

The rates and mechanisms of transformation of a pesticide compound in the hydrologic system are determined by the concentration and structure of the compound of interest (*substrate*), as well as the physical, chemical and biological circumstances under which each reaction takes place—including the concentrations and structures of other chemical species with which the substrate may react. As might be expected, these factors often show complex patterns of interdependence, such as the simultaneous influence of temperature and soil moisture on reaction rates, or the fact that pH can exert both direct and indirect effects on pesticide persistence. The nature of these influences as well as their interactions is discussed below.

9.15.3.4.1 Reactant concentrations

Except for intramolecular reactions and direct photolysis, most transformations of pesticide compounds in the hydrologic system are *bimolecular*, i.e., their rate-limiting step involves a reaction between the substrate and another chemical species. In natural waters, these other species—which may be of biological or abiotic origin, or associated with the surfaces of natural materials—consist primarily of Brønsted acids and bases, nucleophiles, oxidants, reductants and catalysts. Although there are some exceptions (e.g., Hemmamda *et al.*, 1994; Huang and Stone, 2000),the rates of abiotic bimolecular transformation of pesticide compounds in *homogeneous* aqueous solution (i.e., in the absence of a solid phase) have been found to be first-order with respect to the aqueous concentrations of both the substrate and the other reactant—and thus second-order overall—for a wide range of reactions (e.g., Walraevens *et al.*, 1974; Burlinson *et al.*, 1982; Klecka and Gonsior, 1984; Mill and Mabey, 1985; Jafvert and Wolfe, 1987; Haag and Mill, 1988b; Tratnyek and Macalady, 1989; Schwarzenbach *et al.*, 1990; Deeley *et al.*, 1991; Roberts *et al.*, 1992; Curtis and Reinhard, 1994; Wei *et al.*, 2000; Strathmann and Stone, 2001). The widespread use of transformation half-lives for quantifying the persistence of pesticide compounds in the hydrologic system, however (e.g., Mackay *et al.*, 1997), is based on the assumption of *pseudo-first-order kinetics* with respect to the substrate, i.e., that the concentrations of the other reactants are sufficiently high to remain effectively constant during the transformation of the substrate (Moore and Pearson, 1981). For heterogeneous reactions—which may involve the participation of soil, mineral, or biological surfaces—reaction rates may exhibit a different quantitative relation to the concentrations of the substrate (Zepp and Wolfe, 1987) or the other reactants (Kriegman-King and Reinhard, 1994).

Because the rates of biotransformation are influenced by a variety of factors related to the size, growth, and substrate utilization rate of the microbial populations involved, they often exhibit a more complex dependence upon substrate concentration than do the rates of abiotic transformation in homogeneous solution (D'Adamo *et al.*, 1984). Under many circumstances, however, these more complex relations often simplify to being pseudo first-order with respect to substrate concentration, particularly when the latter is substantially lower than that required to support half the maximum rate of growth of the organisms of interest. At concentrations well above this level, transformation rates may be independent of substrate concentration (Paris *et al.*, 1981).

9.15.3.4.2 Structure and properties of the pesticide substrate

The development of synthetic pesticides has always depended upon an understanding of the effects of chemical structure on the toxicity to the target organism(s). (These effects are discussed in a later section.) However, as the deleterious effects of pesticide compounds on non-target organisms became more widely known (e.g., Carson, 1962), it also became increasingly important to learn how the structures of these chemicals control their persistence in the hydrologic system. Indeed, the properties responsible for biological activity are often those that most affect environmental persistence as well (e.g., Smolen and Stone, 1997). In addition to their direct effects on reactivity (discussed below), variations in chemical structure will also influence persistence indirectly if they shift the partitioning among different phases in which reaction rates are substantially different. Because the ways in which the structure of a pesticide compound controls its reactivity depend, in turn, upon reaction mechanism, these effects will be examined separately for each of the major types of reactions.

(i) *Neutral reactions.* As noted previously, more studies have been conducted on neutral reactions (especially hydrolysis and dehydrohalogenation) than on other types of reaction involving pesticide compounds in aqueous solution. Thus, it is not surprising that these reactions have also yielded some of the most extensive understanding of the effects of structure on the reactivity of pesticide compounds in the hydrologic system. Investigations of this topic for hydrolysis and dehydrohalogenation have also provided some of the most statistically significant *quantitative structure-reactivity relations* (QSRRs) observed for pesticide compounds (e.g., Wolfe *et al.*, 1978; Roberts *et al.*, 1993; Schwarzenbach *et al.*, 1993).

Several QSRR studies suggest that among the structural features with the greatest influence over hydrolysis rates are those that affect the pK_a of the leaving group. For example, the initial step in OP hydrolysis, which usually occurs at one of the three phosphate ester linkages, involves the displacement of the leaving group with the lowest pK_a (Smolen and Stone, 1997). As a result, for each of five organothiophosphate ester insecticides investigated by Smolen and Stone (1997), the initial product of hydrolysis was always a phenolate anion (Figure 10). This also explains why dimethyl ester OPs, such as methyl parathion and methyl azinphos, hydrolyze more rapidly than their diethyl ester counterparts (Lartiges and Garrigues, 1995). Similarly, for each of four groups of carbamates (*N*-methyl, *N*-phenyl, *N*,*N*-dimethyl, and *N*-methyl-*N*-phenyl), the second-order reaction rate constants for base-promoted hydrolysis have been found to be inversely correlated with the pK_a values of the displaced leaving groups (Wolfe *et al.*, 1978). Several of the correlation equations devised for predicting pesticide hydrolysis rates from pK_a values have been compiled by Harris (1990a).

Many other structural effects on the rates of nucleophilic substitution reactions of pesticide

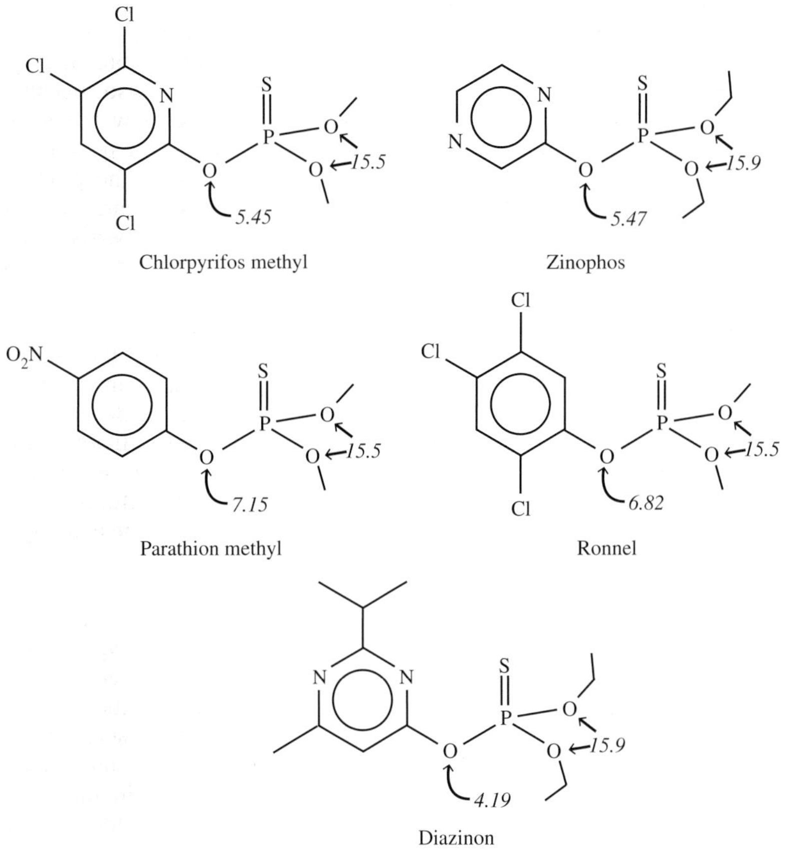

Chlorpyrifos methyl

Zinophos

Parathion methyl

Ronnel

Diazinon

Figure 10 pK_a values for different leaving groups for a series of organothiophosphate insecticides. Leaving group with the lowest pK_a value in each compound is the one first displaced by hydrolysis (Smolen and Stone (1997); reproduced by permission of American Chemical Society from *Environ. Sci. Technol.*, **1997**, *31*, 1664–1673).

compounds have been noted. For example, carbamates with two substituents (other than hydrogen) bound to nitrogen undergo base-promoted hydrolysis more slowly than those with only one (Wolfe *et al.*, 1978), perhaps as a result of *steric hindrance* (i.e., the enhanced restriction of access to the reaction site by the presence of larger substituents near C_α). Steric hindrance is also believed to be responsible for some of the effects of structure on the rates of acid- and base-catalyzed hydrolysis of chloroacetamide herbicides—particularly the size of the ether or alkyl substituent bound to the amide nitrogen (Carlson and Roberts, 2002; Carlson, 2003). Zepp *et al.* (1975) observed that the rates of hydrolysis of 2,4-D esters (2,4-$C_6H_3Cl_2$–O–CH_2COOR) are higher if *R* contains an ether linkage near the ester carboxyl group than if *R* is a hydrocarbon moiety.

The bimolecular nucleophilic (S_{N2}) displacement of halide from haloalkane adjuvants and fumigants is slowed by the presence of alkyl groups or additional halogen atoms bound to the carbon at (C_α), or immediately adjacent to (C_β) the site of attack. Although bromide is a better leaving group than chloride (i.e., it is displaced more rapidly than chloride), trends in the relative rates of displacement of different halogens may also depend upon the nature of the attacking species, as well as other substituents bound to C_α and C_β. These and other effects of structure on the rates of S_N2 displacement of halide from halogenated pesticide compounds have been examined extensively (e.g., Barbash and Reinhard, 1989a; Barbash, 1993; Roberts *et al.*, 1993; Schwarzenbach *et al.*, 1993).

Since bimolecular dehydrohalogenation reactions are initiated by the abstraction of a hydrogen atom (from C_α) by a Brønsted base, they are promoted by structural factors that increase the ease with which this hydrogen atom can be removed from the molecule, such as the presence of halogens or other *electron-withdrawing* substituents bound to C_α. (These are moieties that withdraw electron density from the rest of the molecule, the effect being less pronounced with increasing distance from the substituent.) Because this transformation also involves the departure of halide from the adjacent carbon (C_β), rates of reaction are higher for the loss of HBr than for the loss of HCl (e.g., Burlinson *et al.*, 1982; Barbash and Reinhard, 1992a; Barbash, 1993). Roberts *et al.* (1993) provided a comprehensive discussion of these and other effects of structure on the rates of dehydrohalogenation reactions.

(ii) *Electron-transfer reactions.* Although many types of redox reactions involving pesticide compounds have been investigated (Barbash and Resek, 1996), most of what has been learned about the effects of substrate structure and properties on the rates of these reactions in the hydrologic system has come from studies of reductive dehalogenation (e.g., Peijnenburg *et al.*, 1992a) and nitro group reduction (e.g., Schwarzenbach *et al.*, 1993). Other factors remaining constant, the rates of reductive dehalogenation increase with decreasing strength of the carbon–halogen bond being broken. As a result, the rates of these reactions generally: (i) increase with increasing numbers of halogens on the molecule (e.g., Gantzer and Wackett, 1991), especially at C_α (Mochida *et al.*, 1977; Klecka and Gonsior, 1984; Butler and Hayes, 2000); (ii) decrease among halogens in the order I > Br > Cl (e.g., Kochi and Powers, 1970; Wade and Castro, 1973; Jafvert and Wolfe, 1987; Schwarzenbach *et al.*, 1993); (iii) are higher for the removal of halogen from *polyhaloalkanes* (polyhalogenated hydrocarbons lacking multiple carbon–carbon bonds) than from the corresponding *polyhaloalkenes* (polyhalogenated, nonaromatic hydrocarbons with one or more carbon–carbon double bonds) (Gantzer and Wackett, 1991; Butler and Hayes, 2000); and (iv) are positively correlated with one-electron reduction potentials (e.g., Gantzer and Wackett, 1991; Curtis and Reinhard, 1994; Butler and Hayes, 2000). A positive correlation of reaction rates with reduction potentials has also been reported for the reduction of nitroaromatic compounds (e.g., Schwarzenbach *et al.*, 1990). Systematic relations between the rates of reductive dechlorination of polyhaloethanes and electron densities on carbon have also been observed (Salmon *et al.*, 1981).

Rates of reduction may also be influenced by the manner in which different substituents affect the distribution of electrons in the substrate molecule (*electronic* effects), or hinder access to reaction sites (steric effects). Because the initial, rate-limiting step in most of these reactions involves the addition of an electron to the substrate to form a carbon radical, reaction rates are generally increased by electron-withdrawing groups (e.g., halogens, acetyl, nitro), and decreased by *electron-donating* groups, i.e., substituents such as alkyl groups that donate electron density to the rest of the molecule (e.g., Peijnenburg *et al.*, 1992a). The opposite pattern is expected for rates of oxidation (e.g., Dragun and Helling, 1985). For reactions involving aromatic compounds, electronic effects are substantially more pronounced for substituents located in positions that are either *ortho* or *para* to the substituent being replaced or altered (i.e., either one or three carbons away on the hexagonal aromatic ring) than for those in the *meta* position (i.e., located two carbons away). By contrast, steric effects are most important for substituents in the *ortho* position (Peijnenburg *et al.*, 1992a; Schwarzenbach *et al.*, 1993). These electronic and steric effects on the rates of electron-transfer reactions of pesticide compounds are in agreement with general reactivity theory (March, 1985).

(iii) *Photochemical reactions.* As noted earlier, phototransformations in the hydrologic system occur only for those substrates that either absorb a sufficient amount of light energy within the solar spectrum, or react with photoexcited sensitizers or other photoproduced oxidants. In general, structures that promote light absorption include extensively conjugated hydrocarbon systems (i.e., those possessing alternating double and single bonds), substituents containing unsaturated heteroatoms (e.g., nitro, azo, or carbonyl) and, for substituted aromatics, halo, phenyl, and alkoxyl groups (Mill and Mabey, 1985; Harris, 1990b). As with reductive dehalogenation, the rates at which halogenated aromatics undergo photolysis are correlated with the strength of the carbon–halogen bond being broken, as well as with the tendency of other substituents to withdraw electrons or cause steric hindrance (Peijnenburg *et al.*, 1992b). Photolysis rates have also been found (Katagi, 1992) to be correlated with changes in electron density caused by photo-induced excitation (i.e., the promotion of one or more electrons to higher-energy states following the absorption of light energy). A comprehensive summary of the effects of chemical structure on photoreactivity for a broad range of pesticides was provided by Mill and Mabey (1985).

(iv) *Biotransformations.* Some of the structural features that are associated with lower rates of biotransformation include increased branching and decreased length of hydrocarbon chains, decreased degree of unsaturation, larger numbers of rings in polynuclear aromatic hydrocarbons, the presence of methyl, nitro, amino, or halogen substituents on aromatic rings and, under oxic conditions, increasing numbers of halogens on the molecule (Scow, 1990). As might be expected, many of the ways in which variations in the structure of pesticide compounds affect their rates and mechanisms of biotransformation result from fundamental constraints of structure over chemical reactivity. For example, biotransformation rates among a variety of different pesticides have been found to be linearly correlated with their respective rates of base-catalyzed hydrolysis (Wolfe *et al.*, 1980).

Other effects of chemical structure on biodegradability are related more to biological factors than to purely chemical ones. Examples include features that increase the toxicity of the molecule (e.g., additional halogens) or affect the ease with which the appropriate enzyme(s) can access the reaction site (Bollag, 1982). One illustration of the latter effect is the observation, noted earlier, that biochemical reactions are often enantioselective, leading to variations in biotransformation rates among different *isomers* of the same pesticide compound (e.g., Sakata *et al.*, 1986; Peijnenburg *et al.*, 1992a; Falconer *et al.*, 1995; Wong *et al.*, 2002; Monkiedje *et al.*, 2003). (Isomers are compounds that have the same chemical composition, but slightly different spatial arrangements of atoms. Enantiomers, for example, are isomers that, as noted earlier, are mirror images of one another.) The structural specificity of some biotransformation reactions is also demonstrated by the fact that unequal mixtures of isomers are sometimes produced from the transformation of a single compound (e.g., Parsons *et al.*, 1984).

9.15.3.4.3 Structure and properties of other reactants

The rates and mechanisms of most pesticide transformations in the hydrologic system are, as noted earlier, influenced by the nature of the chemical species reacting with the pesticide compound during the rate-limiting step. These other species participate in the reactions either as Brønsted acids or bases, Lewis acids or bases (electrophiles or nucleophiles, respectively), oxidants, reductants, or catalysts. They may be present in either the aqueous, solid or gaseous phases, and may be of either abiotic or biological origin. In some cases, individual chemical species may react by different mechanisms simultaneously with the same pesticide compound. For example, Kriegman-King and Reinhard (1992) observed that in aqueous solution, hydrogen sulfide can react with tetrachloromethane as both a nucleophile (to form carbon disulfide and carbon dioxide) and a reductant (to form trichloromethane, dichloromethane, carbon monoxide, and other products).

Some reactants may play different roles in their interactions with pesticide compounds, depending on the geochemical conditions. Metals and their complexes, for example, may react with pesticide compounds as hydrolysis catalysts (e.g., Mortland and Raman, 1967; Smolen and Stone, 1997; Huang and Stone, 2000), direct reductants (e.g., Castro and Kray, 1963, 1966; Mochida *et al.*, 1977; Strathmann and Stone, 2001, 2002a,b), or bulk electron donors through electron-transfer agents such as hydroquinones (e.g., Tratnyek and Macalady, 1989; Curtis and Reinhard, 1994). Most of what little information is available on how different metals and their complexes vary in their ability to promote transformations of pesticide compounds, however, focuses on their roles as direct reductants or hydrolysis catalysts (see references cited above); reduced iron appears to be the only metal that has been examined as a potential bulk electron donor in natural systems (Glass, 1972).

A valuable framework for understanding the effects of chemical structure on reactivity is provided by the hard and soft acids and bases (HSAB) model, which classifies Lewis acids and bases as either "hard" or "soft." Hard acids (e.g., H_3O^+, Na^+, Fe^{3+}) and bases (e.g., OH^-, CO_3^{2-}, Cl^-) are relatively small, exhibit low polarizability (and generally high electronegativity), and are more likely to form ionic bonds than covalent ones. Soft acids (e.g., Cd^{2+}, Hg^{2+}, Br_2) and bases (e.g., HS^-, $S_2O_3^{2-}$, I^-, CN^-) have the opposite characteristics (Fleming, 1976). Stated simply, the HSAB model posits that hard acids react most readily with hard bases, while soft acids react most readily with soft bases (Pearson, 1972). Thus, the soft base HS^- reacts with the fumigant 1,2-dichloroethane (Figure 11) almost exclusively through nucleophilic substitution at the (soft) carbon bonded to halogen (Barbash and Reinhard, 1989b), while the harder OH^- attacks the same compound primarily at one of the (hard) hydrogens, leading to dehydrochlorination (Walraevens *et al.*, 1974). Similarly, for nucleophilic substitution reactions involving the thionate OPs (see Figure 10 for compound structures), reaction occurs primarily at one of the (soft) carbon atoms bonded to oxygen in the ester linkages when the attacking species is HS^- (Miah and Jans, 2001), but at the (hard) phosphorus atom when the attacking species is OH^- (Smolen and Stone, 1997). Replacement of (soft) sulfur in thionate OPs with (hard) oxygen in the corresponding oxonate OPs further reduces the electron density on phosphorus, leading to higher rates of hydrolysis (Smolen and Stone, 1997).

Although the HSAB model has been criticized for being insufficiently quantitative (e.g., March, 1985), its predictions have been shown to be consistent with results from frontier molecular orbital calculations for a wide variety of reactions (Klopman, 1968; Fleming, 1976). Such calculations have, in turn, been shown to be useful for elucidating the effects of pesticide structure on reactivity (e.g., Katagi, 1992; Lippa and Roberts, 2002).

In addition to the nature of the *donor atom* (e.g., sulfur in bisulfide versus oxygen in hydroxide), other features that govern the hardness of a nucleophile—and thus the rates and mechanisms of its reaction with pesticide compounds—include the oxidation state of the donor atom

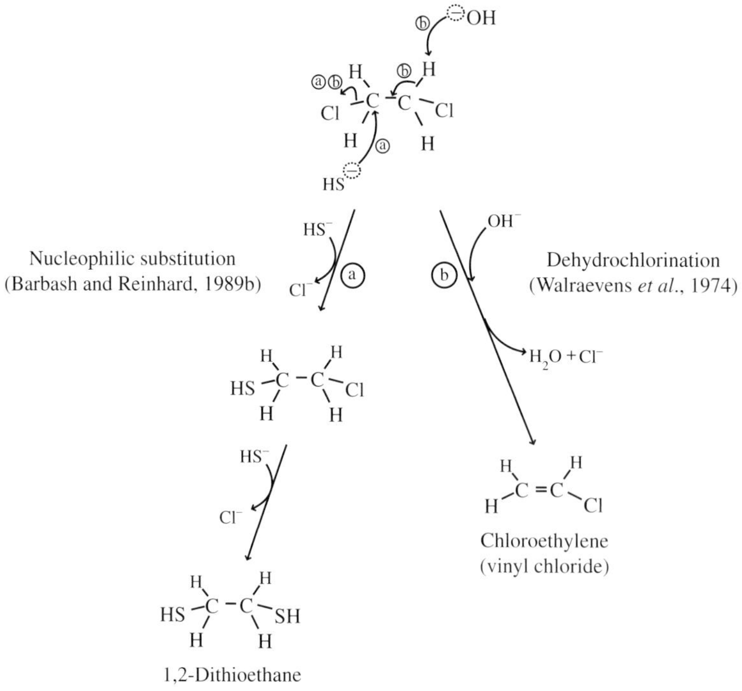

Nucleophilic substitution (Barbash and Reinhard, 1989b)

Dehydrochlorination (Walraevens *et al.*, 1974)

Chloroethylene (vinyl chloride)

1,2-Dithioethane

Figure 11 Principal reactions of the fumigant 1,2-dichloroethane with (a) bisulfide (Barbash and Reinhard, 1989b) and (b) hydroxide (Walraevens *et al.*, 1974) in aqueous solution.

and the hardness, size, and structure of the remainder of the molecule. These patterns have been shown to be evident in reactivity trends observed, for example, among a variety of buffer anions and naturally occurring nucleophiles in their reactions with several haloaliphatic pesticide compounds (Barbash and Reinhard, 1989a; Roberts *et al.*, 1992; Barbash, 1993).

While most applications of the HSAB model have focused on neutral reactions, the model also applies to electron-transfer reactions. The abilities of different transition elements (or other reductants) to reduce a given pesticide compound are generally presumed to increase with their softness—i.e., with their ability to donate electrons, a characteristic that is commonly quantified using one-electron reduction potentials (e.g., Schwarzenbach *et al.*, 1993). Reactivity trends reported among different reduced metal cations, however (e.g., Strathmann and Stone, 2001; Mochida *et al.*, 1977), do not precisely track the order of their reduction potentials (Dean, 1985). The relative propensities of different metals to reduce pesticide compounds are therefore influenced by other factors. Related work suggests that such factors include pH and the nature of the coordinating ligand (Strathmann and Stone, 2001, 2002a, b; Mochida *et al.*, 1977).

Comparisons among different metals with regard to their ability to catalyze the hydrolysis of pesticide compounds appear to be limited (e.g., Mortland and Raman, 1967; Smolen and Stone, 1997; Huang and Stone, 2000). Dramatic decreases in reactivity that have been observed upon precipitation at higher pH suggest, however, that metals catalyze hydrolysis much more effectively as ions in solution than as solid-phase (hydr)oxide surfaces (Huang and Stone, 2000).

Ligands have long been known to exert substantial effects upon the solubility, adsorption, and reactivity of transition elements in the hydrologic system. (Indeed, biochemical evolution has taken great advantage of the latter phenomenon, as shown by the critical roles played by metalloenzymes in many metabolic activities.) Several studies have demonstrated the effects of ligand structure on the rates of reductive transformation of pesticide compounds by transition metals, either as dissolved species (e.g., Singleton and Kochi, 1967; Kochi and Powers, 1970; Mochida *et al.*, 1977; Gantzer and Wackett, 1991; Strathmann and Stone, 2001, 2002a, b) or as part of the solid phase (e.g., Torrents and Stone, 1991). One of the studies of greatest relevance to understanding the effect of ligands on the reactivity of pesticide compounds under natural conditions involved an examination by Klecka and Gonsior (1984) of the reductive dehalogenation of polychloroalkanes by ferrous iron in solution. This work indicated that the rates of these reactions are either wholly dependent upon (chloroform, 1,1,1-trichloroethane), or greatly accelerated by (tetrachloromethane) the coordination of the iron with a porphyrin ring (hematin), relative to the rate when iron is coordinated only to water (Fe^{2+}). Huang and Stone (2000) found that the three transition metal cations known to catalyze the hydrolysis of the carbamate insecticide dimetilan (Cu^{II}, Ni^{II}, and Zn^{II}) exhibit strong affinities for nitrogen- and oxygen-donor ligands. Pb^{II}, which did not catalyze this reaction, showed only a weak affinity for these ligands. As might have been expected, different ligands were found to cause varying degrees of catalytic activity for a given metal.

9.15.3.4.4 Physical factors

(i) *Temperature*. Because both the kinetic energy of molecules and the frequencies of their collisions increase with temperature, so do the rates of thermal reactions, including those that are biologically mediated. By contrast, the rate of photochemical conversion of a molecule from its ground state to an excited state is not dependent upon temperature. Overall rates of photochemical reaction may, however, exhibit either a positive or a negative dependence on temperature, as the net result of competition among activation, quenching and reaction steps (Mill and Mabey, 1985). Nevertheless, the detection in Arctic regions of pesticides that exhibit gas-phase lifetimes of only a few days in the more temperate climates where they were likely to have been applied (Bidleman, 1999) demonstrates the importance of accounting for temperature variations, as well as the rapidity of long-range atmospheric transport (mentioned earlier), in predicting the persistence of these compounds in the atmosphere.

The dependence of thermal reaction rates on temperature is most commonly found to be in accordance with the Arrhenius equation, i.e.,

$$\ln(k) = \ln(A) - \frac{E_a}{RT}$$

where k is the rate constant for the rate-limiting step of the reaction of interest, A is the Arrhenius pre-exponential factor, E_a is the activation energy, R is the universal gas constant, and T is the temperature of reaction (K). As is evident from the form of this equation, the quantities E_a/R and $\ln(A)$ may be estimated from the slope and intercept, respectively, of an Arrhenius plot, i.e., a linear regression of $\ln(k)$ versus $1/T$ (e.g., Figure 12). Arrhenius plots exhibiting a significant departure from linearity are usually interpreted to imply the simultaneous operation of more than one transformation mechanism, with different mechanisms dominating over different temperature ranges. Pesticides for which this phenomenon has been observed include 1,2-dibromo-3-chloropropane (Deeley *et al.*, 1991) triasulfuron and primisulfuron (Dinelli *et al.*, 1998).

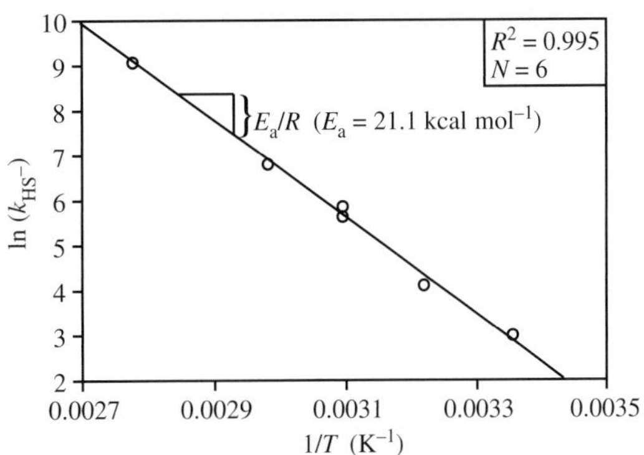

Figure 12 Arrhenius plot for the second-order rate constant (k_{HS^-}) for the nucleophilic displacement of bromide from 1,2-dibromoethane (EDB) by bisulfide ion (HS^-) (after Barbash, 1993).

The Arrhenius relation will not be observed above the temperature at which the decomposition or, as may occur for enzymes, inactivation of one or more of the reactants occurs (T_{max}). Indeed, adherence to this relation at temperatures well above T_{max} for most microorganisms has been used as evidence for an abiotic, rather than a biologically mediated mechanism of transformation (Wolfe and Macalady, 1992). For biotransformations, the Arrhenius equation also fails to describe the temperature dependence of reaction rates below the temperature at which biological functions are inhibited (T_{min}), and above the temperature of maximum transformation rate (T_{opt}). An empirical equation introduced by O'Neill (1968) may be used to estimate the rates of biotransformation as a function of ambient temperature, T_{min}, T_{opt}, T_{max} (in this case, the lethal temperature), E_a, and the maximum biotransformation rate (μ_{max}). Because of the complexity of biochemical systems and the myriad of different structures encompassed by pesticide compounds, T_{min}, T_{opt}, T_{max}, E_a, and μ_{max} are all likely to vary among different compounds, microbial species and geochemical settings (e.g., Gan *et al.*, 1999, 2000). However, Vink *et al.* (1994) demonstrated the successful application of the O'Neill function to describe the temperature dependence of biotransformation for 1,3-dichloropropene and 2,4-D in soils (Figure 13).

Although the dependence of reaction rates on temperature has been well known for over a century (Moore and Pearson, 1981), the temperatures at which transformation half-lives are measured are often not reported, either by the original publications or in data compilations (e.g., Nash, 1988; USDA/ARS, 1995; Pehkonen and Zhang, 2002). Similarly, as noted by Barbash and Resek (1996), the effects of temperature variations on transformation rates are rarely incorporated

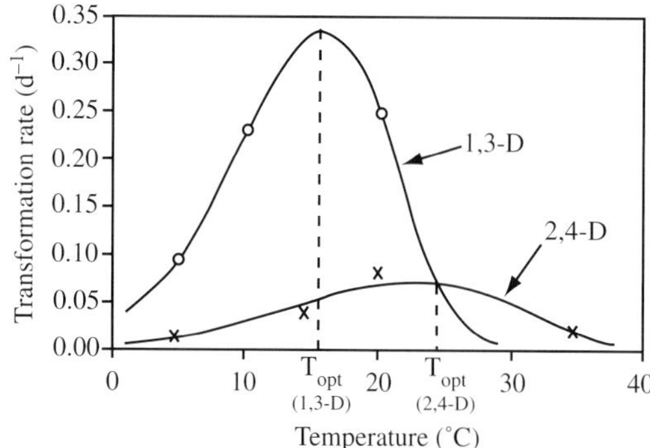

Figure 13 Variations in the rates of biotransformation of 1,3-dichloropropene (1,3-D,circles) and 2,4-D (crosses) with temperature in soils, fitted to the O'Neill function (Vink *et al.* (1944); reproduced by permission of John Wiley & Sons Ltd. on behalf of the SCI from *Pestic. Sci.*, **1994**, *40*, 285–292).

into simulations of pesticide fate in the hydrologic system. Several studies, however, have demonstrated the importance of adjusting transformation rates for seasonal and depth-related variations in temperature when predicting pesticide persistence (Watson, 1977; Padilla *et al.*, 1988; Wu and Nofziger, 1999; Beulke *et al.*, 2000). As is evident from the Arrhenius equation, the adjustment of pesticide transformation rates for variations in temperature requires the availability of data on E_a for the compound and setting of interest. Compilations of E_a values appear to be rare in the literature, but a few summaries are available for the transformations of a variety of herbicides (Nash, 1988; Smith and Aubin, 1993), adjuvants and fumigants (Barbash, 1993), as well as for the activity of soil enzymes known to effect pesticide transformations (Wu and Nofziger, 1999). E_a values—or data from which they may be computed—are currently available for at least 70 pesticide compounds in the published literature (Barbash, unpublished compilation).

(ii) *Soil moisture content.* Rates of pesticide transformation in soils have usually been found to increase with moisture content up to the *field capacity* of the soil (e.g., Burnside and Lavy, 1966; Roeth *et al.*, 1969; Walker, 1974; Anderson and Dulka, 1985; Walker, 1987; Cink and Coats, 1993). (Field capacity is the amount of water remaining in a soil after the rate of gravity drainage becomes negligible.) This dependence has been described through the use of an empirical equation, introduced by Walker (1974), that expresses transformation half-life as an inverse exponential function of soil moisture content (up to field capacity). The Walker equation has been used to quantify the effects of soil moisture on the rates of transformation of atrazine (Rocha and Walker, 1995), propyzamide, napropamide (Walker, 1974), and at least 25 other pesticide compounds (Gottesbüren, 1991). Similarly, Parker and Doxtader (1983) found that the rate of transformation of 2,4-D in a sandy loam increased with decreasing soil moisture tension, but observed an exponential relation between the two variables that was different from that implied by the Walker equation.

The dependence of pesticide transformation rates on soil moisture is commonly assumed to reflect the influence of water content on biotransformation. For example, Parker and Doxtader (1983) attributed the moisture dependence of 2,4-D transformation rates to two different effects of reduced soil moisture on biological processes, i.e., (i) reduced activity of microbial populations and (ii) higher pesticide concentrations, which have been shown by other investigators (e.g., Cink and Coats, 1993) to inhibit microbial degradation. For soil moisture contents above field capacity, however, this relation might not be observed, especially for aerobic transformations that may be inhibited in saturated soils following the consumption of dissolved oxygen. Abiotic factors may also be involved, since a direct relation between pesticide transformation rates and soil moisture has been observed in sterile soils, as well (Anderson and Dulka, 1985). Furthermore, some nonbiological processes may exhibit the opposite effect; both the Lewis and Brønsted acidity of mineral surfaces—and thus their tendency to promote the oxidation or acid-catalyzed reactions of pesticide compounds, respectively—are enhanced with decreasing soil moisture (Voudrias and Reinhard, 1986).

(iii) *Interactions among temperature, moisture and soil texture.* The effects of variations in temperature and moisture content on the rates of pesticide transformation in soil show a complex interdependence, the nature of which may vary among different compounds for the same soil, as well as among different soils for the same compound (Walker, 1974; Nash, 1988; Baer and Calvet, 1999). Equations used to describe the simultaneous effects of temperature and soil moisture on pesticide transformation rates have been reported by Parker and Doxtader (1983) and Dinelli *et al.* (1998). In a comprehensive review of 178 studies that compared observed pesticide concentrations in soils with those predicted using a pesticide persistence model, Beulke *et al.* (2000) demonstrated the importance of accounting for spatial and temporal variations in temperature and moisture in estimating the rates of pesticide transformation. Interactions between abiotic and biological factors are likely to be responsible for the observation that the nature of the relation between soil moisture and the rates of pesticide transformation may, in turn, depend upon soil texture (e.g., Rocha and Walker, 1995; Gan *et al.*, 1999). Rocha and Walker (1995) also observed that

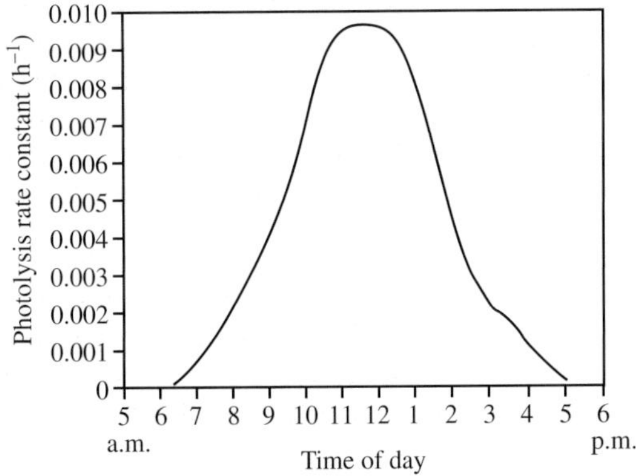

Figure 14 Dependence of the rate constant for the photolysis of 2,4-D butoxyethyl ester (in water at 28 °C) on the time of day in the southern United States (Zepp *et al.* (1975); reproduced by permission of American Chemical Society from *Environ. Sci. Technol.*, **1975**, *9*, 1144–1149).

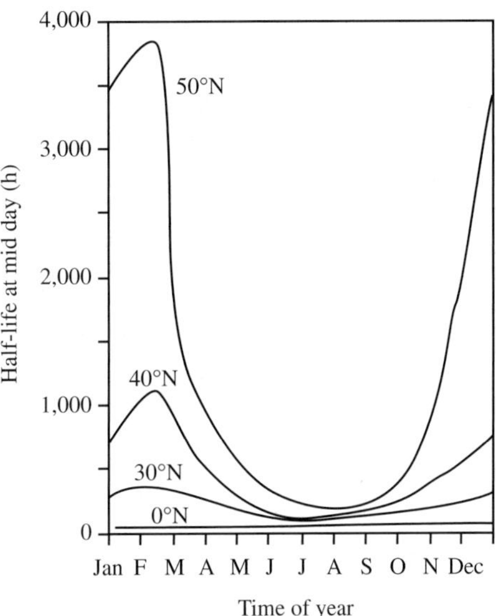

Figure 15 Dependence of the half-life for the photolysis of 2,4-D butoxyethyl ester (in water at 28 °C) on season and northern latitude (Zepp *et al.* (1975); reproduced by permission of American Chemical Society from *Environ. Sci. Technol.*, **1975**, *9*, 1144–1149).

the effect of temperature on atrazine transformation could be a function of soil texture, as well.

(iv) *Solar energy input*. For pesticide compounds that undergo phototransformation in natural waters, the rate of reaction is controlled by a number of factors that affect the amount of light energy reaching the reactant(s) of interest. These include the *irradiance*, or the rate of energy input from the Sun (which, in turn, is controlled in part by latitude, season and time of day) at the wavelengths of maximum light absorption for each reactant, as well as turbidity, color, and depth within the water body of interest. The effects of most of these variables on the rates of photochemical transformation of pesticide compounds are summarized by Mill and Mabey (1985) and Harris (1990b). Zepp *et al.* (1975) presented graphical illustrations of how the rate of 2,4-D butoxyethyl ester photolysis varies with time of day (in Athens, GA, Figure 14), season and latitude in the northern hemisphere (Figure 15).

9.15.3.4.5 Geochemical environment

As is the case for its physical properties, the geochemical characteristics of the reaction medium can also influence the rates and mechanisms of pesticide compound transformation in the hydrologic system, as well as the health and activity of the organisms capable of transforming these compounds. Such characteristics include redox conditions (discussed earlier), pH, ionic strength, the structure and concentrations of any surface-active substances, solvents or ligands that may be present, and the chemical properties of any interfaces with which the reactants may come in contact.

(i) *pH*. Changes in pH can affect the rates of pesticide compound transformation in both direct and indirect ways. Direct effects arise from the fact that shifts in pH reflect changes in the concentrations of two potentially important reactants that are present in all aqueous solutions, i.e., H_3O^+ and OH^-. However, pH changes may also exert substantial indirect effects on pesticide transformations, through their influence on biological activity (e.g., Mallat and Barceló, 1998), the surface properties of reactive natural solids such as manganese oxides (Baldwin *et al.*, 2001) and clays (Voudrias and Reinhard, 1986), and the concentrations of other reactants. For pesticide compounds that are Brønsted acids or bases, such as the sulfonylurea herbicides (Smith and Aubin, 1993), variations in pH may influence transformation rates in solution if the reactivities of the conjugate acid and base are substantially different from one another.

Even if the aqueous concentration of a pesticide compound is independent of pH, however (as is the case for most parent compounds), its transformation rate may still vary with changes in pH if the attacking species is a Brønsted acid or base. In such instances, the pH dependence of the reaction rate will parallel that of the attacking species concentration. For pesticide compounds, this effect has been reported for nucleophilic substitution reactions involving H_3O^+, OH^- (e.g., Mabey and Mill, 1978), and HS^- (Haag and Mill, 1988b; Roberts *et al.*, 1992; Lippa and Roberts, 2002), for reduction (Strathmann and Stone, 2002a,b) or hydrolytic catalysis by divalent metal cations (Huang and Stone, 2000), and for reduction by model compounds employed to mimic the electron-transfer capabilities of hydroquinone moieties found in NOM (Tratnyek and Macalady, 1989; Curtis and Reinhard, 1994). Indeed, the latter observation may explain why the reduction of methyl parathion in anoxic sediments was reported by Wolfe *et al.* (1986) to occur more rapidly under alkaline conditions ($8 < pH < 10$) than under acidic conditions ($2 < pH < 6$). The relative rates among competing photochemical transformations for individual pesticide compounds may also shift with pH. Crosby and Leitis (1973), for example, observed pH-related shifts in the yields of different products from the aqueous photolysis of trifluralin, both in the presence and the absence of soil (Figure 16).

The overall rate at which a pesticide compound undergoes hydrolysis in water represents the sum of the rates of the acid-catalyzed, neutral and base-catalyzed processes (Mabey and Mill, 1978). For those pesticide compounds that react only with H_2O, the rate of hydrolysis is independent of pH (e.g., McCall, 1987; Hong *et al.*, 2001). Some pesticide compounds, however, react with more than one of the three species of interest (H_3O^+, H_2O, and/or OH^-), resulting in a pattern of pH dependence that may display an abrupt shift above or below a particular threshold pH value (Mabey and Mill, 1978; Schwarzenbach *et al.*, 1993). Few compilations of these threshold

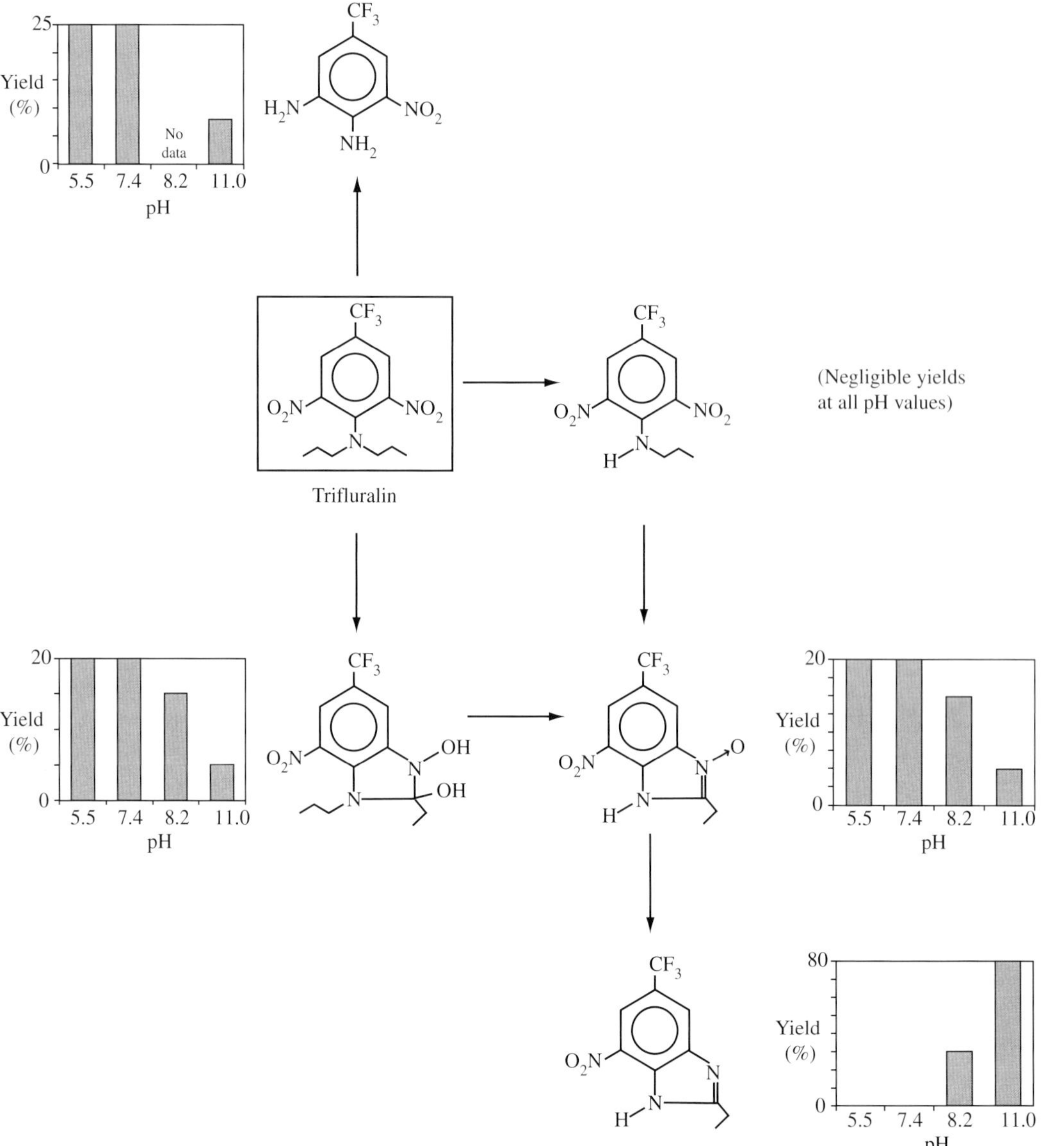

Figure 16 Effects of pH on the yields of different products during the photolysis of trifluralin in summer sunlight (Davis, CA) in water, either with or without soil present. For purposes of display, vertical axis for each product is normalized to the maximum yield for that compound, rather than to 100%. Yields among products at each pH value sum to less than 100% because data for minor products are not shown (after Crosby and Leitis, 1973).

parameters appear to be available for pesticide compounds (Mabey and Mill, 1978; Mabey *et al.*, 1983; Roberts *et al.*, 1993; Schwarzenbach *et al.*, 1993), but individual values may sometimes be inferred from published data (e.g., Zepp *et al.*, 1975; Burlinson *et al.*, 1982; Bank and Tyrrell, 1984; Weintraub *et al.*, 1986; Lightfoot *et al.*, 1987; Cline and Delfino, 1989; Jeffers *et al.*, 1989; Lee *et al.*, 1990; Ngabe *et al.*, 1993). Other results suggest that these threshold parameter values may vary with temperature (Lightfoot *et al.*, 1987).

The dependence of transformation rate on pH is not always consistent among pesticides within the same chemical class. For example, while the hydrolysis reactions of most OP (Konrad and Chesters, 1969; Konrad *et al.*, 1969; Mabey and Mill, 1978) and carbamate insecticides (Wolfe *et al.*, 1978) are primarily base-catalyzed, both diazinon (Konrad *et al.*, 1967) and carbosulfan (Wei *et al.*, 2000) are also subject to the acid-catalyzed reaction. Summaries of the pH dependence of hydrolysis rates for a variety of pesticides have been provided by Mabey and Mill (1978), Bollag (1982), Schwarzenbach *et al.* (1993), and Barbash and Resek (1996).

Several authors have combined the Arrhenius equation with kinetic equations reflecting the pH dependence of reaction rates in order to quantify the simultaneous dependence of pesticide transformation rates on temperature and pH (e.g., Lightfoot *et al.*, 1987; Liqiang *et al.*, 1994; Dinelli *et al.*, 1997). Other work has demonstrated the extent to which E_a values for pesticide hydrolysis vary, if at all, with pH (Lee *et al.*, 1990; Smith and Aubin, 1993; Dinelli *et al.*, 1997, 1998).

Most of what is known about the effects of pH on the rates of pesticide transformation in aqueous solution has been learned from laboratory studies employing buffers to stabilize pH. However, since they are Brønsted acids and bases, pH buffers may also accelerate these reactions—or cause shifts in the relative rates of different transformation mechanisms—compared to what occurs in unbuffered solution (Li and Felbeck, 1972; Burlinson *et al.*, 1982; Barbash and Reinhard, 1989a, 1992b; Deeley *et al.*, 1991; Smolen and Stone, 1997). Such effects, though, are not observed for all pesticide compounds (Hemmamda *et al.*, 1994; Smolen and Stone, 1997; Hong *et al.*, 2001). Indeed, the absence of buffer effects has been used as evidence for unimolecular reaction mechanisms (Bank and Tyrrell, 1984). The varying tendencies of different buffers to accelerate hydrolysis and dehydrohalogenation are consistent with reactivity trends predicted by the HSAB model (Barbash, 1993) and, for hydrolysis, may be predicted quantitatively using a method introduced by Perdue and Wolfe (1983). Relatively few laboratory studies have accounted for buffer effects (e.g., Li and Felbeck, 1972; Burlinson *et al.*, 1982; Miles and Delfino, 1985; Barbash and Reinhard, 1989a, 1992b; Smolen and Stone, 1997), but failing to make such corrections may lead to overestimates in the rates at which these reactions are likely to occur in natural waters.

(ii) *Ionic strength.* Ionic strength is known to influence some chemical reactions, with both the magnitude of the effect and its direction (i.e., acceleration versus inhibition) depending upon the nature of the reaction in question. The effects of ionic strength on pesticide transformations are of particular interest in saline environments such as coastal waters where pesticides are applied for aquaculture, or estuaries—such as the Mississippi River Delta (Pereira *et al.*, 1990; Larson *et al.*, 1995; Clark and Goolsby, 2000) or the Chesapeake Bay (Hainly and Kahn, 1996; Lippa and Roberts, 2002)—into which significant loads of pesticides are discharged by rivers that drain extensive agricultural areas. Transition-state theory predicts that the effects of ionic strength on reaction rates will be smallest for neutral reactants, and become more pronounced with increasing ionic charge on the reactants (Lasaga, 1981). This is consistent with observations regarding the effects of ionic strength (over the range of salinities encountered in most natural waters) on the rates of hydrolysis of pesticide compounds. As predicted by transition-state theory, these effects have been found to be minor at neutral pH for pesticide compounds that are subject only to reaction with H_2O (Barbash and Reinhard, 1992b; Liqiang *et al.*, 1994), but substantial both at higher pH for those subject to base catalysis (Miles and Delfino, 1985) and at lower pH for those subject to acid catalysis (Wei *et al.*, 2000). Also consistent with transition-state theory is the observation by Miles and Delfino (1985) that the effect of salt concentration on the rate of aldicarb sulfone hydrolysis is more pronounced (on a molar basis) for the divalent Ca^{2+} than for the monovalent Na^+.

(iii) *Concentrations and structure of surface-active substances.* Surface-active substances (SAS) that may be present in the hydrologic system—either dissolved in the aqueous phase or as part of the soil—may influence the rates and mechanisms of pesticide transformation. These compounds include fatty acids, polysaccharides, humic substances and other forms of NOM (Thurman, 1985), as well as the surface-active agents (or *surfactants*) commonly used in detergents or as adjuvants in commercial pesticide formulations. All SAS share the common property of *amphiphilicity*, i.e., the ability to associate simultaneously with both polar and nonpolar chemical species in aqueous solution. This property arises from the dual nature of their chemical structure, one portion of the molecule being relatively hydrophilic (and often ionic) and another portion being more hydrophobic. In aqueous solution, these compounds exhibit a tendency to self-associate into assemblages known as *submicellar aggregates* at low concentrations, and *micelles* (Figure 17) above a chemical-specific level (the *critical micelle concentration*). Micelles consist of an inner core comprised of the more hydrophobic portions of the SAS molecules, surrounded by an outer layer comprised of their more hydrophilic portions (J. H. Fendler and E. J. Fendler, 1975).

When SAS are present in solution, pesticide compounds partition between the bulk aqueous solution and the (sub)micellar phase (Figure 17). This partitioning may affect the overall rates and products of transformation of these compounds if their rates of reaction in the (sub)micellar phase are significantly different from those in the aqueous phase (Barbash, 1987; Macalady and Wolfe, 1987; Barbash and Resek, 1996). In some cases, relatively minor variations in SAS structure can have substantial impacts on pesticide transformation rates (Kamiya *et al.*, 1994). Even if reaction rates are not substantially different in the (sub)micellar phase, however, the presence of SAS may modify reaction rates in solution for sparingly soluble pesticide compounds by simply increasing their dissolved concentrations, as may occur in the presence of polar solvents (e.g., Barbash and Reinhard, 1989a; Schwarzenbach *et al.*, 1993; Wei *et al.*, 2000). Rates of biotransformation may be either increased or decreased by partitioning into a (sub)micellar phase, depending upon the effect of the particular SAS on bioavailability, and the nature of the microorganisms involved (Guha and Jaffé, 1996).

The influences of NOM—either in aqueous solution or in the soil phase—on the rates of transformation of pesticide compounds in water are consistent with general principles regarding the effects of SAS on chemical reactions. Because the polar regions of most NOM are either neutral or possess a negative charge, previous research suggests that the presence of NOM in natural waters is likely to accelerate the reactions of pesticide compounds with cations, have a negligible effect on their rates of reaction with neutral species, and inhibit their rates of attack by anions—and that these effects will increase with increasing NOM concentration (Barbash, 1987; Macalady and Wolfe, 1987). These trends are precisely what has been observed for the transformations of pesticide compounds by acid-catalyzed (Li and Felbeck, 1972; Khan, 1978), neutral (Macalady and Wolfe, 1985) and base-catalyzed hydrolysis (Perdue and Wolfe, 1982; Macalady and Wolfe, 1985) in the presence of either dissolved or solid-phase NOM. The anticipated effects of SAS on the rates of other reactions that may transform pesticide compounds in natural waters (i.e., substitution reactions with other nucleophiles,

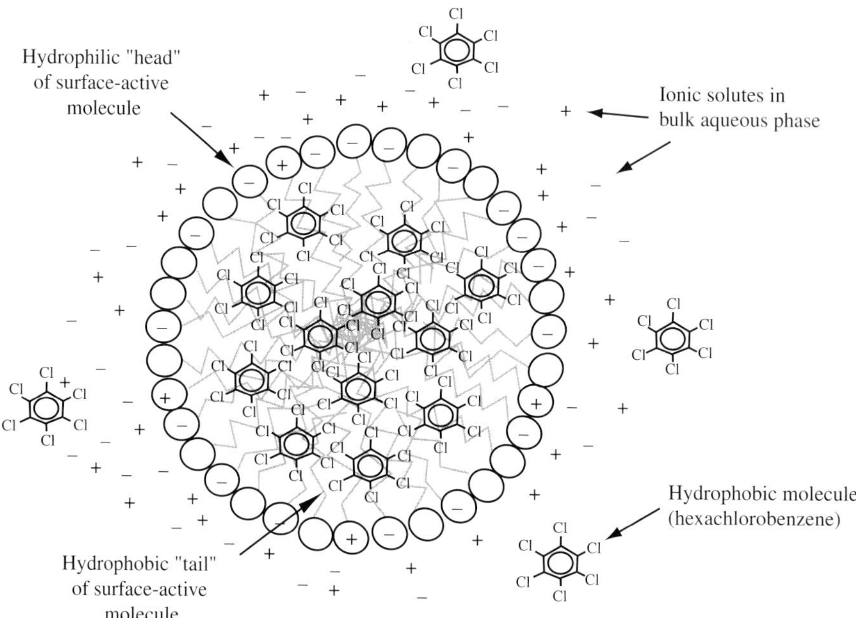

Figure 17 Cross-sectional view of the self-association of surface-active molecules into an idealized micelle. Hydrophobic portion of each surface-active molecule shown as a jagged line; hydrophilic portion shown as a circle. Hydrophobic pesticide molecules, represented by the fungicide hexachlorobenzene, will tend to accumulate within the hydrophobic core of the micelle at a higher concentration than in the bulk aqueous phase. Ionic species, (e.g., hydroxide ion), will show the opposite distribution (after J. H. Fendler and E. J. Fendler, 1975).

dehydrohalogenation, reductive dehalogenation, and free-radical transformations) were reviewed by Barbash (1987).

(iv) *Interfacial effects*. The rates and mechanisms of pesticide compound transformation (both biotic and abiotic) at interfaces in the hydrologic system (air/water, water/solid, and solid/air) are often markedly different from those in the adjacent bulk phases (e.g., Barbash and Resek, 1996). The influence of the air/water interface on these reactions is likely to be most important within the surface microlayer of surface waters, where the more hydrophobic pesticide compounds are likely to concentrate (Zepp *et al.*, 1975)—as alcohols, hydrocarbons, carotenoid pigments, chlorophylls, fatty acids, fatty acid esters, and other naturally derived SAS are known to do (Parsons and Takahashi, 1973). The gas/solid interface is likely to exert the greatest influence on reactivity at the surfaces of dry soils and plants, or in the atmosphere, where surface-catalyzed reactions of volatile halogenated compounds, including the fumigant methyl bromide, can deplete atmospheric ozone. Among the three principal types of interfaces in the hydrologic system, however, the solid/liquid interface appears to have received the most attention regarding this topic, and is therefore the main focus of the following discussion.

The presence of natural solids can significantly modify the rates of transformation in aqueous systems—relative to the rates observed in homogeneous solution—for many pesticide compounds, but may have little effect on others (Barbash and Resek, 1996). Factors that can influence the rates and mechanisms of transformation of pesticide compounds at the water/solid interface include the structure of the compound of interest (e.g., Torrents and Stone, 1991; Baldwin *et al.*, 2001), the composition and surface structure of the mineral phase (e.g., Kriegman-King and Reinhard, 1992; Wei *et al.*, 2001; Carlson *et al.*, 2002), the solid-phase organic-carbon content (e.g., Wolfe and Macalady, 1992), and the characteristics, health, and size of the resident microbial community.

Most of the ways in which natural water/solid interfaces influence the rates of transformation of pesticide compounds, however, are consistent with the various effects of related chemical species on these reactions in homogeneous solution, discussed in preceding sections. Russell *et al.* (1968), for example, observed that the hydrolysis of atrazine at low pH was greatly accelerated in suspensions of montmorillonite clay, relative to the rate of reaction in homogeneous solution at the same pH. The authors attributed this effect to the acid-catalysis of the reaction by the clay surface, which exhibited a pH that was considerably lower (by as much as 3–4 pH units) than that of the bulk solution. As noted previously, the sorption of pesticide compounds to soil organic matter has been observed to influence the rates of their hydrolysis in the same manner as might be expected from the association of these compounds with SAS in solution, leading to an acceleration of the acid-catalyzed process (Li and Felbeck, 1972), a negligible effect on the neutral process, and an inhibition of the base-catalyzed process (Macalady and Wolfe, 1985).

In addition, several naturally occurring metal oxides exhibit semiconducting properties that may catalyze the photochemical production of hydroxyl and hydroperoxyl radicals in aqueous solution—species which, as noted earlier, can react with pesticide compounds (Zepp and Wolfe, 1987). Chemical structures located at the surfaces of natural solids may also participate in pesticide transformation reactions as Brønsted bases, oxidants (e.g., Voudrias and Reinhard, 1986), reductants (e.g., Wolfe *et al.*, 1986), hydrolysis catalysts (Wei *et al.*, 2001), complexing agents (e.g., Torrents and Stone, 1991), or sources of protons for hydrogen bonding (e.g., Armstrong and Chesters, 1968).

Rates of pesticide transformation may be either directly or inversely related to the amount of NOM in the solid phase, depending upon the reaction in question. The common observation of higher transformation rates in more organic-rich soils may result from either enhanced microbial activity in the presence of more abundant NOM (e.g., Veeh *et al.*, 1996) or, for the case of reductive transformations in sterile soils, higher concentrations of biogenic reductants associated with the NOM (Wolfe and Macalady, 1992). By contrast, the inverse relations that have been reported between NOM and pesticide transformation rates are believed to be caused by substantial decreases in the rates of some reactions in the sorbed phase, relative to the dissolved phase (e.g., Graetz *et al.*, 1970; Ogram *et al.*, 1985; Zepp and Wolfe, 1987; Lartiges and Garrigues, 1995; Walse *et al.*, 2002). Indeed, with regard to biotransformations, it is generally assumed that sorption reduces the bioavailability of pesticide compounds (e.g., Zhang *et al.*, 1998). Phototransformations have also been found to occur either faster or more slowly in the sorbed phase, depending upon substrate structure and the nature of the solid phase (Mill and Mabey, 1985).

9.15.3.5 Effects of Transformations on Environmental Transport and Fate

As might be expected, the transformation of a pesticide can generate products with sorption characteristics—and thus aqueous mobilities—that are substantially different from those of the parent compound. (Indeed, one of the principal strategies employed by living organisms for detoxifying xenobiotic compounds is to transform them—either through oxidation or hydrolysis—to products that are more water-soluble, and thus more readily excreted (Coats, 1991).) For example, the herbicide DCPA, a relatively hydrophobic dimethyl ester, undergoes transformation in soil to yield progressively more mobile products as it hydrolyzes first to a "half-acid" ester intermediate, and thence to the diacid, tetrachloroterephthalic acid (Ando, 1992; USEPA, 1992; USDA/ARS, 1995). Similarly, hydrolysis of the moderately hydrophobic fumigant EDB generates ethylene glycol, which is highly mobile (Weintraub *et al.*, 1986; Syracuse Research Corporation, 1988; USDA/ARS, 1995).

Hydrolysis does not always lead to an increase in the mobility of pesticide compounds in the hydrologic system, however. Atrazine, for example, hydrolyzes to form hydroxyatrazine, which is more water soluble than its parent compound, but also exhibits up to twenty times greater affinity for soil organic matter (based on K_{oc}) than atrazine (Moreau and Mouvet, 1997). An early explanation for this phenomenon (Armstrong *et al.*, 1967) suggested that because the replacement of the chlorine atom on atrazine with a hydroxyl group raises the pK_a of the substituted amino groups on the molecule, this also increases its tendency to form hydrogen bonds with soil surfaces.

The oxidation of pesticide compounds usually generates products with aqueous mobilities that are either similar to or greater than that of the parent compound. The oxidation of aldicarb, for example, produces aldicarb sulfoxide and aldicarb sulfone, both of which have lower K_{oc} values than aldicarb (Moye and Miles, 1988). Similarly, because most phototransformations involve either the hydrolysis or oxidation of the parent compound, they yield products that are generally more polar (Mill and Mabey, 1985), and thus more water soluble than the parent compound. Reduction reactions, by contrast, may result in products that are less water soluble than their parent compound. Examples include the reduction of aldicarb sulfoxide to aldicarb (Miles and Delfino, 1985; Lightfoot *et al.*, 1987) and the reduction of phorate sulfoxide to phorate (Coats, 1991). The reactivity of transformation products may be either higher or lower than that of their parent compounds. However, those in the former category (i.e., reactive intermediates) are, of course, much less likely to be detected in the hydrologic system than more stable products.

9.15.4 THE FUTURE

Future progress in understanding the geochemistry and biological effects of pesticide compounds will be fundamentally dependent upon the availability of more detailed information on the temporal and spatial patterns of application and occurrence of these compounds, especially in nonagricultural settings. However, while the spatial and temporal distributions of their sources differ considerably from those of other contaminants, most of the uncertainties that beset our current knowledge of the geochemistry and biological effects of pesticide compounds are essentially identical to those that pertain to industrial solvents, petroleum hydrocarbons, fire-fighting chemicals, personal care products, pharmaceuticals, mining wastes and the other chemical effluvia of modern life. As with these and other anthropogenic contaminants, predictions regarding the transport, effects, and fate of pesticide compounds in natural waters must take more explicit account of a variety of complex phenomena that have received comparatively limited attention in such assessments to date. These include spray drift, large-scale atmospheric transport, preferential transport in the vadose and saturated zones, nonlinear sorption, time-dependent uptake by soils, endocrine disruption, toxicological interactions among pesticides and other contaminants present in mixtures, the effects of temperature and moisture content on transformation rates, and the transport, fate and biological effects of transformation products and adjuvants. Until significant progress is achieved in understanding these phenomena and accounting for the magnitude, timing, and spatial distributions of pesticide and adjuvant use and occurrence, the full implications of the ongoing release of these compounds into the Earth's ecosystems will remain unknown.

ACKNOWLEDGMENTS

Although this chapter officially lists only one author, the assistance of several valued colleagues was indispensable to its production. The author would therefore like to thank both the editor for this volume, Professor Barbara Sherwood-Lollar, and the series editor, Mabel Peterson, for their incredible patience in awaiting its completion. In addition, the author owes an enormous debt of gratitude to two USGS colleagues, Drs. Paul Capel and Mark Sandstrom, for having provided extremely thorough, insightful, and helpful reviews of the manuscript. The author also thanks the staff at the USGS library in Menlo Park, CA, for having provided copies of so many papers in such a timely manner, and to many other colleagues—including Yvonne Roque and Drs. Lynn Roberts, Lisa Nowell, Bill Ball, Markus Flury, and Peter Jeffers—for providing valuable assistance.

REFERENCES

Adrian N. R. and Suflita J. M. (1990) Reductive dehalogenation of a nitrogen heterocyclic herbicide in anoxic aquifer slurries. *Appl. Environ. Microb.* **56**(1), 292–294.

Agertved J., Rügge K., and Barker J. F. (1992) Transformation of the herbicides MCPP and atrazine under natural aquifer conditions. *Ground Water* **30**(4), 500–506.

Ahmed A. E., Kubic V. L., Stevens J. L., and Anders M. W. (1980) Halogenated methanes: metabolism and toxicity. *Federation Proc.* **39**(13), 3150–3155.

Alexander M. (1981) Biodegradation of chemicals of environmental concern. *Science* **211**(4478), 132–138.

Anderson J. J. and Dulka J. J. (1985) Environmental fate of sulfometuron methyl in aerobic soils. *J. Agri. Food Chem.* **33**(4), 596–602.

Ando C. M. (1992) Survey for chlorthal-dimethyl residues in well water of seven California counties. State of California Environmental Protecion Agency, EH-92-01. Sacramento, CA, 60pp.

Armstrong D. E. and Chesters G. (1968) Adsorption catalyzed chemical hydrolysis of atrazine. *Environ. Sci. Technol.* **2**(9), 683–689.

Armstrong D. E., Chesters G., and Harris R. F. (1967) Atrazine hydrolysis in soil. *Soil Sci. Soc. Am. Proc.* **31**, 61–66.

Atkins P. W. (1982) *Physical Chemistry*, 2nd edn. W.H. Freeman, San Francisco, CA, 1095p.

Bacci E., Calamari D., Gaggi C., and Vighi M. (1990) Bioconcentration of organic chemical vapors in plant leaves: experimental measurements and correlation [*sic*]. *Environ. Sci. Technol.* **24**(6), 885–889.

Baer U. and Calvet R. (1999) Fate of soil applied herbicides: experimental data and prediction of dissipation kinetics. *J. Environ. Qual.* **28**, 1765–1777.

Baldwin D. S., Beattie J. K., Coleman L. M., and Jones D. R. (2001) Hydrolysis of an organophosphate ester by manganese dioxide. *Environ. Sci. Technol.* **35**(4), 713–716.

Bank S. and Tyrrell R. J. (1984) Kinetics and mechanism of alkaline and acidic hydrolysis of aldicarb. *J. Agri. Food Chem.* **32**(6), 1223–1232.

Barbash J. E. (1987) The effect of surface-active compounds on chemical reactions of environmental interest in natural waters. *Prep. Ext. Abstr. Am. Chem. Soc. Div. Environ. Chem.* **27**(2), 58–61.

Barbash J. E. (1993) Abiotic reactions of halogenated ethanes and ethylenes with sulfide, nitrate and pH-buffer anions in aqueous solution. PhD Dissertation, Department of Environmental Science and Engineering, Stanford University, Stanford, CA, 417p.

Barbash J. E. and Reinhard M. (1989a) Reactivity of sulfur nucleophiles toward halogenated organic compounds in natural waters. In: *Biogenic Sulfur in the Environment*, ACS Symp. Ser. (eds. E. S. Saltzman and W. J. Cooper). American Chemical Society, Washington, DC, vol. 393, pp. 101–138.

Barbash J. E. and Reinhard M. (1989b) Abiotic dehalogenation of 1,2-dichloroethane and 1,2-dibromoethane in aqueous solution containing hydrogen sulfide. *Environ. Sci. Technol.* **23**(11), 1349–1358.

Barbash J. E. and Reinhard M. (1992a) Abiotic reactions of halogenated ethanes and ethylenes in buffered aqueous solutions containing hydrogen sulfide. *Prep. Ext. Abstr. Am. Chem. Soc. Div. Environ. Chem.* **32**(1), 670–673.

Barbash J. E. and Reinhard M. (1992b) The influence of pH buffers and nitrate concentration on the rate and pathways of abiotic transformation of 1,2-dibromoethane (EDB) in aqueous solution. *Prep. Ext. Abstr. Am. Chem. Soc. Div. Environ. Chem.* **32**(1), 674–677.

Barbash J. E. and Resek E. A. (1996) *Pesticides in Ground Water—Distribution, Trends, and Governing Factors*. CRC Press, Boca Raton, FL, 588p.

Barbash J. E., Thelin G. P., Kolpin D. W., and Gilliom R. J. (1999) Distribution of major herbicides in ground water of the United States. US Geol. Surv. Water-Resources Inv. Report 98-4245, 57p. (http://water.wr.usgs.gov/pnsp/rep/wrir984245/).

Baxter R. M. (1990) Reductive dechlorination of certain chlorinated organic compounds by reduced hematin compared with their behaviour in the environment. *Chemosphere* **21**(4–5), 451–458.

Bell E. M., Hertz-Picciotto I., and Beaumont J. J. (2001) A case-control study of pesticides and fetal death due to congenital anomalies. *Epidemiology* **12**, 148–156.

Betarbet R., Sherer T. B., MacKenzie G., Garcia-Osuna M., Panov A. V., and Greenamyre J. T. (2000) Chronic systemic pesticide exposure reproduces features of *Parkinson's* disease. *Nature Neurosci.* **3**(12), 1301–1306.

Beulke S., Dubus I. G., Brown C. D., and Gottesbüren B. (2000) Simulation of pesticide persistence in the field on the basis of laboratory data—A review. *J. Environ. Qual.* **29**, 1371–1379.

Beynon K. I., Stoydin G., and Wright A. N. (1972) A comparison of the breakdown of the triazine herbicides cyanazine, atrazine and simazine in soils and in maize. *Pestic. Biochem. Physiol.* **2**, 153–161.

Bidleman T. F. (1999) Atmospheric transport and air-surface exchange of pesticides. *Water Air Soil Pollut.* **115**, 115–166.

Bollag J.-M. (1982) Microbial metabolism of pesticides. In *Microbial Transformations of Bioactive Compounds* (ed. J. P. Rosazza). CRC Press, Boca Raton, FL, vol. 2, pp. 126–168.

Bolognesi C., Bonatti S., Degan P., Gallerani E., Peluso M., Rabboni R., Roggieri P., and Abbondanolo A. (1997) Genotoxic activity of glyphosate and its technical formulation Roundup. *J. Agri. Food Chem.* **45**, 1957–1962.

Broholm M. M., Tuxen N., Rügge K., and Bjerg P. L. (2001) Sorption and degradation of the herbicide 2-methyl-4,6-dinitrophenol under aerobic conditions in a sandy aquifer in Vejen, Denmark. *Environ. Sci. Technol.* **35**(24), 4789–4797.

Brusseau M. L. and Rao P. S. C. (1989) The influence of sorbate-organic matter interactions on sorption nonequilibrium. *Chemosphere* **18**(9/10), 1691–1706.

Burlinson N. E., Lee L. A., and Rosenblatt D. H. (1982) Kinetics and products of hydrolysis of 1,2-dibromo-3-chloropropane. *Environ. Sci. Technol.* **16**(9), 627–632.

Burnside O. C. and Lavy T. L. (1966) Dissipation of dicamba. *Weeds* **14**(3), 211–214.

Buser H.-R., Poiger T., and Müller M. D. (2000) Changed enantiomer composition of metolachlor in surface water following the introduction of the enantiomerically enriched product to the market. *Environ. Sci. Technol.* **34**(13), 2690–2696.

Butler E. C. and Hayes K. F. (2000) Kinetics of the transformation of halogenated aliphatic compounds by iron sulfide. *Environ. Sci. Technol.* **34**(3), 422–429.

Capel P. D. (1993) Organic chemical concepts. In *Regional Ground-Water Quality* (ed. W. M. Alley). Van Nostrand Reinhold, New York, NY, pp. 155–179.

Capel P. D., Larson S. J., and Winterstein T. A. (2001) The behaviour of 39 pesticides in surface waters as a function of scale. *Hydrol. Process.* **15**, 1251–1269.

Carlson D. L. (2003) Environmental transformations of chloroacetamide herbicides: hydrolysis and reaction with iron pyrite. PhD Dissertation, Johns Hopkins University, Baltimore, MD.

Carlson D. L. and Roberts A. L. (2002) Acid-and base-catalyzed hydrolysis of chloroacetamide herbicides. Amer. Chem. Soc., Div. Environ. Chem., Nat. Meeting, Symposium on "Chemistry of EPA Contaminant Candidate List Compounds," presented August, 2002, Boston, MA.

Carlson D. L., McGuire M. M., Fairbrother D. H., and Roberts A. L. (2002) Dechlorination of the herbicide alachlor on pyrite (100) surfaces. Amer. Chem. Soc., Div. Environ. Chem., Nat. Meeting, Symposium on "Chemistry of EPA Contaminant Candidate List Compounds," presented August, 2002, Boston, MA.

Carson R. L. (1962) *Silent Spring*, Houghton–Mifflin, Boston, MA, 368p.

Castro C. E. (1977) Biodehalogenation. *Environ. Health Persp.* **21**, 279–283.

Castro C. E. and Kray W. C., Jr. (1963) The cleavage of bonds by low valent transition metal ions: the homogeneous reduction of alkyl halides by chromous sulfate. *J. Am. Chem. Soc.* **85**, 2768–2773.

Castro C. E. and Kray W. C., Jr. (1966) Carbenoid intermediates from polyhalomethanes and chromium(II) the homogeneous reduction of geminal halides by chromous sulfate. *J. Am. Chem. Soc.* **88**, 4447–4455.

Chiou C. T. (1998) Soil sorption of organic pollutants and pesticides. In *Encyclopedia of Environmental Analysis and Remediation* (ed. R. A. Meyers). Wiley, New York, NY, pp. 4517–4554.

Chiu P.-C. and Reinhard M. (1995) Metallocoenzyme-mediated reductive transformation of carbon tetrachloride in titanium(III) citrate aqueous solution. *Environ. Sci. Technol.* **29**(3), 595–603.

Christensen T. H., Bjerg P. L., Banwart S. A., Jakobsen R., Heron G., and Albrechtsen H.-J. (2000) Characterization of redox conditions in groundwater contaminant plumes. *J. Contamin. Hydrol.* **45**, 165–241.

Chu W. and Jafvert C. T. (1994) Photodechlorination of polychlorobenzene congeners in surfactant micelle solutions. *Environ. Sci. Technol.* **28**(13), 2415–2422.

Cink J. H. and Coats J. R. (1993) Effect of concentration, temperature, and soil moisture on the degradation of chlorpyrifos in an urban Iowa soil. In *Pesticides in Urban Environments: Fate and Significance*, ACS Symp. Ser. (eds. K. D. Racke and A. R. Leslie). American Chemical Society, Washington, DC, vol. 522, pp. 62–69.

Clark G. M. and Goolsby D. A. (2000) Occurrence and load of selected herbicides and metabolites in the lower Mississippi river. *Sci. Tot. Environ.* **248**(2–3), 101–113.

Cline P. V. and Delfino J. J. (1989) Transformation kinetics of 1,1,1-trichloroethane to the stable product 1,1-dichloroethene. In *Biohazards of Drinking Water Treatment* (ed. R. A. Larson). Lewis Publishers, New York, NY, pp. 47–56.

Coats J. R. (1991) Pesticide degradation mechanisms and environmental activation. In *Pesticide Transformation Products: Fate and Significance in the Environment*, ACS Symp. Ser. (eds. L. Somasundaram and J. R. Coats). American Chemical Society, Washington, DC, vol. 459, pp. 10–31.

Colborn T., vom Saal S., and Soto A. M. (1993) Developmental effects of endocrine-disrupting chemicals in wildlife and humans. *Environ. Health Persp.* **101**(5), 378–384.

Colborn T., Dumanoski D., and Myers J. P. (1996) *Our Stolen Future*. Penguin, New York, 306p.

Cooper W. J. and Zika R. G. (1983) Photochemical formation of hydrogen peroxide in surface and ground waters exposed to sunlight. *Science* **220**, 711–712.

Crosby D. G. and Leitis E. (1973) The photodecomposition of trifluralin in water. *Bull. Environ. Contamin. Toxicol.* **10**, 237–241.

Curtis G. P. and Reinhard M. (1994) Reductive dehalogenation of hexachloroethane, carbon tetrachloride, and bromoform by anthrahydroquinone disulfonate and humic acid. *Environ. Sci. Technol.* **28**(13), 2393–2401.

Curtis G. P., Reinhard M., and Roberts P. V. (1986) Sorption of hydrophobic organic compounds by sediments. In *Geochemical Processes at Mineral Surfaces*, ACS Symp. Ser. (eds. J. A. Davis and K. F. Hayes). American Chemical Society, Washington, DC, vol. 323, pp. 191–216.

D'Adamo P. D., Rozich A. F., and Gaudy A. F., Jr. (1984) Analysis of growth data with inhibitory carbon sources. *Biotech. Bioeng.* **25**, 397–402.

Dean J. A. (ed.) (1985) *Lange's Handbook of Chemistry*, 13th edn. McGraw-Hill, San Francisco, CA, 1792p.

Deeley G. M., Reinhard M., and Stearns S. M. (1991) Transformation and sorption of 1,2-dibromo-3-chloropropane in subsurface samples collected at Fresno, California. *J. Environ. Qual.* **20**(3), 547–556.

Dilling W. L., Lickly L. C., Lickly T. D., Murphy P. G., and McKellar R. L. (1984) Organic photochemistry: 19. Quantum yields for *O,O*-diethyl-*O*-(3,5,6-trichloro-2-pyridinyl) phosphorothioate (chlorpyrifos) and 3,5,6-trichloro-2-pyridinol in dilute aqueous solutions and their environmental phototransformation rates. *Environ. Sci. Technol.* **18**(7), 540–543.

Dinelli G., Vicari A., Bonetti A., and Catizone P. (1997) Hydrolytic dissipation of four sulfonylurea herbicides. *J. Agri. Food Chem.* **45**(5), 1940–1945.

Dinelli G., Di Martino E., and Vicari A. (1998) Influence of soil moisture and temperature on degradation of three sulfonylurea herbicides in soil. *Agrochimica* **42**(1–2), 50–58.

Donaldson D., Kiely T., and Grube A. (2002) Pesticide industry sales and usage: 1998 and 1999 market estimates. US Environmental Protection Agency, Office of Pesticide Programs, Washington, DC, 33p. [http://www.epa.gov/oppbead1/pestsales/99pestsales/market_estimates1999.pdf].

Dragun J. and Helling C. S. (1985) Physicochemical and structural relationships of organic chemicals undergoing soil- and clay-catalyzed free-radical oxidation. *Soil Sci.* **139**(2), 100–111.

Egli C., Tschan T., Scholtz R., Cook A. M., and Leisinger T. (1988) Transformation of tetrachloromethane to dichloromethane and carbon dioxide by *Acetobacterium woodii*. *Appl. Env. Microb.* **54**(11), 2819–2824.

Ehrenberg L., Osterman-Golkar S., Singh D., and Lundqvist U. (1974) *Radiat. Bot.* **15**, 185–194.

Erickson L. E. and Lee K. H. (1989) Degradation of atrazine and related *s*-triazines. *Crit. Rev. Environ. Cont.* **19**(1), 1–14.

Falconer R. L., Bidleman T. F., Gregor D. J., Semkin R., and Teixeira C. (1995) Enantioselective breakdown of α-hexachlorocyclohexane in a small Arctic lake and its watershed. *Environ. Sci. Technol.* **29**(5), 1297–1302.

Fendler J. H. and Fendler E. J. (1975) *Catalysis in Micellar and Macromolecular Systems*. Academic Press, San Francisco, CA, 545p.

Field J. A. and Thurman E. M. (1996) Glutathione conjugation and contaminant transformation. *Environ. Sci. Tech.* **30**(5), 1413–1417.

Fleming I. (1976) *Frontier Orbitals and Organic Chemical Reactions*. Wiley, New York, 249p.

Flury M. (1996) Experimental evidence of transport of pesticides through field soils—A review. *J. Environ. Qual.* **25**, 25–45.

Gan J., Papiernik S. K., Yates S. R., and Jury W. A. (1999) Temperature and moisture effects on fumigant degradation in soil. *J. Environ. Qual.* **28**(5), 1436–1441.

Gan J., Yates S. R., Knuteson J. A., and Becker J. O. (2000) Transformation of 1,3-dichloropropene in soil by thiosulfate fertilizers. *J. Environ. Qual.* **29**, 1476–1481.

Gantzer C. J. and Wackett L. P. (1991) Reductive dechlorination catalyzed by bacterial transition-metal coenzymes. *Environ. Sci. Technol.* **25**(4), 715–722.

Garry V. F., Harkins M. E., Erickson L. L., Long-Simpson L. K., Holland S. E., and Burroughs B. L. (2002) Birth defects, season of conception, and sex of children born to pesticide applicators living in the Red River Valley of Minnesota, USA. *Environ. Health Persp.* **110**(3), 441–449.

Garrison A. W., Nzengung V. A., Avants J. K., Ellington J. J., Jones W. J., Rennels D., and Wolfe N. L. (2000) Phytodegradation of *p,p'*-DDT and the enantiomers of *o,p'*-DDT. *Environ. Sci. Technol.* **34**(9), 1663–1670.

Gibson S. A. and Suflita J. M. (1986) Extrapolation of biodegradation results to groundwater aquifers: reductive dehalogenation of aromatic compounds. *Appl. Environ. Microb.* **52**(4), 681–688.

Glass B. L. (1972) Relation between the degradation of DDT and the iron redox system in soils. *J. Agri. Food Chem.* **20**(2), 324–327.

Gottesbüren B. (1991) Concept, development and validation of the knowledge-based herbicide advisory system HERBASYS. PhD Dissertation, University of Hannover, FRG.

Graetz D. A., Chesters G., Daniel T. C., Newland L. W., and Lee G. B. (1970) Parathion degradation in lake sediments. *J. Water Pollut. Cont. Fed.* **42**(2), R76–R94.

Guha S. and Jaffé P. R. (1996) Biodegradation kinetics of phenanthrene partitioned into the micellar phase of nonionic surfactants. *Environ. Sci. Technol.* **30**(2), 605–611.

Guillette E. A., Meza M. M., Aquilar M. G., Soto A. D., and Garcia I. E. (1998) An anthropological approach to the evaluation of preschool children exposed to pesticides in Mexico. *Environ. Health Persp.* **106**(6), 347–353.

Haag W. R. and Mill T. (1988a) Effect of a subsurface sediment on hydrolysis of haloalkanes and epoxides. *Environ. Sci. Technol.* **22**(6), 658–663.

Haag W. R. and Mill T. (1988b) Some reactions of naturally occurring nucleophiles with haloalkanes in water. *Environ. Toxicol. Chem.* **7**, 917–924.

Haderlein S. B. and Schwarzenbach R. P. (1993) Adsorption of substituted nitrobenzenes and nitrophenols to mineral surfaces. *Environ. Sci. Technol.* **27**(2), 316–326.

Hainly R. A. and Kahn J. M. (1996) Factors affecting herbicide yields in the Chesapeake Bay watershed, June 1994. *J. Am. Water Resour. Assoc.* **32**(5), 965–984.

Hamaker J. W. and Thompson J. M. (1972) Adsorption. In *Organic Chemicals in the Soil Environment* (eds. C. A. I. Goring and J. W. Hamaker). Dekker, New York, NY, vol. 1, pp. 49–143.

Harner T., Kylin H., Bidleman T. F., and Strachan W. M. (1999) Removal of α- and γ-hexachlorocyclohexane and enantiomers of α-hexachlorocyclohexane in the eastern Arctic Ocean. *Environ. Sci. Technol.* **33**(8), 1157–1164.

Harris C. I. (1967) Fate of 2-chloro-s-triazine herbicides in soil. *J. Agri. Food Chem.* **15**(1), 157–162.

Harris J. C. (1990a) Rate of hydrolysis. In *Handbook of Chemical Property Estimation Methods* (eds. W. J. Lyman, W. F. Reehl, and D. H. Rosenblatt). American Chemical Society, Washington, DC, pp. 7(1)–7(48).

Harris J. C. (1990b) Rate of aqueous photolysis. In *Handbook of Chemical Property Estimation Methods* (eds. W. J. Lyman, W. F. Reehl, and D. H. Rosenblatt). American Chemical Society, Washington, DC, pp. 8(1)–8(43).

Hayes T. B., Collins A., Lee M., Mendoza M., Noriega N., Stuart A. A., and Vonk A. (2002) Hermaphroditic, demasculinized frogs after exposure to the herbicide, atrazine, at low ecologically relevant doses. *Proc. Natl. Acad. Sci. USA* **99**(8), 5476–5480.

Hemmamda S., Calmon M., and Calmon J. P. (1994) Kinetics and hydrolysis mechanism of chlorsulfuron and metsulfuron-methyl. *Pestic. Sci.* **40**, 71–76.

Hippelein M. and McLachlan M. S. (1998) Soil-air partitioning of semivolatile organic compounds: 1. Method development and influence of physical-chemical properties. *Environ. Sci. Technol.* **32**(2), 310–316.

Hong F., Win K. Y., and Pehkonen S. O. (2001) Hydrolysis of terbufos using simulated environmental conditions: rates, mechanisms, and product analysis. *J. Agri. Food Chem.* **49**(12), 5866–5873.

Huang C.-H. and Stone A. T. (2000) Synergistic catalysis of dimetilan hydrolysis by metal ions and organic ligands. *Environ. Sci. Technol.* **34**(19), 4117–4122.

Iwata H., Tanabe S., Sakal N., and Tatsukawa R. (1993) Distribution of persistent organochlorines in the oceanic air and surface seawater and the role of ocean [*sic*] on their global transport and fate. *Environ. Sci. Technol.* **27**(6), 1080–1098.

Jafvert C. T. and Wolfe N. L. (1987) Degradation of selected halogenated ethanes in anoxic sediment-water systems. *Environ. Toxicol. Chem.* **6**, 827–837.

Jeffers P. M. and Wolfe N. L. (1996) On the degradation of methyl bromide in sea water. *Geophys. Res. Lett.* **23**(14), 1773–1776.

Jeffers P. M., Ward L. M., Woytowitch L. M., and Wolfe N. L. (1989) Homogeneous hydrolysis rate constants for selected chlorinated methanes, ethanes, ethenes, and propanes. *Environ. Sci. Technol.* **23**(8), 965–969.

Kamiya M., Nakamura K., and Sasaki C. (1994) Inclusion effects of cyclodextrins on photodegradation rates of parathion and paraoxon in aquatic medium [*sic*]. *Chemosphere* **28**(11), 1961–1966.

Kamrin M. A. (ed.) (1997) *Pesticide Profiles: Toxicity, Environmental Impact, and Fate*. CRC Press, Boca Raton, FL, 676p.

Katagi T. (1992) Quantum chemical estimation of environmental and metabolic fates of pesticides. *J. Pestic. Sci.* **17**(3), S221–S230 (in Japanese).

Katz B. G. (1993) *Biogeochemical and Hydrological Processes Controlling the Transport and Fate of 1,2-Dibromoethane (EDB) in Soil and Ground Water, Central Florida*. US Geological Survey, Water-Supply Paper 2402, Tallahassee, Florida, 35p.

Kearney P. C. and Kaufman D. D. (1972) Microbial degradation of some chlorinated pesticides. In *Degradation of Synthetic Organic Molecules in the Biosphere: Natural, Pesticidal, and Various Other Man-made Compounds*. National Academy of Sciences, pp. 166–189.

Kenaga E. E. (1980) Predicted bioconcentration factors and soil sorption coefficients of pesticides and other chemicals. *Ecotoxicol. Environ. Safety* **4**, 26–38.

Kenneke J. F. and Weber E. J. (2003) Reductive dehalogenation of halomethanes in iron- and sulfate-reducing sediments: 1. Reactivity pattern analysis. *Environ. Sci. Technol.* **37**(4), 713–720.

Khan S. U. (1978) Kinetics of hydrolysis of atrazine in aqueous fulvic acid solution. *Pestic. Sci.* **9**, 39–43.

Klecka G. M. and Gonsior S. J. (1984) Reductive dechlorination of chlorinated methanes and ethanes by reduced iron(II) porphyrins. *Chemosphere* **13**(3), 391–402.

Klecka G. M., Gonsior S. J., and Markham D. A. (1990) Biological transformations of 1,1,1-trichloroethane in subsurface soils and ground water. *Environ. Toxicol. Chem.* **9**, 1437–1451.

Klopman G. (1968) Chemical reactivity and the concept of charge- and frontier-controlled reactions. *J. Am. Chem. Soc.* **90**(2), 223–234.

Kochi J. K. and Powers J. W. (1970) The mechanism of reduction of alkyl halides by chromium (II) complexes: alkylchromium species as intermediates. *J. Am. Chem. Soc.* **92**(1), 137–146.

Kolpin D. W., Barbash J. E., and Gilliom R. J. (2000) Pesticides in ground water of the US, 1992–1996. *Ground Water* **38**(6), 858–863.

Konrad J. G. and Chesters G. (1969) Degradation in soils of ciodrin, an organophosphate insecticide. *J. Agric. Food Chem.* **17**(2), 226–230.

Konrad J. G., Armstrong D. E., and Chesters G. (1967) Soil degradation of diazinon, a phosphorothioate insecticide. *Agron. J.* **59**, 591–594.

Konrad J. G., Chesters G., and Armstrong D. E. (1969) Soil degradation of malathion, a phosphorothioate insecticide. *Proc. Soil Sci. Soc. Am.* **33**, 259–262.

Kray W. C., Jr. and Castro C. E. (1964) The cleavage of bonds by low-valent transition metal ions: the homogeneous dehalogenation of vicinal dihalides by chromous sulfate. *J. Am. Chem. Soc.* **86**, 4603–4608.

Kriegman-King M. and Reinhard M. (1992) Transformation of carbon tetrachloride in the presence of sulfide, biotite and vermiculite. *Environ. Sci. Technol.* **26**(11), 2198–2206.

Kriegman-King M. and Reinhard M. (1994) Transformation of carbon tetrachloride by pyrite in aqueous solution. *Environ. Sci. Technol.* **28**(4), 692–700.

Kuhn E. P. and Suflita J. M. (1989) Dehalogenation of pesticides by anaerobic microorganisms in soils and groundwater—A review. In *Reactions and Movement of*

Organic Chemicals in Soils. Soil Science Society of America, Spl. Publ. No. 22, Soil Science Society of America, Madison, WI, pp. 111–180.

Larson S. J., Capel P. D., Goolsby D. A., Zaugg S. D., and Sandstrom M. W. (1995) Relations between pesticide use and riverine flux in the Mississippi River Basin. *Chemosphere* **31**(5), 3305–3321.

Larson S. J., Capel P. D., and Majewski M. S. (1997) *Pesticides in Surface Waters—Distribution, Trends and Governing Factors*. CRC Press, Boca Raton, FL, 373p.

Larson S. J., Gilliom R. J., and Capel P. D. (1999) Pesticides in streams of the US-initial results from the National Water-Quality Assessment Program. US Geol. Surv. Water Resour. Inv. Report. 98-4222, 92p.

Lartiges S. B. and Garrigues P. P. (1995) Degradation kinetics of organophosphorus and organonitrogen pesticides in different waters under various environmental conditions. *Environ. Sci. Technol.* **29**(5), 1246–1254.

Lasaga A. C. (1981) Transition state theory. In *Kinetics of Geochemical Processes*. Reviews in Mineralogy. (eds. A. C. Lasaga and R. J. Kirkpatrick). Mineralogical Society of America, Washington, DC, vol. 8, pp. 135–169.

Lee P. W., Fukoto J. M., Hernandez H., and Stearns S. M. (1990) Fate of monocrotophos in the environment. *J. Agri. Food Chem.* **38**(2), 567–573.

Li G.-C. and Felbeck G. T., Jr. (1972) Atrazine hydrolysis as catalyzed by humic acids. *Soil Sci.* **114**(3), 201–209.

Lightfoot E. N., Thorne P. S., Jones R. L., Hansen J. L., and Romine R. R. (1987) Laboratory studies on mechanisms for the degradation of aldicarb, aldicarb sulfoxide and aldicarb sulfone. *Environ. Toxicol. Chem.* **6**, 377–394.

Lin N. and Garry V. F. (2000) *In vitro* studies of cellular and molecular developmental toxicity of adjuvants, herbicides and fungicides commonly used in Red River Valley, Minnesota. *J. Toxicol. Environ. Health* **60**, 423–439.

Lippa K. A. and Roberts A. L. (2002) Nucleophilic aromatic substitution reactions of chloroazines with bisulfide (HS^-) and polysulfides (S_n^{2-}). *Environ. Sci. Technol.* **36**(9), 2008–2018.

Liqiang J., Shukui H., Liansheng W., Chao L., and Deben D. (1994) The influential factors of hydrolysis of organic pollutants in environment [*sic*]. *Chemosphere* **28**(10), 1749–1756.

Little C. D., Palumbo A. V., Herbes S. E., Lidstrom M. E., Tyndall R. L., and Gilmer P. J. (1988) Trichloroethylene biodegradation by a methane-oxidizing bacterium. *Appl. Environ. Microb.* **54**, 951–956.

Loch A. R., Lippa K. A., Carlson D. L., Chin Y. P., Traina S. J., and Roberts A. L. (2002) Nucleophilic aliphatic substitution reactions of propachlor, alachlor, and metolachlor with bisulfide (HS^-) and polysulfides (S_n^{2-}). *Environ. Sci. Technol.* **36**(19), 4065–4073.

Loh J. (ed.) (2002) Living Planet Report 2002: Gland, Switzerland, World Wildlife Fund, 37p. (available on the World Wide Web at http://panda.org/news_facts/publications/general/livingplanet/lpr02.cfm).

Lovley D. R., Chapelle F. H., and Woodward J. C. (1994) Use of dissolved hydrogen (H_2) concentrations to determine distribution of microbially catalyzed redox reactions in anoxic groundwater. *Environ. Sci. Technol.* **28**(7), 1205–1210.

Mabey W. R. and Mill T. (1978) Critical review of hydrolysis of organic compounds in water under environmental conditions. *J. Phys. Chem. Ref. Data* **7**(2), 383–425.

Mabey W. R., Barich V., and Mill T. (1983) Hydrolysis of polychlorinated alkanes. American Chemical Society. *Prepr. Ext. Abstr. Am. Chem. Soc., Div. Environ. Chem.* **23**(2), 359–361.

Macalady D. L. and Wolfe N. L. (1985) Effects of sediment sorption and abiotic hydrolyses: 1. Organophosphorothioate esters. *J. Agri. Food Chem.* **33**(2), 167–173.

Macalady D. L. and Wolfe N. L. (1987) Influences of aquatic humic substances on the abiotic hydrolysis of organic contaminants: a critical review. *Prepr. Ext. Abstr. Am. Chem. Soc., Div. Environ. Chem* **27**(1), 12–15.

Mackay D. (1979) Finding fugacity feasible. *Environ. Sci. Technol.* **13**(10), 1218–1223.

Mackay D., Shiu W.-Y., and Ma K.-C. (1997) *Illustrated Handbook of Physical–Chemical Properties and Environmental Fate for Organic Chemicals, Volume V. Pesticide Chemicals*. Lewis Publishers, New York, NY, 812p.

Majewski M. S. and Capel P. D. (1995) *Pesticides in the Atmosphere—Distribution, Trends and Governing Factors*. CRC Press, Boca Raton, FL, 214p.

Majewski M. S., Desjardins R., Rochette P., Pattey E., Seiber J., and Glotfelty D. (1993) Field comparison of an eddy accumulation and an aerodynamic-gradient system for measuring pesticide volatilization fluxes. *Environ. Sci. Technol.* **27**(1), 121–128.

Majewski M. S., Foreman W. T., Goolsby D. A., and Nakagaki N. (1998) Airborne pesticide residues along the Mississippi River. *Environ. Sci. Technol.* **32**(23), 3689–3698.

Mallat E. and Barceló D. (1998) Analysis and degradation study of glyphosate and of aminomethylphosphonic acid in natural waters by means of polymeric and ion-exchange solid-phase extraction columns followed by ion chromatography-post-column derivatization with fluorescence detection. *J. Chromatog. Acta* **823**, 129–136.

Mandelbaum R. T., Wackett L. P., and Allan D. L. (1993) Rapid hydrolysis of atrazine to hydroxyatrazine by soil bacteria. *Environ. Sci. Technol.* **27**(9), 1943–1946.

Mansour M. and Feicht E. A. (1994) Transformation of chemical contaminants by biotic and abiotic processes in water and soil. *Chemosphere* **28**(2), 323–332.

March J. (1985) *Advanced Organic Chemistry: Reactions, Mechanisms and Structure*, 3rd edn. Wiley, New York, NY, 1346p.

Matheron M. (2001) Modes of action for plant disease management chemistries. Presented 6, December 2001 at the 11th Annual Desert Vegetable Crop Workshop (Yuma, AZ). (Accessed online at http://ag.arizona.edu/crops/diseases/papers/dischemistry.html on 9/22/02).

Matsumura F. (1985) *Toxicology of Insecticides*, 2nd edn. Plenum Press, New York, NY, 598p.

McCall P. J. (1987) Hydrolysis of 1,3-dichloropropene in dilute aqueous solution. *Pestic. Sci.* **19**, 235–242.

McCarty P. L. (1972) Energetics of organic matter degradation. In *Water Pollution Microbiology* (ed. R. Mitchell). Wiley-Inter Science, New York, NY, pp. 91–118.

McCarty P. L., Reinhard M., and Rittmann B. E. (1981) Trace organics in groundwater. *Environ. Sci. Technol.* **15**(1), 40–51.

Meijer S. N., Shoeib M., Jantunen L. M. M., Jones K. C., and Harner T. (2003) Air-soil exchange of organochlorine pesticides in agricultural soils: 1. Field measurements using a novel *in situ* sampling device. *Environ. Sci. Technol.* **37**(7), 1292–1299.

Meister T. (ed.) (2000) *Farm Chemicals Handbook, 2000*, Meister Publishing Company, Willoughby, OH, vol. 86, variously paged.

Miah H. M. and Jans U. (2001) Reactions of phosphorothionate triesters with reduced sulfur species. Abstract presented at the 22nd annual meeting of the Society for Environmental Toxicology and Chemistry (November, 2001; Baltimore, MD).

Miles C. J. and Delfino J. J. (1985) Fate of aldicarb, aldicarb sulfoxide, and aldicarb sulfone in Floridan groundwater. *J. Agri. Food Chem.* **33**(3), 455–460.

Mill T. and Mabey W. R. (1985) Photochemical transformations. In *Environmental Exposure From Chemicals* (eds. W. B. Neely and G. Blau). CRC Press, Boca Raton, FL, vol. I, pp. 175–216.

Milligan P. W. and Häggblom M. M. (1999) Biodegradation and biotransformation of dicamba under different reducing conditions. *Environ. Sci. Technol.* **33**(8), 1224–1229.

Mochida I., Noguchi H., Fujitsu H., Seiyama T., and Takeshita K. (1977) Reactivity and selectivity in the reductive elimination of halogen from haloalkanes by chromous, cuprous, and stannous ions. *Can. J. Chem.* **55**, 2420–2425.

Monkiedje A., Spiteller M., and Bester K. (2003) Degradation of racemic and enantiopure metalaxyl in tropical and temperate soils. *Environ. Sci. Technol.* **37**(4), 707–712.

Moore J. W. and Pearson R. G. (1981) *Kinetics and Mechanism,* 3rd edn. Wiley, New York, NY, 455p.

Moreau C. and Mouvet C. (1997) Sorption and desorption of atrazine, deethylatrazine, and hydroxyatrazine by soil and aquifer solids. *J. Environ. Qual.* **26**, 416–424.

Mortland M. M. and Raman K. V. (1967) Catalytic hydrolysis of some organic phosphate pesticides by copper(II). *J. Agri. Food Chem.* **15**(1), 163.

Moye H. A. and Miles C. J. (1988) Aldicarb contamination of groundwater. *Rev. Environ. Cont. Toxicol.* **105**, 99–145.

Murphy S. D. (1986) Toxic effects of pesticides. In *Casarett and Doull's Toxicology: The Basic Science of Poisons*, 3rd edn. (eds. C. D. Klaassen, M. O. Amdur, and J. Doull). Macmillan, New York, NY, pp. 519–581.

Nair D. R. and Schnoor J. L. (1992) Effect of two electron acceptors on atrazine mineralization rates in soil. *Environ. Sci. Technol.* **26**(11), 2298–2300.

Nash R. G. (1988) Dissipation from soil. In *Environmental Chemistry of Herbicides* (ed. R. Grover). CRC Press, Boca Raton, FL, vol. 1, pp. 132–169.

Newland L. W., Chesters G., and Lee G. B. (1969) Degradation of γ-BHC in simulated lake impoundments as affected by aeration. *J. Water Pollut. Cont. Fed.* **41**(5), R174–R188.

Ngabe B., Bidleman T. F., and Falconer R. L. (1993) Base hydrolysis of α- and γ- hexachlorocyclohexanes. *Environ. Sci. Technol.* **27**(9), 1930–1933.

Nicollier G. and Donzel B. (1994) Release of bound ^{14}C-residues from plants and soil using microwave extraction. Poster presented at the 8th International Congress of Pesticide Chemistry, July 4–9, 1994, Washington, D.C., American Chemical Society.

Nowell L. H., Capel P. D., and Dileanis P. D. (1999) *Pesticides in Stream Sediment and Aquatic Biota—Distribution, Trends and Governing Factors*. CRC Press, Boca Raton, FL, 1001p.

Oakes D. J. and Pollak J. K. (1999) Effects of a herbicide formulation, Tordon 75D(R), and its individual components on the oxidative functions of mitochondria. *Toxicol.* **136**, 41–52.

Ogram A. V., Jessup R. E., Ou L. T., and Rao P. S. C. (1985) Effects of sorption on biological degradation rates of (2,4-dichlorophenoxy)acetic acid in soils. *Appl. Environ. Microb.* **49**(3), 582–587.

O'Neill R. V. (1968) Population energetics of the millipede, *Narcus americanus* (*beauvois*). *Ecology* **49**(5), 803–809.

Oremland R. S., Miller L. G., and Strohnaier F. E. (1994) Degradation of methyl bromide in anaerobic sediments. *Environ. Sci. Technol.* **28**(3), 514–520.

Padilla F., LaFrance P., Robert C., and Villeneuve J. (1988) Modeling the transport and the fate of pesticides in the unsaturated zone considering temperature effects. *Ecol. Model.* **44**, 73–88.

Pape-Lindstrom P. A. and Lydy M. J. (1997) Synergistic toxicity of atrazine and organophosphate insecticides contravenes the response addition mixture model. *Environ. Toxicol. Chem.* **16**(11), 2415–2420.

Papiernik S. K. and Spalding R. F. (1998) Atrazine, deethylatrazine, and deisopropylatrazine persistence measured in groundwater *in situ* under low-oxygen conditions. *J. Agri. Food Chem.* **46**(2), 749–754.

Paris D. F., Steen W. C., Baughman G. L., and Barnett J. T., Jr. (1981) Second-order model to predict microbial degradation of organic compounds in natural waters. *Appl. Environ. Microb.* **41**(3), 603–609.

Parker L. W. and Doxtader K. G. (1983) Kinetics of the microbial degradation of 2,4-D in soil: effects of temperature and moisture. *J. Environ. Qual.* **12**(4), 553–558.

Parsons T. R. and Takahashi M. (1973) *Biological Oceanographic Processes*. Pergamon, New York, NY, 186pp.

Parsons F., Wood P. R., and DeMarco J. (1984) Transformations of tetrachloroethene and trichloroethene in microcosms and groundwater. *J. Am. Water Works Assn.*, (Feb), 56–59.

Patlak M. (1996) Estrogens may link pesticides, breast cancer. *Environ. Sci. Technol.* **30**(5), 210A–211A.

Pearson R. G. (1972) The influence of the reagent on organic reactivity. In *Advances in Linear Free Energy Relationships* (eds. N. B. Chapman and J. Shorter). Plenum, New York, NY, pp. 281–319.

Pehkonen S. O. and Zhang Q. (2002) The degradation of organophosphorus pesticides in natural waters: a critical review. *Crit. Rev. Environ. Sci. Technol.* **32**(1), 17–72.

Peijnenburg W. J. G. M., Hart M. J., den Hollander H. A., van de Meent D., Verboom H. H., and Wolfe N. L. (1992a) QSARs for predicting reductive transformation rate constants of halogenated aromatic hydrocarbons in anoxic sediment systems. *Environ. Toxicol. Chem.* **11**, 301–314.

Peijnenburg W. J. G. M., de Beer K. G. M., Haan M. W. A., den Hollander H. A., Stegeman M. H. L., and Verboom H. (1992b) Development of a structure-reactivity relationship for the photohydrolysis of substituted aromatic halides. *Environ. Sci. Technol.* **26**(11), 2116–2121.

Perdue E. J. and Wolfe N. L. (1982) Modification of pollutant hydrolysis kinetics in the presence of humic substances. *Environ. Sci. Technol.* **16**(12), 847–852.

Perdue E. M. and Wolfe N. L. (1983) Prediction of buffer catalysis in field and laboratory studies of pollutant hydrolysis reactions. *Environ. Sci. Technol.* **17**(11), 635–642.

Pereira W. E., Rostad C. E., and Leiker T. J. (1990) Distribution of agrochemicals in the lower Mississippi River and its tributaries. *Sci. Tot. Environ.* **97/98**, 41–53.

Picardal F., Arnold R. G., and Huey B. B. (1995) Effects of electron donor and acceptor conditions on reductive dehalogenation of tetrachloromethane by *Shewanella putrefaciens 200. Appl. Environ. Microb.* **61**(1), 8–12.

Repetto R. and Baliga S. (1996) *Pesticides and the Immune System: the Public Health Risks*. World Resources Institute, Washington, DC, 100p.

Roberts A. L., Sanborn P. N., and Gschwend P. M. (1992) Nucleophilic substitution reactions of dihalomethanes with hydrogen sulfide species. *Environ. Sci. Technol.* **26**(11), 2263–2274.

Roberts A. L., Jeffers P. M., Wolfe N. L., and Gschwend P. M. (1993) Structure-reactivity relationships in dehydrohalogenation reactions of polychlorinated and polybrominated alkanes. *Crit. Rev. Environ. Sci. Technol.* **23**(1), 1–39.

Roberts T. R. and Stoydin G. (1976) The degradation of (Z)- and (E)-1,3-dichloropropenes and 1,2-dichloropropane in soil. *Pestic. Sci.* **7**, 325–335.

Rocha F. and Walker A. (1995) Simulation of the persistence of atrazine in soil at different sites in Portugal. *Weed Res.* **35**(3), 179–186.

Roeth F. W., Lavy T. L., and Burnside O. C. (1969) Atrazine degradation in two soil profiles. *Weed Sci.* **17**, 202–205.

Rügge K., Bjerg P. L., Mosbaek H., and Christensen T. H. (1999) Fate of MCPP and atrazine in an anaerobic landfill leachate plume (Grindsted, Denmark). *Water Res.* **33**(10), 2455–2458.

Russell J. D., Cruz M., White J. L., Bailey G. W., Payne W. R., Jr., Pope J. D., Jr., and Teasley J. I. (1968) Mode of chemical degradation of s-triazines by montmorillonite. *Science* **160**, 1340–1342.

Sakata S., Mikami N., Matsuda T., and Miyamoto J. (1986) Degradation and leaching behavior of the pyrethroid insecticide cypermethrin in soils. *J. Pestic. Sci.* **11**(1), 71–79.

Salmon A. G., Jones R. B., and Mackrodt W. C. (1981) Microsomal dechlorination of chloroethanes: structure-reactivity relationships. *Xenobiotica* **11**(11), 723–734.

Schellenberg K., Leuenberger C., and Schwarzenbach R. P. (1984) Sorption of chlorinated phenols by natural sediments and aquifer materials. *Environ. Sci. Technol.* **18**(9), 652–657.

Scholz N. L., Truelove N. K., French B. L., Berejikian B. A., Quinn T. P., Casillas E., and Collier T. K. (2000) Diazinon disrupts antipredator and homing behaviors in chinook salmon (*Oncorhynchus tsawytscha*). *Can. J. Fish. Aquat. Sci.* **57**, 1911–1918.

Schreinemachers D. M. (2000) Cancer mortality in four northern wheat-producing States. *Environ. Health Persp.* **108**(9), 873–881.

Schwarzenbach R. P. and Westall J. (1981) Transport of nonpolar organic compounds from surface water to groundwater: laboratory sorption studies. *Environ. Sci. Technol.* **15**(11), 1360–1367.

Schwarzenbach R. P., Stierli R., Lanz K., and Zeyer J. (1990) Quinone and iron porphyrin mediated reduction of nitroaromatic compounds in homogeneous aqueous solution. *Environ. Sci. Technol.* **24**(10), 1566–1574.

Schwarzenbach R. P., Gschwend P. M., and Imboden D. M. (1993) *Environmental Organic Chemistry*. Wiley, New York, NY, 681p.

Scow K. M. (1990) Rate of biodegradation. In *Handbook of Chemical Property Estimation Methods* (eds. W. J. Lyman, W. F. Reehl, and D. H. Rosenblatt). American Chemical Society, Washington, DC, pp. 9(1)–9(85).

Shin Y.-O., Chodan J. J., and Wolcott A. R. (1970) Adsorption of DDT by soils, soil fractions, and biological materials. *J. Agri. Food Chem.* **18**(6), 1129–1133.

Singleton D. M. and Kochi J. K. (1967) The mechanism of reductive elimination of vic–dihalides by chromium (II). *J. Am. Chem. Soc.* **89**(25), 6547–6555.

Sirons G. J., Frank R., and Sawyer T. (1973) Residues of atrazine, cyanazine, and their phytotoxic metabolites in a clay loam soil. *J. Ag. Food Chem.* **21**(6), 1016–1020.

Skipper H. D., Gilmour C. M., and Furtick W. R. (1967) Microbial versus chemical degradation of atrazine in soils. *Soil Sci. Soc. Am. Proc.* **31**(5), 653–656.

Smith G. J. (1987) Pesticide use and toxicology in relation to wildlife: organophosphorus and carbamate compounds. *US Fish and Wildlife Service Resource Pub.* **170**, 171p.

Smith A. E. and Aubin A. J. (1993) Degradation of [14-C]amidosulfuron in aqueous buffers and in an acidic soil. *J. Ag. Food Chem.* **41**(12), 2400–2403.

Smolen J. M. and Stone A. T. (1997) Divalent metal ion-catalyzed hydrolysis of phosphorothionate ester pesticides and their corresponding oxonates. *Environ. Sci. Technol.* **31**(6), 1664–1673.

Sparling D. W., Fellers G. M., and McConnell L. L. (2001) Pesticides and amphibian population declines in California, USA. *Environ. Toxicol. Chem.* **20**(7), 1591–1595.

Squillace P. J., Scott J. C., Moran M. J., Nolan B. T., and Kolpin D. W. (2002) VOCs, pesticides, nitrate, and their mixtures in groundwater used for drinking water in the United States. *Environ. Sci. Technol.* **36**(9), 1923–1930.

Strathmann T. J. and Stone A. T. (2001) Reduction of the carbamate pesticides oxamyl and methomyl by dissolved Fe^{II} and Cu^{I}. *Environ. Sci. Technol.* **35**(12), 2461–2469.

Strathmann T. J. and Stone A. T. (2002a) Reduction of the pesticides oxamyl and methomyl by Fe^{II}: effect of pH and inorganic ligands. *Environ. Sci. Technol.* **36**(4), 653–661.

Strathmann T. J. and Stone A. T. (2002b) Reduction of oxamyl and related pesticides by Fe^{II}: influence of organic ligands and natural organic matter. *Environ. Sci. Technol.* **36**(23), 5172–5183.

Stumm W. and Morgan J. J. (1981) *Aquatic Chemistry: An Introduction Emphasizing Chemical Equilibria in Natural Waters*, 2nd edn. Wiley, New York, NY, 780p.

Subramaniam V. and Hoggard P. E. (1988) Metal complexes of glyphosate. *J. Ag. Food Chem.* **36**(6), 1326–1329.

Suntio L. R., Shiu W. Y., Mackay D., Seiber J. N., and Glotfelty D. (1988) Critical review of Henry's Law Constants for pesticides. *Rev. Environ. Cont. Toxicol.* **103**, 1–59.

Syracuse Research Corporation (1988) *Environmental Fate Database*, available on the World Wide Web at http://esc.syrres.com/scripts/CHFcgi.exe (accessed on 4/11/03).

Tesoriero A. J., Löffler F. E., and Hiebscher H. (2001) Fate and origin of 1,2-dichloropropane in an unconfined shallow aquifer. *Environ. Sci. Technol.* **35**(3), 455–461.

Thelin G. P. and Gianessi L. P. (2000) Method for estimating pesticide use for county areas of the conterminous United States. US Geol. Surv. Open-File Report 00-250, 62p.

Thomas R. G. (1990a) Volatilization from water. In *Handbook of Chemical Property Estimation Methods* (eds. W. J. Lyman, W. F. Reehl, and D. H. Rosenblatt). American Chemical Society, Washington, DC, pp. 15(1)–15(34).

Thomas R. G. (1990b) Volatilization from soil. In *Handbook of Chemical Property Estimation Methods* (eds. W. J. Lyman, W. F. Reehl, and D. H. Rosenblatt). American Chemical Society, Washington, DC, pp. 16(1)–16(50).

Thompson H. M. (1996) Interactions between pesticides: a review of reported effects and their implications for wildlife risk assessment. *Ecotoxicology* **5**, 59–81.

Thurman E. M. (1985) *Organic Geochemistry of Natural Waters*. Martinus Nijhoff/Dr. W. Junk, Boston, MA, 497p.

Tominack R. L. (2000) Herbicide formulations. *Clin. Toxicol.* **38**(2), 129–135.

Torrents A. and Stone A. T. (1991) Hydrolysis of phenyl picolinate at the mineral/water interface. *Environ. Sci. Technol.* **25**(1), 143–149.

Tratnyek P. G. and Macalady D. L. (1989) Abiotic reduction of nitro aromatic pesticides in anaerobic laboratory systems. *J. Agri. Food Chem.* **37**(1), 248–254.

United Nations Food and Agriculture Organization (2003) Database on Pesticides Consumption, United Nations Food and Agriculture Organization, Statistics Division. Available on the world wide web at http://www.fao.org/waicent/FAOINFO/economic/pesticid.htm (accessed 4/14/03).

US Department of Agriculture-Agricultural Research Service (USDA/ARS) (1995) Pesticide Properties Database. (Available on-line at http://www.arsusda.govrsml/ppdb2.html [last update May, 1995]).

US Environmental Protection Agency [USEPA] (1992) *Another Look: National Survey of Pesticides in Drinking Water Wells, Phase II Report*. USEPA, EPA 579/09-91-020. Washington, DC, 166p. (+Appendices).

US Environmental Protection Agency (2002) List of inert ingredients contained in products registered for use in the United States (downloaded from http://www.epa.gov/opprd001/inerts/inerts.xls on May 10, 2002).

US General Accounting Office (2001) Agricultural pesticides: management improvements needed to further promote integrated pest management. GAO-01-815, 31p. US General Accounting Office.

US Geological Survey (1998) USGS Pesticide National Synthesis Project—Annual use maps. (Accessed November 11, 2002, on the World-Wide Web at URL http://ca.water.usgs.gov/pnsp/use92/index.html; last update March 20, 1998.).

Van Metre P. C., Wilson J. T., Callender E., and Fuller C. C. (1998) Similar rates of decrease of persistent, hydrophobic and particle-reactive contaminants in riverine systems. *Environ. Sci. Technol.* **32**(21), 3312–3317.

Veeh R. H., Inskeep W. P., and Camper A. K. (1996) Soil depth and temperature effects on microbial degradation of 2,4-D. *J. Environ. Qual.* **25**, 5–12.

Vink J. P. M., Nörtershäuser P., Richter O., Diekkrüger B., and Groen K. P. (1994) Modelling the microbial breakdown of pesticides in soil using a parameter estimation technique. *Pestic. Sci.* **40**, 285–292.

Vogel T. M. and Reinhard M. (1986) Reaction products and rates of disappearance of simple bromoalkanes, 1,

2-dibromopropane, and 1,2-dibromoethane in water. *Environ. Sci. Technol.* **20**(10), 992–997.

Vogel T. M. and McCarty P. L. (1987) Abiotic and biotic transformations of 1,1,1-trichloroethane under methanogenic conditions. *Environ. Sci. Technol.* **21**(12), 1208–1213.

Vogel T. M., Criddle C. S., and McCarty P. L. (1987) Transformations of halogenated aliphatic compounds. *Environ. Sci. Technol.* **21**(8), 722–736.

Voudrias E. A. and Reinhard M. (1986) Abiotic organic reactions at mineral surfaces. In *Geochemical Processes at Mineral Surfaces*, ACS Symp. Ser. (eds. J. A. Davis and K. F. Hayes). American Chemical Society, Washington, DC, vol. 323, pp. 462–486.

Wade R. S. and Castro C. E. (1973) Oxidation of iron(II) porphyrins by alkyl halides. *J. Am. Chem. Soc.* **95**(1), 226–230.

Walker A. (1974) A simulation model for prediction of herbicide persistence. *J. Environ. Qual.* **3**(4), 396–401.

Walker A. (1987) Herbicide persistence in soil. *Rev. Weed Sci.* **3**, 1–17.

Walraevens R., Trouillet P., and Devos A. (1974) Basic elimination of HCl from chlorinated ethanes. *Int. J. Chem. Kin.* **VI**, 777–786.

Walse S. S., Shimizu K. D., and Ferry J. L. (2002) Surface-catalyzed transformations of aqueous endosulfan. *Environ. Sci. Technol.* **36**(22), 4846–4853.

Wang Q., Gan J., Papiernik S. K., and Yates S. R. (2000) Transformation and detoxification of halogenated fumigants by ammonium thiosulfate. *Environ. Sci. Technol.* **34**(17), 3717–3721.

Wang W., Liszewski M., Buchmiller R., and Cherryholmes K. (1995) Occurrence of active and inactive herbicide ingredients at selected sites in Iowa. *Water Air Soil Pollut.* **83**, 21–35.

Wania F. (2003) Assessing the potential of persistent organic chemicals for long-range transport and accumulation in polar regions. *Environ. Sci. Technol.* **37**(7), 1344–1351.

Washington J. W. (1995) Hydrolysis rates of dissolved volatile organic compounds: principles, temperature effects, and literature review. *Ground Water* **33**(3), 415–424.

Watson J. R. (1977) Seasonal variation in the biodegradation of 2,4-D in river water. *Water Res.* **11**, 153–157.

Wauchope R. D. (1978) The pesticide content of surface water draining from agricultural fields—a review. *J. Environ. Qual.* **7**(4), 459–472.

Weber J. B. (1990) Behavior of dinitroaniline herbicides in soils. *Weed Technol.* **4**, 394–406.

Wei J., Furrer G., and Schulin R. (2000) Kinetics of carbosulfan degradation in the aqueous phase in the presence of a cosolvent. *J. Environ. Qual.* **29**, 1481–1487.

Wei J., Furrer G., Kaufmann S., and Schulin R. (2001) Influence of clay minerals on the hydrolysis of carbamate pesticides. *Environ. Sci. Technol.* **35**(11), 2226–2232.

Weintraub R. A., Jex G. W., and Moye H. A. (1986) Chemical and microbial degradation of 1,2-dibromoethane (EDB) in Florida ground water, soil, and sludge. In *Evaluation of Pesticides in Ground Water*, ACS Symp. Ser. (eds. W. Y. Garner, R. C. Honeycutt, and H. N. Nigg). American Chemical Society, Washington, DC, vol. 315, pp. 294–310.

Widmer S. K. and Spalding R. F. (1996) Assessment of herbicide transport and persistence in groundwater: a review. In *Herbicide Metabolites in Surface Water and Groundwater*, ACS Symp. Ser. (eds. M. T. Meyer and E. M. Thurman). American Chemical Society, Washington, DC, vol. 630, pp. 271–287.

William R. D., Burrill L. C., Ball D., Miller T. L., Parker R., Al-Khatib K., Callihan R. H., Eberlein C., and Morishita D. W. (1995) 1995 Pacific Northwest Weed Control Handbook. Oregon State University Extension Service, Corvallis, OR, 358p.

Wolfe N. L. and Macalady D. L. (1992) New perspectives in aquatic redox chemistry: abiotic transformations of pollutants in groundwater and sediments. *J. Cont. Hydrol.* **9**(1–2), 17–34.

Wolfe N. L., Zepp R. G., and Paris D. F. (1978) Use of structure-reactivity relationships to estimate hydrolytic persistence of carbamate pesticides. *Water Res.* **12**, 561–563.

Wolfe N. L., Paris D. F., Steen W. C., and Baughman G. L. (1980) Correlation of microbial degradation rates with chemical structure. *Environ. Sci. Technol.* **14**(9), 1143–1144.

Wolfe N. L., Kitchens B. E., Macalady D. L., and Grundl T. J. (1986) Physical and chemical factors that influence the anaerobic degradation of methyl parathion in sediment systems. *Environ. Toxicol. Chem.* **5**, 1019–1026.

Wong C. S., Capel P. D., and Nowell L. H. (2001) National-scale, field-based evaluation of the biota-sediment accumulation factor model. *Environ. Sci. Technol.* **35**(9), 1709–1715.

Wong C. S., Lau F., Clark M., Mabury S. A., and Muir D. C. G. (2002) Rainbow trout (*Oncorhynchus mykiss*) can eliminate chiral organochlorine compounds enantioselectively. *Environ. Sci. Technol.* **36**(6), 1257–1262.

Woodrow J. E., Seiber J. N., and Baker L. W. (1997) Correlation techniques for estimating pesticide volatilization flux and downwind concentrations. *Environ. Sci. Technol.* **31**(2), 523–529.

Wu J. and Nofziger D. L. (1999) Incorporating temperature effects on pesticide degradation into a management model. *J. Environ. Qual.* **28**, 92–100.

Xu J. M., Gan J., Papiernik S. K., Becker J. O., and Yates S. R. (2003) Incorporation of fumigants into soil organic matter. *Environ. Sci. Technol.* **37**(7), 1288–1291.

Zepp R. G. (1988) Sunlight-induced oxidation and reduction of organic xenobiotics in water. EPA/600/D-88/033. Athens, GA. US Environmental Protection Agency Office of Research and Development, 26p.

Zepp R. G. and Wolfe N. L. (1987) Abiotic transformation of organic chemicals at the particle-water interface. In *Aquatic Surface Chemistry: Chemical Processes at the Particle–Water Interface* (ed. W. Stumm). Wiley, New York, NY, pp. 423–455.

Zepp R. G., Wolfe N. L., Gordon J. A., and Baughman G. L. (1975) Dynamics of 2,4-D esters in surface waters—hydrolysis, photolysis, and vaporization. *Environ. Sci. Technol.* **9**(13), 1144–1149.

Zepp R. G., Schlotzhauer P. F., Simmons M. S., Miller G. C., Baughman G. L., and Wolfe N. L. (1984) Dynamics of pollutant photoreactions in the hydrosphere. *Fresenius Z. Anal. Chem.* **319**, 119–125.

Zhang W., Bouwer E. J., and Ball W. P. (1998) Bioavailability of hydrophobic organic contaminants: effects and implications of sorption-related mass transfer on bioremediation. *Ground Water Mon. Remed.* **18**(1), 126–138.

9.16
The Groundwater Geochemistry of Waste Disposal Facilities

P. L. Bjerg, H.-J. Albrechtsen, P. Kjeldsen, and T. H. Christensen
Technical University of Denmark, Lyngby, Denmark

and

I. Cozzarelli
US Geological Survey, Reston, VA, USA

9.16.1 INTRODUCTION

Landfills of solid waste are abundant sources of groundwater pollution. The potential for generating strongly contaminated leachate from landfill waste is very substantial. Even for small landfills the timescale can be measured in decades or centuries. This indicates that waste dumps with no measures to control leachate entrance into the groundwater may constitute a source of groundwater contamination long after dumping has ceased. In addition to these dumps, engineered landfills with liners and leachate collection systems may *also* constitute a source of groundwater contamination due to inadequate design, construction, and maintenance, resulting in the leakage of leachate.

Landfills may pose several environmental problems (explosion hazards, vegetation damage, dust and air emissions, etc.), but groundwater pollution by leachate is considered to be the most important one and the focus of this chapter. Landfills differ significantly depending on the waste they receive: mineral waste landfills for combustion ashes, hazardous waste landfills, specific industrial landfills serving a single industry, or municipal waste landfills receiving a mixture of municipal waste, construction, and demolition waste, waste from small industries and minor quantities of hazardous waste. The latter type of landfill (termed "old landfills" in this chapter) is very common all over the world. Municipal landfills are characterized by a high content of organic waste that affects the biogeochemical processes in the landfill body and the generation of strongly anaerobic leachate with a high content of dissolved organic carbon, salts, ammonium, and organic compounds and metals released from the waste.

This chapter describes the biogeochemistry of a landfill leachate plume as it emerges from the bottom of a landfill and migrates in an aquifer. The landfill hydrology, source composition, and spreading of contaminants are described in introductory sections. The focus of this chapter is on investigating the biogeochemical processes associated with the natural attenuation of organic contaminants in a leachate plume. Studies have shown that microbial processes and geochemical conditions change over time and distance in contaminant plumes, resulting in different rates of degradation (biotic and abiotic). The availability of electron acceptors, such as iron oxides or dissolved sulfate, is an important factor for evaluating the efficacy and sustainability of natural attenuation as a remedy for leachate plumes. Understanding the complex environments developing in leachate plumes is important in assessing the risk to groundwater resources and for developing cost-effective remediation strategies.

9.16.2 SOURCE AND LEACHATE COMPOSITION

Landfills as pollution sources have three key characteristics:

- they are large often heterogeneous sources both in volume and area;
- they host a mixture of inorganic and organic pollutants; and
- they have an expected pollution potential lasting for decades to centuries.

The area of landfills typically ranges from a few hectares up to more than 50 ha (Christensen *et al.*, 2001). The amount of waste can be enormous ($1\times10^5-5\times10^6\ m^3$).

The presence of a mixture of contaminants in the landfill body affects the overall behavior of the pollution plume, and the interaction between various contaminants. The natural setting of landfills is crucial for understanding the biogeochemistry of landfill leachate plumes. In comparison with other groundwater pollution sources, the complexity of the source means that many different compounds (inorganic and organic) and processes (geochemical and microbial) will be interacting in a landfill leachate plume.

Landfill leachate is generated by excess rainwater percolating through landfilled waste layers. Combined physical, chemical, and microbial processes in the waste transfer pollutants from the waste material to the percolating water. Leachates from common types of landfill that receive a mixture of municipal, commercial, and mixed industrial waste, but excluding significant amounts of concentrated specific chemical waste, may be characterized as a water-based solution containing four groups of pollutants (Christensen *et al.*, 1994).

- Dissolved organic matter, expressed as chemical oxygen demand (COD) or total organic carbon (TOC), including methane, volatile fatty acids (in particular in the acid phase of the waste stabilization; Christensen and Kjeldsen (1989), and more refractory compounds, e.g., fulvic-like and humic-like compounds.
- Inorganic macrocomponents: calcium (Ca^{2+}), magnesium (Mg^{2+}), sodium (Na^{+}), potassium (K^{+}), ammonium (NH_4^{+}), iron (Fe^{2+}), manganese (Mn^{2+}), chloride (Cl^{-}), sulfate (SO_4^{2+}), and bicarbonate (HCO_3^{-}).
- Heavy metals: cadmium (Cd^{2+}), chromium (Cr^{3+}), copper (Cu^{2+}), lead (Pb^{2+}), nickel (Ni^{2+}), and zinc (Zn^{2+}).
- Xenobiotic organic compounds (XOCs) originating from household or industrial chemicals and present in relatively low concentrations in the leachate (usually less than $1\ mg\ L^{-1}$ of individual compounds). These compounds include, among others, a variety of aromatic hydrocarbons, phenols, chlorinated aliphatic hydrocarbons, and pesticides.

Other compounds may be found in leachate from landfills, e.g., borate, sulfide, arsenate, selenate, barium, lithium, mercury, and cobalt. In general, however, these components are not often measured; when they are measured, they are usually present in very low concentrations and are considered only of secondary importance.

Leachate composition varies significantly among landfills, depending on waste composition, waste age, and landfill technology. The waste can be divided into three main groups: household waste, demolition waste, and chemical and hazardous waste.

In nonengineered landfills, the different types of waste are randomly distributed at the site. Historical records are often nonexistent, and investigations are often based on interviews, aerial photos (over time), and excavations. Experiences from Danish sites have shown that combining these methods is an efficient way to develop a good description of the composition of the landfilled waste (Kjeldsen, 1993; Kjeldsen *et al.*, 1998a,b).

Investigations of leachate composition have often been based on only one or a few leachate samples from each landfill. In the context of groundwater pollution, the spatial distribution of the leachate quality must be appreciated. This requires a large number of sampling points (Kjeldsen *et al.*, 1998a). For instance, significant spatial variability in leachate concentrations was observed in wells at the 10 ha Grindsted Landfill (DK) (Kjeldsen *et al.*, 1998a). Very low concentrations of almost all parameters (including XOCs) were found in areas covering ~60–70% of the landfill. A "hot spot," occupying ~10% of the landfill area, was found with concentrations 20–1,000 times higher than in the low-concentration area. Especially for large landfills, information on spatial variability in leachate concentrations is very important as a basis for locating the main sources of the groundwater pollution plume and for selecting cost-effective remedial actions.

Table 1 presents ranges of common leachate parameters compiled from data reported in the literature. The table is based mainly on data from newer landfills (leachate less than 25 yr old). Data from older uncontrolled landfills may exhibit lower values than the minimum values given in the table (Assmuth and Strandberg, 1993; Kjeldsen and Christophersen, 2001). Landfill leachates may contain very high concentrations of dissolved organic matter and inorganic macrocomponents. The concentrations of these components may typically be up to a factor of 10–500 higher than groundwater concentrations in pristine aquifers; however, significant variability is common.

Leachate also contains a broad range of XOCs (Table 2). Since the composition of chemical

Table 1 Composition of landfill leachate. Values in $mg\ L^{-1}$ unless otherwise stated.

Parameter	*Range*
pH (no unit)	4.5–9
Spec. cond. ($\mu S\ cm^{-1}$)	2,500–35,000
Total solids	2,000–60,000
Organic matter	
TOC	30–29,000
Biological oxygen demand (BOD_5)	20–57,000
COD	140–152,000
BOD_5/COD (ratio)	0.02–0.80
Organic nitrogen	14–2,500
Inorganic macrocomponents	
Total phosphorus	0.1–23
Chloride	150–4,500
Sulphate	8–7,750
Hydrogencarbonate	610–7,320
Sodium	70–7,700
Potassium	50–3,700
Ammonium-N	50–2,200
Calcium	10–7,200
Magnesium	30–15,000
Iron	3–5,500
Manganese	0.03–1,400
Heavy metals	
Arsenic	0.01–1
Cadmium	0.0001–0.4
Chromium	0.02–1.5
Cobalt	0.005–1.5
Copper	0.005–10
Lead	0.001–5
Mercury	0.00005–0.16
Nickel	0.015–13
Zinc	0.03–1,000

Source: Kjeldsen *et al.* (2002).

Table 2 The most frequently observed XOCs in landfill leachates. Only pollutants that have been observed in more than three independent investigations are included.

Compound	*Range* ($\mu g\ L^{-1}$)
Aromatic hydrocarbons	
Benzene	0.2–1,630
Toluene	1–12,300
Xylenes	0.8–3,500
Ethylbenzene	0.2–2,300
Trimethylbenzenes	0.3–250
Naphthalene	0.1–260
Halogenated hydrocarbons	
Chlorobenzene	0.1–110
1,2-Dichlorobenzene	0.1–32
1,4-Dichlorobenzene	0.1–16
1,1-Dichloroethane	0.6–46
1,2-Dichloroethane	<6
1,1,1-Trichloroethane	0.1–3,810
Trichloroethylene	0.05–750
Tetrachloroethylene	0.01–250
Dichloromethane	1.0–827
Trichloromethane	1.0–70
Phenols	
Phenol	0.6–1,200
Cresols	1–2,100
Pesticides	
Mecoprop[a]	0.38–150
Phtalates	
Diethylphthalate	10–660
Di-(2-ethylexyl)phtalate	0.6–240
Di-*n*-butylphthalate	0.1–70
Butylbenzyl phtalate	0.2–8
Phosphonates	
Tri-*n*-butylphosphate	1.2–360
Miscellaneous	
Acetone	6–4,400
Camphor[b]	20–260
Fenchone	7–80
Tetrahydrofuran	9–430

Source: Kjeldsen *et al.* (2002).
[a] 2-(2-methyl-4-chlorophenoxy) propionic acid (MCPP).
[b] 1,7,7-trimethyl-bicyclo(-bicyclo[2.2.1]-heptane-2-one.

wastes is highly variable, the compounds identified and their concentration ranges are difficult to summarize in general terms. An important subgroup is that of volatile organic compounds (VOCs); these are organic compounds that tend to vaporize at room temperature and pressure. Typically VOCs are an important component of the compounds in gasoline, lubricants, and solvents. Some VOCs are highly toxic and/or carcinogenic. Aromatic hydrocarbons and chlorinated aliphatic compounds are the most frequently found compounds in landfill leachate. They are also the most common compounds included in analytical programs. Phenols, pharmaceuticals, and pesticides have been found as well, and other ionic or polar compounds (e.g., phthalates and aromatic sulfonates) have been identified in more recent investigations (e.g., Schwarzbauer *et al.*, 2002; Paxeus, 2000).

Several parameters change dramatically as the organic waste in the landfill degrades. During the initial fermentation (acid phase), leachates may have low pH values and high concentrations of many compounds, in particular high concentrations of easily degradable organic compounds such as volatile fatty acids. Later, when the fermentation products are converted effectively to methane and carbon dioxide (methane phase), the pH increases and the degradability of the organic carbon in the leachate decreases. A detailed discussion of the phases which landfills experience is given in Kjeldsen *et al.* (2002).

The data in Tables 1 and 2 are based on leachates from landfills that are younger than 25 yr. The values are difficult to extrapolate beyond the first 25 years of a landfill's life. Belevi and Baccini (1992) estimated by using leaching tests on municipal solid waste that leachate from landfills contains significant concentrations of several compounds for centuries. The leachates contain nitrogen and organic carbon in significant concentrations for several centuries as discussed in detail by Kjeldsen *et al.* (2002).

In summary, leachates contain a variety of compounds (dissolved organic matter, inorganic compounds, heavy metals, and XOCs), due to the mixed nature of the waste in landfills. The spatial variability of the leachate quality is significant, and this can affect the leaching pattern from the landfill and the resulting plumes. A landfill should be seen as a complex source, which is expected to last for decades or even centuries.

9.16.3 SPREADING OF POLLUTANTS IN GROUNDWATER

All compounds in leachate entering an aquifer will be subject to advection and dispersion (dilution) as the leachate mixes with groundwater (Freeze and Cherry, 1979). For nonreactive components, of which chloride is the dominant component, dilution is the only attenuation mechanism. Dilution is the interaction of the leachate flow in the aquifer with the flow of groundwater. Leachate migration should be seen in terms of a three-dimensional plume developing in a three-dimensional geological structure, where gradients, permeabilities, and physical boundaries (geological strata, infiltration, rivers, abstraction wells, etc.) determine the position and migration velocity of the plume. Dilution is governed by macroscopic dispersion and molecular diffusion, but can also be affected by local vertical gradients, leachate density, and, to some extent, viscosity.

Dispersion is the mathematical term in the solute transport equation (Freeze and Cherry, 1979) accounting for dilution or mixing according to concentration gradients. The dispersivity has a longitudinal component (in the flow direction), a vertical component, and a horizontal, transverse component. The longitudinal dispersivity is important only for concentrations at the front of leachate plumes. Of greater interest is the magnitude of the transverse dispersivities, which govern the transverse spreading of the plume. Data from field experiments (review by Gelhar *et al.* (1992), Jensen *et al.* (1993), Petersen (2000), and Adams and Gelhar (1992)) showed small transverse dispersivities (from millimeters to a few centimeters) indicating limited transverse spreading of pollution plumes. Unfortunately, field data showing detailed horizontal delineations of landfill leachate plumes have not been presented since the early investigations of the Borden Landfill (Figure 1). The review of Gelhar *et al.* (1992) also indicates that vertical transverse dispersivities are extremely small, which implies very limited vertical mixing due to dispersion alone. This has been confirmed in a number of landfill studies (e.g., Nicholson *et al.*, 1983; Barker *et al.*, 1986; Lyngkilde and Christensen, 1992b; Bjerg *et al.*, 1995), where steep vertical concentration gradients of chloride have been observed (Figures 1 and 2).

The flow of leachate may, as mentioned, differ physically from that of groundwater at least with respect to the following three aspects.

Local water table gradients below and around the landfill will most likely differ from the general gradients, because the landfill will usually have a different hydrology/hydrogeology compared to the surrounding area. Local water table mounds have been observed at the Borden Landfill (CAN) (MacFarlane *et al.*, 1983), Vejen Landfill (DK) (Kjeldsen, 1993), the Noordwijk Landfill (NL) (van Duijvenbooden and Kooper, 1981), and at the Grindsted Landfill (DK) (Kjeldsen *et al.*, 1998b). The reasons for water mounds are not fully understood (see later discussion in Section 9.16.7.1).

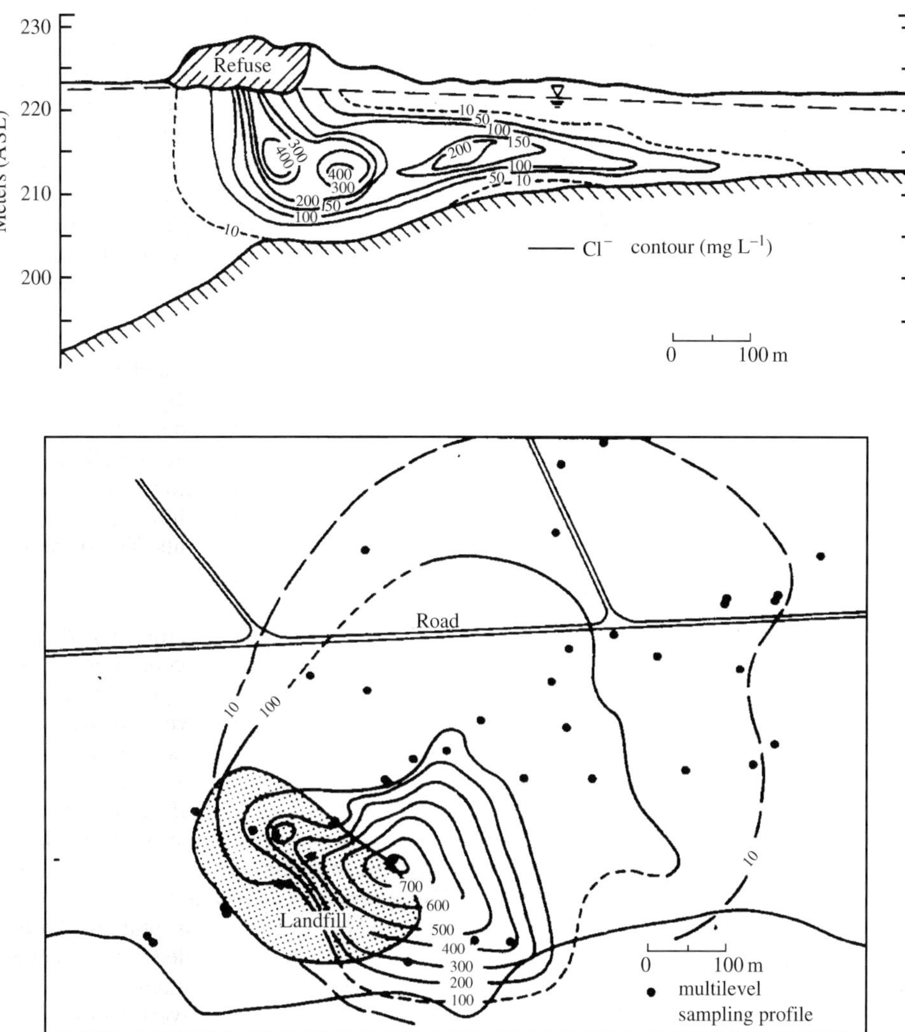

Figure 1 Horizontal and vertical transect contours of Cl^- ($mg\ L^{-1}$) in the Borden Landfill (CAN) (source MacFarlane *et al.*, 1983).

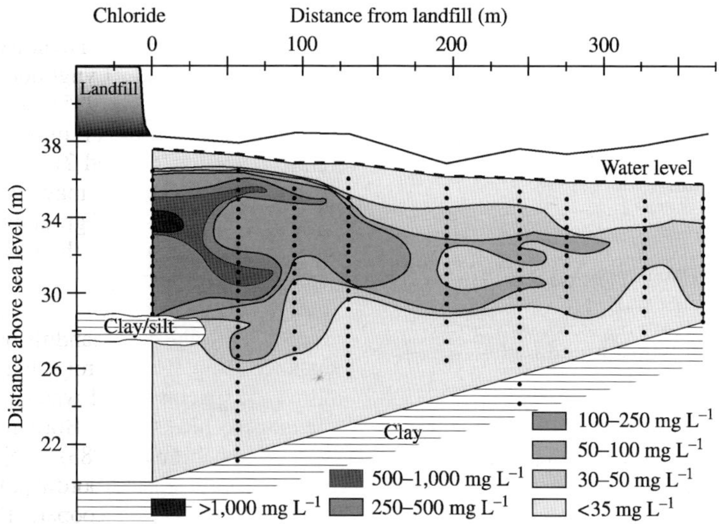

Figure 2 Vertical transect contours of Cl^- ($mg\ L^{-1}$) in the Vejen Landfill (DK) (source Lyngkilde and Christensen, 1992b).

Local mounding effects are enhanced lateral spreading of the leachate plume, and to downward directed hydraulic gradients in the groundwater zone beneath the landfill. The latter can cause an unexpected spreading pattern despite homogeneous aquifer conditions and limited density difference between leachate and the ambient groundwater. The enhanced lateral spreading of the plume may increase the volume of contaminated groundwater and its spatial extent, but provides increased dilution of contaminants.

The viscosity of the leachate may differ from that of the groundwater. A higher viscosity should lead to lower flow velocities, which could influence the dilution of the leachate plume; however, actual field evidence is scarce (Christensen *et al.*, 2001).

The density of the leachate is a function of the temperature and the concentration of dissolved solids. Leachate with a total dissolved solid concentration of 2×10^4 mg L^{-1} is not uncommon (see Table 1); the density of such a leachate is >1% higher than the groundwater density. Density differences may significantly affect the vertical positioning of the plume just below the landfill. Field observations on the downward movement of the plume are often difficult to separate from the effect of local water table mounds (Christensen *et al.*, 2001). A better understanding of the effects of higher leachate densities in field situations is needed. Density effects could be the major cause of vertical leachate spreading in aquifers since "normal" vertical dispersion is usually very small.

In summary, transport and spreading of dissolved landfill leachate pollutants in aquifers are governed by advection/dispersion. The local hydrogeology at the site in terms of water table mounds and seasonal variation in flow field may enhance the spreading horizontally and vertically. Density effects due to high concentrations of inorganic compounds may increase vertical transport, but the understanding of density transport in leachate plumes is poor. Density transport is likely to be most important close to landfills but less so as dilution of the plume increases.

9.16.4 BIOGEOCHEMISTRY OF LANDFILL LEACHATE PLUMES

9.16.4.1 Redox Environments and Redox Buffering

The entry of strongly reduced landfill leachate into a pristine, often oxidized, aquifer, leads to the creation of very complex redox environments. Important processes include organic matter biodegradation, biotic and abiotic redox processes, dissolution/precipitation of minerals, complexation, ion exchange, and sorption. The resulting redox environments strongly influence both the inorganic and organic biogeochemistry of the aquifer, and create the chemical framework for understanding the attenuation processes in the plume.

Characterization of redox environments in pollution plumes has been reviewed by Christensen *et al.* (2000b). A range of different approaches have been used in addressing redox conditions in pollution plumes:

- redox potential,
- redox-sensitive compounds in groundwater samples,
- hydrogen concentrations in groundwater,
- concentrations of volatile fatty acids,
- sediment characteristics, and
- microbial tools.

However, it should be noted that the approaches are not standard, and the value of each approach is still a matter of debate (Christensen *et al.*, 2000b). In addition, the redox conditions in contaminant plumes are not only spatially variable but also temporally variable (McGuire *et al.*, 2000).

In the following, we discuss the redox environments observed in landfill leachate plumes, based on the simplified presentation of redox conditions as given in Figure 3. Later in this chapter it is shown for two case histories (Sections 9.16.5 and 9.16.6) that the actual redox conditions may be somewhat more complex. In an aquifer with a continuous leachate release, a methanogenic zone evolves close to the landfill. Within this zone and downgradient from it, sulfate reduction may take place. Iron reduction takes place further downgradient, where the conditions are less reducing. Zones of manganese and nitrate reduction have been observed, sometimes overlapping the iron-reducing zone. Finally, aerobic conditions may exist in the outskirts of the reduced plume, if the pristine aquifer is oxidized and contains significant amounts of dissolved oxygen (>1 mg L^{-1}).

As illustrated in Figure 3, the content of reduced species (organic matter and ammonium) in groundwater decreases along the flow lines. The redox potential increases with distance. Close to the landfill, dissolved electron acceptors such as oxygen, nitrate, and sulfate are depleted or lowered in concentration. Sulfide may be present due to sulfate-reducing processes. At some distance from the source, the content of reduced dissolved species such as sulfide, ferrous iron, and manganese peaks as a result of redox processes. The composition of the minerals in the aquifer changes with distance, as discussed below. The pollutants leaving the landfill will, unless they are removed from the water, migrate through a series of redox zones, and over time migrate into more oxidizing environments.

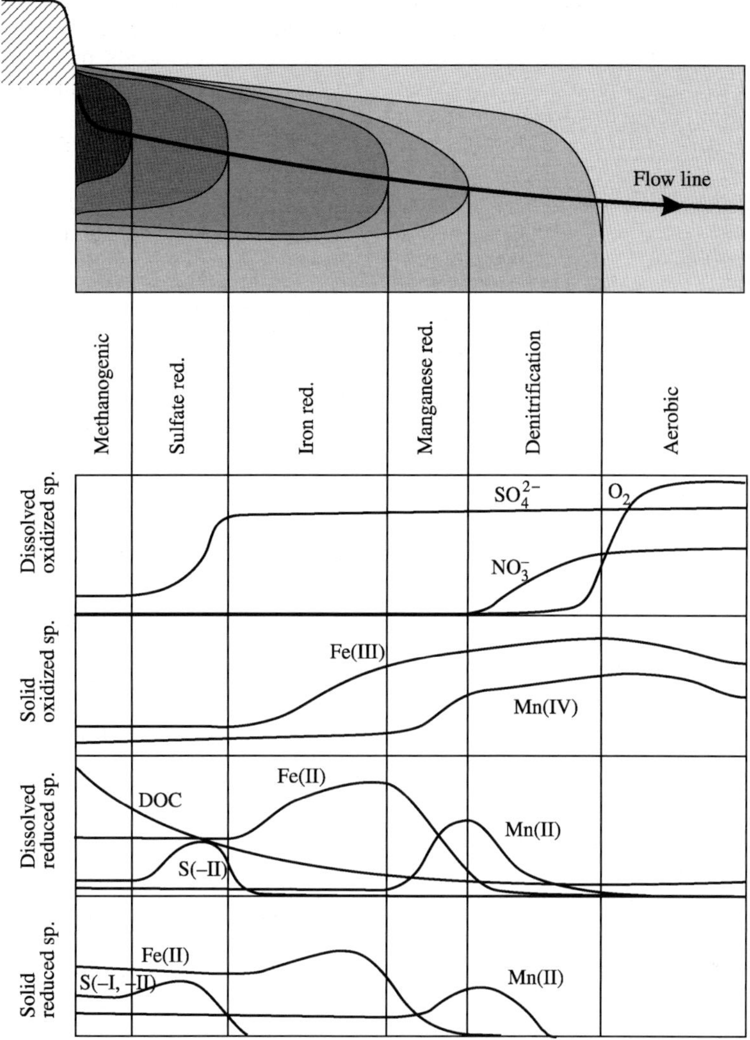

Figure 3 Schematic redox zonation in an originally aerobic aquifer downgradient from a landfill, and the distribution of redox-sensitive species along a streamline in the plume. The axes are not to scale (source Christensen *et al.*, 2001).

Leachate from landfills typically is strongly reduced, rich in organic matter and ammonium, and may be seen as infiltrating water with a great capacity for donating electrons (reduction capacity) during redox reactions. The produced electrons must be accepted by dissolved or solid aquifer electron acceptors. The capacity of the aquifer to accept electrons is called the oxidation capacity (OXC; Scott and Morgan, 1990; Heron *et al.*, 1994a).

The set of reactions that creates the complex redox environments of landfill leachate plumes consists of combinations of two half-reactions: oxidation half-reaction and reduction half-reaction. Table 3 presents the most prominent overall redox reactions, along with their calculated Gibbs free energy change under standard conditions ($\Delta G_0(W)$). The lower (the more negative) the $\Delta G_0(W)$, the more energy is gained, and the more readily the reaction will proceed. Considering the processes of organic matter oxidation, it is evident that when all electron acceptors are present, oxygen will be used first, followed by nitrate, manganese, iron, and sulfate. Finally, methanogenesis and fermentation reactions dominate, when the most favorable electron acceptors are depleted.

Organic matter dominates the reduction capacity of typical leachates (Christensen *et al.*, 2001). Ammonium and methane may also contribute significantly, showing that the fate of these inorganic compounds in the aquifer may also affect the formation of redox environments. The aquifer OXC may be dominated by iron oxides, when calculated for an aquifer volume including aquifer material and groundwater (Table 4). This is caused by the low aqueous solubility of oxygen, and the relatively low nitrate

Table 3 Most prominent redox reactions in landfill leachate plumes. Dissolved organic matter is represented by the model compound CH_2O.

Reaction	*Process*	ΔG_0 *(W)* (kcal mol^{-1})
Methanogenic/fermentative organic matter mineralization	$2CH_2O \rightarrow CH_3COOH \rightarrow CH_4 + CO_2$	−22
Sulfate reduction/OMO	$2CH_2O + SO_4^{2-} + H^+ \rightarrow 2CO_2 + HS^- + 2H_2O$	−25
Iron reduction/OMO	$CH_2O + 4Fe(OH)_3 + 8H^+ \rightarrow CO_2 + 4Fe^{2+} + 11H_2O$	−28
Maganese reduction/OMO	$CH_2O + 2MnO_2 + 4H^+ \rightarrow CO_2 + 2Mn^{2+} + 3H_2O$	−81
Denitrification/OMO	$5CH_2O + 4NO_3^- + 4H^+ \rightarrow CO_2 + 2N_2 + 7H_2O$	−114
Aerobic respiration/OMO	$CH_2O + O_2 \rightarrow CO_2 + H_2O$	−120
CO_2 reduction	$HCO_3^- + H^+ + 4H_2 \rightarrow CH_4 + 3H_2O$	−55
Ammonium oxidation	$NH_4^+ + 2O_2 \rightarrow NO_3^- + 2H^+ + H_2O$	−72
Methane oxidation	$CH_4 + 2O_2 \rightarrow HCO_3^- + H^+ + H_2O$	−196

Source: Christensen *et al.* (2001).
OMO is short for the reaction organic matter oxidation.

Table 4 Oxidation capacity (OXC, milli equivalent per liter of aquifer) calculated for oxidized species in two aerobic aquifers.

Species	*Reduction half-reaction*	*Vejen (DK)*		*Sand Ridge (Illinois, USA)*	
		Content	*OXC* (meq L^{-1})	*Content*	*OXC* (meq L^{-1})
O_2	$O_2 + 4H^+ + 4e^- \rightarrow 2H_2O$	10 mg L^{-1}	0.44	9 mg L^{-1}	0.39
NO_3^-	$NO_3^- + 6H^+ + 5e^- \rightarrow 1/2N_2 + 3H_2O$	15 mg L^{-1}	1.9	0.95 mg L^{-1}	0.12
Mn(IV) (sediment)	$MnO_2 + 4H^+ + 2e^- \rightarrow Mn^{2+} + 2H_2O$	0.1 mg g^{-1}	6	0.39 mg g^{-1}	23
Fe(III) (sediment)	$FeOOH + 3H^+ + e^- \rightarrow Fe^{2+} + 3H_2O$	2 mg g^{-1}	60	6.8 mg g^{-1}	200
SO_4^{2-}	$SO_4^{2-} + 9H^+ + 8e^- \rightarrow HS^- + 4H_2O$	40 mg L^{-1}	1.2	36 mg L^{-1}	1.1

The calculations are based on the shown contents of oxidized species, the proposed reduction half reactions, and assumed physical parameters: Porosities of 0.35 and bulk densities of 1.6 kg L^{-1}. The potential contributions from CO_2 and natural organic matter were not evaluated.
Source: Christensen *et al.* (2001).

and sulfate content of aquifers. The actual importance of the dissolved electron acceptors can, however, not be evaluated solely from an aquifer volume. In a flow system the mixing of electron acceptors at the fringes of the plume will be critical as well. Mixing at the fringes will be governed by the transverse dispersivities (transverse and vertical). Seasonal recharge may play a role as well; this was discussed by McGuire *et al.* (2000) for a mixed contaminant plume.

The relative importance of fringe and core processes will also depend on the degradability of the electron donor. In a phenol plume, Thornton *et al.* (2001) showed that the consumption of aqueous oxidants greatly exceeded that of the mineral oxidants. This was in part because very high phenol concentrations limited degradation inside the plume and partly because the iron reduction potential was small in the sandstone aquifer. This is not expected to be the case in a landfill leachate plume in sandy aquifers, as the major electron donor (organic matter) is degradable by iron reduction and there will probably be a large iron reduction capacity. Also, solid manganese oxides contribute to the OXC, as they can be reduced to dissolved manganese. However, when long-term aquifer changes are in question, iron reduction is likely to dominate, since iron concentrations typically are 20–50 times higher than manganese concentrations in aerobic glaciofluvial sediments (Heron, 1994).

Not all iron oxides are available for reduction. Some iron minerals are solid crystals or even entire iron grains, which makes them resistant to microbial reduction (Lovley, 1991; Postma, 1993; Heron *et al.*, 1994b). Other iron oxides or hydroxides are amorphous and readily reducible. Over time, even some crystalline minerals such as goethite and hematite may be reduced in the complex environment in leachate (Heron and Christensen, 1995). This indicates that the importance of iron as a redox buffer controlling the size of plumes is not given just by the amount of iron oxides present. The composition and microbial availability of iron for reduction are key parameters. Methods for the actual quantification of the microbial iron reduction capacity have, however, not been developed.

Reactive fractions have been addressed by mild chemical extractions (hydrochloric acid or ascorbic acid), but this is only an operationally defined quantity of easily dissolved, oxidized minerals. The actual pool of microbial available iron reduction capacity may be better determined using microbial assays.

The reduction of iron oxides and precipitation of the reduced metals as carbonates or sulfides changes the composition of the solids along a flow line (Figure 3). Overall, the mineral-bound iron oxides are reduced into dissolved ferrous iron, which partly precipitate and partly migrate downgradient into more oxidized zones. When it meets oxygen, and perhaps also manganese oxides, ferrous iron is oxidized and precipitates as amorphous iron hydroxides. The newly precipitated hydroxides form a very reactive and accessible electron acceptor. The migrating part of the reduced iron thus contributes to a regeneration of OXC further away from the landfill. This may be essential in controlling the size of the reduced zones, especially if the plume expands. The substantial buffering by iron oxides, thus, is related to consumption of OXC and the buildup of reduced species in the strongly reduced part of the plume. Overall, iron acts to minimize the size of the plume by the redox buffering reactions, thus greatly retarding the migration of the reduced leachate and associated problematic compounds (Heron, 1994).

9.16.4.2 Microbial Activity and Redox Processes

Inside a landfill leachate plume the environment is characterized by the presence of reduced species and high concentrations of dissolved organic matter. This environment is partly due to the composition of the leachate from the landfill and partly due to microbial processes in the plume. Since this environment is very different from that of uncontaminated, oligotrophic, often aerobic aquifers surrounding the plume, the composition of the microbial population of the plume is dramatically different from the indigenous microbial population of the uncontaminated aquifer.

Microbial populations in landfill leachate-contaminated aquifers are dominated by bacteria (eubacteria and archaea), as shown by analysis of the phospho lipid fatty acids (PLFAs) (Ludvigsen *et al.*, 1999). The total number of bacteria reported in landfill leachate plumes are in the range of $4 \times 10^4 - 1.5 \times 10^9$ cells g^{-1} dry weight (dw) and the number of colony-forming units, living cells, are in the range of $60 - 10^7$ CFU g^{-1} dw (Christensen *et al.*, 2001). However, the large variation caused by different analytical methods and the fact that different types of aquifers were studied mask the difference in the number of bacteria inside and outside the plume. The total number of bacteria in the aquifer downgradient from the Grindsted Landfill (DK) was fairly constant with distance from the landfill, and the ATP content (an estimate of living cells) showed no significant trend (Ludvigsen *et al.*, 1999). In contrast, the number of living bacteria estimated by the PLFA concentration was higher close to the landfill than further away from the landfill. From the measurements of ATP and PLFA, the viable biomass ranged from 10^4 viable cell g^{-1} dw to 10^6 viable cell g^{-1} dw (Ludvigsen *et al.*, 1999), clearly demonstrating the presence of a significantly viable population.

The microbial community structure in the water phase in the landfill leachate plume is clearly different from the community structure outside the plume as shown in the Banisveld aquifer in the Netherlands (Röling *et al.*, 2001) by 16S ribosomal DNA-based denaturing gradient gel electrophoresis (DGGE). Members of the β-subclass of the class *Proteobacteria* dominated upstream of the landfill, but this group was not encountered beneath the landfill where gram-positive bacteria dominated. Further downstream, where the effect of contamination decreased, the community structure shifted partly back; the contribution of the gram-positive bacteria decreased and the β-proteobacteria reappeared. However, the contribution of δ-protoebacteria also increased strongly, and the β-proteobacteria found here (*Acidovorax*, *Rhodoferax*) differed considerably from those found upstream (*Gallionella*, *Azoarcus*) so the community structure continued to be affected by the contamination. Surprisingly, this relationship was not evident in sediment samples, where the major part of the microbial population is present, either because leachate has had little impact on the microorganisms associated with the $10^4 - 10^5$ yr old sediments (Röling *et al.*, 2001), or because the microorganisms in the water phase of the plume were mainly derived from the landfill.

The ability of microbial populations to use different organic substrates under anaerobic conditions was investigated by Röling *et al.* (2000). The contaminated samples were able to use a higher number of substrates than the samples from upstream and downstream. This pattern was observed in water samples as well as in sediment samples, but the populations in sediment samples were able to use up to three times more substrate than populations in water samples (Röling *et al.*, 2000).

Microbial redox processes. Different metabolic types of bacteria (denitrifiers, manganese reducers, iron reducers, sulfate reducers, and methanogens) occur at landfill sites. Some of these metabolic types (sulfate reducers and methanogens) can be separated into physiological groups which use different carbon substrates as observed in the Norman Landfill (USA) leachate-contaminated aquifer (Beeman and Suflita, 1987, 1990; Harris *et al.*, 1999). The microbial population changed in composition throughout the plume at the Grindsted Landfill (DK). Methanogens and sulfate reducers were abundant close to the landfill, but their numbers decreased in the more distant parts of the plume (Ludvigsen *et al.*, 1999). The iron-, manganese-, and nitrate reducers constituted a surprisingly high fraction of the total cell numbers and their abundance varied little with distance. The ubiquitous presence of these groups of redox specific populations provided the aquifer with a substantial potential for the several redox processes. Therefore, the dominance of one redox process reflects mere the environment and the available electron acceptors than the composition of the microbial population.

The potential for microbially mediated redox processes has been documented in several landfill leachate plumes (Acton and Barker, 1992; Albrechtsen and Christensen, 1994; Nielsen *et al.*, 1995a; Ludvigsen *et al.*, 1998; Cozzarelli *et al.*, 2000). Bioassays of unamended groundwater and sediment samples verified the presence of the following metabolic activities in the Grindsted Landfill (DK) plume: denitrification, iron reduction, manganese reduction, sulfate reduction, and methane production (Ludvigsen *et al.*, 1998). Iron reduction, sulfate reduction, and methane production were also observed at the Norman Landfill (Cozzarelli *et al.*, 2000). Examples of bioassays in different locations within the leachate plume at Grindsted Landfill are shown in Figure 4. The rates for the different processes can be estimated from such incubations, but representative rates cannot be selected. This is due to the fact that only few data exist for landfill sites

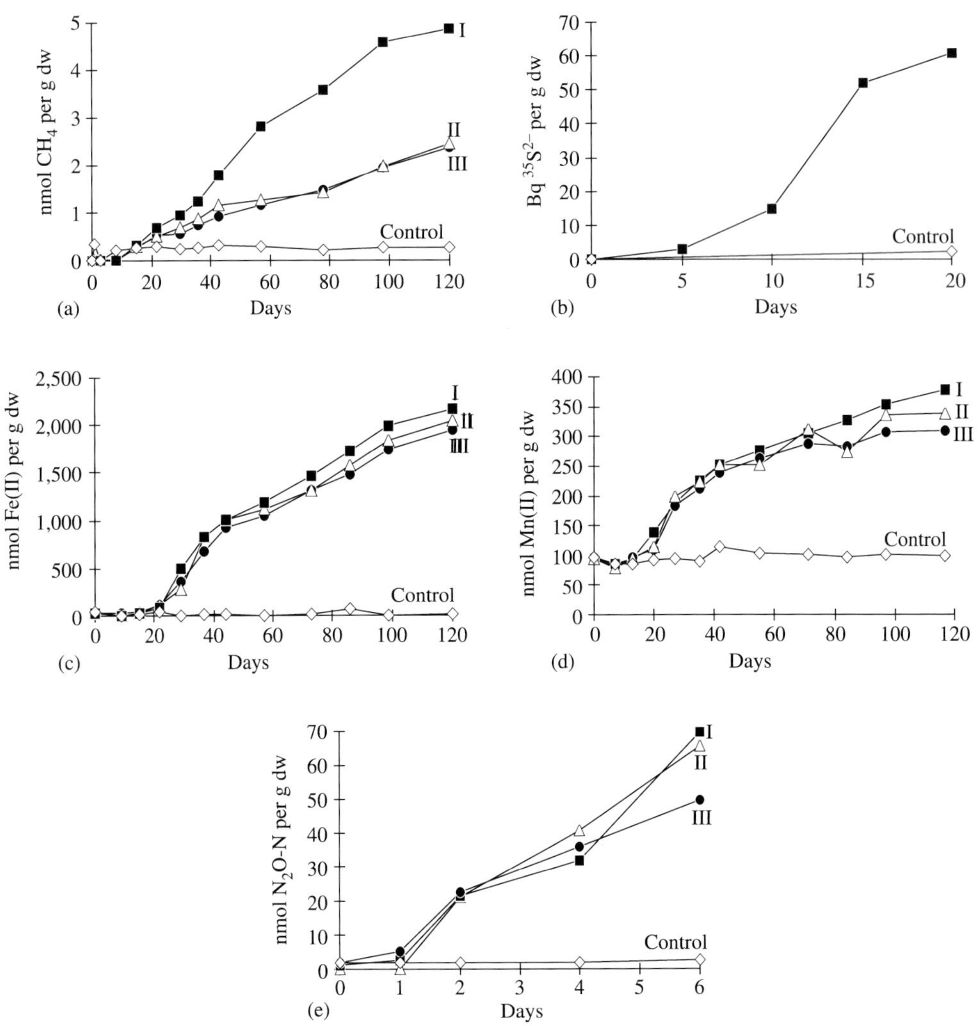

Figure 4 Unamended bioassays showing evidence of each redox reaction from the Grindsted Landfill (DK). The bioassays are performed with sample material from different locations in the plume representing different redox conditions (source Ludvigsen *et al.*, 1998).

(Ludvigsen *et al.*, 1998). That very large variations exist has also been pointed out by McGuire *et al.* (2002), who compared rates in different environments. Several different microbially mediated redox processes occur concomitantly in such microbial assays (Ludvigsen *et al.*, 1998; Cozzarelli *et al.*, 2000), and microbially mediated redox processes do not exclude each other. This somewhat conflicts with a simple thermodynamic model based on Gibbs free energy (Table 3). However, in each sample one electron-accepting process accounted for more than 70% of the equivalent carbon conversion when the measured rates of the electron-accepting processes were used to calculate the carbon conversion of organic matter to carbon dioxide (assuming oxidation level zero of the carbon in the organic matter) (Ludvigsen *et al.*, 1998). This suggests that the concept of redox zones makes sense in terms of dominating redox levels, but that other redox processes may be taking place simultaneously (to be further discussed in Section 9.16.7.2). This may have further implications for the potential of a redox zone to degrade trace amounts of organic chemicals (Rügge *et al.*, 1999a; Albrechtsen *et al.*, 1999).

These microbially mediated redox processes utilize electron acceptors and produce reduced species. This will generate more reduced environments as long as there are electron donors available. The microbial population thus strongly affects their environment in the core of the plume. At the boundaries of the plume, complex microbial communities may exist, and steep redox gradients are created when dissolved electron acceptors are consumed. In addition, reoxidation of sulfides or ferrospecies by oxygen diffusing into the plume may increase the concentration of sulfate and ferric iron, which can stimulate sulfate and iron reduction in these zones as observed at Norman Landfill (Cozzarelli *et al.*, 2000).

In summary, significant numbers of bacteria, detected with several different approaches, are present in landfill leachate plumes. Methanogens, sulfate reducers, iron reducers, manganese reducers, and denitrifiers are believed to be widespread in leachate plumes. Microbial activity seems to occur throughout leachate plumes, although the actual activity (as measured by ATP, PLFA, and redox processes) is low compared to activity in topsoil. Several redox processes can take place in the same samples adding additional diversity to the concept of redox zones illustrated in Figure 3.

9.16.5 OVERVIEW OF PROCESSES CONTROLLING FATE OF LANDFILL LEACHATE COMPOUNDS

The four major compound classes in landfill leachates are dissolved organic carbon, inorganic compound, heavy metals, and XOCs. Typical levels in leachate are reported in Tables 1 and 2. In this section, the fate of these compounds in landfill leachate plumes is summarized. The emphasis is on XOCs, but in order to provide an overview all compound classes are included. A detailed discussion of the fate of individual compounds in landfill leachate plumes can be found in Christensen *et al.* (1994, 2001). A comprehensive investigation of the distribution and geochemistry of inorganic macrocomponents at the Borden Landfill (CAN) is presented by Nicholson *et al.* (1983).

9.16.5.1 Dissolved Organic Matter, Inorganic Macrocomponents, and Heavy Metals

Dissolved organic carbon. Volatile fatty acids constituting a substantial fraction of the dissolved organic carbon in acid-phase leachates are easily degraded according to reported laboratory studies. The dissolved organic matter dominating the methanogenic leachate does not sorb to any substantial degree onto aquifer material and seems fairly recalcitrant with respect to microbial degradation as seen in laboratory experiments (Kjeldsen, 1986). However, with respect to degradation of recalcitrant organic matter, laboratory experiments with short retention times, unstable redox conditions, and limited time for microbial adaptation may fail to simulate the conditions in a leachate polluted aquifer. Observations of actual leachate plumes are usually too limited to provide insight into the fate of the dissolved organic matter. One exception is the report by Lyngkilde and Christensen (1992a) that demonstrated substantial degradation of dissolved organic matter in the anaerobic part of the leachate plume at Vejen Landfill (DK). The observations by DeWalle and Chian (1981) and Rügge *et al.* (1995) may support this, indicating that dissolved organic matter in methanogenic leachate is degradable to a large extent. Brun *et al.* (2002) quantified the degradation of organic carbon in the Vejen Landfill leachate plume. They found half-lives of 100 d in the anaerobic parts of the plume and 1–2 d in the aerobic zone. The anaerobic degradation rate is comparable to that of Sykes *et al.* (1982), who estimated an anaerobic half-life at the Borden Landfill (CAN) of 400 d.

Inorganic macrocomponents. Anions in leachate plumes are mainly important due to their ability to form complexes, take part in dissolution/precipitation processes, and their role as electron acceptors. The formation of complexes may increase the mobility of cations and heavy metals. In addition, many reactions are influenced by pH, which is governed, to a large extent, by the carbonic acid components, in particular HCO_3^-.

The sulfur compounds involved in the sulfate-reduction process are of interest, but the prevalence of sulfate reduction in leachate plumes is not very well understood (Murray *et al.*, 1981). Nitrate and perhaps sulfate as well may be depleted in the core of the plume, but they are certainly significant players at plume boundaries.

The attenuation of cations is primarily governed by cation-exchange processes in addition to dilution (Nicholson *et al.*, 1983). Calcium and magnesium are also influenced by complexation and dissolution/precipitation processes. The attenuation of ammonium and potassium due to cation-exchange processes is significant, while sodium takes little part in cation-exchange processes. Calcium and in some cases also magnesium, typically dominate the cation-exchange complex, can be released and moved to the front of the leachate plume (Kehew *et al.*, 1984).

Ammonium seems to be significantly attenuated in the anaerobic part of the plume as indicated by the detailed investigation of the Grindsted Landfill (DK) leachate plume. The ammonium plume is of limited extent. It is followed by zones of increasing concentration of nitrate and dinitrogen oxide, but the attenuation mechanisms of ammonium are not understood. This issue deserves further research, as ammonium may be seen as one of the critical compounds in landfill leachate plumes (Christensen *et al.*, 2000a).

Dissolved iron and manganese in the leachate are subject to precipitation as sulfides or carbonates, ion exchange, oxidation, and dilution (Nicholson *et al.*, 1983). These processes tend to lower the aqueous concentration of iron and manganese along the flow lines, but reduction of sediment-associated iron and manganese oxides may increase their concentrations further out in the plume (Albrechtsen and Christensen, 1994). Sometimes they are often supersaturated with respect to carbonates (Jensen *et al.*, 2002). Organic complexation of iron and manganese seems only of modest importance. Further down-gradients in the landfill redox potentials are higher, so that iron and manganese may reprecipitate as oxides.

Heavy metals. The behavior of heavy metals in a landfill leachate plume is simultaneously controlled by sorption, precipitation, and complexation, and proper evaluations of metal attenuation must account for this complex system. Generally, heavy metals do not constitute a groundwater pollution problem at landfills (Arneth *et al.*, 1989), because landfill leachates usually contain only modest heavy metal concentrations, and the metals are subject to strong attenuation by sorption and precipitation in the landfill itself (Kjeldsen *et al.*, 2002). Sulfide-producing conditions result in extremely low solubility of the heavy metals (Bisdom *et al.*, 1983). The presence of colloidal as well as organically complexed metals enhances solubilities and mobilities (Christensen *et al.*, 1996), but apparently not to the extent that the metals migrate appreciably in leachate plumes.

9.16.5.2 Xenobiotic Organic Compounds

Sorption. In aquifers characterized by low organic carbon content, most of the XOCs found in leachate plumes are only weakly attenuated by sorption. This applies to the aromatic hydrocarbons, chlorinated hydrocarbons, and the polar compounds. Very few detailed sorption studies involving landfill leachate have been reported (Kjeldsen *et al.*, 1990; Larsen *et al.*, 1992). Preliminary evidence suggests that the presence of leachate, in particular in terms of dissolved organic carbon, does not affect the sorption of XOCs significantly and as such the traditional methods for estimating retardation in aquifers are valid.

The chlorinated aliphatic hydrocarbons are frequently identified in landfill leachate. Adriaens *et al.* (see Chapter 9.14) reviewed their presence and fate in the environment. The chlorinated aliphatic hydrocarbons generally degrade by reductive dechlorination under anaerobic conditions (Vogel *et al.*, 1987). The understanding of this process has increased during the last few years, and it is now generally accepted that the chlorinated compounds can act as electron acceptors (halorespiration; Holliger *et al.* (1999)). This also means that the reductive dechlorination process requires an electron donor such as naturally occurring carbon, petroleum hydrocarbons, or organic carbon in landfill leachate. The greater availability of naturally occurring carbon likely contributes to the extremely rapid reductive dechlorination of chlorinated ethanes and ethenes that was observed in wetland sediments compared to sand aquifers (Lorah and Olsen, 1999a,b; Lorah *et al.*, in press). Degradation of higher chlorinated aliphatic compounds such as tetrachloroethylene (PCE) and trichloroethylene (TCE) under aerobic condition has not been demonstrated, but degradation products such as dichloroethylene (DCE) isomers (primarily *cis*-1,2-DCE) can be oxidized to carbon dioxide. Recent microcosm studies with aquifer and streambed sediments indicate that oxidation of *cis*-DCE and vinyl chloride (VC) to carbon dioxide may also be possible under manganese and iron-reducing conditions (see a review by Bradley (2000)), although the occurrence and significance of this reaction in contaminant plumes has not been demonstrated. In microcosm and enrichment experiments with wetland sediments, Jones *et al.* (2002) found that addition of Fe(III) either as amorphous FeOOH or as Fe(III)NTA slowed dechlorination of chlorinated ethanes and inhibited degradation of *cis*-DCE, *trans*-DCE, and VC. Degradation of the chlorinated ethanes and ethenes was most rapid under methanogenic conditions (Jones *et al.*, 2002; Lorah *et al.*, in press).

Information obtained from different plumes is in accordance with observations from landfill leachate plumes, where the expectation is that PCE and TCE will be reductively dechlorinated. This is supported by observations of lower chlorinated compounds, DCE and VC, in leachate plumes or degradation experiments. The transformation of chlorinated ethenes has been observed under various redox conditions ranging from methanogenic to nitrate-reducing conditions (Nielsen *et al.*, 1995a; Johnston *et al.*, 1996; Bradley, 2000; Rügge *et al.*, 1999a). Tetrachloromethane will rapidly degrade in landfill plumes by reduction with sediment-associated iron and organic carbon (Rügge *et al.*, 1999a; Pecher *et al.*, 1997). The transformation of 1,1,1-TCA in anaerobic environments is rapid and seems to be affected by both abiotic and biotic degradation processes (Bjerg *et al.*, 1999; Nielsen *et al.*, 1995b). Studies of full-scale landfill plumes have shown a significant degradation of TCE (Chapelle and Bradley, 1998) and 1,1-DCA (Ravi *et al.*, 1998). However, in the case of 1,1-DCA, complete dechlorination to ethane was not shown. In summary, the information available suggests that landfill plumes host redox environments, microorganisms, and/or geochemical processes that can effectively attenuate chlorinated aliphatic hydrocarbons.

The aromatic hydrocarbons generally degrade readily under aerobic conditions, but anaerobic degradation by pure bacterial cultures has also been recognized (see Chapter 9.12; Heider *et al.*, 1999). The vast amount of data from natural attenuation studies of petroleum hydrocarbon plumes generally supports anaerobic degradation, especially for benzene, toluene, ethylbenzene, xylenes (BTEX) under field conditions. The first-order degradation rates observed under unspecified anaerobic conditions (Suarez and Rifai, 1999) are typically one or two orders of magnitude lower than rates reported under aerobic conditions (Nielsen *et al.*, 1996).

Detailed observations in landfill leachate plumes have indicated the degradation mainly of toluene, xylenes, and C3–C5 benzenes (Barker *et al.*, 1986; Lyngkilde and Christensen, 1992b; Rügge *et al.*, 1995; Eganhouse *et al.*, 2001). These studies use tracers or compound ratios to rule out dilution and sorption; however, direct proof of degradation has not been provided. Isotopic ratios have been introduced as a powerful tool for identification of degradation. The Vejen Landfill (DK) site was revisited after 10 yr (Baun *et al.*, 2003; Richnow *et al.*, 2003) and evidence of degradation was provided for ethylbenzene and *m*/*p*-xylene by using the isotopic ratio $^{13}C/^{12}C$ in aromatic hydrocarbons. A specific degradation product, benzyl succinic acid (see a review by Beller (2000)), was observed. This documents the degradation of toluene in the plume. Positive results regarding the degradation of benzene are few. Baun *et al.* (2003) concluded that benzene was persistent in the anaerobic part of the Vejen Landfill (DK) plume. Ravi *et al.* (1998) proved that benzene degradation took place in the very long plume at West KL Landfill (USA).

The degradability of toluene and xylenes observed in plumes has been supported by experimental evidence from field and laboratory experiments (Acton and Barker, 1992; Bjerg *et al.*, 1999; Nielsen *et al.*, 1995a; Johnston *et al.*, 1996; Rügge *et al.*, 1999a), while the recalcitrance of benzene has been shown in experiments by the same authors. This adds to the belief that benzene is less readily degradable than most of the other BTEXs under strongly anaerobic conditions in landfill leachate plumes.

The phenolic compounds generally degrade readily under aerobic conditions. Information for anaerobic conditions is mixed, and no distinct pattern has emerged. Studies indicate persistence of phenol, *o*-cresol, 2,4-dichlorophenol, and 2,6-dichlorophenol under iron-reducing and nitrate-reducing conditions (Nielsen *et al.*, 1995a). Grbic-Galic (1990) reviewed the methanogenic transformation of phenolic and aromatic compounds in aquifers in more general terms, and reported the transformation of several phenols.

The pesticides are another important group of pollutants (see Chapter 9.15). Many different herbicides have been identified in landfill leachate; however, very little is known about pesticide degradation potentials in leachate plumes. Mecoprop is frequently observed in leachates (Table 2). At the Vejen Landfill (DK), mecoprop was observed in the plume 130 m downgradient of the landfill at a concentration of 95 $\mu g\ L^{-1}$ (Lyngkilde and Christensen, 1992b). Baun *et al.* (2003) showed in a revisit to the site that MCPP was recalcitrant in the anaerobic part of the plume up to 135 m from the landfill. In an injection experiment in the Grindsted Landfill (DK) plume Rügge *et al.* (1999b) found that atrazine and mecoprop were recalcitrant under strongly anaerobic conditions. Anaerobic dechlorination of phenoxy acids has been proposed. Rügge *et al.* (1995) identified phenoxy acids resembling known herbicides, though, without the chlorine atoms attached in landfill leachate-affected groundwater. Tuxen *et al.* (in press) proposed that a significant part of the phenoxy acids at the Sjoelund Landfill (DK) disappeared due to degradation in the interphase between the anaerobic leachate plume and the surrounding aerobic aquifer. This is consistent with the expected aerobic degradation of phenoxy acids (Broholm *et al.*, 2001). In conclusion, studies on pesticide degradation in different anaerobic environments are few, and due to their general recalcitrance in groundwater environments (Albrechtsen *et al.*, 2001), pesticides may turn out to be critical compounds in landfill leachate plumes.

Recently, information on XOC degradation in different landfill plume redox environments has been expanded. As more results become available, more XOCs are found to be degradable in the intermediate redox zones dominated by sulfate, iron, and nitrate reduction. Transformation of chlorinated aliphatics seems to occur not only under methanogenic conditions, but also in less reducing zones. Aromatic hydrocarbons degrade readily in aerobic environments, but only slowly in reducing environments. Benzene may, in particular, be recalcitrant under strongly reducing conditions. As our ability to perform degradation studies has improved significantly since the 1990s, more detailed information on complex degradation patterns and compound interactions has been revealed. Also, the use of isotopic ratios and the identification of specific degradation products has improved our ability to demonstrate degradation under field conditions. Several compounds have been shown to disappear in plumes, but direct evidence of microbial degradation has only been established for some of these XOCs. Finally, the frequent occurrence of pesticides and the presence of more polar compounds also call for a greater focus on these compounds.

9.16.6 NORMAN LANDFILL (USA)

The Norman Landfill Research Site is a closed municipal solid waste landfill, formerly operated by the city of Norman, Oklahoma. The site is a research site of the Toxic Substances Hydrology Program of the US Geological Survey (USGS). Scientists from the USGS, the University of Oklahoma, the US EPA, and numerous universities have installed wells and instruments to investigate the chemical, biological, and hydrologic processes in groundwater and surface water affected by landfill leachate. At Norman Landfill, a combined geochemical and microbiological approach has been used to identify the important biogeochemical processes occurring in the aquifer contaminated by leachate from the Norman Landfill (Cozzarelli *et al.*, 2000; Harris *et al.*, 1999).

9.16.6.1 Source, Geology and Hydrogeology

The field site is located in the alluvium of the Canadian River. The landfill has received solid waste since 1922. The waste was dumped into trenches that were ~3 m deep and that contained water to a depth of 1.5–2.5 m because of the shallow water table; the waste was subsequently covered with 15 cm of sand. No restrictions were placed on the type of material dumped at the landfill. In 1985 the landfill was closed, and the mounds, which had reached a maximum height greater than 12 m, were covered with local clay and silty-sand material. The Canadian River alluvium is 10–12 m thick and consists predominantly of sand and silty sand, with interbedded mud and gravel. Near the landfill the water table is less than 2 m below the land surface. The hydraulic conductivity of subsurface materials near the landfill, which was measured using slug tests, ranged from 7.3×10^{-2} m d^{-1} to 24 m d^{-1} (Scholl *et al.*, 1999). A discontinuous, low hydraulic-conductivity interval that consisted of silt and clay was detected ~3–4 m below the water table along the transect (A–A′) where most data have been collected (Figure 5). A high hydraulic-conductivity layer containing coarse sand and gravel is located near the base of the alluvium at depth of ~10–12 m. Low-permeability shales and siltstones of the Hennessey Group of Permian age underlie the alluvium, and act as a boundary to vertical groundwater flow. A shallow stream with areas ponded by beaver dams (referred to here as sloughs), which is ~0.75 m deep, lies roughly parallel to the landfill and ~100 m to the southwest (Figure 5). Groundwater flows from the landfill toward the slough and Canadian River.

9.16.6.2 Landfill Leachate Plume

Biogeochemistry of plume. Chloride and nonvolatile dissolved organic carbon (NVOC) profiles along the well 35–80 transect confirm that the plume extends through the entire thickness of the alluvium between the landfill and the slough and has migrated beneath the slough (Figures 6(a) and (b)). Groundwater downgradient from the landfill has high concentrations of NVOC compared to groundwater collected upgradient and northeast of the landfill, where the average concentration is <0.2 mM (Cozzarelli *et al.*, 2000). The high NVOC concentrations result from the dissolution and partial degradation of organic waste in the landfill. The maximum concentration of NVOC (17 mM) was measured close to the landfill edge.

Degradation of organic compounds to inorganic carbon compounds is shown by the increase in

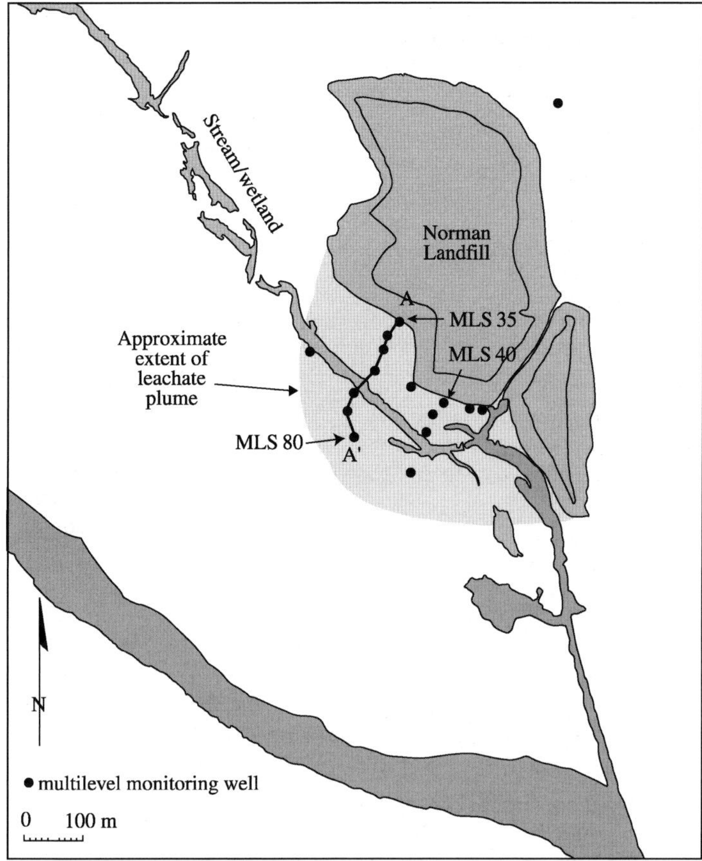

Figure 5 Map showing the Norman Landfill (USA) site. Transect A–A′ runs from MLS 35 to MLS 80 in the direction of groundwater flow. The approximate extent of the plume was determined from geophysical measurements and specific conductance measurements.

alkalinity (up to 7 times the background concentration) of groundwater downgradient from the landfill resulting in an alkalinity plume similar in shape to the NVOC plume (Figure 6(c)). Background water pH values ranged from 6.9 to 7.3, whereas values throughout the high chloride plume along transect A–A′ (Figure 5) ranged from 6.7 to 7.3. Degradation processes in the leachate-contaminated aquifer have resulted in the depletion of oxidized chemical species (such as O_2 and SO_4^{2-}) and the accumulation of reduced products (such as Fe^{2+}) in the groundwater (Figure 6(d)). The entire area that was sampled between the landfill and the slough is anaerobic (<5 μM dissolved oxygen). Aerobic respiration is of limited importance in most of the contaminated aquifer. High levels of ammonium are present in the groundwater immediately downgradient from the landfill (>5 mM, Figure 6(e)) and nitrate levels in the contaminated aquifer and in the background water are very low (<0.1 mM). The large NH_4^+ plume downgradient from the landfill is probably due to the transport of NH_4^+ produced during the fermentation of organic matter within the refuse mounds. At the boundary of the plume the concentration of NH_4^+ decreases sharply. Ammonium bound by cation exchange onto sediment or oxidation along the plume boundary may control the transport of ammonium. Periodically, nitrate concentrations as high as 2.7 mM were measured in 1999 during monthly sampling of water table wells. NH_4^+ oxidation probably occurs in this zone.

High concentrations of Fe^{2+} in the groundwater (>0.25 mM, Figure 6(d)) are consistent with microbial dissolution of iron oxides in the aquifer. Reactive iron oxides were depleted from sediments within the contaminated region of the aquifer (Harris *et al.*, 1999). Hematite grain coatings remained on all samples collected within the anaerobic plume (Breit *et al.*, 1996) and may be available as a source of electron acceptors for iron-reducing bacteria. Nevertheless, the persistence of hematite relative to iron oxyhydroxides in the Norman Landfill plume suggests that the microbial reduction of hematite, if it occurs at all, is very slow.

Sulfate reduction in the plume has resulted in the depletion of SO_4^{2-} from the center of the plume (Figure 6(f)). Dissolved sulfate at the

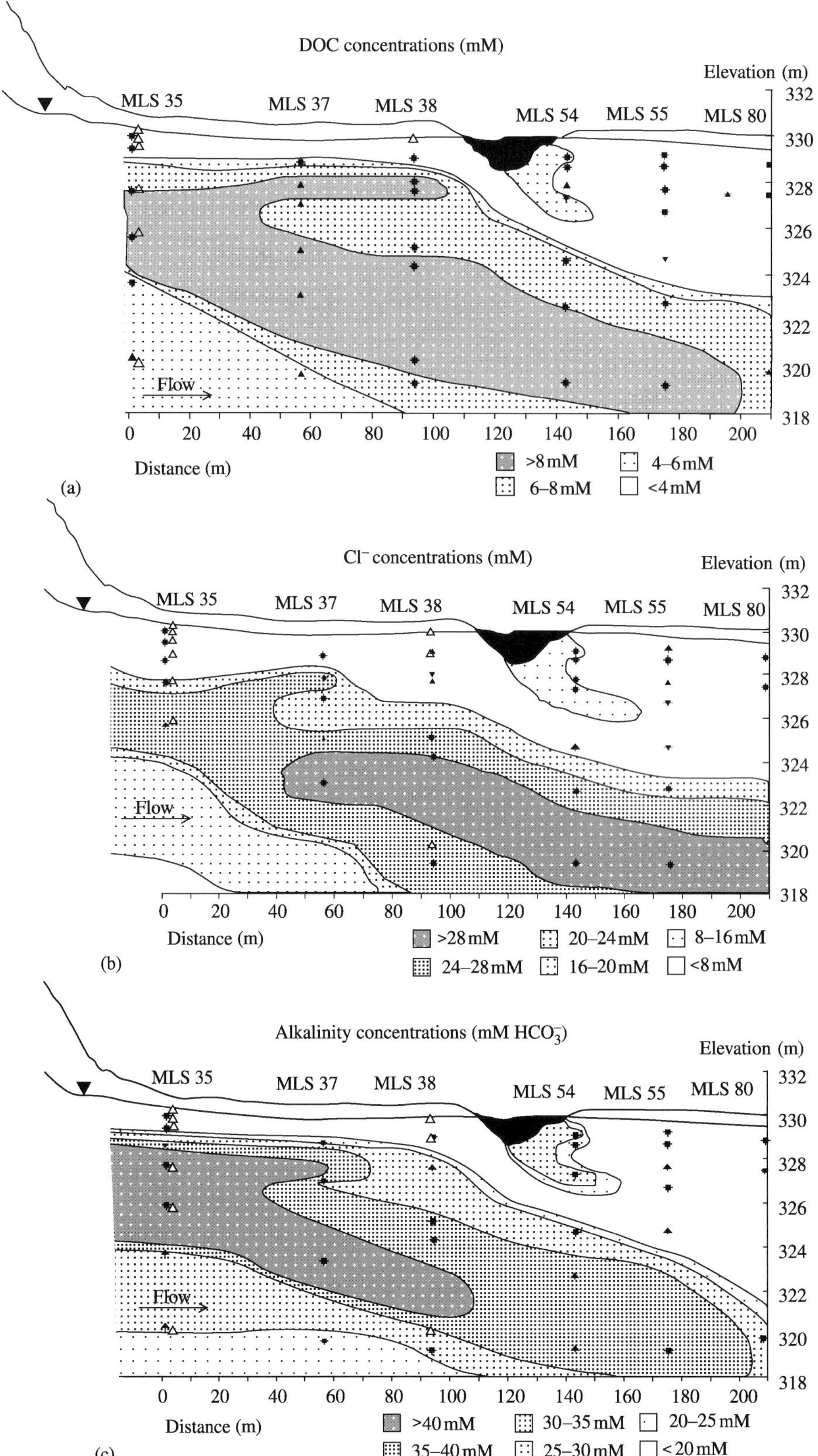

Figure 6 Distribution of: (a) NVOC concentrations (in mM); (b) chloride (Cl^-) concentrations (in mM); (c) alkalinity values (in mM HCO_3^-); (d) ferrous iron concentrations (in mM); (e) ammonium concentrations (in mM); (f) sulfate concentrations (in mM) and methane concentrations (in mM, shown in the plot as individual values at sites 35 and 38), in groundwater downgradient from the Norman Landfill along a transect from site 35 to 80 (after Cozzarelli *et al.*, 2000).

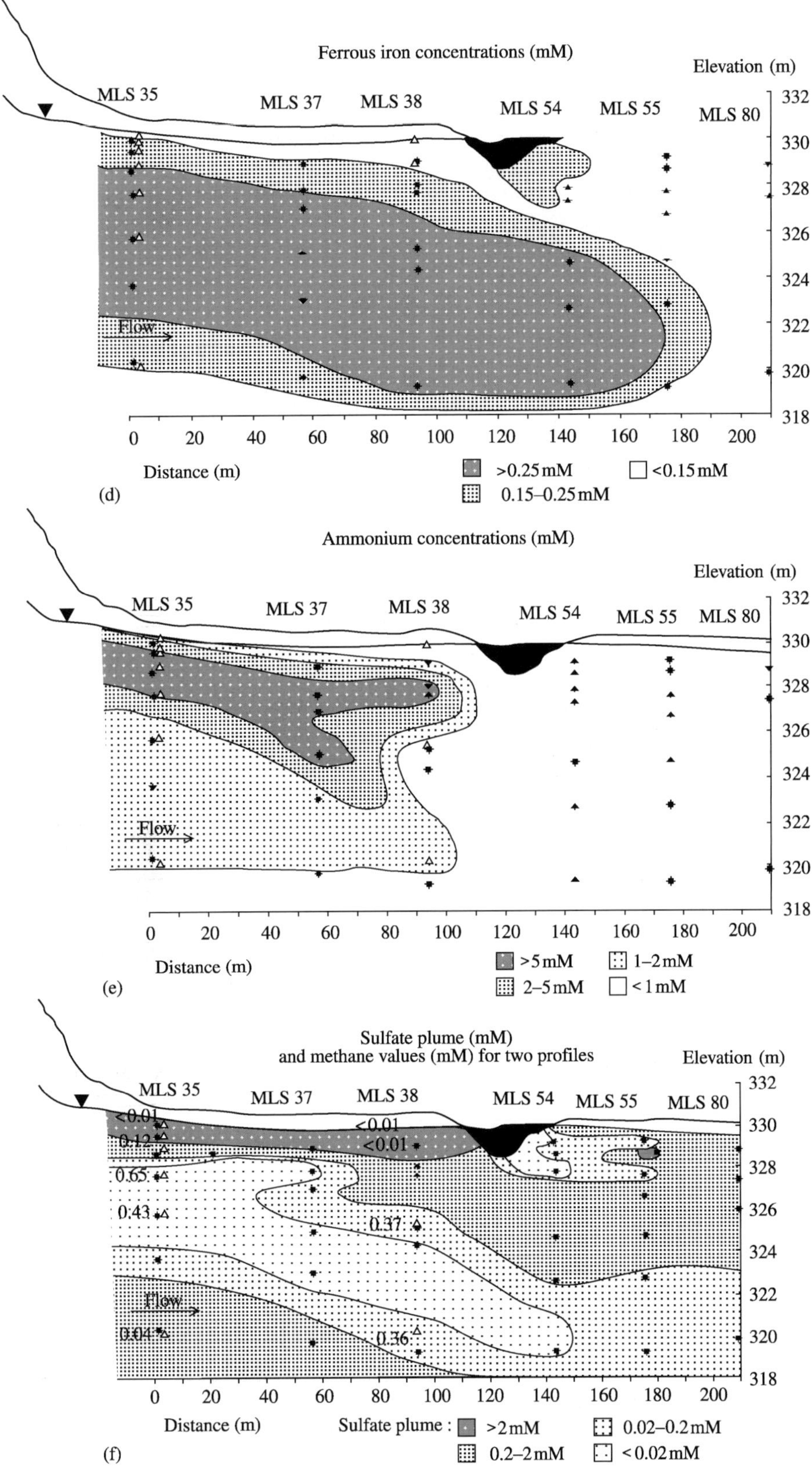

Figure 6 (continued).

edge of the landfill (site 35) was enriched in ^{34}S. At the water table (1.0 m below land surface (bls)), the δ^{34}S of SO_4^{2-} was 33‰; 1.6 m bls, the δ^{34}S of SO_4^{2-} was 67‰. In contrast, sulfur-isotope analyses of SO_4^{2-} in groundwater from uncontaminated alluvium showed that the δ^{34}S of SO_4^{2-} was −5‰ at 1.0 m bls. Although dissolved sulfide concentrations in the plume were low (0.3–5.0 μM), analysis of the sediment cores indicated that iron sulfides have accumulated in the aquifer. The highest concentrations of iron sulfides were detected just beneath the water table where increased sulfate reduction rates were measured (Figure 7).

Concentrations of CH_4 are highest in the center of the anoxic plume. The highest CH_4 concentrations occurred slightly beneath the top of the sulfate-depleted zone (Figure 6(f)). The δD values for groundwater downgradient from the landfill indicate that the groundwater is enriched in deuterium (Cozzarelli *et al.*, 2000). Hackley *et al.* (1996) reported 30–60‰ deuterium enrichment in leachate from three landfills in Illinois, and speculated that most of the enrichment was a result of methanogenesis, with some enrichment resulting from isotopic exchange with hydrogen sulfide. The greatest enrichment in deuterium at the Norman Landfill was measured in the center of the plume where δD of H_2O values ranged from −10.6‰ to −3.4‰, compared to background values of −45.8‰ to −27.9‰. The samples that contained the greatest enrichment in deuterium also had total inorganic carbon (TIC) values enriched in ^{13}C. The δ^{13}C of TIC in the most contaminated groundwater downgradient from the landfill was as heavy as 11.9‰. This indicated a significant enrichment in ^{13}C compared to typical δ^{13}C values of shallow groundwater from the Central Oklahoma aquifer, which ranged from −17.8‰ to −12.5‰ (Parkhurst *et al.*, 1993). This large shift in isotopic composition to enriched values in the groundwater most likely results from biogenic CH_4 production. The high concentrations of CH_4 combined with the heavy δD of H_2O values at the edge of the landfill at Norman, Oklahoma indicate that methanogenesis probably occurs within and underneath the landfill, and that the products of this process are transported in groundwater.

Availability of electron acceptors. The fate of NVOC was investigated in the plume (Figure 6). The NVOC concentrations show little change with distance, indicating that NVOC is not efficiently degraded in this zone. Most of the degradation occurs at the boundaries of the plume where electron acceptors are available (Figure 7). Direct rate measurements made in the laboratory combined with field observations suggest that sulfate reduction and methanogenesis are the most important microbial reactions that affect aquifer geochemistry downgradient from the Norman Landfill. Although the core of the plume is strictly anaerobic and supports both sulfate reduction and methanogenesis, the edges of the plume appear to support iron reduction and, to a greater extent, sulfate reduction, due to the increased availability of readily reactive electron acceptors at these boundaries. The nonuniform availability of electron acceptors and the mixing of the contaminant plume with oxygenated water at the plume boundaries have a significant effect on biogeochemical processes.

Soluble and solid-phase geochemical investigations coupled with laboratory rate experiments

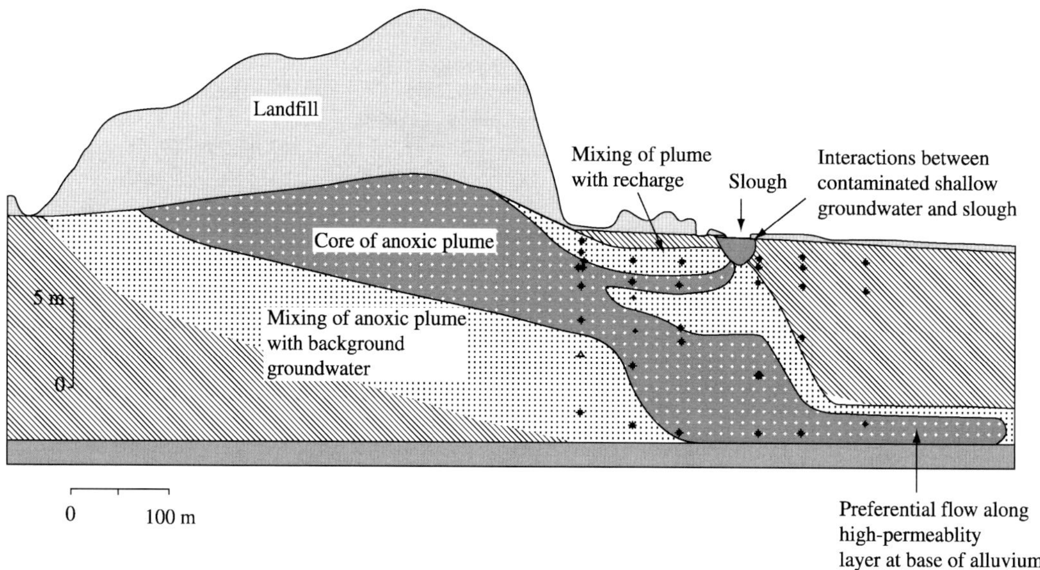

Figure 7 Conceptual model of redox (transport and reaction) zones at the Norman Landfill (USA).

were used to evaluate the factors that control the availability of electron acceptors in this system (Cozzarelli *et al.*, 2000; Ulrich *et al.*, 2003). The sources of electron acceptors vary significantly over relatively small spatial scales (on the order of meters). At least three sources of sulfate that support sulfate-reducing activity in the leachate-impacted aquifer have been identified. Each of these sources supplies sulfate to a different region of the aquifer.

First, the oxidation of iron sulfides to sulfate and/or dissolution of sulfate during recharge events are important in shallow regions near the water table where high rates of sulfate reduction have been measured. *In situ* evidence for aerobic iron sulfide oxidation is found at the water table, where despite relatively high rates of sulfate reduction, the concentration of iron sulfide is comparatively low (Figure 8). Further, sulfate concentrations are highest near the water table and decrease rapidly with depth. Experiments conducted with sediment collected from the study site have shown that hydrogen sulfide leaves the solution rapidly as a constituent of immobile iron sulfide minerals in the sediments. Such minerals were not easily oxidized to sulfate under anaerobic conditions in the presence of a variety of potential electron acceptors (Ulrich *et al.*, 2003). However, when aerobic conditions prevail, as is expected to occur seasonally at the water table, the iron sulfides readily oxidize and the sulfate concentration increases. The observation that iron sulfide oxidation near the water table contributes to the supply of SO_4^{2-} is analogous to findings from an uncontaminated aquifer in the Yegua formation of East Central Texas (Ulrich *et al.*, 1998).

Sulfate reduction within the core of the plume (at intermediate aquifer depths) is limited by the availability of sulfate, which is supplied by the slow process of barite ($BaSO_4$) dissolution. Under conditions of low dissolved sulfate (<10 μM), as is the case in the center of the anaerobic leachate plume, barite is undersaturated and dissolves, releasing both barium and sulfate to solution. Barite grains within the core of the plume show dissolution features (Figure 9). Rapid dissolution of barite is unlikely given its low solubility; however, the amount of barite present in the sediment is sufficient to impact the sulfate budget of the aquifer (Ulrich *et al.*, 2003).

Another source of sulfate in the aquifer is advection just above the confining layer at the bottom of the aquifer. Hydraulic conductivity is relatively high (Scholl and Christenson, 1998) in this depth interval, where coarse-grained sands and gravel are the dominant sediment types. The lower chloride concentration and the lower specific conductance of groundwater in deeper portions of the aquifer relative to the leachate plume are consistent with mixing of leachate with uncontaminated groundwater. Dissolved sulfate in this interval approaches background concentrations and is important in maintaining rates of sulfate reduction. Investigations of plume biogeochemistry (Cozzarelli *et al.*, 2000) indicate that the influx of electron acceptors by mixing with

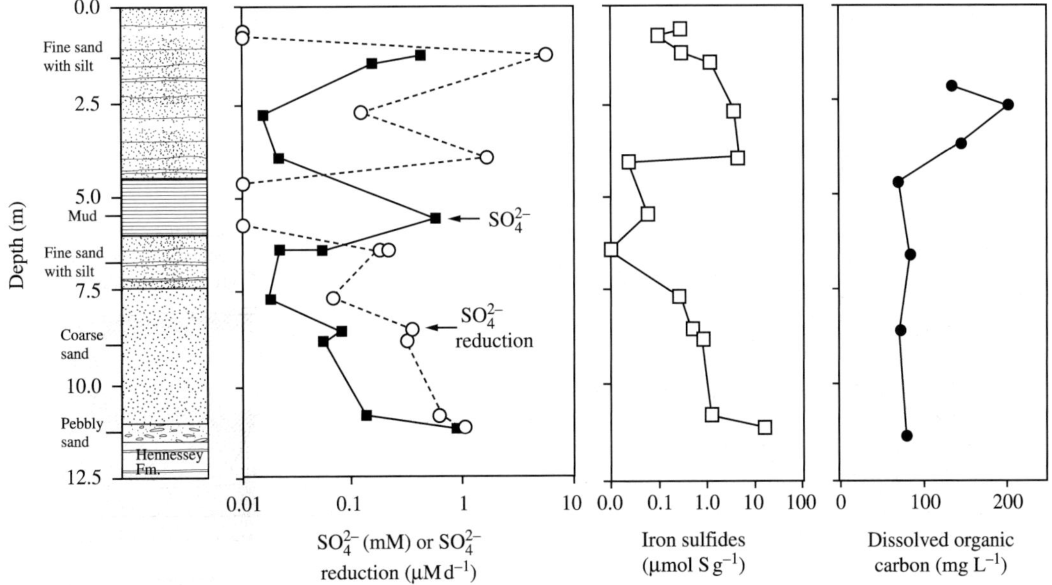

Figure 8 Depth profile (meters below land surface) of the stratigraphy, sulfate reduction rates, sulfate concentration, iron sulfide content of sediments, and dissolved organic carbon in groundwater obtained from an area adjacent to well 40 (source Ulrich *et al.*, 2003).

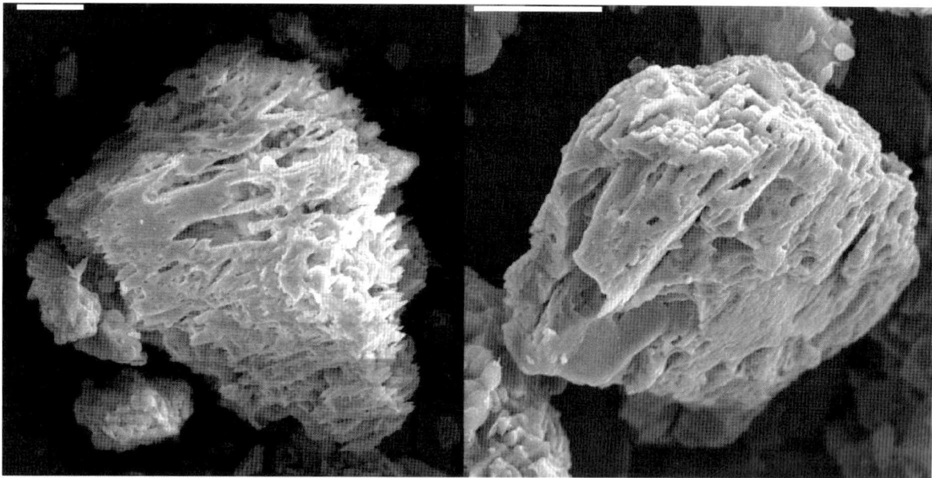

Figure 9 SEM micrographs contrasting textures of detrital barite grains. Grain on the left with dissolution textures was collected from sediment exposed to leachate; nearby pore water contains <10 mg L^{-1} sulfate. Grain on the right was collected from sediment unexposed to leachate and containing 100 mg L^{-1} sulfate. Bar scale is 10 μm (source Ulrich *et al.*, 2003).

recharge or upgradient groundwater is limited to the boundaries of the plume.

Fate of XOCs. Investigation of the distribution of VOCs has provided evidence of natural attenuation of several priority pollutants (Eganhouse *et al.*, 2001). Although VOCs make up <1% of the mass of organic carbon in the Norman Landfill leachate plume, they are useful indicators of biodegradation in the leachate plume. Eganhouse *et al.* (2001) compared the concentrations of two isomers of benzene, isopropylbenzene, and *n*-propylbenzene, in landfill leachate. Isomers of benzene have the same number and type of atoms, but the molecules have slightly different structures. These isomers of benzene have similar physical properties; they should, therefore, be affected similarly by volatilization, dilution, and sorption. Thus, comparisons of the concentration of this pair of compounds, which changes with the distance from the source, allow for the elaboration of the processes responsible for the observed attenuation. The basis for this analysis is that compounds having identical or very similar physico-chemical properties should have the same distribution within a given region of the plume unless differences in compound-specific biodegradation rates cause them to differ (Reinhard *et al.*, 1984; Eganhouse *et al.*, 1996, 2001; Allen-King *et al.*, 1996). The concentration of *n*-propylbenzene decreases much faster as leachate flows away from the landfill than the concentration of isopropylbenzene. This rapid decrease in the concentration of *n*-propylbenzene at Norman Landfill is caused by biodegradation. These techniques can be applied at sites with contaminants other than landfill leachate.

In situ field experiments of microbial processes in zones with different chemical and physical properties have been conducted at Norman Landfill using push–pull test technology and small-scale tracer tests (Scholl *et al.*, 2001; Senko *et al.*, 2002). Push–pull tests are single-well injection-withdrawal tests (Istok *et al.*, 1997). During the injection phase of the test, a solution consisting of groundwater spiked with tracers, electron donors, or electron acceptors is injected or "pushed" into the aquifer. During the extraction phase, the test solution is pumped ("pulled") from the same location and the concentrations of tracers, reactants, and possible reaction products are measured as a function of time in order to construct breakthrough curves and to compute mass balances for each solute. Reaction rate coefficients are computed from the mass of reactant consumed and/or product formed. These tests can be conducted anywhere in an aquifer, making it possible to investigate processes and rates in different geologic and geochemical environments.

At Norman Landfill investigators are using these field injection techniques to investigate how biodegradation rates vary with aquifer permeability (Scholl *et al.*, 2001). Push–pull tracer tests were conducted to measure *in situ* biodegradation rates of simple organic acids in the leachate plume. Replicate wells were placed in three layers: medium sand, silt/clay lenses in sand, and poorly sorted gravel. The *in situ* biodegradation rates of two simple organic acids, formate and lactate, were compared in three different permeability zones within the anoxic leachate plume at the site. These organic acids were used, because they degrade at different rates depending on the dominant microbial processes. The results show that there are differences in biodegradation rates in areas of different permeability. These may be related to differences in microbial community structure, sediment chemistry, and water flow regime.

The conceptual model of biogeochemical zones developed for the Norman Landfill study (Figure 7) provides a framework for understanding the transport of organic contaminants and provides insight into the natural attenuation of the concentration of leachate compounds in the aquifer. This type of approach to assessing the active microbial processes and the availability of electron acceptors can be applied at other sites contaminated with leachate. Once the biogeochemical framework of a system is established, detailed experiments on the rates of processes and the fate and transport of the compounds of concern can be undertaken.

9.16.7 GRINDSTED LANDFILL SITE (DK)

The Grindsted Landfill site in Denmark has been subjected to a number of investigations since 1992. The site has been investigated using a multidisciplinary approach by a large group of researchers of different background (environmental engineering, environmental chemistry, ecotoxicology, geology, and microbiology) from the Technical University of Denmark. The work has been done in co-operation with researchers from universities and research institutions all over the world.

9.16.7.1 Source, Geology and Hydrogeology

The site is located in the western part of Denmark on top of the original land surface (Figure 10). Disposal of waste took place between 1930 and 1977. Approximately, 3×10^5 t of waste were deposited there, mainly between 1960 and 1970 (Kjeldsen *et al.*, 1998a). The waste consists of: municipal solid waste (20%); bulky waste, garden waste, and street sweepings (5%); industrial waste (20%); sewage treatment sludge (30%); and demolition waste (25%). The spatial variability of the leachate quality was investigated by sampling (31 wells) below the landfill in the uppermost groundwater (a small unsaturated zone exists beneath the landfill). The results revealed a significant spatial variability in the leachate composition. Based on these results, landfill could be divided into four main areas (Figure 11). The average concentrations of ammonium, chloride, and NVOC in the strong leachate were typically 20–40 times higher than in the weak leachate. The strong leachate was located in the northern part of the landfill and originated from the dumping of industrial waste. The large heterogeneities in leachate quality may generate plumes with different properties and call for different remedial actions mainly directed towards the industrial hot spot area.

Geology. The Grindsted Landfill is located on a glacial outwash plain. The upper 10–12 m of the unconfined aquifer consist of an upper Quaternary sandy layer and a lower Tertiary sandy layer, locally separated by discontinuous silt and clay layers (Heron *et al.*, 1998). Investigations of the

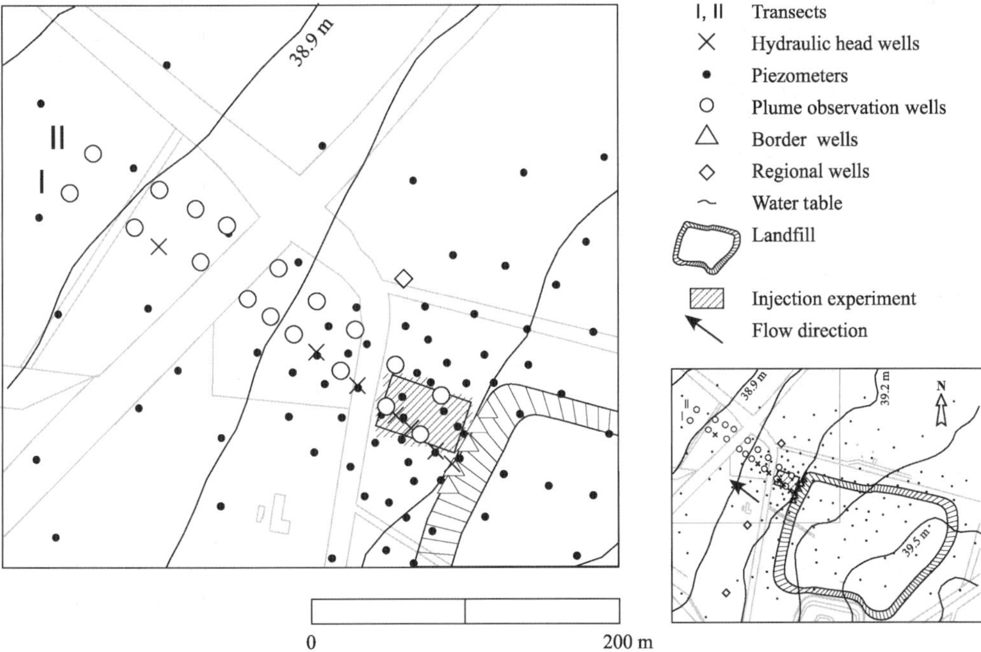

Figure 10 Location of the Grindsted Landfill (DK) site, overview of wells, and investigated transects.

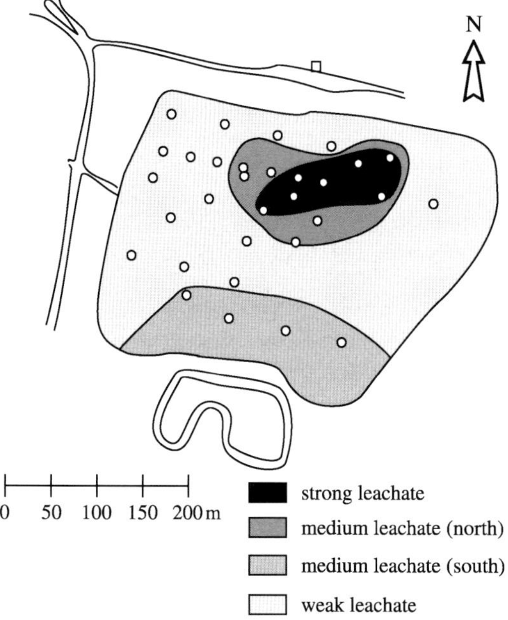

Figure 11 Division of the landfill into four areas with different leachate strength. The division was primarily based on chloride, ammonium, and NVOC. The open circles indicate sampling wells. (source Kjeldsen *et al.*, 1998a).

hydraulic conductivity and hydraulic gradients showed that pore flow velocities were approximately 50 m yr^{-1} and 10 m yr^{-1} for the Quaternary and Tertiary sandy layers, respectively. A clay layer ~1 m thick located 12 m below ground surface extends over a large part of the landfill. Below this layer there is a more regional micaceous sandy layer ~65 m thick. This layer is vertically limited by another low-permeable clay layer ~80 m below ground surface.

The overall groundwater flow direction is north westerly, but the isopotential contours are semicircular, indicating a diverging flow. Locally, inside and close to the landfill the flow field shows significant seasonal variation (Figure 12). The reasons for this mounding are not fully understood, but Kjeldsen *et al.* (1998b) suggest three possibilities: (i) higher infiltration in this part of the landfill; (ii) lower hydraulic conductivity in the aquifer underlying this part of the landfill, possibly due to differences in geology, bacterial growth, precipitates, or gas bubbles of methane and carbon dioxide; or (iii) higher infiltration in the borders of the mounding area. The effects of this local mounding are enhanced lateral spreading of the plume, and downward directed hydraulic gradients in the groundwater below the landfill. The latter

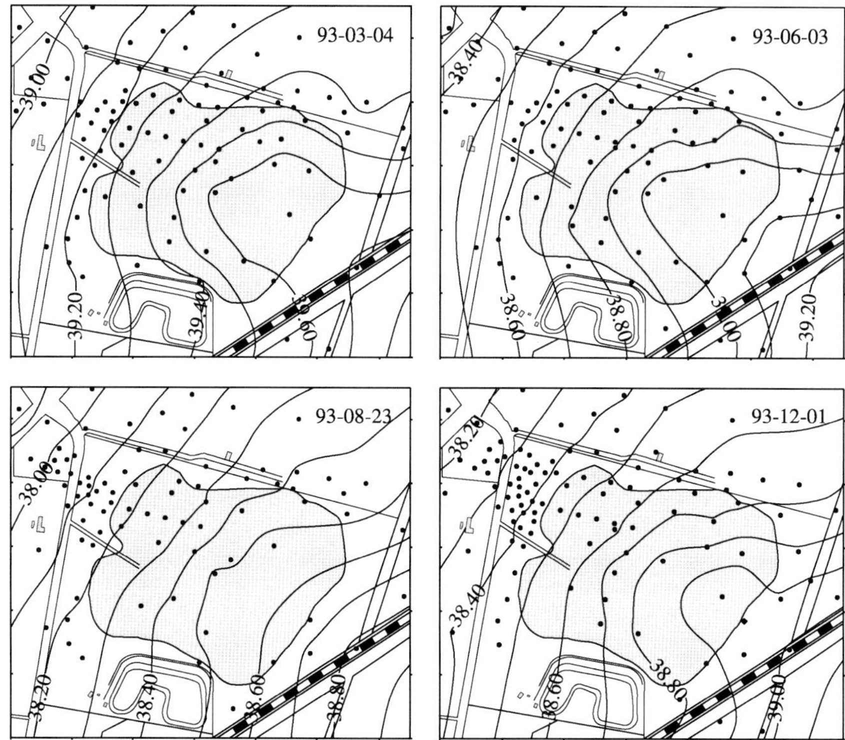

Figure 12 Mounding of the leachate-groundwater table below Grindsted Landfill (DK) shown as isopotential contours for the landfill and surrounding area at four different seasons during 1993 (after Kjeldsen *et al.*, 1998b).

can cause an unexpected vertical spreading pattern, while the enhanced spreading affects the dilution of the plume. Degradation could be increased by enhanced mixing of electron acceptors into the anaerobic parts of the plume. The seasonal variations in the flow field are also important for designing the monitoring network and interpreting the time series monitoring data.

The groundwater quality along the downgradient border of the landfill was mapped in order to illustrate the spreading of leachate into the upper aquifer. Figure 13 shows a three-dimensional sketch of the distribution of chloride, ammonium, and NVOC along the northern and western borders of the landfill as well as the leachate concentrations beneath the landfill.

9.16.7.2 Landfill Leachate Plume

The investigations of the landfill leachate plume at the Grindsted Landfill have been restricted to

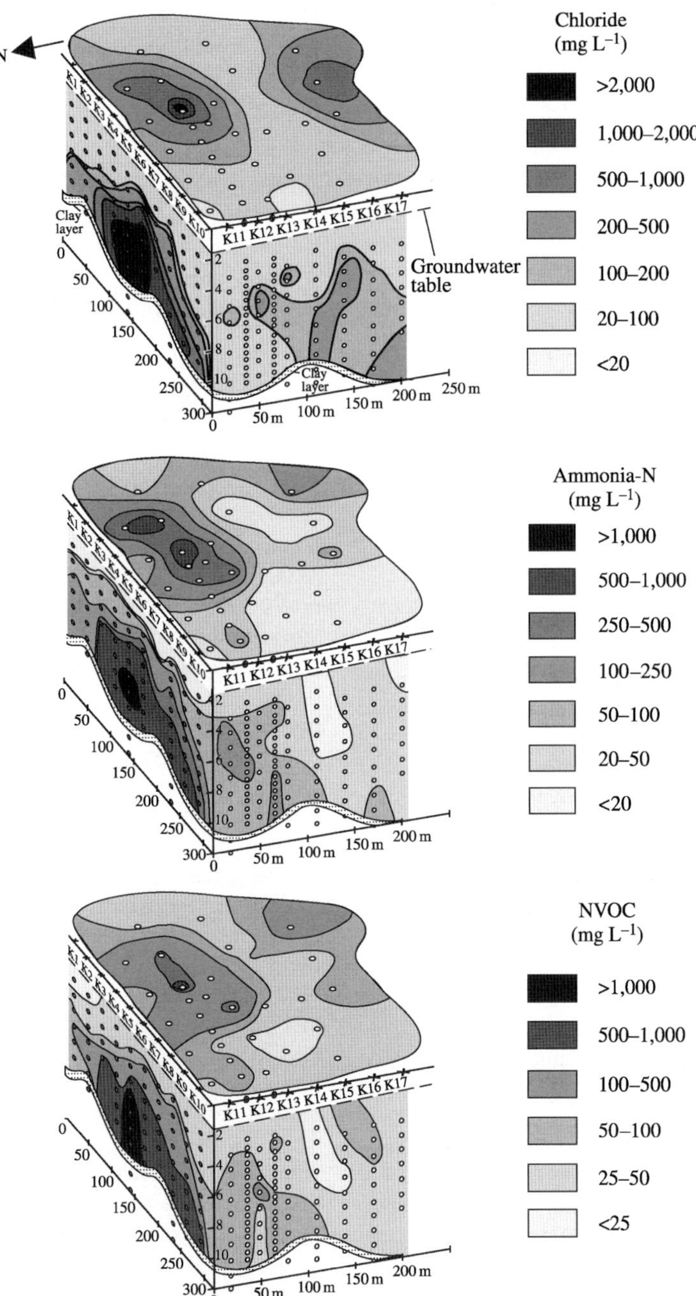

Figure 13 Three-dimensional sketch of leaching pattern for chloride, ammonium, and NVOC from Grindsted Landfill (DK) (source Kjeldsen *et al.*, 1998b).

the upper aquifer (10–12 m) in two transects starting at the western border of the landfill. Transect I (Figure 10) has been characterized in great detail with respect to:

- geology and hydrogeology (Bjerg *et al.*, 1995; Heron *et al.*, 1998; Petersen, 2000);
- inorganic and/or redox-sensitive compounds (Bjerg *et al.*, 1995; Jensen *et al.*, 1998);
- hydrogen levels and *in situ* energetics (Bjerg *et al.*, 1997; Jakobsen *et al.*, 1998);
- aquifer solid composition (Heron *et al.*, 1998);
- microbiology and microbial redox processes (Ludvigsen *et al.*, 1997, 1998, 1999);
- distribution of XOCs (Rügge *et al.*, 1995; Holm *et al.*, 1995); and
- toxicity related to XOCs (Baun *et al.*, 1999, 2000).

Redox environments. At the Grindsted Landfill site, the redox environments were addressed in terms of dissolved redox-sensitive species (Bjerg *et al.*, 1995), aquifer solid compositions (Heron *et al.*, 1998), activity of microorganisms performing each electron-accepting reaction (Ludvigsen *et al.*, 1997), and the concentration of dissolved hydrogen in the groundwater (Jakobsen *et al.*, 1998).

The distribution of dissolved redox-sensitive species (example given in Figure 14) showed that the redox zones were somewhat different in the two parallel vertical transects, which are separated by only 30 m (Figure 15). The overall sequence was consistent with that depicted in Figure 3, although the water chemistry of the Grindsted Landfill (DK) leachate plume suggested that several of the redox zones overlapped. Certainly, the methanogenic and the sulfate-reducing zones overlap (Figure 15). Iron, manganese, and nitrate reducing zones overlap in some cases, but were separated in others. The size of the zones varied considerably between neighboring transects (Bjerg *et al.*, 1995), which is interesting in light of the conclusions drawn earlier for sites for which there are much less data. In this plume methane and ammonium migrate farther than dissolved organic matter (Figure 14), and thus are the dominant reductants at distances greater than 100–150 m from the landfill.

The detailed geological and geochemical description of the aquifer sediment led to an improved understanding of the distribution of iron species in the plume (Heron *et al.*, 1998). The majority of the aquifer consists of mineral-poor fine sands, low in organic matter and iron oxides. Iron and manganese reduction is less important here than at the Vejen Landfill (DK) which is located in the same geographical area, even though very high concentrations of dissolved Fe^{2+} and Mn^{2+} were observed in the Grindsted Landfill (DK) leachate plume. Both a lower initial iron oxide content and a different iron mineralogy (presumably solid grains of crystalline iron oxides) indicated lower iron reactivity.

Bioassays (microbially active, unamended incubations of aquifer solids, and groundwater; Ludvigsen *et al.*, 1998) monitored all the redox processes that occurred and allowed estimates to be made of the rates of the individual redox processes as shown in Figure 16. The rates were fairly low for many of the redox processes. The bioassays also showed that in several samples more than one redox process was significant. However, in most cases one electron acceptor dominated in terms of equivalent rates of organic matter oxidation. Altogether the rates balance fairly well the observed degradation of dissolved organic carbon in the plume. However, the rates determined for denitrification exceeded the dissolved carbon available in that part of the plume suggesting that other electron donors also played a role, perhaps ammonium. It was furthermore demonstrated that low-permeability layers can lead to unexpected redox activities. This was exemplified by the high sulfate reduction activity 170 m from the landfill was caused by localized sulfate and organic-matter-rich deposits, not by leaching from the landfill.

Measurements of dissolved hydrogen have been used to characterize redox levels according to criteria based on the competitive exclusion of terminal electron acceptors for hydrogen oxidation (Lovley and Goodwin, 1988; Chapelle *et al.*, 1995). In the Grindsted Landfill plume, the variations in hydrogen concentration (52 sampling points) were limited, and the values were low (0.004–0.88 nmol L^{-1}) indicating, according to previous criteria, iron-reducing conditions in most of the anaerobic part of the plume (Jakobsen *et al.*, 1998). This was surprising, since the microbial assays and the geochemistry have indicated other active redox processes in the plume (see above). This suggested a need for refining the use of hydrogen in identifying terminal electron acceptors in complex plumes. The hydrogen measurements were used together with the measurements of groundwater chemistry in thermodynamic calculations of free energies of the redox reactions at the temperature of the plume (11 °C). These calculations showed that both sulfate reduction and iron reduction could occur in the plume, since in several places ΔG_r values for each reaction were below a proposed threshold value of -7 kJ mol^{-1} of H_2. Methanogenesis (by CO_2 reduction) showed higher ΔG_r values in all samples, suggesting that methanogenesis only occurs in stagnant pore water, where more reducing conditions may prevail. The small differences in calculated ΔG_r values actually suggested that sulfate reduction and iron

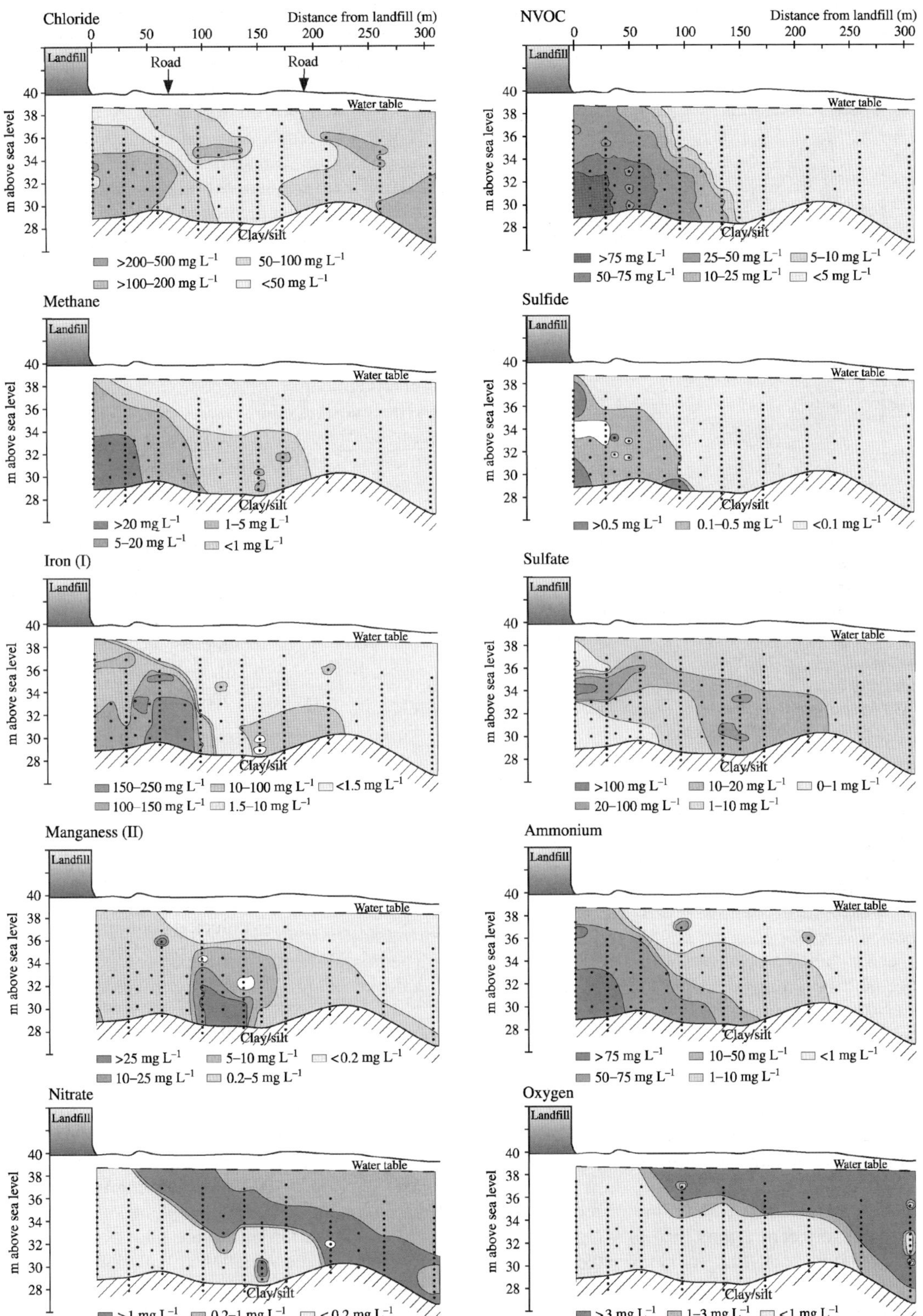

Figure 14 Distribution of Cl^- and dissolved redox-sensitive compounds in Transect 1 downgradient of Grindsted Landfill (DK) (source Bjerg *et al.*, 1995).

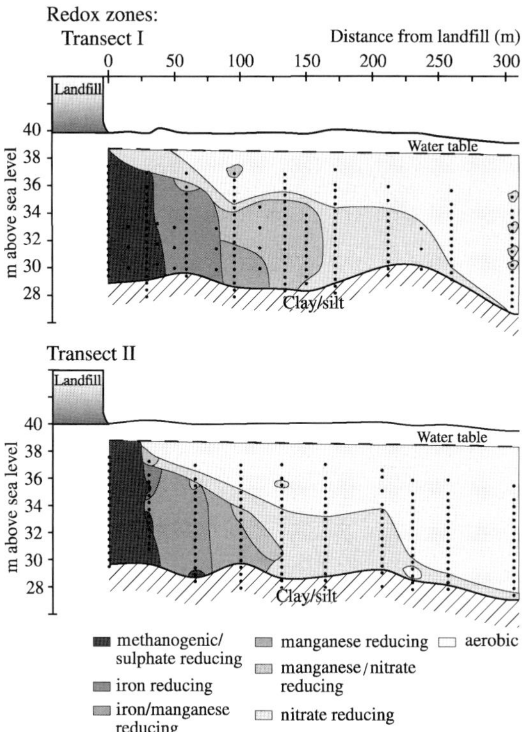

Figure 15 Proposed redox zone distribution in two parallel transect downgradients of Grindsted Landfill (DK) based on the observed groundwater chemistry (source Bjerg *et al.*, 1995).

reduction could take place simultaneously in the same sample.

Overall, this refined use of hydrogen concentrations supported the results of the bioassays and the complex system of redox zones inferred from the distribution of dissolved redox-sensitive species. The Grindsted Landfill (DK) plume is host to all of the proposed redox reactions, but also to secondary oxidation–reduction reactions involving ammonium, methane, manganese oxides, ferrous iron, and sulfides.

Fate of XOCs. The distribution of xenobiotic compounds was mapped in Transect I (Figure 17). More than 15 different organic compounds were identified close to the landfill, with the BTEX compounds dominating. Concentrations of BTEX in the range of 0–222 $\mu g\ L^{-1}$ were observed close to the landfill, with single observations of *o*-xylene concentrations up to 1,550 $\mu g\ L^{-1}$ (Rügge *et al.*, 1995). No chlorinated aliphatic compounds were present in this part of the pollution plume. Most of the XOCs were no longer detectable ~60 m from the edge of the landfill site. Since dilution and sorption could not account for the disappearance of the xenobiotic compounds, it was proposed that the majority of the xenobiotic compounds in the leachate were transformed under methanogenic/sulfate-reducing or iron-reducing conditions in the aquifer. It should be emphasized that the apparent attenuation close to the landfill was not based on direct proof, but on a comparison of the actual distribution to the leaching period, dilution, and sorption in the plume. This was convincing for a number of compounds, and was also supported by reactive solute transport modeling (Petersen, 2000). However, benzene shows a distribution that can indicate recalcitrance in the most reduced parts of the plume, but fast disappearance in more oxidized environments (see discussion on benzene in Petersen (2000)). Final conclusions based on field observations may, therefore, be difficult to reach, and may only be feasible where substantial data from different disciplines are available.

The degradation of xenobiotic compounds can also be investigated by microcosm/column experiments and field injection experiments. At the Grindsted Landfill site the degradation of a mixture of xenobiotic compounds (seven aromatic hydrocarbons, four chlorinated aliphatic hydrocarbons, five nitroaromatic hydrocarbons, and two pesticides) was studied using *in situ* microcosms (ISMs) and laboratory microcosms (LMs) (Bjerg *et al.*, 1999; Rügge *et al.*, 1998, 1999a,b). The data from degradation experiments were compared to the field observation data of the aromatic hydrocarbons and we will limit the discussion to these compounds.

An anaerobic stock solution of the XOCs was injected along with bromide as a tracer into five injection wells, installed 15 m downgradient of the landfill. The amount of water injected in the natural gradient experiment was ~5% of the groundwater flux passing the injection wells, yielding approximate concentrations of 75–330 $\mu g\ L^{-1}$ of the XOCs and 100 mg L^{-1} of bromide immediately downgradient from the injection wells.

The migration of the compounds was monitored in a dense sampling network consisting of a total of 140 multilevel samplers (1,030 sampling points). Over a period of 924 days, samples were collected from ~70 sampling points in the central part of the cloud for the determination of breakthrough curves (BTCs). Degradation and sorption could be determined from the BTCs. After end of the injection, seven cloud snapshots were established covering ~400–700 sampling points each time. From the snapshot, moment analysis provided an evaluation of the mass loss of solute in the system and the spatial distribution of the cloud in the aquifer. The redox conditions in the studied area of the plume were determined by analysis of water-soluble redox-sensitive compounds sampled every 6–8 weeks during the experimental period and by groundwater and sediment characterizations combined with microbial assays conducted on sediment and groundwater sampled 800 d after the start of the injection (Albrechtsen *et al.*, 1999).

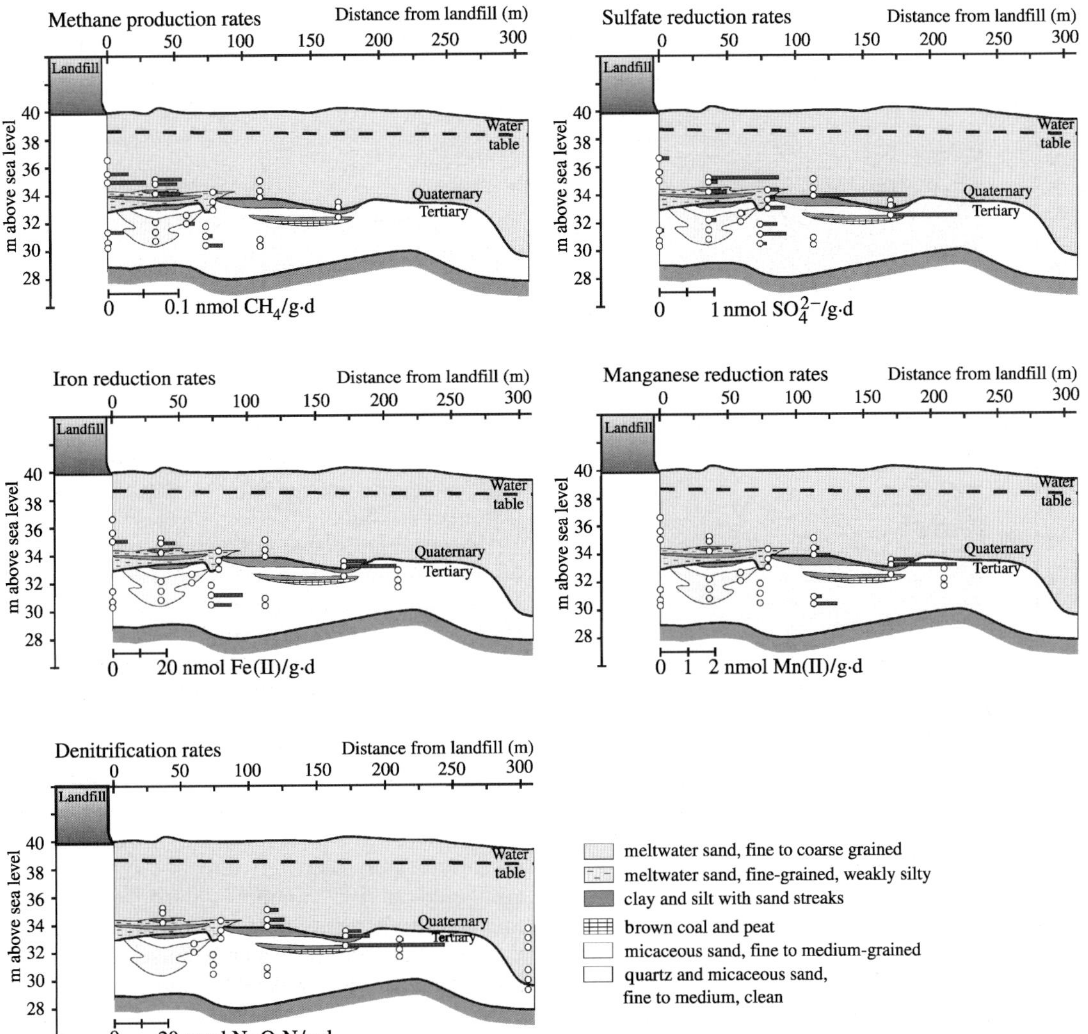

Figure 16 Rates observed by unamended bioassays for individual redox processes at several locations downgradient of Grindsted Landfill (DK) (Ludvigsen *et al.*, 1998).

In the injection experiment, complete degradation of toluene was observed within the most reduced part of the aquifer where iron reduction, sulfate reduction, and methanogenesis occurred. Partial degradation of *o*-xylene was observed. The degradation of *o*-xylene was not initiated before the cloud had reached the part of the aquifer where iron reduction was predominant. Examples of the cloud movement after 649 d are given in Figure 18. Benzene was not degraded within the experimental period of 924 d. Due to highly varying background concentrations, it was not possible to determine whether any degradation of the compounds ethylbenzene and *m*/*p*-xylene had occurred.

In parallel with the anaerobic field injection experiment, ISMs were installed at five locations downgradient of the landfill (see Nielsen *et al.* (1996) for description of this technique). Also, laboratory batch experiments (LBs) were conducted with sediment and groundwater from the corresponding locations. The experimental periods of the ISMs and the LBs were up to 220 d and 537 d, respectively. Both systems involved the same mixture of 18 compounds (Bjerg *et al.*, 1999). For the aromatic compounds only toluene was degraded in the ISM, while also *o*- and *m*/*p*-xylene was degraded in the LB. Benzene, ethylbenzene and naphthalene were not degraded in either the ISM or the LB.

In general, good agreement was observed between the results obtained in the injection experiment, the ISMs, and the LBs; however, a few differences were found, as shown in Table 5. These differences were mainly due to the different experimental periods, 924 d in the injection experiment, and up to 210 d and 537 d in the ISMs and LBs, respectively. However, differences between the static and flow systems also influenced the results. This comparison indicated

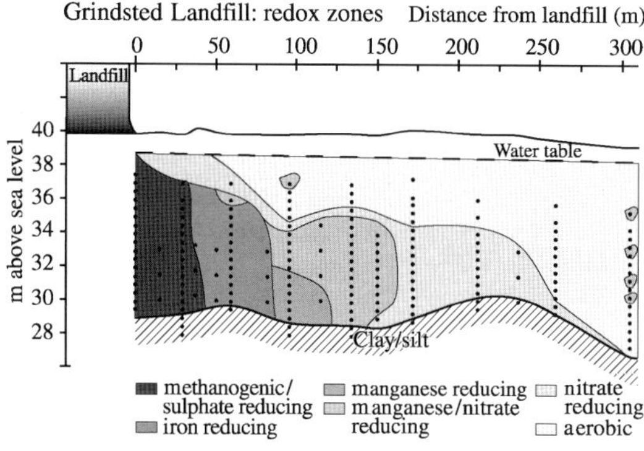

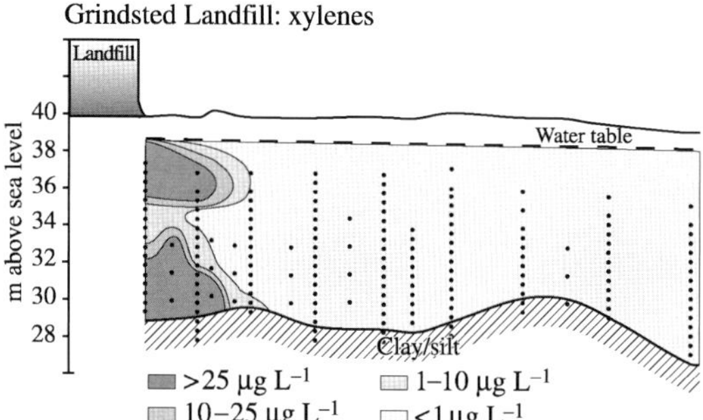

Figure 17 Proposed redox zonation, and distribution of xylenes in Transect I at the downgradient of Grindsted Landfill (DK) (after Bjerg *et al.*, 1995; Rügge *et al.*, 1995).

that the ISM is a good method for studying degradation in the field for compounds with lag periods shorter than 50–100 d. LBs are also a useful and cost-saving approach for studying degradation. The batch setup allows for very long experimental periods. Therefore, the LBs are useful for mixtures of compounds with both shorter and longer lag periods as well as for compounds with varying degradation rates. The results on degradation, however, have been obtained for rather simple compounds, and it is not known how well the batch experiments mimic the field situation for more complex compounds.

The experiments carried out in the anaerobic part of the leachate plume indicate that natural attenuation of toluene and *o*-xylene takes place close to landfill. However, the results for ethylbenzene do not agree with the plume observations. In the case of benzene the degradation may take place at a longer distance from the landfill in more oxidized environments (Petersen, 2000).

In summary, a description of natural attenuation processes for XOCs in a landfill leachate is not an easy task and may require a number of approaches including field observations, experiments in the laboratory and field and also reactive solute transport modeling. The latter is an important tool for integrating data bearing on geology and hydrogeology, redox conditions, distribution of xenobiotic compounds, and degradability/degradation rates.

9.16.8 ENVIRONMENTAL RESTORATION

9.16.8.1 Monitored Natural Attenuation

Natural attenuation refers to processes that naturally transform contaminants to less harmful forms or immobilize contaminants so that they are less of a threat to the environment; see *Natural Attenuation for Groundwater Remediation* by the National Research Council (2000). This includes: dispersion/dilution, sorption, volatilization, and degradation (abiotic, biotic).

Degradation is the most interesting process, because the contaminants can be transformed into less harmful products (carbon dioxide and water). Engineered application of natural attenuation processes as a remedy is termed monitored natural attenuation (MNA, see US EPA). This involves a monitoring component in addition to an evaluation of the natural attenuation processes.

The results from the landfill sites reviewed here indicate a significant potential for natural attenuation of XOCs at landfills sites. There are as of early 2000s, however, only a few examples, compared to the number of landfills around the world. Additional well-documented field examples are badly needed. Experience from more landfill sites will also help in developing procedures for demonstrating natural attenuation, which is not a straightforward procedure. The experience presented in this chapter and the evaluation of natural attenuation as a remedy at landfill sites by Christensen *et al.* (2000a) suggest that five points will be critical:

- local hydrogeological conditions in the landfill area may affect the spreading of the contaminants;
- the size of the landfill and the heterogeneity of the source may create a variable leaching pattern and multiple plumes;
- the complexity of leachate plumes with respect to compounds (inorganic, XOCs) and biogeochemical processes may be an obstacle;
- the time frame of leaching from a landfill site is apt to be very long and will call for long-term evaluation of the attenuation capacity; and
- demonstration of natural attenuation in terms of mass reduction at the field scale is difficult.

The importance of each issue will depend on the actual landfill. It should also be emphasized that even though these problems may be difficult to solve, very few alternatives exist for remediation at landfill sites (see next section).

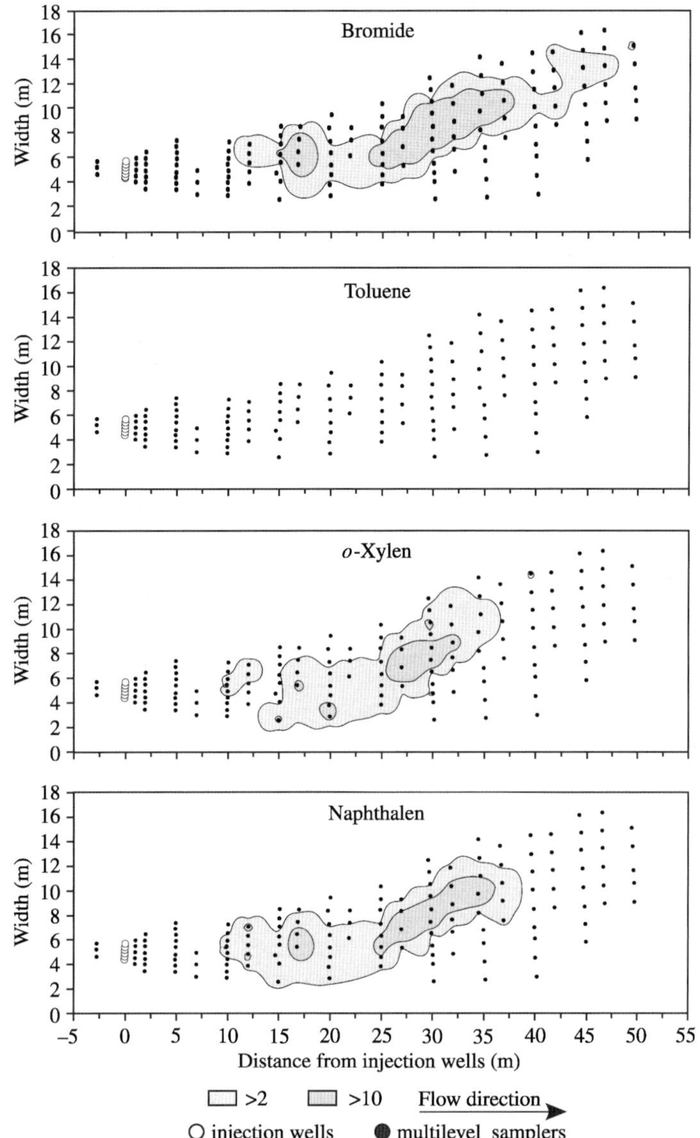

Figure 18 Observed clouds of bromide, toluene, *o*-xylene, naphthalene after 649 d. Bromide in g m^{-2}, xenobiotics in mg m^{-2} (source Rügge *et al.*, 1999a).

9.16.8.2 Engineered or Enhanced Remediation

Remediation of landfill leachate plumes has, in most cases, focused on source control. This is due to the fact that landfills are so large that the mass of pollutants contained in the landfill body would be able to sustain the leachate plume for centuries making a remedy focusing on the plume a long and costly approach. Additional reasons for focusing on the source are that the landfill also may be a "time bomb" containing chemical waste in drums subject to corrosion, landfill gas emission needs to be controlled, improvements of landscape value is requested by the community, and potentially new landfilling capacity is in demand.

Source controls may involve the following.

- Excavation in order to remove drums and highly polluted waste. If the most contaminated waste in the landfill body can be identified, excavation may be an efficient remediation technique. Partly corroded drums may be difficult to handle without contaminating the waste material; this is for precautions and care. The excavated material may be removed for off-site treatment or may be re-landfilled in landfill sections established with landfill liners, leachate drainage collection, and treatment systems.
- Sheet piling in order to control groundwater migration into and out of the waste if waste

Table 5 Potential for degradation of the aromatic compounds in anaerobic leachate affected groundwater at Grindsted Landfill (DK).

Grindsted Landfill	*Benzene*	*Toluene*	*Ethyl-benzene*	*m/p-Xylene*	*o-Xylene*	*Naphthalene*
(1) Field Injection 15–45 m	−	+	?	?	−	−
(1) Field Injection 45–65 m	−	+	?	?	+	−
(2) ISM and LB						
LB (I) 15 m	−	+	−	−	−	−
LB (II) 15 m	−	+	−	−	−	−
LB (II) 25 m	−	+	−	−	−	−
ISM 25 m	−	−	−	−	−	−
LB (II) 35 m	−	−	−	−	−	−
ISM 35 m	−	−	−	−	−	−
LB (II) 45 m	−	+	−	−	−	−
ISM 45 m	−	−	−	−	−	−
LB (II) 55 m	−	+	−	+	−	−
ISM 55 m	−	+	−	−	−	−
LB (I) 60 m	−	+	−	+	+	−

(1) Rügge *et al.* (1999a) and (2) Bjerg *et al.* (1999). +, degradation observed. −, no degradation observed. ?, not possible to determine due to highly varying background conditions.

exists below the groundwater table. The sheet piles should reach low-permeable strata. This approach is, therefore, limited to fairly shallow aquifers. Sheeting piling can effectively reduce the exchange of leachate-contaminated groundwater with the surrounding flow of groundwater. However, sufficient control can only be obtained with this approach if it is combined with continuous removal of leachate-contaminated groundwater from the sheet-piled area. This will create groundwater seepage into the leachate-contaminated area instead of leachate migration out of the area.

- Impermeable top covers in terms of plastic liners or clay membranes can be installed in order to prevent water infiltration into the waste and hence cease leachate generation for waste landfilled above the seasonal water table. Top covers are relatively expensive to install and maintenance must be expected in perpetuity. Adequate surface water removal is important to limit the infiltration through cracks, fractures, or pinholes in the liners. Vegetation is often introduced to improve landscape value but requires special measures to avoid root penetration of the membranes.
- In-landfill treatments have been used in a few experimental cases including aerating the landfill body through vertical wells installed in the central part of the landfill body and extracting the off-gases through collection wells placed in the periphery of the landfill (Heyer *et al.*, 2001). The approach is a combined stripping system for methane and volatile organics and a composting system providing oxygen to the microbial processes in the landfill body. The approach may reduce the contaminant level of the leachate, but most likely will not eliminate leachate generation when aeration is terminated.
- Pump-and-treat systems can be installed to remove contaminated groundwater from areas, beneath the landfill. This will control the migration of the leachate away from the landfill but the pump-and-treat system must be maintained as long as the landfill is a threat to the groundwater. Pump and treat at source is not very different than that in the leachate plume except that the volume is likely to be less and the concentrations of contaminants are likely to be higher when used for source control.

Engineered remediations of leachate plumes are few and mainly involve pump-and-treat schemes. Since leachate plumes may contain a broad spectrum of contaminants potentially with an uneven distribution in the plume, any remediation scheme must be broadly targeted and robust. This supports pump and treat as a suitable candidate.

The number of landfill plumes that have been remediated are very few compared to, for example, plumes with chlorinated solvents and hydrocarbons, and the remediation of leachate plumes has usually involved a combination of approaches focusing on the source remediation and monitored natural attenuation for the plume.

9.16.9 FUTURE CHALLENGES AND RESEARCH TOPICS

The number of detailed landfill studies has been quite limited. A few sites have been characterized to a high degree (including Banisveld Landfill, Borden Landfill, Grindsted Landfill, Norman Landfill, and Vejen Landfill) by using multidisciplinary approaches. These studies consistently show the importance of integrated studies in order to understand the biogeochemical processes in landfill leachate plumes. The results presented in this chapter have revealed a number of topics where research using similar approaches could be beneficial. Suggestions for future challenges and research topics are listed below:

- *Source and spreading of landfill leachate.* Density flow is most likely at landfill sites, but the current understanding of instability effects and our ability to predict the impact of density flow under field conditions is poor. Also, the reasons for the often observed mounding of the groundwater table beneath landfills are not fully understood and need more attention.
- *Biogeochemical processes.* A good conceptual understanding of the biogeochemical processes in landfill leachate plumes exists and the overall understanding of redox processes is good. The challenge is to quantify rates and capacity, so that the relative importance of core processes (e.g., microbial iron reduction capacity) and fringe processes (mixing of electron acceptors) can be evaluated in order to predict the long-term behavior of the plume. Application of stable isotope as well as molecular techniques may be useful in such studies, but simple model/mass balance approaches or advanced multidimensional reactive solute transport models are also needed to integrate flow, transport, and reactive processes.
- *Fate of XOCs.* Significant degradation of XOC has been demonstrated at most landfill sites, but discrepancies between field observations and experimental work have been observed. Lack of knowledge of the expected fate is mainly related to "new" compounds such as pesticides, pharmaceuticals, and phthalates. However, the discrepancies also reveal that the methods used for documentation of natural attenuation processes are poor in landfill leachate plumes. Recent studies have suggested using probe compounds as tracers, degradation products, stable isotopes, and enantiomeric ratios, but these methods are all quite new and need to be improved before they can be applied in practice for documenting natural attenuation. Flux approaches have been suggested for quantification of mass removal for XOCs; however, current applications indicate that this is a costly and complex approach, which needs refinement.
- *Remediation of landfill leachate* plumes has been performed only in very few cases. The most promising technique is monitored natural attenuation, but guidelines and practical experience are necessary to ensure proper implementation. The need for monitoring and the length of the monitoring period of landfills of different lifetimes *in situ* very much a matter of discussion.

REFERENCES

Acton D. W. and Barker J. F. (1992) *In situ* biodegradation potential of aromatic hydrocarbons in anaerobic groundwater. *J. Contamin. Hydrol.* **9**, 325–352.

Adams E. E. and Gelhar L. W. (1992) Field study in a heterogeneous aquifer: 2. Spatial moments analysis. *Water Resour. Res.* **28**, 3293–3307.

Albrechtsen H.-J. and Christensen T. H. (1994) Evidence for microbial iron reduction in a landfill leachate-polluted aquifer (Vejen, Denmark). *Appl. Environ. Microbiol.* **60**, 3920–3925.

Albrechtsen H.-J., Bjerg P. L., Ludvigsen L., Rügge K., and Christensen T. H. (1999) An anaerobic field injection experiment in a landfill leachate plume (Grindsted, Denmark): 2. Deduction of anaerobic (methanogenic, sulfate and Fe(III)reducing) redox conditions. *Water Resour. Res.* **35**, 1247–1256.

Albrechtsen H.-J., Mills M., Aamand J., and Bjerg P. L. (2001) Degradation of herbicides in shallow Danish aquifers—an integrated laboratory and field study. *Pest Manage. Sci.* **57**, 341–350.

Allen-King R. M., Gillham R. W., and Barker J. F. (1996) Fate of dissolved toluene during steady infiltration through unsaturated soil: I. Method emphasizing chloroform as a volatile, sorptive, and recalcitrant tracer. *J. Environ. Qual.* **25**, 279–286.

Arneth J.-D., Milde G., Kerndorff H., and Schleyer R. (1989) Waste deposit influences on groundwater quality as a tool for waste type and site selection for final storage quality. In *The Landfill*, Lecture Notes in Earth Sciences (ed. P. Baccini). Springer, Berlin, vol. 20, pp. 399–424.

Assmuth T. W. and Strandberg T. (1993) Ground-water contamination at Finnish landfills. *Water Air Soil. Pollut.* **69**, 179–199.

Barker J. F., Tessmann J. S., Plotz P. E., and Reinhard M. (1986) The organic geochemistry of a sanitary landfill leachate plume. *J. Contamin. Hydrol.* **1**, 171–189.

Baun A., Kløft L., Bjerg P. L., and Nyholm N. (1999) Toxicity testing of organic chemicals in groundwater polluted with landfill leachate. *Environ. Toxicol. Chem.* **18**, 2046–2056.

Baun A., Jensen S. D., Bjerg P. L., Christensen T. H., and Nyholm N. (2000) Toxicity of organic chemical pollution in groundwater downgradient of a landfill (Grindsted, Denmark). *Environ. Sci. Technol.* **34**, 1647–1652.

Baun A., Ask L., Ledin A., Christensen T. H., and Bjerg P. L. (2003) Natural attenuation of xenobiotic organic compounds in a landfill leachate plume (Vejen, Denmark). *J. Contamin. Hydrol.* **65**, 269–291.

Beeman R. E. and Suflita J. M. (1987) Microbial ecology of a shallow unconfined ground water aquifer polluted by municipal landfill leachate. *Microb. Ecol.* **14**, 39–54.

Beeman R. E. and Suflita J. M. (1990) Environmental factors influencing methanogenesis in a shallow anoxic aquifer: a field and a laboratory study. *J. Ind. Microb.* **5**, 45–58.

Belevi H. and Baccini P. (1992) Long-term leachate emissions from municipal solid waste landfills. In *Landfilling of Waste: Leachate* (eds. T. H. Christensen, R. Cossu, and R. Stegmann). Elsevier, London, pp. 431–440.

Beller H. (2000) Metabolic indicators for detecting *in situ* anaerobic alkaylbenzene degradation. *Biodegradation* **11**, 125–139.

Bisdom E. B. A., Boekestein A., Curmi P., Legas P., Letsch A. C., Loch J. P. G., Nauta R., and Wells C. B. (1983) Submicroscopy and chemistry of heavy metal contaminated precipitates from column experiments simulating conditions in a soil beneath a landfill. *Geoderma* **30**, 1–20.

Bjerg P. L., Rügge K., Pedersen J. K., and Christensen T. H. (1995) Distribution of redox sensitive groundwater quality parameters downgradient of a landfill (Grindsted, Denmark). *Environ. Sci. Technol.* **29**, 1387–1394.

Bjerg P. L., Jakobsen R., Bay H., Rasmussen M., Albrechtsen H.-J., and Christensen T. H. (1997) Effects of sampling well construction on H_2 measurements made for characterization of redox conditions in a contaminated aquifer. *Environ. Sci. Technol.* **31**, 3029–3031.

Bjerg P. L., Rügge K., Cortsen J., Nielsen P. H., and Christensen T. H. (1999) Degradation of aromatic and chlorinated aliphatic hydrocarbons in the anaerobic part of the Grindsted Landfill leachate plume: *in situ* microcosm and laboratory batch experiments. *Ground Water* **37**, 113–121.

Bradley P. M. (2000) Microbial degradation of chloroethenes in groundwater systems. *Hydrogeol. J.* **8**, 104–111.

Breit G. N., Cozzarelli I. M., Johnson R. D., and Norvell J. S. (1996) Interaction of alluvial sediments and a leachate plume from a landfill near Norman Oklahoma. *Geol. Soc. Am. Abstr. Prog.* **28**(7), A258.

Broholm M. M., Rügge K., Tuxen N., Højbjerg A. L., Mosbæk H., and Bjerg P. L. (2001) Fate of herbicides in a shallow aerobic aquifer: a continuous field injection experiment (Vejen, Denmark). *Water Resour. Res.* **37**, 3163–3176.

Brun A., Engesgaard P., Christensen T. H., and Rosbjerg D. (2002) Modelling of transport and biogeochemical processes in pollution plumes: Vejen Landfill, Denmark. *J. Hydrol.* **256**, 228–248.

Chapelle F. H. and Bradley P. M. (1998) Selecting remediation goals by assessing the natural attenuation capacity of groundwater systems. *Bioremed. J.* **2**, 227–238.

Chapelle F. H., McMahon P. B., Dubrovsky N. M., Fujii R. F., Oaksford E. T., and Vroblesky D. A. (1995) Deducing the distribution of terminal electron-accepting processes in hydrologically diverse groundwater systems. *Water Resour. Res.* **31**, 359–371.

Christensen T. H. and Kjeldsen P. (1989) Basic biochemical processes in landfills. In *Sanitary Landfilling: Process, Technology and Environmental Impact* (eds. T. H. Christensen, R. Cossu, and R. Stegmann). Academic Press, London, chap. 2.1, pp. 29–49.

Christensen T. H., Kjeldsen P., Albrechtsen H.-J., Heron G., Nielsen P. H., Bjerg P. L., and Holm P. E. (1994) Attenuation of landfill leachate pollutants in aquifers. *Crit. Rev. Environ. Sci. Technol.* **24**, 119–202.

Christensen J. B., Jensen D. L., and Christensen T. H. (1996) Effect of dissolved organic carbon on the mobility of cadmium, nickel, and zinc in leachate polluted groundwater. *Water Res.* **30**, 3037–3049.

Christensen T. H., Bjerg P. L., and Kjeldsen P. (2000a) Natural attenuation: a feasible approach to remediation of groundwater pollution at landfills? *GWMR* **20**(1), 69–77.

Christensen T. H., Bjerg P. L., Banwart S., Jakobsen R., Heron G., and Albrechtsen H.-J. (2000b) Characterization of redox conditions in groundwater contaminant plumes. *J. Contamin. Hydrol.* **45**, 165–241.

Christensen T. H., Kjeldsen P., Bjerg P. L., Jensen D. L., Christensen J. B., Baun A., Albrechtsen H.-J., and Heron G. (2001) Biogeochemistry of landfill leachate plumes. *Appl. Geochem.* **16**, 659–718.

Cozzarelli I. M., Suflita J. M., Ulrich G. A., Harris S. H., Scholl M. A., Schlottmann J. L., and Christenson S. (2000) Geochemical and microbiological methods for evaluating anaerobic processes in an aquifer contaminated by landfill leachate. *Environ. Sci. Technol.* **34**, 4025–4033.

DeWalle F. B. and Chian S. K. (1981) Detection of trace organics in well near solid waste landfill. *AWWA J.* **73**, 206–211.

Eganhouse R. P., Dorsey T. F., Phinney C. S., and Westcott A. M. (1996) Processes affecting the fate of monoaromatic hydrocarbons in an aquifer contaminated by crude oil. *Environ. Sci. Technol.* **30**, 3304–3312.

Eganhouse R. P., Cozzarelli I. M., Scholl M. A., and Matthews L. L. (2001) Natural attenuation of volatile organic compounds (VOCs) in the leachate plume of a municipal landfill: using alkylbenzenes as a process probe. *Ground Water* **39**, 192–202.

Freeze R. A. and Cherry J. A. (1979) *Groundwater*. Prentice Hall, Englewood Cliffs, NJ.

Gelhar L. W., Welty C., and Rehfeldt K. R. (1992) A critical review of data on field-scale dispersion in aquifers. *Water Resour. Res.* **28**, 1955–1974.

Grbic-Galic D. (1990) Methanogenic transformation of aromatic hydrocarbons and phenols in groundwater aquifers. *Geomicrobiol. J.* **8**, 167–200.

Hackley K. C., Liu C. L., and Coleman D. D. (1996) Environmental isotope characteristics of landfill leachates and gases. *Ground Water* **34**, 827–836.

Harris S. H., Ulrich G. A., and Suflita J. M. (1999) *Dominant Terminal Electron Accepting Processes Occuring at a Landfill Leachate-impacted Site as Indicated by Field and Laboratory Measurements*. US Geological Survey Water Resources Investigations Report #99-4018C, Reston, VA.

Heider J., Spormann A. M., Beller H. R., and Widdel F. (1999) Anaerobic bacterial metabolism of hydrocarbons. *FEMS Microbiol. Rev.* **22**, 459–473.

Heron G. (1994) Redox buffering in landfill leachate contaminated aquifers. PhD Thesis, Department of Environmental Science and Engineering, Technical University of Denmark, Lyngby.

Heron G. and Christensen T. H. (1995) Impact of sediment-bound iron on redox buffering in a landfill leachate polluted aquiffer (Vejen, Denmark). *Environ. Sci. Technol.* **29**, 187–192.

Heron G., Christensen T. H., and Tjell J. C. (1994a) Oxidation capacity of aquifer sediments. *Environ. Sci. Technol.* **28**, 153–158.

Heron G., Crouzet C., Bourg A. C. M., and Christensen T. H. (1994b) Speciation of Fe(II) and Fe(III) in contaminated aquifer sediments using chemical extraction techniques. *Environ. Sci. Technol.* **28**, 1698–1705.

Heron G., Bjerg P. L., Gravesen P., Ludvigsen L., and Christensen T. H. (1998) Geology and sediment geochemistry of a landfill leachate contaminated aquifer (Grindsted, Denmark). *J. Contamin. Hydrol.* **29**, 301–317.

Heyer K.-U., Hupe K., Ritzkowski M., and Stegmann R. (2001) Technical implementation an operation of the low pressure aeration of landfills. In *Sardinia 2001: 8th International Waste Management and Landfill Symposium, October 1–5, Sardinia, Italy*, Proceedings Vol. IV (eds. T. H. Christensen, R. Cossu, and R. Stegmann). CISA, Cagliari, Italy.

Holliger C., Wohkfahrt G., and Diekert G. (1999) Reductive dechlorination in the energy metabolism of anaerobic bacteria. *FEMS Microbiol. Rev.* **22**, 383–398.

Holm J. V., Rügge K., Bjerg P. L., and Christensen T. H. (1995) Occurrence and distribution of pharmaceutical organic compounds in the groundwater downgradient of a landfill (Grindsted, Denmark). *Environ. Sci. Technol.* **29**, 1415–1420.

Istok J. D., Humphrey M. D., Schroth M. H., Hyman M. R., and O'Reilly K. T. (1997) Single-well "push–pull" test for *in situ* determination of microbial activities. *Ground Water* **35**, 619–631.

Jakobsen R., Albrechtsen H.-J., Rasmussen M., Bay H., Bjerg P. L., and Christensen T. H. (1998) H_2 concentrations in a landfill leachate plume (Grindsted, Denmark): *in situ* energetics of terminal electron acceptor processes. *Environ. Sci. Technol.* **32**, 2142–2148.

Jensen K. H., Bitsch K., and Bjerg P. L. (1993) Large-scale dispersion experiments in a sandy aquifer in Denmark: observed tracer movements and numerical analysis. *Water Resour. Res.* **29**, 673–696.

Jensen D. L., Boddum J. K., Redemann S., and Christensen T. H. (1998) Speciation of dissolved iron(II) and manganese(II) in a groundwater pollution plume. *Environ. Sci. Technol.* **32**, 2657–2664.

Jensen D. L., Boddum J. K., Tjell J. C., and Christensen T. H. (2002) The solubility of rhodochrosite ($MnCO_3$) and siderite ($FeCO_3$) in anaerobic aquatic environments. *Appl. Geochem.* **17**, 503–511.

Johnston J. J., Borden R. C., and Barlaz M. A. (1996) Anaerobic biodegradation of alkylbenzene and trichloroetylene in aquifer sediment downgradient of a sanitary landfill. *J. Contamin. Hydrol.* **23**, 263–283.

Jones E. J. P., Deal A., Lorah M. M., Kirshtein J. D., and Voytek M. A. (2002) The effect of Fe(III) on microbial degradation of chlorinated volatile organic compounds in a contaminated freshwater wetland. In *102nd General Meeting of the American Society for Microbiology, Salt Lake City, UT, May 19–23*. American Society for Microbiology, Washington, DC.

Kehew A. E., Schwindt F. J., and Brown D. J. (1984) Hydrogeochemical interaction between a municipal waste stabilization lagoon and a shallow aquifer. *Groundwater* **22**, 746–754.

Kjeldsen P. (1986) Attenuation of landfill leachate in soil and aquifer material. PhD Thesis, Department of Environmental Engineering, Technical University of Denmark, Lyngby.

Kjeldsen P. (1993) Groundwater pollution source characterization of an old landfill. *J. Hydrol.* **142**, 349–371.

Kjeldsen P. and Christophersen M. (2001) Composition of leachate from old landfills in Denmark. *Waste Manage. Res.* **19**, 249–256.

Kjeldsen P., Kjølholt J., Schultz B., Christensen T. H., and Tjell J. C. (1990) Sorption and degradation of chlorophenols, nitrophenols and organophosphorus pesticides in the subsoil under landfills—laboratory studies. *J. Contamin. Hydrol.* **6**, 165–184.

Kjeldsen P., Grundtvig A., Winther P., and Andersen J. S. (1998a) Characterization of an old municipal landfill (Grindsted, Denmark) as a groundwater pollution source: landfill history and leachate composition. *Waste Manage. Res.* **16**, 3–13.

Kjeldsen P., Bjerg P. L., Rügge K., Christensen T. H., and Pedersen J. K. (1998b) Characterization of an old municipal landfill (Grindsted, Denmark) as a groundwater pollution source: landfill hydrology and leachate migration. *Waste Manage. Res.* **16**, 14–22.

Kjeldsen P., Barlaz M. A., Rooker A. P., Baun A., Ledin A., and Christensen T. H. (2002) Present and long term composition of MSW landfill leachate—a review. *Crit. Rev. Environ. Sci. Technol.* **32**, 297–336.

Larsen T., Christensen T. H., Pfeffer F. M., and Enfield C. G. (1992) Landfill leachate effects on sorption of organic micropollutants onto aquifer materials. *J. Contamin. Hydrol.* **9**, 307–324.

Lorah M. M. and Olsen L. D. (1999a) Degradation of 1,1,2,2-tetrachloroethane in a freshwater tidal wetland: field and laboratory evidence. *Environ. Sci. Technol.* **33**, 227–234.

Lorah M. M. and Olsen L. D. (1999b) Natural attenuation of chlorinated volatile organic compounds in a freshwater tidal wetland: field evidence of anaerobic biodegradation. *Water Resour. Res.* **35**, 3811–3827.

Lorah M. M., Voytek M. A., Kirshtein J. D., and Jones E. J. *Anaerobic Degradation of 1,1,2,2-Tetrachloroethane and Association with Microbial Communities in a Freshwater Tidal Wetland, Aberdeen Proving Ground, Maryland: Laboratory Experiments and Comparisons to Field Data.* US Geological Survey Water-Resources Investigations Report 02-4157 (in press).

Lovley D. R. (1991) Dissimilatory Fe(III) and Mn(IV) reduction. *Microbiol. Rev.* **55**, 259–287.

Lovley D. R. and Goodwin S. (1988) Hydrogen concentrations as an indicator of the predominant terminal electron-accepting reactions in aquatic sediments. *Geochim. Cosmochim. Acta* **52**, 2993–3003.

Ludvigsen L., Albrechtsen H.-J., Holst H., and Christensen T. H. (1997) Correlating phospholipid fatty acids (PLFA) in a landfill leachate polluted aquifer with biogeochemical factors by multivariate statistical methods. *FEMS Microbiol. Rev.* **20**, 447–460.

Ludvigsen L., Albrechtsen H.-J., Heron G., Bjerg P. L., and Christensen T. H. (1998) Anaerobic microbial redox processes in a landfill leachate contaminated aquifer (Grindsted, Denmark). *J. Contamin. Hydrol.* **33**, 273–291.

Ludvigsen L., Albrechtsen H.-J., Ringelberg D. B., Ekelund F., and Christensen T. H. (1999) Composition and distribution of mirobial biomass in a landfill leachate contaminated aquifer (Grindsted, Denmark). *Microbiol. Ecol.* **37**, 197–207.

Lyngkilde J. and Christensen T. H. (1992a) Fate of organic contaminants in the redox zones of a landfill leachate pollution plume (Vejen, Denmark). *J. Contamin. Hydrol.* **10**, 291–307.

Lyngkilde J. and Christensen T. H. (1992b) Redox zones of a landfill leachate pollution plume (Vejen, Denmark). *J. Contamin. Hydrol.* **10**, 273–289.

MacFarlane D. S., Cherry J. A., Gillham R. W., and Sudicky E. A. (1983) Migration of contaminants in groundwater at a landfill: a case study. 1. Groundwater flow and plume delineation. *J. Hydrol.* **63**, 1–29.

McGuire J. T., Smith E. W., Long D. T., Hyndman D. W., Haack S. K., Klug M. J., and Velbel M. A. (2000) Temporal variations in parameters reflecting terminal-electron-accepting processes in an aquifer contaminated with waste fuel and chlorinated solvents. *Chem. Geol.* **169**, 471–485.

McGuire J. T., Long D. T., Klug M. J., Haack S. K., and Hyndman D. W. (2002) Evaluating behavior of oxygen, nitrate, and sulfate during recharge and quantifying reduction rates in a contaminated aquifer. *Environ. Sci. Technol.* **36**, 2693–2700.

Murray J., Rouse J. V., and Carpenter A. B. (1981) Groundwater contamination by sanitary landfill leachate and domestic wastewater in carbonate terrain: principal source diagnosis, chemical transport characteristics and design implications. *Water Res.* **15**, 745–757.

National Research Council (2000) *Natural Attenuation for Groundwater Remediation*. National Academy Press, Washington, DC, 274pp.

Nicholson R. V., Cherry J. A., and Reardon E. J. (1983) Migration of contaminants in groundwater at a landfill: a case study. 6. Hydrogeochemistry. *J. Hydrol.* **63**, 131–176.

Nielsen P. H., Albrechtsen H.-J., Heron G., and Christensen T. H. (1995a) *In situ* and laboratory studies on the fate of specific organic compounds in an anaerobic landfill leachate plume: I. Experimental conditions and fate of phenolic compounds. *J. Contamin. Hydrol.* **20**, 27–50.

Nielsen P. H., Bjarnadottir H., Winter P. L., and Christensen T. H. (1995b) *In situ* and laboratory studies on the fate of specific organic compounds in an anaerobic landfill leachate plume: II. Fate of aromatic and chlorinated aliphatic compounds. *J. Contamin. Hydrol.* **20**, 51–66.

Nielsen P. H., Bjerg P. L., Nielsen P., Smith P., and Christensen T. H. (1996) *In situ* and laboratory determined first-order degradation rate constants of specific organic compounds in an aerobic aquifer. *Environ. Sci. Technol.* **30**, 31–37.

Parkhurst D. L., Christenson S. C., and Breit G. N. (1993) *Ground-water Quality Assessment of the Central Oklahoma Aquifer, Oklahoma: Geochemical and Geohydrologic Investigations.* US Geological Survey Open File Report 92-642, US Geological Survey, Oklahoma City, OK.

Paxeus N. (2000) Organic compounds in municipal landfill leachates. *Water Sci. Technol.* **42**(7/8), 323–333.

Pecher K., Haderlein S. B., and Schwarzenbach R. P. (1997) Transformation of polyhalogenated alkanes in suspensions of ferrous iron and iron oxides. In *213th ASC National Meeting, April 13–17, 1997.* ASC Division of Environmental Chemistry, Washington, DC, pp. 185–187.

Petersen M. J. (2000) Modeling of groundwater flow and reactive transport in a landfill leachate plume. PhD Thesis, Department of Hydrodynamics and Hydraulic Engineering, Technical University of Denmark, Lyngby (ISVA Series Paper No. 73).

Postma D. (1993) The reactivity of iron oxides in sediments: a kinetic approach. *Geochim. Cosmochim. Acta* **57**, 5027–5034.

Ravi V., Chen J.-S., Wilson J. T., Johnson J. A., Gierke W., and Murdie L. (1998) Evaluation of natural attenuation of benzene and dichloroethane at the KL landfill. *Bioremed. J.* **2**, 239–258.

Reinhard M., Goodman N. L., and Barker J. F. (1984) Occurrence and distribution of organic chemicals in two landfill leachate plumes. *Environ. Sci. Technol.* **18**, 953–961.

Richnow H. H., Meckenstock R. U., Ask L., Baun A., Ledin A., and Christensen T. H. (2003) *In situ* biodegradation determined by carbon isotope fractionation of aromatic hydrocarbons in an anaerobic landfill leachate plume (Vejen, Denmark). *J. Contamin. Hydrol.* **64**, 59–72.

Rügge K., Bjerg P. L., and Christensen T. H. (1995) Distribution of organic compounds from municipal solid waste in the groundwater downgradient of a landfill (Grindsted, Denmark). *Environ. Sci. Technol.* **29**, 1395–1400.

Rügge K., Hofstetter T. B., Haderlein S. B., Bjerg P. L., Knudsen S., Zraunig C., Mosbæk H., and Christensen T. H. (1998) Characterization of predominant reductants in an anaerobic leachate-affected aquifer by nitroaromatic probe compounds. *Environ. Sci. Technol.* **32**, 23–31.

Rügge K., Bjerg P. L., Pedersen J. K., Mosbæk H., and Christensen T. H. (1999a) An anaerobic field injection experiment in a landfill leachate plume (Grindsted, Denmark): 1. Site description, experimental setup, tracer movement and fate of aromatic and chlorinated aliphatic compounds. *Water Resour. Res.* **35**, 1231–1246.

Rügge K., Bjerg P. L., Mosbæk H., and Christensen T. H. (1999b) Fate of MCPP and atrazine in an anaerobic landfill leachate plume (Grindsted, Denmark). *Water Res.* **33**, 2455–2458.

Röling W. F. M., van Breukelen B. M., Braster M., Groen J., and van Verseveld H. W. (2000) Analysis of microbial communities in a landfill leachate polluted aquifer using a new method for anaerobic physiological profiling and 16S rDNA based fingerprinting. *Microbiol. Ecol.* **40**, 177–188.

Röling W. F. M., van Breukelen B. M., Braster M., Lin B., and van Verseveld H. W. (2001) Relationship between microbial community structure and hydrochemistry in a landfill leachate-polluted aquifer. *Appl. Environ. Microbiol.* **67**, 4619–4629.

Scholl M. A. and Christenson S. C. (1998) *Spatial Variation in Hydraulic Conductivity Determined by Slug Tests in the Canadian River Alluvium near the Norman Landfill, Norman, Oklahoma*. US Geological Survey Water-Resources Investigations Report 97-4292, US Oklahoma City, OK.

Scholl M. A., Cozzarelli I. M., Christenson S. C., Breit G. N., and Schlottmann J. L. (1999) Aquifer heterogeneity at Norman Landfill and its effect on observations of biodegradation processes. In *US Geological Survey Toxic Substances Hydrology Program: Proceedings of the Technical Meeting, Charleston, South Carolina, March 8–12, 1999 Subsurface Contamination from Point Sources* (eds. D. W. Morganwalp and H. T. Buxton). US Geological Survey Water-Resources Investigations Report 99-4018C, West Trenton, NJ, vol. 3, pp. 557–568.

Scholl M. A., Cozzarelli I. M., Christenson S. C., Istok J., Jaeschke J., Ferree D. M., and Senko J. (2001) Measuring variability of *in-situ* biodegradation rates in a heterogeneous aquifer contaminated by landfill leachate. *EOS, Trans., AGU* **82**(20), 146.

Schwarzbauer J., Heim S., Brinker S., and Littke R. (2002) Occurrence and alteration of organic contaminants in seepage and leakage water from a waste deposit landfill. *Water Res.* **36**, 2275–2287.

Scott M. J. and Morgan J. J. (1990) Energetics and conservative properties of redox systems. *ACS Symp. Ser.* **416**, 368–378.

Senko J. M., Istok J. D., Suflita J. M., and Krumholz L. R. (2002) *In-situ* evidence for uranium immobilization and remobilization. *Environ. Sci. Technol.* **36**, 1491–1496.

Suarez M. P. and Rifai H. S. (1999) Biodegradation rates for fuel hydrocarbons and chlorinated solvents in groundwater. *Bioremed. J.* **3**, 337–362.

Sykes J. F., Soupak S., and Farquhar G. J. (1982) Modeling of leachate organic migration in groundwater below sanitary landfills. *Water Resour. Res.* **18**, 135–145.

Thornton S. F., Lerner D. N., and Banwart S. A. (2001) Assessing the natural attenuation of organic contamainants in aquifers using plume-scale electron carbon balances: model development with analysis and parameters sensitivity. *J. Contamin. Hydrol.* **53**, 199–232.

Tuxen N., Ejlskov P., Albrechtsen H.-J., Reitzel L. A., Pedersen J. K., and Bjerg P. L. Application of natural attenuation to ground water contaminated by phenoxy acid herbicides at an old landfill (Sjoelund, Denmark). *Ground Water Monit. R.* (in press).

Ulrich G. A., Martino D., Burger K., Routh J., Grossman E. L., Ammerman J. W., and Suflita J. M. (1998) Sulfur cycling in the terrestrial subsurface: commensal interactions, spatial scales, and microbial heterogeneity. *Microbiol. Ecol.* **36**, 141–151.

Ulrich G. A., Breit G. N., Cozzarelli I. M., and Suflita J. M. (2003) Sources of sulfate supporting anaerobic metabolism in a contaminated aquifer. *Environ. Sci. Technol.* **37**, 1093–1099.

van Duijvenbooden W. and Kooper W. F. (1981) Effects on groundwater flow and groundwater quality of a waste disposal site in Noordwijk, The Netherlands. *Quality of Groundwater: Proc. Int. Symp. Noordwijkerhout, March 1981*. Studies in Environmental Science,. Elsevier, Amsterdam, vol. 17, pp. 253–260.

Vogel T. M., Criddle C. S., and McCarty P. L. (1987) Transformations of halogenated aliphatic compounds. *Environ. Sci. Technol.* **21**, 722–736.

Subject Index

The index is in letter-by-letter order, whereby hyphens and spaces within index headings are ignored in the alphabetization (e.g. Arabian–Nubian Shield precedes Arabian Sea). Terms in parentheses are excluded from the initial alphabetization. In line with normal materials science practice, compound names are not inverted but are filed under substituent prefixes.

The index is arranged in set-out style, with a maximum of three levels of heading. Location references refer to the page number. Major discussion of a subject is indicated by bold page numbers. Page numbers suffixed by *f* or *t* refer to figures or tables.